FRACTIONALIZATION OF PARTICLES IN PHYSICS

This book explores the fractionalization of particles in physics, how interactions between individual particles and their background can modify their fundamental quantum states. Covering a large breadth of topics with an example-driven approach, this comprehensive text explains why phases of matter must be described in terms of both symmetries and their topology. The majority of important results are derived in full with explanations provided, while exercises at the end of each section allow readers to extend and develop their understanding of key topics.

The first part presents polyacetylene as the paradigmatic material in which electric charge can be fractionalized, while the second part introduces the notion of invertible topological phases of matter. The final part is devoted to the "tenfold way", a classification of topological insulators or superconductors. The text requires a solid understanding of quantum mechanics and is a valuable resource for graduate students and researchers in physics.

CHRISTOPHER MUDRY is Head of the Condensed Matter Theory Group at the Paul Scherrer Institute. He earned his PhD in Theoretical Physics from the University of Illinois at Urbana-Champaign, and his research is focused on condensed matter physics.

CLAUDIO CHAMON is Professor of Physics at Boston University. He earned his PhD in Physics from MIT, and his research interests are in condensed matter physics and in quantum and classical computation.

Fractionalization of Particles in Physics

Invertible Topological Phases of Matter

CHRISTOPHER MUDRY
Paul Scherrer Institute

CLAUDIO CHAMON
Boston University

CAMBRIDGE
UNIVERSITY PRESS

CAMBRIDGE
UNIVERSITY PRESS

Shaftesbury Road, Cambridge CB2 8EA, United Kingdom

One Liberty Plaza, 20th Floor, New York, NY 10006, USA

477 Williamstown Road, Port Melbourne, VIC 3207, Australia

314–321, 3rd Floor, Plot 3, Splendor Forum, Jasola District Centre, New Delhi – 110025, India

103 Penang Road, #05–06/07, Visioncrest Commercial, Singapore 238467

Cambridge University Press is part of Cambridge University Press & Assessment, a department of the University of Cambridge.

We share the University's mission to contribute to society through the pursuit of education, learning and research at the highest international levels of excellence.

www.cambridge.org
Information on this title: www.cambridge.org/9781107009547

DOI: 10.1017/9780511841828

When citing this work, please include a reference to the DOI 10.1017/9780511841828

First published 2025

A catalogue record for this publication is available from the British Library.

A Cataloging-in-Publication data record for this book is available from the Library of Congress.

ISBN 978-1-107-00954-7 Hardback

To

André, Cristy, Josette, Louis,

and

Flavia, Felipe, Henrique

Denise, and Nagib (in memory)

Contents

Preface

It was in 2010 that we had agreed with Simon Capelin from Cambridge University Press to write a book entitled *Fractionalization of Particles in Physics*. That it took 14 years to complete this project has two main explanations. First, the writing of the book was supposed not to impede on our ability to do research and to fulfill our teaching and administrative responsibilities in two academic institutions separated by six time zones. Second, the collective understanding of the subject matter has been dramatically evolving since 2010. In retrospect, we are amazed that Neil W. Ashcroft and N. David Mermin needed only seven years to complete their iconic book *Solid State Physics*.

The compromise that we made with regard to the pace at which the field has been progressing is to limit the scope of this book to the study of invertible topological phases of matter, a unifying terminology that treats on equal footing electronic topological insulators and superconductors, on the one hand, and bosonic symmetry protected topological phases of matter, on the other hand. We have promised Cambridge University Press a sequel that will cover the topics of topological and fracton order, hopefully on a shorter time scale.

This book is large in several aspects. Although it contains eight chapters, each one is very detailed in an attempt to make the book as self-contained as may be. Some results are derived in several different and complementary ways. The fractional value of the electric charge bound to a soliton in polyacetylene is computed differently in eight sections in Chapter 3. In doing so, we introduce theoretical tools that are needed in later chapters and in the sequel to this book. Some chapters include material that has only appeared in research papers (Chapters 4-8) to the best of our knowledge. A broad range of concepts in physics and mathematics is thus introduced, hopefully in a didactic way. Chapter 7 introduces random matrix theory in $d = 0$-dimensional space and the theory of localization in $d = 1$-dimensional space from a physics perspective or, from a mathematical perspective, Brownian motion on symmetric spaces. The title of this book refers to the fractionalization of quantum numbers. To measure a fractional charge, one may localize it in space and thus break translation invariance. This fractionalization is understood from the point of view of index theorems in Chapter 3, a tool marrying analysis and topology in mathematics. We then shift to using tools from algebraic topology in Chapters 5–8 to sharpen and generalize what is meant by fractionalization in physics.

Whereas it is impossible to teach the entire book in one semester, it is possible to cover selected parts of each one of the eight chapters in one semester. Every chapter ends with a section devoted to exercises. Most exercises fill intermediary steps, offer alternative interpretations, or provide some background material.

We aimed at writing a textbook, not a review. The papers and textbooks that we cite are not meant to cover in an exhaustive way the relevant literature. They were influential on the way we have developed our understanding of the field. As we do not shy away from repetition in this book, references are simultaneously cited as footnotes and indexed at the end of the book. The former choice is meant to satisfy readers who are curious to know who did what and when without the disruption of searching the reference at the end of the book. The latter choice gives a quick overview of the literature arranged alphabetically.

Acknowledgments

We are grateful to Eduardo Fradkin and Roman Jackiw, who encouraged us to undertake the writing of this book. We deeply regret that Roman could not see the end product before his passing.

Two generations of PhD students, Jyong-Hao Chen and Ömer Aksoy, were kind enough to read critically early versions of the lecture notes for the lectures at ETHZ and EPFL covering the material from this book. They pruned out many mistakes. The thesis of Ömer also shaped Chapters 5 and 6 in a conceptual way.

Over 25 years, many of our collaborators had an influence on our understanding of the material presented in this book. They are Adolfo Grushin, Akira Furusaki, Alexander Altland, Alexei Tsvelik, Andreas Schnyder, Apoorv Tiwari, Ben Simons, Claudio Castelnovo, Chang-Yu Hou, Denise Freed, Gordon Semenoff, Ilya Gruzberg, Luis Santos, Nancy Sandler, Pedro Gomes, Pierre Pujol, Piet Brouwer, Po-Hao Huang, Prashant Sharma, Shinsei Ryu, Shivaji Sondhi, So-Young Pi, Stefanos Kourtis, Steve Kivelson, Takahiro Morimoto, Titus Neupert, Tom Iadecola, Xiao-Gang Wen, and Zhi-Cheng Yang.

Lastly, we are indebted to Maurizio Storni, who proof-read the entire manuscript with his wonderful attention and patience. We benefited from his mastery of physics and mathematics. In particular, Maurizio greatly improved Section 7.3.11.

1
Introduction and Overview

1.1 What Will Be Covered

In this book, we are interested in building blocks of matter made out of atoms. Atoms are themselves made of nuclei and electrons. The characteristic microscopic length scale of such matter is the average distance between the nucleus and the electron in an hydrogen atom, that is, the Bohr radius. On this length scale, nuclei and electrons are point particles. On the one hand, nuclei and electrons have intrinsic properties such as mass, electric charge, linear momentum, angular momentum, and energy that are shared with point particles in classical mechanics. On the other hand, nuclei and electrons have wave-like attributes, discrete quantum numbers, and internal degrees of freedom that can only be described within the realm of quantum mechanics. For distances of the order of or larger than the Bohr radius, their dominant interactions are electromagnetic.

Because the matter of interest in this book is made of an astronomically large number of nuclei and electrons, it must be described within the framework of many-body physics. The ground state is a many-body ground state and so are the excitations built from it. Now, the quantum many-body ground state and its many-body excited states can be very different from their counterparts when all (electromagnetic) interactions between nuclei and electrons are turned off. The simplest treatment of interactions between nuclei and electrons is to assume that the nuclei are localized on the sites of a lattice (not necessarily periodic) and that their interactions with the electrons is approximated by a one-body classical potential for, otherwise, noninteracting electrons. Even in this limit, the many-body ground state for the electrons can acquire exotic properties in the thermodynamic limit. The main goal of this book is to study the conditions under which electronic many-body excitations carry fractions of the quantum numbers of an isolated electron. This exploration will deliver the main concept of this book, namely, that of *invertible topological phases of matter.*

The aim of equilibrium condensed matter physics is to identify what phases of matter are realized in a material. A phase of matter is characterized by a list of attributes. The electrical resistance decreases with decreasing temperature in a metal, a crystal supports sharp collective excitations in the form of phonons, a magnet supports sharp collective excitations in the form of magnons. A phase diagram is the outcome of the theoretical and experimental study of a material in thermodynamic equilibrium. The control parameters in a phase diagram are typically the temperature, chemical potential, pressure, and magnetic field. Regions of the phase diagram correspond to different phases of matter if they are separated by boundaries along which sharp transitions take place. When the parametric boundary between two phases of matter realizes a continuous phase transition, the parametric boundary is then characterized by attributes of a universal character, that is, they only depend on a countable set of dimensionless parameters. Some of the phases of matter can themselves be characterized by universal dimensionless parameters, say the quantized value of the Hall conductivity in the *quantum Hall effect.*

Many of the distinguishing attributes of phases of matter in thermodynamic equilibrium, at sufficiently low temperatures, and in the presence of translation symmetry are inherited from the presence or absence of a spectral gap between the ground and excited states. For example, there is no gap between the ground and excited states in a metal, whereas there is one in an insulator. In 1975, known examples of gapless phases of matter were as follows:

1. The Coulomb phase of relativistic U(1) gauge theories[1]
2. Fermi-liquid theory[2]
3. The non-Fermi-liquid fixed point induced by the current-current interaction between electrons[3,4]
4. Whenever there are Nambu-Goldstone bosons associated with the spontaneous breaking of a continuous symmetry provided space (spacetime) is larger than two[5]
5. Continuous (quantum) critical points
6. The algebraic phase when a Berezinskii–Kosterlitz–Thouless transition takes place (an example of a Luttinger liquid in two-dimensional spacetime).[6]

In 1975, known examples of (partially) gapped phases of matter were the following:

[1] The Coulomb phase of quantum electrodynamics in the electroweak standard model of high-energy physics is defined by demanding that the two-point function for the exponential of the U(1) field-strength tensor decays algebraically fast in four-dimensional Euclidean spacetime.

[2] L. D. Landau, Soviet Physics JETP-USSR **3**(6), 920–925 (1957); **5**(1), 101–108 (1957); **8**(1), 70–74 (1959). [Landau (1957a,b, 1959)].

[3] T. Holstein, R. E. Norton, and P. Pincus, Phys. Rev. B **8**(6), 2649–2656 (1973). [Holstein et al. (1973)].

[4] Although Fermi-liquid theory is robust to the long-range Coulomb interaction because of screening, it is not robust to current-current interactions for the latter are not screened because of gauge invariance. However, the smallness of the ratio between the Fermi velocity and the speed of light in metals renders this instability unobservable as this instability of Fermi-liquid theory is pre-empted by gaping instabilities for all practical purposes.

[5] Y. Nambu, Phys. Rev. **117**(3), 648–663 (1960); Y. Nambu and G. Jona-Lasinio, Phys. Rev. **122**(1), 345–358 (1961); J. Goldstone, Il Nuovo Cimento **19**(1), 154–164 (1961). [Nambu (1960); Nambu and Jona-Lasinio (1961); Goldstone (1961)].

[6] V. J. Emery, in *Highly Conducting One-Dimensional Solids,* pages 247–303, edited by J. T. Devreese, R. P. Evrard, V. E. van Doren, Springer, Boston, 1979. [Emery (1979)].

1. Bloch band insulators
2. Long-range ordered phases breaking spontaneously a discrete symmetry (the Ising model for localized spins, charge-density waves for interacting electrons, or bond-density waves arising from the electron-phonon coupling)
3. Charge Mott insulators such as the Hubbard model at half-filling for any bipartite lattice[6]
4. The confining phases in gauge theories[7]
5. Higgs phases of matter (nonrelativistic superconductors or the weak sector of the standard model of particle physics).[8]

After 1975, it was progressively understood that all gapped phases of matter are not equal in that they can be distinguished by dimensionless and quantized quantum numbers. One such family of gapped phases of matter is now called invertible topological phases of matter. They have a nondegenerate ground state that is separated from all excitations by a gap when space is any closed manifold and translation symmetry holds. They owe their topological attributes to the fact that they can also support (symmetry) protected gapless boundary states when open boundary conditions are selected.

Invertible topological phases of matter display excitations that carry fractional values of the quantum numbers carried by the elementary building blocks of matter in condensed matter physics, say the electric charge when the total electric charge is conserved. Although the emphasis of this book will be on the tenfold classification of strong topological insulators or superconductors, which are fermionic examples of invertible topological phases of matter, we shall also describe bosonic invertible topological phases of matter.

Conceptual discoveries are rarely punctual in time. Even if a paradigm changing concept can be attributed to one or two papers, these papers were not written out of thin air, they were motivated by earlier works. Bearing in mind this caveat, the concepts that will be explored in this book crystallized in the mid 1970s.[9,10] The 1970s saw a convergence of interest between theorists in high-energy and condensed matter physics, on the one hand, and mathematicians, on the other hand, that culminated with the realization that index theorems and quantum anomalies were related and had potentially observable consequences in physics. Predictions deriving from applications of the index theorem in one-dimensional space were made for polyacetylene. However, it is the experimental discovery of the integer[11]

[7] J. Schwinger, Phys. Rev. **128**(5), 2425–2429 (1962). [Schwinger (1962)].
[8] P. W. Anderson, Phys. Rev. **130**(1), 439–442 (1963); P. W. Higgs, Phys. Lett. **12**(2), 132–133 (1964); P. W. Higgs, Phys. Rev. Lett. **13**(16), 508–509 (1964). [Anderson (1963); Higgs (1964b,a)].
[9] R. Jackiw and C. Rebbi, Phys. Rev. D **13**(12), 3398–3409 (1976). [Jackiw and Rebbi (1976)].
[10] W. P. Su, J. R. Schrieffer, and A. J. Heeger, Phys. Rev. Lett. **42**(25), 1698–1701 (1979); Phys. Rev. B **22**(4), 2099–2111 (1980). [Su et al. (1979, 1980)].
[11] K. v. Klitzing, G. Dorda, and M. Pepper, Phys. Rev. Lett. **45**(6), 494–497 (1980). [Klitzing et al. (1980)].

and fractional[12] quantum Hall effects in 1980 and 1982, respectively, that opened the Pandora's box for those phenomena in condensed matter physics that require concepts from topology to be understood.

Paradoxically, after the nexus of common interest in the 1970s between condensed matter physics, high-energy physics, and algebraic topology, the discovery of high-temperature superconductivity on the one hand and progress in string theory on the other hand saw the three communities go their separate ways between 1990 and 2005. The discovery of graphene[13] in 2004 and the prediction by Kane and Mele[14] in 2005 that polyacetylene and the quantum Hall effect were not the only playgrounds for band topology in physics triggered a renewed convergence of interest between these three communities with topology as the focal point. The topological attributes of band insulators or band superconductors by which (symmetry-protected) topological boundary states can evade Anderson localization transcend condensed matter physics as they have classical counterparts in many different mediums supporting the propagation of waves that are of relevance to optics, mechanical engineering, or electrical engineering. Applied science has paid due notice to the possibility of improving the efficiency of data transmission by encoding information into delocalized (symmetry-protected) topological boundary states.

This book is organized into eight Chapters. Chapters 2–4 are dedicated to the phenomenon of charge fractionalization in polyacetylene. The material presented in Chapters 2 and 4 was understood over the decade from 1976 to 1986. Chapters 5 and 6 introduce the notion of invertible topological phases of matter, of which polyacetylene was the first example. The time period covered by Chapters 5 and 6 starts in 1964 and extends to today. Whereas polyacetylene is a fermionic example of an invertible topological phase of matter in one-dimensional space, Chapters 5 and 6 cover bosonic examples of symmetry-protected topological phases of matter in one-dimensional space and their relation to fermionic examples of invertible topological phases of matter. Finally, we return to fermionic invertible topological phases of matter in Chapters 7 and 8, for which we give an exhaustive classification in any dimension of space called the tenfold way. This classification presumes a combination of three discrete symmetries: time-reversal symmetry, charge-conjugation symmetry, or chiral symmetry that is realized by identical point particles obeying the Pauli exclusion principle (fermions). In particular, this classification does not presume any space-group symmetry.

[12] D. C. Tsui, H. L. Stormer, and A. C. Gossard, Phys. Rev. Lett. **48**(22), 1559–1562 (1982). [Tsui et al. (1982)].

[13] K. S. Novoselov, A. K. Geim, S. V. Morozov, et al., Science **306**, 666—669 (2004). [Novoselov et al. (2004)].

[14] C. L. Kane and E. J. Mele, Phys. Rev. Lett. **95**(14), 146802 (2005); Phys. Rev. Lett. **95**(22), 226801 (2005). [Kane and Mele (2005a,b)].

1.2 What Will Not Be Covered

Fermionic invertible topological phases of matter in two or higher dimensions of space and at zero temperature are realized by materials with a vanishing bulk thermal conductance, while their boundaries support a nonvanishing thermal conductance. From the point of view of the bulk, they realize an insulating state. From the point of view of their boundaries, they realize a thermal metal. What makes many fermionic invertible topological phases of matter remarkable is that their ability to transport energy along their boundaries survives the thought experiment by which the electron-electron interaction is adiabatically turned off, that is the existence of gapless (symmetry-protected) topological boundary states is not driven by electron-electron interactions in many fermionic invertible topological phases of matter. However, this need not necessarily be so.

For example, it is possible to construct a local electronic interaction such that the charge and thermal Hall conductance are independently quantized,[15] thereby breaking the Wiedemann-Franz law according to which the thermal and charge conductivity tensor must be proportional. As the Wiedemann-Franz law is a property of noninteracting electrons, it follows that the decoupling of the charge and thermal Hall conductance is a property driven by strong interactions. Although the terminology of fermionic invertible topological phase of matter still applies to this situation, we will not discuss such examples of physics driven by strong interactions in this book.

Another example is the fractional quantum Hall effect (FQHE), for which the value taken by the Hall conductance in units of e^2/h is quantized to a rational value that is not an integer. Hereto, electron-electron interactions are essential for the fractional quantum Hall effect. The manifestation of topology in the fractional quantum Hall effect is of a qualitatively different nature than that governing fermionic invertible topological phases of matter. The fractional quantum Hall effect is an example of a phase of matter displaying *topological order*.[16,17] Topological order is not possible in one-dimensional space when the quantum dynamics is a local one for the observable degrees of freedom. Topological order can only be realized when the dimensionality of space is larger than one.

The study of fractionalization that requires strong interactions will be treated in a separate book that covers bosonic and fermionic topological order. Another arena for fractionalization to be covered in such a book is that of fractons.[18,19,20,21]

15 T. Neupert, C. Chamon, C. Mudry, and R. Thomale, Phys. Rev. B **90**(20), 205101 (2014). [Neupert et al. (2014)].

16 X.-G. Wen, International Journal of Modern Physics B **04**(02), 239–271 (1990). [Wen (1990)].

17 X.-G. Wen, *Quantum Field Theory of Many-Body Systems: From the Origin of Sound to an Origin of Light and Electrons,* Oxford University Press, New York 2007. [Wen (2007)].

18 C. Chamon, Phys. Rev. Lett. **94**(4), 040402 (2005). [Chamon (2005)].

19 S. Bravyi, B. Leemhuis, and B. M. Terhal, Ann. Phys. (N.Y.) **326**(4), 839–866 (2011). [Bravyi et al. (2011)].

20 J. Haah, Phys. Rev. A **83**(4), 042330 (2011). [Haah (2011)].

21 S. Vijay, J. Haah, and L. Fu, Phys. Rev. B **92**(23), 235136 (2015). [Vijay et al. (2015)].

1.3 Why One, Two, and More d-Dimensional Space

It is standard practice when teaching quantum mechanics to solve the Schrödinger equation when space is one-, two-, and so on dimensional. This allows to add stepwise the complexity brought about by angular degrees of freedom, say.

The same is true with classical and quantum statistical physics. Exact solutions of the Ising model for one- and two-dimensional lattices have played a very important role in the study of phase transitions, mean-field theory, the renormalization group, and so on.

In condensed matter physics, quantum confinement has allowed to realize experimentally models for which space is effectively zero-dimensional (quantum dots[22]), one-dimensional (carbon nanotubes[23]), or two-dimensional (graphene[13]). The same is possible by trapping cold atoms in an optical lattice. Conversely, by adiabatic tuning of some couplings entering a Hamiltonian (quantum pumping[24]), it is possible to study an effective Hamiltonian acting on a d greater than three-dimensional space.

The essence of what are invertible topological phases of matter is most easily explained in one-dimensional space. In fact, both fermionic and bosonic invertible topological phases of matter were first predicted theoretically for effectively one-dimensional space. In the case of bosonic invertible topological phases of matter, the first experimental discovery thereof was that of a one-dimensional spin-1 Heisenberg antiferromagnetic quantum magnet. In the case of a fermionic invertible topological phases of matter, the first experimental discovery thereof was that of a two-dimensional gas of electrons displaying the integer quantum Hall effect.

[22] S. M. Reimann and M. Manninen, Rev. Mod. Phys. **74**(4), 1283–1342 (2002); R. Hanson, L. P. Kouwenhoven, J. R. Petta, S. Tarucha, and L. M. K. Vandersypen, Rev. Mod. Phys. **79**(4), 1217–1265 (2007). [Reimann and Manninen (2002); Hanson et al. (2007)].

[23] M. S. Dresselhaus, G. Dresselhaus, and P. Avouris, *Carbon Nanotubes: Synthesis, Structure, Properties, and Applications*, Springer, Berlin Heidelberg, 2001. [Dresselhaus et al. (2001)].

[24] D. J. Thouless, Phys. Rev. B **27**(10), 6083–6087 (1983). [Thouless (1983)].

2

Modeling Polyacetylene

2.1 Introduction

Polyacetylene[1] is a semiconductor[2] with an energy gap of the order of 1.8 eV (as measured from the reflection coefficient[3]). The semiconducting behavior of polyacetylene is unexpected from the point of view of quantum chemistry, as quantum chemistry predicts that polyacetylene should be a metal with the Fermi level crossing the conduction band at half filling thereof. However, polyacetylene is an example of a quasi-one-dimensional metal undergoing a Peierls transition that is triggered by the electron-phonon coupling. At sufficiently low temperatures, the elastic energy cost for a lattice distortion is compensated by the gain in electronic energy through the opening of a gap at the Fermi energy. This lattice distortion (dimerization) breaks spontaneously the point-group symmetry of pristine polyacetylene above the ordering temperature. Chapter 2 is devoted to modeling the phenomenon of dimerization in polyacetylene.

Polyacetylene films have the remarkable property that their conductivity can increase up to seven orders of magnitude in a fully reversible way when exposed to vapors of electron-accepting compounds.[4] The discovery of the conducting properties of polyacetylene upon doping helped to launch the field of organic conductive polymers, even though polyacetylene has not found commercial applications so far. Alan J. Heeger, Alan G. MacDiarmid and Hideki Shirakawa were jointly rewarded

[1] K. Ziegler, E. Holzkamp, H. Breil, and H. Martin, Angew. Chem. **67**(19-20), 541–547 (1955); G. Natta, P. Pino, P. Corradini et al., Moraglio, J. Am. Chem. Soc. **77**(6), 1708–1710 (1955); G. Natta, G. Mazzanti, and P. Corradini, Atti Accad. Naz. Lincei, Rend. Cl. Sci. Fis. Mat. e Nat. **25**, 3–12 (1958). [Ziegler et al. (1955); Natta et al. (1955, 1958)].

[2] M. Hatano, S. Kambara, and S. Okamoto, J. Polym. Sci. **51**(156), S26–S29 (1961); W. H. Watson, W. C. McMordie, and L. G. Lands, J. Polym. Sci. **55**(161), 137–144 (1961); K. Shimamura, M. Hatano, S. Kanbara, and I. Nakada, J. Phys. Soc. Jpn. **23**(3), 578–581 (1967); A. Matsui and K. Nakamura, Jpn. J. Appl. Phys. **6**(12), 1468–1469 (1967); D. J. Berets and D. S. Smith, Trans. Faraday Soc. **64**, 823–828 (1968); F. D. Kleist and N. R. Byrd, J. Polym. Sci. Part A-1: Polym. Chem. **7**(12), 3419–3425 (1969). [Hatano et al. (1961); Watson Jr. et al. (1961); Shimamura et al. (1967); Matsui and Nakamura (1967); Berets and Smith (1968); Kleist and Byrd (1969)].

[3] D. Moses, A. Feldblum, E. Ehrenfreund, A. J. Heeger, T. -C. Chung, and A. G. MacDiarmid, Phys. Rev. B **26**(6), 3361–3369 (1982). [Moses et al. (1982)].

[4] C. K. Chiang, S. C. Gau, C. R. Fincher Jr. et al, Appl. Phys. Lett. **33**(1), 18–20 (1978). [Chiang et al. (1978)].

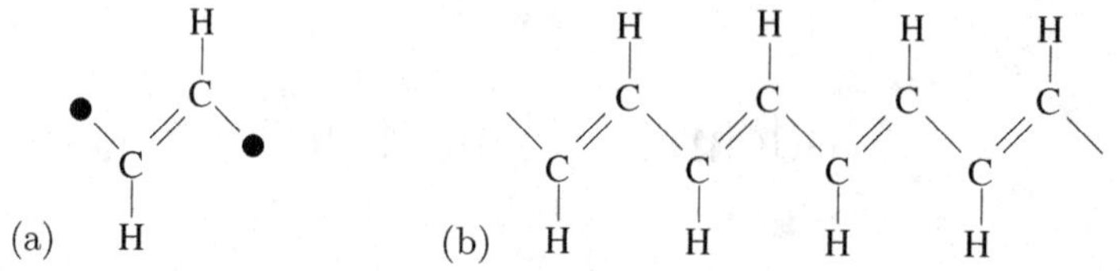

Figure 2.1 The first example and the paradigm for a semiconducting polymer is trans-polyacetylene (that we will abbreviate with polyacetylene in this book). Polyacetylene is a linear polymer in that it is made of weakly coupled chains of CH units forming a quasi-one-dimensional lattice. (a) Cartoon for the acetylene molecule C_2H_2, the repeat unit cell of polyacetylene. Acetylene is made of two CH monomers. The lines between the C and H atoms are both guides to the eyes and pictorial representations of the σ bonds between the C and H atoms resulting from the sp^2 hybridization of three of the four valence electrons of C originating from $2s$ and $2p$ atomic orbitals. The angles between any pair of the three bonds originating from a C atom is $2\pi/3$ in polyacetylene. The filled circle represents the embedding of acetylene in the linear chain that defines polyacetylene. The fourth valence electron of C has the symmetry of a $2p_z$ atomic orbital. It delivers π bonds that are half-filled in polyacetylene. (b) Cartoon for polyacetylene.

with the Nobel Prize in Chemistry 2000 "for the discovery and development of conductive polymers."

However, it is the fact that semiconducting polyacetylene was the first physical platform that was proposed for the realization of a phase of matter that is now called an invertible topological phase of matter, which motivates devoting Chapters 2–4 exclusively to polyacetylene at half-filling. The concept of an invertible topological phase of matter will be characterized by the property that some of its excitations can carry fractional quantum numbers as will be shown in Chapters 3–4 in the special case of polyacetylene.

2.2 Some Chemistry

Polyacetylene belongs to a class of organic materials for which the acetylene molecule depicted in Figure 2.1(a) serves as the elementary building block. Acetylene molecules can bond to form long chains. These chains are examples of conjugated polymers in chemistry and are depicted in Figure 2.1(b). Conjugated polymers can form bundles that, in turn, arrange in a complicated three-dimensional array, as is shown in the electron micrograph of polyacetylene in Figure 2.2(a). However, these bundles can be aligned along a common direction by stretching, as is illustrated in Figure 2.2(b).

The one-dimensional chain of carbon atoms shown in Figure 2.1(b) conveys, through the chemical bonds denoted by the line segments, that each carbon atom is bound to two other carbon atoms and one hydrogen atom by sp^2 orbitals. There are additional chemical bonds that have been omitted because of their relative

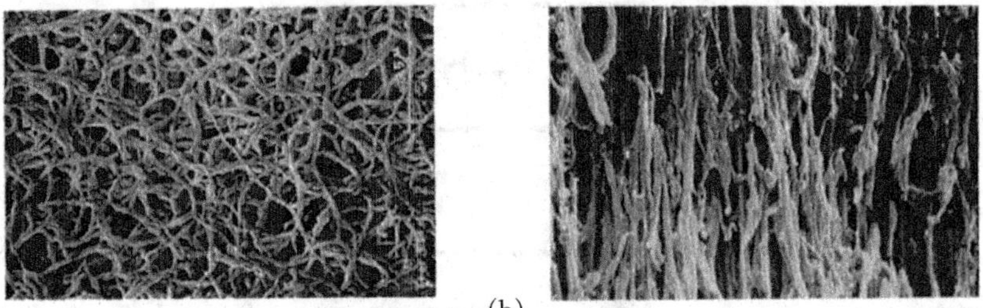

Figure 2.2 Electron micrograph of unstretched (a) and stretched (b) polyacetylene, after Figure 3 from Heeger et al. (1988).

weakness compared to the marked ones, among which one finds inter-chain bonds. Interchain bonds are much weaker than the intrachain bonds. In addition to the sp^2 orbitals that hold the chain together, there is a π orbital that hosts carbon's fourth electron.

It is the itinerant nature of the electron in the π orbital that is responsible for the strongly anisotropic electronic transport properties of polyacetylene. Characterizing electronic transport in polyacetylene requires the tools of experimental and theoretical physics. The schematic of the resulting bands of π conjugated polymers are shown in Figure 2.3. A calculation within band theory of the one-dimensional bands along the chain direction is presented in Figure 2.4.

2.3 Some Phenomenology

Polyacetylene is a semiconductor. First, its dc conductivity increases with increasing temperature away from zero temperature, see Figure 2.5. Second, impurity doping can increase its dc conductivity by orders of magnitudes, see Figure 2.6. Moreover, at sufficiently low temperatures, the dc conductivity along the chains of polyacetylene is orders of magnitude larger than the dc conductivity in the directions transverse to the chains as is illustrated in Figure 2.7. Because of this large anisotropy, electronic transport is said to be quasi-one-dimensional in polyacetylene. As we are going to explain, electronic transport in polyacetylene can be modeled with a one-dimensional tight-binding Hamiltonian in the approximation that neglects all electronic transport in the directions transverse to the chains of polyacetylene.

The tight-binding method is a way to study the electronic structure of a material starting from the orbital wave functions of isolated atoms, and superimposing these orbitals so as to construct the wave functions for the itinerant electrons. It is useful to imagine assembling the solid by bringing isolated atoms closer and closer together. When the electronic orbitals sitting at different atoms start to overlap, they hybridize and form the electronic bands of the solid.

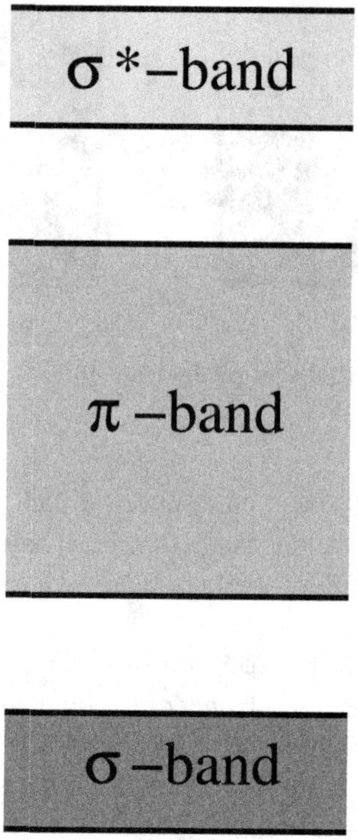

Figure 2.3 There are three schematic bands for π-conjugated polymers. There is the narrow fully-filled band of σ bonding character. There is the narrow fully-empty band of σ^* antibonding character. There is the broad half-filled band of π character.

In tight-binding models, the matrix element for an electron to tunnel between two atoms is known as the hopping amplitude. The hopping of the electron between the π orbitals of a pair of carbon sites in polyacetylene gives rise to a π-orbital band, while the sp^2 bonds hold the chains together. The overlap between the π orbitals of two carbon atoms on the same polymer chain is much larger in magnitude than that for atoms belonging to different chains. Since the atomic wave functions are exponentially localized near the atoms, with a characteristic length scale of the order of the Bohr radius, the overlap between two orbitals in far-away atoms is exponentially small. The strongly anisotropic dc conductivity in polyacetylene is explained by having much larger leading intrachain hopping amplitudes than leading interchain ones.

To a first approximation, we are going to ignore the hopping of the electrons in π orbitals belonging to carbon atoms sitting on different chains. We shall simply assume that electrons in π orbitals are constrained to tunnel between atoms belonging to the same chain, that is, the electronic motion in polyacetylene is predominantly one-dimensional (1D).

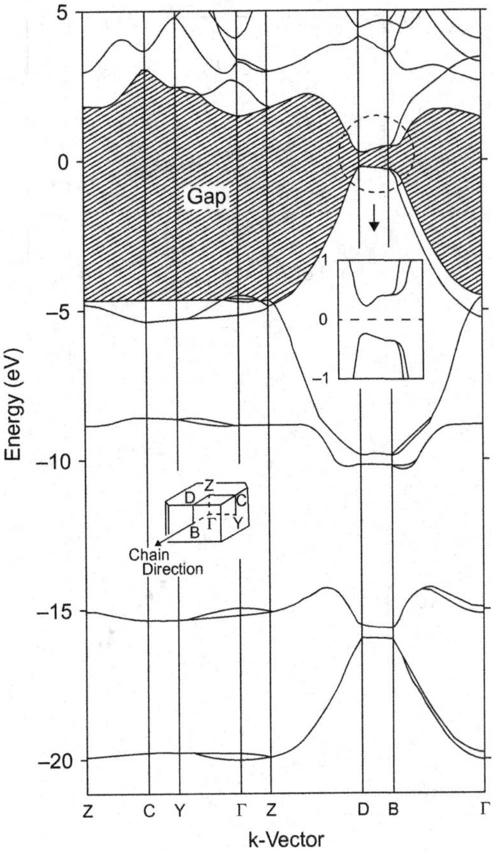

Figure 2.4 Calculated band structure of trans-$(CH)_x$, after Figure 3 from Grant and Batra (1983).

2.4 Tight-Binding Model along a Chain

We shall enumerate the carbon atoms in a chain of N atoms by an index i that runs from $i=1$ at one end of the chain to $i=N$ at the other end. The physical position of the carbon atom is at $\boldsymbol{r}_i$ in three-dimensional (3D) space. The most important tunneling matrix element is between two adjacent atoms in the same chain. The dependence of this hopping amplitude, denoted t_i, on the distance $|\boldsymbol{r}_i - \boldsymbol{r}_{i+1}|$ between these consecutive atoms is governed by the rule

$$t_i = t\, e^{-c\,|\boldsymbol{r}_i - \boldsymbol{r}_{i+1}|/\mathfrak{a}}, \tag{2.1}$$

where $t > 0$ is the characteristic energy scale that controls the band width for the itinerant electrons, c is a nonuniversal (that is it is material dependent) positive number of order unity, and $\mathfrak{a}$ is the average atomic distance (which would be the lattice spacing if the atoms formed a 1D lattice). The magnitude t for the hopping matrix element between adjacent atoms can either be deduced from a first-principle

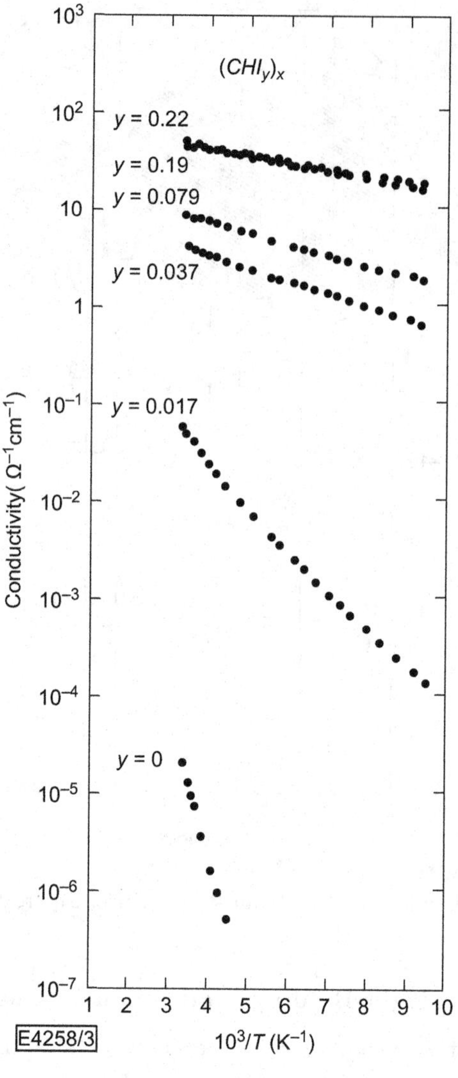

Figure 2.5 Dependence of the logarithm of the dc conductivity of polyacetylene as a function of $1/T$ with T the temperature, after Figure 3 from Chiang et al. (1979).

calculation (see Figure 2.4) or from spectroscopic data (by measuring the energy dispersion relation of charge and magnetic excitations) from a phenomenological point of view. For polyacetylene, $4t \sim 12\,\text{eV}$.

In this way, we arrive at the 1D quantum Hamiltonian[5]

$$\widehat{H} := -\sum_{i=1}^{N} \sum_{\sigma=\uparrow,\downarrow} \left(t_i\, \hat{c}^{\dagger}_{i,\sigma}\, \hat{c}_{i+1,\sigma} + \text{H.c.} \right). \tag{2.2a}$$

[5] The notation H.c. denotes Hermitean conjugation of all preceding terms.

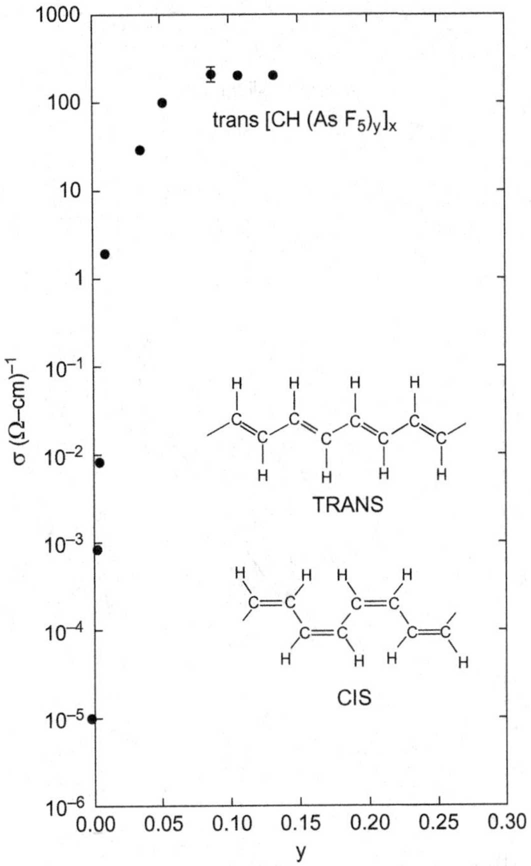

Figure 2.6 Effect of impurity doping y on the logarithm of the dc conductivity of polyacetylene, after Figure 1 from Chiang et al. (1977).

The index i labels a carbon site along the chain depicted in Figure 2.1(b). The index σ labels the spin quantum number of an electron along some quantization axis in spin space. The site-dependent number $t_i \in \mathbb{C}$ is the tunneling amplitude between two consecutive sites along a one-dimensional chain. With open boundary conditions, it is always possible to choose $t_i \in \mathbb{R}$. A hat over a Latin letter emphasizes that this Latin letter describes a mathematical object that is an operator as opposed to being a mere $\mathbb{C}$ number. The operators $\hat{c}^{\dagger}_{i,\sigma}$ and $\hat{c}_{i,\sigma}$ create and annihilate an electron in the orbital that is exponentially localized in position space around $\boldsymbol{r}_i$ with the spin quantum number σ, respectively.[6] Pairs of these operators satisfy the fermionic anticommutation relations:[7]

[6] We are denoting the Cartesian basis of $\mathbb{R}^3$ by $\boldsymbol{e}_1$, $\boldsymbol{e}_2$, and $\boldsymbol{e}_3$.

[7] The notation $\delta_{\bullet,\bullet'} \equiv \delta_{\bullet\bullet'}$, where the pair $\bullet$ and $\bullet'$ stands for a pair of indices taking discrete values, is reserved for the Kronecker delta symbol in this book, that is $\delta_{\bullet,\bullet'} = 1$ if $\bullet = \bullet'$, while $\delta_{\bullet,\bullet'} = 0$ if $\bullet \neq \bullet'$.

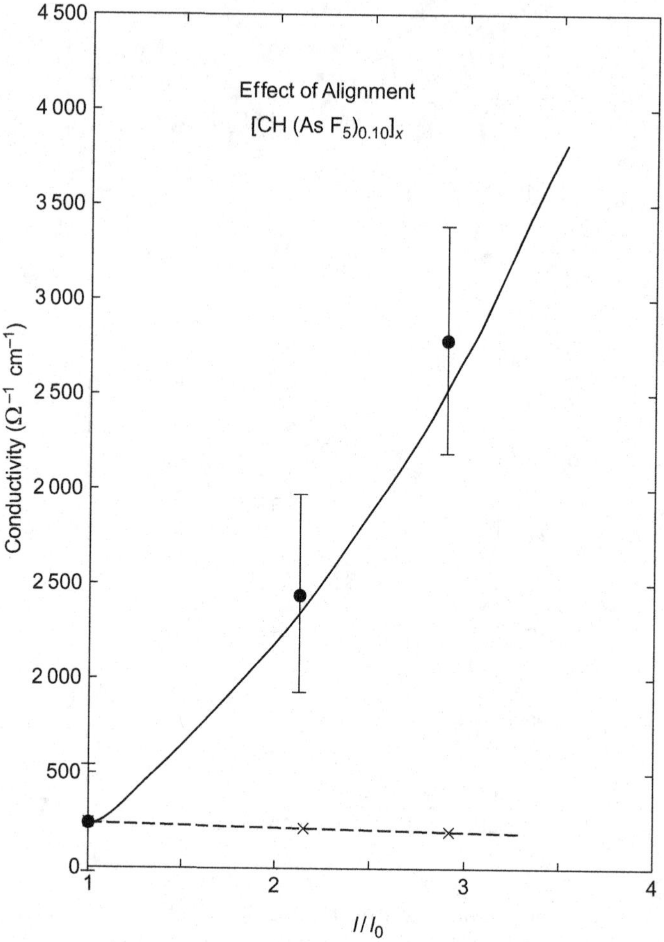

Figure 2.7 Effect of the stretching (l/l_0) on the dc conductivity of strongly doped polyacetylene when measured parallel (•) as opposed to perpendicular (×) to the stretching direction, after Figure 8 from Chiang et al. (1979).

$$\{\hat{c}_{i,\sigma}, \hat{c}^{\dagger}_{i',\sigma'}\} = \delta_{i,i'}\,\delta_{\sigma,\sigma'}, \qquad \{\hat{c}_{i,\sigma}, \hat{c}_{i',\sigma'}\} = \{\hat{c}^{\dagger}_{i',\sigma'}, \hat{c}^{\dagger}_{i,\sigma}\} = 0. \tag{2.2b}$$

Hamiltonian (2.2a) acts on the Fock space $\mathfrak{F}$ with the orthonormal basis obtained by any application of products of creation operators to the empty state $|0\rangle$ that is annihilated by all the $\hat{c}_{i,\sigma}$'s. More precisely, the Fock space $\mathfrak{F}$ is

$$\mathfrak{F} := \operatorname{span}\left\{\prod_{i=1}^{N}\prod_{\sigma=\uparrow,\downarrow}\left(\hat{c}^{\dagger}_{i,\sigma}\right)^{n_{i,\sigma}}|0\rangle, \quad n_{i,\sigma}=0,1, \quad \hat{c}_{i,\sigma}|0\rangle = 0\right\}. \tag{2.2c}$$

It is often useful to choose periodic boundary conditions so as to identify site $i = N+1$ with site $i = 1$. Thereby, the chain is turned into a ring as is depicted in Figure 2.8(a). More generically, one can impose twisted boundary conditions

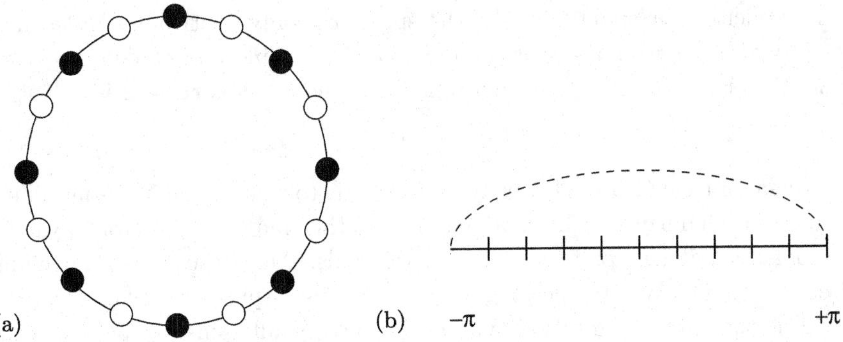

Figure 2.8 (a) One-dimensional regular lattice Λ with the lattice spacing $\mathfrak{a}$ and the topology of a ring imposed by the use of periodic boundary conditions. The length of the ring is L with $L = N\,\mathfrak{a}$. (b) One-dimensional Brillouin zone $\Lambda^{\star}_{\mathrm{BZ}}$ defined by Eq. (2.10c) with the lattice spacing $2\pi/L$ and the topology of a ring imposed by the use of periodic boundary conditions. The wave numbers in the Brillouin zone are depicted by vertical lines. The topology of the ring is implied by the dashed line. The length of the ring is $2\pi/\mathfrak{a}$ in momentum space. The lattice spacing $\mathfrak{a}$ has been set to unity, and $N = 10$ was chosen.

$$\hat{c}^{\dagger}_{i+N,\sigma} = \hat{c}^{\dagger}_{i,\sigma}\, e^{-\mathrm{i}\phi_{\sigma}}, \qquad \hat{c}_{i+N,\sigma} = e^{+\mathrm{i}\phi_{\sigma}}\, \hat{c}_{i,\sigma}, \tag{2.2d}$$

where the twisting parameters $0 \le \phi_{\uparrow,\downarrow} < 2\pi$ have the same effect as (spin dependent) magnetic fluxes through the ring. The choice $\phi_{\uparrow,\downarrow} = 0$ corresponds to the periodic boundary conditions

$$\hat{c}^{\dagger}_{i+N,\sigma} = \hat{c}^{\dagger}_{i,\sigma}, \qquad \hat{c}_{i+N,\sigma} = \hat{c}_{i,\sigma}. \tag{2.2e}$$

2.4.1 Symmetries

Hamiltonian (2.2a) with real-valued nearest-neighbor hopping amplitudes t_i has the following symmetries: (i) time-reversal symmetry (TRS), (ii) particle-hole symmetry (PHS), (iii) chiral symmetry (CHS), and (iv) global charge and spin-1/2 gauge symmetry. In order to establish what is meant by these symmetries, let a and b be any pair of complex numbers, let $i = 1, \cdots, N$ be any site from the chain, and let $\sigma =\uparrow,\downarrow$ be any spin quantum number. We shall choose N to be an even integer.

(i) **TRS** is the property that the Hamiltonian (2.2a) with real-valued nearest-neighbor hoppings t_i and the boundary conditions (2.2e) are unchanged under the antilinear and on-site transformation defined by[8] (**Exercise** 2.1)

$$a\,\hat{c}^{\dagger}_{i,\sigma} + b\,\hat{c}_{i,\sigma} \mapsto (-1)^{\sigma}\left(a^{*}\,\hat{c}^{\dagger}_{i,-\sigma} + b^{*}\,\hat{c}_{i,-\sigma}\right). \tag{2.3a}$$

The multiplicative factor $(-1)^{\sigma} = +1$ if $\sigma =\uparrow$ while it is $(-1)^{\sigma} = -1$ if $\sigma =\downarrow$. It is essential to implement reversal of time through Eq. (2.3a) whenever spin-orbit coupling is present. However, when the full spin-1/2 rotation symmetry is neither

[8] Complex conjugation is here denoted by the use of $*$ as an upper index. Moreover, if $\sigma =\uparrow,\downarrow$, then $-\sigma \equiv\downarrow,\uparrow$.

broken explicitly nor spontaneously, the spin label only contributes a degeneracy factor of two. If so, we may treat the index $\sigma =\uparrow,\downarrow$ as inert when defining reversal of time, that is, we may implement reversal of time by (**Exercise** 2.1)

$$a\,\hat{c}^{\dagger}_{i,\sigma} + b\,\hat{c}_{i,\sigma} \mapsto a^{*}\,\hat{c}^{\dagger}_{i,\sigma} + b^{*}\,\hat{c}_{i,\sigma}. \tag{2.3b}$$

The reversal of time (2.3b) effectively corresponds to that of a spinless fermion. An alternative mechanism by which we may ignore the spin of an electron is when the spin is polarized along a preferred quantization axis. This is the case in the quantum Hall effect due to a very large magnetic field or in a spin-triplet superconductor. As long as the spin-1/2 rotation symmetry holds or all spins are polarized along some quantization axis, we adopt the spinless realization (2.3b) of time reversal. Otherwise, we adopt the realization (2.3a) of time reversal. Throughout Chapter 2, we assume that the spin-1/2 rotation symmetry holds and we represent reversal of time by Eq. (2.3b).

(ii) **PHS** is a consequence of the fact that the many-body Hamiltonian (2.2a) anticommutes with the antilinear and on-site transformation defined by

$$a\,\hat{c}^{\dagger}_{i,\sigma} + b\,\hat{c}_{i,\sigma} \mapsto a^{*}\,\hat{c}_{i,\sigma} + b^{*}\,\hat{c}^{\dagger}_{i,\sigma}. \tag{2.4a}$$

Because N is chosen to be even so that $(-1)^{N} = 1$, PHS is implemented as a symmetry of the many-body Hamiltonian (2.2a) and the boundary conditions (2.2e) through the antilinear and on-site transformation defined by

$$a\,\hat{c}^{\dagger}_{i,\sigma} + b\,\hat{c}_{i,\sigma} \mapsto (-1)^{i}\left(a^{*}\,\hat{c}_{i,\sigma} + b^{*}\,\hat{c}^{\dagger}_{i,\sigma}\right), \tag{2.4b}$$

for the antilinear transformation (2.4b) commutes with the many-body Hamiltonian (2.2a) and leaves the boundary conditions (2.2e) unchanged.

(iii) **CHS** is a consequence of the fact that the many-body Hamiltonian (2.2a) anticommutes with the linear and on-site transformation defined by[9]

$$a\,\hat{c}^{\dagger}_{i,\sigma} + b\,\hat{c}_{i,\sigma} \mapsto (-1)^{i}\left(a\,\hat{c}^{\dagger}_{i,\sigma} + b\,\hat{c}_{i,\sigma}\right). \tag{2.5a}$$

CHS is implemented as a symmetry of the many-body Hamiltonian (2.2a) and the boundary conditions (2.2e) through the linear and on-site transformation

$$a\,\hat{c}^{\dagger}_{i,\sigma} + b\,\hat{c}_{i,\sigma} \mapsto (-1)^{i}\left(a\,\hat{c}_{i,\sigma} + b\,\hat{c}^{\dagger}_{i,\sigma}\right), \tag{2.5b}$$

for the linear transformation (2.5b) commutes with the many-body Hamiltonian (2.2a) and leaves the boundary conditions (2.2e) unchanged.

(iv) **Global charge and spin-1/2 gauge symmetry** is the property that Hamiltonian (2.2a) and boundary conditions (2.2e) are unchanged by the linear and continuous transformation

[9] Hereto, we make use of the assumption that N is even.

$$a\,\hat{c}^{\dagger}_{i,\sigma} + b\,\hat{c}_{i,\sigma} \mapsto a\,\hat{c}^{\dagger}_{i,\sigma'}\,U^{*}_{\sigma\sigma'} + b\,U_{\sigma\sigma'}\,\hat{c}_{i,\sigma'} \tag{2.6}$$

for any matrix U, whose four matrix elements $U_{\sigma\sigma'}$ are labeled by $\sigma, \sigma' =\uparrow, \downarrow$, that belongs to the group U(2) of unitary 2×2 matrices.

One verifies that transformations (2.3b), (2.4b), (2.5b), and (2.6) commute pairwise (**Exercise** 2.2). Correspondingly, the many-body eigenstates of Hamiltonian (2.2a) subject to the boundary conditions (2.2e) can be chosen to be simultaneous eigenstates of the operators representing reversal of time, exchange of particle and holes, the chiral transformation, the global U(1) gauge rotation associated with the conservation of the electron number (charge conservation), and the global SU(2) gauge rotation associated with the conservation of the spin-1/2.

We close Section 2.4 with the following two observations.

1. Any two of the three many-body symmetries (2.3b), (2.4b), and (2.5b), if composed pairwise, yield the third symmetry.
2. Either the particle-hole or the chiral single-particle spectral symmetries implies that all single-particle eigenstates of Hamiltonian (2.2a) with nonvanishing eigenvalues can be arranged in pairs of opposite sign, whereby either a particle-hole transformation or a chiral transformation maps one member of this pair into the other (**Exercise** 2.3).

2.5 Fermiology on the Lattice

Many properties of Hamiltonian (2.2a) depend sensitively on the dependence of the hopping amplitude t_i on the site index i. For example, if Hamiltonian (2.2a) obeys periodic boundary conditions, it then acquires a translation symmetry in the uniform limit $t_i = t$ for $i = 1, \cdots, N$, that is, Hamiltonian (2.2a) is then invariant under the transformation

$$a\,\hat{c}^{\dagger}_{i,\sigma} + b\,\hat{c}_{i,\sigma} \mapsto a\,\hat{c}^{\dagger}_{i+n,\sigma} + b\,\hat{c}_{i+n,\sigma}, \qquad a, b \in \mathbb{C}, \tag{2.7}$$

for any integer n (not only N as dictated by periodic boundary conditions). As we shall see, a model that explains the observed spectral properties of polyacetylene is one for which the hopping amplitudes are staggered, that is,

$$t_i = t + (-1)^i\,\delta t, \tag{2.8}$$

where the dimensionless positive number $|\delta t/t|$ quantifies the degree of dimerization relative to the band width, while the sign $\mathrm{sgn}\,(\delta t/t)$ selects how the translation symmetry in the uniform limit $\delta t = 0$ is broken by one lattice spacing.

We shall proceed in two steps. First, we consider the case $\delta t = 0$ of a uniform hopping. Second, we consider the staggered case $\delta t \neq 0$. The case $\delta t = 0$ of a uniform hopping is solved by going to momentum space. This solution can be presented in a way useful for solving the generic staggered case $\delta t \neq 0$.

2.5.1 Uniform Nearest-Neighbor Hopping along a Chain

When the hopping is uniform, that is, when $t_i = t > 0$ for all sites i, Hamiltonian (2.2a) becomes

$$\widehat{H} := -t \sum_{i=1}^{N} \sum_{\sigma=\uparrow,\downarrow} \left(\hat{c}^{\dagger}_{i,\sigma} \, \hat{c}_{i+1,\sigma} + \text{H.c.} \right). \tag{2.9}$$

We impose the periodic boundary conditions (2.2e). Hence, invariance under the discrete translations (2.7) holds.

Hamiltonian (2.9) can be diagonalized with the help of the unitary Fourier transformation

$$\hat{c}^{\dagger}_{i,\sigma} = \frac{1}{\sqrt{N}} \sum_{k \in \Lambda^{\star}_{\text{BZ}}} \hat{c}^{\dagger}_{k,\sigma} \, e^{-ik\,i}, \qquad \hat{c}_{i,\sigma} = \frac{1}{\sqrt{N}} \sum_{k \in \Lambda^{\star}_{\text{BZ}}} e^{+ik\,i} \, \hat{c}_{k,\sigma}, \tag{2.10a}$$

where

$$\hat{c}^{\dagger}_{k,\sigma} := \frac{1}{\sqrt{N}} \sum_{i=1}^{N} \hat{c}^{\dagger}_{i,\sigma} \, e^{+ik\,i}, \qquad \hat{c}_{k,\sigma} := \frac{1}{\sqrt{N}} \sum_{i=1}^{N} e^{-ik\,i} \, \hat{c}_{i,\sigma}. \tag{2.10b}$$

In order to enforce the boundary conditions, we must choose k from a set called the Brillouin zone. The Brillouin zone that is compatible with the periodic boundary conditions (2.2e) [**Exercise** 2.4, see also Figure 2.8(b)] is the set[10]

$$\Lambda^{\star}_{\text{BZ}} := \left\{ k = \frac{2\pi}{N} n, \; n = -\left\lfloor \frac{N}{2} \right\rfloor + 1, -\left\lfloor \frac{N}{2} \right\rfloor + 2, \cdots, -\left\lfloor \frac{N}{2} \right\rfloor + N \right\}. \tag{2.10c}$$

The operators $\hat{c}^{\dagger}_{k,\sigma}$ and $\hat{c}_{k,\sigma}$ create and annihilate a single-particle electronic state with quasi-momentum k and spin quantum number σ, respectively.[11] They obey the anticommuting relations (**Exercise** 2.5)

$$\{\hat{c}_{k,\sigma}, \hat{c}^{\dagger}_{k',\sigma'}\} = \delta_{k,k'} \, \delta_{\sigma,\sigma'}, \qquad \{\hat{c}_{k,\sigma}, \hat{c}_{k',\sigma'}\} = \{\hat{c}^{\dagger}_{k',\sigma'}, \hat{c}^{\dagger}_{k,\sigma}\} = 0. \tag{2.10d}$$

Working in momentum space yields the more economic representation (**Exercise** 2.6)

$$\widehat{H} = \sum_{k \in \Lambda^{\star}_{\text{BZ}}} \sum_{\sigma=\uparrow,\downarrow} \varepsilon_k \, \hat{c}^{\dagger}_{k,\sigma} \, \hat{c}_{k,\sigma} \tag{2.11a}$$

of Hamiltonian (2.9) obeying the periodic boundary conditions (2.2e). The single-particle energy dispersion relation ε_k is here twofold degenerate. In the thermodynamic limit $N \to \infty$, it is given by

$$\varepsilon_k = -2t \, \cos k, \qquad |k| \leq \pi. \tag{2.11b}$$

[10] The floor function $\lfloor \cdot \rfloor : \mathbb{R} \to \mathbb{Z}$ returns with $\lfloor x \rfloor$ the largest integer smaller than or equal to the real number x.

[11] We are using units with $\hbar = \mathfrak{a} = 1$, where $\hbar$ is the Planck constant and $\mathfrak{a}$ is the lattice spacing.

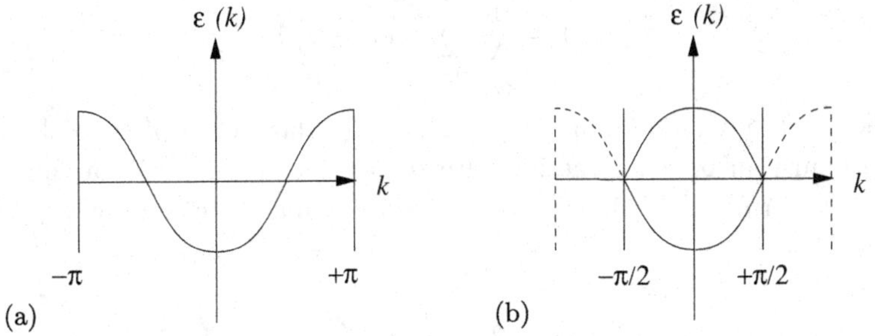

Figure 2.9 (a) The single-particle energy dispersion (2.11b) of the one-dimensional nearest-neighbor hopping Hamiltonian with uniform hopping amplitude t defined in Eq. (2.11a) over the Brillouin zone defined in Eq. (2.10c) in the thermodynamic limit $N \to \infty$. (b) The same dispersion but folded over half the Brillouin zone as is done in Eq. (2.27).

The single-particle energy dispersion (2.11b) is displayed in Figure 2.9(a). The minimum and maximum energies are $-2t$ and $+2t$, respectively, hence the bandwidth, the range of the energies in the band, is $4t$.

At zero temperature, the many-body ground state for $1 \leq N_{\mathrm{e}} \leq 2N$ electrons is obtained by filling the N_{e} single-particle states with the lowest possible energies. These filled states form the Fermi sea. For example, if

$$N = 1 + 2n, \qquad N_{\mathrm{e}} = 2 + 4n_{\mathrm{e}} \tag{2.12a}$$

with $n \geq n_{\mathrm{e}} > 0$ integers, then the ground state (the Fermi sea)

$$|\Psi_{\mathrm{FS}}\rangle \equiv \prod_{k\in\Lambda^{\star}_{\mathrm{BZ}}}^{|k|\leq k_{\mathrm{F}}} \hat{c}^{\dagger}_{+k,\uparrow}\,\hat{c}^{\dagger}_{-k,\downarrow}\,|0\rangle \tag{2.12b}$$

of Hamiltonian (2.11) is obtained by filling successively all $N_{\mathrm{e}}/2$ Kramers' degenerate pairs

$$|k,\sigma\rangle = \hat{c}^{\dagger}_{k,\sigma}|0\rangle, \qquad |-k,-\sigma\rangle = \hat{c}^{\dagger}_{-k,-\sigma}|0\rangle, \tag{2.12c}$$

of single-particle states starting from the bottom of the single-particle energy dispersion (2.11b) all the way up to the Fermi energy

$$\varepsilon_{\mathrm{F}} := -2t\,\cos k_{\mathrm{F}}, \qquad k_{\mathrm{F}} := \frac{2\pi}{N} \times n_{\mathrm{e}}. \tag{2.12d}$$

The many-body ground-state energy of the Fermi sea (2.12b) is (**Exercise** 2.7)

$$E_{\mathrm{FS}} = 2 \sum_{k\in\Lambda^{\star}_{\mathrm{BZ}}}^{|k|\leq k_{\mathrm{F}}} \varepsilon_{k}. \tag{2.13}$$

It is common practice to introduce the single-particle density of states per site (**Exercise** 2.7)

$$\nu(\varepsilon) := \frac{1}{N} \sum_{k \in \Lambda^{\star}_{\mathrm{BZ}}} \delta(\varepsilon - \varepsilon_k). \tag{2.14}$$

Either the thermodynamic limit $N \to \infty$ or a regularization of the delta-function (say by a Gaussian or a Lorentzian) are required to render $\nu(\varepsilon)$ a smooth function of ε. In terms of $\nu(\varepsilon)$, the many-body ground-state energy E_{FS} becomes (**Exercise** 2.7)

$$E_{\mathrm{FS}} = 2\,N \int\limits_{-\infty}^{\varepsilon_{\mathrm{F}}} \mathrm{d}\varepsilon\, \nu(\varepsilon)\, \varepsilon. \tag{2.15}$$

The Fermi sea (2.12b) is an eigenstate of the many-body momentum operator

$$\widehat{P} := \sum_{k \in \Lambda^{\star}_{\mathrm{BZ}}} \sum_{\sigma=\uparrow,\downarrow} k\, \hat{c}^{\dagger}_{k,\sigma}\, \hat{c}_{k,\sigma} \tag{2.16}$$

with vanishing total momentum and an eigenstate of the many-body spin operator[12]

$$\widehat{\boldsymbol{S}} := \frac{1}{2} \sum_{k \in \Lambda^{\star}_{\mathrm{BZ}}} \sum_{\sigma,\sigma'=\uparrow,\downarrow} \hat{c}^{\dagger}_{k,\sigma}\, \boldsymbol{\sigma}_{\sigma\sigma'} \hat{c}_{k,\sigma'} \tag{2.17}$$

with vanishing total spin, that is, a spin singlet. This conclusion is independent of the choice (2.12a) in the thermodynamic limit $N \to \infty$, $N_{\mathrm{e}} \to \infty$, with the ratio $N_{\mathrm{e}}/N \leq 2$ held fixed (**Exercise** 2.8).

The Fermi sea (2.12) describes two distinct states of matter depending on the value taken by the filling fraction

$$\nu_{\mathrm{e}} := \frac{N_{\mathrm{e}}}{2 \times N}. \tag{2.18}$$

When $\nu_{\mathrm{e}} = 0$ or $\nu_{\mathrm{e}} = 1$, the Fermi sea either fails to fill any single-particle state or completely fills all single-particle states, respectively, as is shown for the latter case in Figure 2.10(a). The action of removing one electron with momentum k and spin σ and creating one electron with momentum $k' \neq k$ or spin $\sigma' \neq \sigma$ annihilates this Fermi sea. Consequently, this Fermi sea does not couple to any perturbation that interacts with the electrons through the operator $\hat{c}^{\dagger}_{k',\sigma'}\, \hat{c}_{k,\sigma}$. In turn, this rules out any interaction between this Fermi sea and photons, muons, or neutrons, that is, this Fermi sea is an insulating and magnetically inert ground state in that it is not susceptible to supporting either charge or spin excitations at temperature zero. At the level of single-particle physics, the insulating character of this Fermi sea follows from the fact that the single-particle Bloch state at the bottom or maximum of the band is nondegenerate with respect to the momentum quantum number.

When $0 < \nu_{\mathrm{e}} < 1$, the Fermi sea partially fills the available single-particle Bloch states, as is illustrated at half-filling in Figure 2.10(b). At the corresponding Fermi

[12] The vector $\boldsymbol{\sigma}$ is made of the three Pauli matrices

$$\sigma_1 = \begin{pmatrix} 0 & 1 \\ 1 & 0 \end{pmatrix}, \quad \sigma_2 = \begin{pmatrix} 0 & -\mathrm{i} \\ +\mathrm{i} & 0 \end{pmatrix}, \quad \sigma_3 = \begin{pmatrix} +1 & 0 \\ 0 & -1 \end{pmatrix}.$$

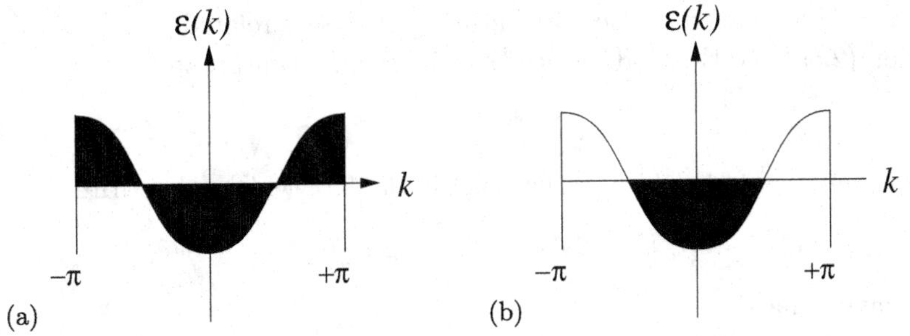

Figure 2.10 (a) Insulating Fermi sea when the electronic density per spin (2.18) is unity. (b) Metallic Fermi sea when the electronic density per spin (2.18) is 1/2.

energy (2.12d), the single-particle Bloch states are fourfold degenerate, whereby a degeneracy of two originates from the electronic spin and another degeneracy of two originates from the periodicity of the single-particle energy dispersion over the Brillouin zone [recall Eq. (2.12c)]. The action of removing one electron with momentum k and spin σ and creating one electron with momentum $k' \neq k$ or spin $\sigma' \neq \sigma$ does not necessarily annihilate this Fermi sea. This Fermi sea is a metallic ground state, for it is susceptible to supporting charge or spin excitations at temperature zero when perturbed by photons, muons, or neutrons for examples. Moreover, the susceptibilities of the Fermi sea at zero temperature are solely controlled by the gradient of the single-particle dispersion evaluated at the Fermi surface, whereby the Fermi surface is defined by the set of all momenta from the Brillouin zone whose single-particle energy equals the Fermi energy (2.12d).[13]

The model (2.2) predicts a metallic ground state at any partial filling of the band depicted in Figure 2.9(a). This is not consistent with the fact that polyacetylene is a semiconductor, that is, it is an insulator at zero temperature, with an energy gap of order 1.8 eV that is smaller than the π-band width of order 12 eV. However, we shall see that it is possible to modify Hamiltonian (2.2a) by the introduction of a second energy scale besides the π-band width $4t \sim 12\,\text{eV}$ that opens a single-particle energy gap $2\Delta \sim 1.8\,\text{eV}$ within the band depicted in Figure 2.9(a).

2.5.2 Sublattice Grading and Spectral Folding

Before we identify those perturbations of Hamiltonian (2.11a) that open a single-particle energy gap and which ones are compatible with the energy gap measured in polyacetylene, we first observe that the dispersion (2.11b) has the mirror symmetry

$$\varepsilon_k = -\varepsilon_{k\pm\pi}. \tag{2.19}$$

[13] This definition of a Fermi surface is independent of the dimensionality d of the Brillouin zone. However, the terminology of Fermi surface is biased in favor of a metal such as copper. For a d-dimensional Brillouin zone, the Fermi surface is an n-dimensional sub-manifold of the Brillouin zone with the integer n ranging from zero when the Fermi surface is made of a set of isolated points, as it must always be when $d = 1$, to $d - 1$.

Second, we observe that the momentum π is closely related to the chiral transformation (2.5a). To this end, notice that for the momentum π,

$$e^{-\mathrm{i}\pi\, i} = (-1)^i. \tag{2.20}$$

We may then recast the chiral transformation (2.5a), which we rewrite

$$\hat{c}^\dagger_{i,\sigma} \mapsto \hat{c}'^\dagger_{i,\sigma} := (-1)^i\, \hat{c}^\dagger_{i,\sigma}, \qquad \hat{c}_{i,\sigma} \mapsto \hat{c}'_{i,\sigma} := (-1)^i\, \hat{c}_{i,\sigma}, \tag{2.21}$$

as the transformation

$$\hat{c}'^\dagger_{k+\pi,\sigma} = \hat{c}^\dagger_{k,\sigma}, \qquad \hat{c}'_{k+\pi,\sigma} = \hat{c}_{k,\sigma}, \tag{2.22}$$

in momentum space.

In turn, insertions of Eqs. (2.22) and (2.19) into Hamiltonian (2.11a) deliver

$$\begin{aligned}
\widehat{H} &= \sum_{k\in\Lambda^\star_{\mathrm{BZ}}} \sum_{\sigma=\uparrow,\downarrow} \varepsilon_k\, \hat{c}^\dagger_{k,\sigma}\, \hat{c}_{k,\sigma} \\
&= \sum_{k\in\Lambda^\star_{\mathrm{BZ}}} \sum_{\sigma=\uparrow,\downarrow} \varepsilon_k\, \hat{c}'^\dagger_{k+\pi,\sigma}\, \hat{c}'_{k+\pi,\sigma} \\
&= \sum_{k\in\Lambda^\star_{\mathrm{BZ}}} \sum_{\sigma=\uparrow,\downarrow} \varepsilon_{k-\pi}\, \hat{c}'^\dagger_{k,\sigma}\, \hat{c}'_{k,\sigma} \\
&= \sum_{k\in\Lambda^\star_{\mathrm{BZ}}} \sum_{\sigma=\uparrow,\downarrow} (-\varepsilon_k)\, \hat{c}'^\dagger_{k,\sigma}\, \hat{c}'_{k,\sigma} \\
&= - \sum_{k\in\Lambda^\star_{\mathrm{BZ}}} \sum_{\sigma=\uparrow,\downarrow} \varepsilon_k\, \hat{c}^\dagger_{k-\pi,\sigma}\, \hat{c}_{k-\pi,\sigma}.
\end{aligned} \tag{2.23}$$

Equation (2.23) is the representation in momentum space of the fact that Hamiltonian (2.11a) anticommutes with the chiral transformation (2.21).

To explore a useful consequence of Eq. (2.23), we introduce the operator-valued two-component creation

$$\hat{\chi}^\dagger_{k,\sigma} := \begin{pmatrix} \hat{c}^\dagger_{k,\sigma} \\ \hat{c}^\dagger_{k-\pi,\sigma} \end{pmatrix}^{\mathsf{T}} \equiv \begin{pmatrix} \hat{c}^\dagger_{k,\sigma} & \hat{c}^\dagger_{k-\pi,\sigma} \end{pmatrix}, \tag{2.24a}$$

and annihilation

$$\hat{\chi}_{k,\sigma} := \begin{pmatrix} \hat{c}_{k,\sigma} \\ \hat{c}_{k-\pi,\sigma} \end{pmatrix}, \tag{2.24b}$$

spinors for each momentum k and for each spin quantum number σ, respectively. Adding Eqs. (2.11a) and (2.23) and dividing by two then delivers

$$\widehat{H} = \sum_{k\in\Lambda^\star_{\mathrm{BZ}}}^{|k|\leq\pi/2} \sum_{\sigma=\uparrow,\downarrow} \varepsilon_k\, \hat{\chi}^\dagger_{k,\sigma}\, \tau_3\, \hat{\chi}_{k,\sigma}. \tag{2.25}$$

Here, we have absorbed the division by two by restricting the Brillouin zone through the halving condition $|k| \leq \pi/2$, while the addition of the first and last lines of Eq. (2.23) is accounted for by the introduction of the 2×2 Pauli matrix τ_3. This matrix is independent of the momentum k and of the spin σ.[14] Hamiltonian (2.25) is of the form

$$\widehat{H} = \sum_{k\in\Lambda^{\star}_{\mathrm{BZ}}}^{|k|\leq\pi/2} \sum_{\sigma=\uparrow,\downarrow} \hat{\chi}^{\dagger}_{k,\sigma}\,\mathcal{H}_k\,\hat{\chi}_{k,\sigma}, \tag{2.26a}$$

with the 2×2 momentum-resolved single-particle Hermitean matrix

$$\mathcal{H}_k = \varepsilon_k\,\tau_3. \tag{2.26b}$$

Its eigenvalues are obtained by diagonalizing the matrix $\mathcal{H}_k$. There are two bands with opposite energies, because the matrix τ_3 has two eigenvalues ± 1. The energies of these two bands, labeled by $\pm$, in the thermodynamic limit ($N\to\infty$) are

$$\varepsilon_{k,\pm} = \pm|\varepsilon_k|, \qquad |k| \leq \pi/2. \tag{2.27}$$

The energy bands, for the halved Brillouin zone, are shown in Figure 2.9(b). For each single-particle eigenstate with energy $+\varepsilon$, there is a mirror single-particle eigenstate with energy $-\varepsilon$. This spectral symmetry is a consequence of the fact that the chiral transformation (2.5a), which is represented by (**Exercise** 2.9)

$$\hat{\chi}_{k,\sigma} \mapsto \tau_1\,\hat{\chi}_{k,\sigma} \tag{2.28a}$$

in the basis (2.24), obeys the anticommutation law (**Exercise** 2.10)

$$\mathcal{H}_k \mapsto \tau_1\,\mathcal{H}_k\,\tau_1 = -\mathcal{H}_k. \tag{2.28b}$$

We are now going to relate explicitly the spectral symmetry (2.28) to the existence of two interpenetrating sublattices of the linear chain and tie it to the chiral transformation (2.5a).

The one-dimensional chain

$$\Lambda := \{\boldsymbol{r}_i := i\,\boldsymbol{e}_1, \quad i = 1,\cdots,N\} \tag{2.29a}$$

with N even (**Exercise** 2.11) can be divided equally into the sublattice A defined by

$$\Lambda^{A} := \{\boldsymbol{r}_i := i\,\boldsymbol{e}_1, \quad i = 1,3,\cdots,N-1\} \tag{2.29b}$$

and the sublattice B defined by

$$\Lambda^{B} := \{\boldsymbol{r}_i := i\,\boldsymbol{e}_1, \quad i = 2,4,\cdots,N\}. \tag{2.29c}$$

[14] We may introduce a new set of Pauli matrices for each new two-valued degree of freedom (or grading). Together with the identity matrix, there are four matrices for each grading. The gradings are combined using tensor products, for example one may define 16 4×4 Hermitean matrices of the form $\sigma_\mu\otimes\tau_\nu$ with $\mu,\nu = 0, 1, 2, 3$, where σ_μ acts on the spin degrees of freedom and τ_μ acts on the two band indices $\pm$ that have been introduced by folding the dispersion on half of the Brillouin zone $\Lambda^{\star}_{\mathrm{BZ}}$.

Sublattices Λ^A and Λ^B each have the same number of sites $N/2$, do not intersect as sets, and their union as sets yields Λ. In a ring geometry, each site of sublattice Λ^A has two sites from sublattice Λ^B as nearest neighbors. We define the Brillouin zone associated with the sublattice Λ^A by (recall footnote 10)

$$\Lambda^{A\star}_{\mathrm{BZ}} := \left\{ k = \frac{2\pi}{N} n, \quad n = -\left\lfloor \frac{N}{4} \right\rfloor + 1, -\left\lfloor \frac{N}{4} \right\rfloor + 2, \cdots, -\left\lfloor \frac{N}{4} \right\rfloor + \frac{N}{2} \right\}. \tag{2.30}$$

Any $k \in \Lambda^{A\star}_{\mathrm{BZ}}$ corresponds to a $k \in \Lambda^{\star}_{\mathrm{BZ}}$ in the range $|k| \leq \pi/2$. In other words, $\Lambda^{A\star}_{\mathrm{BZ}}$ is the halved Brillouin zone shown in Figure 2.9(b).

If we introduce the notation

$$\hat{a}_{j,\sigma} := \hat{c}_{2j-1,\sigma}, \qquad \hat{b}_{j,\sigma} := \hat{c}_{2j,\sigma}, \tag{2.31a}$$

for $j = 1, \cdots, N/2$ and take advantage of the periodic boundary conditions, we can write

$$\widehat{H} = -t \sum_{j=1}^{N/2} \sum_{\sigma=\uparrow,\downarrow} \left[\hat{a}^{\dagger}_{j,\sigma} \left(\hat{b}_{j,\sigma} + \hat{b}_{j-1,\sigma} \right) + \hat{b}^{\dagger}_{j,\sigma} \left(\hat{a}_{j+1,\sigma} + \hat{a}_{j,\sigma} \right) \right]. \tag{2.31b}$$

To diagonalize this Hamiltonian, we introduce the unitary Fourier transformations

$$\hat{a}_{j,\sigma} := \sqrt{\frac{2}{N}} \sum_{k \in \Lambda^{A\star}_{\mathrm{BZ}}} e^{+\mathrm{i}k\,2j}\, \hat{a}_{k,\sigma}, \qquad \hat{a}_{k,\sigma} := \sqrt{\frac{2}{N}} \sum_{j=1}^{N/2} e^{-\mathrm{i}k\,2j}\, \hat{a}_{j,\sigma}, \tag{2.32a}$$

and

$$\hat{b}_{j,\sigma} := \sqrt{\frac{2}{N}} \sum_{k \in \Lambda^{A\star}_{\mathrm{BZ}}} e^{+\mathrm{i}k\,2j}\, \hat{b}_{k,\sigma}, \qquad \hat{b}_{k,\sigma} := \sqrt{\frac{2}{N}} \sum_{j=1}^{N/2} e^{-\mathrm{i}k\,2j}\, \hat{b}_{j,\sigma}. \tag{2.32b}$$

The presence of $2j$ instead of j in the phase factors reflects the fact that the lattice spacing of sublattice Λ^A is twice as large as that of lattice Λ. Now,

$$\begin{aligned} \widehat{H} &= -t \sum_{j=1}^{N/2} \sum_{\sigma=\uparrow,\downarrow} \left[\hat{b}^{\dagger}_{j,\sigma} \left(\hat{a}_{j,\sigma} + \hat{a}_{j+1,\sigma} \right) + \mathrm{H.c.} \right] \\ &= -t \sum_{k \in \Lambda^{A\star}_{\mathrm{BZ}}} \sum_{\sigma=\uparrow,\downarrow} \left[\left(1 + e^{+2\mathrm{i}k}\right) \hat{b}^{\dagger}_{k,\sigma} \hat{a}_{k,\sigma} + \left(1 + e^{-2\mathrm{i}k}\right) \hat{a}^{\dagger}_{k,\sigma} \hat{b}_{k,\sigma} \right]. \end{aligned} \tag{2.33}$$

Hamiltonian (2.33) is brought to the representation

$$\widehat{H} := \sum_{k \in \Lambda^{A\star}_{\mathrm{BZ}}} \sum_{\sigma=\uparrow,\downarrow} \hat{\psi}^{\dagger}_{k,\sigma}\, \widetilde{\mathcal{H}}_k\, \hat{\psi}_{k,\sigma}, \tag{2.34a}$$

where the sublattice structure can be encoded by two degrees of freedom (a 2×2 grading) with the help of the two-component spinors

$$\hat{\psi}^{\dagger}_{k,\sigma} := \begin{pmatrix} \hat{a}^{\dagger}_{k,\sigma} & \hat{b}^{\dagger}_{k,\sigma} \end{pmatrix}, \qquad \hat{\psi}_{k,\sigma} := \begin{pmatrix} \hat{a}_{k,\sigma} \\ \hat{b}_{k,\sigma} \end{pmatrix}, \tag{2.34b}$$

and the 2×2 momentum-resolved single-particle Hermitean matrix is

$$\widetilde{\mathcal{H}}_k := -t\,[1+\cos(2k)]\,\tau_1 - t\,\sin(2k)\,\tau_2. \tag{2.34c}$$

The range of allowed k entering Hamiltonian (2.25) is identical to that entering Hamiltonian (2.34a). The single-particle energies in Eqs. (2.25) and (2.34a) must also be identical, for Eqs. (2.25) and (2.34a) only differ by a choice of basis in the sublattice grading. This claim is verified explicitly in **Exercise** 2.12 where it is shown that the eigenvalues of the single-particle Hamiltonian (2.34c) are

$$\varepsilon_{k,\pm} = \pm 2t\cos k, \qquad k \in \Lambda^{A\star}_{\mathrm{BZ}}. \tag{2.34d}$$

For each single-particle eigenstate with energy $+\varepsilon$, there is a mirror single-particle eigenstate with $-\varepsilon$. This spectral symmetry is a consequence of the fact that the chiral transformation (2.5a), which is represented by

$$\hat{\psi}_{k,\sigma} \mapsto (-\tau_3)\,\hat{\psi}_{k,\sigma} \tag{2.35a}$$

in the basis (2.34b), obeys the anticommutation law

$$\widetilde{\mathcal{H}}_k \mapsto (-\tau_3)\,\widetilde{\mathcal{H}}_k\,(-\tau_3) = -\widetilde{\mathcal{H}}_k. \tag{2.35b}$$

Moreover, this spectral symmetry is a consequence of the many-body chiral symmetry

$$\widehat{U}_{\mathrm{ch}}\,\widehat{H}\,\widehat{U}^{\dagger}_{\mathrm{ch}} = \widehat{H}, \tag{2.36a}$$

under the unitary transformation

$$\widehat{U}_{\mathrm{ch}}\left(c\,\hat{b}_{k,\sigma} + d\,\hat{a}_{k,\sigma}\right)\widehat{U}^{\dagger}_{\mathrm{ch}} = c\,\hat{a}^{\dagger}_{k,\sigma} - d\,\hat{b}^{\dagger}_{k,\sigma} \tag{2.36b}$$

for any pair of complex-valued numbers c and d.

2.5.3 Single-Particle Gap

We are equipped to analyze how it is possible to open a single-particle energy gap in the dispersion (2.27) while lowering the discrete translation symmetry (2.7) to that under

$$a\,\hat{c}^{\dagger}_{i,\sigma} + b\,\hat{c}_{i,\sigma} \mapsto a\,\hat{c}^{\dagger}_{i\pm 2n,\sigma} + b\,\hat{c}_{i\pm 2n,\sigma}, \qquad a,b\in\mathbb{C}, \qquad n\in\mathbb{Z}, \tag{2.37}$$

for any $i \in \Lambda$.[15] We begin from the noninteracting, electronic, and two-orbital Hamiltonian

$$\widehat{H} := \sum_{k\in\Lambda^{A\star}_{\mathrm{BZ}}}\sum_{\sigma=\uparrow,\downarrow} \hat{\psi}^{\dagger}_{k,\sigma}\,\mathcal{H}_k\,\hat{\psi}_{k,\sigma} \tag{2.39a}$$

[15] Be aware that symmetry under translation by two lattice spacings of Λ is the same as symmetry under translation by one lattice spacing of Λ^{A}, that is it generates the symmetry under

$$c\,\hat{a}^{\dagger}_{j,\sigma} + d\,\hat{a}_{j,\sigma} \mapsto c\,\hat{a}^{\dagger}_{j\pm n,\sigma} + d\,\hat{a}_{j\pm n,\sigma}, \qquad c\,\hat{b}^{\dagger}_{j,\sigma} + d\,\hat{b}_{j,\sigma} \mapsto c\,\hat{b}^{\dagger}_{j\pm n,\sigma} + d\,\hat{b}_{j\pm n,\sigma}, \tag{2.38}$$

for any $c,d\in\mathbb{C}$, $n\in\mathbb{Z}$, and $j = 1,\cdots,N/2$.

with the generic 2×2 Hermitean matrix

$$\mathcal{H}_k := h_{k,\mu}\,\tau_\mu \equiv h_{k,0}\,\tau_0 + h_{k,1}\,\tau_1 + h_{k,2}\,\tau_2 + h_{k,3}\,\tau_3 \tag{2.39b}$$

in the basis (2.34b), that is, the components of the operator-valued spinors obey the fermion algebra

$$\begin{aligned}\{\hat{\psi}_{k,\sigma,\alpha}, \hat{\psi}^{\dagger}_{k',\sigma',\alpha'}\} &= \delta_{k,k'}\,\delta_{\sigma,\sigma'}\,\delta_{\alpha,\alpha'},\\ \{\hat{\psi}_{k,\sigma,\alpha}, \hat{\psi}_{k',\sigma',\alpha'}\} &= \{\hat{\psi}^{\dagger}_{k',\sigma',\alpha'}, \hat{\psi}^{\dagger}_{k,\sigma,\alpha}\} = 0,\end{aligned} \tag{2.39c}$$

for $k, k' \in \Lambda^{A\star}_{\mathrm{BZ}}$, $\sigma, \sigma' =\uparrow, \downarrow$, and $\alpha, \alpha' = A, B$. In this basis, the chiral transformation is diagonal and represented by Eq. (2.35a). Hamiltonian $\widehat{H}$ can be thought of as an operator-valued functional of the four independent real-valued functions $h_{k,0}$, $h_{k,1}$, $h_{k,2}$, and $h_{k,3}$ that are periodic across the Brillouin zone $\Lambda^{A\star}_{\mathrm{BZ}}$. Charge conservation is satisfied by $\widehat{H}$ since a creation and an annihilation operator are always paired in $\widehat{H}$. Spin-rotation symmetry is satisfied by $\widehat{H}$ since we are implicitly using the spin-SU(2) invariant Kronecker delta symbol $\delta_{\sigma,\sigma'}$ when contracting the spin quantum numbers of the electron. Conversely, model (2.39) is the most general noninteracting electronic single-orbital tight-binding model with translation by two lattice spacings of lattice Λ defined in Eq. (2.29a), charge, and spin-rotation symmetries, for the group SU(2) has only one independent quadratic Casimir invariant, namely, the Kronecker delta symbol $\delta_{\sigma,\sigma'}$.[16] Locality of $\widehat{H}$ restricts the choice of the functions $h_{k,0}$, $h_{k,1}$, $h_{k,2}$, and $h_{k,3}$ to those varying sufficiently slowly over the Brillouin zone $\Lambda^{A\star}_{\mathrm{BZ}}$ for their Fourier transforms to be exponentially peaked on the lattice Λ^{A}.

We briefly pause before proceeding with the choice of the functions $h_{k,0}$, $h_{k,1}$, $h_{k,2}$, and $h_{k,3}$ that are compatible with transport and spectroscopy in polyacetylene, as we will regularly encounter single-particle Bloch Hamiltonians of the form of $\mathcal{H}_k$ in this book. The single-particle Hamiltonian $\mathcal{H}_k$ is a Hermitean 2×2 matrix that can be diagonalized by elementary means, that is, by solving the eigenvalue equation $\mathcal{H}_k\,|k\rangle = \varepsilon_k\,|k\rangle$ for the eigenenergy ε_k and the eigenstate $|k\rangle$. However, there is a more direct method to solve this eigenvalue problem that takes full advantage of the fact that the Hamiltonian is a 2×2 Hermitean matrix.

To this end, one observes the following two facts. First, the 2×2 unit matrix τ_0 commutes with the three Pauli matrices τ_1, τ_2, and τ_3. Hence,

[16] The group SU(2) is an example of a compact simple Lie group $\mathfrak{G}$. A simple Lie group is a group whose elements are uncountable as a set with some additional structure. The structure in question is that it is possible to continuously deform any simple Lie group element into another one. For a simple Lie group, this smooth deformation is encoded by the fact that any group element is the exponential of a linear combination of the generators of the simple Lie algebra $\mathfrak{g}$ associated with the simple Lie group $\mathfrak{G}$. If the simple Lie group is compact, the structure constants of the simple Lie algebra must obey some additional constraints. For example, any element of $U \in \mathrm{SU}(2)$ can be written as $U = e^{\mathrm{i}\boldsymbol{\theta}\cdot\hat{\boldsymbol{S}}/\hbar}$ where the generators $\hat{\boldsymbol{S}} = (\hat{S}_a)$ of the simple Lie algebra are defined by their commutator $[\hat{S}_a, \hat{S}_b] = \mathrm{i}\,\hbar\,\epsilon_{abc}\,\hat{S}_c$ (the spin algebra) with $a, b, c = 1, 2, 3$ and the summation convention is implied over the repeated Latin indices, while the real numbers $\boldsymbol{\theta} = (\theta_a)$ are arbitrary. The Levi-Civita symbol represents the structure constants of the $\mathfrak{su}(2)$ simple Lie algebra. The quadratic Casimir of the $\mathfrak{su}(2)$ simple Lie algebra is proportional to $\hat{\boldsymbol{S}}^2$.

$$\mathcal{H}'_k := h_{k,0}\,\tau_0 \tag{2.40}$$

can be simultaneously diagonalized with

$$\mathcal{H}''_k := h_{k,1}\,\tau_1 + h_{k,2}\,\tau_2 + h_{k,3}\,\tau_3. \tag{2.41}$$

Second, the three Pauli matrices anticommute pairwise and square to the unit matrix, that is,

$$\tau_a\,\tau_b = \delta_{ab} + \mathrm{i}\epsilon_{abc}\,\tau_c, \tag{2.42}$$

where $a, b, c = 1, 2, 3$, summation convention over the repeated Latin indices is implied, and the symbol ϵ_{abc} denotes the Levi-Civita fully antisymmetric rank-3 tensor with the convention $\epsilon_{123} = 1$. Hence,

$$\mathcal{H}''^{\,2}_k := \left(h^2_{k,1} + h^2_{k,2} + h^2_{k,3}\right)\tau_0, \tag{2.43}$$

that is the eigenvalues of $\mathcal{H}''^{\,2}_k$ are doubly degenerate and are given by

$$h^2_{k,1} + h^2_{k,2} + h^2_{k,3}. \tag{2.44}$$

In turn, the eigenvalues of $\mathcal{H}''_k$ are nondegenerate and are given by

$$\pm\sqrt{h^2_{k,1} + h^2_{k,2} + h^2_{k,3}}. \tag{2.45}$$

The eigenvalues of $\mathcal{H}_k$ are thus

$$\varepsilon_{k,\pm} = h_{k,0} \pm |\boldsymbol{h}_k|, \qquad \boldsymbol{h}_k := \begin{pmatrix} h_{k,1} \\ h_{k,2} \\ h_{k,3} \end{pmatrix} = |\boldsymbol{h}_k| \begin{pmatrix} \cos\phi_k\,\sin\theta_k \\ \sin\phi_k\,\sin\theta_k \\ \cos\theta_k \end{pmatrix}. \tag{2.46a}$$

Moreover, if one defines the two orthonormal states

$$|k,+\rangle := \begin{pmatrix} e^{-\mathrm{i}\phi_k/2}\,\cos(\theta_k/2) \\ +e^{+\mathrm{i}\phi_k/2}\,\sin(\theta_k/2) \end{pmatrix}, \qquad |k,-\rangle := \begin{pmatrix} e^{-\mathrm{i}\phi_k/2}\,\sin(\theta_k/2) \\ -e^{+\mathrm{i}\phi_k/2}\,\cos(\theta_k/2) \end{pmatrix}, \tag{2.46b}$$

one verifies the spectral decomposition (**Exercise** 2.13)

$$\mathcal{H}_k = \varepsilon_{k,+}\,|k,+\rangle\langle k,+| + \varepsilon_{k,-}\,|k,-\rangle\langle k,-|. \tag{2.46c}$$

In other words, $|k,\pm\rangle$ is an eigenstate of $\mathcal{H}_k$ with eigenvalue $\varepsilon_{k,\pm}$.

There is a geometric interpretation to this algebraic solution of the eigenvalue problem $\mathcal{H}_k\,|k\rangle = \varepsilon_k\,|k\rangle$. Define the U(2) matrix

$$U_k := \begin{pmatrix} e^{-\mathrm{i}\phi_k/2}\cos(\theta_k/2) & e^{-\mathrm{i}\phi_k/2}\sin(\theta_k/2) \\ e^{+\mathrm{i}\phi_k/2}\sin(\theta_k/2) & -e^{+\mathrm{i}\phi_k/2}\cos(\theta_k/2) \end{pmatrix} \tag{2.47}$$

with the components of $|k,+\rangle$ as the first column and the components of $|k,-\rangle$ as the second column. This is the unitary transformation that diagonalizes the 2×2 matrix $\mathcal{H}_k$ by rotating the unit vector $\boldsymbol{h}_k/|\boldsymbol{h}_k|$ with the inclination θ_k and

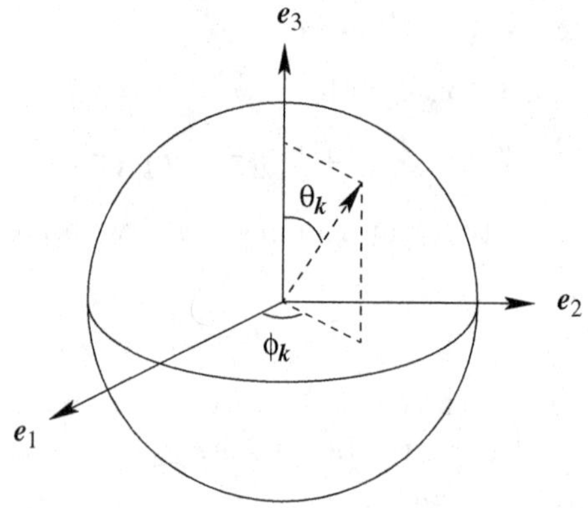

Figure 2.11 The unit (Bloch) sphere in Pauli space. The unit vector $\boldsymbol{h}_k/|\boldsymbol{h}_k|$ is represented by the polar angle θ_k and the azimuthal angle ϕ_k according to Eq. (2.46a).

azimuth ϕ_k to the unit vector $(0,0,1)^{\mathsf{T}}$ in the Pauli space spanned by the three Pauli matrices τ_1, τ_2, and τ_3 that is depicted in Figure 2.11, for the identity

$$U_k^{\dagger}\,\frac{\boldsymbol{h}_k}{|\boldsymbol{h}_k|}\cdot\boldsymbol{\tau}\,U_k=\tau_3 \tag{2.48}$$

holds.

Equipped with these insights, we return to model (2.39), which we diagonalize in the χ basis of the Fock space defined by[17]

$$\hat{\psi}^{\dagger}_{k,\sigma}=:\hat{\chi}^{\dagger}_{k,\sigma}\,U^{\dagger}_{k},\qquad \hat{\psi}_{k,\sigma}=:U_k\,\hat{\chi}_{k,\sigma}, \tag{2.49}$$

where the momentum-resolved unitary 2×2 matrix U_k is given in Eq. (2.47). The orbital basis (ψ) in momentum space is thus related to the Bloch basis (χ) through the unitary transformation (2.47). In the Bloch basis (2.46), model (2.39) becomes

$$\widehat{H}:=\sum_{k\in\Lambda^{A\star}_{\mathrm{BZ}}}\sum_{\sigma=\uparrow,\downarrow}\sum_{\mathsf{a}=\pm}\hat{\chi}^{\dagger}_{k,\sigma,\mathsf{a}}\,\varepsilon_{k,\mathsf{a}}\,\hat{\chi}_{k,\sigma,\mathsf{a}}, \tag{2.50a}$$

[17] We reserve as label of the two (sublattice) components of the operator $\hat{\psi}_{k,\sigma}$ that annihilates an electron with the momentum k and spin σ quantum numbers the Greek alphabet, that is, $\hat{\psi}_{k,\sigma}\equiv(\hat{\psi}_{k,\sigma,\alpha})$ with $\alpha=A,B$. We reserve as label of the two (band) components of the operator $\hat{\chi}_{k,\sigma}$ that annihilates a Bloch state with the momentum k and spin σ quantum numbers the Latin alphabet in modern sans serif fonts, that is, $\hat{\chi}_{k,\sigma}\equiv(\hat{\chi}_{k,\sigma,\mathsf{a}})$ with $\mathsf{a}=\pm$. Equation (2.49) thus reads

$$\hat{\psi}^{\dagger}_{k,\sigma,\alpha}=\sum_{\mathsf{a}=\pm}\hat{\chi}^{\dagger}_{k,\sigma,\mathsf{a}}\left(U^{\dagger}_{k}\right)_{\mathsf{a}\alpha},\qquad \hat{\psi}_{k,\sigma,\alpha}=\sum_{\mathsf{a}=\pm}\left(U_k\right)_{\alpha\mathsf{a}}\,\hat{\chi}_{k,\sigma,\mathsf{a}},$$

for $\alpha=A,B$.

where the components of the operator-valued spinors $\hat{\chi}$ obey the fermion algebra

$$\begin{aligned}\{\hat{\chi}_{k,\sigma,\mathrm{a}}, \hat{\chi}^{\dagger}_{k',\sigma',\mathrm{a}'}\} &= \delta_{k,k'}\,\delta_{\sigma,\sigma'}\,\delta_{\mathrm{a},\mathrm{a}'},\\ \{\hat{\chi}_{k,\sigma,\mathrm{a}}, \hat{\chi}_{k',\sigma',\mathrm{a}'}\} &= \{\hat{\chi}^{\dagger}_{k',\sigma',\mathrm{a}'}, \hat{\chi}^{\dagger}_{k,\sigma,\mathrm{a}}\} = 0.\end{aligned} \tag{2.50b}$$

The Fourier transforms

$$\hat{\chi}^{\dagger}_{j,\sigma} := \sqrt{\frac{2}{N}} \sum_{k\in\Lambda^{A\star}_{\mathrm{BZ}}} \hat{\chi}^{\dagger}_{k,\sigma}\, e^{-\mathrm{i}k\,2j}, \qquad \hat{\chi}_{j,\sigma} := \sqrt{\frac{2}{N}} \sum_{k\in\Lambda^{A\star}_{\mathrm{BZ}}} e^{+\mathrm{i}k\,2j}\,\hat{\chi}_{k,\sigma}, \tag{2.51}$$

define the Wannier basis in position space of the Fock space. The Wannier basis (2.51) and the orbital basis

$$\hat{\psi}^{\dagger}_{j,\sigma} := \sqrt{\frac{2}{N}} \sum_{k\in\Lambda^{A\star}_{\mathrm{BZ}}} \hat{\psi}^{\dagger}_{k,\sigma}\, e^{-\mathrm{i}k\,2j}, \qquad \hat{\psi}_{j,\sigma} := \sqrt{\frac{2}{N}} \sum_{k\in\Lambda^{A\star}_{\mathrm{BZ}}} e^{+\mathrm{i}k\,2j}\,\hat{\psi}_{k,\sigma}, \tag{2.52}$$

can be different. They are mathematically related by a convolution in position space with a kernel that is determined by the dependence on k of the 2×2 unitary matrix U_k defined in Eq. (2.47). The orbital basis is built from single-particle states that are exponentially localized in position space. The Wannier basis is built from the Fourier transform of Bloch single-particle states. Some attributes of the Bloch single-particle states can prevent the Wannier single-particle states from being exponentially localized in position space.

Imposing on $\widehat{H}$ defined in Eq. (2.39) invariance under the reversal of time (2.3b) demands that

$$\begin{aligned} h_{-k,0} &= +h_{+k,0}, & h_{-k,1} &= +h_{+k,1},\\ h_{-k,2} &= -h_{+k,2}, & h_{-k,3} &= +h_{+k,3}.\end{aligned} \tag{2.53}$$

Imposing on $\widehat{H}$ defined in Eq. (2.39) invariance under the chiral transformation (2.5b) demands that

$$h_{k,0} = 0, \qquad h_{k,3} = 0. \tag{2.54}$$

Only the channel associated with the Pauli matrix τ_2 in the sublattice grading is available to open a single-particle energy gap. Together, the conditions for TRS (2.53) and CHS (2.54) also imply that Hamiltonian $\widehat{H}$ commutes with the particle-hole transformation defined by composing reversal of time (2.3b) with the chiral transformation (2.5b). The chiral spectral symmetry (2.35) implies that the single-particle energy eigenstate

$$|k,+,\sigma\rangle = \frac{1}{\sqrt{2}}\begin{pmatrix} e^{-\mathrm{i}\phi_k/2}\\ +e^{+\mathrm{i}\phi_k/2}\end{pmatrix}, \qquad \tan\phi_k = \frac{h_{k,2}}{h_{k,1}}, \tag{2.55a}$$

of $\widehat{H}$ with the twofold degenerate positive eigenenergy

$$\varepsilon_{k,+} := +\sqrt{h^2_{k,1} + h^2_{k,2}} \tag{2.55b}$$

is related to the single-particle energy eigenstate

$$|k,-,\sigma\rangle = \frac{1}{\sqrt{2}}\begin{pmatrix} e^{-\mathrm{i}\phi_k/2} \\ -e^{+\mathrm{i}\phi_k/2} \end{pmatrix}, \qquad \tan\phi_k = \frac{h_{k,2}}{h_{k,1}}, \tag{2.56a}$$

of $\widehat{H}$ with the twofold degenerate negative eigenenergy

$$\varepsilon_{k,-} := -\sqrt{h_{k,1}^2 + h_{k,2}^2} \tag{2.56b}$$

through the chiral transformation (**Exercise** 2.3) that is represented by $-\tau_3$ in the orbital basis ($\hat{\psi}_{k,\sigma}$) and by $-U_k^\dagger\,\tau_3\,U_k$ in the Bloch basis ($\hat{\chi}_{k,\sigma}$).

2.5.4 A Model for Polyacetylene

The crudest model for electronic transport in polyacetylene is one dimensional and introduces a second characteristic energy scale, the staggered nearest-neighbor hopping amplitude $\delta t \in \mathbb{R}$, in addition to the uniform nearest-neighbor hopping amplitude $t > 0$, through the choice

$$h_{k,0} = h_{k,3} = 0, \qquad h_{k,1} = -t_1 - t_2\cos(2k), \qquad h_{k,2} = -t_2\sin(2k), \tag{2.57a}$$

where

$$t_1 := t - \delta t, \qquad t_2 := t + \delta t, \tag{2.57b}$$

in $\widehat{H}$ defined by Eq. (2.39). This gives the effective electronic one-dimensional non-interacting tight-binding Hamiltonian

$$\begin{aligned}\widehat{H}_{\mathrm{polya}} &:= -\sum_{k\in\Lambda^{A\star}_{\mathrm{BZ}}}\sum_{\sigma=\uparrow,\downarrow}\hat{\psi}^\dagger_{k,\sigma}\,\{[t_1 + t_2\cos(2k)]\,\tau_1 + t_2\,\sin(2k)\,\tau_2\}\,\hat{\psi}_{k,\sigma} \\ &= -\sum_{j=1}^{N/2}\sum_{\sigma=\uparrow,\downarrow}\left\{\hat{a}^\dagger_{j,\sigma}\left[(t-\delta t)\,\hat{b}_{j,\sigma} + (t+\delta t)\,\hat{b}_{j-1,\sigma}\right] + \mathrm{H.c.}\right\}\end{aligned} \tag{2.57c}$$

as a model of polyacetylene that depends on three empirical energy scales, $t > 0$, $\delta t \in \mathbb{R}$, and the Fermi energy ε_{F}, see Figure 2.12.

An energy eigenstate $|k,+,\sigma\rangle$ of $\widehat{H}_{\mathrm{polya}}$ given by Eq. (2.55a) with the twofold degenerate positive eigenenergy

$$\varepsilon_{k,+} := +2\sqrt{t^2\cos^2 k + (\delta t)^2\sin^2 k} \tag{2.58a}$$

can always be paired with an energy eigenstate $|k,-,\sigma\rangle$ of $\widehat{H}_{\mathrm{polya}}$ given by Eq. (2.56a) with the twofold degenerate negative eigenenergy

$$\varepsilon_{k,-} := -2\sqrt{t^2\cos^2 k + (\delta t)^2\sin^2 k} \tag{2.58b}$$

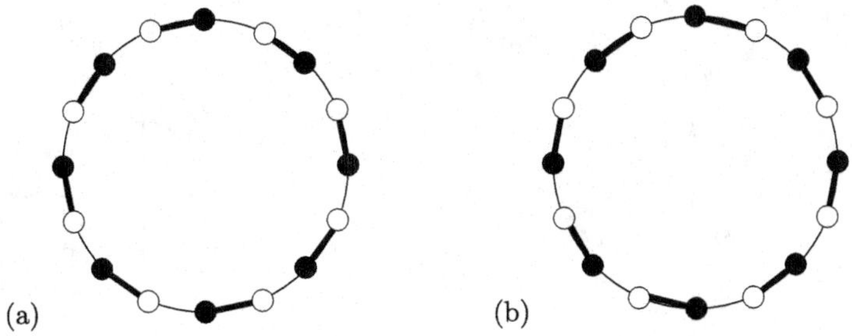

Figure 2.12 Dimerized ring with hopping amplitudes $t + \delta t$ on strong nearest-neighbor bonds (thick line) and $t - \delta t$ on weak nearest-neighbor bonds (thin line). (a) Strong bonds connect empty to filled circles in a clockwise direction. (b) Strong bonds connect filled to empty circles in a clockwise direction.

through the chiral transformation (2.35a) (**Exercise** 2.3). The maximum of the lower branch $\varepsilon_{k,-}$ is separated from the minimum of the upper branch $\varepsilon_{k,+}$ by the single-particle gap

$$2\Delta := \begin{cases} \min\limits_{k\in\Lambda^{A\star}_{\mathrm{BZ}}} \varepsilon_{k,+} - \max\limits_{k\in\Lambda^{A\star}_{\mathrm{BZ}}} \varepsilon_{k,-} = 4\,|\delta t|, & \text{if } |\delta t|/t < 1, \\ \\ \min\limits_{k\in\Lambda^{A\star}_{\mathrm{BZ}}} \varepsilon_{k,+} - \max\limits_{k\in\Lambda^{A\star}_{\mathrm{BZ}}} \varepsilon_{k,-} = 4\,t, & \text{if } |\delta t|/t \geq 1. \end{cases} \tag{2.59}$$

The gap (2.59) is a direct one, that is, the maximum of the lower band and the minimum of the upper band occur either at the momentum $k = \pi/2 \bmod \pi$ when $|\delta t| < t$, or at the momentum $k = 0 \bmod \pi$ when $|\delta t| > t$. This gap is plotted in Figure 2.13(a). The dispersions for different ratios of $|\delta t|/t$ are shown in Figures 2.13(b)-2.13(d). When $|\delta t| = t$, the lower and upper bands are flat, that is independent of the momentum k. When $|\delta t| = t$, the electrons can, but need not, be thought of as Bloch waves since the chain breaks down into independent dimers as shown in the panel of Figure 2.13(c). Phenomenology dictates that $4t \sim 12\,\mathrm{eV}$, $\delta t \sim 0.5\,\mathrm{eV}$, and $\varepsilon_{\mathrm{F}} = 0$ for polyacetylene.

2.6 Fermiology in the Continuum

The sign of the staggered hopping amplitude δt determines if it is the hopping amplitude between the pair of nearest-neighbor sites i and $i+1$ from the lattice Λ that is larger or smaller in magnitude than the hopping amplitude between the pair of nearest-neighbor sites $i+1$ and $i+2$. In the thermodynamic limit, at half-filling, and at vanishing temperature, the point $\delta t/t = 0$ along the one-dimensional parameter space with the dimensionless coordinate $\delta t/t$ is a quantum critical point that separates two gapped phases. It signals a singularity at $\delta t/t = 0$ in the dependence

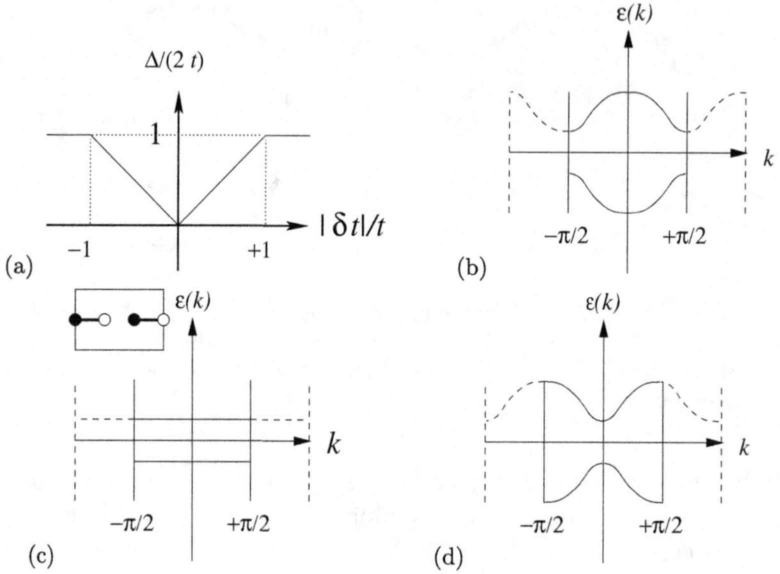

Figure 2.13 (a) Gap (2.59) as a function of $\delta t/t$. (b) Dispersion of the dimerized chain when $|\delta t| \ll t$. (c) Dispersion of the dimerized chain when $|\delta t| = t$, in which case the chain breaks down into $N/2$ disjoint dimers. (d) Dispersion of the dimerized chain when $|\delta t| \gg t$.

on $\delta t/t$ of the partition function at half-filling and at zero temperature, that is, the partition function[18]

$$Z_{\text{polya}}(\beta, \delta t/t, \nu_{\text{e}}) := \text{Tr}_{\mathfrak{F}}\left(e^{-\beta\,\widehat{H}_{\text{polya}}}\right), \qquad \beta := \frac{1}{T}, \tag{2.60}$$

displays a singularity in the thermodynamic limit, at $T = 0$, $\delta t/t = 0$, and half-filling, see Figure 2.14. Each gapped phase breaks the discrete translation invariance by the lattice spacing $\mathfrak{a}$ of the quantum critical point at $\delta t = 0$ in one of two ways.

The staggered hopping amplitude δt is also called the characteristic energy for dimerization. There are two distinct dimerized phases in that they break in one of two ways the discrete translation symmetry by the lattice spacing $\mathfrak{a}$.

The approach as a function of $\delta t/t$ of the quantum critical point at $\delta t/t = 0$, half-filling, and vanishing temperature is characterized by the diverging correlation length

$$\xi \sim \frac{1}{\Delta^{z}} \tag{2.61a}$$

in units in which the Planck constant and the characteristic speed have been set to unity. The critical exponent

$$z = 1 \tag{2.61b}$$

[18] We have set the Boltzmann constant k_{B} to unity.

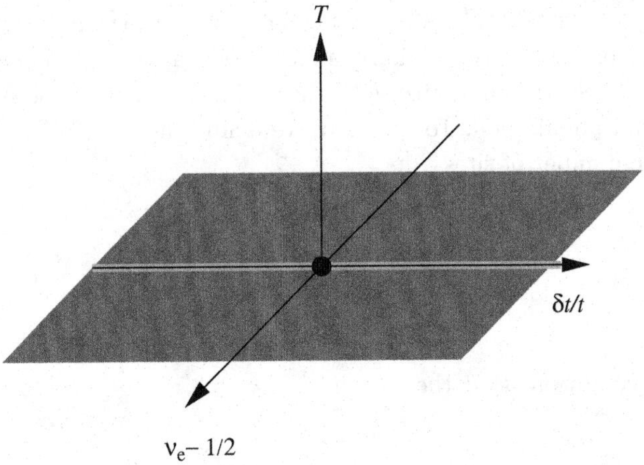

Figure 2.14 Three-dimensional phase diagram derived from the partition function (2.60) with the shifted filling fraction $-1/2 < \nu_{\rm e} - 1/2 < +1/2$, the ratio of coupling constants $-\infty < \delta t/t < +\infty$, and the temperature $0 \leq T < +\infty$ as axis. The point at $\nu_{\rm e} = 1/2$, $\delta t/t = 0$, $T = 0$ is a quantum critical point. It separates the gapped phase $\nu_{\rm e} = 1/2$, $\delta t/t < 0$, $T = 0$ from the gapped phase $\nu_{\rm e} = 1/2$, $\delta t/t > 0$, $T = 0$. The stripe $1/2 > \nu_{\rm e} - 1/2 > 0$, $\delta t/t \in \mathbb{R}$, $T = 0$ together with the stripe $-1/2 < \nu_{\rm e} - 1/2 < 0$, $\delta t/t \in \mathbb{R}$, $T = 0$ have metallic ground states.

is called the dynamical exponent. It relates a change of length scale to a change of energy scale. The correlation length is finite on each side of the quantum critical point along the line at half-filling and vanishing temperature in Figure 2.14. Correspondingly, it is possible to define a local order parameter $m(\delta t/t)$ that vanishes at the quantum critical point $\delta t/t = 0$ along the line at half-filling and vanishing temperature in Figure 2.14 and grows in a continuous fashion as

$$|m(\delta t/t)| \sim |\delta t/t|^{\beta} \tag{2.62a}$$

with the critical exponent

$$\beta = 1 \tag{2.62b}$$

in the vicinity of $\delta t/t = 0$. This order parameter can be chosen to be

$$m(\delta t/t) := \frac{1}{2} \sum_{\sigma=\uparrow,\downarrow} \langle \Psi_{\rm FS} | \hat{\psi}^{\dagger}_{q,\sigma} \, \tau_2 \, \hat{\psi}_{q,\sigma} | \Psi_{\rm FS} \rangle, \tag{2.63}$$

whereby the momentum q is the one associated with the direct gap, that is $q = \pi/2$ for $|\delta t|/t \ll 1$.

The values $z = \beta = 1$ in Eqs. (2.61) and (2.62) follow from the fact that the quantum critical point at $T = 0$, $\delta t/t = 0$, and half-filling, is noninteracting in the electron basis, as we now explain.

For this purpose, we take advantage of the fact that in the vicinity of this quantum critical point, a continuum description is a meaningful approximation for the Fourier component of the electron operator with the momentum k very close to either one of the two Fermi points $\pm k_{\mathrm{F}}$. To this end, we define the scaling limit of the chain Λ by which the number of sites

$$N \to \infty \tag{2.64a}$$

and the lattice spacing

$$\mathfrak{a} \to 0 \tag{2.64b}$$

under the two conditions that the length

$$L := N\,\mathfrak{a} \tag{2.64c}$$

and the filling fraction

$$0 < \nu_{\mathrm{e}} < 1 \tag{2.64d}$$

defined in Eq. (2.18) are held fixed. In the scaling limit (2.64), the ultra-violet momentum cutoff

$$\frac{\Lambda_{\mathrm{uv}}}{2} := \frac{\pi}{2\mathfrak{a}} \tag{2.64e}$$

set by the Brillouin zone $\Lambda^{A\star}_{\mathrm{BZ}}$ diverges. Next, we proceed in two steps.

First, in the scaling limit (2.64), we make the identifications

$$\left\{ j = 1, \cdots, \frac{N}{2}, \qquad \Delta j = 1 \right\} \to \left\{ -\frac{L}{2} \leq x \leq +\frac{L}{2}, \qquad \frac{\mathrm{d}x}{2\mathfrak{a}} = 1 \right\}, \tag{2.65a}$$

and

$$\left\{ k = \frac{2\pi}{N}, \cdots, \pi, \quad \Delta k = \frac{2\pi}{N} \right\} \to \left\{ -\frac{\pi}{2\mathfrak{a}} \leq p \leq +\frac{\pi}{2\mathfrak{a}}, \quad \frac{\mathrm{d}p}{2\pi/L} = 1 \right\}. \tag{2.65b}$$

Sums over the sublattice Λ^{A} or its Brillouin zone $\Lambda^{A\star}_{\mathrm{BZ}}$ can then be identified with the integrals

$$\left\{ \sum_{j=1}^{N/2} = \sum_{j=1}^{N/2} \frac{2\mathfrak{a}}{2\mathfrak{a}} \right\} \to \left\{ \frac{1}{2\mathfrak{a}} \int\limits_{-L/2}^{+L/2} \mathrm{d}x \right\} \tag{2.66a}$$

or

$$\left\{ \sum_{k\in\Lambda^{A\star}_{\mathrm{BZ}}} = \sum_{k\in\Lambda^{A\star}_{\mathrm{BZ}}} \frac{2\pi/L}{2\pi/L} \right\} \to \left\{ \frac{1}{2\pi/L} \int\limits_{-\Lambda_{\mathrm{uv}}/2}^{+\Lambda_{\mathrm{uv}}/2} \mathrm{d}p \right\}, \tag{2.66b}$$

respectively.

Application of Eq. (2.66b) to Hamiltonian (2.39a) suggests absorbing the multiplicative factor $(2\pi/L)^{-1}$ by the rescaling

$$\hat{\psi}_{k,\sigma} \to \sqrt{2\pi/L}\,\hat{\psi}_{\sigma}(p) \Longleftrightarrow \hat{\psi}_{\sigma}(p) \to \frac{1}{\sqrt{2\pi/L}}\,\hat{\psi}_{k,\sigma}, \tag{2.67}$$

when relating lattice operators to field operators for the electron. If we extend the 2×2 matrix-valued $\mathcal{H}_k$ in Eq. (2.39b) from $\Lambda^{A\star}_{\mathrm{BZ}}$ to $[-\pi/2,+\pi/2[$ by demanding that the 2×2 matrix-valued function $\mathcal{H}(p)$ equals $\mathcal{H}_k$ whenever $\mathfrak{a}\,p = k$, that is,

$$\mathcal{H}(p) := h_0(p)\,\tau_0 + h_1(p)\,\tau_1 + h_2(p)\,\tau_2 + h_3(p)\,\tau_3, \tag{2.68a}$$

we infer the identification

$$\widehat{H} \to \sum_{\sigma=\uparrow,\downarrow}\int\limits_{-\Lambda_{\mathrm{uv}}/2}^{+\Lambda_{\mathrm{uv}}/2} \mathrm{d}p\;\hat{\psi}^{\dagger}_{\sigma}(p)\,\mathcal{H}(p)\,\hat{\psi}_{\sigma}(p), \tag{2.68b}$$

in the scaling limit (2.64). Here,

$$\begin{aligned}
\{\hat{\psi}_{\sigma,\alpha}(p),\hat{\psi}^{\dagger}_{\sigma',\alpha'}(p')\} &= \delta_{\sigma,\sigma'}\,\delta_{\alpha,\alpha'}\,\delta(p-p'),\\
\{\hat{\psi}_{\sigma,\alpha}(p),\hat{\psi}_{\sigma',\alpha'}(p')\} &= \{\hat{\psi}^{\dagger}_{\sigma',\alpha'}(p'),\hat{\psi}^{\dagger}_{\sigma,\alpha}(p)\} = 0,
\end{aligned} \tag{2.68c}$$

is the algebra obeyed by the field operators $\hat{\psi}^{\dagger}_{\sigma,\alpha}(p)$ and $\hat{\psi}_{\sigma,\alpha}(p)$ that create and annihilate an electron with spin $\sigma=\uparrow,\downarrow$ on sublattice $\alpha = A, B$, and with momentum p, respectively. It is instructive to compare the algebra (2.68c) to the algebra (2.39c). The bookkeeping between these fermion algebras is provided by

$$\frac{L}{2\pi}\,\delta_{k,k'} \to \delta(p-p') \tag{2.69}$$

in the scaling limit (2.64). For completeness, observe that the Fourier transforms (2.32) are to be replaced by

$$\hat{\psi}_{\sigma,\alpha}(p) = \frac{1}{\sqrt{2\pi}}\int\limits_{-L/2}^{+L/2} \mathrm{d}x\,e^{-\mathrm{i}px}\,\hat{\psi}_{\sigma,\alpha}(x) \tag{2.70a}$$

in the scaling limit (2.64). Hence,

$$\begin{aligned}
&\{\hat{\psi}_{\sigma,\alpha}(x),\hat{\psi}^{\dagger}_{\sigma',\alpha'}(x')\} = \delta_{\sigma,\sigma'}\,\delta_{\alpha,\alpha'}\,\delta(x-x'),\\
&\{\hat{\psi}_{\sigma,\alpha}(x),\hat{\psi}_{\sigma',\alpha'}(x')\} = \{\hat{\psi}^{\dagger}_{\sigma',\alpha'}(x'),\hat{\psi}^{\dagger}_{\sigma,\alpha}(x)\} = 0,\\
&\widehat{H} \to \sum_{\sigma=\uparrow,\downarrow}\int\limits_{-L/2}^{+L/2}\mathrm{d}x\int\limits_{-L/2}^{+L/2}\mathrm{d}x'\,\hat{\psi}^{\dagger}_{\sigma}(x)\,\mathcal{H}(x-x')\,\hat{\psi}_{\sigma}(x'),
\end{aligned} \tag{2.70b}$$

whereby

$$\frac{1}{2\mathfrak{a}}\,\delta_{j,j'} \to \delta(x-x'), \qquad \frac{1}{\sqrt{2\mathfrak{a}}}\,\hat{\psi}_{j,\sigma} \to \hat{\psi}_{\sigma}(x), \tag{2.70c}$$

and

$$\mathcal{H}(x-x') = \int\limits_{-\Lambda_{\mathrm{uv}}/2}^{+\Lambda_{\mathrm{uv}}/2} \frac{\mathrm{d}p}{2\pi}\, e^{+\mathrm{i}\,p\,(x-x')}\, \mathcal{H}(p). \tag{2.70d}$$

Second, we use the fact that the filling fraction is held fixed in the scaling limit (2.64). This means that the magnitude of the dimensionless Fermi momentum

$$k_{\mathrm{F}} \equiv \mathfrak{a}\,|p_{\mathrm{F},\pm}| \tag{2.71a}$$

defined in Eq. (2.12d) is fixed in the scaling limit (2.64). The subscript $\pm$ distinguishes in

$$p_{\mathrm{F},\pm} := \pm k_{\mathrm{F}}/\mathfrak{a} \tag{2.71b}$$

the two Fermi points under the assumption that each branch of the single-particle dispersions

$$\varepsilon_{k,\pm} \to \varepsilon_{\pm}(p) \tag{2.72}$$

from Eq. (2.46a) is a strictly monotonous function from the zone center $p = 0$ to either one of the zone boundaries $p = \pm\pi/(2\mathfrak{a})$ in addition to being periodic over the Brillouin zone $[-\pi/(2\mathfrak{a}), +\pi/(2\mathfrak{a})[$. This assumption is verified for the dispersions (2.58) used to model electronic transport in polyacetylene if and only if $|\delta t| \neq t$. We need to distinguish the case when the filling fraction $0 < \nu_{\mathrm{e}} < 1$ of all available single-particle states is not equal to $1/2$ from the case when $\nu_{\mathrm{e}} = 1/2$.

When $\nu_{\mathrm{e}} \neq 0, 1/2, 1$, there are two distinct Fermi points around which we can always select from the Brillouin zone $[-\pi/(2\mathfrak{a}), +\pi/(2\mathfrak{a})[$ in the scaling limit (2.64) the interval $U_{\Omega,+}$ defined by those momenta $p_{+} = k_{+}/\mathfrak{a}$,

$$\mathfrak{a}\,p = +k_{\mathrm{F}} + k_{+}, \qquad |k_{+}| < \mathfrak{a}\,\Omega/2 \ll k_{\mathrm{F}}, \tag{2.73a}$$

of magnitude less than $\Omega/2$ centered about the Fermi point $p_{\mathrm{F},+}$ and the interval $U_{\Omega,-}$ defined by those momenta $p_{-} = k_{-}/\mathfrak{a}$,

$$\mathfrak{a}\,p = -k_{\mathrm{F}} + k_{-}, \qquad |k_{-}| < \mathfrak{a}\,\Omega/2 \ll k_{\mathrm{F}}, \tag{2.73b}$$

of magnitude less than $\Omega/2$ centered about the Fermi point $p_{\mathrm{F},-}$, as is illustrated in Figure 2.15. We shall reserve the label $\varrho = \pm$ to distinguish the two distinct Fermi points (2.71b). We do the Taylor expansion

$$\varepsilon_{\mathsf{a}_{\mathrm{F}}}(p_{\mathrm{F},\varrho} + p_{\varrho}) = \varepsilon_{\mathrm{F}} + \left(\frac{\mathrm{d}\,\varepsilon_{\mathsf{a}_{\mathrm{F}}}}{\mathrm{d}\,p_{\varrho}}\right)(p_{\mathrm{F},\varrho})\,p_{\varrho} + \cdots \tag{2.74a}$$

for that band a_{F} from either the lower ($\mathsf{a}_{\mathrm{F}} = -$) or upper ($\mathsf{a}_{\mathrm{F}} = +$) bands (2.72) for which

$$\varepsilon_{\mathsf{a}_{\mathrm{F}}}(p_{\mathrm{F},\varrho}) = \varepsilon_{\mathrm{F}}, \qquad \varrho = \pm. \tag{2.74b}$$

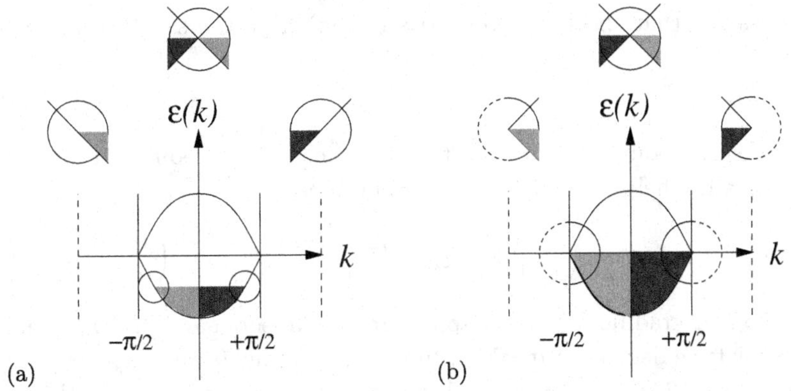

Figure 2.15 The low-energy and long-wavelength effective theory of Hamiltonian (2.39a) is obtained by selecting a small interval of momenta centered around the Fermi point(s). (a) There are two distinct Fermi points if the filling fraction ν_{e} defined in Eq. (2.18) is neither zero, nor half-filling, nor unity. Each Fermi point has one branch of excitations. (b) At half-filling the Fermi points sit at the zone boundaries. Hence, they are not distinct, for they differ by the reciprocal momentum π. This Fermi point has two branches of excitations.

If we introduce the shorthand notations

$$\hat{\chi}^{\dagger}_{\mathrm{F},\sigma,\varrho}(p_{\varrho}) := \hat{\chi}^{\dagger}_{\sigma,\mathrm{a}_{\mathrm{F}}}(p_{\mathrm{F},\varrho} + p_{\varrho}), \qquad \hat{\chi}_{\mathrm{F},\sigma,\varrho}(p_{\varrho}) := \hat{\chi}_{\sigma,\mathrm{a}_{\mathrm{F}}}(p_{\mathrm{F},\varrho} + p_{\varrho}), \tag{2.75a}$$

and the Fermi velocities

$$v_{\mathrm{F},\varrho} := \left(\frac{\mathrm{d}\,\varepsilon_{\mathrm{a}_{\mathrm{F}}}}{\mathrm{d}\,p_{\varrho}}\right)(p_{\mathrm{F},\varrho}), \qquad \varrho = \pm, \tag{2.75b}$$

we then capture the low-energy and long-wavelength sector of Hamiltonian (2.39a) with the effective Hamiltonian

$$\widehat{H}_{\mathrm{eff}} := \sum_{\sigma=\uparrow,\downarrow}\sum_{\varrho=\pm}\int\limits_{-\Omega/2}^{+\Omega/2} \mathrm{d}p_{\varrho}\, \hat{\chi}^{\dagger}_{\mathrm{F},\sigma,\varrho}(p_{\varrho})\,\left(\varepsilon_{\mathrm{F}} + v_{\mathrm{F},\varrho}\,p_{\varrho}\right)\hat{\chi}_{\mathrm{F},\sigma,\varrho}(p_{\varrho}), \tag{2.75c}$$

whereby

$$\begin{aligned} \{\hat{\chi}_{\mathrm{F},\sigma,\varrho}(p), \hat{\chi}^{\dagger}_{\mathrm{F},\sigma',\varrho'}(p')\} &= \delta_{\sigma,\sigma'}\,\delta_{\varrho,\varrho'}\,\delta(p-p'), \\ \{\hat{\chi}_{\mathrm{F},\sigma,\varrho}(p), \hat{\chi}_{\mathrm{F},\sigma',\varrho'}(p')\} &= \{\hat{\chi}^{\dagger}_{\mathrm{F},\sigma',\varrho'}(p'), \hat{\chi}^{\dagger}_{\mathrm{F},\sigma,\varrho}(p)\} = 0, \end{aligned} \tag{2.75d}$$

in the scaling limit (2.64). This is the generic metallic fixed point for a single band of noninteracting electrons in one-dimensional position space. Consequently, any deviation from half-filling in the model (2.57c) for polyacetylene implies metallic behavior of the conductivity.

If we assume that the Fermi velocities (2.75b) have the same magnitude v_{F},[19] that is

$$v_{\mathrm{F},\varrho} := \varrho\, v_{\mathrm{F}} \tag{2.76a}$$

at the two Fermi points (2.71b), it is then customary to introduce yet another 2×2 grading with the help of the basis of 2×2 matrices

$$\varrho_0 = \begin{pmatrix} 1 & 0 \\ 0 & 1 \end{pmatrix}, \quad \varrho_1 = \begin{pmatrix} 0 & 1 \\ 1 & 0 \end{pmatrix}, \quad \varrho_2 = \begin{pmatrix} 0 & -\mathrm{i} \\ +\mathrm{i} & 0 \end{pmatrix}, \quad \varrho_3 = \begin{pmatrix} 1 & 0 \\ 0 & -1 \end{pmatrix}, \tag{2.76b}$$

besides the 2×2 grading (σ) for the spin degrees of freedom and the 2×2 grading (τ) for the sublattice degrees of freedom. In the basis of right- and left-moving spinors defined by Eq. (2.76), the effective low-energy and long-wavelength Hamiltonian (2.75c) simplifies to

$$\begin{aligned} \widehat{H}_{\mathrm{eff}} &= \sum_{\sigma=\uparrow,\downarrow} \int\limits_{-\Omega/2}^{+\Omega/2} \mathrm{d}p\, \hat{\chi}^{\dagger}_{\mathrm{F},\sigma}(p)\, (\varepsilon_{\mathrm{F}}\, \varrho_0 + v_{\mathrm{F}}\, p\, \varrho_3)\, \hat{\chi}_{\mathrm{F},\sigma}(p) \\ &= \sum_{\sigma=\uparrow,\downarrow} \int\limits_{-L/2}^{+L/2} \mathrm{d}x\, \hat{\chi}^{\dagger}_{\mathrm{F},\sigma}(x) \left(\varepsilon_{\mathrm{F}}\, \varrho_0 - \mathrm{i}\, v_{\mathrm{F}} \frac{\mathrm{d}}{\mathrm{d}x}\, \varrho_3 \right) \hat{\chi}_{\mathrm{F},\sigma}(x), \end{aligned} \tag{2.77a}$$

where the Fourier transforms[20]

$$\hat{\chi}_{\mathrm{F},\sigma}(x) := \frac{1}{\sqrt{2\pi}} \int\limits_{-\Omega/2}^{+\Omega/2} \mathrm{d}p\, e^{+\mathrm{i}\, p\, x}\, \hat{\chi}_{\mathrm{F},\sigma}(p) = e^{-\mathrm{i} p_{\mathrm{F}}\, x}\, \hat{\chi}_{\sigma}(x) \tag{2.77b}$$

for each spin component σ define the (spinor) Wannier basis of the Fock space. If we introduce the shorthand notation

$$\hat{\psi}^{\dagger}_{\mathrm{F},\sigma,\varrho}(p_\varrho) := \hat{\psi}^{\dagger}_{\sigma}(p_{\mathrm{F},\varrho} + p_\varrho), \qquad \hat{\psi}_{\mathrm{F},\sigma,\varrho}(p_\varrho) := \hat{\psi}_{\sigma}(p_{\mathrm{F},\varrho} + p_\varrho), \tag{2.78a}$$

with the convention for the Fourier transformation given by[20]

$$\hat{\psi}_{\mathrm{F},\sigma,\varrho}(x) := \frac{1}{\sqrt{2\pi}} \int\limits_{-\Omega/2}^{+\Omega/2} \mathrm{d}p\, e^{+\mathrm{i}\, p\, x}\, \hat{\psi}_{\mathrm{F},\sigma,\varrho}(p) = e^{-\mathrm{i} p_{\mathrm{F}}\, x}\, \hat{\psi}_{\sigma,\varrho}(x), \tag{2.78b}$$

then the single-particle states in the orbital basis (2.78b) are related to the single-particle states in the Wannier basis (2.77b) by a convolution. As was the case on

[19] Either time-reversal symmetry or inversion symmetry (symmetry under the linear transformation by which $k \to -k$) implies that $|v_{\mathrm{F},+}| = |v_{\mathrm{F},-}| = v_{\mathrm{F}}$.

[20] It is important to remember that the momentum cutoff Ω defined in Eq. (2.73) is much smaller than the momentum cutoff Λ_{uv} set by the Brillouin zone $\Lambda^{\star}_{\mathrm{BZ}}$ [see Eq. (2.64e)]. Hence, we are factorizing the dependence on x into a first component $e^{\mathrm{i}\, p_{\mathrm{F}}\, x}$ that undergoes a period of oscillation on the characteristic length scale of $1/p_{\mathrm{F}}$ and a second component $\hat{\chi}_{\mathrm{F},\sigma}(x)$ that undergoes a significant change on the much larger characteristic length scale $1/\Omega$ according to Eq. (2.73).

the lattice, the orbital basis delivers exponentially localized single-particle states, but this need not be the case for the Wannier basis.

When $\nu_{\mathsf{e}} = 1/2$, the periodicity of the energy eigenvalues (2.46a) across the Brillouin zone and the assumption that they are monotonous functions of the momentum from the center to either one of the two boundaries of the Brillouin zone dictate that the Fermi momentum is either located at $p_{\mathrm{F}} = 0$ or at $p_{\mathrm{F}} = -\pi/(2\mathfrak{a})$, the only points of the Brillouin zone that are invariant under the inversion $k \to -k$ up to the reciprocal momentum π. Hence, there is only one Fermi point p_{F} with the Fermi energy ε_{F} at half-filling. We denote with $U_{\mathrm{F},\Omega}$ the interval of size $\Omega \ll \Lambda_{\mathrm{uv}}$ centered about this Fermi point. We also introduce the shorthand notations

$$\hat{\psi}^{\dagger}_{\mathrm{F},\sigma}(p) := \hat{\psi}^{\dagger}_{\sigma}(p_{\mathrm{F}} + p), \qquad \hat{\psi}_{\mathrm{F},\sigma}(p) := \hat{\psi}_{\sigma}(p_{\mathrm{F}} + p), \tag{2.79a}$$

with the convention for the Fourier transformation given by[20]

$$\hat{\psi}_{\mathrm{F},\sigma}(x) := \frac{1}{\sqrt{2\pi}} \int\limits_{-\Omega/2}^{+\Omega/2} \mathrm{d}p\, e^{+\mathrm{i}\,p\,x}\, \hat{\psi}_{\mathrm{F},\sigma}(p) = e^{-\mathrm{i}\,p_{\mathrm{F}}\,x}\, \hat{\psi}_{\sigma}(x), \tag{2.79b}$$

and

$$\hat{\chi}^{\dagger}_{\mathrm{F},\sigma}(p) := \hat{\chi}^{\dagger}_{\sigma}(p_{\mathrm{F}} + p), \qquad \hat{\chi}_{\mathrm{F},\sigma}(p) := \hat{\chi}_{\sigma}(p_{\mathrm{F}} + p), \tag{2.80a}$$

with the convention for the Fourier transformation given by

$$\hat{\chi}_{\mathrm{F},\sigma}(x) := \frac{1}{\sqrt{2\pi}} \int\limits_{-\Omega/2}^{+\Omega/2} \mathrm{d}p\, e^{+\mathrm{i}\,p\,x}\, \hat{\chi}_{\mathrm{F},\sigma}(p) = e^{-\mathrm{i}\,p_{\mathrm{F}}\,x}\, \hat{\chi}_{\sigma}(x), \tag{2.80b}$$

for the operator-valued spinors in the orbital and Bloch basis, respectively. We then capture the low-energy and long-wavelength sector of Hamiltonian (2.39a) with the effective Hamiltonian

$$\begin{aligned}
\widehat{H}_{\mathrm{eff}} &:= \sum_{\sigma=\uparrow,\downarrow} \int\limits_{-\Omega/2}^{+\Omega/2} \mathrm{d}p\, \hat{\psi}^{\dagger}_{\mathrm{F},\sigma}(p)\, \mathcal{H}(p_{\mathrm{F}} + p)\, \hat{\psi}_{\mathrm{F},\sigma}(p) \\
&= \sum_{\sigma=\uparrow,\downarrow} \sum_{\mathsf{a}=\pm} \int\limits_{-\Omega/2}^{+\Omega/2} \mathrm{d}p\, \hat{\chi}^{\dagger}_{\mathrm{F},\sigma,\mathsf{a}}(p)\, \varepsilon_{\mathsf{a}}(p_{\mathrm{F}} + p)\, \hat{\chi}_{\mathrm{F},\sigma,\mathsf{a}}(p),
\end{aligned} \tag{2.81}$$

where it is understood that either the single-particle 2×2 Hermitean matrix $\mathcal{H}(p_{\mathrm{F}} + p)$ or the energy eigenvalues $\varepsilon_{\mathsf{a}}(p_{\mathrm{F}} + p)$ have to be expanded in powers of p up to the first nonvanishing order beyond the order zero.

For the model (2.57c) of polyacetylene at half-filling, with $|\delta t| < t$, its low-energy and long-wavelength sector is captured by the effective Hamiltonian (**Exercise** 2.14)

$$\widehat{H}^{\mathrm{eff}}_{\mathrm{polya}} = \sum_{\sigma=\uparrow,\downarrow} \int\limits_{-\Omega/2}^{+\Omega/2} \mathrm{d}p\, \hat{\psi}^{\dagger}_{\mathrm{F},\sigma}(p) \left(v_{\mathrm{F}}\, p\, \tau_2 + m_{\mathrm{F}}\, v_{\mathrm{F}}^2\, \tau_1\right) \hat{\psi}_{\mathrm{F},\sigma}(p) \tag{2.82a}$$

in the orbital basis of the Fock space or

$$\widehat{H}^{\mathrm{eff}}_{\mathrm{polya}} = \sum_{\sigma=\uparrow,\downarrow} \int\limits_{-\Omega/2}^{+\Omega/2} \mathrm{d}p\, \hat{\chi}^{\dagger}_{\mathrm{F},\sigma}(p) \left[(v_{\mathrm{F}}\, p)^2 + m_{\mathrm{F}}^2\, v_{\mathrm{F}}^4\right]^{1/2} \tau_3\, \hat{\chi}_{\mathrm{F},\sigma}(p) \tag{2.82b}$$

in the Bloch basis of the Fock space. The Fermi speed is[21]

$$v_{\mathrm{F}} := 2\mathfrak{a}\, t \tag{2.82c}$$

and the "relativistic rest energy" is

$$m_{\mathrm{F}}\, v_{\mathrm{F}}^2 := 2\delta t. \tag{2.82d}$$

Observe that the role of the right- and left-moving grading away from half-filling is played by the sublattice grading at half-filling.

Hamiltonian (2.82a) represents two independent copies, each of which is labeled by the spin quantum number $\sigma =\uparrow$ or $\downarrow$, of the simplest example of a relativistic fermionic quantum field theory in (1+1)-dimensional spacetime. This is the noninteracting quantum field theory for a fermion obeying the Dirac equation

$$\mathrm{i}\hbar \left(\frac{\partial \Psi}{\partial t}\right)(t,x) = \left(\alpha\, c\, \frac{\hbar}{\mathrm{i}} \frac{\partial}{\partial x} + \beta\, m\, c^2\right) \Psi(t,x) \tag{2.83a}$$

with the choices

$$\alpha \equiv \tau_2, \qquad \beta \equiv \tau_1, \tag{2.83b}$$

for the 2×2 Dirac matrices α and β in the notation of Dirac and the choices

$$\hbar \equiv 1, \qquad c \equiv v_{\mathrm{F}}, \qquad m \equiv m_{\mathrm{F}}, \tag{2.83c}$$

for the Planck constant, characteristic speed, and Dirac mass, respectively. The Dirac wave function $\Psi(t,x)$ at the coordinate x in one-dimensional space and the coordinate t in time represents a two-component vector field taking values in $\mathbb{C}^2$ with the components $\Psi_1(t,x)$ and $\Psi_2(t,x)$. The Dirac wave function Ψ is also called a spinor field in view of the transformation law that it obeys under a Poincaré transformation (2.88) of Minkowski space (2.85) as we show here.

The Dirac equation (2.83) can be derived from the classical Lagrangian density

$$\mathcal{L}_{\mathrm{D}}(t,x) := \Psi^{\dagger}(t,x) \left(\mathrm{i}\hbar\, \partial_t - \alpha\, c\, \frac{\hbar}{\mathrm{i}}\, \partial_x - \beta\, m\, c^2\right) \Psi(t,x) \tag{2.84}$$

by variation of $\mathcal{L}_{\mathrm{D}}(t,x)$ with respect to the components of the row vector $\Psi^{\dagger}(t,x)$.

[21] We have set the Planck constant $\hbar$ to unity.

To establish the relativistic invariance of the Dirac equation (2.83) with the classical Lagrangian density (2.84), we first multiply it from the left by the 2×2 matrix β. Second, we introduce the standard relativistic notation $(\mu, \nu = 0, 1)$

$$\mathsf{x} \equiv (\mathsf{x}^{\mu}) = (ct, x), \qquad \mathsf{x}^{\mu} = g^{\mu\nu}\,\mathsf{x}_{\nu}, \qquad g^{\mu\nu} = \begin{cases} +1, & \text{if } \mu = \nu = 0, \\ 0, & \text{if } \mu \neq \nu, \\ -1, & \text{if } \mu = \nu = 1, \end{cases} \tag{2.85}$$

for the coordinate x in one-dimensional space and the coordinate t in time. Here, the Einstein summation convention over repeated indices is understood. We trade the Dirac matrices for the gamma matrices defined by

$$\gamma^0 \equiv \beta, \qquad \gamma^1 \equiv \beta\,\alpha \Longrightarrow \{\gamma^{\mu}, \gamma^{\nu}\} = 2g^{\mu\nu}, \qquad \mu, \nu = 0, 1. \tag{2.86}$$

By convention, the 2×2 identity matrix is implicitly understood to multiply $2g^{\mu\nu}$ (twice the Minkowski metric) on the right-hand side of Eq. (2.86). The relativistic invariance of the Dirac equation (2.83) with its classical Lagrangian density (2.84) is then manifest in its covariant representation

$$\left(\mathrm{i}\hbar\,c\,\gamma^{\mu}\partial_{\mu} - m\,c^2\right)\Psi(\mathsf{x}) = 0, \tag{2.87a}$$

with its classical Lagrangian density

$$\mathcal{L}_{\mathrm{D}}(\mathsf{x}) = \overline{\Psi}(\mathsf{x})\left(\mathrm{i}\hbar\,c\,\gamma^{\mu}\,\partial_{\mu} - m\,c^2\right)\Psi(\mathsf{x}), \tag{2.87b}$$

where

$$\overline{\Psi}(\mathsf{x}) := \Psi^{\dagger}(\mathsf{x})\,\gamma_0. \tag{2.87c}$$

Indeed, Eq. (2.87a) is form invariant under the Poincaré transformation[22]

$$\begin{aligned} &\mathsf{x}^{\mu} =: \omega_{\nu}^{\ \mu}\,\mathsf{x}'^{\nu} + a^{\mu}, && \omega_{\nu}^{\ \mu}\,g_{\mu\sigma}\,\omega_{\lambda}^{\ \sigma} = g_{\nu\lambda}, && a^{\mu} \in \mathbb{R}, \\ &\omega^{\mu}_{\ \lambda}\,\gamma^{\lambda} =: \Lambda^{-1}\,\gamma^{\mu}\,\Lambda, && \Lambda^{\dagger} = \gamma^0\,\Lambda^{-1}\,\gamma^0, && \Lambda^{*} = U_{\star}^{\dagger}\,\Lambda\,U_{\star}, \\ &\Psi'(\mathsf{x}') := \Lambda\,\Psi(\mathsf{x}), \end{aligned} \tag{2.88}$$

where the real-valued 2×2 matrix ω with the components $\omega_{\nu}^{\ \mu}$ represents a Lorentz transformation of (1+1)-dimensional spacetime and $U_{\star}$ is any 2×2 unitary matrix such that $\gamma^{\mu *} = U_{\star}^{\dagger}\,\gamma^{\mu}\,U_{\star}$.

The Poincaré covariance of the Dirac equation (2.83) is not a symmetry of the equations of motion of the electron operator in the tight-binding model (2.57c). It is an emergent symmetry whose accuracy is set by the validity of the gradient expansion, an expansion in powers of the momentum p measured relative to the Fermi point p_{F}. This expansion becomes exact in the scaling limit (2.64) by which the Fermi point can be approached to arbitrary accuracy (due to the limit $N \to \infty$) and the momentum cutoff Ω, above which the corrections in powers of p/p_{F} of the Taylor expansions of the lattice dispersions (2.58) beyond the leading orders

22 A. Messiah, *Quantum Mechanics*, Dover, New York 2017. [Messiah (2017)].

become sizable, diverges (due to the limit $\mathfrak{a} \to 0$). The point-group symmetry of the lattice Λ^A that we built into the model (2.57c) for polyacetylene is enhanced to the Poincaré group in the scaling limit (2.64) for which we have derived the Dirac equation.

A second example of an emergent symmetry encoded by the Dirac equation (2.83) is the axial symmetry of the Dirac equation. To define an axial transformation, we observe that the Hermitean matrix

$$\gamma^5 := \gamma^0 \gamma^1 \equiv -\gamma_5 \tag{2.89a}$$

squares to the unit matrix and anticommutes with both γ^0 and γ^1. This is a sole consequence of the Clifford algebra obeyed by those matrices labeled by the space-time index μ whose pairwise anticommutator is the unit matrix multiplied by $2\,g^{\mu\nu}$, see Eq. (2.86). We define the transformation laws

$$\overline{\Psi}(\mathsf{x}) =: \overline{\Psi}_{\theta_{\mathrm{e}},\theta_5}(\mathsf{x})\, e^{-\mathrm{i}(\theta_{\mathrm{e}}-\theta_5\,\gamma_5)} \Longleftrightarrow \Psi(\mathsf{x}) =: e^{+\mathrm{i}(\theta_{\mathrm{e}}+\theta_5\,\gamma_5)}\, \Psi_{\theta_{\mathrm{e}},\theta_5}(\mathsf{x}) \tag{2.89b}$$

and

$$m =: e^{-2\mathrm{i}\theta_5\,\gamma_5} m_{\theta_5}. \tag{2.89c}$$

Equations (2.89b) and (2.89c) dictate the transformation laws obeyed by the Dirac spinor and the Dirac mass under the continuous transformation from the group $\mathrm{U}(1) \times \mathrm{U}(1) \equiv \mathrm{U}_{\mathrm{e}}(1) \times \mathrm{U}_5(1)$ since $0 \le \theta_{\mathrm{e}} < 2\pi$ and $0 \le \theta_5 < 2\pi$. The first continuous parameter θ_{e} is associated with a global gauge transformation that is responsible for the conservation of the electric charge when it is a symmetry of the quantum field theory. The second continuous parameter θ_5 is associated with a global axial gauge transformation that is responsible for the conservation of the axial charge when it is a symmetry of the quantum field theory. Evidently, the Dirac spinor Ψ, its dual $\overline{\Psi}$, and the Dirac mass obey different transformation rules under an electric and an axial gauge transformation. The Dirac spinor Ψ carries both a positive unit electric charge and a positive unit axial charge. Its dual $\overline{\Psi}$ carries a negative unit electric charge but a positive unit axial charge because of the fact that γ_0 anticommutes with γ_5. The Dirac mass carries no electric charge but two negative units of the axial charge. Under the combined transformations (2.89b) and (2.89c), the Lagrangian density (2.87b) is form invariant, that is,

$$\mathcal{L}_{\mathrm{D}}(\mathsf{x}) = \overline{\Psi}_{\theta_{\mathrm{e}},\theta_5}(\mathsf{x}) \left(\mathrm{i}\,\hbar\, c\, \gamma^\mu\, \partial_\mu - m_{\theta_{\mathrm{e}},\theta_5}\, c^2\right) \Psi_{\theta_{\mathrm{e}},\theta_5}(\mathsf{x}) \tag{2.90}$$

for any choice of $0 \le \theta_{\mathrm{e}} < 2\pi$ and $0 \le \theta_5 < 2\pi$. The symmetry (2.90) presumes that m can be treated as a dynamical field. However, if the Dirac mass m is to be treated as a fixed number, the form invariance (2.90) is only to be achieved if $m = 0$. The microscopic origin of the Dirac mass is the staggered nearest-neighbor hopping in the model (2.57c) of polyacetylene, which we shall soon interpret as an effect of the ions on the quantum motion of the electrons from the π band of polyacetylene. A staggered nearest-neighbor hopping that is quantified by the dimerization δt defined in Eq. (2.57) breaks translation symmetry by the lattice spacing $\mathfrak{a}$ of the chain Λ. This discrete translation symmetry of the tight-binding (lattice) Hamiltonian becomes the continuous symmetry with the symmetry group $U_5(1)$ of the massless Dirac Hamiltonian (2.82a) that follows from taking the scaling limit (2.64) and performing a gradient expansion up to leading (linear) order, for the emergent $U_5(1)$ symmetry is broken by the Dirac mass m_{F}, where m_{F} is proportional to the dimerization δt according to Eq. (2.82d). Hereto, it is the sign of m_{F} that selects whether it is the strong or the weak nearest-neighbor hopping amplitude that starts from sublattice A to end in sublattice B in clockwise direction in the lattice Hamiltonian. The pattern of explicit breaking of the $U_5(1)$ symmetry selected by the sign of m_{F} thus delivers two gapped ground states, which we may interpret as the two ways there are to break translation symmetry by one lattice spacing but preserving translation symmetry by two lattice spacings from decorating (strong or weak) nearest-neighbor bonds along a chain (**Exercise** 2.15). A third example of an emerging symmetry encoded by the Dirac equation (2.83) is the scale invariance when $m = 0$.

We restore the units in which $\hbar = c = 1$ and define the dimensionless action

$$S_{\mathrm{D}} := \int \mathrm{d}^2\mathsf{x}\, \mathcal{L}_{\mathrm{D}}(\mathsf{x}). \tag{2.91}$$

This action is covariant under the scale transformation

$$\begin{aligned} \mathsf{x}^{\mu} &=: \kappa\, \mathsf{x}'^{\mu}, & m &=: \kappa^{-1}\, m', \\ \overline{\Psi}(\mathsf{x}) &=: \overline{\Psi}(\mathsf{x}')\, \kappa^{-1/2}, & \Psi(\mathsf{x}) &=: \kappa^{-1/2}\, \Psi(\mathsf{x}'), \end{aligned} \tag{2.92}$$

for any real number $\kappa > 0$.

When $m \neq 0$, define the length scale

$$\xi := 1/|m|, \tag{2.93a}$$

the Feynman scalar propagator

$$\Delta_{\mathrm{F}}(\mathsf{x};\xi) := -\frac{1}{(2\pi)^2} \int \mathrm{d}^2\mathsf{p}\, e^{-\mathrm{i}\mathsf{p}\cdot\mathsf{x}} \frac{1}{\mathsf{p}^2 - m^2 + \mathrm{i}0^+}, \tag{2.93b}$$

and the Feynman spinor propagator

$$S_{\mathrm{F}}(\mathsf{x};\xi) := \left(\mathrm{i}\gamma^{\mu}\frac{\partial}{\partial \mathsf{x}^{\mu}} + m\right) \Delta_{\mathrm{F}}(\mathsf{x};\xi). \tag{2.93c}$$

We are using the shorthand notation $\mathsf{a}\cdot\mathsf{b} \equiv \mathsf{a}_{\mu}\,\mathsf{b}^{\mu} \equiv \mathsf{a}_{\mu}\, g^{\mu\nu}\, \mathsf{b}_{\nu}$ for the scalar product in the (1+1)-dimensional Minkowski space (2.85). The notation $\mathrm{i}0^+$ stands for an infinitesimal positive imaginary number needed to regulate the pole of the integrand. The multiplicative prefactor $-(2\pi)^{-2}$ is convention. There are different possible options to regulate the poles of $1/(\mathsf{p}^2 - m^2)$. The choice made in Eqs (2.93b) and (2.93c) is named after Feynman. This choice is inconsequential with regard to the consequence of the scale covariance of the free (quadratic) action (2.91) for the function (2.93b), and hence for the function (2.93c). Now, scale covariance of the free (quadratic) action (2.91) is equivalent to the fact that

$$\Delta_{\mathrm{F}}(\mathsf{x};\xi)=\Delta_{\mathrm{F}}(\mathsf{x}/\xi;1). \tag{2.94}$$

Indeed, Eq. (2.94) follows from the change of integration variable defined by the relation $\mathsf{p}=:|m|\,\mathsf{k}$ and under which

$$\Delta_{\mathrm{F}}(\mathsf{x}/\xi;1)=-\frac{1}{(2\pi)^2}\int\mathrm{d}^2\mathsf{k}\,e^{-\mathrm{i}\mathsf{k}\cdot\mathsf{x}/\xi}\frac{1}{\mathsf{k}^2-1+\mathrm{i}0^+}. \tag{2.95}$$

One verifies (**Exercise** 2.16) that, when $\mathsf{x}=(0,x)$, the Feynman scalar propagator decays like

$$\Delta_{\mathrm{F}}(\mathsf{x}/\xi;1)\sim e^{-|x|/\xi} \tag{2.96}$$

for $|x|$ much larger than ξ, and so does the Feynman spinor propagator by Eq. (2.93c). If we identify the gap Δ in Eq. (2.61) with the Dirac mass m, we have derived the scaling relation encoded by the dynamical exponent $z=1$ between the mass gap and the diverging correlation length in the proximity of the quantum critical point at $m=0$.

Scale invariance of the free (quadratic) action (2.91) holds only at the quantum critical point $m=0$. Indeed, one verifies (**Exercise** 2.16) that, precisely at the critical point $m=1/\xi=0$ when $\mathsf{x}=(0,x)$, the Feynman scalar propagator grows logarithmically

$$\Delta_{\mathrm{F}}(\mathsf{x},\xi=\infty)\sim-\frac{\mathrm{i}}{2\pi}\ln\left(\frac{|x|}{\mathfrak{a}}\right), \tag{2.97}$$

for $|x|$ much larger than $\mathfrak{a}$, while the Feynman spinor propagator decays like

$$S_{\mathrm{F}}(\mathsf{x},\xi=\infty)\sim\frac{\mathfrak{a}}{|x|} \tag{2.98}$$

because of Eq. (2.93c).

2.7 The Peierls Instability for Polyacetylene

2.7.1 The Su-Schrieffer-Heeger (SSH) Model

It is time to investigate the origin for the staggering of the electronic nearest-neighbor hopping defined by Eq. (2.8). To this end, we introduce the simplest possible one-dimensional lattice model that couples the electrons from the π band of polyacetylene to the acetylene molecules entering the chemical formula of polyacetylene. This model is called the Su-Schrieffer-Heeger (SSH) model.[23]

[23] W.-P. Su, J. R. Schrieffer, and A. J. Heeger, Phys. Rev. Lett. **42**(25), 1698–1701 (1979). [Su et al. (1979)].

First, we consider elastic deformations of the polyacetylene chain in Figure 2.1. We neglect torsion of the chain, that is, we consider deformations such that the sp^2 orbitals lie on a plane, with the π orbital perpendicular to this plane. In this case, the overlap of neighboring π orbitals only depends on the distance between neighboring carbon atoms. The i-th from N carbon atoms is located at position $r_i = R_i + u_i \in \mathbb{R}$, where $R_i = i\,\mathfrak{a}$ is the equilibrium location of the atom, and u_i is its displacement relative to this equilibrium position. The other degree of freedom of the i-th carbon atom of mass M is its momentum $p_i \in \mathbb{R}$ along the one-dimensional chain.

Second, we associate with Figure 2.1 the quantum Hamiltonian $\widehat{H}_{\text{phonons}}$ for a harmonic linear chain, whereby $\widehat{H}_{\text{phonons}}$ is defined by

$$\widehat{H}_{\text{phonons}} := \sum_{i=1}^{N} \left[\frac{\hat{p}_i^2}{2M} + \frac{\kappa}{2} \left(\hat{u}_i - \hat{u}_{i+1} \right)^2 \right] \tag{2.99a}$$

with the equal-time canonical commutation relations

$$\left[\hat{u}_i, \hat{p}_{i'} \right] := \mathrm{i}\hbar\, \delta_{i,i'} \tag{2.99b}$$

for any pair $i, i' = 1, \cdots, N$ and with the equal-time periodic boundary conditions

$$\hat{u}_i = \hat{u}_{i+N}, \qquad \hat{p}_i = \hat{p}_{i+N}. \tag{2.99c}$$

The dimensionfull constant $\kappa > 0$ is the elastic or spring constant. It originates implicitly from the σ-bond electrons in Figures 2.1, 2.3, and 2.4.

The normal modes of $\widehat{H}_{\text{phonons}}$ are called phonons. In the Heisenberg representation, they are given by (**Exercise** 2.17)

$$\hat{u}_i(t) = \frac{1}{\sqrt{N}} \sum_{l=1}^{N} \sqrt{\frac{\hbar}{2\,M\,\varpi_l}} \left[\hat{a}_l\, e^{+\mathrm{i}(q_l\, i - \varpi_l\, t)} + \hat{a}_l^{\dagger}\, e^{-\mathrm{i}(q_l\, i - \varpi_l\, t)} \right], \tag{2.100a}$$

$$\hat{p}_i(t) = \frac{-\mathrm{i}}{\sqrt{N}} \sum_{l=1}^{N} \sqrt{\frac{\hbar\, M\, \varpi_l}{2}} \left[\hat{a}_l\, e^{+\mathrm{i}(q_l\, i - \varpi_l\, t)} - \hat{a}_l^{\dagger}\, e^{-\mathrm{i}(q_l\, i - \varpi_l\, t)} \right], \tag{2.100b}$$

for $i = 1, \cdots, N$. Their dispersion is given by

$$\varpi_l := \sqrt{\frac{2\kappa}{M} (1 - \cos q_l)}, \qquad q_l := \frac{2\pi}{N} l, \tag{2.100c}$$

given some initial conditions specified by the N pairs of operators $\hat{a}_l^{\dagger}$ and $\hat{a}_l$. In turn, these operators obey the boson algebra

$$\left[\hat{a}_l, \hat{a}_{l'}^{\dagger} \right] = \delta_{l,l'}, \qquad \left[\hat{a}_l^{\dagger}, \hat{a}_{l'}^{\dagger} \right] = \left[\hat{a}_l, \hat{a}_{l'} \right] = 0, \tag{2.100d}$$

for $l,l'=1,\cdots,N$ and span the bosonic Fock space

$$\mathfrak{F}_{\text{phonons}} := \text{span}\left\{\prod_{l=1}^{N}\frac{\left(\hat{a}_l^{\dagger}\right)^{n_l}}{\sqrt{n_l!}}|0\rangle, \quad n_l=0,1,\cdots, \quad \hat{a}_l|0\rangle=0\right\}. \tag{2.100e}$$

These phonons are gapless with the linear dispersion $\varpi_l=\sqrt{\kappa/M}\,q_l+\mathcal{O}(q_l^2)$, that is $\sqrt{\kappa/M}$ plays the role of a characteristic speed.

Third, the only electrons that we shall explicitly retain from Figures 2.1, 2.3, and 2.4 are the π electrons. We postulate that their quantum dynamics, in the limit in which they are decoupled from the molecules they originate from, is governed by Hamiltonian (2.9) that we rename

$$\widehat{H}_{\text{electrons}} := -t\sum_{i=1}^{N}\sum_{\sigma=\uparrow,\downarrow}\left(\hat{c}_{i,\sigma}^{\dagger}\,\hat{c}_{i+1,\sigma}+\text{H.c.}\right)-\mu\sum_{i=1}^{N}\sum_{\sigma=\uparrow,\downarrow}\hat{c}_{i,\sigma}^{\dagger}\,\hat{c}_{i,\sigma}. \tag{2.101}$$

The creation and annihilation operators for the electrons obey the algebra (2.2b) and the periodic boundary conditions (2.2e). They span the Fock space (2.2c) that we rename $\mathfrak{F}_{\text{electrons}}$. Here, we have also introduced the real-valued chemical potential μ.

Fourth, the SSH model postulates an interaction between the phonons and the π electrons that is governed by the Hamiltonian

$$\widehat{H}_{\text{e-p}} := -\alpha\sum_{i=1}^{N}\left(\hat{u}_i-\hat{u}_{i+1}\right)\sum_{\sigma=\uparrow,\downarrow}\left(\hat{c}_{i,\sigma}^{\dagger}\,\hat{c}_{i+1,\sigma}+\text{H.c.}\right) \tag{2.102a}$$

with the electron-phonon coupling α carrying the dimension of energy per unit length. This Hamiltonian acts on the tensor product

$$\mathfrak{F}_{\text{SSH}} := \mathfrak{F}_{\text{phonons}}\otimes\mathfrak{F}_{\text{electrons}}. \tag{2.102b}$$

All together, we arrive at the SSH Hamiltonian

$$\widehat{H}_{\text{SSH}} := \widehat{H}_{\text{phonons}}+\widehat{H}_{\text{electrons}}+\widehat{H}_{\text{e-p}} \tag{2.103a}$$

acting on the Fock space $\mathfrak{F}_{\text{SSH}}$. The partition function is defined by [recall Eq. (2.60)]

$$Z_{\text{SSH}} := \lim_{N\to\infty}\text{Tr}_{\mathfrak{F}_{\text{SSH}}}\left(e^{-\beta\widehat{H}_{\text{SSH}}}\right), \qquad \beta=\frac{1}{k_{\text{B}}T}. \tag{2.103b}$$

Here, the thermodynamic limit $N\to\infty$ is taken holding the filling fraction per spin

$$\nu_{\text{e}} := \frac{1}{2N\beta}\frac{\partial Z_{\text{SSH}}}{\partial\mu} \tag{2.103c}$$

fixed to any number between 0 and 1. Equation (2.103) defines the SSH model in thermodynamic equilibrium at the inverse temperature $\beta=1/(k_{\text{B}}T)$.

We have already solved the SSH model (2.103) in the limit $\alpha = 0$, for which the phonons decouple from the electrons. The SSH model is not exactly solvable (integrable) when $\alpha \neq 0$ and $0 \leq M < \infty$. Integrability for any nonvanishing electron-phonon coupling $\alpha \neq 0$ can be, however, recovered if the phonons do not undergo a zero-point motion, that is, in the classical limit $M \to \infty$ for the phonons. This limit was solved in closed form within a saddle-point approximation by SSH on the lattice and by Takayama, Lin-Liu, and Maki[24] in the scaling limit defined in Section 2.6.

The classical limit for the phonons can be obtained by comparing the zero point motion of the ions to the lattice spacing. This exercise yields the condition (**Exercise** 2.18)

$$c\,\frac{\hbar}{\sqrt{M\,\kappa}} < \mathfrak{a}^2 \tag{2.104}$$

with c a numerical constant of order 1. This condition is trivially satisfied in the limit when $\hbar \to 0$; it is also satisfied in the limit of heavy acetylene masses, $M \to \infty$. In the latter case, we may replace the operators $\hat{u}_i$ by u_i, any real-valued eigenvalue of $\hat{u}_i$. In doing so, one selects the subspace

$$\mathfrak{F}^{\text{s-c}}_{\text{SSH}} \subset \mathfrak{F}_{\text{SSH}} \tag{2.105}$$

that is isomorphic to $\mathfrak{F}_{\text{electrons}}$. This limit allows us to replace the phonon Hamiltonian (2.99a) by the classical Hamiltonian (with all hats removed from the operators)

$$H_{\text{phonons}} := \sum_{i=1}^{N} \left[\frac{\kappa}{2}\left(u_i - u_{i+1}\right)^2\right], \tag{2.106}$$

We may also replace the electron-phonon Hamiltonian (2.102a) by the semiclassical Hamiltonian

$$\widehat{H}^{\text{s-c}}_{\text{e-p}} := -\sum_{i=1}^{N} \sum_{\sigma=\uparrow,\downarrow} \alpha\left(u_i - u_{i+1}\right)\left(\hat{c}^{\dagger}_{i,\sigma}\,\hat{c}_{i+1,\sigma} + \text{H.c.}\right). \tag{2.107}$$

Hamiltonian $\widehat{H}^{\text{s-c}}_{\text{e-p}}$ is nothing but Hamiltonian (2.2a) with the identification $t_i \to \alpha\left(u_i - u_{i+1}\right)$. Thus, the semiclassical limit (2.104) of the SSH model (2.103) is

$$\widehat{H}^{\text{s-c}}_{\text{SSH}} := H_{\text{phonons}} + \widehat{H}_{\text{electrons}} + \widehat{H}^{\text{s-c}}_{\text{e-p}} \tag{2.108a}$$

with the partition function

$$Z^{\text{s-c}}_{\text{SSH}} := \lim_{N\to\infty} \text{Tr}_{\mathfrak{F}^{\text{s-c}}_{\text{SSH}}}\left(e^{-\beta \widehat{H}^{\text{s-c}}_{\text{SSH}}}\right), \qquad \beta = \frac{1}{k_{\text{B}}\,T}, \tag{2.108b}$$

whereby the thermodynamic limit $N \to \infty$ is taken holding the filling fraction per spin (2.103c) fixed to any number between 0 and 1.

[24] H. Takayama, Y. R. Lin-Liu, and K. Maki, Phys. Rev. B **21**(6), 2388—2393 (1980). [Takayama et al. (1980)].

Even though

$$\widehat{H}_{\text{electrons}} + \widehat{H}^{\text{s-c}}_{\text{e-p}} = -\sum_{i=1}^{N} \sum_{\sigma=\uparrow,\downarrow} \Big\{ \left[t + \alpha\left(u_i - u_{i+1}\right)\right] \times \left(\hat{c}^{\dagger}_{i,\sigma}\, \hat{c}_{i+1,\sigma} + \text{H.c.}\right) + \mu\, \hat{c}^{\dagger}_{i,\sigma}\, \hat{c}_{i,\sigma} \Big\} \tag{2.109}$$

is considerably simpler than $\widehat{H}_{\text{electrons}} + \widehat{H}_{\text{e-p}}$, it cannot be solved in closed form for arbitrary choices of $u_1, \cdots, u_N$. However, the staggered choice [recall Eq. (2.8)]

$$u_i - u_{i+1} = (-1)^i\, u \Longleftrightarrow t_i := t + (-1)^i\, \delta t, \qquad \delta t := \alpha\, u, \tag{2.110}$$

with u any real-valued number, delivers the closed solution of Section 2.5.4. Moreover, this choice is self-consistent at half-filling, that is, $\nu_{\text{e}} = 1/2$, as we now show.

2.7.2 Mean-Field Solution for the SSH Model at Half-Filling

We choose the filling fraction $\nu_{\text{e}} = 1/2$. We define the continuous family of states

$$|\Psi(u)\rangle := |u\rangle \otimes |\Psi_{\text{FS}}\rangle \in \mathfrak{F}_{\text{SSH}}, \tag{2.111a}$$

which we label by the real-valued number u, to be the direct product between the simultaneous (many-body) eigenstate of all the position operators for the acetylene molecules, whereby

$$\hat{u}_i\, |u\rangle = u_i\, |u\rangle, \qquad u_i = (-1)^i\, \frac{u}{2}, \qquad u \in \mathbb{R}, \tag{2.111b}$$

and the (many-body) Fermi sea

$$|\Psi_{\text{FS}}\rangle := \prod_{k\in\Lambda^{A\star}_{\text{BZ}}}^{|k|<\pi/2} \prod_{\sigma=\uparrow,\downarrow} \hat{\chi}^{\dagger}_{k,\sigma,-}\, |0\rangle \tag{2.111c}$$

shown in Figure 2.16, where the creation operators $\hat{\chi}^{\dagger}_{k,\sigma,-}$ are defined by Eqs. (2.50) and (2.58b). By construction, $|\Psi(u)\rangle$ is an eigenstate of $\widehat{H}^{\text{s-c}}_{\text{SSH}}$ defined in Eq. (2.108a) with the many-body eigenvalue

$$E(u) = \frac{1}{2}\, \kappa\, u^2\, N - 2 \times 2 \sum_{k\in\Lambda^{A\star}_{\text{BZ}}}^{|k|<\pi/2} \sqrt{t^2\, \cos^2 k + (\alpha\, u)^2\, \sin^2 k}. \tag{2.112a}$$

This many-body energy eigenvalue is the difference between the energy cost

$$E^{\text{s-c}}_{\text{phonons}}(u) := \frac{1}{2}\, \kappa\, u^2\, N \tag{2.112b}$$

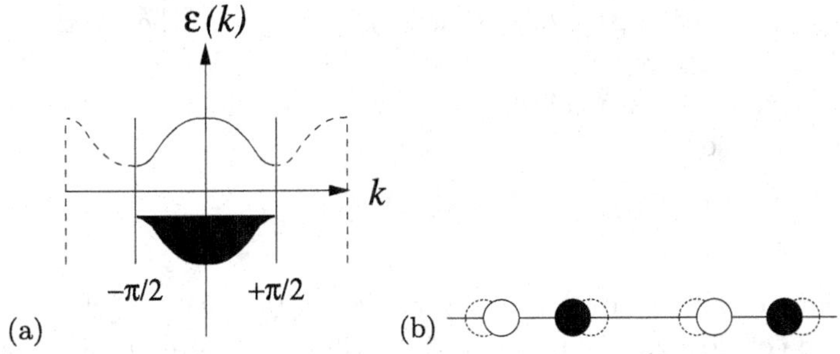

Figure 2.16 Filled Fermi sea at half-filling (a) for polyacetylene with one (b) of the two possible weak dimerization patterns for the CH molecules that we here represent by alternating empty or filled circles. (The dotted circles represent the equilibrium positions of the CH molecules along a chain with the lattice spacing $\mathfrak{a}$.)

incurred when distorting the acetylene molecules away from their equilibrium positions by the staggered displacement $\pm u/2$ and the energy gain

$$E^{\text{s-c}}_{\text{e-p}}(u) := -2 \times 2 \sum_{k \in \Lambda^{A\star}_{\text{BZ}}}^{|k|<\pi/2} \sqrt{t^2 \cos^2 k + (\alpha\, u)^2 \sin^2 k} \tag{2.112c}$$

resulting from opening a gap at the Fermi surface (one of the two multiplicative factors of 2 originates from the spin-1/2 twofold degeneracy). (The latter gain follows from the fact that the last occupied single-particle energy is lowered relative to the chemical potential $\mu = 0$ at half-filling with the opening of the electronic single-particle gap.)

We are going to locate the minimum of the function $E(u)$ defined by Eq. (2.112) in the thermodynamic limit $N \to \infty$ holding $\mathfrak{a}$ fixed. To this end, we need to evaluate the right-hand side of the electronic energy gain (2.112c).

In the thermodynamic limit $N \to \infty$ holding the lattice spacing fixed, we may replace the sum over the wave numbers k from the reduced Brillouin zone by the integral

$$E^{\text{s-c}}_{\text{e-p}}(u) = -\,4 \int\limits_{-\pi/2}^{+\pi/2} \frac{\mathrm{d}k}{2\pi/N} \sqrt{t^2 \cos^2 k + (\alpha\, u)^2 \sin^2 k}. \tag{2.113}$$

Since the integrand is an even function of the wave number, we may write

$$\frac{E(u)}{N} = \frac{1}{2}\kappa\, u^2 - \frac{8\,t}{2\pi} \int\limits_{0}^{\pi/2} \mathrm{d}k \sqrt{1 - \left[1 - \left(\frac{\alpha\, u}{t}\right)^2\right] \sin^2 k}, \tag{2.114}$$

to leading order in an expansion in powers of $1/N$. All single-particle electronic energies have here been expressed in units of t.

When the dimerization is strong, that is $1 \ll (\alpha\, u/t)^2$, the elastic energy cost dominates over the electronic energy gain so that $E(u)/N \sim u^2$.

When the dimerization is weak, that is $1 \gg (\alpha\, u/t)^2$, it is useful to introduce the dimensionless function

$$-1 \ll z(u) := \frac{\alpha\, u}{t} \ll +1 \tag{2.115}$$

of the staggered displacement that is parameterized by u. We may then express the right-hand side of Eq. (2.114) in terms of the elliptic integral of the second kind

$$E\left(\varphi|x^2\right) := \int\limits_0^{\varphi} \mathrm{d}\theta\, \sqrt{1 - x^2 \sin^2\theta}. \tag{2.116}$$

Thus, the many-body eigenstate (2.111) has the many-body energy eigenvalue

$$\frac{E(u)}{N} = \frac{1}{2}\kappa\, u^2 - \frac{8\,t}{2\pi} E\left(\frac{\pi}{2}\middle| 1 - z^2(u)\right), \tag{2.117}$$

to leading order in an expansion in powers of $1/N$. With the help of the expansion

$$E\left(\frac{\pi}{2}\middle| 1 - z^2\right) = 1 + \left(\frac{z}{2}\right)^2 \left[\ln\left(\frac{16}{z^2}\right) - 1\right], \qquad z^2 \ll 1, \tag{2.118}$$

for the complete elliptic integral of the second kind,

$$\frac{E(u)}{N} = -\frac{8\,t}{2\pi} - \left\{\frac{2\,t}{2\pi}\left(\frac{\alpha}{t}\right)^2 \left[\ln\left(\left(\frac{4\,t}{\alpha\, u}\right)^2\right) - 1\right] - \frac{1}{2}\kappa\right\} u^2, \tag{2.119}$$

to leading order in an expansion in powers of $1/N$ and to the first three leading orders in an expansion in powers of $\alpha\, u/t$. Expansion (2.119) implies that $E(u = 0)/N = -8t/(2\pi)$ is a local maximum as the bracketed term on the right-hand side of Eq. (2.119) is always positive for sufficiently small relative displacement u.

We conclude that the many-body eigenstate (2.111) has the many-body energy eigenvalue (2.114). The latter is an even function of u that grows quadratically with u when $\alpha\, u/t \gg 1$, reaches a local maximum at $u = 0$, and has thus at least two absolute minima at $\pm|u_{\mathrm{MF}}| \neq 0$. The dependence on u of the many-body energy eigenvalue (2.114) can be evaluated numerically. It delivers the two absolute minima $\pm|u_{\mathrm{MF}}| \neq 0$ as shown in Figure 2.17. These minima are the roots of

$$\kappa = \frac{8\,\alpha^2}{2\pi\, t} \int\limits_0^{\pi/2} \mathrm{d}k \frac{\sin^2 k}{\sqrt{\cos^2 k + (\alpha\, u/t)^2 \sin^2 k}}. \tag{2.120}$$

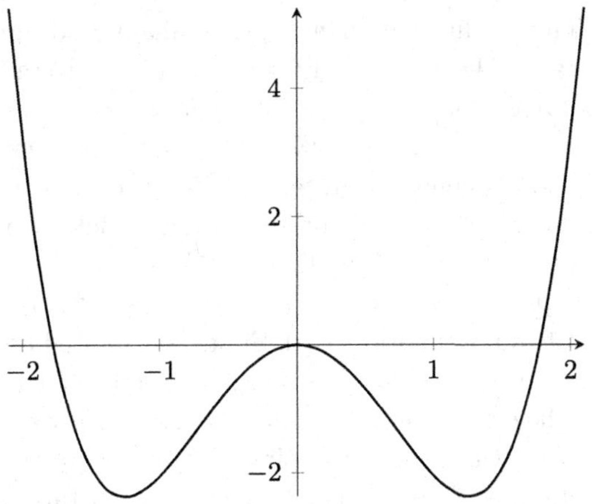

Figure 2.17 The qualitative dependence on u defined in Eq. (2.110) of the many-body energy eigenvalue (2.114) in the semiclassical approximation (2.108a).

In the weak coupling limit $(\alpha\, u)^2/t^2 \ll 1$, the estimate[25]

$$\int_0^{\pi/2} \mathrm{d}k\, \frac{\sin^2 k}{\sqrt{\cos^2 k + (\alpha\, u/t)^2 \sin^2 k}} \approx \int_0^{\pi/2} \mathrm{d}k\, \frac{1}{\sqrt{k^2 + (\alpha\, u/t)^2}} \approx \ln\left(\frac{\pi}{|\alpha\, u/t|}\right) \tag{2.121}$$

follows for the integral on the right-hand side. This gives the roots

$$u_{\mathrm{MF}} \approx \pm \frac{\pi\, t}{|\alpha|} \exp\left(-\frac{2\pi\, \kappa\, t}{8\,\alpha^2}\right). \tag{2.122}$$

The dependence of these two roots on the electron-phonon coupling is singular at $\alpha = 0$. The quantum critical point $\alpha = 0$ at which the phonons decouple from the electrons represents an essential singularity of the dependence on α of the mean-field electronic gap

$$2\,|\alpha\, u_{\mathrm{MF}}| \approx 2\,\pi\, t \exp\left(-\frac{2\pi\, \kappa\, t}{8\,\alpha^2}\right) \tag{2.123}$$

(**Exercise** 2.19).

The mean-field solution (2.123) to the SSH model shows that the nature of the many-body ground state of an harmonic chain can be changed qualitatively by coupling the phonons to a nonvanishing density of electrons for an arbitrarily small

[25] The first approximation follows from identifying the pole of the integrand at $k = \pi/2$ in the limit $\alpha = 0$. The second approximation follows from $\frac{\mathrm{d}}{\mathrm{d}x} \ln\left(x + \sqrt{x^2 + a^2}\right) = \frac{1}{\sqrt{x^2+a^2}}$ with $\sqrt{1+x^2} = 1 + \frac{x^2}{2} + \cdots$ and $\ln(1+x) = x + \cdots$.

coupling strength. The qualitative change in the many-body ground state arises in this example because the many-body ground state selected by the electron-phonon interaction breaks spontaneously a translation symmetry by one lattice spacing down to a translation symmetry by two lattice spacings, whenever the electronic filling fraction is commensurate with the discrete translation symmetry group of the SSH model. The justification for the mean-field approximation can be traced to the smallness of the dimensionless ratio between the mass of the electron and the mass of the proton in vacuum. The smallness of this ratio suggests the following hierarchy of approximations. (i) First, the many-body ground state is approximated by the direct product of a many-body state for the phonons with a many-body state for the electrons. (ii) Second, the many-body state for the phonons is devoid of quantum fluctuations (zero-point motion). If so, the many-body state for the electrons is obtained from the single-particle eigenstates and eigenvalues of the electrons subjected to a classical static background encoding a classical many-body state for the phonons. Quantum fluctuations in the many-body state for the phonons can be accounted for by replacing the mean-field approximation by the so-called random-phase approximation. This is achieved by relaxing approximation (ii). However, if the electron-phonon coupling becomes sufficiently strong, the very first approximation (i) can break down. Nevertheless, as crude as it is, the mean-field approximation for the SSH model is intrinsically of a non-perturbative nature. We are going to pursue in Chapter 3 the study of the SSH model within approximations (i) and (ii) (that is Fermiology) with the twist that we allow for space defects in the static dimerization pattern.

The mean-field treatment of the SSH model is very similar to that of one-dimensional commensurate charge-density waves (**Exercise** 2.20).

2.8 Exercises

2.1 In classical mechanics, a free point-like particle with mass m and charge e located at the position $\boldsymbol{x}(t) \in \mathbb{R}^3$ at time t obeys Newton equations

$$\ddot{\boldsymbol{x}} = 0. \tag{2.124a}$$

This second-order differential equation in time is invariant under reversal of time

$$t \mapsto t' := -t. \tag{2.124b}$$

Under reversal of time, momentum

$$\boldsymbol{p} := m\,\dot{\boldsymbol{x}} \tag{2.124c}$$

and angular momentum

$$\boldsymbol{L} := \boldsymbol{x} \wedge m\,\dot{\boldsymbol{x}} \tag{2.124d}$$

are odd. In nonrelativistic quantum mechanics for a spinless and a spin-1/2 particle, Newton equation is replaced by the Schrödinger equation

$$\mathrm{i}\hbar\frac{\partial}{\partial t}\psi(\boldsymbol{x},t)=\left\{\frac{1}{2m}\left[\hat{\boldsymbol{p}}-\frac{e}{c}\,\boldsymbol{A}(\boldsymbol{x},t)\right]^2+e\,A_0(\boldsymbol{x},t)+V(\boldsymbol{x},t)\right\}\psi(\boldsymbol{x},t) \quad (2.125)$$

and the Pauli equation

$$\begin{aligned}\mathrm{i}\hbar\frac{\partial}{\partial t}\Psi(\boldsymbol{x},t)=&\left\{\frac{1}{2m}\left[\boldsymbol{\sigma}\cdot\left(\hat{\boldsymbol{p}}-\frac{e}{c}\,\boldsymbol{A}(\boldsymbol{x},t)\right)\right]^2+\sigma_0\,e\,A_0(\boldsymbol{x},t)\right.\\&\left.+\sigma_0\,V(\boldsymbol{x},t)\right\}\Psi(\boldsymbol{x},t),\end{aligned} \quad (2.126)$$

respectively. Here, $\hat{\boldsymbol{p}}\equiv\hbar\partial/\mathrm{i}\partial\boldsymbol{x}$, $\boldsymbol{A}$ is a background vector potential, A_0 is a background scalar potential, and V is a potential, while σ_0 is the unit 2×2 matrix acting on the spin-1/2 degrees of freedom and $\boldsymbol{\sigma}$ is the associated vector of Pauli matrices. Denote by K the operation of complex conjugation. Verify the following.

2.1.1 K and $\sigma_2\,\mathsf{K}$ act antilinearly on the Hilbert spaces for spinless and spin-1/2 particles, respectively.

2.1.2 $\mathsf{K}^2=1$ and $(\sigma_2\,\mathsf{K})^2=-1$.

2.1.3 $\mathsf{K}\,\hat{\boldsymbol{x}}\,\mathsf{K}=+\hat{\boldsymbol{x}}, \qquad \mathsf{K}\,\hat{\boldsymbol{p}}\,\mathsf{K}=-\hat{\boldsymbol{p}}, \qquad \sigma_2\,\mathsf{K}\,\boldsymbol{\sigma}\,\mathsf{K}\,\sigma_2=-\boldsymbol{\sigma}.$

2.1.4 The Schrödinger and Pauli equations are invariant under the operation of time reversal that is defined by

$$\begin{aligned}&\boldsymbol{x}\mapsto\boldsymbol{x}':=+\boldsymbol{x}, \qquad \boldsymbol{t}\mapsto\boldsymbol{t}':=-\boldsymbol{t},\\&\boldsymbol{A}(\boldsymbol{x},t)\mapsto\boldsymbol{A}'(\boldsymbol{x}',t'):=-\boldsymbol{A}(\boldsymbol{x}',-t'),\\&A_0(\boldsymbol{x},t)\mapsto A_0'(\boldsymbol{x}',t'):=+A_0(\boldsymbol{x}',-t'),\\&V(\boldsymbol{x},t)\mapsto V'(\boldsymbol{x}',t'):=+V(\boldsymbol{x}',-t'),\\&\psi(\boldsymbol{x},t)\mapsto\psi'(\boldsymbol{x}',t'):=\mathsf{K}\,\psi(\boldsymbol{x}',-t'),\\&\Psi(\boldsymbol{x},t)\mapsto\Psi'(\boldsymbol{x}',t'):=\sigma_2\,\mathsf{K}\,\Psi(\boldsymbol{x}',-t').\end{aligned}$$

2.2 For any pair of the transformations (2.3a), (2.3b), (2.4b), (2.5b), and (2.6) compute their commutator.

2.3 Let $\mathcal{K}$ be either a unitary or antiunitary transformation of the Hilbert space on which the single-particle Hermitean Hamiltonian $\mathcal{H}$ acts. Assume that $\mathcal{K}^\dagger\,\mathcal{H}\,\mathcal{K}=-\mathcal{H}$, that is $\mathcal{K}$ anticommutes with $\mathcal{H}$. Show that if Ψ is an eigenstate of $\mathcal{H}$ with eigenvalue $\varepsilon\neq 0$, then $\mathcal{K}^\dagger\,\Psi$ is an eigenstate of $\mathcal{H}$ with eigenvalue $-\varepsilon\neq 0$.

2.4 Verify that the Brillouin zone (2.10c) is compatible with periodic boundary conditions. Enumerate all the momenta from the Brillouin zone (2.10c) when $N=2,3,4,5$. This counting is useful to motivate the choice (2.12a).

2.5 Verify Eq. (2.10d) by applying Eq. (2.10a) to Eq. (2.2b).

2.6 Verify Eq. (2.11) by applying Eq. (2.10) to Eq. (2.9).

2.7 In the thermodynamic limit, the summations in Eqs. (2.13) and (2.14) can be expressed as integrals. Calculate these integrals in the thermodynamic limit $N \to \infty$, $N_{\mathrm{e}} \to \infty$, holding the ratio $0 \leq N_{\mathrm{e}}/(2N) \leq 1$ fixed. Verify that the two expressions you have obtained obey the relation (2.15).

2.8 Verify that the total momentum operator (2.16) and the total spin operator (2.17) both commute with Hamiltonian (2.9). Verify that the Fermi sea is annihilated by both operators in the thermodynamic limit in two ways. [*Hint*: First, take the thermodynamic limit and evaluate the Fermi sea eigenvalues of the total momentum and total spin by performing the corresponding Riemann integrations over the Brillouin zone. Second, consider all three cases for which a finite number $N_{\mathrm{e}} \gg 6$ of electrons fails to satisfy the choice (2.12a). For each case, estimate the correction to the total momentum and total spin of the Fermi sea compared to the situation when Eq. (2.12a) holds.]

2.9 Verify Eq. (2.28a).

2.10 Verify Eq. (2.28b).

2.11 Single-particle energy eigenvalues of Hamiltonian (2.2a) are discrete if the number N of lattice sites is finite. According to Exercise 2.3, the chiral symmetry of Hamiltonian (2.2a) implies a pairing of single-particle energy eigenstates provided their single-particle energy eigenvalues are nonvanishing. What can be said of zero modes, the single-particle energy eigenstates with vanishing single-particle energy eigenvalue, as a function of the parity (evenness or oddness) of N? To answer this question, write down the single-particle Hamiltonian corresponding to Hamiltonian (2.2a) as a $2N \times 2N$ Hermitean matrix in the orbital basis with (i) open boundary conditions (with $t_N = 0$) and (ii) periodic boundary conditions (with $t_N \neq 0$). What is the fate of a (the) zero mode(s)

2.11.1 if translation invariance of Hamiltonian (2.2a) with $t_i = t$ for $i = 1, \cdots, N-1$ and $t_N = 0$ is broken by the additive perturbation

$$\delta\widehat{H} := -\sum_{i=1}^{N} \sum_{\sigma=\uparrow,\downarrow} \delta t_i \left(\hat{c}^{\dagger}_{i,\sigma}\, \hat{c}_{i+1,\sigma} + \mathrm{H.c.}\right), \tag{2.127}$$

where $\delta t_i \in \mathbb{R}$ for $i = 1, \cdots, N-1$ and $\delta t_N = 0$ under open boundary conditions?

2.11.2 if time-reversal symmetry and spin-rotation symmetry of Hamiltonian (2.2a) is broken by the additive perturbation

$$\delta\widehat{H} := -\delta B \sum_{i=1}^{N-1} \sum_{\sigma=\uparrow,\downarrow} (-1)^{\sigma} \hat{c}^{\dagger}_{i,\sigma}\, \hat{c}_{i+1,\sigma}, \tag{2.128}$$

where $\delta B \in \mathbb{R}$ and $(-1)^{\sigma} = +1$ if $\sigma =\uparrow$, while $(-1)^{\sigma} = -1$ if $\sigma =\downarrow$ under open boundary conditions?

2.11.3 if chiral symmetry of Hamiltonian (2.2a) is broken by the additive perturbation

$$\delta\widehat{H} := -\delta\mu \sum_{i=1}^{N} \sum_{\sigma=\uparrow,\downarrow} \hat{c}^{\dagger}_{i,\sigma}\,\hat{c}_{i,\sigma}, \tag{2.129}$$

where $\delta\mu \in \mathbb{R}$ under open boundary conditions?

2.12 Construct the 2×2 unitary transformation relating the $\hat{\chi}_{k,\sigma}$ basis used in Eq. (2.25) to the $\hat{\psi}_{k,\sigma}$ basis used in Eq. (2.34a) for any k in the folded band and for any σ. Verify Eq. (2.34d).

2.13 Verify that the right-hand side of Eq. (2.46c) equals the right-hand side of Eq. (2.39b).

2.14 We assume that polyacetylene is approximated by the model (2.57c). For simplicity, we ignore the spin index. Thus, we start from the single-particle Hamiltonian

$$\mathcal{H}(p) := -\left[t_1 + t_2 \cos\Big((p_{\mathrm{F}} + p)\,2\mathfrak{a}\Big)\right]\tau_1 - t_2 \sin\Big((p_{\mathrm{F}} + p)\,2\mathfrak{a}\Big)\,\tau_2, \tag{2.130a}$$

where

$$t_1 := t - \delta t, \qquad t_2 := t + \delta t, \qquad |\delta t| < t, \tag{2.130b}$$

and

$$p_{\mathrm{F}} := \frac{\pi}{2\mathfrak{a}}. \tag{2.130c}$$

Assume that δt is of the same order as p. Verify that, to leading order in powers of p or δt,

$$\mathcal{H}(p) \approx v_{\mathrm{F}}\,p\,\tau_2 + m_{\mathrm{F}}\,v_{\mathrm{F}}^2\,\tau_1, \qquad v_{\mathrm{F}} := 2\mathfrak{a}\,t, \qquad m_{\mathrm{F}}\,v_{\mathrm{F}}^2 := 2\delta t. \tag{2.131}$$

This representation is not unique, however. To see this, do the gauge transformation [recall Eq. (2.31a)]

$$\hat{a}_j \mapsto \mathrm{i}^{2j-1}\,\hat{a}_j, \qquad \hat{b}_j \mapsto \mathrm{i}^{2j}\,\hat{b}_j \tag{2.132a}$$

and verify that

$$\mathcal{H}(p) \approx v_{\mathrm{F}}\,p\,\tau_1 + m_{\mathrm{F}}\,v_{\mathrm{F}}^2\,\tau_2, \qquad v_{\mathrm{F}} := 2\mathfrak{a}\,t, \qquad m_{\mathrm{F}}\,v_{\mathrm{F}}^2 := 2\delta t, \tag{2.132b}$$

that is the gauge transformation is implemented by an interchange of τ_1 and τ_2 after linearization.

2.15 How do the eigenvalues of the Dirac Hamiltonian (2.82) change if the term

$$m'\,\tau_3 \tag{2.133}$$

with m' a constant real number is added to the Dirac Hamiltonian (2.82)? Add to the lattice Hamiltonian (2.57c) describing polyacetylene a one-body contribution that produces the term (2.133) in the Dirac Hamiltonian obtained after linearization (the answer is not unique). What are the lattice symmetries that are broken by the one-body contribution from which m' originates? Conversely, what symmetries are broken by m' that are not broken by m_{F}?

2.16 Define the following scalar functions by the values they take at a point x in the (1+1)-dimensional Minkowski space defined in Eq. (2.85):

$$\Delta_{\mathrm{F}}(\mathsf{x}) := -\int \frac{d^2\mathsf{p}}{(2\pi)^2}\, e^{-\mathrm{i}\mathsf{p}\cdot\mathsf{x}}\, \frac{1}{\mathsf{p}^2 - m^2 + \mathrm{i}0^+}, \tag{2.134a}$$

$$\Delta_{+}(\mathsf{x}) := +\int \frac{d^2\mathsf{p}}{(2\pi)^2}\, e^{-\mathrm{i}\mathsf{p}\cdot\mathsf{x}}\, \Theta(\mathsf{p}^0)\, \delta(\mathsf{p}^2 - m^2), \tag{2.134b}$$

$$\Delta(\mathsf{x}) := -\mathrm{i}\int \frac{d^2\mathsf{p}}{(2\pi)^2}\, e^{-\mathrm{i}\mathsf{p}\cdot\mathsf{x}}\, \mathrm{sgn}(\mathsf{p}^0)\, \delta(\mathsf{p}^2 - m^2). \tag{2.134c}$$

Here, 0^+ is a positive infinitesimal number, 0^- is a negative infinitesimal number, Θ is the Heaviside function that vanishes when its argument is negative and is unity otherwise, and $\mathrm{sgn}(\mathsf{p}^0) = \Theta(+\mathsf{p}^0) - \Theta(-\mathsf{p}^0)$. Show that

$$\Delta_{\mathrm{F}}(\mathsf{x}) = \frac{\mathrm{i}}{2\pi}\, K_0(|m|\sqrt{-\mathsf{x}^2 + \mathrm{i}\,0^+}), \tag{2.135a}$$

$$\Delta_{+}(\mathsf{x}) = \frac{1}{(2\pi)^2}\, K_0(|m|\sqrt{-\mathsf{x}^2 + \mathrm{i}\,\mathsf{x}^0\,0^+}), \tag{2.135b}$$

$$\Delta(\mathsf{x}) = -\frac{1}{4\pi}\, \Theta(\mathsf{x}^2)\, \mathrm{sgn}(\mathsf{x}^0)\, J_0(|m|\sqrt{\mathsf{x}^2}), \tag{2.135c}$$

where, for any integer $n = 0, 1, 2, \cdots$,

$$J_n(z) := \frac{1}{2\pi}\int_{-\pi}^{+\pi} d\theta\, e^{-\mathrm{i}\,(n\,\theta - z\,\sin\theta)}, \qquad z \in \mathbb{C}, \tag{2.136}$$

is the integral representation of the Bessel function of the first kind and

$$K_n(z) := \int_{0}^{+\infty} dy\, e^{-z\cosh y}\, \cosh(n\,y), \qquad |\arg z| < \pi/2, \tag{2.137}$$

is the integral representation of the modified Bessel function of the second kind. Use Eq. (2.135) to prove Eq. (2.97).

2.17 Show that Hamiltonian (2.99) for a quantum harmonic chain takes the diagonal form

$$\widehat{H}_{\mathrm{phonons}} := \sum_{l=1}^{N} \hbar\,\varpi_l \left(\hat{a}_l^{\dagger}\,\hat{a}_l + \frac{1}{2}\right) \tag{2.138}$$

in terms of the phonon operators defined in Eq. (2.100).

2.18 Prove the inequality (2.104). *Hint 1:* Start from the uncertainity relation

$$\Delta x\,\Delta p \geq \frac{\hbar}{2} \Longleftrightarrow \Delta p \geq \frac{\hbar}{2\,\Delta x}. \tag{2.139}$$

Choose Δx to be the lattice spacing $\mathfrak{a}$. Needed is an estimate of Δp. To this end, compute the second cumulant of the momentum operator in the state annihilated by any $\hat{a}_l$ in Eq. (2.100b). *Hint 2:* Consider a single harmonic oscillator for a particle of mass M and with the spring constant κ. Compute the second cumulant of the position operator in the ground state of the Harmonic oscillator and demand that this cumulant is much smaller than $\mathfrak{a}^2$. Compare the numerical constants in the inequality (2.104) obtained using these two hints. Are they the same and if they are not, why?

2.19 Derive the counterparts to Eqs. (2.122) and (2.123) when using the effective Dirac Hamiltonian defined in Eq. (2.82) instead of the full lattice model for the electrons in the SSH model.

2.20 We are going to modify the SSH Hamiltonian (2.103) by replacing the electron-phonon coupling (2.102) with

$$\widehat{H}^{\mathrm{CDW}}_{\text{e-p}} := -\alpha_{\mathrm{CDW}} \sum_{i=1}^{N} \left(\hat{u}_i - \hat{u}_{i+1}\right) \sum_{\sigma=\uparrow,\downarrow} \hat{c}^{\dagger}_{i,\sigma}\, \hat{c}_{i,\sigma}. \tag{2.140}$$

This Hamiltonian is the simplest toy model for a commensurate one-dimensional charge-density wave.[26]

2.20.1 Repeat all the counterparts to the steps that led to Eq. (2.123).

2.20.2 What are the symmetries obeyed by the mean-field fermionic Hamiltonian that describes the charge-density wave instability?

2.20.3 What is the charge-density wave counterpart to the effective Dirac Hamiltonian (2.82)? (*Hint:* The answer can be guessed without any calculation using symmetry arguments.)

[26] G. Grüner, Rev. Mod. Phys. **60**(4), 1129–1181 (1988). [Grüner (1988)].

3

Fractionalization in Polyacetylene

3.1 Introduction

The phenomenon of fractionalization in polyacetylene was predicted theoretically by Su, Schrieffer, and Heeger in 1979.[1] It relies on the interplay between three quantum numbers. First, there is the electric charge of electrons, a local attribute. Second, there is the spin-1/2 of the electron, another local attribute. Third, there is the topological charge that characterizes the asymptotic behavior of an order parameter, a bond-density wave. The meaning of fractionalization in polyacetylene is that the charge and spin quantum numbers of the many-body electronic ground state when the bond-density wave, treated as a classical background, is topologically trivial, differ from those of the many-body electronic ground state when the bond-density wave is topologically nontrivial.

A seminal theoretical analysis with the same outcome had been made independently in the setting of quantum field theory by Jackiw and Rebbi in 1976.[2] Jackiw and Rebbi showed that the massive Dirac equation in (1+1)-dimensional space and time admits a zero mode when the mass term supports a single domain wall. They also showed that upon second quantization of the Dirac equation in the presence of this zero mode, the spectrum of the many-body fermion charge operator is shifted by 1/2 relative to that of the many-body fermion charge operator for a constant mass term.

We will follow this narrative. We begin with the analysis of Su, Schrieffer, and Heeger that applies to an approximation of polyacetylene in terms of one-dimensional tight-binding model for spinless fermions (we ignore the conserved spin-1/2 degrees of freedom of electrons in polyacetylene for simplicity) in Sections 3.2, 3.3, and 3.4.

We then move to the approach favored by Jackiw and Rebbi, namely, one based on quantum field theory. We compute the conserved fermion charge induced by an order parameter, the background, that supports point-like defects with their own conserved topological charges using scattering theory in Section 3.5, supersymmetry

[1] W. P. Su, J. R. Schrieffer, and A. J. Heeger, Phys. Rev. Lett. **42**(25), 1698–1701 (1979); Phys. Rev. B **22**(4), 2099–2111 (1980). [Su et al. (1979, 1980)].

[2] R. Jackiw and C. Rebbi, Phys. Rev. D **13**(12), 3398–3409 (1976). [Jackiw and Rebbi (1976)].

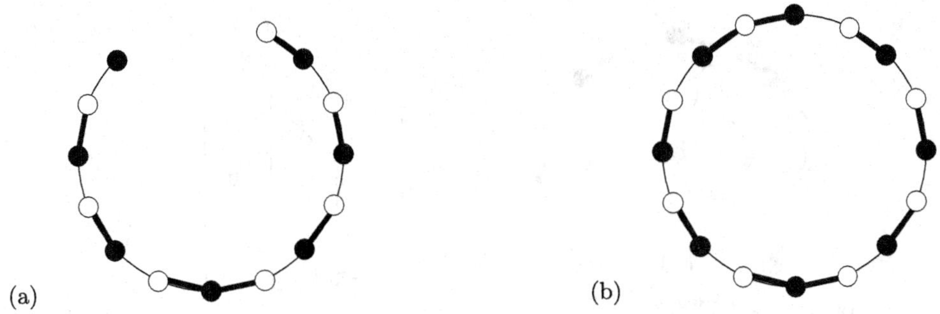

Figure 3.1 (a) Dimerization for an open chain with one point-defect at which two strong bonds meet. (b) Dimerization for a closed chain with two point-defects at which two strong bonds meet.

in Section 3.6, a gradient expansion in Section 3.7, that is an adiabatic approximation as shown in Section 3.8, functional bosonization in Section 3.9, and Abelian bosonization in Section 3.10. Section 3.11 covers the computation of the conserved fermion charge induced by defects in the order parameter treated as a background at any nonvanishing temperature. We close with the effects of short-range quartic interactions on the fractionalization of the charge when stacking several copies of polyacetylene in Section 3.12.

3.2 Point Defects in the Dimerization

We continue to work in the semiclassical limit (2.104), in which we may treat the phonons as a classical background to the electrons. However, instead of assuming the background (2.110) that breaks the translation symmetry of the SSH Hamiltonian (2.103) by one lattice spacing, we assume the existence of a profile in the dependence of u_i on the position i of the acetylene molecule that interpolates between the two mean-field solutions (2.122). Such a profile can be associated with a point-like defect at which two strong (weak) bonds meet. A single defect of this type is pictured on an open chain in Figure 3.1(a). For a closed chain (with the topology of a ring and an even number of sites), only an even number of defects is compatible with periodic boundary conditions, as is illustrated in Figure 3.1(b).

For any given periodic background $u_i = u_{i+N}$ with $i = 1, \cdots, N$, the many-body ground-state energy eigenvalue $E[u_i]$ of $\widehat{H}^{\text{s-c}}_{\text{SSH}}$ defined by Eq. (2.108) is obtained from the sum of two terms. There is the many-body elastic energy stored in u_i with $i = 1, \cdots, N$. There is the many-body electronic energy stored in the Fermi sea for the electrons. Because translation symmetry of sublattice Λ^A [recall definition (2.29)] is broken by the defect, the label for the single-particle electronic energies need not be the wave numbers from the reduced Brillouin zone $\Lambda^{A\star}_{\text{BZ}}$. We shall label

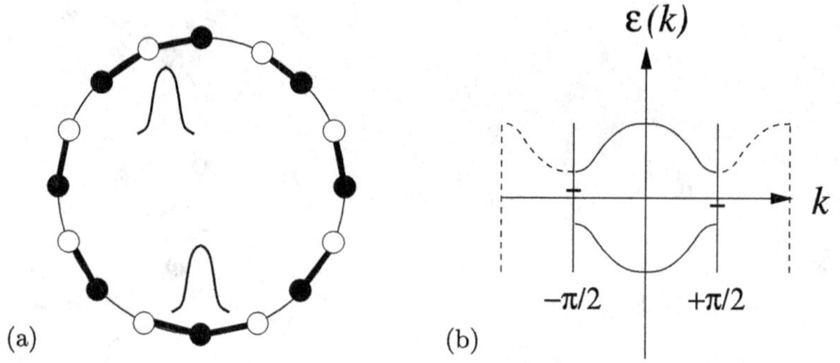

Figure 3.2 (a) Bound states localized at the point defects in the dimerization pattern. (b) Electronic single-particle spectrum in the presence of a pair of soliton and antisoliton that are far apart.

the single-particle energy eigenvalues of $\widehat{H}_{\text{electrons}} + \widehat{H}^{\text{s-c}}_{\text{e-p}}$ defined by Eq. (2.108) with the index $\iota = 1, \cdots, N$. Each single-particle energy eigenvalue remains twofold degenerate owing to the spin-1/2 twofold degeneracy. Hence,

$$\frac{E[u_i]}{N} = \frac{1}{2N}\sum_{i=1}^{N}\kappa\left(u_i - u_{i+1}\right)^2 + \frac{2}{N}\sum_{\iota=1,\cdots,N}^{\varepsilon_\iota \leq 0}\varepsilon_\iota. \tag{3.1}$$

Although this energy density cannot be evaluated in closed form in general, it can be evaluated numerically and shown to remain finite in the thermodynamic limit $N \to \infty$ holding the number of defects finite. In turn, the profile of a single defect at site j in an open chain can be optimized so as to deliver an absolute (degenerate) minimum for $E[u_i^{(j)}]/N$ if the boundary conditions

$$u_i^{(j)} \sim u_{\text{MF}} \tanh\left(\frac{i-j}{\zeta_{\text{s}}}\right), \qquad |i-j| \gg 1, \tag{3.2}$$

are met. Here, the label j is the position at which the two uniform mean-field vacua $\pm|u_{\text{MF}}|$ meet and ζ_{s} is the characteristic size for the region over which u_i interpolates between $+|u_{\text{MF}}|$ and $-|u_{\text{MF}}|$. The profile (3.2) that minimizes $E[u_i]/N$ is called a soliton. Its characteristic size ζ_{s} is model dependent, for it results from the competition of the semiclassical phonon energy and the adiabatic electronic energy, each of which depends on the microscopics (through κ, t, etc.). The soliton in Eq. (3.2) can be assigned the charge

$$Q_{\text{s}} := \text{sgn}(u_{\text{MF}}) \in \{-1, +1\}. \tag{3.3}$$

By convention, a soliton has charge $+1$, while an antisoliton has charge -1.

3.3 Zero Modes Bound to Topological Defects

The uniform dimerization (2.110) opens a gap in the electronic single-particle spectrum of the semiclassical SSH Hamiltonian (2.108) at half-filling. We are going to show that solitons defined in Section 3.2 bind single-particle bound states, each of which has a single-particle energy within the gap and an amplitude that decays exponentially fast away from the location of the domain wall, as is illustrated in Figure 3.2. In the limit in which domain walls are infinitely far apart, the single-particle energy eigenvalues of these bound states are precisely zero.

3.3.1 Zero Modes from the Lattice

We start from the spinless version of Hamiltonian (2.2), which we rewrite as

$$\widehat{H} := -\sum_{i,j=1}^{N} \hat{c}_i^{\dagger}\,\mathcal{H}_{ij}\,\hat{c}_j, \qquad \mathcal{H}_{ij} = t_i\,\delta_{i,j-1} + t_j^{*}\,\delta_{i,j+1}. \tag{3.4a}$$

Open boundary conditions are implemented by setting

$$t_N = 0, \tag{3.4b}$$

periodic ones by setting (**Exercise** 3.1)

$$\hat{c}_{N+1} = \hat{c}_1 \tag{3.4c}$$

as an operator identity.

In first quantization, the stationary Schrödinger equation for Hamiltonian (3.4) reads

$$\cdots, \qquad t_{i-1}^{*}\,\psi_{i-1} + t_i\,\psi_{i+1} = -\varepsilon\,\psi_i, \qquad \cdots, \tag{3.5}$$

where $\psi_i = \langle i|\psi\rangle \in \mathbb{C}$ is the amplitude of a single-particle wave function at the site $i = 1, \cdots, N$ of the chain (**Exercise** 3.2).

For any solution to Eq. (3.5) at the nonvanishing single-particle energy ε, the amplitudes on the odd sites are related to the amplitudes on the even sites. Moreover, any solution to Eq. (3.5) at a nonvanishing single-particle energy ε delivers the solution to Eq. (3.5) at the nonvanishing single-particle energy $-\varepsilon$ through the orthogonal transformation

$$\psi_i \mapsto (-1)^i\,\psi_i, \qquad i = 1, \cdots, N, \tag{3.6}$$

whereby N must be even if periodic boundary conditions are imposed. We call the transformation (3.6) a chiral transformation (**Exercise** 3.3).

Any solution $\psi^{(\mathrm{zm})}$ to Eq. (3.5) at the vanishing single-particle energy $\varepsilon = 0$, a zero mode, if it exists, is very special. First, the amplitudes on the even sites decouple from the amplitudes on the odd sites for a zero mode, since, for a zero mode, Eq. (3.5) reduces to

$$\cdots, \qquad t^{*}_{i-1}\,\psi^{(\mathrm{zm})}_{i-1}+t_{i}\,\psi^{(\mathrm{zm})}_{i+1}=0, \qquad \cdots. \tag{3.7}$$

Second, if we choose a site $1 \ll j_{\mathrm{s}} \ll N$ and a positive integer l such that $1 \ll j_{\mathrm{s}}-2\,l$ while $j_{\mathrm{s}}+2\,l \ll N$, then the amplitude $\psi^{(\mathrm{zm})}_{j_{\mathrm{s}}+2l}$ of a zero mode is recursively related to the amplitude $\psi^{(\mathrm{zm})}_{j_{\mathrm{s}}}$ by

$$\psi^{(\mathrm{zm})}_{j_{\mathrm{s}}+2l}=\left(-\frac{t^{*}_{j_{\mathrm{s}}+2l-2}}{t_{j_{\mathrm{s}}+2l-1}}\right)\cdots\left(-\frac{t^{*}_{j_{\mathrm{s}}+2}}{t_{j_{\mathrm{s}}+3}}\right)\left(-\frac{t^{*}_{j_{\mathrm{s}}}}{t_{j_{\mathrm{s}}+1}}\right)\psi^{(\mathrm{zm})}_{j_{\mathrm{s}}}, \tag{3.8a}$$

while the amplitude $\psi^{(\mathrm{zm})}_{j_{\mathrm{s}}-2l}$ of a zero mode is recursively related to the amplitude $\psi^{(\mathrm{zm})}_{j_{\mathrm{s}}}$ by

$$\psi^{(\mathrm{zm})}_{j_{\mathrm{s}}-2l}=\left(-\frac{t_{j_{\mathrm{s}}-2l+1}}{t^{*}_{j_{\mathrm{s}}-2l}}\right)\cdots\left(-\frac{t_{j_{\mathrm{s}}-3}}{t^{*}_{j_{\mathrm{s}}-4}}\right)\left(-\frac{t_{j_{\mathrm{s}}-1}}{t^{*}_{j_{\mathrm{s}}-2}}\right)\psi^{(\mathrm{zm})}_{j_{\mathrm{s}}}. \tag{3.8b}$$

The amplitude $\psi^{(\mathrm{zm})}_{j_{\mathrm{s}}}$ is the seed value of the recursive relation (3.8). Of course, the existence of a zero mode is only guaranteed if the recursive relations (3.8) are compatible with the chosen boundary and normalization conditions. Third, if a zero mode $\psi^{(\mathrm{zm})}$ is an eigenstate of the chiral transformation (3.6), then it must obey

$$\pm\,\psi^{(\mathrm{zm})}_{i}=(-1)^{i}\,\psi^{(\mathrm{zm})}_{i}, \qquad i=1,\cdots,N. \tag{3.9}$$

This condition enforces that a zero mode, if it exists and if it is an eigenstate of the chiral transformation, must vanish on either the even or the odd sites of the chain (**Exercise** 3.4).

We now apply Eq. (3.8) to the (weak) dimerization Ansatz

$$t^{(\mathrm{MF})}_{i}:=t+(-1)^{i}\,\delta t, \qquad 0<|\delta t|<t, \tag{3.10}$$

where the dimerization δt is real-valued and assumed to be smaller in magnitude than $t>0$. If periodic boundary conditions are imposed, there is no zero mode. This is not true anymore for an open chain with the even number N of sites in the presence of either a single soliton or a single antisoliton. Let $1 \ll j \ll N$ be a defective site in that the hopping amplitude interpolates between the two allowed patterns of dimerization according to the asymptotics

$$t^{(j)}_{i}\sim\begin{cases} t+(-1)^{i}\,\delta t, & j\ll i\ll N,\\ \\ t-(-1)^{i}\,\delta t, & 1\ll i\ll j.\end{cases} \tag{3.11}$$

Here, we shall use the convention that the case (3.11) with $\delta t>0$ is called a soliton, while the case (3.11) with $\delta t<0$ is called an antisoliton [compare with Eq. (3.2)].

Next, we apply Eq. (3.8) to the Ansatz (3.11) and require that this zero mode vanishes as

$$\psi_i^{(\mathrm{zm})} \sim e^{-|i-j|/\xi}, \qquad |i-j| \gg 1, \tag{3.12}$$

where $\xi \equiv 1/\Delta$ is the characteristic length scale set by the dimerization, after the thermodynamic limit $N \to \infty$ has been taken. Normalizability of Eq. (3.8) applied to the Ansatz (3.11) hinges on whether the dimensionless ratio $-t_i^{(j)}/t_{i+1}^{(j)}$ has magnitude larger or smaller than unity (recall that $t > 0$):

- If Eq. (3.11) is a soliton ($\delta t > 0$), the recursive zero mode (3.8) is supported on the odd sites,

$$\lim_{\substack{i \gg j \\ i\,\mathrm{odd}}} \left| \frac{t_i^{(j)}}{t_{i+1}^{(j)}} \right| = \left| \frac{t - \delta t}{t + \delta t} \right| < 1, \qquad \lim_{\substack{i \ll j \\ i\,\mathrm{odd}}} \left| \frac{t_{i-1}^{(j)}}{t_{i-2}^{(j)}} \right| = \left| \frac{t - \delta t}{t + \delta t} \right| < 1. \tag{3.13a}$$

- If Eq. (3.11) is an antisoliton ($\delta t < 0$), the recursive zero mode (3.8) is supported on the even sites,

$$\lim_{\substack{i \gg j \\ i\,\mathrm{even}}} \left| \frac{t_i^{(j)}}{t_{i+1}^{(j)}} \right| = \left| \frac{t + \delta t}{t - \delta t} \right| < 1, \qquad \lim_{\substack{i \ll j \\ i\,\mathrm{even}}} \left| \frac{t_{i-1}^{(j)}}{t_{i-2}^{(j)}} \right| = \left| \frac{t + \delta t}{t - \delta t} \right| < 1. \tag{3.13b}$$

Conditions (3.13) ensure that a zero mode decays exponentially fast away from the location of the point defect (the seed value), a sufficient condition for normalizability in the thermodynamic limit $N \to \infty$. Of course, for any finite N, there is a pair of zero modes. The first zero mode is supported on one sublattice. The second zero mode is supported on the other sublattice. One of these zero modes decays exponentially fast away from its seed value; the other zero mode grows exponentially fast away from its seed value. It is the sign of the dimerization that selects which of the two finite-size solutions to the recursive equation (3.8) remains normalizable in the thermodynamic limit $N \to \infty$.

3.3.2 Zero Modes from the Continuum

The most general 2×2 Dirac Hamiltonian in one-dimensional space that is compatible with translation symmetry takes the form

$$\mathcal{H}(p) := \alpha\, p + \beta\, m + \mathrm{i}\beta\, \gamma_5\, m_5 \equiv \tau_2\, p + \tau_1\, \phi_\infty + \tau_3\, \mu_{\mathrm{s}}, \tag{3.14a}$$

where we have chosen the 2×2 dimensional representation of the Dirac matrices to be ($\boldsymbol{\tau}$ are a triplet of Pauli matrices)

$$\beta \equiv \gamma^0 = \tau_1, \quad \alpha \equiv \gamma^0\, \gamma^1 = \tau_2, \quad \gamma_5 \equiv -\gamma^0\, \gamma^1 = -\tau_2, \quad \mathrm{i}\beta\, \gamma_5 = \tau_3. \tag{3.14b}$$

The real-valued constants $m \equiv \phi_\infty$ and $m_5 \equiv \mu_{\mathrm{s}}$ carry the dimensions of energy when the units are chosen so that velocities are dimensionless.[3] The energy scale $m \equiv \phi_\infty$ encodes in the Dirac Hamiltonian a microscopic dimerization. The energy scale $m_5 \equiv \mu_{\mathrm{s}}$ encodes in the Dirac Hamiltonian a microscopic staggered chemical potential, that is an energy penalty (gain) for spinless electrons to be localized on the sites of sublattice Λ^A (Λ^B), say. Eigenstates of the Dirac Hamiltonian (3.14) are plane waves that, according to Eq. (2.46), may be chosen to be represented with the pairs of two-component spinors

$$\Psi_+(p) = \begin{pmatrix} e^{-\mathrm{i}\varphi(p)/2} \cos\big(\theta(p)/2\big) \\ +e^{+\mathrm{i}\varphi(p)/2} \sin\big(\theta(p)/2\big) \end{pmatrix} \tag{3.15a}$$

and

$$\Psi_-(p) = \begin{pmatrix} e^{-\mathrm{i}\varphi(p)/2} \sin\big(\theta(p)/2\big) \\ -e^{+\mathrm{i}\varphi(p)/2} \cos\big(\theta(p)/2\big) \end{pmatrix} \tag{3.15b}$$

with the energy eigenvalues

$$\varepsilon_\pm(p) = \pm\sqrt{p^2 + m^2 + m_5^2} \equiv \pm\sqrt{p^2 + \phi_\infty^2 + \mu_{\mathrm{s}}^2}, \tag{3.15c}$$

respectively. Here,

$$\begin{pmatrix} \phi_\infty \\ p \\ \mu_{\mathrm{s}} \end{pmatrix} = \sqrt{p^2 + \phi_\infty^2 + \mu_{\mathrm{s}}^2} \begin{pmatrix} \cos\varphi(p)\,\sin\theta(p) \\ \sin\varphi(p)\,\sin\theta(p) \\ \cos\theta(p) \end{pmatrix}. \tag{3.15d}$$

Any dependence $\phi(x)$ and $\mu_{\mathrm{s}}(x)$ on the position x breaks translation symmetry. However, when

$$\mu_{\mathrm{s}}(x) = 0, \tag{3.16}$$

any zero mode of ($\hbar = 1$)

$$\mathcal{H}_\phi(x) := -\mathrm{i}\tau_2 \frac{\mathrm{d}}{\mathrm{d}x} + \tau_1\,\phi(x) \tag{3.17}$$

for some special profile of the function $\phi(x)$ is robust to smooth changes in this profile because of the chiral spectral symmetry

$$\mathcal{H}_\phi(x) = -\tau_3\,\mathcal{H}_\phi(x)\,\tau_3, \tag{3.18}$$

as we are going to explain. Here, the chiral transformation

$$a\,\Psi(x) + a'\,\Psi'(x) \mapsto a\,\tau_3\,\Psi(x) + a'\,\tau_3\,\Psi'(x) \tag{3.19a}$$

[3] The notation m_5 is the natural one in particle physics. The notation μ_{s} is the natural one in condensed matter physics. The rational for the notation ϕ_∞ will become more transparent as we proceed.

for any pair of $\mathbb{C}$ numbers a and a' and any pair of complex-valued spinors $\Psi(x)$ and $\Psi'(x)$, an orthogonal transformation that squares to the identity, maps the eigenstate satisfying

$$\mathcal{H}_{\phi}(x)\,\Psi_{\varepsilon}(x) = \varepsilon\,\Psi_{\varepsilon}(x) \tag{3.19b}$$

into the eigenstate satisfying

$$\mathcal{H}_{\phi}(x)\,\Psi_{-\varepsilon}(x) = -\varepsilon\,\Psi_{-\varepsilon}(x), \tag{3.19c}$$

where

$$\Psi_{-\varepsilon}(x) = \tau_3\,\Psi_{\varepsilon}(x). \tag{3.19d}$$

In particular, any simultaneous eigenstate of the chiral transformation τ_3 and of $\mathcal{H}_{\phi}(x) = -\tau_3\,\mathcal{H}_{\phi}(x)\,\tau_3$ is a zero mode:

$$\left.\begin{array}{l}\tau_3\,\Psi(x) = \pm\Psi(x)\\ \mathcal{H}_{\phi}\,\Psi(x) = \varepsilon\,\Psi(x)\end{array}\right\} \Longrightarrow \mathcal{H}_{\phi}(x)\,\Psi(x) = 0. \tag{3.20}$$

The two components of the zero mode defined by Eq. (3.20) must satisfy the real-valued first-order differential equations

$$\begin{pmatrix} 0 & -\frac{\mathrm{d}}{\mathrm{d}x} + \phi(x) \\ +\frac{\mathrm{d}}{\mathrm{d}x} + \phi(x) & 0 \end{pmatrix} \begin{pmatrix} u(x) \\ v(x) \end{pmatrix} = 0, \tag{3.21}$$

whose formal solution, given some initial condition $u(0)$ and $v(0)$, follows from the integration

$$\left[+\frac{\mathrm{d}}{\mathrm{d}x} + \phi(x)\right] u(x) = 0 \iff \qquad u(x) = u(0)\,e^{-\int_0^x \mathrm{d}x'\,\phi(x')}, \tag{3.22a}$$

$$\left[-\frac{\mathrm{d}}{\mathrm{d}x} + \phi(x)\right] v(x) = 0 \iff \qquad v(x) = v(0)\,e^{+\int_0^x \mathrm{d}x'\,\phi(x')}. \tag{3.22b}$$

We impose the boundary conditions that the components of the zero modes vanish at infinity. If the asymptotic value of the order parameter $|\phi(\pm\infty)| \neq 0$, these boundary conditions can only be met in the following two cases.

- If the function $\phi(x)$ supports a soliton defined by the convention $\phi(-\infty) < 0$ and $\phi(+\infty) > 0$, then $u(x)$ vanishes exponentially fast as $x \to \pm\infty$ for any $u(0)$, while we must impose the choice $v(0) = 0$ to meet the boundary conditions at $x = \pm\infty$.

- If the function $\phi(x)$ supports an antisoliton defined by the convention $\phi(-\infty) > 0$ and $\phi(+\infty) < 0$, then $v(x)$ vanishes exponentially fast as $x \to \pm\infty$ for any $v(0)$, while we must impose the choice $u(0) = 0$ to meet the boundary conditions at $x = \pm\infty$.

In either case, there is only one zero mode that satisfies the boundary conditions at $x = \pm\infty$. That only one of the solutions $u(x)$ or $v(x)$ meets the boundary conditions is consistent with the fact that zero modes are eigenstates of the chiral transformation (**Exercise** 3.5).

The two ordinary differential equations defined by Eq. (3.21) are examples of the eigenvalue problem

$$\begin{pmatrix} 0 & \mathcal{D} \\ \mathcal{D}^\dagger & 0 \end{pmatrix} \begin{pmatrix} f \\ g \end{pmatrix} = 0, \tag{3.23}$$

given some initial conditions on f and g (and their derivatives if necessary) supplemented by boundary conditions. Here, $\mathcal{D}$ is some differential operator and $\mathcal{D}^\dagger$ its adjoint. The off-diagonal block structure (3.23) used to embed the pair of operators $\mathcal{D}$ and $\mathcal{D}^\dagger$ is a generalization of the chiral symmetry under the transformations (3.6) or (3.19a). The counterpart for the eigenvalue problem (3.23) to the solutions (3.22) occurs when the eigenvalue problem (3.23) only admits zero modes whereby either

$$f = 0, \qquad \mathcal{D} g = 0, \tag{3.24a}$$

or

$$\mathcal{D}^\dagger f = 0, \qquad g = 0. \tag{3.24b}$$

It turns out that for a class of operators $\mathcal{D}$ and $\mathcal{D}^\dagger$ and boundary conditions, the number of admissible zero modes for the eigenvalue problem (3.23) is given by the index (see Section 3.6.4)

$$\mathrm{Index}(\mathcal{D}) := \dim \mathrm{Ker}\left(\mathcal{D}^\dagger\right) - \dim \mathrm{Ker}\left(\mathcal{D}\right). \tag{3.25}$$

This index is the difference between the number of linear independent solutions to the equations $\mathcal{D}^\dagger f = 0$ and $\mathcal{D} g = 0$. There are powerful theorems from mathematics that relates this difference, the analytical index, to another index, the topological index. The topological index is an attribute that quantifies a topological property of the background entering the operator $\mathcal{D}$.[4] For polyacetylene, the topological index is $+1$ for the soliton background, and -1 for the antisoliton background.

To summarize, we found the normalized zero mode

$$\Psi^{\mathrm{s}}_{\varepsilon=0}(x) := \frac{1}{\sqrt{\mathcal{N}}} \begin{pmatrix} e^{-\int_0^x \mathrm{d}x' \, \phi(x')} \\ 0 \end{pmatrix}, \qquad \mathcal{N} := \int_{-\infty}^{\infty} \mathrm{d}x \, e^{-2\int_0^x \mathrm{d}x' \, \phi(x')}, \tag{3.26a}$$

for the soliton background and the normalized zero mode

$$\Psi^{\bar{\mathrm{s}}}_{\varepsilon=0}(x) := \frac{1}{\sqrt{\mathcal{N}}} \begin{pmatrix} 0 \\ e^{+\int_0^x \mathrm{d}x' \, \phi(x')} \end{pmatrix}, \qquad \mathcal{N} := \int_{-\infty}^{\infty} \mathrm{d}x \, e^{+2\int_0^x \mathrm{d}x' \, \phi(x')}, \tag{3.26b}$$

for the antisoliton background when solving the eigenvalue problem (3.21).

[4] When space is a compact manifold, this class of theorems are known as the Atiyah-Singer index theorem. When space is non-compact and odd-dimensional one refers to the Callias index theorem.

3.4 A First Run at Charge Fractionalization

3.4.1 Overview

One of the remarkable consequences of zero modes in polyacetylene that are bounded to solitons or antisolitons is that they can be assigned a fractional charge. Here, we shall present counting arguments to support the notion of charge fractionalization. We are going to present two counting arguments that lead to the conclusion that the charge bound to a soliton is 1/2 of the electron charge. The first argument is based on quantum mechanics. The second argument is a heuristic one based on a dimer picture. It can be attributed to Su and Schrieffer.[5]

3.4.2 Induced Charge Taking the Value 1/2

Let $|i\rangle$ be the single-particle state with the (spinless) electron localized at site $i = 1, \cdots, N$ of the chain. These single-particle states form a complete orthonormal basis:

$$1 = \sum_{i=1}^{N} |i\rangle\langle i|, \qquad \langle i|i'\rangle = \delta_{i,i'}, \qquad i, i' = 1, \cdots, N. \tag{3.27}$$

Let $|\varepsilon_\iota\rangle$ denote a single-particle energy eigenstate for some Hamiltonian defined on the Hilbert space spanned by the lattice basis (3.27), that is, one single-particle eigenstate from the orthonormal basis $\{|\varepsilon_\iota\rangle,\ \iota = 1, \cdots, N\}$ that satisfies the completeness relation

$$1 = \sum_{\iota=1}^{N} |\varepsilon_\iota\rangle\langle\varepsilon_\iota|, \qquad \langle\varepsilon_\iota|\varepsilon_{\iota'}\rangle = \delta_{\iota,\iota'}, \qquad \iota, \iota' = 1, \cdots, N. \tag{3.28}$$

From now on, we drop the explicit reference to the countable index ι of the single-particle energy eigenstates. For a given site i, we may combine Eqs. (3.27) and (3.28) to find the local identity

$$\begin{aligned} 1 &= \langle i|i\rangle \\ &= \sum_{\varepsilon} \langle i|\varepsilon\rangle\langle\varepsilon|i\rangle \\ &\equiv \sum_{\varepsilon} \left|\psi_{i,\varepsilon}\right|^2 \\ &= \sum_{\varepsilon} \nu_{i,\varepsilon}\,. \end{aligned} \tag{3.29a}$$

Here,

$$\nu_{i,\varepsilon} := \left|\psi_{i,\varepsilon}\right|^2 = |\langle i|\varepsilon\rangle|^2 \tag{3.29b}$$

[5] W. P. Su and J. R. Schrieffer, Phys. Rev. Lett. **46**(11), 738–741 (1981). [Su and Schrieffer (1981)].

is the dimensionless density of states at the site i and the (discrete) energy ε. The local identity (3.29a) is a sum rule that applies to any single-particle lattice Hamiltonian.

Let $\nu^{\infty}_{i,\varepsilon}$ denote the position- and energy-resolved dimensionless density of states at site i and energy ε for spinless electrons undergoing nearest-neighbor hopping along a chain made of N sites with pristine dimerization. Let $\nu^{\mathrm{s}\bar{\mathrm{s}}}_{i,\varepsilon}$ denote the position- and energy-resolved dimensionless density of states at site i and energy ε for spinless electrons undergoing nearest-neighbor hopping along a chain made of N sites with a defective dimerization supporting a soliton centered around i_{s} and an antisoliton centered around $i_{\bar{\mathrm{s}}}$. The sum rule (3.29a) thus implies the local identity

$$\sum_{\varepsilon} \nu^{\mathrm{s}\bar{\mathrm{s}}}_{i,\varepsilon} = \sum_{\varepsilon} \nu^{\infty}_{i,\varepsilon} \tag{3.30}$$

for any site $i = 1, \cdots, N$.

We assume that the separation

$$r_{\mathrm{s}\bar{\mathrm{s}}} := |i_{\mathrm{s}} - i_{\bar{\mathrm{s}}}| \tag{3.31}$$

between the centers of the soliton and the antisoliton is much larger than the characteristic length

$$\xi := 1/\Delta \tag{3.32}$$

(in units for which $\hbar$ and the characteristic velocity are both one),

$$r_{\mathrm{s}\bar{\mathrm{s}}} \gg \xi, \tag{3.33}$$

where Δ is the characteristic energy scale for the dimerization.

The position- and energy-resolved dimensionless density of states $\nu^{\infty}_{i,\varepsilon}$ is controlled solely by the quasi-continuum of Bloch states in the valence (lower) and the conduction (upper) bands.[6] In the limit $N \gg 1$, the position- and energy-resolved dimensionless density of states $\nu^{\mathrm{s}\bar{\mathrm{s}}}_{i,\varepsilon}$ consists (by assumption) of both a quasi-continuum of Bloch states in the valence (lower) and the conduction (upper) bands and two mid-gap states (quasi-zero modes) ψ_{i,ε_-} and ψ_{i,ε_+} with the single-particle energies

$$\varepsilon_- = -\varepsilon_+ < 0, \qquad \psi_{i,\varepsilon_-} = (-1)^i\, \psi_{i,\varepsilon_+}, \tag{3.34a}$$

respectively,[7] whereby [by assumption (3.33)]

$$|\varepsilon_- - \varepsilon_+| \approx \Delta\, e^{-r_{\mathrm{s}\bar{\mathrm{s}}}/\xi}. \tag{3.34b}$$

We choose the chemical potential to be above the last state from the valence quasi-continuum and below ε_-. We may now rewrite the local sum rule (3.30) as

[6] The terminology of a quasi-continuous spectrum accounts for the fact that the dimensionality N of the single-particle Hilbert space is finite before taking the thermodynamic limit $N \to \infty$. Hence, the energy eigenvalue spectrum is discrete. The defining property of the quasi-continuous spectrum is that the spacing between any two consecutive energies belonging to the quasi-continuous spectrum decreases with increasing N.

[7] We made use of the spectral symmetry under the chiral transformation (3.6).

$$\sum_{\varepsilon<\varepsilon_-} \nu^{s\bar{s}}_{i,\varepsilon} + \left|\psi_{i,\varepsilon_-}\right|^2 + \left|\psi_{i,\varepsilon_+}\right|^2 + \sum_{\varepsilon>\varepsilon_+} \nu^{s\bar{s}}_{i,\varepsilon} = \sum_{\varepsilon<0} \nu^{\infty}_{i,\varepsilon} + \sum_{\varepsilon>0} \nu^{\infty}_{i,\varepsilon}. \tag{3.35}$$

The chiral symmetry implies that the position- and energy-resolved dimensionless density of states is an even function of the energy, for the chiral transformation $\psi_{i,-|\varepsilon|} \mapsto (-1)^i \, \psi_{i,+|\varepsilon|}$ implies that $|\psi_{i,-|\varepsilon|}| \mapsto |\psi_{i,+|\varepsilon|}|$ for the quasi-continuum states. Hence, the local sum rule (3.35) becomes

$$2\sum_{\varepsilon<\varepsilon_-} \nu^{s\bar{s}}_{i,\varepsilon} + \left|\psi_{i,\varepsilon_-}\right|^2 + \left|\psi_{i,\varepsilon_+}\right|^2 = 2\sum_{\varepsilon<0} \nu^{\infty}_{i,\varepsilon}. \tag{3.36}$$

Both the energy-integrated but position-resolved dimensionless densities of states

$$\nu^{s\bar{s}}_{i} := \sum_{\varepsilon<\varepsilon_-} \nu^{s\bar{s}}_{i,\varepsilon} \tag{3.37}$$

and

$$\nu^{\infty}_{i} := \sum_{\varepsilon<0} \nu^{\infty}_{i,\varepsilon} = \sum_{\varepsilon<\varepsilon_-} \nu^{\infty}_{i,\varepsilon} \tag{3.38}$$

are potentially divergent quantities in the thermodynamic limit $N \to \infty$ for unbounded spectra such as Dirac spectra. However, their difference

$$\delta\nu^{s\bar{s}-\infty}_{i} := \nu^{s\bar{s}}_{i} - \nu^{\infty}_{i} \tag{3.39}$$

remains finite in the thermodynamic limit $N \to \infty$ and so does the regularized representation

$$\delta\nu^{s\bar{s}-\infty}_{i} = -\frac{1}{2}\left(\left|\psi_{i,\varepsilon_-}\right|^2 + \left|\psi_{i,\varepsilon_+}\right|^2\right) \tag{3.40}$$

of the local sum rule (3.36). Subtraction of infinities are common practice in quantum field theory. The regulated local sum rule (3.40) is akin to performing a normal ordering with respect to the Fermi sea in the context of interacting fermions at a nonvanishing fermionic density.

The (quasi) zero modes ψ_{i,ε_-} and ψ_{i,ε_+} are, to a very good approximation when $r_{s\bar{s}}/\xi \gg 1$, the bonding and antibonding linear combinations of the zero mode ψ^{s}_{i} obtained for an open chain supporting a single soliton and the zero mode $\psi^{\bar{s}}_{i}$ obtained for an open chain supporting a single antisoliton,

$$\psi_{i,\varepsilon_\mp} \approx \frac{1}{\sqrt{2}}\left(\psi^{s}_{i} \pm \psi^{\bar{s}}_{i}\right). \tag{3.41}$$

With the help of (3.41), the local sum rule (3.40) becomes

$$\delta\nu^{s\bar{s}-\infty}_{i} \approx -\frac{1}{2}\left|\psi^{s}_{i}\right|^2 - \frac{1}{2}\left|\psi^{\bar{s}}_{i}\right|^2. \tag{3.42}$$

This approximation is exact in the limit $r_{s\bar{s}}/\xi \to \infty$.

We *define* the space- and energy-integrated charge $Q^{s\bar{s}-\infty}$ bound by the pair of soliton and antisoliton to be

$$Q^{s\bar{s}-\infty} := \sum_i \delta\nu_i^{s\bar{s}-\infty}. \tag{3.43}$$

The charge (3.43) is measured relative to a reference charge, the charge of the many-body Fermi sea at half-filling with a pristine dimerization. It is given by

$$Q^{s\bar{s}-\infty} = -1, \tag{3.44}$$

if we normalize the zero modes to one.

What does (3.44) tell us? That we have a deficit of exactly one electron. This is because we chose not to fill the single-particle states at $\varepsilon_- < 0$ and $\varepsilon_+ > 0$ by picking the Fermi level right below the value ε_-. So we are one state away from half-filling, and therefore we are one spinless electron shy of having the same charge for the system with topological defects as compared to the one without defects. The total charge with or without defects has to be an integer, and so is the difference between the charge for the two cases. So far everything is standard. Where are the fractional charges then?

The fractional charges sit near the soliton and the antisoliton. By computing the total charge of the system we cannot see them. We need to focus our attention to a region surrounding either the soliton or the antisoliton, but not both in the same field of view. We cannot probe the system in its entirety.

To make this idea precise, we define the weighted sum over the charge densities

$$Q_f^{s\bar{s}-\infty} := \sum_i f_i\, \delta\nu_i^{s\bar{s}-\infty} \approx -\frac{1}{2}\sum_i f_i\, |\psi_i^{s}|^2 - \frac{1}{2}\sum_i f_i\, |\psi_i^{\bar{s}}|^2, \tag{3.45}$$

where $f\colon \{1,\cdots,N\} \to [0,1], i \mapsto f_i$ is any test function taking the positive values f_i. The choice f for the test function defines our field of view. If we want to probe the charge bound to the soliton centered at i_s, we choose the test function f^s that takes the value $f_i^s = 1$ for i in a neighborhood of size no less than ξ centered about i_s, while it vanishes otherwise, say $f_i^s = 0$ for i in a neighborhood of size of order ξ centered about $i_{\bar{s}}$. Conversely, if we want to probe the charge bound to the antisoliton centered at $i_{\bar{s}}$, we choose the test function $f^{\bar{s}}$ that takes the value $f_i^{\bar{s}} = 1$ for i in a neighborhood of size no less than ξ centered about $i_{\bar{s}}$, while it vanishes otherwise, say $f_i^{\bar{s}} = 0$ for i in a neighborhood of size of order ξ centered about i_s. With these two test functions, one finds

$$Q_{f^s} \approx -\frac{1}{2}, \tag{3.46a}$$

and

$$Q_{f^{\bar{s}}} \approx -\frac{1}{2}, \tag{3.46b}$$

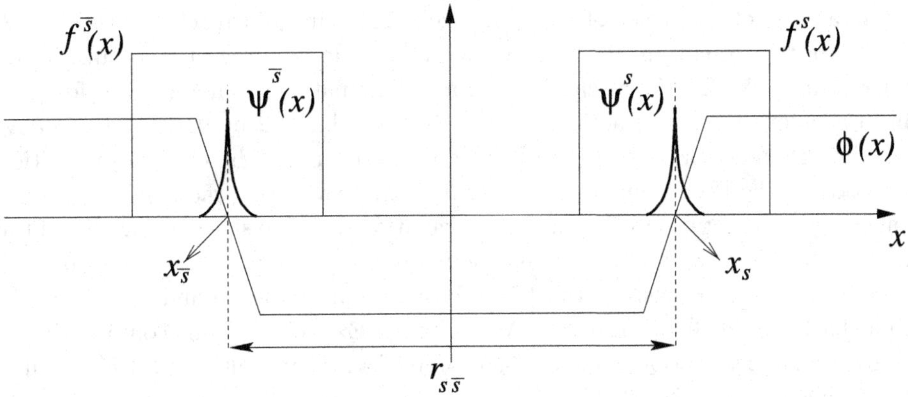

Figure 3.3 The characteristic length scales entering the counting formula (3.45) obey the hierarchy (3.47).

respectively. These approximations become exact in the limit $r_{\mathrm{s}\bar{\mathrm{s}}}/\xi \to \infty$. We conclude that there is a particle deficit of 1/2 of an electron in a neither too-small nor too-large neighborhood of a topological defect, relative to the pristine dimerized phase of polyacetylene at half-filling. In a closed system one cannot have a global deficit of a fraction of a particle, but one can have this deficit locally, where locality is set by the characteristic length scale ξ. The hierarchy

$$r_{\mathrm{s}\bar{\mathrm{s}}} \gg \sup f^{\mathrm{s}} = \sup f^{\bar{\mathrm{s}}} \gg \sup \psi^{\mathrm{s}} = \sup \psi^{\bar{\mathrm{s}}} \approx \xi \gg \sup \partial_x \phi^{\mathrm{s}} = \sup \partial_x \phi^{\bar{\mathrm{s}}} \tag{3.47}$$

between the length scales $r_{\mathrm{s}\bar{\mathrm{s}}}$ (the separation between an antisoliton and a soliton), the size of the support of the field of view $\sup f^{\mathrm{s}} = \sup f^{\bar{\mathrm{s}}}$, the size of the support of the bound states $\sup \psi^{\mathrm{s}} = \sup \psi^{\bar{\mathrm{s}}} \approx \xi$, and the size of the support of the soliton (antisoliton) $\sup \partial_x \phi^{\mathrm{s}} = \sup \partial_x \phi^{\mathrm{s}}$ is depicted in Figure 3.3. The last inequality prohibits the proliferation of mid-gap states in addition to the zero modes arising from the index theorem.

The method used by Schrieffer to compute the fractional charge involves computing the expectation values of two many-body ground states and then subtracting one expectation value, that of a reference many-body ground state free of defect, from the expectation value for the many-body ground state with a point-like defect. In the thermodynamic limit, a limit which is required for the exponentially small splitting in energy between the two bound states induced by the soliton and antisoliton pair to be strictly vanishing, the subtraction is of the $\infty - \infty$ type. Subtracting one ∞ from another one does not have to give a unique answer. The order of limits taken to reach these infinities matters. It is thus desirable to use an alternate derivation of the fractional charge 1/2 that gives a complementary understanding of the order of limits invoked (implicitly) to reach the result 1/2. The methods of quantum field theory, with the help of which Jackiw and Rebbi first obtained the fractional charge 1/2, are illuminating in that regard, for they pin in a unique way the fractional charge to the value 1/2 by defining the many-body charge operator in such a way that it is odd under the charge-conjugation symmetry of the Hamiltonian (**Exercise** 3.6). Moreover, the fractional charge 1/2 is not a mere expectation value in their calculation. It is the sharp eigenvalue of the many-body charge operator provided the many-body charge operator is expanded in the single-particle basis of the Dirac Hamiltonian with one and only one soliton as background (**Exercise** 3.6). We outline in **Exercise** 3.7 the computation of the fractional expectation value of the charge of a soliton (antisoliton) for a Dirac Hamiltonian in the presence of a pair of soliton and antisoliton.

In view of the essential role played by the symmetry under charge conjugation to pin the fractional charge 1/2 bound to a soliton, we anticipate that the fractional charge bound to a soliton can vary continuously (but not arbitrarily if the orders of limits are agreed on) as a function of a weak breaking of the charge-conjugation symmetry.

3.4.3 Induced Charges Taking the Values $\nu = 1/n$, $n = 1, 2, \cdots$

We first review an alternative heuristic derivation of the fractional charge 1/2 bound to a soliton in polyacetylene. We then generalize this heuristic argument to treat the case of defects in n-merized pattern of symmetry breaking that bind a rational charge of $1/n$ with n any nonvanishing positive integer.

3.4.3.1 Dimerization at Half-Filling

The Hamiltonian is given by Eq. (3.4a) at the strong dimerization point

$$t_i := t\left[1 - (-1)^i\right], \qquad i = 1, \cdots, N-1, \tag{3.48}$$

corresponding to the dimerization $\delta t = -t$. Periodic boundary conditions (3.4c) are chosen, that is, N is necessarily even. Finally, the filling fraction is

$$\nu := \frac{N_{\mathrm{e}}}{N} = \frac{1}{2}, \tag{3.49}$$

where N_{e} is the number of spinless electrons.

At the strong dimerization point (3.48), Hamiltonian (3.4) reduces to the direct sum

$$\widehat{H} = -2t \sum_{j=1}^{N/2} \begin{pmatrix} \hat{c}^{\dagger}_{2j-1} & \hat{c}^{\dagger}_{2j} \end{pmatrix} \begin{pmatrix} 0 & 1 \\ 1 & 0 \end{pmatrix} \begin{pmatrix} \hat{c}_{2j-1} \\ \hat{c}_{2j} \end{pmatrix}. \tag{3.50}$$

Hence, the spectral decomposition of Hamiltonian (3.50) is

$$\begin{aligned}\widehat{H} =& -2t\sum_{j=1}^{N/2}\begin{pmatrix}\hat{\chi}^{\dagger}_{2j-1,+} & \hat{\chi}^{\dagger}_{2j-1,-}\end{pmatrix}\begin{pmatrix}+1 & 0\\ 0 & -1\end{pmatrix}\begin{pmatrix}\hat{\chi}_{2j-1,+}\\ \hat{\chi}_{2j-1,-}\end{pmatrix}\\ =& -\sum_{\mathrm{a}=\pm}\varepsilon_{\mathrm{a}}\sum_{j=1}^{N/2}\hat{\chi}^{\dagger}_{2j-1,\mathrm{a}}\,\hat{\chi}_{2j-1,\mathrm{a}},\end{aligned} \tag{3.51a}$$

where

$$\hat{\chi}^{\dagger}_{2j-1,\pm} := \frac{1}{\sqrt{2}}\left(\hat{c}^{\dagger}_{2j-1} \pm \hat{c}^{\dagger}_{2j}\right), \qquad \hat{\chi}_{2j-1,\pm} := \frac{1}{\sqrt{2}}\left(\hat{c}_{2j-1} \pm \hat{c}_{2j}\right), \tag{3.51b}$$

and

$$\varepsilon_{\pm} = \pm 2t. \tag{3.51c}$$

At the strong dimerization point (3.48), Hamiltonian (3.51a) can be thought of as made of two flat bands consisting of $N/2$ degenerate single-particle states

$$|2j-1,\pm\rangle := \hat{\chi}^{\dagger}_{2j-1,\pm}|0\rangle, \qquad j = 1,\cdots,N/2, \tag{3.52}$$

with the single-particle energies (3.51c) independent of the label for the sublattice Λ^A, respectively. The single-particle state $|2j-1,+\rangle$ represents the bonding linear combination to have an electron either on site $2j-1$ or on site $2j$. The single-particle state $|2j-1,-\rangle$ represents the antibonding linear combination to have an electron either on site $2j-1$ or on site $2j$. Single-particle energy eigenstates within one band can be localized within the unit cell of sublattice Λ^A, but they cannot be localized on a single site of the lattice $\Lambda = \Lambda^A \cup \Lambda^B$. The two flat bands of Bloch states

$$|k,\pm\rangle := \frac{1}{\sqrt{N/2}}\sum_{j=1}^{N/2} e^{\mathrm{i}k\,(2j-1)}\,|2j-1,\pm\rangle, \qquad k = \frac{2\pi}{N}n, \qquad n = 1,\cdots,\frac{N}{2}, \tag{3.53}$$

in units for which the lattice spacing $\mathfrak{a}$ of lattice Λ has been set to unity, can equally well be chosen for the spectral decomposition of the Hamiltonian at the strong dimerization point.

At the strong dimerization point (3.48) and at half-filling, we are going to choose the representation

$$\left|\text{dimer}, \nu = \frac{1}{2}\right\rangle := \prod_{j=1}^{N/2}\hat{\chi}^{\dagger}_{2j-1,+}|0\rangle \tag{3.54}$$

for the fermionic many-body ground state of a chain made of the even number N of sites. The many-body ground state (3.54) is a product state. It is because of this property of the many-body ground state that it is advantageous to choose the single-particle basis (3.52). Indeed, the many-body ground state (3.54) can then be represented by a classical bond (dimer) covering of the chain with bonds (dimers)

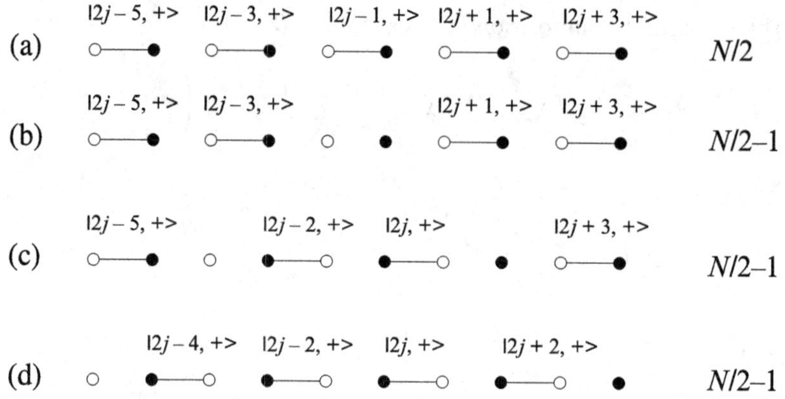

Figure 3.4 (a) Product-state representation (3.54) of the fermionic many-body ground state at half-filling as a valence-bond (dimer) covering. (b) Product-state representation (3.55) of one of the $N/2$ degenerate fermionic many-body ground states at one single-particle state short of half-filling as a valence-bond (dimer) covering with two monomers. (c) Product-state representation (3.57) of a many-body state with two monomers a distance $r_{s\bar{s}} = 5$ apart in units of the lattice spacing of Λ. (d) Product-state representation (3.57) of a many-body state with two monomers a distance $r_{s\bar{s}} = 9$ apart in units of the lattice spacing of Λ.

joining nearest-neighbor sites following the rules that (i) any given odd site $2j-1$ of the chain is the starting of one and only one dimer, (ii) any bond (dimer) labeled by the odd site $2j-1$ on its left end is a pictorial representation of the single-particle state $|2j-1,+\rangle$, and (iii) the covering represents a direct product of the single-particle states $|2j-1,+\rangle$ with $j = 1,\cdots,N/2$. Figure 3.4(a) shows this correspondence (**Exercise** 3.8).

At the strong dimerization point (3.48) and at the filling fraction corresponding to one spinless electron less than half-filling, the fermionic many-body ground state of a chain made of the even number N of sites is $N/2$-fold degenerate (periodic boundary conditions are assumed) with the representative

$$\left|\text{dimer}, \nu = \frac{1}{2} - \frac{1}{N}\right\rangle := \hat{\chi}^{\dagger}_{1,+}\cdots\hat{\chi}^{\dagger}_{2j-3,+}\ \ 1\ \ \hat{\chi}^{\dagger}_{2j+1,+}\cdots\hat{\chi}^{\dagger}_{N-1,+}|0\rangle. \tag{3.55}$$

Figure 3.4(b) is the valence-bond (dimer) covering that represents the many-body ground state (3.55) with

$$N_{\mathrm{e}} = \frac{N}{2} - 1 \tag{3.56}$$

spinless electrons. Sites $2j-1$ and $2j$ are not anymore the end points of a bond. Site $2j-1$ is called a monomer from sublattice Λ^{A}. Site $2j$ is called a monomer from sublattice Λ^{B}. The charge deficit of the many-body state $\left|\text{dimer}, \nu = \frac{1}{2} - \frac{1}{N}\right\rangle$ is one spinless electron relative to the many-body state $\left|\text{dimer}, \nu = \frac{1}{2}\right\rangle$. The two monomers in Figure 3.4(b) shared equally one spinless electron in the reference state 3.4(a). Now, they share equally the deficit of one spinless electron.

We proceed by defining the many-body state

$$\begin{aligned}\left|\text{dimer}, \nu = \frac{1}{2} - \frac{1}{N}, r_{\mathrm{s}\bar{\mathrm{s}}} := 4n+1\right\rangle := \hat{\chi}^{\dagger}_{1,+} \cdots \hat{\chi}^{\dagger}_{2(j-n)-3,+} \times \\ \hat{\chi}^{\dagger}_{2(j-n),+} \cdots \hat{\chi}^{\dagger}_{2(j+n)-2,+} \times \\ \hat{\chi}^{\dagger}_{2(j+n)+1,+} \cdots \hat{\chi}^{\dagger}_{N-1,+} |0\rangle \end{aligned} \tag{3.57a}$$

that supports a monomer located on the site $2(j-n)-1$ of sublattice Λ^A and a monomer located on the site $2(j+n)$ of sublattice Λ^B. Here, we have defined

$$\hat{\chi}^{\dagger}_{2j,\pm} := \frac{1}{\sqrt{2}} \left(\hat{c}^{\dagger}_{2j} \pm \hat{c}^{\dagger}_{2j+1} \right), \qquad \hat{\chi}_{2j,\pm} := \frac{1}{\sqrt{2}} \left(\hat{c}_{2j} \pm \hat{c}_{2j+1} \right) \tag{3.57b}$$

for $j = 1, \cdots, N/2$. The case $n=1$ is shown in Figure 3.4(c). The case $n=2$ is shown in Figure 3.4(d). What we have done, classically, is to translate (slide) the bond (dimer) covering by the lattice spacing of $\Lambda = \Lambda^A \cup \Lambda^B$ from the left to the right between the site $2j-1-2n$ and the site $2j-1$ of sublattice Λ^A in Figure 3.4(b) and to translate (slide) the bond (dimer) covering by the lattice spacing of $\Lambda = \Lambda^A \cup \Lambda^B$ from the right to the left between the site $2j+2n$ and the site $2j$ of sublattice Λ^B in Figure 3.4(b). In doing so, the pair of monomers is now a distance $4n+1$ in units of the lattice spacing of $\Lambda = \Lambda^A \cup \Lambda^B$ apart. The valence-bond (dimer) covering changes as one passes the monomers. It skips a step at the monomer. It goes from connecting (reading from left to right) an odd-to-even site, to connecting an even-to-odd site in between the monomers, to again connecting an odd-to-even site past the second monomer. The two patterns of dimerization correspond to the $\mathbb{Z}_2$ arbitrariness when choosing the sign of the dimerization with the magnitude $|\delta t| = t$ in Eq. (3.48).

The many-body state (3.57) is not anymore an eigenstate of the Hamiltonian with a pristine dimerization pattern at the filling fraction (3.56). On the other hand, the many-body state (3.57) still has a deficit of one occupied single-particle state relative to the reference many-body ground state (3.54). The missing spinless electron is equally shared between the two monomers. So there is a deficit of 1/2 an electron around each monomer. This is a classical explanation for the fractional charge in polyacetylene. To turn this cartoon explanation into a proof, it must be shown that the many-body state (3.57) becomes arbitrarily close to the ground state of the dimerized Hamiltonian supporting a pair of soliton and antisoliton defects a distance $r_{\mathrm{s}\bar{\mathrm{s}}}$ apart in the grand-canonical ensemble and after the limit $N \to \infty$ followed by the limit $r_{\mathrm{s}\bar{\mathrm{s}}}/\xi \to \infty$ have been taken. As before, ξ is the characteristic size of the support of the pair of zero modes bound to the pair of soliton and antisoliton.

3.4.3.2 n-Merization away from Half-Filling

Let $n = 2, 3, 4, \cdots$ be a positive integer, let t_1, t_2, $\cdots$, t_{n-1}, t_n be n complex numbers, let N be a multiple of n, and let $i = 1, \cdots, N$ be a site from the linear chain Λ. The many-body Hamiltonian is defined by Eq. (3.4a) with the choice

$$t_i = t_{i+N} := \begin{cases} t_1, & \text{if } i = n\,(j-1)+1, \\ t_2, & \text{if } i = n\,(j-1)+2, \\ \vdots & \\ t_{n-1}, & \text{if } i = n\,(j-1)+n-1, \\ t_n, & \text{if } i = n\,(j-1)+n, \end{cases} \tag{3.58}$$

for $j = 1, \cdots, N/n$. A generic choice (3.58) breaks the translation symmetry of the lattice Λ down to all translations generated by the symmetry under

$$i \mapsto i+n, \qquad i = 1, \cdots, N. \tag{3.59}$$

The filling fraction that we shall consider is

$$\nu := \frac{N_{\mathrm{e}}}{N} = \frac{1}{n}, \tag{3.60}$$

where N_{e} is the number of spinless electrons. The strong n-merization point in the parameter space spanned by $t_1 \in \mathbb{C}$, $t_2 \in \mathbb{C}$, $\cdots$, $t_{n-1} \in \mathbb{C}$, $t_n \in \mathbb{C}$ is the choice

$$t_n = 0. \tag{3.61}$$

At the strong n-merization point (3.61), Hamiltonian (3.4a) with the hopping amplitudes (3.58) reduces to the direct sum over n flat bands with the N/n-fold degenerate single-particle energies

$$\varepsilon_1 \leq \varepsilon_2 \leq \cdots \leq \varepsilon_n. \tag{3.62}$$

We shall assume that the choice made for the hopping amplitudes (3.58) implies the existence of the nonvanishing gap

$$\Delta := \varepsilon_2 - \varepsilon_1 > 0. \tag{3.63}$$

We can always introduce the single-particle basis

$$|j, m\rangle := \hat{\chi}^{\dagger}_{j,m}|0\rangle, \qquad j = 1, \cdots, \frac{N}{n}, \qquad m = 1, \cdots, n, \tag{3.64a}$$

such that

$$\widehat{H} = \sum_{m=1}^{n} \varepsilon_m \sum_{j=1}^{N/n} \hat{\chi}^{\dagger}_{j,m}\, \hat{\chi}_{j,m}, \tag{3.64b}$$

where the pair of operators $\hat{\chi}^{\dagger}_{j',m'}$ and $\hat{\chi}_{j,m}$ obey the canonical fermion algebra

$$\{\hat{\chi}_{j,m}, \hat{\chi}^{\dagger}_{j',m'}\} = \delta_{j,j'}\,\delta_{m,m'} \tag{3.64c}$$

for $j, j' = 1, \cdots, N/n$ and $m, m' = 1, \cdots, n$ and with all other anticommutators vanishing. The ground state at the filling fraction (3.60) is then the product state

$$|n\text{-mer}, \nu = 1/n\rangle = \prod_{j=1}^{N/n} \hat{\chi}^{\dagger}_{j,1}\,|0\rangle, \tag{3.64d}$$

where $|0\rangle$ is annihilated by $\hat{\chi}_{j,m}$ for any $j = 1, \cdots, N/n$ and $m = 1, \cdots, n$.

For any given pair $j = 1, \cdots, N/n$ and $m = 1, \cdots, n$, the wave function

$$\langle i|j, m\rangle = \begin{cases} \psi^{(m)}_{m'}, & \text{if } i = n\,(j-1) + m', \qquad m' = 1, \cdots, n, \\ 0, & \text{otherwise.} \end{cases} \tag{3.65}$$

returns the amplitude at site $i = 1, \cdots, N$ for the single-particle eigenstate (3.64a). This amplitude is nonvanishing if and only if the integer $m' = 1, \cdots, n$ exists such that

$$i = n\,(j-1) + m', \tag{3.66}$$

in which case it only depends on the band m and the position m' within the repeat unit cell indexed by j.

We leave it to the reader to (i) generalize Figure 3.4(a) to the case of a generic strong n-merization pattern of translation symmetry breaking, (ii) to remove a spinless electron supported within a repeat unit cell of size n in units of the lattice spacing of Λ as was done in Figure 3.4(b), (iii) to translate the strong n-merization pattern so as to effectively separate the n monomers obtained in step (ii) as was done in Figures 3.4(c) and 3.4(d), (iv) to identify the n possible domain walls defined by the well-separated monomers, and (v) to argue that each isolated domain wall labeled by the integer $i_{\text{dw}} = n\,(j_{\text{dw}} - 1) + m_{\text{dw}}$ with $j_{\text{dw}} = 1, \cdots, N/n$ and $m_{\text{dw}} = 1, \cdots, n$ (the acronym "dw" is a shorthand for domain wall) binds a charge of magnitude

$$|Q_{m_{\text{dw}}}| = \left|\psi^{(1)}_{m_{\text{dw}}}\right|^2 \tag{3.67}$$

for a generic strong n-merization pattern (**Exercise** 3.9). When a spinless electron is spread uniformly between the n sites of a repeat unit cell, that is, when $|\psi^{(1)}_{m_{\text{dw}}}| = \sqrt{1/n}$, then

$$|Q_{m_{\text{dw}}}| = \frac{1}{n}. \tag{3.68}$$

The calculation of the fractional charge (3.46) relies in an essential way on the chiral spectral symmetry. This is not so for the derivation of the fractional charges (3.67) and (3.68). We are going to use complementary approaches in Sections 3.5,

3.6, 3.7, 3.9, and 3.10 to compute, at zero temperature, the fractional charge bound to point-like defects in polyacetylene that do not rely on the chiral spectral symmetry.

3.5 Fractionalization from Scattering Theory

3.5.1 Overview

We are going to treat the effect of an explicit breaking of the chiral spectral symmetry on the charge bound to a domain-wall defect in the dimerization pattern of polyacetylene from the point of view of scattering theory. The use of scattering theory is done in the same spirit as the Friedel sum rule that explains how a classical point charge is screened in a three-dimensional Fermi liquid.[8] The calculation of the fractional charge from scattering theory was shown by Yamagishi[9] to be an application of the Levinson theorem that relates the difference between the phase shifts of scattering states at zero and infinite energies, respectively, to the number of states bound to a scattering potential.[10] However, unlike with the dynamical screening of a classical point charge by a Fermi sea, the induced fermion number is here a sharp quantum number.

To this end we first express the phenomenon of charge fractionalization in polyacetylene as a boundary effect in one-dimensional position space that is captured by Eq. (3.73). We then evaluate Eq. (3.73) using the scattering Ansatz (3.99) to obtain the remarkable dependence (3.109) on the staggered chemical potential of the charge bound to a domain-wall defect in polyacetylene.

This method was pioneered by Jackiw.[11] We will follow the exposition of Jackiw and Semenoff.[12]

3.5.2 Definitions

Our starting point is the single-particle 2×2 Dirac Hamiltonian

$$\mathcal{H}_{\mathrm{s}}(x) := \tau_2\, \hat{p} + \tau_1\, \phi(x) + \tau_3\, \mu_{\mathrm{s}} \tag{3.69a}$$

acting on the vector space of functions spanned by the plane-wave basis on the real line tensored into $\mathbb{C}^2$. The momentum operator is represented by

$$\hat{p} := \frac{\mathrm{d}}{\mathrm{i}\mathrm{d}x} \tag{3.69b}$$

[8] J. Friedel, London Edinburgh Dublin Philos. Mag. & J. Sci. **43**(337), 153–189 (1952). [Friedel (1952)].

[9] H. Yamagishi, Phys. Rev. Lett. **50**(6), 458–458 (1983); Phys. Rev. D **27**(10), 2383–2396 (1983). [Yamagishi (1983a,b)].

[10] N. Levinson, Kgl. Danske Videnskab. Selskab. Mat. Fys. Medd. **25**(9), 1–29 (1949). [Levinson (1949)].

[11] R. Jackiw, in: *Quantum Structure of Space and Time,* pages 169–184, edited by M. J. Duff and C. J. Isham, Cambridge University Press, New York 1982. [Jackiw (1982)].

[12] R. Jackiw and G. Semenoff, Phys. Rev. Lett. **50**(6), 439–442 (1983). [Jackiw and Semenoff (1983)].

in the position basis. We impose the boundary conditions

$$\lim_{x\to\pm\infty} \phi(x) = \pm\phi_{\infty}, \qquad \phi_{\infty} > 0, \qquad x \in \mathbb{R}, \tag{3.69c}$$

on the real-valued soliton profile $\phi(x)$. The staggered chemical potential $\mu_{\rm s}$ is real-valued and breaks the chiral symmetry generated by τ_3. We also define the reference single-particle 2×2 Dirac Hamiltonian

$$\mathcal{H}_{\infty} := \tau_2\,\hat{p} + \tau_1\,\phi_{\infty} + \tau_3\,\mu_{\rm s} \tag{3.69d}$$

acting on the same Hilbert space as the single-particle Hamiltonian (3.69a).

We introduce the two complete sets of orthogonal eigenfunctions

$$\mathcal{H}_{\infty}\,\Psi_{\infty}(\varepsilon,x) = \varepsilon\,\Psi_{\infty}(\varepsilon,x), \qquad \varepsilon \in \mathbb{R}, \tag{3.70a}$$

and

$$\mathcal{H}_{\rm s}(x)\,\Psi(\varepsilon,x) = \varepsilon\,\Psi(\varepsilon,x), \qquad \varepsilon \in \mathbb{R}, \tag{3.70b}$$

for any $x \in \mathbb{R}$. For any $x \in \mathbb{R}$, the local densities of the single-particle level ε are

$$\rho_{\Psi_{\infty}}(\varepsilon) := \Psi^{\dagger}_{\infty}(\varepsilon,x)\,\Psi_{\infty}(\varepsilon,x) \tag{3.71a}$$

[by translation invariance, $\rho_{\Psi_{\infty}}(\varepsilon)$ is independent of x] and

$$\rho_{\Psi}(\varepsilon,x) := \Psi^{\dagger}(\varepsilon,x)\,\Psi(\varepsilon,x), \tag{3.71b}$$

respectively.

We also define the density of states ν_{∞} and $\nu_{\rm s}$ to be the measures (distributions) such that $\nu_{\infty}(\varepsilon)\,{\rm d}\varepsilon$ and $\nu_{\rm s}(\varepsilon)\,{\rm d}\varepsilon$ are the numbers of eigenstates of $\mathcal{H}_{\infty}$ and of $\mathcal{H}_{\rm s}$ in the interval $[\varepsilon, \varepsilon + {\rm d}\varepsilon]$, respectively.

The local densities of the many-body ground states are obtained by integrating over all occupied single-particle states below the chemical potential μ:

$$\Upsilon_{\infty}(x) := \int\limits_{-\infty}^{\mu} {\rm d}\varepsilon\,\nu_{\infty}(\varepsilon)\,\rho_{\Psi_{\infty}}(\varepsilon) \tag{3.72a}$$

and

$$\Upsilon(x) := \int\limits_{-\infty}^{\mu} {\rm d}\varepsilon\,\nu_{\rm s}(\varepsilon)\,\rho_{\Psi}(\varepsilon,x), \tag{3.72b}$$

respectively. The pair of functions $\Upsilon_{\infty}(x)$ and $\Upsilon(x)$ denote the local density of states integrated in energy over all occupied single-particle states without and with a soliton profile, respectively. Because of the translation symmetry of $\mathcal{H}_{\infty}$, the reference local density $\Upsilon_{\infty}(x)$ is independent of x, that is,

$$\Upsilon_{\infty}(x) = \Upsilon_{\infty}. \tag{3.72c}$$

The value Υ_∞ is none but the average fermion density at the chemical potential μ. The global charge in the soliton background is measured relative to that without the soliton background by integrating over space the difference $\Upsilon(x) - \Upsilon_\infty(x)$ between the local densities of states integrated in energy over all occupied single-particle states. In other words,

$$Q := \int_{\mathbb{R}} \mathrm{d}x \, [\Upsilon(x) - \Upsilon_\infty]. \tag{3.73}$$

The subtraction by Υ_∞ in the integrand of the right-hand side can be interpreted as the subtraction by a local background charge that guarantees charge neutrality.

3.5.3 Evaluation of the Induced Charge

We are going to prove the following three facts. In order to evaluate Eq. (3.73), we note first that the spectrum of $\mathcal{H}_\infty$ depends continuously on the momentum k through the condition

$$\varepsilon^2(k) = k^2 + \phi_\infty^2 + \mu_\mathrm{s}^2, \qquad \forall k \in \mathbb{R}, \tag{3.74}$$

on the eigenenergy $\varepsilon(k)$ of $\mathcal{H}_\infty$. Second,

$$\Theta_{\mu_\mathrm{s}}(x) := \begin{pmatrix} \exp\left(-\int\limits_0^x \mathrm{d}y \, \phi(y)\right) \\ 0 \end{pmatrix} \tag{3.75}$$

is the normalizable eigenstate with energy eigenvalue μ_s of $\mathcal{H}_\mathrm{s}(x)$, a bound state. Bound states are here eigenstates of $\mathcal{H}_\mathrm{s}(x)$ with eigenenergies bounded in magnitude between $|\mu_\mathrm{s}|$ and $\sqrt{\phi_\infty^2 + \mu_\mathrm{s}^2}$. Bound states are necessarily normalizable. As we shall explain in Section 3.6, bound states whose energy eigenvalues are larger than $|\mu_\mathrm{s}|$ in magnitude do not contribute to the global charge (3.73). Hence, for simplicity but without loss of generality, we shall assume that the bound state (3.75) is unique (up to a multiplicative phase factor). This amounts to requiring that the soliton profile around its localization center is sufficiently sharp (**Exercise** 3.7). Third,

$$\Psi(k,x) := \sqrt{\frac{\varepsilon(k) + \mu_\mathrm{s}}{2\,\varepsilon(k)}} \begin{pmatrix} 1 \\ \frac{1}{\varepsilon(k)+\mu_\mathrm{s}} \left[\partial_x + \phi(x)\right] \end{pmatrix} u(k,x), \tag{3.76a}$$

where $u(k,x)$ is a solution to the stationary Schrödinger equation

$$\left[-\partial_x^2 + \phi^2(x) - \phi'(x)\right] u(k,x) = \left(k^2 + \phi_\infty^2\right) u(k,x), \qquad k, x \in \mathbb{R}, \tag{3.76b}$$

and is an eigenstate of $\mathcal{H}_\mathrm{s}(x)$ with the energy eigenvalue $\varepsilon(k)$ obeying

$$\varepsilon^2(k) := k^2 + \phi_\infty^2 + \mu_\mathrm{s}^2, \tag{3.76c}$$

and plane-wave asymptotics as $x \to \pm\infty$. Fourth, it follows that for any ε satisfying $|\varepsilon| > \sqrt{\phi_\infty^2 + \mu_\mathrm{s}^2}$,

$$\mathrm{d}\varepsilon\, \nu_\infty(\varepsilon) = \mathrm{d}\varepsilon\, \nu_\mathrm{s}(\varepsilon) \equiv \mathrm{d}\varepsilon\, \nu(\varepsilon). \tag{3.77}$$

Proof Observe that

$$\begin{aligned}\mathcal{H}_{\mathrm{s}}^{2}(x) &= \left[\tau_{2}\,\hat{p}+\tau_{1}\,\phi(x)+\tau_{3}\,\mu_{\mathrm{s}}\right]^{2}\\ &= \tau_{0}\left[\hat{p}^{2}+\phi^{2}(x)+\mu_{\mathrm{s}}^{2}\right]-\mathrm{i}\tau_{3}\left[\hat{p},\phi(x)\right]\\ &= \tau_{0}\left[-\partial_{x}^{2}+\phi^{2}(x)+\mu_{\mathrm{s}}^{2}\right]-\tau_{3}\,\phi'(x),\end{aligned} \tag{3.78a}$$

where the shorthand notation

$$\phi'(x):=(\partial_{x}\,\phi)(x),\qquad x\in\mathbb{R}, \tag{3.78b}$$

was introduced, and is a nonnegative definite Hermitean operator. When

$$\phi(x)=\phi_{\infty},\qquad \forall x\in\mathbb{R}, \tag{3.79}$$

Eq. (3.74) follows.

The interpretation of the spinor (3.75) as the bound state of $\mathcal{H}_{\mathrm{s}}(x)$ with the lowest eigenenergy follows from

$$\begin{aligned}\mathcal{H}_{\mathrm{s}}(x)\,\Theta_{\mu_{\mathrm{s}}}(x) &= \left(\tau_{2}\,\frac{\partial}{\mathrm{i}\partial x}+\tau_{1}\,\phi(x)+\tau_{3}\,\mu_{\mathrm{s}}\right)\begin{pmatrix}\exp\left(-\int_{0}^{x}\mathrm{d}y\,\phi(y)\right)\\ 0\end{pmatrix}\\ &= \mu_{\mathrm{s}}\,\Theta_{\mu_{\mathrm{s}}}(x)\end{aligned} \tag{3.80}$$

and from the fact that $\mathcal{H}_{\mathrm{s}}^{2}$ is the sum over the two nonnegative definite Hermitean operators $\tau_{0}\left[-\partial_{x}^{2}+\phi^{2}(x)\right]-\tau_{3}\,\phi'(x)$ and $\tau_{0}\,\mu_{\mathrm{s}}^{2}$, respectively.

We now turn to the continuous spectrum. Denote with

$$\Psi(\varepsilon,x):=\mathcal{N}\begin{pmatrix}u(\varepsilon,x)\\ v(\varepsilon,x)\end{pmatrix},\qquad \forall x\in\mathbb{R}, \tag{3.81a}$$

a solution to the eigenvalue problem

$$\mathcal{H}_{\mathrm{s}}(x)\,\Psi(\varepsilon,x)=\varepsilon\,\Psi(\varepsilon,x) \tag{3.81b}$$

with the asymptotics of a plane wave, that is, with the normalization condition that

$$\lim_{x\to\pm\infty}|\Psi(\varepsilon,x)|^{2}=1. \tag{3.81c}$$

It follows from Eq. (3.81b) that

$$\mathcal{H}_{\mathrm{s}}^{2}(x)\,\Psi(\varepsilon,x)=\varepsilon^{2}\,\Psi(\varepsilon,x),\qquad \forall x\in\mathbb{R}. \tag{3.82}$$

Combining Eqs. (3.78a), (3.81a), and (3.81b) yields

$$\left[-\partial_{x}^{2}+\phi^{2}(x)-\phi'(x)+\mu_{\mathrm{s}}^{2}\right]\,u(\varepsilon,x)=\varepsilon^{2}\,u(\varepsilon,x) \tag{3.83}$$

for all $x\in\mathbb{R}$. Combining Eqs. (3.69a), (3.70b), and (3.81a) yields

$$v(\varepsilon,x)=\frac{1}{\varepsilon+\mu_{\mathrm{s}}}\,\left[\partial_{x}+\phi(x)\right]\,u(\varepsilon,x) \tag{3.84}$$

for all $x\in\mathbb{R}$ (**Exercise** 3.10).

The stationary Schrödinger equation (3.83) describes a single-particle on the line subject to the static potential

$$V(x) := \phi^2(x) - \phi'(x) + \mu_{\mathrm{s}}^2 \tag{3.85}$$

that decays[13] to the constant value $\phi_\infty^2 + \mu_{\mathrm{s}}^2$ as $x \to \pm\infty$. Hence, the eigenenergy $\varepsilon(k)$ from the continuous spectrum must obey

$$\varepsilon^2(k) = k^2 + \phi_\infty^2 + \mu_{\mathrm{s}}^2, \qquad \forall k \in \mathbb{R}, \tag{3.86}$$

with the continuum density of states

$$\nu(\varepsilon) = \frac{1}{2\pi}\left|\frac{\mathrm{d}k}{\mathrm{d}\varepsilon}\right| = \frac{1}{2\pi}\frac{|\varepsilon|}{k(\varepsilon)}, \qquad k(\varepsilon) = \sqrt{\varepsilon^2 - \phi_\infty^2 - \mu_{\mathrm{s}}^2}. \tag{3.87}$$

In order to fix the normalization constant $\mathcal{N}$ on the right-hand side of Eq. (3.81a), we demand that $u(\varepsilon, x)$ be a plane wave normalized to the number 1 as $x \to \pm\infty$, that is,

$$\lim_{x \to \pm\infty} |u(\varepsilon, x)|^2 = 1. \tag{3.88}$$

This implies the asymptotics

$$\lim_{x \to \pm\infty} v(\varepsilon, x) = \frac{1}{\varepsilon + \mu_{\mathrm{s}}} (\mathrm{i}k \pm \phi_\infty)\, u(\varepsilon, \pm\infty), \tag{3.89a}$$

$$\lim_{x \to \pm\infty} |v(\varepsilon, x)|^2 = \frac{\varepsilon^2 - \mu_{\mathrm{s}}^2}{(\varepsilon + \mu_{\mathrm{s}})^2}, \tag{3.89b}$$

$$\lim_{x \to \pm\infty} \left[|u(\varepsilon, x)|^2 + |v(\varepsilon, x)|^2\right] = 1 + \frac{\varepsilon - \mu_{\mathrm{s}}}{\varepsilon + \mu_{\mathrm{s}}}. \tag{3.89c}$$

The desired normalization (3.81c) follows when

$$\mathcal{N} = \sqrt{\frac{\varepsilon + \mu_{\mathrm{s}}}{2\,\varepsilon}}, \qquad \varepsilon^2 \geq \phi_\infty^2 + \mu_{\mathrm{s}}^2, \tag{3.90}$$

is chosen in Eq. (3.81a). □

The local density (3.71b) of the single-particle level (3.76a) with the energy eigenvalue ε obeying

$$\varepsilon^2 = k^2 + \phi_\infty^2 + \mu_{\mathrm{s}}^2 \tag{3.91a}$$

is

$$\rho_\Psi(\varepsilon, x) = \frac{\varepsilon + \mu_{\mathrm{s}}}{2\,\varepsilon}\,|u(\varepsilon, x)\,|^2 + \frac{1}{2\,\varepsilon\,(\varepsilon + \mu_{\mathrm{s}})}\,|[\partial_x + \phi(x)]\, u(\varepsilon, x)|^2. \tag{3.91b}$$

[13] This decay is exponential with the choice $\phi(x) := \phi_\infty \tanh(x/\zeta)$ for the soliton profile.

We rewrite

$$
\begin{aligned}
\left|(\partial_x+\phi)\,u\right|^2 &= \left|\partial_x\,u\right|^2+\phi^2\left|u\right|^2+\phi\,(\partial_x\,u^*)\,u+\phi\,u^*\,(\partial_x\,u)\\
&= \underline{u^*\left(-\partial_x^2+\phi^2-\phi'\right)u}+\partial_x\left(u^*\,\partial_x\,u+\phi\,u^*\,u\right)\\
&= \underline{(\varepsilon+\mu_{\mathrm{s}})(\varepsilon-\mu_{\mathrm{s}})u^*\,u}+\partial_x\left(u^*\,\partial_x\,u+\phi\,u^*\,u\right).
\end{aligned}
\tag{3.92}
$$

Equation (3.83) relates here the underlined terms. Inserting Eq. (3.92) into Eq. (3.91b) gives, for any ε such that $\varepsilon^2=k^2+\phi_\infty^2+\mu_{\mathrm{s}}^2$,

$$
\begin{aligned}
\rho_\Psi(\varepsilon,x) &= \left|u(\varepsilon,x)\right|^2\\
&\quad+\frac{1}{2\,\varepsilon\,(\varepsilon+\mu_{\mathrm{s}})}\partial_x\left[u^*(\varepsilon,x)\,\partial_x\,u(\varepsilon,x)+\phi(x)\,u^*(\varepsilon,x)\,u(\varepsilon,x)\right].
\end{aligned}
\tag{3.93}
$$

When $\phi(x)=\phi_\infty$ for all $x\in\mathbb{R}$, $u_\infty^*(\varepsilon,x)\,\partial_x\,u_\infty(\varepsilon,x)$ and $|u_\infty(\varepsilon,x)|^2$ are both independent of x so that

$$
\rho_{\Psi_\infty}(\varepsilon)\equiv|u_\infty(\varepsilon,x)|^2,\qquad \varepsilon^2=k^2+\phi_\infty^2+\mu_{\mathrm{s}}^2. \tag{3.94}
$$

When the chemical potential is just below the staggered chemical potential, that is, when $\mu=\mu_{\mathrm{s}}+0^-$, the global charge (3.73) is now

$$
\begin{aligned}
Q=&\int_{\mathbb{R}}\mathrm{d}x\int_{-\infty}^{-\sqrt{\phi_\infty^2+\mu_{\mathrm{s}}^2}}\mathrm{d}\varepsilon\,\nu(\varepsilon)\,\left[\left|u(\varepsilon,x)\right|^2-\left|u_\infty(\varepsilon,x)\right|^2\right]\\
&+\int_{-\infty}^{-\sqrt{\phi_\infty^2+\mu_{\mathrm{s}}^2}}\mathrm{d}\varepsilon\,\nu(\varepsilon)\,\frac{1}{2\,\varepsilon\,(\varepsilon+\mu_{\mathrm{s}})}\\
&\qquad\times\left[u^*(\varepsilon,x)\,\partial_x\,u(\varepsilon,x)+\phi(x)\,u^*(\varepsilon,x)\,u(\varepsilon,x)\right]_{x=-\infty}^{x=+\infty}.
\end{aligned}
\tag{3.95}
$$

Here, the number of continuum states in the interval $[\varepsilon,\varepsilon+\mathrm{d}\varepsilon]$ is $\mathrm{d}\varepsilon\,\nu(\varepsilon)$ according to Eq. (3.77). The double integral (over x and over ε) can be evaluated by assuming that the basis of eigenstates defined by the eigenvalue problem (3.83) are complete for both a background with a soliton profile or one without (the translation invariant reference state). If so, the eigenstates $u(\varepsilon,x)$ with the energy eigenvalues $\varepsilon^2\geq\phi_\infty^2+\mu_{\mathrm{s}}^2$ are one short of being complete so that

$$
\int_{\mathbb{R}}\mathrm{d}x\int_{-\infty}^{-\sqrt{\phi_\infty^2+\mu_{\mathrm{s}}^2}}\mathrm{d}\varepsilon\,\nu(\varepsilon)\,\left[\left|u(\varepsilon,x)\right|^2-\left|u_\infty(\varepsilon,x)\right|^2\right]=-1. \tag{3.96}
$$

The global charge is thus solely determined by the boundary term in

$$Q = -1 + \int\limits_{-\infty}^{-\sqrt{\phi_\infty^2+\mu_s^2}} \mathrm{d}\varepsilon\, \nu(\varepsilon)\, \frac{1}{2\,\varepsilon\,(\varepsilon+\mu_s)} \times \left[u^*(\varepsilon,x)\,\partial_x\, u(\varepsilon,x) + \phi(x)\, u^*(\varepsilon,x)\, u(\varepsilon,x)\right]_{x=-\infty}^{x=+\infty}. \tag{3.97}$$

The boundary contribution to the global charge (3.73) is the integral

$$\int\limits_{-\infty}^{-\sqrt{\phi_\infty^2+\mu_s^2}} \mathrm{d}\varepsilon\, \nu(\varepsilon)\, \frac{1}{2\,\varepsilon\,(\varepsilon+\mu_s)} \times \left[u^*(\varepsilon,x)\,\partial_x\, u(\varepsilon,x) + \phi(x)\, u^*(\varepsilon,x)\, u(\varepsilon,x)\right]_{x=-\infty}^{x=+\infty}. \tag{3.98a}$$

There is the contribution

$$\int\limits_{-\infty}^{-\sqrt{\phi_\infty^2+\mu_s^2}} \mathrm{d}\varepsilon\, \nu(\varepsilon)\, \frac{\phi_\infty}{\varepsilon\,(\varepsilon+\mu_s)} = \int\limits_{-\infty}^{-\sqrt{\phi_\infty^2+\mu_s^2}} \mathrm{d}\varepsilon\, \nu(\varepsilon) \times \frac{\phi_\infty}{2\,\varepsilon\,(\varepsilon+\mu_s)} \left[\mathrm{sgn}(x)\,|u(\varepsilon,x)|^2\right]_{x=-\infty}^{x=+\infty} \tag{3.98b}$$

that arises from the density of the occupied continuum states at $x = \pm\infty$ [the normalization conditions (3.88) were used to reach the left-hand side from the right-hand side]. There is the contribution

$$\int\limits_{-\infty}^{-\sqrt{\phi_\infty^2+\mu_s^2}} \mathrm{d}\varepsilon\, \nu(\varepsilon)\, \frac{1}{2\,\varepsilon\,(\varepsilon+\mu_s)} \left[u^*(\varepsilon,x)\,\partial_x\, u(\varepsilon,x)\right]_{x=-\infty}^{x=+\infty} = 0 \tag{3.98c}$$

that arises from the current density of the occupied continuum states at $x = \pm\infty$. This is a vanishing contribution as we now verify.

Proof To prove Eq. (3.98c), we make the scattering Ansatz

$$u(\varepsilon,x) = \begin{cases} T\, e^{+ikx}, & x \to +\infty, \\ e^{+ikx} + R\, e^{-ikx}, & x \to -\infty, \end{cases} \tag{3.99a}$$

for the function $u(\varepsilon,x)$ defined in Eq. (3.81):

$$|u(\varepsilon,x)|^2 = \begin{cases} |T|^2, & x \to +\infty, \\ 1 + |R|^2 + 2|R|\, \cos\left(2\,k\,x + \arg R^*\right), & x \to -\infty, \end{cases} \tag{3.99b}$$

for its squared amplitude, and

$$u^*(\varepsilon,x)\,\partial_x\,u(\varepsilon,x)=\mathrm{i}k\,|T|^2 \tag{3.99c}$$

as $x\to+\infty$, while

$$u^*(\varepsilon,x)\,\partial_x\,u(\varepsilon,x)=\mathrm{i}k\left[1-|R|^2+2\,\mathrm{i}|R|\,\sin\left(2\,k\,x+\arg R^*\right)\right] \tag{3.99d}$$

as $x\to-\infty$.

This scattering Ansatz describes a plane wave with the single-particle energy $\varepsilon=-\sqrt{k^2+\phi_\infty^2+\mu_{\mathrm{s}}^2}$ that is incoming from the left onto the soliton profile centered at the origin of the real line. This incoming plane wave is either transmitted or reflected by the soliton profile with amplitudes T or R, respectively. Conservation of flux implies

$$1=|R|^2+|T|^2. \tag{3.100}$$

Hence, subtracting Eq. (3.99d) from Eq. (3.99c) only leaves the oscillatory term from Eq. (3.99d).

To treat the oscillatory term $k\,\sin\left(2\,k\,x+\arg R^*\right)$ with $x\to\pm\infty$, we shall demand that $k\,x$ remains finite as $x\to\pm\infty$ for it to be well defined. If so, we must effectively set $k=0$ in the argument of the oscillatory term. However, this implies that the oscillatory term is multiplied by $k=0$ so that it does not contribute to the global charge. □

We are then left with

$$\begin{aligned}Q=&-1+\int\limits_{-\infty}^{-\sqrt{\phi_\infty^2+\mu_{\mathrm{s}}^2}}\mathrm{d}\varepsilon\,\nu(\varepsilon)\,\frac{\phi_\infty}{\varepsilon\,(\varepsilon+\mu_{\mathrm{s}})}\\=&-1+\phi_\infty\int\limits_{\mathbb{R}}\frac{\mathrm{d}k}{2\pi}\,\frac{1}{\sqrt{k^2+\phi_\infty^2+\mu_{\mathrm{s}}^2}\left(\sqrt{k^2+\phi_\infty^2+\mu_{\mathrm{s}}^2}-\mu_{\mathrm{s}}\right)}.\end{aligned} \tag{3.101}$$

The integral

$$I:=\phi_\infty\int\limits_{\mathbb{R}}\frac{\mathrm{d}k}{2\pi}\,\frac{1}{\sqrt{k^2+\phi_\infty^2+\mu_{\mathrm{s}}^2}\left(\sqrt{k^2+\phi_\infty^2+\mu_{\mathrm{s}}^2}-\mu_{\mathrm{s}}\right)} \tag{3.102}$$

can be evaluated as follows. First, we do the change of integration variable

$$k:=|\mu_{\mathrm{s}}|\,k' \tag{3.103a}$$

in terms of which

$$\begin{aligned}I=\frac{\phi_\infty/|\mu_{\mathrm{s}}|}{2\pi}\int\limits_{\mathbb{R}}\mathrm{d}k'\,&\frac{1}{\sqrt{1+(\phi_\infty/\mu_{\mathrm{s}})^2+k'^2}}\\&\times\frac{1}{\sqrt{1+(\phi_\infty/\mu_{\mathrm{s}})^2+k'^2}-\mathrm{sgn}(\mu_{\mathrm{s}})}.\end{aligned} \tag{3.103b}$$

This step is followed by the change of variable

$$k'' := k'/\lambda, \qquad \lambda := +\sqrt{1+(\phi_\infty/\mu_{\mathrm{s}})^2}, \tag{3.104a}$$

in terms of which

$$\begin{aligned} I &= \frac{\phi_\infty/|\mu_{\mathrm{s}}|}{2\pi}\int\limits_{\mathbb{R}}\frac{\mathrm{d}k'}{\lambda}\,\frac{1}{\sqrt{1+(k'/\lambda)^2}\left[\lambda\sqrt{1+(k'/\lambda)^2}-\mathrm{sgn}(\mu_{\mathrm{s}})\right]} \\ &= \frac{\phi_\infty/|\mu_{\mathrm{s}}|}{2\pi}\int\limits_{\mathbb{R}}\mathrm{d}k''\,\frac{1}{\sqrt{1+(k'')^2}\left[\lambda\sqrt{1+(k'')^2}-\mathrm{sgn}(\mu_{\mathrm{s}})\right]}. \end{aligned} \tag{3.104b}$$

Second, we do another change of integration variable,

$$k'' := \tan\theta, \qquad \mathrm{d}\tan\theta = \frac{\mathrm{d}\theta}{\cos^2\theta} = (1+\tan^2\theta)\,\mathrm{d}\theta, \tag{3.105}$$

in terms of which

$$\begin{aligned} I &= \frac{\phi_\infty/|\mu_{\mathrm{s}}|}{2\pi}\int\limits_{-\pi/2}^{+\pi/2}\frac{\mathrm{d}\tan\theta}{\sqrt{1+\tan^2\theta}\left(\lambda\sqrt{1+\tan^2\theta}-\mathrm{sgn}(\mu_{\mathrm{s}})\right)} \\ &= \frac{\phi_\infty/|\mu_{\mathrm{s}}|}{2\pi}\int\limits_{-\pi/2}^{+\pi/2}\frac{\mathrm{d}\theta}{\lambda-\mathrm{sgn}(\mu_{\mathrm{s}})\cos\theta}. \end{aligned} \tag{3.106}$$

Third, we use the identity

$$\int\frac{\mathrm{d}x}{a+b\cos x} = \frac{2}{\sqrt{a^2-b^2}}\arctan\left(\frac{\sqrt{a^2-b^2}\,\tan x/2}{a+b}\right) \tag{3.107}$$

that is valid when $a^2 > b^2 \geq 0$, with the identifications $a \to \lambda > 1$ and $b \to -\mathrm{sgn}(\mu_{\mathrm{s}})$. When $\phi_\infty/|\mu_{\mathrm{s}}| > 0$, we then obtain (**Exercise** 3.11)

$$\begin{aligned} I &= \frac{2}{\pi}\arctan\left(\frac{\phi_\infty/|\mu_{\mathrm{s}}|}{\sqrt{1+(\phi_\infty/\mu_{\mathrm{s}})^2}-\mathrm{sgn}(\mu_{\mathrm{s}})}\right) \\ &= \begin{cases} 1-\frac{1}{\pi}\arctan\left(\frac{\phi_\infty}{|\mu_{\mathrm{s}}|}\right), & \text{if } \mu_{\mathrm{s}}>0, \\ \frac{1}{\pi}\arctan\left(\frac{\phi_\infty}{|\mu_{\mathrm{s}}|}\right), & \text{if } \mu_{\mathrm{s}}<0. \end{cases} \end{aligned} \tag{3.108}$$

When $\mu = \mu_{\mathrm{s}} + 0^-$, the global charge (3.73) has the final representation

$$\begin{aligned} Q &= -1+I \\ &= \begin{cases} -\frac{1}{\pi}\arctan\left(\frac{\phi_\infty}{|\mu_{\mathrm{s}}|}\right), & \text{if } \mu_{\mathrm{s}}>0, \\ -1+\frac{1}{\pi}\arctan\left(\frac{\phi_\infty}{|\mu_{\mathrm{s}}|}\right), & \text{if } \mu_{\mathrm{s}}<0. \end{cases} \end{aligned} \tag{3.109}$$

The global charge (3.109) only depends on the values $\pm\phi_\infty$ with $\phi_\infty > 0$ taken by the soliton profile at $x = \pm\infty$, respectively. The global charge is insensitive to any smooth deformation of the soliton profile as long as it leaves the asymptotic values $\pm\phi_\infty$ with $\phi_\infty > 0$ unchanged. This insensitivity to local but smooth changes to the soliton profile is a topological signature. Finally, we conclude that the charge $Q = -1/2$ if the limit $\mu_{\mathrm{s}} \to 0$ is taken with $\phi_\infty > 0$.

3.6 Fractionalization from Supersymmetry

3.6.1 Overview

The Dirac Hamiltonian (3.69) has a special property that allows the exact computation of the conserved induced fermion charge. There is a hidden supersymmetry obeyed by a Hamiltonian that is linearly related to the square of the Dirac Hamiltonian (3.69). Niemi[14] took advantage of this relation to give an integral representation of the conserved induced fermion charge in terms of the heat-kernel regularization of the Witten index from supersymmetric quantum mechanics.[15] This connection to supersymmetric quantum mechanics is the subject matter of Section 3.6, for which we follow the presentation given by Niemi and Semenoff.[16]

3.6.2 Definitions

The features of the Dirac Hamiltonian (3.69) that lead to the fractionalization of the charge (3.109) are shared by the following generalizations of the Dirac Hamiltonian (3.69). Introduce the $N \times N$ complex-valued matrices P_i obeying the conditions

$$P_i^\dagger\, P_j + P_j^\dagger\, P_i = 2\,\delta_{ij}, \qquad P_i\, P_j^\dagger + P_j\, P_i^\dagger = 2\,\delta_{ij}, \tag{3.110a}$$

for any pair $i, j = 1, \cdots, d$. To any point x in d-dimensional Euclidean space $\mathbb{R}^d$, we associate the $N \times N$ complex-valued matrix $\phi(x)$. The d constant matrices P_i and the matrix field $\phi(x)$ are combined into the pair of first-order differential operators

$$\widehat{D}^\dagger := P_i^\dagger\, \mathrm{i}\partial_i + \phi^\dagger(x), \qquad \widehat{D} := P_i\, \mathrm{i}\partial_i + \phi(x), \qquad \partial_i \equiv \frac{\partial}{\partial x_i}, \tag{3.110b}$$

with the summation convention over the repeated Latin indices. This pair of operators acts on the vector space spanned by the plane waves in d-dimensional Euclidean space. We shall assume that the dependence of the matrix field ϕ on $x \in \mathbb{R}^d$ is smooth. Let μ_{s} be a real number. We define the single-particle Hamiltonian

$$\mathcal{H} := \begin{pmatrix} +\mu_{\mathrm{s}} & \widehat{D} \\ \widehat{D}^\dagger & -\mu_{\mathrm{s}} \end{pmatrix}, \tag{3.110c}$$

14 A. J. Niemi, Phys. Lett. B **146**(3), 213–216 (1984); Nucl. Phys. B **253**, 14–46 (1985). [Niemi (1984, 1985)].

15 E. Witten, Nucl. Phys. B **202**, 253–316 (1982). [Witten (1982)].

16 A. J. Niemi and G. W. Semenoff, Phys. Rep. **135**(3), 99–193 (1986). [Niemi and Semenoff (1986)].

where we follow the physicist convention by which an entry such as μ_{s} is multiplying implicitly the $N \times N$ identity matrix.

By construction, $\mathcal{H}$ is Hermitean. It is of the Dirac type since

$$\mathcal{H} = \mathcal{H}_{\phi} + \mu_{\mathrm{s}}\,\Gamma_5, \tag{3.111a}$$

where

$$\mathcal{H}_{\phi} := \Gamma_i\,\mathrm{i}\partial_i + \Phi(x), \tag{3.111b}$$

and the $2N \times 2N$ matrices

$$\Gamma_i := \begin{pmatrix} 0 & P_i \\ P_i^{\dagger} & 0 \end{pmatrix}, \tag{3.111c}$$

anticommute pairwise,

$$\{\Gamma_i, \Gamma_j\} = 2\,\delta_{ij}, \qquad i,j = 1,\cdots,d, \tag{3.111d}$$

because of Eq. (3.110a). These d matrices and the Hermitean matrix-valued field

$$\Phi(x) := \begin{pmatrix} 0 & \phi(x) \\ \phi^{\dagger}(x) & 0 \end{pmatrix} \tag{3.111e}$$

all anticommute with

$$\Gamma_5 := \begin{pmatrix} +1 & 0 \\ 0 & -1 \end{pmatrix}. \tag{3.111f}$$

The Dirac nature of Hamiltonian (3.111a) implies that its square is a positive Hamiltonian, whose eigenvalues are bounded from below by μ_{s}^2,

$$\mathcal{H}^2 = \mathcal{H}_{\phi}^2 + \mu_{\mathrm{s}}^2 \geq \mu_{\mathrm{s}}^2. \tag{3.112}$$

Moreover, according to the representation (3.110c), the upper and lower components of any eigenstate

$$\psi \equiv \begin{pmatrix} u \\ v \end{pmatrix} \tag{3.113}$$

of $\mathcal{H}$ with eigenvalue ε must obey the coupled equations

$$\widehat{D}^{\dagger}\,u = (\varepsilon + \mu_{\mathrm{s}})\,v, \tag{3.114a}$$

$$\widehat{D}\,v = (\varepsilon - \mu_{\mathrm{s}})\,u. \tag{3.114b}$$

Iteration of this pair of equations implies that

$$\widehat{D}\,\widehat{D}^{\dagger}\,u = \left(\varepsilon^2 - \mu_{\mathrm{s}}^2\right)\,u, \tag{3.115a}$$

$$\widehat{D}^{\dagger}\,\widehat{D}\,v = \left(\varepsilon^2 - \mu_{\mathrm{s}}^2\right)\,v. \tag{3.115b}$$

We are going to deduce from Eqs. (3.114) and (3.115) that the spectrum of $\mathcal{H}$ is qualitatively different at the spectral threshold

$$\varepsilon^2 = \mu_{\mathrm{s}}^2 \tag{3.116}$$

than it is above this spectral threshold, that is, when

$$\varepsilon^2 > \mu_s^2. \tag{3.117}$$

On the one hand, for any nonvanishing solution u to the stationary Schrödinger equation (3.115a) above the spectral threshold (3.116), there is the nonvanishing solution $\widehat{D}^\dagger u$ to the stationary Schrödinger equation (3.115b) with the same non-vanishing eigenvalue $\varepsilon^2 - \mu_s^2 > 0$, for

$$\widehat{D}^\dagger\,\widehat{D}\,(\widehat{D}^\dagger\,u) = \widehat{D}^\dagger\,(\widehat{D}\,\widehat{D}^\dagger\,u) = \widehat{D}^\dagger\,(\varepsilon^2 - \mu_s^2)\,u = (\varepsilon^2 - \mu_s^2)\,\widehat{D}^\dagger\,u. \tag{3.118}$$

Similarly, for any nonvanishing solution v to the stationary Schrödinger equation (3.115b) above the spectral threshold (3.116), there is the nonvanishing solution $\widehat{D}\,v$ to the stationary Schrödinger equation (3.115a) with the same nonvanishing eigenvalue $\varepsilon^2 - \mu_s^2 > 0$, for

$$\widehat{D}\,\widehat{D}^\dagger\,(\widehat{D}\,v) = \widehat{D}\,(\widehat{D}^\dagger\,\widehat{D}\,v) = \widehat{D}\,(\varepsilon^2 - \mu_s^2)\,v = (\varepsilon^2 - \mu_s^2)\,\widehat{D}\,v. \tag{3.119}$$

This observation is nothing but the fact that the two representations

$$\psi = \begin{pmatrix} u \\ v \end{pmatrix} = \begin{pmatrix} u \\ \frac{\widehat{D}^\dagger\,u}{\varepsilon+\mu_s} \end{pmatrix} \tag{3.120a}$$

and

$$\psi = \begin{pmatrix} u \\ v \end{pmatrix} = \begin{pmatrix} \frac{\widehat{D}\,v}{\varepsilon-\mu_s} \\ v \end{pmatrix} \tag{3.120b}$$

of Eqs. (3.114) merely differ by the (arbitrary) choice to impose a normalization condition on either u or v that is dictated by solving Eq. (3.115a) or Eq. (3.115b), respectively, above the spectral threshold (3.116). A normalization convention that is suited for computing the fractional charge based on scattering theory is either

$$\psi(\varepsilon,x) := \sqrt{\frac{\varepsilon+\mu_s}{2\,\varepsilon}}\begin{pmatrix} u(\varepsilon,x) \\ v(\varepsilon,x) \end{pmatrix}, \qquad v(\varepsilon,x) = \frac{1}{\varepsilon+\mu_s}\left(\widehat{D}^\dagger\,u\right)(\varepsilon,x), \tag{3.121a}$$

if we impose the plane-wave asymptotics $\lim_{|x|\to\infty}|u(\varepsilon,x)| = 1$ on the solution u to Eq. (3.115a), or

$$\psi(\varepsilon,x) := \sqrt{\frac{\varepsilon-\mu_s}{2\,\varepsilon}}\begin{pmatrix} u(\varepsilon,x) \\ v(\varepsilon,x) \end{pmatrix}, \qquad u(\varepsilon,x) = \frac{1}{\varepsilon-\mu_s}\left(\widehat{D}\,v\right)(\varepsilon,x), \tag{3.121b}$$

if we impose the plane-wave asymptotics $\lim_{|x|\to\infty}|v(\varepsilon,x)| = 1$ on the solution v to Eq. (3.115b), respectively, above the spectral threshold (3.116).

On the other hand, among all possible solutions to the eigenvalue problem (3.114) at the spectral threshold (3.116), there are the solutions

$$\psi = \begin{pmatrix} u \\ 0 \end{pmatrix}, \qquad \widehat{D}^\dagger\,u = 0, \tag{3.122a}$$

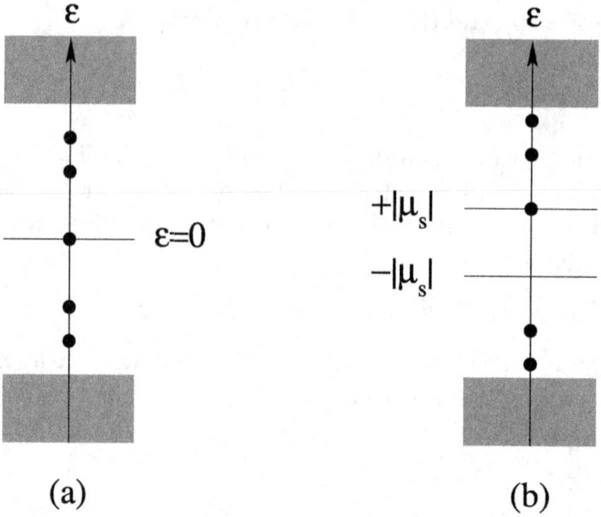

Figure 3.5 (a) Decomposition of the spectrum of $\mathcal{H}_\phi$ defined in Eq. (3.111b) into a point spectrum and a continuous spectrum. The unitary spectral symmetry $\{\mathcal{H}_\phi, \Gamma_5\} = 0$ enforces a pairing of all single-particle energy eigenstates with nonvanishing energy eigenvalues into pairs with opposite single-particle eigenenergies. If ψ is annihilated by $\mathcal{H}_\phi$, so is $\Gamma_5\,\psi$. Hence, zero modes of $\mathcal{H}_\phi$ can always be chosen to be eigenstates of Γ_5. (b) Decomposition of the spectrum of $\mathcal{H}$ defined in Eq. (3.111a) into a point spectrum and a continuous spectrum. Any zero mode of $\mathcal{H}_\phi$ defined in Eq. (3.111b) becomes a threshold state at the threshold energies $\pm|\mu_{\mathrm{s}}|$. The non-unitary spectral symmetry $\{\mathcal{H}, \mathcal{A}\} = 0$ with $\mathcal{A} = \mathcal{H}_\phi\,\Gamma_5$ enforces the mirror symmetry under $\varepsilon \mapsto -\varepsilon$ for the non-threshold spectrum of $\mathcal{H}$. The non-threshold spectrum of $\mathcal{H}$ is the set of all eigenenergies in $\mathcal{H}$ with the single-particle energies $\varepsilon^2 > \mu_{\mathrm{s}}^2$.

at $\varepsilon = +\mu_{\mathrm{s}}$ and there are the solutions

$$\psi = \begin{pmatrix} 0 \\ v \end{pmatrix}, \qquad \widehat{D}\, v = 0, \tag{3.122b}$$

at $\varepsilon = -\mu_{\mathrm{s}}$, respectively. At the spectral threshold (3.116), Eqs. (3.115a) and (3.115b) are met by the conditions $\widehat{D}^\dagger\, u = 0$ and $\widehat{D}\, v = 0$, respectively. At the spectral threshold (3.116), Eqs. (3.115a) and (3.115b) leave open the possibility that $\widehat{D}\, u \neq 0$ and $\widehat{D}^\dagger\, v \neq 0$, even though $\widehat{D}^\dagger\, u = 0$ and $\widehat{D}\, v = 0$, respectively. Hence, the possibility[17]

$$\dim\left(\mathrm{Ker}\,\widehat{D}^\dagger\right) \neq \dim\left(\mathrm{Ker}\,\widehat{D}\right) \tag{3.123}$$

arises.

When $\mu_{\mathrm{s}} = 0$, the spectral threshold collapses to the band center, see Figure 3.5. In this limit, the single-particle Dirac Hamiltonian (3.111a) acquires a unitary

[17] More precise definitions of ker and Ker are reviewed in Section 3.6.4.

spectral symmetry generated by Γ_5 in that Γ_5 anticommutes with $\mathcal{H}$. Any single-particle eigenstate ψ of $\mathcal{H}$ with the nonvanishing single-particle eigenvalue ε is mapped unitarily into the single-particle eigenstate $\Gamma_5\,\psi$ of $\mathcal{H}$ with the nonvanishing single-particle energy eigenvalue $-\varepsilon$ when $\mu_{\mathrm{s}} = 0$. Zero modes of $\mathcal{H}$ taking the form (3.122a) and (3.122b) are eigenstates of Γ_5 with the eigenvalues $+1$ and -1, respectively, when $\mu_{\mathrm{s}} = 0$.

When $\mu_{\mathrm{s}} \neq 0$, no single-particle eigenstate within the energy window $-|\mu_{\mathrm{s}}| < \varepsilon < +|\mu_{\mathrm{s}}|$ can be found for the single-particle Dirac Hamiltonian (3.111a). Any zero mode of $\mathcal{H}_{\phi}$ defined in Eq. (3.111b) that is an eigenstate of Γ_5 has been moved to either the lower threshold $-\mu_{\mathrm{s}}$ or the upper threshold $+\mu_{\mathrm{s}}$. The unitary matrix Γ_5 does not anticommute with the single-particle Dirac Hamiltonian (3.111a). However, one verifies that the non-unitary operator

$$\mathcal{A} := \frac{1}{2}\,[\mathcal{H},\Gamma_5] = \frac{1}{2}\,[\mathcal{H}_{\phi},\Gamma_5] = \mathcal{H}_{\phi}\,\Gamma_5 = \begin{pmatrix} 0 & -\widehat{D} \\ +\widehat{D}^{\dagger} & 0 \end{pmatrix} \tag{3.124}$$

anticommutes with $\mathcal{H}$,

$$\{\mathcal{H},\mathcal{A}\} = 0, \tag{3.125}$$

and annihilates any threshold state of the form (3.122). Hence, the spectrum of Hamiltonian (3.111a) that is not annihilated by $\mathcal{A}$ is symmetric with respect to $\varepsilon = 0$. For example, any scattering state $\psi(+\varepsilon,x)$ of the form (3.121a) is related to the scattering state $\psi(-\varepsilon,x)$ of the form (3.121a) by (**Exercise** 3.12)

$$\psi(-\varepsilon,x) = -\frac{\operatorname{sgn}(\varepsilon)}{\sqrt{\varepsilon^2-\mu_{\mathrm{s}}^2}}\,(\mathcal{A}\,\psi)\,(\varepsilon,x). \tag{3.126}$$

Hereto, if we assume the continuum normalization

$$\int\limits_{\mathbb{R}^d} \mathrm{d}^d x\,\psi^{\dagger}(\varepsilon,x)\,\psi(\varepsilon',x) = \delta(\varepsilon-\varepsilon'), \tag{3.127}$$

we infer that

$$\int\limits_{\mathbb{R}^d} \mathrm{d}^d x\,\left(\mathcal{A}\,\psi\right)^{\dagger}(\varepsilon,x)\,\left(\mathcal{A}\,\psi\right)(\varepsilon',x) = \left(\varepsilon^2-\mu_{\mathrm{s}}^2\right)\,\delta(\varepsilon-\varepsilon'). \tag{3.128}$$

Correspondingly, the number of states in the interval $[|\varepsilon|,|\varepsilon|+\mathrm{d}\varepsilon]$ from the continuous spectrum of $\mathcal{H}$ with positive single-particle energy eigenvalues need not equal the number of states in the interval $[-|\varepsilon|-\mathrm{d}\varepsilon,-|\varepsilon|]$ from the continuous spectrum of $\mathcal{H}$ with negative single-particle energy eigenvalues.

The ground-state expectation value of the many-body fermion number operator is defined by (**Exercise** 3.6)

$$Q := \frac{1}{2}\int\limits_{\mathbb{R}^d} \mathrm{d}^d x\,\left\langle \mathrm{FS}\left|\left[\widehat{\Psi}^{\dagger}(x),\widehat{\Psi}(x)\right]\right|\mathrm{FS}\right\rangle. \tag{3.129a}$$

The ground state $|\mathrm{FS}\rangle$ is the many-body state that is annihilated by all the annihilation operators in the mode expansions

$$\begin{aligned}\widehat{\Psi}^{\dagger}(x) := & \sum_{n_+} \hat{a}^{\dagger}_{n_+}\, \psi^{\dagger}_{n_+,+}(x) + \sum_{n_-} \hat{b}_{n_-}\, \psi^{\dagger}_{n_-,-}(x) \\ & + \sum_{n} \left[\hat{c}^{\dagger}_{n}\, \psi^{\dagger}_{n,+}(x) + \hat{d}_{n}\, \psi^{\dagger}_{n,-}(x)\right] \\ & + \int\limits_{\mathbb{R}^d} \mathrm{d}^d p \left[\hat{c}^{\dagger}(p)\, \psi^{\dagger}_{+}(p,x) + \hat{d}(p)\, \psi^{\dagger}_{-}(p,x)\right]\end{aligned} \qquad (3.129\text{b})$$

and

$$\begin{aligned}\widehat{\Psi}(x) := & \sum_{n_+} \psi_{n_+,+}(x)\, \hat{a}_{n_+} + \sum_{n_-} \psi_{n_-,-}(x)\, \hat{b}^{\dagger}_{n_-} \\ & + \sum_{n} \left[\psi_{n,+}(x)\, \hat{c}_{n} + \psi_{n,-}(x)\, \hat{d}^{\dagger}_{n}\right] \\ & + \int\limits_{\mathbb{R}^d} \mathrm{d}^d p \left[\psi_{+}(p,x)\, \hat{c}(p) + \psi_{-}(p,x)\, \hat{d}^{\dagger}(p)\right].\end{aligned} \qquad (3.129\text{c})$$

The mode expansion has been organized as follows, see Figure 3.5. The label n_+ is reserved for any threshold eigenstate of $\mathcal{H}$ with the positive single-particle eigenvalue $+|\mu_{\mathrm{s}}|$ that is represented by the wave function $\psi_{n_+,+}(x)$. The label n_- is reserved for any threshold eigenstate of $\mathcal{H}$ with the negative single-particle eigenvalue $-|\mu_{\mathrm{s}}|$ that is represented by the wave function $\psi_{n_-,-}(x)$. The discrete label n is reserved for any pair of bound states with the mirror symmetric single-particle eigenvalues $\varepsilon_{n,+} > +|\mu_{\mathrm{s}}|$ and $\varepsilon_{n,-} = -\varepsilon_{n,+}$ that are represented by the wave functions $\psi_{n,+}(x)$ and $\psi_{n,-}(x) \propto \mathcal{A}\,\psi_{n,+}(x)$, respectively. All these single-particle eigenstates are pairwise orthonormal. The momentum label p is reserved for any pair of states with the mirror symmetric single-particle eigenvalues $\varepsilon_{+}(p)$ and $\varepsilon_{-}(p) = -\varepsilon_{+}(p)$ that are represented by the wave functions with plane-wave asymptotics $\psi_{+}(p,x)$ and $\psi_{-}(p,x) \propto \mathcal{A}\,\psi_{+}(p,x)$, respectively. Distinct continuum states are orthogonal and they are continuum normalized. Finally, we postulate the sole nonvanishing anticommutators

$$\left\{\hat{a}_{n_+}, \hat{a}^{\dagger}_{n'_+}\right\} = \delta_{n_+,n'_+}, \qquad \left\{\hat{b}_{n_-}, \hat{b}^{\dagger}_{n'_-}\right\} = \delta_{n_-,n'_-}, \qquad (3.129\text{d})$$

for the threshold sector

$$\left\{\hat{c}_{n}, \hat{c}^{\dagger}_{n'}\right\} = \delta_{n,n'}, \qquad \left\{\hat{d}_{n}, \hat{d}^{\dagger}_{n'}\right\} = \delta_{n,n'} \qquad (3.129\text{e})$$

for the sector of bound states in between the threshold and continuum, and

$$\left\{\hat{c}(p), \hat{c}^{\dagger}(p')\right\} = \left\{\hat{d}(p), \hat{d}^{\dagger}(p')\right\} = \delta(p-p') \qquad (3.129\text{f})$$

for the continuum sector. Insertion of the mode expansions (3.129b) and (3.129c) into the many-body fermion number operator present in Eq. (3.129a) gives

$$
\begin{aligned}
Q = & -\frac{1}{2}\int\limits_{\mathbb{R}^d} \mathrm{d}^d x \left[\sum_{n_+} \psi^\dagger_{n_+,+}(x)\,\psi_{n_+,+}(x) - \sum_{n_-} \psi^\dagger_{n_-,-}(x)\,\psi_{n_-,-}(x)\right] \\
& -\frac{1}{2}\int\limits_{\mathbb{R}^d} \mathrm{d}^d x \int\limits_{\mathbb{R}^d} \mathrm{d}^d p \left[\psi^\dagger_+(p,x)\,\psi_+(p,x) - \psi^\dagger_-(p,x)\,\psi_-(p,x)\right].
\end{aligned}
\tag{3.130}
$$

The integration can be performed before the summations over n_+ and n_- for the first line on the right-hand side of Eq. (3.130), since the threshold states are normalized to unity. The first line on the right-hand side of Eq. (3.130) then reduces to the difference in the number of threshold states with the single-particle energies $+|\mu_{\mathrm{s}}|$ and $-|\mu_{\mathrm{s}}|$. This number is proportional to the index of the single-particle Hamiltonian $\mathcal{H}_\phi$, which is defined by (see Section 3.6.4)

$$
\operatorname{Index}\mathcal{H}_\phi := \dim\left(\operatorname{Ker}\widehat{D}^\dagger\right) - \dim\left(\operatorname{Ker}\widehat{D}\right).
\tag{3.131}
$$

There is no contribution from the bound states in between the threshold sector and the continuum sector. This is so because these bound states are normalized to unity, so that the integral over x can safely be performed before summing over their discrete single-particle energies, and they come in pair with opposite eigenenergies. The second line on the right-hand side of Eq. (3.130) arises from the continuum. The order of integration matters greatly, for the continuum single-particle state are not normalizable. However, the integral over momentum can be converted into an integral over the single-particle energies from the continuum over the spatially resolved single-particle density of states per unit energy

$$
\nu(\varepsilon,x) :=
\begin{cases}
\int\limits_{\mathbb{R}^d} \mathrm{d}^d p\,\delta\Big(\varepsilon - \varepsilon_+(p)\Big)\,\psi^\dagger_+(p,x)\,\psi_+(p,x), & \text{if } \varepsilon > 0, \\
\\
\int\limits_{\mathbb{R}^d} \mathrm{d}^d p\,\delta\Big(\varepsilon - \varepsilon_-(p)\Big)\,\psi^\dagger_-(p,x)\,\psi_-(p,x), & \text{if } \varepsilon < 0,
\end{cases}
\tag{3.132a}
$$

multiplying the sign function of the single-particle energy. In other words, we may write

$$
Q = -\frac{1}{2}\operatorname{sgn}(\mu_{\mathrm{s}})\operatorname{Index}\mathcal{H}_\phi - \frac{1}{2}\int\limits_{\mathbb{R}^d}\mathrm{d}^d x \int\limits_{\mathbb{R}} \mathrm{d}\varepsilon\,\nu(\varepsilon,x)\operatorname{sgn}(\varepsilon).
\tag{3.132b}
$$

Hence, if $\nu(\varepsilon,x)$ is an even function of energy, the second line on the right-hand side of Eq. (3.130) vanishes. This is what happens when $\mu_{\mathrm{s}} = 0$ owing to the unitary chiral spectral symmetry by which Γ_5 anticommutes with $\mathcal{H}_\phi$ defined in Eq. (3.111b) and thus with $\mathcal{H}$ since $\mathcal{H} = \mathcal{H}_\phi$ when $\mu_{\mathrm{s}} = 0$. When $\mu_{\mathrm{s}} \neq 0$, the spectral symmetry induced by the fact that the non-unitary $\mathcal{A}$ anticommutes with $\mathcal{H}$ is not sufficient to enforce that $\nu(\varepsilon, x)$ is an even function of energy. The second line of Eq. (3.130) can be nonvanishing when $\mu_{\mathrm{s}} \neq 0$, as we saw explicitly when evaluating the second term on the right-hand side of Eq. (3.132b) in Section 3.5 for the case $d = N = 1$ and showing that it reduces to a boundary (that is topological) contribution.

3.6.3 Supersymmetry

The spectral symmetry (3.125) is the signature of a hidden supersymmetry as we now explain.

We are given the Dirac Hamiltonian $\mathcal{H}$ for which we choose the representation (3.110) with its explicit 2×2 grading instead of the more abstract representation (3.111) belonging to a Clifford algebra. We define the triplet of operators

$$\mathbf{Q}^{\dagger} := \begin{pmatrix} 0 & 0 \\ +\mathrm{i}\widehat{D}^{\dagger} & 0 \end{pmatrix}, \qquad \mathbf{Q} := \begin{pmatrix} 0 & -\mathrm{i}\widehat{D} \\ 0 & 0 \end{pmatrix}, \qquad \mathbf{H} := \frac{1}{2} \begin{pmatrix} \widehat{D}\,\widehat{D}^{\dagger} & 0 \\ 0 & \widehat{D}^{\dagger}\,\widehat{D} \end{pmatrix}. \tag{3.133}$$

The triplet of operators $\mathbf{Q}^{\dagger}$, $\mathbf{Q}$, and $\mathbf{H}$ act on the same Hilbert space as the one on which the Dirac Hamiltonian (3.110) is defined.

One verifies that the triplet (3.133) closes the algebra

$$\mathbf{Q}^{\dagger}\,\mathbf{Q}^{\dagger} = 0, \qquad \mathbf{Q}\,\mathbf{Q} = 0, \qquad \{\mathbf{Q}, \mathbf{Q}^{\dagger}\} = 2\,\mathbf{H}. \tag{3.134}$$

Any triplet of operators $\{\mathbf{Q}^{\dagger}, \mathbf{Q}; \mathbf{H}\}$ that satisfies the algebra (3.134) realizes a $N = 2$ supersymmetric (SUSY) algebra.[18] We also define the pair of Hermitean operators

$$\mathbf{W} := \begin{pmatrix} +1 & 0 \\ 0 & -1 \end{pmatrix}, \qquad \mathbf{F} := \begin{pmatrix} 0 & 0 \\ 0 & 1 \end{pmatrix}. \tag{3.136}$$

Again, these operators act on the same Hilbert space as the one on which the Dirac Hamiltonian (3.110) is defined. One verifies that

$$[\mathbf{W}, \mathbf{H}] = 0, \quad \{\mathbf{W}, \mathbf{Q}^{\dagger}\} = \{\mathbf{W}, \mathbf{Q}\} = 0, \qquad \mathbf{W}^2 = 1, \quad \mathbf{W} = (-1)^{\mathbf{F}}. \tag{3.137}$$

A Hermitean operator that commutes with a $N = 2$ SUSY Hamiltonian, anti-commutes with the complex supercharges, and defines a unitary involution on the Hilbert space, is called the Witten parity operator (the Witten operator).

[18] The set of operators $\{\mathbf{Q}_1, \cdots, \mathbf{Q}_N; \mathbf{H}\}$ defines a supersymmetric (SUSY) algebra of order N if there exists a strictly positive real number $\lambda > 0$ such that

$$\mathbf{Q}_i = \mathbf{Q}_i^{\dagger}, \qquad \mathbf{H} = \mathbf{H}^{\dagger}, \qquad \left\{\mathbf{Q}_i, \mathbf{Q}_j\right\} = \lambda\,\mathbf{H}\,\delta_{i,j}, \qquad i, j = 1, \cdots, N. \tag{3.135}$$

The N Hermitean operators $\mathbf{Q}_i$ are called supercharges. The Hermitean operator $\mathbf{H}$ is called a SUSY Hamiltonian. The $N = 2$ SUSY algebra (3.134) follows from the identifications $\lambda \to 1$, $\mathbf{Q}_1 \to (\mathbf{Q} + \mathbf{Q}^{\dagger})/2$, and $\mathbf{Q}_2 \to (\mathbf{Q} - \mathbf{Q}^{\dagger})/(2\mathrm{i})$.

One verifies the following.

1. Each supercharge $\mathbf{Q}^{\dagger}$ and $\mathbf{Q}$ commutes with the $N=2$ SUSY Hamiltonian $\mathbf{H}$. They are the generators of the $N=2$ SUSY symmetry.

Proof This is a direct consequence of the $N=2$ SUSY algebra (3.134). For example,

$$[\mathbf{Q},\mathbf{H}]=\frac{1}{2}\left[\mathbf{Q},\left\{\mathbf{Q},\mathbf{Q}^{\dagger}\right\}\right]=\frac{1}{2}\left(\mathbf{Q}^{2}\mathbf{Q}^{\dagger}-\mathbf{Q}^{\dagger}\mathbf{Q}^{2}\right)=0. \qquad \square$$

2. The eigenvalues of the $N=2$ SUSY Hamiltonian $\mathbf{H}$ are always nonnegative.

Proof If ε_{+} is an eigenvalue of the Hermitean operator $\mathbf{Q}\,\mathbf{Q}^{\dagger}$ with eigenstate ψ_{+}, then $\varepsilon_{+}\,|\psi_{+}|^{2}=|\mathbf{Q}^{\dagger}\,\psi_{+}|^{2}\geq 0$, that is, $\varepsilon_{+}\geq 0$. If ε_{-} is an eigenvalue of the Hermitean operator $\mathbf{Q}^{\dagger}\,\mathbf{Q}$ with eigenstate ψ_{-}, then $\varepsilon_{-}\,|\psi_{-}|^{2}=|\mathbf{Q}\,\psi_{-}|^{2}\geq 0$, that is, $\varepsilon_{-}\geq 0$. Hence, $\mathbf{H}=(1/2)(\mathbf{Q}\,\mathbf{Q}^{\dagger}+\mathbf{Q}^{\dagger}\,\mathbf{Q})$ is the sum of two positive semi-definite operators. $\square$

3. Any zero mode of the $N=2$ SUSY Hamiltonian $\mathbf{H}$ is annihilated by both $\mathbf{Q}^{\dagger}$ and $\mathbf{Q}$. Thus, any zero mode of the $N=2$ SUSY Hamiltonian $\mathbf{H}$ provides a one-dimensional (trivial) representation of the SUSY.

Proof If ψ is an eigenstate of $\mathbf{H}$ with eigenvalue ε, then $\varepsilon\geq 0$ and

$$\varepsilon\,|\psi|^{2}=\frac{1}{2}\left(|\mathbf{Q}\,\psi|^{2}+|\mathbf{Q}^{\dagger}\,\psi|^{2}\right)\geq 0, \tag{3.138}$$

that is $\mathbf{Q}\,\psi=0$ and $\mathbf{Q}^{\dagger}\,\psi=0$ for $\varepsilon=0$. $\square$

4. Any nonvanishing eigenenergy $\varepsilon>0$ of the $N=2$ SUSY Hamiltonian $\mathbf{H}$ is at least twofold degenerate. Moreover, $\mathbf{Q}\,\mathbf{Q}^{\dagger}$ and $\mathbf{Q}^{\dagger}\,\mathbf{Q}$ share the same nonvanishing eigenvalue $2\,\varepsilon>0$, that is,

$$\operatorname{spec}\left(\mathbf{Q}\,\mathbf{Q}^{\dagger}\right)\setminus\{0\}=\operatorname{spec}\left(\mathbf{Q}^{\dagger}\,\mathbf{Q}\right)\setminus\{0\}. \tag{3.139}$$

The spectra of $\mathbf{Q}\,\mathbf{Q}^{\dagger}$ and $\mathbf{Q}^{\dagger}\,\mathbf{Q}$ are said to be essential isospectral.

Proof The Witten parity operator $\mathbf{W}$ commutes with the $N=2$ SUSY Hamiltonian $\mathbf{H}$. The Witten parity operator allows to define a resolution of the identity in terms of the pair of orthogonal projectors

$$\mathbf{P}_{\pm}=\frac{1}{2}\,(1\pm\mathbf{W})=\frac{1}{2}\left[1\pm(-1)^{\mathbf{F}}\right], \tag{3.140}$$

each of which is a constant of the motion (the supercharges are independent of time by assumption). If $\varepsilon>0$ is a strictly positive eigenvalue of $\mathbf{H}$ with the eigenstate ψ, then so are the pair of orthogonal states $\psi_{\pm}:=\mathbf{P}_{\pm}\,\psi$. Moreover, ψ_{+} is an eigenstate of $\mathbf{Q}\,\mathbf{Q}^{\dagger}$ with eigenvalue $2\,\varepsilon>0$, while ψ_{-} is an eigenstate of $\mathbf{Q}^{\dagger}\,\mathbf{Q}$ with eigenvalue $2\,\varepsilon>0$.

Next, we prove Eq. (3.139). Let ψ be an eigenstate of $\mathbf{Q}\,\mathbf{Q}^{\dagger}$ with the nonvanishing eigenvalue $2\,\varepsilon>0$. If we define $\varphi:=\mathbf{Q}^{\dagger}\,\psi$, then

$$\mathbf{Q}^{\dagger}\,\mathbf{Q}\,\varphi=\mathbf{Q}^{\dagger}\left(\mathbf{Q}\,\mathbf{Q}^{\dagger}\,\psi\right)=2\,\varepsilon\,\mathbf{Q}^{\dagger}\,\psi=2\,\varepsilon\,\varphi,$$

that is φ is an eigenstate of $\mathbf{Q}^\dagger\,\mathbf{Q}$ with the eigenvalue $2\,\varepsilon > 0$. Conversely, if φ is an eigenstate of $\mathbf{Q}^\dagger\,\mathbf{Q}$ with the nonvanishing eigenvalue $2\,\varepsilon > 0$, then $\psi := \mathbf{Q}\,\varphi$ is an eigenstate of $\mathbf{Q}\,\mathbf{Q}^\dagger$ with the eigenvalue $2\,\varepsilon > 0$. □

5. To any strictly positive energy eigenvalue $\varepsilon > 0$ of $\mathbf{H}$, there corresponds an at least two-dimensional eigenspace spanned by the orthogonal basis

$$\psi_- = -\mathbf{W}\,\psi_- = \frac{1}{\sqrt{2\,\varepsilon}}\,\mathbf{Q}^\dagger\,\psi_+, \qquad \psi_+ = +\mathbf{W}\,\psi_+ = \frac{1}{\sqrt{2\,\varepsilon}}\,\mathbf{Q}\,\psi_-. \tag{3.141}$$

Proof Since $[\mathbf{H},\mathbf{W}] = 0$, the operators $\mathbf{H}$ and $\mathbf{W} \equiv (-1)^{\mathbf{F}}$ can be simultaneously diagonalized. Because $\mathbf{W}$ is Hermitean and squares to the identity, its eigenvalues are ± 1. Let ψ_+ be a simultaneous eigenstate of $\mathbf{H}$ and $\mathbf{W}$ with the eigenvalues $\varepsilon > 0$ and $+1$, respectively, that is,

$$\mathbf{H}\,\psi_+ = \varepsilon\,\psi_+, \qquad \mathbf{W}\,\psi_+ = +\psi_+. \tag{3.142}$$

We define the state

$$\psi_- := \frac{1}{\sqrt{2\,\varepsilon}}\,\mathbf{Q}^\dagger\,\psi_+. \tag{3.143}$$

Since $[\mathbf{H},\mathbf{Q}^\dagger] = 0$, ψ_- is also an eigenstate of $\mathbf{H}$ with energy eigenvalue $\varepsilon > 0$. Moreover, since $\{\mathbf{W},\mathbf{Q}^\dagger\} = 0$, ψ_- is an eigenstate of $\mathbf{W}$ with eigenvalue -1. We have thus shown that

$$\mathbf{H}\,\psi_- = \varepsilon\,\psi_-, \qquad \mathbf{W}\,\psi_- = -\psi_-. \tag{3.144}$$

In turn, this implies the orthogonality

$$\psi_+^\dagger\psi_- = 0. \tag{3.145}$$

The state ψ_- shares the same normalization as ψ_+, for

$$|\psi_-|^2 = \frac{1}{2\,\varepsilon}\,\psi_+^\dagger\,\underbrace{\mathbf{Q}\,\mathbf{Q}^\dagger\,\psi_+}_{2\,\varepsilon\,\psi_+} = |\psi_+|^2. \tag{3.146}$$

Finally, the state ψ_- fulfills

$$\frac{1}{\sqrt{2\varepsilon}}\,\mathbf{Q}\,\psi_- = \frac{1}{2\,\varepsilon}\,\mathbf{Q}\,\mathbf{Q}^\dagger\,\psi_+ = \psi_+. \tag{3.147}$$

□

The triplet $\{\mathbf{Q}^\dagger,\mathbf{Q};\mathbf{H}\}$ obeying the $N = 2$ SUSY algebra (3.134) is said to have a good SUSY if its ground-state energy eigenvalue vanishes. The SUSY is said to be broken if the ground-state energy eigenvalue is strictly positive, in which case the ground state is at least twofold degenerate. Examples of SUSY spectra are shown in Figure 3.6.

The Dirac Hamiltonian $\mathcal{H}$ defined in Eq. (3.110) is related to the SUSY Hamiltonian $\mathbf{H}$ defined in Eq. (3.133) by

$$\mathcal{H}^2 = 2\,\mathbf{H} + \mu_{\rm s}^2 \Longleftrightarrow \mathbf{H} = \frac{1}{2}\left(\mathcal{H}^2 - \mu_{\rm s}^2\right). \tag{3.148}$$

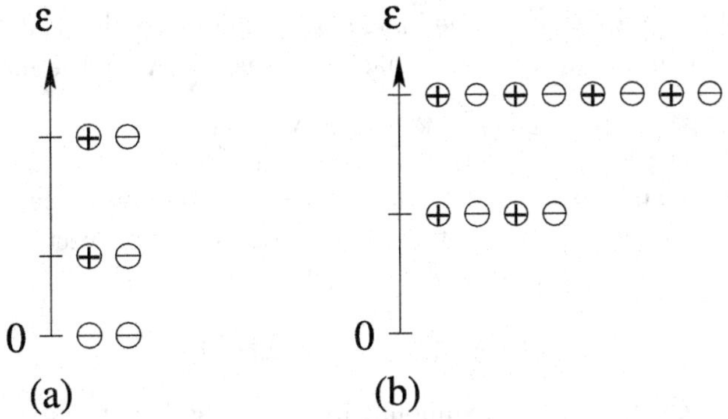

Figure 3.6 Examples of spectra for a SUSY Hamiltonian $\mathbf{H}$, with the caveat that periodic boundary conditions are imposed so that the spectrum is discrete. The Witten parity of eigenstates is denoted by a circle enclosing the Witten parity eigenvalue. (a) Spectrum with good SUSY. (b) Spectrum with broken SUSY.

Moreover, the non-unitary operator $\mathcal{A}$ defined by Eq. (3.124) that anticommutes with the Dirac Hamiltonian $\mathcal{H}$ defined by Eq. (3.110) is related to the complex supercharges $\mathbf{Q}^{\dagger}$ and $\mathbf{Q}$ by

$$\mathcal{A} = -\mathrm{i}\left(\mathbf{Q} + \mathbf{Q}^{\dagger}\right) \Longleftrightarrow \mathbf{Q} + \mathbf{Q}^{\dagger} = +\mathrm{i}\mathcal{A}. \tag{3.149}$$

The eigenstates of the Dirac Hamiltonian $\mathcal{H}$ defined by Eq. (3.110) are related by Eq. (3.148) to the eigenstates of the SUSY Hamiltonian $\mathbf{H}$ defined by Eq. (3.133). If the triplet $\{\mathbf{Q}^{\dagger}, \mathbf{Q}; \mathbf{H}\}$ realizes an unbroken SUSY, the threshold states of the Dirac Hamiltonian (3.110) are the zero modes of the SUSY Hamiltonian (3.133). The threshold states of the Dirac Hamiltonian (3.110) are thus the zero modes of $\mathbf{Q}^{\dagger}$ and $\mathbf{Q}$ and they provide a one-dimensional representation of the SUSY. Even though non-threshold eigenstates of the Dirac Hamiltonian (3.110) are not eigenstates of the Witten parity operator (as a result of taking the "square root" of $\mathcal{H}^2$), the SUSY signature of $\mathbf{H}$ that is encoded by Eq. (3.141) is realized by the pairing (3.126) for the non-threshold eigenstates of $\mathcal{H}$.

3.6.4 Witten Index

The pair of first-order differential operators $\widehat{D}$ and $\widehat{D}^{\dagger}$ entering in the definition of the Dirac Hamiltonian (3.110) takes center stage in the analysis that follows. We are going to relate the index (3.131) of this pair of operators to the Witten index (soon to be defined) of the triplet $\{\mathbf{Q}^{\dagger}, \mathbf{Q}; \mathbf{H}\}$ defined in Eq. (3.133). Before doing so, we need to recall the following mathematical concepts.

Let f denote a linear map between two finite-dimensional vector spaces X and Y over the field $\mathbb{K}$, that is for any pair $x, x' \in X$ and for any pair $\alpha, \alpha' \in \mathbb{K}$,

$$f\colon X \to Y, \qquad f(\alpha\, x + \alpha'\, x') = \alpha\, f(x) + \alpha'\, f(x'). \tag{3.150a}$$

The domain of the map f is X. The image of the map f, which is denoted $\operatorname{im} f$, is the set of vectors in Y which are the image of some $x \in X$ by the map f:

$$\operatorname{im} f := \{y \in Y \mid \exists x \in X, f(x) = y\}\,. \tag{3.150b}$$

The image of the map f is also called the codomain of the linear map f. The kernel of the linear map f, which is denoted $\ker f$, is the set of all vectors from X that maps under f to the null vector of Y

$$\ker f := \{x \in X \mid f(x) = 0\}\,. \tag{3.150c}$$

The kernel $\ker f$ characterizes the amount by which the linear map fails to be one-to-one (injective). The cokernel of the linear map f, which is denoted $\operatorname{coker} f$, is the quotient space $Y/\operatorname{im} f$ obtained by identifying all vectors of Y that differ by a vector from the codomain $\operatorname{im} f$ of f,

$$\operatorname{coker} f := \{[y] := \{y' \in Y \mid y - y' = f(x) \text{ for some } x \in X\} \mid y \in Y\}\,. \tag{3.150d}$$

We generalize the finite-dimensional spaces X and Y to infinite-dimensional ones V and W, respectively, whereby we assume that each vector space is equipped with a norm and is complete in the sense that any Cauchy sequence has a well defined limit. Such vector spaces are called Banach spaces. The kernel, image (codomain), and cokernel of a linear map $\widehat{F}\colon V \to W$ are denoted $\operatorname{Ker}\widehat{F}$, $\operatorname{Im}\widehat{F}$, and $\operatorname{Coker}\widehat{F}$, respectively. If the linear operator $\widehat{F}\colon V \to W$ is bounded (that is continuous) and if both $\dim\operatorname{Ker}(\widehat{F})$ and $\dim\operatorname{Coker}(\widehat{F})$ are finite, then $\widehat{F}$ is called a Fredholm operator and the integer

$$\operatorname{Index}(\widehat{F}) := \dim\operatorname{Ker}(\widehat{F}) - \dim\operatorname{Coker}(\widehat{F}) \tag{3.151}$$

is called the Fredholm index of $\widehat{F}$.

Furthermore, if V and W are Hilbert spaces (that is Banach spaces whose norm obey the parallelogram law so that their norm can define a scalar product), then

$$\dim\operatorname{Coker}(\widehat{F}) = \dim\operatorname{Ker}(\widehat{F}^{\dagger}). \tag{3.152}$$

We are going to demonstrate that, for any pair of Hilbert spaces V and W related by the linear map $\widehat{F}$,

$$\dim\operatorname{Ker}(\widehat{F}) = \dim\operatorname{Ker}(\widehat{F}^{\dagger}\,\widehat{F}). \tag{3.153a}$$

Proof Let $\psi \in V$ with V the Hilbert space with the scalar product $(\cdot,\cdot)_V$. Let $\phi \in W$ with W the Hilbert space with the scalar product $(\cdot,\cdot)_W$. The pair of mappings

$$\begin{aligned} \widehat{F}\colon V &\to W, \\ \psi &\mapsto \widehat{F}\,\psi \end{aligned}$$

and

$$\begin{aligned}\widehat{F}^{\dagger}\colon\ & W \to V,\\ & \phi \mapsto \widehat{F}^{\dagger}\,\phi\end{aligned}$$

are related by the defining condition

$$(\phi,\widehat{F}\,\psi)_{W} = (\widehat{F}^{\dagger}\,\phi,\psi)_{V}, \qquad \forall\psi\in V, \qquad \forall\phi\in W.$$

We have that

$$\begin{aligned}\widehat{F}^{\dagger}\,\widehat{F}\colon\ & V \to V,\\ & \psi \mapsto \widehat{F}^{\dagger}\,\widehat{F}\,\psi\end{aligned}$$

and

$$\begin{aligned}\widehat{F}\,\widehat{F}^{\dagger}\colon\ & W \to W,\\ & \phi \mapsto \widehat{F}\,\widehat{F}^{\dagger}\,\phi.\end{aligned}$$

If $\psi\in\mathrm{Ker}\,(\widehat{F})$, then $\widehat{F}\psi=0$. In turn, $\widehat{F}^{\dagger}\,\widehat{F}\,\psi=0$ must hold so that

$$\mathrm{Ker}\,(\widehat{F})\subset\mathrm{Ker}\,(\widehat{F}^{\dagger}\,\widehat{F}).$$

To prove that $\mathrm{Ker}\,(\widehat{F}^{\dagger}\,\widehat{F})\subset\mathrm{Ker}\,(\widehat{F})$, it suffices to show that it is impossible to find a state ψ such that $\widehat{F}\,\psi\neq0$ but $\widehat{F}^{\dagger}\,\widehat{F}\,\psi=0$. Suppose that we have found a ψ such that $\widehat{F}\,\psi\neq0$, while $\widehat{F}^{\dagger}\,\widehat{F}\,\psi=0$. If so, $\widehat{F}\,\psi\neq0$ implies that $(\widehat{F}\,\psi,\widehat{F}\,\psi)_{W}>0$, while $\widehat{F}^{\dagger}\,\widehat{F}\,\psi=0$ implies that $(\psi,\widehat{F}^{\dagger}\,\widehat{F}\,\psi)_{V}=0$. However, the defining condition dictates that $(\psi,\widehat{F}^{\dagger}\,\widehat{F}\,\psi)_{V}=(\widehat{F}\,\psi,\widehat{F}\,\psi)_{W}$. As it is impossible to satisfy simultaneously $(\widehat{F}\,\psi,\widehat{F}\,\psi)_{W}>0$ and $(\widehat{F}\,\psi,\widehat{F}\,\psi)_{W}=0$, we must have that

$$\mathrm{Ker}\,(\widehat{F}^{\dagger}\,\widehat{F})\subset\mathrm{Ker}\,(\widehat{F}). \qquad \square$$

Similarly,

$$\dim\mathrm{Ker}\,(\widehat{F}^{\dagger})=\dim\mathrm{Ker}\,(\widehat{F}\,\widehat{F}^{\dagger}). \tag{3.153b}$$

We may then define the pair of integer-valued indices

$$\Xi_{\widehat{F}}:=\dim\mathrm{Ker}\,(\widehat{F})+\dim\mathrm{Ker}\,(\widehat{F}^{\dagger})=\dim\mathrm{Ker}\,(\widehat{F}^{\dagger}\,\widehat{F})+\dim\mathrm{Ker}\,(\widehat{F}\,\widehat{F}^{\dagger}) \tag{3.154}$$

and

$$\Delta_{\widehat{F}}:=\dim\mathrm{Ker}\,(\widehat{F})-\dim\mathrm{Ker}\,(\widehat{F}^{\dagger})=\dim\mathrm{Ker}\,(\widehat{F}^{\dagger}\,\widehat{F})-\dim\mathrm{Ker}\,(\widehat{F}\,\widehat{F}^{\dagger}). \tag{3.155}$$

The integer $\Xi_{\widehat{F}}$ counts all the linearly independent zero modes of $\widehat{F}$ and $\widehat{F}^{\dagger}$. The integer $\Delta_{\widehat{F}}$, the Fredholm index of the operator $\widehat{F}$, counts the difference between the number of the linearly independent zero modes of $\widehat{F}$ and the number of the linearly independent zero modes of $\widehat{F}^{\dagger}$.

We assume that we may identify the Fredholm operator $\widehat{F}$ with $\widehat{D}^{\dagger}$ or $\widehat{D}$ defined in Eq. (3.110). Let the positive integer n_{+} denote the total number of linearly independent zero modes of $\widehat{D}\,\widehat{D}^{\dagger}$, that is,

$$n_{+} := \dim \operatorname{Ker}(\widehat{D}\,\widehat{D}^{\dagger}). \tag{3.156a}$$

Let the positive integer n_{-} denote the total number of linearly independent zero modes of $\widehat{D}^{\dagger}\,\widehat{D}$, that is,

$$n_{-} := \dim \operatorname{Ker}(\widehat{D}^{\dagger}\,\widehat{D}). \tag{3.156b}$$

Because $\widehat{D}\,\widehat{D}^{\dagger}$ and $\widehat{D}^{\dagger}\,\widehat{D}$ are differential operators acting on infinite-dimensional Hilbert spaces, any one of the positive integers n_{+} and n_{-} could be infinite. However, if $\widehat{D}^{\dagger}$ and $\widehat{D}$ are Fredholm operators, we may then define the pair of two integers:

$$\Xi := n_{+} + n_{-}, \qquad \Delta := n_{+} - n_{-}. \tag{3.156c}$$

As $|\Delta| < \Xi$, we infer that SUSY is good whenever $\Delta \neq 0$. When $\Delta = 0$, no conclusion can be drawn on whether SUSY is good or broken.

In general, $\widehat{D}$ and $\widehat{D}^{\dagger}$ are not Fredholm operators, that is n_{+} and n_{-} are ill-defined numbers. Theoretical physicists are masters at extracting finite numbers out of ill-defined numbers. For example, the Witten index for the triplet $\{\mathbf{Q}^{\dagger}, \mathbf{Q}; \mathbf{H}\}$ is defined to be

$$\begin{aligned}
\Delta :=& \lim_{\beta\to\infty} \operatorname{Tr}\left[(-1)^{\mathbf{F}}\, e^{-\beta\,\mathbf{H}}\right] \\
=& \lim_{\beta\to\infty} \left[\operatorname{Tr}|_{+}\left(e^{-\beta\,\widehat{D}\,\widehat{D}^{\dagger}/2}\right) - \operatorname{Tr}|_{-}\left(e^{-\beta\,\widehat{D}^{\dagger}\,\widehat{D}/2}\right)\right] \\
\equiv& \operatorname{Index}\mathcal{H}_{\phi}.
\end{aligned} \tag{3.157}$$

The Witten index is nothing but the difference in the number of linearly independent zero modes with opposite Witten parity according to the first equality. In this equality, the trace is over the space of suitable wave functions on the real line tensored with the vector space $\mathbb{C}^2$. According to the second equality, the Witten index is obtained by computing two partition functions at nonvanishing temperature $1/(k_{\mathrm{B}}\,\beta)$ (with k_{B} denoting the Boltzmann constant and in units in which $k_{\mathrm{B}} = 1$) followed by taking the zero temperature limit $1/\beta \to 0$. The first partition function is defined by tracing over the Hilbert space obtained by applying the projector $\mathbf{P}_{+}$ over the Hilbert space on which $\mathbf{H}$ acts. The second partition function is defined by tracing over the Hilbert space obtained by applying the projector $\mathbf{P}_{-}$ over the Hilbert space on which $\mathbf{H}$ acts. These two subspaces need not share the same dimensionality. They are both infinite dimensional, but may differ by a finite amount, the number of their zero modes, owing to the spectra of $\mathbf{Q}\,\mathbf{Q}^{\dagger}$ and $\mathbf{Q}^{\dagger}\,\mathbf{Q}$ being essential isospectral. The Witten index is nonvanishing when the pair $\widehat{D}$ and $\widehat{D}^{\dagger}$ can be thought of as "rectangular" and "non-square" matrices with an infinite number of columns and rows that are related by transposition and complex conjugation. The Witten index is a regularization of the analytical index, see Eqs. (3.25) and (3.131), according to the third equality. (The third equality follows from the fact that the index of any two Hamiltonian differing by a multiplicative factor is, by construction, the same.)

This regularization is called the heat-kernel regularization. It was introduced by Atiyah, Bott, and Patodi as an alternative proof of the Atiyah-Singer index theorem. Regularizations need not be unique. Correspondingly, it can be shown that

$$\Delta = \lim_{\beta\to\infty} \Delta_{\mathrm{HK}}(\beta) = \lim_{z\to 0} \Delta_{\mathrm{RK}}(z) = \lim_{\varepsilon\to 0} \Delta_{\mathrm{IDOS}}(\varepsilon), \tag{3.158a}$$

where the index HK in the regularization ($\beta > 0$)

$$\Delta_{\mathrm{HK}}(\beta) := \mathrm{Tr}\,\left[(-1)^{\mathbf{F}}\, e^{-\beta\mathbf{H}}\right] \tag{3.158b}$$

stands for heat kernel, the index RK in the regularization ($z < 0$)

$$\Delta_{\mathrm{RK}}(z) := \mathrm{Tr}\,\left[(-1)^{\mathbf{F}}\,\frac{z}{\mathbf{H}-z}\right] \tag{3.158c}$$

stands for resolvent kernel, and the index IDOS in the regularization ($\varepsilon > 0$)

$$\Delta_{\mathrm{IDOS}}(\varepsilon) := \mathrm{Tr}\,\left[(-1)^{\mathbf{F}}\,\Theta(\varepsilon - \mathbf{H})\right] \tag{3.158d}$$

stands for integrated density of states (Θ denotes the Heaviside step function).

What is the dependence of the regularized Witten index (3.158) if we assume that $\phi(x)$ entering the Dirac Hamiltonian (3.110) is changed in some continuous fashion?

To answer this question unambiguously, we assume that periodic boundary conditions have been imposed so that the spectra of $\mathcal{H}$, $\widehat{D}\,\widehat{D}^{\dagger}$, $\widehat{D}^{\dagger}\,\widehat{D}$, and $\mathbf{H}$ are all discrete (see Figure 3.7). We denote with $\varepsilon_{\mathrm{min}}(g)$ the parametric dependence of the eigenvalue of $\mathbf{H}(g)$ on some interval $a \le g \le b$ such that (i) $\varepsilon_{\mathrm{min}}(g)$ is the smallest nonvanishing eigenvalue on the interval $a \le g < g'$, (ii) $\varepsilon_{\mathrm{min}}(g) = 0$ on the interval $g' \le g \le g''$, and (iii) $\varepsilon_{\mathrm{min}}(g)$ is the smallest nonvanishing eigenvalue on the interval $g'' < g \le b$.[19] Whenever $\varepsilon_{\mathrm{min}}(g)$ is nonvanishing, it is at least doubly degenerate because of the SUSY. We shall assume that it is doubly degenerate on the interval $a \le g < g'$ and $2n$-fold degenerate on the interval $g'' < g \le b$. At $g = g'$, both the integer numbers $n_{+}(g') = n_{+}(a) + 1$ and $n_{-}(g') = n_{-}(a) + 1$ have increased by one. Correspondingly, $\Xi(g') = \Xi(a) + 2$ while $\Delta(g') = \Delta(a)$. For $g > g''$, both the integer numbers $n_{+}(g) = n_{+}(g') - n$ and $n_{-}(g) = n_{-}(g') - n$ have decreased by the integer $n > 0$. Correspondingly, for $g > g''$, $\Xi(g) = \Xi(g') - 2n$ while $\Delta(g) = \Delta(g')$. Even though $\Xi(g)$ undergoes two discontinuous jumps, $\Delta(g)$ is constant, that is,

$$\left(\frac{\mathrm{d}\Delta}{\mathrm{d}g}\right)(g) = 0, \tag{3.159}$$

[19] The choice of strict inequalities is dictated by the fact that the interval on which a continuous function vanishes is a closed interval.

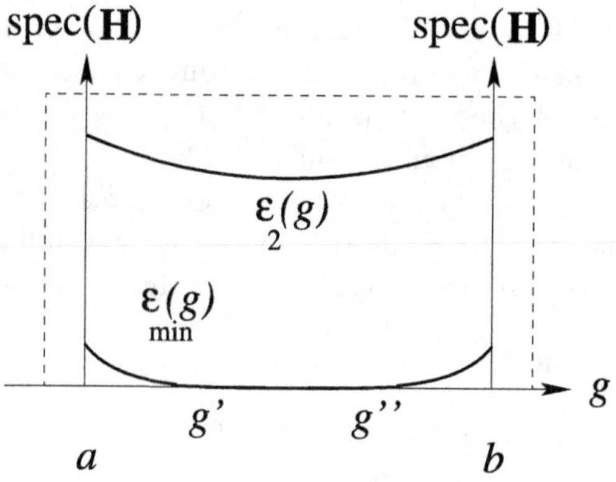

Figure 3.7 Adiabatic change in the spectrum of the SUSY Hamiltonian **H** defined in Eq. (3.133), with the caveat that periodic boundary conditions are imposed, as a function of some parameter $a \leq g \leq b$. SUSY is assumed not to be broken on the interval $a \leq g \leq b$. The dashed frame represents a window of energy during the adiabatic process defined by the interval $a \leq g \leq b$ that contains the three lowest energy eigenvalues of $\mathbf{H}(g)$. The dimensionality of the subspace spanned by all the linearly independent eigenstates with the first three ordered energy eigenvalues $0 \leq \varepsilon_{\mathrm{min}}(g) < \varepsilon_2(g)$ is unchanged in the interval $a \leq g \leq b$. Whereas the number of zero modes increases at g' and decreases at g'', the difference in the number of zero modes with opposite Witten parity is constant in the interval $a \leq g \leq b$.

under the assumed adiabatic dependence of $\varepsilon_{\mathrm{min}}(g)$ on the interval $a \leq g \leq b$. The property that the Witten index is unchanged under an adiabatic parametric dependence of the SUSY triplet $\{\mathbf{Q}^\dagger, \mathbf{Q}; \mathbf{H}\}$ is attributed to it being of topological character. The Witten index

$$\Delta = n_+ - n_- \equiv \mathrm{Tr}\,(-1)^{\mathbf{F}}, \tag{3.160}$$

that is the Fredholm index of a SUSY Hamiltonian, is a topological index, as opposed to the pair of indices

$$\Xi = n_+ + n_-, \qquad \Delta = n_+ - n_- \tag{3.161}$$

of a generic Fredholm operator.

Under what situation would the adiabatic assumption break down and with it the topological attribute (3.159) of the Witten index? Under an adiabatic parametric change, eigenvalues of **H** are not allowed to suddenly appear or disappear from some sufficiently small window of energies. A sufficient condition for a parametric change of **H** to be adiabatic is that the dimensionality of the effective low-energy Hilbert spaces prior and after the parametric change are identical, that is, no level crossing has occurred in the relevant window of energies. By switching on or off terms in **H**, that is, by making changes to **H** that are of the same order as the terms contained in **H**, this condition can be violated. A first example occurs at a gap closing transition.

As a second example, we recall that the boundary conditions in Eq. (3.69c) originate from ϕ minimizing a potential with two degenerate absolute minima at $\pm\phi_\infty$ (recall Figure 2.17). If this potential is changed parametrically so that it acquires a single minimum at $\phi = 0$ beyond a critical value g_c, the spectrum of $\mathbf{H}(g)$ must change in a nonadiabatic way by exchanging low-energy states with states at "infinite energy" upon crossing the value g_c. Such break-downs of adiabaticity are signaled by a dependence on g of the Witten index that is nonanalytic at g_c, that is, a jump of the Witten index.

3.6.5 Path-Integral Representation of the Witten Index

The heat-kernel regularization of the Witten index in SUSY quantum mechanics is represented by a path integral over classical paths and over Grassmann variables. This path integral is then shown to reduce to a Riemann integral over saddle points.

3.6.5.1 Path-Integral Representation

Motivated by Eq. (3.133), we define

$$\mathbf{H} := \frac{1}{2}\{\mathbf{Q}^\dagger, \mathbf{Q}\}, \tag{3.162a}$$

where[20]

$$\mathbf{Q}^\dagger := -\left[\hat{p} - \mathrm{i}\phi(\hat{q})\right] \otimes \hat{c}^\dagger, \qquad \mathbf{Q} := -\left[\hat{p} + \mathrm{i}\phi(\hat{q})\right] \otimes \hat{c}, \tag{3.162b}$$

with the Hermitean operators

$$\hat{p} := -\mathrm{i}\frac{\mathrm{d}}{\mathrm{d}x}, \qquad \hat{q} := x, \tag{3.162c}$$

acting on the vector space spanned by plane waves in one dimension and the pair of adjoint operators

$$\hat{c}^\dagger := \begin{pmatrix} 0 & 0 \\ 1 & 0 \end{pmatrix}, \qquad \hat{c} := \begin{pmatrix} 0 & 1 \\ 0 & 0 \end{pmatrix}, \tag{3.162d}$$

acting on $\mathbb{C}^2$. There follows the algebra

$$[\hat{q}, \hat{q}] = [\hat{p}, \hat{p}] = 0, \qquad [\hat{q}, \hat{p}] = \mathrm{i}, \tag{3.163a}$$

$$\{\hat{c}, \hat{c}\} = \{\hat{c}^\dagger, \hat{c}^\dagger\} = 0, \qquad \{\hat{c}, \hat{c}^\dagger\} = 1, \tag{3.163b}$$

$$[\hat{q}, \hat{c}] = [\hat{p}, \hat{c}] = [\hat{q}, \hat{c}^\dagger] = [\hat{p}, \hat{c}^\dagger] = 0, \tag{3.163c}$$

and

$$\mathbf{H} = \frac{1}{2}\left[\hat{p}^2 + \phi^2(\hat{q})\right] \otimes \left\{\hat{c}, \hat{c}^\dagger\right\} + \frac{\mathrm{i}}{2}\left[\hat{p}, \phi(\hat{q})\right] \otimes \left[\hat{c}^\dagger, \hat{c}\right]. \tag{3.163d}$$

[20] The minus sign on the right-hand side of the expressions defining $\mathbf{Q}^\dagger$ and $\mathbf{Q}$ is a matter of convention. Here, it is chosen to be consistent with Eqs. (3.110c) and (3.134).

If we introduce the notation

$$\phi'(x) := \left(\frac{\mathrm{d}\phi}{\mathrm{d}x}\right)(x), \tag{3.164a}$$

and if we drop the explicit use of the tensor product, we arrive at the representation[21]

$$\mathbf{H} = \frac{1}{2}\left[\hat{p}^2 + \phi^2(\hat{q})\right] + \frac{1}{2}\phi'(\hat{q})\left[\hat{c}^\dagger, \hat{c}\right] \tag{3.164b}$$

for a point particle of unit mass that (i) moves along the real line $x \in \mathbb{R}$ subject to the potential $\phi^2(x)/2$ and (ii) is coupled to a two-level system through the position-dependent coupling $\phi'(x)$. The quantum Lagrangian corresponding to the Hamiltonian (3.164b) is[22]

$$\mathbf{L} := \frac{1}{2}\left(\frac{\partial\hat{q}}{\partial t}\right)^2 - \frac{1}{2}\phi^2(\hat{q}) + \hat{c}^\dagger\,\mathrm{i}\left(\frac{\partial\hat{c}}{\partial t}\right) - \frac{1}{2}\phi'(\hat{q})\left[\hat{c}^\dagger, \hat{c}\right]. \tag{3.164c}$$

The point particle of unit mass decouples from the two-level system when the potential ϕ is a linear function of the eigenvalues $x \in \mathbb{R}$ of $\hat{q}$, that is when

$$\phi(x) = a + b\,x \tag{3.165}$$

for a and b real-valued. The special case $a = 0$ and $b = 1$ is called the SUSY harmonic oscillator. Indeed, if

$$\phi(x) = x \tag{3.166a}$$

is inserted into the right-hand side of Eq. (3.164b), one then obtains

$$\begin{aligned}\mathbf{H}^{\mathrm{SUSY}}_{\mathrm{HarO}} &:= \frac{1}{2}\left(\hat{p}^2 + \hat{q}^2\right) + \left(\hat{c}^\dagger\,\hat{c} - \frac{1}{2}\right)\\ &= \hat{a}^\dagger\,\hat{a} + \hat{c}^\dagger\,\hat{c},\end{aligned} \tag{3.166b}$$

where we have performed the unitary transformation

$$\left.\begin{aligned}\hat{q} &=: \tfrac{1}{\sqrt{2}}\left(\hat{a} + \hat{a}^\dagger\right)\\ \hat{p} &=: \tfrac{-\mathrm{i}}{\sqrt{2}}\left(\hat{a} - \hat{a}^\dagger\right)\end{aligned}\right\} \Longrightarrow \frac{1}{2}\left(\hat{p}^2 + \hat{q}^2\right) = \hat{a}^\dagger\,\hat{a} + \frac{1}{2}, \qquad [\hat{a}, \hat{a}^\dagger] = 1. \tag{3.166c}$$

Alternatively, we could have defined the supercharges

$$Q^\dagger_{\mathrm{HarO}} := -\mathrm{i}\sqrt{2}\,\hat{a}\,\hat{c}^\dagger \tag{3.167a}$$

and

$$Q_{\mathrm{HarO}} := +\mathrm{i}\sqrt{2}\,\hat{a}^\dagger\,\hat{c} \tag{3.167b}$$

[21] The rational for the multiplicative factor 1/2 on the right-hand side of Eq. (3.162a) is to recover the standard normalization for the kinetic energy in quantum mechanics.

[22] The momentum $\hat{p}$ that is canonically conjugate to $\hat{q}$ is $\partial_t\hat{q}$. It must obey the commutator $[\hat{q}, \hat{p}] = \mathrm{i}$. The momentum that is canonically conjugate to $\hat{c}$ is $\mathrm{i}\hat{c}^\dagger$. It must obey the anticommutator $\{\hat{c}, \mathrm{i}\hat{c}^\dagger\} = \mathrm{i}$. Hence, the algebra (3.163a)–(3.163c) follows.

to obtain

$$\mathbf{H}^{\mathrm{SUSY}}_{\mathrm{HarO}} = \frac{1}{2}\left\{\boldsymbol{Q}^{\dagger}_{\mathrm{HarO}}, \boldsymbol{Q}_{\mathrm{HarO}}\right\}. \tag{3.167c}$$

The orthonormal basis of eigenstates for the SUSY harmonic oscillator (3.166b) is given by

$$|n_{\mathrm{b}}, n_{\mathrm{f}}\rangle := \left(\prod_{n_{\mathrm{b}}=0}^{\infty} \frac{\left(\hat{a}^{\dagger}\right)^{n_{\mathrm{b}}}}{\sqrt{n_{\mathrm{b}}!}}\right)\left(\prod_{n_{\mathrm{f}}=0,1} \left(\hat{c}^{\dagger}\right)^{n_{\mathrm{f}}}\right)|0\rangle, \tag{3.168a}$$

whereby the nondegenerate ground state

$$|n_{\mathrm{b}} = 0, n_{\mathrm{f}} = 0\rangle \equiv |0\rangle \tag{3.168b}$$

is annihilated by both $\hat{a}$ and $\hat{c}$ and, thus, has vanishing energy. The energy spectrum

$$\varepsilon_{n_{\mathrm{b}},n_{\mathrm{f}}} := n_{\mathrm{b}} + n_{\mathrm{f}} \tag{3.169}$$

of $\mathbf{H}^{\mathrm{SUSY}}_{\mathrm{HarO}}$ is discrete with the level spacing one between any two consecutive energy eigenvalues (the unit of energy has been set to one in $\mathbf{H}^{\mathrm{SUSY}}_{\mathrm{HarO}}$). Any nonvanishing energy eigenstate is twofold degenerate, for

$$\varepsilon_{n_{\mathrm{b}},0} = \varepsilon_{n_{\mathrm{b}}-1,1}, \qquad n_{\mathrm{b}} = 1, 2, \cdots. \tag{3.170a}$$

Here, the corresponding pair of degenerate energy eigenstates are called bosonic when

$$|n_{\mathrm{b}}, n_{\mathrm{f}} = 0\rangle = +(-1)^{\hat{c}^{\dagger}\hat{c}}|n_{\mathrm{b}}, n_{\mathrm{f}} = 0\rangle \tag{3.170b}$$

and fermionic when

$$|n_{\mathrm{b}} - 1, n_{\mathrm{f}} = 1\rangle = -(-1)^{\hat{c}^{\dagger}\hat{c}}|n_{\mathrm{b}} - 1, n_{\mathrm{f}} = 1\rangle. \tag{3.170c}$$

For any pair $n_{\mathrm{b}} = 0, 1, 2, \cdots$ and $n_{\mathrm{f}} = 0, 1$,

$$\boldsymbol{Q}^{\dagger}_{\mathrm{HarO}}|n_{\mathrm{b}}, n_{\mathrm{f}}\rangle = -\mathrm{i}\sqrt{2}\sqrt{n_{\mathrm{b}}}\,\delta_{n_{\mathrm{f}},0}|n_{\mathrm{b}} - 1, n_{\mathrm{f}} + 1\rangle \tag{3.171a}$$

and

$$\boldsymbol{Q}_{\mathrm{HarO}}|n_{\mathrm{b}}, n_{\mathrm{f}}\rangle = +\mathrm{i}\sqrt{2}\sqrt{n_{\mathrm{b}} + 1}\,\delta_{n_{\mathrm{f}},1}|n_{\mathrm{b}} + 1, n_{\mathrm{f}} - 1\rangle. \tag{3.171b}$$

Remarkably, the values taken by the heat-kernel regularization of the Witten index for the SUSY harmonic oscillator $\mathbf{H}^{\mathrm{SUSY}}_{\mathrm{HarO}}$ is one for any value of $\beta > 0$,

$$\Delta_{\mathrm{HK}}(\beta) = 1, \qquad \forall\beta > 0. \tag{3.172}$$

Proof The trace on the right-hand side of Eq. (3.158b) factorizes into the product of two traces. There is the trace over the bosonic Fock space

$$\mathfrak{F}_{\mathrm{b}} := \mathrm{span}\left\{\frac{\left(\hat{a}^{\dagger}\right)^{n}}{\sqrt{n!}}|0\rangle \;\middle|\; n = 0, 1, 2, \cdots, \qquad \hat{a}\,|0\rangle = 0\right\} \tag{3.173a}$$

that is spanned by the action of the bosonic creation operator $\hat{a}^\dagger$ on the state annihilated by $\hat{a}$,

$$\begin{aligned}
\mathrm{Tr}\big|_{\mathfrak{F}_\mathrm{b}} \left(e^{-\beta\,\hat{a}^\dagger\,\hat{a}}\right) &= \sum_{n=0}^{\infty} e^{-\beta\,n} \\
&= \frac{1}{1-e^{-\beta}} \\
&= \frac{e^{+\beta}}{e^{+\beta}-1}.
\end{aligned} \tag{3.173b}$$

This geometric series is closely related to the Bose-Einstein distribution.[23] There is the trace over the fermionic Fock space

$$\mathfrak{F}_\mathrm{f} := \mathrm{span}\ \{|0\rangle,\ |1\rangle \mid \hat{c}\,|0\rangle = 0\} \tag{3.176a}$$

that is spanned by the action of the fermionic creation operator $\hat{c}^\dagger$ on the state annihilated by $\hat{c}$,

$$\begin{aligned}
\mathrm{tr}\big|_{\mathfrak{F}_\mathrm{f}} \left[(-1)^{\hat{c}^\dagger\,\hat{c}}\, e^{-\beta\,\hat{c}^\dagger\,\hat{c}}\right] &= \mathrm{tr}\big|_{\mathbb{C}^2} \left(e^{\mathrm{i}\pi\,\hat{c}^\dagger\,\hat{c}}\, e^{-\beta\,\hat{c}^\dagger\,\hat{c}}\right) \\
&= \mathrm{tr}\big|_{\mathbb{C}^2} \left(e^{-(\beta-\mathrm{i}\pi)\,\hat{c}^\dagger\,\hat{c}}\right) \\
&= 1-e^{-\beta} \\
&= \frac{e^{+\beta}-1}{e^{+\beta}}.
\end{aligned} \tag{3.176b}$$

Observe that, had we taken the fermionic trace

$$\mathrm{tr}\big|_{\mathfrak{F}_\mathrm{f}} \left(e^{-\beta\,\hat{c}^\dagger\,\hat{c}}\right) = 1+e^{-\beta} \tag{3.177}$$

instead, we would have obtained a generating function closely related to the Fermi-Dirac distribution (see footnote 23). □

A less elementary way to understand Eq. (3.172), that will be seen to be extremely powerful, is the path-integral representation [the path integral over the bosons is derived in **Exercise** 3.13, while that over the fermions is derived in **Exercise** 3.14]

[23] In the grand canonical ensemble (volume V, inverse temperature β, and the chemical potential μ), the Bose-Einstein and Fermi-Dirac distributions of the single-particle energy ε are

$$f_\mathrm{BE}(\varepsilon) := \frac{1}{e^{\beta\,(\varepsilon-\mu)}-1}, \qquad f_\mathrm{FD}(\varepsilon) := \frac{1}{e^{\beta\,(\varepsilon-\mu)}+1}, \tag{3.174}$$

respectively. They follow from $-1/\beta$ times the logarithmic derivative with respect to ε of the partition functions

$$Z_\mathrm{BE}(V,\beta,\mu) := \prod_{\varepsilon'} \frac{1}{1-e^{-\beta\,(\varepsilon'-\mu)}}, \qquad Z_\mathrm{FD}(V,\beta,\mu) := \prod_{\varepsilon'} \left[1+e^{-\beta\,(\varepsilon'-\mu)}\right], \tag{3.175}$$

respectively.

$$\Delta_{\mathrm{HK}}(\beta) := \mathrm{Tr}\big|_{\mathfrak{F}_{\mathrm{b}}\otimes\mathfrak{F}_{\mathrm{f}}}\left[(-1)^{\hat{c}^{\dagger}\,\hat{c}}\,e^{-\beta\,(\hat{a}^{\dagger}\,\hat{a}+\hat{c}^{\dagger}\,\hat{c})}\right]$$
$$= \int \mathcal{D}[\varphi^{*},\varphi] \int \mathcal{D}[\psi^{*},\psi] e^{-\int_{0}^{\beta} \mathrm{d}\tau [\varphi^{*}\,(\partial_{\tau}+1)\,\varphi+\psi^{*}\,(\partial_{\tau}+1)\,\psi]}, \qquad (3.178a)$$

where $\mathcal{D}[\varphi^{*},\varphi]$ is the measure for the pair of independent real-valued functions $(\varphi+\varphi^{*})/2 \equiv \mathrm{Re}\,\varphi$ and $(\varphi-\varphi^{*})/(2\mathrm{i}) \equiv \mathrm{Im}\,\varphi$ that obey the periodic boundary conditions

$$\mathrm{Re}\,\varphi(\tau+\beta) = \mathrm{Re}\,\varphi(\tau), \qquad \mathrm{Im}\,\varphi(\tau+\beta) = \mathrm{Im}\,\varphi(\tau), \qquad 0 \le \tau < \beta, \qquad (3.178b)$$

and $\mathcal{D}[\psi^{*},\psi]$ is the measure for the pair of independent Grassmann-valued functions ψ^{*} and ψ that obey the periodic boundary conditions

$$\psi^{*}(\tau+\beta) = \psi^{*}(\tau), \qquad \psi(\tau+\beta) = \psi(\tau), \qquad 0 \le \tau < \beta. \qquad (3.178c)$$

The result (3.172) is a consequence of the identities (3.693) and (3.718), according to which

$$\Delta_{\mathrm{HK}}(\beta) = \frac{\mathrm{Det}\,(\partial_{\tau}+1)}{\mathrm{Det}\,(\partial_{\tau}+1)} = 1. \qquad (3.179)$$

As a sanity check, we do one by one the bosonic and fermionic path integrals on the right-hand side of Eq. (3.178a). The boundary conditions (3.178b) and (3.178c) imply the existence of the Fourier expansions

$$\varphi^{*}(\tau) = \frac{1}{\sqrt{\beta}} \sum_{\varpi_l} a^{*}_{\varpi_l} e^{+\mathrm{i}\varpi_l\,\tau}, \qquad \varphi(\tau) = \frac{1}{\sqrt{\beta}} \sum_{\varpi_l} a_{\varpi_l} e^{-\mathrm{i}\varpi_l\,\tau}, \qquad (3.180a)$$

and

$$\psi^{*}(\tau) = \frac{1}{\sqrt{\beta}} \sum_{\varpi_l} c^{*}_{\varpi_l} e^{+\mathrm{i}\varpi_l\,\tau}, \qquad \psi(\tau) = \frac{1}{\sqrt{\beta}} \sum_{\varpi_l} c_{\varpi_l} e^{-\mathrm{i}\varpi_l\,\tau}, \qquad (3.180b)$$

where the frequencies

$$\varpi_l = \frac{2\,\pi}{\beta}\,l, \qquad l \in \mathbb{Z}, \qquad (3.180c)$$

are the so-called bosonic Matsubara frequencies.[24] The real and imaginary parts of the complex-valued expansion coefficients $a^{*}_{\varpi_l}$ and a_{ϖ_l} define the bosonic measure

[24] The fermionic Matsubara frequencies

$$\omega_n = \frac{\pi}{\beta}\,(2n+1), \qquad n \in \mathbb{Z}, \qquad (3.181)$$

follow from imposing antiperiodic boundary conditions in imaginary time.

$$\begin{aligned}\int\mathcal{D}[\varphi^*,\varphi]\,e^{-\int_0^\beta\varphi^*(\tau)\,(\partial_\tau+1)\,\varphi\,(\tau)} &= \prod_{l\in\mathbb{Z}}\int_{\mathbb{C}}\frac{\mathrm{d}a^*_{\varpi_l}\,\mathrm{d}a_{\varpi_l}}{2\pi\mathrm{i}}\,e^{-a^*_{\varpi_l}\,(-\mathrm{i}\varpi_l+1)\,a_{\varpi_l}}\\ &=\prod_{l\in\mathbb{Z}}\frac{1}{-\mathrm{i}\varpi_l+1}\\ &\equiv\frac{1}{\operatorname{Det}\,(\partial_\tau+1)}.\end{aligned} \tag{3.182}$$

The Grassmann-valued expansion coefficients $c^*_{\varpi_l}$ and c_{ϖ_l} define the fermionic measure

$$\begin{aligned}\int\mathcal{D}[\psi^*,\psi]\,e^{-\int_0^\beta\psi^*(\tau)\,(\partial_\tau+1)\,\psi\,(\tau)} &= \prod_{l\in\mathbb{Z}}\int\mathrm{d}c^*_{\varpi_l}\,\mathrm{d}c_{\varpi_l}\,e^{-c^*_{\varpi_l}\,(-\mathrm{i}\varpi_l+1)\,c_{\varpi_l}}\\ &=\prod_{l\in\mathbb{Z}}(-\mathrm{i}\varpi_l+1)\\ &\equiv\operatorname{Det}\,(\partial_\tau+1)\,.\end{aligned} \tag{3.183}$$

A generalization of the representation (3.178) of the Witten index for the SUSY harmonic oscillator in terms of a path integral is a useful one needed to tackle the generic case when the second derivative with respect to x of ϕ, that is ϕ'', is nonvanishing for some $x\in\mathbb{R}$ in the SUSY Hamiltonian (3.164b). The coupling between the point particle of unit mass and the two-level system can be rewritten as

$$\frac{1}{2}\,\phi'(\hat{q})\,\left(2\,\hat{c}^\dagger\,\hat{c}-1\right)=\frac{1}{2}\,\phi'(\hat{q})\,\left(|1\rangle\langle 1|-|0\rangle\langle 0|\right). \tag{3.184}$$

The coupling between the point particle of unit mass and the two level system is the same as the Zeeman coupling between the spin-1/2 of an electron of unit charge and the (not necessarily uniform) magnetic field (in units with $\hbar=1$ and $c=1$) $B(\hat{q})\,\boldsymbol{e}_3=\phi'(\hat{q})\,\boldsymbol{e}_3$ aligned along the spin-1/2 quantization axis $\boldsymbol{e}_3$. Because the pair of operators $\{\hat{c},\hat{c}^\dagger\}$ and $[\hat{c},\hat{c}^\dagger]$ commute, the operator

$$\begin{aligned}\hat{\sigma} &:=[\hat{c}^\dagger,\hat{c}]\\ &=2\,\hat{c}^\dagger\,\hat{c}-1\\ &=|1\rangle\langle 1|-|0\rangle\langle 0|\end{aligned} \tag{3.185}$$

has the nonvanishing eigenvalues

$$\sigma:=\pm 1 \tag{3.186}$$

and commutes with $\mathbf{H}$.

If so, we may label any eigenstate of $\mathbf{H}$ with the nonvanishing eigenvalue $\varepsilon>0$ by the ket $|\varepsilon,\sigma\rangle$ with σ, the eigenvalue of $\hat{\sigma}$. The spectral decomposition

$$\mathbf{H}=\sum_{\sigma=\pm 1}\int_0^\infty\mathrm{d}\varepsilon\,\nu(\varepsilon)\,\varepsilon\,|\varepsilon,\sigma\rangle\langle\varepsilon,\sigma| \tag{3.187a}$$

with the resolution of the identity

$$1=\sum_{\iota}|\iota\rangle\langle\iota|+\sum_{\sigma=\pm1}\int_{0}^{\infty}\mathrm{d}\varepsilon\,\nu(\varepsilon)\,|\varepsilon,\sigma\rangle\langle\varepsilon,\sigma| \tag{3.187b}$$

and the orthogonality conditions

$$\langle\iota|\iota'\rangle=\delta_{\iota,\iota'},\qquad \langle\varepsilon,\sigma|\varepsilon',\sigma'\rangle=N_{\varepsilon,\sigma}\,\delta_{\varepsilon,\varepsilon'}\,\delta_{\sigma,\sigma'},\qquad \langle\iota|\varepsilon,\sigma\rangle=0, \tag{3.187c}$$

follow. The zero modes $\{|\iota\rangle\}$ of $\mathbf{H}$, if they exist, are necessarily normalizable. Eigenstates of $\mathbf{H}$ with nonvanishing eigenvalues need not be normalizable, as is implied by the normalization factor $N_{\varepsilon,\sigma}$. This happens when the density of states $\nu(\varepsilon)$ is a smooth function of ε and $\int\mathrm{d}\varepsilon\,\nu(\varepsilon)\,N_{\varepsilon,\sigma}\,\delta_{\varepsilon,\varepsilon'}=1$ over any energy interval that contains ε', while $\int\mathrm{d}\varepsilon\,\nu(\varepsilon)\,N_{\varepsilon,\sigma}\,\delta_{\varepsilon,\varepsilon'}=0$ otherwise.

The unitary evolution operator

$$\mathbf{U}(t,t_0):=e^{-\mathrm{i}\mathbf{H}\,(t-t_0)} \tag{3.188a}$$

evolves a solution to the Schrödinger equation

$$\mathrm{i}\partial_t\,|\Psi;t\rangle=\mathbf{H}\,|\Psi;t\rangle \tag{3.188b}$$

with the initial condition

$$|\Psi;t_0\rangle\equiv|\Psi_0\rangle \tag{3.188c}$$

according to the rule

$$|\Psi;t\rangle=\mathbf{U}(t,t_0)\,|\Psi_0\rangle. \tag{3.188d}$$

Its matrix elements in the orthogonal basis

$$|x,\sigma\rangle:=|x\rangle\otimes|\sigma\rangle,\qquad \hat{q}\,|x\rangle=x\,|x\rangle,\qquad \hat{\sigma}\,|\sigma\rangle=\sigma\,|\sigma\rangle, \tag{3.189a}$$

are block diagonal

$$\langle x',\sigma'|\mathbf{U}(t,t_0)|x,\sigma\rangle=\delta_{\sigma,\sigma'}\,\langle x',\sigma|\mathbf{U}(t,t_0)|x,\sigma\rangle. \tag{3.189b}$$

In turn, the standard Feynman path-integral representation of quantum mechanics in one-dimensional space, that is,

$$\langle x',\sigma|\,\mathbf{U}(t,t_0)\,|x,\sigma\rangle=\int_{x}^{x'}\mathcal{D}[q]\,\exp\left(+\mathrm{i}\int_{t_0}^{t}\mathrm{d}t'\,L(q)\right), \tag{3.189c}$$

where

$$L(q):=\frac{1}{2}\left(\frac{\partial q}{\partial t}\right)^2-\frac{1}{2}\,\phi^2(q)-\frac{\sigma}{2}\,\phi'(q), \tag{3.189d}$$

holds.

The representation (3.189) as a Feynman path integral of the SUSY unitary evolution operator (3.188) is a first step toward a path-integral representation of the heat-kernel regularization of the Witten index. This path-integral representation must involve an integration over all the classical paths that evolve in time between the same initial and final states, since a trace enters in the definition (3.157) of the Witten index. However, this evolution is not quite in time. Rather it takes place in imaginary time owing to the heat-kernel regularization. Moreover, the path-integral representation that we seek should correspond to taking the trace on the right-hand side of the first equality in Eq. (3.157), rather than the Feynman path integral corresponding to the right-hand side of the second equality in Eq. (3.157) in the spirit of what we have done so far. We shall achieve this goal by representing the two-level degrees of freedom by a path integral, as we did for the SUSY harmonic oscillator.

The trace on the right-hand side of Eq. (3.158b) includes tracing over all suitable wave functions defined on the real line $x \in \mathbb{R}$ as well as tracing over the vector space $\mathbb{C}^2$. Tracing over all suitable wave functions defined on the real line $x \in \mathbb{R}$ is to be done with the Feynman path integral in imaginary time over all classical paths obeying periodic boundary conditions. Needed is a representation for the trace over the vector space $\mathbb{C}^2$ as a path integral. This is done exactly in the same way as we did for the SUSY harmonic oscillator (3.166b), except for one caveat. We should not forget that deriving a representation of a partition function or of the unitary time evolution in terms of coherent states requires the Hamiltonian to be normal ordered. After normal ordering the energy levels of the resulting Hamiltonian are shifted by a constant term relative to the energy levels of the original Hamiltonian. This brings about a multiplicative normalization factor of the partition function or unitary time evolution to be represented by a path integral. Hence, the alternating trace that defines the heat-kernel regularization of the Witten index has the path-integral representation

$$\Delta_{\mathrm{HK}}(\beta) = \mathcal{N} \int \mathcal{D}[q] \int \mathcal{D}[\psi^*, \psi]\, e^{-\int_0^\beta \mathrm{d}\tau\, \mathcal{L}(q,\psi^*,\psi)}, \tag{3.190a}$$

where the paths q and the Grassmann-valued fields ψ^* and ψ obey periodic boundary conditions in imaginary time,

$$q(\tau+\beta) = q(\tau), \qquad \psi^*(\tau+\beta) = \psi^*(\tau), \qquad \psi(\tau+\beta) = \psi(\tau), \tag{3.190b}$$

and the Lagrangian density is given by

$$\mathcal{L}(q,\psi^*,\psi) = \frac{1}{2}\left[\left(\frac{\partial q}{\partial \tau}\right)^2 + \phi^2(q)\right] + \psi^* \left(\frac{\partial}{\partial \tau} + \phi'(q)\right)\psi. \tag{3.190c}$$

The multiplicative normalization $\mathcal{N}$ arose when normal ordering the Hamiltonian in preparation for the path-integral representation. We shall fix this normalization in due course by demanding that the Witten index is unity when specializing to the SUSY harmonic oscillator.

The representation (3.190) of the heat-kernel regularization of the Witten index can be simplified further.

3.6.5.2 Integrating the Fermions Exactly

We first perform the path integral over the Grassmann integration variables. This gives

$$\Delta_{\mathrm{HK}}(\beta)=\mathcal{N}'\int\mathcal{D}[q]\left|\mathrm{Det}\left(\frac{\partial}{\partial\tau}+\phi'(q)\right)\right|e^{-\int\limits_0^\beta \mathrm{d}\tau\,\mathcal{L}(q)}, \tag{3.191a}$$

where the path q obeys periodic boundary conditions in imaginary time,

$$q(\tau+\beta)=q(\tau) \tag{3.191b}$$

and the Lagrangian density is given by

$$\mathcal{L}(q):=\frac{1}{2}\left[\left(\frac{\partial q}{\partial\tau}\right)^2+\phi^2(q)\right]. \tag{3.191c}$$

We assume that there exists a function $V(q)$ with $\phi(q)=V'(q)$, such that Eq. (3.191c) takes the form

$$\mathcal{L}(q):=\frac{1}{2}\left[\left(\frac{\partial q}{\partial\tau}\right)^2+\left(\frac{\partial V}{\partial q}\right)^2(q)\right]. \tag{3.192}$$

This assumption is always verified when $\phi(q)$ is a polynomial in q. It also holds for trigonometric and exponential functions of q. This assumption brings about the following useful identities. First,

$$\begin{aligned}\mathcal{L}(q)&=\frac{1}{2}\left[\frac{\partial q}{\partial\tau}+V'(q)\right]^2-V'(q)\left(\frac{\partial q}{\partial\tau}\right)\\&=\frac{1}{2}\left[\frac{\partial q}{\partial\tau}+V'(q)\right]^2-\frac{\mathrm{d}V(q)}{\mathrm{d}\tau}.\end{aligned} \tag{3.193a}$$

We used the chain rule for derivatives to reach the last equality. The periodic boundary conditions (3.191b) imply that the total time derivative $\mathrm{d}V(q)/\mathrm{d}\tau$ drops out from the Euclidean action

$$\int\limits_0^\beta \mathrm{d}\tau\,\mathcal{L}(q)=\int\limits_0^\beta \mathrm{d}\tau\,\frac{1}{2}\left[\frac{\partial q}{\partial\tau}+V'(q)\right]^2. \tag{3.193b}$$

Second, define the path

$$Q:=\frac{\partial q}{\partial\tau}+V'(q). \tag{3.194}$$

The path Q inherits the periodic boundary conditions (3.191b),

$$Q(\tau+\beta)=Q(\tau). \tag{3.195}$$

If the path q is varied by the infinitesimal δq, the path Q is varied by the infinitesimal

$$\delta Q=\left(\frac{\partial}{\partial\tau}+V''(q)\right)\delta q. \tag{3.196}$$

The Jacobian

$$\left|\frac{\delta Q}{\delta q}\right|=\left|\mathrm{Det}\left(\frac{\partial}{\partial\tau}+V''(q)\right)\right| \tag{3.197}$$

follows. This Jacobian is none but the absolute value of the fermionic determinant in the integrand on the right-hand side of Eq. (3.191a). By changing the integration variable from q to Q in Eq. (3.191a), we arrive at the desired exact path-integral representation of heat-kernel regularization of the Witten index,

$$\Delta_{\mathrm{HK}}(\beta)=\mathcal{N}'\int\mathcal{D}[Q]\,e^{-\int_0^\beta d\tau\,\frac{1}{2}Q^2}, \tag{3.198a}$$

where the path Q obeys periodic boundary conditions in imaginary time,

$$Q(\tau+\beta)=Q(\tau). \tag{3.198b}$$

3.6.5.3 Integrating the Fermions in the Background of Static Paths Only

We return to the representation (3.190) of the heat-kernel regularization of the Witten index. We observe that, in the limit $\beta\to\infty$, the heat-kernel regularization of the Witten index (3.190) is dominated by the static bosonic paths for which $(\partial q/\partial\tau)^2=0$. If we do the same static approximation for the fermionic paths, we find the estimate

$$\Delta_{\mathrm{HK}}(\beta)\approx\mathcal{N}\int\limits_{-\infty}^{+\infty}dq\int d\psi^*\int d\psi\,e^{-\beta\left[\frac{1}{2}\phi^2(q)+\psi^*\phi'(q)\psi\right]}. \tag{3.199}$$

In Eq. (3.199), any one of the three integrations is done over one integration variable. The integration over q is a one-dimensional Riemann integration over the real line. The integrations over ψ^* and ψ are two independent Grassmann integrations. The two Grassmann integrations deliver

$$\Delta_{\mathrm{HK}}(\beta)\approx\mathcal{N}\beta\int\limits_{-\infty}^{+\infty}dq\,\phi'(q)\,e^{-\frac{\beta}{2}\phi^2(q)}. \tag{3.200}$$

With the identification of $\mathcal{N}'$ with $\mathcal{N}\beta$, the same static approximation follows from Eqs. (3.194) and (3.198a).

If we assume that ϕ is injective on $\mathbb{R}$, the manipulation

$$\mathrm{d}q\,\phi'(q) = \mathrm{d}q\,\frac{\mathrm{d}\phi}{\mathrm{d}q} = \mathrm{d}\phi, \qquad \phi_{\pm} := \lim_{q\to\pm\infty} \phi(q), \tag{3.201a}$$

suggests the change of integration variables

$$\Delta_{\mathrm{HK}}(\beta) \approx \mathcal{N}\,\beta \int\limits_{\phi_-}^{\phi_+} \mathrm{d}\phi\, e^{-\frac{\beta}{2}\phi^2}. \tag{3.201b}$$

The final change of variable that we perform is

$$y := \sqrt{\frac{\beta}{2}}\,\phi, \qquad y_{\pm} := \sqrt{\frac{\beta}{2}}\,\phi_{\pm}, \qquad \mathcal{N}' = \mathcal{N}\sqrt{2\,\beta}, \tag{3.202a}$$

and

$$\Delta_{\mathrm{HK}}(\beta) \approx \mathcal{N}' \int\limits_{\sqrt{\phi_-^2\,\beta/2}}^{\sqrt{\phi_+^2\,\beta/2}} \mathrm{d}y\, e^{-y^2}. \tag{3.202b}$$

The normalization $\mathcal{N}'$ is fixed by demanding that the integral

$$\mathcal{N}' \int\limits_{-\infty}^{+\infty} \mathrm{d}y\, e^{-y^2} = \sqrt{\pi}\,\mathcal{N}' \tag{3.203}$$

corresponding to the static approximation (3.202) applied to the SUSY harmonic oscillator [for which $\phi(q) = q$ implies $\phi_{\pm} = \pm\infty$] is unity. The final representation for the saddle-point approximation to the heat-kernel regularization of the Witten index is

$$\Delta_{\mathrm{HK}}(\beta) \approx \frac{1}{\sqrt{\pi}} \int\limits_{\sqrt{\phi_-^2\,\beta/2}}^{\sqrt{\phi_+^2\,\beta/2}} \mathrm{d}y\, e^{-y^2}. \tag{3.204}$$

Equation (3.204) will be used to derive the exact value of the fractional charge in polyacetylene in Section 3.6.7, even though it is an approximation for $\beta < \infty$.

3.6.6 Integral Representation of the Induced Charge

For the Dirac Hamiltonian $\mathcal{H}$ defined in Eq. (3.110), we have exploited the hidden SUSY to establish the relation (3.157) relating the index of the contribution $\mathcal{H}_{\phi}$ to $\mathcal{H}$ to the regularized Witten index. We are going to show that the induced fermion number defined in Eq. (3.129a), and whose spectral representation is given by Eq. (3.132b), has the integral representation

$$Q = -\mu_{\mathrm{s}} \int\limits_{0}^{\infty} \frac{\mathrm{d}\omega}{2\pi} \int\limits_{0}^{\infty} \mathrm{d}\beta\, e^{-(\omega^2+\mu_{\mathrm{s}}^2)\,\beta/2}\, \Delta_{\mathrm{HK}}(\beta)$$
$$= -\frac{1}{2} \int\limits_{0}^{\infty} \frac{\mathrm{d}u\, e^{-u/2}}{\sqrt{2\pi\, u}}\, \Delta_{\mathrm{HK}}(u/\mu_{\mathrm{s}}^2). \tag{3.205a}$$

Here, the integrand $\Delta_{\mathrm{HK}}(\beta)$ on the right-hand side of the first equality is the heat-kernel regularization of the Witten index,

$$\Delta_{\mathrm{HK}}(\beta) := \mathrm{Tr}\, \left[(-1)^{\mathbf{F}}\, e^{-\beta\mathbf{H}}\right], \tag{3.205b}$$

where the Dirac Hamiltonian $\mathcal{H}$ is related to the SUSY Hamiltonian $\mathbf{H}$ by Eq. (3.148). The computation of the induced fermion number for any Dirac Hamiltonian $\mathcal{H}$ of the form (3.110) thus reduces to computing the integral over the heat-kernel regularization of the corresponding Witten index weighted by a Gaussian measure.

Proof The proof of Eq. (3.205) relies on comparing the number of eigenvalues

$$N_{\widehat{D}\,\widehat{D}^\dagger}(\lambda, \lambda + \mathrm{d}\lambda) := \nu_{\widehat{D}\,\widehat{D}^\dagger}(\lambda)\,\mathrm{d}\lambda \tag{3.206}$$

of the operator $\widehat{D}\,\widehat{D}^\dagger$ in the interval $[\lambda, \lambda + \mathrm{d}\lambda]$ against the number of eigenvalues

$$N_{\widehat{D}^\dagger\,\widehat{D}}(\lambda, \lambda + \mathrm{d}\lambda) := \nu_{\widehat{D}^\dagger\,\widehat{D}}(\lambda)\,\mathrm{d}\lambda \tag{3.207}$$

of the operator $\widehat{D}^\dagger\,\widehat{D}$ in the interval $[\lambda, \lambda+\mathrm{d}\lambda]$. The (formal) definitions (3.206) and (3.207) make no distinction between the contributions arising from point-like spectra and continuum spectra, that is, $\nu_{\widehat{D}\,\widehat{D}^\dagger}(\lambda)$ and $\nu_{\widehat{D}^\dagger\,\widehat{D}}(\lambda)$ are to be interpreted as distributions or measures, for it is only after integration that they return numbers. Because the spectra of $\widehat{D}\,\widehat{D}^\dagger$ and $\widehat{D}^\dagger\,\widehat{D}$ are essential isospectral, this comparison is to be done for the range $0 \leq \lambda < \infty$. However, even though the spectra of $\widehat{D}\,\widehat{D}^\dagger$ and $\widehat{D}^\dagger\,\widehat{D}$ are essential isospectral, their density of states $\nu_{\widehat{D}\,\widehat{D}^\dagger}(\lambda)$ and $\nu_{\widehat{D}^\dagger\,\widehat{D}}(\lambda)$ need not be equal given the non-unitary relation (3.141).

The strategy of the proof is to manipulate the spectral representation of the Witten index so as to relate it to the spectral representation of the induced fermion charge.

We begin with the spectral representation of the heat-kernel regularization of the Witten index,

$$\Delta_{\mathrm{HK}}(\beta) := \mathrm{Tr}|_{+}\, e^{-\beta\,\widehat{D}\,\widehat{D}^\dagger/2} - \mathrm{Tr}|_{-}\, e^{-\beta\,\widehat{D}^\dagger\,\widehat{D}/2}$$
$$\equiv \int\limits_{0}^{\infty} \mathrm{d}\lambda\, e^{-\beta\,\lambda/2} \left[\nu_{\widehat{D}\,\widehat{D}^\dagger}(\lambda) - \nu_{\widehat{D}^\dagger\,\widehat{D}}(\lambda)\right]. \tag{3.208}$$

It is useful to perform the Laplace transformation

$$
\begin{aligned}
(\mathcal{L}\Delta_{\mathrm{HK}})(z/2) &:= \int_0^\infty \mathrm{d}\beta\, e^{-\beta z/2}\, \Delta_{\mathrm{HK}}(\beta) \\
&= 2\int_0^\infty \mathrm{d}\lambda\, \left[\nu_{\widehat{D}\,\widehat{D}^\dagger}(\lambda) - \nu_{\widehat{D}^\dagger\,\widehat{D}}(\lambda)\right] \frac{1}{\lambda + z}
\end{aligned}
\tag{3.209}
$$

for any complex number z with $\mathrm{Re}\, z > 0$. In particular, we may choose $z = \omega^2 + \mu_{\mathrm{s}}^2$, in which case

$$
(\mathcal{L}\Delta_{\mathrm{HK}})\left(\frac{\omega^2+\mu_{\mathrm{s}}^2}{2}\right) = 2\int_0^\infty \mathrm{d}\lambda\, \left[\nu_{\widehat{D}\,\widehat{D}^\dagger}(\lambda) - \nu_{\widehat{D}^\dagger\,\widehat{D}}(\lambda)\right] \frac{1}{\lambda + \omega^2 + \mu_{\mathrm{s}}^2}. \tag{3.210}
$$

The change of integration variable

$$
0 \leq \lambda =: \varepsilon^2 - \mu_{\mathrm{s}}^2 < \infty, \qquad \mathrm{d}\lambda = 2\,\varepsilon\,\mathrm{d}\varepsilon, \qquad |\mu_{\mathrm{s}}| \leq \varepsilon < \infty, \tag{3.211}
$$

delivers the integral representation

$$
\begin{aligned}
(\mathcal{L}\Delta_{\mathrm{HK}})\left(\frac{\omega^2+\mu_{\mathrm{s}}^2}{2}\right) = 4\int_0^\infty \mathrm{d}\varepsilon\, &\left[\nu_{\widehat{D}\,\widehat{D}^\dagger}(\varepsilon^2-\mu_{\mathrm{s}}^2) - \nu_{\widehat{D}^\dagger\,\widehat{D}}(\varepsilon^2-\mu_{\mathrm{s}}^2)\right] \\
&\times \frac{\varepsilon}{\omega^2+\varepsilon^2}.
\end{aligned}
\tag{3.212}
$$

[The lower bound of integration was extended from $|\mu_{\mathrm{s}}| \leq \varepsilon$ to $0 < \varepsilon$ by demanding that $\nu_{\widehat{D}\,\widehat{D}^\dagger}(\lambda) = \nu_{\widehat{D}^\dagger\,\widehat{D}}(\lambda) = 0$ when $\lambda < 0$.]

With the help of the residue theorem, we have the identity

$$
\int_0^\infty \frac{\mathrm{d}\omega}{2\pi} \frac{f(\omega)}{\omega^2+\varepsilon^2} = \frac{f(\mathrm{i}|\varepsilon|)}{4\,|\varepsilon|} \tag{3.213}
$$

if the function $f(z) = f(-z)$ is analytic and such that the line integral over a semicircle of infinite radius vanishes. In turn, we may write

$$
\int_0^\infty \frac{\mathrm{d}\omega}{2\pi}\, (\mathcal{L}\Delta_{\mathrm{HK}})\left(\frac{\omega^2+\mu_{\mathrm{s}}^2}{2}\right) = \int_0^\infty \mathrm{d}\varepsilon\, \left[\nu_{\widehat{D}\,\widehat{D}^\dagger}(\varepsilon^2-\mu_{\mathrm{s}}^2) - \nu_{\widehat{D}^\dagger\,\widehat{D}}(\varepsilon^2-\mu_{\mathrm{s}}^2)\right]. \tag{3.214}
$$

We now turn to the spectral representation of the induced fermion number. Let

$$
N_{\mathcal{H}}(\varepsilon, \varepsilon + \mathrm{d}\varepsilon) := \nu_{\mathcal{H}}(\varepsilon)\,\mathrm{d}\varepsilon \tag{3.215}
$$

be the number of eigenvalues of the Dirac Hamiltonian $\mathcal{H}$ defined by Eq. (3.110) in the interval $[\varepsilon, \varepsilon + \mathrm{d}\varepsilon]$. As was the case with Eqs. (3.206) and (3.207), no distinction is made between the contributions arising from point-like spectra and continuum spectra. We may then rewrite Eq. (3.132b) as

$$Q = -\frac{1}{2} \int\limits_{-\infty}^{+\infty} \mathrm{d}\varepsilon\, \nu_{\mathcal{H}}(\varepsilon)\, \mathrm{sgn}(\varepsilon). \tag{3.216}$$

This suggests doing the formal additive decomposition of the spectral measure

$$\nu_{\mathcal{H}}(\varepsilon) = \frac{1}{2}\left[\nu_{\mathcal{H}}(+\varepsilon) + \nu_{\mathcal{H}}(-\varepsilon)\right] + \frac{1}{2}\left[\nu_{\mathcal{H}}(+\varepsilon) - \nu_{\mathcal{H}}(-\varepsilon)\right] \tag{3.217a}$$

into the even part of the spectral measure denoted

$$\nu^{(\mathrm{e})}_{\mathcal{H}}(\varepsilon) := \frac{1}{2}\left[\nu_{\mathcal{H}}(+\varepsilon) + \nu_{\mathcal{H}}(-\varepsilon)\right] \tag{3.217b}$$

and the odd part of the spectral measure denoted

$$\nu^{(\mathrm{o})}_{\mathcal{H}}(\varepsilon) := \frac{1}{2}\left[\nu_{\mathcal{H}}(+\varepsilon) - \nu_{\mathcal{H}}(-\varepsilon)\right]. \tag{3.217c}$$

Equipped with this decomposition,

$$Q = -\int\limits_{0}^{\infty} \mathrm{d}\varepsilon\, \nu^{(\mathrm{o})}_{\mathcal{H}}(\varepsilon). \tag{3.218}$$

We seek an identity between integrals over spectral measures that relates the measures defined by Eqs. (3.206) and (3.207) on the one hand and Eqs. (3.215) and (3.217c) on the other hand. To this end, for any complex number z^2 that is outside of the spectrum of $\mathcal{H}$ (the latter being a subset of the real axis of the complex plane), we do the following two manipulations. First,

$$\begin{aligned}\int\limits_{-\infty}^{+\infty} \mathrm{d}\varepsilon\, \nu^{(\mathrm{o})}_{\mathcal{H}}(\varepsilon)\, \frac{\varepsilon}{\varepsilon^2 + z^2} &= 2 \int\limits_{|\mu_{\mathrm{s}}|}^{+\infty} \mathrm{d}\varepsilon\, \nu^{(\mathrm{o})}_{\mathcal{H}}(\varepsilon)\, \frac{\varepsilon}{\varepsilon^2 + z^2} \\ &= \mathrm{Tr}\left(\frac{\mathcal{H}}{\mathcal{H}^2 + z^2}\right).\end{aligned} \tag{3.219}$$

Second,

$$\mathrm{Tr}\left(\frac{\mathcal{H}}{\mathcal{H}^2 + z^2}\right) = \mathrm{Tr}|_{+}\, \frac{\mu_{\mathrm{s}}}{\widehat{D}\,\widehat{D}^{\dagger} + \mu_{\mathrm{s}}^2 + z^2} - \mathrm{Tr}|_{-}\, \frac{\mu_{\mathrm{s}}}{\widehat{D}^{\dagger}\,\widehat{D} + \mu_{\mathrm{s}}^2 + z^2}. \tag{3.220}$$

The traces $\mathrm{Tr}|_{+}$ and $\mathrm{Tr}|_{-}$ can be performed using the densities of states $\nu_{\widehat{D}\,\widehat{D}^{\dagger}}$ and $\nu_{\widehat{D}^{\dagger}\,\widehat{D}}$, respectively, that is,

$$\operatorname{Tr}\left(\frac{\mathcal{H}}{\mathcal{H}^2+z^2}\right)=\int\limits_0^{+\infty}\mathrm{d}\lambda\,\left[\nu_{\widehat{D}\,\widehat{D}^\dagger}(\lambda)-\nu_{\widehat{D}^\dagger\,\widehat{D}}(\lambda)\right]\frac{\mu_{\mathrm{s}}}{\lambda+\mu_{\mathrm{s}}^2+z^2}. \tag{3.221}$$

With the change of variable (3.211),

$$\operatorname{Tr}\left(\frac{\mathcal{H}}{\mathcal{H}^2+z^2}\right)=\int\limits_{-\infty}^{+\infty}\mathrm{d}\varepsilon\,\mu_{\mathrm{s}}\operatorname{sgn}(\varepsilon)\,\left[\nu_{\widehat{D}\,\widehat{D}^\dagger}(\varepsilon^2-\mu_{\mathrm{s}}^2)-\nu_{\widehat{D}^\dagger\,\widehat{D}}(\varepsilon^2-\mu_{\mathrm{s}}^2)\right]\frac{\varepsilon}{\varepsilon^2+z^2} \tag{3.222}$$

can be combined with Eq. (3.219). One thus obtains

$$\begin{aligned}&\int\limits_{-\infty}^{+\infty}\mathrm{d}\varepsilon\,\nu^{(\mathrm{o})}_{\mathcal{H}}(\varepsilon)\,\frac{\varepsilon}{\varepsilon^2+z^2}=\\&\qquad\int\limits_{-\infty}^{+\infty}\mathrm{d}\varepsilon\,\mu_{\mathrm{s}}\operatorname{sgn}(\varepsilon)\,\left[\nu_{\widehat{D}\,\widehat{D}^\dagger}(\varepsilon^2-\mu_{\mathrm{s}}^2)-\nu_{\widehat{D}^\dagger\,\widehat{D}}(\varepsilon^2-\mu_{\mathrm{s}}^2)\right]\frac{\varepsilon}{\varepsilon^2+z^2}.\end{aligned} \tag{3.223}$$

We may then identify the two measures

$$\nu^{(\mathrm{o})}_{\mathcal{H}}(\varepsilon)=\mu_{\mathrm{s}}\operatorname{sgn}(\varepsilon)\,\left[\nu_{\widehat{D}\,\widehat{D}^\dagger}(\varepsilon^2-\mu_{\mathrm{s}}^2)-\nu_{\widehat{D}^\dagger\,\widehat{D}}(\varepsilon^2-\mu_{\mathrm{s}}^2)\right]. \tag{3.224}$$

The induced fermion number (3.218) can now be expressed in terms of the measures (3.206) and (3.207) according to

$$Q=-\mu_{\mathrm{s}}\int\limits_0^{\infty}\mathrm{d}\varepsilon\,\left[\nu_{\widehat{D}\,\widehat{D}^\dagger}(\varepsilon^2-\mu_{\mathrm{s}}^2)-\nu_{\widehat{D}^\dagger\,\widehat{D}}(\varepsilon^2-\mu_{\mathrm{s}}^2)\right]. \tag{3.225}$$

After comparison of Eq. (3.225) with Eq. (3.214), we infer that

$$Q=-\mu_{\mathrm{s}}\int\limits_0^{\infty}\frac{\mathrm{d}\omega}{2\pi}\,(\mathcal{L}\Delta_{\mathrm{HK}})\left(\frac{\omega^2+\mu_{\mathrm{s}}^2}{2}\right). \tag{3.226}$$

This is nothing but the first equality in Eq. (3.205a). The second equality follows from doing the ω integration, on the right-hand side of the first equality in Eq. (3.205a). Indeed, the Gaussian ω integration can be safely interchanged with the β integration. It delivers the multiplicative factor $(1/2)\times\sqrt{2\pi/\beta}$:

$$Q=-\frac{\mu_{\mathrm{s}}}{2}\int\limits_0^{\infty}\frac{\mathrm{d}\beta}{\sqrt{2\pi\,\beta}}\,e^{-\mu_{\mathrm{s}}^2\,\beta/2}\,\Delta_{\mathrm{HK}}(\beta). \tag{3.227}$$

The change of variable $\beta=:u/\mu_{\mathrm{s}}^2$ delivers the second equality in Eq. (3.205a). □

3.6.7 Derivation of the Fractional Charge (3.109)

We start from

$$Q=-\frac{1}{2}\int_{0}^{\infty}\frac{\mathrm{d}u\,e^{-u/2}}{\sqrt{2\pi\,u}}\,\Delta_{\mathrm{HK}}(u/\mu_{\mathrm{s}}^{2}) \tag{3.228a}$$

for $0<\mu_{\mathrm{s}}^{2}$. According to Eq. (3.204), the heat-kernel regularization of the Witten index is, in the static approximation, given by

$$\Delta_{\mathrm{HK}}^{\mathrm{sta}}(\beta):=\frac{1}{\sqrt{\pi}}\left(\int_{0}^{\sqrt{\phi_{+}^{2}\,\beta/2}}\mathrm{d}y\,e^{-y^{2}}-\int_{0}^{\sqrt{\phi_{-}^{2}\,\beta/2}}\mathrm{d}y\,e^{-y^{2}}\right). \tag{3.228b}$$

Here, $\phi_{\pm}\neq 0$ is a pair of nonvanishing real-valued numbers and $\beta\geq 0$. We are going to derive the fractional charge (3.109) as an exercise in the manipulation of hypergeometric functions. In preparation for this task, we shall make use of the following definitions. For any complex number a and for any nonnegative integer $n=0,1,2,3,\cdots$

$$(a)_{0}\equiv 1,\qquad (a)_{n}\equiv a\,(a+1)\cdots(a+n-1). \tag{3.229}$$

For any pair of positive integers $p,q=1,2,3\cdots$, we define the generalized hypergeometric series

$${}_{p}F_{q}\left(a_{1},\cdots,a_{p};c_{1},\cdots,c_{q};z\right):=\sum_{n=0}^{\infty}\frac{(a_{1})_{n}\cdots(a_{p})_{n}}{(c_{1})_{n}\cdots(c_{q})_{n}}\frac{z^{n}}{n!}. \tag{3.230}$$

With the help of the error function (see Eq. **7.1.1** from Abramowitz and Stegun[25])

$$\operatorname{erf}(z):=\frac{2}{\sqrt{\pi}}\int_{0}^{z}\mathrm{d}t\,e^{-t^{2}}=-\operatorname{erf}(-z), \tag{3.231}$$

$$\Delta_{\mathrm{HK}}^{\mathrm{sta}}(\beta)=\frac{1}{2}\left[\operatorname{erf}\left(\operatorname{sgn}(\phi_{+})\sqrt{\frac{\phi_{+}^{2}\,\beta}{2}}\right)-\operatorname{erf}\left(\operatorname{sgn}(\phi_{-})\sqrt{\frac{\phi_{-}^{2}\,\beta}{2}}\right)\right], \tag{3.232}$$

we do the approximation

$$\begin{aligned}Q&\approx-\frac{1}{4}\int_{0}^{\infty}\frac{\mathrm{d}u\,e^{-u/2}}{\sqrt{2\pi\,u}}\left[\operatorname{erf}\left(\operatorname{sgn}(\phi_{+})\sqrt{\phi_{+}^{2}u/2\,\mu_{\mathrm{s}}^{2}}\right)-\operatorname{erf}\left(\operatorname{sgn}(\phi_{-})\sqrt{\phi_{-}^{2}\,u/2\,\mu_{\mathrm{s}}^{2}}\right)\right]\\&=-\frac{1}{4}\frac{1}{\sqrt{2\pi}}\sum_{\pm}(\pm)\sqrt{\frac{2\,\mu_{\mathrm{s}}^{2}}{\phi_{\pm}^{2}}}\int_{0}^{\infty}\mathrm{d}t\,t^{-1/2}\,e^{-\frac{\mu_{\mathrm{s}}^{2}}{\phi_{\pm}^{2}}t}\operatorname{erf}\left(\operatorname{sgn}(\phi_{\pm})\sqrt{t}\right).\end{aligned} \tag{3.233}$$

25 M. Abramowitz and I. A. Stegun, *Handbook of Mathematical Functions,* Dover, New York 1970. [Abramowitz and Stegun (1970)].

To do this integral, we trade the error function for the Kummer function through the relation (see Eq. **7.1.21** from Abramowitz and Stegun)

$$\operatorname{erf}(z) := \frac{2\,z\,e^{-z^2}}{\sqrt{\pi}}\;{}_1F_1\left(1;\frac{3}{2};z^2\right) = -\operatorname{erf}(-z) \tag{3.234a}$$

with the Kummer function defined by the power series

$${}_1F_1\,(a;b;z) := \sum_{n=0}^{\infty}\frac{(a)_n}{(b)_n}\frac{z^n}{n!}. \tag{3.234b}$$

The charge now reads

$$Q \approx -\frac{1}{2}\frac{1}{\sqrt{2}\,\pi}\sum_{\pm}(\pm)\operatorname{sgn}(\phi_\pm)\sqrt{\frac{2\,\mu_{\mathrm{s}}^2}{\phi_\pm^2}}\int\limits_0^\infty \mathrm{d}t\, e^{-\left(1+\frac{\mu_{\mathrm{s}}^2}{\phi_\pm^2}\right)t}\;{}_1F_1\left(1;\frac{3}{2};t\right). \tag{3.235}$$

According to Eq. **7.621.4** from Gradshteyn and Ryzhik[26], the Kummer function ${}_1F_1\,(a;c;kt)$ is related to the Gauss hypergeometric series

$${}_2F_1\,(a,b;c;z) := \sum_{n=0}^{\infty}\frac{(a)_n\,(b)_n}{(c)_n}\frac{z^n}{n!} = {}_2F_1\,(b,a;c;z) \tag{3.236}$$

through the integral relation

$$\int\limits_0^\infty \mathrm{d}t\, e^{-st}\,t^{b-1}\;{}_1F_1\,(a;c;kt) = \Gamma(b)\,s^{-b}\;{}_2F_1\left(a,b;c;\frac{k}{s}\right) \tag{3.237}$$

provided $|s| > |k|$, $\operatorname{Re} b > 0$, and $\operatorname{Re} s > \max(0, \operatorname{Re} k)$. It is the choice $a = b = k = 1$, $c = 3/2$, and $s = 1 + \frac{\mu_{\mathrm{s}}^2}{\phi_\pm^2} > 1$ that applies, in which case the left-hand side of Eq. (3.237) is the Laplace transform of ${}_1F_1\left(1;\frac{3}{2};t\right)$. The charge is given by

$$Q \approx -\frac{1}{2}\frac{1}{\sqrt{2}\,\pi}\sum_{\pm}(\pm)\operatorname{sgn}(\phi_\pm)\sqrt{\frac{2\,\mu_{\mathrm{s}}^2}{\phi_\pm^2}}\,\frac{\phi_\pm^2}{\mu_{\mathrm{s}}^2+\phi_\pm^2}\;{}_2F_1\left(1,1;\frac{3}{2};\frac{\phi_\pm^2}{\mu_{\mathrm{s}}^2+\phi_\pm^2}\right). \tag{3.238}$$

By applying the transformation formula (see Eq. **9.131.1** from Gradshteyn and Ryzhik)

$${}_2F_1\,(a,b;c;z) = \frac{1}{(1-z)^a}\;{}_2F_1\left(a,c-b;c;\frac{z}{z-1}\right) \tag{3.239a}$$

[26] I. S. Gradshteyn, I. M. Ryzhik, and A, Jeffrey, *Table of Integrals, Series, and Products*, Academic Press, London, 1994. [Gradshteyn and Ryzhik (1994)].

with $a = b = 1$, $c = 3/2$, $c - b = 1/2$, and

$$1 > z = \frac{\phi_\pm^2}{\mu_s^2 + \phi_\pm^2} > 0, \tag{3.239b}$$

$$1 > 1 - z = \frac{\mu_s^2}{\mu_s^2 + \phi_\pm^2} > 0, \tag{3.239c}$$

$$\frac{1}{1-z} = \frac{\mu_s^2 + \phi_\pm^2}{\mu_s^2} > 1, \tag{3.239d}$$

$$\frac{z}{z-1} = -\frac{\phi_\pm^2}{\mu_s^2} < 0, \tag{3.239e}$$

we find

$$Q \approx -\frac{1}{2}\frac{1}{\pi}\sum_\pm (\pm)\,\mathrm{sgn}(\phi_\pm)\sqrt{\frac{\mu_s^2}{\phi_\pm^2}\frac{\phi_\pm^2}{\mu_s^2}}\;{}_2F_1\left(1,\frac{1}{2};\frac{3}{2};-\frac{\phi_\pm^2}{\mu_s^2}\right). \tag{3.240}$$

By applying the representation (see Eq. **9.121.27** from Gradshteyn and Ryzhik)

$$z^2\,{}_2F_1\left(1,\frac{1}{2};\frac{3}{2};-z^2\right) = z^2\,{}_2F_1\left(\frac{1}{2},1;\frac{3}{2};-z^2\right) = z\ \arctan\ z, \tag{3.241}$$

we find

$$\begin{aligned} Q \approx & -\frac{1}{2}\frac{1}{\pi}\sum_\pm (\pm)\,\mathrm{sgn}(\phi_\pm)\sqrt{\frac{\mu_s^2}{\phi_\pm^2}}\sqrt{\frac{\phi_\pm^2}{\mu_s^2}}\ \arctan\left(\sqrt{\frac{\phi_\pm^2}{\mu_s^2}}\right) \\ = & -\frac{1}{2}\frac{1}{\pi}\sum_\pm (\pm)\,\mathrm{sgn}(\phi_\pm)\ \arctan\left(\sqrt{\frac{\phi_\pm^2}{\mu_s^2}}\right). \end{aligned} \tag{3.242}$$

Equation (3.109) follows with the identification $\phi_\pm^2 \equiv \phi_\infty^2$ and $\phi_\pm = \pm\sqrt{\phi_\infty^2}$. Remarkably, the static approximation (3.204) of the heat-kernel regularization for the Witten index suffices to recover the exact fractional charge induced by a soliton.

3.7 Fractionalization from the Gradient Expansion

3.7.1 Overview

We have made extensive use of spectral symmetries, the chiral spectral symmetry in Section 3.4 and the hidden SUSY in Sections 3.5 and 3.6, to compute exactly the conserved induced fermion number in the presence of the static background $\Phi(x)$ from the Clifford algebra defined by Eq. (3.111e). Goldstone and Wilczek relaxed this assumption in 1981 and presented a systematic expansion, the gradient expansion, of the conserved fermion current that delivers the *exact fractional part* of the conserved induced fermion number.[27] We are going to reproduce this result following the paper by Midorikawa.[28]

3.7.2 Definitions

Our starting point is the single-particle time-dependent Dirac Hamiltonian

$$\mathcal{H}(t,x) := \tau_2 \frac{\partial}{\mathrm{i}\partial x} + \tau_1 \phi_1(t,x) + \tau_3 \phi_2(t,x) \tag{3.243a}$$

acting on the vector space of functions spanned by the plane-wave basis on the real line tensored into $\mathbb{C}^2$. We shall impose the condition on the real-valued scalar fields ϕ_1 and ϕ_2 that the mass

$$m := \lim_{|x|\to\infty} \sqrt{\phi_1^2(t,x) + \phi_2^2(t,x)} \equiv \lim_{|x|\to\infty} |\phi(t,x)| \tag{3.243b}$$

is independent of time $t \in \mathbb{R}$ and position $x \in \mathbb{R}$ for large $|x|$, while

$$|\phi(t,x)| > 0 \tag{3.243c}$$

for all times and positions. [The time-dependent single-particle Hamiltonian (3.243a) should be compared with the static single-particle Hamiltonian (3.69a).] One verifies that the Dirac Hamiltonian (3.243a) is of the form (3.110c) with the identifications

$$d \to 1, \quad N \to 1, \quad P_1 \to \mathrm{i}, \quad \phi \to \phi_1(t,x), \quad \mu_{\mathrm{s}} \to \phi_2(t,x), \quad \Gamma_5 \to \tau_3. \tag{3.244}$$

Because of the t dependence of $\phi_1(t,x)$ and $\phi_2(t,x)$, energy is not conserved. Because of the x dependence of $\phi_2(t,x)$, Eqs. (3.78a) and (3.125) become

$$\mathcal{H}^2(t,x) = \tau_0 \left[-\frac{\partial^2}{\partial x^2} + \phi_1^2(t,x) + \phi_2^2(t,x) \right] - \tau_3 \left(\frac{\partial \phi_1}{\partial x} \right)(t,x) + \tau_1 \left(\frac{\partial \phi_2}{\partial x} \right)(t,x) \tag{3.245}$$

and

$$\{\mathcal{H}(t,x), \mathcal{A}\} = \begin{pmatrix} 0 & -\left(\frac{\partial \phi_2}{\partial x}\right)(t,x) \\ +\left(\frac{\partial \phi_2}{\partial x}\right)(t,x) & 0 \end{pmatrix}, \tag{3.246}$$

respectively. Hence, the hidden SUSY from Sections 3.5 and 3.6 is not of any direct use now. We need to consider the time-dependent Dirac equation.

The corresponding time-dependent Dirac equation can be written

$$\begin{aligned} 0 &= \left(\tau_1 \mathrm{i}\frac{\partial}{\partial t} + \tau_1 \tau_2 \mathrm{i}\frac{\partial}{\partial x} - \phi_1(t,x) - \tau_1 \tau_3 \phi_2(t,x) \right) \Psi(t,x) \\ &\equiv \left(\gamma^0 \mathrm{i}\frac{\partial}{\partial t} + \gamma^1 \mathrm{i}\frac{\partial}{\partial x} - \phi_1(t,x) - \mathrm{i}\gamma_5 \phi_2(t,x) \right) \Psi(t,x), \end{aligned} \tag{3.247a}$$

[27] J. Goldstone and F. Wilczek, Phys. Rev. Lett. **47**(14), 986–989 (1981). [Goldstone and Wilczek (1981)].

[28] S. Midorikawa, Prog. Theor. Phys. **69**(6), 1831–1834 (1983). [Midorikawa (1983)].

where the pair of gamma matrices

$$\gamma^0 := \tau_1, \qquad \gamma^1 := \tau_1\,\tau_2 = \mathrm{i}\tau_3, \tag{3.247b}$$

obeys the Clifford algebra

$$\{\gamma^\mu, \gamma^\nu\} = 2\,g^{\mu\nu} := \begin{cases} +2, & \text{if } \mu = \nu = 0, \\ -2, & \text{if } \mu = \nu = 1, \\ 0, & \text{otherwise,} \end{cases} \tag{3.247c}$$

and transforms like two vectors do under Lorentz transformations, while the Hermitean 2×2 matrix

$$\gamma_5 := \gamma_0\,\gamma_1 = -\gamma^0\,\gamma^1 = -\tau_2 \tag{3.247d}$$

transforms like a pseudo-scalar under Lorentz transformations (**Exercise** 3.5).

3.7.3 Symmetries

The single-particle two-current

$$j^\mu_\Psi(t,x) := \left(\overline{\Psi}\,\gamma^\mu\,\Psi\right)(t,x), \qquad \mu = 0,1, \qquad \overline{\Psi} := \Psi^\dagger\gamma^0, \tag{3.248}$$

obeys the continuity equation

$$\partial_\mu\, j^\mu_\Psi(t,x) = 0. \tag{3.249}$$

As is verified in **Exercise** 3.15, this is a consequence of the fact that the Lagrangian density

$$\mathcal{L}_\Psi(t,x) := \overline{\Psi}(t,x)\left[\mathrm{i}\gamma^\mu\,\partial_\mu - \Phi(t,x)\right]\Psi(t,x), \tag{3.250a}$$

where we have introduced the shorthand notation

$$\Phi(t,x) := \phi_1(t,x) + \mathrm{i}\gamma_5\,\phi_2(t,x) \tag{3.250b}$$

and the summation convention over repeated indices is implied, is form invariant under the global gauge transformation

$$\overline{\Psi}(t,x) =: \overline{\Xi}(t,x)\,e^{-\mathrm{i}\alpha}, \qquad \Psi(t,x) =: e^{+\mathrm{i}\alpha}\,\Xi(t,x), \tag{3.251}$$

for any $0 \leq \alpha < 2\pi$ independent of time t and space x.

Another symmetry is that of the action

$$S_\Psi := \int \mathrm{d}y^0 \int \mathrm{d}y^1\,\mathcal{L}_\Psi(y^0,y^1) \tag{3.252a}$$

under the scaling transformation

$$y^0 \mapsto \kappa\,y^0, \quad y^1 \mapsto \kappa\,y^1, \quad \Phi \mapsto \kappa^{-1}\,\Phi, \quad \overline{\Psi} \mapsto \overline{\Psi}\,\kappa^{-1/2}, \quad \Psi \mapsto \kappa^{-1/2}\,\Psi, \tag{3.252b}$$

for any dimensionless nonvanishing real-valued $\kappa > 0$.

We are also going to take advantage of the following transformation laws of the conserved two-current (3.248) and the Lagrangian density (3.250). With the help of the time- and space-dependent polar angle

$$\varphi(t,x) := \arctan\left(\frac{\phi_2(t,x)}{\phi_1(t,x)}\right), \tag{3.253a}$$

we have the polar representation

$$\Phi(t,x) = |\phi(t,x)|\, e^{+\mathrm{i}\gamma_5\,\varphi(t,x)}. \tag{3.253b}$$

Next, we define the time- and space-dependent unitary transformation of the single-particle Hilbert space through

$$\overline{\Psi}(t,x) =: \overline{\Xi}(t,x)\, e^{-\mathrm{i}\gamma_5\,\varphi(t,x)/2}, \qquad \Psi(t,x) =: e^{-\mathrm{i}\gamma_5\,\varphi(t,x)/2}\,\Xi(t,x). \tag{3.254}$$

We made use of the fact that γ^0 anticommutes with γ_5 to deduce the transformation law obeyed by $\overline{\Psi}(t,x)$, given that obeyed by $\Psi(t,x)$. The conserved two-current (3.248) is form invariant under the local unitary transformation (3.254):

$$\begin{aligned} j^{\mu}_{\Psi}(t,x) =& \overline{\Xi}(t,x)\,\gamma^{\mu}\,\Xi(t,x) \\ \equiv& j^{\mu}_{\Xi}(t,x), \qquad \mu = 0,1, \end{aligned} \tag{3.255}$$

as a consequence of the fact that γ^{μ} anticommutes with γ_5. This is not so for the Lagrangian density (3.250), which turns into

$$\begin{aligned} \mathcal{L}_{\Psi}(t,x) =& \overline{\Xi}(t,x)\Bigg[\gamma^0\left(\mathrm{i}\partial_t - \frac{1}{2}(\partial_x\,\varphi)(t,x)\right) + \gamma^1\left(\mathrm{i}\partial_x - \frac{1}{2}(\partial_t\,\varphi)(t,x)\right) \\ & - |\phi(t,x)|\Bigg]\Xi(t,x) \\ \equiv& \mathcal{L}_{\Xi}(t,x) \end{aligned} \tag{3.256}$$

under the local unitary transformation (3.254). If we introduce the gauge field a_{μ} with the covariant components

$$a_0(t,x) := -\frac{1}{2}(\partial_x\,\varphi)(t,x), \qquad a_1(t,x) := -\frac{1}{2}(\partial_t\,\varphi)(t,x), \tag{3.257a}$$

we may write

$$\mathcal{L}_{\Xi} = \overline{\Xi}\left[\gamma^{\mu}\left(\mathrm{i}\partial_{\mu} + a_{\mu}\right) - |\phi|\right]\Xi. \tag{3.257b}$$

The local unitary transformation (3.254) thus rotates the linear combination of Dirac masses (3.253) from the Clifford algebra along the unital direction from the Clifford algebra (**Exercise** 3.5 for the definition of the unital element of a Clifford algebra) at every time t and every position x at the cost of introducing the gauge field (3.257a) through the minimal coupling (3.257b). This gauge field is the source of the electrical field

$$\begin{aligned} E(t,x) :=& -\left(\partial_x a^0\right)(t,x) - \partial_t a^1(t,x) \\ =& +\frac{1}{2}\left(\partial_x^2 \varphi\right)(t,x) - \frac{1}{2}\left(\partial_t^2 \varphi\right)(t,x). \end{aligned} \tag{3.258}$$

The continuity equation (3.249) implies that the total single-particle charge

$$\mathcal{Q}_{\Psi}(t) := \int \mathrm{d}x\, j^0_{\Psi}(t,x) \tag{3.259}$$

is time-independent if boundary terms are to be dropped, that is,

$$\left(\frac{\mathrm{d}\mathcal{Q}_{\Psi}}{\mathrm{d}t}\right)(t) = 0. \tag{3.260}$$

We now assume that any instantaneous single-particle eigenstate $\Psi_{\varepsilon(t)}$ with the instantaneous single-particle energy eigenvalue $\varepsilon(t)$ of $\mathcal{H}(t,x)$ has a smooth dependence on t. Under this assumption, we may try the intuitive guess $\sum_{\Psi_{\varepsilon(t)}}^{\varepsilon(t)<0} \mathcal{Q}_{\Psi_{\varepsilon(t)}}$ for the many-body generalization of the conserved single-particle charge (3.259). In fact, there are several formalisms that deliver the many-body version $\hat{\jmath}^{\mu}$ of the single-particle two-current j^{μ}_{Ψ} defined in Eq. (3.248). We may use second quantization as is done in **Exercise** 3.6. Alternatively, we may use the path-integral representation of quantum field theory as advocated by Feynman. We will follow the former option for now. The latter option will be used in Section 3.9.

3.7.4 Gradient Expansion for the Current

For the calculation that follows, we adopt the standard relativistic notation from high-energy physics. We shall use x and y to denote the two vectors with the contravariant components x^{μ} and y^{μ} in two-dimensional Minkowski spacetime $\mathbb{M}_{1,1}$. We shall also use the two vectors p and k with the covariant components p_{μ} and k_{μ} in two-dimensional Minkowski energy-momentum $\mathbb{M}_{1,1}$. Lorentz contraction of two vectors is

$$\mathsf{p}\cdot\mathsf{x} \equiv \mathsf{p}_{\mu}\,\mathsf{x}^{\mu} \equiv \mathsf{p}^0\,\mathsf{x}^0 - \mathsf{p}^1\,\mathsf{x}^1. \tag{3.261a}$$

The Feynman slash is the contraction

$$\not{\mathsf{p}} \equiv \gamma^{\mu}\,\mathsf{p}_{\mu}, \tag{3.261b}$$

an element from the Clifford algebra that squares to $\mathsf{p}^2\,\tau_0$.

Let $\hat{\not{\mathsf{p}}} - \hat{\Phi}$ denote the operator with the representation $\mathrm{i}\not{\partial} - \Phi(\mathsf{x})$ in the position basis. The propagator $\hat{S}_{\mathrm{F}}$ is the operator defined by

$$\left(\hat{\not{\mathsf{p}}} - \hat{\Phi}\right)\hat{S}_{\mathrm{F}} = \hat{S}_{\mathrm{F}}\left(\hat{\not{\mathsf{p}}} - \hat{\Phi}\right) = \tau_0 \otimes \hat{1}, \tag{3.262}$$

where $\hat{1}$ is the unit operator acting on the vector space spanned by the plane waves and the 2×2 unit matrix τ_0 is the unit operator acting on $\mathbb{C}^2$. It is convenient to define the related pair of operators through the pair of 2×2 matrices $S_{\mathrm{F}}^{\mathrm{r}}(\mathsf{x},\mathsf{y})$ and $S_{\mathrm{F}}^{\mathrm{l}}(\mathsf{x},\mathsf{y})$ by

$$\left[\mathrm{i}\overrightarrow{\partial\!\!\!/}_{\mathsf{x}} - \Phi(\mathsf{x})\right] S_{\mathrm{F}}^{\mathrm{r}}(\mathsf{x},\mathsf{y}) := \tau_0\, \delta(\mathsf{x}-\mathsf{y}) \tag{3.263a}$$

and

$$S_{\mathrm{F}}^{\mathrm{l}}(\mathsf{x},\mathsf{y}) \left[\mathrm{i}\overleftarrow{\partial\!\!\!/}_{\mathsf{y}} - \Phi(\mathsf{y})\right] := \tau_0\, \delta(\mathsf{x}-\mathsf{y}), \tag{3.263b}$$

respectively. We need both right and left propagators in order to regularize the product of two quantum fields $\widehat{O}_1$ and $\widehat{O}_2$ acting at the same point $\mathsf{x} \in \mathbb{M}_{1,1}$ (this is called point splitting) in a symmetric way, that is,

$$\widehat{O}_1(\mathsf{x})\,\widehat{O}_2(\mathsf{x}) \equiv \lim_{y\to 0} \frac{1}{2} \left[\widehat{O}_1(\mathsf{x})\,\widehat{O}_2(\mathsf{x}+\mathsf{y}) + \widehat{O}_1(\mathsf{x}+\mathsf{y})\,\widehat{O}_2(\mathsf{x})\right]. \tag{3.264}$$

For $\mu = 0, 1$, we *define* the many-body ground-state expectation value of the many-body two-current $\hat{j}^{\mu}(\mathsf{x})$ to be

$$\langle \hat{j}^{\mu}(\mathsf{x}) \rangle := -\frac{\mathrm{i}}{2} \operatorname{tr} \left[\gamma^{\mu}\, S_{\mathrm{F}}^{\mathrm{r}}(\mathsf{x},\mathsf{x}) + \gamma^{\mu}\, S_{\mathrm{F}}^{\mathrm{l}}(\mathsf{x},\mathsf{x})\right], \tag{3.265}$$

where the symbol tr is reserved for tracing over the Clifford algebra. We shall revisit the calculation that follows using path-integral methods. In doing so we will gain a deeper intuition for the definition (3.265). That this is a meaningful definition in that it delivers a conserved current obeying the expected symmetries will be born out by what follows.

If we insert the Fourier expansions

$$\delta(\mathsf{x}-\mathsf{y}) = \int \frac{\mathrm{d}^2 \mathsf{p}}{(2\pi)^2}\, e^{-\mathrm{i}\mathsf{p}\cdot(\mathsf{x}-\mathsf{y})}, \tag{3.266a}$$

$$S_{\mathrm{F}}^{\mathrm{r}}(\mathsf{x},\mathsf{y}) =: \int \frac{\mathrm{d}^2 \mathsf{p}}{(2\pi)^2}\, e^{-\mathrm{i}\mathsf{p}\cdot(\mathsf{x}-\mathsf{y})}\, S_{\mathrm{F}}^{\mathrm{r}}(\mathsf{x},\mathsf{p}), \tag{3.266b}$$

$$S_{\mathrm{F}}^{\mathrm{l}}(\mathsf{x},\mathsf{y}) =: \int \frac{\mathrm{d}^2 \mathsf{p}}{(2\pi)^2}\, e^{-\mathrm{i}\mathsf{p}\cdot(\mathsf{x}-\mathsf{y})}\, S_{\mathrm{F}}^{\mathrm{l}}(\mathsf{p},\mathsf{y}), \tag{3.266c}$$

into Eqs. (3.263a) and (3.263b), we obtain

$$S_{\mathrm{F}}^{\mathrm{r}}(\mathsf{x},\mathsf{p}) = +\frac{\tau_0}{\not{\mathsf{p}} - \Phi(\mathsf{x}) + \mathrm{i}\overrightarrow{\partial\!\!\!/}_{\mathsf{x}}} \tag{3.267a}$$

and

$$S_{\mathrm{F}}^{\mathrm{l}}(\mathsf{p},\mathsf{y}) = -\frac{\tau_0}{\not{\mathsf{p}} + \Phi(\mathsf{y}) - \mathrm{i}\overleftarrow{\partial\!\!\!/}_{\mathsf{y}}}, \tag{3.267b}$$

respectively. The gradient expansion of (3.265) follows from inserting the expansions

$$S^{\mathrm{r}}_{\mathrm{F}}(\mathsf{x},\mathsf{p}) = +\frac{\tau_0}{\not{\mathsf{p}}-\Phi(\mathsf{x})} + \frac{\tau_0}{\not{\mathsf{p}}-\Phi(\mathsf{x})}\left(-\mathrm{i}\overrightarrow{\not{\partial}}_{\mathsf{x}}\right)\frac{\tau_0}{\not{\mathsf{p}}-\Phi(\mathsf{x})} + \cdots \tag{3.268a}$$

and

$$S^{\mathrm{l}}_{\mathrm{F}}(\mathsf{p},\mathsf{y}) = -\frac{\tau_0}{\not{\mathsf{p}}+\Phi(\mathsf{y})} - \frac{\tau_0}{\not{\mathsf{p}}+\Phi(\mathsf{y})}\left(+\mathrm{i}\overleftarrow{\not{\partial}}_{\mathsf{y}}\right)\frac{\tau_0}{\not{\mathsf{p}}+\Phi(\mathsf{y})} - \cdots \tag{3.268b}$$

into

$$\langle \hat{\jmath}^{\mu}(\mathsf{x})\rangle = -\frac{\mathrm{i}}{2}\int\frac{\mathrm{d}^2\mathsf{p}}{(2\pi)^2}\,\mathrm{tr}\,\left[\gamma^{\mu}\,S^{\mathrm{r}}_{\mathrm{F}}(\mathsf{x},\mathsf{p}) + \gamma^{\mu}\,S^{\mathrm{l}}_{\mathrm{F}}(\mathsf{p},\mathsf{x})\right], \tag{3.268c}$$

up to the desired order. This expansion is free of any infra-red divergence by assumption (3.243c). The first nonvanishing contribution to the gradient expansion of (3.265) is of first order in the gradient expansion. It is given by (**Exercise** 3.16)

$$\begin{aligned}
\left\langle \hat{\jmath}^{\mu}(\mathsf{x})\right\rangle =& -\frac{\mathrm{i}}{2}\int\frac{\mathrm{d}^2\mathsf{p}}{(2\pi)^2}\,\mathrm{tr}\,\left[\gamma^{\mu}\,S^{\mathrm{r}}_{\mathrm{F}}(\mathsf{x},\mathsf{p}) + \gamma^{\mu}\,S^{\mathrm{l}}_{\mathrm{F}}(\mathsf{p},\mathsf{x})\right]\\
=& -\mathrm{i}\int\frac{\mathrm{d}^2\mathsf{p}}{(2\pi)^2}\,\mathrm{tr}\left[\gamma^{\mu}\frac{\tau_0}{\not{\mathsf{p}}-\Phi(\mathsf{x})}\left(-\mathrm{i}\not{\partial}_{\mathsf{x}}\right)\frac{\tau_0}{\not{\mathsf{p}}-\Phi(\mathsf{x})}\right] + \cdots\\
=& -2\mathrm{i}\,\epsilon^{\mu\nu}\,\epsilon_{ab}\,\phi_a(\mathsf{x})(\partial_{\nu}\,\phi_b)(\mathsf{x})\int\frac{\mathrm{d}^2\mathsf{p}}{(2\pi)^2}\,\frac{1}{\left(\mathsf{p}^2-|\boldsymbol{\phi}(\mathsf{x})|^2\right)^2} + \cdots\\
=& +\frac{1}{2\pi}\,\epsilon^{\mu\nu}\,\epsilon_{ab}\left(\frac{\phi_a\partial_{\nu}\phi_b}{|\boldsymbol{\phi}|^2}\right)(\mathsf{x}) + \cdots\\
=& +\frac{1}{2\pi}\,\epsilon^{\mu\nu}\,(\partial_{\nu}\varphi)\,(\mathsf{x}) + \cdots.
\end{aligned} \tag{3.269}$$

This is a remarkable result that is uniquely determined, up to the proportionality constant $1/(2\pi)$, by demanding that the right-hand side (i) transforms like the left-hand side under the scaling transformation (3.252b), (ii) is a bilinear form in terms of the two components of the normalized unit vector $\boldsymbol{\phi}/|\boldsymbol{\phi}|$ that is odd under the transformation $\phi_1 \to +\phi_1$ and $\phi_2 \to -\phi_2$, (iii) is of first order in the derivative expansion, (iv) transforms like a two-vector with respect to Lorentz transformations, and (v) obeys the continuity equation. The higher-order corrections in the gradient expansion implied by the term $\cdots$ are of order $1/|\boldsymbol{\phi}(\mathsf{x})|$ and need not only depend on the Levi-Civita tensor

$$\epsilon^{\mu\nu} = -\epsilon_{\mu\nu} = \begin{cases} +1, & \text{if } \mu=0 \text{ and } \nu=1,\\ -1, & \text{if } \mu=1 \text{ and } \nu=0,\\ 0, & \text{otherwise.}\end{cases} \tag{3.270}$$

Unlike the leading contribution to the right-hand side of Eq. (3.269), the subleading contributions $\cdots$ may also depend on the metric (3.247c) of Minkowski space $\mathbb{M}_{1,1}$ and are thus not merely topological in character.

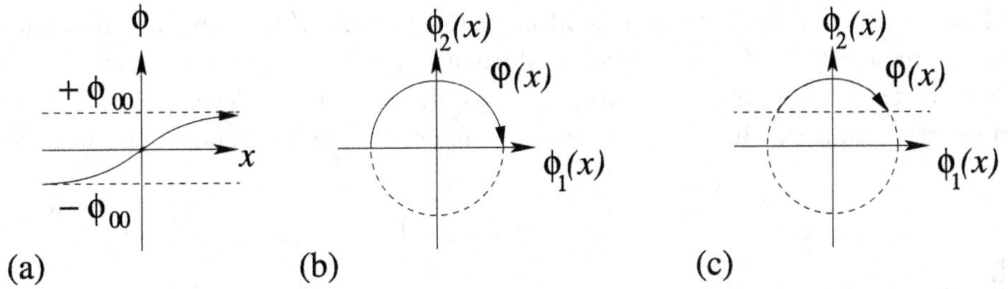

Figure 3.8 (a) Soliton profile (3.271). (b) Deformation of the soliton profile (3.271) into the profile (3.272). (c) Profile (3.275).

3.7.5 Applications to Polyacetylene

The proportionality constant $1/(2\pi)$ on the right-hand side of Eq. (3.269) follows from the fact that a soliton induces the charge $1/2$, as we have seen in Section 3.4.2. To prove this claim, we choose the static Ansatz

$$\phi_{\mathrm{s}}(t,x) := \begin{pmatrix} \phi(x) \\ 0 \end{pmatrix} \tag{3.271a}$$

with the boundary conditions

$$\lim_{x\to\pm\infty} \phi(x) = \pm\phi_\infty \equiv \pm m. \tag{3.271b}$$

We may deform the soliton profile (3.271) with the singular point in position space at which $\phi(x)=0$, see Figure 3.8(a), into the static profile

$$\Phi_{\mathrm{s}}(t,x) = m\, e^{\mathrm{i}\gamma_5\,\varphi(x)} \tag{3.272a}$$

with the boundary conditions

$$\lim_{x\to\pm\infty} \varphi(x) = \begin{cases} 0, & \text{if } x\to+\infty, \\ \pi, & \text{if } x\to-\infty. \end{cases} \tag{3.272b}$$

The soliton profile (3.272) has the nonvanishing and constant mass

$$m \equiv \sqrt{\det[\Phi_{\mathrm{s}}(t,x)]} \tag{3.272c}$$

[recall the definition of γ_5 in Eq. (3.247d)] for all $x\in\mathbb{R}$, see Figure 3.8(b). According to Eq. (3.269), the induced expectation value in the Fermi sea of the many-body charge operator is proportional to the integration over space of $\left\langle \hat{j}^0(t,x)\right\rangle$, that is, it is proportional to

$$\int\limits_{\mathbb{R}} \mathrm{d}x\, (\partial_x\varphi)(x) = \varphi(+\infty) - \varphi(-\infty) = -\pi. \tag{3.273}$$

The normalization $1/(2\pi)$ follows if this induced charge equals $1/2$ in magnitude.

The sign of the charge for the soliton profile (3.272) is ambiguous. This ambiguity is removed by lifting the twofold degeneracy of the many-body ground state when the chemical potential is precisely zero. To this end, we break the sublattice symmetry through a time- and space-independent staggered chemical potential μ_{s},

$$\phi_{\mathrm{s}}(t,x) = \begin{pmatrix} \phi(x) \\ \mu_{\mathrm{s}} \end{pmatrix} \tag{3.274a}$$

with the boundary conditions

$$\lim_{x\to\pm\infty} \phi(x) = \pm\phi_{\infty}. \tag{3.274b}$$

The corresponding polar decomposition of the mass in the Clifford algebra is [see Figure 3.8(c)]

$$\Phi_{\mathrm{s}}(t,x) = \sqrt{\phi_{\infty}^{2} + \mu_{\mathrm{s}}^{2}}\, e^{\mathrm{i}\gamma_5\,\varphi(x)} \tag{3.275a}$$

with the boundary conditions

$$\lim_{x\to\pm\infty} \varphi(x) = \begin{cases} \arctan\frac{\mu_{\mathrm{s}}}{\phi_{\infty}} = \frac{\pi}{2} - \arctan\frac{\phi_{\infty}}{\mu_{\mathrm{s}}}, & \text{if } x\to+\infty, \\ \\ \pi - \arctan\frac{\mu_{\mathrm{s}}}{\phi_{\infty}} = \frac{\pi}{2} + \arctan\frac{\phi_{\infty}}{\mu_{\mathrm{s}}}, & \text{if } x\to-\infty. \end{cases} \tag{3.275b}$$

There follows, by Eqs. (3.269) and (3.270), the induced charge

$$\begin{aligned} Q :=& +\frac{1}{2\pi}\int\limits_{\mathbb{R}} \mathrm{d}x\,(\partial_x\varphi)(x) \\ =& -\frac{1}{\pi}\arctan\frac{\phi_{\infty}}{\mu_{\mathrm{s}}} \\ =& \begin{cases} -\frac{1}{\pi}\arctan\frac{\phi_{\infty}}{|\mu_{\mathrm{s}}|}, & \text{if } \mu_{\mathrm{s}}>0, \\ \\ -1+\frac{1}{\pi}\arctan\frac{\phi_{\infty}}{|\mu_{\mathrm{s}}|}, & \text{if } \mu_{\mathrm{s}}<0, \end{cases} \end{aligned} \tag{3.276}$$

in agreement with Eq. (3.109). The limit $\mu_{\mathrm{s}} \to 0$ can then be safely taken so as to define the charge assignment of the soliton profile shown in Figure 3.8(a) in a unique way.

Equation (3.269) allows to define a quantum pump as follows (see Figure 3.9). We make the adiabatic Ansatz

$$\phi_{\mathrm{pump}}(t,x) = m\begin{pmatrix} \cos(\omega\,t) \\ \sin(\omega\,t) \end{pmatrix} \tag{3.277}$$

for some frequency $|\omega| \ll m$. The corresponding polar decomposition of the mass in the Clifford algebra is [recall Eq. (3.253)]

$$\Phi_{\mathrm{pump}}(t,x) = m\,e^{\mathrm{i}\gamma_5\,\varphi_{\mathrm{pump}}(t)}, \qquad \varphi_{\mathrm{pump}}(t) = \omega\,t. \tag{3.278}$$

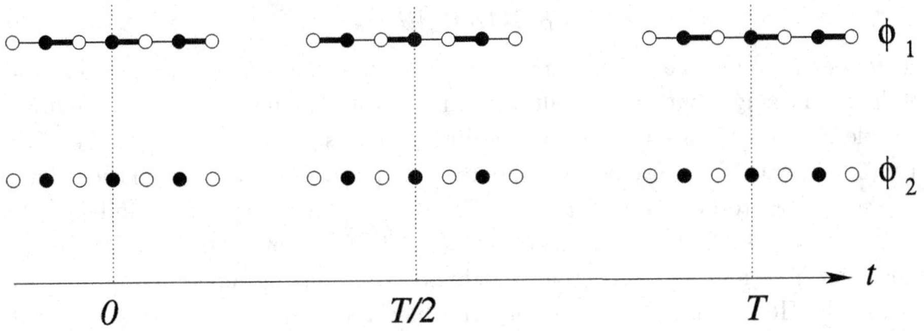

Figure 3.9 The time evolution during one period $T = 2\pi/\omega$ of a quantum pump, by which one electron flows from the left to the right of an arbitrary point in position space, is monitored with the time dependence of the bond-density wave, $\phi_1(t)$, and the charge-density wave, $\phi_2(t)$. Each order breaks the translation symmetry of the lattice by one lattice spacing. Sites of the lattice are shown as circles. Their coloring distinguishes the two sublattices. Bond ordering is implied by modulating the thickness of the bonds. Charge ordering is implied by modulating the radius of the circles. Bond ordering is maximal (vanishing) at $t = 0, T/2, T$ ($t = T/4, 3T/4$). Charge ordering is vanishing (maximal) at $t = 0, T/2, T$ ($t = T/4, 3T/4$).

Hence,

$$\partial_x \varphi_{\text{pump}}(t) = 0, \qquad \partial_t \varphi_{\text{pump}}(t) = \omega. \tag{3.279}$$

According to Eq. (3.269), an induced dc current is flowing that is proportional to $\omega/(2\pi)$. The charge I that is pumped through the period T is proportional to $\omega T/(2\pi)$, that is, one electron is pumped through the period $T = 2\pi/\omega$:

$$I = \epsilon^{10}. \tag{3.280}$$

3.8 Fractionalization from the Adiabatic Expansion

3.8.1 Overview

The derivation of the induced fermion number that follows from the gradient expansion (3.269) of the conserved fermion two-current is an example of an adiabatic approximation. We stated in the introduction of Section 3.7 that this adiabatic approximation delivers the *exact fractional part* of the conserved induced fermion number of a Dirac Hamiltonian. We now justify this claim. We proceed in two steps. We first establish Eq. (3.294) without any specific reference to Dirac Hamiltonians. We then specialize to the case of Dirac Hamiltonians to establish Eq. (3.300). Our claim is then justified by comparing Eqs. (3.294) and (3.300).

3.8.2 Definitions

Let $\mathcal{H}_{\mathrm{i}}$ be the reference single-particle Hamiltonian. The index i refers to initial. Let $\mathcal{H}_{\mathrm{f}}$ be the single-particle Hamiltonian of interest. The index f refers to final. For example, we may choose to work in one-dimensional space ($d=1$), $\mathcal{H}_{\mathrm{f}}$ to be of the form (3.243) with the static background $\Phi(x)\equiv\tau_1\,\phi_1(x)+\tau_3\,\phi_2(x)$ from the Clifford algebra, and $\mathcal{H}_{\mathrm{i}}$ to be $\lim_{x\to\infty}\mathcal{H}_{\mathrm{f}}$. Both $\mathcal{H}_{\mathrm{i}}$ and $\mathcal{H}_{\mathrm{f}}$ are assumed time-independent and act on the same Hilbert space. Let $\widehat{H}_{\mathrm{i}}$ and $\widehat{H}_{\mathrm{f}}$ denote the second-quantized extensions of $\mathcal{H}_{\mathrm{i}}$ and $\mathcal{H}_{\mathrm{f}}$, respectively, acting on the same fermionic Fock space. We work in the Heisenberg picture of quantum mechanics in which operators that do not depend explicitly on time are endowed with a time (t) dependence inherited by the unitary evolution generated by the explicitly time-independent Hamiltonians $\widehat{H}_{\mathrm{i}}$ and $\widehat{H}_{\mathrm{f}}$. We assume the existence of a current density $\hat{\boldsymbol{j}}_{\mathrm{i}}$ such that the continuity equation

$$\frac{\mathrm{d}\hat{j}^0_{\mathrm{i}}}{dt}+\boldsymbol{\nabla}\cdot\hat{\boldsymbol{j}}_{\mathrm{i}}=0,\qquad \frac{\mathrm{d}\hat{j}^0_{\mathrm{i}}}{dt}:=-\mathrm{i}\left[\hat{j}^0_{\mathrm{i}},\widehat{H}_{\mathrm{i}}\right], \tag{3.281}$$

where $\hat{j}^0_{\mathrm{i}}$ is the fermion number density operator, holds for the reference Hamiltonian $\widehat{H}_{\mathrm{i}}$. We assume the existence of a current density $\hat{\boldsymbol{j}}_{\mathrm{f}}$ and of a function

$$\begin{aligned}\mathcal{A}_{\mathrm{f}}\colon \mathbb{R}^d &\to \mathbb{R},\\ \boldsymbol{x}&\mapsto \mathcal{A}_{\mathrm{f}}(\boldsymbol{x}),\end{aligned} \tag{3.282a}$$

such that the *anomalous* continuity equation

$$\frac{\mathrm{d}\hat{j}^0_{\mathrm{f}}}{dt}+\boldsymbol{\nabla}\cdot\hat{\boldsymbol{j}}_{\mathrm{f}}=\mathcal{A}_{\mathrm{f}},\qquad \frac{\mathrm{d}\hat{j}^0_{\mathrm{f}}}{dt}:=-\mathrm{i}\left[\hat{j}^0_{\mathrm{f}},\widehat{H}_{\mathrm{f}}\right], \tag{3.282b}$$

holds for the Hamiltonian of interest $\widehat{H}_{\mathrm{f}}$. The real-valued function $\mathcal{A}_{\mathrm{f}}$ quantifies the anomaly locally. Even though $\mathcal{A}_{\mathrm{f}}=0$ when $\mathcal{H}_{\mathrm{f}}$ is of the form (3.243), we shall encounter an anomalous continuity equation of this kind in Section 3.10 in one-dimensional space ($d=1$) when treating chiral Dirac fermions, that is, when the effective one-dimensional model describes the boundary states of two-dimensional Chern insulators. The origin of this local anomaly is traced to Eq. (3.484). Hence, we include the possibility of an anomalous continuity equation for the sake of generality.

Let the ket $|\mathrm{GS}_{\mathrm{f}}\rangle$ denote the many-body ground state of $\widehat{H}_{\mathrm{f}}$ obtained by occupying all single-particle eigenstates of $\mathcal{H}_{\mathrm{f}}$ with eigenvalues less than the chemical potential μ. By assumption,[29]

$$Q_{\mathrm{f}}\equiv\left\langle \mathrm{GS}_{\mathrm{f}}\middle|\widehat{Q}_{\mathrm{f}}\middle|\mathrm{GS}_{\mathrm{f}}\right\rangle:=\int\limits_{\mathbb{R}^d}\mathrm{d}^d\boldsymbol{x}\,\left\langle \mathrm{GS}_{\mathrm{f}}\middle|\hat{j}^0_{\mathrm{f}}(t,\boldsymbol{x})\middle|\mathrm{GS}_{\mathrm{f}}\right\rangle\equiv\int\limits_{\mathbb{R}^d}\mathrm{d}^d\boldsymbol{x}\;j^0_{\mathrm{f}}(t,\boldsymbol{x}) \tag{3.283}$$

is independent of the unitary time (t) evolution generated by $\widehat{H}_{\mathrm{f}}$. Similarly, if the ket $|\mathrm{GS}_{\mathrm{i}}\rangle$ denotes the many-body ground state of $\widehat{H}_{\mathrm{i}}$ obtained by occupying all

[29] We are thus assuming that the integral over space of both the ground-state expectation value of $\boldsymbol{\nabla}\cdot\hat{\boldsymbol{j}}_{\mathrm{f}}$ and $\mathcal{A}_{\mathrm{f}}$ vanish.

single-particle eigenstates of $\mathcal{H}_{\mathrm{i}}$ with eigenvalues less than the chemical potential μ, then

$$Q_{\mathrm{i}} \equiv \left\langle \mathrm{GS}_{\mathrm{i}} \left| \widehat{Q}_{\mathrm{i}} \right| \mathrm{GS}_{\mathrm{i}} \right\rangle := \int\limits_{\mathbb{R}^d} \mathrm{d}^d \boldsymbol{x} \, \left\langle \mathrm{GS}_{\mathrm{i}} \left| \hat{j}^0_{\mathrm{i}}(t, \boldsymbol{x}) \right| \mathrm{GS}_{\mathrm{i}} \right\rangle \equiv \int\limits_{\mathbb{R}^d} \mathrm{d}^d \boldsymbol{x} \, j^0_{\mathrm{i}}(t, \boldsymbol{x}) \qquad (3.284)$$

is independent of the unitary time (t) evolution generated by $\widehat{H}_{\mathrm{i}}$. We seek the answer to the following question. Is it possible to compute Q_{f} from computing smooth deformations of Q_{i}? To answer this question, we observe that a family of Hamiltonians $\{\widehat{H}(\tau)\}$ on the interval $\tau_{\mathrm{i}} \leq \tau \leq \tau_{\mathrm{f}}$ that interpolates between the reference Hamiltonian $\widehat{H}_{\mathrm{i}}$ at the initial τ_{i} and the Hamiltonian of interest $\widehat{H}_{\mathrm{f}}$ at the final τ_{f} can always be constructed. For example, we may choose the family of single-particle Hamiltonians

$$\mathcal{H}(\tau) := \frac{\tau - \tau_{\mathrm{f}}}{\tau_{\mathrm{i}} - \tau_{\mathrm{f}}} \, \mathcal{H}_{\mathrm{i}} + \frac{\tau - \tau_{\mathrm{i}}}{\tau_{\mathrm{f}} - \tau_{\mathrm{i}}} \, \mathcal{H}_{\mathrm{f}} \qquad (3.285\mathrm{a})$$

with the family of many-body ground states $|\mathrm{GS}(\tau)\rangle$ to the family of many-body Hamiltonians

$$\widehat{H}(\tau) := \frac{\tau - \tau_{\mathrm{f}}}{\tau_{\mathrm{i}} - \tau_{\mathrm{f}}} \, \widehat{H}_{\mathrm{i}} + \frac{\tau - \tau_{\mathrm{i}}}{\tau_{\mathrm{f}} - \tau_{\mathrm{i}}} \, \widehat{H}_{\mathrm{f}} \qquad (3.285\mathrm{b})$$

obtained by occupying all single-particle eigenstates of $\mathcal{H}(\tau)$ with eigenvalues less than the chemical potential μ. The answer to this question depends on certain conditions to be fulfilled by the family of Hamiltonians $\{\widehat{H}(\tau)\}$.

3.8.3 Absence of Spectral Flow

Equations (3.294) and (3.300) will be derived assuming the absence of spectral flow.

The conditions that we impose are the following. We assume the existence of a many-body gap between $|\mathrm{GS}_{\mathrm{i}}\rangle$ and any many-body excited state of the reference Hamiltonian $\widehat{H}_{\mathrm{i}}$. We do the same for the Hamiltonian of interest $\widehat{H}_{\mathrm{f}}$. Finally, we assume that the many-body ground state $|\mathrm{GS}(\tau)\rangle$ of the interpolating Hamiltonian $\widehat{H}(\tau)$ also supports a many-body gap to all excited many-body states.

Because we are doing Fermiology, that is, treating noninteracting fermions subjected to different backgrounds, these conditions are equivalent to the chemical potential μ lying in a spectral gap for the entire family $\{\mathcal{H}(\tau)\}$ of single-particle Hamiltonians. Any single-particle state that crosses the value of μ at some $\tau_{\mathrm{i}} < \tau' < \tau_{\mathrm{f}}$ violates the latter condition. We call such a violation a *single-particle spectral flow*. The condition that we impose on the interval $\tau_{\mathrm{i}} \leq \tau \leq \tau_{\mathrm{f}}$ is thus that of the absence of a single-particle spectral flow.

The absence of spectral flow on the interval $\tau_{\mathrm{i}} \leq \tau \leq \tau_{\mathrm{f}}$ implies that the unitary time evolution

$$|\mathrm{GS}(\tau)\rangle := \left[\widehat{\mathcal{T}}_{\tau'}\exp\left(-\mathrm{i}\int\limits_{\tau_{\mathrm{i}}}^{\tau}\mathrm{d}\tau'\,\widehat{H}(\tau')\right)\right]|\mathrm{GS}_{\mathrm{i}}\rangle \tag{3.286}$$

that follows from solving the Schrödinger equation

$$\mathrm{i}\partial_{\tau}|\Psi(\tau)\rangle = \widehat{H}(\tau)\,|\Psi(\tau)\rangle \tag{3.287a}$$

with the initial condition

$$|\Psi(\tau=\tau_{\mathrm{i}}\rangle = |\mathrm{GS}_{\mathrm{i}}\rangle \tag{3.287b}$$

is proportional, up to a phase factor $e^{\mathrm{i}\Theta(\tau)}$, to the instantaneous ground state $|\mathrm{GS}_{\mathrm{adia}}(\tau)\rangle$ of $\widehat{H}(\tau)$, that is,

$$|\mathrm{GS}(\tau)\rangle = e^{\mathrm{i}\Theta(\tau)}\,|\mathrm{GS}_{\mathrm{adia}}(\tau)\rangle. \tag{3.288}$$

Hence, $|\mathrm{GS}(\tau)\rangle$ interpolates between $|\mathrm{GS}_{\mathrm{i}}\rangle$ and $|\mathrm{GS}_{\mathrm{f}}\rangle$.

We shall assume the existence of the anomalous continuity equation

$$\frac{\mathrm{d}\hat{J}^0}{\mathrm{d}\tau} + \boldsymbol{\nabla}\cdot\hat{\boldsymbol{J}} = \mathcal{A}, \qquad \frac{\mathrm{d}\hat{J}^0}{\mathrm{d}\tau} := -\mathrm{i}\left[\hat{J}^0, \widehat{H}(\tau)\right], \tag{3.289a}$$

in the Heisenberg picture with the unitary time (τ) evolution generated by $\widehat{H}(\tau)$ on the *full* interval $\tau_{\mathrm{i}} \le \tau \le \tau_{\mathrm{f}}$. Here, the $\mathbb{R}$-valued function

$$\begin{aligned}\mathcal{A}\colon [\tau_{\mathrm{i}},\tau_{\mathrm{f}}]\times\mathbb{R}^d &\to \mathbb{R},\\ (\tau,\boldsymbol{x}) &\mapsto \mathcal{A}(\tau,\boldsymbol{x}),\end{aligned} \tag{3.289b}$$

is chosen to obey the boundary conditions

$$\mathcal{A}(\tau_{\mathrm{i}},\boldsymbol{x}) = 0, \qquad \mathcal{A}(\tau_{\mathrm{f}},\boldsymbol{x}) = \mathcal{A}_{\mathrm{f}}(\boldsymbol{x}), \tag{3.289c}$$

at the ends of the interval $\tau_{\mathrm{i}} \le \tau \le \tau_{\mathrm{f}}$ in such a way that the identifications

$$(\hat{J}^0,\hat{\boldsymbol{J}})(\tau_{\mathrm{i}},\boldsymbol{x}) = (\hat{j}^0_{\mathrm{i}},\hat{\boldsymbol{j}}_{\mathrm{i}})(\boldsymbol{x}), \qquad (\hat{J}^0,\hat{\boldsymbol{J}})(\tau_{\mathrm{f}},\boldsymbol{x}) = (\hat{j}^0_{\mathrm{f}},\hat{\boldsymbol{j}}_{\mathrm{f}})(\boldsymbol{x}) \tag{3.289d}$$

are meaningful.

Define the instantaneous expectation value

$$J^0_{\mathrm{adia}}(\tau,\boldsymbol{x}) := \left\langle \mathrm{GS}_{\mathrm{adia}}(\tau)\left|\hat{J}^0(\tau=0,\boldsymbol{x})\right|\mathrm{GS}_{\mathrm{adia}}(\tau)\right\rangle \tag{3.290a}$$

of the local fermion density and the instantaneous expectation value

$$\boldsymbol{J}_{\mathrm{adia}}(\tau,\boldsymbol{x}) := \left\langle \mathrm{GS}_{\mathrm{adia}}(\tau)\left|\hat{\boldsymbol{J}}(\tau=0,\boldsymbol{x})\right|\mathrm{GS}_{\mathrm{adia}}(\tau)\right\rangle \tag{3.290b}$$

of the local fermion current density. Define

$$J^0(\tau,\boldsymbol{x}) := \left\langle \mathrm{GS}(\tau)\left|\hat{J}^0(\tau=0,\boldsymbol{x})\right|\mathrm{GS}(\tau)\right\rangle \tag{3.291a}$$

to be the expectation of the local fermion density in the many-body ground state $|\mathrm{GS}(\tau)\rangle$ and

$$\boldsymbol{J}(\tau,\boldsymbol{x}) := \left\langle \mathrm{GS}(\tau) \middle| \hat{\boldsymbol{J}}(\tau=0,\boldsymbol{x}) \middle| \mathrm{GS}(\tau) \right\rangle \tag{3.291b}$$

to be the expectation value of the local fermion current density in the many-body ground state $|\mathrm{GS}(\tau)\rangle$. As we are assuming that

$$J^{\mu}_{\mathrm{adia}}(\tau,\boldsymbol{x}) = J^{\mu}(\tau,\boldsymbol{x}), \tag{3.292}$$

it follows that the continuity and differentiability of $J^0(\tau,\boldsymbol{x})$ and $\boldsymbol{J}(\tau,\boldsymbol{x})$ on the interval $\tau_{\mathrm{i}} \leq \tau \leq \tau_{\mathrm{f}}$ is equivalent to that of $J^0_{\mathrm{adia}}(\tau,\boldsymbol{x})$ and $\boldsymbol{J}_{\mathrm{adia}}(\tau,\boldsymbol{x})$. Consequently, we may do the manipulations

$$\begin{aligned} Q_{\mathrm{f}} - Q_{\mathrm{i}} &= \int\limits_{\mathbb{R}^d} \mathrm{d}^d\boldsymbol{x}\; j^0_{\mathrm{f}}(\boldsymbol{x}) - \int\limits_{\mathbb{R}^d} \mathrm{d}^d\boldsymbol{x}\; j^0_{\mathrm{i}}(\boldsymbol{x}) \\ &= \int\limits_{\tau_{\mathrm{i}}}^{\tau_{\mathrm{f}}} \mathrm{d}\tau \frac{\mathrm{d}}{\mathrm{d}\tau} \int\limits_{\mathbb{R}^d} \mathrm{d}^d\boldsymbol{x}\; J^0(\tau,\boldsymbol{x}). \end{aligned} \tag{3.293}$$

We are in position to take advantage of the anomalous continuity equation (3.289a) obeyed at the operator level locally in space $\boldsymbol{x}$ and adiabatic time τ by applying it to Eq. (3.293) on the full interval $\tau_{\mathrm{i}} \leq \tau \leq \tau_{\mathrm{f}}$,

$$Q_{\mathrm{f}} - Q_{\mathrm{i}} = -\int\limits_{\tau_{\mathrm{i}}}^{\tau_{\mathrm{f}}} \mathrm{d}\tau \oint\limits_{\partial\,\mathbb{R}^d} \mathrm{d}\boldsymbol{a}\cdot \boldsymbol{J}(\tau,\boldsymbol{x}) + \int\limits_{\tau_{\mathrm{i}}}^{\tau_{\mathrm{f}}} \mathrm{d}\tau \int\limits_{\mathbb{R}^d} \mathrm{d}^d\boldsymbol{x}\; \mathcal{A}(\tau,\boldsymbol{x}). \tag{3.294}$$

Here, $\mathrm{d}\boldsymbol{a} \in \mathbb{R}^d$ is the infinitesimal vector that is normal to the $(d-1)$-dimensional boundary $\partial\,\mathbb{R}^d$ of $\mathbb{R}^d$. The first term on the right-hand side of Eq. (3.294) is manifestly a boundary term. For all the cases of interest in this book, the second term on the right-hand side of Eq. (3.294) is also a boundary term. This means that $Q_{\mathrm{f}} - Q_{\mathrm{i}}$ is invariant under any smooth and local changes of $\boldsymbol{J}(\tau,\boldsymbol{x})$ and $\mathcal{A}(\tau,\boldsymbol{x})$ that preserves the asymptotic values of $\lim_{|\boldsymbol{x}|\to\infty} \boldsymbol{J}(\tau,\boldsymbol{x})$ and $\lim_{|\boldsymbol{x}|\to\infty} \mathcal{A}(\tau,\boldsymbol{x})$. Such smooth and local changes of $\boldsymbol{J}(\tau,\boldsymbol{x})$ and $\mathcal{A}(\tau,\boldsymbol{x})$ originate from smooth deformations of the family $\{\mathcal{H}(\tau)\}$ that do not produce a spectral flow. On the other hand, as we shall verify by way of explicit examples, smooth changes of the family $\{\mathcal{H}(\tau)\}$ that are parameterized by τ can induce spectral flows that invalidate Eq. (3.294).

We now specialize to any pair of Hamiltonians $\widehat{H}_{\mathrm{i}}$ and $\widehat{H}_{\mathrm{f}}$ that are both of the Dirac type. Most importantly, we assume that the chemical potential μ is vanishing, that is, we work at an effectively vanishing density of fermion matter (no Fermi surface).

The single-particle Dirac Hamiltonian $\mathcal{H}(\tau)$ that interpolates between the reference single-particle Dirac Hamiltonian $\mathcal{H}_{\mathrm{i}}$ when $\tau = \tau_{\mathrm{i}}$ and the single-particle Dirac

Hamiltonian of interest $\mathcal{H}_{\mathrm{f}}$ when $\tau = \tau_{\mathrm{f}}$ has the spectral measure (density of states) $\nu_{\mathcal{H}(\tau)}(\varepsilon)$, that is

$$N_{\varepsilon,\varepsilon+\mathrm{d}\varepsilon} := \nu_{\mathcal{H}(\tau)}(\varepsilon)\,\mathrm{d}\varepsilon \tag{3.295a}$$

returns the number of single-particle eigenstates of $\mathcal{H}(\tau)$ in the interval $[\varepsilon, \varepsilon + \mathrm{d}\varepsilon]$.

Furthermore, in line with Eqs. (3.130) and (3.626), we define the adiabatic charge operator

$$\widehat{Q}^{\mathrm{adia}}(\tau) := \int \mathrm{d}\kappa\, \nu(\kappa)\, \left[\hat{c}^{\dagger}(\kappa)\,\hat{c}(\kappa) \;-\hat{d}^{\dagger}(\kappa)\,\hat{d}(\kappa)\right] - \frac{1}{2}\,\eta^{\mathrm{adia}}_{\mathcal{H}(\tau)}. \tag{3.295b}$$

This definition is unique if we demand that the Dirac charge operator is antisymmetric under charge conjugation at vanishing chemical potential. The reason why this operator is qualified as being adiabatic is the following. Even though the collective label κ could label any basis of the single-particle Hilbert space, here it labels the quantum numbers of the single-particle eigenstate of the reference Hamiltonian $\mathcal{H}_{\mathrm{i}}$ out of which $\mathcal{H}_{\mathrm{f}}$ evolves through the τ unitary evolution adiabatically, that is the labels κ remain good quantum labels during the entire τ evolution. In the definition (3.295b), the only nonvanishing equal-time anticommutators are

$$\left\{\hat{c}(\kappa), \hat{c}^{\dagger}(\kappa')\right\} = \delta(\kappa - \kappa'), \qquad \left\{\hat{d}(\kappa), \hat{d}^{\dagger}(\kappa')\right\} = \delta(\kappa - \kappa'). \tag{3.295c}$$

Hereto, the number

$$N_{\kappa,\kappa+\mathrm{d}\kappa} := \nu(\kappa)\,\mathrm{d}\kappa \tag{3.295d}$$

counts all the quantum numbers in the interval $[\kappa, \kappa + \mathrm{d}\kappa]$, and

$$1 =: \delta(\kappa)\,\mathrm{d}\kappa. \tag{3.295e}$$

Finally, the τ dependence in the definition (3.295b) arises from the dependence on the $\mathbb{C}$ number

$$\eta^{\mathrm{adia}}_{\mathcal{H}(\tau)} := \int\limits_{-\infty}^{+\infty} \mathrm{d}\varepsilon\, \nu_{\mathcal{H}(\tau)}(\varepsilon)\,\mathrm{sgn}(\varepsilon) \tag{3.295f}$$

that measures the spectral asymmetry of $\mathcal{H}(\tau)$.

We make two observations. First, Eq. (3.295b) implies that the eigenvalues

$$\mathrm{spec}\left(\widehat{Q}^{\mathrm{adia}}(\tau) + \frac{1}{2}\,\eta^{\mathrm{adia}}_{\mathcal{H}(\tau)}\right) \subset \mathbb{Z} \tag{3.296}$$

are integers. It is only through the spectral asymmetry $\eta^{\mathrm{adia}}_{\mathcal{H}(\tau)}$ that $Q^{\mathrm{adia}}(\tau)$ may change continuously. Second, if we introduce the even and odd parts of the measure (density of states) $\nu_{\mathcal{H}(\tau)}(\varepsilon)$ through

$$\left[\nu_{\mathcal{H}(\tau)}(\varepsilon)\right]_{\mathrm{even}} := \frac{1}{2}\left[\nu_{\mathcal{H}(\tau)}(+\varepsilon) + \nu_{\mathcal{H}(\tau)}(-\varepsilon)\right] \tag{3.297a}$$

and

$$\left[\nu_{\mathcal{H}(\tau)}(\varepsilon)\right]_{\text{odd}} := \frac{1}{2}\left[\nu_{\mathcal{H}(\tau)}(+\varepsilon) - \nu_{\mathcal{H}(\tau)}(-\varepsilon)\right], \tag{3.297b}$$

then

$$\begin{aligned}\eta^{\text{adia}}_{\mathcal{H}(\tau)} &= \int\limits_{-\infty}^{+\infty} \mathrm{d}\varepsilon \left[\nu_{\mathcal{H}(\tau)}(\varepsilon)\right]_{\text{odd}} \operatorname{sgn}(\varepsilon)\\ &= 2\int\limits_{0}^{\infty} \mathrm{d}\varepsilon \left[\nu_{\mathcal{H}(\tau)}(\varepsilon)\right]_{\text{odd}} \operatorname{sgn}(\varepsilon).\end{aligned} \tag{3.297c}$$

By construction,

$$\begin{aligned}Q_{\text{i}} &:= \left\langle \text{GS}_{\text{i}} \middle| \widehat{Q}_{\text{i}} \middle| \text{GS}_{\text{i}} \right\rangle\\ &= \left\langle \text{GS}_{\text{adia}}(\tau_{\text{i}}) \middle| \widehat{Q}^{\text{adia}}(\tau_{\text{i}}) \middle| \text{GS}_{\text{adia}}(\tau_{\text{i}}) \right\rangle\\ &= -\frac{1}{2}\,\eta^{\text{adia}}_{\mathcal{H}(\tau_{\text{i}})},\end{aligned} \tag{3.298}$$

for the annihilation operator $\hat{c}(\kappa)$ that annihilates a Dirac fermion with positive energy eigenvalue ε_κ of $\mathcal{H}_{\text{i}}$ annihilates $|\text{GS}_{\text{i}}\rangle$, while the annihilation operator $\hat{d}(\kappa)$ that creates a Dirac fermion with negative energy eigenvalue ε_κ of $\mathcal{H}_{\text{i}}$ annihilates $|\text{GS}_{\text{i}}\rangle$. By construction,

$$Q^{\text{adia}}(\tau) := \left\langle \text{GS}_{\text{adia}}(\tau) \middle| \widehat{Q}^{\text{adia}}(\tau) \middle| \text{GS}_{\text{adia}}(\tau) \right\rangle = -\frac{1}{2}\,\eta^{\text{adia}}_{\mathcal{H}(\tau)} \tag{3.299}$$

is a smooth function of τ, for we have ruled out the scenario by which the sign of a single-particle eigenvalue of $\mathcal{H}(\tau)$ changes on the interval $\tau_{\text{i}} \le \tau \le \tau_{\text{f}}$. [If a single-particle eigenvalue of $\mathcal{H}(\tau)$ changes sign at τ', then the relation (3.288) breaks down just beyond τ', while $\eta^{\text{adia}}_{\mathcal{H}(\tau)}$ undergoes a discontinuous jump of two in magnitude upon crossing τ'.] Thus, we have the identity

$$Q_{\text{f}} - Q_{\text{i}} = -\frac{1}{2}\left(\eta^{\text{adia}}_{\mathcal{H}_{\text{f}}} - \eta^{\text{adia}}_{\mathcal{H}_{\text{i}}}\right) = -\frac{1}{2}\int\limits_{\tau_{\text{i}}}^{\tau_{\text{f}}} \mathrm{d}\tau\, \frac{\mathrm{d}}{\mathrm{d}\tau}\,\eta^{\text{adia}}_{\mathcal{H}(t)}. \tag{3.300}$$

Comparison of Eqs. (3.294) and (3.300) gives

$$\eta^{\text{adia}}_{\mathcal{H}_{\text{f}}} - \eta^{\text{adia}}_{\mathcal{H}_{\text{i}}} = 2\left[\int\limits_{\tau_{\text{i}}}^{\tau_{\text{f}}} \mathrm{d}\tau \oint\limits_{\partial\mathbb{R}^d} \mathrm{d}\boldsymbol{a}\cdot\boldsymbol{J}(\tau,\boldsymbol{x}) - \int\limits_{\tau_{\text{i}}}^{\tau_{\text{f}}} \mathrm{d}\tau \int\limits_{\mathbb{R}^d} \mathrm{d}^d\boldsymbol{x}\, \mathcal{A}(\tau,\boldsymbol{x})\right]. \tag{3.301}$$

We conclude that the difference between the adiabatic spectral asymmetry of the Dirac Hamiltonian $\mathcal{H}_{\text{f}}$ and that of the reference Dirac Hamiltonian $\mathcal{H}_{\text{i}}$ inherits the invariance of $Q_{\text{f}} - Q_{\text{i}}$ under any smooth and local changes of $\boldsymbol{J}(\tau,\boldsymbol{x})$ and $\mathcal{A}(\tau,\boldsymbol{x})$ that preserve the asymptotic values of $\lim_{|\boldsymbol{x}|\to\infty}\boldsymbol{J}(\tau,\boldsymbol{x})$ and $\lim_{|\boldsymbol{x}|\to\infty}\mathcal{A}(\tau,\boldsymbol{x})$.

The qualifier adiabatic is synonymous to the condition (3.288) for the absence of a spectral flow holding on the interval $\tau_{\mathrm{i}} \leq \tau \leq \tau_{\mathrm{f}}$.

By taking advantage of the identity,

$$\begin{aligned}\mathrm{sgn}(\varepsilon) =& \frac{\varepsilon}{|\varepsilon|}\\ =& 2 \times \int\limits_{-\infty}^{+\infty} \frac{\mathrm{d}z}{2\pi\mathrm{i}} \frac{\mathrm{i}\varepsilon}{z^2+\varepsilon^2}\\ =& \frac{2}{\pi} \times \int\limits_{0}^{+\infty} \mathrm{d}z \, \frac{\varepsilon}{z^2+\varepsilon^2}\end{aligned} \tag{3.302}$$

for any $\varepsilon \neq 0$, we choose to combine the spectral representation (3.299) with Eqs. (3.295f) and (3.297c) into

$$\begin{aligned}Q(\tau) =& Q^{\mathrm{adia}}(\tau)\\ =& -\frac{1}{2}\eta^{\mathrm{adia}}_{\mathcal{H}(\tau)}\\ =& -\frac{2}{\pi}\int\limits_0^\infty \mathrm{d}\varepsilon \left[\nu_{\mathcal{H}(\tau)}(\varepsilon)\right]_{\mathrm{odd}} \int\limits_0^\infty \mathrm{d}z \, \frac{\varepsilon}{\varepsilon^2+z^2}\\ =& -\frac{1}{\pi}\int\limits_0^\infty \mathrm{d}z \int\limits_{-\infty}^{+\infty} \mathrm{d}\varepsilon \left[\nu_{\mathcal{H}(\tau)}(\varepsilon)\right]_{\mathrm{odd}} \frac{\varepsilon}{\varepsilon^2+z^2}.\end{aligned} \tag{3.303}$$

We have assumed that we can safely interchange the order of integrations to reach the last equality. Finally, we can undo the spectral representation in favor of reinstating the representation of the single-particle Hamiltonian $\mathcal{H}(\tau)$ in the position basis,

$$Q(\tau) = -\frac{1}{\pi}\int\limits_0^\infty \mathrm{d}z \int\limits_{\mathbb{R}^d} \mathrm{d}^d\boldsymbol{x}\,\mathrm{tr}\left\langle \boldsymbol{x} \left| \frac{\mathcal{H}(\tau)}{\mathcal{H}^2(\tau)+z^2} \right| \boldsymbol{x} \right\rangle. \tag{3.304}$$

Here, the trace tr is to be performed over the Clifford algebra as we are dealing with a Dirac Hamiltonian.

The existence of a single-particle gap

$$\Delta := \inf\{\Delta(\tau) | \tau_{\mathrm{i}} \leq \tau \leq \tau_{\mathrm{f}}\} > 0 \tag{3.305}$$

for the family $\{\mathcal{H}(\tau)\}$ over the entire interval $\tau_{\mathrm{i}} \leq \tau \leq \tau_{\mathrm{f}}$ allows to define the characteristic length (in units with $\hbar = c = 1$)

$$\xi = \frac{1}{\Delta}. \tag{3.306}$$

By dimensional analysis, we can write

$$\operatorname{tr}\left\langle \boldsymbol{x} \left| \frac{\mathcal{H}(\tau)}{\mathcal{H}^2(\tau)+z^2} \right| \boldsymbol{x} \right\rangle = \Delta^{d-1} \times F(\boldsymbol{x}, z, \tau), \tag{3.307}$$

where the real-valued function F is dimensionless. For simplicity, we assume that the anomaly $\mathcal{A}$ vanishes. The dependence on τ of F is then the signature of the functional dependence of $\mathcal{H}(\tau)$ on the background. If this background, we call it Φ, varies slowly on the characteristic length scale ξ, we may expand F in powers of the space derivatives of Φ. Each derivative must come multiplied by $\xi = 1/\Delta$, since F is dimensionless. Hence, the small parameter in this expansion is $1/\Delta$. According to the first term on the right-hand side of Eq. (3.294), this gradient expansion is that of the divergence of a current density ($\mathcal{A} = 0$ by assumption). Thus, we expect that re-summation over all subleading terms from this gradient expansion after the z and $\boldsymbol{x}$ integrals have been performed delivers exactly zero in the absence of any spectral flow (an integer otherwise).

To illustrate the inner workings of such a gradient expansion, we make the choices $d = 1$ for the dimension of space,

$$\mathcal{H}_{\mathrm{f}}(x) := -\mathrm{i}\tau_2\, \partial_x + \tau_1\, \phi_1(x) + \tau_3\, \phi_2(x) \tag{3.308a}$$

for the single-particle Dirac Hamiltonian of interest, and

$$\mathcal{H}_{\mathrm{i}} := \lim_{x\to\infty} \mathcal{H}_{\mathrm{f}}(x) \tag{3.308b}$$

for the reference single-particle Dirac Hamiltonian [recall Eq. (3.243)]. With the choice $\mathcal{H}_{\mathrm{i}}$ made in Eq. (3.308b), we have

$$Q_{\mathrm{i}} = 0. \tag{3.309}$$

To organize the gradient expansion of Eq. (3.307), we observe that, for any $z \in \mathbb{C}$,

$$\mathcal{H}^2_{\mathrm{f}}(x) + \tau_0\, z^2 = \left(-\partial^2_x + |\phi(x)|^2 + z^2\right)\, \tau_0 - \tau_3\, \phi'_1(x) + \tau_1\, \phi'_2(x), \tag{3.310}$$

where $\phi'_i \equiv \partial_x \phi_i$ is the one-dimensional gradient of the background real-valued scalar field ϕ_i with $i = 1, 2$. We use the shorthand notation

$$\mathcal{H}'_{\mathrm{f}} := -\tau_3\, \phi'_1 + \tau_1\, \phi'_2. \tag{3.311}$$

We also notice that we may perform a Taylor expansion around $x = \infty$ on the operator multiplied by τ_0 on the right-hand side of Eq. (3.310), thereby obtaining $\mathcal{H}^2_{\mathrm{i}} + \tau_0\, z^2 + \cdots$ where the $\cdots$ contain derivatives of the background fields,

$$\mathcal{H}^2_{\mathrm{f}} + \tau_0\, z^2 = \left(\mathcal{H}^2_{\mathrm{i}} + \tau_0\, z^2 + \cdots\right) + \mathcal{H}'_{\mathrm{f}}. \tag{3.312}$$

Second, we do the geometrical expansion

$$\frac{1}{\mathcal{H}^2_{\mathrm{f}} + z^2} = \frac{1}{(\mathcal{H}^2_{\mathrm{i}} + z^2 + \cdots)} - \frac{1}{(\mathcal{H}^2_{\mathrm{i}} + z^2 + \cdots)}\, \mathcal{H}'_{\mathrm{f}}\, \frac{1}{(\mathcal{H}^2_{\mathrm{i}} + z^2 + \cdots)} + \cdots, \tag{3.313}$$

where the unital element τ_0 from the Clifford algebra is implicit. Third, we compute the trace tr over the Clifford algebra. To the zero-th order in the gradient expansion,

$$\operatorname{tr}\left\langle x\left|\frac{\mathcal{H}_{\mathrm{f}}}{\mathcal{H}_{\mathrm{f}}^2+z^2}\right|x\right\rangle = 0. \tag{3.314a}$$

To the first order in the gradient expansion,

$$\begin{aligned}\operatorname{tr}\left\langle x\left|\frac{\mathcal{H}_{\mathrm{f}}}{\mathcal{H}_{\mathrm{f}}^2+z^2}\right|x\right\rangle =& -\operatorname{tr}\left\langle x\left|\frac{1}{\mathcal{H}_{\mathrm{i}}^2+z^2}\phi_2'\frac{1}{\mathcal{H}_{\mathrm{i}}^2+z^2}\phi_1\right|x\right\rangle \\ &+\operatorname{tr}\left\langle x\left|\frac{1}{\mathcal{H}_{\mathrm{i}}^2+z^2}\phi_1'\frac{1}{\mathcal{H}_{\mathrm{i}}^2+z^2}\phi_2\right|x\right\rangle.\end{aligned} \tag{3.314b}$$

Fourth, we insert the resolution of the identity in position space twice: first to the right of the first propagator $(\mathcal{H}_{\mathrm{i}}^2+z^2)^{-1}$ and second to the left of the second propagator $(\mathcal{H}_{\mathrm{i}}^2+z^2)^{-1}$. From the translation symmetry of $\mathcal{H}_{\mathrm{i}}^2+z^2$,

$$\left\langle x\left|\frac{\tau_0}{\mathcal{H}_{\mathrm{i}}^2+z^2}\right|y\right\rangle = \int\limits_{-\infty}^{+\infty}\frac{\mathrm{d}k}{2\pi}\,e^{+\mathrm{i}k\,(x-y)}\frac{\tau_0}{k^2+|\boldsymbol{\phi}(\infty)|^2+z^2}, \tag{3.315a}$$

$$\left\langle y\left|\frac{\tau_0}{\mathcal{H}_{\mathrm{i}}^2+z^2}\right|x\right\rangle = \int\limits_{-\infty}^{+\infty}\frac{\mathrm{d}p}{2\pi}\,e^{-\mathrm{i}p\,(x-y)}\frac{\tau_0}{p^2+|\boldsymbol{\phi}(\infty)|^2+z^2}. \tag{3.315b}$$

Fifth, we do the gradient expansions $\phi_2'(y)=\phi_2'(x)+\cdots$ and $\phi_1'(y)=\phi_1'(x)+\cdots$ and ignore the $\cdots$ at the order under consideration in the gradient expansion. The y integration then delivers $2\pi\,\delta(k-p)$ to this order in the gradient expansion. After integration over the momentum p, we are left with

$$\operatorname{tr}\left\langle x\left|\frac{\mathcal{H}_{\mathrm{f}}}{\mathcal{H}_{\mathrm{f}}^2+z^2}\right|x\right\rangle = -2\int\limits_{-\infty}^{+\infty}\frac{\mathrm{d}k}{2\pi}\frac{(\phi_1\,\phi_2'-\phi_2\,\phi_1')\,(x)}{(k^2+|\boldsymbol{\phi}(\infty)|^2+z^2)^2}+\cdots. \tag{3.316}$$

The integration over the momentum k gives the multiplicative factor $(|\boldsymbol{\phi}(\infty)|^2+z^2)^{-3/2}/4$,

$$\operatorname{tr}\left\langle x\left|\frac{\mathcal{H}_{\mathrm{f}}}{\mathcal{H}_{\mathrm{f}}^2+z^2}\right|x\right\rangle = -\frac{1}{2}\frac{(\phi_1\,\phi_2'-\phi_2\,\phi_1')\,(x)}{(|\boldsymbol{\phi}(\infty)|^2+z^2)^{3/2}}+\cdots. \tag{3.317}$$

The primitive of the function $(z^2+a^2)^{-3/2}$ is $z(z^2+a^2)^{-1/2}/a^2$. Hence,

$$\begin{aligned}-\frac{1}{\pi}\int\limits_0^{\infty}\mathrm{d}z\operatorname{tr}\left\langle x\left|\frac{\mathcal{H}_{\mathrm{f}}}{\mathcal{H}_{\mathrm{f}}^2+z^2}\right|x\right\rangle =& +\frac{1}{2\pi}\frac{(\phi_1\,\phi_2'-\phi_2\,\phi_1')\,(x)}{|\boldsymbol{\phi}(\infty)|^2}+\cdots \\ =& +\frac{1}{2\pi}\frac{(\phi_1\,\phi_2'-\phi_2\,\phi_1')\,(x)}{|\boldsymbol{\phi}(x)|^2}+\cdots \\ =& +\frac{1}{2\pi}\frac{\mathrm{d}}{\mathrm{d}x}\arctan\left(\frac{\phi_2(x)}{\phi_1(x)}\right)+\cdots.\end{aligned} \tag{3.318}$$

As it should be, the right-hand side is a total derivative. Consequently, the induced fermion charge (3.300) for the Hamiltonian of interest $\widehat{H}_{\mathrm{f}}$ defined by Eq. (3.308) is exactly given by

$$Q_{\mathrm{f}} = +\frac{1}{2\pi}\left[\arctan\left(\frac{\phi_2(+\infty)}{\phi_1(+\infty)}\right) - \arctan\left(\frac{\phi_2(-\infty)}{\phi_1(-\infty)}\right)\right], \tag{3.319}$$

if we assume that integrating over space *all* the terms implied by $\cdots$ in Eq. (3.318) vanishes. Equation (3.319) agrees with Eq. (3.269). Moreover, owing to the identity $\arctan(x) + \arctan(1/x) = \pi/2$, we recover the induced fermion charge (3.109) for the profile

$$\lim_{x\to\pm\infty} \phi_1(x) = \pm\phi_\infty, \qquad \phi_2(x) = \mu_{\mathrm{s}}, \tag{3.320}$$

with $\phi_\infty \in \mathbb{R}$ and the staggered chemical potential $\mu_{\mathrm{s}} > 0$.

3.8.4 Presence of Spectral Flow

The question needing to be addressed is how should Eqs. (3.294) and (3.300) be modified when the parametric dependence on τ of the single-particle Hamiltonian $\mathcal{H}(\tau)$ that interpolates between $\mathcal{H}(\tau_{\mathrm{i}}) \equiv \mathcal{H}_{\mathrm{i}}$ and $\mathcal{H}(\tau_{\mathrm{f}}) \equiv \mathcal{H}_{\mathrm{f}}$ is piece-wise continuous due the presence of n spectral flows at the intermediary times $\tau_1 < \cdots < \tau_n$. We are going to show that the fractional part of $Q_{\mathrm{f}} - Q_{\mathrm{i}}$ is unchanged by any one of these n spectral flows. As a corollary, any instance of spectral flow can only change $Q_{\mathrm{f}} - Q_{\mathrm{i}}$ by an integer.

We assume that, at any one of the n evolution times $\tau_1 < \tau_2 < \cdots < \tau_n$ from the open interval $]\tau_{\mathrm{i}}, \tau_{\mathrm{f}}[$, a number $n^{(i)}_{-\to+}$ ($n^{(i)}_{+\to-}$) of eigenvalues of $\mathcal{H}(\tau_i - 0^+)$ cross the chemical potential $\mu = 0$ from negative to positive values (from positive to negative values), that is, the adiabatic relation (3.286) only holds on the $(n+1)$ open intervals $]\tau_{\mathrm{i}}, \tau_1[$, $]\tau_1, \tau_2[$, $\cdots$, $]\tau_n, \tau_{\mathrm{f}}[$. The function of τ

$$\eta_{\mathcal{H}(\tau)} := \int\limits_{-\infty}^{+\infty} \mathrm{d}\varepsilon\, \nu_{\mathcal{H}(\tau)}(\varepsilon)\, \mathrm{sgn}(\varepsilon) \tag{3.321}$$

is then continuous on any one of the $(n+1)$ open intervals $]\tau_{\mathrm{i}}, \tau_1[$, $]\tau_1, \tau_2[$, $\cdots$, $]\tau_n, \tau_{\mathrm{f}}[$, while it jumps by $+2$ (-2) for each eigenvalue of $\mathcal{H}(\tau)$ that crosses from negative to positive values (from positive to negative values) at τ_i with $i = 1, \cdots, n$. Hence, the function of τ [Θ denotes the Heaviside step function]

$$\eta^{\mathrm{cont}}_{\mathcal{H}(\tau)} := \eta_{\mathcal{H}(\tau)} - 2\sum_{i=1}^{n}\left(n^{(i)}_{-\to+} - n^{(i)}_{+\to-}\right)\Theta(\tau - \tau_i) \tag{3.322}$$

is continuous on the interval $[\tau_{\mathrm{i}}, \tau_{\mathrm{f}}]$. Furthermore, the identity (3.300) is not valid anymore, it must be amended owing to the fact that

$$-\frac{1}{2}\left(\eta_{\mathcal{H}_\mathrm{f}}-\eta_{\mathcal{H}_\mathrm{i}}\right)=-\frac{1}{2}\int\limits_{\tau_\mathrm{i}}^{\tau_\mathrm{f}}\mathrm{d}\tau\,\frac{\mathrm{d}}{\mathrm{d}\tau}\,\eta^{\mathrm{cont}}_{\mathcal{H}(t)}-\sum_{i=1}^{n}\left(n^{(i)}_{-\to+}-n^{(i)}_{+\to-}\right). \tag{3.323}$$

In order to relate the left-hand side of Eq. (3.323) to the induced fermion number, we need to limit the validity of Eq. (3.289a) to each $(n+1)$ intervals $]\tau_\mathrm{i},\tau_1[,\cdots,$ $]\tau_n,\tau_\mathrm{f}[$ separately. Because of the spectral flow taking place at τ_i $(i=1,\cdots,n)$, we introduce $(n+1)$ families of adiabatic charge operators

$$\widehat{Q}^{\mathrm{cont}}_\mathrm{i}(\tau):=\int\mathrm{d}\kappa\,\nu(\kappa)\,\left[\hat{c}^\dagger_\mathrm{i}(\kappa)\,\hat{c}_\mathrm{i}(\kappa)\,-\hat{d}^\dagger_\mathrm{i}(\kappa)\,\hat{d}_\mathrm{i}(\kappa)\right]-\frac{1}{2}\,\eta^{\mathrm{cont}}_{\mathcal{H}(\tau)}, \tag{3.324a}$$

for $\tau_\mathrm{i}<\tau<\tau_1$,

$$\widehat{Q}^{\mathrm{cont}}_1(\tau):=\int\mathrm{d}\kappa\,\nu(\kappa)\,\left[\hat{c}^\dagger_1(\kappa)\,\hat{c}_1(\kappa)\,-\hat{d}^\dagger_1(\kappa)\,\hat{d}_1(\kappa)\right]-\frac{1}{2}\,\eta^{\mathrm{cont}}_{\mathcal{H}(\tau)}, \tag{3.324b}$$

for $\tau_1<\tau<\tau_2$, and so on until

$$\widehat{Q}^{\mathrm{cont}}_n(\tau):=\int\mathrm{d}\kappa\,\nu(\kappa)\,\left[\hat{c}^\dagger_n(\kappa)\,\hat{c}_n(\kappa)\,-\hat{d}^\dagger_n(\kappa)\,\hat{d}_n(\kappa)\right]-\frac{1}{2}\,\eta^{\mathrm{cont}}_{\mathcal{H}(\tau)}, \tag{3.324c}$$

for $\tau_n<\tau<\tau_\mathrm{f}$, instead of the single family of adiabatic charge operators (3.295b). The difference between the charge operator $\widehat{Q}^{\mathrm{cont}}_{i-1}(\tau_i-0^+)$ on the interval $]\tau_{i-1},\tau_i[$ and the charge operator $\widehat{Q}^{\mathrm{cont}}_{i}(\tau_i+0^+)$ on the interval $]\tau_i,\tau_{i+1}[$, where $i=1,\cdots,n$ and $i-1=0$ corresponds to i, reflects the nonadiabatic change in the many-body ground state that occurs at τ_i because of the spectral flow. Indeed, $n^{(i)}_{-\to+}$ $\hat{d}$'s from $\widehat{Q}^{\mathrm{cont}}_{i-1}(\tau_i-0^+)$ on the interval $]\tau_{i-1},\tau_i[$ have been turned into $\hat{c}$'s for $\widehat{Q}^{\mathrm{cont}}_{i}(\tau_i+0^+)$ on the interval $]\tau_i,\tau_{i+1}[$, while $n^{(i)}_{+\to-}$ $\hat{c}$'s from $\widehat{Q}^{\mathrm{cont}}_{i-1}(\tau_i-0^+)$ on the interval $]\tau_{i-1},\tau_i[$ have been turned into $\hat{d}$'s for $\widehat{Q}^{\mathrm{cont}}_{i}(\tau_i+0^+)$ on the interval $]\tau_i,\tau_{i+1}[$. Correspondingly, we may write

$$\begin{aligned}
Q_\mathrm{f}-Q_\mathrm{i}=&+Q_\mathrm{f}-Q_n(\tau_n+0^+)-\left(n^{(n)}_{-\to+}-n^{(n)}_{+\to-}\right)\\
&+Q_{n-1}(\tau_n-0^+)-Q_{n-1}(\tau_{n-1}+0^+)-\left(n^{(n-1)}_{-\to+}-n^{(n-1)}_{+\to-}\right)\\
&\vdots\\
&+Q_1(\tau_2-0^+)-Q_1(\tau_1+0^+)-\left(n^{(1)}_{-\to+}-n^{(1)}_{+\to-}\right)\\
&+Q_\mathrm{i}(\tau_1-0^+)-Q_\mathrm{i}.
\end{aligned} \tag{3.325}$$

By comparing Eqs. (3.323) and (3.325), we deduce that

$$Q_\mathrm{f}-Q_\mathrm{i}=-\frac{1}{2}\left(\eta_{\mathcal{H}_\mathrm{f}}-\eta_{\mathcal{H}_\mathrm{i}}\right) \tag{3.326a}$$

and

$$
\begin{aligned}
-\frac{1}{2}\int\limits_{\tau_{\mathrm{i}}}^{\tau_{\mathrm{f}}} \mathrm{d}\tau\, \frac{\mathrm{d}}{\mathrm{d}\tau}\, \eta^{\mathrm{cont}}_{\mathcal{H}(t)} =\; & Q_{\mathrm{f}} - Q_n(\tau_n + 0^+) \\
& + Q_{n-1}(\tau_n - 0^+) - Q_{n-1}(\tau_{n-1} + 0^+) \\
& \vdots \\
& + Q_1(\tau_2 - 0^+) - Q_1(\tau_1 + 0^+) \\
& + Q_{\mathrm{i}}(\tau_1 - 0^+) - Q_{\mathrm{i}},
\end{aligned}
\qquad (3.326b)
$$

are the extensions of Eqs. (3.294) and (3.300) in the presence of a spectral flow.

For each of the $(n+1)$ differences $Q_{\mathrm{i}}(\tau_1 - 0^+) - Q_{\mathrm{i}}, \cdots, Q_{\mathrm{f}} - Q_n(\tau_n + 0^+)$, an equation similar to Eq. (3.294) holds. Hence, the fractional part of the induced fermion number $Q_{\mathrm{f}} - Q_{\mathrm{i}}$ retains its robustness to local and smooth changes of $\boldsymbol{J}(\tau, \boldsymbol{x})$ and $\mathcal{A}(\tau, \boldsymbol{x})$ that preserve the asymptotic values of $\lim_{|\boldsymbol{x}|\to\infty} \boldsymbol{J}(\tau, \boldsymbol{x})$ and $\lim_{|\boldsymbol{x}|\to\infty} \mathcal{A}(\tau, \boldsymbol{x})$, in spite of the presence of a spectral flow. Spectral flows can only change the integer part of the induced fermion charge $Q_{\mathrm{f}} - Q_{\mathrm{i}}$. Application of Eq. (3.294) to each line on the right-hand side of Eq. (3.326b) captures the fractional part of the induced fermion charge both without or with spectral flow.

3.8.5 Two Examples of Spectral Flows

An example of a spectral flow was obtained by MacKenzie and Wilczek upon continuously changing the asymptotic values obeyed by the background.[30] Choose the family $\{\mathcal{H}(\tau)\}$ of single-particle Dirac Hamiltonians at vanishing chemical potential, each of which is represented by [recall Eq. (3.308)]

$$\mathcal{H}(\tau, x) := -\mathrm{i}\tau_2\, \partial_x + \tau_1\, \phi_1(\tau, x) + \tau_3\, \phi_2(\tau, x) \qquad (3.327a)$$

with

$$\phi_1(\tau, x) := m\, \cos\varphi(\tau, x), \qquad \phi_2(\tau, x) := m\, \sin\varphi(\tau, x), \qquad (3.327b)$$

where $m > 0$ and

$$\varphi(\tau, x) := \tau\, \mathrm{sgn}(x) \qquad (3.327c)$$

for $x \in \mathbb{R}$ and $0 \leq \tau < 2\pi$. The reference single-particle Hamiltonian is

$$\mathcal{H}_{\mathrm{i}} := \mathcal{H}(\tau = 0) = -\mathrm{i}\tau_2\, \partial_x + m\, \tau_1. \qquad (3.328)$$

It is translation invariant, anticommutes with τ_3, displays the plane-wave spectrum

$$\varepsilon_{\pm}(\tau = 0, k) = \pm\sqrt{k^2 + m^2}, \qquad (3.329)$$

supports the spectral gap $|\varepsilon| < m$, and has vanishing total fermion number.

[30] R. MacKenzie and F. Wilczek, Phys. Rev. D **30**(10), 2194–2200 (1984). [MacKenzie and Wilczek (1984)].

On the open interval $0 < \tau < (\pi/2)$, $\mathcal{H}(\tau, x)$ breaks translation invariance, has neither unitary (antiunitary) spectral symmetry nor effective SUSY, has the normalized bound state

$$\psi_{\mathrm{bd}}(\tau, x) = \sqrt{\frac{m\,|\sin\tau|}{2}}\, e^{-m\,(\sin\tau)\,|x|} \begin{pmatrix} +1 \\ -1 \end{pmatrix} \tag{3.330a}$$

with the negative energy eigenvalue

$$\varepsilon_{\mathrm{bd}} = -m\,\cos\tau \tag{3.330b}$$

within the spectral gap $|\varepsilon| < m$ (**Exercise** 3.17), and has the fractional induced fermion number [recall Eq. (3.319)]

$$\begin{aligned} Q_{\mathrm{f}} - Q_{\mathrm{i}} &= \frac{1}{2\pi}\left[\lim_{x\to+\infty}\varphi(\tau,x) - \lim_{x\to-\infty}\varphi(\tau,x)\right] - 0 \\ &= \frac{\tau}{\pi}. \end{aligned} \tag{3.330c}$$

The single-particle Hamiltonian

$$\mathcal{H}_{\mathrm{s}} := \mathcal{H}(\tau = \pi/2 - 0^{+}) \tag{3.331}$$

breaks translation invariance, anticommutes with τ_1, supports a zero mode (mid-gap state) within the spectral gap $|\varepsilon| < m$, and has the fractional induced fermion number $+1/2$ since the mid-gap state is occupied.

The fractional part of the induced fermion number is τ/π on the interval $(\pi/2) < \tau < \pi$. However, the induced fermion number jumps by the integer -1 upon crossing $\tau = \pi/2$, since one occupied single-particle bound state with the negative energy $-m\,\cos\big((\pi/2) - 0^{+}\big)$ turns into one unoccupied single-particle bound state with the positive energy $-m\,\cos\big((\pi/2) + 0^{+}\big)$.

The bound state with the positive single-particle energy $-m\,\cos(\tau)$ on the open interval $(\pi/2) < \tau < \pi$ reaches the threshold $+m$ of the continuum with the single-particle energies $\varepsilon_{+}(k)$ at $\tau = \pi - 0^{+}$. The fractional part of the induced fermion number is $+1$ at $\tau = \pi$ and cancels the integer part -1 of the induced fermion number as is expected from the single-particle Hamiltonian

$$\mathcal{H}(\tau = \pi) = -\mathrm{i}\tau_2\,\partial_x - m\,\tau_1. \tag{3.332}$$

A bound state crosses in similar fashion from the continuum threshold $+m$ to the continuum threshold $-m$ on the interval $\pi < \tau < 2\pi$. The reference single-particle Hamiltonian is recovered at $\tau = 2\pi$.

The spectral flow of the single-particle Dirac Hamiltonians (3.327) at vanishing chemical potential is shown in Figure 3.10.

Another example of spectral flow was obtained by Blankenbecler and Boyanovsky through a continuous local change of the background Φ from the Clifford algebra,

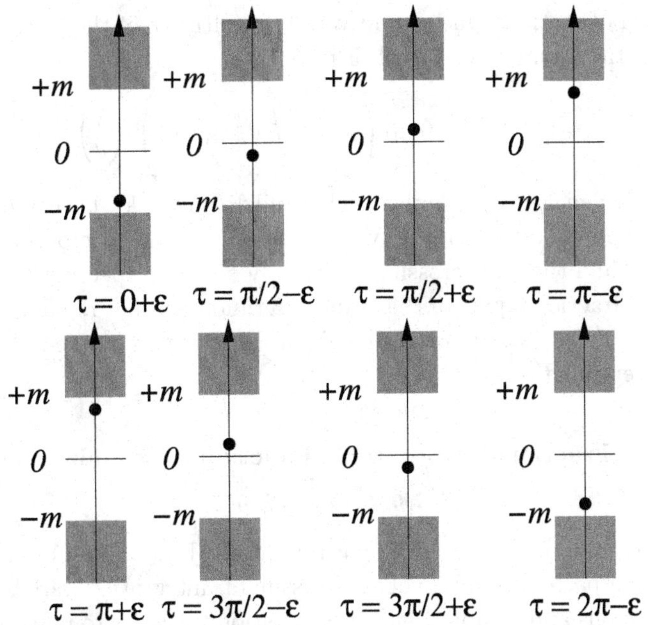

Figure 3.10 Eight still shots from the movie of the spectral flow of the single-particle Dirac Hamiltonians (3.327) at vanishing chemical potential. The two continua are denoted by the intersections between the shaded boxes and the vertical single-particle energy axis. The bound state is denoted by the intersection between the black circle and the vertical single-particle energy axis.

while holding asymptotic values of the background fixed.[31] Choose the family of single-particle Dirac Hamiltonians $\{\mathcal{H}(\tau)\}$ at vanishing chemical potential, each of which is represented by [recall Eq. (3.308)]

$$\mathcal{H}(\tau,x) := -\mathrm{i}\tau_2\,\partial_x + \tau_1\,\phi_1(\tau,x) + \tau_3\,\phi_2(\tau,x) \tag{3.333a}$$

with

$$\phi_1(\tau,x) := \phi_\infty\,\mathrm{sgn}(x), \tag{3.333b}$$

and

$$\phi_2(\tau,x) := \begin{cases} \mu_{\mathrm{s}}, & x < -\tau, \\ 0, & -\tau < x < +\tau, \\ \mu_{\mathrm{s}}, & \tau < x, \end{cases} \tag{3.333c}$$

where $0 < \phi_\infty$, $0 < \mu_{\mathrm{s}}$, and $0 \leq \tau$. The component ϕ_1 is a static (that is independent of τ) step-like soliton of the background $\Phi = \tau_1\,\phi_1 + \tau_3\,\phi_2$. The component ϕ_2 is

[31] R. Blankenbecler and D. Boyanovsky, Phys. Rev. D **31**(8), 2089-2099 (1985). [Blankenbecler and Boyanovsky (1985)].

a potential well of static height μ_s and variable width 2τ of the background Φ. At $\tau = 0$, there is the bound state [recall Eq. (3.75)]

$$\psi(\tau = 0, x) = \exp\left(-\phi_\infty \int_0^x dy\, \mathrm{sgn}(y)\right) \begin{pmatrix} 1 \\ 0 \end{pmatrix} \tag{3.334}$$

with the positive energy μ_s. Blankenbecler and Boyanovsky have shown that the energy of this bound state decreases with increasing τ until it crosses the chemical potential $\mu = 0$ and that this crossing is the only case of a spectral flow on the half line $0 < \tau$. The fractional part of the induced fermion number remains given by the right-hand side of Eq. (3.319), even though the integer part of the induced fermion number is a step function of τ.

3.9 Fractionalization from Functional Bosonization

3.9.1 Overview

Quantization can be implemented by path integrals. In quantum electrodynamics, for example, the path integral involves Grassmann integration variables to represent Dirac electrons and complex-valued integration variables to represent photons. The operation by which the Grassmann integration variables are integrated, either exactly or in a perturbative way, is referred to as functional bosonization.

Functional bosonization was applied in high-energy physics by Weinberg and Coleman[32] and in condensed matter physics by Halperin, Lubensky, and Ma[33] to predict that fluctuations arising from integrating out fermions can drive a phase transition that is continuous in the mean-field approximation into a first-order phase transition.

Fujikawa showed in 1979 how quantum anomalies,[34] that had been derived using the operator formalism starting from the seminal work of Schwinger in 1959,[35] were to be understood as the impossibility to implement simultaneously all symmetries of the action on the Grassmann measure. This paper paved the way for the derivation of bosonization rules,[36] a dictionary between conserved currents in bosonic Fock spaces, on the one hand, and fermionic Fock spaces, on the other hand, with the help of functional bosonization.

Following Schaposnik,[37] we are going to interpret fractional induced fermion numbers of one-dimensional Dirac Hamiltonians at vanishing chemical potential as manifestations of the axial (chiral) gauge anomaly in functional bosonization.

32 S. Coleman and E. Weinberg, Phys. Rev. D **7**(6), 1888–1910 (1973). [Coleman and Weinberg (1973)].
33 B. I. Halperin, T. C. Lubensky, and S. K. Ma, Phys. Rev. Lett. **32**(6), 292–295 (1974). [Halperin et al. (1974)].
34 K. Fujikawa, Phys. Rev. Lett. **42**(18), 1195–1198 (1979). [Fujikawa (1979)].
35 J. Schwinger, Phys. Rev. Lett. **3**(6), 296–297 (1959). [Schwinger (1959)].
36 R. E. Gamboa Saraví, F. A. Schaposnik, and J. E. Solomin, Nucl. Phys. B **185**, 239–253 (1981). [Gamboa Saraví et al. (1981)].
37 F. A. Schaposnik, Z. Phys. C **28**, 127–131 (1985). [Schaposnik (1985)].

3.9.2 Definitions

Define the functional

$$Z[\Phi, \not{A}] := \int \mathcal{D}[\bar{\psi}]\, \mathcal{D}[\psi]\, e^{+\mathrm{i} \int \mathrm{d}^2\mathrm{x}\, \bar{\psi}\, (\mathrm{i}\not{\partial} + \not{A} - \Phi)\, \psi}, \tag{3.335a}$$

of the mass $\Phi = \phi_1 + \mathrm{i}\gamma_5\, \phi_2$ and of the external gauge field $\not{A} \equiv \gamma^{\mu}\, A_{\mu}$, both from the Clifford algebra.

We have chosen units in which the Planck constant $\hbar = 1$ and the characteristic speed $c = 1$. The integration measure $\mathrm{d}^2\mathrm{x}$ stands for $\mathrm{d}x\, \mathrm{d}t$. The symbol $\int \mathcal{D}[\bar{\psi}]\, \mathcal{D}[\psi]$ denotes a linear functional that maps the algebra over the complex numbers generated by the pair of *independent anticommuting numbers* (Grassmann numbers) $\bar{\psi}(t, x)$ and $\psi(t, x)$ into the complex numbers in such a way that[38]

$$Z[\Phi, \not{A}] = \mathrm{Det}\, \left[-\mathrm{i}\left(\mathrm{i}\not{\partial} + \not{A} - \Phi\right)\right] \equiv \mathrm{Det}\, \left[-\mathrm{i}\left(\mathrm{i}\gamma^{\mu}\, \partial_{\mu} + \gamma^{\mu}\, A_{\mu} - \Phi\right)\right]. \tag{3.335b}$$

The reader should not worry about the precise definition of the so-called Grassmann integral $\int \mathcal{D}[\bar{\psi}]\, \mathcal{D}[\psi]$ at this stage. He should only accept as facts that (i) it is possible to formulate second quantization of fermions by means of a partition function defined as a path integral over Grassmann-valued fields (**Exercise** 3.14), and (ii) Grassmann integration of any Gaussian function of a bilinear form with kernel $\mathcal{K}$ delivers the (functional) determinant of the kernel $\mathcal{K}$, that is the (infinite) product of all eigenvalues of $\mathcal{K}$. The kernel $\mathcal{K}$ is here the kernel of the Lagrangian density (3.250) with the addition of the external gauge field $\not{A}$ through the minimal coupling. Even though $Z[\Phi, \not{A}]$ is ill-defined in the thermodynamic limit, for it is then an infinite product over numbers that need not be bounded in magnitude, the ratio $Z[\Phi, \not{A}]/Z[\Phi, 0]$ is a well-defined number if rules (regulation) on how to divide two infinite products are agreed on. Similarly, the ratio

$$\begin{aligned}\langle j^{\mu}(t, x)\rangle &\equiv \left\langle \left(\bar{\psi}\, \gamma^{\mu}\, \psi\right)(t, x)\right\rangle \\ &:= -\mathrm{i} \frac{\delta \log Z[\Phi, \not{A}]}{\delta A_{\mu}(t, x)}\bigg|_{\not{A}=0}\end{aligned} \tag{3.336}$$

with $\mu = 0, 1$ defines the two-current in the background of the mass Φ, provided rules (regulation) on how to divide the infinite products in the ratio

$$\frac{\delta \log Z[\Phi, \not{A}]}{\delta A_{\mu}(t, x)} = \frac{1}{Z[\Phi, \not{A}]} \frac{\delta Z[\Phi, \not{A}]}{\delta A_{\mu}(t, x)} \tag{3.337}$$

are agreed on. The ambiguity that arises when taking ratios of infinite products has the same origin as the ambiguity requiring normal ordering, say as in Eq. (3.39) or in **Exercise** 3.6, in order to extract a finite expectation value for the number operator when the many-body ground state, the filled Fermi sea say, is built out of infinitely many single-particle states. There are at least two ways to proceed.

[38] Because of assumptions (3.243b) and (3.243c), this functional determinant is nonvanishing, although it might not be finite.

3.9.3 Brute-Force Gradient Expansion

First, we may opt to write [from now on, we are using the shorthand notation introduced in Eq. (3.261)]

$$\begin{aligned}\langle j^{\mu}(\mathrm{x})\rangle &= -\mathrm{i}\,\frac{\delta \log \mathrm{Det}\left[\mathrm{i}\partial\!\!\!/ + A\!\!\!/ - \Phi\right]}{\delta A_{\mu}(\mathrm{x})}\Bigg|_{A\!\!\!/=0} \\ &= -\mathrm{i}\,\frac{\delta \operatorname{Tr} \log\left[\mathrm{i}\partial\!\!\!/ + A\!\!\!/ - \Phi\right]}{\delta A_{\mu}(\mathrm{x})}\Bigg|_{A\!\!\!/=0} \\ &= -\mathrm{i}\left\langle \mathrm{x}\left|\operatorname{tr}\left(\gamma^{\mu}\,\frac{1}{\mathrm{i}\partial\!\!\!/ - \Phi}\right)\right|\mathrm{x}\right\rangle. \end{aligned} \tag{3.338}$$

The (functional) trace Tr is done over the functional space over which the kernel $\mathrm{i}\partial\!\!\!/ + A\!\!\!/ - \Phi$ is defined. The ket $|\mathrm{x}\rangle$ is an instantaneous eigenstate of the position operator with eigenvalue x^1 at time x^0 that is selected by the functional differentiation with respect to $A_{\mu}(\mathrm{x}^0,\mathrm{x}^1)$. The trace tr is done over the Clifford algebra. If we make the assumption that the function

$$\Phi^{\dagger}(\mathrm{x})\,\Phi(\mathrm{x}) = |\phi(\mathrm{x})|^2\,\tau_0 \tag{3.339a}$$

is independent of space and time, that is,

$$|\phi(\mathrm{x})|^2 = m^2 \tag{3.339b}$$

for all $\mathrm{x} \in \mathbb{M}_{1,1}$, then the gradient expansion follows from taking advantage of the identities

$$\left(\mathrm{i}\partial\!\!\!/ + \Phi^{\dagger}\right)\,\left(\mathrm{i}\partial\!\!\!/ - \Phi\right) = \left(-\partial^2 - m^2\right)\,\tau_0 - \widehat{\delta\Phi}, \tag{3.340a}$$

where

$$\widehat{\delta\Phi} := \mathrm{i}(\partial\!\!\!/\Phi) + \mathrm{i}\left(\Phi - \Phi^{\dagger}\right)\,\partial\!\!\!/, \tag{3.340b}$$

is a differential operator and

$$\begin{aligned}\left(\mathrm{i}\partial\!\!\!/ - \Phi\right)^{-1} &= \left[\left(\mathrm{i}\partial\!\!\!/ + \Phi^{\dagger}\right)^{-1}\,\left(\mathrm{i}\partial\!\!\!/ + \Phi^{\dagger}\right)\,\left(\mathrm{i}\partial\!\!\!/ - \Phi\right)\right]^{-1} \\ &= \left[\left(-\partial^2 - m^2\right)\,\tau_0 - \widehat{\delta\Phi}\right]^{-1}\,\left(\mathrm{i}\partial\!\!\!/ + \Phi^{\dagger}\right) \\ &= \frac{\tau_0}{\tau_0 - (-\partial^2 - m^2)^{-1}\,\widehat{\delta\Phi}}\,\frac{\tau_0}{(-\partial^2 - m^2)}\,\left(\mathrm{i}\partial\!\!\!/ + \Phi^{\dagger}\right) \\ &= \sum_{n=0}^{\infty}\left[\frac{\tau_0}{(-\partial^2 - m^2)}\,\widehat{\delta\Phi}\right]^{n}\,\frac{\tau_0}{(-\partial^2 - m^2)}\,\left(\mathrm{i}\partial\!\!\!/ + \Phi^{\dagger}\right). \end{aligned} \tag{3.341}$$

It is (**Exercise** 3.18)

$$\begin{aligned}\langle j^\mu(\mathrm{x})\rangle = -\,\mathrm{i}\sum_{n=0}^{\infty}\Big\langle \mathrm{x}\Big|\mathrm{tr}\,\Big\{\gamma^\mu\left[\frac{\tau_0}{-\partial^2-m^2}\,\widehat{\delta\Phi}\right]^n \times \\ \frac{\tau_0}{-\partial^2-m^2}\left(\mathrm{i}\partial\!\!\!/+\Phi^\dagger\right)\Big\}\Big|\mathrm{x}\Big\rangle.\end{aligned} \qquad (3.342)$$

If we relax the condition (3.339) by replacing it with Eqs. (3.243b) and (3.243c),[39] then the gradient expansion is to be performed after point splitting the current in terms of left and right propagators as was done in Eq. (3.268).

3.9.4 Functional Bosonization

Second, we may opt to compute the partition function $Z[\Phi, A\!\!\!/]$ exactly, if possible, or within a gradient expansion otherwise. Either ways, this computation defines functional bosonization. When computing $Z[\Phi, A\!\!\!/]$, a contribution from $\Phi = \phi_1 + \mathrm{i}\gamma_5\,\phi_2$ and $A\!\!\!/ = \gamma^\mu\,A_\mu$ from the Clifford algebra that has decoupled from the fermionic fields $\bar\psi$ and ψ will be shown to arise. This contribution is called the chiral anomaly, for it stems from the Jacobian acquired by the Grassmann measure under the transformation (3.254). It is this contribution that delivers the leading term in the gradient expansion (3.342). This contribution is topological in that it only depends on the asymptotic behavior of Φ when $A\!\!\!/ = 0$ at the space-like boundary $\mathrm{x} \equiv (\mathrm{x}^0, \pm\infty)$ of Minkowski space and not on the detailed profile of Φ as a function of x^1. The chiral anomaly can be computed exactly for $Z[\Phi, A\!\!\!/]$ defined in Eq. (3.335) because of the peculiarities of Minkowski space $\mathbb{M}_{1,1}$ compared to the higher-dimensional Minkowski spaces $\mathbb{M}_{1,d}$ with an odd dimensionality d of space.

The polar decomposition of $\Phi(\mathrm{x})$ is

$$\begin{aligned}\Phi(\mathrm{x}) &= |\phi(\mathrm{x})|\,e^{+\mathrm{i}\gamma_5\,\varphi(\mathrm{x})},\\ |\phi(\mathrm{x})| &:= \sqrt{\phi_1^2(\mathrm{x})+\phi_2^2(\mathrm{x})},\\ \varphi(\mathrm{x}) &:= \arctan\left(\frac{\phi_2(\mathrm{x})}{\phi_1(\mathrm{x})}\right).\end{aligned} \qquad (3.343)$$

We perform the local unitary transformation over the Grassmann integration variables

$$\bar\psi(\mathrm{x}) =: \bar\chi(\mathrm{x})\,e^{-\mathrm{i}\gamma_5\,\varphi(\mathrm{x})/2}, \qquad \psi(\mathrm{x}) =: e^{-\mathrm{i}\gamma_5\,\varphi(\mathrm{x})/2}\,\chi(\mathrm{x}). \qquad (3.344)$$

Under this local axial rotation,[40]

$$\left(\bar\psi\,\psi\right)(\mathrm{x}) = \left(\bar\chi\,e^{-\mathrm{i}\gamma_5\,\varphi}\,\chi\right)(\mathrm{x}), \qquad (3.345)$$

[39] We still demand that (3.243c) holds in order to prevent any infra-red divergence.

[40] The terminology axial stems from the fact that this unitary transformation on the Clifford algebra is generated by

$$\gamma_5 \equiv \gamma_0\,\gamma_1 = -\gamma^0\,\gamma^1 \equiv -\gamma^5.$$

while

$$\bar{\psi}\left(\mathrm{i}\partial\!\!\!/ + A\!\!\!/ - \Phi\right)\psi = \bar{\chi}\left(\mathrm{i}\partial\!\!\!/ + A\!\!\!/ + a\!\!\!/ - |\phi|\,\tau_0\right)\chi, \tag{3.346a}$$

where

$$\begin{aligned} a\!\!\!/(\mathrm{x}) :=& \frac{1}{2}\gamma^{\mu}\,\gamma_5\,(\partial_{\mu}\,\varphi)\,(\mathrm{x}) \\ =& \gamma^0\,(-1)\,\frac{1}{2}\,(\partial_1\,\varphi)\,(\mathrm{x}) + \gamma^1\,(-1)\,\frac{1}{2}\,(\partial_0\,\varphi)\,(\mathrm{x}). \end{aligned} \tag{3.346b}$$

Equation (3.345) implies that a mass term is not invariant under the axial gauge transformation (3.344). Equation (3.346) implies that (i) the two-current is invariant under the axial gauge transformation (3.344) and (ii) the covariant derivative changes by the transformation law $A\!\!\!/ \mapsto A\!\!\!/ + a\!\!\!/$, whereby the induced gauge field $a\!\!\!/$ from the Clifford algebra has the covariant components [compare with Eq. (3.257a)]

$$a_0(\mathrm{x}) := -\frac{1}{2}\,(\partial_1\,\varphi)\,(\mathrm{x}), \qquad a_1(\mathrm{x}) := -\frac{1}{2}\,(\partial_0\,\varphi)\,(\mathrm{x}). \tag{3.347}$$

If we assume that the transformation law of the Grassmann measure is (i) invariant under a local gauge transformation, (ii) while it transforms under the local axial rotation (3.344) as

$$\mathcal{D}[\bar{\psi}]\,\mathcal{D}[\psi] = \mathcal{D}[\bar{\chi}]\,\mathcal{D}[\chi]\,e^{+\log\,J[\Phi,A\!\!\!/]}, \tag{3.348}$$

we conclude that

$$Z[\Phi,A\!\!\!/] = e^{\log\,J[\Phi,A\!\!\!/]+\log\,\mathrm{Det}\,\left[-\mathrm{i}\left(\mathrm{i}\partial\!\!\!/+A\!\!\!/+a\!\!\!/-|\phi|\,\tau_0\right)\right]}, \tag{3.349}$$

and

$$\mathrm{i}\,\langle j^{\mu}(\mathrm{x})\rangle = \left.\frac{\delta\,\log\,J[\Phi,A\!\!\!/]}{\delta\,A_{\mu}(\mathrm{x})}\right|_{A\!\!\!/=0} + \left.\frac{\delta\,\log\,\mathrm{Det}\,\left[\mathrm{i}\partial\!\!\!/ + A\!\!\!/ + a\!\!\!/ - |\phi|\,\tau_0\right]}{\delta\,A_{\mu}(\mathrm{x})}\right|_{A\!\!\!/=0}. \tag{3.350}$$

The second term on the right-hand side of Eq. (3.350) is rewritten

$$\frac{\delta\,\log\,\mathrm{Det}\,\left[\mathrm{i}\partial\!\!\!/ + a\!\!\!/ - |\phi|\,\tau_0\right]}{\delta\,a_{\mu}(\mathrm{x})} = \left\langle \mathrm{x}\left|\mathrm{tr}\,\left\{\gamma^{\mu}\,\left[\mathrm{i}\partial\!\!\!/ - |\phi(\mathrm{x})|\,\tau_0 + a\!\!\!/(\mathrm{x})\right]^{-1}\right\}\right|\mathrm{x}\right\rangle. \tag{3.351}$$

Its gradient expansion is obtained from Eqs. (3.268) with the substitution

$$\Phi(\mathrm{x}) \to |\phi(\mathrm{x})|\,\tau_0 - a\!\!\!/(\mathrm{x}). \tag{3.352}$$

Its contribution to the current is vanishing to first order in the gradient expansion [remember that $a_{\mu}(\mathrm{x})$ is itself of first order in the gradient expansion] (**Exercise** 3.19). If we combine Eqs. (3.269) and (3.350) with this observation, we conclude that

$$\left.\frac{\delta\,\log\,J[\Phi,A\!\!\!/]}{\delta\,A_{\mu}(\mathrm{x})}\right|_{A\!\!\!/=0} = \frac{\mathrm{i}}{2\pi}\,\epsilon^{\mu\nu}\,(\partial_{\nu}\,\varphi)\,(\mathrm{x}) + \cdots, \qquad \mu = 0,1, \tag{3.353}$$

must hold to leading order in the gradient expansion. Indeed, it is shown in **Exercise** 3.20 that

$$\log J[\Phi, A\!\!\!/] = \frac{\mathrm{i}}{2\pi} \int \mathrm{d}^2 \mathrm{x} \Big\{ A_\mu \, \epsilon^{\mu\nu} \, \partial_\nu \, \varphi + \frac{1}{4} \, \partial_\mu \varphi \, \partial^\mu \varphi + \frac{1}{2} \, |\phi|^2 \, [\cos(2\varphi) - 1] \Big\}. \tag{3.354}$$

Equation (3.354) quantifies the chiral (axial) anomaly associated with the Dirac equation in one-dimensional space.[41] It is the chiral (axial) anomaly that is responsible for the fractionalization of the charge induced by static domain walls in polyacetylene or to the spectral flow by which a dc current transports one electric charge per adiabatic cyclic oscillation of the dimerization and staggered chemical potential in polyacetylene.

3.10 Fractionalization from Chiral Abelian Bosonization

3.10.1 Overview

To leading order in the gradient expansion, the expectation value (3.269) for the induced fermionic current depends on only *one* bosonic field $\varphi = \arctan(\phi_2/\phi_1)$, even though the single-particle Hamiltonian depends on two independent bosonic fields ϕ_1 and ϕ_2. This observation is not coincidental. It is closely related to Abelian bosonization in one-dimensional space. We are going to derive the rules for Abelian bosonization, with which help we are going to gain deeper insights into the meaning of Eq. (3.269). To this end, we begin with the study of the most general bosonic Hamiltonian that describes N chiral bosonic modes propagating in one-dimensional space. We will then specialize to the case of $N = 2$ with full relativistic symmetry and show how it relates to a theory of Dirac fermions propagating in one-dimensional space.

Abelian bosonization is attributed to Coleman,[42] Mandelstam,[43] and Luther and Peschel,[44] respectively. However, Abelian bosonization can be traced to the much older works of Kronig,[45] Tomonaga,[46] and Luttinger,[47] in which it is observed that gapless fermions in one-dimensional space are equivalent to gapless phonons. Here, we follow the more general formulation of chiral Abelian bosonization given by Haldane,[48] as it lends itself to a description of one-dimensional quantum effective field theories arising in the low-energy sector along the boundary in space of $(2+1)$-dimensional topological field theories.

[41] The two terminologies of chiral and axial anomalies can be found in the literature.
[42] S. Coleman, Phys. Rev. D **11**(8), 2088–2097 (1975). [Coleman (1975)].
[43] S. Mandelstam, Phys. Rev. D **11**(10), 3026–3030 (1975). [Mandelstam (1975)].
[44] A. Luther and I. Peschel, Phys. Rev. B **12**(9), 3908–3917 (1975). [Luther and Peschel (1975)].
[45] R. de L. Kronig, Physica **2**, 968–980 (1935). [Kronig (1935)].
[46] S. Tomonaga, Prog. Theor. Phys. **5**(4), 544–569 (1950). [Tomonaga (1950)].
[47] J. M. Luttinger, J. Math. Phys. **4**(9), 1154–1162 (1963). [Luttinger (1963)].
[48] F. D. M. Haldane, Phys. Rev. Lett. **74**(11), 2090–2093 (1995). [Haldane (1995)].

3.10.2 Definition

Define the quantum Hamiltonian (in units with the electric charge e, the characteristic speed, and $\hbar$ set to one)

$$\widehat{H} := \int\limits_0^L \mathrm{d}x \left[\frac{1}{4\pi} V_{ij} \left(D_x \hat{u}_i\right) \left(D_x \hat{u}_j\right) + A_0 \left(\frac{1}{2\pi} q_i K^{-1}_{ij} \left(D_x \hat{u}_j\right)\right)\right] (t,x), \tag{3.355a}$$

$$D_x \hat{u}_i(t,x) := \left(\partial_x \hat{u}_i + q_i A_1\right)(t,x).$$

The indices $i,j = 1,\cdots,N$ label the bosonic modes. Summation is implied over repeated indices. The N real-valued quantum fields $\hat{u}_i(t,x)$ obey the equal-time commutation relations

$$\left[\hat{u}_i(t,x), \hat{u}_j(t,y)\right] = \mathrm{i}\pi \left[K_{ij}\, \mathrm{sgn}(x-y) + L_{ij}\right] \tag{3.355b}$$

for any pair $i,j = 1,\cdots,N$. The function $\mathrm{sgn}(x) = -\mathrm{sgn}(-x)$ gives the sign of the real variable x and will be assumed to be periodic with periodicity L. The $N\times N$ matrix K is symmetric, invertible, and integer-valued. Given the pair $i,j = 1,\cdots,N$, any one of its matrix elements thus obeys

$$K_{ij} = K_{ji} \in \mathbb{Z}, \qquad K^{-1}_{ij} = K^{-1}_{ji} \in \mathbb{Q}. \tag{3.355c}$$

The $N\times N$ matrix L is antisymmetric

$$L_{ij} = -L_{ji} = \begin{cases} 0, & \text{if } i=j, \\ \mathrm{sgn}(i-j)\left(K_{ij} + q_i\, q_j\right), & \text{otherwise,} \end{cases} \tag{3.355d}$$

for $i,j = 1,\cdots,N$. The sign function $\mathrm{sgn}(i)$ of any integer i is here not made periodic and taken to vanish at the origin of $\mathbb{Z}$. The external scalar gauge potential $A_0(t,x)$ and vector gauge potential $A_1(t,x)$ are real-valued functions of the time t and space x coordinates. They are also chosen to be periodic under $x \mapsto x+L$. The $N\times N$ matrix V is symmetric and positive definite

$$V_{ij} = V_{ji} \in \mathbb{R}, \qquad v_i\, V_{ij}\, v_j > 0, \qquad i,j = 1,\cdots,N, \tag{3.355e}$$

for any nonvanishing vector $v = (v_i) \in \mathbb{R}^N$. The charges q_i are integer-valued and satisfy

$$(-1)^{K_{ii}} = (-1)^{q_i}, \qquad i = 1,\cdots,N. \tag{3.355f}$$

Finally, we shall impose the boundary conditions

$$\hat{u}_i(t,x+L) = \hat{u}_i(t,x) \bmod 2\pi, \tag{3.355g}$$

and

$$\left(\partial_x \hat{u}_i\right)(t,x+L) = \left(\partial_x \hat{u}_i\right)(t,x), \tag{3.355h}$$

for any $i = 1,\cdots,N$.

3.10.3 Chiral Equations of Motion

For any $i,j=1,\cdots,N$, one verifies with the help of the equal-time commutation relation

$$\left[\hat{u}_i(t,x), D_y\,\hat{u}_j(t,y)\right] = -2\pi\mathrm{i}\,K_{ij}\,\delta(x-y) \tag{3.356}$$

that the equations of motions are (**Exercise** 3.21)

$$\begin{aligned}\mathrm{i}\,(\partial_t\,\hat{u}_i)\,(t,x) :=& \left[\hat{u}_i(t,x), \widehat{H}\right]\\ =& -\mathrm{i}K_{ij}\,V_{jk}\,(\partial_x\,\hat{u}_k + q_k\,A_1)\,(t,x) - \mathrm{i}q_i\,A_0(t,x).\end{aligned} \tag{3.357}$$

Introduce the covariant derivatives

$$D_\mu\,\hat{u}_k := \left(\partial_\mu\,\hat{u}_k + q_k\,A_\mu\right), \qquad \partial_0\equiv\partial_t, \qquad \partial_1\equiv\partial_x, \tag{3.358}$$

for $\mu=0,1$ and $k=1,\cdots,N$. The equations of motion

$$0 = \delta_{ik}\,D_0\,\hat{u}_k + K_{ij}\,V_{jk}\,D_1\,\hat{u}_k \tag{3.359}$$

are chiral. Doing the substitutions $\hat{u}_i\mapsto\hat{v}_i$ and $K\mapsto -K$ everywhere in Eq. (3.355) delivers the chiral equations of motions

$$0 = \delta_{ik}\,D_0\,\hat{v}_k - K_{ij}\,V_{jk}\,D_1\,\hat{v}_k, \tag{3.360}$$

with the opposite chirality. Evidently, the chiral equations of motion (3.359) and (3.360) are first-order differential equations, as opposed to the Klein-Gordon equations of motion obeyed by a relativistic quantum scalar field (**Exercise** 3.22).

3.10.4 Conserved Topological Charges

We turn off the external gauge potentials

$$A_0(t,x) = A_1(t,x) = 0. \tag{3.361}$$

For any $i=1,\cdots,N$, define the operator

$$\begin{aligned}\widehat{\mathcal{N}}_i(t) :=& \frac{1}{2\pi}\int\limits_0^L \mathrm{d}x\;(\partial_x\hat{u}_i)\,(t,x)\\ =& \frac{1}{2\pi}\left[\hat{u}_i(t,L) - \hat{u}_i(t,0)\right].\end{aligned} \tag{3.362}$$

This operator is conserved if and only if

$$(\partial_x\hat{u}_i)\,(t,x) = (\partial_x\hat{u}_i)\,(t,x+L), \qquad 0\le x\le L, \tag{3.363}$$

for (**Exercise** 3.23)

$$\mathrm{i}\left(\partial_t\widehat{\mathcal{N}}_i\right)(t) = -\frac{\mathrm{i}}{2\pi}K_{ik}V_{kl}\left[(\partial_x\hat{u}_l)\,(t,L) - (\partial_x\hat{u}_l)\,(t,0)\right]. \tag{3.364}$$

Now, if we demand that there exists an $n_i \in \mathbb{Z}$ such that

$$\hat{u}_i(t, x+L) = \hat{u}_i(t,x) + 2\pi\, n_i, \tag{3.365}$$

it then follows that

$$\widehat{\mathcal{N}}_i = n_i. \tag{3.366}$$

Condition (3.365) is stronger than condition (3.355g). The latter condition implies no less and no more that the eigenvalues of $\widehat{\mathcal{N}}_i$ are integer-valued,

$$\text{spectrum}\left(\widehat{\mathcal{N}}_i\right) \subset \mathbb{Z}. \tag{3.367}$$

Next, we show that any pair $\widehat{\mathcal{N}}_i$ and $\widehat{\mathcal{N}}_j$ commutes. To this end, for any $i,j = 1,\cdots,N$, we do the brute force manipulations

$$\begin{aligned}\left[\widehat{\mathcal{N}}_i, \widehat{\mathcal{N}}_j\right] =& \frac{1}{2\pi}\int\limits_0^L \mathrm{d}y \left[\widehat{\mathcal{N}}_i, \left(\partial_y \hat{u}_j\right)(t,y)\right] \\ =& \frac{1}{2\pi}\int\limits_0^L \mathrm{d}y\, \partial_y \left[\widehat{\mathcal{N}}_i, \hat{u}_j(t,y)\right].\end{aligned} \tag{3.368}$$

However, according to **Exercise** 3.24,

$$\left[\widehat{\mathcal{N}}_i, \hat{u}_j(y)\right] = \mathrm{i}K_{ij} \tag{3.369}$$

is independent of y. Hence,

$$\left[\widehat{\mathcal{N}}_i, \widehat{\mathcal{N}}_j\right] = 0 \tag{3.370}$$

holds as was announced.

The local counterpart to the global conservation of the topological charge is

$$\partial_t\, \hat{\rho}^{\mathsf{top}}_i + \partial_x\, \hat{\jmath}^{\mathsf{top}}_i = 0, \tag{3.371a}$$

where the local topological density operator is defined by

$$\hat{\rho}^{\mathsf{top}}_i(t,x) := \frac{1}{2\pi}\left(\partial_x \hat{u}_i\right)(t,x) \tag{3.371b}$$

and the local topological current operator is defined by

$$\hat{\jmath}^{\mathsf{top}}_i(t,x) := \frac{1}{2\pi} K_{ik}\, V_{kl}\, \left(\partial_x \hat{u}_l\right)(t,x) \tag{3.371c}$$

for $i = 1,\cdots,N$. The local topological density operator obeys the equal-time algebra

$$\left[\hat{\rho}^{\mathsf{top}}_i(t,x), \hat{\rho}^{\mathsf{top}}_j(t,y)\right] = -\frac{\mathrm{i}}{2\pi} K_{ij}\, \partial_x \delta(x-y) \tag{3.372a}$$

for any $i,j=1,\cdots,N$. The local topological current operator obeys the equal-time algebra

$$\left[\hat{j}^{\mathrm{top}}_{i}(t,x),\hat{j}^{\mathrm{top}}_{j}(t,y)\right]=-\frac{\mathrm{i}}{2\pi}\left(K_{ik}\,V_{kl}\right)\left(K_{jk'}\,V_{k'l'}\right)K_{ll'}\,\partial_{x}\delta(x-y) \tag{3.372b}$$

for any $i,j=1,\cdots,N$. Finally,

$$\left[\hat{\rho}^{\mathrm{top}}_{i}(t,x),\hat{j}^{\mathrm{top}}_{j}(t,y)\right]=-\frac{\mathrm{i}}{2\pi}\left(K_{jk}\,V_{kl}\right)K_{il}\,\partial_{x}\delta(x-y) \tag{3.372c}$$

for any $i,j=1,\cdots,N$.

We also introduce the local charges and currents

$$\hat{\rho}_{i}(t,x):=K^{-1}_{ij}\,\hat{\rho}^{\mathrm{top}}_{j}(t,x) \tag{3.373a}$$

and

$$\hat{j}_{i}(t,x):=K^{-1}_{ij}\,\hat{j}^{\mathrm{top}}_{j}(t,x), \tag{3.373b}$$

respectively, for any $i=1,\cdots,N$. The continuity equation (3.371a) is unchanged under this linear transformation,

$$\partial_{t}\,\hat{\rho}_{i}+\partial_{x}\,\hat{j}_{i}=0, \tag{3.373c}$$

for any $i=1,\cdots,N$. The topological current algebra (3.372) transforms into

$$\left[\hat{\rho}_{i}(t,x),\hat{\rho}_{j}(t,y)\right]=-\frac{\mathrm{i}}{2\pi}\,K^{-1}_{ij}\,\partial_{x}\delta(x-y), \tag{3.374a}$$

$$\left[\hat{j}_{i}(t,x),\hat{j}_{j}(t,y)\right]=-\frac{\mathrm{i}}{2\pi}\,V_{ik}\,V_{jl}\,K_{kl}\,\partial_{x}\delta(x-y), \tag{3.374b}$$

$$\left[\hat{\rho}_{i}(t,x),\hat{j}_{j}(t,y)\right]=-\frac{\mathrm{i}}{2\pi}\,V_{ij}\,\partial_{x}\delta(x-y), \tag{3.374c}$$

for any $i,j=1,\cdots,N$.

At last, if we contract the continuity equation (3.373c) with the integer-valued charge vector, we obtain the flavor-global continuity equation [compare with Eq. (3.772)]

$$\partial_{t}\,\hat{\rho}+\partial_{x}\,\hat{j}=0, \tag{3.375a}$$

where the local flavor-global charge operator is [compare with Eq. (3.771a)]

$$\hat{\rho}(t,x):=q_{i}\,K^{-1}_{ij}\,\hat{\rho}^{\mathrm{top}}_{j}(t,x) \tag{3.375b}$$

and the local flavor-global current operator is [compare with Eq. (3.771b)]

$$\hat{j}(t,x):=q_{i}\,K^{-1}_{ij}\,\hat{j}^{\mathrm{top}}_{j}(t,x). \tag{3.375c}$$

The flavor-resolved current algebra (3.374) turns into the flavor-global current algebra

$$[\hat{\rho}(t,x),\hat{\rho}(t,y)] = -\frac{\mathrm{i}}{2\pi}\left(q_i\,K^{-1}_{ij}\,q_j\right)\,\partial_x\delta(x-y), \tag{3.376a}$$

$$\left[\hat{j}(t,x),\hat{j}(t,y)\right] = -\frac{\mathrm{i}}{2\pi}\left(q_i\,V_{ik}\,K_{kl}\,V_{lj}\,q_j\right)\,\partial_x\delta(x-y), \tag{3.376b}$$

$$\left[\hat{\rho}(t,x),\hat{j}(t,y)\right] = -\frac{\mathrm{i}}{2\pi}\left(q_i\,V_{ij}\,q_j\right)\,\partial_x\delta(x-y). \tag{3.376c}$$

3.10.5 Quasi-Particle and Particle Excitations

When Eq. (3.361) holds, there exist N conserved global topological (that is integer-valued) charges $\widehat{\mathcal{N}}_i$ with $i=1,\cdots,N$ defined in Eq. (3.362) that commute pairwise. Define the N global charges

$$\widehat{Q}_i := \int\limits_0^L \mathrm{d}x\,\hat{\rho}_i(t,x) = K^{-1}_{ij}\,\widehat{\mathcal{N}}_j, \qquad i=1,\cdots,N. \tag{3.377}$$

We shall shortly interpret these charges as the elementary Fermi-Bose charges.

For any $i=1,\cdots,N$, define the pair of vertex operators

$$\widehat{\Psi}^{\dagger}_{\text{q-p},i}(t,x) := e^{-\mathrm{i}K^{-1}_{ij}\,\hat{u}_j(t,x)} \tag{3.378a}$$

and

$$\widehat{\Psi}^{\dagger}_{\text{f-b},i}(t,x) := e^{-\mathrm{i}\delta_{ij}\,\hat{u}_j(t,x)}, \tag{3.378b}$$

respectively. The quasi-particle vertex operator $\widehat{\Psi}^{\dagger}_{\text{q-p},i}(t,x)$ can be multi-valued under a shift by 2π of all $\hat{u}_j(t,x)$ where $j=1,\cdots,N$. The Fermi-Bose vertex operator $\widehat{\Psi}^{\dagger}_{\text{f-b},i}(t,x)$ is always single-valued under a shift by 2π of all $\hat{u}_j(t,x)$, where $j=1,\cdots,N$.

For any pair $i,j=1,\cdots,N$, the commutator (3.369) delivers the identities

$$\left[\widehat{\mathcal{N}}_i,\widehat{\Psi}^{\dagger}_{\text{q-p},j}(t,x)\right] = \delta_{ij}\,\widehat{\Psi}^{\dagger}_{\text{q-p},j}(t,x), \quad \left[\widehat{\mathcal{N}}_i,\widehat{\Psi}^{\dagger}_{\text{f-b},j}(t,x)\right] = K_{ij}\,\widehat{\Psi}^{\dagger}_{\text{f-b},j}(t,x), \tag{3.379}$$

and

$$\left[\widehat{Q}_i,\widehat{\Psi}^{\dagger}_{\text{q-p},j}(t,x)\right] = K^{-1}_{ij}\,\widehat{\Psi}^{\dagger}_{\text{q-p},j}(t,x), \quad \left[\widehat{Q}_i,\widehat{\Psi}^{\dagger}_{\text{f-b},j}(t,x)\right] = \delta_{ij}\,\widehat{\Psi}^{\dagger}_{\text{f-b},j}(t,x), \tag{3.380}$$

respectively. The quasi-particle vertex operator $\widehat{\Psi}^{\dagger}_{\text{q-p},i}(t,x)$ is an eigenstate of the topological number operator $\widehat{\mathcal{N}}_i$ with eigenvalue 1. The Fermi-Bose vertex operator $\widehat{\Psi}^{\dagger}_{\text{f-b},i}(t,x)$ is an eigenstate of the charge number operator $\widehat{Q}_i$ with eigenvalue 1.

The Baker-Campbell-Hausdorff formula implies that

$$e^{\hat{A}}\,e^{\hat{B}} = e^{\hat{A}+\hat{B}}\,e^{+(1/2)[\hat{A},\hat{B}]} = e^{\hat{B}}\,e^{\hat{A}}\,e^{[\hat{A},\hat{B}]} \tag{3.381}$$

whenever two operators $\hat{A}$ and $\hat{B}$ have a $\mathbb{C}$-number as their commutator.

A first application of the Baker-Campbell-Hausdorff formula to any pair of quasi-particle vertex operators at equal time t but two distinct space coordinates $x \neq y$ gives (**Exercise** 3.25)

$$\widehat{\Psi}^{\dagger}_{\text{q-p},i}(t,x)\,\widehat{\Psi}^{\dagger}_{\text{q-p},j}(t,y) = e^{-\mathrm{i}\pi\,\Theta^{\text{q-p}}_{ij}}\,\widehat{\Psi}^{\dagger}_{\text{q-p},j}(t,y)\,\widehat{\Psi}^{\dagger}_{\text{q-p},i}(t,x), \tag{3.382a}$$

where

$$\Theta^{\text{q-p}}_{ij} := K^{-1}_{ji}\,\mathrm{sgn}(x-y) + \left(K^{-1}_{ik}\,K^{-1}_{jl}\,K_{kl} + q_k\,K^{-1}_{ik}\,K^{-1}_{jl}\,q_l\right)\mathrm{sgn}(k-l). \tag{3.382b}$$

Here, it is understood that

$$\mathrm{sgn}(k-l) = 0 \tag{3.383}$$

when $k = l = 1,\cdots,N$. Hence, the quasi-particle vertex operators need not obey either bosonic or fermionic statistics since $K^{-1}_{ij} \in \mathbb{Q}$.

The same exercise applied to the Fermi-Bose vertex operators yields (**Exercise** 3.26)

$$\widehat{\Psi}^{\dagger}_{\text{f-b},i}(t,x)\,\widehat{\Psi}^{\dagger}_{\text{f-b},j}(t,y) = \begin{cases} (-1)^{K_{ii}}\,\widehat{\Psi}^{\dagger}_{\text{f-b},i}(t,y)\,\widehat{\Psi}^{\dagger}_{\text{f-b},i}(t,x), & \text{if } i = j, \\ \\ (-1)^{q_i\,q_j}\,\widehat{\Psi}^{\dagger}_{\text{f-b},j}(t,y)\,\widehat{\Psi}^{\dagger}_{\text{f-b},i}(t,x), & \text{if } i \neq j, \end{cases} \tag{3.384}$$

when $x \neq y$. The self statistics of the Fermi-Bose vertex operators is carried by the diagonal matrix elements $K_{ii} \in \mathbb{Z}$. The mutual statistics of any pair of Fermi-Bose vertex operators labeled by $i \neq j$ is carried by the product $q_i\,q_j \in \mathbb{Z}$ of the integer-valued charges q_i and q_j. Had we not assumed that K_{ij} with $i \neq j$ are integers, the mutual statistics would not be Fermi-Bose because of the nonlocal term $K_{ij}\mathrm{sgn}\,(x-y)$.

A third application of the Baker-Campbell-Hausdorff formula allows to determine the boundary conditions (**Exercise** 3.27)

$$\widehat{\Psi}^{\dagger}_{\text{q-p},i}(t,x+L) = \widehat{\Psi}^{\dagger}_{\text{q-p},i}(t,x)\,e^{-2\pi\mathrm{i}\,K^{-1}_{ij}\widehat{\mathcal{N}}_j}\,e^{-\pi\mathrm{i}\,K^{-1}_{ii}} \tag{3.385}$$

and

$$\widehat{\Psi}^{\dagger}_{\text{f-b},i}(t,x+L) = \widehat{\Psi}^{\dagger}_{\text{f-b},i}(t,x)\,e^{-2\pi\mathrm{i}\widehat{\mathcal{N}}_i}\,e^{-\pi\mathrm{i}\,K_{ii}} \tag{3.386}$$

obeyed by the quasi-particle and Fermi-Bose vertex operators, respectively.

We close this discussion with the following definitions. Introduce the operators

$$\widehat{Q} := q_i\,\widehat{Q}_i, \qquad \widehat{\Psi}^{\dagger}_{\text{q-p},\boldsymbol{m}} := e^{-\mathrm{i}m_i K^{-1}_{ij}\hat{u}_j(t,x)}, \qquad \widehat{\Psi}^{\dagger}_{\text{f-b},\boldsymbol{m}} := e^{-\mathrm{i}m_i\delta_{ij}\hat{u}_j(t,x)}, \tag{3.387}$$

where $\boldsymbol{m} \in \mathbb{Z}^N$ is the vector with the integer-valued components m_i for any $i = 1,\cdots,N$. The N charges q_i with $i = 1,\cdots,N$ that enter Hamiltonian (3.355a) can also be viewed as the components of the vector $\boldsymbol{q} \in \mathbb{Z}^N$. Define the functions

$$\begin{aligned} q\colon \mathbb{Z}^N &\to \mathbb{Z}, \\ \boldsymbol{m} &\mapsto q(\boldsymbol{m}) := q_i\,m_i \equiv \boldsymbol{q}\cdot\boldsymbol{m}, \end{aligned} \tag{3.388a}$$

and

$$\begin{aligned} K\colon \mathbb{Z}^N &\to \mathbb{Z}, \\ \boldsymbol{m} &\mapsto K(\boldsymbol{m}) := m_i\, K_{ij}\, m_j. \end{aligned} \tag{3.388b}$$

On the one hand, for any distinct pair of space coordinates $x \neq y$, we deduce from Eqs. (3.380), (3.382), and (3.385) that

$$\left[\widehat{Q}, \widehat{\Psi}^{\dagger}_{\text{q-p},\boldsymbol{m}}(t,x)\right] = \left(q_i\, K^{-1}_{ij}\, m_j\right) \widehat{\Psi}^{\dagger}_{\text{q-p},\boldsymbol{m}}(t,x), \tag{3.389a}$$

$$\widehat{\Psi}^{\dagger}_{\text{q-p},\boldsymbol{m}}(t,x)\, \widehat{\Psi}^{\dagger}_{\text{q-p},\boldsymbol{n}}(t,y) = e^{-\mathrm{i}\pi\, m_i\, \Theta^{\text{q-p}}_{ij}\, n_j}\, \widehat{\Psi}^{\dagger}_{\text{q-p},\boldsymbol{n}}(t,y)\, \widehat{\Psi}^{\dagger}_{\text{q-p},\boldsymbol{m}}(t,x), \tag{3.389b}$$

$$\widehat{\Psi}^{\dagger}_{\text{q-p},\boldsymbol{m}}(t,x+L) = \widehat{\Psi}^{\dagger}_{\text{q-p},\boldsymbol{m}}(t,x)\, e^{-2\pi\mathrm{i}\, m_i\, K^{-1}_{ij}\, \widehat{\mathcal{N}}_j}\, e^{-\pi\mathrm{i}\, m_i\, K^{-1}_{ij}\, m_j}, \tag{3.389c}$$

respectively. On the other hand, for any distinct pair of space coordinate $x \neq y$, we deduce from Eqs. (3.380), (3.384), and (3.386) that

$$\left[\widehat{Q}, \widehat{\Psi}^{\dagger}_{\text{f-b},\boldsymbol{m}}(t,x)\right] = q(\boldsymbol{m})\, \widehat{\Psi}^{\dagger}_{\text{f-b},\boldsymbol{m}}(t,x), \tag{3.390a}$$

$$\widehat{\Psi}^{\dagger}_{\text{f-b},\boldsymbol{m}}(t,x)\, \widehat{\Psi}^{\dagger}_{\text{f-b},\boldsymbol{n}}(t,y) = e^{-\mathrm{i}\pi\, m_i\, \Theta^{\text{f-b}}_{ij}\, n_j}\, \widehat{\Psi}^{\dagger}_{\text{f-b},\boldsymbol{n}}(t,y)\, \widehat{\Psi}^{\dagger}_{\text{f-b},\boldsymbol{m}}(t,x), \tag{3.390b}$$

$$\widehat{\Psi}^{\dagger}_{\text{f-b},\boldsymbol{m}}(t,x+L) = \widehat{\Psi}^{\dagger}_{\text{f-b},\boldsymbol{m}}(t,x)\, e^{-2\pi\mathrm{i}\, m_i \widehat{\mathcal{N}}_i}\, e^{-\pi\mathrm{i}\, m_i K_{ij} m_j}, \tag{3.390c}$$

respectively, where

$$\Theta^{\text{f-b}}_{ij} := K_{ij}\, \operatorname{sgn}(x-y) + \left(K_{ij} + q_i\, q_j\right)\, \operatorname{sgn}(i-j). \tag{3.390d}$$

The integer quadratic form $K(\boldsymbol{m})$ is thus seen to dictate whether the vertex operator $\widehat{\Psi}^{\dagger}_{\text{f-b},\boldsymbol{m}}(t,x)$ realizes a fermion or a boson. The vertex operator $\widehat{\Psi}^{\dagger}_{\text{f-b},\boldsymbol{m}}(t,x)$ realizes a fermion if and only if

$$K(\boldsymbol{m}) \text{ is an odd integer} \tag{3.391}$$

or a boson if and only if

$$K(\boldsymbol{m}) \text{ is an even integer.} \tag{3.392}$$

Because of assumption (3.355f),

$$(-1)^{K(\boldsymbol{m})} = (-1)^{q(\boldsymbol{m})}. \tag{3.393}$$

Hence, the vertex operator $\widehat{\Psi}^{\dagger}_{\text{f-b},\boldsymbol{m}}(t,x)$ realizes a fermion if and only if

$$q(\boldsymbol{m}) \text{ is an odd integer} \tag{3.394}$$

or a boson if and only if

$$q(\boldsymbol{m}) \text{ is an even integer.} \tag{3.395}$$

3.10.6 Bosonization Rules

We are going to relate the theory of chiral bosons (3.355) without external gauge fields to the massless limit of the Dirac Hamiltonian (3.243a). To this end, we proceed in three steps.

Step 1: We make the choices

$$N = 2, \qquad i,j = 1,2 \equiv -,+, \tag{3.396a}$$

and

$$K := \begin{pmatrix} +1 & 0 \\ 0 & -1 \end{pmatrix}, \qquad V := \begin{pmatrix} +1 & 0 \\ 0 & +1 \end{pmatrix}, \qquad q = \begin{pmatrix} 1 \\ 1 \end{pmatrix}, \tag{3.396b}$$

in Eq. (3.355). With these choices, the free bosonic Hamiltonian on the real line is

$$\widehat{H}_{\mathrm{B}} = \int\limits_{\mathbb{R}} \mathrm{d}x \, \frac{1}{4\pi} \left[(\partial_x \hat{u}_-)^2 + (\partial_x \hat{u}_+)^2 \right], \tag{3.397a}$$

where

$$[\hat{u}_-(t,x), \hat{u}_-(t,y)] = +\mathrm{i}\pi \, \mathrm{sgn}(x-y), \tag{3.397b}$$
$$[\hat{u}_+(t,x), \hat{u}_+(t,y)] = -\mathrm{i}\pi \, \mathrm{sgn}(x-y), \tag{3.397c}$$
$$[\hat{u}_-(t,x), \hat{u}_+(t,y)] = -\mathrm{i}\pi. \tag{3.397d}$$

There follows the chiral equations of motion [recall Eq. (3.357)]

$$\partial_t \hat{u}_- = -\partial_x \hat{u}_-, \qquad \partial_t \hat{u}_+ = +\partial_x \hat{u}_+, \tag{3.398}$$

obeyed by the right-mover $\hat{u}_-$ and the left-mover $\hat{u}_+$ (**Exercise** 3.28), the current algebra [recall Eq. (3.376)]

$$\left[\hat{\rho}(t,x), \hat{\rho}(t,y)\right] = 0, \tag{3.399a}$$
$$\left[\hat{\jmath}(t,x), \hat{\jmath}(t,y)\right] = 0, \tag{3.399b}$$
$$\left[\hat{\rho}(t,x), \hat{\jmath}(t,y)\right] = -\frac{\mathrm{i}}{\pi} \partial_x \delta(x-y), \tag{3.399c}$$

obeyed by the density

$$\hat{\rho} = +\frac{1}{2\pi} \left(\partial_x \hat{u}_- - \partial_x \hat{u}_+ \right) \equiv \hat{\jmath}_- + \hat{\jmath}_+ \tag{3.399d}$$

and the current density[49]

$$\hat{\jmath} = +\frac{1}{2\pi} \left(\partial_x \hat{u}_- + \partial_x \hat{u}_+ \right) \equiv \hat{\jmath}_- - \hat{\jmath}_+, \tag{3.399e}$$

[49] Notice that the chiral equations of motion implies that $\hat{\jmath} = -\frac{1}{2\pi}\left(\partial_t \hat{u}_- - \partial_t \hat{u}_+\right)$.

and the identification of the pair of quasi-particle vertex operators [recall Eq. (3.378a)]

$$\hat{\psi}^{\dagger}_{-} := \sqrt{\frac{1}{4\pi\,\mathfrak{a}}}\, e^{-\mathrm{i}\hat{u}_{-}}, \qquad \hat{\psi}^{\dagger}_{+} := \sqrt{\frac{1}{4\pi\,\mathfrak{a}}}\, e^{+\mathrm{i}\hat{u}_{+}}, \tag{3.400}$$

with a pair of creation operators for fermions. The multiplicative prefactor $1/\sqrt{4\pi}$ is a matter of convention and the constant $\mathfrak{a}$ carries the dimension of length, that is, the fermion fields carries the dimension of $1/\sqrt{\text{length}}$. By construction, the chiral currents

$$\hat{j}_{-} := +\frac{1}{2\pi}\,\partial_x\,\hat{u}_{-}, \qquad \hat{j}_{+} := -\frac{1}{2\pi}\,\partial_x\,\hat{u}_{+} \tag{3.401}$$

obey the chiral equations of motion

$$\partial_t\,\hat{j}_{-} = -\partial_x\,\hat{j}_{-}, \qquad \partial_t\,\hat{j}_{+} = +\partial_x\,\hat{j}_{+}, \tag{3.402}$$

that is they depend solely on $(t-x)$ and $(t+x)$, respectively. As with the chiral fields $\hat{u}_{-}$ and $\hat{u}_{+}$, the chiral currents $\hat{j}_{-}$ and $\hat{j}_{+}$ are right-moving and left-moving solutions, respectively, of the Klein-Gordon equation

$$(\partial_t^2 - \partial_x^2)\, f(t,x) = (\partial_t - \partial_x)\,(\partial_t + \partial_x)\, f(t,x) = 0. \tag{3.403}$$

Step 2: We define the free Dirac Hamiltonian

$$\widehat{H}_{\mathrm{D}} := -\int\limits_{\mathbb{R}} \mathrm{d}x\, \hat{\psi}^{\dagger}_{\mathrm{D}}\,\gamma^0\,\gamma^1\,\mathrm{i}\partial_x\,\hat{\psi}_{\mathrm{D}}, \tag{3.404a}$$

where the equal-time algebra

$$\left\{\hat{\psi}_{\mathrm{D}\,\alpha}(t,x), \hat{\psi}^{\dagger}_{\mathrm{D}\,\beta}(t,y)\right\} = \delta_{\alpha\beta}\,\delta(x-y), \qquad \alpha,\beta = 1,2 \tag{3.404b}$$

between the components of the operator-valued Dirac spinors delivers the only nonvanishing equal-time anticommutators. If we define the chiral projections ($\gamma_5 \equiv -\gamma^5 \equiv -\gamma^0\gamma^1$)

$$\hat{\psi}^{\dagger}_{\mathrm{D}\mp} := \hat{\psi}^{\dagger}_{\mathrm{D}}\,\frac{1}{2}(1 \mp \gamma_5), \qquad \hat{\psi}_{\mathrm{D}\mp} := \frac{1}{2}(1 \mp \gamma_5)\,\hat{\psi}_{\mathrm{D}}, \tag{3.405a}$$

there follows the chiral equations of motion (**Exercise** 3.29)

$$\partial_t\,\hat{\psi}_{\mathrm{D}-} = -\partial_x\,\hat{\psi}_{\mathrm{D}-}, \qquad \partial_t\,\hat{\psi}_{\mathrm{D}+} = +\partial_x\,\hat{\psi}_{\mathrm{D}+}. \tag{3.405b}$$

The annihilation operator $\hat{\psi}_{\mathrm{D}-}$ removes a right-moving fermion. The annihilation operator $\hat{\psi}_{\mathrm{D}+}$ removes a left-moving fermion. Moreover, the Lagrangian density

$$\widehat{\mathcal{L}}_{\mathrm{D}} := \hat{\psi}^{\dagger}_{\mathrm{D}}\,\gamma^0\,\mathrm{i}\gamma^{\mu}\,\partial_{\mu}\,\hat{\psi}_{\mathrm{D}} \tag{3.406}$$

obeys the additive decomposition

$$\widehat{\mathcal{L}}_{\mathrm{D}} = \hat{\psi}^{\dagger}_{\mathrm{D}-}\,\mathrm{i}(\partial_0 + \partial_1)\,\hat{\psi}_{\mathrm{D}-} + \hat{\psi}^{\dagger}_{\mathrm{D}+}\,\mathrm{i}(\partial_0 - \partial_1)\,\hat{\psi}_{\mathrm{D}+} \tag{3.407}$$

with the two independent chiral currents

$$\hat{j}_{\mathrm{D}-} := \hat{\psi}^{\dagger}_{\mathrm{D}-}\,\hat{\psi}_{\mathrm{D}-}, \qquad \hat{j}_{\mathrm{D}+} := \hat{\psi}^{\dagger}_{\mathrm{D}+}\,\hat{\psi}_{\mathrm{D}+}, \tag{3.408a}$$

obeying the independent conservation laws (**Exercise** 3.30)

$$\partial_t\,\hat{j}_{\mathrm{D}-} = -\partial_x\,\hat{j}_{\mathrm{D}-}, \qquad \partial_t\,\hat{j}_{\mathrm{D}+} = +\partial_x\,\hat{j}_{\mathrm{D}+}. \tag{3.408b}$$

Finally, it can be shown that if the chiral currents are normal ordered with respect to the filled Fermi sea with a vanishing chemical potential, then the only nonvanishing equal-time commutators are (**Exercise** 3.31)

$$\left[\hat{j}_{\mathrm{D}-}(t,x),\hat{j}_{\mathrm{D}-}(t,y)\right] = -\frac{\mathrm{i}}{2\pi}\,\partial_x\,\delta(x-y), \tag{3.409a}$$

and

$$\left[\hat{j}_{\mathrm{D}+}(t,x),\hat{j}_{\mathrm{D}+}(t,y)\right] = +\frac{\mathrm{i}}{2\pi}\,\partial_x\,\delta(x-y). \tag{3.409b}$$

Step 3: The Dirac chiral current algebra (3.409) is equivalent to the bosonic chiral current algebra (3.399). This equivalence is interpreted as the fact that (i) the bosonic theory (3.397) is equivalent to the Dirac theory (3.404), and (ii) there is a one-to-one correspondence between the following operators acting on their respective Fock spaces. To establish this one-to-one correspondence, we introduce the pair of bosonic fields

$$\hat{\phi}(\mathrm{x}^0,\mathrm{x}^1) := \hat{u}_-(\mathrm{x}^0-\mathrm{x}^1) + \hat{u}_+(\mathrm{x}^0+\mathrm{x}^1), \tag{3.410a}$$

$$\hat{\theta}(\mathrm{x}^0,\mathrm{x}^1) := \hat{u}_-(\mathrm{x}^0-\mathrm{x}^1) - \hat{u}_+(\mathrm{x}^0+\mathrm{x}^1). \tag{3.410b}$$

Now, the relevant one-to-one correspondence between operators in the Dirac theory for fermions and operators in the chiral bosonic theory is given in **Table** 3.1.

3.10.7 Applications to Polyacetylene

Consider the many-body Dirac Hamiltonian

$$\widehat{H}_{\mathrm{D}} := \widehat{H}_{\mathrm{D}\,0} + \widehat{H}_{\mathrm{D}\,1}, \tag{3.411a}$$

where the massless contribution is

$$\widehat{H}_{\mathrm{D}\,0} := \int\limits_{\mathbb{R}} \mathrm{d}x\,\left(\hat{\psi}^{\dagger}_{\mathrm{D}+}\,\mathrm{i}\partial_x\,\hat{\psi}_{\mathrm{D}+} - \hat{\psi}^{\dagger}_{\mathrm{D}-}\,\mathrm{i}\partial_x\,\hat{\psi}_{\mathrm{D}-}\right), \tag{3.411b}$$

while the mass contribution is

$$\widehat{H}_{\mathrm{D}\,1} := \int\limits_{\mathbb{R}} \mathrm{d}x\,\left[\phi_1\left(\hat{\psi}^{\dagger}_{\mathrm{D}-}\,\hat{\psi}_{\mathrm{D}+} + \hat{\psi}^{\dagger}_{\mathrm{D}+}\,\hat{\psi}_{\mathrm{D}-}\right) + \mathrm{i}\phi_2\left(\hat{\psi}^{\dagger}_{\mathrm{D}-}\,\hat{\psi}_{\mathrm{D}+} - \hat{\psi}^{\dagger}_{\mathrm{D}+}\,\hat{\psi}_{\mathrm{D}-}\right)\right]. \tag{3.411c}$$

Table 3.1 *Abelian bosonization rules*

	Fermions	Bosons
Free gapless Lagrangian density	$\hat{\bar{\psi}}_{\mathrm{D}}\,\mathrm{i}\gamma^{\mu}\,\partial_{\mu}\,\hat{\psi}_{\mathrm{D}}$	$\frac{1}{8\pi}\,(\partial^{\mu}\,\hat{\phi})(\partial_{\mu}\,\hat{\phi})$
Current	$\hat{\bar{\psi}}_{\mathrm{D}}\,\gamma^{\mu}\,\hat{\psi}_{\mathrm{D}}$	$\frac{1}{2\pi}\,\epsilon^{\mu\nu}\,\partial_{\nu}\,\hat{\phi}$
Chiral currents	$\hat{\psi}^{\dagger}_{\mathrm{D}\mp}\,\hat{\psi}_{\mathrm{D}\mp}$	$\pm\frac{1}{2\pi}\partial_{x}\,\hat{u}_{\mp}$
Right and left movers	$\hat{\psi}^{\dagger}_{\mathrm{D}\mp}$	$\sqrt{\frac{1}{4\pi\,a}}\,e^{\mp\mathrm{i}\hat{u}_{\mp}}$
Backward scattering	$\hat{\psi}^{\dagger}_{\mathrm{D}-}\,\hat{\psi}_{\mathrm{D}+}$	$\frac{1}{4\pi\,a}\,e^{-\mathrm{i}\hat{\phi}}$
Cooper pairing	$\hat{\psi}^{\dagger}_{\mathrm{D}-}\,\hat{\psi}^{\dagger}_{\mathrm{D}+}$	$\frac{1}{4\pi\,a}\,e^{-\mathrm{i}\hat{\theta}}$
Scalar mass	$\hat{\psi}^{\dagger}_{\mathrm{D}-}\,\hat{\psi}_{\mathrm{D}+}+\hat{\psi}^{\dagger}_{\mathrm{D}+}\,\hat{\psi}_{\mathrm{D}-}$	$\frac{1}{2\pi\,a}\,\cos\hat{\phi}$
Pseudo-scalar mass	$\hat{\psi}^{\dagger}_{\mathrm{D}-}\,\hat{\psi}_{\mathrm{D}+}-\hat{\psi}^{\dagger}_{\mathrm{D}+}\,\hat{\psi}_{\mathrm{D}-}$	$\frac{-\mathrm{i}}{2\pi\,a}\,\sin\hat{\phi}$

Abelian bosonization rules in two-dimensional Minkowski space. The conventions of relevance to the scalar mass $\hat{\bar{\psi}}\,\hat{\psi}$ and the pseudo-scalar mass $\hat{\bar{\psi}}_{\mathrm{D}}\,\gamma_{5}\,\hat{\psi}_{\mathrm{D}}$ are $\hat{\bar{\psi}}_{\mathrm{D}}=\hat{\psi}^{\dagger}_{\mathrm{D}}\,\gamma^{0}$ with $\hat{\psi}_{\mathrm{D}}$ the operator-valued Dirac spinor adjoint to $\hat{\psi}^{\dagger}_{\mathrm{D}}=(\hat{\psi}^{\dagger}_{\mathrm{D}-},\hat{\psi}^{\dagger}_{\mathrm{D}+})$, whereby $\gamma^{0}=\tau_{1}$ and $\gamma^{1}=-\mathrm{i}\tau_{2}$ so that $\gamma^{5}=-\gamma_{5}=+\gamma^{0}\,\gamma^{1}=\tau_{3}$.

The only nonvanishing equal-time anticommutators are given by Eq. (3.404b).[50]

According to the bosonization rules from **Table** 3.1 and with the help of the polar decomposition

$$\phi_{1}(t,x)=|\boldsymbol{\phi}(t,x)|\,\cos\varphi(t,x),\quad \phi_{2}(t,x)=|\boldsymbol{\phi}(t,x)|\,\sin\varphi(t,x), \tag{3.412}$$

the many-body bosonic Hamiltonian that is equivalent to the many-body Dirac Hamiltonian (3.411) is

$$\widehat{H}_{\mathrm{B}}:=\widehat{H}_{\mathrm{B}\,0}+\widehat{H}_{\mathrm{B}\,1}, \tag{3.413a}$$

where

$$\widehat{H}_{\mathrm{B}\,0}:=\int\limits_{\mathbb{R}}\mathrm{d}x\,\frac{1}{8\pi}\left[\widehat{\Pi}^{2}+\left(\partial_{x}\hat{\phi}\right)^{2}\right], \tag{3.413b}$$

[50] Observe that the representation of the single-particle Dirac Hamiltonian entering Eq. (3.411) follows from performing the unitary transformation $\tau_2\to+\tau_3$, $\tau_1\to+\tau_1$, $\tau_3\to-\tau_2$ on the single-particle Hamiltonian (3.243a).

while

$$\widehat{H}_{\mathrm{B}\,1} := \int_{\mathbb{R}} \mathrm{d}x\, \frac{1}{2\pi\,\mathfrak{a}}\, |\phi|\, \cos\left(\hat{\phi} - \varphi\right). \tag{3.413c}$$

Here, the canonical momentum

$$\widehat{\Pi}(t,x) := \left(\partial_t\,\hat{\phi}\right)(t,x) \tag{3.414a}$$

shares with $\hat{\phi}(t,x)$ the only nonvanishing equal-time commutator

$$\left[\hat{\phi}(t,x), \widehat{\Pi}(t,y)\right] = 4\pi\,\mathrm{i}\delta(x-y). \tag{3.414b}$$

Hamiltonian (3.413) is interacting, and its interaction (3.413c) can be traced to the mass contributions in the noninteracting Dirac Hamiltonian (3.411). The interaction (3.413c) is minimized when the operator identity

$$\hat{\phi}(t,x) = \varphi(t,x) + \pi \tag{3.415}$$

holds. This identity can only be met in the limit

$$|\phi(t,x)| \to \infty \tag{3.416}$$

for all time t and position x in view of the algebra (3.414) and the competition between the contributions (3.413b) and (3.413c).

Close to the limit (3.416), the bosonization formula for the conserved current

$$\hat{\bar{\psi}}_{\mathrm{D}}\,\gamma^{\mu}\,\hat{\psi}_{\mathrm{D}} \to \frac{1}{2\pi}\,\epsilon^{\mu\nu}\,\partial_{\nu}\hat{\phi} \tag{3.417}$$

simplifies to

$$\frac{1}{2\pi}\,\epsilon^{\mu\nu}\,\partial_{\nu}\hat{\phi} \approx \frac{1}{2\pi}\,\epsilon^{\mu\nu}\,\partial_{\nu}\varphi. \tag{3.418}$$

On the one hand, the conserved charge

$$\widehat{Q} := \int_{\mathbb{R}} \mathrm{d}x\, \left(\hat{\bar{\psi}}_{\mathrm{D}}\,\gamma^{0}\,\hat{\psi}_{\mathrm{D}}\right)(t,x) \to \frac{\epsilon^{01}}{2\pi}\,\left[\hat{\phi}(t,x=+\infty) - \hat{\phi}(t,x=-\infty)\right] \tag{3.419}$$

for the static profile $\varphi(x)$ is approximately given by

$$\widehat{Q} \approx \frac{\epsilon^{01}}{2\pi}\,\left[\varphi(x=+\infty) - \varphi(x=-\infty)\right]. \tag{3.420}$$

On the other hand, the number of electrons per period $T = 2\pi/\omega$ that flows across a point x

$$\hat{I} := \int_{0}^{T} \mathrm{d}t\, \left(\hat{\bar{\psi}}_{\mathrm{D}}\,\gamma^{1}\,\hat{\psi}_{\mathrm{D}}\right)(t,x) \to \frac{\epsilon^{10}}{2\pi}\,\left[\hat{\phi}(T,x) - \hat{\phi}(0,x)\right] \tag{3.421}$$

for the uniform profile $\varphi(t)=\omega t$ is approximately given by

$$\hat{I} \approx \frac{\epsilon^{10}}{2\pi}\,\omega\,T = \epsilon^{10}. \tag{3.422}$$

We have thus reproduced the leading-order results (3.276) and (3.280) from the gradient expansion to leading order in the semiclassical expansion about the limit (3.416). Unlike the results (3.276) and (3.280) that are expectation values, results (3.420) and (3.422) are sharp operator identities in the limit (3.416). The small parameter in both expansions is $1/m$, where $m := \lim_{x\to\infty}|\phi(t,x)|$.

3.10.8 U(1) *Gauge Anomaly: The Integer Quantum Hall Effect*

The chiral equations of motion (3.359) and (3.360) are invariant under the local U(1) gauge symmetry

$$\begin{aligned}
\hat{u}_i(t,x) &=: \hat{u}'_i(t,x) + q_i\,\chi(t,x),\\
A_0(t,x) &=: A'_0(t,x) - (\partial_t\,\chi)\,(t,x),\\
A_1(t,x) &=: A'_1(t,x) - (\partial_x\,\chi)\,(t,x),
\end{aligned} \tag{3.423a}$$

for any real-valued function χ that satisfies the periodic boundary conditions

$$\chi(t,x+L) = \chi(t,x). \tag{3.423b}$$

Functional differentiation of Hamiltonian (3.355a) with respect to the gauge potentials allows to define the two-current with the components

$$\begin{aligned}
\hat{\jmath}^{0}(t,x) :&= \frac{\delta\widehat{H}}{\delta A_0(t,x)}\\
&= \frac{1}{2\pi}\,q_i\,K^{-1}_{ij}\,\left(D_1\,\hat{u}_j\right)(t,x)
\end{aligned} \tag{3.424a}$$

and

$$\begin{aligned}
\hat{\jmath}^{1}(t,x) :&= \frac{\delta\widehat{H}}{\delta A_1(t,x)}\\
&= \frac{1}{2\pi}\,q_i\,V_{ij}\,\left(D_1\,\hat{u}_j\right)(t,x) + \frac{1}{2\pi}\,\left(q_i\,K^{-1}_{ij}\,q_j\right)A_0(t,x).
\end{aligned} \tag{3.424b}$$

We introduce the shorthand notation

$$\sigma_{\mathrm{H}} := \frac{1}{2\pi}\,\left(q_i\,K^{-1}_{ij}\,q_j\right) \in \frac{1}{2\pi}\,\mathbb{Q} \tag{3.425}$$

for the second term on the right-hand side of Eq. (3.424b). The subscript stands for Hall as we shall shortly interpret σ_{H} as a dimensionless Hall conductance.

The transformation law of the two-current (3.424) under the local gauge transformation (3.423) is

$$\hat{J}^0(t,x) = \hat{J}^{0\,\prime}(t,x) \tag{3.426a}$$

and

$$\hat{J}^1(t,x) = \hat{J}^{1\,\prime}(t,x) - \sigma_{\mathrm{H}}\,(\partial_t \chi)\,(t,x), \tag{3.426b}$$

where the two-current $\hat{J}^{\mu\,\prime}$ follows from $\hat{J}^{\mu}$ by replacing all operators by primed ones. The two-current (3.424) is only invariant under gauge transformations (3.423) that are static when $\sigma_{\mathrm{H}} \neq 0$.

With the help of

$$\left[D_x\,\hat{u}_i(t,x), D_y\,\hat{u}_j(t,y)\right] = -2\pi\mathrm{i}\,K_{ij}\,\delta'(x-y) \tag{3.427}$$

for $i,j = 1,\cdots,N$, one verifies that the time derivative of $\hat{J}^0(t,x)$ is (**Exercise** 3.32)

$$\begin{aligned}\frac{\partial\,\hat{J}^0}{\partial t} =&\, -\mathrm{i}\left[\hat{J}^0, \widehat{H}\right] + \sigma_H\,\frac{\partial\,A_1}{\partial t}\\ =&\, -\frac{\partial\,\hat{J}^1}{\partial x} + \sigma_H\,\frac{\partial\,A_1}{\partial t}.\end{aligned} \tag{3.428}$$

There follows the continuity equation

$$\partial_\mu\,\hat{J}^\mu = 0 \tag{3.429}$$

provided A_1 is time-independent or $\sigma_H = 0$.

For any nonvanishing σ_{H}, the continuity equation

$$\partial_\mu\,\hat{J}^\mu = \sigma_H\,\frac{\partial\,A_1}{\partial t} \tag{3.430}$$

is anomalous as soon as the vector gauge potential A_1 is time-dependent. The edge theory (3.355) is said to be chiral when $\sigma_{\mathrm{H}} \neq 0$, in which case the continuity equation (3.430) is anomalous. The anomalous continuity equation (3.430) is form covariant under any smooth gauge transformation (3.423). The choice of gauge may be fixed by the condition

$$\frac{\partial\,A_0}{\partial x} = 0, \tag{3.431a}$$

for which the anomalous continuity equation (3.430) then becomes

$$\left(\partial_\mu\,\hat{J}^\mu\right)(t,x) = +\sigma_H\,E(t,x), \tag{3.431b}$$

where

$$E(t,x) := +\left(\frac{\partial\,A_1}{\partial t}\right)(t,x) \equiv -\left(\frac{\partial\,A^1}{\partial t}\right)(t,x) \tag{3.431c}$$

represents the electric field in this gauge.

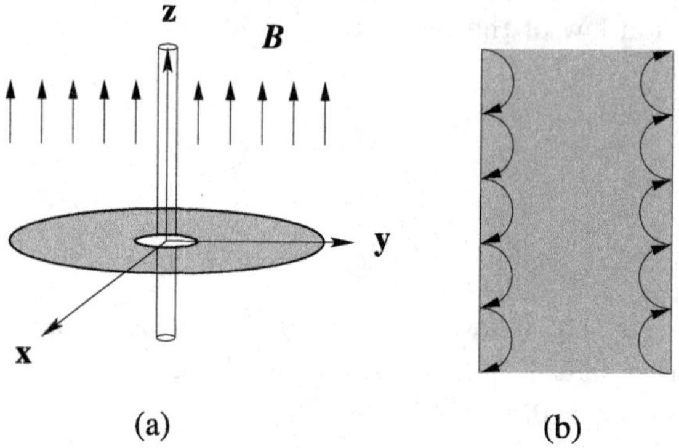

Figure 3.11 (a) A ring of outer radius $R \equiv L/(2\pi)$ and inner radius r in which electrons are confined. A uniform and static magnetic field $\boldsymbol{B}$ normal to the ring is present. The hierarchy $\ell_B \ll r \ll R$ of length scales is assumed, where $\ell_B \equiv \sqrt{\hbar c/|e\,B|}$ is the magnetic length. A time-dependent vector potential $\boldsymbol{A}(t,\boldsymbol{r})$ is induced by a time-dependent flux supported within a solenoid (sln) of radius $r_{\rm sln} \ll r$. This Corbino geometry has a cylindrical symmetry. (b) The classical motion of electrons confined to a plane normal to a uniform static magnetic field is circular. In the limit $R \to \infty$ holding R/r fixed, the Corbino disk turns into a Hall bar. An electron within a magnetic length of the boundary undergoes a classical skipping orbit. Upon quantization, a classical electron undergoing a skipping orbit turns into a chiral electron. Upon bosonization, a chiral electron turns into a chiral boson.

To interpret the anomalous continuity equation (3.431) of the bosonic chiral edge theory (3.355), we recall that x is a compact coordinate because of the periodic boundary conditions (3.355g), (3.355h), and (3.423). For simplicity, we assume

$$E(t,x) = E(t). \tag{3.432}$$

The interval $0 \leq x \leq L$ is thought of as a circle of perimeter L centered at the origin of the three-dimensional Euclidean space, as shown in Figure 3.11. The vector potential $A^1(t)$ and the electric field $E(t) = -\left(\frac{\partial A^1}{\partial t}\right)(t)$ along the circle of radius $R \equiv L/(2\pi)$ are then the polar components of a three-dimensional gauge field $A^\mu(t,\boldsymbol{r}) = (A^0,\boldsymbol{A})(t,\boldsymbol{r})$ and of a three-dimensional electric field $\boldsymbol{E}(t,\boldsymbol{r})$, respectively, in a cylindrical geometry. The electro-magnetic fields are

$$\boldsymbol{E}(t,\boldsymbol{r}) = -(\boldsymbol{\nabla}\,A^0)(t,\boldsymbol{r}) - (\partial_t\,\boldsymbol{A})(t,\boldsymbol{r}), \qquad \boldsymbol{B}(t,\boldsymbol{r}) = (\boldsymbol{\nabla}\wedge\boldsymbol{A})(t,\boldsymbol{r}). \tag{3.433}$$

The dimensionless Hall conductance $\sigma_{\rm H}$ encodes the linear response of spin-polarized electrons confined to move along this circle in the presence of a uniform and static magnetic field normal to the plane that contains this circle. The time-dependent anomalous term on the right-hand side of the anomalous continuity equation (3.431b) is caused by a solenoid of radius $r_{\rm sln} \ll r \ll R$ in a puncture of the plane that contains the solenoid of radius $r_{\rm sln}$ supporting a time-dependent flux. The combination of this time-dependent flux with the uniform static magnetic field exerts a Lorentz force on spin-polarized electrons moving along circles in the ring with the inner edge of radius r and the outer edge of radius R. This Lorentz force causes a net transfer of charge between the inner and outer edges

$$\frac{1}{L}\int\limits_{0}^{T}\mathrm{d}t\,\partial_t Q(t) := \int\limits_{0}^{T}\mathrm{d}t\,\langle\partial_t \hat{J}^0(t)\rangle = \sigma_{\mathrm{H}}\int\limits_{0}^{T}\mathrm{d}t\,E(t) \tag{3.434}$$

during the adiabatic evolution with period T of the normalized many-body ground state of the outer edge, provided we may identify the anomalous continuity equation (3.431b) with that of chiral spin-polarized electrons propagating along the outer edge in Figure 3.11. Hereto, separating the many-body ground state at the outer edge from all spin-polarized electrons supported between the inner and outer edge requires the existence of an energy scale separating it from many-body states in which these bulk spin-polarized electrons participate and by demanding that the inverse of this energy scale, a length scale, is much smaller than $R-r$. This energy scale is brought about by the uniform and static magnetic field $\boldsymbol{B}$ in Figure 3.11. That none of this pumped charge is lost in the shaded region of the ring follows if it is assumed that the spin-polarized electrons are unable to transport (dissipatively) a charge current across any circle of radius less than R and greater than r. The Hall conductance in the Corbino geometry of Figure 3.11 is then a rank 2 antisymmetric tensor proportional to the rank 2 Levi-Civita antisymmetric tensor with σ_{H} the proportionality constant in units of e^2/h. The charge density and current density for the ring obey a continuity equation as full gauge invariance is restored in the ring.

The chiral bosonic theory (3.355) is nothing but a theory for chiral electrons at the outer edge of the Corbino disk, as we still have to demonstrate. Chiral fermions are a fraction of the original fermion (a spin-polarized electron). More precisely, low-energy fermions have been split into one half that propagate on the outer edge and another half that propagate on the inner edge of the Corbino disk. The price for this fractionalization is an apparent breakdown of gauge invariance and charge conservation, when each chiral edge is treated independently from the other. Manifest charge conservation and gauge invariance are only restored if all low-energy degrees of freedom from the Corbino disk are treated on equal footing.

3.10.9 From the Hamiltonian to the Lagrangian Formalism

What is the Minkowski path integral that is equivalent to the quantum theory defined by Eq. (3.355)? In other words, we seek the path integral

$$Z := \int \mathcal{D}[u]\, e^{\mathrm{i}S[u]} \tag{3.435a}$$

with the Minkowski action

$$S[u] := \int_{-\infty}^{+\infty} \mathrm{d}t\, L[u] \equiv \int_{-\infty}^{+\infty} \mathrm{d}t \int_{0}^{L} \mathrm{d}x\, \mathcal{L}[u](t,x) \tag{3.435b}$$

such that

$$H := \int_{0}^{L} \mathrm{d}x \left[\Pi_i\,(\partial_t\, u_i) - \mathcal{L}[u]\right] \tag{3.436}$$

can be identified with $\widehat{H}$ in Eq. (3.355a) after elevating the classical fields

$$u_i(t,x) \tag{3.437a}$$

and

$$\Pi_i(t,x) := \frac{\delta\,\mathcal{L}}{\delta(\partial_t\, u_i)(t,x)} \tag{3.437b}$$

entering $\mathcal{L}[u]$ to the status of quantum fields $\hat{u}_i(t,x)$ and $\hat{\Pi}_j(t,y)$ upon imposing the equal-time commutation relations

$$\left[\hat{u}_i(t,x), \hat{\Pi}_j(t,y)\right] = +\frac{\mathrm{i}}{2}\,\delta_{ij}\,\delta(x-y) \tag{3.437c}$$

for any $i,j = 1,\cdots,N$. The unusual factor $1/2$ (instead of 1) on the right-hand side of the commutator between pairs of canonically conjugate fields arises because each scalar field u_i with $i = 1,\cdots,N$ is chiral, that is, it represents "one-half" of a canonical scalar field.

Without loss of generality, we set $A_0 = A_1 = 0$ in Eq. (3.355a). We try

$$\mathcal{L} := -\frac{1}{4\pi}\left[(\partial_x\, u_i)\, K^{-1}_{ij}\,(\partial_t\, u_j) + (\partial_x\, u_i)\, V_{ij}\,(\partial_x\, u_j)\right] \tag{3.438a}$$

with the chiral equations of motion

$$\begin{aligned}
0 &= \partial_\mu \frac{\delta\,\mathcal{L}}{\delta\,\partial_\mu\, u_i} - \frac{\delta\,\mathcal{L}}{\delta\, u_i} \\
&= \partial_t \frac{\delta\,\mathcal{L}}{\delta\,\partial_t\, u_i} + \partial_x \frac{\delta\,\mathcal{L}}{\delta\,\partial_x\, u_i} - \frac{\delta\,\mathcal{L}}{\delta\, u_i} \\
&= -\frac{1}{4\pi}\left(K^{-1}_{ji}\,\partial_t\,\partial_x + K^{-1}_{ij}\,\partial_x\,\partial_t + 2V_{ij}\,\partial_x\,\partial_x\right) u_j \\
&= -\frac{K^{-1}_{ij}}{2\pi}\,\partial_x\left(\delta_{jl}\,\partial_t + K_{jk}\,V_{kl}\,\partial_x\right) u_l
\end{aligned} \tag{3.438b}$$

for any $i = 1,\cdots,N$. Observe that the term that mixes time t and space x derivatives only becomes imaginary in Euclidean time $\tau = it$.

Proof The canonical momentum Π_i to the field u_i is

$$\Pi_i(t,x) := \frac{\delta\,\mathcal{L}}{\delta\,(\partial_t\,u_i)(t,x)} = -\frac{1}{4\pi}\,K^{-1}_{ij}\,\left(\partial_x\,u_j\right)(t,x) \tag{3.439}$$

for any $i = 1, \cdots, N$ owing to the matrix K being symmetric. The Legendre transform

$$\mathcal{H} := \Pi_i\,(\partial_t\,u_i) - \mathcal{L} \tag{3.440}$$

delivers

$$\mathcal{H} = \frac{1}{4\pi}\,(\partial_x\,u_i)\,V_{ij}\,\left(\partial_x\,u_j\right). \tag{3.441}$$

We now quantize the theory by elevating the classical fields u_i to the status of operators $\hat{u}_i$ obeying the algebra (3.355b). This gives a quantum theory that meets all the demands of the quantum chiral edge theory (3.355) in all compatibility with the canonical quantization rules (3.437c), for

$$\begin{aligned}
\left[\hat{u}_i(t,x), \hat{\Pi}_j(t,y)\right] &= -\frac{1}{4\pi}\,K^{-1}_{jk}\,\partial_y\,[\hat{u}_i(t,x), \hat{u}_k(t,y)] \\
{\scriptstyle \text{Eq. (3.355b)}} &= -\frac{1}{4\pi}\,K^{-1}_{jk}\,(\pi \mathrm{i})K_{ik}\,(-2)\delta(x-y) \\
{\scriptstyle K_{ik}\,=\,K_{ki}} &= +\frac{\mathrm{i}}{2}\,K^{-1}_{jk}\,K_{ki}\,\delta(x-y) \\
&= +\frac{\mathrm{i}}{2}\,\delta_{ij}\,\delta(x-y)
\end{aligned} \tag{3.442}$$

where $i, j = 1, \cdots, N$. □

Finally, analytical continuation to Euclidean time

$$\tau = \mathrm{i}t \tag{3.443a}$$

allows to define the finite-temperature quantum chiral theory through the path integral

$$Z_\beta := \int \mathcal{D}[u]\,\exp\left(-\int_0^\beta \mathrm{d}\tau \int_0^L \mathrm{d}x\,\mathcal{L}\right), \tag{3.443b}$$

$$\mathcal{L} := \frac{1}{4\pi}\,\left[\mathrm{i}\,(\partial_x\,u_i)\,K^{-1}_{ij}\,\left(\partial_\tau\,u_j\right) + (\partial_x\,u_i)\,V_{ij}\,\left(\partial_x\,u_j\right)\right]. \tag{3.443c}$$

3.11 Fractionalization at Nonvanishing Temperature

3.11.1 Overview

Temperature trades quantum fluctuations in favor of thermal fluctuations. The fractionalization of the fermion number presented so far is a many-body quantum effect tied to the charge-conjugation symmetry of the fermion number operator at vanishing Fermi energy. At vanishing temperature, the chemical potential defines the Fermi energy, that we have set to zero by assumption, for noninteracting fermions. Quantum fluctuations merely reflect the fact that the adiabatic deformation of the order parameter, the background, from a topological trivial representative to a topologically nontrivial representative induces a (single-particle) spectral flow across the Fermi energy that is responsible for the fractional part of the induced fermion number in the many-body ground state. At infinite temperature, the chemical potential is negative and scales with the number of fermions for noninteracting fermions. The notion that a characteristic energy fixed by a symmetry principle, charge-conjugation symmetry, is crossed under a flow of single-particle energies is moot in this limit. There cannot be any fractional induced fermion number in this limit. The chemical potential is a decreasing function of temperature for noninteracting fermions. The amount of spectral flow across the chemical potential decreases with increasing temperature. As more excited many-body states participate to the thermal average of the fermion number operator with increasing temperature, the more charge-conjugation symmetry at the Fermi energy is broken by thermal fluctuations. Correspondingly, the fractional part of the induced fermion number at zero temperature, if any, must decrease in magnitude with increasing temperature and vanish at infinite temperature. (The fractional part of the induced fermion number is already exponentially suppressed for temperature larger than the gap to the many-body excited states.)

The smooth crossover for the fractional part of the induced fermion number with increasing temperature was studied initially by Niemi and Semenoff[51] and by Midorikawa.[52] Their analysis was refined by Aitchison and Dunne.[53] We are going to describe the effects of fluctuations induced by temperature on the expectation value of the fermion number operator following Dunne and Rao.[54] We shall answer the following question. Does the fractional part of the induced fermion number at non-vanishing temperature retain its topological character at vanishing temperature?

3.11.2 Definition

Define the zero-temperature partition function

$$Z[\Phi, \not{A}] := \int \mathcal{D}[\bar{\psi}]\, \mathcal{D}[\psi]\, e^{+\mathrm{i}\int \mathrm{d}^2x\, \mathcal{L}[\Phi,\not{A}]} \tag{3.444a}$$

with the Lagrangian density

$$\mathcal{L}[\Phi, \not{A}] := \bar{\psi}\, \left(\mathrm{i}\not{\partial} + \not{A} - \Phi\right)\, \psi, \tag{3.444b}$$

51 A. J. Niemi and G. W. Semenoff, Phys. Lett. B **135**(1), 121–124 (1984). [Niemi and Semenoff (1984)].

52 S. Midorikawa, Prog. Theor. Phys. **69**(6), 1831–1834 (1983); Phys. Rev. D **31**(6), 1499–1502 (1985). [Midorikawa (1983, 1985)].

53 I. J. R. Aitchison and G. V. Dunne, Phys. Rev. Lett. **86**(9), 1690–1693 (2001). [Aitchison and Dunne (2001)].

54 G. V. Dunne and K. Rao, Phys. Rev. D **64**(2), 025003 (2001). [Dunne and Rao (2001)].

the external gauge field

$$\not{A} := \gamma^{\mu} A_{\mu} \tag{3.444c}$$

and the background

$$\Phi := |\phi| \, e^{+\mathrm{i}\gamma_5 \varphi}, \qquad \gamma_5 := \gamma_0 \, \gamma_1, \qquad \phi := |\phi| \begin{pmatrix} \cos\varphi \\ \sin\varphi \end{pmatrix}, \tag{3.444d}$$

both from the Clifford algebra. The integration measure $\mathrm{d}^2\mathrm{x}$ stands for $\mathrm{d}t\,\mathrm{d}x$. The induced fermion charge at zero temperature is defined by

$$Q := \int_{-\infty}^{+\infty} \mathrm{d}x \, \langle j^0(t,x) \rangle \tag{3.445a}$$

with

$$\langle j^{\mu}(t,x) \rangle := -\mathrm{i} \, \frac{\delta \log Z[\Phi, \not{A}]}{\delta A_{\mu}(t,x)} \Bigg|_{\not{A}=0}, \qquad \mu = 0, 1, \tag{3.445b}$$

the conserved charge current.

We seek the induced conserved fermion charge at nonvanishing temperature for two backgrounds:

1. A static, single domain wall (kink) background $\phi(t,x)$ defined by

$$\left(\frac{\mathrm{d}\phi}{\mathrm{d}t}\right)(t,x) = 0 \tag{3.446a}$$

with

$$\phi_1(t,x) \equiv \phi(x), \qquad \phi_2(t,x) \equiv \mu_{\mathrm{s}}, \tag{3.446b}$$

where

$$\lim_{x\to\pm\infty} \phi(x) = \pm\phi_{\infty}, \qquad \phi_{\infty} \in \mathbb{R}. \tag{3.446c}$$

2. A static, single kink in the sigma model background $\phi(t,x)$ defined by

$$\left(\frac{\mathrm{d}\phi}{\mathrm{d}t}\right)(t,x) = 0 \tag{3.447a}$$

with

$$|\phi(t,x)|^2(t,x) = m^2 \geq 0, \tag{3.447b}$$

where

$$\lim_{x\to\pm\infty} \varphi(t,x) = \varphi_{\pm\infty} \in [0, 2\pi[. \tag{3.447c}$$

By performing the analytical continuation to imaginary time

$$\begin{aligned} &t =: -\mathrm{i}\tau \equiv -\mathrm{i}x_2^{\mathrm{E}} \equiv -\mathrm{i}x_{\mathrm{E}}^2, && x \equiv x_1^{\mathrm{E}} \equiv x_{\mathrm{E}}^1, \\ &\gamma^0 = \gamma_0 =: \Gamma_2 \equiv \Gamma^2, && \gamma^1 = -\gamma_1 =: +\mathrm{i}\Gamma_1 \equiv +\mathrm{i}\Gamma^1, \\ &A_0 = A^0 =: +\mathrm{i}A_2^{\mathrm{E}} \equiv +\mathrm{i}A_{\mathrm{E}}^2, && A_1 = -A^1 =: A_1^{\mathrm{E}} \equiv A_{\mathrm{E}}^1, \end{aligned} \tag{3.448}$$

and imposing antiperiodic boundary conditions in imaginary time on the Grassmann integration variables

$$\bar{\psi}(x_1^{\mathrm{E}}, x_2^{\mathrm{E}} + \beta) = -\bar{\psi}(x_1^{\mathrm{E}}, x_2^{\mathrm{E}}), \qquad \psi(x_1^{\mathrm{E}}, x_2^{\mathrm{E}} + \beta) = -\psi(x_1^{\mathrm{E}}, x_2^{\mathrm{E}}), \tag{3.449}$$

we obtain the Euclidean partition function

$$Z_{\mathrm{E}}[\Phi_{\mathrm{E}}, \not{A}_{\mathrm{E}}] := \int \mathcal{D}[\bar{\psi}]\,\mathcal{D}[\psi] \exp\left(- \int\limits_{-\infty}^{+\infty} \mathrm{d}x_1^{\mathrm{E}} \int\limits_0^\beta \mathrm{d}x_2^{\mathrm{E}}\, \mathcal{L}_{\mathrm{E}}[\Phi_{\mathrm{E}}, \not{A}_{\mathrm{E}}] \right). \tag{3.450a}$$

The range of integration in the imaginary time direction x_2^{E} is the interval $[0, \beta[$ as a result of imposing antiperiodic boundary conditions on the Grassmann integration variables. The Lagrangian density in two-dimensional Euclidean space is

$$\mathcal{L}_{\mathrm{E}}[\Phi_{\mathrm{E}}, \not{A}_{\mathrm{E}}] := \bar{\psi}\left(\not{\partial}_{\mathrm{E}} - \mathrm{i}\not{A}_{\mathrm{E}} + \Phi_{\mathrm{E}}\right)\psi, \tag{3.450b}$$

where

$$\not{\partial}_{\mathrm{E}} := \Gamma_\mu\,\partial_\mu, \qquad \not{A}_{\mathrm{E}} := \Gamma_\mu\,A_\mu^{\mathrm{E}}, \qquad \Phi_{\mathrm{E}} := |\phi|\,e^{-\Gamma_1\Gamma_2\varphi}. \tag{3.450c}$$

As the components of the conserved charge current $j^\mu \equiv \bar{\psi}\,\gamma^\mu\,\psi$ have the Euclidean representation

$$j^0 := \bar{\psi}\,\gamma^0\,\psi = \bar{\psi}\,\Gamma_2\,\psi =: j_2^{\mathrm{E}}, \qquad j^1 := \bar{\psi}\,\gamma^1\,\psi = +\mathrm{i}\bar{\psi}\,\Gamma_1\,\psi =: +\mathrm{i}j_1^{\mathrm{E}}, \tag{3.451a}$$

the induced conserved fermion charge at any temperature $T = 1/(k_{\mathrm{B}}\,\beta)$ (k_{B} the Boltzmann constant) is given by

$$Q(\beta) := \int\limits_{-\infty}^{+\infty} \mathrm{d}x\, \langle j_2^{\mathrm{E}}(x, \tau)\rangle \tag{3.451b}$$

with

$$\langle j_\mu^{\mathrm{E}}(x, \tau)\rangle := -\mathrm{i}\left.\frac{\delta \log Z_{\mathrm{E}}[\Phi_{\mathrm{E}}, \not{A}_{\mathrm{E}}]}{\delta A_\mu^{\mathrm{E}}(x, \tau)}\right|_{\not{A}^{\mathrm{E}}=0}, \qquad \mu = 1, 2. \tag{3.451c}$$

3.11.3 Temperature Dependence of the Induced Fermion Charge

The fermion density operator is defined to be

$$\hat{\rho}(x) := \frac{1}{2}\left[\hat{\psi}^\dagger(x), \hat{\psi}(x)\right] \tag{3.452}$$

according to Eq. (3.129a). Here, the operator $\hat{\psi}^\dagger(x)$ that creates a fermion localized at x and its adjoint $\hat{\psi}(x)$ that annihilates a fermion localized at x along the line $-\infty < x < +\infty$ can be expanded in terms of the single-particle eigenstates of the Dirac Hamiltonian $\mathcal{H}$ defined in Eq. (3.243) for $N = 1$, as was done in Eq. (3.129). This definition insures that the local charge is odd under the conjugation of charge by which $\hat{\psi}^\dagger(x) \mapsto \hat{\psi}(x)$ and $\hat{\psi}(x) \mapsto \hat{\psi}^\dagger(x)$ (**Exercise** 3.6). The conserved charge operator is

$$\widehat{Q} := \frac{1}{2} \int\limits_{-\infty}^{+\infty} \mathrm{d}x \, \left[\hat{\psi}^\dagger(x), \hat{\psi}(x)\right]. \tag{3.453}$$

Hereto, the conserved charge operator is odd under charge conjugation. The conserved charge induced by the background Φ is the expectation value

$$Q(\beta) := \frac{\operatorname{Tr}\left(e^{-\beta \widehat{H}} \widehat{Q}\right)}{\operatorname{Tr}\left(e^{-\beta \widehat{H}}\right)}, \tag{3.454a}$$

where

$$\widehat{H} := \int\limits_{-\infty}^{+\infty} \mathrm{d}x \, \hat{\psi}^\dagger(x) \, \mathcal{H} \, \hat{\psi}(x). \tag{3.454b}$$

The sensitivity of the induced conserved charge to the background Φ comes about from the dependence on Φ of the many-body eigenstates of $\widehat{H}$.

The steps that lead to Eqs. (3.132b) and (3.216) can be repeated at any nonvanishing temperature $T = 1/(k_{\mathrm{B}} \beta)$ to deliver

$$Q(\beta) = -\frac{1}{2} \int\limits_{-\infty}^{+\infty} \mathrm{d}\varepsilon \, \nu_{\mathcal{H}}(\varepsilon) \, \tanh\left(\frac{\beta \, \varepsilon}{2}\right), \tag{3.455a}$$

where

$$\nu_{\mathcal{H}}(\varepsilon) \, \mathrm{d}\varepsilon := \frac{1}{\pi} \operatorname{Im} \operatorname{Tr}\left(\frac{1}{\mathcal{H} - \varepsilon - \mathrm{i}0^+}\right) \mathrm{d}\varepsilon \tag{3.455b}$$

returns the number of single-particle eigenstates of $\mathcal{H}$ in the energy interval $[\varepsilon, \varepsilon + \mathrm{d}\varepsilon]$. Hence, $\nu_{\mathcal{H}}(\varepsilon)$ is a measure, the density of states.[55]

The presence of $\tanh(\beta \, \varepsilon/2)$ in the integrand on the right-hand side of Eq. (3.455a) can be traced to the fermion density operator being *defined to be odd under charge conjugation.* As such it should be

$$\lim_{\beta \to \infty} \tanh\left(\frac{\beta \, \varepsilon}{2}\right) = \operatorname{sgn}(\varepsilon). \tag{3.456}$$

[55] A measure is not an ordinary function. It is a distribution. For example, the contribution of a bound state at the energy $\varepsilon_{\mathrm{bs}}$ to $\nu_{\mathcal{H}}(\varepsilon)$ is proportional to $\delta(\varepsilon - \varepsilon_{\mathrm{bs}})$.

Therefore, the expression (3.455) for the fermion charge induced by the background Φ at a nonvanishing temperature is a regularization of the spectral asymmetry (3.132b) or (3.216). Under the transformation $\varepsilon \mapsto -\varepsilon$ that encodes charge conjugation,

$$\tanh\left(\frac{\beta\,\varepsilon}{2}\right) \mapsto -\tanh\left(\frac{\beta\,\varepsilon}{2}\right). \tag{3.457}$$

It follows that, if $\nu_{\mathcal{H}}(\varepsilon)$ is a smooth function of ε, then

$$Q(\beta) = -\frac{1}{4}\int\limits_{-\infty}^{+\infty} \mathrm{d}\varepsilon\, \left[\nu_{\mathcal{H}}(+\varepsilon) - \nu_{\mathcal{H}}(-\varepsilon)\right] \tanh\left(\frac{\beta\,\varepsilon}{2}\right) \tag{3.458}$$

as

$$-\frac{1}{4}\int\limits_{-\infty}^{+\infty} \mathrm{d}\varepsilon\, \left[\nu_{\mathcal{H}}(+\varepsilon) + \nu_{\mathcal{H}}(-\varepsilon)\right] \tanh\left(\frac{\beta\,\varepsilon}{2}\right) = 0. \tag{3.459}$$

In particular, if the spectral measure $\nu_{\mathcal{H}}(\varepsilon)$ is a smooth and even function of ε, then $Q(\beta)$ must vanish for all temperatures. This is precisely what happens for the reference massive Dirac Hamiltonian $\mathcal{H}_{\infty}$ defined in Eq. (3.69d) when the staggered chemical potential $\mu_{\mathrm{s}} = 0$. However, the spectral measure $\nu_{\mathcal{H}}(\varepsilon)$ is not a smooth function of ε when bound states are present. The three identities

$$\tanh\left(\frac{\beta\,\varepsilon}{2}\right) = 1 - 2f_{\mathrm{FD}}(+\varepsilon), \tag{3.460a}$$

$$\tanh\left(\frac{\beta\,\varepsilon}{2}\right) = \mathrm{sgn}(\varepsilon)\, \left[1 - 2f_{\mathrm{FD}}(+|\varepsilon|)\right], \tag{3.460b}$$

$$\tanh\left(\frac{\beta\,\varepsilon}{2}\right) = \mathrm{sgn}(\varepsilon)\, \left[f_{\mathrm{FD}}(-|\varepsilon|) - f_{\mathrm{FD}}(+|\varepsilon|)\right], \tag{3.460c}$$

obeyed by $\tanh(\beta\,\varepsilon/2)$ are useful to account for the contributions of bound states to $Q(\beta)$.

Identities (3.460a) and (3.460b), that is

$$\begin{aligned}
\tanh\left(\frac{\beta\,\varepsilon}{2}\right) &= \frac{e^{+\beta\,\varepsilon/2} - e^{-\beta\,\varepsilon/2}}{e^{+\beta\,\varepsilon/2} + e^{-\beta\,\varepsilon/2}} \\
&= \frac{e^{+\beta\,\varepsilon} - 1}{e^{+\beta\,\varepsilon} + 1} \\
&= \frac{e^{+\beta\,\varepsilon} + 1 - 2}{e^{+\beta\,\varepsilon} + 1} \\
&= 1 - 2f_{\mathrm{FD}}(\varepsilon),
\end{aligned} \tag{3.461}$$

and

$$\begin{aligned}\tanh\left(\frac{\beta\,\varepsilon}{2}\right) =& \operatorname{sgn}(\varepsilon)\tanh\left(\frac{\beta\,|\varepsilon|}{2}\right)\\ =& \operatorname{sgn}(\varepsilon)\left[1-2f_{\mathrm{FD}}(|\varepsilon|)\right],\end{aligned} \tag{3.462}$$

respectively, relate $\tanh(\beta\,\varepsilon/2)$ to the Fermi-Dirac distribution. Identity (3.460c), that is

$$\begin{aligned}\tanh\left(\frac{\beta\,\varepsilon}{2}\right) =& \operatorname{sgn}(\varepsilon)\tanh\left(\frac{\beta\,|\varepsilon|}{2}\right)\\ =& \operatorname{sgn}(\varepsilon)\left(\frac{e^{+\beta\,|\varepsilon|/2}-e^{-\beta\,|\varepsilon|/2}}{e^{+\beta\,|\varepsilon|/2}+e^{-\beta\,|\varepsilon|/2}}\right)\\ =& \operatorname{sgn}(\varepsilon)\left(\frac{e^{+\beta\,|\varepsilon|/2}-e^{-\beta\,|\varepsilon|/2}}{e^{+\beta\,|\varepsilon|/2}+e^{-\beta\,|\varepsilon|/2}}\right)\left(\frac{e^{+\beta\,|\varepsilon|/2}+e^{-\beta\,|\varepsilon|/2}}{e^{+\beta\,|\varepsilon|/2}+e^{-\beta\,|\varepsilon|/2}}\right)\\ =& \operatorname{sgn}(\varepsilon)\left(\frac{1}{e^{-\beta\,|\varepsilon|}+1}-\frac{1}{e^{+\beta\,|\varepsilon|}+1}\right)\\ =& \operatorname{sgn}(\varepsilon)\left[f_{\mathrm{FD}}(-|\varepsilon|)-f_{\mathrm{FD}}(+|\varepsilon|)\right],\end{aligned} \tag{3.463}$$

relates $\tanh(\beta\,\varepsilon/2)$ to the difference between the Fermi-Dirac distribution $f_{\mathrm{FD}}(-|\varepsilon|)$ to occupy a single-particle state at the negative single-particle energy $-|\varepsilon|$ and the Fermi-Dirac distribution $f_{\mathrm{FD}}(+|\varepsilon|)$ to occupy a single-particle state at the positive single-particle energy $+|\varepsilon|$.

These innocuous looking identities help to deduce some far reaching interpretations of the spectral representation (3.455) for the fermion charge induced by the background Φ. For example, we may use Eq. (3.460b) to do the additive decomposition

$$Q(\beta) = Q' + Q''(\beta) \tag{3.464a}$$

of Eq. (3.455a). The contribution at vanishing temperature

$$Q' := -\frac{1}{2}\int\limits_{-\infty}^{+\infty} \mathrm{d}\varepsilon\,\nu_{\mathcal{H}}(\varepsilon)\operatorname{sgn}(\varepsilon), \tag{3.464b}$$

which is solely controlled by the many-body ground state, has been separated from the contribution at nonvanishing temperature

$$Q''(\beta) := +\int\limits_{-\infty}^{+\infty} \mathrm{d}\varepsilon\,\nu_{\mathcal{H}}(\varepsilon)\operatorname{sgn}(\varepsilon)\,f_{\mathrm{FD}}(|\varepsilon|), \tag{3.464c}$$

which encodes all many-body excited eigenstates. These two contributions come with opposite signs.

The contribution at vanishing temperature is either exactly topological, as we have shown in Sections 3.5 and 3.6 for the static kink (3.446), or it is topological to leading order in a gradient expansion (in a semiclassical approximation), as we have shown in Sections 3.7, 3.9, and 3.10 for the static kink in the sigma model background (3.447).

The contribution at nonvanishing temperature is generically non-topological. It is a function of β that vanishes at zero temperature and converges to the value

$$\lim_{\beta\to 0} Q''(\beta) = -Q' \geq 0 \tag{3.465}$$

at infinite temperature. Hence, the fermion charge induced by the background Φ must vanish at infinite temperature. The function $Q''(\beta)$ can be topological in that it only depends on the asymptotic values taken by the static background $\Phi(x)$ as $x \to \pm\infty$ under special circumstances, such as when $\mathcal{H}^2$ is linearly related to a SUSY Hamiltonian as is the case for the static kink (3.446). The function $Q''(\beta)$ is a full functional of $\Phi(x)$ otherwise, say for the static kink of the sigma model (3.447).

3.11.3.1 A Closer Look at the Thermal Corrections When SUSY Holds

It is enlightening to show explicitly how the hidden SUSY for the static kink (3.446) conspires to retain the topological character of the thermal correction (3.464) to the induced fermion charge.

To this end, we rewrite the density of states (3.455b) as the functional trace Tr over the difference between the retarded and advanced single-particle Green functions:

$$\nu_{\mathcal{H}}(\varepsilon) = \lim_{\epsilon\downarrow 0} \frac{1}{2\pi\mathrm{i}} \left[\mathrm{Tr}\left(\frac{1}{\mathcal{H}-\varepsilon-\mathrm{i}\epsilon}\right) - \mathrm{Tr}\left(\frac{1}{\mathcal{H}-\varepsilon+\mathrm{i}\epsilon}\right)\right]. \tag{3.466}$$

The integration over the single-particle energies $\varepsilon \in \mathbb{R}$ in the spectral representation of the induced fermion charge (3.455a) is to be interpreted as the contour integral in the complex energy plane $z \in \mathbb{C}$ along the oriented path [as is shown in Figure 3.12(a)]

$$\Gamma := \Gamma_+ \cup \Gamma_-, \tag{3.467a}$$

where

$$\Gamma_+ := \{z = +\varepsilon + \mathrm{i}\epsilon | \varepsilon \in \mathbb{R}\} \tag{3.467b}$$

and

$$\Gamma_- := \{z = -\varepsilon - \mathrm{i}\epsilon | \varepsilon \in \mathbb{R}\}. \tag{3.467c}$$

The induced fermion charge (3.455a) becomes

$$Q(\beta) = -\frac{1}{2}\int_{\Gamma} \frac{\mathrm{d}z}{2\pi\mathrm{i}}\, \mathrm{Tr}\left(\frac{1}{\mathcal{H}-z}\right) \tanh\left(\frac{\beta z}{2}\right). \tag{3.468}$$

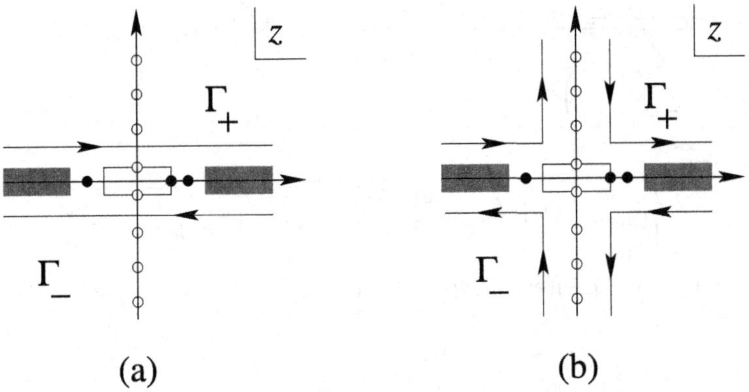

Figure 3.12 (a) Contour $\Gamma = \Gamma_+ \cup \Gamma_-$ on the right-hand side of Eq. (3.468). The spectrum of $\mathcal{H}$ is continuous along the real axis in the gray boxes. The spectrum of $\mathcal{H}$ is discrete and shown as filled circles for $|\mu_{\mathrm{s}}| \leq \varepsilon < \sqrt{\phi_\infty^2 + \mu_{\mathrm{s}}^2}$. There is a threshold eigenstate of $\mathcal{H}$ at $\varepsilon = |\mu_{\mathrm{s}}|$. The white box does not support eigenvalues of $\mathcal{H}$. (b) Deformation of $\Gamma = \Gamma_+ \cup \Gamma_-$ to pick up the poles of $\tanh(\beta\, z/2)$ shown as circles colored in white.

This representation is flexible for it also allows the oriented Γ to be deformed so as to pick up the simple poles of the function $\tanh\left(\frac{\beta\, z}{2}\right)$ [as is shown in Figure 3.12(b)]. In doing so, one trades the integral representation (3.455a) of the induced fermion charge in favor of an infinite sum at the location of the Matsubara poles

$$z_n = \frac{(2n+1)\,\pi\,\mathrm{i}}{\beta}, \qquad n \in \mathbb{Z}. \tag{3.469}$$

An important difference with the spectral representation (3.455a) is that it is the even part

$$\left[\mathrm{Tr}\left(\frac{1}{\mathcal{H}-z}\right)\right]_{\mathrm{even}} := \frac{1}{2}\left[\mathrm{Tr}\left(\frac{1}{\mathcal{H}-z}\right) + \mathrm{Tr}\left(\frac{1}{\mathcal{H}+z}\right)\right] \tag{3.470}$$

that survives the integration along the oriented path Γ weighted by $\tanh\left(\frac{\beta\, z}{2}\right)$, as the change of integration variables $z \mapsto -z$ reverses the orientation of Γ.

An application of the Callias index theorem delivers the remarkable relation

$$\left[\mathrm{Tr}\left(\frac{1}{\mathcal{H}-z}\right)\right]_{\mathrm{even}} = \frac{\mu_{\mathrm{s}}\,\phi_\infty}{(\mu_{\mathrm{s}}^2 - z^2)\sqrt{\phi_\infty^2 + \mu_{\mathrm{s}}^2 - z^2}}. \tag{3.471}$$

Before proving Eq. (3.471), one verifies (**Exercise** 3.33) that Eq. (3.471) together with the identity

$$Q(\beta) = -\frac{1}{2}\int\limits_{\Gamma} \frac{\mathrm{d}z}{2\pi\mathrm{i}} \left[\mathrm{Tr}\left(\frac{1}{\mathcal{H}-z}\right)\right]_{\mathrm{even}} \tanh\left(\frac{\beta\, z}{2}\right), \tag{3.472}$$

deliver the series $[\tilde{\varphi}_{\infty} \equiv \arctan(\phi_{\infty}/\mu_{\mathrm{s}}) = (\pi/2) - \arctan(\mu_{\mathrm{s}}/\phi_{\infty}) = (\pi/2) - \varphi_{\infty}]$

$$
\begin{aligned}
Q(\beta) = & -\frac{2}{\pi}\left(\frac{\mu_{\mathrm{s}}\,\beta}{\pi}\right)^{2}\cos\varphi_{\infty}\times \\
& \sum_{n=0}^{\infty}\frac{1}{\left[(2n+1)^{2}+\left(\frac{\mu_{\mathrm{s}}\,\beta}{\pi}\right)^{2}\right]\sqrt{(2n+1)^{2}\sin^{2}\varphi_{\infty}+\left(\frac{\mu_{\mathrm{s}}\,\beta}{\pi}\right)^{2}}},
\end{aligned}
\tag{3.473}
$$

on the one hand, or the integral representation

$$
\begin{aligned}
Q(\beta) = & -\frac{1}{2}\,\mathrm{sgn}\left(\frac{\pi}{2}-\varphi_{\infty}\right)\tanh\left(\frac{\mu_{\mathrm{s}}\,\beta}{2}\right) \\
& + \frac{\cos\varphi_{\infty}\,\sin\varphi_{\infty}}{\pi}\int\limits_{1}^{\infty}\mathrm{d}u\,\frac{\tanh\left(\frac{\mu_{\mathrm{s}}\,\beta\,u}{2\sin\varphi_{\infty}}\right)}{\sqrt{u^{2}-1}\,\left(u^{2}-\sin^{2}\varphi_{\infty}\right)},
\end{aligned}
\tag{3.474}
$$

on the other hand.

Proof The proof of Eq. (3.471) proceeds in two steps. First, we need a useful representation of Eq. (3.470), namely, Eq. (3.479b). We then need to turn this representation into a boundary term.

Step 1: A useful representation of Eq. (3.470).

The Dirac Hamiltonian (3.69a) has the 2×2 representation

$$
\mathcal{H} := \begin{pmatrix} +\mu_{\mathrm{s}} & -\nabla+\phi(x) \\ +\nabla+\phi(x) & -\mu_{\mathrm{s}} \end{pmatrix}, \qquad \nabla := \frac{\mathrm{d}}{\mathrm{d}x}. \tag{3.475a}
$$

The Dirac Hamiltonian (3.475a) is a special case of Eq. (3.110c) with $N=1$,

$$
\widehat{D} = -\nabla + \phi, \tag{3.475b}
$$

and

$$
\widehat{D}^{\dagger} = +\nabla + \phi. \tag{3.475c}
$$

Its square is the decomposable 2×2 matrix

$$
\begin{aligned}
\mathcal{H}^{2} := & \begin{pmatrix} -\Delta-\phi'(x)+\phi^{2}(x)+\mu_{\mathrm{s}}^{2} & 0 \\ 0 & -\Delta+\phi'(x)+\phi^{2}(x)+\mu_{\mathrm{s}}^{2} \end{pmatrix} \\
\equiv & \,\mathcal{H}_{\phi}^{2}+\mu_{\mathrm{s}}^{2},
\end{aligned}
\tag{3.476}
$$

where $\Delta \equiv \nabla^{2}$ is the Laplace operator, $\phi' \equiv \nabla\phi$, and $\mathcal{H}_{\phi}$ was defined in Eq. (3.111b). The positive semidefinite Schrödinger Hamiltonians

$$
\widehat{D}\,\widehat{D}^{\dagger} = -\Delta - \phi'(x) + \phi^{2}(x) \tag{3.477a}
$$

and

$$
\widehat{D}^{\dagger}\,\widehat{D} = -\Delta + \phi'(x) + \phi^{2}(x) \tag{3.477b}
$$

are essential isospectral and

$$\mathbf{H} := \frac{1}{2}\left(\mathcal{H}^2 - \mu_{\mathrm{s}}^2\right) \tag{3.477c}$$

is a SUSY Hamiltonian.

For any $z \in \mathbb{C} \setminus \mathbb{R}$,

$$\begin{aligned}\frac{1}{\mathcal{H}-z} &= (\mathcal{H}-z)^{-1}\,(\mathcal{H}+z)^{-1}\,(\mathcal{H}+z)\\ &= \begin{pmatrix} \frac{1}{\widehat{D}\,\widehat{D}^{\dagger}+\mu_{\mathrm{s}}^2-z^2} & 0\\ 0 & \frac{1}{\widehat{D}^{\dagger}\,\widehat{D}+\mu_{\mathrm{s}}^2-z^2}\end{pmatrix}\begin{pmatrix} +\mu_{\mathrm{s}}+z & \widehat{D}\\ \widehat{D}^{\dagger} & -\mu_{\mathrm{s}}+z\end{pmatrix}.\end{aligned} \tag{3.478}$$

Upon taking the functional trace, one finds

$$\begin{aligned}\mathrm{Tr}\left(\frac{1}{\mathcal{H}-z}\right) &:= \int\limits_{-\infty}^{+\infty} \mathrm{d}x\,\mathrm{tr}\,\left\langle x\left|(\mathcal{H}-z)^{-1}\right|x\right\rangle\\ &= \left[\mathrm{Tr}\left(\frac{1}{\mathcal{H}-z}\right)\right]_{\text{even}} + \left[\mathrm{Tr}\left(\frac{1}{\mathcal{H}-z}\right)\right]_{\text{odd}},\end{aligned} \tag{3.479a}$$

where

$$\left[\mathrm{Tr}\left(\frac{1}{\mathcal{H}-z}\right)\right]_{\text{even}} := \mathrm{Tr}\begin{pmatrix} \frac{+\mu_{\mathrm{s}}}{\widehat{D}\,\widehat{D}^{\dagger}+\mu_{\mathrm{s}}^2-z^2} & 0\\ 0 & \frac{-\mu_{\mathrm{s}}}{\widehat{D}^{\dagger}\,\widehat{D}+\mu_{\mathrm{s}}^2-z^2}\end{pmatrix} \tag{3.479b}$$

and

$$\left[\mathrm{Tr}\left(\frac{1}{\mathcal{H}-z}\right)\right]_{\text{odd}} := \mathrm{Tr}\begin{pmatrix} \frac{z}{\widehat{D}\,\widehat{D}^{\dagger}+\mu_{\mathrm{s}}^2-z^2} & 0\\ 0 & \frac{z}{\widehat{D}^{\dagger}\,\widehat{D}+\mu_{\mathrm{s}}^2-z^2}\end{pmatrix}. \tag{3.479c}$$

Step 2: Turning Eq. (3.479b) into a boundary term.

The result (3.484) that follows does not hinge on space being one-dimensional. For this reason, we follow the notation of Eq. (3.110) and work in N-dimensional position space $\mathbb{R}^N$. We are after the matrix elements

$$\begin{aligned}\left[\mathrm{tr}\left\langle \boldsymbol{x}\left|\frac{1}{\mathcal{H}-z}\right|\boldsymbol{y}\right\rangle\right]_{\text{even}} &= \mathrm{tr}\left\langle \boldsymbol{x}\left|\Gamma_5\left(\frac{\mu_{\mathrm{s}}}{\mathcal{H}^2-z^2}\right)\right|\boldsymbol{y}\right\rangle\\ &\overset{\text{Eq. (3.112)}}{=} \mathrm{tr}\left\langle \boldsymbol{x}\left|\Gamma_5\left(\frac{\mu_{\mathrm{s}}}{\mathcal{H}_{\phi}^2+\mu_{\mathrm{s}}^2-z^2}\right)\right|\boldsymbol{y}\right\rangle,\end{aligned} \tag{3.480}$$

where $\boldsymbol{x}$ and $\boldsymbol{y}$ denote two points in position space $\mathbb{R}^N$. We introduce the shorthand notation

$$0 \neq \kappa := \sqrt{\mu_{\mathrm{s}}^2 - z^2} \in \mathbb{C}. \tag{3.481}$$

We have

$$
\begin{aligned}
\left[\operatorname{tr}\left\langle \boldsymbol{x}\left|\frac{1}{\mathcal{H}-z}\right|\boldsymbol{y}\right\rangle\right]_{\text{even}}
&=\mu_{\text{s}}\operatorname{tr}\left\langle \boldsymbol{x}\left|\Gamma_5\left(\frac{1}{\left(\mathcal{H}_\phi+\mathrm{i}\kappa\right)\left(\mathcal{H}_\phi-\mathrm{i}\kappa\right)}\right)\right|\boldsymbol{y}\right\rangle\\
&=\frac{\mathrm{i}\mu_{\text{s}}}{\kappa}\operatorname{tr}\left\langle \boldsymbol{x}\left|\Gamma_5\left(\frac{1}{\mathcal{H}_\phi+\mathrm{i}\kappa}\right)\right|\boldsymbol{y}\right\rangle .
\end{aligned}
\tag{3.482}
$$

Proof Because Γ_5 anticommutes with $\mathcal{H}_\phi$, we can write

$$
\begin{aligned}
\operatorname{tr}\left[\Gamma_5\left(\frac{1}{\mathcal{H}_\phi+\mathrm{i}\kappa}\right)\right]
&=-\operatorname{tr}\left[\left(\frac{1}{\mathcal{H}_\phi-\mathrm{i}\kappa}\right)\Gamma_5\right]\\
&=\frac{1}{2}\operatorname{tr}\left\{\Gamma_5\left[\left(\frac{1}{\mathcal{H}_\phi+\mathrm{i}\kappa}\right)-\left(\frac{1}{\mathcal{H}_\phi-\mathrm{i}\kappa}\right)\right]\right\}\\
&=-\mathrm{i}\kappa\operatorname{tr}\left[\Gamma_5\left(\frac{1}{\left(\mathcal{H}_\phi+\mathrm{i}\kappa\right)\left(\mathcal{H}_\phi-\mathrm{i}\kappa\right)}\right)\right].
\end{aligned}
\tag{3.483}
$$

□

We are going to show that the right-hand side is a total derivative in the limit $\boldsymbol{y}\to\boldsymbol{x}$. Integration over $\boldsymbol{x}$ to compute the functional trace Tr then delivers the boundary term on the right-hand side of Eq. (3.471).

The intermediate steps leading to the important result

$$
\begin{aligned}
&\sum_{i=1}^{N}\left(\frac{\partial}{\partial x_i}+\frac{\partial}{\partial y_i}\right)\operatorname{tr}\left\langle \boldsymbol{x}\left|\mathrm{i}\Gamma_i\Gamma_5\left(\frac{1}{\mathcal{H}_\phi+\mathrm{i}\kappa}\right)\right|\boldsymbol{y}\right\rangle=\\
&\qquad 2\mathrm{i}\kappa\operatorname{tr}\left\langle \boldsymbol{x}\left|\Gamma_5\left(\frac{1}{\mathcal{H}_\phi+\mathrm{i}\kappa}\right)\right|\boldsymbol{y}\right\rangle\\
&\qquad -\operatorname{tr}\left\{\left[\Phi(\boldsymbol{x})-\Phi(\boldsymbol{y})\right]\left\langle \boldsymbol{x}\left|\Gamma_5\left(\frac{1}{\mathcal{H}_\phi+\mathrm{i}\kappa}\right)\right|\boldsymbol{y}\right\rangle\right\}
\end{aligned}
\tag{3.484}
$$

are outlined in **Exercise** 3.33. Equation (3.484) is yet another signature of the regularized (through point splitting) axial anomaly of the Dirac Hamiltonian $\mathcal{H}_\phi+\mathrm{i}\kappa$. The left-hand side of Eq. (3.484) is the point-split divergence over an axial-like current. This axial-like current would be conserved if the right-hand side were to vanish. The vanishing of the right-hand side holds if $\kappa=0$, since $\mathcal{H}_\phi$ anticommutes with Γ_5. For a nonvanishing κ, there are two contributions with the potential to spoil the continuity equation obeyed in the limit $\kappa=0$. Both contributions are controlled by the $N\times N$ matrix $\langle\boldsymbol{x}|\Gamma_5/\left(\mathcal{H}_\phi+\mathrm{i}\kappa\right)|\boldsymbol{y}\rangle$. This matrix is multiplied by the $\mathbb{C}$ number κ on the one hand and the $N\times N$ matrix $\Phi(\boldsymbol{x})-\Phi(\boldsymbol{y})$ on the other hand. The latter contribution vanishes in the limit $\boldsymbol{y}\to\boldsymbol{x}$ if and only if the matrix field $\Phi(\boldsymbol{x})$ is a smooth function of $\boldsymbol{x}$ everywhere in $\mathbb{R}^N$. This need not be the case, say if $\Phi(\boldsymbol{x})$ represents a gauge field supporting a monopole for example.

In the case at hand, $N = 1$, $\Phi(x)$ is smooth for all $x \in \mathbb{R}$. If we take the limit $y \to x$ and use the product rule for differentiation, we then infer that Eq. (3.484) simplifies to

$$\mathrm{tr}\left\langle x\left|\Gamma_5\left(\frac{1}{\mathcal{H}_\phi+\mathrm{i}\kappa}\right)\right|x\right\rangle = \frac{1}{2\mathrm{i}\kappa}\nabla\,\mathrm{tr}\left\langle x\left|\mathrm{i}\Gamma_1\Gamma_5\left(\frac{1}{\mathcal{H}_\phi+\mathrm{i}\kappa}\right)\right|x\right\rangle. \tag{3.485}$$

Insertion of Eq. (3.485) on the right-hand side of Eq. (3.482) in the limit $y \to x$ gives

$$\begin{aligned}\left[\mathrm{tr}\left\langle x\left|\frac{1}{\mathcal{H}-z}\right|x\right\rangle\right]_{\text{even}} =& \frac{\mathrm{i}\mu_{\mathrm{s}}}{\kappa}\,\mathrm{tr}\left\langle x\left|\Gamma_5\left(\frac{1}{\mathcal{H}_\phi+\mathrm{i}\kappa}\right)\right|x\right\rangle \\ =& \frac{\mu_{\mathrm{s}}}{2\,\kappa^2}\nabla\,\mathrm{tr}\left\langle x\left|\mathrm{i}\Gamma_1\Gamma_5\left(\frac{1}{\mathcal{H}_\phi+\mathrm{i}\kappa}\right)\right|x\right\rangle,\end{aligned} \tag{3.486}$$

where we recall that $\kappa := \sqrt{\mu_{\mathrm{s}}^2 - z^2}$, $\Gamma_5 = \tau_3$, $\Gamma_1 = -\tau_2$, and $\mathrm{i}\Gamma_1\Gamma_5 = \tau_1$. The desired functional trace is the boundary term

$$\begin{aligned}\left[\mathrm{Tr}\left(\frac{1}{\mathcal{H}-z}\right)\right]_{\text{even}} =& \int\limits_{-\infty}^{+\infty}\mathrm{d}x\,\left[\mathrm{tr}\left\langle x\left|\frac{1}{\mathcal{H}-z}\right|x\right\rangle\right]_{\text{even}} \\ =& \frac{1}{2}\frac{\mu_{\mathrm{s}}}{\kappa^2}\,\mathrm{tr}\left\langle x\left|\frac{\tau_1}{\mathcal{H}_\phi+\mathrm{i}\kappa}\right|x\right\rangle\Bigg|_{x=-\infty}^{x=+\infty},\end{aligned} \tag{3.487}$$

as advertized.

If $\phi_\pm$ denote the asymptotic values of $\phi(x)$ as $x \to \pm\infty$ and taking into account that the gradient $(\nabla\phi)(x) \equiv \phi'(x)$ must vanish asymptotically, $\lim_{x\to\pm\infty}\phi'(x) = 0$, we can do the manipulations [with the help of Eqs. (3.475a) and (3.476)]

$$\begin{aligned}&\mathrm{tr}\left\langle x=\pm\infty\left|\frac{\tau_1}{\mathcal{H}_\phi+\mathrm{i}\kappa}\right|x=\pm\infty\right\rangle = \\ &\quad\mathrm{tr}\left\langle x=\pm\infty\left|\frac{\tau_1}{\mathcal{H}_\phi+\mathrm{i}\kappa}\,\frac{\mathcal{H}_\phi-\mathrm{i}\kappa}{\mathcal{H}_\phi-\mathrm{i}\kappa}\right|x=\pm\infty\right\rangle = \\ &\qquad 2\left\langle x=\pm\infty\left|\frac{\phi(x)}{-\nabla^2+\phi^2(x)+\mu_{\mathrm{s}}^2-z^2}\right|x=\pm\infty\right\rangle = \\ &\qquad\quad 2\,\phi_\pm\int\limits_{-\infty}^{+\infty}\frac{\mathrm{d}k}{2\pi\mathrm{i}}\,\frac{\mathrm{i}}{k^2+\phi_\pm^2+\mu_{\mathrm{s}}^2-z^2}.\end{aligned} \tag{3.488}$$

The integration over the momenta is to be performed with the help of the residue theorem. If, without loss of generality, the first-order pole $\mathrm{i}(\phi_\pm^2 + \mu_{\mathrm{s}}^2 - z^2)^{1/2}$ is chosen, then the residue $1/[2(\phi_{\pm}^2+\mu_{\rm s}^2-z^2)^{1/2}\,{\rm i}]$ follows. Consequently, Eq. (3.487) becomes

$$\left[\mathrm{Tr}\left(\frac{1}{\mathcal{H}-z}\right)\right]_{\rm even}=\frac{1}{2}\frac{\mu_{\rm s}}{\mu_{\rm s}^2-z^2}\sum_{\sigma=\pm}\frac{\sigma\,\phi_\sigma}{\sqrt{\phi_\sigma^2+\mu_{\rm s}^2-z^2}}. \tag{3.489}$$

Equation (3.471) follows from choosing $\phi_{\pm}=\pm\phi_\infty$. □

3.11.3.2 A Closer Look at the Thermal Corrections When SUSY Does Not Hold

When SUSY does not hold, we expect that $Q''(\beta)$ defined in Eq. (3.464c) has no topological character. To illustrate this fact, we consider the case of a static single kink in the sigma model background Φ defined in Eq. (3.447).

We do the local unitary transformation (3.344) and work with the partition function (3.349). The induced fermion charge Q' follows from the contribution (3.353) to the induced fermion current caused by the axial anomaly, while the induced fermion charge $Q''(\beta)$ follows solely from the analytical continuation (3.448) performed on the current defined by Eq. (3.351).

The current defined by Eq. (3.351) is the one induced by a massive Dirac Hamiltonian (with the mass $m>0$) with the static electric field

$$E(x):=-\frac{1}{2}(\Delta\varphi)(x)\equiv-\frac{1}{2}(\varphi'')(x), \tag{3.490}$$

if we combine Eq. (3.347) with Eq. (3.447). The static potential causing this static electric field enters the Dirac Hamiltonian as a space dependent chemical potential

$$\mu(x):=-\frac{1}{2}(\nabla\varphi)(x)\equiv-\frac{1}{2}(\varphi')(x). \tag{3.491}$$

If we assume that

$$|\mu(x)|\ll m, \tag{3.492a}$$

we may then perform the semiclassical approximation of Eq. (3.464c), by which

$$Q''(\beta)\approx\int\limits_{-\infty}^{+\infty}\mathrm{d}x\int\limits_{-\infty}^{+\infty}\frac{\mathrm{d}k}{2\pi}\left[f_{\mathrm{FD}\,+}(x,k)-f_{\mathrm{FD}\,-}(x,k)\right], \tag{3.492b}$$

where we have introduced the pair of semiclassical (local) Fermi-Dirac functions

$$f_{\mathrm{FD}\,\pm}(x,k):=\frac{1}{e^{\beta\left[\sqrt{k^2+m^2}\mp\mu(x)\right]}+1}. \tag{3.492c}$$

The interpretation of this approximation is that the static background field $\Phi(x)$ corresponding to the static kink (3.447) induces a slowly varying local electric charge density

$$\rho_{\mathrm{sc}}(x) := \int\limits_{-\infty}^{+\infty} \frac{\mathrm{d}k}{2\pi} \left[f_{\mathrm{FD}\,+}(x,k) - f_{\mathrm{FD}\,-}(x,k) \right], \tag{3.493}$$

in a semiclassical approximation that neglects the axial anomaly or other quantum effects such as the existence of bound states. In the low temperature limit,

$$\frac{1}{\beta} \ll m, \tag{3.494}$$

we may do the expansions

$$f_{\mathrm{FD}\,\pm}(x,k) = e^{-\beta\sqrt{k^2+m^2}} e^{\pm\beta\,\mu(x)} + \cdots, \tag{3.495a}$$

$$\begin{aligned} f_{\mathrm{FD}\,+}(x,k) - f_{\mathrm{FD}\,-}(x,k) &= +\,2\,e^{-\beta\sqrt{k^2+m^2}} \sinh\left(\beta\,\mu(x)\right) + \cdots \\ &= -\,2\,e^{-\beta\sqrt{k^2+m^2}} \sinh\left(\frac{\beta\,\varphi'(x)}{2}\right) + \cdots, \end{aligned} \tag{3.495b}$$

and

$$\begin{aligned} \int\limits_{-\infty}^{+\infty} \frac{\mathrm{d}k}{2\pi} e^{-\beta\sqrt{k^2+m^2}} &\sim \int\limits_{-\infty}^{+\infty} \frac{\mathrm{d}k}{2\pi} \exp\left(-\beta\,m \left[1 + \frac{1}{2}\left(\frac{k}{m}\right)^2 + \cdots\right]\right) \\ &= \sqrt{\frac{m}{2\pi\beta}}\, e^{-\beta\,m} + \cdots. \end{aligned} \tag{3.495c}$$

This gives the estimate

$$Q''(\beta) \sim -\sqrt{\frac{2m}{\pi\beta}}\, e^{-\beta\,m} \int\limits_{-\infty}^{+\infty} \mathrm{d}x\, \sinh\left(\frac{\beta\,\varphi'(x)}{2}\right) \tag{3.496a}$$

provided

$$\varphi'(x) \ll m, \qquad 1 \ll \beta\,m. \tag{3.496b}$$

As anticipated, the detailed functional shape of φ enters in Eq. (3.496). Any local and smooth change of φ changes the semiclassical value of $Q''(\beta)$.

The nonuniversal semiclassical correction (3.496) is the leading one if and only if there are no isolated bound states with the minimum single-particle energy $0 < |\varepsilon_{\mathrm{min}}| < m$. When this assumption fails,

$$Q''(\beta) \sim \mathrm{sgn}(\varepsilon_{\mathrm{min}})\, e^{-\beta\,|\varepsilon_{\mathrm{min}}|} \tag{3.497a}$$

provided

$$1 \ll \beta\,|\varepsilon_{\mathrm{min}}|. \tag{3.497b}$$

The value of $\varepsilon_{\mathrm{min}}$, unlike that for the threshold single-particle state when SUSY holds, is sensitive to the detailed shape of the function $\varphi(x)$ and has thus no topological character.

3.12 Stability with Respect to Fermion Interactions

3.12.1 Overview

The electrons of polyacetylene carry an electric charge. Hence, they must necessarily interact through the Coulomb interaction, an effect that we have ignored so far.

The Coulomb interaction $V_{\mathrm{Cb}}(\boldsymbol{r}-\boldsymbol{r}')$ between any two classical point charges located at $\boldsymbol{r}$ and $\boldsymbol{r}'$ in three-dimensional space and carrying the electric charges e and e', respectively, is

$$V_{\mathrm{Cb}}(\boldsymbol{r}-\boldsymbol{r}') = e\,\frac{1}{|\boldsymbol{r}-\boldsymbol{r}'|}\,e' \tag{3.498}$$

in Gaussian units. Now, the Coulomb interaction between this pair of test charges is screened by the electrons making up the many-body eigenstates. This is to say that the Fourier component

$$V_{\mathrm{Cb}}(\boldsymbol{q}) = e\,\frac{4\pi}{\boldsymbol{q}^2}\,e' \tag{3.499}$$

of the bare Coulomb interaction (3.498) at the momentum transfer $\boldsymbol{q}$ is turned by the quantum fluctuations originating from the electronic many-body eigenstates into the frequency and momentum dependent screened interaction

$$V^{\mathrm{eff}}_{\mathrm{Cb}}(\omega,\boldsymbol{q}) = \frac{1}{\varepsilon(\omega,\boldsymbol{q})}\,V_{\mathrm{Cb}}(\boldsymbol{q}). \tag{3.500}$$

The deviation of the dielectric function $\varepsilon(\omega,\boldsymbol{q})$ away from unity quantifies the amount by which the many-body eigenstates screen the bare Coulomb interaction (3.498).

Fermions that interact with each others *not too strongly* have a many-body ground state obtained by occupying all the single-particle states up to a single-particle energy, the Fermi energy, that is, fixed by the fermion density. When single-particle states are labeled by (crystal) momenta, it is often the case that the Fermi energy defines a surface at constant energy in the Brillouin zone called the Fermi surface. (In one-dimensional space, the Fermi surface nearly always collapses to Fermi points. In dimensions of space larger than one, there is a qualitative difference between Fermi points and Fermi surfaces.) Interacting fermions with a Fermi surface are called Fermi liquids.

For a Fermi liquid, occupied dressed single-particle states below the Fermi energy make up the metallic ground state. Pairs of dressed single-particle states made of a quasiparticle (an occupied dressed single-particle state above the Fermi energy) and a quasihole (an empty dressed single-particle state below the Fermi energy) make up many-body excited states. Many-body excited states built out of quasiparticle and quasihole excitations can be found with an energy arbitrary close to that of the Fermi sea in a Fermi liquid. Virtual quasiparticle-quasihole excitations in the vicinity of the Fermi surface (quantum fluctuations) determine the screening captured by the asymptotics

$$\lim_{\omega\to 0}\varepsilon(\omega,\boldsymbol{q})\propto\frac{1}{\boldsymbol{q}^2}\tag{3.501}$$

to leading order in an expansion in powers of $\boldsymbol{q}^2$. The behavior (3.501) implies that the effective two-body interaction between quasiparticles and quasiholes from the Fermi liquid decays exponentially fast with their separation in the limit by which $\omega\to 0$ before $\boldsymbol{q}\to\boldsymbol{0}$ owing to the quantum fluctuations (virtual quasiparticle-quasihole excitations). The ability for the electrons with the bare mass m and the density n making up the Fermi sea below the Fermi surface to screen the bare Coulomb interaction extends to all frequencies up to the plasma frequency

$$\varpi_{\mathrm{P}}:=(4\pi\,n\,e^2/m)^{-1/2}.\tag{3.502}$$

For an insulator, the many-body ground state is separated from all many-body excitations by a gap. When the insulating many-body ground state can be thought of as a Fermi sea in the Brillouin zone, it has no Fermi surface. Virtual particle-hole excitations (quantum fluctuations) above this gap determine the screening captured by the asymptotics

$$\lim_{\omega\to 0}\varepsilon(\omega,\boldsymbol{q})\propto 1\tag{3.503}$$

to leading order in an expansion in powers of $\boldsymbol{q}^2$. The behavior (3.503) implies that the effective two-body interaction between quasiparticle and quasiholes remains of the long-ranged form (3.498) in the limit by which $\omega\to 0$ before $\boldsymbol{q}\to\boldsymbol{0}$. However, if the single-particle gap 2Δ is smaller than the plasma energy $\hbar\varpi_{\mathrm{P}}$, then the effective two-body interaction (3.500) for the range of frequencies

$$\frac{2\Delta}{\hbar}\ll\omega\ll\varpi_{\mathrm{P}}\tag{3.504}$$

is screened very much in the same way as if the ground state would be metallic.

Modeling polyacetylene at zero temperature as a one-dimensional noninteracting tight-binding model is reasonable if the predicted insulating band gap is larger than the hopping amplitudes between the one-dimensional polymers bunching into polyacetylene. However, at zero temperature, it would not be consistent to treat the effects of electron-electron interactions in polyacetylene perturbatively, while ignoring the hopping amplitudes between the one-dimensional polymers bunching into polyacetylene.

Modeling the effects of electron-electron interactions in polyacetylene within a one-dimensional interacting tight-binding model for temperatures much larger than the characteristic strength of the hopping amplitude between the one-dimensional polymers bunching into polyacetylene is nevertheless plausible for two reasons.

First, the electron density n in polyacetylene is

$$n \sim 10^{22}\,\mathrm{cm}^{-1/3}. \tag{3.505}$$

Hence, the plasma frequency (3.502) becomes

$$\hbar\varpi_{\mathrm{P}} \sim 3.5\,\mathrm{eV} > 1.8\,\mathrm{eV} \sim 2\Delta \tag{3.506}$$

in polyacetylene.

Second, it has been argued[56] that if the Coulomb interaction in polyacetylene is treated perturbatively, then these perturbative corrections arise from integrals over frequencies and transfer momenta involving the bare Coulomb interaction. As long as such integrals are dominated by the range (3.504) with Eq. (3.506), the effective interaction is screened.

The minimal lattice model that captures the effect of interactions in polyacetylene, for not too low temperatures and for any observable for which the effective interaction is well approximated by a screened interaction, consists of the one-dimensional noninteracting tight-binding model (2.2) perturbed by short-ranged two-body interactions. Such one-dimensional interacting lattice models are then often solved by numerical means.

In the long-wave length limit corresponding to low energies, the qualitative nature of the numerical solutions is distilled by the noninteracting massive Dirac Hamiltonian (2.82) perturbed by short-ranged two-body interactions in the one-dimensional continuum.

The question to be addressed in Section 3.12 is the following. We shall interpret Figure 3.1(b) as a partitioning of the one-dimensional "universe" into two vacua that are distinguished by their dimerization patterns. There are two domain walls separating the two vacua. Both domain walls will be interpreted as the boundaries between a pair of one-dimensional "half-universes." Each boundary supports a bound state. The pair of bound states become degenerate zero modes as the separation between the domain walls is taken to infinity, an operation that defines the thermodynamic limit. From the perspective of any one of the two one-dimensional "half-universes" in the thermodynamic limit, there are two boundary zero modes, one at each end of the "half-universe." Proving their existence and the computation of their quantum numbers was achieved in the noninteracting limit. We shall establish the conditions under which these boundary zero modes are robust to short-ranged many-body interactions between the fermions. For polyacetylene, a necessary and sufficient condition will be that these interactions do not close the single-particle gap. However, following the pioneering work of Fidkowski and Kitaev,[57] it will be shown that this condition is not sufficient for a suitable generalization of polyacetylene. The suitable generalization that we shall construct

[56] S. Barisic, J. Phys. France **44**(2), 185–199 (1983); W.-K. Wu and S. Kivelson, Phys. Rev. B **33**(12), 8546–8557 (1986). [Barisic (1983); Wu and Kivelson (1986)].

[57] L. Fidkowski and A. Kitaev, Phys. Rev. B **81**(13), 134509 (2010). [Fidkowski and Kitaev (2010)].

relies on the observation that the one-dimensional model (3.4) for polyacetylene is an example of a one-dimensional topological insulator or, more generally, of a fermionic one-dimensional invertible topological phase of matter.

More precisely, we are going to explain that polyacetylene, as we have described it so far, is a special case of an insulator in the symmetry class BDI. The symmetry class BDI is characterized by two protecting symmetries, one of which is called a chiral symmetry. The chiral symmetry is shared in a defining way by two more symmetry classes, the symmetry class CII, and the symmetry class AIII. We are going to explain why noninteracting one-dimensional insulators in the chiral symmetry classes BDI, CII, and AIII can each be characterized by the group of integers $\mathbb{Z}$ and why this group is to be associated with a topological attribute. We shall then explain that the group $\mathbb{Z}$ that assigns a topological attribute is not stable to interactions that respect the defining symmetries of the symmetry classes BDI, CII, and AIII. If interaction driven instabilities leading to the breaking of charge conservation are ignored, we will show that the subgroups $\mathbb{Z}_4 \subset \mathbb{Z}$ for the symmetry classes BDI and AIII, and the subgroup $\mathbb{Z}_2 \subset \mathbb{Z}$ for the symmetry class CII are the reductions of the group $\mathbb{Z}$ brought about by interactions that preserve all the defining symmetries of BDI, CII, and AIII, respectively.

3.12.2 Conventions

We will use the following conventions. The operation of complex conjugation will be denoted by K. Linear maps from and to the two-dimensional vector space $\mathbb{C}^2$ shall be represented by 2×2 matrices that we expand in terms of the unit matrix τ_0 and the three Pauli matrices τ_1, τ_2, and τ_3. Linear maps from and to the four-dimensional vector space $\mathbb{C}^4 = \mathbb{C}^2 \otimes \mathbb{C}^2$ will be represented by 4×4 matrices that we expand in terms of the 16 Hermitean matrices

$$X_{\mu\mu'} \equiv \tau_\mu \otimes \sigma_{\mu'}, \qquad \mu, \mu' = 0, 1, 2, 3, \tag{3.507}$$

where σ_ν is a second set of 2×2 matrices comprised of the unit matrix and the three Pauli matrices. Linear maps from and to the 2^n-dimensional vector space $\mathbb{C}^{2^n} = \mathbb{C}^2 \otimes \cdots \otimes \mathbb{C}^2$ will be represented by $2^n \times 2^n$ matrices that we expand in terms of the 4^n Hermitean matrices

$$X_{\mu_1 \cdots \mu_n} \equiv \tau^{(1)}_{\mu_1} \otimes \tau^{(2)}_{\mu_2} \otimes \cdots \otimes \tau^{(n)}_{\mu_n} \tag{3.508}$$

where $\mu_1, \cdots, \mu_n = 0, 1, 2, 3$.

3.12.3 Definition of the Symmetry Class BDI

The electrons in polyacetylene carry two spin-1/2 degrees of freedom. This is reflected by the use of 4×4 Dirac matrices in the one-dimensional Dirac Hamiltonian (2.82). However, we have ignored the spin-1/2 of the electron when discussing domain walls in polyacetylene by trading spinfull electrons for spinless fermions. This was done without loss of generality when the spin-rotation symmetry holds in the noninteracting limit. Ignoring the spin-1/2 of the electron could matter, however, when considering the qualitative effects of many-body fermion interactions. To appreciate the qualitative differences that could be brought about by the spin-1/2 degrees of freedom, we shall thus consider both spinless and spinfull fermions.

Consider the one-dimensional bulk single-particle Dirac Hamiltonian

$$\mathcal{H}^{(0)}(x) := -\mathrm{i}\partial_x\,\tau_3 + m(x)\,\tau_2 \tag{3.509a}$$

with Dirac matrices of dimension (full rank)

$$r = 2 \equiv r_{\mathrm{min}}. \tag{3.509b}$$

This single-particle Hamiltonian is said to belong to the symmetry class BDI if and only if there exist two antiunitary transformations $\mathcal{T}$ and $\mathcal{C}$ such that

$$\mathcal{T}\,\mathcal{H}^{(0)}(x)\,\mathcal{T}^{-1} = +\mathcal{H}^{(0)}(x), \qquad \mathcal{T}^{\mathsf{T}} = +\mathcal{T}, \tag{3.509c}$$

$$\mathcal{C}\,\mathcal{H}^{(0)}(x)\,\mathcal{C}^{-1} = -\mathcal{H}^{(0)}(x), \qquad \mathcal{C}^{\mathsf{T}} = +\mathcal{C}. \tag{3.509d}$$

This is indeed the case as one may choose

$$\mathcal{T} := \tau_1\,\mathsf{K}, \qquad \mathcal{C} := \tau_0\,\mathsf{K}, \tag{3.509e}$$

where we recall that K denotes complex conjugation. The antiunitary transformation $\mathcal{T}$ represents reversal of time. The antiunitary transformation $\mathcal{C}$ represents charge conjugation, that is exchange of particles and holes in a second-quantized setting. They are chosen such that they commute. Remarkably, the unitary generator

$$\mathcal{S} := \mathcal{T}\,\mathcal{C} = \tau_1 \tag{3.509f}$$

for the chiral transformation is recovered by composing $\mathcal{T}$ and $\mathcal{C}$. In the symmetry class BDI, only two of the TRS, PHS, and CHS are independent of the third one.

The Dirac mass matrix τ_2 is here the only one allowed for 2×2 Dirac matrices under the constraints (3.509c) and (3.509d) (**Exercise** 3.34). Hence, we conclude that a one-dimensional Dirac Hamiltonian that is represented by $2{\times}2$ Dirac matrices and belongs to the symmetry class BDI admits, up to a multiplicative sign, a unique Dirac mass matrix that squares to the unit 2×2 matrix τ_0. By construction, halving the dimension (full rank) of the Dirac matrices forbids defining a mass term. This is why the subscript "min" is added to the dimension (full rank) (3.509b).

We recall that when translation symmetry is broken by the mass term supporting the domain wall

$$m(x) = m_{\infty}\,\mathrm{sgn}(x), \qquad m_{\infty} \in \mathbb{R}, \tag{3.510a}$$

at $x=0$, then the zero mode

$$e^{-\mathrm{i}\tau_3\,\tau_2\int\limits_0^x \mathrm{d}x'\,m(x')}\chi = e^{-|m_\infty\,x|}\,\chi, \tag{3.510b}$$

where

$$\tau_1\,\chi = \mathrm{sgn}\,(m_\infty)\,\chi, \tag{3.510c}$$

is the only normalizable state bound to this domain wall. This boundary state is an eigenstate of the single-particle boundary Hamiltonian

$$\mathcal{H}^{(0)}_{\mathrm{bd}} = 0. \tag{3.510d}$$

Suppose that we consider $\nu = 1,2,\cdots$ identical copies of the single-particle Hamiltonian (3.509) by defining

$$\mathcal{H}^{(0)}_{\nu}(x) := \mathcal{H}^{(0)}(x)\otimes \mathbb{1}, \tag{3.511a}$$

and

$$\mathcal{T} := \tau_1\otimes\mathbb{1}\,\mathsf{K}, \qquad \mathcal{C} := \tau_0\otimes\mathbb{1}\,\mathsf{K}, \tag{3.511b}$$

where $\mathbb{1}$ is a $\nu\times\nu$ unit matrix. Observe that $\mathcal{T}$ and $\mathcal{C}$ commute with $\tau_1\otimes\mathbb{1}$ and with each other. The domain wall (3.510a) must then support ν linearly independent boundary zero modes. They are annihilated by the single-particle boundary Hamiltonian

$$\mathcal{H}^{(0)}_{\mathrm{bd}\,\nu} = \mathcal{H}^{(0)}_{\mathrm{bd}}\otimes\mathbb{1} = 0. \tag{3.512}$$

This number ν of zero modes is endowed with the following robustness. First, there is a unique mass matrix of the form $\tau_2\otimes\mathbb{1}$ and, second, any other $2\nu\times 2\nu$ mass matrix allowed in the symmetry class BDI must commute with $\tau_2\otimes\mathbb{1}$. This is to say that the ν zero modes bounded to the domain wall (3.510a) entering $\mathcal{H}^{(0)}_{\nu}(x)$ remain eigenstates in the presence of any other $2\nu\times 2\nu$ mass matrix allowed in the symmetry class BDI.

All $2\nu\times 2\nu$ one-dimensional massive Dirac Hamiltonians in the symmetry class BDI are of the form[58]

$$\mathcal{H}^{(0)}_{M}(x) := -\mathrm{i}\partial_x\,\tau_3\otimes\mathbb{1} + \tau_2\otimes M(x) + \cdots \tag{3.513a}$$

with $M(x)$ a $\nu\times\nu$ Hermitean matrix obeying the symmetry constraint

$$M^{*}(x) = M(x) \tag{3.513b}$$

and $\cdots$ representing any $2\nu\times 2\nu$ Hermitean matrix that does not anticommute with $\tau_3\otimes\mathbb{1}$ and is allowed by the BDI symmetry constraints.

[58] One verifies that the term $\tau_1\otimes M'(x)$ with any Hermitean $\nu\times\nu$ matrix $M'(x)$ is not allowed by the chiral spectral symmetry.

We assume that the matrix elements of $M(x)$ are smooth functions of x. We may then partition the line $x \in \mathbb{R}$ into open intervals Ω_ι defined by the condition

$$\frac{\mathrm{d}}{\mathrm{d}x}\mathrm{sgn}\{\det[M(x)]\} = 0, \qquad \forall x \in \Omega_\iota, \tag{3.514a}$$

and separated by boundaries at which

$$\det[M(x)] = 0. \tag{3.514b}$$

Each connected domain Ω_ι of the line can be indexed by the index

$$\xi_\iota := \frac{1}{2}\,\mathrm{tr}\,\{\mathrm{sgn}\,[U^\dagger(x)\,M(x)\,U(x)]\}, \qquad \forall x \in \Omega_\iota, \tag{3.514c}$$

where $U(x)$ is the unitary matrix that diagonalizes $M(x)$. We take the thermodynamic limit by which the size of any connected domain Ω_ι diverges. The integer number $\nu_{\iota,\iota+1}$ of zero modes that are bounded to a domain wall between two consecutive domains Ω_ι and $\Omega_{\iota+1}$ is then

$$\nu_{\iota,\iota+1} = |\xi_\iota - \xi_{\iota+1}|. \tag{3.514d}$$

This index counts how many eigenvalues of $M(x)$ vanish on a domain wall of $M(x)$. In the limit $\nu \to \infty$, the topological index $\nu_{\iota,\iota+1}$ takes values in $\mathbb{Z}$.

If the Hamiltonian of polyacetylene obeys the spin-rotation symmetry, it reduces to two independent sectors, one for the spin up and one for the spin down. Each of these sectors corresponds to the case of $\nu = 1$ above if described by a single-particle massive Dirac Hamiltonian. However, spin-rotation symmetry can be broken while preserving the time-reversal symmetry. If so, it is necessary to define the operation of time reversal by fully accounting for its action on the spin-1/2 degrees of freedom carried by an electron as we do next.

3.12.4 Definition of the Symmetry Class CII

We start from the following generalization of the one-dimensional continuum model (2.82) for polyacetylene in the chiral basis (high-energy terminology), that is, in the basis of right- and left-movers in the condensed matter terminology. This generalization is indispensable in the noninteracting limit if spin-rotation symmetry is broken while time-reversal symmetry is maintained, say in the presence of spin-orbit coupling.

Consider the one-dimensional bulk single-particle Dirac Hamiltonian

$$\mathcal{H}^{(0)}(x) := -\mathrm{i}\partial_x\,X_{30} + m(x)\,X_{20} \tag{3.515a}$$

with Dirac matrices of dimension (full rank)

$$r = 4 \equiv r_{\min}. \tag{3.515b}$$

This single-particle Hamiltonian is said to belong to the symmetry class CII if and only if there exist two antiunitary transformations $\mathcal{T}$ and $\mathcal{C}$ such that

$$\mathcal{T}\,\mathcal{H}^{(0)}(x)\,\mathcal{T}^{-1} = +\mathcal{H}^{(0)}(x), \qquad \mathcal{T}^{\mathsf{T}} = -\mathcal{T}, \tag{3.515c}$$

$$\mathcal{C}\,\mathcal{H}^{(0)}(x)\,\mathcal{C}^{-1} = -\mathcal{H}^{(0)}(x), \qquad \mathcal{C}^{\mathsf{T}} = -\mathcal{C}. \tag{3.515d}$$

This is indeed the case as one may choose

$$\mathcal{T} := \mathrm{i}X_{12}\,\mathsf{K}, \qquad \mathcal{C} := \mathrm{i}X_{02}\,\mathsf{K}, \tag{3.515e}$$

where we recall that K denotes complex conjugation. The antiunitary transformation $\mathcal{T}$ represents reversal of time. The antiunitary transformation $\mathcal{C}$ represents charge conjugation, that is exchange of particles and holes in a second-quantized setting. The multiplicative phase factor i insures that $\mathcal{T}$ and $\mathcal{C}$ commute. Remarkably, the unitary generator

$$\mathcal{S} := -\mathcal{T}\,\mathcal{C} = X_{10} \tag{3.515f}$$

for the chiral transformation is recovered by composing $\mathcal{T}$ and $\mathcal{C}$. In the symmetry class CII, only two of the TRS, PHS, and CHS are independent of the third one.

The Dirac mass matrix X_{20} is here the only one allowed for 4×4 Dirac matrices under the constraints (3.515c) and (3.515d) (**Exercise** 3.35). Hence, we conclude that a one-dimensional Dirac Hamiltonian that is represented by $4{\times}4$ Dirac matrices and belongs to the symmetry class CII admits, up to a multiplicative sign, a unique Dirac mass matrix that squares to the unit 4×4 matrix $\tau_0 \otimes \sigma_0$. One verifies that halving the dimension (full rank) of the Dirac matrices forbids defining a mass term. This is why the subscript "min" is added to the dimension (full rank) (3.515b).

We recall that when translation symmetry is broken by the mass term supporting the domain wall

$$m(x) = m_{\infty}\,\mathrm{sgn}(x), \qquad m_{\infty} \in \mathbb{R}, \tag{3.516a}$$

at $x = 0$, then the zero mode

$$e^{-\mathrm{i}X_{30}\,X_{20}\int_0^x \mathrm{d}x'\,m(x')}\,\chi = e^{-|m_{\infty}\,x|}\,\chi, \tag{3.516b}$$

where

$$X_{10}\,\chi = \mathrm{sgn}\,(m_{\infty})\,\chi, \tag{3.516c}$$

is a normalizable state bound to this domain wall. This boundary state is an eigenstate of the single-particle boundary Hamiltonian

$$\mathcal{H}^{(0)}_{\mathrm{bd}} = 0. \tag{3.516d}$$

Equation (3.516c) is solved by two linearly independent χ's. This is a consequence of Kramers' theorem (**Exercise** 3.36).[59] Alternatively, one may state that there is a single pair of Kramers' degenerate zero modes.

Suppose that we consider $\nu = 1, 2, \cdots$ identical copies of the single-particle Hamiltonian (3.515) by defining

$$\mathcal{H}^{(0)}_{\nu}(x) := \mathcal{H}^{(0)}(x) \otimes \mathbb{1}, \tag{3.517a}$$

and

$$\mathcal{T} := \mathrm{i}X_{12} \otimes \mathbb{1}\,\mathsf{K}, \qquad \mathcal{C} := \mathrm{i}X_{02} \otimes \mathbb{1}\,\mathsf{K}, \tag{3.517b}$$

where $\mathbb{1}$ is a $\nu \times \nu$ unit matrix. Observe that $\mathcal{T}$ and $\mathcal{C}$ commute with $X_{10} \otimes \mathbb{1}$ and with each other. The domain wall (3.516a) must then support ν linearly independent pairs of Kramers' degenerate zero modes. They are annihilated by the single-particle boundary Hamiltonian

$$\mathcal{H}^{(0)}_{\mathrm{bd}\,\nu} = \mathcal{H}^{(0)}_{\mathrm{bd}} \otimes \mathbb{1} = 0. \tag{3.518}$$

This number $2\,\nu$ of zero modes is endowed with the following robustness. First, there is a unique mass matrix of the form $X_{20} \otimes \mathbb{1}$ and, second, any other $4\nu \times 4\nu$ mass matrix allowed in the symmetry class CII must commute with $X_{20} \otimes \mathbb{1}$. This is to say that the ν pairs of Kramers' degenerate zero modes bounded to the domain wall (3.516a) entering $\mathcal{H}^{(0)}_{\nu}(x)$ remain eigenstates in the presence of any other $4\nu \times 4\nu$ mass matrix allowed in the symmetry class CII.

All $4\nu \times 4\nu$ one-dimensional massive Dirac Hamiltonians in the symmetry class CII are of the form[60]

$$\mathcal{H}^{(0)}_{M}(x) := -\mathrm{i}\partial_x\, X_{30} \otimes \mathbb{1} + X_{20} \otimes M(x) + \cdots, \tag{3.519a}$$

with $M(x)$ a $\nu \times \nu$ Hermitean matrix obeying the symmetry constraint

$$M^{*}(x) = M(x), \tag{3.519b}$$

and $\cdots$ representing any $4\nu \times 4\nu$ Hermitean matrix that does not anticommute with $X_{30} \otimes \mathbb{1}$ and is allowed by the CII symmetry constraints.

We assume that the matrix elements of $M(x)$ are smooth functions of x. We may then partition the line $x \in \mathbb{R}$ into open intervals Ω_{ι} defined by the condition

$$\frac{\mathrm{d}}{\mathrm{d}x}\mathrm{sgn}\{\det\left[M(x)\right]\} = 0, \qquad \forall x \in \Omega_{\iota}, \tag{3.520a}$$

and separated by boundaries at which

$$\det\left[M(x)\right] = 0. \tag{3.520b}$$

[59] Kramers showed in H. A. Kramers, Proceedings of the Royal Netherlands Academy of Arts and Sciences **33**(9), 959–972 (1930) [Kramers (1930)], that, for every energy eigenstate of a time-reversal symmetric Hamiltonian with half-integer total spin, there is at least one more eigenstate with the same energy.

[60] One verifies that the term $X_{10} \otimes M'(x)$ with any Hermitean $\nu \times \nu$ matrix $M'(x)$ is not allowed by the chiral spectral symmetry.

Each connected domain Ω_ι of the line can be indexed by the index

$$\xi_\iota := \frac{1}{2}\,\mathrm{tr}\,\{\mathrm{sgn}\,[U^\dagger(x)\,M(x)\,U(x)]\}, \qquad \forall x \in \Omega_\iota, \tag{3.520c}$$

where $U(x)$ is the unitary matrix that diagonalizes $M(x)$. We take the thermodynamic limit by which the size of any connected domain Ω_ι diverges. The integer number $\nu_{\iota,\iota+1}$ of pairs of Kramers' degenerate zero modes that are bounded to a domain wall between two consecutive domains Ω_ι and $\Omega_{\iota+1}$ is then

$$\nu_{\iota,\iota+1} = |\xi_\iota - \xi_{\iota+1}|. \tag{3.520d}$$

This index counts how many eigenvalues of $M(x)$ vanish on a domain wall of $M(x)$. In the limit $\nu \to \infty$, the topological index $\nu_{\iota,\iota+1}$ takes values in $\mathbb{Z}$.

3.12.5 Definition of the Symmetry Class AIII

Time-reversal symmetry is broken by a magnetic field in two ways. For spinless fermions, we must shift the momentum operator by the vector gauge field whose rotation is the magnetic field. In addition to this orbital effect, a magnetic field couples to the spin-1/2 degrees of freedom of the electron through a Zeeman term. In the symmetry class BDI, the breaking of time-reversal symmetry by an applied magnetic field enters solely through the orbital effect. In the symmetry class CII, a Zeeman term must also be accounted for in the presence of an applied magnetic field. In the case of a strong and uniform magnetic field (larger than the band width without magnetic field), the Zeeman term polarizes all the spin-1/2 degrees of freedom making up the Fermi sea. All fermions participating to the many-body ground state are effectively spinless, the magnetic field manifests itself only through the orbital effect at all energies below the single-particle Zeeman energy gap.

Consider the one-dimensional bulk single-particle Dirac Hamiltonian

$$\mathcal{H}^{(0)}(x) := -\mathrm{i}\partial_x \tau_3 + m(x)\,\tau_2 \tag{3.521a}$$

with Dirac matrices of dimension (full rank)

$$r = 2 \equiv r_{\mathrm{min}}. \tag{3.521b}$$

This single-particle Hamiltonian is said to belong to the symmetry class AIII if and only if there exists one unitary transformation Γ_5 that anticommutes with $\mathcal{H}^{(0)}(x)$:

$$\left\{\Gamma_5, \mathcal{H}^{(0)}(x)\right\} = 0. \tag{3.521c}$$

This is indeed the case as one may choose

$$\Gamma_5 := \tau_1. \tag{3.521d}$$

As was the case in the symmetry class BDI, the Dirac mass matrix τ_2 is here the only one allowed for 2×2 Dirac matrices under the constraint (3.521c) (**Exercise** 3.34). Hence, we conclude that a one-dimensional Dirac Hamiltonian that is represented by 2×2 Dirac matrices and belongs to the symmetry class AIII admits, up to a multiplicative sign, a unique Dirac mass matrix that squares to the unit 2×2 matrix τ_0. By construction, halving the dimension (full rank) of the Dirac matrices forbids defining a mass term. This is why the subscript "min" is added to the dimension (full rank) (3.521b).

The single-particle Hamiltonian (3.521a) supports the zero mode (3.510) at the boundary where it identically vanishes:

$$\mathcal{H}^{(0)}_{\mathrm{bd}}(x) = 0. \tag{3.522}$$

Upon tensoring Hamiltonian (3.521a) and the matrix (3.521d) by the $\nu \times \nu$ unit matrix $\mathbb{1}$, so that

$$\mathcal{H}^{(0)}(x) \mapsto \mathcal{H}^{(0)}_{\nu}(x) := \mathcal{H}^{(0)}(x) \otimes \mathbb{1}, \qquad \tau_1 \mapsto \Gamma_5 := \tau_1 \otimes \mathbb{1}, \tag{3.523}$$

there follows $\nu = 1, 2, 3, \cdots$ boundary zero modes. This number ν of zero modes is endowed with the following robustness. First, there is a unique mass matrix of the form $\tau_2 \otimes \mathbb{1}$ and, second, any other $2\nu \times 2\nu$ mass matrix allowed in the symmetry class AIII must commute with $\tau_2 \otimes \mathbb{1}$. This is to say that the ν zero modes bounded to the domain wall (3.510a) entering $\mathcal{H}^{(0)}_{\nu}(x)$ remain eigenstates in the presence of any other $2\nu \times 2\nu$ mass matrix allowed in the symmetry class AIII.

All $2\nu \times 2\nu$ one-dimensional massive Dirac Hamiltonians in the symmetry class AIII are of the form[61]

$$\mathcal{H}^{(0)}_{M}(x) := -\mathrm{i}\partial_x\, \tau_3 \otimes \mathbb{1} + \tau_2 \otimes M(x) + \cdots \tag{3.524a}$$

with $M(x)$ a $\nu \times \nu$ Hermitean matrix and $\cdots$ representing any $2\nu \times 2\nu$ Hermitean matrix that does not anticommute with $\tau_3 \otimes \mathbb{1}$ and is allowed by the AIII symmetry constraints, that is,

$$\left\{\Gamma_5, \mathcal{H}^{(0)}_{M}\right\} = 0. \tag{3.524b}$$

We assume that the matrix elements of $M(x)$ are smooth functions of x. We may then partition the line $x \in \mathbb{R}$ into open intervals Ω_ι defined by the condition

$$\frac{\mathrm{d}}{\mathrm{d}x}\mathrm{sgn}\{\det\left[M(x)\right]\} = 0, \qquad \forall x \in \Omega_\iota, \tag{3.525a}$$

and separated by boundaries at which

$$\det\left[M(x)\right] = 0. \tag{3.525b}$$

[61] One verifies that the term $\tau_1 \otimes M'(x)$ with any Hermitean $\nu \times \nu$ matrix $M'(x)$ is not allowed by the chiral spectral symmetry.

Each connected domain Ω_ι of the line can be indexed by the index

$$\xi_\iota := \frac{1}{2}\operatorname{tr}\{\operatorname{sgn}[U^\dagger(x)\,M(x)\,U(x)]\}, \qquad \forall x \in \Omega_\iota, \tag{3.525c}$$

where $U(x)$ is the unitary matrix that diagonalizes $M(x)$. We take the thermodynamic limit by which the size of any connected domain Ω_ι diverges. The integer number $\nu_{\iota,\iota+1}$ of zero modes that are bounded to a domain wall between two consecutive domains Ω_ι and $\Omega_{\iota+1}$ is then

$$\nu_{\iota,\iota+1} = |\xi_\iota - \xi_{\iota+1}|. \tag{3.525d}$$

This index counts how many eigenvalues of $M(x)$ vanish on a domain wall of $M(x)$. In the limit $\nu \to \infty$, the topological index $\nu_{\iota,\iota+1}$ takes values in $\mathbb{Z}$.

3.12.6 Short-Range Quartic Interactions: Strategy

We have introduced the three symmetry classes BDI, CII, and AIII by examining single-particle massive Dirac Hamiltonians in one-dimensional space. These symmetries can be realized for lattice models if the lattice is bipartite and the single-particle Hamiltonian is off-diagonal with respect to the sublattice grading, irrespectively of the dimensionality of the bipartite lattice. The symmetry class BDI is the most symmetric one, for the spin-rotation symmetry holds together with the time-reversal symmetry in addition to the chiral symmetry. The symmetry class CII applies when the spin-rotation symmetry for half-integer spins is broken while reversal of time remains a symmetry in addition to the chiral symmetry. The symmetry class AIII is the less symmetric one, for the chiral symmetry holds while reversal of time is not a symmetry anymore.

For example, the tight-binding model (2.2) with open boundary conditions or with periodic boundary conditions for an even number of sites falls into the symmetry class BDI when the hopping amplitudes are all real-valued.[62] When periodic boundary conditions are imposed with an even number of sites and time-reversal symmetry is broken, the tight-binding model (2.2) falls into the symmetry class AIII. Finally, symmetry class CII applies to the generalization

$$\widehat{H} := -\sum_{i=1}^{N} \sum_{\sigma,\sigma'=\uparrow,\downarrow} \left(\hat{c}^\dagger_{i,\sigma}\, t_{i,\sigma\sigma'}\, \hat{c}_{i+1,\sigma'} + \text{H.c.}\right) \tag{3.526}$$

of the tight-binding model (2.2) by which one demands that some of the $t_{i,\sigma\sigma'}$ be purely imaginary, while maintaining time-reversal symmetry.

Any defectuous dimerization pattern for the hopping amplitudes in the tight-binding model (2.2) as depicted in Figure 2.12 binds zero modes to the defect in

[62] When spin-rotation symmetry holds, the single-particle Hamiltonian is decomposable. It is the direct sum of two single-particle Hamiltonians, one for spin up and one for spin down. Each of these irreducible blocks belongs to the symmetry class BDI.

the thermodynamic limit by which the distance between any two consecutive defects is taken to infinity. We generalize the tight-binding model (2.2) by writing

$$\widehat{H} := -\sum_{i=1}^{N}\sum_{\sigma,\sigma'=1}^{2\nu}\left(\hat{c}^{\dagger}_{i,\sigma}\, t_{i,\sigma\sigma'}\, \hat{c}_{i+1,\sigma'} + \text{H.c.}\right), \tag{3.527}$$

where it is now understood that t_i in Eq. (3.4) is replaced by a $2\nu\times2\nu$ Hermitean matrix and $\hat{c}_i \equiv (\hat{c}_{i,\sigma})$ (for given i) is an operator-valued vector whose 2ν components $\hat{c}_{i,\sigma}$ $(\sigma = 1,\cdots,2\nu)$ obey the fermion algebra. We then deduce that a defectuous dimerization pattern on the site i binds at most 2ν zero modes to this site. The two restrictions of Hamiltonian (3.527) to the pair of sets $\{t_i,\, i\in I\} \equiv \{(t_{i,\sigma\sigma'}),\, i\in I\}$ and $\{t_j,\, j\in J\} \equiv \{(t_{j,\sigma\sigma'}),\, j\in J\}$ on two consecutive intervals I and J, whereby any one of I and J is made of consecutive sites from the lattice, are topologically inequivalent if an integer number from the set $\{1,2,\cdots,2\nu\}$ of zero modes is bound to the boundary at which the dimerization pattern separating the two intervals is defectuous. This is to say that any restriction of Hamiltonian (3.527) to a macroscopically large interval of the lattice with a dimerization pattern free of defects can be assigned a topological number that is in one-to-one correspondence with the set of the integers in the limit $\nu\to\infty$.

In one-dimensional space, any suitable family of noninteracting Hamiltonians belonging to one of the symmetry classes BDI, CII, and AIII – say a family of tight-binding Hamiltonians of the form (3.527) with $\nu\to\infty$ that depends parametrically on a set of smooth parameters – can be organized into equivalence classes of Hamiltonians. Two Hamiltonians are here understood to be equivalent through the number of zero modes that they support at the boundary of space if open boundary conditions are used. We shall call these equivalence classes topological classes. Any two members within a topological class can be deformed into each other by a smooth (adiabatic) deformation of their matrix elements without closing the bulk energy gap. These equivalence classes are endowed with an Abelian group structure $\mathfrak{G}$, whereby we have shown that

$$\mathfrak{G} = \mathbb{Z}. \tag{3.528}$$

We consider many-body interactions that are short-ranged and preserve the protecting symmetries of the noninteracting limit explicitly. These many-body interactions may be strong on the boundary, yet are not too strong as measured against the single-particle gap for the bulk states of the insulators.

We shall explain why the noninteracting topological classification with the Abelian group $\mathfrak{G}$ breaks down in the presence of such many-body interactions for the symmetry classes BDI, CII, and AIII. More precisely, an Abelian group $\mathfrak{G}_{\text{int}}$ that encodes the topological classes of gapped ground states for interacting fermions can be smaller than $\mathfrak{G}$ as a group (some quotient group of $\mathfrak{G}$).

Our strategy will be to start from the boundary single-particle Hamiltonian

$$\mathcal{H}^{(0)}_{\mathrm{bd}} := 0 \tag{3.529}$$

supporting ν zero modes that we derived starting from a one-dimensional massive Dirac Hamiltonian in the symmetry classes AIII, BDI, and CII, respectively. We will then add boundary interactions that are compatible with the symmetry classes AIII, BDI, and CII, respectively.

3.12.7 Short-Range Quartic Interactions: Outcomes

To set the stage, we begin with the single-particle Hamiltonian (3.524) in the symmetry class AIII, which we rewrite as

$$\widehat{H}^{(0)} := \int \mathrm{d}x\, \widehat{\Psi}^{\dagger}(x)\left[-\mathrm{i}\partial_x\, \tau_3 \otimes \mathbb{1} + \tau_2 \otimes M(x)\right] \widehat{\Psi}(x) \tag{3.530a}$$

in order to accommodate interactions below. Here,

$$\left\{\widehat{\Psi}_j(x), \widehat{\Psi}^{\dagger}_{j'}(x')\right\} = \delta_{jj'}\, \delta(x-x'), \qquad j,j' = 1,\cdots,2\nu, \tag{3.530b}$$

are the only nonvanishing equal-time anticommutators.

We define the chiral transformation $\widehat{\mathrm{S}}$ in two steps. First, the actions of $\widehat{\mathrm{S}}$ on the creation and annihilation operators follow from

$$\widehat{\mathrm{S}}\, \widehat{\Psi}_j(x)\, \widehat{\mathrm{S}}^{-1} := \sum_{j'=1}^{2\nu} \mathcal{S}_{j'j}\, \widehat{\Psi}^{\dagger}_{j'}(x) \equiv \sum_{j'=1}^{2\nu} \Gamma^5_{j'j}\, \widehat{\Psi}^{\dagger}_{j'}(x), \qquad j = 1,\cdots,2\nu. \tag{3.531a}$$

Second, we *choose* to extend these actions *antilinearly* on any linear combination with complex-valued coefficients of products of creation and annihilation operators. In particular, the first quantized representation of $\widehat{\mathrm{S}}$ is the composition $\mathcal{S}\,\mathsf{K}$ of the unitary transformation $\mathcal{S} \equiv \Gamma^5 = \tau_1 \otimes \mathbb{1}$ with the operation for complex conjugation K. By making use of **Exercise** 3.37, one verifies that

$$\widehat{\mathrm{S}}\, \widehat{H}^{(0)}\, \widehat{\mathrm{S}}^{-1} = \widehat{H}^{(0)}. \tag{3.531b}$$

Moreover, U(1) charge conservation holds since one may freely multiply the creation and annihilation operators by a global phase, that is,

$$\hat{U}(\alpha)\, \widehat{H}^{(0)}\, \hat{U}^{-1}(\alpha) = \widehat{H}^{(0)} \tag{3.532a}$$

for any $\hat{U}(\alpha)$ defined by the linear extension of the rule

$$\begin{aligned} \hat{U}(\alpha)\, \widehat{\Psi}^{\dagger}(x)\, \hat{U}^{-1}(\alpha) &:= e^{-\mathrm{i}\alpha}\, \widehat{\Psi}^{\dagger}(x), \qquad 0 \le \alpha < 2\pi, \\ \hat{U}(\alpha)\, \widehat{\Psi}(x)\, \hat{U}^{-1}(\alpha) &:= e^{+\mathrm{i}\alpha}\, \widehat{\Psi}(x), \qquad 0 \le \alpha < 2\pi. \end{aligned} \tag{3.532b}$$

An example of a bulk interacting Hamiltonian in the symmetry class AIII is

$$\widehat{H} := \widehat{H}^{(0)} + \widehat{H}^{\mathrm{int}} \tag{3.533a}$$

with the noninteracting bulk contribution (3.530) and a quartic local bulk interaction

$$\widehat{H}^{\mathrm{int}} := \int \mathrm{d}x \left[\left(\widehat{\Psi}^{\dagger} \tau_2 \otimes V_2 \widehat{\Psi} \right)^2 (x) + \left(\widehat{\Psi}^{\dagger} \tau_1 \otimes V_1 \widehat{\Psi} \right)^2 (x) \right], \tag{3.533b}$$

where $V_1 = V_1^{\dagger}$ and $V_2 = V_2^{\dagger}$ are any pair of Hermitean $\nu \times \nu$ matrices.

If we define a boundary at $x = 0$ by imposing that $\det M(x)$ is a smooth function of x that changes sign once at $x = 0$ on the real line $x \in \mathbb{R}$ with a zero of order ν, then the noninteracting boundary Hamiltonian

$$\widehat{H}^{(0)}_{\mathrm{bd}} = 0 \tag{3.534}$$

vanishes identically. Along the boundary, $\widehat{H}^{(0)}_{\mathrm{bd}}$ thus supports ν localized zero modes that are all eigenstates of Γ_5 with the same eigenvalue of Γ_5, which we will choose to be $+1$ without loss of generality. We associate to each of these zero modes the operators $\widehat{\Psi}^{\mathsf{a}}_{\mathrm{zm}}$ with $\mathsf{a} = 1, \cdots, \nu$ satisfying the algebra with

$$\left\{ \widehat{\Psi}^{\mathsf{a}}_{\mathrm{zm}}, \widehat{\Psi}^{\mathsf{b}\dagger}_{\mathrm{zm}} \right\} = \delta^{\mathsf{ab}}, \qquad \mathsf{a}, \mathsf{b} = 1, \cdots, \nu, \tag{3.535}$$

the only nonvanishing anticommutators. The Hilbert space on the boundary is the 2^{ν}-dimensional vector space

$$\mathfrak{F}^{\mathrm{AIII}}_{\mathrm{bd}} := \mathrm{span} \left\{ \prod_{\mathsf{a}=1}^{\nu} \left(\widehat{\Psi}^{\mathsf{a}\dagger}_{\mathrm{zm}} \right)^{n^{\mathsf{a}}_{\mathrm{zm}}} |0\rangle \,\middle|\, n^{\mathsf{a}}_{\mathrm{zm}} = 0, 1, \right. \\ \left. \widehat{\Psi}^{\mathsf{a}}_{\mathrm{zm}} |0\rangle = 0, \qquad \mathsf{a} = 1, \cdots, \nu \right\}, \tag{3.536}$$

that is a 2^{ν}-dimensional fermionic Fock space. The action of the bulk chiral transformation on the boundary reduces to the antilinear unitary transformation

$$\widehat{\mathsf{S}}_{\mathrm{bd}} \left(z\, \widehat{\Psi}^{\mathsf{a}}_{\mathrm{zm}} + w\, \widehat{\Psi}^{\mathsf{a}\dagger}_{\mathrm{zm}} \right) \widehat{\mathsf{S}}^{-1}_{\mathrm{bd}} = z^*\, \widehat{\Psi}^{\mathsf{a}\dagger}_{\mathrm{zm}} + w^*\, \widehat{\Psi}^{\mathsf{a}}_{\mathrm{zm}}, \qquad \mathsf{a} = 1, \cdots, \nu, \tag{3.537}$$

for any pair of complex numbers $z, w \in \mathbb{C}$, since there is no distinction anymore between left and right movers on the boundary. The global U(1) gauge symmetry on the boundary is represented by the linear extension of the rule

$$\begin{aligned} \hat{U}_{\mathrm{bd}}(\alpha)\, \widehat{\Psi}^{\mathsf{a}\dagger}_{\mathrm{zm}}\, \hat{U}^{-1}_{\mathrm{bd}}(\alpha) &:= e^{-\mathrm{i}\alpha}\, \widehat{\Psi}^{\mathsf{a}\dagger}_{\mathrm{zm}}, \qquad 0 \le \alpha < 2\pi, \\ \hat{U}_{\mathrm{bd}}(\alpha)\, \widehat{\Psi}^{\mathsf{a}}_{\mathrm{zm}}\, \hat{U}^{-1}_{\mathrm{bd}}(\alpha) &:= e^{+\mathrm{i}\alpha}\, \widehat{\Psi}^{\mathsf{a}}_{\mathrm{zm}}, \qquad 0 \le \alpha < 2\pi. \end{aligned} \tag{3.538}$$

The most general Hermitean interaction on the boundary that preserves the global U(1) symmetry generated by Eq. (3.538) must be of the form

$$\widehat{H}^{\mathrm{int}}_{\mathrm{bd}} := \sum_{\mu=1}^{\nu} \frac{V^{[\mathsf{a}_1\cdots\mathsf{a}_\mu][\mathsf{b}_1\cdots\mathsf{b}_\mu]}_{\mathrm{bd}}}{(\mu!)^2} \epsilon^{\mathsf{a}_1\cdots\mathsf{a}_\mu} \epsilon^{\mathsf{b}_1\cdots\mathsf{b}_\mu} \widehat{\Psi}^{\mathsf{a}_1\dagger}_{\mathrm{zm}} \cdots \widehat{\Psi}^{\mathsf{a}_\mu\dagger}_{\mathrm{zm}} \widehat{\Psi}^{\mathsf{b}_1}_{\mathrm{zm}} \cdots \widehat{\Psi}^{\mathsf{b}_\mu}_{\mathrm{zm}} + \mathrm{H.c.}, \tag{3.539a}$$

where the symbol $[\mathsf{a}_1 \cdots \mathsf{a}_\mu]$ is symmetric under any permutation of $\mathsf{a}_1 \cdots \mathsf{a}_\mu$ and the sum over repeated indices is implied. When $\mu > 1$, we used the rank μ Levi-Civita tensors to account explicitly for the Pauli principle (for $\mu = 1$, $\epsilon^1 \equiv 1$). Imposing the chiral symmetry on

$$\begin{aligned}\widehat{V}^{(\mu)}_{\mathrm{bd}} :=& \frac{V^{[\mathsf{a}_1\cdots\mathsf{a}_\mu][\mathsf{b}_1\cdots\mathsf{b}_\mu]}_{\mathrm{bd}}}{(\mu!)^2}\,\epsilon^{\mathsf{a}_1\cdots\mathsf{a}_\mu}\,\epsilon^{\mathsf{b}_1\cdots\mathsf{b}_\mu}\,\widehat{\Psi}^{\mathsf{a}_1\dagger}_{\mathrm{zm}}\cdots\widehat{\Psi}^{\mathsf{a}_\mu\dagger}_{\mathrm{zm}}\,\widehat{\Psi}^{\mathsf{b}_1}_{\mathrm{zm}}\cdots\widehat{\Psi}^{\mathsf{b}_\mu}_{\mathrm{zm}}\\ &+\frac{V^{[\mathsf{a}_1\cdots\mathsf{a}_\mu][\mathsf{b}_1\cdots\mathsf{b}_\mu]*}_{\mathrm{bd}}}{(\mu!)^2}\,\epsilon^{\mathsf{a}_1\cdots\mathsf{a}_\mu}\,\epsilon^{\mathsf{b}_1\cdots\mathsf{b}_\mu}\,\widehat{\Psi}^{\mathsf{b}_\mu\dagger}_{\mathrm{zm}}\cdots\widehat{\Psi}^{\mathsf{b}_1\dagger}_{\mathrm{zm}}\,\widehat{\Psi}^{\mathsf{a}_\mu}_{\mathrm{zm}}\cdots\widehat{\Psi}^{\mathsf{a}_1}_{\mathrm{zm}}\end{aligned} \tag{3.539b}$$

is solved by demanding that (**Exercise** 3.38)

$$\{\mathsf{a}_1,\cdots,\mathsf{a}_\mu\}\cap\{\mathsf{b}_1,\cdots,\mathsf{b}_\mu\}=\emptyset \tag{3.539c}$$

and that

$$V^{[\mathsf{a}_1\cdots\mathsf{a}_\mu][\mathsf{b}_1\cdots\mathsf{b}_\mu]}_{\mathrm{bd}} = (-1)^\mu\,V^{[\mathsf{a}_1\cdots\mathsf{a}_\mu][\mathsf{b}_1\cdots\mathsf{b}_\mu]}_{\mathrm{bd}}. \tag{3.539d}$$

When $\nu = 1$, the charge-conserving Hamiltonian (3.539a) is

$$\widehat{H}^{\mathrm{int}}_{\mathrm{bd}} = V_{\mathrm{bd}}\,\widehat{\Psi}^{1\dagger}_{\mathrm{zm}}\,\widehat{\Psi}^{1}_{\mathrm{zm}} + \mathrm{H.c.}, \tag{3.540a}$$

where $V_{\mathrm{bd}} \in \mathbb{C}$. Condition (3.539c) is violated and condition (3.539d) implies that $V_{\mathrm{bd}} = 0$. Hence, chiral symmetry implies that

$$\widehat{H}^{\mathrm{int}}_{\mathrm{bd}} = 0. \tag{3.540b}$$

When $\nu = 2$, the charge-conserving Hamiltonian (3.539a) is

$$\widehat{H}^{\mathrm{int}}_{\mathrm{bd}} = V^{\mathsf{ab}}_{\mathrm{bd}}\,\widehat{\Psi}^{\mathsf{a}\dagger}_{\mathrm{zm}}\,\widehat{\Psi}^{\mathsf{b}}_{\mathrm{zm}} + V^{[12][12]}_{\mathrm{bd}}\,\widehat{\Psi}^{1\dagger}_{\mathrm{zm}}\,\widehat{\Psi}^{2\dagger}_{\mathrm{zm}}\,\widehat{\Psi}^{1}_{\mathrm{zm}}\,\widehat{\Psi}^{2}_{\mathrm{zm}} + \mathrm{H.c.}, \tag{3.541a}$$

where $V^{\mathsf{ab}}_{\mathrm{bd}}, V^{[12][12]}_{\mathrm{bd}} \in \mathbb{C}$. Condition (3.539c) is violated by the quartic term on the right-hand side and condition (3.539d) implies that $V^{\mathsf{ab}}_{\mathrm{bd}} = 0$. Hence, chiral symmetry implies that

$$\widehat{H}^{\mathrm{int}}_{\mathrm{bd}} = 0. \tag{3.541b}$$

When $\nu = 3$, the charge-conserving Hamiltonian (3.539a) is

$$\begin{aligned}\widehat{H}^{\mathrm{int}}_{\mathrm{bd}} =& V^{\mathsf{ab}}_{\mathrm{bd}}\,\widehat{\Psi}^{\mathsf{a}\dagger}_{\mathrm{zm}}\,\widehat{\Psi}^{\mathsf{b}}_{\mathrm{zm}} + \mathrm{H.c.}\\ &+\frac{V^{[\mathsf{a}_1\mathsf{a}_2][\mathsf{b}_1\mathsf{b}_2]}_{\mathrm{bd}}}{(2!)^2}\epsilon^{\mathsf{a}_1\mathsf{a}_2}\,\epsilon^{\mathsf{b}_1\mathsf{b}_2}\,\widehat{\Psi}^{\mathsf{a}_1\dagger}_{\mathrm{zm}}\,\widehat{\Psi}^{\mathsf{a}_2\dagger}_{\mathrm{zm}}\,\widehat{\Psi}^{\mathsf{b}_1}_{\mathrm{zm}}\,\widehat{\Psi}^{\mathsf{b}_2}_{\mathrm{zm}} + \mathrm{H.c.}\\ &+\frac{V^{[\mathsf{a}_1\mathsf{a}_2\mathsf{a}_3][\mathsf{b}_1\mathsf{b}_2\mathsf{b}_3]}_{\mathrm{bd}}}{(3!)^2}\epsilon^{\mathsf{a}_1\mathsf{a}_2\mathsf{a}_3}\,\epsilon^{\mathsf{b}_1\mathsf{b}_2\mathsf{b}_3}\,\widehat{\Psi}^{\mathsf{a}_1\dagger}_{\mathrm{zm}}\,\widehat{\Psi}^{\mathsf{a}_2\dagger}_{\mathrm{zm}}\,\widehat{\Psi}^{\mathsf{a}_3\dagger}_{\mathrm{zm}}\,\widehat{\Psi}^{\mathsf{b}_1}_{\mathrm{zm}}\,\widehat{\Psi}^{\mathsf{b}_2}_{\mathrm{zm}}\,\widehat{\Psi}^{\mathsf{b}_3}_{\mathrm{zm}} + \mathrm{H.c.},\end{aligned} \tag{3.542a}$$

where $V^{\mathsf{ab}}_{\mathrm{bd}}, V^{[\mathsf{a}_1\mathsf{a}_2][\mathsf{b}_1\mathsf{b}_2]}_{\mathrm{bd}}, V^{[\mathsf{a}_1\mathsf{a}_2\mathsf{a}_3][\mathsf{b}_1\mathsf{b}_2\mathsf{b}_3]}_{\mathrm{bd}} \in \mathbb{C}$. Condition (3.539c) is violated on the second and third line of the right-hand side. Condition (3.539d) implies that $V^{\mathsf{ab}}_{\mathrm{bd}} = V^{[\mathsf{a}_1\mathsf{a}_2\mathsf{a}_3][\mathsf{b}_1\mathsf{b}_2\mathsf{b}_3]}_{\mathrm{bd}} = 0$. Hence, chiral symmetry implies that

$$\widehat{H}^{\mathrm{int}}_{\mathrm{bd}} = 0. \tag{3.542b}$$

The first value of ν in the symmetry class AIII for which the interaction on the boundary is nonvanishing is $\nu = 4$, in which case it is sufficient to choose the interaction

$$\widehat{H}^{\mathrm{int}}_{\mathrm{bd}} := V_{\mathrm{bd}} \left[\left(\widehat{\Psi}^{1\dagger}_{\mathrm{zm}} \, \widehat{\Psi}^{2}_{\mathrm{zm}} \right) \left(\widehat{\Psi}^{3\dagger}_{\mathrm{zm}} \, \widehat{\Psi}^{4}_{\mathrm{zm}} \right) + \mathrm{H.c.} \right], \qquad V_{\mathrm{bd}} \in \mathbb{R}, \tag{3.543}$$

in order to establish an instability of the noninteracting topological classification. We choose the basis of the Fock space (3.536) when $\nu = 4$ to be spanned by

$$\begin{array}{llllll} |0000\rangle, & & & & & \\ |1000\rangle, & |0100\rangle, & |0010\rangle, & |0001\rangle, & & \\ |1100\rangle, & |1001\rangle, & |0011\rangle, & |0110\rangle, & |1010\rangle, & |0101\rangle, \\ |1110\rangle, & |1101\rangle, & |1011\rangle, & |0111\rangle, & & \\ |1111\rangle. & & & & & \end{array} \tag{3.544}$$

The only elements $|1010\rangle$ and $|0101\rangle$ of this basis that are not annihilated by the interaction (3.543) are mapped into

$$\widehat{H}^{\mathrm{int}}_{\mathrm{bd}} \, |1010\rangle = V_{\mathrm{bd}} \, |0101\rangle, \qquad \widehat{H}^{\mathrm{int}}_{\mathrm{bd}} \, |0101\rangle = V_{\mathrm{bd}} \, |1010\rangle, \tag{3.545}$$

by $\widehat{H}^{\mathrm{int}}_{\mathrm{bd}}$. Hence, the interaction (3.543) lifts the 16fold degeneracy of the Hamiltonian (3.534) into three vector subspaces. There is the one-dimensional vector subspace spanned by the eigenstate with energy $-|V_{\mathrm{bd}}|$. There is the 14-dimensional vector subspace with eigenenergy 0 that is spanned by all the states in the basis (3.544) except $|1010\rangle$ and $|0101\rangle$. There is the one-dimensional vector subspace spanned by the eigenstate with energy $+|V_{\mathrm{bd}}|$. We conclude that the noninteracting topological classification in terms of the group $\mathbb{Z}$ of the one-dimensional massive Dirac Hamiltonian in the symmetry class AIII is unstable to interactions that preserve the global U(1) gauge symmetry associated with charge conservation and the chiral symmetry. If one is given $\nu = 1, 2, 3, \cdots$ noninteracting topological zero modes localized on the boundary of a chain in the symmetry class AIII, switching on a generic local interaction only retains $\nu \bmod 4$ topological gapless boundary modes. Correspondingly, it is postulated that the group $\mathbb{Z}$ that labels the topologically distinct noninteracting insulating topological phases is reduced by such interactions to the subgroup

$$\mathbb{Z}/4\mathbb{Z} =: \mathbb{Z}_4 \subset \mathbb{Z} \tag{3.546}$$

for the symmetry class AIII in one-dimensional space.[63]

[63] The construction of a group structure will be done in Section 6.5.2.

The same line of reasoning applies directly to the symmetry class BDI as the interaction (3.543) is compatible with the TRS of BDI. Hereto, the group $\mathbb{Z}$ of the one-dimensional massive Dirac Hamiltonian in the symmetry class BDI is unstable to interactions that preserve the global U(1) gauge symmetry associated with charge conservation and the chiral symmetry. The group $\mathbb{Z}$ is reduced by such interactions to the subgroup $\mathbb{Z}_4$ for the symmetry class BDI in one-dimensional space.

Because the starting point in the symmetry class CII is a massive bulk Dirac Hamiltonian of minimal rank 4, the case $\nu = 2$ already delivers a quartic interaction of the form (3.543) that is compatible with the global U(1) gauge symmetry associated with charge conservation, spin-1/2 time-reversal symmetry, and the chiral symmetry. The group $\mathbb{Z}$ is reduced by such interactions to the subgroup $\mathbb{Z}_2$ for the symmetry class CII in one-dimensional space.

After we have introduced superconductors in Chapter 7, we shall revisit the stability analysis of topological insulators or topological superconductors in d-dimensional space with $d = 1, 2, \cdots$ using an alternative line of reasoning in Chapter 8.

3.13 Exercises

3.1 Show that any phase ϕ_i in the polar decomposition $t_i = |t_i| e^{\mathrm{i}\phi_i}$ of the hopping amplitudes in Hamiltonian (3.4a) can always be gauged out when open boundary conditions are imposed. Can one always gauge out these phases if periodic boundary conditions are imposed?

3.2 Verify that the Schrödinger equation (3.5) with open boundary conditions can be rewritten as

$$\begin{pmatrix} \psi_i \\ \psi_{i-1} \end{pmatrix} = \mathcal{M}_{i,i'} \begin{pmatrix} \psi_{i'} \\ \psi_{i'-1} \end{pmatrix}, \qquad i' = 2, \cdots, i-1, \tag{3.547a}$$

where

$$\mathcal{M}_{i,i'} = \prod_{j=i'}^{i-1} \begin{pmatrix} -t_j^{-1}\varepsilon & -t_j^{-1} t_{j-1}^{*} \\ 1 & 0 \end{pmatrix}. \tag{3.547b}$$

This representation allows to extract from the Schrödinger equation (3.5) the 2×2 transfer matrix $\mathcal{M}_{N,2}$ that maps the seed values (ψ_2, ψ_1) to the final values (ψ_N, ψ_{N-1}) for a chain made of N sites. Verify that if $i - i'$ is even, then $\mathcal{M}_{i,i'}$ is diagonal when $\varepsilon = 0$.

3.3 Show that the eigenvalues of the orthogonal transformation (3.6) are ± 1.

3.4 Verify Eq. (3.9).

3.5 The goal of this exercise is to generalize the massive Dirac Hamiltonian (3.14) to any dimension d of space. This task is achieved by introducing Clifford algebras,[64,65] the simplest example thereof we have already encountered in the form of the two matrices α and β defined in Eq. (3.14).

Before constructing the Clifford algebras required to generalize the massive Dirac Hamiltonian (3.14) to any dimension d of space, we are going to relate the properties of the 2×2 unit matrix τ_0 and the three Pauli matrices $\boldsymbol{\tau}$ to their Clifford algebras.

3.5.1 Verify that any Hermitean 2×2 matrix can be written as a linear combination with real coefficients of the matrices τ_0, τ_1, τ_2, and τ_3.

3.5.2 Verify that the set of all Hermitean 2×2 matrix is a vector space of dimension four over the field $\mathbb{R}$ of real numbers.

So far, we have only taken advantage of the fact that matrices of fixed order are closed under the addition and the multiplication by the elements of the field in which their matrix elements take values. No use has yet been made of the fact that matrices can be multiplied pairwise. To take advantage of the possibility opened by matrix multiplication, it is useful to depart from the original notation used by Dirac. Let

$$I := \tau_0, \qquad \Gamma_1 := \alpha \equiv \tau_2, \qquad \Gamma_2 := \beta \equiv \tau_1. \tag{3.548}$$

3.5.3 Verify that the triplet (3.548) of Hermitean 2×2 matrices obeys the anticommutation relations

$$\{\Gamma_a, \Gamma_b\} = 2\,\delta_{ab}\, I, \qquad a,b = 1,2. \tag{3.549}$$

Let V be the two-dimensional vector space over the field $\mathbb{R}$ defined by (summation over repeated indices is implied)

$$V := \{v\,|\,v = v^a\,\Gamma_a,\ v^a \in \mathbb{R} \text{ for } a = 1,2\}. \tag{3.550}$$

Define the bilinear map

$$\begin{aligned}(\,\cdot\,,\,\cdot\,)\colon V\times V &\to \mathbb{R},\\ (v,w) &\mapsto \frac{1}{2}\operatorname{tr}(w\,v).\end{aligned} \tag{3.551}$$

3.5.4 Verify that the bilinear map (3.551) defines a scalar product.

3.5.5 Verify that

$$(\Gamma_a, \Gamma_b) = \delta_{ab}, \qquad a,b = 1,2, \tag{3.552}$$

that is Γ_1 and Γ_2 define an orthonormal basis of V equipped with the nondegenerate scalar product (3.551).

[64] Y. Choquet-Bruhat, C. DeWitt-Morette, and M. Dillard-Bleick, *Analysis, Manifolds, and Physics,* Elsevier Science, Amsterdam, 1982; *Lectures on Clifford (Geometric) Algebras and Applications,* edited by R. Ablamowicz and G. Sobczyk, Birkhäuser, Boston, 2004. [Choquet-Bruhat et al. (1982); Ablamowicz and Sobczyk (2004)].

[65] I. Porteous, *Clifford Algebras and the Classical Groups,* Cambridge Studies in Advanced Mathematics, Vol. **50**, Cambridge University Press, Cambridge 1995. [Porteous (1995)].

The set $C\ell_2(V,\mathbb{R})$ is defined by

$$C\ell_2(V,\mathbb{R}) := \{v\,|\,v = v^0\,I + v^1\,\Gamma_1 + v^2\,\Gamma_2 + v^{12}\,\Gamma_1\,\Gamma_2 \\ \text{with } v^0, v^1, v^2, v^{12} \in \mathbb{R}\,\}. \tag{3.553}$$

The following exercises motivate the terminologies algebra, unital algebra, and Clifford algebra.

3.5.6 Verify that $C\ell_2(V,\mathbb{R})$ is isomorphic to the vector space of 2×2 Hermitean matrices.

3.5.7 Verify that $C\ell_2(V,\mathbb{R})$ is closed under matrix multiplication.

3.5.8 Verify that matrix multiplication in $C\ell_2(V,\mathbb{R})$ meets the following conditions. Choose any elements v, w, and z from $C\ell_2(V,\mathbb{R})$. Choose any x and y from $\mathbb{R}$ (elements from the field are also called scalars). Verify the identities

$$(v+w)\,z = v\,z + w\,z \qquad \text{(left distributivity)}, \tag{3.554a}$$

$$v\,(w+z) = v\,w + v\,z \qquad \text{(right distributivity)}, \tag{3.554b}$$

$$(x\,v)(y\,w) = (x\,y)\,(v\,w) \qquad \text{(bilinearity)}. \tag{3.554c}$$

It is because the set $C\ell_2(V,\mathbb{R})$ is a vector space equipped with a binary operation, the matrix multiplication, that is, closed and satisfies the rules (3.554) that it is called an algebra over the field $\mathbb{R}$. The existence of the unit 2×2 matrix, that is, of the element I such that $I\,v = v\,I = v$ for any v from $C\ell_2(V,\mathbb{R})$, makes of $C\ell_2(V,\mathbb{R})$ a unital algebra over the real numbers. As an algebra, $C\ell_2(V,\mathbb{R})$ is distinguished by the fact that it contains and is generated by the vector space V over the field $\mathbb{R}$, whereby V is equipped with the positive definite quadratic form

$$Q\colon V \to \mathbb{R}, \\ v \mapsto Q(v) = (v,v). \tag{3.555}$$

Thus, the subspace V of the vector space $C\ell_2(V,\mathbb{R})$ is privileged in the Clifford algebra $C\ell_2(V,\mathbb{R})$, for any element $v := v^a\,\Gamma_a$ from V satisfies

$$v^2 = Q(v)\,I = \left[(v^1)^2 + (v^2)^2\right]\,I. \tag{3.556}$$

What makes the massive Dirac Hamiltonian (3.14) special, when the axial mass m_5 (staggered chemical potential μ_{s}) vanishes, is that it belongs to V. The following exercises emphasize the special role played by V in $C\ell_2(V,\mathbb{R})$.

3.5.9 Set $m=0$ and $m_5=0$ in the Dirac Hamiltonian (3.14). Show that, up to a multiplicative phase, there are two and only two distinct elements $\Gamma_{\mathcal{T}}$ and $\Gamma_{\mathcal{T}'}$ from $C\ell_2(V,\mathbb{R})$ such that

$$\Gamma_{\mathcal{T}}\,\mathcal{H}^*(-p)\,\Gamma_{\mathcal{T}}^{-1} = \mathcal{H}(p), \qquad \Gamma_{\mathcal{T}'}\,\mathcal{H}^*(-p)\,\Gamma_{\mathcal{T}'}^{-1} = \mathcal{H}(p). \tag{3.557}$$

Explain why either of these two transformations can be interpreted as representing reversal of time. Explain why reversal of time with $\Gamma_{\mathcal{T}}^{\mathsf{T}} = +\Gamma_{\mathcal{T}}$ is suited for a spinless particle. Explain why reversal of time with $\Gamma_{\mathcal{T}'}^{\mathsf{T}} = -\Gamma_{\mathcal{T}'}$ is suited for a spin-1/2 particle.

3.5.10 Set the product $m\,m_5 \neq 0$ in the Dirac Hamiltonian (3.14). Show that

$$\Gamma_{\mathcal{T}}\,\mathcal{H}^{*}(-p)\,\Gamma_{\mathcal{T}}^{-1} = \mathcal{H}(p), \qquad \Gamma_{\mathcal{T}'}\,\mathcal{H}^{*}(-p)\,\Gamma_{\mathcal{T}'}^{-1} \neq \mathcal{H}(p). \tag{3.558}$$

3.5.11 Set $m_5 = 0$ in the Dirac Hamiltonian (3.14). Show that, up to a multiplicative phase, there exists a unique Γ_5 from $C\ell_2(V,\mathbb{R})$ such that

$$\Gamma_5\,\mathcal{H}(p)\,\Gamma_5^{-1} = -\mathcal{H}(p) \tag{3.559}$$

holds for any m. We conclude that V is the subspace of $C\ell_2(V,\mathbb{R})$ selected by the choice of $\Gamma_{\mathcal{T}}$ and the choice of Γ_5 to implement time-reversal symmetry and chiral spectral symmetry, respectively.

3.5.12 Set $m_5 = 0$ in the Dirac Hamiltonian (3.14). Show that

$$\Gamma_5\,\Gamma_{\mathcal{T}}\,\mathcal{H}^{*}(-p)\,\Gamma_{\mathcal{T}}^{-1}\,\Gamma_5^{-1} = -\mathcal{H}(p). \tag{3.560}$$

In particle physics, this is called a spectral conjugation symmetry. In condensed matter physics, this is called a spectral particle-hole symmetry.

3.5.13 Solve for the eigenvalues of the Dirac Hamiltonian (3.14) for arbitrary momentum p, mass m, and axial mass m_5 by squaring $\mathcal{H}(p)$.

We close the discussion of $C\ell_2(V,\mathbb{R})$ by establishing how it is connected to geometry. The basis element

$$\Gamma_1\,\Gamma_2 = -\mathrm{i}\tau_3 \tag{3.561}$$

of $C\ell_2(V,\mathbb{R})$ is the anti-Hermitean generator of U(1) rotations that leaves V invariant as a vector space and leaves Q invariant as a quadratic form.

3.5.14 Let $U(\phi) := \exp(\phi\,\Gamma_1\,\Gamma_2)$ for any $0 \leq \phi < 2\pi$. Compute

$$\mathcal{H}(p,\theta) := (\phi)\,\mathcal{H}(p)\,U^{-1}(\phi). \tag{3.562}$$

3.5.15 Deduce from $\mathcal{H}(p,\theta)$ that the action of the conjugation $U(\phi)\,v\,U^{-1}(\phi)$ on any $v \in V$ is the same as a anticlockwise O(2) rotation by the angle 2ϕ of the two-vector $(v^1, v^2)^{\mathsf{T}}$ from the Euclidean space $\mathbb{R}^2$.

3.5.16 Having established that the basis elements of V transform like the covariant components of a vector in $\mathbb{R}^2$, deduce that I and $\Gamma_1\,\Gamma_2$ transform as a scalar and a pseudo-scalar, respectively, under any O(2) transformation of $\mathbb{R}^2$. *Hint:* Show that I and $\Gamma_1\,\Gamma_2$ transform like the fully symmetric (δ_{ij}) and fully antisymmetric (ϵ_{ij}) rank two tensors in $\mathbb{R}^2$, respectively.

The time-dependent Dirac equation

$$(\mathrm{i}\partial_t + \Gamma_1\,\mathrm{i}\partial_x - m\,\Gamma_2)\,\psi(t,x) = 0 \tag{3.563a}$$

can be recast in the manifestly Lorentz covariant representation

$$\left(\gamma^0\,\mathrm{i}\partial_0+\gamma^1\,\mathrm{i}\partial_1-m\right)\psi(x^0,x^1)=0, \tag{3.563b}$$

where

$$\gamma^0\equiv+\gamma_0:=\Gamma_2=\tau_1, \qquad \gamma^1\equiv-\gamma_1=\Gamma_2\,\Gamma_1:=+\mathrm{i}\tau_3. \tag{3.563c}$$

3.5.17 Verify that

$$\{\gamma^\mu,\gamma^\nu\}=2\,g^{\mu\nu}, \qquad \mu,\nu=0,1, \tag{3.564a}$$

where $g^{\mu\nu}$ is the Lorentz metric, that is,

$$g^{\mu\nu}=\begin{cases}+1, & \text{if } \mu=\nu=0,\\ -1, & \text{if } \mu=\nu=1,\\ 0, & \text{otherwise.}\end{cases} \tag{3.564b}$$

Raising and lowering of the Lorentz indices is done with the help of the Lorentz metric with the signature $(1,-1)$.
The real Clifford algebra $C\ell_{1,1}(V,\mathbb{R})$ is defined by

$$\begin{aligned}C\ell_{1,1}(V,\mathbb{R}):=\big\{v\,\big|\,v=v^0\,\gamma_0+v^1\,\gamma_1+v^{01}\,\gamma_0\,\gamma_1+v^4\,I\\ \text{with } v^0,v^1,v^{01},v^4\in\mathbb{R}\big\},\end{aligned} \tag{3.565a}$$

where the vector space

$$V:=\{v\,|\,v=v^\mu\,\gamma_\mu,\ v^\mu\in\mathbb{R}\text{ for }\mu=0,1\} \tag{3.565b}$$

is equipped with the symmetric bilinear form

$$\begin{aligned}(\,\cdot\,,\,\cdot\,)\colon V\times V&\to\mathbb{R},\\ (v,w)&\mapsto\frac{1}{2}\operatorname{tr}(w\,v),\end{aligned} \tag{3.565c}$$

and the indefinite quadratic form

$$\begin{aligned}Q\colon V&\to\mathbb{R},\\ v&\mapsto Q(v)=(v,v).\end{aligned} \tag{3.565d}$$

Any element $v:=v^\mu\,\gamma_\mu$ of V thus satisfies

$$v^2=Q(v)\,I=\left[(v^0)^2-(v^1)^2\right]\,I. \tag{3.566}$$

Moreover, both V as a vector space and Q as a quadratic form are invariant under the Lorentz boost generated by $\gamma_0\,\gamma_1$.

3.5.18 Show that, for any $\chi\in\mathbb{R}$, $\exp(\chi\,\gamma_0\,\gamma_1)=\exp(-\chi\,\tau_2)$ is a Lorentz boost with V as its invariant vector space and Q as its invariant quadratic form.

3.5.19 Show that I and $\gamma_0\,\gamma_1$ transform as a scalar and a pseudo-scalar, respectively, under any Lorentz transformation [an element of $\mathrm{O}(1,1)$] of two-dimensional Minkowski space ($\mathbb{M}^2$). *Hint:* Show that I and $\gamma_0\,\gamma_1$ transform like the fully symmetric ($\delta^{\mu}{}_{\nu}$) and fully antisymmetric ($\epsilon_{\mu\nu}=\epsilon^{\mu\nu}$) rank two tensors in $\mathbb{M}^2$, respectively.

We are ready to increase dimensionality by one, that is, $d=2$. The translation-invariant Dirac Hamiltonian is

$$\mathcal{H}(\boldsymbol{p})=\alpha_1\,p_1+\alpha_2\,p_2+\beta\,m, \tag{3.567a}$$

where m is the mass. Without loss of generality, we choose the representation

$$\alpha_1:=\tau_2,\qquad \alpha_2:=\tau_3,\qquad \beta:=\tau_1. \tag{3.567b}$$

In effect, we have used $\mathrm{i}\beta\,\gamma_5$ of Hamiltonian (3.14) to complete the kinetic term in the massive Dirac equation (3.567).

3.5.20 Construct the unique (up to a multiplicative phase) representation for time reversal when $m=0$ in Hamiltonian (3.567). Is it even or odd under transposition?

3.5.21 Construct the unique (up to a multiplicative phase) representation for the chiral spectral symmetry when $m=0$ in Hamiltonian (3.567).

3.5.22 Construct the unique (up to a multiplicative phase) representation for the charge conjugation spectral symmetry when $m=0$ in Hamiltonian (3.567). Is it even or odd under transposition?

3.5.23 What is the fate of these symmetries when $m\neq 0$.

3.5.24 Assume that translation symmetry is broken by a dependence of the mass m on the coordinate x and assume that m interpolates between the negative value $-m$ at $x=-\infty$ to the positive value $+m$ at $x=+\infty$ with a single zero at $x=0$. Solve for the zero mode whose amplitude remains bounded everywhere and compute its chirality. Repeat these calculations when m is negative instead of positive.

It is impossible to represent a massive Dirac Hamiltonian that preserves time-reversal symmetry in two-dimensional space with 2×2 Dirac (γ) matrices. Conversely, if the low-energy and long-wavelength effective theory is captured by a single 2×2 Dirac Hamiltonian in two-dimensional space and if time-reversal symmetry is neither broken explicitly nor spontaneously, then the low-energy and long-wavelength effective theory is necessarily gapless.

3.5.25 Define $\mathcal{H}'(p)$ by applying reversal of time on the massive Dirac Hamiltonian (3.567a). Define the 4×4 massive Dirac Hamiltonian $\mathcal{H}_{\mathrm{TRS}}(p):=\mathcal{H}(p)\oplus\mathcal{H}'(p)$. How is reversal of time represented with respect to the chosen representation of $\mathcal{H}_{\mathrm{TRS}}(p)$? Verify that $\mathcal{H}_{\mathrm{TRS}}(p)$ is time-reversal symmetric. Find all 4×4 Hermitean matrices that anticommute with the kinetic energy and identify those that are even (odd) under reversal of time.

Let $d = 3, 5, 7, \cdots$ be an odd integer. Define the translation-invariant Dirac Hamiltonian (summation convention over repeated indices)

$$\mathcal{H}(p) := \Gamma_i\, p^i + \Gamma_{d+1}\, m, \tag{3.568a}$$

where

$$\{\Gamma_a, \Gamma_b\} = 2\,\delta_{ab}\, I, \qquad a, b = 1, \cdots, d+1, \tag{3.568b}$$

for any d-momentum with the components $p^i \in \mathbb{R}$ for $i = 1, \cdots, d$ and the mass $m \in \mathbb{R}$. Here, I is the unital element defined by

$$I\,\Gamma_a = \Gamma_a\, I = \Gamma_a \tag{3.569}$$

for $a = 1, \cdots, d+1$. The Dirac Hamiltonian (3.568) is a point in the vector space

$$V := \{v \,|\, v = v^a\, \Gamma_a,\ v^a \in \mathbb{R} \text{ for } a = 1, \cdots, d+1\}. \tag{3.570}$$

This vector space V generates the Clifford algebra

$$\begin{aligned} C\ell_{d+1}(V, \mathbb{R}) := \big\{ v \,\big|\, v = {} & v^0\, I + v^1\, \Gamma_1 + \cdots + v^{d+1}\, \Gamma_{d+1} \\ & + v^{12}\, \Gamma_1\, \Gamma_2 + \cdots + v^{(d-1)d}\, \Gamma_{d-1}\, \Gamma_d \\ & + v^{123}\, \Gamma_1\, \Gamma_2\, \Gamma_3 + \cdots + v^{(d-2)(d-1)d}\, \Gamma_{d-2}\, \Gamma_{d-1}\, \Gamma_d \\ & + \cdots + v^{1\cdots(d+1)}\, \Gamma_1 \cdots \Gamma_{d+1} \\ & \text{with } v^0, \cdots, v^{1\cdots(d+1)} \in \mathbb{R} \big\}. \end{aligned} \tag{3.571}$$

The basis of $C\ell_{d+1}(V, \mathbb{R})$, when interpreted as a vector space, consists of any product $\Gamma_{a_1} \cdots \Gamma_{a_k}$ with $1 \leq a_1 < a_2 < \cdots < a_k \leq d+1$ and $0 \leq k \leq d+1$ (the case $k = 0$, that corresponds to the empty product, is defined as the multiplicative identity element I). The dimensionality of $C\ell_{d+1}(V, \mathbb{R})$, when interpreted as a vector space, is

$$\sum_{k=0}^{d+1} \binom{d+1}{k} = 2^{d+1}. \tag{3.572}$$

As a vector space, $C\ell_{d+1}(V, \mathbb{R})$ is isomorphic to $\mathbb{R}^{2^{d+1}}$. In turn, $\mathbb{R}^{2^{d+1}}$ is isomorphic to the vector space of $2^{(d+1)/2} \times 2^{(d+1)/2}$ Hermitean matrices. We conclude that we may represent V by choosing as a basis $d+1$ suitable linearly independent Hermitean matrices of order $2^{(d+1)/2}$, while we may represent $C\ell_{d+1}(V, \mathbb{R})$ by choosing as a basis 2^{d+1} suitable linearly independent matrices of order $2^{(d+1)/2}$. For concreteness, we shall make the choice

$$\Gamma_1 := \underbrace{\tau_2 \otimes \tau_3 \otimes \tau_3 \otimes \cdots \otimes \tau_3 \otimes \tau_3}_{(d+1)/2-\text{times}}, \tag{3.573a}$$

$$\Gamma_2 := \underbrace{\tau_1 \otimes \tau_3 \otimes \tau_3 \otimes \cdots \otimes \tau_3 \otimes \tau_3}_{(d+1)/2-\text{times}}, \tag{3.573b}$$

$$\Gamma_3 := \underbrace{\tau_0 \otimes \tau_2 \otimes \tau_3 \otimes \cdots \otimes \tau_3 \otimes \tau_3}_{(d+1)/2-\text{times}}, \tag{3.573c}$$

$$\Gamma_4 := \underbrace{\tau_0 \otimes \tau_1 \otimes \tau_3 \otimes \cdots \otimes \tau_3 \otimes \tau_3}_{(d+1)/2-\text{times}}, \tag{3.573d}$$

$$\vdots$$

$$\Gamma_d := \underbrace{\tau_0 \otimes \tau_0 \otimes \tau_0 \otimes \cdots \otimes \tau_0 \otimes \tau_2}_{(d+1)/2-\text{times}}, \tag{3.573e}$$

$$\Gamma_{d+1} := \underbrace{\tau_0 \otimes \tau_0 \otimes \tau_0 \otimes \cdots \otimes \tau_0 \otimes \tau_1}_{(d+1)/2-\text{times}}, \tag{3.573f}$$

for the basis of V. Here, there is one distinct quadruplet τ_0 and $\boldsymbol{\tau}$ acting on each one of the two-dimensional subspaces in the direct product decomposition of the spinor vector space, the latter being isomorphic to $\bigotimes_{n=1}^{(d+1)/2} \mathbb{C}^2$.

3.5.26 Verify that

$$\Gamma_1 \cdots \Gamma_{d+1} = (-\mathrm{i})^{(d+1)/2} \underbrace{\tau_3 \otimes \tau_3 \otimes \tau_3 \otimes \cdots \otimes \tau_3 \otimes \tau_3}_{(d+1)/2-\text{times}}. \tag{3.574}$$

3.5.27 Verify that

$$\left(\Gamma_1 \cdots \Gamma_{d+1}\right)^\dagger = (-1)^{(d+1)/2}\, \Gamma_1 \cdots \Gamma_{d+1} \tag{3.575}$$

in two different ways. First, use the basis (3.573). Second, use the anticommutators (3.568b).

3.5.28 Verify that

$$\{\Gamma_a, \Gamma_1 \cdots \Gamma_{d+1}\} = 0, \qquad a = 1, \cdots, d+1, \tag{3.576}$$

in two different ways. First, use the basis (3.573). Second, use the anticommutators (3.568b).

3.5.29 Verify that the matrix $\Gamma_{\mathcal{T}}$ from $C\ell_{d+1}(V, \mathbb{R})$ that satisfies the algebra

$$\Gamma_{\mathcal{T}}\, \mathcal{H}^*(-p)\, \Gamma_{\mathcal{T}}^{-1} = \mathcal{H}(p) \tag{3.577a}$$

with the Dirac Hamiltonian (3.568) is unique (up to a multiplicative phase) and given by

$$\Gamma_{\mathcal{T}} = \begin{cases} \tau_0, & \text{for } d = 1, \\ \tau_2 \otimes \tau_0, & \text{for } d = 3, \\ \tau_1 \otimes \tau_2 \otimes \tau_0, & \text{for } d = 5, \\ \tau_2 \otimes \tau_1 \otimes \tau_2 \otimes \tau_0, & \text{for } d = 7, \\ \tau_1 \otimes \tau_2 \otimes \tau_1 \otimes \tau_2 \otimes \tau_0, & \text{for } d = 9, \end{cases} \tag{3.577b}$$

and so on for $d = 11, 13, \cdots$. Explain why the transformation law (3.577a) implements reversal of time.

3.5.30 Define

$$\widetilde{\Gamma}_5 := (+\mathrm{i})^{d(d+1)/2}\, \Gamma_1 \cdots \Gamma_{d+1} \tag{3.578}$$

for $d = 1, 3, \cdots$ an odd integer. Verify that

$$\Gamma_{\mathcal{T}}\, \widetilde{\Gamma}^*_5\, \Gamma^{-1}_{\mathcal{T}} = (-1)^{d(d+3)/2}\, \widetilde{\Gamma}_5, \qquad \widetilde{\Gamma}^2_5 = I, \tag{3.579}$$

and

$$\widetilde{\Gamma}_5\, \mathcal{H}(p)\, \widetilde{\Gamma}^{-1}_5 = -\mathcal{H}(p) \tag{3.580}$$

hold.

3.5.31 Define

$$\Gamma_{\mathcal{C}} := \widetilde{\Gamma}_5\, \Gamma_{\mathcal{T}} \tag{3.581}$$

and verify that

$$\Gamma_{\mathcal{C}}\, \mathcal{H}^*(-p)\, \Gamma^{-1}_{\mathcal{C}} = -\mathcal{H}(p). \tag{3.582}$$

3.5.32 Show that Γ_{d+1} and $\widetilde{\Gamma}_5$ are the only two matrices (up to multiplication by a complex number) from $C\ell_{d+1}(V, \mathbb{R})$ that anticommute with Γ_i for $i = 1, \cdots, d$, that is with the kinetic energy of the Dirac Hamiltonian (3.568). *Hint:* Use the relation between $C\ell_{d+1}(V, \mathbb{R})$, the orthogonal group $\mathrm{O}(d+1)$, and the two invariant tensors δ_{ab} and $\epsilon_{a_1 \cdots a_{d+1}}$ on $\mathbb{R}^{d+1} \otimes \mathbb{R}^{d+1}$ and $\mathbb{R}^{d+1} \otimes \cdots \otimes \mathbb{R}^{d+1}$, respectively.

We are ready to generalize the massive translation-invariant Dirac Hamiltonian (3.14) to

$$\mathcal{H}(p) := \Gamma_i\, p^i + \Gamma_{d+1}\, m + \widetilde{\Gamma}_5\, m_5 \tag{3.583}$$

in any odd-dimensional space. The mass $m \in \mathbb{R}$ sets the energy scale for a single-particle contribution to the Dirac Hamiltonian that transforms like a scalar under $\mathrm{O}(d+1)$ transformations. The mass $m_5 \in \mathbb{R}$ sets the energy scale for a single-particle contribution to the Dirac Hamiltonian that transforms like a pseudo-scalar under $\mathrm{O}(d+1)$ transformations, since it may change its sign under those $\mathrm{O}(d+1)$ transformations with a negative determinant. In high-energy physics, m_5 is called an axial mass. A nonvanishing

m_5 breaks the spectral symmetries (3.580) and (3.582). It is time-reversal symmetric for $d = 1, 5, 9, \cdots = 4n - 3$, while it breaks time-reversal symmetry for $d = 3, 7, 11, \cdots = 4n - 1$, for $n = 1, 2, \cdots$.

3.5.33 Break the translation symmetry of the massive Dirac Hamiltonian (3.583) in odd-dimensional space with $m_5 = 0$ by assuming that m is independent of the contravariant coordinates $x^2, \cdots, x^d$ but varies from $-m$ to $+m$ with $m(x^1 = 0) = 0$ as x^1 runs over the real line from $-\infty$ to $+\infty$, respectively. Show that

$$\psi(x_1, \cdots, x_d) = \chi(x_2, \cdots, x_d) \exp\left(-\mathrm{i}\Gamma_1 \Gamma_{d+1} \int\limits_0^{x_1} \mathrm{d}y\, m(y) \right), \qquad (3.584a)$$

where $\chi(x_2, \cdots, x_d) \in \mathbb{C}^n$ with $n = 2^{(d+1)/2}$ obeys the massless Dirac equation

$$\Gamma_i \frac{\partial}{\mathrm{i}\partial x^i} \chi = 0 \qquad (3.584b)$$

in $\mathbb{R}^{d-1}$, is simultaneously a zero mode of

$$\mathcal{H}_{\mathrm{dw}} := \Gamma_i \frac{\partial}{\mathrm{i}\partial x^i} + \Gamma_{d+1}\, m(x^1) \qquad (3.584c)$$

and an eigenstate of $\widetilde{\Gamma}_5$ that decays exponentially fast away from the $(d-1)$-dimensional domain wall $x^1 = 0$. (The domain wall $x^1 = 0$ should be thought of as an embedding of $\mathbb{R}^{d-1}$ in $\mathbb{R}^d$.)

3.5.34 What happens to the zero mode (3.584a) if a constant mass m_5 of small magnitude in that $0 < |m_5| \ll m$ is switched on?

3.5.35 We now choose to study the massive Dirac Hamiltonian in the even dimension $d + 1$ of space with the caveat that it must be constructed from the Dirac matrices in their $2^{(d+1)/2} \times 2^{(d+1)/2}$-dimensional representation. (i) Give two ways of doing this by inspection of the Dirac Hamiltonian (3.583). (ii) In both cases, give the operation of reversal of time by inspection of the kinetic energy and show that a nonvanishing mass necessarily breaks time-reversal symmetry when the space dimension is $d + 1 = 4n - 2$ with $n = 1, 2, \cdots$; while a nonvanishing mass does not break time-reversal symmetry when the space dimension is $d + 1 = 4n$ with $n = 1, 2, \cdots$. (iii) Enlarge the dimensionality of the representation of the Dirac matrix by tensoring with a new set of matrices τ_0 and $\boldsymbol{\tau}$ so that the new massive Dirac Hamiltonian is time-reversal symmetric.

3.6 In this exercise, we are going to follow the original derivation by Jackiw and Rebbi[66] of the charge 1/2 bound to a domain wall in the Dirac mass of the Dirac equation in one-dimensional space.

[66] R. Jackiw and C. Rebbi, Phys. Rev. **D** 13(12), 3398–3409 (1976). [Jackiw and Rebbi (1976)].

Jackiw and Rebbi consider the following Dirac Hamiltonians defined on the real line. There are the translation invariant Dirac Hamiltonians

$$\mathcal{H}_{\infty} := -\mathrm{i}\tau_2 \frac{\mathrm{d}}{\mathrm{d}x} + \tau_1 \phi_{\infty}, \tag{3.585a}$$

and

$$\tau_2 \mathcal{H}_{\infty} \tau_2 := -\mathrm{i}\tau_2 \frac{\mathrm{d}}{\mathrm{d}x} - \tau_1 \phi_{\infty}, \tag{3.585b}$$

where we choose $\phi_{\infty} > 0$ without loss of generality. There are the Dirac Hamiltonians

$$\mathcal{H}_{\mathrm{s}} := -\mathrm{i}\tau_2 \frac{\mathrm{d}}{\mathrm{d}x} + \tau_1 \phi_{\mathrm{s}}(x), \tag{3.586a}$$

and

$$\mathcal{H}_{\bar{\mathrm{s}}} := -\mathrm{i}\tau_2 \frac{\mathrm{d}}{\mathrm{d}x} - \tau_1 \phi_{\mathrm{s}}(x) = \tau_2 \mathcal{H}_{\mathrm{s}} \tau_2, \tag{3.586b}$$

where, for concreteness, we choose

$$\phi_{\mathrm{s}}(x) := \phi_{\infty} \tanh\left(\phi_{\infty}(x - x_{\mathrm{s}})\right). \tag{3.586c}$$

Hamiltonian $\mathcal{H}_{\mathrm{s}}$ has one soliton as background. Hamiltonian $\mathcal{H}_{\bar{\mathrm{s}}}$ has one antisoliton as background. The soliton and antisoliton are centered at x_{s} on the real line.

3.6.1 With the help of dimensional analysis, deduce that ϕ_{∞} carries the units of inverse length.

3.6.2 Enumerate the distinct antiunitary transformations $\mathcal{U}_{\mathcal{T}}\colon \mathbb{C}^2 \to \mathbb{C}^2$ that leave all four Dirac Hamiltonians invariant, that is say

$$\mathcal{U}_{\mathcal{T}} \mathcal{H}_{\infty} \mathcal{U}_{\mathcal{T}}^{-1} = +\mathcal{H}_{\infty}. \tag{3.587}$$

Two antiunitary transformations are not distinct if they differ by a phase factor. (*Hint*: Decompose $\mathcal{U}$ into the composition of the operation of complex conjugation denoted K in this book and into a unitary 2×2 matrix.)

3.6.3 Enumerate all distinct transformations $\mathcal{U}\colon \mathbb{C}^2 \to \mathbb{C}^2$ (unitary or antiunitary) under which the four Dirac Hamiltonians are odd, that is say

$$\mathcal{U} \mathcal{H}_{\infty} \mathcal{U}^{-1} = -\mathcal{H}_{\infty}. \tag{3.588}$$

Two such transformations are not distinct if they differ by a phase factor. Is there any relation between the solutions to (3.587) and (3.588)?

Equipped with the single-particle Dirac Hamiltonians (3.585) and (3.586), we want to move to a quantum field theory that naturally accounts for a Dirac sea (a many-body ground state). We do this first for the translation-invariant Hamiltonians (3.585). We then repeat the same exercise for the Hamiltonians (3.586).

3.6.4 Construct explicitly all the linearly independent solutions to the eigenvalue problem

$$\mathcal{H}_{\infty}\,\psi_{\infty,\pm}(p,x)=\varepsilon_{\infty,\pm}(p)\,\psi_{\infty,\pm}(p,x), \tag{3.589a}$$

$$\varepsilon_{\infty,\pm}(p):=\pm\sqrt{p^2+\phi^2_{\infty}}\equiv\pm\varepsilon_{\infty}(p), \tag{3.589b}$$

and show that they can be chosen such that

$$\int \mathrm{d}p\,\psi^{\dagger}_{\infty,\sigma}(p,x)\,\psi_{\infty,\sigma'}(p,x')=\delta_{\sigma,\sigma'}\,\delta(x-x'), \tag{3.589c}$$

$$\int \mathrm{d}x\,\psi^{\dagger}_{\infty,\sigma}(p,x)\,\psi_{\infty,\sigma'}(p',x)=\delta_{\sigma,\sigma'}\,\delta(p-p'). \tag{3.589d}$$

Define the second-quantized (many-body) Hamiltonian

$$\widehat{H}_{\infty}:=\int \mathrm{d}x\,\widehat{\Psi}^{\dagger}_{\infty}(t,x)\,\mathcal{H}_{\infty}\widehat{\Psi}_{\infty}(t,x). \tag{3.590a}$$

Here,

$$\widehat{\Psi}^{\dagger}_{\infty}(t,x):=\sum_{\sigma=\pm}\int \mathrm{d}p\,e^{+\mathrm{i}\varepsilon_{\infty,\sigma}(p)\,t}\,\psi^{\dagger}_{\infty,\sigma}(p,x)\,\hat{c}^{\dagger}_{\sigma}(p) \tag{3.590b}$$

and

$$\widehat{\Psi}_{\infty}(t,x):=\sum_{\sigma=\pm}\int \mathrm{d}p\,e^{-\mathrm{i}\varepsilon_{\infty,\sigma}(p)\,t}\,\psi_{\infty,\sigma}(p,x)\,\hat{c}_{\sigma}(p) \tag{3.590c}$$

are spinor-valued quantum fields, whereby their spinor indices are carried by the pair of single-particle eigenstates $\psi_{\infty,\sigma}(p,x)\in\mathbb{C}^2$ with $\sigma=\pm$, while their quantum nature is encoded by the fermion algebra with the only nonvanishing anticommutators

$$\left\{\hat{c}_{\sigma}(p),\hat{c}^{\dagger}_{\sigma'}(p')\right\}=\delta_{\sigma,\sigma'}\,\delta(p-p'). \tag{3.590d}$$

3.6.5 Show that

$$\left\{\widehat{\Psi}_{\infty,\alpha}(t,x),\widehat{\Psi}^{\dagger}_{\infty,\alpha'}(t,x')\right\}=\delta_{\alpha,\alpha'}\,\delta(x-x') \tag{3.591}$$

are the only nonvanishing equal-time anticommutators for any pair of spinor indices $\alpha,\alpha'=1,2$ and for any pair x and x' on the real line.

3.6.6 Show that

$$\widehat{H}_{\infty}:=\int \mathrm{d}p\,\varepsilon_{\infty}(p)\left[\hat{c}^{\dagger}_{+}(p)\,\hat{c}_{+}(p)-\hat{c}^{\dagger}_{-}(p)\,\hat{c}_{-}(p)\right]. \tag{3.592}$$

3.6.7 Deduce from Eq. (3.592) that the many-body ground state is the Fermi sea obtained by filling all single-particle states with negative energy eigenvalues,

$$|\mathrm{FS}\rangle:=\prod_{p}\hat{c}^{\dagger}_{-}(p)\,|0\rangle \tag{3.593}$$

with $|0\rangle$ annihilated by $\hat{c}_\sigma(p)$ for any p and any σ, when the Fermi energy is within the single-particle gap.

3.6.8 Define the many-body momentum operator by

$$\widehat{P} := \int \mathrm{d}p\, p \left[\hat{c}^\dagger_+(p)\,\hat{c}_+(p) + \hat{c}^\dagger_-(p)\,\hat{c}_-(p)\right] \tag{3.594}$$

and one gives its representation in the basis defined by the representation (3.590).

3.6.9 Define the many-body charge operator by

$$\widehat{Q} := \int \mathrm{d}p \left[\hat{c}^\dagger_+(p)\,\hat{c}_+(p) + \hat{c}^\dagger_-(p)\,\hat{c}_-(p)\right] \tag{3.595}$$

and one gives its representation in the basis defined by the representation (3.590).

3.6.10 Show that the many-body Hamiltonian (3.592), the many-body momentum operator (3.594), and the many-body charge operator (3.595), commute pairwise and that the many-body momentum operator (3.594) has the many-body ground-state eigenvalue zero.

3.6.11 Show that, under the antilinear transformation defined by

$$\begin{aligned} a^*_+\,\hat{c}^\dagger_+(p) + b_+\,\hat{c}_+(p) + a^*_-\,\hat{c}^\dagger_-(p) + b_-\,\hat{c}_-(p) \mapsto& \\ a_+\,\hat{c}_-(-p) + b^*_+\,\hat{c}^\dagger_-(-p) + a_-\,\hat{c}_+(-p) + b^*_-\,\hat{c}^\dagger_+(-p)& \end{aligned} \tag{3.596}$$

for any complex numbers a_+, b_+, a_-, b_-, (i) the fermion algebra (3.590d) is preserved, (ii) the many-body Hamiltonian (3.592) transforms like

$$\widehat{H}_\infty \mapsto +\widehat{H}_\infty, \tag{3.597}$$

(iii) the many-body momentum operator (3.594) transforms like

$$\widehat{P} \mapsto +\widehat{P} - 2\,\delta(q=0) \int \mathrm{d}p\, p, \tag{3.598}$$

and (iv) the many-body charge operator (3.595) transforms like

$$\widehat{Q} \mapsto -\widehat{Q} + 2\,\delta(q=0) \int \mathrm{d}p\, 1. \tag{3.599}$$

The transformation law (3.596) implements the single-particle transformation (called charge conjugation)[67]

$$\mathcal{H}_\infty \mapsto \tau_3\, \mathcal{H}^*_\infty\, \tau_3 \tag{3.600}$$

at the many-body level.

It is convenient to perform a transformation that leaves the algebra (3.590d) unchanged but in terms of which the many-body ground state is annihilated

[67] The terminology of particle-hole transformation is the one used in the context of superconductivity.

by all annihilation operators. It is also convenient to choose this transformation such that the many-particle momentum operator (3.594) is odd under exchange of the new operators.

3.6.12 Show that the transformation

$$\begin{aligned} & a_+^* \, \hat{c}_+^\dagger(p) + b_+ \, \hat{c}_+(p) + a_-^* \, \hat{c}_-^\dagger(p) + b_- \, \hat{c}_-(p) \mapsto \\ & \qquad a_+^* \, \hat{c}^\dagger(+p) + b_+ \, \hat{c}(+p) + a_- \, \hat{d}(-p) + b_-^* \, \hat{d}^\dagger(-p) \end{aligned} \tag{3.601}$$

for any complex numbers a_+, b_+, a_-, b_- achieves these goals and gives the representations

$$\widehat{H}_\infty = + \int \mathrm{d}p \, \varepsilon_\infty(p) \left[\hat{c}^\dagger(p) \, \hat{c}(p) - \hat{d}(p) \, \hat{d}^\dagger(p) \right], \tag{3.602}$$

$$\widehat{P} = \int \mathrm{d}p \, p \left[\hat{c}^\dagger(p) \, \hat{c}(p) - \hat{d}(p) \, \hat{d}^\dagger(p) \right], \tag{3.603}$$

and

$$\widehat{Q} = + \int \mathrm{d}p \left[\hat{c}^\dagger(p) \, \hat{c}(p) + \hat{d}(p) \, \hat{d}^\dagger(p) \right]. \tag{3.604}$$

It is desirable to bring all annihilation operators to the right of creation operators to avoid any ordering ambiguity. This might come at the cost of $\mathbb{C}$ numbers, possibly infinite ones, when this operation is performed on bilinear forms of creation and annihilation operators. To dispose of such undesirable $\mathbb{C}$ numbers, mere rigid shifts of many-body spectra of bilinear operators, the procedure of normal ordering is used. Let $(\cdots)$ represents any polynomial in the creation and annihilation operators obeying the canonical fermion algebra. Normal ordering of $(\cdots)$ is denoted by $:(\cdots):$ and consists in moving all creation operators to the left of the annihilation operators as if they were mere anticommuting numbers (that is Grassmann numbers) and not operators.

3.6.13 Show that

$$\begin{aligned} :\widehat{H}_\infty: &= + \int \mathrm{d}p \, \varepsilon_\infty(p) \left[\hat{c}^\dagger(p) \, \hat{c}(p) + \hat{d}^\dagger(p) \, \hat{d}(p) \right] \\ &= \widehat{H}_\infty + \delta(q=0) \int \mathrm{d}p \, \varepsilon_\infty(p), \end{aligned} \tag{3.605}$$

$$\begin{aligned} :\widehat{P}: &= \int \mathrm{d}p \, p \left[\hat{c}^\dagger(p) \, \hat{c}(p) + \hat{d}^\dagger(p) \, \hat{d}(p) \right] \\ &= \widehat{P} + \delta(q=0) \int \mathrm{d}p \, p, \end{aligned} \tag{3.606}$$

and

$$
\begin{aligned}
:\widehat{Q}: &= + \int \mathrm{d}p \left[\hat{c}^{\dagger}(p)\,\hat{c}(p) - \hat{d}^{\dagger}(p)\,\hat{d}(p)\right] \\
&= \widehat{Q} - \delta(q=0) \int \mathrm{d}p\, 1.
\end{aligned}
\tag{3.607}
$$

3.6.14 Show that the normal-ordered Hamiltonian (3.605) and the normal-ordered momentum (3.606) are even under the conjugation of charge defined by $\hat{c} \leftrightarrow \hat{d}$, while the normal-ordered charge (3.607) is odd. The oddness of the normal-ordered charge (3.607) is desired from the intuition that the conjugation of charge defined by $\hat{c} \leftrightarrow \hat{d}$ exchanges particle and antiparticles. An antiparticle carries the opposite charge assignment to the one of its particle.[68]

3.6.15 Show that the normal-ordered charge (3.607) may be represented, following Dirac, as

$$
\begin{aligned}
:\widehat{Q}: &\equiv \frac{1}{2} \int \mathrm{d}p \left\{\left[\hat{c}^{\dagger}(p), \hat{c}(p)\right] - \left[\hat{d}^{\dagger}(p), \hat{d}(p)\right]\right\} \\
&= \frac{1}{2} \int \mathrm{d}x \sum_{\alpha=1,2} \left[\widehat{\Psi}^{\dagger}_{\infty,\alpha}(t,x), \widehat{\Psi}_{\infty,\alpha}(t,x)\right].
\end{aligned}
\tag{3.608}
$$

3.6.16 Show that

$$
:\widehat{H}_{\infty}: |\mathrm{FS}\rangle = :\widehat{P}: |\mathrm{FS}\rangle = :\widehat{Q}: |\mathrm{FS}\rangle = 0. \tag{3.609}
$$

We are now ready to attack the problem of constructing the quantum field theory for the single-particle Hamiltonian (3.586a). To this end, we seek all the eigenstates with their eigenvalues of $\mathcal{H}_{\mathrm{s}}$ that are either normalizable on the real line or have the asymptotics of plane waves as $x \to \pm\infty$. To this end, we shall posit the existence of two cases. The eigenfunction $\psi_{\mathrm{s},\pm}(x)$ of $\mathcal{H}_{\mathrm{s}}$ with the nonvanishing eigenvalue $\varepsilon_{\mathrm{s},\pm}$ must satisfy

$$
\mathcal{H}_{\mathrm{s}}\, \psi_{\mathrm{s},\pm}(x) = \varepsilon_{\mathrm{s},\pm}\, \psi_{\mathrm{s},\pm}(x), \tag{3.610a}
$$

where

$$
\varepsilon_{\mathrm{s},\pm} = \pm|\varepsilon_{\mathrm{s}}| \tag{3.610b}
$$

and

$$
\psi_{\mathrm{s},\pm}(x) := \begin{pmatrix} u_{\mathrm{s},\pm}(x) \\ v_{\mathrm{s},\pm}(x) \end{pmatrix} \in \mathbb{C}^2. \tag{3.610c}
$$

[68] At the single-particle level, we may understand this observation by considering the minimal couplings $\mathrm{i}\partial_t \to (\mathrm{i}\partial_t + e\,A_0)$ and $-\mathrm{i}\boldsymbol{\nabla} \to (-\mathrm{i}\boldsymbol{\nabla} + e\,\boldsymbol{A}/c)$ to an electro-magnetic gauge field $(A_0, \boldsymbol{A})$. Consistency under complex conjugation and reversal of time implies that $+e\,A_0 \to +e\,A_0$ and $e\,\boldsymbol{A}/c \to -e\,\boldsymbol{A}/c$ while consistency under complex conjugation alone implies $+e\,A_0 \to -e\,A_0$ and $e\,\boldsymbol{A}/c \to -e\,\boldsymbol{A}/c$. For consistency, the transformation laws $e \to +e$, $A_0 \to +A_0$, and $\boldsymbol{A} \to -\boldsymbol{A}$ are used under complex conjugation and reversal of time, while $e \to -e$, $A_0 \to +A_0$, and $\boldsymbol{A} \to +\boldsymbol{A}$ are used under complex conjugation alone.

The eigenfunction $\psi_{s,0}(x)$ of $\mathcal{H}_s$ is a nondegenerate zero mode of $\mathcal{H}_s$ if

$$\mathcal{H}_s\,\psi_{s,0}(x)=0. \tag{3.610d}$$

3.6.17 Even though translation symmetry is broken by the soliton profile (3.586c), show that (i) $\mathcal{H}_s$ is even under composition of the inversion

$$(x-x_s)\mapsto -(x-x_s) \tag{3.611}$$

about the soliton center x_s with the conjugation

$$\mathcal{H}_s\mapsto \tau_3\,\mathcal{H}_s\,\tau_3, \tag{3.612}$$

and (ii)

$$\begin{cases}\psi_{s,-}(x)=\tau_3\,\psi^*_{s,+}(x), & \text{if } |\varepsilon_{s,\pm}|=|\varepsilon_s|>0,\\[2ex] \tau_3\,\psi_{s,0}(x)=\pm\psi_{s,0}(x), & \text{otherwise.}\end{cases} \tag{3.613}$$

3.6.18 Show that the eigenvalue problem (3.610) is equivalent to solving the one-dimensional Schrödinger equation

$$\left[-\frac{d^2}{dx^2}-\left(\frac{d\phi_s}{dx}\right)(x)+\phi_s^2(x)\right]u_{s,\pm}(x)=|\varepsilon_s|^2\,u_{s,\pm}(x) \tag{3.614a}$$

supplemented with

$$v_{s,\pm}(x)=\frac{1}{\varepsilon_{s,\pm}}\left[\frac{d}{dx}+\phi_s(x)\right]u_{s,\pm}(x), \tag{3.614b}$$

whenever $|\varepsilon_{s,\pm}|=|\varepsilon_s|>0$, and

$$u_{s,0}(x)=\exp\left(-\int\limits_{x_s}^{x}dx'\,\phi_s(x')\right),\qquad v_{s,0}(x)=0, \tag{3.614c}$$

otherwise.

3.6.19 Verify that the spectrum of the Dirac Hamiltonian (3.610) contains the continuum

$$\varepsilon_{s,\pm}(p):=\pm\sqrt{p^2+\phi_\infty^2}\equiv\pm\varepsilon_\infty(p) \tag{3.615a}$$

with the nondegenerate eigenfunctions

$$\psi_{s,-}(p,x)=\tau_3\,\psi^*_{s,+}(p,x) \tag{3.615b}$$

specified by

$$u_{s,\pm}(p,x)=-\frac{\phi_\infty}{\phi_\infty+ip}\left[\tanh\left(\phi_\infty\,(x-x_s)\right)-\frac{ip}{\phi_\infty}\right]e^{ip\,(x-x_s)} \tag{3.615c}$$

and the nondegenerate zero eigenvalue with the normalized eigenfunction (N is the normalization factor)

$$\psi_{\mathrm{s},0}(x) = N \begin{pmatrix} u_{\mathrm{s},0}(x) \\ 0 \end{pmatrix}. \tag{3.615d}$$

3.6.20 Show that the normal-mode expansions with the single-particle basis (3.615) for the second-quantized spinor-field operators are, after making use of Eq. (3.601), given by

$$\begin{aligned}\widehat{\Psi}^{\dagger}_{\mathrm{s}}(t,x) = \psi^{\dagger}_{\mathrm{s},0}(x)\,\hat{b}^{\dagger} + \int \mathrm{d}p \Big[& e^{+\mathrm{i}\varepsilon_{\infty}(p)\,t}\,\psi^{\dagger}_{\mathrm{s},+}(p,x)\,\hat{c}^{\dagger}(p) \\ & + e^{-\mathrm{i}\varepsilon_{\infty}(p)\,t}\,\psi^{\dagger}_{\mathrm{s},+}(p,x)\,\tau_3\,\hat{d}(p)\Big]\end{aligned} \tag{3.616a}$$

and

$$\begin{aligned}\widehat{\Psi}_{\mathrm{s}}(t,x) = \psi_{\mathrm{s},0}(x)\,\hat{b} + \int \mathrm{d}p \Big[& e^{-\mathrm{i}\varepsilon_{\infty}(p)\,t}\,\psi_{\mathrm{s},+}(p,x)\,\hat{c}(p) \\ & + e^{+\mathrm{i}\varepsilon_{\infty}(p)\,t}\,\tau_3\,\psi_{\mathrm{s},+}(p,x)\,\hat{d}^{\dagger}(p)\Big],\end{aligned} \tag{3.616b}$$

where we impose the fermion algebra with the only nonvanishing anticommutators

$$\left\{\hat{c}(p),\hat{c}^{\dagger}(p')\right\} = \left\{\hat{d}(p),\hat{d}^{\dagger}(p')\right\} = \delta(p-p') \tag{3.616c}$$

for the continuum and

$$\left\{\hat{b},\hat{b}^{\dagger}\right\} = 1 \tag{3.616d}$$

for the zero mode.

3.6.21 If we use the definitions

$$\widehat{H}_{\mathrm{s}} := \int \mathrm{d}x\,\widehat{\Psi}^{\dagger}_{\mathrm{s}}(t,x)\,\mathcal{H}_{\mathrm{s}}\,\widehat{\Psi}_{\mathrm{s}}(t,x) \tag{3.617}$$

and

$$\widehat{Q} := \int \mathrm{d}x\,\widehat{\Psi}^{\dagger}_{\mathrm{s}}(t,x)\,\widehat{\Psi}_{\mathrm{s}}(t,x), \tag{3.618}$$

verify that (i)

$$\widehat{H}_{\mathrm{s}} \mapsto \widehat{H}_{\mathrm{s}}, \tag{3.619}$$

while

$$\widehat{Q} \mapsto -\widehat{Q} + 2\,\delta(y=0)\int \mathrm{d}x\,1, \tag{3.620}$$

under the antilinear transformation (this mapping is understood to hold component wise)

$$a^{*}\,\widehat{\Psi}^{\dagger}_{\mathrm{s}}(t,x) + b\,\widehat{\Psi}_{\mathrm{s}}(t,x) \mapsto a\,\widehat{\Psi}_{\mathrm{s}}(t,x)\,\tau_3 + b^{*}\,\tau_3\,\widehat{\Psi}^{\dagger}_{\mathrm{s}}(t,x) \tag{3.621}$$

for any complex number a and b. Show that (ii)

$$
\begin{aligned}
:\widehat{H}_{\mathrm{s}}: &= + \int \mathrm{d}p\, \varepsilon_{\infty}(p) \left[\hat{c}^{\dagger}(p)\,\hat{c}(p) + \hat{d}^{\dagger}(p)\,\hat{d}(p)\right] \\
&= \widehat{H}_{\mathrm{s}} + \delta(q=0) \int \mathrm{d}p\, \varepsilon_{\infty}(p),
\end{aligned}
\tag{3.622}
$$

and

$$
\begin{aligned}
:\widehat{Q}: &= \hat{b}^{\dagger}\,\hat{b} + \int \mathrm{d}p \left[\hat{c}^{\dagger}(p)\,\hat{c}(p) - \hat{d}^{\dagger}(p)\,\hat{d}(p)\right] \\
&= \widehat{Q} - \delta(q=0) \int \mathrm{d}p\, 1.
\end{aligned}
\tag{3.623}
$$

Verify (iii) that $:\widehat{H}_{\mathrm{s}}:$ commutes with $:\widehat{Q}:$ and that (iv) the many-body ground states of $:\widehat{H}_{\mathrm{s}}:$ are twofold degenerate when the chemical potential lies exactly at the single-particle energy 0, while $:\widehat{Q}:$ has the eigenvalue 0 (1) if the chemical potential is just below (above) the single-particle energy 0. Correspondingly, show that (v)

$$
:\widehat{H}_{\mathrm{s}}: \mapsto :\widehat{H}_{\mathrm{s}}:, \qquad :\widehat{Q}: \mapsto -:\widehat{Q}: +1, \tag{3.624}
$$

under

$$
\hat{c} \leftrightarrow \hat{d} \tag{3.625a}
$$

and

$$
\hat{b}^{\dagger} \mapsto \hat{b}, \qquad \hat{b} \mapsto \hat{b}^{\dagger}. \tag{3.625b}
$$

Hence, mere normal ordering is not sufficient to define a many-body charge operator that is odd under charge conjugation in the presence of a soliton. This observation motivates the definition

$$
\widehat{Q}_{\mathrm{s}} := \int \mathrm{d}x\, \frac{1}{2} \sum_{\alpha=1,2} \left[\widehat{\Psi}^{\dagger}_{\mathrm{s},\alpha}(x), \widehat{\Psi}_{\mathrm{s},\alpha}(x)\right] = :\widehat{Q}: - \frac{1}{2} \tag{3.626}
$$

with the desired property that it is odd under both representations (3.621) and (3.625) of charge conjugation.

3.6.22 Define the many-body ground states

$$
|\mathrm{s},+\rangle := \hat{b}^{\dagger}|\mathrm{FS}\rangle \tag{3.627}
$$

and

$$
|\mathrm{s},-\rangle := |\mathrm{FS}\rangle, \tag{3.628}
$$

where $|\mathrm{FS}\rangle$ was defined in Eq. (3.593). Show that

$$
:\widehat{H}_{\mathrm{s}}:\, |\mathrm{s},\pm\rangle = 0, \qquad \widehat{Q}_{\mathrm{s}}\, |\mathrm{s},\pm\rangle = \pm\frac{1}{2}\, |\mathrm{s},\pm\rangle. \tag{3.629}
$$

The many-body soliton ground state is doubly degenerate with each state carrying a fractional fermion number $\pm 1/2$, if we demand that the fermion number be odd under charge conjugation. The fermion numbers of all states obtained by adding occupied single-particle momenta p with the single-particle positive energies $\sqrt{p^2+\phi_\infty^2}$ to either one of the degenerate many-body soliton ground states $|\mathrm{s},-\rangle$ or $|\mathrm{s},+\rangle$ are then necessarily half-integers.

3.7 This exercise is inspired by the treatment of charge fractionalization made by Rajaraman and Bell.[69] It differs from **Exercise** 3.6 in that the fractional charge is computed in the presence of a pair of soliton and antisoliton.

First, we seek the solutions to the stationary Dirac equation

$$\mathcal{H}_{\mathrm{s}\bar{\mathrm{s}}}(\hat{p},x)\,\psi(x)=\varepsilon\,\psi(x), \tag{3.630a}$$

$$\mathcal{H}_{\mathrm{s}\bar{\mathrm{s}}}(\hat{p},x):=-\mathrm{i}\tau_2\,\frac{\mathrm{d}}{\mathrm{d}x}+\tau_1\,\phi_{\mathrm{s}\bar{\mathrm{s}}}(x), \tag{3.630b}$$

$$\phi_{\mathrm{s}\bar{\mathrm{s}}}(x):=\phi_\infty\,\mathrm{sgn}(x), \tag{3.630c}$$

on the interval

$$-\frac{L}{2}\leq x\leq+\frac{L}{2} \tag{3.630d}$$

subject to the periodic boundary conditions

$$\psi(-L/2)=\psi(+L/2). \tag{3.630e}$$

The conventions for the Pauli matrices τ_2 and τ_1 are the usual ones and $\phi_\infty>0$.

3.7.1 Verify that there is a soliton at $x=0$ and an antisoliton at $L/2\equiv-L/2$ so that the distance $r_{\mathrm{s}\bar{\mathrm{s}}}$ between the soliton and antisoliton is half the size L over which periodic boundary conditions are imposed.

3.7.2 Verify that the stationary Dirac equation (3.630) has the eigenspinors

$$\psi_{k_n}(x):=\frac{1}{\sqrt{L}}\begin{pmatrix}\cos(k_n\,|x|+\delta_{k_n})\\ -\sin(k_n\,x)\end{pmatrix} \tag{3.631a}$$

and

$$\tau_1\,\psi_{-k_n}(x):=\frac{1}{\sqrt{L}}\begin{pmatrix}\sin(k_n\,x)\\ \cos(k_n\,|x|-\delta_{k_n})\end{pmatrix} \tag{3.631b}$$

with the twofold degenerate positive eigenenergy

$$\varepsilon_{n,+}:=+\sqrt{k_n^2+\phi_\infty^2},\qquad k_n=\frac{2\pi}{L}\,n,\qquad n=1,2,3,\cdots, \tag{3.631c}$$

and the phase shift

$$\delta_{k_n}=\arctan\left(\phi_\infty/|k_n|\right) \tag{3.631d}$$

[69] R. Rajaraman and J. S. Bell, Phys. Lett. B **116**(2), 151–154 (1982). [Rajaraman and Bell (1982)].

as a consequence of the identity

$$\tau_1\,\mathcal{H}_{\mathrm{s\bar{s}}}(-\hat{p},x)\,\tau_1=\mathcal{H}_{\mathrm{s\bar{s}}}(\hat{p},x). \tag{3.631e}$$

3.7.3 Verify that the stationary Dirac equation (3.630) has the eigenspinors

$$\widetilde{\psi}_{k_n}(x):=\tau_3\,\psi_{+k_n}(x) \tag{3.632a}$$

and

$$\tau_1\,\widetilde{\psi}_{-k_n}(x):=\tau_1\,\tau_3\,\psi_{-k_n}(x) \tag{3.632b}$$

with the twofold degenerate negative eigenenergy

$$\varepsilon_{n,-}:=-\sqrt{k_n^2+\phi_\infty^2},\qquad k_n=\frac{2\pi}{L}\,n,\qquad n=1,2,3,\cdots. \tag{3.632c}$$

What is the origin of the spectral symmetry between positive and negative eigenenergies?

3.7.4 Verify that the stationary Dirac equation (3.630) has the eigenspinors

$$\psi_0(x):=\sqrt{\frac{\phi_\infty}{1-e^{-\phi_\infty L}}}\begin{pmatrix}e^{-\phi_\infty\,|x|}\\0\end{pmatrix} \tag{3.633a}$$

and

$$\widetilde{\psi}_0(x):=\sqrt{\frac{\phi_\infty}{1-e^{-\phi_\infty L}}}\begin{pmatrix}0\\e^{+\phi_\infty\,[|x|-(L/2)]}\end{pmatrix} \tag{3.633b}$$

with the twofold degenerate eigenenergy

$$\varepsilon_0=0. \tag{3.633c}$$

What are the localization centers of these zero modes?

3.7.5 Verify that the eigenstates (3.631), (3.632), and (3.633) are pairwise orthonormal and complete, that is, they realize a resolution of the identity.

3.7.6 For any $n\in\mathbb{Z}$, we introduce the notations

$$\eta_n(x):=\begin{cases}\psi_{k_n}(x), & \text{if } n>0,\\ \psi_0(x), & \text{if } n=0,\\ \tau_1\,\psi_{-k_{|n|}}(x), & \text{if } n<0,\end{cases} \tag{3.634}$$

and

$$\widetilde{\eta}_n(x):=\begin{cases}\widetilde{\psi}_{k_n}(x), & \text{if } n>0,\\ \widetilde{\psi}_0(x), & \text{if } n=0,\\ \tau_1\,\widetilde{\psi}_{-k_{|n|}}(x), & \text{if } n<0.\end{cases} \tag{3.635}$$

Define the (spinor) field operator

$$\widehat{\Psi}(t,x):=\sum_{n\in\mathbb{Z}}\left[\eta_n(x)\,e^{-\mathrm{i}\varepsilon_{n,+}\,t}\,\hat{c}_n+\widetilde{\eta}_n(x)\,e^{+\mathrm{i}\varepsilon_{n,+}\,t}\,\hat{d}_n^\dagger\right]. \tag{3.636}$$

Verify the following equivalence. Demanding that

$$\left\{\widehat{\Psi}_{\alpha}(t,x),\widehat{\Psi}^{\dagger}_{\alpha'}(t,x')\right\}=\delta_{\alpha,\alpha'}\,\delta(x-x') \tag{3.637}$$

is the only nonvanishing equal-time anticommutators for any pair of spinor indices $\alpha,\alpha'=1,2$ and for any pair x,x' from the interval $[-(L/2),(L/2)]$ is equivalent to demanding that

$$\left\{\hat{c}_{n},\hat{c}^{\dagger}_{n'}\right\}=\left\{\hat{d}_{n},\hat{d}^{\dagger}_{n'}\right\}=\delta_{n,n'} \tag{3.638}$$

is the only nonvanishing anticommutators for any pair $n,n'\in\mathbb{Z}$. Denote with $|0\rangle$ the many-body state that is annihilated by $\hat{c}_n$ and $\hat{d}_n$ for all $n\in\mathbb{Z}$. This many-body state has all the single-particle states $\widetilde{\eta}_n(x)$ with $n\in\mathbb{Z}$ occupied.

3.7.7 Define the local density operator

$$\hat{\rho}(t,x):=\frac{1}{2}\sum_{\alpha=1,2}\left[\widehat{\Psi}^{\dagger}_{\alpha}(t,x),\widehat{\Psi}_{\alpha}(t,x)\right]\equiv\frac{1}{2}\left[\widehat{\Psi}^{\dagger}(t,x),\widehat{\Psi}\,(t,x)\right]. \tag{3.639}$$

To shorten the notation, we suppress the summation over the spinor indices from now on. Verify that, at $t=0$,

$$\begin{aligned}\hat{\rho}(t=0,x)=&-\frac{1}{2}\left[\eta^{\dagger}_{0}(x)\,\eta_{0}(x)-\widetilde{\eta}^{\dagger}_{0}(x)\,\widetilde{\eta}_{0}(x)\right]\\&+\sum_{m,n\in\mathbb{Z}}\left[\eta^{\dagger}_{m}(x)\,\eta_{n}(x)\,\hat{c}^{\dagger}_{m}\,\hat{c}_{n}-\widetilde{\eta}^{\dagger}_{m}(x)\,\widetilde{\eta}_{n}(x)\,\hat{d}^{\dagger}_{n}\,\hat{d}_{m}\right]\\&+\sum_{m,n\in\mathbb{Z}}\left[\eta^{\dagger}_{m}(x)\,\widetilde{\eta}_{n}(x)\,\hat{c}^{\dagger}_{m}\,\hat{d}^{\dagger}_{n}+\widetilde{\eta}^{\dagger}_{m}(x)\,\eta_{n}(x)\,\hat{d}_{m}\,\hat{c}_{n}\right].\end{aligned} \tag{3.640}$$

3.7.8 Verify that the total fermion number operator

$$\widehat{Q}:=\int\limits_{-L/2}^{+L/2}\mathrm{d}x\,\hat{\rho}(t,x) \tag{3.641}$$

is independent of time and explicitly given by

$$\widehat{Q}=\sum_{n\in\mathbb{Z}}\left(\hat{c}^{\dagger}_{n}\,\hat{c}_{n}-\hat{d}^{\dagger}_{n}\,\hat{d}_{n}\right). \tag{3.642}$$

Deduce that the total fermion number operator has integer-valued eigenvalues and that $|0\rangle$ is an eigenstate with eigenvalue zero to the total fermion number operator.

3.7.9 Verify that the four many-body states

$$|0\rangle,\qquad \hat{c}^{\dagger}_{0}\,|0\rangle,\qquad \hat{d}^{\dagger}_{0}\,|0\rangle,\qquad \hat{c}^{\dagger}_{0}\,\hat{d}^{\dagger}_{0}\,|0\rangle, \tag{3.643}$$

are degenerate in energy (they span all the ground states) and are eigenstates of the total fermion number operator with the eigenvalues

$$0, \qquad +1, \qquad -1, \qquad 0, \tag{3.644}$$

respectively.

3.7.10 Let $\Theta(x)$ denote the Heaviside function equal to one if $x > 0$ and zero otherwise. Define the box function

$$f_\ell(x) := \Theta\Big(+(\ell/2) - x\Big) - \Theta\Big(-(\ell/2) - x\Big) \tag{3.645}$$

taking the values one on the interval $]-(\ell/2), +(\ell/2)[$ and zero otherwise. Assume that

$$L \gg \ell \gg 1/\phi_\infty. \tag{3.646}$$

Define the boxed fermion number operator

$$\widehat{Q}_{f_\ell}(t) := \int\limits_{-L/2}^{+L/2} \mathrm{d}x\, f_\ell(x)\, \hat{\rho}(t,x) \equiv \int\limits_{-\ell/2}^{+\ell/2} \mathrm{d}x\, \hat{\rho}(t,x). \tag{3.647}$$

Verify that

$$\begin{aligned} \langle 0|\, \widehat{Q}_{f_\ell}(t=0)\, |0\rangle &\approx \langle 0|\, \hat{d}_0\, \widehat{Q}_{f_\ell}(t=0)\, \hat{d}^\dagger_0\, |0\rangle \approx -\frac{1}{2}, \\ \langle 0|\, \hat{c}_0\, \widehat{Q}_{f_\ell}(t=0)\, \hat{c}^\dagger_0\, |0\rangle &\approx \langle 0|\, \hat{d}_0\, \hat{c}_0\, \widehat{Q}_{f_\ell}(t=0)\, \hat{c}^\dagger_0\, \hat{d}^\dagger_0\, |0\rangle \approx +\frac{1}{2}. \end{aligned} \tag{3.648}$$

How small are the terms as a function of ℓ/L and $(1/\phi_\infty)/\ell$ that have been ignored on the right-hand sides?

3.7.11 Define

$$\operatorname{var} \widehat{Q}_{f_\ell}\Big|_{|0\rangle} := \langle 0|\, \Big[\widehat{Q}_{f_\ell} - \langle 0|\widehat{Q}_{f_\ell}\, |0\rangle\Big]^2\, |0\rangle. \tag{3.649}$$

Verify that this variance is

$$\operatorname{var} \widehat{Q}_{f_\ell}\Big|_{|0\rangle} = \sum_{m,n\in\mathbb{Z}} \left| \int\limits_{-\ell/2}^{+\ell/2} \mathrm{d}x\, \eta^\dagger_m(x)\, \widetilde{\eta}_n(x) \right|^2 \tag{3.650}$$

and that it originates from particle-hole excitations. How would Eq. (3.650) change if there were no pair of soliton and antisoliton? We shall devote Chapter 4 to interpret this result and to show that it does not contradict the notion of a sharp fractional charge.

3.8 Repeat the construction of the ground state at half-filling for the choice $\delta t = +t$ instead of the choice $\delta t = -t$ made in Eq. (3.48).

3.9 This exercise is dedicated to the case of strong n-merization. Let Λ be a chain with N sites and $n = 2, 3, 4, \ldots$ such that n divides N.

3.9.1 Show that if $i \in \Lambda$, then there exists a unique $j = 1, \cdots, N/n$ and a unique $m = 1, \cdots, n$ such that $i = n\,(j-1) + m$.

3.9.2 Show that, at the strong n-merization point (3.61), the matrix in Hamiltonian (3.4) can be written as

$$\mathcal{H} = \bigoplus_{j=1}^{N/n} \mathcal{H}^{(n\,j)}, \tag{3.651}$$

where the $n \times n$ irreducible Hermitean matrix $\mathcal{H}^{(n\,j)}$ is defined by the matrix elements

$$\left(\mathcal{H}^{(n\,j)}\right)_{m,l} = t_{n(j-1)+m}\,\delta_{m,l-1} + t^{*}_{n(j-1)+l}\,\delta_{m,l+1} \tag{3.652}$$

for $m, l = 1, \cdots, n$.

3.9.3 Show that the matrix elements of $\mathcal{H}^{(n\,j)}$ are independent of $j = 1, \cdots, N/n$.

3.9.4 Verify the spectral representation (3.64b).

3.9.5 Diagonalize explicitly the $n \times n$ matrix (3.652) when $n = 3$ that corresponds to the strong trimerization pattern of translation-symmetry breaking. Generalize Figure 3.4 to the case of $n = 3$ and derive the fractional charges (3.68) and (3.67), respectively.

3.10 Verify explicitly that upon inserting Eq. (3.84) into

$$\begin{pmatrix} +\mu_{\mathrm{s}} & -\partial_x + \phi \\ +\partial_x + \phi & -\mu_{\mathrm{s}} \end{pmatrix} \begin{pmatrix} u \\ v \end{pmatrix} = \begin{pmatrix} \mu_{\mathrm{s}}\, u - (\partial_x - \phi)\, v \\ (\partial_x + \phi)\, u - \mu_{\mathrm{s}}\, v \end{pmatrix}, \tag{3.653}$$

there follows

$$\begin{pmatrix} +\mu_{\mathrm{s}} & -\partial_x + \phi \\ +\partial_x + \phi & -\mu_{\mathrm{s}} \end{pmatrix} \begin{pmatrix} u \\ v \end{pmatrix} = \varepsilon \begin{pmatrix} u \\ v \end{pmatrix}. \tag{3.654}$$

3.11 Let $x > 0$ be a strictly positive real-valued number. Verify the identities

$$\frac{x}{\sqrt{1+x^2} \mp 1} = \frac{\sqrt{1+x^2} \pm 1}{x}. \tag{3.655}$$

Define

$$y := \arctan\left(\frac{x}{\sqrt{1+x^2}+1}\right) = \arctan\left(\frac{\sqrt{1+x^2}-1}{x}\right). \tag{3.656}$$

Verify that

$$x = \frac{2\tan y}{1-\tan^2 y} = \tan(2y), \tag{3.657}$$

that is

$$y = \frac{1}{2}\arctan x. \tag{3.658}$$

Deduce the case $\mu_{\mathrm{s}} < 0$ in Eq. (3.108). Use the identity

$$\arctan\left(\frac{1}{x}\right) = \frac{\pi}{2} - \arctan x \tag{3.659}$$

to prove the case $\mu_{\mathrm{s}} > 0$ in Eq. (3.108).

3.12 Verify Eq. (3.126).

3.13 This exercise reviews what bosonic coherent states are. Bosonic coherent states are the tools required to construct the path-integral representation of the quantum mechanics built out of bosonic degrees of freedom.

Let $\mathfrak{F}$ denote the Hilbert space of the harmonic oscillator, which we interpret as the Fock space for the boson whose occupation numbers label the eigenstates of the harmonic oscillator. Define the uncountable set of coherent states for the harmonic oscillator, in short bosonic coherent states, by

$$|\alpha\rangle_{\mathrm{cs}} := e^{\alpha\,\hat{a}^{\dagger}}|0\rangle := \sum_{n=0}^{\infty}\frac{\alpha^{n}}{\sqrt{n!}}|n\rangle, \qquad \alpha\in\mathbb{C}. \tag{3.660a}$$

The adjoint set is (α^{*} denotes the complex-conjugate of $\alpha \in \mathbb{C}$)

$${}_{\mathrm{cs}}\langle\alpha| := \langle 0|e^{\hat{a}\,\alpha^{*}} := \sum_{n=0}^{\infty}\langle n|\frac{(\alpha^{*})^{n}}{\sqrt{n!}}, \qquad \alpha\in\mathbb{C}. \tag{3.660b}$$

Prove the following properties of bosonic coherent states.

3.13.1 Coherent state $|\alpha\rangle_{\mathrm{cs}}$ is a right eigenstate with eigenvalue α of the annihilation operator $\hat{a}$,[70]

$$\hat{a}|\alpha\rangle_{\mathrm{cs}} = \alpha|\alpha\rangle_{\mathrm{cs}}. \tag{3.661}$$

3.13.2 Coherent state ${}_{\mathrm{cs}}\langle\alpha|$ is a left eigenstate with eigenvalue α^{*} of the creation operator $\hat{a}^{\dagger}$.

$${}_{\mathrm{cs}}\langle\alpha|\hat{a}^{\dagger} = {}_{\mathrm{cs}}\langle\alpha|\alpha^{*}. \tag{3.662}$$

3.13.3 The action of creation operator $\hat{a}^{\dagger}$ on coherent state $|\alpha\rangle_{\mathrm{cs}}$ is differentiation with respect to α,

$$\hat{a}^{\dagger}|\alpha\rangle_{\mathrm{cs}} = \frac{\mathrm{d}}{\mathrm{d}\alpha}|\alpha\rangle_{\mathrm{cs}}. \tag{3.663}$$

3.13.4 The action of creation operator $\hat{a}$ on coherent state ${}_{\mathrm{cs}}\langle\alpha|$ is differentiation with respect to α^{*},

$${}_{\mathrm{cs}}\langle\alpha|\hat{a} = \frac{\mathrm{d}}{\mathrm{d}\alpha^{*}}\,{}_{\mathrm{cs}}\langle\alpha|. \tag{3.664}$$

3.13.5 The overlap ${}_{\mathrm{cs}}\langle\alpha|\beta\rangle_{\mathrm{cs}}$ between two coherent states is $\exp(\alpha^{*}\beta)$,

$${}_{\mathrm{cs}}\langle\alpha|\beta\rangle_{\mathrm{cs}} = e^{\alpha^{*}\beta}. \tag{3.665}$$

[70] Non-Hermitean operators need not have the same left and right eigenstates.

3.13.6 There exists a resolution of the identity in terms of bosonic coherent states,

$$\begin{aligned}\mathbb{1} =& \int \frac{\mathrm{d}z^*\mathrm{d}z}{2\pi\mathrm{i}}\; e^{-z^* z}\; |z\rangle_{\mathrm{cs}\;\mathrm{cs}}\langle z| \\ \equiv& \frac{1}{\pi}\int\limits_{-\infty}^{+\infty} \mathrm{dRe}\, z \int\limits_{-\infty}^{+\infty} \mathrm{dIm}\, z\; e^{-z^* z}\; |z\rangle_{\mathrm{cs}\;\mathrm{cs}}\langle z|.\end{aligned} \tag{3.666}$$

Proof Write

$$\widehat{O} := \int \frac{\mathrm{d}z^*\mathrm{d}z}{2\pi\mathrm{i}}\; e^{-z^* z}\; |z\rangle_{\mathrm{cs}\;\mathrm{cs}}\langle z|. \tag{3.667}$$

By construction, $\widehat{O}$ belongs to the algebra of operators generated by $\hat{a}$ and $\hat{a}^\dagger$.

3.13.6.1 Step 1: With the help of Eqs. (3.661) and (3.664) show that

$$[\hat{a}, |z\rangle_{\mathrm{cs}\;\mathrm{cs}}\langle z|] = \left(z - \frac{\mathrm{d}}{\mathrm{d}z^*}\right) |z\rangle_{\mathrm{cs}\;\mathrm{cs}}\langle z|. \tag{3.668}$$

Hence, after making use of integration by parts,

$$\begin{aligned}[\hat{a}, \widehat{O}] =& \int \frac{\mathrm{d}z^*\mathrm{d}z}{2\pi\mathrm{i}}\; e^{-z^* z} \left(z - \frac{\mathrm{d}}{\mathrm{d}z^*}\right) |z\rangle_{\mathrm{cs}\;\mathrm{cs}}\langle z| \\ =& 0.\end{aligned} \tag{3.669}$$

3.13.6.2 Step 2: By taking the adjoint of Eq. (3.669), $[\hat{a}^\dagger, \widehat{O}] = 0$.

3.13.6.3 Step 3: Show that

$$\langle 0|\widehat{O}|0\rangle = 1. \tag{3.670}$$

3.13.6.4 Step 4: Any linear operator from $\mathfrak{F}$ to $\mathfrak{F}$ belongs to the algebra generated by $\hat{a}$ and $\hat{a}^\dagger$. Since $\widehat{O}$ commutes with both $\hat{a}$ and $\hat{a}^\dagger$ by steps 1 and 2, $\widehat{O}$ commutes with all linear operators from $\mathfrak{F}$ to $\mathfrak{F}$. By Schur's Lemma, $\widehat{O}$ must be proportional to the identity operator. By Step 3, the proportionality factor is 1.

□

3.13.7 For any operator $\widehat{O}$ that is some linear combination of products of $\hat{a}$'s and $\hat{a}^\dagger$'s, show that

$$\mathrm{Tr}\,\widehat{O} := \int \frac{\mathrm{d}z^*\mathrm{d}z}{2\pi\mathrm{i}}\; e^{-z^* z}\; {}_{\mathrm{cs}}\langle z|\widehat{O}|z\rangle_{\mathrm{cs}}. \tag{3.671}$$

3.13.8 For any operator $\widehat{O}$ that is some linear combination of products of $\hat{a}$'s and $\hat{a}^\dagger$'s, normal ordering of $\widehat{O}$, which is denoted $:\!\widehat{O}\!:$, is the operation of moving all creation operators to the left of annihilation operators as if all operators were to commute. For example,

$$\begin{aligned}&\widehat{O} := \hat{a}^\dagger\,\hat{a}\,\hat{a}\,\hat{a}^\dagger + \hat{a}^\dagger\,\hat{a}\,\hat{a}^\dagger\\ &\qquad\Longrightarrow :\widehat{O}: = \hat{a}^\dagger\,\hat{a}^\dagger\,\hat{a}\,\hat{a} + \hat{a}^\dagger\,\hat{a}^\dagger\,\hat{a} = \widehat{O} - 2\hat{a}^\dagger\hat{a} - \hat{a}^\dagger.\end{aligned} \tag{3.672}$$

Show that the matrix element of any normal ordered operator $:\widehat{O}(\hat{a}^\dagger,\hat{a}):$ between any two coherent states ${}_{\mathrm{cs}}\langle z|$ and $|z'\rangle_{\mathrm{cs}}$ is given by

$${}_{\mathrm{cs}}\langle z|\,:\widehat{O}(\hat{a}^\dagger,\hat{a}):\,|z'\rangle_{\mathrm{cs}} = e^{z^* z'}\;:O(z^*,z'):\,. \tag{3.673}$$

Here, $:O(z^*,z'):$ is the complex-valued function obtained from the normal ordered operator $:\widehat{O}(\hat{a}^\dagger,\hat{a}):$ by substituting $\hat{a}^\dagger$ for the complex number z^* and $\hat{a}$ for the complex number z'.

3.13.9 Define the continuous family of unitary operators

$$D(\alpha) := e^{\alpha\hat{a}^\dagger - \alpha^*\hat{a}}, \qquad \alpha\in\mathbb{C}. \tag{3.674}$$

From Glauber formula[71]

$$D(\alpha) = e^{-\frac{|\alpha|^2}{2}}\,e^{+\alpha\hat{a}^\dagger}\,e^{-\alpha^*\hat{a}}, \tag{3.676}$$

show that

$$D(\alpha)|0\rangle = e^{-\frac{|\alpha|^2}{2}}\,|\alpha\rangle_{\mathrm{cs}}. \tag{3.677}$$

Hence, $D(\alpha)$ is the unitary transformation that rotates the vacuum $|0\rangle$ into the coherent state $|\alpha\rangle_{\mathrm{cs}}$, up to a proportionality constant.

3.13.10 Define the anharmonic oscillator of order $n = 2, 3, 4, \cdots$ by

$$\begin{aligned}&\widehat{H} := \widehat{H}_0 + \widehat{H}_n,\\ &\widehat{H}_0 := \hbar\omega\left(\hat{a}^\dagger\hat{a} + \frac{1}{2}\right), \qquad \widehat{H}_n := \sum_{m=3}^{2n}\lambda_m\left(\hat{a}^\dagger + \hat{a}\right)^m.\end{aligned} \tag{3.678}$$

Of the real-valued parameters λ_m, $m = 3, 4, \cdots, 2n$, it is only required that $\lambda_{2n} > 0$. This insures that there exists a vacuum $|0\rangle$ annihilated by $\hat{a}$. With the help of the bosonic algebra, it is possible to move all annihilation operators to the right of the creation operators in the interaction $\widehat{H}_n$. This action generates many terms that can be grouped by ascending order in the combined number of creation and annihilation operators. The monomials of largest order are all contained in $:\widehat{H}_n:$. For example,

$$:(\hat{a}^\dagger + \hat{a})^3: = \left(\hat{a}^\dagger\,\hat{a}^\dagger\,\hat{a}^\dagger + 3\hat{a}^\dagger\,\hat{a}^\dagger\,\hat{a} + \mathrm{H.c.}\right). \tag{3.679}$$

[71] Let A and B be two operators that both commute with their commutator $[A,B]$. Then,

$$e^{A}\,e^{B} = e^{A+B}\,e^{\frac{1}{2}[A,B]}. \tag{3.675}$$

Evidently, $:\widehat{H}_n:$ cannot be written anymore as a polynomial in $\hat{x} \propto (\hat{a}^\dagger + \hat{a})$ of degree $2n$.
After normal ordering of $\widehat{H}$, the canonical partition function on the Hilbert space $\mathfrak{F}$ for the harmonic oscillator becomes

$$Z(\beta) := e^{-\beta E_0}\,\mathrm{Tr}\, e^{-\beta :\widehat{H}:} = e^{-\beta\, E_0} \sum_{n=0}^{\infty} \langle n| e^{-\beta :\widehat{H}:} |n\rangle, \tag{3.680}$$

where E_0 is the normal ordering energy, that is, the expectation value $\langle 0|\widehat{H}|0\rangle$. We will now give an alternative representation of the canonical partition function that relies on the use of coherent states. We begin with the trace formula (3.671)

$$Z(\beta) = \exp(-\beta\, E_0) \int \frac{\mathrm{d}\varphi_0^* \mathrm{d}\varphi_0}{2\pi\mathrm{i}}\, e^{-\varphi_0^*\varphi_0}\, {}_{\mathrm{cs}}\langle \varphi_0| \exp(-\beta :\widehat{H}:) |\varphi_0\rangle_{\mathrm{cs}}. \tag{3.681}$$

For M a large positive integer, write

$$\begin{aligned} \exp(-\beta :\widehat{H}:) &= \exp\left(-\frac{\beta}{M} \sum_{j=0}^{M-1} :\widehat{H}: \right) \\ &= 1 - \frac{\beta}{M} \sum_{j=0}^{M-1} :\widehat{H}: + \mathcal{O}\left[(\beta/M)^2 \right]. \end{aligned} \tag{3.682}$$

3.13.11 To the same order of accuracy, verify that

$$e^{-\beta :\widehat{H}:} = \left(e^{-\beta :\widehat{H}: /M} \right)^M . \tag{3.683}$$

Insert the resolution of identity (3.666) $(M-1)$-times,

$$e^{-\beta :\widehat{H}:} = e^{-\frac{\beta}{M} :\widehat{H}:} \left(\prod_{j=M-1}^{1} \int \frac{\mathrm{d}\varphi_j^* \mathrm{d}\varphi_j}{2\pi\mathrm{i}}\, e^{-\varphi_j^*\varphi_j}\, |\varphi_j\rangle_{\mathrm{cs}}\, {}_{\mathrm{cs}}\langle \varphi_j|\, e^{-\frac{\beta}{M} :\widehat{H}:} \right). \tag{3.684}$$

3.13.12 Verify that Eq. (3.673) together with Eq. (3.682) gives

$$\begin{aligned} {}_{\mathrm{cs}}\langle \varphi_0|\, e^{-\frac{\beta}{M} :\widehat{H}:}\, |\varphi_{M-1}\rangle_{\mathrm{cs}} =& e^{+\varphi_0^*\varphi_{M-1} - \frac{\beta}{M} :H(\varphi_0^*, \varphi_{M-1}):} \\ &+ \mathcal{O}\left[(\beta/M)^2 \right], \end{aligned} \tag{3.685a}$$

and

$$\begin{aligned} {}_{\mathrm{cs}}\langle \varphi_j|\, e^{-\frac{\beta}{M} :\widehat{H}:}\, |\varphi_{j-1}\rangle_{\mathrm{cs}} =& e^{+\varphi_j^*\varphi_{j-1} - \frac{\beta}{M} :H(\varphi_j^*, \varphi_{j-1}):} \\ &+ \mathcal{O}\left[(\beta/M)^2 \right], \end{aligned} \tag{3.685b}$$

for $j = M-1, M-2, \cdots, 1$. The operator-valued function $:\widehat{H}:$ of $\hat{a}$ and $\hat{a}^\dagger$ has been replaced by a complex-valued function $:H:$ of φ and φ^*, respectively.

3.13.13 Verify that, altogether, a M-dimensional integral representation of the partition function has been found,

$$
\begin{aligned}
Z(\beta) = \exp(-\beta\,E_0) \int &\left(\prod_{j=0}^{M-1} \frac{\mathrm{d}\varphi_j^* \mathrm{d}\varphi_j}{2\pi\mathrm{i}}\right) \\
&\times \exp\left(-\sum_{j=1}^{M}\left[\varphi_j^*\left(\varphi_j - \varphi_{j-1}\right) + \frac{\beta}{M}:H(\varphi_j^*, \varphi_{j-1}):\right]\right) \\
&+ \mathcal{O}\left[(\beta/M)^2\right],
\end{aligned}
\tag{3.686a}
$$

whereby

$$
\varphi_M := \varphi_0, \qquad \varphi_M^* := \varphi_0^*. \tag{3.686b}
$$

It is customary to write, in the limit $M \to \infty$, the functional path-integral representation of the partition function

$$
Z(\beta) = e^{-\beta\,E_0} \int \mathcal{D}[\varphi^*, \varphi] e^{-S_{\mathrm{E}}[\varphi^*, \varphi]}, \tag{3.687a}
$$

where the so-called Euclidean action $S_{\mathrm{E}}[\varphi^*, \varphi]$ is given by

$$
S_{\mathrm{E}}[\varphi^*, \varphi] = \int_0^{\beta} \mathrm{d}\tau \left\{\varphi^*(\tau)\partial_\tau\varphi(\tau) + :H[\varphi^*(\tau), \varphi(\tau)]:\right\}, \tag{3.687b}
$$

and the complex-valued fields $\varphi^*(\tau)$ and $\varphi(\tau)$ obey the periodic boundary conditions

$$
\varphi^*(\tau) = \varphi^*(\tau + \beta), \qquad \varphi(\tau) = \varphi(\tau + \beta). \tag{3.687c}
$$

Hence, their Fourier transform are

$$
\varphi^*(\tau) = \frac{1}{\beta}\sum_{l\in\mathbb{Z}} \varphi_l^*\, e^{+\mathrm{i}\varpi_l\tau}, \qquad \varphi(\tau) = \frac{1}{\beta}\sum_{l\in\mathbb{Z}} \varphi_l\, e^{-\mathrm{i}\varpi_l\tau}. \tag{3.688a}
$$

The frequencies

$$
\varpi_l := \frac{2\pi}{\beta} l, \qquad l \in \mathbb{Z}, \tag{3.688b}
$$

are the so-called bosonic Matsubara frequencies.
Convergence of the (functional) integral representing the partition function is guaranteed by the contribution $\lambda_{2n}(\varphi^* + \varphi)^{2n}$ to the interaction $:H_n(\varphi^*, \varphi):$. Thus, convergence of an integral is the counterpart in a path-integral representation to the existence of a ground state in operator language.

3.13.14 Verify that quantum mechanics at zero temperature is recovered from the partition function after performing the analytical continuation (also called a Wick rotation)

$$\tau = +\mathrm{i}t, \qquad \mathrm{d}\tau = +\mathrm{i}\mathrm{d}t, \qquad \partial_\tau = -\mathrm{i}\partial_t, \tag{3.689}$$

under which

$$\begin{aligned} -S_{\mathrm{E}} &\to +\mathrm{i}S \\ &= +\mathrm{i}\int\limits_{-\infty}^{+\infty} \mathrm{d}t \left\{\varphi^*(t)\mathrm{i}\partial_t\varphi(t) - :H[\varphi^*(t),\varphi(t)]:\right\}. \end{aligned} \tag{3.690}$$

The path-integral representation of the anharmonic oscillator relies solely on two properties of bosonic coherent states: Equations (3.666) and (3.673). Raising, $\hat{a}^\dagger$, and lowering, $\hat{a}$, operators are not unique to bosons. One can also associate raising and lowering operators to fermions. Raising and lowering operators are also well known to be involved in the theory of the angular momentum. In general, raising and lowering operators appear whenever a finite (infinite) set of operators obey a finite (infinite) dimensional Lie algebra. Coherent states are those states that are eigenstates of lowering operators in the Lie algebra and they obey extensions of Eqs. (3.666) and (3.673). Hence, it is possible to generalize the path-integral representation of the partition function for the anharmonic oscillator to Hamiltonians expressed in terms of operators obeying a fermion, spin, or any type of Lie algebra. Due to the nonvanishing overlap of coherent states, a first-order imaginary-time derivative term always appears in the action. This term is called a Berry phase when it yields a pure phase in an otherwise real-valued Euclidean action as is the case, say, when dealing with spin Hamiltonians.[72] It is the first-order imaginary-time derivative term that encodes quantum mechanics in the path-integral representation of the partition function.

3.13.15 Show that if $H[\varphi^*,\varphi]$ is a quadratic form with the kernel $\mathcal{H}$, then

$$Z(\beta) = \frac{1}{\mathrm{Det}\,(\partial_\tau + \mathcal{H})}, \qquad Z = \frac{1}{\mathrm{Det}\,[-\mathrm{i}\,(\mathrm{i}\partial_t - \mathcal{H})]}. \tag{3.693}$$

[72] By writing

$$\varphi(\tau) = \sqrt{\frac{1}{2}}\,[x(\tau) + \mathrm{i}p(\tau)], \qquad \varphi^*(\tau) = \sqrt{\frac{1}{2}}\,[x(\tau) - \mathrm{i}p(\tau)], \tag{3.691}$$

we can derive the path-integral representation of the (an)harmonic oscillator in terms of the coordinate and momentum of the single particle of unit mass $m = 1$, unit characteristic frequency $\omega = 1$, and with $\hbar = 1$. The first-order partial derivative term becomes purely imaginary

$$\int\limits_0^\beta \mathrm{d}\tau (\varphi^*\partial_\tau\varphi)(\tau) = \mathrm{i}\int\limits_0^\beta \mathrm{d}\tau (x\partial_\tau p)(\tau). \tag{3.692}$$

3.14 This exercise reviews what fermionic coherent states are. Fermionic coherent states are the tools required to construct the path-integral representation of the quantum mechanics built out of fermionic degrees of freedom.

Consider the pair of creation and annihilation operator $\hat{c}^{\dagger}$ and $\hat{c}$ that obey the anticommutation relations

$$\{\hat{c},\hat{c}^{\dagger}\}=1,\qquad \{\hat{c}^{\dagger},\hat{c}^{\dagger}\}=\{\hat{c},\hat{c}\}=0. \tag{3.694}$$

Define $|0\rangle$ to be the state obeying the condition

$$\hat{c}\,|0\rangle=0. \tag{3.695}$$

This state is assumed normalized to the number 1. It is thus unique up to a multiplicative phase. The state $|0\rangle$ is called the vacuum in physics. In group theory, it is the highest weight state. The vector space $\mathfrak{F}$ over the field of complex numbers defined by

$$\mathfrak{F}:=\operatorname{span}\{|0\rangle,\hat{c}^{\dagger}\,|0\rangle\}\equiv\operatorname{span}\{|0\rangle,|1\rangle\} \tag{3.696}$$

is two-dimensional. It is isomorphic to $\mathbb{C}^2$.

3.14.1 Verify the last two statements.

The vector space $\mathfrak{F}$ is an example of a Fock space in physics.

The Fock space $\mathfrak{F}$ is also an example of an exterior algebra. To see this, let V be a one-dimensional vector space over the field of complex numbers with v its basis vector. Let $\wedge$ denote the wedge (antisymmetric) product.

3.14.2 Verify the isomorphism

$$\mathfrak{F}\cong\bigoplus_{n=0,1}(\wedge V)^n, \tag{3.697}$$

where $(\wedge V)^0=\mathbb{C}$ and $(\wedge V)^1=V$ by convention.

3.14.3 How different is $\bigoplus_{n=0,1}(\wedge V)^n$ from $\bigoplus_{n=0}^{q}(\wedge V)^n$ for q an integer larger than one?

3.14.4 For m, n, and q integers, what is the dimensionality of $\bigoplus_{n=0}^{q}(\wedge V)^n$ if V is m-dimensional?

The Fock space $\mathfrak{F}$ is an example of a Grassmann algebra. Indeed, the Fock space $\mathfrak{F}$ is also isomorphic to the vector space $\mathfrak{G}$ over the field of complex numbers defined to be the set of all possible linear combination with complex-valued coefficients of all monomials that can be built out of the pair of objects I and c obeying the multiplication rules

$$I\,c=c\,I=c,\qquad c^2=0. \tag{3.698}$$

3.14.5 Prove this statement.

3.14.6 Verify that $\mathfrak{G}$ satisfies Eq. (3.554) and is thus an unital algebra.

In physics, thermodynamic equilibrium at the temperature $1/(k_{\mathrm{B}}\,\beta)$ (k_{B} the Boltzmann constant) is encoded quantitatively by the single-particle partition function

$$Z(\beta) := \mathrm{tr}|_{\mathfrak{F}}\, e^{-\beta\,\hat{c}^{\dagger}\,\varepsilon\,\hat{c}} = 1 + e^{-\beta\,\varepsilon} \tag{3.699}$$

for the canonical ensemble. For example, the "internal energy," that is the thermal average over the single-particle energies 0 for the state $|0\rangle$ and ε for the state $|1\rangle$, is

$$U(\beta) := -\frac{\partial \ln Z(\beta)}{\partial \beta} = \frac{\varepsilon\, e^{-\beta\,\varepsilon}}{1 + e^{-\beta\,\varepsilon}} = \sum_{\varepsilon_{\iota}=0,\varepsilon} \varepsilon_{\iota}\, f_{\mathrm{FD}}(\varepsilon_{\iota}), \tag{3.700a}$$

where

$$f_{\mathrm{FD}}(\varepsilon_{\iota}) := \frac{1}{e^{\beta\,\varepsilon_{\iota}} + 1} \tag{3.700b}$$

is the Fermi-Dirac distribution in the canonical ensemble. We are going to show that it is possible to represent (3.699) as an integral over Grassmann-valued integration variables.

The key idea is to introduce a basis of states for the Fock space $\mathfrak{F}$ such that the states from this basis are (i) over-complete, (ii) eigenstates of the annihilation operator $\hat{c}$, (iii) provide a resolution of the identity. The condition (i) for over-completeness is necessary to achieve the condition (ii). Condition (iii) will be achieved by defining a functional that maps elements of a Grassmann algebra to the field over which it is defined in such a way that it respects key properties of the Riemann integral on the functional space of integrable functions, say linearity.

Over-completeness is brought about as follows. We associate to the annihilation operator $\hat{c}$ the Grassmann number c. We associate to the annihilation operator $\hat{c}^{\dagger}$ the Grassmann number c^{*}. We impose that c and c^{*} pair-wise anticommute so that we may define the four-dimensional Grassmann algebra over the field of complex numbers

$$\mathfrak{G} := \mathrm{span}\,\{I,\; c^{*},\; c,\; c^{*}\,c\}, \tag{3.701a}$$

where

$$\begin{aligned} &I\,c = c\,I = c, \qquad I\,c^{*} = c^{*}\,I = c^{*}, \qquad I\,c^{*}\,c = c^{*}\,c\,I = c^{*}\,c,\\ &c^{*}\,c = -c\,c^{*}, \qquad c^{*2} = c^{2} = 0. \end{aligned} \tag{3.701b}$$

The unital element I will often be implicit, identified with the real number 1, denoted $1_{\mathcal{F}}$, and so on.

Define the Grassmann Fock space $\mathcal{F}_{\mathfrak{G}}$ to be the vector space spanned by $|0\rangle$ and $|1\rangle$ over two copies of the Grassmann algebra (3.701), that is

$$\begin{aligned}\mathcal{F}_{\mathfrak{G}} := \Big\{ |\mathfrak{a}\,,\mathfrak{b}\,\rangle = &\,(a_1 + a_2\, c + a_3\, c^* + a_4\, c^*\, c)\,|0\rangle \\ &+ (b_1 + b_2\, c + b_3\, c^* + b_4\, c^*\, c)\,|1\rangle, \quad a_i, b_i \in \mathbb{C}\Big\},\end{aligned} \tag{3.702a}$$

whereby the consistency rule that Grassmann numbers c and c^* anticommute with fermion annihilation $\hat{c}$ and creation $\hat{c}^\dagger$ operators,

$$\{c, \hat{c}\} = \{c, \hat{c}^\dagger\} = \{c^*, \hat{c}\} = \{c^*, \hat{c}^\dagger\} = 0, \tag{3.702b}$$

must be imposed.
Define the pair of fermionic coherent states $|a_2\, c\rangle$ and $|a_3^*\, c^*\rangle$ from the Grassmann Fock space $\mathcal{F}_{\mathfrak{G}}$ by

$$|a_2\, c\rangle := e^{-a_2\, c\, \hat{c}^\dagger}\,|0\rangle = \left(1 - a_2\, c\, \hat{c}^\dagger\right)|0\rangle, \tag{3.703a}$$

and

$$|a_3^*\, c^*\rangle := e^{-a_3^*\, c^*\, \hat{c}^\dagger}\,|0\rangle = \left(1 - a_3^*\, c^*\, \hat{c}^\dagger\right)|0\rangle, \tag{3.703b}$$

respectively. The corresponding pair of adjoint fermionic coherent states $\langle a_2\, c|$ and $\langle a_3^*\, c^*|$ from the Grassmann Fock space $\mathcal{F}_{\mathfrak{G}}$ are defined by

$$\langle a_2\, c| := \langle 0|e^{-\hat{c}\, c^*\, a_2^*} = \langle 0|\,(1 - \hat{c}\, c^*\, a_2^*)\,, \tag{3.704a}$$

and

$$\langle a_3^*\, c^*| := \langle 0|e^{-\hat{c}\, c\, a_3} = \langle 0|\,(1 - \hat{c}\, c\, a_3)\,, \tag{3.704b}$$

respectively.

3.14.7 Verify that

$$\hat{c}\,|a_2\, c\,\rangle = a_2\, c\,|a_2\, c\rangle, \qquad \langle a_2\, c\,|\,\hat{c}^\dagger = \langle a_2\, c\,|\, a_2^*\, c^*, \tag{3.705a}$$

and

$$\hat{c}\,|a_3^*\, c^*\rangle = a_3^*\, c^*\,|a_3^*\, c^*\rangle, \qquad \langle a_3^*\, c^*|\,\hat{c}^\dagger = \langle a_3^*\, c^*|\, a_3\, c\,. \tag{3.705b}$$

3.14.8 Verify that fermionic coherent states are neither normalized nor orthogonal,

$$\langle c|c\,\rangle = e^{+c^*\, c}, \quad \langle c^*|c^*\rangle = e^{-c^*\, c}, \quad \langle c\,|c^*\rangle = 1, \quad \langle c^*|c\,\rangle = 1. \tag{3.706}$$

3.14.9 Verify that the expectation value of any normal-ordered operator $\widehat{O}(\hat{c}^\dagger, \hat{c})$ in the fermionic coherent state $|c\rangle$ is

$$\langle c|\widehat{O}(\hat{c}^\dagger, \hat{c})|c\rangle = e^{+c^*\, c}\, O(c^*, c). \tag{3.707}$$

Here, the Grassmann-valued function $O(c^*, c)$ is obtained from the operator $\widehat{O}(\hat{c}^\dagger, \hat{c})$ by replacing $\hat{c}^\dagger$ with c^* and $\hat{c}$ with c.

Fermionic coherent states allow for path-integral representations of partition functions for fermions. To see this, we still need a resolution of the identity, which, in turn, demands the notion of Grassmann integration. Grassmann integrations $\int \mathrm{d}c$ and $\int \mathrm{d}c^*$ are multilinear mappings from the Grassmann algebra spanned by I and c and I and c^*, respectively, to the complex numbers. They are defined by linear extension of the rules

$$\begin{aligned}
&0 =: \int \mathrm{d}c\, 1, \quad 0 =: \int \mathrm{d}c\, c^* = -c^* \int \mathrm{d}c\, 1, \quad 1 =: \int \mathrm{d}c\, c, \\
&0 =: \int \mathrm{d}c^*\, 1, \quad 0 =: \int \mathrm{d}c^*\, c = -c \int \mathrm{d}c^*\, 1, \quad 1 =: \int \mathrm{d}c^*\, c^*, \\
&\int \mathrm{d}c^* \int \mathrm{d}c\, (\cdots) = - \int \mathrm{d}c \int \mathrm{d}c^*\, (\cdots) \equiv \int \mathrm{d}c^* \mathrm{d}c\, (\cdots).
\end{aligned} \tag{3.708}$$

Thus,

$$\begin{aligned}
&\int \mathrm{d}c\, (a_1 + a_2 c + a_3 c^* + a_4 c^*\, c) = a_2 - a_4 c^*, \\
&\int \mathrm{d}c^*\, (a_1 + a_2 c + a_3 c^* + a_4 c^*\, c) = a_3 + a_4 c, \\
&\int \mathrm{d}c^* \int \mathrm{d}c\, (a_1 + a_2 c + a_3 c^* + a_4 c^*\, c) = -a_4.
\end{aligned} \tag{3.709}$$

Grassmann integration over the Grassmann Fock space $\mathcal{F}_{\mathfrak{G}}$ in Eq. (3.702) is the same as Grassmann integration over the Grassmann algebra with the caveat that $\hat{c}$ and $\hat{c}^\dagger$ anticommute with $\int \mathrm{d}c$ and $\int \mathrm{d}c^*$.

3.14.10 We are now ready to establish a resolution of the identity for fermionic coherent states. Indeed, with the help of the resolution of the identity

$$|0\rangle\langle 0| + |1\rangle\langle 1| =: 1_{\mathcal{F}} \tag{3.710}$$

in $\mathcal{F}$, verify that

$$\int \mathrm{d}c^* \int \mathrm{d}c\, e^{-c^*\, c}\, |c\rangle\langle c| = 1_{\mathcal{F}}. \tag{3.711}$$

3.14.11 Moreover, verify the trace formula

$$\int \mathrm{d}c^* \int \mathrm{d}c\, e^{-c^*\, c} \langle -c|\widehat{O}| + c\rangle = \mathrm{tr}_{\mathcal{F}}\, \widehat{O}, \tag{3.712}$$

for any linear operator $\widehat{O} : \mathcal{F} \to \mathcal{F}$. It is the asymmetry in the sign of c entering the bra relative to that entering the ket in the trace formula (3.712) that delivers the Fermi-Dirac distribution function (3.700b).

3.14.12 Verify with the help of the overlap, resolution of the identity, and trace formula in Eqs. (3.706), (3.711), and (3.712), respectively, that it is possible to represent the partition function (3.699) as the Grassmann path integral,

$$Z(\beta) = \lim_{M\to\infty}\left(\prod_{j=0}^{M-1}\int \mathrm{d}c_j^*\int \mathrm{d}c_j\right) e^{-\sum_{j=1}^{M}\left[c_j^*\,(c_j-c_{j-1})+\frac{\beta}{M}\,H(c_j^*,c_{j-1})\right]}$$
$$\equiv \int \mathcal{D}[c^*,c]\, e^{-\int_0^\beta \mathrm{d}\tau\,\{c^*(\tau)\,(\partial_\tau\, c)(\tau)+H[c^*(\tau),c(\tau)]\}}, \tag{3.713a}$$

where

$$\begin{aligned} c_M &\equiv -c_0 \Longrightarrow c(\tau+\beta) = -c(\tau),\\ c_M^* &\equiv -c_0^* \Longrightarrow c^*(\tau+\beta) = -c^*(\tau). \end{aligned} \tag{3.713b}$$

3.14.13 Verify that

$$\begin{aligned} 1-e^{-\beta\,\varepsilon} &= Z\left(\beta+\mathrm{i}\frac{\pi}{\varepsilon}\right)\\ &= \mathrm{tr}|_{\mathcal{F}}\left[(-1)^{\hat{c}^\dagger\,\hat{c}}\, e^{-\beta\,\hat{c}^\dagger\,\varepsilon\,\hat{c}}\right]. \end{aligned} \tag{3.714}$$

3.14.14 Verify the alternating trace formula

$$\int \mathrm{d}c^*\int \mathrm{d}c\, e^{-c^*\,c}\,\langle +c|\widehat{O}|+c\rangle = \mathrm{tr}_{\mathcal{F}}\left[(-1)^{\hat{c}^\dagger\,\hat{c}}\,\widehat{O}\right], \tag{3.715}$$

for any linear operator $\widehat{O}:\mathcal{F}\to\mathcal{F}$.

3.14.15 Verify with the help of the overlap, resolution of the identity, and trace formula in Eqs. (3.706), (3.711), and (3.715), respectively, that it is possible to represent the trace on the right-hand side of the last equality in Eq. (3.714) as the Grassmann path integral

$$Z\left(\beta+\frac{\mathrm{i}\pi}{\varepsilon}\right) = \lim_{M\to\infty}\left(\prod_{j=0}^{M-1}\int \mathrm{d}c_j^*\int \mathrm{d}c_j\right) e^{-\sum_{j=1}^{M}\left[c_j^*\,(c_j-c_{j-1})+\frac{\beta}{M}\,H(c_j^*,c_{j-1})\right]}$$
$$\equiv \int \mathcal{D}[c^*,c]\, e^{-\int_0^\beta \mathrm{d}\tau\,\{c^*(\tau)\,(\partial_\tau\, c)(\tau)+H[c^*(\tau),c(\tau)]\}}, \tag{3.716a}$$

where

$$c^*(\tau+\beta) = +c^*(\tau), \qquad c(\tau+\beta) = +c(\tau). \tag{3.716b}$$

3.14.16 Show that the analytical continuation $\tau = it$ after the limit $\beta\to\infty$ has been taken gives

$$Z = \int \mathcal{D}[c^*,c]\, e^{+\mathrm{i}\int_{\mathbb{R}} \mathrm{d}t\,\{c^*(t)\,(\mathrm{i}\partial_t\, c)(t)-H[c^*(t),c(t)]\}}. \tag{3.717}$$

3.14.17 Show that if $H[c^*,c]$ is a quadratic form with the kernel $\mathcal{H}$, then

$$Z(\beta) = \mathrm{Det}\,(\partial_\tau+\mathcal{H}), \qquad Z = \mathrm{Det}\,[-\mathrm{i}\,(\mathrm{i}\partial_t-\mathcal{H})]. \tag{3.718}$$

3.14.18 Let A be any complex number and let c^* and c denote two independent Grassmann variables. Because the Grassmann integral is constructed such that

$$\int \mathrm{d}c^* \int \mathrm{d}c\, e^{-c^*\,A\,c} = A, \tag{3.719a}$$

there follows that

$$A = \int \mathrm{d}c^* \int \mathrm{d}c\, e^{-c^*\,A\,c} \tag{3.719b}$$

$$= A \int \frac{\mathrm{d}c^*}{\sqrt{A}} \int \frac{\mathrm{d}c}{\sqrt{A}}\, e^{-(c^*\,\sqrt{A})\,(\sqrt{A}\,c)}$$

$$= A \int \mathrm{d}\zeta^* \int \mathrm{d}\zeta\, e^{-\zeta^*\,\zeta}. \tag{3.719c}$$

This means that the transformation

$$\zeta^* =: c^*\,\sqrt{A}, \qquad \zeta =: \sqrt{A}\,c \tag{3.720a}$$

implies the Jacobian

$$J \equiv \frac{\mathrm{d}c^*}{\mathrm{d}\zeta^*}\,\frac{\mathrm{d}c}{\mathrm{d}\zeta} = A \tag{3.720b}$$

that relates the two Grassmann measures. What are the counterparts to Eqs. (3.719) and (3.720) if A is a positive real-valued number, and we replace c^* by the complex number z^*, c by the complex-conjugate z of z^*, and the Grassmann integrals by the Riemann integral over the complex plane with the measure $\mathrm{d}z^*\,\mathrm{d}z/(2\pi\mathrm{i})$?

3.14.19 Denote with $\bar{\psi}$ and ψ two independent Grassmann-valued fields belonging to some functional space equipped with the scalar product $(\cdot,\cdot)$. Let K denote a linear map on this functional space. Define the invertible linear maps

$$\bar{\psi} =: \bar{\psi}'\,\bar{U}, \tag{3.721a}$$

and

$$\psi =: U\,\psi', \tag{3.721b}$$

respectively. Define

$$K' := \bar{U}\,K\,U. \tag{3.721c}$$

Show that the generalization of Eq. (3.719) is

$$\int \mathcal{D}[\bar{\psi}] \int \mathcal{D}[\psi]\, e^{-(\bar{\psi},K\,\psi)} = \int \mathcal{D}[\bar{\psi}'] \int \mathcal{D}[\psi']\, J\, e^{-(\bar{\psi}',K'\,\psi')}, \tag{3.722a}$$

where the Jacobian J is

$$J \equiv \frac{\mathcal{D}[\bar{\psi}\,]\,\mathcal{D}[\psi\,]}{\mathcal{D}[\bar{\psi}']\,\mathcal{D}[\psi']} = \left(\mathrm{Det}\,\bar{U}^{-1}\right)\,\left(\mathrm{Det}\,U^{-1}\right). \tag{3.722b}$$

3.15 Verify the continuity equation (3.249) in two ways. First, take the time derivative of j^0 defined by Eq. (3.248) and make use of the Dirac equation (3.247). Second, derive Noether's theorem from the fact that the Lagrangian density (3.250) is invariant under the global U(1) gauge transformation (3.251) by which $\Psi(t,x)$ changes by a multiplicative phase.

3.16 Let $\Phi(\mathsf{x})$ be defined by Eq. (3.250b) [recall that (t,x) was identified with the two-vector $\mathsf{x}\in\mathbb{M}_{1,1}$]

3.16.1 Verify that

$$\Phi(\mathsf{x})\,\Phi^{\dagger}(\mathsf{x})=|\phi(\mathsf{x})|^2\,\tau_0. \tag{3.723}$$

3.16.2 Verify that

$$\left[\not{\mathsf{p}}\mp\Phi(\mathsf{x})\right]\left[\not{\mathsf{p}}\pm\Phi^{\dagger}(\mathsf{x})\right]=\left(\mathsf{p}^2-|\phi(\mathsf{x})|^2\right)\,\tau_0. \tag{3.724}$$

3.16.3 Verify that

$$\frac{\tau_0}{\not{\mathsf{p}}\mp\Phi(\mathsf{x})}=\frac{\not{\mathsf{p}}\pm\Phi^{\dagger}(\mathsf{x})}{\mathsf{p}^2-|\phi(\mathsf{x})|^2}. \tag{3.725}$$

3.16.4 Perform the traces and integrals leading to Eq. (3.269).

3.17 In order to verify Eqs. (3.330a) and (3.330b), we start from

$$\mathcal{H}(\tau,x)=\begin{pmatrix}+\phi_2(\tau,x) & -\partial_x+\phi_1(\tau,x)\\ +\partial_x+\phi_1(\tau,x) & -\phi_2(\tau,x)\end{pmatrix}. \tag{3.726}$$

3.17.1 Show that, if

$$\psi_{\mathrm{bd}}(\tau,x):=\sqrt{\frac{|N(\tau)|}{2}}\,e^{-N(\tau)\,f(x)}\begin{pmatrix}+1\\-1\end{pmatrix} \tag{3.727a}$$

with the real-valued function $N(\tau)$ and the positive-valued function $f(x)$ obeying the first-order differential equation

$$N(\tau)\left(\frac{\partial f}{\partial x}\right)(x)=\phi_2(\tau,x), \tag{3.727b}$$

then

$$\mathcal{H}(\tau,x)\,\psi_{\mathrm{bd}}(\tau,x)=-\phi_1(\tau,x)\,\psi_{\mathrm{bd}}(\tau,x). \tag{3.727c}$$

Verify that the choices

$$f(x)=|x|,\qquad \phi_2(\tau,x)=N(\tau)\,\mathrm{sgn}(x), \tag{3.728}$$

solve Eq. (3.727b).

3.17.2 Verify that the choices

$$N(\tau)=m\,\sin\tau,\qquad \phi_1(\tau,x)=m\,\cos\tau, \tag{3.729}$$

give the bound state (3.330a) with the eigenenergy (3.330b) to the Dirac Hamiltonian (3.327).

3.18 Verify that Eq. (3.342) agrees with the gradient expansion (3.269) to leading order.

3.19 Assume for simplicity that $|\phi(\mathsf{x})| = m$ is nonvanishing and verify explicitly that the contribution from Eq. (3.351) to the current (3.350) is vanishing to first order in the gradient expansion. Furthermore, show that it is proportional to $\epsilon^{\mu\nu}\,\partial_\nu\,\Box\,\varphi$, where $\Box = \partial_\mu\,\partial^\mu$ is the d'Alembert operator, to the first subleading order in the gradient expansion.

3.20 We are going to derive Eq. (3.354) with the help of the identity (3.722). Recall that $\mathsf{x} = (\mathsf{x}^0, \mathsf{x}^1) \in \mathbb{M}_{1,1}$. We shall introduce a parameter $0 \le t \le 1$, the adiabatic time, that should not be confused with the time x^0 from Minkowski space $\mathbb{M}_{1,1}$.

Define the Dirac kernel

$$K := -\mathrm{i}\left(\mathrm{i}\partial\!\!\!/ + A\!\!\!/ - \Phi\right) \tag{3.730}$$

that enters in the partition function defined in Eq. (3.335) as

$$Z[\Phi, A\!\!\!/] = \mathrm{Det}\left[-\mathrm{i}\left(\mathrm{i}\partial\!\!\!/ + A\!\!\!/ - \Phi\right)\right] = \mathrm{Det}\,K. \tag{3.731}$$

Motivated by Eq. (3.344), we define the family of local unitary (axial) transformations $\{\bar{U}_t(\mathsf{x})\}$ and $\{U_t(\mathsf{x})\}$ that are indexed by the real-valued adiabatic parameter $t \in [0,1]$ through

$$\bar{\psi}(\mathsf{x}) =: \bar{\psi}_t(\mathsf{x})\,\bar{U}_t(\mathsf{x}), \qquad \bar{U}_t(\mathsf{x}) := e^{-\mathrm{i}\gamma_5\,\varphi(\mathsf{x})\,t/2}, \tag{3.732a}$$

and

$$\psi(\mathsf{x}) =: U_t(\mathsf{x})\,\psi_t(\mathsf{x}), \qquad U_t(\mathsf{x}) := e^{-\mathrm{i}\gamma_5\,\varphi(\mathsf{x})\,t/2}, \tag{3.732b}$$

respectively.

3.20.1 Verify that

$$\left(\bar{\psi}\,\psi\right)(\mathsf{x}) = \left(\bar{\psi}_t\,\bar{U}_t\,U_t\,\psi_t\right)(\mathsf{x}) = \left(\bar{\psi}_t\,e^{-\mathrm{i}\gamma_5\,\varphi\,t}\,\psi_t\right)(\mathsf{x}), \tag{3.733}$$

and

$$\bar{\psi}\,K\,\psi = \bar{\psi}_t\,K_t\,\psi_t, \tag{3.734a}$$

where

$$K_t := \bar{U}_t\,K\,U_t = -\mathrm{i}\left(\mathrm{i}\partial\!\!\!/ + A\!\!\!/ + a\!\!\!/\,t - |\phi|\,e^{\mathrm{i}\gamma_5\,(1-t)\,\varphi}\right) \tag{3.734b}$$

with

$$\begin{aligned} a\!\!\!/(\mathsf{x}) :=& \frac{1}{2}\gamma^\mu\,\gamma_5\,\left(\partial_\mu\,\varphi\right)(\mathsf{x}) \\ =& \gamma^0\,(-1)\,\frac{1}{2}\,\left(\partial_1\,\varphi\right)(\mathsf{x}) + \gamma^1\,(-1)\,\frac{1}{2}\,\left(\partial_0\,\varphi\right)(\mathsf{x}). \end{aligned} \tag{3.734c}$$

Needed is the explicit dependence on Φ and $\not{A}$ of the Jacobian

$$J[\Phi,\not{A}] := \lim_{N\to\infty}\prod_{q=1}^{N}\frac{\mathcal{D}\left[\bar{\psi}_{(q-1)\,\mathrm{d}t}\right]\mathcal{D}\left[\psi_{(q-1)\,\mathrm{d}t}\right]}{\mathcal{D}\left[\bar{\psi}_{q\,\mathrm{d}t}\right]\mathcal{D}\left[\psi_{q\,\mathrm{d}t}\right]} \tag{3.735}$$

with the convention that $\bar{\psi}_0\equiv\bar{\psi}$, $\psi_0\equiv\psi$, and $\mathrm{d}t\equiv 1/N$. For this purpose, we need to relate the Grassmann measures $\mathcal{D}[\bar{\psi}_t]\,\mathcal{D}[\psi_t]$ and $\mathcal{D}[\bar{\psi}_{t+\mathrm{d}t}]\,\mathcal{D}[\psi_{t+\mathrm{d}t}]$ when $\mathrm{d}t$ is infinitesimal.
Assume that we may write

$$\begin{aligned}
\mathcal{D}[\bar{\psi}_t]\,\mathcal{D}[\psi_t] &= \left(\prod_m \mathrm{d}\bar{b}_{t;m}\right)\left(\prod_n \mathrm{d}b_{t;n}\right),\\
\bar{\psi}_t(\mathrm{x}) &= \sum_m \xi^{\dagger}_{t;m}(\mathrm{x})\,\bar{b}_{t;m},\\
\psi_t(\mathrm{x}) &= \sum_n b_{t;n}\,\xi_{t;n}(\mathrm{x}),
\end{aligned} \tag{3.736a}$$

where $\{\xi_{t;n}\}$ is the complete set of orthogonal instantaneous eigenspinors with eigenvalues $\lambda_{t;n}$ of K_t, that is,

$$\begin{aligned}
&K_t\,\xi_{t;n}(\mathrm{x}) = \lambda_{t;n}\,\xi_{t;n}(\mathrm{x}),\\
&\sum_n \xi_{t;n}(\mathrm{x})\,\xi^{\dagger}_{t;n}(\mathrm{y}) = \tau_0\,\delta(\mathrm{x}-\mathrm{y}),\\
&\int \mathrm{d}^2\mathrm{x}\,\xi^{\dagger}_{t;m}(\mathrm{x})\,\xi_{t;n}(\mathrm{x}) = 0, \qquad m\neq n.
\end{aligned} \tag{3.736b}$$

Similarly,

$$\begin{aligned}
\mathcal{D}[\bar{\psi}_{t+\mathrm{d}t}]\,\mathcal{D}[\psi_{t+\mathrm{d}t}] &= \left(\prod_m \mathrm{d}\bar{b}_{t+\mathrm{d}t;m}\right)\left(\prod_n \mathrm{d}b_{t+\mathrm{d}t;n}\right),\\
\bar{\psi}_{t+\mathrm{d}t}(\mathrm{x}) &= \sum_{m'} \xi^{\dagger}_{t+\mathrm{d}t;m'}(\mathrm{x})\,\bar{b}_{t+\mathrm{d}t;m'},\\
\psi_{t+\mathrm{d}t}(\mathrm{x}) &= \sum_{n'} b_{t+\mathrm{d}t;n'}\,\xi_{t+\mathrm{d}t;n'}(\mathrm{x}).
\end{aligned} \tag{3.737}$$

3.20.2 Use Eq. (3.732) to show that

$$\begin{aligned}
\bar{\psi}_{t+\mathrm{d}t}(\mathrm{x}) &= \sum_m \xi^{\dagger}_{t;m}(\mathrm{x})\,\bar{b}_{t+\mathrm{d}t;m}\,e^{+\mathrm{i}\gamma_5\frac{\varphi(\mathrm{x})}{2}\mathrm{d}t},\\
\psi_{t+\mathrm{d}t}(\mathrm{x}) &= \sum_n b_{t+\mathrm{d}t;n}\,e^{+\mathrm{i}\gamma_5\frac{\varphi(\mathrm{x})}{2}\mathrm{d}t}\xi_{t;n}(\mathrm{x}).
\end{aligned} \tag{3.738}$$

3.20.3 Verify that, to first order in $\mathrm{d}t$,

$$\begin{aligned}\bar{b}_{t+\mathrm{d}t;m'} &= \sum_{m} \left(\bar{U}_{t+\mathrm{d}t,t}\right)_{m',m} \bar{b}_{t;m}, \\ \left(\bar{U}_{t+\mathrm{d}t,t}\right)_{m',m} &= \int \mathrm{d}^2\mathrm{x}\, \xi^{\dagger}_{t;m}(\mathrm{x}) \left[1 + \mathrm{i}\gamma_5 \frac{\varphi(\mathrm{x})}{2}\, \mathrm{d}t\right] \xi_{t;m'}(\mathrm{x}),\end{aligned} \tag{3.739}$$

and

$$\begin{aligned}b_{t+\mathrm{d}t;n'} &= \sum_{n} \left(U_{t+\mathrm{d}t,t}\right)_{n',n} b_{t;n}, \\ \left(U_{t+\mathrm{d}t,t}\right)_{n',n} &= \int \mathrm{d}^2\mathrm{x}\, \xi^{\dagger}_{t;n'}(\mathrm{x}) \left[1 + \mathrm{i}\gamma_5 \frac{\varphi(\mathrm{x})}{2}\, \mathrm{d}t\right] \xi_{t;n}(\mathrm{x}).\end{aligned} \tag{3.740}$$

3.20.4 Deduce that, to first order in $\mathrm{d}t$,

$$\begin{aligned}\operatorname{Det} \bar{U}_{t+\mathrm{d}t,t} &= \operatorname{Det} U_{t+\mathrm{d}t,t} \\ &= e^{+\sum_{m} \int \mathrm{d}^2\mathrm{x}\, \xi^{\dagger}_{t;m}(\mathrm{x})\, \mathrm{i}\gamma_5 \frac{\varphi(\mathrm{x})}{2}\, \mathrm{d}t\, \xi_{t;m}(\mathrm{x})}.\end{aligned} \tag{3.741}$$

3.20.5 Deduce that (recall that $\mathrm{d}t = 1/N$)

$$\begin{aligned}J[\Phi, \not{A}] &= \lim_{N\to\infty} \prod_{q=1}^{N} \operatorname{Det} \bar{U}_{q\,\mathrm{d}t,(q-1)\,\mathrm{d}t} \operatorname{Det} U_{q\,\mathrm{d}t,(q-1)\,\mathrm{d}t} \\ &= e^{+2\times\sum_{m} \int_0^1 \mathrm{d}t \int \mathrm{d}^2\mathrm{x}\, \xi^{\dagger}_{t;m}(\mathrm{x})\, \mathrm{i}\gamma_5 \frac{\varphi(\mathrm{x})}{2}\, \xi_{t;m}(\mathrm{x})}.\end{aligned} \tag{3.742}$$

We have reduced the computation of the Jacobian $J[\Phi, \not{A}]$ to the computation of the local chiral (axial) anomaly [recall that $\mathrm{x} = (\mathrm{x}^0, \mathrm{x}^1) \in \mathbb{M}_{1,1}$]

$$J[\phi, \not{A}] = e^{\int \mathrm{d}^2\mathrm{x}\, \mathcal{A}_5(\mathrm{x})}, \tag{3.743a}$$

where

$$\begin{aligned}\mathcal{A}_5(\mathrm{x}) &\equiv 2 \times \sum_{m} \int_0^1 \mathrm{d}t\, \xi^{\dagger}_{t;m}(\mathrm{x})\, \mathrm{i}\gamma_5 \frac{\varphi(\mathrm{x})}{2}\, \xi_{t;m}(\mathrm{x}) \\ &\equiv 2 \times \int_0^1 \mathrm{d}t \operatorname{Tr} \left[|x\rangle_{tt}\langle \mathrm{x}|\, \mathrm{i}\gamma_5 \frac{\varphi(\mathrm{x})}{2}\, |x\rangle_{tt}\langle \mathrm{x}|\right].\end{aligned} \tag{3.743b}$$

The functional trace Tr is to be performed with the instantaneous eigenstates of the kernel K_{t}. This trace is an ill-conditioned infinite sum. To make sense of it, we need to introduce a regularization that insures that the regularized trace is convergent. The choice of regularization is not unique. We choose a regularization that preserves the local gauge symmetry, for we want the total fermion number to be a good quantum number. This choice is

$$
\begin{aligned}
\mathcal{A}_5(\mathsf{x}) :=& 2\times \lim_{M\to\infty}\sum_m \int_0^1 \mathrm{d}t\, \xi^{\dagger}_{t;m}(\mathsf{x})\, \mathrm{i}\gamma_5\, \frac{\varphi(\mathsf{x})}{2} e^{-\lambda^2_m/M^2}\, \xi_{t;m}(\mathsf{x}) \\
=& 2\times \lim_{M\to\infty}\sum_m \int_0^1 \mathrm{d}t\, \xi^{\dagger}_{t;m}(\mathsf{x})\, \mathrm{i}\gamma_5\, \frac{\varphi(\mathsf{x})}{2} e^{-K^2_t/M^2}\, \xi_{t;m}(\mathsf{x}).
\end{aligned}
\tag{3.744}
$$

3.20.6 To justify the choice $e^{-K^2_t/M^2}$ for the regulator in Eq. (3.744), consider the limit $A\!\!\!/ = \Phi = 0$. Show that the Wick rotation $\tau := \mathrm{i}x^0$ turn the indefinite Hermitean operator K^2_t into a positive definite Hermitean operator.

3.20.7 Verify that the regulation (3.744) is gauge invariant.

3.20.8 With the help of Eqs. (3.247), (3.270), and $\epsilon^{01} = -\epsilon^{10} = 1$, verify that

$$
\operatorname{tr}\gamma_5 = 0, \qquad \operatorname{tr}\gamma_5^2 = 2, \tag{3.745a}
$$

$$
\operatorname{tr}\left(\gamma_5\,\gamma^{\mu}\right) = 0, \qquad \operatorname{tr}\left(\gamma_5\,\gamma^{\mu}\,\gamma_5\right) = 0, \tag{3.745b}
$$

$$
\operatorname{tr}\left(\gamma_5\,\gamma^{\mu}\,\gamma^{\nu}\right) = -2\,\epsilon^{\mu\nu}, \qquad \operatorname{tr}\left(\gamma_5\,\gamma^{\mu}\,\gamma^{\nu}\,\gamma_5\right) = 2\,g^{\mu\nu}, \tag{3.745c}
$$

where $\mu,\nu = 0,1$, and tr denotes the trace over the Clifford algebra.

3.20.9 With the help of Eq. (3.734b), verify that

$$
\begin{aligned}
-\operatorname{tr}\left(\mathrm{i}\gamma_5\,\frac{\varphi}{2}\,K^2_t\right) =& \varphi\,\left(\epsilon^{\mu\nu}\,\partial_{\mu}\,A_{\nu}\right) + t\,\varphi\,\left(\epsilon^{\mu\nu}\,\partial_{\mu}\,a_{\nu}\right) \\
&- \varphi\,|\phi|^2\,\sin\Big(2\,(1-t)\,\varphi\Big).
\end{aligned}
\tag{3.746}
$$

Integration over the adiabatic time t gives

$$
\begin{aligned}
-\int_0^1 \mathrm{d}t\,\operatorname{tr}\left(\mathrm{i}\gamma_5\,\frac{\varphi}{2}\,K^2_t\right) =& \varphi\,\left(\epsilon^{\mu\nu}\,\partial_{\mu}\,A_{\nu}\right) + \frac{1}{2}\,\varphi\,\left(\epsilon^{\mu\nu}\,\partial_{\mu}\,a_{\nu}\right) \\
&+ \frac{1}{2}\,|\phi|^2\,\left[\cos(2\,\varphi) - 1\right].
\end{aligned}
\tag{3.747}
$$

3.20.10 We assume that, after regularization, we are free to choose any basis of the functional space to represent the local chiral (axial) anomaly. If so, we can trade the instantaneous eigenbasis of K_t in favor of the same plane-wave basis for every adiabatic time $0 \le t \le 1$. Verify that the representation in terms of the plane-wave basis of the local chiral (axial) anomaly is

$$
\mathcal{A}_5(\mathsf{x}) = 2\times \lim_{M\to\infty}\int \frac{\mathrm{d}^2\mathsf{k}\,e^{-\mathsf{k}^2/M^2}}{(2\pi\,M)^2}\int_0^1 \mathrm{d}t\,\operatorname{tr}\left(-\mathrm{i}\gamma_5\,\frac{\varphi(\mathsf{x})}{2}\,K^2_t\right). \tag{3.748}
$$

3.20.11 Verify that

$$
\int \frac{\mathrm{d}^2\mathsf{k}\,e^{-\mathsf{k}^2/M^2}}{(2\pi\,M)^2} = +\frac{\mathrm{i}}{4\pi}. \tag{3.749}
$$

Hint: Do the analytical continuation $k^1 \to \mathrm{i}k^1$.

3.20.12 Verify that

$$\begin{aligned}\mathcal{A}_5(\mathsf{x}) = & -\frac{\mathrm{i}}{2\pi}\,(\partial_\mu\varphi)\,(\mathsf{x})\,\epsilon^{\mu\nu}\left[A_\nu(\mathsf{x})+\frac{1}{2}\,a_\nu(\mathsf{x})\right] \\ & +\frac{\mathrm{i}}{4\pi}\,|\phi(\mathsf{x})|^2\left[\cos\left(2\,\varphi(\mathsf{x})\right)-1\right] \\ & +\frac{\mathrm{i}}{2\pi}\,\partial_\mu\left\{\varphi(\mathsf{x})\,\epsilon^{\mu\nu}\left[A_\nu(\mathsf{x})+\frac{1}{2}\,a_\nu(\mathsf{x})\right]\right\}.\end{aligned} \tag{3.750}$$

3.20.13 Verify that Eq. (3.354) follows from combining Eq. (3.743) with Eq. (3.750).

3.21 Verify Eqs. (3.357) and (3.356).

3.22 Take the continuum limit of the Hamiltonian defined in Eq. (2.99) and contrast its equations of motion to the chiral equations of motion (3.359) and (3.360) when the order N of the matrices K and V is $N=1$. Show that the free massless relativistic quantum field theory for a neutral particle propagating in Minkowski space $\mathbb{M}_{1,1}$ is equivalent to choosing $N=2$, $K_{ij}=(-1)^{i+1}\,\delta_{ij}$, and $V_{ij}=\delta_{ij}$ in Eq. (3.355), when the external gauge fields are vanishing and the units $c=1=\hbar=1$ have been chosen.

3.23 Verify Eq. (3.364).

3.24 Verify Eq. (3.369).

3.25 Verify Eq. (3.382).

3.26 Verify Eq. (3.384).

3.27 Verify Eqs. (3.385) and (3.386).

3.28 Justify from Eq. (3.398) the terminology of right and left movers. *Hint:* Try the plane-wave Ansatz $e^{\mathrm{i}(p\,x-\omega\,t)}$.

3.29 After expressing the Dirac Hamiltonian (3.404a) in the chiral basis (3.405a), derive the chiral equations of motion (3.405b).

3.30 Verify the equations of motion (3.408b) by computing $-\mathrm{i}[\hat{J}_{\mathrm{D}\mp},\widehat{H}_{\mathrm{D}}]$.

3.31 Start from

$$\widehat{H}_{\mathrm{D}} := \int\limits_0^L \mathrm{d}x\,\left[\hat{\psi}^\dagger_{\mathrm{D}-}(x)\,\hat{p}\,\hat{\psi}_{\mathrm{D}-}(x)-\hat{\psi}^\dagger_{\mathrm{D}+}(x)\,\hat{p}\,\hat{\psi}_{\mathrm{D}+}(x)\right],\qquad \hat{p}:=-\mathrm{i}\partial_x, \tag{3.751a}$$

with the only nonvanishing equal-time anticommutators given by

$$\left\{\hat{\psi}_{\mathrm{D}\alpha}(x),\hat{\psi}^\dagger_{\mathrm{D}\beta}(y)\right\}=\delta_{\alpha,\beta}\,\delta(x-y),\qquad \alpha,\beta=-,+, \tag{3.751b}$$

and the periodic boundary conditions

$$\hat{\psi}^\dagger_{\mathrm{D}\alpha}(x+L)=\hat{\psi}^\dagger_{\mathrm{D}\alpha}(x),\qquad \hat{\psi}_{\mathrm{D}\alpha}(x+L)=\hat{\psi}_{\mathrm{D}\alpha}(x),\qquad \alpha=-,+. \tag{3.751c}$$

For any one of $\alpha=-,+$, define the current

$$\hat{\jmath}_{\mathrm{D}\alpha}(x):=\hat{\psi}^\dagger_{\mathrm{D}\alpha}(x)\,\hat{\psi}_{\mathrm{D}\alpha}(x). \tag{3.752}$$

We are going to prove that

$$\left[\hat{\jmath}_{\mathrm{D}\alpha}(x),\hat{\jmath}_{\mathrm{D}\alpha'}(x')\right]=\alpha\,\delta_{\alpha,\alpha'}\,\frac{\mathrm{i}}{2\pi}\,\partial_x\delta(x-x') \tag{3.753}$$

for any $\alpha,\alpha'=-,+$.

3.31.1 Step 1: Verify that

$$\begin{aligned}\left[\hat{\psi}^{\dagger}_{\mathrm{D}\alpha}(x)\,\hat{\psi}_{\mathrm{D}\alpha}(y),\hat{\psi}^{\dagger}_{\mathrm{D}\alpha'}(x')\,\hat{\psi}_{\mathrm{D}\alpha'}(y')\right]=&\\ \delta_{\alpha,\alpha'}\,\delta(y-x')\,\hat{\psi}^{\dagger}_{\mathrm{D}\alpha}(x)\,\hat{\psi}_{\mathrm{D}\alpha'}(y')&\\ -\,\delta_{\alpha,\alpha'}\,\delta(x-y')\,\hat{\psi}^{\dagger}_{\mathrm{D}\alpha'}(x')\,\hat{\psi}_{\mathrm{D}\alpha}(y).&\end{aligned} \tag{3.754}$$

We have to be careful when taking the limit $y\to x$ and $y'\to x'$ holding $x'-x\neq 0$ fixed. We shall be considering the limit

$$y=x+\epsilon,\qquad y'=x'+\epsilon \tag{3.755}$$

with $\epsilon>0$ infinitesimal.

3.31.2 Step 2: One verifies that if one performs the Fourier expansion

$$\hat{\psi}^{\dagger}_{\mathrm{D}\alpha}(x)=\frac{1}{\sqrt{L}}\sum_{p\in\mathbb{Z}}e^{-\mathrm{i}\frac{2\pi}{L}\,p\,x}\,\hat{\psi}^{\dagger}_{\mathrm{D}\alpha,p},\qquad \hat{\psi}_{\mathrm{D}\alpha}(x)=\frac{1}{\sqrt{L}}\sum_{p\in\mathbb{Z}}e^{+\mathrm{i}\frac{2\pi}{L}\,p\,x}\,\hat{\psi}_{\mathrm{D}\alpha,p}, \tag{3.756}$$

for $\alpha=-,+$, then

$$\widehat{H}_{\mathrm{D}}:=\frac{2\pi}{L}\sum_{p\in\mathbb{Z}}p\left(\hat{\psi}^{\dagger}_{\mathrm{D}-,p}\,\hat{\psi}_{\mathrm{D}-,p}-\hat{\psi}^{\dagger}_{\mathrm{D}+,p}\,\hat{\psi}_{\mathrm{D}+,p}\right). \tag{3.757}$$

The state $|0\rangle$ defined by

$$\hat{\psi}_{\mathrm{D}\alpha,p}|0\rangle=0 \tag{3.758}$$

for any $\alpha=-,+$ and any $p\in\mathbb{Z}$ is not the ground state of $\widehat{H}_{\mathrm{D}}$. The ground state of $\widehat{H}_{\mathrm{D}}$ is the Dirac sea

$$|\mathrm{DS}\rangle:=\left(\prod_{p\in\mathbb{Z}}^{p\leq 0}\hat{\psi}^{\dagger}_{\mathrm{D}-,p}\right)\left(\prod_{p\in\mathbb{Z}}^{p\geq 0}\hat{\psi}^{\dagger}_{\mathrm{D}+,p}\right)|0\rangle. \tag{3.759}$$

3.31.3 Step 3: One verifies that

$$\left\langle \mathrm{DS}\middle|\,\hat{\psi}^{\dagger}_{\mathrm{D}\alpha,p}\,\hat{\psi}_{\mathrm{D}\alpha',q}\,\middle|\mathrm{DS}\right\rangle=\delta_{\alpha,\alpha'}\,\delta_{p,q}\,\Theta_{\alpha\,p},\qquad \Theta_{\alpha\,p}:=\begin{cases}1, & \alpha\,p\geq 0,\\ 0, & \alpha\,p<0,\end{cases} \tag{3.760}$$

for any $p,q\in\mathbb{Z}$ and $\alpha=-,+$.

3.31.4 Step 4: One verifies that

$$\left\langle \mathrm{DS} \middle| \hat{\psi}^{\dagger}_{\mathrm{D}\alpha}(x)\, \hat{\psi}_{\mathrm{D}\alpha'}(y') \middle| \mathrm{DS} \right\rangle = \frac{\delta_{\alpha,\alpha'}}{L} \sum_{p\in\mathbb{Z}} e^{\mathrm{i}\frac{2\pi}{L}p\,(y'-x)}\, \Theta_{\alpha p} = \frac{\delta_{\alpha,\alpha'}}{L} \sum_{p=0}^{\infty} \left(e^{\alpha \mathrm{i}\frac{2\pi}{L}(y'-x)} \right)^{p} \tag{3.761}$$

for $\alpha = -,+$. To make sense of the right-hand side, we do the regularization

$$y' - x \to y' - x + \mathrm{i}\alpha\, 0^{+} \tag{3.762}$$

for which the right-hand side becomes the geometrical series

$$\left\langle \mathrm{DS} \middle| \hat{\psi}^{\dagger}_{\mathrm{D}\alpha}(x)\, \hat{\psi}_{\mathrm{D}\alpha'}(y') \middle| \mathrm{DS} \right\rangle = \frac{\delta_{\alpha,\alpha'}}{L} \frac{1}{1 - e^{\alpha \mathrm{i}\frac{2\pi}{L}(y'-x+\mathrm{i}\alpha\, 0^{+})}}. \tag{3.763a}$$

Similarly,

$$\left\langle \mathrm{DS} \middle| \hat{\psi}^{\dagger}_{\mathrm{D}\alpha'}(x')\, \hat{\psi}_{\mathrm{D}\alpha}(y) \middle| \mathrm{DS} \right\rangle = \frac{\delta_{\alpha,\alpha'}}{L} \frac{1}{1 - e^{\alpha \mathrm{i}\frac{2\pi}{L}(y-x'+\mathrm{i}\alpha\, 0^{+})}}. \tag{3.763b}$$

3.31.5 Step 5: One verifies that if $y = x+\epsilon$ and $y' = x'+\epsilon$ with $\epsilon > 0$ infinitesimal, then, in the thermodynamic limit $L \to \infty$ holding $x' - x$ fixed,

$$\left\langle \mathrm{DS} \middle| \hat{\psi}^{\dagger}_{\mathrm{D}\alpha}(x)\, \hat{\psi}_{\mathrm{D}\alpha'}(x'+\epsilon) \middle| \mathrm{DS} \right\rangle = \delta_{\alpha,\alpha'} \frac{\alpha\,\mathrm{i}}{2\pi\,(x'-x+\epsilon)}, \tag{3.764a}$$

$$\left\langle \mathrm{DS} \middle| \hat{\psi}^{\dagger}_{\mathrm{D}\alpha'}(x')\, \hat{\psi}_{\mathrm{D}\alpha}(x+\epsilon) \middle| \mathrm{DS} \right\rangle = \delta_{\alpha,\alpha'} \frac{\alpha\,\mathrm{i}}{2\pi\,(x-x'+\epsilon)}. \tag{3.764b}$$

3.31.6 Step 6: For any one of $\alpha = -,+$, define the current

$$\hat{\jmath}_{\mathrm{D}\alpha}(x) := \lim_{\epsilon\to 0} \hat{\psi}^{\dagger}_{\mathrm{D}\alpha}(x)\, \hat{\psi}_{\mathrm{D}\alpha}(x+\epsilon) \tag{3.765}$$

by careful point splitting. For any operator $\widehat{O}$, define the operation of normal ordering with respect to the Fermi sea as the subtraction by the $\mathbb{C}$ number

$$:\widehat{O}: := \widehat{O} - \left\langle \mathrm{DS} \middle| \widehat{O} \middle| \mathrm{DS} \right\rangle. \tag{3.766}$$

We may then write

$$\hat{\jmath}_{\mathrm{D}\alpha}(x) := :\hat{\psi}^{\dagger}_{\mathrm{D}\alpha}(x)\, \hat{\psi}_{\mathrm{D}\alpha}(x): + \lim_{\epsilon\to 0} \left\langle \mathrm{DS} \middle| \hat{\psi}^{\dagger}_{\mathrm{D}\alpha}(x)\, \hat{\psi}_{\mathrm{D}\alpha}(x+\epsilon) \middle| \mathrm{DS} \right\rangle \tag{3.767}$$

as it is understood that the limit

$$:\hat{\psi}^{\dagger}_{\mathrm{D}\alpha}(x)\, \hat{\psi}_{\mathrm{D}\alpha}(x): := \lim_{\epsilon\to 0} :\hat{\psi}^{\dagger}_{\mathrm{D}\alpha}(x)\, \hat{\psi}_{\mathrm{D}\alpha}(x+\epsilon): \tag{3.768}$$

is well defined. If so, verify that Eqs. (3.754) and (3.755) give

$$
\begin{aligned}
&\left[\hat{\psi}^{\dagger}_{\mathrm{D}\alpha}(x)\,\hat{\psi}_{\mathrm{D}\alpha}(x+\epsilon),\hat{\psi}^{\dagger}_{\mathrm{D}\alpha'}(x')\,\hat{\psi}_{\mathrm{D}\alpha'}(x'+\epsilon)\right]=\\
&\qquad\delta_{\alpha,\alpha'}\,\delta(x-x'+\epsilon)\,\left\langle \mathrm{DS}\left|\,\hat{\psi}^{\dagger}_{\mathrm{D}\alpha}(x)\,\hat{\psi}_{\mathrm{D}\alpha'}(x'+\epsilon)\right|\mathrm{DS}\right\rangle\\
&\qquad-\delta_{\alpha,\alpha'}\,\delta(x-x'-\epsilon)\,\left\langle \mathrm{DS}\left|\,\hat{\psi}^{\dagger}_{\mathrm{D}\alpha'}(x')\,\hat{\psi}_{\mathrm{D}\alpha}(x+\epsilon)\right|\mathrm{DS}\right\rangle.
\end{aligned}
\tag{3.769}
$$

With the help of the identity $\delta(x-a)\,f(x)=\delta(x-a)\,f(a)$, verify that

$$
\begin{aligned}
&\left[\hat{\psi}^{\dagger}_{\mathrm{D}\alpha}(x)\,\hat{\psi}_{\mathrm{D}\alpha}(x+\epsilon),\hat{\psi}^{\dagger}_{\mathrm{D}\alpha'}(x')\,\hat{\psi}_{\mathrm{D}\alpha'}(x'+\epsilon)\right]=\\
&\qquad\delta_{\alpha,\alpha'}\,\delta(x-x'+\epsilon)\,\frac{\alpha\,\mathrm{i}}{2\pi\,2\epsilon}\\
&\qquad-\delta_{\alpha,\alpha'}\,\delta(x-x'-\epsilon)\,\frac{\alpha\,\mathrm{i}}{2\pi\,2\epsilon}.
\end{aligned}
\tag{3.770}
$$

3.31.7 Step 7: Verify that Eq. (3.753) holds.

3.32 First, verify Eq. (3.427). Second, verify (3.428) by computing $-\mathrm{i}[\hat{J}^0,\widehat{H}]$ with the help of Eq. (3.427). Alternatively, introduce the density

$$
\hat{\rho}(t,x):=\frac{1}{2\pi}\,q_i\,K^{-1}_{ij}\,\left(\partial_x\,\hat{u}_j\right)(t,x)
\tag{3.771a}
$$

and the current density

$$
\hat{\jmath}(t,x):=\frac{1}{2\pi}\,q_i\,V_{ij}\,\left(\partial_x\,\hat{u}_j\right)(t,x).
\tag{3.771b}
$$

First, verify that taking the divergence over $\hat{J}^{\mu}$ gives

$$
\begin{aligned}
\partial_{\mu}\,\hat{J}^{\mu}\equiv&\,\partial_t\,\hat{J}^0+\partial_x\,\hat{J}^1\\
=&\,\partial_t\,\hat{\rho}+\sigma_{\mathrm{H}}\,\partial_t\,A_1\\
&+\partial_x\,\hat{\jmath}+\underline{\underline{\frac{1}{2\pi}\left(q_i\,V_{ij}\,q_j\right)\partial_x\,A_1}}+\underline{\sigma_{\mathrm{H}}\,\partial_x\,A_0}.
\end{aligned}
\tag{3.772}
$$

Second, verify with the help of the chiral equations of motion (3.357) that

$$
\partial_t\,\hat{\rho}+\partial_x\,\hat{\jmath}=\underline{\underline{-\frac{1}{2\pi}\left(q_i\,V_{ij}\,q_j\right)\partial_x\,A_1}}-\underline{\sigma_{\mathrm{H}}\,\partial_x\,A_0}.
\tag{3.773}
$$

3.33 This exercise is dedicated to filling some intermediate steps when deriving Eq. (3.471) and applying it to the fractionalization of charge at a nonvanishing temperature.

3.33.1 Combine Eqs. (3.471) and (3.472) to derive Eqs. (3.473) and (3.474).

3.33.2 Show that, for $0 \leq \varphi_\infty \leq \pi$,

$$\int\limits_1^\infty \mathrm{d}u \, \frac{1}{\sqrt{u^2-1}\,(u^2-\sin^2\varphi_\infty)} = \frac{1}{|\cos\varphi_\infty|\,\sin\varphi_\infty}\left(\frac{\pi}{2}-\left|\frac{\pi}{2}-\varphi_\infty\right|\right) \tag{3.774}$$

and

$$\lim_{\beta\to\infty} Q(\beta) = \lim_{\beta\to\infty}\left[-\frac{1}{\pi}\left(\frac{\pi}{2}-\varphi_\infty\right)+\operatorname{sgn}\left(\frac{\pi}{2}-\varphi_\infty\right)e^{-\mu_{\mathrm{s}}\beta}+\cdots\right]. \tag{3.775}$$

The following intermediate steps are useful to derive Eq. (3.484). All manipulations apply to the Dirac Hamiltonian defined in Eq. (3.110). The notation used here is defined in the beginning of Section 3.6.2.

3.33.3 Let $\boldsymbol{\nabla}$ denote the gradient operator with the N components ∇_i with $i = 1,\cdots,N$. The summation convention over repeated indices will be implied. For any distinct pair $\boldsymbol{x},\boldsymbol{y}\in\mathbb{R}^N$, let

$$F(\boldsymbol{x},\boldsymbol{y}) := \operatorname{tr}\left\langle \boldsymbol{x}\left|\left[\mathrm{i}\Gamma_i\nabla_i,\Gamma_5\frac{1}{\mathcal{H}_\phi+\mathrm{i}\kappa}\right]\right|\boldsymbol{y}\right\rangle, \tag{3.776}$$

where we recall that

$$\mathcal{H} := \mathcal{H}_\phi + \mu_{\mathrm{s}}\Gamma_5, \qquad \mathcal{H}_\phi := \mathrm{i}\Gamma_i\nabla_i + \Phi, \tag{3.777}$$

in the notation of Section 3.6.2. Show the identity

$$\begin{aligned} F(\boldsymbol{x},\boldsymbol{y}) =& \operatorname{tr}\left\langle \boldsymbol{x}\left|\left[\underline{\mathcal{H}_\phi}-\Phi(\boldsymbol{x})+\mathrm{i}\kappa-\underline{\mathrm{i}\kappa}\right]\Gamma_5\frac{1}{\mathcal{H}_\phi+\mathrm{i}\kappa}\right|\boldsymbol{y}\right\rangle \\ &-\operatorname{tr}\left\langle \boldsymbol{x}\left|\Gamma_5\frac{1}{\mathcal{H}_\phi+\mathrm{i}\kappa}\left[\underline{\mathcal{H}_\phi}-\Phi(\boldsymbol{y})+\underline{\mathrm{i}\kappa}-\mathrm{i}\kappa\right]\right|\boldsymbol{y}\right\rangle. \end{aligned} \tag{3.778}$$

3.33.4 Verify that the terms that have been underlined in Eq. (3.778) cancel upon performing the trace tr over the $N\times N$ matrices so that

$$\begin{aligned} F(\boldsymbol{x},\boldsymbol{y}) =& \operatorname{tr}\left\langle \boldsymbol{x}\left|\left[-\Phi(\boldsymbol{x})+\underline{\underline{\mathrm{i}\kappa}}\right]\Gamma_5\frac{1}{\mathcal{H}_\phi+\mathrm{i}\kappa}\right|\boldsymbol{y}\right\rangle \\ &-\operatorname{tr}\left\langle \boldsymbol{x}\left|\Gamma_5\frac{1}{\mathcal{H}_\phi+\mathrm{i}\kappa}\left[-\Phi(\boldsymbol{y})-\underline{\underline{\mathrm{i}\kappa}}\right]\right|\boldsymbol{y}\right\rangle. \end{aligned} \tag{3.779}$$

3.33.5 Verify that the terms that have been underlined twice in Eq. (3.779) give

$$2\mathrm{i}\kappa\operatorname{tr}\left\langle \boldsymbol{x}\left|\Gamma_5\frac{1}{\mathcal{H}_\phi+\mathrm{i}\kappa}\right|\boldsymbol{y}\right\rangle. \tag{3.780}$$

3.33.6 Take advantage of the cyclicity of the trace tr to verify that

$$\begin{aligned}F(\boldsymbol{x},\boldsymbol{y}) =& 2\mathrm{i}\kappa\,\mathrm{tr}\left\langle \boldsymbol{x}\left|\Gamma_5\,\frac{1}{\mathcal{H}_\phi+\mathrm{i}\kappa}\right|\boldsymbol{y}\right\rangle \\ &-\mathrm{tr}\left\{[\Phi(\boldsymbol{x})-\Phi(\boldsymbol{y})]\left\langle \boldsymbol{x}\left|\Gamma_5\,\frac{1}{\mathcal{H}_\phi+\mathrm{i}\kappa}\right|\boldsymbol{y}\right\rangle\right\}.\end{aligned} \tag{3.781}$$

3.33.7 By virtue of the fact that $\frac{\partial}{\partial \boldsymbol{x}}\langle \boldsymbol{x}|\boldsymbol{p}\rangle = -\frac{\partial}{\partial \boldsymbol{x}}\langle \boldsymbol{p}|\boldsymbol{x}\rangle$ for any position $\boldsymbol{x}\in\mathbb{R}^N$ and any momentum $\boldsymbol{p}\in\mathbb{R}^N$ and with the help of the cyclicity of the trace tr, verify that

$$F(\boldsymbol{x},\boldsymbol{y}) = \left(\frac{\partial}{\partial x_i}+\frac{\partial}{\partial y_i}\right)\mathrm{tr}\left\langle \boldsymbol{x}\left|\mathrm{i}\Gamma_i\,\Gamma_5\,\frac{1}{\mathcal{H}_\phi+\mathrm{i}\kappa}\right|\boldsymbol{y}\right\rangle. \tag{3.782}$$

3.33.8 Verify Eq. (3.484).

3.34 What is the most general form of the space and time-independent one-dimensional Dirac Hamiltonian when represented in terms of 2×2 Dirac matrices? What restrictions follow from imposing the BDI symmetry constraints (3.509c), (3.509d), (3.509e)?

3.35 What is the most general form of the space and time-independent one-dimensional Dirac Hamiltonian when represented in terms of 4×4 Dirac matrices? What restrictions follow from imposing the CII symmetry constraints (3.515c), (3.515d), (3.515e)?

3.36 Show that for every energy eigenstate of a time-reversal symmetric Hamiltonian with half-integer total spin, there is at least one more eigenstate with the same energy. *Hint:* Assume that the Hamiltonian $\mathcal{H}$ commutes with the operation $\mathcal{T}$ implementing reversal of time, whereby $\mathcal{T}$ squares to minus the identity.

3.37 Define the many-body quadratic form

$$\widehat{H} = \int \mathrm{d}^d\boldsymbol{x}\int \mathrm{d}^d\boldsymbol{y}\sum_{ij}\hat{\psi}^\dagger_i(t,\boldsymbol{x})\,\mathcal{H}_{ij}(\boldsymbol{x},\boldsymbol{y})\,\hat{\psi}_j(t,\boldsymbol{y}), \tag{3.783a}$$

where

$$\mathcal{H}_{ij}(\boldsymbol{x},\boldsymbol{y}) = \mathcal{H}^*_{ji}(\boldsymbol{y},\boldsymbol{x}), \tag{3.783b}$$

and

$$\{\hat{\psi}_i(t,\boldsymbol{x}),\hat{\psi}^\dagger_j(t,\boldsymbol{y})\} = \delta_{ij}\,\delta(\boldsymbol{x}-\boldsymbol{y}) \tag{3.783c}$$

are the only nonvanishing equal-time anticommutators.

3.37.1 *Time-reversal symmetry.* Let K denote complex conjugation. Define the time-reversal transformation by the antiunitary transformation

$$\widehat{\mathrm{T}} := \widehat{\mathcal{T}}\,\mathsf{K} \tag{3.784a}$$

that reverses time but leaves space unchanged by demanding that

$$\widehat{\mathcal{T}}^{-1} = \widehat{\mathcal{T}}^{\dagger} \tag{3.784b}$$

and

$$\widehat{\mathrm{T}}\,\hat{\psi}_j(t,\boldsymbol{y})\,\widehat{\mathrm{T}}^{-1} = \sum_{j'} \mathcal{T}^{*}_{j'j}\,\hat{\psi}_{j'}(-t,\boldsymbol{y}). \tag{3.784c}$$

Show that

$$\widehat{\mathrm{T}}\,\widehat{H}\,\widehat{\mathrm{T}}^{-1} = \widehat{H} \tag{3.785a}$$

if and only if

$$\sum_{ij} \mathcal{T}_{i'i}\,\mathcal{H}^{*}_{ij}(\boldsymbol{x},\boldsymbol{y})\,\mathcal{T}^{-1}_{jj'} = \mathcal{H}_{i'j'}(\boldsymbol{x},\boldsymbol{y}). \tag{3.785b}$$

3.37.2 *Particle-hole (charge-conjugation) symmetry.* Assume that

$$\int \mathrm{d}^d\boldsymbol{x} \sum_i \mathcal{H}_{ii}(\boldsymbol{x},\boldsymbol{x}) = 0. \tag{3.786}$$

Define the particle-hole transformation by the unitary transformation

$$\widehat{\mathrm{C}} := \widehat{\mathcal{C}} \tag{3.787a}$$

that reverses the sign of the fermion number

$$\hat{n}_i(x) - \frac{1}{2}\,\delta(\boldsymbol{x}=0) := \hat{\psi}^{\dagger}_i(\boldsymbol{x})\,\hat{\psi}_i(\boldsymbol{x}) - \frac{1}{2}\,\delta(\boldsymbol{x}=0) \tag{3.787b}$$

measured relative to the background of the fermion density 1/2 but leaves space unchanged by demanding that

$$\widehat{\mathcal{C}}^{-1} = \widehat{\mathcal{C}}^{\dagger} \tag{3.787c}$$

and

$$\widehat{\mathrm{C}}\,\hat{\psi}_j(t,\boldsymbol{y})\,\widehat{\mathrm{C}}^{-1} = \sum_{j'} \mathcal{C}_{j'j}\,\hat{\psi}^{\dagger}_{j'}(t,\boldsymbol{y}). \tag{3.787d}$$

Show that

$$\widehat{\mathrm{C}}\,\widehat{H}\,\widehat{\mathrm{C}}^{-1} = \widehat{H} \tag{3.788a}$$

if and only if

$$\sum_{ij} \mathcal{C}_{i'i}\,\mathcal{H}^{*}_{ij}(\boldsymbol{x},\boldsymbol{y})\,\mathcal{C}^{-1}_{jj'} = -\mathcal{H}_{i'j'}(\boldsymbol{x},\boldsymbol{y}). \tag{3.788b}$$

3.37.3 *Chiral symmetry.* Assume that

$$\int \mathrm{d}^d\boldsymbol{x} \sum_i \mathcal{H}_{ii}(\boldsymbol{x},\boldsymbol{x}) = 0. \tag{3.789}$$

Define the chiral transformation by the antiunitary transformation

$$\widehat{\mathrm{S}} := \widehat{\mathcal{S}}\,\mathsf{K} \tag{3.790a}$$

that reverses the sign of the fermion number

$$\hat{n}_i(x) - \frac{1}{2}\,\delta(\boldsymbol{x}=0) := \hat{\psi}^{\dagger}_i(\boldsymbol{x})\,\hat{\psi}_i(\boldsymbol{x}) - \frac{1}{2}\,\delta(\boldsymbol{x}=0) \tag{3.790b}$$

measured relative to the background of the fermion density 1/2 but leaves space unchanged by demanding that

$$\widehat{\mathcal{S}}^{-1} = \widehat{\mathcal{S}}^{\dagger} \tag{3.790c}$$

and

$$\widehat{\mathrm{S}}\,\hat{\psi}_j(t,\boldsymbol{y})\,\widehat{\mathrm{S}}^{-1} = \sum_{j'} \mathcal{S}_{j'j}\,\hat{\psi}^{\dagger}_{j'}(t,\boldsymbol{y}). \tag{3.790d}$$

Show that

$$\widehat{\mathrm{S}}\,\widehat{H}\,\widehat{\mathrm{S}}^{-1} = \widehat{H} \tag{3.791a}$$

if and only if

$$\sum_{ij} \mathcal{S}_{i'i}\,\mathcal{H}_{ij}(\boldsymbol{x},\boldsymbol{y})\,\mathcal{S}^{-1}_{jj'} = -\mathcal{H}_{i'j'}(\boldsymbol{x},\boldsymbol{y}). \tag{3.791b}$$

The unitary symmetry under $\widehat{\mathrm{C}}$ is called charge conjugation symmetry or particle-hole symmetry (PHS). The antiunitary symmetry under $\widehat{\mathrm{S}}$ is called the chiral symmetry (CHS). The antiunitary symmetry under $\widehat{\mathrm{T}}$ is called time-reversal symmetry (TRS).

3.38 Show that demanding the invariance of Eq. (3.539b) under the chiral transformation (3.537) delivers conditions (3.539c) and (3.539d). *Hint:* First, assume that if Eq. (3.539c) holds, then Eq. (3.539d) must follow. Second, show that Eq. (3.539c) must hold.

4

Sharpness of the Fractional Charge

4.1 Introduction

We presented in Chapter 3 several ways to compute the expectation value of the charge operator for the many-body noninteracting ground states of one-dimensional fermions in the presence of classical domain walls. We showed under what conditions this expectation value differs from that of a reference many-body noninteracting ground state of one-dimensional fermions by a fractional number. Moreover, it was shown in **Exercise** 3.6 that, in the presence of a single domain wall in the thermodynamic limit of a Dirac Hamiltonian, the fractional charge was the eigenvalue of the many-body noninteracting ground state with respect of a properly defined charge operator, that is, a charge operator that is odd under charge conjugation. For the latter reason, the phenomenon of charge fractionalization investigated in Chapter 3 was interpreted as the fractionalization of the quantum number for the conserved fermion-number (-charge) operator. On the other hand, we saw in **Exercise** 3.7 that there can be quantum fluctuations associated with measuring the charge localized around either the soliton or the antisoliton in the background of a pair of soliton and antisoliton that are far apart.

Chapter 4 is devoted to revisiting the issue of the quantum fluctuations associated with the measurement process of the fractional charge. To this end we will follow the papers by Kivelson and Schrieffer,[1] Rajaraman and Bell,[2] Bell and Rajaraman,[3] and Jackiw et al.[4]

We will first show in Section 4.2 that no charge fractionalization can be measured for a single quantum particle subjected to a double delta-function potential well. We will then show in Section 4.3 that charge fractionalization is measurable for the SSH model in the presence of a solition-antisoliton pair when approximated by a Dirac Hamiltonian. A lattice generalization of the SSH model with a staggered chemical potential taking the values q and $1-q$ with $0<q<1$ on the even and

[1] S. Kivelson and J. R. Schrieffer, Phys. Rev. B, **25**(10), 6447–6451 (1982). [Kivelson and Schrieffer (1982)].
[2] R. Rajaraman and J. S. Bell, Phys. Lett. B **116**(2), 151–154 (1982). [Rajaraman and Bell (1982)].
[3] J. S. Bell and R. Rajaraman, Nucl. Phys. B **220**[FS8], 1–12 (1983). [Bell and Rajaraman (1983)].
[4] R. Jackiw, A. K. Kerman, I. Klebanov, and G. Semenoff, Nucl. Phys. B **225**[FS9], 233–246 (1983). [Jackiw et al. (1983)].

odd sites of the lattice, respectively, is solved at the strong dimerization point in Section 4.4. Hereto, it is shown that charge fractionalization is measurable. Finally, in Section 4.5 we give the conditions under which a fractional charge is measurable in any many-body, noninteracting, and local lattice Hamiltonian.

There were two ingredients in Chapter 3 for the fractionalization of the charge to the values $\pm 1/2$. First, a many-body ground state (a Fermi sea) is separated from all many-body excited states in the spectrum of noninteracting spinless fermions by a band gap. Second, there are isolated and localized single-particle mid-gap states close to the middle of the gap between particle-hole symmetric valence and conduction bands. The goal of this chapter is to argue that, in the thermodynamic limit, there exists a conserved operator $\widehat{Q}_{\mathrm{s}}$ associated with any one (indexed by the label s) of these localized single-particle mid-gap states such that (i) its expectation value in the ground state of the Hamiltonian is $\pm 1/2$ *with vanishing quantum fluctuations*, and (ii) this expectation value can be interpreted as a conserved electric charge in that $\widehat{Q}_{\mathrm{s}}$ couples with an externally applied electrical field $\boldsymbol{E}$ through the Lorentz force $\widehat{Q}_{\mathrm{s}}\,\boldsymbol{E}$. To appreciate the nontrivial nature of this claim, we present the following counter example to a sharp fractionalization of the fermion number.

4.2 Example 1: The Double-Well Potential

Consider the single-particle Hamiltonian describing a nonrelativistic quantum particle propagating along the line in the presence of a pair of delta-function potential wells a distance r apart, see Figure 4.1. (As a warm up, the solution to the single delta-function potential is reviewed in **Exercise** 4.1.) Let the mass of this particle be set to unity, together with $\hbar$ and the characteristic speed, that is,

$$\mathcal{H}_{\mathrm{dw}} := -\frac{1}{2}\frac{\mathrm{d}^2}{\mathrm{d}x^2} - v\left[\delta\left(x+\frac{r}{2}\right)+\delta\left(x-\frac{r}{2}\right)\right] \tag{4.1}$$

is the single-particle Hamiltonian (the index "dw" stands for double well). The strength of the delta-function potential wells is $v>0$. Observe that Hamiltonian (4.1) is invariant under complex conjugation. This symmetry is none but TRS for a spinless point particle.

The single-particle spectrum consists of two bound states $|+\rangle$ (the bonding state) and $|-\rangle$ (the antibonding state) and a continuum of states $|p\rangle$ indexed by the momentum $p\in\mathbb{R}$.

The bound states are normalizable on the real line. They are constructed as follows. Let $W(x)$ denote the Lambert W function defined implicitly by

$$x = W(x)\,e^{W(x)} \tag{4.2a}$$

with the Taylor expansion

$$W(x) = \sum_{n=1}^{\infty}\frac{(-n)^{n-1}}{n!}\,x^n \tag{4.2b}$$

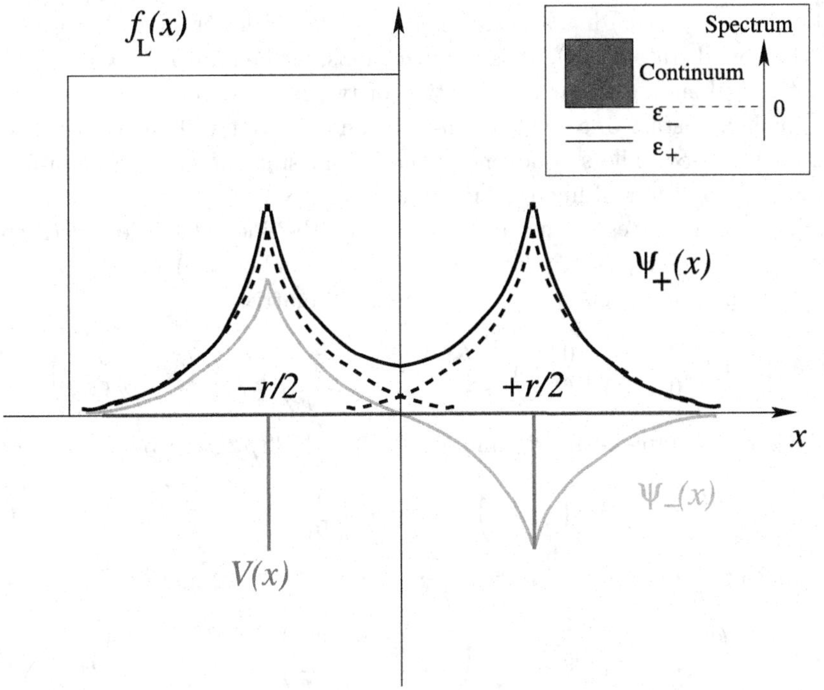

Figure 4.1 The function $V(x)$ depicts a pair of delta-function potential wells at $\mp r/2$. The bound state with the lowest energy $\varepsilon_+ < 0$ is the eigenfunction $\psi_+(x) = +\psi_+(-x)$. The bound state with the largest energy $\varepsilon_- < 0$ is the eigenfunction $\psi_-(x) = -\psi_-(-x)$. The eigenfunction $\psi_+(x)$ is a bonding linear combination of two functions $\psi_{\mathrm{L},+}$ and $\psi_{\mathrm{R},+}$ that are exponentially localized about the left and right delta-function potential wells, respectively. They are depicted by dashed lines. The eigenfunction $\psi_-(x)$ is an antibonding linear combination of two functions $\psi_{\mathrm{L},-}$ and $\psi_{\mathrm{R},-}$ that are exponentially localized about the left and right delta-function potential wells, respectively. They are not shown. The continuum has the dispersion $\varepsilon(p) = p^2/2$. The function $f_{\mathrm{L}}(x)$ depicts a box test function centered at $x = -r/2$ with $\omega = \ell$. The spectrum is shown in the inset.

about $x = 0$. The pair of bound states are orthonormal on the line,

$$\mathcal{H}_{\mathrm{dw}}\,\psi_\pm(x) = \varepsilon_\pm\,\psi_\pm(x), \qquad \psi_\pm(x) \equiv \langle x|\pm\rangle, \tag{4.2c}$$

where

$$\varepsilon_+ \equiv -\frac{\kappa_+^2}{2} < \varepsilon_- \equiv -\frac{\kappa_-^2}{2}, \qquad \kappa_\pm = v + \frac{1}{r}\,W\left(\pm v\,r\,e^{-v\,r}\right), \tag{4.2d}$$

and

$$\psi_\pm(x) \propto e^{-\kappa_\pm\,|x+r/2|} \pm e^{-\kappa_\pm\,|x-r/2|}, \qquad \int \mathrm{d}x\,\psi_+(x)\,\psi_-(x) = 0. \tag{4.2e}$$

The bound state $\psi_+(x)$ has an eigenenergy ε_+ that is lower than the eigenenergy ε_- of the bound state $\psi_-(x)$. This is expected as, in the limit $r \to \infty$, $\psi_+(x)$ is the symmetric or bonding linear combination of two isolated bound states. The first one is the bound state of a single delta-function potential well on the real line that is centered at $-\infty$. The second one is the bound state of a single delta-function potential well on the real line that is centered at $+\infty$.

The continuum states are not normalizable on the line. They are constructed as follows. For any $p \in \mathbb{R}$, choose the complex-valued number $A(p)$ and $B(p)$. Associate to the delta-potential well at $x = -r/2$ the 2×2 matrix

$$M_{\mathrm{L}}(p) := \begin{pmatrix} e^{-\mathrm{i}p\,r/2} & 0 \\ 0 & e^{+\mathrm{i}p\,r/2} \end{pmatrix}^{-1} \begin{pmatrix} 1 + \frac{\mathrm{i}v}{p} & +\frac{\mathrm{i}v}{p} \\ -\frac{\mathrm{i}v}{p} & 1 - \frac{\mathrm{i}v}{p} \end{pmatrix} \begin{pmatrix} e^{-\mathrm{i}p\,r/2} & 0 \\ 0 & e^{+\mathrm{i}p\,r/2} \end{pmatrix} \tag{4.3a}$$

and the pair of complex-valued numbers $A'(p)$ and $B'(p)$ given by

$$\begin{pmatrix} A'(p) \\ B'(p) \end{pmatrix} := M_{\mathrm{L}}(p) \begin{pmatrix} A(p) \\ B(p) \end{pmatrix}. \tag{4.3b}$$

Associate to the delta-potential well at $x = +r/2$ the 2×2 matrix

$$M_{\mathrm{R}}(p) := \begin{pmatrix} e^{+\mathrm{i}p\,r/2} & 0 \\ 0 & e^{-\mathrm{i}p\,r/2} \end{pmatrix}^{-1} \begin{pmatrix} 1 + \frac{\mathrm{i}v}{p} & +\frac{\mathrm{i}v}{p} \\ -\frac{\mathrm{i}v}{p} & 1 - \frac{\mathrm{i}v}{p} \end{pmatrix} \begin{pmatrix} e^{+\mathrm{i}p\,r/2} & 0 \\ 0 & e^{-\mathrm{i}p\,r/2} \end{pmatrix} \tag{4.3c}$$

and the pair of complex-valued numbers $A''(p)$ and $B''(p)$ given by

$$\begin{pmatrix} A''(p) \\ B''(p) \end{pmatrix} := M_{\mathrm{R}}(p) \begin{pmatrix} A'(p) \\ B'(p) \end{pmatrix}. \tag{4.3d}$$

One verifies that (**Exercise** 4.2)

$$\mathcal{H}_{\mathrm{dw}}\, \psi(p,x) = \varepsilon(p)\, \psi(p,x), \qquad \psi(p,x) \equiv \langle x|p\rangle, \tag{4.3e}$$

where

$$\varepsilon(p) = \frac{p^2}{2}, \qquad \psi(p,x) = \begin{cases} A(p)\, e^{+\mathrm{i}p\,x} + B(p)\, e^{-\mathrm{i}p\,x}, & x < -\frac{r}{2}, \\ A'(p)\, e^{+\mathrm{i}p\,x} + B'(p)\, e^{-\mathrm{i}p\,x}, & -\frac{r}{2} < x < +\frac{r}{2}, \\ A''(p)\, e^{+\mathrm{i}p\,x} + B''(p)\, e^{-\mathrm{i}p\,x}, & +\frac{r}{2} < x, \end{cases} \tag{4.3f}$$

and

$$\int \mathrm{d}x\, \psi^*(p,x)\, \psi(q,x) \propto \delta(p-q), \qquad \int \mathrm{d}x\, \psi^*(p,x)\, \psi_\pm(x) = 0. \tag{4.3g}$$

The TRS of Hamiltonian (4.1) implies that

$$\psi(p,x) = \psi^*(-p,x) \Longleftrightarrow \alpha(p) = \alpha^*(-p),\ \beta(p) = \beta^*(-p),\ M_{\mathrm{W}}(p) = M^*_{\mathrm{W}}(-p), \tag{4.4}$$

for $\alpha(p) = A(p), A'(p), A''(p)$, $\beta(p) = B(p), B'(p), B''(p)$, and W=L,R, respectively. We use the unit 2×2 matrix τ_0 and the vector of Pauli matrices $\boldsymbol{\tau}$ as the basis for the vector space of 2×2 matrices to which the so-called transfer matrix $M_{\mathrm{W}}(p)$ belongs. One verifies (**Exercise** 4.3) that

$$M_{\mathrm{W}}^{\dagger}(p)\,\tau_3\,M_{\mathrm{W}}(p) = \tau_3, \tag{4.5a}$$

that is flux conservation in the representation

$$|A(p)|^2 - |B(p)|^2 = |A'(p)|^2 - |B'(p)|^2 = |A''(p)|^2 - |B''(p)|^2 \tag{4.5b}$$

holds.

As with the case of a soliton and antisoliton pair from Chapter 3, we have (i) two bound states whose wave fun ctions decay in magnitude exponentially fast away from a pair of localization centers (the delta-function potential wells), (ii) become degenerate when the separation r between the delta-function potential wells is taken to infinity, and (iii) there exists a continuum of states. We place the chemical potential μ between ε_+ and ε_-:

$$\varepsilon_+ < \mu < \varepsilon_-. \tag{4.6}$$

The ground state is then a nondegenerate single-particle one for the double well of delta-function potentials separated by the distance $r < \infty$. This is a crucial difference with the soliton and antisoliton pair from Chapter 3 for which the ground state was a many-body one.

As the limit $r \to \infty$ is taken, Eq. (4.2d) implies that

$$\lim_{r\to\infty} \varepsilon_+ = \lim_{r\to\infty} \mu = \lim_{r\to\infty} \varepsilon_- = -\frac{v^2}{2}. \tag{4.7}$$

In order to compare the physics of the double-well potential to that of the soliton and antisoliton pair from Chapter 3 for which the ground state was a many-body one, we construct the four-dimensional fermionic Fock space associated with the two single-particle bound states with the single-particle energies $\varepsilon_{\pm}$. In this four-dimensional Fock space, having no fermions, one fermion, or two fermions occupying the single-particle states with the single-particle energies $\varepsilon_{\pm} - (-v^2/2)$ costs the same fourfold degenerate vanishing energy in the singular limit (4.7) if we measure single-particle energies relative to $(-v^2/2)$.

We thus define the pair of field operators [compare with Eq. (3.616)]

$$\widehat{\Psi}^{\dagger}(x,t) := \sum_{\pm} e^{+\mathrm{i}\varepsilon_{\pm} t}\,\psi_{\pm}(x)\,\hat{b}^{\dagger}_{\pm} + \int \mathrm{d}p\, e^{+\mathrm{i}\varepsilon(p)\,t}\,\psi^{*}(p,x)\,\hat{c}^{\dagger}(p), \tag{4.8a}$$

and

$$\widehat{\Psi}(x,t) := \sum_{\pm} e^{-\mathrm{i}\varepsilon_{\pm} t}\,\psi_{\pm}(x)\,\hat{b}_{\pm} + \int \mathrm{d}p\, e^{-\mathrm{i}\varepsilon(p)\,t}\,\psi(p,x)\,\hat{c}(p), \tag{4.8b}$$

where the only nonvanishing anticommutators are

$$\{\hat{b}_+, \hat{b}^{\dagger}_+\} = 1, \qquad \{\hat{b}_-, \hat{b}^{\dagger}_-\} = 1, \qquad \{\hat{c}(p), \hat{c}^{\dagger}(q)\} = \delta(p-q). \tag{4.8c}$$

The second-quantized Hamiltonian and total fermion number are

$$\widehat{H}_{\rm dw} = \sum_{\pm} \varepsilon_{\pm}\, \hat{b}^{\dagger}_{\pm}\, \hat{b}_{\pm} + \int \mathrm{d}p\, \varepsilon(p)\, \hat{c}^{\dagger}(p)\, \hat{c}(p), \tag{4.9}$$

and

$$\widehat{Q}_{\rm dw} = \sum_{\pm} \hat{b}^{\dagger}_{\pm}\, \hat{b}_{\pm} + \int \mathrm{d}p\, \hat{c}^{\dagger}(p)\, \hat{c}(p), \tag{4.10}$$

respectively.

The single-particle ground state

$$|1,0\rangle \equiv |n_{+} = 1, n_{-} = 0\rangle := \hat{b}^{\dagger}_{+}\, |0\rangle, \tag{4.11a}$$

where $|0\rangle$ is the state annihilated by all annihilation operators, is nondegenerate for $0 < r < \infty$. It becomes degenerate with the first single-particle excited state

$$|0,1\rangle \equiv |n_{+} = 0, n_{-} = 1\rangle := \hat{b}^{\dagger}_{-}\, |0\rangle, \tag{4.11b}$$

as well as with the vacuum

$$|0,0\rangle \equiv |n_{+} = 0, n_{-} = 0\rangle := |0\rangle, \tag{4.11c}$$

or the two fermion state

$$|1,1\rangle \equiv |n_{+} = 1, n_{-} = 1\rangle := \hat{b}^{\dagger}_{+}\, \hat{b}^{\dagger}_{-}\, |0\rangle, \tag{4.11d}$$

of

$$\widehat{H}_{\rm dw} - \mu\, \widehat{Q}_{\rm dw} \tag{4.12}$$

in the limit $r \to \infty$.

The total fermion number operator has integer-valued eigenvalues and the single-particle ground state is an eigenstate of $\widehat{Q}_{\rm dw}$ with eigenvalue 1:

$$\widehat{Q}_{\rm dw}\, |\mathrm{GS}\rangle = |\mathrm{GS}\rangle, \qquad |\mathrm{GS}\rangle := |1,0\rangle, \tag{4.13}$$

when $0 < r < \infty$. In the limit $r \to \infty$, the total fermion number of the normalized ground state

$$|\mathrm{GS}\rangle := \sum_{n_{+},n_{-}=0}^{1} a_{n_{+},n_{-}}\, |n_{+}, n_{-}\rangle, \qquad \sum_{n_{+},n_{-}=0}^{1} |a_{n_{+},n_{-}}|^{2} = 1, \tag{4.14}$$

need not be sharp. The jump of degeneracy of the ground states from one to four (in the four-dimensional fermionic Fock space) in the limit $r \to \infty$ renders this limit singular, as will be explained shortly.

To mimic the computation of the fractional charge from Chapter 3 we assume that r is the largest length scale pertaining to the single-particle Hamiltonian (4.1), that is $r \gg \kappa_{\mp}^{-1}$. Let ω and ℓ be two intermediary length scales obeying

$$r \gg \omega \gg \ell \gg \kappa_{\mp}^{-1} \tag{4.15a}$$

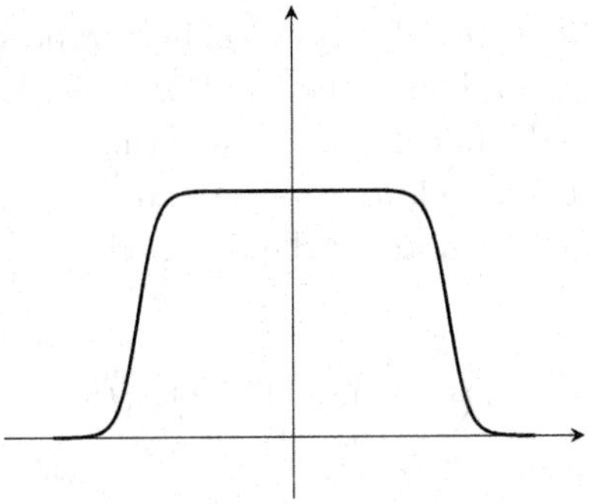

Figure 4.2 The function (4.15c) centered at $x=-r/2$.

and let

$$f_{\mathrm{L}}:\mathbb{R}\to[0,1],\qquad x\mapsto f_{\mathrm{L}}(x), \tag{4.15b}$$

be a test function taking values in the closed interval $[0,1]$ such that it decreases away from its maximum at $x=-r/2$ like the function (see Figure 4.2)

$$\begin{cases} e^{-[x+(r/2)+(\ell/2)]^{2n}/[(\omega-\ell)/2]^{2n}}, & x\leq-(r/2)-(\ell/2),\\ 1, & -(r/2)-(\ell/2)\leq x\leq-(r/2)+(\ell/2),\\ e^{-[x+(r/2)-(\ell/2)]^{2n}/[(\omega-\ell)/2]^{2n}}, & x\geq-(r/2)+(\ell/2), \end{cases} \tag{4.15c}$$

with $n=1,2,\cdots$ any positive integer if we demand that its first $2n-1$ derivatives are continuous at $x=-(r/2)\mp(\ell/2)$. In other words, the test function f_{L} is centered about the left well, takes the value 1 over the region where $\psi_{+}+\psi_{-}$ is significantly different from zero, while it is exponentially suppressed over the region where $\psi_{+}-\psi_{-}$ is significantly different from zero. To compute the fermion number in the interval $[-(r/2)-(\omega/2),-(r/2)+(\omega/2)]$, we define the smeared number operator

$$\widehat{Q}_{f_{\mathrm{L}}}(t):=\int\mathrm{d}x\,f_{\mathrm{L}}(x)\,\widehat{\Psi}^{\dagger}(x,t)\,\widehat{\Psi}(x,t). \tag{4.15d}$$

The order in which the three limits $r\to\infty$, $\omega\to\infty$, and $\ell\to\infty$ are taken matters.

If the limit $r\to\infty$ is taken after $\omega\to\infty$ and $\ell\to\infty$, one obtains the limit at $r=\infty$ of the total fermion number operator (4.10),

$$\lim_{r\to\infty}\widehat{Q}_{\mathrm{dw}}=\lim_{r\to\infty}\lim_{f_{\mathrm{L}}\to1}\widehat{Q}_{f_{\mathrm{L}}}(t), \tag{4.16}$$

with the caveat that the boundary conditions at $x=\pm\infty$ are selected by how the limits $\omega\to\infty$ and $\ell\to\infty$ are taken. For example, the sharp box limit $\omega=\ell\to\infty$

selects sharp boundaries due to $\mathrm{d}f_{\mathrm{L}}(x)/\mathrm{d}x$ being proportional to a difference of delta-functions. On the other hand, the limit $\omega, \ell \to \infty$ with $\ell/\omega \to 0$ selects smooth boundaries due to $\mathrm{d}f_{\mathrm{L}}(x)/\mathrm{d}x$ being smooth.

If the limit $r \to \infty$ is taken before $\omega \to \infty$ and $\ell \to \infty$ with $\ell/\omega \to 0$ (that is $f_{\mathrm{L}} \to 1$ in a smooth way), one obtains the smeared fermion number operator for the left delta-function well

$$\widehat{Q}_{\mathrm{L}}(t) = \lim_{f_{\mathrm{L}} \to 1} \lim_{r \to \infty} \widehat{Q}_{f_{\mathrm{L}}}(t). \tag{4.17}$$

The first question we want to address is what is the fate of the time dependence of $\widehat{Q}_{f_{\mathrm{L}}}(t)$ in the limit (4.17). Is it time-dependent or does it become time-independent as in Eq. (4.16)?

If we insert the mode expansion (4.8) on the right-hand side of Eq. (4.15d), we identify the contributions

$$\begin{aligned}\int \mathrm{d}x\, f_{\mathrm{L}}(x) \Big[& \psi_{+}^{2}(x)\, \hat{b}_{+}^{\dagger}\, \hat{b}_{+} + \psi_{-}^{2}(x)\, \hat{b}_{-}^{\dagger}\, \hat{b}_{-} \\ & + e^{+\mathrm{i}(\varepsilon_{+}-\varepsilon_{-})\,t}\, \psi_{+}(x)\, \psi_{-}(x)\, \hat{b}_{+}^{\dagger}\, \hat{b}_{-} + e^{+\mathrm{i}(\varepsilon_{-}-\varepsilon_{+})\,t}\, \psi_{-}(x)\, \psi_{+}(x)\, \hat{b}_{-}^{\dagger}\, \hat{b}_{+} \Big]\end{aligned} \tag{4.18a}$$

between the bound states, the contributions

$$\begin{aligned}\int \mathrm{d}x\, f_{\mathrm{L}}(x) \int \mathrm{d}p \Big[& e^{+\mathrm{i}[\varepsilon_{+}-\varepsilon(p)]\,t}\, \psi_{+}(x)\, \psi(p,x)\, \hat{b}_{+}^{\dagger}\, \hat{c}(p) \\ & + e^{+\mathrm{i}[\varepsilon_{-}-\varepsilon(p)]\,t}\, \psi_{-}(x)\, \psi(p,x)\, \hat{b}_{-}^{\dagger}\, \hat{c}(p) + \mathrm{H.c.} \Big]\end{aligned} \tag{4.18b}$$

between the bound and continuum states, and the contributions

$$\int \mathrm{d}x\, f_{\mathrm{L}}(x) \int \mathrm{d}p \int \mathrm{d}q\, e^{+\mathrm{i}[\varepsilon(p)-\varepsilon(q)]\,t}\, \psi^{*}(p,x)\, \psi(q,x)\, \hat{c}^{\dagger}(p)\, \hat{c}(q) \tag{4.18c}$$

between the continuum states. By inspection we see that there are explicit time-dependent phase factors. Correspondingly, the identity

$$\left[\widehat{A}\,\widehat{B}, \widehat{C}\,\widehat{D}\right] = \widehat{A}\left\{\widehat{B}, \widehat{C}\right\}\widehat{D} - \widehat{A}\,\widehat{C}\left\{\widehat{B}, \widehat{D}\right\} + \left\{\widehat{A}, \widehat{C}\right\}\widehat{D}\,\widehat{B} - \widehat{C}\left\{\widehat{A}, \widehat{D}\right\}\widehat{B} \tag{4.19}$$

applied to the time evolution

$$\frac{\mathrm{d}}{\mathrm{d}t}\widehat{Q}_{f_{\mathrm{L}}} = -\mathrm{i}\left[\widehat{Q}_{f_{\mathrm{L}}}, \widehat{H}\right] \tag{4.20}$$

delivers the contributions

$$-\mathrm{i}\int \mathrm{d}x\, f_{\mathrm{L}}(x)\, e^{+\mathrm{i}(\varepsilon_{+}-\varepsilon_{-})\,t}\, \psi_{+}(x)\, \psi_{-}(x)\, (\varepsilon_{-}-\varepsilon_{+})\, \hat{b}_{+}^{\dagger}\, \hat{b}_{-} + \mathrm{H.c.} \tag{4.21a}$$

from the bound states, the mixed contributions

$$
\begin{aligned}
&-\mathrm{i}\int \mathrm{d}x\, f_{\mathrm{L}}(x) \int \mathrm{d}p\, e^{+\mathrm{i}[\varepsilon_{+}-\varepsilon(p)]\,t}\, \psi_{+}(x)\, \psi(p,x)\, [\varepsilon(p)-\varepsilon_{+}]\, \hat{b}^{\dagger}_{+}\, \hat{c}(p) \\
&-\mathrm{i}\int \mathrm{d}x\, f_{\mathrm{L}}(x) \int \mathrm{d}p\, e^{+\mathrm{i}[\varepsilon_{-}-\varepsilon(p)]\,t}\, \psi_{-}(x)\, \psi(p,x)\, [\varepsilon(p)-\varepsilon_{-}]\, \hat{b}^{\dagger}_{-}\, \hat{c}(p) \\
&+\mathrm{H.c.}
\end{aligned}
\tag{4.21b}
$$

from the bound and continuum states, and the contributions

$$
\begin{aligned}
&-\mathrm{i}\int \mathrm{d}x\, f_{\mathrm{L}}(x) \int \mathrm{d}p \int \mathrm{d}q\, e^{+\mathrm{i}[\varepsilon(p)-\varepsilon(q)]\,t}\, \psi^{*}(p,x)\, \psi(q,x)\, [\varepsilon(q)-\varepsilon(p)] \\
&\times \hat{c}^{\dagger}(p)\, \hat{c}(q)
\end{aligned}
\tag{4.21c}
$$

from the continuum states.

The continuity equation (4.20) turns into the conservation law

$$
\frac{\mathrm{d}}{\mathrm{d}t}\left(\lim_{r\to\infty}\widehat{Q}_{\mathrm{dw}}\right)=0 \tag{4.22}
$$

for the total fermion number in the limit (4.16), owing to the orthogonality of the single-particle eigenstates of Hamiltonian (4.1).

The limit (4.17) also delivers a conservation law, namely,

$$
\frac{\mathrm{d}}{\mathrm{d}t}\widehat{Q}_{\mathrm{L}}=0, \tag{4.23}
$$

but for different reasons.

For any pair of bound states $\sigma,\sigma'=\pm$ and any pair p and p' from the continuum, define the (complex-valued) functions

$$
j(\sigma,\sigma';x):=\frac{1}{2\mathrm{i}}\left[\psi_{\sigma}(x)\frac{\partial}{\partial x}\psi_{\sigma'}(x)-\psi_{\sigma'}(x)\frac{\partial}{\partial x}\psi_{\sigma}(x)\right], \tag{4.24a}
$$

$$
j(\sigma,p;x):=\frac{1}{2\mathrm{i}}\left[\psi_{\sigma}(x)\frac{\partial}{\partial x}\psi(p,x)-\psi(p,x)\frac{\partial}{\partial x}\psi_{\sigma}(x)\right], \tag{4.24b}
$$

$$
j(p,\sigma;x):=\frac{1}{2\mathrm{i}}\left[\psi^{*}(p,x)\frac{\partial}{\partial x}\psi_{\sigma}(x)-\psi_{\sigma}(x)\frac{\partial}{\partial x}\psi^{*}(p,x)\right], \tag{4.24c}
$$

$$
j(p,p',x):=\frac{1}{2\mathrm{i}}\left[\psi^{*}(p,x)\frac{\partial}{\partial x}\psi(p',x)-\psi(p',x)\frac{\partial}{\partial x}\psi^{*}(p,x)\right]. \tag{4.24d}
$$

These functions are the matrix elements of the conserved current density operator for the Schrödinger equation in one-dimensional space. One verifies (**Exercise** 4.4) that Eqs. (4.21a), (4.21b), and (4.21c) are nothing but

$$
\int \mathrm{d}x \left(\frac{\mathrm{d}f_{\mathrm{L}}}{\mathrm{d}x}\right)(x)\, e^{+\mathrm{i}(\varepsilon_{+}-\varepsilon_{-})\,t}\, j(+,-;x)\, \hat{b}^{\dagger}_{+}\, \hat{b}_{-} + \mathrm{H.c.} \tag{4.25a}
$$

for the bound states,

$$\begin{aligned}
&\int \mathrm{d}x \left(\frac{\mathrm{d}f_{\mathrm{L}}}{\mathrm{d}x}\right)(x) \int \mathrm{d}p\, e^{+\mathrm{i}[\varepsilon_{+}-\varepsilon(p)]\,t}\, \jmath(+,p;x)\, \hat{b}^{\dagger}_{+}\, \hat{c}(p) \\
&\int \mathrm{d}x \left(\frac{\mathrm{d}f_{\mathrm{L}}}{\mathrm{d}x}\right)(x) \int \mathrm{d}p\, e^{+\mathrm{i}[\varepsilon_{-}-\varepsilon(p)]\,t}\, \jmath(-,p;x)\, \hat{b}^{\dagger}_{-}\, \hat{c}(p) \\
&+\mathrm{H.c.}
\end{aligned} \tag{4.25b}$$

for the bound and continuum states, and

$$\int \mathrm{d}x \left(\frac{\mathrm{d}f_{\mathrm{L}}}{\mathrm{d}x}\right)(x) \int \mathrm{d}p \int \mathrm{d}q\, e^{+\mathrm{i}[\varepsilon(p)-\varepsilon(q)]\,t}\, \jmath(p,q;x)\, \hat{c}^{\dagger}(p)\, \hat{c}(q) \tag{4.25c}$$

for the continuum states, respectively.

The limit (4.17) is defined so that $f_{\mathrm{L}} \to 1$ with $(\partial_x f_{\mathrm{L}}) \to 0$ uniformly. Hence, the identity

$$\lim_{f_{\mathrm{L}}\to 1} \int_a^b \mathrm{d}x \left(\frac{\mathrm{d}f_{\mathrm{L}}}{\mathrm{d}x}\right)(x)\, g(x) = \int_a^b \mathrm{d}x \lim_{f_{\mathrm{L}}\to 1} \left(\frac{\mathrm{d}f_{\mathrm{L}}}{\mathrm{d}x}\right)(x)\, g(x) = 0 \tag{4.26}$$

holds on any finite interval $[a,b]$ of the real line provided the definite integral

$$\int_a^b \mathrm{d}x \left(\frac{\mathrm{d}f_{\mathrm{L}}}{\mathrm{d}x}\right)(x)\, g(x) \tag{4.27}$$

is well defined for some arbitrarily chosen function g and for all members of the sequences f_{L} that converge to 1 with $(\partial_x f_{\mathrm{L}}) \to 0$ converging to 0 uniformly on $[a,b]$. If the function g is chosen such that the improper integral

$$\int_{\mathbb{R}} \mathrm{d}x\, g(x) \tag{4.28}$$

is well defined, then the identity

$$\lim_{f_{\mathrm{L}}\to 1} \int_{\mathbb{R}} \mathrm{d}x \left(\frac{\mathrm{d}f_{\mathrm{L}}}{\mathrm{d}x}\right)(x)\, g(x) = \int_{\mathbb{R}} \mathrm{d}x \lim_{f_{\mathrm{L}}\to 1} \left(\frac{\mathrm{d}f_{\mathrm{L}}}{\mathrm{d}x}\right)(x)\, g(x) = 0 \tag{4.29}$$

holds as well. In Eqs. (4.25a) and (4.25b), the role of the function g is taken by the functions $\jmath(\sigma,\sigma';x)$ or $\jmath(\sigma,p;x)$, defined in Eq. (4.24), for very large r.

This logic breaks down for Eq. (4.25c) as the functions $\jmath(p,p';x)$ are not integrable over the real line. This is so because Eq. (4.25c) arises from the contributions from the continuum to the time derivative of the smeared charge operator. Plane waves are not normalizable functions and any integral such as Eq. (4.3g) delivers a generalized function (distribution). We must rely on a different strategy to prove that Eq. (4.25c) vanishes in the limit (4.17).

In Eq. (4.25c), the function

$$\begin{aligned} j(p,q;x) = &+\frac{1}{2}\,(p+q)\,\alpha^{*}(p)\,\alpha(q)\,e^{-\mathrm{i}(p-q)x} \\ &+\frac{1}{2}\,(p-q)\,\alpha^{*}(p)\,\beta(q)\,e^{-\mathrm{i}(p+q)x} \\ &-\frac{1}{2}\,(p-q)\,\beta^{*}(p)\,\alpha(q)\,e^{+\mathrm{i}(p+q)x} \\ &-\frac{1}{2}\,(p+q)\,\beta^{*}(p)\,\beta(q)\,e^{+\mathrm{i}(p-q)x} \end{aligned} \tag{4.30a}$$

is a linear combination of plane waves with the coefficients

$$\alpha(p) = \begin{cases} A(p), & x < -r/2, \\ A'(p), & -r/2 < x < +r/2, \\ A''(p), & +r/2 < x, \end{cases} \tag{4.30b}$$

$$\beta(p) = \begin{cases} B(p), & x < -r/2, \\ B'(p), & -r/2 < x < +r/2, \\ B''(p), & +r/2 < x, \end{cases} \tag{4.30c}$$

obeying

$$\alpha(p) = \alpha^{*}(-p), \qquad \beta(p) = \beta^{*}(-p), \tag{4.30d}$$

because of TRS.

For concreteness, we choose

$$f_{\mathrm{L}}(x) := e^{-\eta\,|x+(r/2)|}, \qquad (\partial_x f_{\mathrm{L}})(x) = -\eta\,\mathrm{sgn}\big(x+(r/2)\big)\,e^{-\eta\,|x+(r/2)|}, \tag{4.31}$$

where η is a positive real-valued parameter. Hence, the uniform limits $f \to 1$ and $(\partial_x f_{\mathrm{L}}) \to 0$ on the real line amounts to the limit $\eta \to 0$. Now, we can do the integral

$$\begin{aligned} I(p,q;\eta,r) :=& \int \mathrm{d}x\,(\partial_x f_{\mathrm{L}})(x)\,j(p,q;x) \\ =& +\frac{(p+q)\,\eta}{2}\left[\frac{A^{*}(p)\,A(q)}{\eta-\mathrm{i}(p-q)} + \frac{A'^{*}(p)\,A'(q)}{-\eta-\mathrm{i}(p-q)}\right] e^{+\mathrm{i}(p-q)\,r/2} \\ &+\frac{(p-q)\,\eta}{2}\left[\frac{A^{*}(p)\,B(q)}{\eta-\mathrm{i}(p+q)} + \frac{A'^{*}(p)\,B'(q)}{-\eta-\mathrm{i}(p+q)}\right] e^{+\mathrm{i}(p+q)\,r/2} \\ &-\frac{(p-q)\,\eta}{2}\left[\frac{B^{*}(p)\,A(q)}{\eta+\mathrm{i}(p+q)} + \frac{B'^{*}(p)\,A'(q)}{-\eta+\mathrm{i}(p+q)}\right] e^{-\mathrm{i}(p+q)\,r/2} \\ &-\frac{(p+q)\,\eta}{2}\left[\frac{B^{*}(p)\,B(q)}{\eta+\mathrm{i}(p-q)} + \frac{B'^{*}(p)\,B'(q)}{-\eta+\mathrm{i}(p-q)}\right] e^{-\mathrm{i}(p-q)\,r/2} \\ &+\mathcal{O}(e^{-\eta\,r/2}). \end{aligned} \tag{4.32}$$

For any given, p, q, and η, the limit $r\to\infty$ removes the contribution from the delta-function potential well at $x=+r/2$ to $I(p,q;\eta,r)$. Even though the limit $r\to\infty$ of $I(p,q;\eta,r)$ is not convergent because of the phase factors containing the phases $(p\pm q)r/2$, its magnitude can be bounded from above. This is all that is needed to establish that Eq. (4.25c) vanishes in the limit $r\to\infty$ followed by $\eta\to 0$. After the limit $r\to\infty$ of $I(p,q;\eta,r)$ has been taken, the limit $\eta\to 0$ is manifestly convergent to the value 0 for any pair of momenta p and q satisfying $p\neq\pm q$. When $p=q$ (forward scattering), we must invoke the conservation of flux (4.5) to deduce that the limit $r\to\infty$ followed by the limit $\eta\to 0$ of $I(p,p;\eta,r)$ is 0. When $p=-q$ (backward scattering), we must invoke the symmetry under reversal of time (4.4) to deduce that the limit $r\to\infty$ followed by the limit $\eta\to 0$ of $I(p,-p;\eta,r)$ is 0.

We seek the generic conditions on the sequence of functions f_{L} for the integral

$$I(p,q;f_{\mathrm{L}},r):=\int \mathrm{d}x\left(\frac{\mathrm{d}f_{\mathrm{L}}}{\mathrm{d}x}\right)(x)\,j(p,q;x), \tag{4.33}$$

where the matrix elements $j(p,q;x)$ of the conserved current density are given by Eq. (4.30), to vanish when the uniform limit $f_{\mathrm{L}}\to 1$ and $(\partial_x f_{\mathrm{L}})\to 0$ is taken after $r\to\infty$. To this end, we make use of the Heaviside function

$$\Theta(x):=\begin{cases}0, & x<0,\\ 1, & x>0,\end{cases} \tag{4.34a}$$

through the definitions

$$F_{\mathrm{L},<}(x):=(\partial_x f_{\mathrm{L}})(x)\,\Theta\big(-x-(r/2)\big), \tag{4.34b}$$

$$F_{\mathrm{L},>}(x):=(\partial_x f_{\mathrm{L}})(x)\,\Theta\big(x+(r/2)\big). \tag{4.34c}$$

We shall use the conventions

$$\begin{aligned}F_{\mathrm{L},<}(p)&:=\int \mathrm{d}x\,e^{-\mathrm{i}p\,x}\,F_{\mathrm{L},<}(x)=F^{*}_{\mathrm{L},<}(-p),\\ F_{\mathrm{L},<}(p=0)&=+f_{\mathrm{L}}(-r/2),\end{aligned} \tag{4.34d}$$

$$\begin{aligned}F_{\mathrm{L},>}(p)&:=\int \mathrm{d}x\,e^{-\mathrm{i}p\,x}\,F_{\mathrm{L},>}(x)=F^{*}_{\mathrm{L},>}(-p),\\ F_{\mathrm{L},>}(p=0)&=-f_{\mathrm{L}}(-r/2),\end{aligned} \tag{4.34e}$$

for the Fourier transformations. Combining Eqs. (4.30), (4.33), and (4.34) replaces Eq. (4.32) with

$$
\begin{aligned}
I(p,q;f_{\mathrm{L}},r) :=& \int \mathrm{d}x\,(\partial_x f_{\mathrm{L}})(x)\,\jmath(p,q;x) \\
=& + \frac{(p+q)}{2}\left[\frac{A^*(p)\,A(q)}{1/F_{\mathrm{L},<}(p-q)} + \frac{A'^*(p)\,A'(q)}{1/F_{\mathrm{L},>}(p-q)}\right] \\
&+ \frac{(p-q)}{2}\left[\frac{A^*(p)\,B(q)}{1/F_{\mathrm{L},<}(p+q)} + \frac{A'^*(p)\,B'(q)}{1/F_{\mathrm{L},>}(p+q)}\right] \\
&- \frac{(p-q)}{2}\left[\frac{B^*(p)\,A(q)}{1/F^*_{\mathrm{L},<}(p+q)} + \frac{B'^*(p)\,A'(q)}{1/F^*_{\mathrm{L},>}(p+q)}\right] \\
&- \frac{(p+q)}{2}\left[\frac{B^*(p)\,B(q)}{1/F^*_{\mathrm{L},<}(p-q)} + \frac{B'^*(p)\,B'(q)}{1/F^*_{\mathrm{L},<}(p-q)}\right] \\
&+ \mathcal{O}\Big(f_{\mathrm{L}}(+r/2)\Big).
\end{aligned}
\tag{4.35}
$$

Flux conservation, symmetry under reversal of time, and the fact that $F_{\mathrm{L},<}(p = 0) = -F_{\mathrm{L},>}(p = 0)$ imply that this integral vanishes if either $p = q$ or $p = -q$. Hence, it is sufficient to demand that

$$\lim_{r\to\infty} f_{\mathrm{L}}(+r/2) = 0, \tag{4.36a}$$

$$\lim_{f_{\mathrm{L}}\to 1}\lim_{r\to\infty} F_{\mathrm{L},<}(p) = 0, \tag{4.36b}$$

$$\lim_{f_{\mathrm{L}}\to 1}\lim_{r\to\infty} F_{\mathrm{L},>}(p) = 0 \tag{4.36c}$$

hold for any $p \neq 0$ in order for an upper bound on $\lim_{f_{\mathrm{L}}\to 1}\lim_{r\to\infty}|I(p,q;f_{\mathrm{L}},r)|$ to vanish. Conditions (4.36) are met for the test functions (4.15c). This is why Eq. (4.25c) vanishes in the limit (4.17).

Having established that

$$\widehat{Q}_{\mathrm{L}}(t) = \widehat{Q}_{\mathrm{L}}, \tag{4.37}$$

that is $\widehat{Q}_{\mathrm{L}}$ is time-independent, we want to perform the integrals over x in Eqs. (4.18a), (4.18b), and (4.18c) in the limit (4.17).

To this end, we reiterate that, in the limit by which $r \to \infty$ before $\omega \to \infty$ and $\ell \to \infty$ with $\ell/\omega \to 0$ [that is $f_{\mathrm{L}} \to 1$ and $(\partial_x f_{\mathrm{L}}) \to 0$ in a smooth way], the single-particle ground state

$$|1,0\rangle \equiv \hat{b}^{\dagger}_{+}\,|0\rangle := \frac{1}{\sqrt{2}}\left(|\mathrm{L},+\rangle + |\mathrm{R},+\rangle\right), \tag{4.38a}$$

where

$$\psi_{\mathrm{L},+}(x) \equiv \langle x|\mathrm{L},+\rangle \propto e^{-\kappa_{+}|x+r/2|} \tag{4.38b}$$

and

$$\psi_{\mathrm{R},+}(x) \equiv \langle x|\mathrm{R},+\rangle \propto e^{-\kappa_{+}|x-r/2|}, \tag{4.38c}$$

becomes degenerate with the first single-particle excited state

$$|0,1\rangle \equiv \hat{b}^{\dagger}_{-}\,|0\rangle := \frac{1}{\sqrt{2}}\left(|\mathrm{L},-\rangle - |\mathrm{R},-\rangle\right), \tag{4.39a}$$

where

$$\psi_{\mathrm{L},-}(x) \equiv \langle x|\mathrm{L},-\rangle \propto e^{-\kappa_{-}|x+r/2|} \tag{4.39b}$$

and

$$\psi_{\mathrm{R},-}(x) \equiv \langle x|\mathrm{R},-\rangle \propto e^{-\kappa_{-}|x-r/2|}, \tag{4.39c}$$

as well as with the vacuum

$$|0,0\rangle \equiv |0\rangle \tag{4.40}$$

and the two-particle state

$$|1,1\rangle \equiv \hat{b}^{\dagger}_{+}\,\hat{b}^{\dagger}_{-}\,|0\rangle. \tag{4.41}$$

Hence, the integrals entering contribution (4.18a) from the bound states are

$$\begin{aligned}
\frac{1}{2} =& \frac{1}{2}\lim_{r\to\infty}\int \mathrm{d}x\,\psi^{2}_{\mathrm{L},+}(x)\\
=& \frac{1}{2}\lim_{r\to\infty}\int \mathrm{d}x\,\psi^{2}_{\mathrm{L},-}(x)\\
=& \lim_{f_{\mathrm{L}}\to 1}\lim_{r\to\infty}\int \mathrm{d}x\,f_{\mathrm{L}}(x)\,\psi^{2}_{+}(x)\\
=& \lim_{f_{\mathrm{L}}\to 1}\lim_{r\to\infty}\int \mathrm{d}x\,f_{\mathrm{L}}(x)\,\psi^{2}_{-}(x)\\
=& \lim_{f_{\mathrm{L}}\to 1}\lim_{r\to\infty}\int \mathrm{d}x\,f_{\mathrm{L}}(x)\,\psi_{+}(x)\,\psi_{-}(x)
\end{aligned} \tag{4.42}$$

in the limit (4.17). The four contributions from the bound states to $\widehat{Q}_{\mathrm{L}}$ define the quadratic from

$$\begin{aligned}
\frac{1}{2}\left(\hat{b}^{\dagger}_{+}\,\hat{b}_{+} + \hat{b}^{\dagger}_{-}\,\hat{b}_{-} + \hat{b}^{\dagger}_{+}\,\hat{b}_{-} + \hat{b}^{\dagger}_{-}\,\hat{b}_{+}\right) =& \begin{pmatrix}\hat{b}^{\dagger}_{+} & \hat{b}^{\dagger}_{-}\end{pmatrix}\frac{1}{2}\begin{pmatrix}1 & 1\\ 1 & 1\end{pmatrix}\begin{pmatrix}\hat{b}_{+}\\ \hat{b}_{-}\end{pmatrix}\\
=& \begin{pmatrix}\hat{b}^{\dagger}_{\mathrm{L}} & \hat{b}^{\dagger}_{\mathrm{R}}\end{pmatrix}\begin{pmatrix}1 & 0\\ 0 & 0\end{pmatrix}\begin{pmatrix}\hat{b}_{\mathrm{L}}\\ \hat{b}_{\mathrm{R}}\end{pmatrix},
\end{aligned} \tag{4.43a}$$

where

$$\hat{b}_{\mathrm{L}} := \frac{1}{\sqrt{2}}\left(\hat{b}_{+} + \hat{b}_{-}\right), \qquad \hat{b}_{\mathrm{R}} := \frac{1}{\sqrt{2}}\left(\hat{b}_{+} - \hat{b}_{-}\right). \tag{4.43b}$$

We are going to show in two steps that the eigenvalues 1 and 0 of the quadratic form (4.43) can be identified with the pair of eigenvalues of $\widehat{Q}_{\mathrm{L}}$ associated with the state bound to the left delta-function potential well.

First, the mixed contributions (4.18b) to $\widehat{Q}_{\mathrm{L}}$ vanish in the limit (4.17), as follows from multiplying $e^{+\mathrm{i}[\varepsilon_{+}-\varepsilon(p)]\,t}$ in Eq. (4.18b) with $\frac{\varepsilon_{+}-\varepsilon(p)}{\varepsilon_{+}-\varepsilon(p)}$ while multiplying $e^{+\mathrm{i}[\varepsilon_{-}-\varepsilon(p)]\,t}$ in Eq. (4.18b) with $\frac{\varepsilon_{-}-\varepsilon(p)}{\varepsilon_{-}-\varepsilon(p)}$ and proceeding as we did to establish that the contributions to (4.21b) vanish in the limit (4.17).

Second, the contributions (4.18c) to $\widehat{Q}_{\mathrm{L}}$ from the continuum simplify in the limit (4.17) to

$$\int \mathrm{d}p\ \left[T_{\mathrm{L}}\,\hat{c}^{\dagger}(p)\,\hat{c}(p) + R_{\mathrm{L}}\,\hat{c}^{\dagger}(p)\,\hat{c}(-p)\right], \tag{4.44}$$

where T_{L} is real-valued and R_{L} is complex-valued. The pair T_{L} and R_{L} are the counterpart to the number 1/2 in Eq. (4.42) that results from effectively projecting onto the left half of the real line through the limit (4.17). The proof of Eq. (4.44) proceeds in two steps. First, one may multiply $e^{+\mathrm{i}[\varepsilon(p)-\varepsilon(q)]\,t}$ with $\frac{\varepsilon(p)-\varepsilon(q)}{\varepsilon(p)-\varepsilon(q)}$ when $\varepsilon(p) \neq \varepsilon(q)$ to proceed as we did to establish that the contributions to (4.21c) vanish in the limit (4.17). The only nonvanishing contributions to the triple integrals (4.18c) occurs when $\varepsilon(p) = \varepsilon(q)$, that is, when $p = \pm q$ for the assumed quadratic dispersion,

$$\lim_{f_{\mathrm{L}}\to 1}\lim_{r\to\infty} \int \mathrm{d}x\, f_{\mathrm{L}}(x)\,\psi^{*}(p,x)\,\psi(q,x) \propto \delta\Big(\varepsilon(p)-\varepsilon(q)\Big). \tag{4.45}$$

This is why there are two terms in the integrand of Eq. (4.44).

To recap, we have shown that

$$\begin{aligned}\widehat{Q}_{\mathrm{L}} =& \frac{1}{2}\left(\hat{b}^{\dagger}_{+}\,\hat{b}_{+} + \hat{b}^{\dagger}_{-}\,\hat{b}_{-} + \hat{b}^{\dagger}_{+}\,\hat{b}_{-} + \hat{b}^{\dagger}_{-}\,\hat{b}_{+}\right)\\ &+ \int \mathrm{d}p\ \left[T_{\mathrm{L}}\,\hat{c}^{\dagger}(p)\,\hat{c}(p) + R_{\mathrm{L}}\,\hat{c}^{\dagger}(p)\,\hat{c}(-p)\right]\end{aligned} \tag{4.46a}$$

can be brought to the diagonal representation

$$\widehat{Q}_{\mathrm{L}} = \hat{b}^{\dagger}_{\mathrm{L}}\,\hat{b}_{\mathrm{L}} + \int \mathrm{d}p\ \left[T_{\mathrm{L}}\,\hat{c}^{\dagger}(p)\,\hat{c}(p) + R_{\mathrm{L}}\,\hat{c}^{\dagger}(p)\,\hat{c}(-p)\right]. \tag{4.46b}$$

Three important properties follow. First, one verifies with the help of the identity (4.19) that the smeared fermion number operator for the left delta-function well (4.46a) commutes with the total fermion number operator (4.10) for any r. Second, one verifies with the help of the identity (4.19) that the smeared fermion number operator for the left delta-function well (4.46a) commutes with the Hamiltonian for the double delta-function (4.9) in the limit (4.17). Third, the smeared fermion number operator for the left delta-function well (4.46b) has the four eigenstates (in the fermion Fock space) [recall Eqs. (4.38) and (4.39)]

$$\begin{aligned}
&|0,0\rangle,\\
&\hat{b}^{\dagger}_{\mathrm{L}}\,|0,0\rangle \equiv \frac{1}{\sqrt{2}}\left(|1,0\rangle + |0,1\rangle\right), \qquad \hat{b}^{\dagger}_{\mathrm{R}}\,|0,0\rangle \equiv \frac{1}{\sqrt{2}}\left(|1,0\rangle - |0,1\rangle\right),\\
&\hat{b}^{\dagger}_{\mathrm{L}}\,\hat{b}^{\dagger}_{\mathrm{R}}\,|0,0\rangle \equiv \hat{b}^{\dagger}_{-}\,\hat{b}^{\dagger}_{+}\,|0,0\rangle,
\end{aligned} \tag{4.47}$$

with the four integer-valued eigenvalues

$$0, \qquad 1, \qquad 0, \qquad 1, \tag{4.48}$$

respectively.

There are thus two ways to select the ground states in the limit (4.17). On the one hand, if we select the ground state in the limit (4.17) to be the eigenstate

$$|\mathrm{GS}\rangle_{\mathrm{L}} := \hat{b}^{\dagger}_{\mathrm{L}}\,|0\rangle \tag{4.49}$$

of the smeared fermion number operator for the left delta-function well (4.46b), then this ground state is an eigenstate of $\widehat{Q}_{\mathrm{L}}$ with the integer-valued eigenvalue 1. This is the ground state that would be selected if an electric field E were to contribute to the Hamiltonian (4.9) for the double delta-function potential well in the limit (4.17) (with the chemical potential always obeying the single-particle condition $\varepsilon_{+} < \mu < \varepsilon_{-} < 0$) by the addition of the term $E\,\widehat{Q}_{\mathrm{L}}$, before being switched off. On the other hand, if no external electric field is applied and if the limit (4.17) is taken with the chemical potential always obeying the single-particle condition $\varepsilon_{+} < \mu < \varepsilon_{-} < 0$, then it is the ground state

$$|\mathrm{GS}\rangle \equiv \hat{b}^{\dagger}_{+}\,|0\rangle \tag{4.50}$$

of Hamiltonian (4.9) that is selected. If so, $\langle\mathrm{GS}|\,\widehat{Q}_{\mathrm{L}}|\mathrm{GS}\rangle = 1/2$ with a nonvanishing variance of $1/4$ as we now explain.

Indeed, the limit by which $r \to \infty$ before $\omega \to \infty$ and $\ell \to \infty$ with $\ell/\omega \to 0$ (that is $f_{\mathrm{L}} \to 1$ in a smooth way) gives the limiting value

$$\begin{aligned}
\langle\mathrm{GS}|\,\widehat{Q}_{\mathrm{L}}|\mathrm{GS}\rangle &= \lim_{f_{\mathrm{L}}\to 1}\lim_{r\to\infty}\langle\mathrm{GS}|\,\widehat{Q}_{f_{\mathrm{L}}}|\mathrm{GS}\rangle\\
&= \lim_{f_{\mathrm{L}}\to 1}\lim_{r\to\infty}\int \mathrm{d}x\, f_{\mathrm{L}}(x)\,\psi^{2}_{+}(x)\\
&= \lim_{f_{\mathrm{L}}\to 1}\lim_{r\to\infty}\int \mathrm{d}x\, f_{\mathrm{L}}(x)\,\frac{1}{2}\psi^{2}_{\mathrm{L},+}(x)\\
&= \frac{1}{2}
\end{aligned} \tag{4.51}$$

for the ground-state expectation value of the smeared charge operator centered about the left potential well.

Of course, this is not to say that we have realized a quantum state with the value 1/2 for the fractional charge. What we have shown is that the probability to find the fermion in the left potential well is 1/2. Indeed, this statistical interpretation for the value 1/2 of the charge in the left potential well is a consequence of the variance

$$\begin{aligned}\operatorname{var}\widehat{Q}_{f_{\mathrm{L}}}\Big|_{\mathrm{GS}} &:= \left\langle \mathrm{GS}\left|\left(\widehat{Q}_{f_{\mathrm{L}}}-\langle\mathrm{GS}|\,\widehat{Q}_{f_{\mathrm{L}}}|\mathrm{GS}\rangle\right)^2\right|\mathrm{GS}\right\rangle \\ &= \langle\mathrm{GS}|\,\widehat{Q}^2_{f_{\mathrm{L}}}|\mathrm{GS}\rangle-\left(\langle\mathrm{GS}|\,\widehat{Q}_{f_{\mathrm{L}}}|\mathrm{GS}\rangle\right)^2\end{aligned} \tag{4.52}$$

taking the nonvanishing value 1/4 in the limit by which the separation between the two potential wells is taken to infinity, $r\to\infty$, before the limit $\omega\to\infty$ and $\ell\to\infty$ with $\ell/\omega\to 0$ (that is $f_{\mathrm{L}}\to 1$ in a smooth way).

Proof Equation (4.51) gives

$$\left(\langle\mathrm{GS}|\,\widehat{Q}_{\mathrm{L}}|\mathrm{GS}\rangle\right)^2=\lim_{f_{\mathrm{L}}\to 1}\lim_{r\to\infty}\left(\langle\mathrm{GS}|\,\widehat{Q}_{f_{\mathrm{L}}}|\mathrm{GS}\rangle\right)^2=\frac{1}{4}. \tag{4.53}$$

It suffices to show that $\langle\mathrm{GS}|\,\widehat{Q}^2_{\mathrm{L}}|\mathrm{GS}\rangle=1/2$. This is a consequence of the fact that [recall Eqs. (4.18), (4.38), (4.39), and set $|\mathrm{FES}\rangle\equiv\hat{b}^{\dagger}_{-}|0\rangle$ for the first excited state of $\mathcal{H}_{\mathrm{dw}}$]

$$\begin{aligned}\frac{1}{2} &= \lim_{f_{\mathrm{L}}\to 1}\lim_{r\to\infty}\int \mathrm{d}x\, f_{\mathrm{L}}(x)\,\langle\mathrm{GS}|\,e^{+\mathrm{i}(\varepsilon_{+}-\varepsilon_{-})\,t}\,\psi_{+}(x)\,\psi_{-}(x)\,\hat{b}^{\dagger}_{+}\,\hat{b}_{-}\,|\mathrm{FES}\rangle \\ &= \lim_{f_{\mathrm{L}}\to 1}\lim_{r\to\infty}\langle\mathrm{GS}|\,\widehat{Q}_{f_{\mathrm{L}}}|\mathrm{FES}\rangle,\end{aligned} \tag{4.54}$$

that is the states $\widehat{Q}_{f_{\mathrm{L}}}|\mathrm{FES}\rangle$ and $|\mathrm{GS}\rangle$ remain non-orthogonal in the limit by which $r\to\infty$ before $\omega\to\infty$ and $\ell\to\infty$ with $\ell/\omega\to 0$ (that is $f_{\mathrm{L}}\to 1$ in a smooth way), as we now show. Indeed, upon insertion of the resolution of the identity in terms of the single-particle eigenstates of $\mathcal{H}_{\mathrm{dw}}$,

$$\begin{aligned}\langle\mathrm{GS}|\,\widehat{Q}^2_{f_{\mathrm{L}}}|\mathrm{GS}\rangle &= \langle\mathrm{GS}|\,\widehat{Q}_{f_{\mathrm{L}}}|\mathrm{GS}\rangle\langle\mathrm{GS}|\,\widehat{Q}_{f_{\mathrm{L}}}|\mathrm{GS}\rangle \\ &\quad+\langle\mathrm{GS}|\,\widehat{Q}_{f_{\mathrm{L}}}|\mathrm{FES}\rangle\langle\mathrm{FES}|\widehat{Q}_{f_{\mathrm{L}}}|\mathrm{GS}\rangle \\ &\quad+\langle\mathrm{GS}|\,\widehat{Q}_{f_{\mathrm{L}}}\int\mathrm{d}p\,|p\rangle\langle p|\,\widehat{Q}_{f_{\mathrm{L}}}|\mathrm{GS}\rangle.\end{aligned} \tag{4.55}$$

In the limit by which $r\to\infty$ before $\omega\to\infty$ and $\ell\to\infty$ with $\ell/\omega\to 0$ (that is $f_{\mathrm{L}}\to 1$ and $(\partial_x f_{\mathrm{L}})\to 0$ in a smooth way),

$$
\begin{aligned}
\langle p|\,\widehat{Q}_{\mathrm{L}}|\mathrm{GS}\rangle =& \lim_{f_{\mathrm{L}}\to 1}\lim_{r\to\infty}\langle p|\,\widehat{Q}_{f_{\mathrm{L}}}|\mathrm{GS}\rangle \\
{\scriptstyle \text{Eq. (4.18b)}}\ =& \lim_{f_{\mathrm{L}}\to 1}\lim_{r\to\infty} e^{-\mathrm{i}[\varepsilon_{+}-\varepsilon(p)]\,t}\int \mathrm{d}x\, f_{\mathrm{L}}(x)\,\psi_{+}(x)\,\psi^{*}(p,x) \\
& \times \langle p|\,\hat{c}^{\dagger}(p)\,\hat{b}_{+}\,|\mathrm{GS}\rangle \\
{\scriptstyle \text{Eq. (4.25b)}}\ =& \lim_{f_{\mathrm{L}}\to 1}\lim_{r\to\infty} \mathrm{i}\,\frac{e^{+\mathrm{i}[\varepsilon(p)-\varepsilon_{+}]\,t}}{\varepsilon_{+}-\varepsilon(p)}\int \mathrm{d}x\left(\frac{\mathrm{d}f_{\mathrm{L}}}{\mathrm{d}x}\right)(x)\,j(p,+;x) \\
=& \,0,
\end{aligned}
\tag{4.56}
$$

so that

$$
\begin{aligned}
\langle \mathrm{GS}|\,\widehat{Q}^{2}_{\mathrm{L}}|\mathrm{GS}\rangle =& \lim_{f_{\mathrm{L}}\to 1}\lim_{r\to\infty}\langle \mathrm{GS}|\,\widehat{Q}^{2}_{f_{\mathrm{L}}}|\mathrm{GS}\rangle \\
=& \lim_{f_{\mathrm{L}}\to 1}\lim_{r\to\infty}\left[\left|\langle \mathrm{GS}|\,\widehat{Q}_{f_{\mathrm{L}}}|\mathrm{GS}\rangle\right|^{2}+\left|\langle \mathrm{GS}|\,\widehat{Q}_{f_{\mathrm{L}}}|\mathrm{FES}\rangle\right|^{2}\right] \\
=& \frac{1}{4}+\frac{1}{4} \\
=& \frac{1}{2}.
\end{aligned}
\tag{4.57}
$$

At last,

$$
\begin{aligned}
\operatorname{var}\widehat{Q}_{\mathrm{L}}\Big|_{\mathrm{GS}} =& \langle \mathrm{GS}|\,\widehat{Q}^{2}_{\mathrm{L}}|\mathrm{GS}\rangle - \Big(\langle \mathrm{GS}|\,\widehat{Q}_{\mathrm{L}}|\mathrm{GS}\rangle\Big)^{2} \\
=& \frac{1}{2}-\frac{1}{4} \\
=& \frac{1}{4}.
\end{aligned}
$$

This variance remains nonvanishing in the limit $r\to\infty$ before $\omega\to\infty$ and $\ell\to\infty$ with $\ell/\omega\to 0$ (that is $f_{\mathrm{L}}\to 1$ in a smooth way), for the states $|\mathrm{GS}\rangle$ and $\widehat{Q}_{\mathrm{L}}\,|\mathrm{FES}\rangle$ remain non-orthogonal in this limit, that is, the smeared number operator $\widehat{Q}_{\mathrm{L}}$ when applied to the ground state $|\mathrm{GS}\rangle$ can reach the state $|\mathrm{FES}\rangle$ that becomes degenerate with the ground state as $r\to\infty$. □

We conclude that the single-particle ground state of the double delta-function potential well in the limit (4.17) can be chosen to be either an eigenstate of the smeared number operator $\widehat{Q}_{\mathrm{L}}$ with the integer-valued eigenvalue 1, or to provide $\widehat{Q}_{\mathrm{L}}$ with the expectation value 1/2 and the variance 1/4. No charge fractionalization occurs for the double delta-function potential well.

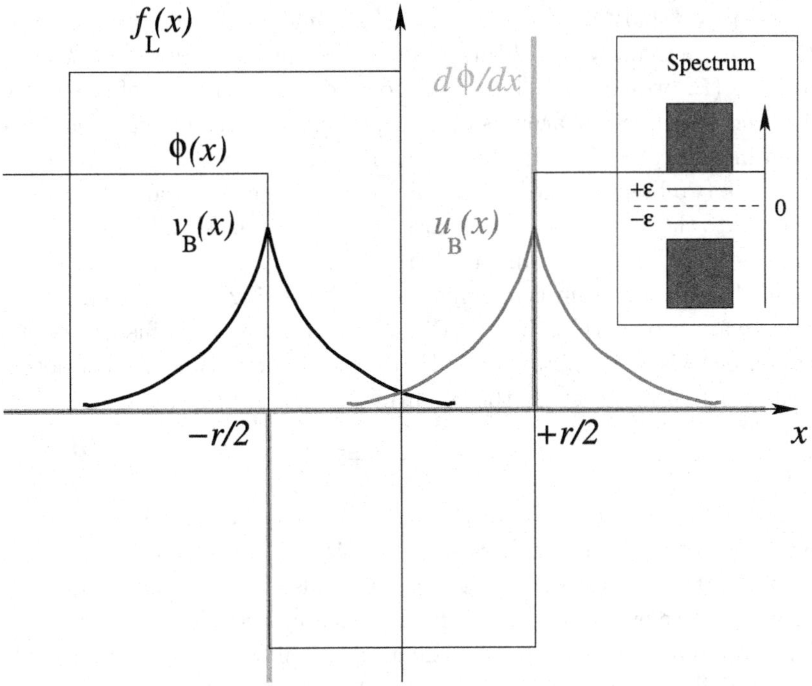

Figure 4.3 The function $\phi(x)$ depicts a pair of domain walls (soliton and antisoliton) centered at $\mp r/2$, respectively. Its derivative $\mathrm{d}\phi/\mathrm{d}x$ has a delta-function well at $-(r/2)$ and delta-function wall at $+(r/2)$. The Dirac spinor bound state $\psi_{\mathrm{B}}^{\mathsf{T}}(x) = \big(u_{\mathrm{B}}(x), v_{\mathrm{B}}(x)\big)$ has the single-particle energy $+\varepsilon > 0$ with its upper component $u_{\mathrm{B}}(x)$ exponentially localized about $x = +(r/2)$ and its lower component $v_{\mathrm{B}}(x)$ exponentially localized about $x = -(r/2)$. The Dirac spinor bound state $\tau_3\, \psi_{\mathrm{B}}(x)$ has the single-particle energy $-\varepsilon < 0$. The continuum consists of a valence band unbounded from below and of a conduction band unbounded from above. The function $f_{\mathrm{L}}(x)$ depicts a box test function centered at $x = -r/2$ with $\omega = \ell$. The spectrum in the inset is particle-hole (chiral) symmetric.

4.3 Example 2: The Pair of Soliton and Antisoliton

We turn our attention to the case of the Dirac Hamiltonian [recall Eqs. (3.14) and (3.17)]

$$\mathcal{H}_{\mathrm{s\bar{s}}} := \tau_2\, \frac{\mathrm{d}}{\mathrm{i}\mathrm{d}x} + \tau_1\, \phi(x) \tag{4.58a}$$

with the soliton-antisoliton profile

$$\phi(x) = \begin{cases} +\phi_\infty, & x < -r/2, \\ -\phi_\infty, & -r/2 < x < +r/2, \\ +\phi_\infty, & r/2 < x. \end{cases} \tag{4.58b}$$

The choice of two sharp domain walls simplifies analytical treatments and guarantees that no more than two bound states are present in the spectrum at the energies $-\varepsilon < 0 < +\varepsilon$ (**Exercise** 4.5). This choice does not result in a loss of generality with regard to the issue of the sharpness of the fractional charge. The threshold energies to the continua are at $\pm\phi_{\infty}$ with the convention that $\phi_{\infty} > 0$. Figure 4.3 depicts the profiles of the two bound states. These profiles should be compared to those of the states bound to the delta-function potential wells in Figure 4.1. There is one crucial difference between Figures 4.3 and 4.1. The Dirac Hamiltonian (4.58) anticommutes with τ_3 so that its spectrum is symmetric about the single-particle energy 0. This is why there are two continua for the Dirac Hamiltonian (4.58) instead of only one for the nonrelativistic Hamiltonian (4.1). Hence, pinning the chemical potential μ to the value 0 suffices to satisfy the condition

$$-\varepsilon < \mu < +\varepsilon, \tag{4.59}$$

for any positive eigenvalue $\varepsilon > 0$ of the Dirac Hamiltonian (4.58).

The ground state $|\mathrm{GS}\rangle$ of the Dirac Hamiltonian (4.58) with $\mu = 0$ has all single-particle eigenstates with negative eigenenergies filled and all single-particle eigenstates with positive eigenenergies empty. This is a many-body ground state, unlike that for Hamiltonian (4.1) with the chemical potential satisfying $\varepsilon_{+} < \mu < \varepsilon_{-}$ for the bound-state eigenenergies $\varepsilon_{\pm}$ defined in Eq. (4.2d). The many-body excited states with fixed number of fermions (canonical ensemble) are obtained from the ground state $|\mathrm{GS}\rangle$ by promoting fermions occupying negative single-particle eigenenergy levels to positive single-particle eigenenergy levels. In the limit $r \to \infty$, and if we only fix the number of fermions on the average (grand-canonical ensemble), the ground state $|\mathrm{GS}\rangle$ becomes degenerate with three other many-body eigenstates. A basis for the fourfold degenerate ground states in the grand-canonical ensemble and at $r = \infty$ is denoted by

$$|n_{+}, n_{-}\rangle_{\mathrm{s\bar{s}}}, \qquad n_{+}, n_{-} = 0, 1. \tag{4.60a}$$

The many-body state $|n_{+}, n_{-}\rangle$ has all single-particle states from the continua with strictly positive (negative) eigenenergies empty (occupied), while the two single-particle mid-gap states at the eigenenergies $\pm\varepsilon$ for $r < \infty$ have occupancy $n_{\pm}$, respectively. In this notation, the nondegenerate many-body ground state for $r < \infty$ is

$$|\mathrm{GS}\rangle_{\mathrm{s\bar{s}}} \equiv |0, 1\rangle_{\mathrm{s\bar{s}}}. \tag{4.60b}$$

We are going to perform a particle-hole transformation on the creation operator associated with single-particle eigenstates with negative eigenenergies. Under this particle-hole transformation, $|0, 1\rangle_{\mathrm{s\bar{s}}}$ will be identified as the vacuum $|0\rangle$.

We recall that the upper (u) and lower (v) components of a solution to the stationary Dirac equation

$$\mathcal{H}_{\mathrm{s}\bar{\mathrm{s}}}\,\Psi=\varepsilon\,\Psi,\qquad \Psi=\begin{pmatrix}u\\ v\end{pmatrix}\Longleftrightarrow\begin{cases}-\frac{\mathrm{d}}{\mathrm{d}x}v+\phi\,v=\varepsilon\,u,\\ \\ +\frac{\mathrm{d}}{\mathrm{d}x}u+\phi\,u=\varepsilon\,v,\end{cases}\tag{4.61a}$$

must obey the stationary Schrödinger equations

$$\left[-\frac{\mathrm{d}^2}{\mathrm{d}x^2}+\left(\phi^2-\frac{\mathrm{d}\phi}{\mathrm{d}x}\right)\right]u=\varepsilon^2\,u\tag{4.61b}$$

and

$$\left[-\frac{\mathrm{d}^2}{\mathrm{d}x^2}+\left(\phi^2+\frac{\mathrm{d}\phi}{\mathrm{d}x}\right)\right]v=\varepsilon^2\,v,\tag{4.61c}$$

respectively.

We use the notation

$$\mathcal{H}_{\mathrm{s}\bar{\mathrm{s}}}\left[\tau_3\Psi_{\mathrm{B}}(x)\right]=-\varepsilon\left[\tau_3\Psi_{\mathrm{B}}(x)\right],\qquad \mathcal{H}_{\mathrm{s}\bar{\mathrm{s}}}\,\Psi_{\mathrm{B}}(x)=+\varepsilon\,\Psi_{\mathrm{B}}(x),\tag{4.62a}$$

for the pair of bound states with eigenenergies $-\varepsilon<0<+\varepsilon$, respectively, and choose the normalization conditions

$$\int\mathrm{d}x\,u_{\mathrm{B}}^2(x)=\int\mathrm{d}x\,v_{\mathrm{B}}^2(x)=\frac{1}{2},\qquad \Psi_{\mathrm{B}}(x)=\begin{pmatrix}u_{\mathrm{B}}(x)\\ v_{\mathrm{B}}(x)\end{pmatrix},\tag{4.62b}$$

for the real-valued upper and lower components of $\Psi_{\mathrm{B}}(x)$. The upper component u_{B} satisfies the stationary Schrödinger equation in the potential

$$V_{\mathrm{R}}(x)=\phi_\infty^2+2\,\phi_\infty\,\delta\big(x+(r/2)\big)-2\,\phi_\infty\,\delta\big(x-(r/2)\big).\tag{4.63}$$

Hence, u_{B} is exponentially localized around the domain wall at $x=+(r/2)$, where it experiences a delta-function potential well (see Figure 4.3). The lower component v_{B} satisfies the stationary Schrödinger equation in the potential

$$V_{\mathrm{L}}(x)=\phi_\infty^2-2\,\phi_\infty\,\delta\big(x+(r/2)\big)+2\,\phi_\infty\,\delta\big(x-(r/2)\big).\tag{4.64}$$

Hence, v_{B} is exponentially localized around the domain wall at $x=-(r/2)$, where it experiences a delta-function potential well (see Figure 4.3).

We use the notation

$$\mathcal{H}_{\mathrm{s}\bar{\mathrm{s}}}\left[\tau_3\,\Psi^*(p,x)\right]=-\varepsilon(p)\left[\tau_3\,\Psi^*(p,x)\right],\qquad \mathcal{H}_{\mathrm{s}\bar{\mathrm{s}}}\,\Psi(p,x)=+\varepsilon(p)\,\Psi(p,x),\tag{4.65a}$$

where

$$\varepsilon(p):=\sqrt{p^2+\phi_\infty^2}\tag{4.65b}$$

for the pair of continuum states with eigenenergies $-\varepsilon(p)\leq-\phi_\infty<0<+\phi_\infty\leq+\varepsilon(p)$, respectively, and choose the normalization conditions

$$\int\mathrm{d}x\,u^*(p,x)\,u(q,x)=\int\mathrm{d}x\,v^*(p,x)\,v(q,x)=\frac{1}{2}\,\delta(p-q)\tag{4.65c}$$

for the upper and lower components of

$$\Psi(p,x)=\begin{pmatrix}u(p,x)\\ v(p,x)\end{pmatrix}=\begin{pmatrix}u^{*}(-p,x)\\ v^{*}(-p,x)\end{pmatrix}=\Psi^{*}(-p,x). \tag{4.65d}$$

We define the pair of field operators [compare with Eq. (3.616) and recall that $\Psi^{*}_{\mathrm{B}}(x)=\Psi_{\mathrm{B}}(x)$]

$$\begin{aligned}\widehat{\Psi}^{\dagger}_{\mathrm{s}\bar{\mathrm{s}}}(x,t):=&\,e^{+\mathrm{i}\varepsilon\,t}\,\Psi^{\dagger}_{\mathrm{B}}(x)\,\hat{c}^{\dagger}+e^{-\mathrm{i}\varepsilon\,t}\,\left[\tau_{3}\,\Psi_{\mathrm{B}}(x)\right]^{\dagger}\,\hat{d}\\ &+\int\mathrm{d}p\,\left[e^{+\mathrm{i}\varepsilon(p)\,t}\,\Psi^{\dagger}(p,x)\,\hat{c}^{\dagger}(p)+e^{-\mathrm{i}\varepsilon(p)\,t}\,\Psi^{\mathsf{T}}(p,x)\,\tau_{3}\,\hat{d}(p)\right],\end{aligned} \tag{4.66a}$$

and

$$\begin{aligned}\widehat{\Psi}_{\mathrm{s}\bar{\mathrm{s}}}(x,t):=&\,e^{-\mathrm{i}\varepsilon\,t}\,\Psi_{\mathrm{B}}(x)\,\hat{c}+e^{+\mathrm{i}\varepsilon\,t}\,\tau_{3}\,\Psi_{\mathrm{B}}(x)\,\hat{d}^{\dagger}\\ &+\int\mathrm{d}p\,\left[e^{-\mathrm{i}\varepsilon(p)\,t}\,\Psi(p,x)\,\hat{c}(p)+e^{+\mathrm{i}\varepsilon(p)\,t}\,\tau_{3}\,\Psi^{*}(p,x)\,\hat{d}^{\dagger}(p)\right],\end{aligned} \tag{4.66b}$$

where the only nonvanishing anticommutators are

$$\{\hat{c},\hat{c}^{\dagger}\}=1,\quad\{\hat{d},\hat{d}^{\dagger}\}=1,\quad\{\hat{c}(p),\hat{c}^{\dagger}(q)\}=\{\hat{d}(p),\hat{d}^{\dagger}(q)\}=\delta(p-q). \tag{4.66c}$$

The second-quantized Hamiltonian and total fermion number are

$$\begin{aligned}\widehat{H}_{\mathrm{s}\bar{\mathrm{s}}}:=&\int\mathrm{d}x\,\widehat{\Psi}^{\dagger}_{\mathrm{s}\bar{\mathrm{s}}}(x,t)\,\mathcal{H}_{\mathrm{s}\bar{\mathrm{s}}}\widehat{\Psi}_{\mathrm{s}\bar{\mathrm{s}}}(x,t)\\ =&\,\varepsilon\left(\hat{c}^{\dagger}\,\hat{c}-\hat{d}\,\hat{d}^{\dagger}\right)+\int\mathrm{d}p\,\varepsilon(p)\,\left[\hat{c}^{\dagger}(p)\,\hat{c}(p)-\hat{d}(p)\,\hat{d}^{\dagger}(p)\right]\end{aligned} \tag{4.67}$$

and

$$\int\mathrm{d}x\,\widehat{\Psi}^{\dagger}_{\mathrm{s}\bar{\mathrm{s}}}(x,t)\,\widehat{\Psi}_{\mathrm{s}\bar{\mathrm{s}}}(x,t)=\left(\hat{c}^{\dagger}\,\hat{c}+\hat{d}\,\hat{d}^{\dagger}\right)+\int\mathrm{d}p\,\left[\hat{c}^{\dagger}(p)\,\hat{c}(p)+\hat{d}(p)\,\hat{d}^{\dagger}(p)\right] \tag{4.68}$$

respectively. The many-body ground state at the chemical potential $\mu=0$ is

$$|\mathrm{GS}\rangle_{\mathrm{s}\bar{\mathrm{s}}}:=|0\rangle, \tag{4.69}$$

where $|0\rangle$ is the state annihilated by all annihilation operators $\hat{c}$, $\hat{d}$, $\hat{c}(p)$, and $\hat{d}(p)$. It is nondegenerate for $0<r<\infty$. It becomes degenerate with the many-body excited states

$$\hat{c}^{\dagger}\,|\mathrm{GS}\rangle_{\mathrm{s}\bar{\mathrm{s}}},\qquad\hat{d}^{\dagger}\,|\mathrm{GS}\rangle_{\mathrm{s}\bar{\mathrm{s}}},\qquad\hat{c}^{\dagger}\,\hat{d}^{\dagger}\,|\mathrm{GS}\rangle_{\mathrm{s}\bar{\mathrm{s}}}, \tag{4.70}$$

in the limit $r\to\infty$. The jump of degeneracy of the ground states from one to four in the limit $r\to\infty$ renders this limit singular as we are going to see shortly.

As we did in Eq. (3.626), we adopt a definition of the fermion number that is odd under the particle-hole transformation

$$\begin{aligned}&a\,\left(\widehat{\Psi}_{\mathrm{s}\bar{\mathrm{s}}}\right)_{\sigma}(x,t)+b^{*}\,\left(\widehat{\Psi}^{\dagger}_{\mathrm{s}\bar{\mathrm{s}}}\right)_{\sigma}(x,t)\mapsto\\ &\qquad a^{*}\,\left(\widehat{\Psi}^{\dagger}_{\mathrm{s}\bar{\mathrm{s}}}\right)_{\sigma'}(x,t)\,(\tau_{3})_{\sigma\sigma'}+b\,(\tau_{3})_{\sigma'\sigma}\left(\widehat{\Psi}_{\mathrm{s}\bar{\mathrm{s}}}\right)_{\sigma'}(x,t),\end{aligned} \tag{4.71}$$

where the spinor index $\sigma = 1, 2$ is given, the summation convention over the repeated lower index $\sigma' = 1, 2$ is implied, and a and b are complex numbers. To this end, we define the smeared fermion number

$$\begin{aligned}\widehat{Q}_{\mathrm{s\bar{s}}\,f}(t) &:= \int \mathrm{d}x\, f(x)\, \frac{1}{2} \sum_{\sigma=1,2} \left[\left(\widehat{\Psi}^{\dagger}_{\mathrm{s\bar{s}}}\right)_{\sigma}(x,t), \left(\widehat{\Psi}_{\mathrm{s\bar{s}}}\right)_{\sigma}(x,t)\right] \\ &\equiv \int \mathrm{d}x\, f(x)\, \frac{1}{2} \left[\widehat{\Psi}^{\dagger}_{\mathrm{s\bar{s}}}(x,t), \widehat{\Psi}_{\mathrm{s\bar{s}}}(x,t)\right]\end{aligned} \tag{4.72}$$

for an arbitrary test function f taking values in the interval $[0, 1]$. To avoid having to always carry a cumbersome distinction between bound states and continuum states during intermediary steps, we introduce the notation

$$\begin{aligned}&\sum\!\!\!\!\!\!\int \mathrm{d}p\, \varepsilon(p) \left[\hat{c}^{\dagger}(p)\, \hat{c}(p) + \hat{d}^{\dagger}(p)\, \hat{d}(p)\right] \equiv \\ &\qquad \varepsilon\left(\hat{c}^{\dagger}\, \hat{c} + \hat{d}^{\dagger}\, \hat{d}\right) + \int \mathrm{d}p\, \varepsilon(p) \left[\hat{c}^{\dagger}(p)\, \hat{c}(p) + \hat{d}^{\dagger}(p)\, \hat{d}(p)\right].\end{aligned} \tag{4.73}$$

One verifies that (**Exercise** 4.6)

$$\widehat{Q}_{\mathrm{s\bar{s}}\,f}(t) = \widehat{Q}^{(1)}_{\mathrm{s\bar{s}}\,f}(t) + \widehat{Q}^{(2)}_{\mathrm{s\bar{s}}\,f}(t) + \widehat{Q}^{(3)}_{\mathrm{s\bar{s}}\,f}(t), \tag{4.74a}$$

where

$$\begin{aligned}\widehat{Q}^{(1)}_{\mathrm{s\bar{s}}\,f}(t) &= \int \mathrm{d}x\, f(x) \sum\!\!\!\!\!\!\int \mathrm{d}p \sum\!\!\!\!\!\!\int \mathrm{d}q\, e^{+\mathrm{i}[\varepsilon(p)-\varepsilon(q)]\,t} \\ &\quad \times \left[u^{*}(p,x)\, u(q,x) + v^{*}(p,x)\, v(q,x)\right]\left[\hat{c}^{\dagger}(p)\, \hat{c}(q) - \hat{d}^{\dagger}(p)\, \hat{d}(q)\right],\end{aligned} \tag{4.74b}$$

$$\begin{aligned}\widehat{Q}^{(2)}_{\mathrm{s\bar{s}}\,f}(t) &= \int \mathrm{d}x\, f(x) \sum\!\!\!\!\!\!\int \mathrm{d}p \sum\!\!\!\!\!\!\int \mathrm{d}q\, e^{+\mathrm{i}[\varepsilon(p)+\varepsilon(q)]\,t} \\ &\quad \times \left[u^{*}(p,x)\, u(-q,x) - v^{*}(p,x)\, v(-q,x)\right]\, \hat{c}^{\dagger}(p)\, \hat{d}^{\dagger}(q),\end{aligned} \tag{4.74c}$$

$$\begin{aligned}\widehat{Q}^{(3)}_{\mathrm{s\bar{s}}\,f}(t) &= \int \mathrm{d}x\, f(x) \sum\!\!\!\!\!\!\int \mathrm{d}p \sum\!\!\!\!\!\!\int \mathrm{d}q\, e^{-\mathrm{i}[\varepsilon(p)+\varepsilon(q)]\,t} \\ &\quad \times \left[u(p,x)\, u^{*}(-q,x) - v(p,x)\, v^{*}(-q,x)\right]\, \hat{d}(p)\, \hat{c}(q).\end{aligned} \tag{4.74d}$$

In the uniform limit $f \to 1$ and $(\partial_x f) \to 0$, one may use the orthogonality between eigenstates and their normalizations to infer that the total fermion number

$$\begin{aligned}\widehat{Q}_{\mathrm{s\bar{s}}} &:= \lim_{f\to 1} \widehat{Q}_{\mathrm{s\bar{s}}\,f}(t) \\ &= \hat{c}^{\dagger}\, \hat{c} - \hat{d}^{\dagger}\, \hat{d} + \int \mathrm{d}p \left[\hat{c}^{\dagger}(p)\, \hat{c}(p) - \hat{d}^{\dagger}(p)\, \hat{d}(p)\right]\end{aligned} \tag{4.75}$$

is time-independent and has integer-valued eigenvalues.

We consider again a test function f_{L} of the form (4.15c) as depicted in Figure 4.2 that is centered about the domain wall at $x = -(r/2)$. We define the smeared fermion number

$$\widehat{Q}_{\mathrm{s}\bar{\mathrm{s}}\, f_{\mathrm{L}}}(t) := \int \mathrm{d}x\, f_{\mathrm{L}}(x)\, \frac{1}{2} \left[\widehat{\Psi}^{\dagger}_{\mathrm{s}\bar{\mathrm{s}}}(x,t), \widehat{\Psi}_{\mathrm{s}\bar{\mathrm{s}}}(x,t) \right] \tag{4.76}$$

[where the sum over the components is implicitly assumed, as in Eq. (4.72)] together with the limit by which $r \to \infty$ before $\omega \to \infty$ and $\ell \to \infty$ with $\ell/\omega \to 0$ [that is $f_{\mathrm{L}} \to 1$ and $(\partial_x f_{\mathrm{L}}) \to 0$ in a smooth way],

$$\widehat{Q}_{\mathrm{s}\bar{\mathrm{s}}\,\mathrm{L}} := \lim_{f_{\mathrm{L}} \to 1} \lim_{r \to \infty} \widehat{Q}_{\mathrm{s}\bar{\mathrm{s}}\, f_{\mathrm{L}}}(t). \tag{4.77}$$

We have anticipated that this limit is time-independent, that is, $\widehat{Q}_{\mathrm{s}\bar{\mathrm{s}}\,\mathrm{L}}$ commutes with Hamiltonian (4.67). The verification of this claim follows the same lines as the proof that (4.23) holds. After computing the commutator between $\widehat{Q}_{\mathrm{s}\bar{\mathrm{s}}\,\mathrm{L}}$ and $\widehat{H}_{\mathrm{s}\bar{\mathrm{s}}}$ a multiplicative factor involving the difference between the single-particle energies of the time-dependent phase factor appears. The product of this difference with the pair of single-particle eigenfunctions it multiplies can be converted into the gradient of the matrix elements of the conserved Dirac current density in one-dimensional space. Partial integration moves this gradient to the test function. The space derivative of the test function converges to 0 as $f_{\mathrm{L}} \to 1$ uniformly by definition of the limit (4.77). We may safely interchange the limit $(\partial_x f_{\mathrm{L}}) \to 0$ with the space integral entering $\mathrm{d}\widehat{Q}_{\mathrm{s}\bar{\mathrm{s}}\,\mathrm{L}}/\mathrm{d}\,t$ when considering the matrix elements of the conserved Dirac current density involving at least one single-particle bound state. Interchanging the limit $(\partial_x f_{\mathrm{L}}) \to 0$ with the space integral entering $\mathrm{d}\widehat{Q}_{\mathrm{s}\bar{\mathrm{s}}\,\mathrm{L}}/\mathrm{d}\,t$ is not legal when considering the matrix elements of the conserved Dirac current density involving only Dirac plane waves. One must rely on flux conservation, TRS, and the smoothness of the function $(\partial_x f_{\mathrm{L}}) \to 0$ to show that the Dirac counterpart to Eq. (4.35) also vanishes in the uniform limit $f_{\mathrm{L}} \to 1$ and $(\partial_x f_{\mathrm{L}}) \to 0$. Along the same lines, one verifies that (**Exercise** 4.7)

$$\begin{aligned} \widehat{Q}_{\mathrm{s}\bar{\mathrm{s}}\,\mathrm{L}} =& \frac{1}{2} \left(\hat{c}^{\dagger}\, \hat{c} - \hat{d}^{\dagger}\, \hat{d} - \hat{c}^{\dagger}\, \hat{d}^{\dagger} - \hat{d}\, \hat{c} \right) \\ &+ \int \mathrm{d}p\, T_{\mathrm{L}} \left[\hat{c}^{\dagger}(p)\, \hat{c}(p) - \hat{d}^{\dagger}(p)\, \hat{d}(p) \right] \\ &+ \int \mathrm{d}p\, R_{\mathrm{L}} \left[\hat{c}^{\dagger}(p)\, \hat{c}(-p) - \hat{d}^{\dagger}(p)\, \hat{d}(-p) \right], \end{aligned} \tag{4.78a}$$

where T_{L} is real-valued and R_{L} is complex-valued. We observe that we can diagonalize the zero-mode sector by introducing the linear combinations

$$\hat{d}^{\dagger}_{\mathrm{s}\bar{\mathrm{s}}} := \frac{1}{\sqrt{2}} \left(\hat{d}^{\dagger} - \hat{c} \right) \qquad \hat{d}_{\mathrm{s}\bar{\mathrm{s}}} := \frac{1}{\sqrt{2}} \left(\hat{d} - \hat{c}^{\dagger} \right), \tag{4.78b}$$

and in terms of which

$$\begin{aligned}\widehat{Q}_{\mathrm{s}\bar{\mathrm{s}}\,\mathrm{L}} =& \frac{1}{2} - \hat{d}^{\dagger}_{\mathrm{s}\bar{\mathrm{s}}}\,\hat{d}_{\mathrm{s}\bar{\mathrm{s}}} \\ &+ \int \mathrm{d}p\, T_{\mathrm{L}} \left[\hat{c}^{\dagger}(p)\,\hat{c}(p) - \hat{d}^{\dagger}(p)\,\hat{d}(p)\right] \\ &+ \int \mathrm{d}p\, R_{\mathrm{L}} \left[\hat{c}^{\dagger}(p)\,\hat{c}(-p) - \hat{d}^{\dagger}(p)\,\hat{d}(-p)\right].\end{aligned} \tag{4.78c}$$

The states

$$|0,0\rangle_{\mathrm{s}\bar{\mathrm{s}}} \equiv |\mathrm{GS}\rangle_{\mathrm{s}\bar{\mathrm{s}}}, \tag{4.79a}$$

$$|0,\bar{1}\rangle_{\mathrm{s}\bar{\mathrm{s}}} \equiv \hat{d}^{\dagger}\,|\mathrm{GS}\rangle_{\mathrm{s}\bar{\mathrm{s}}}, \tag{4.79b}$$

$$|1,0\rangle_{\mathrm{s}\bar{\mathrm{s}}} \equiv \hat{c}^{\dagger}\,|\mathrm{GS}\rangle_{\mathrm{s}\bar{\mathrm{s}}}, \tag{4.79c}$$

$$|1,\bar{1}\rangle_{\mathrm{s}\bar{\mathrm{s}}} \equiv \hat{c}^{\dagger}\,\hat{d}^{\dagger}\,|\mathrm{GS}\rangle_{\mathrm{s}\bar{\mathrm{s}}}, \tag{4.79d}$$

are the fourfold degenerate ground states at $r=\infty$ obtained from tracking the four lowest energy eigenstates of Hamiltonian (4.67) in the limit $r\to\infty$. In particular, the ground state $|0,0\rangle_{\mathrm{s}\bar{\mathrm{s}}} \equiv |\mathrm{GS}\rangle_{\mathrm{s}\bar{\mathrm{s}}}$ is the one selected by tracking the state of Hamiltonian (4.67) by filling all the single-particle levels below the chemical potential $\mu=0$ in the limit $r\to\infty$. This basis for all the ground states at $r=\infty$ diagonalizes the Hamiltonian (4.67) and the total number operator (4.68), but not the smeared left charge number operator (4.78). In particular, the eigenvalues in the basis (4.79) for the operator obtained after normal ordering the total number operator (4.68) are the integers

$$0, \qquad -1, \qquad +1, \qquad 0, \tag{4.79e}$$

respectively. Because the Hamiltonian (4.67) commutes with the smeared left charge number operator (4.78) in the limit (4.77), both can be simultaneously diagonalized with the help of the Bogoliubov transformation

$$\widehat{U}_{\mathrm{s}\bar{\mathrm{s}}} := \exp\left(-\frac{\pi}{4}\left(\hat{c}\,\hat{d} + \hat{c}^{\dagger}\,\hat{d}^{\dagger}\right)\right), \qquad \hat{d}_{\mathrm{s}\bar{\mathrm{s}}} = \widehat{U}_{\mathrm{s}\bar{\mathrm{s}}}\,\hat{d}\,\widehat{U}^{\dagger}_{\mathrm{s}\bar{\mathrm{s}}}. \tag{4.80a}$$

The basis for all the ground states that diagonalizes simultaneously Hamiltonian (4.67), the total number operator (4.68) after normal ordering, and the smeared left charge number operator (4.78) in the limit (4.77) is

$$\begin{aligned}|\bar{0}\rangle_{\mathrm{s}\bar{\mathrm{s}}} &:= \widehat{U}_{\mathrm{s}\bar{\mathrm{s}}}\,|0,0\rangle_{\mathrm{s}\bar{\mathrm{s}}} = \frac{1}{\sqrt{2}}\left(|0,0\rangle_{\mathrm{s}\bar{\mathrm{s}}} - |1,\bar{1}\rangle_{\mathrm{s}\bar{\mathrm{s}}}\right), \\ |-1\rangle_{\mathrm{s}\bar{\mathrm{s}}} &:= \widehat{U}_{\mathrm{s}\bar{\mathrm{s}}}\,|0,\bar{1}\rangle_{\mathrm{s}\bar{\mathrm{s}}} = |0,\bar{1}\rangle_{\mathrm{s}\bar{\mathrm{s}}}, \\ |+1\rangle_{\mathrm{s}\bar{\mathrm{s}}} &:= \widehat{U}_{\mathrm{s}\bar{\mathrm{s}}}\,|1,0\rangle_{\mathrm{s}\bar{\mathrm{s}}} = |1,0\rangle_{\mathrm{s}\bar{\mathrm{s}}}, \\ |0\rangle_{\mathrm{s}\bar{\mathrm{s}}} &:= \widehat{U}_{\mathrm{s}\bar{\mathrm{s}}}\,|1,\bar{1}\rangle_{\mathrm{s}\bar{\mathrm{s}}} = \frac{1}{\sqrt{2}}\left(|0,0\rangle_{\mathrm{s}\bar{\mathrm{s}}} + |1,\bar{1}\rangle_{\mathrm{s}\bar{\mathrm{s}}}\right).\end{aligned} \tag{4.80b}$$

The eigenvalues in the basis (4.80b) of the smeared left charge number operator (4.78) are the rational numbers

$$+\frac{1}{2}, \qquad -\frac{1}{2}, \qquad +\frac{1}{2}, \qquad -\frac{1}{2}, \tag{4.80c}$$

respectively. They carry the normal-ordered total number operator eigenvalues

$$0, \qquad -1, \qquad +1, \qquad 0, \tag{4.80d}$$

respectively.

The smeared left charge number operator (4.78) is not diagonal in the basis (4.79),

$$\begin{aligned}
\widehat{Q}_{\mathrm{s\bar{s}\,L}}\,|0,0\rangle_{\mathrm{s\bar{s}}} &= -\frac{1}{2}\,|1,\bar{1}\rangle_{\mathrm{s\bar{s}}},\\
\widehat{Q}_{\mathrm{s\bar{s}\,L}}\,|0,\bar{1}\rangle_{\mathrm{s\bar{s}}} &= -\frac{1}{2}\,|0,\bar{1}\rangle_{\mathrm{s\bar{s}}},\\
\widehat{Q}_{\mathrm{s\bar{s}\,L}}\,|1,0\rangle_{\mathrm{s\bar{s}}} &= +\frac{1}{2}\,|1,0\rangle_{\mathrm{s\bar{s}}},\\
\widehat{Q}_{\mathrm{s\bar{s}\,L}}\,|1,\bar{1}\rangle_{\mathrm{s\bar{s}}} &= -\frac{1}{2}\,|0,0\rangle_{\mathrm{s\bar{s}}}.
\end{aligned} \tag{4.81}$$

The matrix elements of the smeared left charge number operator (4.78) in the basis (4.79) are

$$\begin{pmatrix}
0 & 0 & 0 & -1/2\\
0 & -1/2 & 0 & 0\\
0 & 0 & +1/2 & 0\\
-1/2 & 0 & 0 & 0
\end{pmatrix}. \tag{4.82}$$

The computation of the variance associated with ${}_{\mathrm{s\bar{s}}}\langle 0,0|\,\widehat{Q}_{\mathrm{s\bar{s}\,L}}\,|0,0\rangle_{\mathrm{s\bar{s}}}$ is now straightforward:

$$\begin{aligned}
{}_{\mathrm{s\bar{s}}}\langle 0,0|\,\widehat{Q}^{2}_{\mathrm{s\bar{s}\,L}}\,|0,0\rangle_{\mathrm{s\bar{s}}} &= \frac{1}{2}\left({}_{\mathrm{s\bar{s}}}\langle\bar{0}| + {}_{\mathrm{s\bar{s}}}\langle 0|\right)\widehat{Q}^{2}_{\mathrm{s\bar{s}\,L}}\left(|\bar{0}\rangle_{\mathrm{s\bar{s}}} + |0\rangle_{\mathrm{s\bar{s}}}\right)\\
&= \frac{1}{2}\left(\frac{1}{4}+\frac{1}{4}\right)\\
&= \frac{1}{4}
\end{aligned} \tag{4.83}$$

and ${}_{\mathrm{s\bar{s}}}\langle 0,0|\,\widehat{Q}_{\mathrm{s\bar{s}\,L}}\,|0,0\rangle_{\mathrm{s\bar{s}}} = 0$ implies

$$\begin{aligned}
\operatorname{var}\widehat{Q}_{\mathrm{s\bar{s}\,L}}\Big|_{|0,0\rangle_{\mathrm{s\bar{s}}}} &= {}_{\mathrm{s\bar{s}}}\langle 0,0|\,\widehat{Q}^{2}_{\mathrm{s\bar{s}\,L}}\,|0,0\rangle_{\mathrm{s\bar{s}}} - \left({}_{\mathrm{s\bar{s}}}\langle 0,0|\,\widehat{Q}_{\mathrm{s\bar{s}\,L}}\,|0,0\rangle_{\mathrm{s\bar{s}}}\right)^{2}\\
&= \frac{1}{4}.
\end{aligned} \tag{4.84a}$$

Similarly, (**Exercise** 4.8)

$$
\begin{aligned}
\operatorname{var}\widehat{Q}_{\mathrm{s\bar{s}\,L}}\Big|_{|0,\bar{1}\rangle_{\mathrm{s\bar{s}}}} &= {}_{\mathrm{s\bar{s}}}\langle 0,\bar{1}|\,\widehat{Q}^{2}_{\mathrm{s\bar{s}\,L}}\,|0,\bar{1}\rangle_{\mathrm{s\bar{s}}} - \left({}_{\mathrm{s\bar{s}}}\langle 0,\bar{1}|\,\widehat{Q}_{\mathrm{s\bar{s}\,L}}\,|0,\bar{1}\rangle_{\mathrm{s\bar{s}}}\right)^{2} \\
&= 0, \qquad (4.84b) \\
\operatorname{var}\widehat{Q}_{\mathrm{s\bar{s}\,L}}\Big|_{|1,0\rangle_{\mathrm{s\bar{s}}}} &= {}_{\mathrm{s\bar{s}}}\langle 1,0|\,\widehat{Q}^{2}_{\mathrm{s\bar{s}\,L}}\,|1,0\rangle_{\mathrm{s\bar{s}}} - \left({}_{\mathrm{s\bar{s}}}\langle 1,0|\,\widehat{Q}_{\mathrm{s\bar{s}\,L}}\,|1,0\rangle_{\mathrm{s\bar{s}}}\right)^{2} \\
&= 0, \qquad (4.84c) \\
\operatorname{var}\widehat{Q}_{\mathrm{s\bar{s}\,L}}\Big|_{|1,\bar{1}\rangle_{\mathrm{s\bar{s}}}} &= {}_{\mathrm{s\bar{s}}}\langle 1,\bar{1}|\,\widehat{Q}^{2}_{\mathrm{s\bar{s}\,L}}\,|1,\bar{1}\rangle_{\mathrm{s\bar{s}}} - \left({}_{\mathrm{s\bar{s}}}\langle 1,\bar{1}|\,\widehat{Q}_{\mathrm{s\bar{s}\,L}}\,|1,\bar{1}\rangle_{\mathrm{s\bar{s}}}\right)^{2} \\
&= \frac{1}{4}. \qquad (4.84d)
\end{aligned}
$$

These variances are to be contrasted with the variances

$$0 = {}_{\mathrm{s\bar{s}}}\langle\bar{0}|\,\widehat{Q}^{2}_{\mathrm{s\bar{s}\,L}}\,|\bar{0}\rangle_{\mathrm{s\bar{s}}} - \left({}_{\mathrm{s\bar{s}}}\langle\bar{0}|\,\widehat{Q}_{\mathrm{s\bar{s}\,L}}\,|\bar{0}\rangle_{\mathrm{s\bar{s}}}\right)^{2}, \qquad (4.85a)$$

$$0 = {}_{\mathrm{s\bar{s}}}\langle +1|\,\widehat{Q}^{2}_{\mathrm{s\bar{s}\,L}}\,|+1\rangle_{\mathrm{s\bar{s}}} - \left({}_{\mathrm{s\bar{s}}}\langle +1|\,\widehat{Q}_{\mathrm{s\bar{s}\,L}}\,|+1\rangle_{\mathrm{s\bar{s}}}\right)^{2}, \qquad (4.85b)$$

$$0 = {}_{\mathrm{s\bar{s}}}\langle -1|\,\widehat{Q}^{2}_{\mathrm{s\bar{s}\,L}}\,|-1\rangle_{\mathrm{s\bar{s}}} - \left({}_{\mathrm{s\bar{s}}}\langle -1|\,\widehat{Q}_{\mathrm{s\bar{s}\,L}}\,|-1\rangle_{\mathrm{s\bar{s}}}\right)^{2}, \qquad (4.85c)$$

$$0 = {}_{\mathrm{s\bar{s}}}\langle 0|\,\widehat{Q}^{2}_{\mathrm{s\bar{s}\,L}}\,|0\rangle_{\mathrm{s\bar{s}}} - \left({}_{\mathrm{s\bar{s}}}\langle 0|\,\widehat{Q}_{\mathrm{s\bar{s}\,L}}\,|0\rangle_{\mathrm{s\bar{s}}}\right)^{2}. \qquad (4.85d)$$

We compare in **Table** 4.1 the main results of Section 4.2 with those of Section 4.3. In either one of the two scaling limits (4.17) and (4.77), the Hamiltonian, total charge number operator, and the smeared left charge number operator can be diagonalized simultaneously.

To facilitate this comparison, we do the following particle-hole transformation on the generators of the Fock space for the double delta-function well. Let

$$\hat{c} := \hat{b}_{-}, \qquad \hat{d} := \hat{b}^{\dagger}_{+}. \qquad (4.86)$$

Under this particle-hole transformation, the non-diagonal representation of the left smeared charge operator (4.46a) becomes

$$
\begin{aligned}
\widehat{Q}_{\mathrm{dw\,L}} = &\frac{1}{2}\left(\hat{c}^{\dagger}\,\hat{c} - \hat{d}^{\dagger}\,\hat{d} + \hat{c}^{\dagger}\,\hat{d}^{\dagger} + \hat{d}\,\hat{c} + 1\right) \\
&+ \int \mathrm{d}p\,\left[T_{\mathrm{L}}\,\hat{c}^{\dagger}(p)\,\hat{c}(p) + R_{\mathrm{L}}\,\hat{c}^{\dagger}(p)\,\hat{c}(-p)\right].
\end{aligned}
\qquad (4.87)
$$

If we define the basis

$$
\begin{aligned}
|0,0\rangle_{\mathrm{dw}} &\equiv |1,0\rangle, \\
|0,\bar{1}\rangle_{\mathrm{dw}} &\equiv |0,0\rangle, \\
|1,0\rangle_{\mathrm{dw}} &\equiv |1,1\rangle, \\
|1,\bar{1}\rangle_{\mathrm{dw}} &\equiv |0,1\rangle,
\end{aligned}
\qquad (4.88a)
$$

where the basis $\{|0,0\rangle, |1,0\rangle, |0,1\rangle, |1,1\rangle\}$ for the fourfold degenerate ground states of the double delta-function potential well in the limit $r \to \infty$ was defined in Eq. (4.11), then the matrix elements of the smeared left charge number operator (4.87) in the basis (4.88a) are

$$\begin{pmatrix} 1/2 & 0 & 0 & 1/2 \\ 0 & 0 & 0 & 0 \\ 0 & 0 & 1 & 0 \\ 1/2 & 0 & 0 & 1/2 \end{pmatrix}. \tag{4.88b}$$

The smeared left charge number operator (4.87) has the integer-valued eigenvalues

$$0, \qquad 0, \qquad 1, \qquad 1, \tag{4.89a}$$

in the basis

$$\begin{aligned} |\bar{0}\rangle_{\mathrm{dw}} &:= \frac{1}{\sqrt{2}}\left(|0,0\rangle_{\mathrm{dw}} - |1,\bar{1}\rangle_{\mathrm{dw}}\right), \\ |-1\rangle_{\mathrm{dw}} &:= |0,\bar{1}\rangle_{\mathrm{dw}}, \\ |+1\rangle_{\mathrm{dw}} &:= |1,0\rangle_{\mathrm{dw}}, \\ |0\rangle_{\mathrm{dw}} &:= \frac{1}{\sqrt{2}}\left(|0,0\rangle_{\mathrm{dw}} + |1,\bar{1}\rangle_{\mathrm{dw}}\right). \end{aligned} \tag{4.89b}$$

What renders the case of the soliton and antisoliton ($\mathrm{s}\bar{\mathrm{s}}$) remarkable in comparison to the case of the double delta-function potential well (dw) is that the smeared left charge number operator acquires fractional eigenvalues in the former case. Hence, given the value $\mu = 0$ for the chemical potential, it is possible to select a ground state among the four degenerate ground states for the pair of soliton and antisoliton that is an eigenstate of the smeared left charge number operator carrying a fractional-valued quantum number as follows. It suffices to add the coupling $E\,\widehat{Q}_{\mathrm{s}\bar{\mathrm{s}}\,f_{\mathrm{L}}}$ to an applied external electric field E prior to taking the limit (4.77) and switching off this coupling after the limit (4.77) has been taken. In contrast, performing the same exercise for the dw case yields a ground state among the four degenerate ground states for the pair of delta-function wells that is an eigenstate of the smeared left charge number operator carrying an integer-valued quantum number.

According to Eqs. (4.78a) and (4.87), this difference arises from an apparently innocuous spectral shift by $-1/2$ for the $\mathrm{s}\bar{\mathrm{s}}$ case relative to the dw case up to a permutation of the four eigenvalues. This shift is all but trivial, however. This shift is only possible when the ground state supports the finite density

$$\nu_{\mathrm{e}} = \lim_{N,N_{\mathrm{e}}\to\infty} \frac{N_{\mathrm{e}}}{N} = \frac{1}{2} \tag{4.90}$$

of spinless fermions (electrons) for some underlying lattice regularization of the continuum where N is the number of lattice sites. Indeed, for the Dirac Hamiltonian at the chemical potential $\mu = 0$, the Fermi sea $|0,0\rangle_{\mathrm{s}\bar{\mathrm{s}}} \equiv |\mathrm{GS}\rangle_{\mathrm{s}\bar{\mathrm{s}}}$ in Eq. (4.79)

Table 4.1 *Measuring the charge*

	$:\widehat{H}_{s\bar{s}}:$	$:\widehat{Q}_{s\bar{s}}:$	$:\widehat{Q}_{s\bar{s}\,L}:$	$\widehat{H}_{dw} - \mu\,\widehat{Q}_{dw}$	$\widehat{Q}_{dw}$	$\widehat{Q}_{dw\,L}$
$\|\bar{0}\rangle_{s\bar{s}}$	0	0	+1/2			
$\|-1\rangle_{s\bar{s}}$	0	−1	−1/2			
$\|+1\rangle_{s\bar{s}}$	0	+1	+1/2			
$\|0\rangle_{s\bar{s}}$	0	0	−1/2			
$\|\bar{0}\rangle_{dw}$				0	1	0
$\|-1\rangle_{dw}$				0	0	0
$\|+1\rangle_{dw}$				0	2	1
$\|0\rangle_{dw}$				0	1	1

Comparison between the case of a double delta-function well (dw) and the case of a pair of soliton and antisoliton (s$\bar{\text{s}}$). For the dw, the second-quantized Hamiltonian, the total number operator, and the smeared charge operator centered on the left well are found in Eqs. (4.9), (4.10), and (4.46b), respectively. For the s$\bar{\text{s}}$, the second-quantized Hamiltonian, the total number operator, and the smeared charge operator centered on the left soliton are found in Eqs. (4.67), (4.68), and (4.78), respectively. All operators are acting on the Fock space for the grand-canonical ensemble and a suitable normal ordering is performed in order to remove infinities for the s$\bar{\text{s}}$ case. The chemical potential is always in between the two bound states localized about the double delta-function well for the dw case. The chemical potential is at zero energy for the s$\bar{\text{s}}$ case. The limits (4.17) and (4.77) are implied, respectively.

is the reference many-body state with respect to which the Hamiltonian, the total fermion number, and the smeared fermions numbers are normal ordered. This normal ordering delivers the prescription (4.72) that defines the smeared fermion number, as opposed to the prescription (4.68). This normal ordering dictates that the smeared fermion number be measured relative to the number "$N\,\nu_{\mathrm{e}}$" with the nonvanishing value $\nu_{\mathrm{e}} = 1/2$. For the double delta-function well, the density of spinless fermions in the state $|0,0\rangle_{\mathrm{dw}} = \hat{b}^{\dagger}_{+}\,|0\rangle$ is vanishing (we are considering at most two spinless electrons on the entire real line). The vanishing density $\nu_{\mathrm{e}} = 0$ for the double delta-function well is the origin for the additive number 1 in the sector of the bound states on the right-hand side of Eq. (4.87).

This difference between the s$\bar{\text{s}}$ and dw cases has the following interpretation in terms of the spectral flows induced by interpolating the separation between the localization centers of the bound states from zero to infinity that are depicted in Figure 4.4.

By decreasing the separation $r_{s\bar{s}}$ between the soliton and antisoliton localization centers, the two single-particle mid-gap states $\hat{b}^{\dagger}_{+}\,|0\rangle$ and $\hat{b}^{\dagger}_{-}\,|0\rangle$ migrate toward

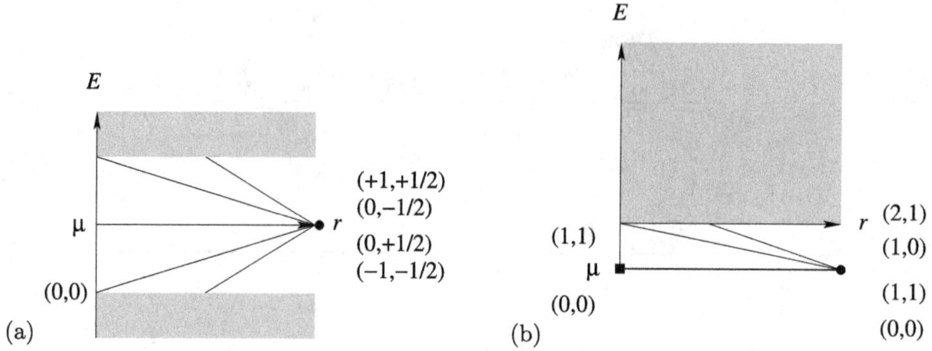

Figure 4.4 (a) One- and two-particles spectral flows E as a function of the separation r between two defects in the Dirac mass of a one-dimensional Dirac Hamiltonian of rank 2 obeying TRS and CHS, whereby the defects are domain walls. (b) Zero-, one- and two-particles spectral flows E as a function of the separation r between two delta-function potential wells for nonrelativistic spinless fermions. The chemical potential μ vanishes for (a). The chemical potential μ tracks the lowest eigenenergy of the single-particle bound states for (b). The two axis cross at $E = r = 0$ in cases (a) and (b). The black circle along the r axis represents the fourfold degenerate many-body ground states in the limit $r = \infty$ in cases (a) and (b). The black square along the r axis represents the twofold degenerate many-body ground states in the limit $r = 0$ in case (b). The continuum of single-particle excitations is depicted by the gray area. The quantum numbers for the total fermion number and the left smeared fermion number, respectively, are presented at $r = \infty$ and $r = 0$ as row vectors. In case (a), the total fermion number and the left smeared fermion number are normal ordered.

the continua of single-particle delocalized states with negative and positive single-particle eigenenergies, respectively. In the limit $r_{s\bar{s}} \to 0$, the soliton and antisoliton pair annihilate as is implied by the topological charge of opposite signs that they carry. The physical process for this annihilation is that the localization centers of the soliton and antisoliton meet precisely when the mid-gap states merge into the two single-particle continua separated by the pristine Dirac mass gap of the translation invariant state obtained by filling all single-particle states below the gap. Both the expectation value of the charge operator and of the left smeared charge operator is precisely $1/2$ in this reference state. This reference state could have equally well been chosen for the normal ordering. The fractional charges of magnitude $1/2$ for the smeared left charge operator when the soliton and antisoliton pair are infinitely apart can thus be interpreted as resulting from the spectral flow between the limiting spectra of two Dirac Hamiltonians with chiral symmetry, one having as nondegenerate ground state the nondegenerate ground state for the Dirac Hamiltonian with a pristine mass at half-filling, the other having the four-fold degenerate ground states from **Table** 4.1 for the Dirac Hamiltonian with a mass supporting two domain walls infinitely far apart.

Topology plays no role for the double well potential. In the limit $r \to 0$ when the separation r between the double delta-function potential wells is vanishing, the two delta-function potential wells merge into a single delta-function potential well with an interaction strength twice as large. The single-particle antibonding bound state merges into the continuum of plane waves precisely when $r = 0$. There remains one single-particle bound state and two degenerate many-body ground states when the chemical potential tracks the value of the bound state with the lowest single-particle eigenenergy. Both the total fermion number operator and the smeared charge operator centered about the localization center of the bound state that remains at $r = 0$ have the integer-valued eigenvalues 0 and 1 for these twofold degenerate ground states.

We deduce from the s$\bar{\text{s}}$ and dw examples that the two necessary conditions for the smeared left charge number operator to carry a fractional-valued quantum number are that (i) the ground state is a many-body ground state obtained by filling infinitely many single-particle states in the thermodynamic limit prior to taking the scaling limit (4.77) and (ii) that this ground state becomes degenerate with a finite number of excited states after the scaling limit (4.77) has been taken. Although the case of the dw meets condition (ii), it fails to meet condition (i) as the ground state has no more than two occupied single-particle states.

4.4 Example 3: The Dimer State

We proceed with a third example that is solved on the lattice. Let the pair of sites $i, i' = 1, \cdots, 2N$ from the lattice Λ be ordered, that is, $i < i'$. We can draw a (directed) bond between this pair of ordered sites as shown in Figure 4.5(a). Let q be any real number in the interval $[0,1] \subset \mathbb{R}$. We define the pair of (directed) "bond" operators [compare with Eq. (3.51b)]

$$\begin{aligned} \hat{\chi}^{\dagger}_{i,i';+}(q) &:= \left(\sqrt{q}\,\hat{c}^{\dagger}_{i} + \sqrt{1-q}\,\hat{c}^{\dagger}_{i'}\right) = +\hat{\chi}^{\dagger}_{i',i;+}(1-q), \\ \hat{\chi}^{\dagger}_{i,i';-}(q) &:= \left(\sqrt{1-q}\,\hat{c}^{\dagger}_{i} - \sqrt{q}\,\hat{c}^{\dagger}_{i'}\right) = -\hat{\chi}^{\dagger}_{i',i;-}(1-q). \end{aligned} \tag{4.91a}$$

When $q = 1/2$, $\hat{\chi}^{\dagger}_{i,i';+}(q)$ $[\hat{\chi}^{\dagger}_{i,i';-}(q)]$ creates a bonding (an antibonding) single-particle state shared by the sites i and i'. Operators such as $\hat{\chi}^{\dagger}_{i,i';\pm}(q)$ act on the 2^{2N}-dimensional Fock space $\mathfrak{F}_{\Lambda}$ spanned by the spinless fermion algebra

$$\{\hat{c}_{i}, \hat{c}^{\dagger}_{i'}\} = \delta_{i,i'}, \qquad \{\hat{c}^{\dagger}_{i}, \hat{c}^{\dagger}_{i'}\} = \{\hat{c}_{i}, \hat{c}_{i'}\} = 0, \tag{4.91b}$$

out of the vacuum state $|0\rangle$ annihilated by all $\hat{c}_i$'s. We shall also impose the periodic boundary conditions

$$\hat{c}_{i} = \hat{c}_{i+2N}. \tag{4.91c}$$

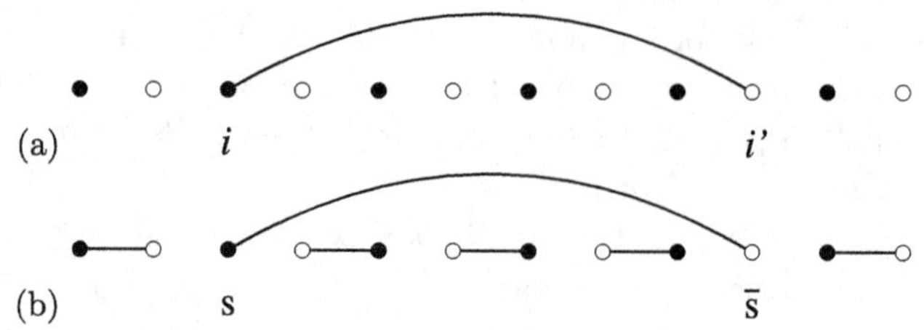

Figure 4.5 (a) A linear chain Λ made of the even number $2N$ of sites. Odd sites are colored in black. Even sites are colored in white. The ordered pair of sites $i < i' \in \Lambda$ is connected by a bond. (b) Pictorial representation of an itinerant quantum dimer model. A linear chain made of the even number $2N$ of sites along which fermions hop. The bonds connecting pair of sites represent the nonvanishing hopping amplitudes $t_{ii'}$ defined in Eq. (4.93). The odd site s (colored in black) represents a soliton. The even site $\bar{\mathrm{s}}$ (colored in white) represents an antisoliton.

We seek a noninteracting many-body Hamiltonian such that its normalized many-body ground state at half-filling can be written as the product state

$$|\Psi(q)\rangle := \prod_{j=1}^{N} \hat{\chi}^{\dagger}_{2j-1,2j;+}(q)\,|0\rangle, \qquad \forall q \in [0,1]. \tag{4.91d}$$

A solution is the many-body noninteracting Hamiltonian

$$\begin{aligned}\widehat{H}(q) =& -2t\sum_{j=1}^{N}\left[\hat{\chi}^{\dagger}_{2j-1,2j;+}(q)\;\hat{\chi}_{2j-1,2j;+}(q) - \hat{\chi}^{\dagger}_{2j-1,2j;-}(q)\;\hat{\chi}_{2j-1,2j;-}(q)\right]\\ =& -2t\sum_{j=1}^{N}\left[(2q-1)\,\hat{c}^{\dagger}_{2j-1}\,\hat{c}_{2j-1} - (2q-1)\,\hat{c}^{\dagger}_{2j}\,\hat{c}_{2j}\right]\\ & -2t\sum_{j=1}^{N}2\sqrt{q(1-q)}\left(\hat{c}^{\dagger}_{2j-1}\,\hat{c}_{2j} + \hat{c}^{\dagger}_{2j}\,\hat{c}_{2j-1}\right)\end{aligned} \tag{4.92}$$

for any $t>0$. For $q=1/2$, Hamiltonian (4.92) reduces to Hamiltonian (3.4) at the strong dimerization point (3.48), that is the direct sum (3.50). Hamiltonian (3.50) is diagonalized in Eq. (3.51). Its many-body ground state at half-filling is (3.54) when the number of sites entering in the definition of Hamiltonian (3.4) is even. Figure 3.4(a) gives a pictorial representation of the valence bond (dimer) ground state (4.91d). The same pictorial representation applies equally well to the many-body state (4.91d) which is the ground state of Hamiltonian (4.92).

Our next task is to give a quantum mechanical interpretation in terms of four degenerate many-body ground states labeled by the occupancy or vacancy of the monomers of Figures 3.4(b-d) when they are pulled infinitely apart. To this end, we work in the grand-canonical ensemble with the chemical potential μ set to zero, that is, the ground states are obtained by filling all strictly negative single-particle energy eigenstates. We choose two ordered site $i_{\mathrm{s}} < i_{\bar{\mathrm{s}}}$ from Λ such that i_{s} is odd (in units of the lattice spacing $\mathfrak{a}$ that we set to unity)

$$i_{\mathrm{s}} = 2\,j_{\mathrm{s}} - 1, \qquad j_{\mathrm{s}} = 1, \cdots, N, \tag{4.93a}$$

while $i_{\bar{\mathrm{s}}}$ is even:

$$i_{\bar{\mathrm{s}}} = 2\,j_{\bar{\mathrm{s}}}, \qquad j_{\bar{\mathrm{s}}} = 1, \cdots, N. \tag{4.93b}$$

The distance $r_{\mathrm{s}\bar{\mathrm{s}}}$ between the even site $i_{\bar{\mathrm{s}}}$ and the odd site i_{s} is an odd number

$$r_{\mathrm{s}\bar{\mathrm{s}}} := i_{\bar{\mathrm{s}}} - i_{\mathrm{s}} \equiv 2\,r' + 1, \qquad r' \equiv j_{\bar{\mathrm{s}}} - j_{\mathrm{s}} = 0, 1, \cdots, N-1. \tag{4.93c}$$

For $q = 1/2$, we define the noninteracting many-body Hamiltonian

$$\widehat{H}_{\mathrm{s}\bar{\mathrm{s}}}(1/2) := -\sum_{i<i'\in\Lambda} \left(\hat{c}^{\dagger}_{i}\, t_{ii'}\, \hat{c}_{i'} + \mathrm{H.c.}\right), \tag{4.93d}$$

where, for any ordered pair $i < i'$ from Λ,

$$t_{ii'} := \begin{cases} 2\,t\,\delta_{i',i+1}, & i = 2j-1, \quad j = 1, \cdots, j_{\mathrm{s}} - 1, \\ 0, & i = 2j, \quad j = 1, \cdots, j_{\mathrm{s}} - 1, \\ 2\,t\,e^{-8\,t\,r_{\mathrm{s}\bar{\mathrm{s}}}}\,\delta_{i',i+r_{\mathrm{s}\bar{\mathrm{s}}}}, & i = 2\,j_{\mathrm{s}} - 1, \\ 0, & i = 2j-1, \quad j = j_{\mathrm{s}} + 1, \cdots, j_{\bar{\mathrm{s}}}, \\ 2\,t\,\delta_{i',i+1}, & i = 2j, \quad j = j_{\mathrm{s}}, \cdots, j_{\bar{\mathrm{s}}} - 1, \\ 0, & i = 2\,j, \quad j = j_{\bar{\mathrm{s}}}, \cdots, N-1, \\ 2\,t\,\delta_{i',i+1}, & i = 2j-1, \quad j = j_{\bar{\mathrm{s}}} + 1, \cdots, N, \\ 0, & i = 2N, \end{cases} \tag{4.93e}$$

with $t > 0$. Each nonvanishing $t_{ii'}$ is assigned a bond in Figure 4.5(b). All bonds are between two sites of opposite parity (belonging to different sublattices). All bonds but one are between consecutive sites. The exception is the bond between site i_{s} and site $i_{\bar{\mathrm{s}}}$. We leave it as an excercise to the reader to generalize this construction so that the many-body Hamiltonian $\widehat{H}_{\mathrm{s}\bar{\mathrm{s}}}(q)$ at half filling is defined such that its many-body ground state is the product of all the bond operators $\hat{\chi}^{\dagger}_{i,i';+}(q)$ defined in Eq. (4.91a) that are labeled by the distinct ordered pair $i < i'$ corresponding to the bonds in Figure 4.5(b), whereby this product multiplies from the left the state annihilated by all $\hat{c}_i$ and $\hat{c}_{i'}$ with $i, i' = 1, \cdots, 2N$.

We seek four many-body eigenstates of $\widehat{H}_{\mathrm{s}\bar{\mathrm{s}}}(q)$ in the fermionic Fock space that become degenerate in the limit $r_{\mathrm{s}\bar{\mathrm{s}}} \to \infty$. We order those four states in ascending order with respect to their eigenenergies. These states are

$$
\begin{aligned}
|0,0\rangle_{\mathrm{s}\bar{\mathrm{s}}} &:= \hat{\chi}^{\dagger}_{i_{\mathrm{s}},i_{\bar{\mathrm{s}}};+}(q)\,|\Psi_{\mathrm{s}\bar{\mathrm{s}}}(q)\rangle, &(4.94\mathrm{a})\\
|0,\bar{1}\rangle_{\mathrm{s}\bar{\mathrm{s}}} &:= |\Psi_{\mathrm{s}\bar{\mathrm{s}}}(q)\rangle, &(4.94\mathrm{b})\\
|1,0\rangle_{\mathrm{s}\bar{\mathrm{s}}} &:= \hat{\chi}^{\dagger}_{i_{\mathrm{s}},i_{\bar{\mathrm{s}}};-}(q)\,\hat{\chi}^{\dagger}_{i_{\mathrm{s}},i_{\bar{\mathrm{s}}};+}(q)\,|\Psi_{\mathrm{s}\bar{\mathrm{s}}}(q)\rangle, &(4.94\mathrm{c})\\
|1,\bar{1}\rangle_{\mathrm{s}\bar{\mathrm{s}}} &:= \hat{\chi}^{\dagger}_{i_{\mathrm{s}},i_{\bar{\mathrm{s}}};-}(q)\,|\Psi_{\mathrm{s}\bar{\mathrm{s}}}(q)\rangle, &(4.94\mathrm{d})
\end{aligned}
$$

where

$$
\begin{aligned}
|\Psi_{\mathrm{s}\bar{\mathrm{s}}}(q)\rangle := &\left(\prod_{j=1}^{j_{\mathrm{s}}-1}\hat{\chi}^{\dagger}_{2j-1,2j;+}(q)\right)\\
&\times\left(\prod_{j=j_{\mathrm{s}}}^{j_{\mathrm{s}}-1+r'}\hat{\chi}^{\dagger}_{2j,2j+1;+}(q)\right)\\
&\times\left(\prod_{j=j_{\bar{\mathrm{s}}}+1}^{N}\hat{\chi}^{\dagger}_{2j-1,2j;+}(q)\right)|0\rangle
\end{aligned}
\tag{4.94e}
$$

is the many-body state with any one of the pair of sites i_{s} and $i_{\bar{\mathrm{s}}}$ being unoccupied with probability one. We have omitted to make explicit the q dependence of these four states on the left-hand side of Eqs. (4.94a)-(4.94d) since observables such as their eigenenergies or their total fermion number are independent of $q \in [0,1]$. The four many-body states (4.94a)-(4.94d) are product states. They are the lattice counterparts to the four states in Eq. (4.79). Their energy eigenvalues, measured relative to that of the ground state $|0,0\rangle_{\mathrm{s}\bar{\mathrm{s}}}$, are the single-particle and two-particle energies

$$
0, \qquad +4\,t\,e^{-8\,t\,r_{\mathrm{s}\bar{\mathrm{s}}}}, \qquad +4\,t\,e^{-8\,t\,r_{\mathrm{s}\bar{\mathrm{s}}}}, \qquad +8\,t\,e^{-8\,t\,r_{\mathrm{s}\bar{\mathrm{s}}}}, \tag{4.95}
$$

respectively. These all converge to zero exponentially fast as $r_{\mathrm{s}\bar{\mathrm{s}}} \to \infty$. The four states (4.94a)-(4.94d) are eigenstates of the total number operator

$$
\widehat{Q} := \sum_{i\in\Lambda}\hat{n}_i, \qquad \hat{n}_i := \hat{c}^{\dagger}_i\,\hat{c}_i. \tag{4.96}
$$

Their eigenvalues measured relative to N are

$$
0, \qquad -1, \qquad +1, \qquad 0, \tag{4.97}
$$

respectively. These eigenvalues agree with the eigenvalues (4.79e).

Let

$$
\begin{aligned}
f\colon \Lambda &\to [0,1],\\
i &\mapsto f_i,
\end{aligned}
\tag{4.98a}
$$

and

$$\begin{aligned} f^{s\bar{s}} : \Lambda &\to [0,1], \\ i &\mapsto f_i - \delta_{i,i_s} f_{i_s} - \delta_{i,i_{\bar{s}}} f_{i_{\bar{s}}}, \end{aligned} \tag{4.98b}$$

be a pair of test functions and let

$$\widehat{Q}_f := \sum_{i\in\Lambda} f_i \, \hat{n}_i = \widehat{Q}_{f^{s\bar{s}}} + \sum_{i\in\{i_s, i_{\bar{s}}\}} f_i \, \hat{n}_i \tag{4.98c}$$

with

$$\widehat{Q}_{f^{s\bar{s}}} := \sum_{i\in\Lambda} f^{s\bar{s}}_i \, \hat{n}_i = \sum_{i\in\Lambda\backslash\{i_s, i_{\bar{s}}\}} f_i \, \hat{n}_i \tag{4.98d}$$

be the corresponding pair of smeared charge operators. With the help of

$$|0,0\rangle_{s\bar{s}} = \left(\sqrt{q}\, \hat{c}^{\dagger}_{i_s} + \sqrt{1-q}\, \hat{c}^{\dagger}_{i_{\bar{s}}}\right) |\Psi_{s\bar{s}}(q)\rangle, \tag{4.99a}$$

$$|0,\bar{1}\rangle_{s\bar{s}} = |\Psi_{s\bar{s}}(q)\rangle, \tag{4.99b}$$

$$|1,0\rangle_{s\bar{s}} = \hat{c}^{\dagger}_{i_s} \hat{c}^{\dagger}_{i_{\bar{s}}} |\Psi_{s\bar{s}}(q)\rangle, \tag{4.99c}$$

$$|1,\bar{1}\rangle_{s\bar{s}} = \left(\sqrt{1-q}\, \hat{c}^{\dagger}_{i_s} - \sqrt{q}\, \hat{c}^{\dagger}_{i_{\bar{s}}}\right) |\Psi_{s\bar{s}}(q)\rangle, \tag{4.99d}$$

one verifies that

$$\begin{aligned} \widehat{Q}_f \, |0,0\rangle_{s\bar{s}} = &\left(\sqrt{q}\, \hat{c}^{\dagger}_{i_s} + \sqrt{1-q}\, \hat{c}^{\dagger}_{i_{\bar{s}}}\right) \widehat{Q}_{f^{s\bar{s}}} \, |\Psi_{s\bar{s}}(q)\rangle \\ &+ \left(\sqrt{q}\, f_{i_s} \hat{c}^{\dagger}_{i_s} + \sqrt{1-q}\, f_{i_{\bar{s}}} \hat{c}^{\dagger}_{i_{\bar{s}}}\right) |\Psi_{s\bar{s}}(q)\rangle, \end{aligned} \tag{4.100a}$$

$$\widehat{Q}_f \, |0,\bar{1}\rangle_{s\bar{s}} = \widehat{Q}_{f^{s\bar{s}}} \, |\Psi_{s\bar{s}}(q)\rangle, \tag{4.100b}$$

$$\widehat{Q}_f \, |1,0\rangle_{s\bar{s}} = + \hat{c}^{\dagger}_{i_s} \hat{c}^{\dagger}_{i_{\bar{s}}} \widehat{Q}_{f^{s\bar{s}}} \, |\Psi_{s\bar{s}}(q)\rangle + (f_{i_s} + f_{i_{\bar{s}}})\, \hat{c}^{\dagger}_{i_s} \hat{c}^{\dagger}_{i_{\bar{s}}} \, |\Psi_{s\bar{s}}(q)\rangle, \tag{4.100c}$$

$$\begin{aligned} \widehat{Q}_f \, |1,\bar{1}\rangle_{s\bar{s}} = &\left(\sqrt{1-q}\, \hat{c}^{\dagger}_{i_s} - \sqrt{q}\, \hat{c}^{\dagger}_{i_{\bar{s}}}\right) \widehat{Q}_{f^{s\bar{s}}} \, |\Psi_{s\bar{s}}(q)\rangle \\ &+ \left(\sqrt{1-q}\, f_{i_s} \hat{c}^{\dagger}_{i_s} - \sqrt{q}\, f_{i_{\bar{s}}} \hat{c}^{\dagger}_{i_{\bar{s}}}\right) |\Psi_{s\bar{s}}(q)\rangle, \end{aligned} \tag{4.100d}$$

is the action of the smeared charge operator on the four lowest energy eigenstates (4.99). The two-dimensional subspace spanned by the orthonormal basis (4.99a) and (4.99d) is mixed under the action of $\widehat{Q}_f$ since states (4.100a) and (4.100d) are not orthogonal for a generic test function.

A nice consequence of Eq. (4.100) is that the expectation value of the smeared charge operator on the four lowest energy eigenstates is the sum of the following two terms,

$$
\begin{aligned}
{}_{\mathrm{s}\bar{\mathrm{s}}}\langle 0,0|\,\widehat{Q}_f\,|0,0\rangle_{\mathrm{s}\bar{\mathrm{s}}} &= \mathcal{N}_{f\mathrm{s}\bar{\mathrm{s}}} + \left[q\,f_{i_{\mathrm{s}}} + (1-q)\,f_{i_{\bar{\mathrm{s}}}}\right], && (4.101\mathrm{a})\\
{}_{\mathrm{s}\bar{\mathrm{s}}}\langle 0,\bar{1}|\,\widehat{Q}_f\,|0,\bar{1}\rangle_{\mathrm{s}\bar{\mathrm{s}}} &= \mathcal{N}_{f\mathrm{s}\bar{\mathrm{s}}}, && (4.101\mathrm{b})\\
{}_{\mathrm{s}\bar{\mathrm{s}}}\langle 1,0|\,\widehat{Q}_f\,|1,0\rangle_{\mathrm{s}\bar{\mathrm{s}}} &= \mathcal{N}_{f\mathrm{s}\bar{\mathrm{s}}} + (f_{i_{\mathrm{s}}} + f_{i_{\bar{\mathrm{s}}}}), && (4.101\mathrm{c})\\
{}_{\mathrm{s}\bar{\mathrm{s}}}\langle 1,\bar{1}|\,\widehat{Q}_f\,|1,\bar{1}\rangle_{\mathrm{s}\bar{\mathrm{s}}} &= \mathcal{N}_{f\mathrm{s}\bar{\mathrm{s}}} + \left[(1-q)\,f_{i_{\mathrm{s}}} + q\,f_{i_{\bar{\mathrm{s}}}}\right], && (4.101\mathrm{d})
\end{aligned}
$$

where the shorthand notation

$$
\mathcal{N}_{f\mathrm{s}\bar{\mathrm{s}}} := \langle \Psi_{\mathrm{s}\bar{\mathrm{s}}}(q)|\,\widehat{Q}_{f\mathrm{s}\bar{\mathrm{s}}}\,|\Psi_{\mathrm{s}\bar{\mathrm{s}}}(q)\rangle \tag{4.101e}
$$

is used and

$$
\begin{aligned}
\mathcal{N}_{f\mathrm{s}\bar{\mathrm{s}}} = {} & \sum_{j=1}^{j_{\mathrm{s}}-1} \left[q\,f_{2j-1} + (1-q)\,f_{2j}\right] \\
& + \sum_{j=j_{\mathrm{s}}}^{j_{\mathrm{s}}-1+r'} \left[q\,f_{2j} + (1-q)\,f_{2j+1}\right] \\
& + \sum_{j=j_{\bar{\mathrm{s}}}+1}^{N} \left[q\,f_{2j-1} + (1-q)\,f_{2j}\right].
\end{aligned} \tag{4.101f}
$$

Define normal ordering of $\widehat{Q}_f$ by subtracting from $\widehat{Q}_f$ its expectation value in the ground state $|0,0\rangle_{\mathrm{s}\bar{\mathrm{s}}}$, that is,

$$
:\widehat{Q}_f:\equiv \widehat{Q}_f - {}_{\mathrm{s}\bar{\mathrm{s}}}\langle 0,0|\,\widehat{Q}_f\,|0,0\rangle_{\mathrm{s}\bar{\mathrm{s}}}. \tag{4.102a}
$$

With this definition,

$$
\begin{aligned}
{}_{\mathrm{s}\bar{\mathrm{s}}}\langle 0,0|\,:\widehat{Q}_f:\,|0,0\rangle_{\mathrm{s}\bar{\mathrm{s}}} &= 0, && (4.102\mathrm{b})\\
{}_{\mathrm{s}\bar{\mathrm{s}}}\langle 0,\bar{1}|\,:\widehat{Q}_f:\,|0,\bar{1}\rangle_{\mathrm{s}\bar{\mathrm{s}}} &= -\left[q\,f_{i_{\mathrm{s}}} + (1-q)\,f_{i_{\bar{\mathrm{s}}}}\right], && (4.102\mathrm{c})\\
{}_{\mathrm{s}\bar{\mathrm{s}}}\langle 1,0|\,:\widehat{Q}_f:\,|1,0\rangle_{\mathrm{s}\bar{\mathrm{s}}} &= +\left[(1-q)\,f_{i_{\mathrm{s}}} + q\,f_{i_{\bar{\mathrm{s}}}}\right], && (4.102\mathrm{d})\\
{}_{\mathrm{s}\bar{\mathrm{s}}}\langle 1,\bar{1}|\,:\widehat{Q}_f:\,|1,\bar{1}\rangle_{\mathrm{s}\bar{\mathrm{s}}} &= +(1-2\,q)\left(f_{i_{\mathrm{s}}} - f_{i_{\bar{\mathrm{s}}}}\right). && (4.102\mathrm{e})
\end{aligned}
$$

As it should be, we recover Eq. (4.97) when $f_{i_{\mathrm{s}}} = f_{i_{\bar{\mathrm{s}}}}$ for any value of $0 \leq q \leq 1$. The diagonal matrix elements of the matrix (4.82) are recovered if $q = 1/2$, $f_{i_{\mathrm{s}}} = 1$, and $f_{i_{\bar{\mathrm{s}}}} = 0$. When $q = 1/2$, the normal ordering (4.102a) is equivalent to subtracting the local background charge $\nu_{\mathrm{e}} = 1/2$ from $\hat{n}_i$ in the definitions (4.96) and (4.98). When $0 \leq q \leq 1$, the normal ordering (4.102a) is equivalent to subtracting (i) the staggered local background charges q from $\hat{n}_{2j-1}$ and $(1-q)$ from $\hat{n}_{2j}$ for $j = 1, \cdots, j_{\mathrm{s}} - 1$ and $j = j_{\bar{\mathrm{s}}} + 1, \cdots, N$; (ii) the staggered local background charges $(1-q)$ from $\hat{n}_{2j-1}$ and q from $\hat{n}_{2j}$ for $j = j_{\mathrm{s}}, \cdots, j_{\mathrm{s}} - 1 + r' \equiv j_{\bar{\mathrm{s}}} - 1$; and (iii) the staggered local background charges q from $\hat{n}_{i_{\mathrm{s}}}$ and $(1-q)$ from $\hat{n}_{\bar{\mathrm{s}}}$ in the definitions (4.96) and (4.98).

For completeness, we give the off-diagonal matrix elements of $\widehat{Q}_f$ in the basis (4.99). We note that the many-body states (4.99b) and (4.99c) are orthonormal, for they differ through their total fermion number. They also differ through their total fermion number with the many-body states (4.99a) and (4.99d). Hence, the only nonvanishing off-diagonal matrix elements of $\widehat{Q}_f$ are between the many-body states (4.99a) and (4.99d). However, the bonding $\hat{\chi}^{\dagger}_{i_{\mathrm{s}},i_{\bar{\mathrm{s}}};+}(q)\,|0\rangle$ and antibonding $\hat{\chi}^{\dagger}_{i_{\mathrm{s}},i_{\bar{\mathrm{s}}};-}(q)\,|0\rangle$ single-particle states are orthogonal. Hence, we find no dependence on $\mathcal{N}_{f^{\mathrm{s}\bar{\mathrm{s}}}}$ in the off-diagonal matrix elements

$$
{}_{\mathrm{s}\bar{\mathrm{s}}}\langle 0,0|\,\widehat{Q}_f\,|1,\bar{1}\rangle_{\mathrm{s}\bar{\mathrm{s}}} = \langle 1,\bar{1}|\,\widehat{Q}_f\,|0,0\rangle_{\mathrm{s}\bar{\mathrm{s}}} = \sqrt{q\,(1-q)}\,\left(f_{i_{\mathrm{s}}} - f_{i_{\bar{\mathrm{s}}}}\right). \tag{4.103}
$$

They vanish when $f_{i_{\mathrm{s}}} = f_{i_{\bar{\mathrm{s}}}}$, as it should be.

We choose the left test function

$$
\begin{aligned}
f_{\mathrm{L}} : \Lambda &\to [0,1],\\
i &\mapsto f_{\mathrm{L}\,i},
\end{aligned} \tag{4.104a}
$$

such that

$$
f_{\mathrm{L}\,i_{\mathrm{s}}} = 1, \qquad f_{\mathrm{L}\,i_{\bar{\mathrm{s}}}} = 0. \tag{4.104b}
$$

The corresponding left smeared charge operator is

$$
\widehat{Q}_{f_{\mathrm{L}}} := \sum_{i\in\Lambda} f_{\mathrm{L}\,i}\,\hat{n}_i \equiv \widehat{Q}_{f^{\mathrm{s}\bar{\mathrm{s}}}_{\mathrm{L}}} + \hat{n}_{i_{\mathrm{s}}}. \tag{4.104c}
$$

With the help of (4.100), the action of the left smeared charge operator on the four lowest energy eigenstates (4.99) is

$$
\begin{aligned}
\widehat{Q}_{f_{\mathrm{L}}}\,|0,0\rangle_{\mathrm{s}\bar{\mathrm{s}}} &= \left(\sqrt{q}\,\hat{c}^{\dagger}_{i_{\mathrm{s}}} + \sqrt{1-q}\,\hat{c}^{\dagger}_{i_{\bar{\mathrm{s}}}}\right)\widehat{Q}_{f^{\mathrm{s}\bar{\mathrm{s}}}_{\mathrm{L}}}\,|\Psi_{\mathrm{s}\bar{\mathrm{s}}}(q)\rangle + \sqrt{q}\,\hat{c}^{\dagger}_{i_{\mathrm{s}}}\,|\Psi_{\mathrm{s}\bar{\mathrm{s}}}(q)\rangle, && (4.105\mathrm{a})\\
\widehat{Q}_{f_{\mathrm{L}}}\,|0,\bar{1}\rangle_{\mathrm{s}\bar{\mathrm{s}}} &= \widehat{Q}_{f^{\mathrm{s}\bar{\mathrm{s}}}_{\mathrm{L}}}\,|\Psi_{\mathrm{s}\bar{\mathrm{s}}}(q)\rangle, && (4.105\mathrm{b})\\
\widehat{Q}_{f_{\mathrm{L}}}\,|1,0\rangle_{\mathrm{s}\bar{\mathrm{s}}} &= \hat{c}^{\dagger}_{i_{\mathrm{s}}}\,\hat{c}^{\dagger}_{i_{\bar{\mathrm{s}}}}\,\widehat{Q}_{f^{\mathrm{s}\bar{\mathrm{s}}}_{\mathrm{L}}}\,|\Psi_{\mathrm{s}\bar{\mathrm{s}}}(q)\rangle + \hat{c}^{\dagger}_{i_{\mathrm{s}}}\,\hat{c}^{\dagger}_{i_{\bar{\mathrm{s}}}}\,|\Psi_{\mathrm{s}\bar{\mathrm{s}}}(q)\rangle, && (4.105\mathrm{c})\\
\widehat{Q}_{f_{\mathrm{L}}}\,|1,\bar{1}\rangle_{\mathrm{s}\bar{\mathrm{s}}} &= \left(\sqrt{1-q}\,\hat{c}^{\dagger}_{i_{\mathrm{s}}} - \sqrt{q}\,\hat{c}^{\dagger}_{i_{\bar{\mathrm{s}}}}\right)\widehat{Q}_{f^{\mathrm{s}\bar{\mathrm{s}}}_{\mathrm{L}}}\,|\Psi_{\mathrm{s}\bar{\mathrm{s}}}(q)\rangle + \sqrt{1-q}\,\hat{c}^{\dagger}_{i_{\mathrm{s}}}\,|\Psi_{\mathrm{s}\bar{\mathrm{s}}}(q)\rangle. && (4.105\mathrm{d})
\end{aligned}
$$

The first application of Eq. (4.105) delivers the matrix elements of $\widehat{Q}_{f_{\mathrm{L}}}$ in the basis (4.99). They are [compare with Eqs. (4.101) and (4.103)]

$$
\begin{pmatrix}
\mathcal{N}_{f^{\mathrm{s}\bar{\mathrm{s}}}_{\mathrm{L}}} + q & 0 & 0 & +\sqrt{q\,(1-q)}\\
0 & \mathcal{N}_{f^{\mathrm{s}\bar{\mathrm{s}}}_{\mathrm{L}}} & 0 & 0\\
0 & 0 & \mathcal{N}_{f^{\mathrm{s}\bar{\mathrm{s}}}_{\mathrm{L}}} + 1 & 0\\
+\sqrt{q\,(1-q)} & 0 & 0 & \mathcal{N}_{f^{\mathrm{s}\bar{\mathrm{s}}}_{\mathrm{L}}} + 1 - q
\end{pmatrix} \tag{4.106a}
$$

with

$$\mathcal{N}_{f_{\mathrm{L}}^{\mathrm{s}\bar{\mathrm{s}}}} := \langle\Psi_{\mathrm{s}\bar{\mathrm{s}}}(q)|\,\widehat{Q}_{f_{\mathrm{L}}^{\mathrm{s}\bar{\mathrm{s}}}}\,|\Psi_{\mathrm{s}\bar{\mathrm{s}}}(q)\rangle. \tag{4.106b}$$

Normal ordering with respect to $\mathcal{N}_{f_{\mathrm{L}}^{\mathrm{s}\bar{\mathrm{s}}}}+q$ removes any dependence on the left test function from the matrix elements of $:\widehat{Q}_{f_{\mathrm{L}}}:$ in the basis (4.94):

$$\begin{pmatrix} 0 & 0 & 0 & +\sqrt{q\,(1-q)} \\ 0 & -q & 0 & 0 \\ 0 & 0 & 1-q & 0 \\ +\sqrt{q\,(1-q)} & 0 & 0 & 1-2\,q \end{pmatrix}. \tag{4.107}$$

If we do the gauge transformation

$$\hat{\chi}^{\dagger}_{i_{\mathrm{s}},i_{\bar{\mathrm{s}}};+}(q)\mapsto\hat{\chi}^{\dagger}_{i_{\mathrm{s}},i_{\bar{\mathrm{s}}};+}(q),\qquad \hat{\chi}^{\dagger}_{i_{\mathrm{s}},i_{\bar{\mathrm{s}}};-}(q)\mapsto-\hat{\chi}^{\dagger}_{i_{\mathrm{s}},i_{\bar{\mathrm{s}}};-}(q), \tag{4.108}$$

and choose $q=1/2$, we find the same matrix as in Eq. (4.82). The basis that diagonalizes the representation (4.107) of $:\widehat{Q}_{f_{\mathrm{L}}}:$ also diagonalizes the Hamiltonian defined by Eq. (4.93) after the limit $r_{\mathrm{s}\bar{\mathrm{s}}}\to\infty$ has been taken. The eigenvalues of the matrix (4.107) are

$$1-q,\qquad -q,\qquad 1-q\qquad -q, \tag{4.109a}$$

respectively. The eigenvectors of the matrix (4.107) are

$$\begin{pmatrix} \sqrt{q} \\ 0 \\ 0 \\ +\sqrt{1-q} \end{pmatrix},\quad \begin{pmatrix} 0 \\ 1 \\ 0 \\ 0 \end{pmatrix},\quad \begin{pmatrix} 0 \\ 0 \\ 1 \\ 0 \end{pmatrix},\quad \begin{pmatrix} \sqrt{1-q} \\ 0 \\ 0 \\ -\sqrt{q} \end{pmatrix}, \tag{4.109b}$$

respectively.

The second application of Eq. (4.105) delivers the variance of $\widehat{Q}_{f_{\mathrm{L}}}$ in the basis (4.94). These four variances can all be expressed as the sum of two contributions, the first one being the variance

$$\begin{aligned} \operatorname{var}\widehat{Q}_{f_{\mathrm{L}}^{\mathrm{s}\bar{\mathrm{s}}}}\Big|_{\Psi_{\mathrm{s}\bar{\mathrm{s}}}(q)} &:= \langle\Psi_{\mathrm{s}\bar{\mathrm{s}}}(q)|\left(\widehat{Q}_{f_{\mathrm{L}}^{\mathrm{s}\bar{\mathrm{s}}}}-\mathcal{N}_{f_{\mathrm{L}}^{\mathrm{s}\bar{\mathrm{s}}}}\right)^{2}|\Psi_{\mathrm{s}\bar{\mathrm{s}}}(q)\rangle \\ &= \langle\Psi_{\mathrm{s}\bar{\mathrm{s}}}(q)|\,\widehat{Q}_{f_{\mathrm{L}}^{\mathrm{s}\bar{\mathrm{s}}}}\,\widehat{Q}_{f_{\mathrm{L}}^{\mathrm{s}\bar{\mathrm{s}}}}\,|\Psi_{\mathrm{s}\bar{\mathrm{s}}}(q)\rangle-\left(\mathcal{N}_{f_{\mathrm{L}}^{\mathrm{s}\bar{\mathrm{s}}}}\right)^{2}, \end{aligned} \tag{4.110}$$

where $\mathcal{N}_{f_{\mathrm{L}}^{\mathrm{s}\bar{\mathrm{s}}}}$ was defined in Eq. (4.106b). The variances of the left smeared charge operator in the basis (4.94) are

$$\operatorname{var}\widehat{Q}_{f_{\mathrm{L}}}\Big|_{|0,0\rangle_{\mathrm{s\bar{s}}}} := {}_{\mathrm{s\bar{s}}}\langle 0,0|\left(\widehat{Q}_{f_{\mathrm{L}}} - {}_{\mathrm{s\bar{s}}}\langle 0,0|\,\widehat{Q}_{f_{\mathrm{L}}}\,|0,0\rangle_{\mathrm{s\bar{s}}}\right)^2 |0,0\rangle_{\mathrm{s\bar{s}}}$$
$$= \operatorname{var}\widehat{Q}_{f^{\mathrm{s\bar{s}}}_{\mathrm{L}}}\Big|_{\Psi_{\mathrm{s\bar{s}}}(q)} + q\,(1-q), \tag{4.111a}$$

$$\operatorname{var}\widehat{Q}_{f_{\mathrm{L}}}\Big|_{|0,\bar{1}\rangle_{\mathrm{s\bar{s}}}} := {}_{\mathrm{s\bar{s}}}\langle 0,\bar{1}|\left(\widehat{Q}_{f_{\mathrm{L}}} - {}_{\mathrm{s\bar{s}}}\langle 0,\bar{1}|\,\widehat{Q}_{f_{\mathrm{L}}}\,|0,\bar{1}\rangle_{\mathrm{s\bar{s}}}\right)^2 |0,\bar{1}\rangle_{\mathrm{s\bar{s}}}$$
$$= \operatorname{var}\widehat{Q}_{f^{\mathrm{s\bar{s}}}_{\mathrm{L}}}\Big|_{\Psi_{\mathrm{s\bar{s}}}(q)}, \tag{4.111b}$$

$$\operatorname{var}\widehat{Q}_{f_{\mathrm{L}}}\Big|_{|1,0\rangle_{\mathrm{s\bar{s}}}} := {}_{\mathrm{s\bar{s}}}\langle 1,0|\left(\widehat{Q}_{f_{\mathrm{L}}} - {}_{\mathrm{s\bar{s}}}\langle 1,0|\,\widehat{Q}_{f_{\mathrm{L}}}\,|1,0\rangle_{\mathrm{s\bar{s}}}\right)^2 |1,0\rangle_{\mathrm{s\bar{s}}}$$
$$= \operatorname{var}\widehat{Q}_{f^{\mathrm{s\bar{s}}}_{\mathrm{L}}}\Big|_{\Psi_{\mathrm{s\bar{s}}}(q)}, \tag{4.111c}$$

$$\operatorname{var}\widehat{Q}_{f_{\mathrm{L}}}\Big|_{|1,\bar{1}\rangle_{\mathrm{s\bar{s}}}} := {}_{\mathrm{s\bar{s}}}\langle 1,\bar{1}|\left(\widehat{Q}_{f_{\mathrm{L}}} - {}_{\mathrm{s\bar{s}}}\langle 1,\bar{1}|\,\widehat{Q}_{f_{\mathrm{L}}}\,|1,\bar{1}\rangle_{\mathrm{s\bar{s}}}\right)^2 |1,\bar{1}\rangle_{\mathrm{s\bar{s}}}$$
$$= \operatorname{var}\widehat{Q}_{f^{\mathrm{s\bar{s}}}_{\mathrm{L}}}\Big|_{\Psi_{\mathrm{s\bar{s}}}(q)} + (1-q)\,q, \tag{4.111d}$$

respectively. In turn, all four variances of the left smeared charge operator in the basis that diagonalizes the representation (4.107) of $:\widehat{Q}_{f_{\mathrm{L}}}:$ are equal to the variance (4.110) of the smeared charge operator with respect to the dimer state for a chain with $2(N-1)$ sites. By subtracting the variance (4.110) from Eq. (4.111) and selecting $q = 1/2$, we recover the same (normal-ordered) variances (4.84) as for the treatment we made of the left smeared charge operator for the one-dimensional Dirac Hamiltonian with a pair of soliton and antisoliton.

We need to evaluate (4.110) in order to complete the computation of the variances (4.111). We shall do this exercise for a generic test function (4.98a). If we insert the definition (4.98c) for the smeared charge operator together with

$$\operatorname{var}\widehat{Q}_{f^{\mathrm{s\bar{s}}}}|_{\Psi_{\mathrm{s\bar{s}}}(q)} = \operatorname{var}\widehat{Q}_{f}|_{\Psi_{\mathrm{s\bar{s}}}(q)}, \tag{4.112}$$

we obtain

$$\begin{aligned}\operatorname{var}\widehat{Q}_{f}\Big|_{\Psi_{\mathrm{s\bar{s}}}(q)} = \sum_{i,i'\in\Lambda}\Big(& f_i\, f_{i'}\,\langle\Psi_{\mathrm{s\bar{s}}}(q)|\,\hat{n}_i\,\hat{n}_{i'}|\Psi_{\mathrm{s\bar{s}}}(q)\rangle \\ & - f_i\, f_{i'}\,\langle\Psi_{\mathrm{s\bar{s}}}(q)|\,\hat{n}_i|\Psi_{\mathrm{s\bar{s}}}(q)\rangle\,\langle\Psi_{\mathrm{s\bar{s}}}(q)|\,\hat{n}_{i'}|\Psi_{\mathrm{s\bar{s}}}(q)\rangle\Big).\end{aligned} \tag{4.113}$$

The variance (4.113) requires computing the connected correlators

$$\begin{aligned}\langle\,\hat{n}_i\,\hat{n}_{i'}\rangle_{\mathrm{c}} &\equiv \langle\,\hat{n}_i\,\hat{n}_{i'}\rangle - \langle\,\hat{n}_i\rangle\,\langle\,\hat{n}_{i'}\rangle \\ &\equiv \langle\Psi_{\mathrm{s\bar{s}}}(q)|\,\hat{n}_i\,\hat{n}_{i'}|\Psi_{\mathrm{s\bar{s}}}(q)\rangle - \langle\Psi_{\mathrm{s\bar{s}}}(q)|\,\hat{n}_i|\Psi_{\mathrm{s\bar{s}}}(q)\rangle\,\langle\Psi_{\mathrm{s\bar{s}}}(q)|\,\hat{n}_{i'}|\Psi_{\mathrm{s\bar{s}}}(q)\rangle\end{aligned} \tag{4.114}$$

for any pair of sites $i \leq i' \in \Lambda$. To this end, useful identities are

$$\hat{n}_i^2 = \hat{n}_i, \tag{4.115a}$$
$$\hat{n}_i\,|0\rangle = 0, \tag{4.115b}$$
$$\hat{n}_i\,\hat{\chi}^{\dagger}_{j,j';+}(q)|0\rangle = \delta_{i,j}\,\sqrt{q}\,\hat{c}^{\dagger}_{j}\,|0\rangle + \delta_{i,j'}\,\sqrt{1-q}\,\hat{c}^{\dagger}_{j'}\,|0\rangle, \tag{4.115c}$$
$$\hat{n}_i\,\hat{\chi}^{\dagger}_{i,i';-}(q)|0\rangle = \delta_{i,j}\,\sqrt{1-q}\,\hat{c}^{\dagger}_{j}\,|0\rangle - \delta_{i,j'}\,\sqrt{q}\,\hat{c}^{\dagger}_{j'}\,|0\rangle, \tag{4.115d}$$

for any pair of ordered sites $j < j' \in \Lambda$ (recall that $\hat{n}_i$ has the eigenvalues $n_i = 0, 1$). The connected correlators (4.114) vanish unless (i) $i = i' \in \Lambda\backslash\{i_{\mathrm{s}}, i_{\bar{\mathrm{s}}}\}$ or (ii) the pair i and i' belong to a dimer from the dimer covering that defines the state $|\Psi_{\mathrm{s}\bar{\mathrm{s}}}(q)\rangle$. In case (i),

$$\begin{aligned}\langle\,\hat{n}_i\,\hat{n}_i\rangle_{\mathrm{c}} &= \langle\,\hat{n}_i^2\,\rangle - (\langle\,\hat{n}_i\,\rangle)^2\\ &= \langle\,\hat{n}_i\,\rangle - (\langle\,\hat{n}_i\,\rangle)^2\\ &= \langle\,\hat{n}_i\,\rangle\,(1 - \langle\,\hat{n}_i\,\rangle)\\ &= q\,(1-q),\end{aligned} \tag{4.116}$$

since $\langle\,\hat{n}_i\,\rangle$ for a dimer is either q or $1-q$, while we have ruled out the monomer cases, that is, $\langle\,\hat{n}_{i_{\mathrm{s}}}\,\rangle = \langle\,\hat{n}_{i_{\bar{\mathrm{s}}}}\,\rangle = 0$, through the restriction $i = i' \in \Lambda \setminus \{i_{\mathrm{s}}, i_{\bar{\mathrm{s}}}\}$. Observe that the right-hand side is independent of $i \in \Lambda \setminus \{i_{\mathrm{s}}, i_{\bar{\mathrm{s}}}\}$. In case (ii), let $i' = \bar{i}$, where the un-directed dimer $(i, \bar{i})$ matches one of the directed dimers from the dimer covering that defines the state $|\Psi_{\mathrm{s}\bar{\mathrm{s}}}(q)\rangle$,

$$\begin{aligned}\langle\,\hat{n}_i\,\hat{n}_{\bar{i}}\,\rangle_{\mathrm{c}} &= \langle\,\hat{n}_i\,(1-\hat{n}_i)\,\rangle_{\mathrm{c}}\\ &= \langle\,\hat{n}_i\,(1-\hat{n}_i)\,\rangle - \langle\,\hat{n}_i\,\rangle\,\langle\,(1-\hat{n}_i)\,\rangle\\ &= -\,\langle\,\hat{n}_i\,\rangle\,\langle\,(1-\hat{n}_i)\,\rangle\\ &= -\,q\,(1-q).\end{aligned} \tag{4.117}$$

To reach the third equality, we took advantage of the fact that $\hat{n}_i$ has the eigenvalues 0 and 1. To reach the last equality, we used the fact that the many-body expectation value reduces to the expectation value in the properly directed single dimer state corresponding to $(i, \bar{i})$. Observe that the right-hand side is independent of whether i is to the left or right of $\bar{i}$. The manipulations

$$\begin{aligned}\operatorname{var}\widehat{Q}_f\Big|_{\Psi_{\mathrm{s}\bar{\mathrm{s}}}(q)} &= \sum_{i,i'\in\Lambda} f_i\,f_{i'}\,\langle\,\hat{n}_i\,\hat{n}_{i'}\rangle_{\mathrm{c}}\\ &= \frac{1}{2}\sum_{i,i'\in\Lambda}\Big(f_i\,f_{i'}\,\langle\,\hat{n}_i\,\hat{n}_{i'}\rangle_{\mathrm{c}} + f_{i'}\,f_i\,\langle\,\hat{n}_{i'}\,\hat{n}_i\,\rangle_{\mathrm{c}}\Big)\end{aligned} \tag{4.118}$$

on the variance (4.113) are independent of the dimer state $|\Psi_{\mathrm{s}\bar{\mathrm{s}}}(q)\rangle$. They apply to any state in the fermionic Fock space. For the state $|\Psi_{\mathrm{s}\bar{\mathrm{s}}}(q)\rangle$, the monomers at i_{s} and $i_{\bar{\mathrm{s}}}$ at which the soliton and antisoliton bind zero modes if one fermion is added must be excluded from the sum over diagonal terms in the unrestricted double sum over $i,i'\in\Lambda$, since their contributions is vanishing, that is,

$$\begin{aligned}\operatorname{var}\widehat{Q}_f\Big|_{\Psi_{\mathrm{s}\bar{\mathrm{s}}}(q)} &= \frac{1}{2}\sum_{i\in\Lambda\setminus\{i_{\mathrm{s}},i_{\bar{\mathrm{s}}}\}}\left[f_i^2\,\langle\,\hat{n}_i^2\,\rangle_{\mathrm{c}} + f_{\bar{i}}^2\,\langle\,\hat{n}_{\bar{i}}^2\,\rangle_{\mathrm{c}} + 2\,f_i\,f_{\bar{i}}\,\langle\,\hat{n}_i\,\hat{n}_{\bar{i}}\,\rangle_{\mathrm{c}}\right]\\ &= \frac{1}{2}\,q\,(1-q)\sum_{i\in\Lambda\setminus\{i_{\mathrm{s}},i_{\bar{\mathrm{s}}}\}}(f_i-f_{\bar{i}})^2.\end{aligned} \tag{4.119}$$

It is important to emphasize that the choice of the test function f greatly matters when evaluating the variances of the smeared charge operator. To illustrate this fact, suppose that the (box) test function f takes the values

$$f_i = f_{i+2N} = \begin{cases} 0, & i = 1,\cdots,i_{\mathrm{s}}-1,\\ 1, & i = i_{\mathrm{s}},\cdots,i_{\bar{\mathrm{s}}},\\ 0, & i = i_{\bar{\mathrm{s}}}+1,\cdots,2N,\end{cases} \tag{4.120}$$

obeying periodic boundary conditions, see Figure 4.6. If so,

$$(f_i - f_{i+1})^2 = 0, \qquad i\in\Lambda\setminus\{i_{\mathrm{s}},i_{\bar{\mathrm{s}}}\}, \tag{4.121}$$

implies that the variance (4.119) vanishes. Another choice for the test function f that implies a vanishing variance is to impose the weaker condition

$$(f_i - f_{\bar{i}})^2 = 0 \tag{4.122}$$

for every un-directed dimer (i,i') from the dimer covering that defines the state $|\Psi_{\mathrm{s}\bar{\mathrm{s}}}(q)\rangle$; see Figure 4.6. With the choice (4.122), the normal ordering of the smeared charge operator (4.102a) is the same as that with respect to the pristine dimerized state for any $0\leq q\leq 1$ (**Exercise** 4.9). Finally, if we choose the test function f to be the function f_{L} that is symmetric about the site i_{s} where it takes the value 1, strictly decreasing away from i_{s}, and exponentially suppressed over the length scale ω away from i_{s}, we may take the continuum limit assuming a d-dimensional space to obtain the estimate

$$\operatorname{var}\widehat{Q}_{f_{\mathrm{L}}}\Big|_{\Psi_{\mathrm{s}\bar{\mathrm{s}}}(q)} \sim \frac{1}{2}\,q\,(1-q)\,\omega^{d-2}\int\frac{\mathrm{d}^d\boldsymbol{x}}{\omega^d}\,\omega^2\left(\frac{\partial f_{\mathrm{L}}}{\partial\boldsymbol{x}}\right)^2(|\boldsymbol{x}|/\omega). \tag{4.123}$$

Hence, if it were possible to generalize the soliton and antisoliton point defects to dimer coverings of higher-dimensional lattices, we would conclude that the variance of the left smeared charge operator vanishes in the limit by which $r_{\mathrm{s}\bar{\mathrm{s}}}\to\infty$ before $\omega\to\infty$ if and only if $d=1$. The variance (4.123) in the continuum limit of the dimer model is of order 1 when $d=2$ and diverges like ω^{d-2} when $d>2$. A generic property of a completely filled band of Bloch electrons when the dimensionality of space d is larger than one is that the variance (4.123) is nonvanishing.

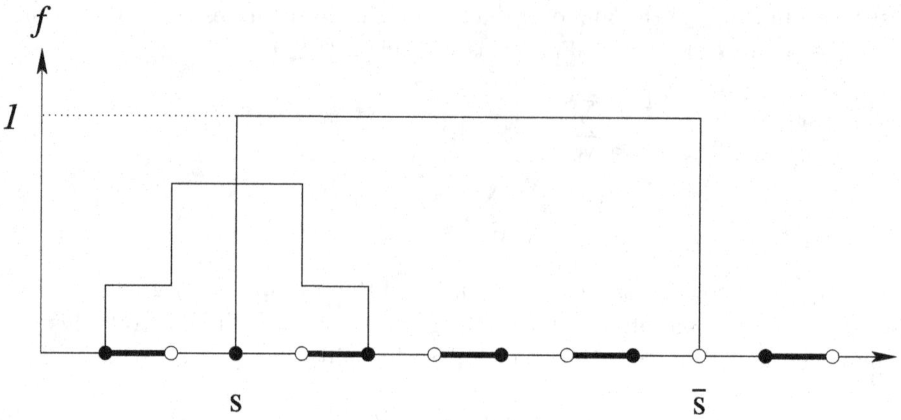

Figure 4.6 Box and step test functions (4.120) and (4.122), respectively, and the dimer state (4.94a).

4.5 Definition of Point-Like Fractional Charges

Having established what is meant by a sharp fractional charge in the context of both the one-dimensional Dirac Hamiltonian and at the strong dimerization point of a lattice Hamiltonian, we proceed with a definition of point-like fractional charges in any many-body but noninteracting lattice Hamiltonian for fermions. This definition is independent of the dimensionality of space.

A prerequisite to discussing the sharpness of a fractional quantum number is that the quantum number in question must be conserved. Throughout Chapter 4, the fermion number (charge) is assumed to be conserved. As with the previous chapters, we are concerned with noninteracting fermions. For simplicity but without loss of generality, we consider spinless fermions. To discuss charge conservation in this context, we consider a lattice Λ made of sites that we shall denote by the letters i and j when they come in pairs. If the cardinality of Λ is denoted $|\Lambda|$, then the Fock space for the fermions hopping on this lattice is of dimensionality $2^{|\Lambda|}$. We assume the Hamiltonian

$$\widehat{H} := -\frac{1}{2}\sum_{i,j\in\Lambda}\left(\hat{c}_i^{\dagger}\,t_{ij}\,\hat{c}_j + \text{H.c.}\right), \quad t_{ij} = t_{ji}^{*}, \quad \{\hat{c}_i, \hat{c}_j^{\dagger}\} = \delta_{ij}, \quad \{\hat{c}_i, \hat{c}_j\} = 0 \tag{4.124a}$$

in the orbital basis. For any pair of sites i and j, the hopping amplitude t_{ij} is complex-valued. Its magnitude decays no slower then exponentially fast beyond the characteristic length ζ as a function of the distance $|i-j|$ between the pair of sites i and j, that is,

$$|t_{ij}| \sim e^{-|i-j|/\zeta}, \qquad |i-j| \gg \zeta, \tag{4.124b}$$

in order to establish locality in the orbital basis. For any site i, the real-valued diagonal element

$$t_{ii} \equiv \mu_i \tag{4.124c}$$

represents a local chemical potential μ_i. The multiplicative factor $1/2$ compensates for the unconstrained double sum over the lattice sites. As usual, the ket $|0\rangle$ denotes the many-body state annihilated by $\hat{c}_i$ for all lattice sites.

The global symmetry under

$$\hat{c}_i^\dagger \mapsto e^{-\mathrm{i}\phi}\,\hat{c}_i^\dagger, \qquad \hat{c}_i \mapsto e^{+\mathrm{i}\phi}\,\hat{c}_i, \qquad \phi \in [0, 2\pi[\tag{4.125}$$

of Hamiltonian (4.124) is the reason for which the total fermion charge operator

$$\widehat{Q} := \sum_{i\in\Lambda} \hat{c}_i^\dagger\,\hat{c}_i = \widehat{Q}^\dagger \tag{4.126}$$

is conserved, that is,

$$\frac{\mathrm{d}\,\widehat{Q}}{\mathrm{d}t} = 0. \tag{4.127}$$

Proof To prove this global conservation law, it is useful to derive the following continuity equation. For any site i, we introduce the local fermion number operator

$$\hat{n}_i := \hat{c}_i^\dagger\,\hat{c}_i = \hat{n}_i^\dagger. \tag{4.128}$$

It obeys the local equation of motion (in units for which $\hbar = 1$)

$$\begin{aligned}\frac{\mathrm{d}\,\hat{n}_i}{\mathrm{d}t} &= -\mathrm{i}\left[\hat{n}_i, \widehat{H}\right]\\ &= -\sum_{j\in\Lambda}\left(\hat{c}_j^\dagger\,\mathrm{i}t_{ij}^*\,c_i - \hat{c}_i^\dagger\,\mathrm{i}t_{ij}\,c_j\right).\end{aligned} \tag{4.129}$$

One then verifies that the global charge (4.126) is indeed conserved (**Exercise** 4.10). □

To proceed, we rewrite Eq. (4.129) as

$$\frac{\mathrm{d}\,\hat{n}_i}{\mathrm{d}t} = -\sum_{j\in\Lambda\setminus\{i\}}\left(\hat{c}_j^\dagger\,\mathrm{i}t_{ji}\,c_i - \hat{c}_i^\dagger\,\mathrm{i}t_{ij}\,c_j\right). \tag{4.130}$$

To reach the last equality, we used the fact that $t_{ij} = t_{ji}^*$ and that the difference in the parenthesis on the right-hand side of Eq. (4.129) vanishes when $j = i$. The Hermitean operator

$$\hat{J}_i := \sum_{j\in\Lambda\setminus\{i\}}\left(\hat{c}_j^\dagger\,\mathrm{i}t_{ji}\,c_i - \hat{c}_i^\dagger\,\mathrm{i}t_{ij}\,c_j\right) = \hat{J}_i^\dagger \tag{4.131}$$

on the right-hand side of Eq. (4.129) is the net current at the site $i \in \Lambda$. It is related by the continuity equation

$$\frac{\mathrm{d}\,\hat{n}_i}{\mathrm{d}t} + \hat{J}_i = 0 \tag{4.132}$$

to the time derivative of the local fermion number operator (4.128) (**Exercise** 4.11). It is also desirable to introduce a directed current operator, the summand on the right-hand side of Eq. (4.131),

$$\hat{J}_{i\to j} := \hat{c}^{\dagger}_{j}\,\mathrm{i}t_{ji}\,\hat{c}_{i} - \hat{c}^{\dagger}_{i}\,\mathrm{i}t_{ij}\,\hat{c}_{j} = \hat{J}^{\dagger}_{i\to j}. \tag{4.133}$$

This is a Hermitean operator that is odd under the interchange of the pair of sites i and j,

$$\hat{J}_{i\to j} = -\hat{J}_{j\to i}. \tag{4.134}$$

The continuity equation (4.132) becomes

$$\frac{\mathrm{d}\,\hat{n}_i}{\mathrm{d}t} + \sum_{j\in\Lambda\setminus\{i\}} \hat{J}_{i\to j} = 0 \tag{4.135}$$

when expressed in terms of the directed current operator.

The global gauge symmetry (4.125) implies both the conservation law (4.127) and the continuity equation (4.135). It is possible to interpolate between both equations by introducing a family of Hermitean and positive semidefinite operators with the representative

$$\widehat{Q}_f := \sum_{i\in\Lambda} f_i\,\hat{c}^{\dagger}_{i}\,\hat{c}_{i} = \widehat{Q}^{\dagger}_{f} \tag{4.136a}$$

indexed by the test function

$$\begin{aligned} f\colon \Lambda &\to [0,1],\\ i &\mapsto f_i. \end{aligned} \tag{4.136b}$$

We shall call $\widehat{Q}_f$ a smeared charge operator. Indeed, the choice

$$\begin{aligned} f\colon \Lambda &\to [0,1],\\ i &\mapsto 1, \end{aligned} \tag{4.137}$$

by which all local fermion number operators are weighted by the number 1, delivers Eq. (4.126), while the choice

$$\begin{aligned} f\colon \Lambda &\to [0,1],\\ j &\mapsto \delta_{j,i}, \end{aligned} \tag{4.138}$$

by which all local fermion number operators are weighted by the number 0 except for the local fermion number operators at i that is weighted by the number 1, delivers Eq. (4.128). As desired,

$$\frac{\mathrm{d}\widehat{Q}_f}{\mathrm{d}t}+\frac{1}{2}\sum_{i,j\in\Lambda}\left(f_i-f_j\right)\hat{J}_{i\to j}=0 \tag{4.139}$$

interpolates between the conservation law (4.127) and the continuity equation (4.135).

For any test function f defined as in Eq. (4.136b), the smeared charge operator (4.136a) has the eigenstates

$$|\{n_i=0,1\}\rangle:=\prod_{i\in\Lambda}\left(\hat{c}_i^{\dagger}\right)^{n_i}|0\rangle \tag{4.140a}$$

with the eigenvalues

$$Q_f:=\sum_{i\in\Lambda}f_i\,n_i. \tag{4.140b}$$

If we restrict the smeared charge operator (4.136a) to any test function taking values in the set $\{0,1\}$,

$$\begin{aligned} f\colon \Lambda &\to \{0,1\},\\ i &\mapsto f_i, \end{aligned} \tag{4.141}$$

then the smeared charge operator (4.136a) has integer-valued eigenvalues. Evidently, the total charge number operator (4.126) and the local fermion number operator (4.128) have integer-valued eigenvalues for any finite cardinality of the lattice Λ. A necessary condition for the smeared charge operator (4.136a) to possess noninteger eigenvalues is that its test function (4.136b) takes values in the open interval $]0,1[$.

Let L be the linear size of the lattice Λ. This is the largest length scale. After the lattice spacing $\mathfrak{a}$, the second smallest length scale is, without loss of generality, the length scale ζ defined in Eq. (4.124b) that quantifies the locality of the many-body noninteracting Hamiltonian (4.124a). We say that a many-body noninteracting ground state of Hamiltonian (4.124) at a nonvanishing filling fraction ν_{e} supports a point-like fractional charge q_i at site $i\in\Lambda$ if it can be chosen as the eigenstate of a suitable smeared charge operator $\widehat{Q}_i$ in a suitable scaling limit, that is

1. if a single-particle eigenstate of Hamiltonian (4.124), whose support is concentrated in the subset $\Xi_i\subset\Lambda$ centered about i in the sense that this single-particle eigenstate decays no slower than exponentially fast in magnitude away from site i with the characteristic decay length ξ (the linear size of Ξ_i is thus of order ξ where $\mathfrak{a}\leq\zeta\leq\xi$), is responsible for a many-body ground-state degeneracy depending on whether this single-particle state is occupied or not in the scaling limit to be defined below,
2. if, upon increasing L, it is possible to construct a sequence of subsets $\Omega_i\subset\Lambda$ centered about i with $\Xi_i\subset\Omega_i$, each of which contains no other point-like

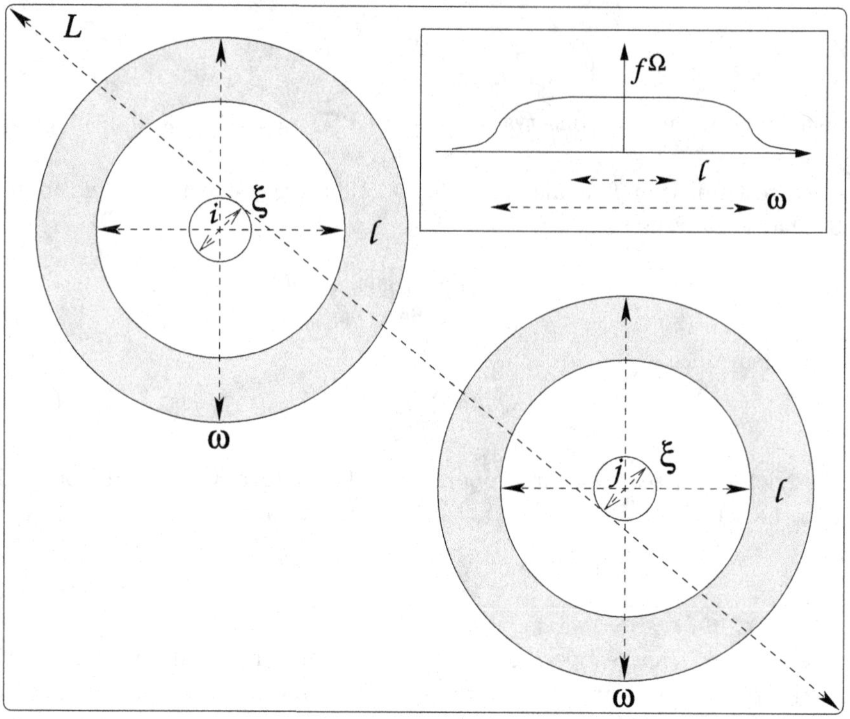

Figure 4.7 The lattice Λ of linear size L is pictured by the large square with rounded corners. Two single-particle states bound to the sites i and j are pictured by the circles of radius ξ centered at i and j, respectively, for their amplitudes both decay exponentially fast with the characteristic length ξ away from their maxima at i and j, respectively. Sites i and j are assigned the subsets Ω_i and Ω_j, respectively, each of which is characterized by two length scales. There is the length ℓ over which the test functions f^{Ω_i} and f^{Ω_j} take values very close to unity (see inset). There is the length ω over which the test functions f^{Ω_i} and f^{Ω_j} away from their maximal value of 1 around i and j, respectively, take a value of order $\exp(-\omega/\ell)$ with $\ell \ll \omega \ll L$ (see inset). If $|i-j|$ is the distance between i and j, the thermodynamic limit is then defined by the limit $\ell, \omega, |i-j|, L \to \infty$ with $\xi/\ell \to 0$, $\ell/\omega \to 0$, $\omega/|i-j| \to 0$, and $|i-j|/L \to 0$.

fractional charge, together with a sequence of test functions f^{Ω_i}, each of which is characterized by two length scales ℓ and ω, whereby ℓ with $\xi \ll \ell$ is the linear size of the region centered about i for which f^{Ω_i} is close to the value 1 while ω with $\ell \ll \omega \ll L$ is the linear size of Ω_i over which the values taken by f^{Ω_i} have all decreased to an exponentially small value of order $\exp(-\omega/\ell)$ (see Figure 4.7),

3. if the operator norm

$$\lim_{\xi/\ell\to 0} \lim_{\ell/\omega\to 0} \lim_{\omega/L\to 0} \left\| \frac{1}{2} \sum_{i',j'\in\Lambda} \left(f^{\Omega_i}_{i'} - f^{\Omega_i}_{j'} \right) \hat{J}_{i'\to j'} \right\| \to 0 \tag{4.142a}$$

of the current operator entering the continuity equation (4.139) vanishes in the scaling limit $\ell, \omega, L \to \infty$ with $\xi/\ell \to 0$, $\ell/\omega \to 0$, and $\omega/L \to 0$ so that the smeared charge operator

$$\widehat{Q}_i(t) := \lim_{\xi/\ell\to 0}\lim_{\ell/\omega\to 0}\lim_{\omega/L\to 0} \widehat{Q}_{f^{\Omega_i}}(t) \tag{4.142b}$$

centered at i becomes time-independent,

$$\widehat{Q}_i(t) = \widehat{Q}_i \iff \left[\widehat{Q}_i, \widehat{H}\right] = 0, \tag{4.142c}$$

4. and, finally, if one of the degenerate many-body ground states selected by $\widehat{H} + E\,\widehat{Q}_i$, with E an infinitesimal electric field, is an eigenstate of $\widehat{Q}_i$ with the eigenvalue Q_i defined by the limiting value

$$Q_i := \lim_{\xi/\ell\to 0}\lim_{\ell/\omega\to 0}\lim_{\omega/L\to 0} Q_{f^{\Omega_i}}, \qquad \widehat{Q}_{f^{\Omega_i}}|Q_{f^{\Omega_i}}\rangle = Q_{f^{\Omega_i}}|Q_{f^{\Omega_i}}\rangle, \tag{4.143a}$$

such that (i) the number

$$q_i := \lim_{\xi/\ell\to 0}\lim_{\ell/\omega\to 0}\lim_{\omega/L\to 0}\left(Q_{f^{\Omega_i}} - \left\lfloor Q_{f^{\Omega_i}}\right\rfloor\right) \tag{4.143b}$$

is not an integer[5] and (ii)

$$\nu_{\mathrm{e}} = \lim_{\xi/\ell\to 0}\lim_{\ell/\omega\to 0}\lim_{\omega/L\to 0}\frac{\left\lfloor Q_{f^{\Omega_i}}\right\rfloor}{|\Lambda|} \tag{4.143c}$$

is the filling fraction.

Remark (1) There are three microscopic characteristic lengths, $\mathfrak{a} \leq \zeta \leq \xi$. There are three macroscopic characteristic lengths $\ell \ll \omega \ll L$ that scale to infinity. The relative growths (that is the order of limits) of the three characteristic lengths ℓ, ω, and L define how the smeared charge is to be measured in the scaling limit (4.142).

Remark (2) The existence of the length scale ξ implies a single-particle spectral gap of order $1/\xi$ (in units with $\hbar = c = 1$) in the thermodynamic limit $L \to \infty$ within which one finds a bound state whose wave function decays exponentially fast in amplitude away from a point in space with the characteristic decay length ξ (see Figure 4.7). The limit $\xi/\ell \to 0$ justifies identifying a mid-gap state with a local object (a particle).

Remark (3) The limit $\ell/\omega \to 0$ has the following purposes. First, it guarantees that any pair of mid-gap states are so far apart that their hybridization is negligible up to an exponential accuracy (see Figure 4.7). Second, the length scale $\omega - \ell$ can be interpreted as the linear width of a boundary defined by equating $\langle\widehat{Q}_{f^{\Omega}}\rangle$ with the expectation value ${}_{\mathrm{OBC}}\langle\widehat{Q}\rangle_{\mathrm{OBC}}$ over the ground state when open boundary

[5] The floor function $\lfloor\cdot\rfloor : \mathbb{R} \to \mathbb{Z}$ returns with $\lfloor x\rfloor$ the largest integer smaller than or equal to the real number x.

conditions (OBC) are imposed in a subset $\Omega \subset \Lambda$ of linear size ω in which no more than one mid-gap state is localized (see Figure 4.7). Hence, if this mid-gap state is associated with a point-like fractional charge of magnitude $|q| < 1$, then the complementary charge of magnitude $1 - |q|$ is localized along the smooth boundary of Ω. If so, the scaling limit $\ell, \omega, L \to \infty$ with $\ell/\omega \to 0$ and $\omega/L \to 0$ moves the complementary charges of all point-like fractional charges outside of all field of views probed by the test functions f^{Ω_i}, leaving only the point-like fractional part q_i to be measured by the smeared operator $\widehat{Q}_{f^{\Omega_i}}$ within linear-response theory.

4.6 Exercises

4.1 We seek the eigenvalues and eigenfunctions of a single delta-function potential located at the origin of the real line. Let

$$\mathcal{H}_{\mathrm{sw}}(x) := -\frac{1}{2}\frac{\mathrm{d}^2}{\mathrm{d}x^2} + v\,\delta(x). \tag{4.144}$$

The shorthand acronym "sw" stands for "single well" and $v \in \mathbb{R}$.

4.1.1 Verify that

$$\begin{aligned} \psi(x) = {} & \psi_{\mathrm{inc}}(x) \\ & + \int \mathrm{d}x' \int \frac{\mathrm{d}p'}{2\pi} \frac{e^{+\mathrm{i}p'\,(x-x')}}{\varepsilon - (p'^2/2) + \mathrm{i}0^+}\, v\,\delta(x')\,\psi(x'), \end{aligned} \tag{4.145a}$$

is a solution with energy $\varepsilon > 0$ to

$$\mathcal{H}_{\mathrm{sw}}(x)\,\psi(x) = \varepsilon\,\psi(x) \tag{4.145b}$$

with the homogeneous ($v = 0$) solution

$$\psi_{\mathrm{inc}}(x) = e^{+\mathrm{i}\sqrt{2\,\varepsilon}\,x} \tag{4.145c}$$

and with the outgoing boundary conditions

$$\lim_{x\to\pm\infty} [\psi(x) - \psi_{\mathrm{inc}}(x)] \sim \lim_{x\to\pm\infty} \frac{1}{\mathrm{i}\sqrt{2\,\varepsilon}}\, e^{+\mathrm{i}\sqrt{2\,\varepsilon}\,|x|}. \tag{4.145d}$$

Hint: The integrand on the right-hand side of Eq. (4.145a) has poles at $p'_{\pm} = \pm\sqrt{2\,\varepsilon}$. Show that adding the positive infinitesimal $\mathrm{i}0^+$ in the numerator of the integrand on the right-hand side of Eq. (4.145a) shifts the pole at p'_- just below the real axis, while it shifts the pole at p'_+ just above the real axis. When $x - x' > 0$ ($x - x' < 0$), draw an integration path in the p' complex plane that picks up the contribution from the pole at p'_+ (p'_-) but not that from the pole p'_- (p'_+).

4.1.2 Give an interpretation of the multiplicative factor $1/(\mathrm{i}\sqrt{2\,\varepsilon})$ entering on the right-hand side of Eq. (4.145d). *Hint:* Compute the current density of the right-hand side of Eq. (4.145d) and compare it with that of the incoming state (4.145c).

4.1.3 Do the integrals over x' first and then over p' in Eq. (4.145a) to show that

$$\psi(x) = \psi_{\mathrm{inc}}(x) + \frac{v}{\mathrm{i}p}\, e^{+\mathrm{i}p\,|x|}\, \psi(x=0), \qquad p = \sqrt{2\,\varepsilon} > 0. \tag{4.146}$$

4.1.4 Show that the choice made for the normalization of $\psi_{\mathrm{inc}}(x)$ in Eq. (4.145c) gives

$$\psi(x) = e^{+\mathrm{i}p\,x} + \frac{v}{\mathrm{i}p - v}\, e^{+\mathrm{i}p\,|x|}, \qquad p = \sqrt{2\,\varepsilon} > 0. \tag{4.147}$$

4.1.5 Let the acronym "trans" stand for "transmitted." Define

$$\psi_{\mathrm{trans}}(x) := \psi(x), \qquad x > 0. \tag{4.148a}$$

Show that

$$\psi_{\mathrm{trans}}(x) = \frac{\mathrm{i}p}{\mathrm{i}p - v} e^{+\mathrm{i}p\,x}, \qquad x > 0. \tag{4.148b}$$

Define

$$t(p) := \frac{\mathrm{i}p}{\mathrm{i}p - v}. \tag{4.148c}$$

4.1.6 Let the acronym "reflec" stand for "reflected." Define

$$\psi_{\mathrm{reflec}}(x) := \psi(x) - e^{+\mathrm{i}p\,x}, \qquad x < 0. \tag{4.149a}$$

Show that

$$\psi_{\mathrm{reflec}}(x) = \frac{v}{\mathrm{i}p - v} e^{-\mathrm{i}p\,x}, \qquad x < 0. \tag{4.149b}$$

Define

$$r(p) := \frac{v}{\mathrm{i}p - v}. \tag{4.149c}$$

4.1.7 Explain why it is meaningful to interpret $t(p)$ as the transmission amplitude from the left across the delta-function potential. Explain why it is meaningful to interpret $r(p)$ as the reflection amplitude from the left across the delta-function potential. Verify the identity

$$|t(p)|^2 + |r(p)|^2 = 1 \tag{4.150}$$

that implements flux conservation.

4.1.8 Write

$$t(p) = |t(p)|\, e^{+\mathrm{i}\delta(p)}, \quad |t(p)| = \sqrt{\frac{p^2}{p^2 + v^2}}, \quad \delta(p) = -\arctan\left(\frac{v}{p}\right). \tag{4.151}$$

Explain why $\delta(p)$ can be interpreted as a phase shift.

4.1.9 Deduce from the pole of the transmission amplitude at $p=-\mathrm{i}\,v$ the existence of the bound state with the energy

$$\varepsilon=-\frac{v^2}{2} \tag{4.152}$$

and the normalized eigenstate

$$\psi(x)=\sqrt{|v|}\,e^{-|v\,x|} \tag{4.153}$$

if and only if $v<0$.

4.2 Show that the pair of coefficients $A''(p)\in\mathbb{C}$ and $B''(p)\in\mathbb{C}$ are linearly related by Eq. (4.3c) to the pair of coefficients $A'(p)\in\mathbb{C}$ and $B'(p)\in\mathbb{C}$ from demanding that $\psi(p,x)$ is continuous at $+r/2$, while $(\partial_x\psi)(p,x)$ jumps by $-2v\,\psi(+r/2)$ across $+r/2$. Similarly, show that the pair of coefficients $A'(p)\in\mathbb{C}$ and $B'(p)\in\mathbb{C}$ are linearly related by Eq. (4.3a) to the pair of coefficients $A\,(p)\in\mathbb{C}$ and $B\,(p)\in\mathbb{C}$. The pair of coefficients $A\,(p)\in\mathbb{C}$ and $B\,(p)\in\mathbb{C}$ are fixed by the boundary conditions on $\psi(p,x)$ and $(\partial_x\psi)(p,x)$ at $x=-\infty$ say.

4.3 Verify Eq. (4.5) and construct the 2×2 scattering matrix S_{W} that maps incoming into outgoing plane waves at the potential well W, where $\mathrm{W}=\mathrm{L},\mathrm{R}$.

4.4 Verify Eqs. (4.25a), (4.25b), and (4.25c).

4.5 Given is the Dirac Hamiltonian (4.58). It depends parametrically on the asymptotic value $\phi_\infty>0$ of the profile $\phi(x)$ that supports the pair of soliton and antisoliton that are a distance $r>0$ apart on the real line [say at $x=-(r/2)$ and $x=+(r/2)$]. We seek the solutions to the stationary Dirac equation (4.61).

4.5.1 Let $\kappa>0$ be the positive solution of the nonlinear equation

$$0<\kappa^2=\phi^2_\infty\left(1-e^{-2\,\kappa\,r}\right)<\phi^2_\infty. \tag{4.154a}$$

As it should be, $\kappa^2=\phi^2_\infty$ in the limit $r\to\infty$. Let ε be the positive single-particle energy defined by

$$\varepsilon^2:=\phi^2_\infty-\kappa^2=\phi^2_\infty\,e^{-2\,\kappa\,r}. \tag{4.154b}$$

As it shoud be, $\varepsilon^2=0$ in the limit $r\to\infty$. Let $\mathcal{N}>0$ be the normalization

$$\mathcal{N}:=\sqrt{\frac{\kappa}{c^2}},\qquad c^2=4\left(1+\frac{\kappa}{\phi_\infty}\right)\left[1-\phi_\infty\,r\left(\frac{\phi_\infty}{\kappa}-\frac{\kappa}{\phi_\infty}\right)\right]. \tag{4.154c}$$

In the limit $r\to\infty$, $c^2=8$. Let

$$y:=x+(r/2). \tag{4.154d}$$

Verify that

$$u_{\mathrm{B}}(x)=\mathcal{N}\begin{cases} e^{-\kappa\,r}\,e^{+\kappa\,y}, & y<0,\\ e^{-\kappa\,r}\left[\left(1+\frac{\phi_\infty}{\kappa}\right)e^{+\kappa\,y}-\frac{\phi_\infty}{\kappa}\,e^{-\kappa\,y}\right], & 0<y<r,\\ \left(\frac{\phi_\infty+\kappa}{\phi_\infty-\kappa}\right)^{1/2}\;e^{-\kappa\,y}, & r<y, \end{cases} \tag{4.154e}$$

and

$$v_{\mathrm{B}}(x)=\mathcal{N}\begin{cases} \left(1+\frac{\kappa}{\phi_\infty}\right)e^{+\kappa\,y}, & y<0,\\ \left(\frac{\kappa}{\phi_\infty}-\frac{\phi_\infty}{\kappa}\right)e^{+\kappa\,y}+\left(1+\frac{\phi_\infty}{\kappa}\right)e^{-\kappa\,y}, & 0<y<r,\\ e^{-\kappa\,y}, & r<y, \end{cases} \tag{4.154f}$$

solve Eq. (4.62).

4.5.2 For any $p\in\mathbb{R}$, choose the complex-valued numbers $A(p)=A^{*}(-p)$ and $B(p)=B^{*}(-p)]$. Let $\varepsilon(p)>0$ be the positive single-particle energy defined by

$$\varepsilon^{2}(p):=p^{2}+\phi_\infty^{2}. \tag{4.155a}$$

Define the phase shift $\delta(p)$ by

$$e^{\mathrm{i}\delta(p)}:=\sqrt{\frac{\phi_\infty+\mathrm{i}p}{\phi_\infty-\mathrm{i}p}}. \tag{4.155b}$$

Let

$$M_{\mathrm{L}}(p):=\begin{pmatrix} 1-\mathrm{i}\frac{\phi_\infty}{p} & -\mathrm{i}\frac{\phi_\infty}{p}\\ +\mathrm{i}\frac{\phi_\infty}{p} & 1+\mathrm{i}\frac{\phi_\infty}{p} \end{pmatrix} \tag{4.155c}$$

and

$$M_{\mathrm{R}}(p):=\begin{pmatrix} 1+\mathrm{i}\frac{\phi_\infty}{p} & +\mathrm{i}\frac{\phi_\infty}{p}\,e^{-2\mathrm{i}p\,r}\\ -\mathrm{i}\frac{\phi_\infty}{p}\,e^{+2\mathrm{i}p\,r} & 1-\mathrm{i}\frac{\phi_\infty}{p} \end{pmatrix} \tag{4.155d}$$

define the transfer matrices at $x=-(r/2)$ and $x=+(r/2)$, respectively. Let

$$\begin{pmatrix} A'(p)\\ B'(p) \end{pmatrix}:=M_{\mathrm{L}}(p)\begin{pmatrix} A(p)\\ B(p) \end{pmatrix} \tag{4.155e}$$

and let

$$\begin{pmatrix} A''(p)\\ B''(p) \end{pmatrix}:=M_{\mathrm{R}}(p)\begin{pmatrix} A'(p)\\ B'(p) \end{pmatrix}. \tag{4.155f}$$

Let

$$y:=x+(r/2). \tag{4.155g}$$

Verify that

$$u(p,x)=\begin{cases}A(p)\,e^{+\mathrm{i}p\,y}+B(p)\,e^{-\mathrm{i}p\,y}, & y<0,\\ A'(p)\,e^{+\mathrm{i}p\,y}+B'(p)\,e^{-\mathrm{i}p\,y}, & 0<y<r,\\ A''(p)\,e^{+\mathrm{i}p\,y}+B''(p)\,e^{-\mathrm{i}p\,y}, & r<y,\end{cases}\tag{4.155h}$$

and

$$v(p,x)=\begin{cases}A(p)\,e^{+\mathrm{i}[p\,y+\delta(p)]}+B(p)\,e^{-\mathrm{i}[p\,y+\delta(p)]}, & y<0,\\ A'(p)\,e^{+\mathrm{i}[p\,y-\delta(p)+\pi]}+B'(p)\,e^{-\mathrm{i}[p\,y-\delta(p)+\pi]}, & 0<y<r,\\ A''(p)\,e^{+\mathrm{i}[p\,y+\delta(p)]}+B''(p)\,e^{-\mathrm{i}[p\,y+\delta(p)]}, & r<y,\end{cases}\tag{4.155i}$$

solve Eq. (4.65).

4.6 Verify Eq. (4.74). Verify that the right-hand sides in Eq. (4.74) are odd under the transformation by which the c creation and annihilation operators are exchanged for the d creation and annihilation operators.

4.7 Verify Eq. (4.78). To this end show that

$$\lim_{f_{\mathrm{L}}\to 1}\lim_{r\to\infty}\int \mathrm{d}x\, f_{\mathrm{L}}(x)\,\left[u_{\mathrm{B}}^2(x)+v_{\mathrm{B}}^2(x)\right]=+\frac{1}{2},\tag{4.156a}$$

$$\lim_{f_{\mathrm{L}}\to 1}\lim_{r\to\infty}\int \mathrm{d}x\, f_{\mathrm{L}}(x)\,\left[u_{\mathrm{B}}^2(x)-v_{\mathrm{B}}^2(x)\right]=-\frac{1}{2},\tag{4.156b}$$

for the bound states,

$$\begin{aligned}&\int \mathrm{d}x\, f_{\mathrm{L}}(x)\,\left[u(p,x)\,u(q,x)-v(p,x)\,v(q,x)\right]=\\ &\qquad\frac{1}{\varepsilon(p)+\varepsilon(q)}\int \mathrm{d}x\,\left(\frac{\mathrm{d}f_{\mathrm{L}}}{\mathrm{d}x}\right)(x)\,\left[u(p,x)\,v(q,x)+v(p,x)\,u(q,x)\right]\end{aligned}\tag{4.156c}$$

for any pair of continuum states, and

$$\begin{aligned}&\int \mathrm{d}x\, f_{\mathrm{L}}(x)\,\left[u^*(p,x)\,u(q,x)+v^*(p,x)\,v(q,x)\right]=\\ &\qquad\frac{1}{\varepsilon(p)-\varepsilon(q)}\int \mathrm{d}x\,\left(\frac{\mathrm{d}f_{\mathrm{L}}}{\mathrm{d}x}\right)(x)\,\left[v^*(p,x)\,u(q,x)-u^*(p,x)\,v(q,x)\right]\end{aligned}\tag{4.156d}$$

for any pair of continuum states with unequal single-particle eigenenergies. The limit $f_{\mathrm{L}}\to 1$, $\partial_x f_{\mathrm{L}}\to 0$ uniformly on the real line cannot be interchanged with the integral over the real line for any pair of continuum states, as was explained when computing the contribution (4.25c). The Dirac counterparts to the integral (4.32) must be evaluated with the help of the wave functions derived in **Exercise** 4.5.

4.8 We are going to verify Eq. (4.84b) from a direct calculation that makes use of the eigenfunctions constructed in **Exercise** 4.5 prior to taking the scaling limit (4.77). We start from

$$\operatorname{var}\widehat{Q}_{\mathrm{s}\bar{\mathrm{s}}\,\mathrm{L}}\Big|_{|0,\bar{1}\rangle_{\mathrm{s}\bar{\mathrm{s}}}} = {}_{\mathrm{s}\bar{\mathrm{s}}}\langle 0,\bar{1}|\,\widehat{Q}^{2}_{\mathrm{s}\bar{\mathrm{s}}\,\mathrm{L}}\,|0,\bar{1}\rangle_{\mathrm{s}\bar{\mathrm{s}}} - \left({}_{\mathrm{s}\bar{\mathrm{s}}}\langle 0,\bar{1}|\,\widehat{Q}_{\mathrm{s}\bar{\mathrm{s}}\,\mathrm{L}}\,|0,\bar{1}\rangle_{\mathrm{s}\bar{\mathrm{s}}}\right)^{2}. \tag{4.157}$$

4.8.1 With the help of Eqs. (4.74) and (4.79b), verify that

$$\begin{aligned}
\operatorname{var}\widehat{Q}_{\mathrm{s}\bar{\mathrm{s}}\,f_{\mathrm{L}}}\Big|_{|0,\bar{1}\rangle_{\mathrm{s}\bar{\mathrm{s}}}} &= \\
&+\int \mathrm{d}p\left|\int \mathrm{d}x\, f_{\mathrm{L}}(x)\,[u(p,x)\,u_{\mathrm{B}}(x)+v(p,x)\,v_{\mathrm{B}}(x)]\right|^{2}\\
&+\int \mathrm{d}p\left|\int \mathrm{d}x\, f_{\mathrm{L}}(x)\,[u_{\mathrm{B}}(x)\,u(p,x)-v_{\mathrm{B}}(x)\,v(p,x)]\right|^{2}\\
&+\int \mathrm{d}p\int \mathrm{d}q\left|\int \mathrm{d}x\, f_{\mathrm{L}}(x)\,[u(p,x)\,u(q,x)-v(p,x)\,v(q,x)]\right|^{2}.
\end{aligned} \tag{4.158}$$

4.8.2 Use the completeness relations

$$\begin{aligned}
\int \mathrm{d}p\,\left[u^{*}(p,x)\,u\,(p,x')+u_{\mathrm{B}}(x)\,u_{\mathrm{B}}(x')\right] &= \frac{1}{2}\,\delta(x-x'),\\
\int \mathrm{d}p\,\left[v^{*}(p,x)\,v\,(p,x')+v_{\mathrm{B}}(x)\,v_{\mathrm{B}}(x')\right] &= \frac{1}{2}\,\delta(x-x')
\end{aligned} \tag{4.159}$$

to eliminate the overlaps between bound and continuum states, that is

$$\begin{aligned}
\operatorname{var}\widehat{Q}_{\mathrm{s}\bar{\mathrm{s}}\,f_{\mathrm{L}}}\Big|_{|0,\bar{1}\rangle_{\mathrm{s}\bar{\mathrm{s}}}} &= \\
&+\int \mathrm{d}x\, f_{\mathrm{L}}^{2}(x)\,\left[u_{\mathrm{B}}^{2}(x)+v_{\mathrm{B}}^{2}(x)\right]\\
&-\left\{\int \mathrm{d}x\, f_{\mathrm{L}}(x)\,\left[u_{\mathrm{B}}^{2}(x)-v_{\mathrm{B}}^{2}(x)\right]\right\}^{2}\\
&-\left\{\int \mathrm{d}x\, f_{\mathrm{L}}(x)\,\left[u_{\mathrm{B}}^{2}(x)+v_{\mathrm{B}}^{2}(x)\right]\right\}^{2}\\
&+\int \mathrm{d}p\int \mathrm{d}q\left|\int \mathrm{d}x\, f_{\mathrm{L}}(x)\,[u(p,x)\,u(q,x)-v(p,x)\,v(q,x)]\right|^{2}.
\end{aligned} \tag{4.160}$$

4.8.3 Verify that (use **Exercise** 4.7)

$$\lim_{f_{\mathrm{L}}\to 1}\lim_{r\to\infty}\int \mathrm{d}x\, f_{\mathrm{L}}^{2}(x)\,\left[u_{\mathrm{B}}^{2}(x)+v_{\mathrm{B}}^{2}(x)\right] = \frac{1}{2}, \tag{4.161a}$$

$$\lim_{f_{\mathrm{L}}\to 1}\lim_{r\to\infty}\left\{\int \mathrm{d}x\, f_{\mathrm{L}}(x)\,\left[u_{\mathrm{B}}^{2}(x)-v_{\mathrm{B}}^{2}(x)\right]\right\}^{2} = \left(-\frac{1}{2}\right)^{2} = \frac{1}{4}, \tag{4.161b}$$

$$\lim_{f_{\mathrm{L}}\to 1}\lim_{r\to\infty}\left\{\int \mathrm{d}x\, f_{\mathrm{L}}(x)\,\left[u_{\mathrm{B}}^{2}(x)+v_{\mathrm{B}}^{2}(x)\right]\right\}^{2} = \left(+\frac{1}{2}\right)^{2} = \frac{1}{4}, \tag{4.161c}$$

so that

$$\operatorname{var}\widehat{Q}_{\mathrm{s}\bar{\mathrm{s}}\,\mathrm{L}} = \lim_{f_{\mathrm{L}}\to 1}\lim_{r\to\infty} \operatorname{var}\widehat{Q}_{\mathrm{s}\bar{\mathrm{s}}\,f_{\mathrm{L}}}\Big|_{|0,\bar{1}\rangle_{\mathrm{s}\bar{\mathrm{s}}}} = \lim_{f_{\mathrm{L}}\to 1}\lim_{r\to\infty} \int \mathrm{d}p \int \mathrm{d}q \times \left| \int \mathrm{d}x\, f_{\mathrm{L}}(x)\,[u(p,x)\,u(q,x) - v(p,x)\,v(q,x)] \right|^2 . \tag{4.161d}$$

4.8.4 Verify that (use **Exercise** 4.7)

$$\begin{aligned}\operatorname{var}\widehat{Q}_{\mathrm{s}\bar{\mathrm{s}}\,\mathrm{L}} =& \lim_{f_{\mathrm{L}}\to 1}\lim_{r\to\infty} \operatorname{var}\widehat{Q}_{\mathrm{s}\bar{\mathrm{s}}\,f_{\mathrm{L}}}\Big|_{|0,\bar{1}\rangle_{\mathrm{s}\bar{\mathrm{s}}}}\\ =& \lim_{f_{\mathrm{L}}\to 1}\lim_{r\to\infty} \int \mathrm{d}p \int \mathrm{d}q \frac{1}{[\varepsilon(p)+\varepsilon(q)]^2}\\ &\times \left| \int \mathrm{d}x \left(\frac{\mathrm{d}f_{\mathrm{L}}}{\mathrm{d}x}\right)(x)\,[u(p,x)\,v(q,x) + v(p,x)\,u(q,x)] \right|^2 .\end{aligned} \tag{4.162}$$

4.8.5 Choose f_{L} to be a box test function. Verify that

$$\operatorname{var}\widehat{Q}_{\mathrm{s}\bar{\mathrm{s}}\,\mathrm{L}} = \lim_{f_{\mathrm{L}}\to 1}\lim_{r\to\infty} \operatorname{var}\widehat{Q}_{\mathrm{s}\bar{\mathrm{s}}\,f_{\mathrm{L}}}\Big|_{|0,\bar{1}\rangle_{\mathrm{s}\bar{\mathrm{s}}}} \propto \ln\Lambda \tag{4.163a}$$

diverges logarithmically with the ultraviolet momentum cutoff

$$\phi_\infty^2 \ll p^2 + q^2 < \Lambda^2. \tag{4.163b}$$

4.8.6 Choose $f_{\mathrm{L}}(x) = e^{-\alpha|x+(r/2)|}$ with $\alpha > 0$. Verify that

$$\operatorname{var}\widehat{Q}_{\mathrm{s}\bar{\mathrm{s}}\,\mathrm{L}} = \lim_{\alpha\to 0}\lim_{r\to\infty} \operatorname{var}\widehat{Q}_{\mathrm{s}\bar{\mathrm{s}}\,f_{\mathrm{L}}}\Big|_{|0,\bar{1}\rangle_{\mathrm{s}\bar{\mathrm{s}}}} \propto \lim_{\alpha\to 0}\alpha^2 = 0 \tag{4.164}$$

by making explicit use of the scattering eigenstates of **Exercise** 4.5.

4.9 Let the state

$$|0,0\rangle_{\mathrm{s}\bar{\mathrm{s}}} := \hat{\chi}^{\dagger}_{i_{\mathrm{s}},i_{\bar{\mathrm{s}}};+}(q)\,|\Psi_{\mathrm{s}\bar{\mathrm{s}}}(q)\rangle \tag{4.165a}$$

be defined by Eqs. (4.94a) and (4.94e). Verify that the expectation value of $\widehat{Q}_f$ in the state $|0,0\rangle_{\mathrm{s}\bar{\mathrm{s}}}$ is

$$\begin{aligned}&[f_1\,q + f_2\,(1-q)] + \cdots + \left[f_{i_{\mathrm{s}}-2}\,q + f_{i_{\mathrm{s}}-1}\,(1-q)\right]\\ &+ \left[f_{i_{\mathrm{s}}}\,q \qquad\qquad\qquad\qquad \right]\\ &+ \left[f_{i_{\mathrm{s}}+1}\,q + f_{i_{\mathrm{s}}+2}\,(1-q)\right] + \cdots + \left[f_{i_{\bar{\mathrm{s}}-2}}\,q + f_{i_{\bar{\mathrm{s}}-1}}\,(1-q)\right]\\ &+ \left[\qquad\qquad f_{i_{\bar{\mathrm{s}}}}\,(1-q)\right]\\ &+ \left[f_{i_{\bar{\mathrm{s}}}+1}\,q + f_{i_{\bar{\mathrm{s}}}+2}\,(1-q)\right] + \cdots + [f_{2N-1}\,q + f_{2N}\,(1-q)] .\end{aligned} \tag{4.165b}$$

Let the state

$$|0,\bar{1}\rangle_{\mathrm{s}\bar{\mathrm{s}}} := |\Psi_{\mathrm{s}\bar{\mathrm{s}}}(q)\rangle \tag{4.166a}$$

be defined by Eq. (4.94e). Verify that the expectation value of $\widehat{Q}_f$ in the state $|0,\bar{1}\rangle_{\mathrm{s}\bar{\mathrm{s}}}$ is

$$\begin{aligned} &[f_1\,q+f_2\,(1-q)]+\cdots+\left[f_{i_{\mathrm{s}}-2}\,q+f_{i_{\mathrm{s}}-1}\,(1-q)\right] \\ &+\left[f_{i_{\mathrm{s}}+1}\,q+f_{i_{\mathrm{s}}+2}\,(1-q)\right]+\cdots+\left[f_{i_{\bar{\mathrm{s}}-2}}\,q+f_{i_{\bar{\mathrm{s}}-1}}\,(1-q)\right] \\ &+\left[f_{i_{\bar{\mathrm{s}}}+1}\,q+f_{i_{\bar{\mathrm{s}}}+2}\,(1-q)\right]+\cdots+\left[f_{2N-1}\,q+f_{2N}\,(1-q)\right]. \end{aligned} \tag{4.166b}$$

Verify that

$${}_{\mathrm{s}\bar{\mathrm{s}}}\langle 0,0|\,\widehat{Q}_f\,|0,0\rangle_{\mathrm{s}\bar{\mathrm{s}}}={}_{\mathrm{s}\bar{\mathrm{s}}}\langle 0,\bar{1}|\,\widehat{Q}_f\,|0,\bar{1}\rangle_{\mathrm{s}\bar{\mathrm{s}}}+f_{i_{\mathrm{s}}}\,q+f_{i_{\bar{\mathrm{s}}}}\,(1-q). \tag{4.167}$$

Let

$$|\Psi_{\mathrm{dimer}}(q)\rangle:=\prod_{j=1}^{N}\hat{\chi}^{\dagger}_{2j-1,2j;+}(q)\,|0\rangle. \tag{4.168a}$$

Verify that the expectation value of $\widehat{Q}_f$ in the state $|\Psi_{\mathrm{dimer}}(q)\rangle$ is

$$\begin{aligned} &[f_1\,q+f_2\,(1-q)]+\cdots+\left[f_{i_{\mathrm{s}}-2}\,q+f_{i_{\mathrm{s}}-1}\,(1-q)\right] \\ &+\left[f_{i_{\mathrm{s}}}\,q+f_{i_{\mathrm{s}}+1}\,(1-q)\right] \\ &+\left[f_{i_{\mathrm{s}}+2}\,q+f_{i_{\mathrm{s}}+3}\,(1-q)\right]+\cdots+\left[f_{i_{\bar{\mathrm{s}}-3}}\,q+f_{i_{\bar{\mathrm{s}}-2}}\,(1-q)\right] \\ &+\left[f_{i_{\bar{\mathrm{s}}-1}}\,q+f_{i_{\bar{\mathrm{s}}}}\,(1-q)\right] \\ &+\left[f_{i_{\bar{\mathrm{s}}}+1}\,q+f_{i_{\bar{\mathrm{s}}}+2}\,(1-q)\right]+\cdots+\left[f_{2N-1}\,q+f_{2N}\,(1-q)\right]. \end{aligned} \tag{4.168b}$$

Verify that

$$\langle\Psi_{\mathrm{s}\bar{\mathrm{s}}}(q)|\,\hat{\chi}_{i_{\mathrm{s}},i_{\bar{\mathrm{s}}};+}(q)\,\widehat{Q}_f\,\hat{\chi}^{\dagger}_{i_{\mathrm{s}},i_{\bar{\mathrm{s}}};+}(q)\,|\Psi_{\mathrm{s}\bar{\mathrm{s}}}(q)\rangle=\langle\Psi_{\mathrm{dimer}}(q)|\,\widehat{Q}_f\,|\Psi_{\mathrm{dimer}}(q)\rangle \tag{4.169}$$

if

$$q=1/2 \tag{4.170}$$

for any test function f or if

$$0=f_{i_{\mathrm{s}}+1}-f_{i_{\mathrm{s}}+2}+f_{i_{\mathrm{s}}+3}-f_{i_{\mathrm{s}}+4}+\cdots+f_{i_{\bar{\mathrm{s}}}-4}-f_{i_{\bar{\mathrm{s}}}-3}+f_{i_{\bar{\mathrm{s}}}-2}-f_{i_{\bar{\mathrm{s}}}-1}, \tag{4.171}$$

for any $0\le q\le 1$. *Hint:* Verify that subtracting Eq. (4.165b) from Eq. (4.168b) gives $(1-2\,q)$ times

$$\left(f_{i_{\mathrm{s}}+1}-f_{i_{\mathrm{s}}+2}\right)+\left(f_{i_{\mathrm{s}}+3}-f_{i_{\mathrm{s}}+4}\right)+\cdots+\left(f_{i_{\bar{\mathrm{s}}}-2}-f_{i_{\bar{\mathrm{s}}}-1}\right). \tag{4.172}$$

4.10 Use Eq. (4.129) to prove that the global charge (4.126) is time-independent. [*Hint:* Consider the sum over all sites $i \in \Lambda$ for both sides of Eq. (4.129).]

4.11 Assume the continuity equation

$$(\partial_t \rho)(t, \boldsymbol{r}) + (\boldsymbol{\nabla} \cdot \boldsymbol{j})(t, \boldsymbol{r}) = 0 \tag{4.173a}$$

in the continuum. Let $U_\epsilon(\boldsymbol{r})$ be a small ball centered at $\boldsymbol{r}$. Introduce two symbols $n\big(t, U_\epsilon(\boldsymbol{r})\big)$ and $J\big(t, U_\epsilon(\boldsymbol{r})\big)$ and assume that they are related by the equation

$$(\partial_t n)(t, U_\epsilon(\boldsymbol{r})) + J(t, U_\epsilon(\boldsymbol{r})) = 0. \tag{4.173b}$$

How is n linearly related to ρ and how is $\boldsymbol{\nabla} \cdot \boldsymbol{j}$ linearly related to J if Eq. (4.173a) is to follow from Eq. (4.173b)?

5

From Spin-1/2 Cluster c Chains to Majorana c Chains

5.1 Introduction

Thus far, we have mostly focused on the zero modes that appear on domain walls (solitons) in the middle of a one-dimensional lattice on which the electrons of polyacetylene hop. The same physics is at play at the endpoints of a finite (or semi-infinite) chain. To see the connection, consider explicitly the lattice solution

$$\psi_{2l}^{(\mathrm{zm})} = \left(-\frac{t'}{t}\right)^{l} \psi_{0}^{(\mathrm{zm})} \tag{5.1}$$

of Eq. (3.8a) for a semi-infinite chain starting at site $j = 0$ with dimerized real-valued hopping hoppings t and $t' = t - \delta t$. There are two cases to be considered: (i) $t' < t$, in which case one obtains the wave function amplitudes $\psi_{2l}^{(\mathrm{zm})}$ that decay exponentially with $l > 0$, away from the boundary at $l = 0$; and (ii) $t' > t$, in which case the zero mode solution is non-normalizable and, consequently, discarded. Therefore, we encounter the two situations depicted in Figure 5.1 (b and c) that depend on the dimerization pattern near the left boundary, respectively.

It is with the theoretical study of soliton-like excitations in polyacetylene that the physics community encountered the first examples of what are now called topological band insulators, that is noninteracting fermionic examples of phases of matter with (i) a nondegenerate gapped ground state if periodic boundary conditions are imposed and (ii) protected boundary zero modes if open boundary conditions are imposed.

We shall present a formalism that will allow us to understand when a system that is gapped in the bulk if periodic boundary conditions are considered, as is the case for the dimerized polyacetelene chain, can yield zero modes on the boundary if the chain is opened. The two situations in Figure 5.1 correspond to two distinct phases in the classification scheme of this section. Tuning the counterpart to the dimensionless ratio t'/t for polyacetylene allows one to switch from one situation to the other.

Moreover, we will also introduce examples of fractionalization beyond fermionic systems, which comprised the focus thus far. We present solvable models of quantum

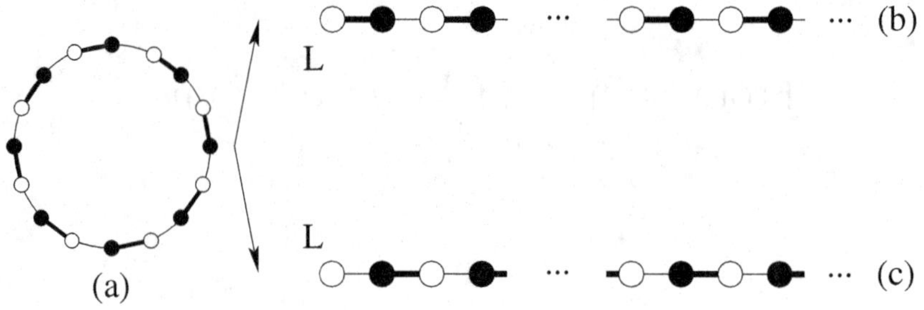

Figure 5.1 (a) Polyacetylene in a ring geometry with one of two possible dimerization pattern [strong (weak) hopping amplitude is the thick (thin) bond]. (b) Semi-infinite dimerized polyacetylene obtained by removing a weak bond in panel (a). (c) Semi-infinite dimerized polyacetylene obtained by removing a strong bond in panel (a). In panel (b) and (c) there is one and only one boundary, denoted L, with the coordinate 0. Figure 5.1 should be compared with Figure 6.6.

spin-1/2 Hamiltonians that, much as the example of polyacetelene above, have protected zero modes at the boundary. These examples illustrate the workings of a general framework introduced by Gu and Wen[1] that (i) is capable of encapsulating bosonic systems, (ii) is extendable to fermionic systems, and (iii) classifies what that they called symmetry-protected topological (SPT) phases of matter. The ground states of these SPT phases of matter have the following properties.

(i) They are gapped and nondegenerate if periodic boundary conditions are imposed.
(ii) They are degenerate if open boundary conditions are imposed owing to the existence of boundary zero modes protected by symmetries.
(iii) They can be deformed adiabatically (that is without closing the gap) to a product of single-particle states in the absence of the protecting symmetries if periodic boundary conditions are imposed.

The adjective topological is motivated here by the following intuition. First, eigenvalue degeneracies that originate from symmetries are not expected to change with either the system size or the choice of symmetry-preserving boundary conditions. Second, two distinct ground-state degeneracies for a given SPT phase (when imposing open boundary conditions) cannot be changed smoothly into one another by local perturbations that preserve the protecting symmetries. Interpolation without violating the protecting symmetries between two SPT phases that are only distinct by their ground-state degeneracies (when imposing open boundary conditions) requires the closing of the gap in the bulk, that is a quantum phase transition.

The first historical example of SPT phase of matter in a quantum spin system is the nearest-neighbor antiferromagnetic quantum spin-1 chain. Even though this

[1] Z.-C. Gu and X.-G. Wen, Phys. Rev. B **80**(15), 155131 (2009). [Gu and Wen (2009)].

lattice Hamiltonian had been studied since the early days of quantum mechanics, the recognition that it is an example of a (symmetry-protected) topological phase occurred much later, namely, between 1983 and 1989. This theoretical discovery was initiated by the conjecture of Haldane that the nearest-neighbor antiferromagnetic quantum spin-S chain obeying periodic boundary conditions is gapless when S is a half integer but gapful when S is an integer.[2] The second milestone was achieved by Affleck, Kennedy, Lieb, and Tasaki (AKLT).[3] Affleck, Kennedy, Lieb, and Tasaki added to the nearest-neighbor antiferromagnetic quantum spin-1 chain a specific biquadratic nearest-neighbor interaction and proved that, in the thermodynamic limit, (i) the ground state is nondegenerate and separated from all excited states by a gap when periodic boundary conditions hold and (ii) the ground state is fourfold degenerate due to one boundary zero mode per edge carrying the spin-quantum number $1/2$ when open boundary conditions hold. This spectral sensitivity to boundary conditions is a topological signature, as was the third milestone, the discovery by den Nijs and Rommelse[4] that a signature of the Haldane phase is that it is characterized by a nonvanishing expectation value for a nonlocal (string-like) order parameter.

Topological band insulators are not quite compatible with the terminology SPT of Gu and Wen, for fermions obey the Pauli principle. Indeed, even in the atomic limit for which the position of a fermion becomes a good quantum number, the many-body ground state, a Slater determinant, is entangled relative to the single-particle product state. Moreover, the protection of the boundary states of topological band insulators cannot always be ascribed to a symmetry that can be switched on and off on demand, for this protection can stem solely from the conservation of the fermion parity, which is postulated to always hold in physics. A terminology that allows to treat bosonic SPT phases of matter on equal footing with the strong topological insulators or superconductors is that of invertible topological phases of matter. The rational for using the adjective "invertible" will be developed in Chapters 5 and 6. Fermionic invertible topological phases of matter occur whenever the gapped fermionic many-body ground states are nondegenerate if periodic boundary conditions are imposed, while the fermionic many-body ground states are degenerate if open boundary conditions are imposed.

The purpose of this chapter is to construct and study a family of local spin-$1/2$ Hamiltonians on a chain in Sections 5.2 and 5.3, respectively, that are related to both the physics of polyacetylene and the physics of the AKLT Hamiltonian as will be shown in Sections 5.4–5.7. Hereto, we shall make use of a very important transformation called the Jordan-Wigner transformation in Section 5.4. This is a

[2] F. D. M. Haldane, Phys. Lett. A **93**(9), 464–468 (1983); Phys. Rev. Lett. **50**(15), 1153–1156 (1983). [Haldane (1983a,b)].

[3] I. Affleck, T. Kennedy, E. H. Lieb, and H. Tasaki, Phys. Rev. Lett. **59**(7), 799–802 (1987); Commun. Math. Phys. **115**(3), 477–528 (1988). [Affleck et al. (1987, 1988)].

[4] M. den Nijs and K. Rommelse, Phys. Rev. B **40**(7), 4709–4734 (1989). [den Nijs and Rommelse (1989)].

nonlocal transformation that converts products of spin-1/2 operators along strings into local Majorana operators and, conversely, converts products of Majorana operators along strings into local spin-1/2 operators.

This family of local spin-1/2 Hamiltonians on a chain will be called spin-1/2 cluster c chains and will be labeled by the integer $c \in \mathbb{Z}$. By design, a spin-1/2 cluster c chain is mapped with the help of a Jordan-Wigner transformation introduced in Section 5.4 into a local quadratic form of Majorana operators that we shall call a Majorana c chain. Majorana c chains will be studied in Sections 5.5–5.7.

The Majorana $c=0$ chain is nothing but a chain of decoupled sites, each of which supports two quantum levels. It realizes the atomic limit of a topologically trivial fermionic band insulator. The Majorana $c = 1$ chain is much more interesting. It is a special case of a one-dimensional tight-binding Hamiltonian realizing a Bogoliubov-de-Gennes superconductor called a Kitaev chain.[5] Kitaev was motivated by quantum information for which Majorana modes might play an important role for performing quantum computations that are robust to quantum decoherence. Kitaev designed a superconducting ring that is gapped but supports one Majorana zero mode per end when it is cut open. The Majorana $c > 1$ chain is nothing but a stacking of c Majorana 1 chains (with the $c < 0$ chain resulting from an inversion in space of the $c > 0$ chain). Hence, it supports $|c|$ Majorana zero modes per ends. As was the case when stacking identical copies of polyacetylene, the number of zero modes can be used as a topological index that distinguishes distinct topological gapped phases of noninteracting Majorana matter. The Majorana character of the zero modes implies an even-odd effect depending on whether the number of Majorana states per boundary is even or odd, as will be discussed in detail in Section 5.7 when relating the Hilbert space of closed Majorana c chains to the ground states of open Majorana c chains. This even-odd effect originates from the fact that any spin-1/2 cluster c chain with c an odd integer breaks spontaneously the global time-reversal symmetry for spin-1/2 degrees of freedom, a symmetry that is absent when c is even.

A natural question is then what is the fate of these zero modes when the Majorana c chain is perturbed by terms of order higher than two in the Majorana operators? This question was answered by Fidkowski and Kitaev[6] and Turner, Pollmann, and Berg.[7] Both groups found that the topological classification of Majorana c chains by the integer $c \in \mathbb{Z}$ is reduced to the finite group $\mathbb{Z}_8$ if Majorana c chains are perturbed by interactions that preserve the protecting symmetries of Majorana c chains. In fact, we have already encountered examples in Section 3.12 for which interactions change the set of topological indices characterizing a noninteracting

[5] A. Kitaev, Phys. Usp. **44**(10S), 131–136 (2001). [Kitaev (2001)].

[6] L. Fidkowski and A. Kitaev, Phys. Rev. B **81**(13), 134509 (2010); Phys. Rev. B **83**(7), 075103 (2011). [Fidkowski and Kitaev (2010, 2011)].

[7] A. M. Turner, F. Pollmann, and E. Berg, Phys. Rev. B **83**(7), 075102 (2011). [Turner et al. (2011)].

invertible topological phases of fermions. Our strategy to establish this breakdown is inspired by the 2011 paper of Fidkowski and Kitaev (see footnote 6).

Undoing the Jordan-Wigner transformation that maps the spin-1/2 cluster c chain to the Majorana c chain suggests that the ground-state degeneracies of open spin-1/2 cluster c chains are also unstable to generic local interactions that commute with the protecting symmetries of the spin-1/2 cluster c chains. It turns out that this stability analysis can always be reduced to the cases for which c is an even integer as is explained in Section 5.3. Indeed, we will show using the notion of symmetry fractionalization in Section 5.3 that the topological classification of spin-1/2 cluster $c = 2m$ chains is reduced to the finite group $\mathbb{Z}_4$ if the integrability of the spin-1/2 cluster $c = 2m$ chains is broken by interactions that preserve the protecting symmetry of spin-1/2 cluster $c = 2m$ chains. On the other hand, spin-1/2 cluster $c = 2m + 1$ chains, unlike the spin-1/2 cluster $c = 2m$ chains, are symmetric under reversal of time. This implies a Kramers' degeneracy of the energy eigenvalues that is lifted in the thermoydnamic limit by the mechanism of spontaneous symmetry breaking. The effective Hamiltonian governing the quantum dynamics of a spin-1/2 cluster $c = 2m + 1$ chain, once spontaneous symmetry breaking has taken place, is the same as that for a spin-1/2 cluster $c = 2m$ chain.

One important lesson that will be drawn from this chapter is that, whereas the symmetry-protected degeneracy of the ground states of an open Majorana c chain is solely to be attributed to the phenomenon of symmetry fractionalization, the symmetry-protected degeneracy of the ground states of an open spin-1/2 cluster c chain originates either from symmetry fractionalization or from spontaneous symmetry breaking or from both depending on the value of c. This is so because of the nonlocal nature of the Jordan-Wigner transformation relating a spin-1/2 cluster c chain to a Majorana c chain. Other important lessons are the following.

1. The Majorana $c = 1$ chain can be fermionized (locally) into a special case of a superconducting Kitaev chain (see Section 5.5.2).
2. Linear combinations of the Majorana $c = 0$ and $c = 2$ chains are unitarily equivalent to a spinless SSH Hamiltonian through a two-site unitary transformation (which is thus local) (see Section 5.5.3).
3. Linear combinations of the Majorana $c = 0$ and $c = 4$ chains are unitarily equivalent to a spinfull SSH Hamiltonian through a four-site unitary transformation (which is thus local) (see Section 5.5.4).

Drawing on the (nonlocal) Jordan-Wigner transformation, we infer the following.

1′. A special case of a superconducting Kitaev chain is unitarily equivalent to the Ising model (the spin-1/2 cluster $c = 1$ chain).

2′. A spinless SSH Hamiltonian is unitarily equivalent to a linear combination of spin-1/2 cluster $c = 0$ and $c = 2$ chains.

3′. A spinfull SSH Hamiltonian is unitarily equivalent to a linear combination of spin-1/2 cluster $c = 0$ and $c = 4$ chains.

5.2 Spin-1/2 Cluster c Chains

It was realized by Suzuki in 1971[8] that a family of classical statistical models on two-dimensional lattices were related to the family of spin-1/2 cluster c chains. Susuki also realized that a spin-1/2 cluster c chain could be diagonalized by mapping it into noninteracting fermions hopping on a chain.[9] Fast forward to 2011 when Son et al. (who were unaware of the work done by Suzuki 40 years earlier) showed that the spin-1/2 cluster $c = 2$ chain realizes an invertible topological phase protected by a $\mathbb{Z}_2 \times \mathbb{Z}_2$ symmetry.[10] As the spin-1/2 cluster $c = 0$ chain realizes a featureless paramagnetic phase, while the spin-1/2 cluster $c = 1$ chain realizes the one-dimensional Ising model, the goal of Section 5.2 is to establish in systematic fashion the interplay between spontaneous symmetry breaking (SSB) and topological symmetry protection (TSP) as a function of $c \in \mathbb{Z}$. We will see that SSB and TSP can coexist when $|c| > 1$ is odd, whereas only TSP is possible when $|c| > 0$ is even.

5.2.1 *Definition, Symmetries, and Energy Spectra*

5.2.1.1 *Hilbert Space*

Let

$$\Lambda := \{j \,|\, j = 1, \cdots, N\} \tag{5.2a}$$

denote a chain of sites with two consecutive sites a distance (the lattice spacing) $\mathfrak{a}$ apart. We allow the strictly positive integer N to be either even or odd in Section 5.2. We shall then choose N to be even. To each site $j \in \Lambda$, we assign the four Hermitean 2×2 matrices

$$\begin{aligned} I_j &:= \begin{pmatrix} 1 & 0 \\ 0 & 1 \end{pmatrix}, & Z_j &:= \begin{pmatrix} +1 & 0 \\ 0 & -1 \end{pmatrix}, \\ X_j &:= \begin{pmatrix} 0 & 1 \\ 1 & 0 \end{pmatrix}, & Y_j &:= \begin{pmatrix} 0 & -\mathrm{i} \\ +\mathrm{i} & 0 \end{pmatrix}. \end{aligned} \tag{5.2b}$$

These four matrices act on the "local" vector space

$$\mathfrak{h}_j \cong \mathbb{C}^2. \tag{5.2c}$$

[8] M. Suzuki, Phys. Lett. A **34**(6), 338–339 (1971). [Suzuki (1971a)].
[9] M. Suzuki, Prog. Theor. Phys. **46**(5), 1337–1359 (1971). [Suzuki (1971b)].
[10] W. Son, L. Amico, R. Fazio, et al., Europhys. Lett. **95**(5), 50001 (2011). [Son et al. (2011)].

The "global" Hilbert space under consideration is

$$\mathfrak{H} := \bigotimes_{j=1}^{N} \mathfrak{h}_j \cong \mathbb{C}^{2^N}. \tag{5.3a}$$

On this Hilbert space, we define the operators

$$\begin{aligned} \widehat{\mathbb{1}} &:= \bigotimes_{j'\in\Lambda} I_{j'}, & \widehat{Z}_j &:= \left(\bigotimes_{j'\in\Lambda}^{j'\neq j} I_{j'}\right) \otimes Z_j, \\ \widehat{X}_j &:= \left(\bigotimes_{j'\in\Lambda}^{j'\neq j} I_{j'}\right) \otimes X_j, & \widehat{Y}_j &:= \left(\bigotimes_{j'\in\Lambda}^{j'\neq j} I_{j'}\right) \otimes Y_j, \end{aligned} \tag{5.3b}$$

where, for any $j \in \Lambda$, we shall impose the conditions

$$I_j \equiv I_{j+N}, \qquad X_j \equiv X_{j+N}, \qquad Y_j \equiv Y_{j+N}, \qquad Z_j \equiv Z_{j+N}. \tag{5.3c}$$

With conditions (5.3c), one can readily impose periodic boundary conditions.

Before we define the family of spin-1/2 cluster c chains, we review some useful transformations on the operators defined in Eq. (5.3).

The transformation

$$\widehat{X}_j \mapsto e^{+\mathrm{i}\pi\widehat{X}_j/4}\,\widehat{X}_j\,e^{-\mathrm{i}\pi\widehat{X}_j/4} = \widehat{X}_j, \tag{5.4a}$$

$$\widehat{Y}_j \mapsto e^{+\mathrm{i}\pi\widehat{X}_j/4}\,\widehat{Y}_j\,e^{-\mathrm{i}\pi\widehat{X}_j/4} = -\widehat{Z}_j, \tag{5.4b}$$

$$\widehat{Z}_j \mapsto e^{+\mathrm{i}\pi\widehat{X}_j/4}\,\widehat{Z}_j\,e^{-\mathrm{i}\pi\widehat{X}_j/4} = +\widehat{Y}_j \tag{5.4c}$$

realizes a local clockwise rotation by $\pi/2$ about the axis $\widehat{X}_j$. The transformation

$$\widehat{X}_j \mapsto e^{+\mathrm{i}\pi\widehat{Y}_j/4}\,\widehat{X}_j\,e^{-\mathrm{i}\pi\widehat{Y}_j/4} = +\widehat{Z}_j, \tag{5.5a}$$

$$\widehat{Y}_j \mapsto e^{+\mathrm{i}\pi\widehat{Y}_j/4}\,\widehat{Y}_j\,e^{-\mathrm{i}\pi\widehat{Y}_j/4} = \widehat{Y}_j, \tag{5.5b}$$

$$\widehat{Z}_j \mapsto e^{+\mathrm{i}\pi\widehat{Y}_j/4}\,\widehat{Z}_j\,e^{-\mathrm{i}\pi\widehat{Y}_j/4} = -\widehat{X}_j \tag{5.5c}$$

realizes a local clockwise rotation by $\pi/2$ about the axis $\widehat{Y}_j$. The transformation

$$\widehat{X}_j \mapsto e^{+\mathrm{i}\pi\widehat{Z}_j/4}\,\widehat{X}_j\,e^{-\mathrm{i}\pi\widehat{Z}_j/4} = -\widehat{Y}_j, \tag{5.6a}$$

$$\widehat{Y}_j \mapsto e^{+\mathrm{i}\pi\widehat{Z}_j/4}\,\widehat{Y}_j\,e^{-\mathrm{i}\pi\widehat{Z}_j/4} = +\widehat{X}_j, \tag{5.6b}$$

$$\widehat{Z}_j \mapsto e^{+\mathrm{i}\pi\widehat{Z}_j/4}\,\widehat{Z}_j\,e^{-\mathrm{i}\pi\widehat{Z}_j/4} = \widehat{Z}_j \tag{5.6c}$$

realizes a local clockwise rotation by $\pi/2$ about the axis $\widehat{Z}_j$. The corresponding global clockwise rotations by $\pi/2$ follow from conjugation with the unitary operators

$$\widehat{R}^{X}_{\pi/2} := \prod_{j=1}^{N} e^{+\mathrm{i}\pi\,\widehat{X}_j/4}, \qquad \left(\widehat{R}^{X}_{\pi/2}\right)^2 = \widehat{R}^{X}_{\pi}, \tag{5.7a}$$

$$\widehat{R}^{Y}_{\pi/2} := \prod_{j=1}^{N} e^{+\mathrm{i}\pi\,\widehat{Y}_j/4}, \qquad \left(\widehat{R}^{Y}_{\pi/2}\right)^2 = \widehat{R}^{Y}_{\pi}, \tag{5.7b}$$

$$\widehat{R}^{Z}_{\pi/2} := \prod_{j=1}^{N} e^{+\mathrm{i}\pi\,\widehat{Z}_j/4}, \qquad \left(\widehat{R}^{Z}_{\pi/2}\right)^2 = \widehat{R}^{Z}_{\pi}, \tag{5.7c}$$

respectively.

Define the global unitary transformation

$$\widehat{X} := \prod_{j=1}^{N} \widehat{X}_j \tag{5.8a}$$

that squares to unity

$$\left(\widehat{X}\right)^2 = \widehat{1}. \tag{5.8b}$$

The transformation

$$\widehat{X}_j \mapsto \widehat{X}\,\widehat{X}_j\,\widehat{X} = \widehat{X}_j, \tag{5.9a}$$

$$\widehat{Y}_j \mapsto \widehat{X}\,\widehat{Y}_j\,\widehat{X} = -\widehat{Y}_j, \tag{5.9b}$$

$$\widehat{Z}_j \mapsto \widehat{X}\,\widehat{Z}_j\,\widehat{X} = -\widehat{Z}_j \tag{5.9c}$$

realizes the global rotation by π about the X axis $(x,y,z) \mapsto (x,-y,-z)$ on the operators defined in Eq. (5.3).

Define the global unitary transformation

$$\widehat{Y} := \prod_{j=1}^{N} \widehat{Y}_j \tag{5.10a}$$

that squares to unity

$$\left(\widehat{Y}\right)^2 = \widehat{1}. \tag{5.10b}$$

The transformation

$$\widehat{X}_j \mapsto \widehat{Y}\,\widehat{X}_j\,\widehat{Y} = -\widehat{X}_j, \tag{5.11a}$$

$$\widehat{Y}_j \mapsto \widehat{Y}\,\widehat{Y}_j\,\widehat{Y} = \widehat{Y}_j, \tag{5.11b}$$

$$\widehat{Z}_j \mapsto \widehat{Y}\,\widehat{Z}_j\,\widehat{Y} = -\widehat{Z}_j \tag{5.11c}$$

realizes the global rotation by π about the Y axis $(x,y,z) \mapsto (-x,y,-z)$ on the operators defined in Eq. (5.3).

Define the global unitary transformation

$$\widehat{Z} := \prod_{j=1}^{N} \widehat{Z}_j \tag{5.12a}$$

that squares to unity

$$\left(\widehat{Z}\right)^2 = \widehat{\mathbb{1}}. \tag{5.12b}$$

The transformation

$$\widehat{X}_j \mapsto \widehat{Z}\,\widehat{X}_j\,\widehat{Z} = -\widehat{X}_j, \tag{5.13a}$$

$$\widehat{Y}_j \mapsto \widehat{Z}\,\widehat{Y}_j\,\widehat{Z} = -\widehat{Y}_j, \tag{5.13b}$$

$$\widehat{Z}_j \mapsto \widehat{Z}\,\widehat{Z}_j\,\widehat{Z} = \widehat{Z}_j \tag{5.13c}$$

realizes the global rotation by π about the Z axis $(x,y,z) \mapsto (-x,-y,z)$ on the operators defined in Eq. (5.3).

Let K denote complex conjugation. Define the global antiunitary transformation

$$\widehat{T} := \widehat{Y}\,\mathsf{K} \equiv \left(\prod_{j=1}^{N} \widehat{Y}_j\right)\mathsf{K} \tag{5.14a}$$

that squares to minus unity when the number N of sites is odd:

$$\left(\widehat{T}\right)^2 = (-1)^N\,\widehat{\mathbb{1}}. \tag{5.14b}$$

The transformation

$$\widehat{X}_j \mapsto \widehat{T}\,\widehat{X}_j\,\widehat{T}^{-1} = -\widehat{X}_j, \tag{5.15a}$$

$$\widehat{Y}_j \mapsto \widehat{T}\,\widehat{Y}_j\,\widehat{T}^{-1} = -\widehat{Y}_j, \tag{5.15b}$$

$$\widehat{Z}_j \mapsto \widehat{T}\,\widehat{Z}_j\,\widehat{T}^{-1} = -\widehat{Z}_j \tag{5.15c}$$

realizes the global inversion $(x,y,z) \mapsto (-x,-y,-z)$ on the operators defined in Eq. (5.3). This is the transformation law expected from reversal of time for a quantum spin-1/2 degree of freedom.

The global mirror transformations

$$\begin{pmatrix}\widehat{X}_j\\ \widehat{Y}_j\\ \widehat{Z}_j\end{pmatrix}\mapsto\begin{pmatrix}\widehat{X}_j\\ \widehat{Y}_j\\ -\widehat{Z}_j\end{pmatrix}, \tag{5.16a}$$

$$\begin{pmatrix}\widehat{X}_j\\ \widehat{Y}_j\\ \widehat{Z}_j\end{pmatrix}\mapsto\begin{pmatrix}\widehat{X}_j\\ -\widehat{Y}_j\\ \widehat{Z}_j\end{pmatrix}, \tag{5.16b}$$

$$\begin{pmatrix}\widehat{X}_j\\ \widehat{Y}_j\\ \widehat{Z}_j\end{pmatrix}\mapsto\begin{pmatrix}-\widehat{X}_j\\ \widehat{Y}_j\\ \widehat{Z}_j\end{pmatrix} \tag{5.16c}$$

for any $j \in \Lambda$ are realized by conjugation with the global antiunitary transformations

$$\widehat{M}^{XY} := \widehat{Z}\,\widehat{T}, \qquad \left(\widehat{M}^{XY}\right)^2 = \widehat{\mathbb{1}}, \tag{5.17a}$$

$$\widehat{M}^{ZX} := \widehat{Y}\,\widehat{T}, \qquad \left(\widehat{M}^{ZX}\right)^2 = \widehat{\mathbb{1}}, \tag{5.17b}$$

$$\widehat{M}^{YZ} := \widehat{X}\,\widehat{T}, \qquad \left(\widehat{M}^{YZ}\right)^2 = \widehat{\mathbb{1}}, \tag{5.17c}$$

respectively.

Define the global antiunitary transformation

$$\widehat{T}' := \mathsf{K} \tag{5.18a}$$

that squares to unity

$$\left(\widehat{T}'\right)^2 = \widehat{\mathbb{1}}. \tag{5.18b}$$

The transformation

$$\widehat{X}_j \mapsto \widehat{T}'\,\widehat{X}_j\,\widehat{T}' = \widehat{X}_j, \tag{5.19a}$$

$$\widehat{Y}_j \mapsto \widehat{T}'\,\widehat{Y}_j\,\widehat{T}' = -\widehat{Y}_j, \tag{5.19b}$$

$$\widehat{Z}_j \mapsto \widehat{T}'\,\widehat{Z}_j\,\widehat{T}' = \widehat{Z}_j \tag{5.19c}$$

realizes the global mirror transformation $(x, y, z) \mapsto (x, -y, z)$ on the operators defined in Eq. (5.3).

Define the global antiunitary transformation

$$\widehat{\Pi}^{XY} := \left(\prod_{j=1}^{N} e^{-\mathrm{i}\pi\,\widehat{Z}_j/4}\right)\mathsf{K} \tag{5.20a}$$

that squares to unity

$$\left(\widehat{\Pi}^{XY}\right)^2 = \widehat{\mathbb{1}}. \tag{5.20b}$$

The transformation

$$\widehat{X}_j \mapsto \widehat{\Pi}^{XY}\,\widehat{X}_j\,\widehat{\Pi}^{XY} = \widehat{Y}_j, \tag{5.21a}$$

$$\widehat{Y}_j \mapsto \widehat{\Pi}^{XY}\,\widehat{Y}_j\,\widehat{\Pi}^{XY} = \widehat{X}_j, \tag{5.21b}$$

$$\widehat{Z}_j \mapsto \widehat{\Pi}^{XY}\,\widehat{Z}_j\,\widehat{\Pi}^{XY} = \widehat{Z}_j \tag{5.21c}$$

realizes the global permutation $(x,y,z) \mapsto (y,x,z)$ on the operators defined in Eq. (5.3).

Similarly, the antiunitary transformation

$$\widehat{\Pi}^{YZ} := \left(\prod_{j=1}^{N} e^{+\mathrm{i}\pi\,\widehat{X}_j/4}\right)\mathsf{K} \tag{5.22a}$$

with

$$\left(\widehat{\Pi}^{YZ}\right)^2 = \widehat{\mathbb{1}} \tag{5.22b}$$

realizes the global permutation $(x,y,z) \mapsto (x,z,y)$ on the operators defined in Eq. (5.3).

At last, the transformation

$$\widehat{\Pi}^{ZX} := \left(\prod_{j=1}^{N} e^{-\mathrm{i}\pi\,\widehat{Y}_j/4}\right)\widehat{Z}\,\mathsf{K} \tag{5.23a}$$

with

$$\left(\widehat{\Pi}^{ZX}\right)^2 = \widehat{\mathbb{1}} \tag{5.23b}$$

realizes the global permutation $(x,y,z) \mapsto (z,y,x)$ on the operators defined in Eq. (5.3).

The relations

$$\widehat{\Pi}^{YZ}\,\widehat{T}'\,\widehat{\Pi}^{YZ} = (+\mathrm{i})^N\,\widehat{X}\,\widehat{T}' \tag{5.24a}$$

with

$$\left(\widehat{\Pi}^{YZ}\,\widehat{T}'\,\widehat{\Pi}^{YZ}\right)^2 = \widehat{\mathbb{1}}, \tag{5.24b}$$

and

$$\widehat{\Pi}^{YZ}\left(\widehat{X}\,\widehat{T}'\right)\widehat{\Pi}^{YZ} = (+\mathrm{i})^N\,\widehat{T}' \tag{5.24c}$$

with

$$\left(\widehat{\Pi}^{YZ}\left(\widehat{X}\,\widehat{T}'\right)\widehat{\Pi}^{YZ}\right)^2 = \widehat{\mathbb{1}} \tag{5.24d}$$

will allow to relate pairwise the spin-1/2 cluster c chains that we now introduce.

Table 5.1 *Symmetries of the spin-1/2 cluster c chains*

$c\in\mathbb{Z}$	$\widehat{H}_c^{(b_c)} := -\frac{\hbar\omega}{2}\sum_{j=1}^{N-b_c}\widehat{\mathrm{C}}_j$	$\widehat{X},\widehat{T}'$	$\widehat{Y},\widehat{Z},\widehat{T}$	$\widehat{\Pi}^{YZ}$
$c<0$	$\widehat{\mathrm{C}}_j=\widehat{Y}_j\widehat{X}_{j+1}\cdots\widehat{X}_{j+\lvert c\rvert-1}\widehat{Y}_{j+\lvert c\rvert}$	$\widehat{H}_c^{(b_c)}$	$(-1)^{\lvert c\rvert+1}\,\widehat{H}_c^{(b_c)}$	$\widehat{H}_{-c}^{(b_{-c})}$
$c=0$	$\widehat{\mathrm{C}}_j=\widehat{X}_j$	$\widehat{H}_0^{(b_0)}$	$-\widehat{H}_0^{(b_0)}$	$\widehat{H}_0^{(b_0)}$
$c>0$	$\widehat{\mathrm{C}}_j=\widehat{Z}_j\widehat{X}_{j+1}\cdots\widehat{X}_{j+c-1}\widehat{Z}_{j+c}$	$\widehat{H}_c^{(b_c)}$	$(-1)^{c+1}\,\widehat{H}_c^{(b_c)}$	$\widehat{H}_{-c}^{(b_{-c})}$

Transformation laws obeyed by the spin-1/2 cluster c chain $\widehat{H}_c^{(b_c)}$ defined in Eq. (5.25) under the conjugation $\widehat{H}_c^{(b_c)}\mapsto\widehat{U}\,\widehat{H}_c^{(b_c)}\,\widehat{U}^{-1}$ with $\widehat{U}=\widehat{X},\widehat{T}',\widehat{Y},\widehat{Z},\widehat{T},\widehat{\Pi}^{YZ}$.

5.2.1.2 Hamiltonians

Let $\hbar\omega>0$ define a characteristic energy scale. Let $c\in\mathbb{Z}$ be an integer and let $N\geq|c|+1$ be another integer. Let $b_c=0$ if periodic boundary conditions (PBC) are imposed or $b_c=|c|$ if open boundary conditions (OPC) are imposed. If $c<-1$, define

$$\widehat{H}_{c<0}^{(b_c)} := -\frac{\hbar\omega}{2}\sum_{j=1}^{N-b_c}\widehat{Y}_j\,\widehat{X}_{j+1}\cdots\widehat{X}_{j+|c|-1}\,\widehat{Y}_{j+|c|}. \tag{5.25a}$$

If $c=-1,0,+1$, define

$$\widehat{H}_c^{(b_c)} := \begin{cases} -\frac{\hbar\omega}{2}\sum_{j=1}^{N-b_c}\widehat{Y}_j\,\widehat{Y}_{j+1}, & c=-1,\\ -\frac{\hbar\omega}{2}\sum_{j=1}^{N}\widehat{X}_j, & c=0,\\ -\frac{\hbar\omega}{2}\sum_{j=1}^{N-b_c}\widehat{Z}_j\,\widehat{Z}_{j+1}, & c=+1. \end{cases} \tag{5.25b}$$

If $c>1$, define

$$\widehat{H}_{c>0}^{(b_c)} := -\frac{\hbar\omega}{2}\sum_{j=1}^{N-b_c}\widehat{Z}_j\,\widehat{X}_{j+1}\cdots\widehat{X}_{j+c-1}\,\widehat{Z}_{j+c}. \tag{5.25c}$$

We call this family of spin-1/2 Hamiltonians the family of spin-1/2 cluster c chains.

Because any spin-1/2 cluster c chain (5.25) is represented by a real-valued Hamiltonian, it commutes with $\widehat{T}'$ defined in Eq. (5.18):

$$\left[\widehat{T}', \widehat{H}_c^{(b_c)}\right] = 0. \tag{5.26}$$

Hence, $\widehat{T}'$ generates the symmetry group $\mathbb{Z}_2$, that is, we can identify the set

$$\{\widehat{T}', (\widehat{T}')^2\} \tag{5.27a}$$

with the cyclic group

$$\mathbb{Z}_2 := \{e, g \,|\, e\, g = g\, e, \quad g^2 = e\} \tag{5.27b}$$

made of two elements through the group isomorphism

$$e \mapsto \left(\widehat{T}'\right)^2, \qquad g \mapsto \widehat{T}'. \tag{5.27c}$$

Because any spin-1/2 cluster c chain (5.25) is a quadratic form in the $\widehat{Z}$'s when $c > 0$, is a quadratic form in the $\widehat{Y}$'s when $c < 0$, and depends solely on the $\widehat{X}$'s when $c = 0$, it commutes with $\widehat{X}$ defined in Eq. (5.8):

$$\left[\widehat{X}, \widehat{H}_c^{(b_c)}\right] = 0. \tag{5.28}$$

Hence, $\widehat{X}$ generates the symmetry group $\mathbb{Z}_2$.

Either the unitary transformation (5.7a) or the antiunitary transformation (5.22) realizes the map

$$\widehat{H}_{c>0}^{(b_c)} \mapsto \widehat{H}_{c<0}^{(b_c)}, \qquad \widehat{H}_{c<0}^{(b_c)} \mapsto \widehat{H}_{c>0}^{(b_c)}. \tag{5.29}$$

Any spin-1/2 cluster c chain (5.25) obeys the transformation law

$$\widehat{H}_c^{(b_c)} \mapsto (-1)^{c+1}\, \widehat{H}_c^{(b_c)} \tag{5.30}$$

under any one of the global unitary transformations $\widehat{Y}$ defined in Eq. (5.10) or $\widehat{Z}$ defined in Eq. (5.12) as well as under the antiunitary transformation $\widehat{T}$ defined in Eq. (5.14). Hence, any one of $\widehat{Y}$, $\widehat{Z}$, or $\widehat{T}$ generates the symmetry group $\mathbb{Z}_2$ when c is an odd integer.

Table 5.1 summarizes the transformations obeyed by the family of Hamiltonians (5.25).

There are two more identities that are sensitive to the parity of c in the spin-1/2 cluster c chain $\widehat{H}_c^{(b_c)}$.

First, for any $c>1$ and $N\geq c+1$,

$$
\begin{aligned}
&\prod_{j=1}^{N}\widehat{Z}_j\,\widehat{X}_{j+1}\cdots\widehat{X}_{j+c-1}\,\widehat{Z}_{j+c}\\
&\quad=\begin{array}{llllll}
\widehat{Z}_1 & \widehat{X}_2 & \cdots & \widehat{X}_c & \underline{\widehat{Z}_{c+1}} & \\
 & \widehat{Z}_2 & \widehat{X}_3 & \cdots & \underline{\underline{\widehat{X}_{c+1}}} & \widehat{Z}_{c+2}\\
 & & \ddots & \ddots & \ddots & \\
 & & & \widehat{Z}_c & \underline{\underline{\widehat{X}_{c+1}}}\cdots\widehat{X}_{2c-1}\,\widehat{Z}_{2c} & \\
 & & & & \underline{\widehat{Z}_{c+1}}\,\widehat{X}_{c+2}\cdots\widehat{X}_{2c}\widehat{Z}_{2c+1} & \\
 & & & & \ddots\quad\ddots\quad\ddots &
\end{array}\\
&\quad=\left[(-1)^N\prod_{j=1}^{N}\widehat{X}_j\right]^{c-1}=
\begin{cases}
(-1)^N\,\widehat{X}, & \text{if } c \text{ is even},\\[2ex]
\widehat{\mathbb{1}}, & \text{if } c \text{ is odd}.
\end{cases}
\end{aligned}
\tag{5.31}
$$

[The multiplicative factor of $(-1)^{N(c-1)}$ in the penultimate equality arises because, reading vertically, (i) there are $c-1$ $\widehat{X}$'s that are underlined twice between a pair of $\widehat{Z}$'s that are underlined once with all $\widehat{X}$'s and $\widehat{Z}$'s sharing the same lattice label and (ii) there are $N-c$ such vertical strings of letters.] For $c=1$, the product of the clusters $\widehat{\mathrm{C}}_j:=\widehat{Z}_j\,\widehat{Z}_{j+1}$ with $j=1,\cdots,N\geq 2$ is the identity. For $c=0$, the product of the clusters $\widehat{\mathrm{C}}_j:=\widehat{X}_j$ with $j=1,\cdots,N$ is $\widehat{X}$ as defined in Eq. (5.8a). For $c=-1$, the product of the clusters $\widehat{\mathrm{C}}_j:=\widehat{Y}_j\,\widehat{Y}_{j+1}$ with $j=1,\cdots,N\geq 2$ is the identity. For $c<-1$, the product of the clusters $\widehat{\mathrm{C}}_j$ with $j=1,\cdots,N\geq |c|+1$ is also given by the right-hand side of Eq. (5.31). Hence, when c is odd and periodic boundary conditions are chosen, taking the product over all clusters $\widehat{\mathrm{C}}_j$ with $j=1,\cdots,N$ acts as a constraint.

Second, for any odd $c\in\mathbb{Z}$, we have the factorization

$$
\widehat{W}_j\,\widehat{X}_{j+1}\cdots\widehat{X}_{j+|c|-1}\,\widehat{W}_{j+|c|}=\widehat{M}^{(c)}_j\,\widehat{M}^{(c)}_{j+1},\qquad j=1,\cdots,N,
\tag{5.32a}
$$

with $\widehat{W}_j=\widehat{Y}_j$ if $c<0$, $\widehat{W}_j=\widehat{Z}_j$ if $c>0$, and $\widehat{M}^{(c)}_j$ a monomial of order $|c|$ in the local Pauli operators labeled by $j,\cdots,j+|c|-1$.

Proof On the one hand, when $c=1$

$$
\widehat{M}^{(1)}_j:=\widehat{Z}_j=-\widehat{T}\,\widehat{M}^{(1)}_j\,\widehat{T}^{-1},\qquad j=1,\cdots,N,
\tag{5.32b}
$$

while when $c=3,5,7,\cdots$,

$$
\begin{aligned}
\widehat{M}^{(c)}_j&:=\widehat{Z}_j\,\widehat{Y}_{j+1}\,\widehat{Z}_{j+2}\cdots\widehat{Y}_{j+c-2}\,\widehat{Z}_{j+c-1}\\
&=-\,\widehat{T}\,\widehat{M}^{(c)}_j\,\widehat{T}^{-1},\qquad j=1,\cdots,N.
\end{aligned}
\tag{5.32c}
$$

On the other hand, when $c=-1$

$$\widehat{M}_j^{(-1)} := \widehat{Y}_j = -\widehat{T}\,\widehat{M}_j^{(-1)}\,\widehat{T}^{-1}, \qquad j=1,\cdots,N, \tag{5.32d}$$

while $c=-3,-5,-7,\cdots$

$$\begin{aligned}\widehat{M}_j^{(c)} :=& \widehat{Y}_j\,\widehat{Z}_{j+1}\,\widehat{Y}_{j+2}\cdots\widehat{Z}_{j+|c|-2}\,\widehat{Y}_{j+|c|-1}\\ =& -\widehat{T}\,\widehat{M}_j^{(c)}\,\widehat{T}^{-1}, \qquad j=1,\cdots,N.\end{aligned} \tag{5.32e}$$

□

Observe that

$$\left\{\widehat{X},\widehat{M}_j^{(c)}\right\}=0, \qquad j=1,\cdots,N, \tag{5.32f}$$

since $\widehat{M}_j^{(c)}$ is a product of an odd number of operators that each anticommute with $\widehat{X}$.

We are going to show that any spin-1/2 cluster c chain (5.25) has a spectral gap $\hbar\omega$ that separates the ground states from the excited states. We are also going to establish under what conditions the ground states are degenerate when periodic boundary conditions are imposed and how sensitive this ground-state degeneracy is to changing from periodic to open boundary conditions. According to the results summarized in **Table** 5.2, the ground state is nondegenerate for periodic boundary conditions when the label $c\in\mathbb{Z}$ for the cluster is even, while it is twofold degenerate for periodic boundary conditions when the label $c\in\mathbb{Z}$ for the cluster is odd. The absence or presence of this twofold degeneracy extends to all energy eigenstates. This absence or this presence originates in the fact that the spin-1/2 cluster c chains break time-reversal symmetry when the label $c\in\mathbb{Z}$ for the cluster is even, while they are time-reversal symmetric when the label $c\in\mathbb{Z}$ for the cluster is odd, respectively. The ground-state degeneracy increases by the factor $2^{2\lfloor|c/2|\rfloor}$ when open boundary conditions are chosen. (The floor function $\lfloor x\rfloor$ returns the largest integer smaller than or equal to $x\in\mathbb{R}$.) This sensitivity is attributed to the existence of boundary states costing no energy when open boundary conditions are selected.

5.2.1.3 Energy Spectra

For any $j\in\Lambda$ and for any integer c, we define the Hermitean cluster operator

$$\widehat{C}_j := \begin{cases} \widehat{Y}_j\,\widehat{X}_{j+1}\cdots\widehat{X}_{j+|c|-1}\,\widehat{Y}_{j+|c|}, & \text{for } c<-1,\\[2ex] \widehat{Y}_j\,\widehat{Y}_{j+1}, & \text{for } c=-1,\\[2ex] \widehat{X}_j, & \text{for } c=0,\\[2ex] \widehat{Z}_j\,\widehat{Z}_{j+1}, & \text{for } c=+1,\\[2ex] \widehat{Z}_j\,\widehat{X}_{j+1}\cdots\widehat{X}_{j+c-1}\,\widehat{Z}_{j+c}, & \text{for } c>+1.\end{cases} \tag{5.33}$$

Table 5.2 *Spectral degeneracies of the spin-1/2 cluster c chains*

Spin-1/2 cluster c chain	BC	GSD	SD E_n
$\widehat{H}_0^{(b_0)}$	Any	1	$\frac{N!}{n!(N-n)!}$
$\widehat{H}_{2m}^{(b_{2m})}$, $m \neq 0$	PBC	1	$\frac{N!}{n!(N-n)!}$
	OBC	$2^{\lvert 2m\rvert}$	$2^{\lvert 2m\rvert} \times \frac{(N-\lvert 2m\rvert)!}{n!(N-\lvert 2m\rvert-n)!}$
$\widehat{H}_{2m+1}^{(b_{2m+1})}$	PBC	2	$2 \times \frac{(N-1)!}{n!(N-1-n)!}$
	OBC	$2 \times 2^{\lvert 2m+1\rvert-1}$	$2 \times 2^{\lvert 2m+1\rvert-1} \times \frac{(N-\lvert 2m+1\rvert)!}{n!(N-\lvert 2m+1\rvert-n)!}$

The spectral degeneracies of any spin-1/2 cluster c chain $\widehat{H}_c^{(b_c)}$ of length $N \geq |c|+1$ defined in Eq. (5.25) depend on the parity of $c \in \mathbb{Z}$ and on the choice $b_c = 0$ corresponding to periodic boundary conditions or the choice $b_c = |c|$ corresponding to open boundary conditions when $c \neq 0$. We abbreviate boundary conditions by BC, ground-state degeneracy by GSD, and spectral degeneracy by SD.

By inspection of the right-hand side in the first equality in Eq. (5.31), one deduces that

$$\left[\widehat{\mathrm{C}}_j, \widehat{\mathrm{C}}_{j'}\right] = 0 \tag{5.34}$$

for any $j, j' \in \Lambda$. Moreover,

$$\left(\widehat{\mathrm{C}}_j\right)^2 = \widehat{\mathbb{1}} \tag{5.35}$$

for any $j \in \Lambda$. Hence, the spin-1/2 cluster c chain (5.25) is the sum

$$\widehat{H}_c^{(b_c)} = -\frac{\hbar\omega}{2} \sum_{j=1}^{N-b_c} \widehat{\mathrm{C}}_j \tag{5.36}$$

over pairwise mutually commuting Hermitean operators, each of which squares to unity.

The eigenvalues of any Hermitean cluster operator $\widehat{\mathrm{C}}_j$ defined in Eq. (5.33) are

$$\mathrm{c}_j = \pm 1, \tag{5.37a}$$

according to Eq. (5.35). The simultaneous eigenstates of all Hermitean cluster operators (5.33) are defined by the conditions

$$\widehat{\mathrm{C}}_j \left|\mathrm{c}_1, \cdots, \mathrm{c}_j, \cdots, \mathrm{c}_N\right\rangle = \mathrm{c}_j \left|\mathrm{c}_1, \cdots, \mathrm{c}_j, \cdots, \mathrm{c}_N\right\rangle \tag{5.37b}$$

for any $j \in \Lambda$ and

$$\left\langle \mathrm{c}_1, \cdots, \mathrm{c}_j, \cdots, \mathrm{c}_N \,\middle|\, \mathrm{c}'_1, \cdots, \mathrm{c}'_j, \cdots, \mathrm{c}'_N \right\rangle = \prod_{j=1}^{N} \delta_{\mathrm{c}_j, \mathrm{c}'_j}. \tag{5.37c}$$

The simultaneous eigenstates of all Hermitean cluster operators (5.33) realize an orthonormal basis of the Hilbert space $\mathfrak{H}$ that was defined in Eq. (5.3).

According to Eqs. (5.31) and (5.32), we have two important identities that are sensitive to the parity of $c \in \mathbb{Z}$. First,

$$\prod_{j=1}^{N} \widehat{\mathrm{C}}_j = \begin{cases} (-1)^{N\,(1-\delta_{c,0})}\,\widehat{X}, & \text{if } c \text{ is even,} \\ \widehat{1\!\!1}, & \text{if } c \text{ is odd.} \end{cases} \tag{5.38}$$

Second, for odd c, we have the factorization [recall Eq. (5.32)]

$$\widehat{\mathrm{C}}_j = \widehat{M}^{(c)}_j\,\widehat{M}^{(c)}_{j+1}, \tag{5.39a}$$

in terms of two monomials of order $|c|$ that are odd under reversal of time:

$$\widehat{T}\,\widehat{M}^{(c)}_j\,\widehat{T}^{-1} = -\widehat{M}^{(c)}_j. \tag{5.39b}$$

5.2.1.4 The Case of Even Spin-1/2 Cluster $c = 2m$ Chains

For any even spin-1/2 cluster $2m$ chain

$$\widehat{H}^{(b_{2m})}_{2m} := -\frac{\hbar\omega}{2} \sum_{j=1-|m|+\frac{b_{2m}}{2}}^{N-|m|-\frac{b_{2m}}{2}} \widehat{\mathrm{C}}_j = -\widehat{T}\,\widehat{H}^{(b_{2m})}_{2m}\,\widehat{T}^{-1}, \qquad m \in \mathbb{Z}, \tag{5.40}$$

we shall use the orthonormal basis of the Hilbert space (5.3a) defined by the conditions

$$\begin{aligned} &\widehat{\mathrm{C}}_j \left| \mathrm{c}_{1-|m|}, \cdots, \underline{\mathrm{c}_1, \cdots, \mathrm{c}_{N-|2m|}}, \cdots, \mathrm{c}_{N-|m|} \right\rangle \\ &\quad = \mathrm{c}_j \left| \mathrm{c}_{1-|m|}, \cdots, \underline{\mathrm{c}_1, \cdots, \mathrm{c}_{N-|2m|}}, \cdots, \mathrm{c}_{N-|m|} \right\rangle \end{aligned} \tag{5.41a}$$

for any

$$j \in \left\{ 1-|m|, \cdots, \underline{1, \cdots, N-|2m|}, \cdots, N-|m| \right\} \cong \Lambda \tag{5.41b}$$

and

$$\left\langle \cdots, \mathrm{c}_j, \cdots \middle| \cdots, \mathrm{c}'_j, \cdots \right\rangle = \prod_{j=1-|m|}^{N-|m|} \delta_{\mathrm{c}_j, \mathrm{c}'_j}. \tag{5.41c}$$

The rational for ordering the sites on the ring Λ ($j \in \Lambda \Rightarrow j + N \equiv j$) as is done in Eq. (5.41) is to distinguish the bulk (underlined sites) from the boundaries (sites that are to the left or to the right of the underlined sites) as will become transparent when we trade periodic for open boundary conditions.

On the one hand, if we impose periodic boundary conditions, $b_{2m} = 0$, then

$$\begin{aligned}
\widehat{H}^{(0)}_{2m} \left| \mathrm{c}_{1-|m|}, \cdots, \underline{\mathrm{c}_1, \cdots, \mathrm{c}_{N-|2m|}}, \cdots, \mathrm{c}_{N-|m|} \right\rangle &= \\
-\frac{\hbar\omega}{2} \sum_{j=1-|m|}^{N-|m|} \widehat{\mathrm{C}}_j \left| \mathrm{c}_{1-|m|}, \cdots, \underline{\mathrm{c}_1, \cdots, \mathrm{c}_{N-|2m|}}, \cdots, \mathrm{c}_{N-|m|} \right\rangle &= \\
-\frac{\hbar\omega}{2} \sum_{j=1-|m|}^{N-|m|} \mathrm{c}_j \left| \mathrm{c}_{1-|m|}, \cdots, \underline{\mathrm{c}_1, \cdots, \mathrm{c}_{N-|2m|}}, \cdots, \mathrm{c}_{N-|m|} \right\rangle &\equiv \\
E_n \left| \mathrm{c}_{1-|m|}, \cdots, \underline{\mathrm{c}_1, \cdots, \mathrm{c}_{N-|2m|}}, \cdots, \mathrm{c}_{N-|m|} \right\rangle&,
\end{aligned} \tag{5.42a}$$

where

$$E_n = -\frac{\hbar\omega}{2} (N - 2n) \tag{5.42b}$$

with $n = 0, 1, \cdots, N$ the number of $\mathrm{c}_j = -1$ for $j \in \Lambda$. The degeneracy d_n of the eigenvalue E_n is the number of ways to choose n out of N, namely,

$$d_n = \frac{N!}{n!\,(N-n)!}, \qquad n = 0, 1, \cdots, N. \tag{5.42c}$$

On the other hand, if we impose open boundary conditions, $b_{2m} = |2m|$, then

$$\begin{aligned}
\widehat{H}^{(|2m|)}_{2m} \left| \mathrm{c}_{1-|m|}, \cdots, \underline{\mathrm{c}_1, \cdots, \mathrm{c}_{N-|2m|}}, \cdots, \mathrm{c}_{N-|m|} \right\rangle &= \\
-\frac{\hbar\omega}{2} \sum_{j=1}^{N-|2m|} \widehat{\mathrm{C}}_j \left| \mathrm{c}_{1-|m|}, \cdots, \underline{\mathrm{c}_1, \cdots, \mathrm{c}_{N-|2m|}}, \cdots, \mathrm{c}_{N-|m|} \right\rangle &= \\
-\frac{\hbar\omega}{2} \sum_{j=1}^{N-|2m|} \mathrm{c}_j \left| \mathrm{c}_{1-|m|}, \cdots, \underline{\mathrm{c}_1, \cdots, \mathrm{c}_{N-|2m|}}, \cdots, \mathrm{c}_{N-|m|} \right\rangle &\equiv \\
E_n \left| \mathrm{c}_{1-|m|}, \cdots, \underline{\mathrm{c}_1, \cdots, \mathrm{c}_{N-|2m|}}, \cdots, \mathrm{c}_{N-|m|} \right\rangle&,
\end{aligned} \tag{5.43a}$$

where

$$E_n = -\frac{\hbar\omega}{2} (N - |2m| - 2n) \tag{5.43b}$$

with $n = 0, \cdots, N - |2m|$ the number of $c_j = -1$ for $j = 1, \cdots, N - |2m|$. The degeneracy d_n of the eigenvalue E_n is the number of ways to choose n out of $N-|2m|$ times a factor of 2 for each entry on the boundary set

$$\begin{aligned} \Lambda_{\mathrm{bd}} &\equiv \Lambda_m^{\mathrm{L}} \cup \Lambda_m^{\mathrm{R}}, \\ \Lambda_m^{\mathrm{L}} &:= \{1 - |m|, 1 - |m| + 1, \cdots, 0\}, \\ \Lambda_m^{\mathrm{R}} &:= \{N - |2m| + 1, \cdots, N - |m|\}, \end{aligned} \tag{5.43c}$$

namely,

$$d_n = 2^{|2m|} \times \frac{(N - |2m|)!}{n!\,(N - |2m| - n)!}, \qquad n = 0, \cdots, N - |2m|. \tag{5.43d}$$

The spectrum of any even spin-1/2 cluster $2m$ chain $\widehat{H}^{(b_{2m})}_{2m}$, for both periodic and open boundary conditions, is discrete with any two consecutive energy eigenvalues separated by the gap $\hbar\omega$. The ground state is nondegenerate with periodic boundary conditions. The degeneracy of the ground states increases by the multiplicative factor $2^{|2m|}$ when changing from periodic to open boundary conditions. This change of degeneracy is attributed to a set of $|m|$ independent two-level degrees of freedom decoupling from the Hamiltonian on the left boundary Λ_m^{L} defined in Eq. (5.43c) and to another set of $|m|$ independent two-level degrees of freedom decoupling from the Hamiltonian on the right boundary Λ_m^{R} defined in Eq. (5.43c). For $2m = 0$, the ground state is special as it remains nondegenerate for both periodic and open boundary conditions.

5.2.1.5 The Case of Odd Spin-1/2 Cluster $c = 2m + 1$ *Chains*

For any odd spin-1/2 cluster $2m + 1$ chain

$$\widehat{H}^{(b_{2m+1})}_{2m+1} := -\frac{\hbar\omega}{2} \sum_{j=1}^{N-b_{2m+1}} \widehat{\mathrm{C}}_j, \qquad m \in \mathbb{Z}, \tag{5.44}$$

we take advantage of three properties. First, $\widehat{H}^{(b_{2m+1})}_{2m+1}$ is even under reversal of time,

$$\widehat{H}^{(b_{2m+1})}_{2m+1} = +\widehat{T}\,\widehat{H}^{(b_{2m+1})}_{2m+1}\,\widehat{T}^{-1}, \qquad m \in \mathbb{Z}, \tag{5.45}$$

as all the pairwise commuting cluster operators $\widehat{\mathrm{C}}_j$ are even under reversal of time. Second, only $N - 1$ out of the N cluster operators $\widehat{\mathrm{C}}_j$ are independent owing to the constraint

$$\prod_{j=1}^{N} \widehat{\mathrm{C}}_j = \widehat{\mathbb{1}}. \tag{5.46}$$

Third, $\widehat{\mathrm{C}}_j$ factorizes into two cluster operators, each of which is odd under reversal of time according to Eq. (5.32). Hence,

$$\widehat{H}^{(b_{2m+1})}_{2m+1} := -\frac{\hbar\omega}{2} \sum_{j=1}^{N-b_{2m+1}} \widehat{M}^{(2m+1)}_{j} \widehat{M}^{(2m+1)}_{j+1}, \tag{5.47a}$$

where the Hermitean cluster operator

$$\widehat{M}^{(2m+1)}_{j} = -\widehat{T}\,\widehat{M}^{(2m+1)}_{j}\,\widehat{T}^{-1} \tag{5.47b}$$

was defined in Eq. (5.32). Furthermore, the identities

$$\left(\widehat{M}^{(2m+1)}_{j}\right)^2 = \widehat{\mathbb{1}}, \tag{5.48a}$$

$$\left[\widehat{M}^{(2m+1)}_{j}, \widehat{M}^{(2m+1)}_{j'}\right] = 0, \tag{5.48b}$$

$$\left[\widehat{M}^{(2m+1)}_{j}, \widehat{\mathrm{C}}_{j'}\right] = 0, \tag{5.48c}$$

hold for any $j, j' \in \Lambda$.

Proof Equation (5.48a) follows from the Pauli algebra.
To prove Eq. (5.48b), we observe that $\widehat{M}^{(2m+1)}_{j}$ is a monomial of order $|2m|+\mathrm{sgn}(m)$.

It then suffices to consider the commutators when $j' = j+1$ and $j' = j+2$, as will become evident shortly.
We may interpret $\widehat{M}^{(2m+1)}_{j}$ as a string of letters labeled by the sites of Λ with the

letters I on all sites except for the site j, which is assigned the letter Z (Y), $j+1$, which is assigned the letter Y (Z), and so on until the site $j + |2m| + \mathrm{sgn}(m) - 1$, which is assigned the letter Z (Y) when $m > 0$ ($m < 0$).
The operator $\widehat{M}^{(2m+1)}_{j+1}$ is obtained from the operator $\widehat{M}^{(2m+1)}_{j}$ by shifting the

strings of letters unequal to I by one site to the right.
Hence, the number of site-resolved mismatches between the letters Z and Y, when comparing $\widehat{M}^{(2m+1)}_{j}$ and $\widehat{M}^{(2m+1)}_{j+1}$, is the even number $|2m|+\mathrm{sgn}(m)-1$. Exchanging $\widehat{M}^{(2m+1)}_{j}$ and $\widehat{M}^{(2m+1)}_{j+1}$ costs the factor $(-1)^{|2m|} = +1$. The operator $\widehat{M}^{(2m+1)}_{j+2}$ is obtained from the operator $\widehat{M}^{(2m+1)}_{j}$ by shifting the strings of letters unequal to I by two sites to the right. The only site-resolved mismatches between the letters in $\widehat{M}^{(2m+1)}_{j}$ and $\widehat{M}^{(2m+1)}_{j+2}$ is limited to the letter I with the mismatched letter Z or to the letter I with the mismatched letter Y. Exchanging $\widehat{M}^{(2m+1)}_{j}$ and $\widehat{M}^{(2m+1)}_{j+2}$ costs the factor $+1$. Equation (5.48c) follows from Eqs. (5.48b) and (5.39). □

The eigenvalues of the Hermitean operator $\widehat{M}^{(2m+1)}_{j}$ defined in Eq. (5.39) are

$$\mathrm{m}_j = \pm 1 \tag{5.49a}$$

according to Eq. (5.48a). The simultaneous eigenstates of $\widehat{M}_1^{(2m+1)}$, $\widehat{\mathrm{C}}_1, \cdots, \widehat{\mathrm{C}}_{N-1}$ are defined by the conditions

$$\begin{aligned}\widehat{M}_1^{(2m+1)} & \left|\mathrm{m}_1, \mathrm{c}_1, \cdots, \mathrm{c}_j, \cdots, \mathrm{c}_{N-1}\right\rangle \\ & = \mathrm{m}_1 \left|\mathrm{m}_1, \mathrm{c}_1, \cdots, \mathrm{c}_j, \cdots, \mathrm{c}_{N-1}\right\rangle,\end{aligned} \tag{5.49b}$$

$$\begin{aligned}\widehat{\mathrm{C}}_j & \left|\mathrm{m}_1, \mathrm{c}_1, \cdots, \mathrm{c}_j, \cdots, \mathrm{c}_{N-1}\right\rangle \\ & = \mathrm{c}_j \left|\mathrm{m}_1, \mathrm{c}_1, \cdots, \mathrm{c}_j, \cdots, \mathrm{c}_{N-1}\right\rangle,\end{aligned} \tag{5.49c}$$

for any $j = 1, \cdots, N-1$, and

$$\begin{aligned}&\left\langle \mathrm{m}_1, \mathrm{c}_1, \cdots, \mathrm{c}_j, \cdots, \mathrm{c}_{N-1} \,\middle|\, \mathrm{m}_1', \mathrm{c}_1', \cdots, \mathrm{c}_j', \cdots, \mathrm{c}_{N-1}' \right\rangle \\ &\qquad = \delta_{\mathrm{m}_1, \mathrm{m}_1'} \prod_{j=1}^{N-1} \delta_{\mathrm{c}_j, \mathrm{c}_j'}.\end{aligned} \tag{5.49d}$$

These eigenstates realize an orthonormal basis of the Hilbert space $\mathfrak{H}$ that was defined in Eq. (5.3).

The operator $\widehat{\mathrm{C}}_j$ and a fortiori $\widehat{H}_{2m+1}^{(b_{2m+1})}$ are even, while $\widehat{M}_j^{(2m+1)}$ is odd, under conjugation by $\widehat{T}$ (reversal of time). It follows that all eigenstates of $\widehat{H}_{2m+1}^{(b_{2m+1})}$ are at least twofold degenerate, as they are independent of the two values taken by the eigenvalue m_1 of $\widehat{M}_1^{(2m+1)}$. To see this explicitly, we treat the cases of periodic and open boundary conditions successively.

On the one hand, if we impose periodic boundary conditions, $b_{2m+1} = 0$, then

$$\begin{aligned}\widehat{H}_{2m+1}^{(0)} \left|\mathrm{m}_1, \mathrm{c}_1, \cdots, \mathrm{c}_{N-1}\right\rangle &= -\frac{\hbar\omega}{2} \sum_{j=1}^{N} \widehat{\mathrm{C}}_j \left|\mathrm{m}_1, \mathrm{c}_1, \cdots, \mathrm{c}_{N-1}\right\rangle \\ &= -\frac{\hbar\omega}{2} \sum_{j=1}^{N} \mathrm{c}_j \left|\mathrm{m}_1, \mathrm{c}_1, \cdots, \mathrm{c}_{N-1}\right\rangle \\ &\equiv E_n \left|\mathrm{m}_1, \mathrm{c}_1, \cdots, \mathrm{c}_{N-1}\right\rangle,\end{aligned} \tag{5.50a}$$

where

$$E_n = -\frac{\hbar\omega}{2} \left(N - 2n - 2\,\delta_{\mathrm{c}_N, -1}\right), \qquad \mathrm{c}_N := \prod_{j=1}^{N-1} \mathrm{c}_j, \tag{5.50b}$$

with $n = 0, 1, \cdots, N-1$ the number of $\mathrm{c}_j = -1$ for $j = 1, \cdots, N-1$. The degeneracy d_n of the eigenvalue E_n is two (accounting for $\mathrm{m}_1 = \pm 1$) times the number of ways to choose n out of $N-1$, namely,

$$d_n = 2 \times \frac{(N-1)!}{n!\,(N-1-n)!}, \qquad n = 0, 1, \cdots, N-1. \tag{5.50c}$$

On the other hand, if we impose open boundary conditions, $b_{2m+1}=|2m+1|$, then

$$\begin{aligned}\widehat{H}^{(|2m+1|)}_{2m+1}\,|\mathrm{m}_1,\mathrm{c}_1,\cdots,\mathrm{c}_{N-1}\rangle &= -\frac{\hbar\omega}{2}\sum_{j=1}^{N-|2m+1|}\widehat{\mathrm{C}}_j\,|\mathrm{m}_1,\mathrm{c}_1,\cdots,\mathrm{c}_{N-1}\rangle\\ &= -\frac{\hbar\omega}{2}\sum_{j=1}^{N-|2m+1|}\mathrm{c}_j\,|\mathrm{m}_1,\mathrm{c}_1,\cdots,\mathrm{c}_{N-1}\rangle\\ &\equiv E_n\,|\mathrm{m}_1,\mathrm{c}_1,\cdots,\mathrm{c}_{N-1}\rangle,\end{aligned} \tag{5.51a}$$

where

$$E_n=-\frac{\hbar\omega}{2}\left(N-|2m+1|-2n\right), \tag{5.51b}$$

with $n=0,\cdots,N-|2m+1|$ the number of $\mathrm{c}_j=-1$ for $j=1,\cdots,N-|2m+1|$. The degeneracy d_n of the eigenvalue E_n is the factor $2^{|2m+1|}$ (that accounts for the fact that E_n does not depend on m_1, $\mathrm{c}_{N-|2m+1|+1}$, $\cdots$, c_{N-1}) times the number of ways to choose n out of $N-|2m+1|$, namely,

$$d_n=2\times 2^{|2m+1|-1}\times\frac{(N-|2m+1|)!}{n!\,(N-|2m+1|-n)!},\qquad n=0,\cdots,N-|2m+1|. \tag{5.51c}$$

The spectrum of any odd spin-1/2 cluster $2m+1$ chain $\widehat{H}^{(b_{2m+1})}_{2m+1}$, for both periodic and open boundary conditions, is discrete with any two consecutive energy eigenvalues separated by the gap $\hbar\omega$. The ground state is twofold degenerate with periodic boundary conditions. The degeneracy of the ground states increases by the multiplicative factor $2^{|2m+1|-1}$ when changing from periodic to open boundary conditions. This change of degeneracy is attributed to a set of $(|2m+1|-1)/2$ independent two-level degrees of freedom decoupling from the Hamiltonian on the left boundary $\Lambda^{\mathrm{L}}_{m'}$ defined in Eq. (5.43c) and to another set of $(|2m+1|-1)/2$ independent two-level degrees of freedom decoupling from the Hamiltonian on the right boundary $\Lambda^{\mathrm{R}}_{m'}$ defined in Eq. (5.43c), whereby $m'\equiv(|2m+1|-1)/2$. We observe that the multiplicity

$$\frac{(N-|2m+1|)!}{n!\,(N-|2m+1|-n)!} \tag{5.52}$$

for a spin-1/2 cluster $2m+1$ chain made of N sites on the right-hand side of Eq. (5.51c) is the same as the multiplicity

$$\frac{(N'-2|m'|)!}{n!\,(N'-2|m'|-n)!} \tag{5.53}$$

for a spin-1/2 cluster $2m'$ chain made of N' sites on the right-hand side of Eq. (5.43d) provided $N'\equiv N-1$ and $m'\equiv(|2m+1|-1)/2$.

In the thermodynamic limit $N \to \infty$, the $\mathbb{Z}_2$ symmetry generated by reversal of time is spontaneously broken. We can then make the mean-field approximation

$$\widehat{H}^{(b_{2m+1})}_{2m+1} \approx \widehat{H}^{(b_{|2m+1|-1})}_{2m+1\,\mathrm{MF}}, \tag{5.54a}$$

where

$$\widehat{H}^{(b_{|2m+1|-1})}_{2m+1\,\mathrm{MF}} := -\frac{\hbar\omega}{2} \sum_{j=1}^{N-b_{|2m+1|-1}} B_j\, \widehat{M}^{(2m+1)}_j \tag{5.54b}$$

and

$$B_j := \frac{1}{2}\left\langle\left(\widehat{M}^{(2m+1)}_j + \widehat{M}^{(2m+1)}_{j+1}\right)\right\rangle \tag{5.54c}$$

is the expectation value of $\widehat{M}^{(2m+1)}_j$ (averaged over two consecutive sites) in the mean-field ground state. The mean-field Hamiltonian is the sum of pairwise commuting monomials of odd order $|2m+1|$. This is why we use $b_{|2m+1|-1} = 0$ for periodic boundary conditions and $b_{|2m+1|-1} = |2m+1| - 1$ for open boundary conditions. The mean-field spectrum is discrete and proportional to that of the even spin-1/2 cluster $(|2m+1|-1)$ chain $\widehat{H}^{(b_{|2m+1|-1})}_{|2m+1|-1}$ if B_j is independent of $j \in \Lambda$. Hence, the mean-field ground states are either nondegenerate or $2^{|2m+1|-1}$-fold degenerate when periodic or open boundary conditions are imposed, respectively. This change of degeneracy is attributed to a set of $(|2m+1|-1)/2$ independent two-level degrees of freedom decoupling from the mean-field Hamiltonian on the left boundary $\Lambda^{\mathrm{L}}_{m'}$ defined in Eq. (5.43c) and to another set of $(|2m+1|-1)/2$ independent two-level degrees of freedom decoupling from the mean-field Hamiltonian on the right boundary $\Lambda^{\mathrm{R}}_{m'}$ defined in Eq. (5.43c) with $m' = (|2m+1|-1)/2$.

5.3 Symmetry Fractionalization of Spin-1/2 Cluster c Chains

5.3.1 Overview

Common to all spin-1/2 cluster c chains is the fact that the ground states are always gapped and their degeneracies are given by $2^{|c|}$ when open boundary conditions are imposed. The spin-1/2 cluster c chains nevertheless fall into two families depending on the parity of c. When periodic boundary conditions are imposed, the ground states are nondegenerate and gapped for even c, whereas the ground states are twofold degenerate and gapped for odd c. This parity effect is reflected by the fact that time-reversal symmetry is broken explicitly for even c, whereas it is present for odd c.

Because time-reversal symmetry is broken spontaneously in the thermodynamic limit for odd c, it is always possible to reduce the study of the low-lying excited states of a spin-1/2 cluster $c = 2m+1$ chain to that of a spin-1/2 cluster $c = 2m^*$ chain with $m^* = m$ if $m > 0$ or $m^* = (m+1)$ if $m < 0$. Without loss of generality, one may then ask if the ground-state degeneracy $2^{|2m|}$ of a spin-1/2 cluster $2m$ chain obeying open boundary conditions is stable to local perturbations. We will give a definite answer to this question under the assumption that such local perturbations are symmetric and spontaneous symmetry breaking is precluded.

To this end, we need to revisit the symmetries of the spin-1/2 cluster c chains, and how they are realized on the boundaries when open boundary conditions are selected. First, we are going to select the symmetries that shall be imposed. Second, we are going to show that, whereas the protecting symmetries are represented by a group homomorphism when periodic boundary conditions are imposed, they can be represented projectively when open boundary conditions are imposed. The terminology of symmetry fractionalization in the title of Section 5.3 is synonymous to realizing projectively the symmetries on the boundaries when they are realized by a group homomorphism in the bulk.

We will then show that, because the protecting symmetry is represented projectively on the boundaries of a spin-1/2 cluster $2m$ chain, there exists an integer-valued function

$$D(2m) = D(2m + 2p) \tag{5.55a}$$

that (i) is periodic with periodicity p as function of m and (ii) gives the minimal degeneracy

$$1 \leq D(2m) \leq 2^{2|m|} \tag{5.55b}$$

of the ground states of an open spin-1/2 cluster $2m$ chain that is stable to any local symmetric perturbation as long as the protecting symmetry is not broken spontaneously. The periodicity p depends on the protecting symmetries. Here, it is

$$p = 4, \tag{5.55c}$$

as is dictated by the symmetry group Eq. (5.57).

5.3.2 On-Site Symmetries

We shall demand in Section 5.3 that all global on-site symmetries that were defined in Section 5.2.1 each square to the identity and all commute pairwise, in order to prevent global degeneracies arising from symmetry. This is achieved by demanding that the cardinality of the chain Λ is the even integer $2N$ with N an integer.

We consider the family of Hamiltonians

$$\widehat{H}_c^{(b_c)} := -\frac{\hbar\omega}{2} \sum_{j=1}^{2N-b_c} \widehat{C}_j, \tag{5.56a}$$

where $b_c = 0, |c|$ specifies the periodic ($b_c = 0$) or open ($b_c = |c|$) boundary conditions and we have defined

$$\widehat{C}_j := \begin{cases} \widehat{Y}_j \widehat{X}_{j+1} \cdots \widehat{X}_{j+|c|-1} \widehat{Y}_{j+|c|}, & \text{if } c < -1, \\ \widehat{Y}_j \widehat{Y}_{j+1}, & \text{if } c = -1, \\ \widehat{X}_j, & \text{if } c = 0, \\ \widehat{Z}_j \widehat{Z}_{j+1}, & \text{if } c = +1, \\ \widehat{Z}_j \widehat{X}_{j+1} \cdots \widehat{X}_{j+c-1} \widehat{Z}_{j+c}, & \text{if } c > +1, \end{cases} \tag{5.56b}$$

with $\widehat{X}_j$, $\widehat{Y}_j$, and $\widehat{Z}_j$ the Pauli spin operators at site j realizing the spin-1/2 representation of the $\mathfrak{su}(2)$ Lie algebra. The local terms $\widehat{C}_j$ are pairwise commuting and each has eigenvalues ± 1. Therefore, the Hamiltonian (5.56) is a sum of pairwise commuting terms. We set the total number of sites to be even, that is, $2N$. This choice guarantees that the global operation of time reversal squares to the identity according to Eq. (5.14b) and commutes with the global rotation (5.9) by the angle π around the X axis in spin-1/2 space.

We shall distinguish the cases of c odd and c even. For each case, we shall single out an on-site[11] symmetry group of the spin-1/2 cluster c chain.

5.3.2.1 Global On-Site Symmetry Group When c Is Even

When c is an even integer, Hamiltonian (5.56) is invariant under the global symmetry group

$$\mathrm{G}_{\mathrm{e}} := \mathbb{Z}_2^{T'} \times \mathbb{Z}_2. \tag{5.57a}$$

The group

$$\mathbb{Z}_2^{T'} := \{e, t'\} \tag{5.57b}$$

is cyclic of order 2 with the product rule $e = t'\,t'$. The superscript "T'" refers to the generator t' being represented by an antiunitary operator as it implements the composition of reversal of time, by which the direction of any spin-1/2 is reversed, with a rotation of π around the Y axis in spin space. The group

$$\mathbb{Z}_2 := \{e, g\} \tag{5.57c}$$

is also cyclic of order 2 with the product rule $e = g\,g$. The generator g implements a rotation by π around the X axis in the spin space. The generators t' and g have the global representations

$$\widehat{U}_{\mathrm{e}}(t') := \widehat{\mathbb{1}}\,\mathsf{K}, \qquad \widehat{U}_{\mathrm{e}}(g) := \prod_{j=1}^{2N} \widehat{X}_j, \tag{5.57d}$$

[11] Inversion of a spatial coordinate is not an on-site operation in space. It is thus ruled out as an on-site symmetry. Reversal of time, although it leaves all space coordinates invariant, reverses the time axis. Thus, reversal of time is an on-site symmetry but cannot be an internal symmetry in spacetime. In particular, reversal of time is not considered an internal symmetry in any relativistic theory, classical or quantum.

where $\widehat{\mathbb{1}}$ is the unit operator acting on the Hilbert space $\mathbb{C}^{2^{2N}}$ and K denotes complex conjugation, respectively. The actions of these generators on the local spin-1/2 degrees of freedom are

$$\widehat{U}_{\mathrm{e}}(t') \begin{pmatrix} \widehat{X}_j & \widehat{Y}_j & \widehat{Z}_j \end{pmatrix}^{\mathsf{T}} \widehat{U}^{\dagger}_{\mathrm{e}}(t') = \begin{pmatrix} +\widehat{X}_j & -\widehat{Y}_j & +\widehat{Z}_j \end{pmatrix}^{\mathsf{T}}, \tag{5.57e}$$

$$\widehat{U}_{\mathrm{e}}(g) \begin{pmatrix} \widehat{X}_j & \widehat{Y}_j & \widehat{Z}_j \end{pmatrix}^{\mathsf{T}} \widehat{U}^{\dagger}_{\mathrm{e}}(g) = \begin{pmatrix} +\widehat{X}_j & -\widehat{Y}_j & -\widehat{Z}_j \end{pmatrix}^{\mathsf{T}}, \tag{5.57f}$$

for $j = 1, \cdots, 2N$. The map[12]

$$\widehat{U}_{\mathrm{e}} \colon \mathrm{G}_{\mathrm{e}} \to \mathrm{GL}(\mathbb{C}^{2^{2N}}) \tag{5.58}$$

defined by (5.57) is a group homomorphism.

5.3.2.2 Global On-Site Symmetry Group When c Is Odd

When c is an odd integer, Hamiltonian (5.56) is invariant under the global symmetry group

$$\mathrm{G}_{\mathrm{o}} := \mathbb{Z}_2^{T} \times \mathbb{Z}_2^{T'} \times \mathbb{Z}_2. \tag{5.59a}$$

The groups $\mathbb{Z}_2^{T'}$ and $\mathbb{Z}_2$ were defined in Eqs. (5.57b) and (5.57c). The group

$$\mathbb{Z}_2^{T} := \{e, t\} \tag{5.59b}$$

is cyclic of order 2 with the product rule $e = t\,t$. The superscript "T" refers to the generator t being represented by an antiunitary operator as it implements reversal of time by which the direction of any spin-1/2 is reversed. These generators have the global representations:

$$\widehat{U}_{\mathrm{o}}(t) := \left(\prod_{j=1}^{2N} \widehat{Y}_j\right) \mathsf{K}, \qquad \widehat{U}_{\mathrm{o}}(t') := \widehat{\mathbb{1}}\,\mathsf{K}, \qquad \widehat{U}_{\mathrm{o}}(g) := \prod_{j=1}^{2N} \widehat{X}_j. \tag{5.59c}$$

The actions of these generators on the local spin-1/2 degrees of freedom are

$$\widehat{U}_{\mathrm{o}}(t) \begin{pmatrix} \widehat{X}_j & \widehat{Y}_j & \widehat{Z}_j \end{pmatrix}^{\mathsf{T}} \widehat{U}^{\dagger}_{\mathrm{o}}(t) = \begin{pmatrix} -\widehat{X}_j & -\widehat{Y}_j & -\widehat{Z}_j \end{pmatrix}^{\mathsf{T}}, \tag{5.59d}$$

$$\widehat{U}_{\mathrm{o}}(t') \begin{pmatrix} \widehat{X}_j & \widehat{Y}_j & \widehat{Z}_j \end{pmatrix}^{\mathsf{T}} \widehat{U}^{\dagger}_{\mathrm{o}}(t') = \begin{pmatrix} +\widehat{X}_j & -\widehat{Y}_j & +\widehat{Z}_j \end{pmatrix}^{\mathsf{T}}, \tag{5.59e}$$

$$\widehat{U}_{\mathrm{o}}(g) \begin{pmatrix} \widehat{X}_j & \widehat{Y}_j & \widehat{Z}_j \end{pmatrix}^{\mathsf{T}} \widehat{U}^{\dagger}_{\mathrm{o}}(g) = \begin{pmatrix} +\widehat{X}_j & -\widehat{Y}_j & -\widehat{Z}_j \end{pmatrix}^{\mathsf{T}}, \tag{5.59f}$$

for $j = 1, \cdots, 2N$. The map

$$\widehat{U}_{\mathrm{o}} \colon \mathrm{G}_{\mathrm{o}} \to \mathrm{GL}(\mathbb{C}^{2^{2N}}) \tag{5.60}$$

defined by Eq. (5.59) is a group homomorphism.

[12] Here, $\mathrm{GL}(\mathbb{C}^{2^{2N}})$ denotes the group of all invertible linear transformations of $\mathbb{C}^{2^{2N}}$.

5.3.2.3 Spectra of Spin-1/2 Cluster c Chains

We have derived in Section 5.2 the spectra of Hamiltonians (5.56) when periodic boundary conditions are imposed, that is, $b_c = 0$. For any even integer c, Hamiltonian $\widehat{H}_c^{(b_c=0)}$ has a nondegenerate gapped ground state. For any odd integer c, Hamiltonian $\widehat{H}_c^{(b_c=0)}$ has a gapped ground state with twofold degeneracy. The twofold degeneracy of the odd-c cluster Hamiltonians is associated with the spontaneous breaking of the symmetry group $\mathrm{G}_{\mathrm{o}} = \mathbb{Z}_2^T \times \mathbb{Z}_2^{T'} \times \mathbb{Z}_2$ to a subgroup of order 4 which is the Cartesian product of two cyclic groups of order 2. We are going to construct this subgroup explicitly. In doing so, we will have established that, from the point of view of the notion of symmetry fractionalization, the case of odd c can always be reduced to that of an even integer c' such that $|c - c'| = 1$.

Spectra of Spin-1/2 Cluster c-Even Chains

Operators $\widehat{C}_j$ in Hamiltonian (5.56) are pairwise commuting and square to the identity, that is,

$$\left[\widehat{C}_j, \widehat{C}_{j'}\right] = 0, \qquad \widehat{C}_j^2 = \widehat{\mathbb{1}}, \tag{5.61}$$

for any $j, j' = 1, \cdots, 2N$. As the $2N$ Hermitean operators $\widehat{C}_j$ can be simultaneously diagonalized, all eigenstates of Hamiltonian (5.56) can be chosen to be pairwise orthogonal and labeled by the eigenvalues ± 1 of the operators $\widehat{C}_j$, that is

$$\widehat{C}_j \,|\mathrm{c}_1, \cdots, \mathrm{c}_{2N}\rangle = \mathrm{c}_j \,|\mathrm{c}_1, \cdots, \mathrm{c}_{2N}\rangle, \qquad \mathrm{c}_j = \pm 1, \qquad j = 1, \cdots, 2N, \tag{5.62a}$$

$$\langle \mathrm{c}_1, \cdots, \mathrm{c}_{2N} \,|\, \mathrm{c}'_1, \cdots, \mathrm{c}'_{2N}\rangle = \delta_{\mathrm{c}_1, \mathrm{c}'_1} \cdots \delta_{\mathrm{c}_{2N}, \mathrm{c}'_{2N}}, \quad \mathrm{c}_j, \mathrm{c}'_j = \pm 1, \quad j = 1, \cdots, 2N, \tag{5.62b}$$

$$\widehat{H}_{c\in 2\mathbb{Z}}^{(b_c=0)} \,|\mathrm{c}_1, \mathrm{c}_2, \cdots, \mathrm{c}_{2N}\rangle = -\frac{\hbar\omega}{2}\left(\sum_{j=1}^{2N} \mathrm{c}_j\right) |\mathrm{c}_1, \mathrm{c}_2, \cdots, \mathrm{c}_{2N}\rangle. \tag{5.62c}$$

The orthonormal basis (5.62a) spans the 2^{2N}-dimensional Hilbert space $\mathbb{C}^{2^{2N}}$ on which $\widehat{H}_{c\in 2\mathbb{Z}}^{(b_c=0)}$ is defined. The nondegenerate ground state of $\widehat{H}_{c\in 2\mathbb{Z}}^{(b_c=0)}$ is specified by the eigenvalues $\mathrm{c}_j = 1$ for all $j = 1, \cdots, 2N$. This ground state $|1, 1, \cdots, 1\rangle$ is separated from all excited states by an excitation gap of $\hbar\omega$ when $b_c = 0$. The ground state $|1, 1, \cdots, 1\rangle$ is symmetric under the symmetry group $\mathrm{G}_{\mathrm{e}} = \mathbb{Z}_2^{T'} \times \mathbb{Z}_2$, that is

$$\widehat{U}_{\mathrm{e}}(t') \,|1, 1, \cdots, 1\rangle = e^{\mathrm{i}\varphi(t')} \,|1, 1, \cdots, 1\rangle, \tag{5.63a}$$

$$\widehat{U}_{\mathrm{e}}(g) \,|1, 1, \cdots, 1\rangle = e^{\mathrm{i}\varphi(g)} \,|1, 1, \cdots, 1\rangle, \tag{5.63b}$$

where $\varphi(t') \in [0, 2\pi[$ and $\varphi(g) = 0, \pi$ are arbitrary phase factors.

Spectra of Spin-1/2 Cluster c-Odd Chains

Operators $\widehat{C}_j$ in Hamiltonian (5.56) are pairwise commuting and square to the identity, that is,

$$\left[\widehat{C}_j, \widehat{C}_{j'}\right] = 0, \qquad \widehat{C}_j^2 = \widehat{\mathbb{1}}, \tag{5.64}$$

for any $j, j' = 1, \cdots, 2N$. However, because c is an odd integer, not all $\widehat{C}_j$ are independent of each other. Indeed, the identity

$$\prod_{j=1}^{2N} \widehat{C}_j = \widehat{\mathbb{1}} \tag{5.65}$$

holds. In other words, only $2N-1$ out of $2N$ Hermitean operators $\widehat{C}_j$ are independent. Hence, Hamiltonian (5.56) becomes

$$\widehat{H}^{(b_c=0)}_{c\in 2\mathbb{Z}+1} = -\frac{\hbar\omega}{2}\sum_{j=1}^{2N-1}\widehat{C}_j - \frac{\hbar\omega}{2}\prod_{j=1}^{2N-1}\widehat{C}_j \tag{5.66}$$

when periodic boundary conditions are imposed.

Because c is odd, each $\widehat{C}_j$ can be factorized into

$$\widehat{C}_j = \widehat{M}^{(c)}_j\,\widehat{M}^{(c)}_{j+1}, \tag{5.67a}$$

with the monomial $\widehat{M}^{(c)}_j$ of odd order $|c|$ defined by

$$\widehat{M}^{(c)}_j := \begin{cases} \widehat{Z}_j\,\widehat{Y}_{j+1}\,\widehat{Z}_{j+2}\,\widehat{Y}_{j+3}\,\cdots\,\widehat{Z}_{j+c-3}\,\widehat{Y}_{j+c-2}\,\widehat{Z}_{j+c-1}, & \text{if } c > +1,\\ \widehat{Z}_j, & \text{if } c = +1,\\ \widehat{Y}_j, & \text{if } c = -1,\\ \widehat{Y}_j\,\widehat{Z}_{j+1}\,\widehat{Y}_{j+2}\,\widehat{Z}_{j+3}\,\cdots\,\widehat{Y}_{j+|c|-3}\,\widehat{Z}_{j+|c|-2}\,\widehat{Y}_{j+|c|-1}, & \text{if } c < -1, \end{cases} \tag{5.67b}$$

obeying

$$\left(\widehat{M}^{(c)}_j\right)^2 = \widehat{\mathbb{1}}, \qquad \left[\widehat{M}^{(c)}_j, \widehat{M}^{(c)}_{j'}\right] = 0, \quad \left[\widehat{M}^{(c)}_j, \widehat{C}_{j'}\right] = 0, \qquad j, j' = 0, 1, \cdots, 2N. \tag{5.67c}$$

One observes that the 2^{2N} eigenstates of the Hamiltonian (5.66) can be labeled by the eigenvalues $\mathrm{c}_j = \pm 1$ of the operators $\widehat{C}_j$ for $j = 1, \cdots, 2N-1$ and the eigenvalues $\mathrm{m}_1 = \pm 1$ of the operator $\widehat{M}^{(c)}_1$. It follows that all energy eigenvalues of Hamiltonian (5.66) are at least twofold degenerate.

Alternatively, we can understand this twofold spectral degeneracy by recasting Hamiltonian (5.66) as the Ising Hamiltonian

$$\widehat{H}^{(b_c=0)}_{c\in 2\mathbb{Z}+1} = -\frac{\hbar\omega}{2}\sum_{j=1}^{2N}\widehat{M}^{(c)}_j\,\widehat{M}^{(c)}_{j+1}. \tag{5.68a}$$

We may then diagonalize Hamiltonian (5.68a) in the orthonormal basis

$$\widehat{M}_j^{(c)}\,|\mathrm{m}_1,\mathrm{m}_2,\cdots,\mathrm{m}_{2N}\rangle=\mathrm{m}_j\,|\mathrm{m}_1,\mathrm{m}_2,\cdots,\mathrm{m}_{2N}\rangle,\qquad \mathrm{m}_j=\pm1, \tag{5.68b}$$

$$\langle\mathrm{m}_1,\cdots,\mathrm{m}_{2N}\,|\,\mathrm{m}'_1,\cdots,\mathrm{m}'_{2N}\rangle=\delta_{\mathrm{m}_1,\mathrm{m}'_1}\cdots\delta_{\mathrm{m}_{2N},\mathrm{m}'_{2N}},\quad \mathrm{m}_j,\mathrm{m}'_j=\pm1, \tag{5.68c}$$

where $j=1,\cdots,2N$, and for which

$$\widehat{H}^{(b_c=0)}_{c\in2\mathbb{Z}+1}\,|\mathrm{m}_1,\mathrm{m}_2,\cdots,\mathrm{m}_{2N}\rangle=-\frac{\hbar\omega}{2}\left(\sum_{j=1}^{2N}\mathrm{m}_j\,\mathrm{m}_{j+1}\right)|\mathrm{m}_1,\mathrm{m}_2,\cdots,\mathrm{m}_{2N}\rangle. \tag{5.68d}$$

The minimal twofold degeneracy of the energy spectrum follows from the fact that $\mathrm{m}_j\,\mathrm{m}_{j+1}$ is unchanged if the sign of each $\mathrm{m}_1,\cdots,\mathrm{m}_N$ is reversed simultaneously. In the thermodynamic limit $N\to\infty$ at zero temperature, this global Ising symmetry is broken spontaneously.

Without loss of generality, we consider the symmetry-breaking ground state labeled by the eigenvalues $\mathrm{m}_1=\cdots=\mathrm{m}_N=1$ and use the mean-field Hamiltonian

$$\widehat{H}^{(b_{|c|-1})}_{\mathrm{MF},\,c\in2\mathbb{Z}+1}:=-\frac{\hbar\omega}{2}\sum_{j=1}^{2N-b_{|c|-1}}\widehat{M}_j^{(c)}. \tag{5.69}$$

When $b_{|c|-1}=0$, this ground state is nondegenerate and separated from all excited states by an excitation gap of $\hbar\omega$.

Hamiltonian (5.69) breaks the $\mathbb{Z}_2^T$ symmetry under reversal of time for $\widehat{M}_j^{(c)}$ is odd under simultaneous sign reversal of all spin operators. In fact, we have the transformations laws

$$\widehat{U}_{\mathrm{o}}(t)\,\widehat{M}_j^{(c)}\,\widehat{U}_{\mathrm{o}}^{\dagger}(t)=-\widehat{M}_j^{(c)}, \tag{5.70a}$$

$$\widehat{U}_{\mathrm{o}}(t')\,\widehat{M}_j^{(c)}\,\widehat{U}_{\mathrm{o}}^{\dagger}(t')=(-1)^{\frac{c-1}{2}}\widehat{M}_j^{(c)}, \tag{5.70b}$$

$$\widehat{U}_{\mathrm{o}}(g)\,\widehat{M}_j^{(c)}\,\widehat{U}_{\mathrm{o}}^{\dagger}(g)=-\widehat{M}_j^{(c)}. \tag{5.70c}$$

Nevertheless, by composing pairs of these three operations, it is possible to construct the subgroup

$$\mathrm{G}_{\mathrm{MF}}=\widetilde{\mathbb{Z}}_2^{\widetilde{T}'}\times\widetilde{\mathbb{Z}}_2,\qquad \widetilde{\mathbb{Z}}_2^{\widetilde{T}'}:=\{e,\tilde{t}'\},\qquad \widetilde{\mathbb{Z}}_2:=\{e,\tilde{g}\}, \tag{5.71}$$

with the representations

$$\widehat{U}_{\mathrm{MF}}(\tilde{t}'):=\begin{cases}\widehat{\mathbb{1}}\,\mathsf{K}, & \text{if } (c-1)/2=0 \bmod 2,\\ \left(\prod_{j=1}^{2N}\widehat{Z}_j\right)\mathsf{K}, & \text{if } (c-1)/2=1 \bmod 2,\end{cases} \tag{5.72a}$$

$$\widehat{U}_{\mathrm{MF}}(\tilde{g}):=\begin{cases}\prod_{j=1}^{2N}\widehat{Z}_j, & \text{if } (c-1)/2=0 \bmod 2,\\ \prod_{j=1}^{2N}\widehat{Y}_j, & \text{if } (c-1)/2=1 \bmod 2,\end{cases} \tag{5.72b}$$

such that

$$\widehat{U}_{\mathrm{MF}}(\tilde{t}')\begin{pmatrix}\widehat{X}_j & \widehat{Y}_j & \widehat{Z}_j\end{pmatrix}^{\mathsf{T}}\widehat{U}^{\dagger}_{\mathrm{MF}}(\tilde{t}') = \begin{cases}\begin{pmatrix}+\widehat{X}_j & -\widehat{Y}_j & +\widehat{Z}_j\end{pmatrix}^{\mathsf{T}}, & \text{if } (c-1)/2 = 0 \bmod 2,\\ \begin{pmatrix}-\widehat{X}_j & +\widehat{Y}_j & +\widehat{Z}_j\end{pmatrix}^{\mathsf{T}}, & \text{if } (c-1)/2 = 1 \bmod 2,\end{cases} \tag{5.72c}$$

$$\widehat{U}_{\mathrm{MF}}(\tilde{g})\begin{pmatrix}\widehat{X}_j & \widehat{Y}_j & \widehat{Z}_j\end{pmatrix}^{\mathsf{T}}\widehat{U}^{\dagger}_{\mathrm{MF}}(\tilde{g}) = \begin{cases}\begin{pmatrix}-\widehat{X}_j & -\widehat{Y}_j & +\widehat{Z}_j\end{pmatrix}^{\mathsf{T}}, & \text{if } (c-1)/2 = 0 \bmod 2,\\ \begin{pmatrix}-\widehat{X}_j & +\widehat{Y}_j & -\widehat{Z}_j\end{pmatrix}^{\mathsf{T}}, & \text{if } (c-1)/2 = 1 \bmod 2,\end{cases} \tag{5.72d}$$

and

$$\widehat{M}^{(c)}_j = \widehat{U}_{\mathrm{MF}}(\tilde{t}')\,\widehat{M}^{(c)}_j\,\widehat{U}^{\dagger}_{\mathrm{MF}}(\tilde{t}') = \widehat{U}_{\mathrm{MF}}(\tilde{g})\,\widehat{M}^{(c)}_j\,\widehat{U}^{\dagger}_{\mathrm{MF}}(\tilde{g}), \qquad j = 1,\cdots,2N. \tag{5.72e}$$

It follows that the group (5.71) is a symmetry group of the mean-field Hamiltonian (5.69).

After projecting the spin-1/2 cluster odd-c Hamiltonian onto the mean-field Hamiltonian (5.69), the spectrum and symmetries of the latter are the same and isomorphic, respectively, to those of the spin-1/2 cluster $[c - \mathrm{sgn}(c)]$ Hamiltonian. The consequences of the group cohomology that we study next are the same for a spin-1/2 cluster even-c' Hamiltonian and the mean-field projection of the spin-1/2 cluster c Hamiltonian with $|c - c'| = 1$.

5.3.3 Projective Representations of Symmetries

5.3.3.1 Overview

We are going to compute the indices classifying bosonic invertible topological phases for the spin-1/2 cluster c models defined in Hamiltonian (5.56). To this end, we will construct explicitly the boundary representations of the global symmetries when $c \bmod 8 = 0, 2, 4, 6$, since the cases of odd $c \bmod 8 = 1, 3, 5, 7$ are reduced by the spontaneous symmetry breaking of time reversal to the cases of $c \bmod 8 = 0, 2, 4, 6$, respectively. We shall show that for each $c \bmod 8 = 0, 2, 4, 6$ the algebra obeyed by the boundary representations of the global symmetries allows to define a pair of distinct c-dependent indices.

5.3.3.2 Strategy

Choose

$$c = 2m, \qquad m \in \mathbb{Z}, \tag{5.73a}$$

to be even. The cardinality of the chain

$$\Lambda := \{j = 1, \cdots, |\Lambda|\} \tag{5.73b}$$

is the large even integer

$$|\Lambda| = 2N. \tag{5.73c}$$

We are going to verify the following five key facts that allow the construction of projective representations of the action of the global symmetries on the boundaries.

1. When open boundary conditions are imposed, the cluster operators

$$\widehat{C}_{2N-|2m|+1}, \cdots, \widehat{C}_{2N} \tag{5.74}$$

are not present in Hamiltonian (5.56). Moreover, either on the left boundary [observe the difference with the left boundary defined in Eq. (5.43c)]

$$\Lambda_{\mathrm{L}} := \{j = 1, 2, \cdots, |2m|\} \tag{5.75}$$

or on the right boundary [observe the difference with the right boundary defined in Eq. (5.43c)]

$$\Lambda_{\mathrm{R}} := \{j = 2N - |2m| + 1, \cdots, 2N - 1, 2N\}, \tag{5.76}$$

the set of all operators commuting with $\widehat{H}_c^{(b_c=|c|)}$ is a Clifford algebra with $|2m|$ generators represented by $2^{|2m|}$-dimensional matrices acting on the Hilbert space

$$\mathfrak{h}_{\Lambda_{\mathrm{B}}} := \mathbb{C}^{2^{|2m|}} \tag{5.77}$$

on either the left (B = L) or the right (B = R) boundary, respectively.

2. This Clifford algebra contains the Lie subalgebra

$$\underbrace{\mathfrak{su}(2) \oplus \cdots \oplus \mathfrak{su}(2)}_{|m| \text{ times}}, \tag{5.78}$$

which is represented reducibly with $2^{|2m|}$-dimensional matrices.[13]

3. It is possible to represent the action of the protecting symmetries on either the left or the right boundaries using the generators of the Lie subalgebra (5.78).
4. The same projective representation of the protecting symmetries is realized on either the left or right boundary.
5. All possible projective representations of the protecting symmetries are fixed by the second cohomology group of the protecting symmetry group. As this second cohomology group is a finite-order Abelian group, the recurrence of a projective representation of the protecting symmetries on either the left or right boundary must be periodic as a function of c.

[13] The difference between even and odd c from the perspective of Clifford algebras spanned by $|c|$ generators comes about because their centers span either a one-dimensional or a two-dimensional vector space when $|c|$ is even and odd, respectively. In the case of odd c, the product of all $|c|$ generators is an operator that is linearly independent from the identity and commutes with all the $|c|$ generators of the Clifford algebra.

We now verify these five facts.

The Cases $c = 0, \pm 2, \pm 4$

We treat Hamiltonian (5.56) with open boundary conditions $b_c = |c|$ for $c = 0, \pm 2, +4$ and $c = -4$.

The case of $c = 0$ When $c = 0$, Hamiltonian (5.56) with open boundary conditions becomes

$$\widehat{H}_0^{(b_0)} := -\frac{\hbar\omega}{2}\sum_{j=1}^{2N}\widehat{X}_j, \tag{5.79}$$

which has a nondegenerate and gapped ground state. All $\widehat{X}_j$ with $j = 1, \cdots, 2N$ are present in $\widehat{H}_0^{(0)}$ and commute pairwise. In other words, the set of gapless boundary degrees of freedom is the empty set

$$\mathfrak{H}_{\mathrm{B},0} := \{\}, \qquad \mathrm{B} = \mathrm{L}, \mathrm{R}, \tag{5.80a}$$

on which t' and g have the trivial representations

$$U_{\mathrm{e,B}}(t') := U_{\mathrm{e,B}}(g) := 1. \tag{5.80b}$$

Using the definitions[14]

$$U_{\mathrm{e,B}}(t')\,U_{\mathrm{e,B}}(t') = (-1)^{[\nu_{\mathrm{B}}]}, \tag{5.81a}$$

$$U_{\mathrm{e,B}}(t')\,U_{\mathrm{e,B}}(g)\,U^{\dagger}_{\mathrm{e,B}}(t')\,U_{\mathrm{e,B}}(g) = (-1)^{[\rho_{\mathrm{B}}]} \tag{5.81b}$$

we associate the pair of (trivial) indices

$$([\nu_{\mathrm{B}}], [\rho_{\mathrm{B}}]) = (0, 0), \qquad \mathrm{B} = \mathrm{L}, \mathrm{R}, \tag{5.81c}$$

to the spin-1/2 cluster $c = 0$ chain. The map

$$U_{\mathrm{e,B}} \colon \mathrm{G}_{\mathrm{e}} \to \mathrm{GL}(\mathbb{C}^1) \tag{5.82}$$

defined by Eq. (5.80) is a group homomorphism.

The case $c = +2$ When $c = +2$, the choice of open boundary conditions implies that Hamiltonian

$$\widehat{H}_2^{(b_2=2)} = -\frac{\hbar\omega}{2}\sum_{j=1}^{2N-2}\widehat{Z}_j\,\widehat{X}_{j+1}\,\widehat{Z}_{j+2} \tag{5.83}$$

has a $2^2 = 4$-fold degenerate ground state separated from all excited states by a gap.

[14] The motivation for these definitions and the rational for bracketing the indices ν_{B} and ρ_{B} will be explained in Section 6.5.

Since open boundary conditions are selected, the pair of commuting operators

$$\widehat{Z}_{2N-1}\,\widehat{X}_{2N}\,\widehat{Z}_{1}, \qquad \widehat{Z}_{2N}\,\widehat{X}_{1}\,\widehat{Z}_{2}, \tag{5.84}$$

are present in $\widehat{H}_2^{(b_2=0)}$ but are absent in $\widehat{H}_2^{(b_2=2)}$.

The set $\mathfrak{O}_{\mathrm{L}}$ of all operators that commute with the Hamiltonian (5.83) and have support on the left boundary

$$\Lambda_{\mathrm{L}} := \{j = 1, 2\} \tag{5.85a}$$

is the Clifford algebra spanned by the generators

$$\left\{\widehat{X}_1\,\widehat{Z}_2, \quad \widehat{Z}_1\right\}. \tag{5.85b}$$

This Clifford algebra is a $2^2 = 4$-dimensional vector space.

The Clifford algebra with the generators (5.85b) contains a four-dimensional representation of the $\mathfrak{su}(2)$ Lie algebra. To verify this claim, we make explicit the tensor product that is implicit in the generators (5.85) of the Clifford algebra. If we momentarily denote the unit 2×2 matrix by I_j together with 2×2 Pauli matrices X_j, Y_j, and Z_j acting on site $j = 1, 2$, it then suffices to define the triplet

$$S^{x}_{\mathrm{L}} := \frac{1}{2}\,X_1 \otimes Z_2, \quad S^{y}_{\mathrm{L}} := \frac{1}{2}\,Y_1 \otimes Z_2, \quad S^{z}_{\mathrm{L}} := \frac{1}{2}\,Z_1 \otimes I_2, \tag{5.86a}$$

of 4×4 matrices and verify that the $\mathfrak{su}(2)$ Lie algebra

$$\left[S^{\alpha}_{\mathrm{L}}, S^{\beta}_{\mathrm{L}}\right] = \mathrm{i}\epsilon^{\alpha\beta\gamma}\,S^{\gamma}_{\mathrm{L}}, \qquad \alpha,\beta = x,y,z, \tag{5.86b}$$

where $\epsilon^{\alpha\beta\gamma}$ is the Levi-Civita symbol and the summation convention over the repeated index $\gamma = x,y,z$ is implied, holds.

The representation of the symmetry group $\mathrm{G}_{\mathrm{e}} = \mathbb{Z}_2^{T'} \times \mathbb{Z}_2$ [recall Eq. (5.57)] on the boundary Λ_{L} is achieved in three steps.

First, we deduce the transformation laws

$$U_{\mathrm{e}}(t')\,\begin{pmatrix}S^{x}_{\mathrm{L}} & S^{y}_{\mathrm{L}} & S^{z}_{\mathrm{L}}\end{pmatrix}^{\mathsf{T}} U^{\dagger}_{\mathrm{e}}(t') = \begin{pmatrix}+S^{x}_{\mathrm{L}} & -S^{y}_{\mathrm{L}} & +S^{z}_{\mathrm{L}}\end{pmatrix}^{\mathsf{T}}, \quad U_{\mathrm{e}}(t') := I_1 \otimes I_2\,\mathsf{K}, \tag{5.87a}$$

$$U_{\mathrm{e}}(g)\,\begin{pmatrix}S^{x}_{\mathrm{L}} & S^{y}_{\mathrm{L}} & S^{z}_{\mathrm{L}}\end{pmatrix}^{\mathsf{T}} U^{\dagger}_{\mathrm{e}}(g) = \begin{pmatrix}-S^{x}_{\mathrm{L}} & +S^{y}_{\mathrm{L}} & -S^{z}_{\mathrm{L}}\end{pmatrix}^{\mathsf{T}}, \quad U_{\mathrm{e}}(g) := X_1 \otimes X_2, \tag{5.87b}$$

by combining Eqs. (5.57) and (5.86).

Second, we define the action of t' and g on the left boundary by demanding that

$$U_{\mathrm{e,L}}(t')\,\begin{pmatrix}S^{x}_{\mathrm{L}} & S^{y}_{\mathrm{L}} & S^{z}_{\mathrm{L}}\end{pmatrix}^{\mathsf{T}} U^{\dagger}_{\mathrm{e,L}}(t') := U_{\mathrm{e}}(t')\,\begin{pmatrix}S^{x}_{\mathrm{L}} & S^{y}_{\mathrm{L}} & S^{z}_{\mathrm{L}}\end{pmatrix}^{\mathsf{T}} U^{\dagger}_{\mathrm{e}}(t'), \tag{5.88a}$$

$$U_{\mathrm{e,L}}(g)\,\begin{pmatrix}S^{x}_{\mathrm{L}} & S^{y}_{\mathrm{L}} & S^{z}_{\mathrm{L}}\end{pmatrix}^{\mathsf{T}} U^{\dagger}_{\mathrm{e,L}}(g) := U_{\mathrm{e}}(g)\,\begin{pmatrix}S^{x}_{\mathrm{L}} & S^{y}_{\mathrm{L}} & S^{z}_{\mathrm{L}}\end{pmatrix}^{\mathsf{T}} U^{\dagger}_{\mathrm{e}}(g). \tag{5.88b}$$

Third, we denote by $\mathbb{1}_{\mathrm{L}}$ the unit 4×4 matrix and we define the action K_{L} of complex conjugation on the four-dimensional representation (5.86) of the $\mathfrak{su}(2)$ Lie algebra by demanding that

$$\mathsf{K}_{\mathrm{L}}\, S^{x}_{\mathrm{L}}\, \mathsf{K}_{\mathrm{L}} := +S^{x}_{\mathrm{L}}, \qquad \mathsf{K}_{\mathrm{L}}\, S^{y}_{\mathrm{L}}\, \mathsf{K}_{\mathrm{L}} := -S^{y}_{\mathrm{L}}, \qquad \mathsf{K}_{\mathrm{L}}\, S^{z}_{\mathrm{L}}\, \mathsf{K}_{\mathrm{L}} := +S^{z}_{\mathrm{L}}. \tag{5.89a}$$

One then verifies that Eqs. (5.87), (5.88), and (5.89a) are satisfied by the four-dimensional representation

$$U_{\mathrm{e,L}}(t') := \mathbb{1}_{\mathrm{L}}\, \mathsf{K}_{\mathrm{L}}, \qquad U_{\mathrm{e,L}}(g) := 2\, S^{y}_{\mathrm{L}}. \tag{5.89b}$$

Using the definitions

$$U_{\mathrm{e,L}}(t')\, U_{\mathrm{e,L}}(t') = (-1)^{[\nu_{\mathrm{L}}]}\, \mathbb{1}_{\mathrm{L}}, \tag{5.90a}$$

$$U_{\mathrm{e,L}}(t')\, U_{\mathrm{e,L}}(g)\, U^{\dagger}_{\mathrm{e,L}}(t')\, U_{\mathrm{e,L}}(g) = (-1)^{[\rho_{\mathrm{L}}]}\, \mathbb{1}_{\mathrm{L}} \tag{5.90b}$$

we associate the pair of indices

$$([\nu_{\mathrm{L}}], [\rho_{\mathrm{L}}]) = (0,1) \tag{5.90c}$$

to the spin-1/2 cluster $c=+2$ chain. The map

$$U_{\mathrm{e,L}}\colon \mathrm{G}_{\mathrm{e}} \to \mathrm{GL}(\mathbb{C}^{4})/F^{\star}, \qquad F^{\star} := \{z\,\mathbb{1}_{L}\,|\, z\in\mathbb{C}\setminus\{0\}\} \tag{5.91}$$

defined by Eq. (5.89b) is a projective representation[15] since it is a group homomorphism.

One verifies that the Clifford algebra spanned by the generators

$$\left\{\widehat{Z}_{2N-1}\,\widehat{X}_{2N},\ \ \widehat{Z}_{2N}\right\} \tag{5.92a}$$

for the right boundary

$$\Lambda_{\mathrm{R}} := \{j = 2N-1, 2N\} \tag{5.92b}$$

delivers the pair of indices

$$([\nu_{\mathrm{R}}], [\rho_{\mathrm{R}}]) = (0,1) \tag{5.92c}$$

for the projective representation of t' and g on the right boundary.

The case $c=-2$ The case $c=-2$ is deduced from the case $c=+2$ by interchanging all the $\widehat{Z}_j$ and $\widehat{Y}_j$ operators for $j=1,\cdots,2N$. The set $\mathfrak{O}_{\mathrm{L}}$ of all operators that commute with the Hamiltonian $\widehat{H}^{(b_{-2}=2)}_{-2}$ and have support on the boundary

$$\Lambda_{\mathrm{L}} := \{j = 1, 2\} \tag{5.93a}$$

is the Clifford algebra spanned by the generators

$$\left\{\widehat{X}_1\,\widehat{Y}_2,\ \ \widehat{Y}_1\right\}. \tag{5.93b}$$

[15] A more detailed discussion of what are projective representations will be given in Section 5.3.4.

The Clifford algebra with the generators (5.93b) contains the $\mathfrak{su}(2)$ Lie algebra generated by the 4×4 matrices

$$S^x_{\mathrm{L}} := \frac{1}{2}\,X_1\otimes Y_2, \qquad S^y_{\mathrm{L}} := -\frac{1}{2}\,Z_1\otimes Y_2, \qquad S^z_{\mathrm{L}} := \frac{1}{2}\,Y_1\otimes I_2. \tag{5.94}$$

The representation of the symmetry group $\mathrm{G}_{\mathrm{e}} = \mathbb{Z}_2^{T'}\times\mathbb{Z}_2$ [recall Eq. (5.57)] on the boundary Λ_{L} is achieved in three steps.

First, we deduce the transformation laws

$$U_{\mathrm{e}}(t')\,\begin{pmatrix}S^x_{\mathrm{L}} & S^y_{\mathrm{L}} & S^z_{\mathrm{L}}\end{pmatrix}^{\mathsf{T}} U^{\dagger}_{\mathrm{e}}(t') = \begin{pmatrix}-S^x_{\mathrm{L}} & -S^y_{\mathrm{L}} & -S^z_{\mathrm{L}}\end{pmatrix}^{\mathsf{T}}, \quad U_{\mathrm{e}}(t') := I_1\otimes I_2\,\mathsf{K}, \tag{5.95a}$$

$$U_{\mathrm{e}}(g)\,\begin{pmatrix}S^x_{\mathrm{L}} & S^y_{\mathrm{L}} & S^z_{\mathrm{L}}\end{pmatrix}^{\mathsf{T}} U^{\dagger}_{\mathrm{e}}(g) = \begin{pmatrix}-S^x_{\mathrm{L}} & +S^y_{\mathrm{L}} & -S^z_{\mathrm{L}}\end{pmatrix}^{\mathsf{T}}, \quad U_{\mathrm{e}}(g) := X_1\otimes X_2, \tag{5.95b}$$

by combining Eqs. (5.57) and (5.94).

Second, we define the action of t' and g on the boundary by demanding that

$$U_{\mathrm{e,L}}(t')\,\begin{pmatrix}S^x_{\mathrm{L}} & S^y_{\mathrm{L}} & S^z_{\mathrm{L}}\end{pmatrix}^{\mathsf{T}} U^{\dagger}_{\mathrm{e,L}}(t') := U_{\mathrm{e}}(t')\,\begin{pmatrix}S^x_{\mathrm{L}} & S^y_{\mathrm{L}} & S^z_{\mathrm{L}}\end{pmatrix}^{\mathsf{T}} U^{\dagger}_{\mathrm{e}}(t'), \tag{5.96a}$$

$$U_{\mathrm{e,L}}(g)\,\begin{pmatrix}S^x_{\mathrm{L}} & S^y_{\mathrm{L}} & S^z_{\mathrm{L}}\end{pmatrix}^{\mathsf{T}} U^{\dagger}_{\mathrm{e,L}}(g) := U_{\mathrm{e}}(g)\,\begin{pmatrix}S^x_{\mathrm{L}} & S^y_{\mathrm{L}} & S^z_{\mathrm{L}}\end{pmatrix}^{\mathsf{T}} U^{\dagger}_{\mathrm{e}}(g). \tag{5.96b}$$

Third, we denote by $\mathbb{1}_{\mathrm{L}}$ the unit 4×4 matrix and we define the action K_{L} of complex conjugation on the four-dimensional representation (5.94) of the $\mathfrak{su}(2)$ Lie algebra by demanding that

$$\mathsf{K}_{\mathrm{L}}\,S^x_{\mathrm{L}}\,\mathsf{K}_{\mathrm{L}} := +S^x_{\mathrm{L}}, \qquad \mathsf{K}_{\mathrm{L}}\,S^y_{\mathrm{L}}\,\mathsf{K}_{\mathrm{L}} := -S^y_{\mathrm{L}}, \qquad \mathsf{K}_{\mathrm{L}}\,S^z_{\mathrm{L}}\,\mathsf{K}_{\mathrm{L}} := +S^z_{\mathrm{L}}. \tag{5.97a}$$

One then verifies that Eqs. (5.95), (5.96), and (5.97) are satisfied by the four-dimensional representation

$$U_{\mathrm{e,L}}(t') := 2\,S^y_{\mathrm{L}}\,\mathsf{K}_{\mathrm{L}} \qquad U_{\mathrm{e,L}}(g) := 2\,S^y_{\mathrm{L}}. \tag{5.97b}$$

Using the definitions

$$U_{\mathrm{e,L}}(t')\,U_{\mathrm{e,L}}(t') = (-1)^{[\nu_{\mathrm{L}}]}\,\mathbb{1}_{\mathrm{L}}, \tag{5.98a}$$

$$U_{\mathrm{e,L}}(t')\,U_{\mathrm{e,L}}(g)\,U^{\dagger}_{\mathrm{e,L}}(t')\,U_{\mathrm{e,L}}(g) = (-1)^{[\rho_{\mathrm{L}}]}\,\mathbb{1}_{\mathrm{L}}, \tag{5.98b}$$

we associate the pair of indices

$$([\nu_{\mathrm{L}}],[\rho_{\mathrm{L}}]) = (1,1) \tag{5.98c}$$

to the spin-1/2 cluster $c=-2$ chain. The map

$$U_{\mathrm{e,L}}\colon \mathrm{G}_{\mathrm{e}} \to \mathrm{GL}(\mathbb{C}^4)/F^{\star}, \qquad F^{\star} := \{z\,\mathbb{1}_L \mid z\in\mathbb{C}\setminus\{0\}\} \tag{5.99}$$

defined by Eq. (5.97b) is a projective representation since it is a group homomorphism.

One verifies that the Clifford algebra spanned by the generators

$$\left\{\widehat{Y}_{2N-1}\widehat{X}_{2N}, \quad \widehat{Y}_{2N}\right\} \tag{5.100a}$$

for the right boundary

$$\Lambda_{\mathrm{R}} := \{j = 2N-1, 2N\} \tag{5.100b}$$

delivers the pair of indices

$$([\nu_{\mathrm{R}}], [\rho_{\mathrm{R}}]) = (1, 1) \tag{5.100c}$$

for the projective representation of t' and g on the right boundary.

A complementary point of view on the spin-1/2 $c = \pm 2$ chains is given in **Exercise** 5.1.

The case $c = +4$ When $c = +4$, the choice of open boundary conditions implies that Hamiltonian

$$\widehat{H}_{4}^{(b_4=4)} = -\frac{\hbar\omega}{2}\sum_{j=1}^{2N-4} \widehat{Z}_{j}\,\widehat{X}_{j+1}\,\widehat{X}_{j+2}\,\widehat{X}_{j+3}\,\widehat{Z}_{j+4} \tag{5.101}$$

has a $2^4 = 16$-fold degenerate ground state separated from all excited states by a gap.

Since open boundary conditions are selected, the quadruplet of pairwise commuting operators

$$\begin{gathered}\widehat{Z}_{2N-3}\,\widehat{X}_{2N-2}\,\widehat{X}_{2N-1}\,\widehat{X}_{2N}\,\widehat{Z}_{1}, \qquad \widehat{Z}_{2N-2}\,\widehat{X}_{2N-1}\,\widehat{X}_{2N}\,\widehat{X}_{1}\,\widehat{Z}_{2}, \\ \widehat{Z}_{2N-1}\,\widehat{X}_{2N}\,\widehat{X}_{1}\,\widehat{X}_{2}\,\widehat{Z}_{3}, \qquad \widehat{Z}_{2N}\,\widehat{X}_{1}\,\widehat{X}_{2}\,\widehat{X}_{3}\,\widehat{Z}_{4}\end{gathered} \tag{5.102}$$

are present in $\widehat{H}_{4}^{(b_4=0)}$ but are absent in $\widehat{H}_{4}^{(b_4=4)}$.

The set $\mathfrak{O}_{\mathrm{L}}$ of all operators that commute with Hamiltonian (5.101) and have support on the boundary

$$\Lambda_{\mathrm{L}} := \{j = 1, 2, 3, 4\} \tag{5.103a}$$

is the Clifford algebra spanned by the generators

$$\left\{\widehat{X}_{1}\,\widehat{X}_{2}\,\widehat{X}_{3}\,\widehat{Z}_{4}, \quad \widehat{X}_{1}\,\widehat{X}_{2}\,\widehat{Z}_{3}, \quad \widehat{X}_{1}\,\widehat{Z}_{2}, \quad \widehat{Z}_{1}\right\}. \tag{5.103b}$$

This Clifford algebra is a $2^4 = 16$-dimensional vector space.

The Clifford algebra with the generators (5.103b) contains a 16-dimensional representation of the $\mathfrak{su}(2) \oplus \mathfrak{su}(2)$ Lie algebra. To verify this claim, we make explicit the tensor product that is implicit in the generators (5.103) of the Clifford algebra. If we momentarily denote the unit 2×2 matrix by I_j together with 2×2 Pauli matrices X_j, Y_j, and Z_j acting on site $j = 1,2,3,4$, it then suffices to define the pair of triplets

$$S^x_{\mathrm{L}} := \frac{1}{2} X_1 \otimes X_2 \otimes X_3 \otimes Z_4, \qquad S^y_{\mathrm{L}} := -\frac{1}{2} X_1 \otimes X_2 \otimes Z_3 \otimes I_4,$$
$$S^z_{\mathrm{L}} := \frac{1}{2} I_1 \otimes I_2 \otimes Y_3 \otimes Z_4, \tag{5.104a}$$

$$\widetilde{S}^x_{\mathrm{L}} := \frac{1}{2} X_1 \otimes Z_2 \otimes Y_3 \otimes Z_4, \qquad \widetilde{S}^y_{\mathrm{L}} := -\frac{1}{2} Z_1 \otimes I_2 \otimes Y_3 \otimes Z_4,$$
$$\widetilde{S}^z_{\mathrm{L}} := \frac{1}{2} Y_1 \otimes Z_2 \otimes I_3 \otimes I_4, \tag{5.104b}$$

of 16×16 matrices and verify that the $\mathfrak{su}(2) \oplus \mathfrak{su}(2)$ Lie algebra

$$\left[S^\alpha_{\mathrm{L}}, S^\beta_{\mathrm{L}}\right] = \mathrm{i}\epsilon^{\alpha\beta\gamma}\, S^\gamma_{\mathrm{L}}, \qquad \left[\widetilde{S}^\alpha_{\mathrm{L}}, \widetilde{S}^\beta_{\mathrm{L}}\right] = \mathrm{i}\epsilon^{\alpha\beta\gamma}\, \widetilde{S}^\gamma_{\mathrm{L}}, \qquad \left[S^\alpha_{\mathrm{L}}, \widetilde{S}^\beta_{\mathrm{L}}\right] = 0, \tag{5.104c}$$

where $\alpha, \beta = x, y, z$, $\epsilon^{\alpha\beta\gamma}$ is the Levi-Civita symbol, and the summation convention over the repeated index $\gamma = x, y, z$ is implied, holds.

The representation of the symmetry group $\mathrm{G}_{\mathrm{e}} = \mathbb{Z}_2^{T'} \times \mathbb{Z}_2$ [recall Eq. (5.57)] on the boundary Λ_{L} is achieved in three steps.

First, we deduce the transformation laws

$$U_{\mathrm{e}}(t')\,\left(S^x_{\mathrm{L}} \quad S^y_{\mathrm{L}} \quad S^z_{\mathrm{L}}\right)^{\mathsf{T}} U^\dagger_{\mathrm{e}}(t') = \left(+S^x_{\mathrm{L}} \quad +S^y_{\mathrm{L}} \quad -S^z_{\mathrm{L}}\right)^{\mathsf{T}},$$
$$U_{\mathrm{e}}(t') := I_1 \otimes \cdots \otimes I_4\, \mathsf{K}, \tag{5.105a}$$

$$U_{\mathrm{e}}(g)\,\left(S^x_{\mathrm{L}} \quad S^y_{\mathrm{L}} \quad S^z_{\mathrm{L}}\right)^{\mathsf{T}} U^\dagger_{\mathrm{e}}(g) = \left(-S^x_{\mathrm{L}} \quad -S^y_{\mathrm{L}} \quad +S^z_{\mathrm{L}}\right)^{\mathsf{T}},$$
$$U_{\mathrm{e}}(g) := X_1 \otimes \cdots \otimes X_4, \tag{5.105b}$$

and

$$U_{\mathrm{e}}(t')\,\left(\widetilde{S}^x_{\mathrm{L}} \quad \widetilde{S}^y_{\mathrm{L}} \quad \widetilde{S}^z_{\mathrm{L}}\right)^{\mathsf{T}} U^\dagger_{\mathrm{e}}(t') = \left(-\widetilde{S}^x_{\mathrm{L}} \quad -\widetilde{S}^y_{\mathrm{L}} \quad -\widetilde{S}^z_{\mathrm{L}}\right)^{\mathsf{T}},$$
$$U_{\mathrm{e}}(t') := I_1 \otimes \cdots \otimes I_4\, \mathsf{K}, \tag{5.105c}$$

$$U_{\mathrm{e}}(g)\,\left(\widetilde{S}^x_{\mathrm{L}} \quad \widetilde{S}^y_{\mathrm{L}} \quad \widetilde{S}^z_{\mathrm{L}}\right)^{\mathsf{T}} U^\dagger_{\mathrm{e}}(g) = \left(-\widetilde{S}^x_{\mathrm{L}} \quad -\widetilde{S}^y_{\mathrm{L}} \quad +\widetilde{S}^z_{\mathrm{L}}\right)^{\mathsf{T}},$$
$$U_{\mathrm{e}}(g) := X_1 \otimes \cdots \otimes X_4, \tag{5.105d}$$

by combining Eqs. (5.57) and (5.104).

Second, we define the action of t' and g on the left boundary by demanding that

$$U_{\mathrm{e,L}}(t')\,\left(S^x_{\mathrm{L}} \quad S^y_{\mathrm{L}} \quad S^z_{\mathrm{L}}\right)^{\mathsf{T}} U^\dagger_{\mathrm{e,L}}(t') := U_{\mathrm{e}}(t')\,\left(S^x_{\mathrm{L}} \quad S^y_{\mathrm{L}} \quad S^z_{\mathrm{L}}\right)^{\mathsf{T}} U^\dagger_{\mathrm{e}}(t'), \tag{5.106a}$$
$$U_{\mathrm{e,L}}(g)\,\left(S^x_{\mathrm{L}} \quad S^y_{\mathrm{L}} \quad S^z_{\mathrm{L}}\right)^{\mathsf{T}} U^\dagger_{\mathrm{e,L}}(g) := U_{\mathrm{e}}(g)\,\left(S^x_{\mathrm{L}} \quad S^y_{\mathrm{L}} \quad S^z_{\mathrm{L}}\right)^{\mathsf{T}} U^\dagger_{\mathrm{e}}(g), \tag{5.106b}$$

and

$$U_{\mathrm{e,L}}(t')\begin{pmatrix}\widetilde{S}^x_{\mathrm{L}} & \widetilde{S}^y_{\mathrm{L}} & \widetilde{S}^z_{\mathrm{L}}\end{pmatrix}^{\mathsf{T}} U^{\dagger}_{\mathrm{e,L}}(t') := U_{\mathrm{e}}(t')\begin{pmatrix}\widetilde{S}^x_{\mathrm{L}} & \widetilde{S}^y_{\mathrm{L}} & \widetilde{S}^z_{\mathrm{L}}\end{pmatrix}^{\mathsf{T}} U^{\dagger}_{\mathrm{e}}(t'), \tag{5.106c}$$

$$U_{\mathrm{e,L}}(g)\begin{pmatrix}\widetilde{S}^x_{\mathrm{L}} & \widetilde{S}^y_{\mathrm{L}} & \widetilde{S}^z_{\mathrm{L}}\end{pmatrix}^{\mathsf{T}} U^{\dagger}_{\mathrm{e,L}}(g) := U_{\mathrm{e}}(g)\begin{pmatrix}\widetilde{S}^x_{\mathrm{L}} & \widetilde{S}^y_{\mathrm{L}} & \widetilde{S}^z_{\mathrm{L}}\end{pmatrix}^{\mathsf{T}} U^{\dagger}_{\mathrm{e}}(g). \tag{5.106d}$$

Third, we denote by $\mathbb{1}_{\mathrm{L}}$ the unit 16×16 matrix and we define the action K_{L} of complex conjugation on the 16-dimensional representation (5.104) of the $\mathfrak{su}(2)\oplus\mathfrak{su}(2)$ Lie algebra by demanding that

$$\mathsf{K}_{\mathrm{L}}\, S^x_{\mathrm{L}}\, \mathsf{K}_{\mathrm{L}} := +S^x_{\mathrm{L}}, \qquad \mathsf{K}_{\mathrm{L}}\, S^y_{\mathrm{L}}\, \mathsf{K}_{\mathrm{L}} := -S^y_{\mathrm{L}}, \qquad \mathsf{K}_{\mathrm{L}}\, S^z_{\mathrm{L}}\, \mathsf{K}_{\mathrm{L}} := +S^z_{\mathrm{L}}, \tag{5.107a}$$

$$\mathsf{K}_{\mathrm{L}}\, \widetilde{S}^x_{\mathrm{L}}\, \mathsf{K}_{\mathrm{L}} := -\widetilde{S}^x_{\mathrm{L}}, \qquad \mathsf{K}_{\mathrm{L}}\, \widetilde{S}^y_{\mathrm{L}}\, \mathsf{K}_{\mathrm{L}} := +\widetilde{S}^y_{\mathrm{L}}, \qquad \mathsf{K}_{\mathrm{L}}\, \widetilde{S}^z_{\mathrm{L}}\, \mathsf{K}_{\mathrm{L}} := +\widetilde{S}^z_{\mathrm{L}}. \tag{5.107b}$$

One then verifies that Eqs. (5.105), (5.106), (5.107a), and (5.107b) are satisfied by the 16-dimensional representation

$$U_{\mathrm{e,L}}(t') := 4\, S^x_{\mathrm{L}}\, \widetilde{S}^x_{\mathrm{L}}\, \mathsf{K}_{\mathrm{L}}, \qquad U_{\mathrm{e,L}}(g) := 4\, S^z_{\mathrm{L}}\, \widetilde{S}^z_{\mathrm{L}}. \tag{5.107c}$$

Using the definitions

$$U_{\mathrm{e,L}}(t')\, U_{\mathrm{e,L}}(t') = (-1)^{[\nu_{\mathrm{L}}]}\, \mathbb{1}_{\mathrm{L}}, \tag{5.108a}$$

$$U_{\mathrm{e,L}}(t')\, U_{\mathrm{e,L}}(g)\, U^{\dagger}_{\mathrm{e,L}}(t')\, U_{\mathrm{e,L}}(g) = (-1)^{[\rho_{\mathrm{L}}]}\, \mathbb{1}_{\mathrm{L}}, \tag{5.108b}$$

we associate the pair of indices

$$([\nu_{\mathrm{L}}],[\rho_{\mathrm{L}}]) = (1,0) \tag{5.108c}$$

to the spin-1/2 cluster $c=+4$ chain. The map

$$U_{\mathrm{e,L}}\colon \mathrm{G_e} \to \mathrm{GL}(\mathbb{C}^{16})/F^{\star}, \qquad F^{\star} := \{z\,\mathbb{1}_{L}\,|\, z\in\mathbb{C}\setminus\{0\}\} \tag{5.109}$$

defined by Eq. (5.107c) is a projective representation since it is a group homomorphism.

One verifies that the Clifford algebra spanned by the generators

$$\left\{\widehat{Z}_{2N-3}\,\widehat{X}_{2N-2}\,\widehat{X}_{2N-1}\,\widehat{X}_{2N},\quad \widehat{Z}_{2N-2}\,\widehat{X}_{2N-1}\,\widehat{X}_{2N},\quad \widehat{Z}_{2N-1}\,\widehat{X}_{2N},\quad \widehat{Z}_{2N}\right\} \tag{5.110a}$$

for the right boundary

$$\Lambda_{\mathrm{R}} := \{j = 2N-3, 2N-2, 2N-1, 2N\} \tag{5.110b}$$

delivers the pair of indices

$$([\nu_{\mathrm{R}}],[\rho_{\mathrm{R}}]) = (1,0) \tag{5.110c}$$

for the projective representation of t' and g on the right boundary.

The case $c=-4$ The case $c=-4$ is deduced from the case $c=+4$ by interchanging all the $\widehat{Z}_j$ and $\widehat{Y}_j$ operators for $j=1,\cdots,2N$. The set $\mathfrak{O}_{\mathrm{L}}$ of all operators that commute with the Hamiltonian $\widehat{H}^{(b_{-4}=4)}_{-4}$ and have support on the boundary

$$\Lambda_{\mathrm{L}} := \{j=1,2,3,4\} \tag{5.111a}$$

is the Clifford algebra spanned by the generators

$$\{\widehat{Y}_1,\quad \widehat{X}_1\widehat{Y}_2,\quad \widehat{X}_1\widehat{X}_2\widehat{Y}_3,\quad \widehat{X}_1\widehat{X}_2\widehat{X}_3\widehat{Y}_4\}. \tag{5.111b}$$

The Clifford algebra with the generators (5.111b) contains the $\mathfrak{su}(2)\oplus\mathfrak{su}(2)$ Lie algebra generated by the 16×16 matrices

$$\begin{aligned} S^x_{\mathrm{L}} &:= \frac{1}{2}X_1\otimes X_2\otimes X_3\otimes Y_4, \qquad S^y_{\mathrm{L}} := \frac{1}{2}X_1\otimes X_2\otimes Y_3\otimes I_4,\\ S^z_{\mathrm{L}} &:= \frac{1}{2}I_1\otimes I_2\otimes Z_3\otimes Y_4, \end{aligned} \tag{5.112a}$$

$$\begin{aligned} \widetilde{S}^x_{\mathrm{L}} &:= \frac{1}{2}X_1\otimes Y_2\otimes Z_3\otimes Y_4, \qquad \widetilde{S}^y_{\mathrm{L}} := \frac{1}{2}Y_1\otimes I_2\otimes Z_3\otimes Y_4,\\ \widetilde{S}^z_{\mathrm{L}} &:= \frac{1}{2}Z_1\otimes Y_2\otimes I_3\otimes I_4. \end{aligned} \tag{5.112b}$$

The representation of the symmetry group $\mathrm{G}_{\mathrm{e}}=\mathbb{Z}_2^{T'}\times\mathbb{Z}_2$ [recall Eq. (5.57)] on the boundary Λ_{L} is achieved in three steps.

First, we deduce the transformation laws

$$\begin{aligned} U_{\mathrm{e}}(t')\left(S^x_{\mathrm{L}}\ \ S^y_{\mathrm{L}}\ \ S^z_{\mathrm{L}}\right)^{\mathsf{T}}U^{\dagger}_{\mathrm{e}}(t') &= \left(-S^x_{\mathrm{L}}\ \ -S^y_{\mathrm{L}}\ \ -S^z_{\mathrm{L}}\right)^{\mathsf{T}},\\ U_{\mathrm{e}}(t') &:= I_1\otimes\cdots\otimes I_4\,\mathsf{K}, \end{aligned} \tag{5.113a}$$

$$\begin{aligned} U_{\mathrm{e}}(g)\left(S^x_{\mathrm{L}}\ \ S^y_{\mathrm{L}}\ \ S^z_{\mathrm{L}}\right)^{\mathsf{T}}U^{\dagger}_{\mathrm{e}}(g) &= \left(-S^x_{\mathrm{L}}\ \ -S^y_{\mathrm{L}}\ \ +S^z_{\mathrm{L}}\right)^{\mathsf{T}},\\ U_{\mathrm{e}}(g) &:= X_1\otimes\cdots\otimes X_4, \end{aligned} \tag{5.113b}$$

and

$$\begin{aligned} U_{\mathrm{e}}(t')\left(\widetilde{S}^x_{\mathrm{L}}\ \ \widetilde{S}^y_{\mathrm{L}}\ \ \widetilde{S}^z_{\mathrm{L}}\right)^{\mathsf{T}}U^{\dagger}_{\mathrm{e}}(t') &= \left(+\widetilde{S}^x_{\mathrm{L}}\ \ +\widetilde{S}^y_{\mathrm{L}}\ \ -\widetilde{S}^z_{\mathrm{L}}\right)^{\mathsf{T}},\\ U_{\mathrm{e}}(t') &:= I_1\otimes\cdots\otimes I_4\,\mathsf{K}, \end{aligned} \tag{5.113c}$$

$$\begin{aligned} U_{\mathrm{e}}(g)\left(\widetilde{S}^x_{\mathrm{L}}\ \ \widetilde{S}^y_{\mathrm{L}}\ \ \widetilde{S}^z_{\mathrm{L}}\right)^{\mathsf{T}}U^{\dagger}_{\mathrm{e}}(g) &= \left(-\widetilde{S}^x_{\mathrm{L}}\ \ -\widetilde{S}^y_{\mathrm{L}}\ \ +\widetilde{S}^z_{\mathrm{L}}\right)^{\mathsf{T}},\\ U_{\mathrm{e}}(g) &:= X_1\otimes\cdots\otimes X_4, \end{aligned} \tag{5.113d}$$

by combining Eqs. (5.57) and (5.112).

Second, we define the action of t' and g on the boundary by demanding that

$$U_{\mathrm{e,L}}(t')\left(S^x_{\mathrm{L}}\ \ S^y_{\mathrm{L}}\ \ S^z_{\mathrm{L}}\right)^{\mathsf{T}}U^{\dagger}_{\mathrm{e,L}}(t') := U_{\mathrm{e}}(t')\left(S^x_{\mathrm{L}}\ \ S^y_{\mathrm{L}}\ \ S^z_{\mathrm{L}}\right)^{\mathsf{T}}U^{\dagger}_{\mathrm{e}}(t'), \tag{5.114a}$$

$$U_{\mathrm{e,L}}(g)\left(S^x_{\mathrm{L}}\ \ S^y_{\mathrm{L}}\ \ S^z_{\mathrm{L}}\right)^{\mathsf{T}}U^{\dagger}_{\mathrm{e,L}}(g) := U_{\mathrm{e}}(g)\left(S^x_{\mathrm{L}}\ \ S^y_{\mathrm{L}}\ \ S^z_{\mathrm{L}}\right)^{\mathsf{T}}U^{\dagger}_{\mathrm{e}}(g), \tag{5.114b}$$

and

$$U_{\mathrm{e,L}}(t')\begin{pmatrix}\widetilde{S}^x_{\mathrm{L}} & \widetilde{S}^y_{\mathrm{L}} & \widetilde{S}^z_{\mathrm{L}}\end{pmatrix}^{\mathsf{T}} U^\dagger_{\mathrm{e,L}}(t') := U_{\mathrm{e}}(t')\begin{pmatrix}\widetilde{S}^x_{\mathrm{L}} & \widetilde{S}^y_{\mathrm{L}} & \widetilde{S}^z_{\mathrm{L}}\end{pmatrix}^{\mathsf{T}} U^\dagger_{\mathrm{e}}(t'), \tag{5.114c}$$

$$U_{\mathrm{e,L}}(g)\begin{pmatrix}\widetilde{S}^x_{\mathrm{L}} & \widetilde{S}^y_{\mathrm{L}} & \widetilde{S}^z_{\mathrm{L}}\end{pmatrix}^{\mathsf{T}} U^\dagger_{\mathrm{e,L}}(g) := U_{\mathrm{e}}(g)\begin{pmatrix}\widetilde{S}^x_{\mathrm{L}} & \widetilde{S}^y_{\mathrm{L}} & \widetilde{S}^z_{\mathrm{L}}\end{pmatrix}^{\mathsf{T}} U^\dagger_{\mathrm{e}}(g). \tag{5.114d}$$

Third, we denote by $\mathbb{1}_{\mathrm{L}}$ the unit 16×16 matrix and we define the action K_{L} of complex conjugation on the 16-dimensional representation (5.112) of the $\mathfrak{su}(2)\oplus\mathfrak{su}(2)$ Lie algebra by demanding that

$$\mathsf{K}_{\mathrm{L}}\,S^x_{\mathrm{L}}\,\mathsf{K}_{\mathrm{L}} := -S^x_{\mathrm{L}}, \qquad \mathsf{K}_{\mathrm{L}}\,S^y_{\mathrm{L}}\,\mathsf{K}_{\mathrm{L}} := +S^y_{\mathrm{L}}, \qquad \mathsf{K}_{\mathrm{L}}\,S^z_{\mathrm{L}}\,\mathsf{K}_{\mathrm{L}} := +S^z_{\mathrm{L}}, \tag{5.115a}$$

and

$$\mathsf{K}_{\mathrm{L}}\,\widetilde{S}^x_{\mathrm{L}}\,\mathsf{K}_{\mathrm{L}} := +\widetilde{S}^x_{\mathrm{L}}, \qquad \mathsf{K}_{\mathrm{L}}\,\widetilde{S}^y_{\mathrm{L}}\,\mathsf{K}_{\mathrm{L}} := -\widetilde{S}^y_{\mathrm{L}}, \qquad \mathsf{K}_{\mathrm{L}}\,\widetilde{S}^z_{\mathrm{L}}\,\mathsf{K}_{\mathrm{L}} := +\widetilde{S}^z_{\mathrm{L}}. \tag{5.115b}$$

One then verifies that Eqs. (5.113), (5.114), (5.115a), and (5.115b) are satisfied by the 16-dimensional representation

$$U_{\mathrm{e,L}}(t') := 4\,S^x_{\mathrm{L}}\,\widetilde{S}^x_{\mathrm{L}}\,\mathsf{K}_{\mathrm{L}}, \qquad U_{\mathrm{e,L}}(g) := 4\,S^z_{\mathrm{L}}\,\widetilde{S}^z_{\mathrm{L}}. \tag{5.115c}$$

Using the definitions

$$U_{\mathrm{e,L}}(t')\,U_{\mathrm{e,L}}(t') = (-1)^{[\nu_{\mathrm{L}}]}\,\mathbb{1}_{\mathrm{L}}, \tag{5.116a}$$

$$U_{\mathrm{e,L}}(t')\,U_{\mathrm{e,L}}(g)\,U^\dagger_{\mathrm{e,L}}(t')\,U_{\mathrm{e,L}}(g) = (-1)^{[\rho_{\mathrm{L}}]}\,\mathbb{1}_{\mathrm{L}}, \tag{5.116b}$$

we associate the pair of indices

$$([\nu_{\mathrm{L}}],[\rho_{\mathrm{L}}]) = (1,0) \tag{5.116c}$$

to the spin-1/2 cluster $c=-4$ chain. This is the same pair of indices as in Eq. (5.108c). The map

$$U_{\mathrm{e,L}}\colon \mathrm{G}_{\mathrm{e}} \to \mathrm{GL}(\mathbb{C}^{16})/F^\star, \qquad F^\star := \{z\,\mathbb{1}_L \mid z\in\mathbb{C}\setminus\{0\}\} \tag{5.117}$$

defined by Eq. (5.115c) is a projective representation since it is a group homomorphism.

One verifies that the Clifford algebra spanned by the generators

$$\left\{\widehat{Y}_{2N-3}\,\widehat{X}_{2N-2}\,\widehat{X}_{2N-1}\,\widehat{X}_{2N},\quad \widehat{Y}_{2N-2}\,\widehat{X}_{2N-1}\,\widehat{X}_{2N},\quad \widehat{Y}_{2N-1}\,\widehat{X}_{2N},\quad \widehat{Y}_{2N}\right\} \tag{5.118a}$$

for the right boundary

$$\Lambda_{\mathrm{R}} := \{j = 2N-3, 2N-2, 2N-1, 2N\} \tag{5.118b}$$

delivers the pair of indices

$$([\nu_{\mathrm{R}}],[\rho_{\mathrm{R}}]) = (1,0) \tag{5.118c}$$

for the projective representation of t' and g on the right boundary. This is the same pair of indices as in Eq. (5.110c).

A complementary point of view on the spin-1/2 $c = \pm 4$ chains is given in **Exercise** 5.2.

The Cases $c = \pm 1, \pm 3$

We treat Hamiltonian (5.69) with open boundary conditions $b_c = |c|$ for $c = \pm 1, \pm 3$ for which we verify explicitly the rule

$$\begin{aligned}&\text{The case of spin-1/2 cluster } (2m+1)\text{-odd chains is brought to}\\ &\text{the case of spin-1/2 cluster } [(2m+1) - (-1)^{\lfloor|\frac{2m+1}{4}|\rfloor}\,\mathrm{sgn}(2m+1)]\text{-even chains.}\end{aligned} \tag{5.119}$$

The cases $c = \pm 1$ When $c = \pm 1$, the mean-field Hamiltonian (5.69) becomes

$$\widehat{H}^{(b_0=0)}_{\mathrm{MF},\pm 1} = \begin{cases} -\frac{\hbar\omega}{2}\sum\limits_{j=1}^{2N}\widehat{Z}_j, & \text{if } c = +1,\\[2ex] -\frac{\hbar\omega}{2}\sum\limits_{j=1}^{2N}\widehat{Y}_j, & \text{if } c = -1, \end{cases} \tag{5.120}$$

which is Hamiltonian (5.79) with $\widehat{X}_j$ replaced by $\widehat{Z}_j$ for $c = +1$ or $\widehat{Y}_j$ for $c = -1$.

Consequently, the set of gapless boundary degrees of freedom is empty, that is,

$$\mathfrak{D}_{\mathrm{B},\pm 1} := \{\}, \qquad \mathrm{B} = \mathrm{L}, \mathrm{R}. \tag{5.121}$$

By convention, we associate the trivial indices

$$([\nu_{\mathrm{B}}], [\rho_{\mathrm{B}}]) = (0, 0), \qquad \mathrm{B} = \mathrm{L}, \mathrm{R}, \tag{5.122}$$

to the spin-1/2 cluster $c = \pm 1$ chains.

The case $c = +3$ When $c = +3$, the mean-field Hamiltonian (5.69) with open boundary conditions becomes

$$\widehat{H}^{(b_{+2}=2)}_{\mathrm{MF},+3} = -\frac{\hbar\omega}{2}\sum_{j=1}^{2N-2}\widehat{Z}_j\,\widehat{Y}_{j+1}\,\widehat{Z}_{j+2}. \tag{5.123}$$

This mean-field Hamiltonian follows from Hamiltonian (5.83) upon replacing $\widehat{X}_j$ with $\widehat{Y}_j$.

One verifies the existence of a Clifford algebra on the left boundary

$$\Lambda_{\mathrm{L}} = \{j = 1, 2\} \tag{5.124}$$

that (i) commutes with Hamiltonian (5.123) and (ii) contains an $\mathfrak{su}(2)$ Lie algebra whose generators can be represented by the four-dimensional matrices

$$S^x_{\mathrm{L}} := \frac{1}{2}\,Y_1 \otimes Z_2, \qquad S^y_{\mathrm{L}} := -\frac{1}{2}\,X_1 \otimes Z_2, \qquad S^z_{\mathrm{L}} := \frac{1}{2}\,Z_1 \otimes I_2. \tag{5.125}$$

The representation of the symmetry group $\mathrm{G}_{\mathrm{MF}} = \widetilde{\mathbb{Z}}_2^{\tilde{T}'} \times \widetilde{\mathbb{Z}}_2$ [recall Eq. (5.72)] on the boundary Λ_{L} is achieved in three steps.

First, we deduce the transformation laws

$$U_{\mathrm{MF}}(\tilde{t}')\,\begin{pmatrix}S^x_{\mathrm{L}} & S^y_{\mathrm{L}} & S^z_{\mathrm{L}}\end{pmatrix}^{\mathsf{T}}\,U^{\dagger}_{\mathrm{MF}}(\tilde{t}') = \begin{pmatrix}+S^x_{\mathrm{L}} & -S^y_{\mathrm{L}} & +S^z_{\mathrm{L}}\end{pmatrix}^{\mathsf{T}}, \qquad U_{\mathrm{MF}}(\tilde{t}') := Z_1 \otimes Z_2\,\mathsf{K}, \tag{5.126a}$$

$$U_{\mathrm{MF}}(\tilde{g})\,\begin{pmatrix}S^x_{\mathrm{L}} & S^y_{\mathrm{L}} & S^z_{\mathrm{L}}\end{pmatrix}^{\mathsf{T}}\,U^{\dagger}_{\mathrm{MF}}(\tilde{g}) = \begin{pmatrix}-S^x_{\mathrm{L}} & +S^y_{\mathrm{L}} & -S^z_{\mathrm{L}}\end{pmatrix}^{\mathsf{T}}, \qquad U_{\mathrm{MF}}(\tilde{g}) := Y_1 \otimes Y_2, \tag{5.126b}$$

by combining Eqs. (5.72) and (5.125).

Second, we define the action of $\tilde{t}'$ and $\tilde{g}$ on the left boundary by demanding that

$$U_{\mathrm{MF,L}}(\tilde{t}')\,\begin{pmatrix}S^x_{\mathrm{L}} & S^y_{\mathrm{L}} & S^z_{\mathrm{L}}\end{pmatrix}^{\mathsf{T}}\,U^{\dagger}_{\mathrm{MF,L}}(\tilde{t}') := U_{\mathrm{MF}}(\tilde{t}')\,\begin{pmatrix}S^x_{\mathrm{L}} & S^y_{\mathrm{L}} & S^z_{\mathrm{L}}\end{pmatrix}^{\mathsf{T}}\,U^{\dagger}_{\mathrm{MF}}(\tilde{t}'), \tag{5.127a}$$

$$U_{\mathrm{MF,L}}(\tilde{g})\,\begin{pmatrix}S^x_{\mathrm{L}} & S^y_{\mathrm{L}} & S^z_{\mathrm{L}}\end{pmatrix}^{\mathsf{T}}\,U^{\dagger}_{\mathrm{MF,L}}(\tilde{g}) := U_{\mathrm{MF}}(\tilde{g})\,\begin{pmatrix}S^x_{\mathrm{L}} & S^y_{\mathrm{L}} & S^z_{\mathrm{L}}\end{pmatrix}^{\mathsf{T}}\,U^{\dagger}_{\mathrm{MF}}(\tilde{g}). \tag{5.127b}$$

Third, we denote by $\mathbb{1}_{\mathrm{L}}$ the unit 4×4 matrix and we define the action K_{L} of complex conjugation on the four-dimensional representation (5.125) of the $\mathfrak{su}(2)$ Lie algebra by demanding that

$$\mathsf{K}_{\mathrm{L}}\,S^x_{\mathrm{L}}\,\mathsf{K}_{\mathrm{L}} := -S^x_{\mathrm{L}}, \qquad \mathsf{K}_{\mathrm{L}}\,S^y_{\mathrm{L}}\,\mathsf{K}_{\mathrm{L}} := +S^y_{\mathrm{L}}, \qquad \mathsf{K}_{\mathrm{L}}\,S^z_{\mathrm{L}}\,\mathsf{K}_{\mathrm{L}} := +S^z_{\mathrm{L}}. \tag{5.128a}$$

One then verifies that Eqs. (5.126), (5.127), and (5.128a) are satisfied by the four-dimensional representation

$$U_{\mathrm{MF,L}}(\tilde{t}') := 2\,S^z_{\mathrm{L}}\,\mathsf{K}_{\mathrm{L}}, \qquad U_{\mathrm{MF,L}}(\tilde{g}) := 2\,S^y_{\mathrm{L}}. \tag{5.128b}$$

Using the definitions

$$U_{\mathrm{MF,L}}(\tilde{t}')\,U_{\mathrm{MF,L}}(\tilde{t}') = (-1)^{[\nu_{\mathrm{L}}]}\,\mathbb{1}_{\mathrm{L}}, \tag{5.129a}$$

$$U_{\mathrm{MF,L}}(\tilde{t}')\,U_{\mathrm{MF,L}}(\tilde{g})\,U^{\dagger}_{\mathrm{MF,L}}(\tilde{t}')\,U_{\mathrm{MF,L}}(\tilde{g}) = (-1)^{[\rho_{\mathrm{L}}]}\,\mathbb{1}_{\mathrm{L}}, \tag{5.129b}$$

we associate the pair of indices

$$([\nu_{\mathrm{L}}],[\rho_{\mathrm{L}}]) = (0,1) \tag{5.129c}$$

to the spin-1/2 cluster $c=+3$ chain. This is the same pair of indices as in Eq. (5.90c). The map

$$U_{\mathrm{MF,L}}\colon \mathrm{G}_{\mathrm{MF}} \to \mathrm{GL}(\mathbb{C}^4)/F^{\star}, \qquad F^{\star} := \{z\,\mathbb{1}_L \,|\, z\in\mathbb{C}\setminus\{0\}\} \tag{5.130}$$

defined by Eq. (5.128b) is a projective representation since it is a group homomorphism.

One verifies that the pair of indices

$$([\nu_{\mathrm{R}}],[\rho_{\mathrm{R}}]) = (0,1) \tag{5.131}$$

is found for the projective representation of $\tilde{t}'$ and $\tilde{g}$ on the left boundary

$$\Lambda_{\mathrm{R}} = \{2N-1, 2N\}. \tag{5.132}$$

The case $c = -3$ The case of $c = -3$ is deduced from the case of $c = +3$ by interchanging all the $\widehat{Z}_j$ and $\widehat{Y}_j$ operators for $j = 1, \cdots, 2N$ in the mean-field Hamiltonian (5.123).

One verifies the existence of a Clifford algebra on the left boundary

$$\Lambda_{\mathrm{L}} = \{j = 1, 2\} \tag{5.133}$$

that (i) commutes with the mean-field Hamiltonian

$$\widehat{H}^{(b_{-2}=2)}_{\mathrm{MF},-3} = -\frac{\hbar\omega}{2}\sum_{j=1}^{2N-2} \widehat{Y}_j\,\widehat{Z}_{j+1}\,\widehat{Y}_{j+2} \tag{5.134}$$

and (ii) contains an $\mathfrak{su}(2)$ Lie algebra whose generators can be represented by the four-dimensional matrices

$$S^x_{\mathrm{L}} := \frac{1}{2} Z_1 \otimes Y_2, \qquad S^y_{\mathrm{L}} := \frac{1}{2} X_1 \otimes Y_2, \qquad S^z_{\mathrm{L}} := \frac{1}{2} Y_1 \otimes I_2. \tag{5.135}$$

The representation of the symmetry group $G_{\mathrm{MF}} = \widetilde{\mathbb{Z}}_2^{\widetilde{T}'} \times \widetilde{\mathbb{Z}}_2$ [recall Eq. (5.72)] on the boundary Λ_{L} is achieved in three steps.

First, we deduce the transformation laws

$$\begin{aligned} U_{\mathrm{MF}}(\tilde{t}')\,\left(S^x_{\mathrm{L}}\ \ S^y_{\mathrm{L}}\ \ S^z_{\mathrm{L}}\right)^{\mathsf{T}} U^{\dagger}_{\mathrm{MF}}(\tilde{t}') &= \left(-S^x_{\mathrm{L}}\ \ -S^y_{\mathrm{L}}\ \ -S^z_{\mathrm{L}}\right)^{\mathsf{T}}, \\ U_{\mathrm{MF}}(\tilde{t}') &:= \mathsf{K}, \end{aligned} \tag{5.136a}$$

$$\begin{aligned} U_{\mathrm{MF}}(\tilde{g})\,\left(S^x_{\mathrm{L}}\ \ S^y_{\mathrm{L}}\ \ S^z_{\mathrm{L}}\right)^{\mathsf{T}} U^{\dagger}_{\mathrm{MF}}(\tilde{g}) &= \left(-S^x_{\mathrm{L}}\ \ +S^y_{\mathrm{L}}\ \ -S^z_{\mathrm{L}}\right)^{\mathsf{T}}, \\ U_{\mathrm{MF}}(\tilde{g}) &:= Z_1 \otimes Z_2, \end{aligned} \tag{5.136b}$$

by combining Eqs. (5.72) and (5.135).

Second, we define the action of $\tilde{t}'$ and $\tilde{g}$ on the left boundary by demanding that

$$U_{\mathrm{MF,L}}(\tilde{t}')\,\left(S^x_{\mathrm{L}}\ \ S^y_{\mathrm{L}}\ \ S^z_{\mathrm{L}}\right)^{\mathsf{T}} U^{\dagger}_{\mathrm{MF,L}}(\tilde{t}') := U_{\mathrm{MF}}(\tilde{t}')\,\left(S^x_{\mathrm{L}}\ \ S^y_{\mathrm{L}}\ \ S^z_{\mathrm{L}}\right)^{\mathsf{T}} U^{\dagger}_{\mathrm{MF}}(\tilde{t}'), \tag{5.137a}$$

$$U_{\mathrm{MF,L}}(\tilde{g})\,\left(S^x_{\mathrm{L}}\ \ S^y_{\mathrm{L}}\ \ S^z_{\mathrm{L}}\right)^{\mathsf{T}} U^{\dagger}_{\mathrm{MF,L}}(\tilde{g}) := U_{\mathrm{MF}}(\tilde{g})\,\left(S^x_{\mathrm{L}}\ \ S^y_{\mathrm{L}}\ \ S^z_{\mathrm{L}}\right)^{\mathsf{T}} U^{\dagger}_{\mathrm{MF}}(\tilde{g}). \tag{5.137b}$$

Third, we denote by $\mathbb{1}_{\mathrm{L}}$ the unit 4×4 matrix and we define the action K_{L} of complex conjugation on the four-dimensional representation (5.135) of the $\mathfrak{su}(2)$ Lie algebra by demanding that

$$\mathsf{K}_{\mathrm{L}}\,S^{x}_{\mathrm{L}}\,\mathsf{K}_{\mathrm{L}} := -S^{x}_{\mathrm{L}}, \qquad \mathsf{K}_{\mathrm{L}}\,S^{y}_{\mathrm{L}}\,\mathsf{K}_{\mathrm{L}} := +S^{y}_{\mathrm{L}}, \qquad \mathsf{K}_{\mathrm{L}}\,S^{z}_{\mathrm{L}}\,\mathsf{K}_{\mathrm{L}} := +S^{z}_{\mathrm{L}}. \tag{5.138a}$$

One then verifies that Eqs. (5.136), (5.137), and (5.138a) are satisfied by the four-dimensional representation

$$U_{\mathrm{MF,L}}(\tilde{t}') := 2\,S^{x}_{\mathrm{L}}\,\mathsf{K}_{\mathrm{L}}, \qquad U_{\mathrm{MF,L}}(\tilde{g}) := 2\,S^{y}_{\mathrm{L}}. \tag{5.138b}$$

Using the definitions

$$U_{\mathrm{MF,L}}(\tilde{t}')\,U_{\mathrm{MF,L}}(\tilde{t}') = (-1)^{[\nu_{\mathrm{L}}]}\,\mathbb{1}_{\mathrm{L}}, \tag{5.139a}$$

$$U_{\mathrm{MF,L}}(\tilde{t}')\,U_{\mathrm{MF,L}}(\tilde{g})\,U^{\dagger}_{\mathrm{MF,L}}(\tilde{t}')\,U_{\mathrm{MF,L}}(\tilde{g}) = (-1)^{[\rho_{\mathrm{L}}]}\,\mathbb{1}_{\mathrm{L}}, \tag{5.139b}$$

we associate the pair of indices

$$([\nu_{\mathrm{L}}],[\rho_{\mathrm{L}}]) = (1,1) \tag{5.139c}$$

to the spin-1/2 cluster $c=-3$ chain. This is the same pair of indices as in Eq. (5.98c). The map

$$U_{\mathrm{MF,L}}\colon G_{\mathrm{MF}} \to \mathrm{GL}(\mathbb{C}^{4})/F^{\star}, \qquad F^{\star} := \{z\,\mathbb{1}_{L}\,|\,z\in\mathbb{C}\setminus\{0\}\} \tag{5.140}$$

defined by Eq. (5.138b) is a projective representation since it is a group homomorphism.

One verifies that the pair of indices

$$([\nu_{\mathrm{R}}],[\rho_{\mathrm{R}}]) = (1,1) \tag{5.141}$$

is found for the projective representation of $\tilde{t}'$ and $\tilde{g}$ on the left boundary

$$\Lambda_{\mathrm{R}} = \{2N-1, 2N\}. \tag{5.142}$$

5.3.4 Group Cohomology of the Protecting Symmetry Group

The ground states of Hamiltonians (5.56) obeying open boundary conditions are $2^{|c|}$-fold degenerate as is reported in **Table** 5.3. This is a consequence of their integrability, for any spin-1/2 cluster c chain is the sum of pairwise commuting Hermitean operators with each one squaring to the identity. Without loss of generality, we consider the case of even

$$c = 2m, \tag{5.143}$$

for which we have constructed the $2^{|c|}$-dimensional projective representation

$$U_{\mathrm{B}}(t')\,U_{\mathrm{B}}(t') = (-1)^{[\nu_{\mathrm{B}}]}\,\mathbb{1}_{\mathrm{B}}, \tag{5.144a}$$

$$U_{\mathrm{B}}(t')\,U_{\mathrm{B}}(g)\,U^{\dagger}_{\mathrm{B}}(t')\,U_{\mathrm{B}}(g) = (-1)^{[\rho_{\mathrm{B}}]}\,\mathbb{1}_{\mathrm{B}} \tag{5.144b}$$

of the symmetry group

Table 5.3 *Cohomological indices for the spin-1/2 cluster c chains*

c	$([\nu_{\mathrm{L}}],[\rho_{\mathrm{L}}])$	$([\nu_{\mathrm{R}}],[\rho_{\mathrm{R}}])$	$\dim \mathfrak{h}^{(c)}_{\mathrm{B,min}}$	$D(c)$	$\dim \mathfrak{h}^{(c)}_{\mathrm{gs}}$
0	$(0,0)$	$(0,0)$	1	1	1
1	$(0,0)$	$(0,0)$	1	1	$2=2^1$
2	$(0,1)$	$(0,1)$	2	4	$4=2^2$
3	$(0,1)$	$(0,1)$	2	4	$8=2^3$
4	$(1,0)$	$(1,0)$	2	4	$16=2^4$
5	$(1,1)$	$(1,1)$	2	4	$32=2^5$
6	$(1,1)$	$(1,1)$	2	4	$64=2^6$
7	$(0,0)$	$(0,0)$	1	1	$128=2^7$
8	$(0,0)$	$(0,0)$	1	1	$256=2^8$

The doublet $([\nu_{\mathrm{B}}],[\rho_{\mathrm{B}}])$ on the left and right boundaries of a spin-1/2 cluster c chain that defines the projective representation of the symmetry group (5.57) for even c or (5.59) and (5.71) for odd c that is realized on the left (B=L) or right (B=R) boundaries of an open chain. It is assumed that time-reversal symmetry is broken spontaneously when c is odd. The column $\dim \mathfrak{h}^{(c)}_{\mathrm{B,min}}$ is the dimensionality of the smallest Hilbert space $\mathfrak{h}^{(c)}_{\mathrm{B,min}}$ for which it is possible to realize the projective algebra on either one of the left or right boundary. Squaring this number gives the topologically protected degeneracy $D(c)$ of the ground states of a spin-1/2 cluster c open chain. The column $\dim \mathfrak{h}^{(c)}_{\mathrm{gs}}$ is the degeneracy of the ground state of the spin-1/2 cluster c open chain that is protected by its integrability if spontaneous symmetry breaking is precluded.

$$G_{\mathrm{e}} \cong \mathbb{Z}_2^{T'} \times \mathbb{Z}_2 \equiv \{e,t'\} \times \{e,g\} \tag{5.144c}$$

on any one of the left (B = L) or right (B = R) boundaries (5.75) and (5.76), respectively. We found that the pairs of symmetry indices $([\nu_{\mathrm{B}}],[\rho_{\mathrm{B}}])$ display a periodicity of four when measured by m, for

$$\begin{aligned} ([\nu_{\mathrm{B}}],[\rho_{\mathrm{B}}]) &= \begin{cases} (0,0), & \text{for } m=0,\\ (0,1), & \text{for } m=+1,\\ (1,0), & \text{for } m=+2,\\ (1,1), & \text{for } m=-1, \end{cases} \\ \iff ([\nu_{\mathrm{B}}],[\rho_{\mathrm{B}}]) &= \begin{cases} (0,0), & \text{for } m=0,\\ (0,1), & \text{for } m=+1,\\ (1,0), & \text{for } m=+2,\\ (1,1), & \text{for } m=+3. \end{cases} \end{aligned} \tag{5.145}$$

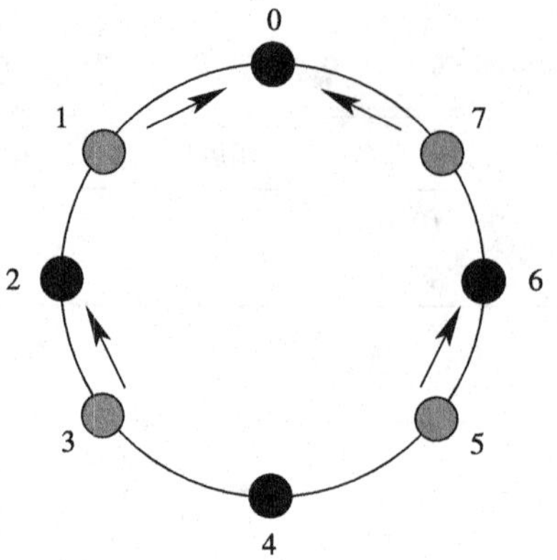

Figure 5.2 If $c = 1, 3, 5, 7$ and $c' = 0, 2, 4, 6$ label anticlockwise the eight roots of unity on a circle as shown, then the directed pair $\langle c, c' \rangle$ that satisfies condition (5.146) can be represented by four arrows as shown.

We report in **Table** 5.3 the doublets $([\nu_{\mathrm{L}}], [\rho_{\mathrm{B}}])$ on the left $(\mathrm{B} = \mathrm{L})$ and right $(\mathrm{B} = \mathrm{R})$ boundaries of a spin-1/2 cluster c open chain for $c = 0, 1, 2, 3, 4, 5, 6, 7, 8$. The periodicity of four for the even open chains $c = 0, 2, 4, 6, 8$ and for the odd open chains $c = 1, 3, 5, 7$ is manifest. The symmetry between the left and right boundaries is also manifest. This symmetry is a consequence of the fact that all elements of the second cohomology group (5.155) are their own inverse. One verifies that the rule by which the group cohomology for a spin-1/2 cluster even-c' Hamiltonian is the same as that of the mean-field projection of the spin-1/2 cluster odd-c Hamiltonian holds if c' is related to c by

$$c' = c - (-1)^{\lfloor |\frac{c}{4}| \rfloor} \operatorname{sgn}(c). \tag{5.146}$$

Figure 5.2 gives an interpretation to the factor $(-1)^{\lfloor |\frac{c}{4}| \rfloor}$.

We also present in **Table** 5.3 the minimal dimensionality needed to represent the projective algebra (5.144). The square of this dimensionality is the minimal degeneracy of the ground states of a spin-1/2 cluster spin c open chain when perturbed by local interactions that neither break explicitly nor spontaneously the symmetry group (5.57) for even c or (5.59) and (5.71) for odd c and do not close the gap between the ground states and all excited states. The degeneracy of two for the boundary B with $([\nu_{\mathrm{B}}], [\rho_{\mathrm{B}}]) = (0, 1)$ arises from the fact that the symmetry operation g, say, is represented on the boundary by an operator that anticommutes with the representation of the symmetry operation t', say. The degeneracy of two for the boundary B with $([\nu_{\mathrm{B}}], [\rho_{\mathrm{B}}]) = (1, 0)$ arises from a Kramers' degeneracy as the operator representing t', say, squares to minus the identity operator on boundary B. A boundary B for which $([\nu_{\mathrm{B}}], [\rho_{\mathrm{B}}]) = (1, 1)$ delivers the same minimal degeneracy of two as that of a boundary B for which $([\nu_{\mathrm{B}}],[\rho_{\mathrm{B}}])=(0,1)$. Indeed, if the product of the representations of t' and g, say, anticommutes while the representation of t' squares to minus the identity, then the product of t' and g is necessarily represented by an operator that squares to the identity.

We denote an element of the group $\mathbb{Z}_2\times\mathbb{Z}_2$ by

$$([\nu_{\mathrm{B}}],[\rho_{\mathrm{B}}]), \qquad [\nu_{\mathrm{B}}],[\rho_{\mathrm{B}}]=0,1. \tag{5.147}$$

The origin of this group in Eq. (5.145) can be understood from the point of view of group cohomology.

A representation U of a symmetry group G is a group homomorphism

$$\begin{aligned} U\colon \mathrm{G} &\to \mathrm{GL}(\mathbb{C}^n),\\ g &\mapsto U(g) \end{aligned} \tag{5.148}$$

that assigns to a symmetry operation g from the symmetry group G a complex-valued $n\times n$ matrix $U(g)$ from the general linear group $\mathrm{GL}(\mathbb{C}^n)$ of complex-valued $n\times n$ matrices.

A projective representation $\widetilde{U}$ of a symmetry group G is a group homomorphism

$$\begin{aligned} \widetilde{U}\colon \mathrm{G} &\to \mathrm{GL}(\mathbb{C}^n)\\ g &\mapsto \widetilde{U}(g) \end{aligned} \tag{5.149}$$

up to some multiplicative constant, that is, the function

$$\begin{aligned} c\colon \mathrm{G}\times\mathrm{G} &\to \mathbb{C},\\ (g,h) &\mapsto c(g,h) \end{aligned} \tag{5.150}$$

exists such that the rule

$$\widetilde{U}(g)\,\widetilde{U}(h)=c(g,h)\,\widetilde{U}(g\,h) \tag{5.151}$$

is compatible with the existence of a neutral element and the associativity in G.

Any representation U of a symmetry group G induces a projective representation $\widetilde{U}$ by composing it with the quotient map

$$\mathrm{GL}(\mathbb{C}^n)\to\mathrm{PGL}(\mathbb{C}^n)\equiv\mathrm{GL}(\mathbb{C}^n)/F^{\star}, \tag{5.152}$$

where $F^{\star}$ is the center of $\mathrm{GL}(\mathbb{C}^n)$, that is the subgroup made of all nonvanishing matrices that are proportional to the unit $n\times n$ matrix.

The projective general linear group $\mathrm{PGL}(\mathbb{C}^n)$ allows to give an alternative definition of a projective representation of the symmetry group G. A projective representation $\widetilde{U}$ is a collection of elements $\widetilde{U}(g)\in\mathrm{PGL}(\mathbb{C}^n)$ labeled by the elements $g\in\mathrm{G}$ such that the rule

$$\widetilde{U}(g)\,\widetilde{U}(h)=\widetilde{U}(g\,h) \tag{5.153}$$

holds for any $g, h \in \mathrm{G}$. A projective representation can be lifted to a group homomorphism if it is possible to choose in any equivalence class $\widetilde{U} \in \mathrm{PGL}(\mathbb{C}^n)$ an $U(g) \in \mathrm{GL}(\mathbb{C}^n)$ so that the rule

$$U(g)\,U(h) = U(g\,h) \tag{5.154}$$

holds for any $g, h \in \mathrm{G}$. It is the second cohomology group $H^2(\mathrm{G}, F^\star)$ that dictates when a projective representation $\widetilde{U}$ of G can be lifted to a linear representation U. A precise definition of the second cohomology group $H^2(\mathrm{G}, F^\star)$ will be given in Section 6.3. For the spin-1/2 cluster spin c chains, it is given by[16]

$$H^2(\mathbb{Z}_2^{T'} \times \mathbb{Z}_2, \mathrm{U}(1)_{\mathfrak{c}}) = \mathbb{Z}_2 \times \mathbb{Z}_2 \tag{5.155a}$$

when c is even and

$$H^2(\widetilde{\mathbb{Z}}_2^{\widetilde{T}'} \times \widetilde{\mathbb{Z}}_2, \mathrm{U}(1)_{\mathfrak{c}}) = \mathbb{Z}_2 \times \mathbb{Z}_2 \tag{5.155b}$$

when c is odd. For example, the four pair of indices in Eq. (5.145) are the four elements of the second cohomology group (5.155a).

When two blocks of matter that each support a bosonic invertible topological phase are brought adiabatically into contact so as to form a new block of matter, the resulting invertible topological phases is bosonic and acquires the indices dictated by the group composition rule of the relevant second group cohomology, as we shall explain in Section 6.5. This operation is called stacking and the group composition rules of the relevant second group cohomology define the stacking rules. For example, for two spin-1/2 cluster $2m$ chains, one with the pair of indices $([\nu_1], [\rho_1])$, and the other one with the pair of indices $([\nu_2], [\rho_2])$, their stacking results in a topological phase with the pair of indices

$$([\nu_\otimes], [\rho_\otimes]) \equiv ([\nu_1], [\rho_1]) \otimes ([\nu_2], [\rho_2]) := ([\nu_1] + [\nu_2], [\rho_1] + [\rho_2])\,. \tag{5.156}$$

In particular, we can obtain all four bosonic invertible topological phases with the pair of indices given in Eq. (5.145) by stacking the phases $(0,1)$ and $(1,0)$ with themselves or with each other.

5.4 Jordan-Wigner Transformation to the Majorana Representation

Let $j \in \Lambda \equiv \{1, \cdots, N\}$ with N any positive integer larger than the nonnegative integer $|c|$. Following Jordan and Wigner,[17] define the pair of Hermitean operators (K denotes the operator implementing complex conjugation)

$$\hat{\xi}_j := \widehat{X}_1 \cdots \widehat{X}_{j-1}\,\widehat{Y}_j = \hat{\xi}_j^\dagger = -\mathsf{K}\,\hat{\xi}_j\,\mathsf{K}, \tag{5.157a}$$

$$\hat{\eta}_j := \widehat{X}_1 \cdots \widehat{X}_{j-1}\,\widehat{Z}_j = \hat{\eta}_j^\dagger = +\mathsf{K}\,\hat{\eta}_j\,\mathsf{K}, \tag{5.157b}$$

16 The subscript $\mathfrak{c}$ in $\mathrm{U}(1)_{\mathfrak{c}}$ arises when the symmetry group is represented using antilinear operators. The precise meaning of this notation will be given in Section 6.3.

17 P. Jordan and E. Wigner, Z. Phys. **47**, 631–651 (1928). [Jordan and Wigner (1928)].

acting on the Hilbert space (5.3a). In view of the Pauli algebra,[18]

$$\hat{\xi}_j^2 = 1, \qquad \hat{\eta}_j^2 = 1, \qquad -\mathrm{i}\hat{\xi}_j\,\hat{\eta}_j = \widehat{X}_j. \tag{5.158}$$

Another consequence of the Pauli algebra is that these N pairs of operators realize the Clifford algebra

$$\left\{\hat{\xi}_j, \hat{\xi}_{j'}\right\} = \left\{\hat{\eta}_j, \hat{\eta}_{j'}\right\} = 2\delta_{j,j'}, \qquad \left\{\hat{\xi}_j, \hat{\eta}_{j'}\right\} = 0, \qquad j, j' \in \Lambda, \tag{5.159}$$

(**Exercise** 5.3). Other useful identities are

$$\begin{aligned}
\mathrm{i}\hat{\xi}_j\,\hat{\eta}_{j+|c|} =&\ \mathrm{i}\widehat{X}_1 \cdots \widehat{X}_{j-1}\,\widehat{Y}_j \\
&\times \widehat{X}_1 \cdots \widehat{X}_{j-1}\,\widehat{X}_j\,\widehat{X}_{j+1} \cdots \widehat{X}_{j+|c|-1}\,\widehat{Z}_{j+|c|} \\
=&\ \widehat{Z}_j\,\widehat{X}_{j+1} \cdots \widehat{X}_{j+|c|-1}\,\widehat{Z}_{j+|c|},
\end{aligned} \tag{5.160a}$$

for $j = 1, \cdots, N - |c|$ and

$$\begin{aligned}
\mathrm{i}\hat{\xi}_j\,\hat{\eta}_{j-|c|} =&\ \mathrm{i}\widehat{X}_1 \cdots \widehat{X}_{j-|c|-1}\,\widehat{X}_{j-|c|}\,\widehat{X}_{j-|c|+1} \cdots \widehat{X}_{j-1}\,\widehat{Y}_j \\
&\times \widehat{X}_1 \cdots \widehat{X}_{j-|c|-1}\,\widehat{Z}_{j-|c|} \\
=&\ \widehat{Y}_{j-|c|}\,\widehat{X}_{j-|c|+1} \cdots \widehat{X}_{j-1}\,\widehat{Y}_j,
\end{aligned} \tag{5.160b}$$

for $j = 1 + |c|, \cdots, N$. We thus arrive at the identities

$$-\sum_{j=1+|c|}^{N} \mathrm{i}\hat{\xi}_j\,\hat{\eta}_{j-|c|} = -\sum_{j=1+|c|}^{N} \widehat{Y}_{j-|c|}\,\widehat{X}_{j-|c|+1} \cdots \widehat{X}_{j-1}\,\widehat{Y}_j, \tag{5.161a}$$

$$-\sum_{j=1}^{N} \mathrm{i}\hat{\xi}_j\,\hat{\eta}_j = +\sum_{j=1}^{N} \widehat{X}_j, \tag{5.161b}$$

$$-\sum_{j=1}^{N-|c|} \mathrm{i}\hat{\xi}_j\,\hat{\eta}_{j+|c|} = -\sum_{j=1}^{N-|c|} \widehat{Z}_j\,\widehat{X}_{j+1} \cdots \widehat{X}_{j+|c|-1}\,\widehat{Z}_{j+|c|}. \tag{5.161c}$$

The last identities that we will need are the inverse relations

$$\widehat{X}_j = -\mathrm{i}\hat{\xi}_j\,\hat{\eta}_j, \tag{5.162a}$$

$$\widehat{Y}_j = \left(-\mathrm{i}\hat{\xi}_{j-1}\,\hat{\eta}_{j-1}\right) \cdots \left(-\mathrm{i}\hat{\xi}_1\,\hat{\eta}_1\right)\,\hat{\xi}_j, \tag{5.162b}$$

$$\widehat{Z}_j = \left(-\mathrm{i}\hat{\xi}_{j-1}\,\hat{\eta}_{j-1}\right) \cdots \left(-\mathrm{i}\hat{\xi}_1\,\hat{\eta}_1\right)\,\hat{\eta}_j. \tag{5.162c}$$

We define the Jordan-Wigner transformation as the map on the Pauli algebra

$$\widehat{X}_j^2 = 1, \qquad \widehat{X}_j\,\widehat{Y}_j = -\widehat{Y}_j\,\widehat{X}_j = \mathrm{i}\widehat{Z}_j, \qquad \left[\widehat{X}_j, \widehat{Y}_{j'}\right] = 0, \tag{5.163}$$

[18] When equating an operator to a mere $\mathbb{C}$ number, it is implied that the latter is multiplying the identity operator acting on the relevant Hilbert space.

(up to cyclic permutations of the letters X, Y, Z and for any pair $j \neq j' \in \Lambda$) that is spanned by the operators $\widehat{X}_j$, $\widehat{Y}_j$, and $\widehat{Z}_j$, by which $\widehat{X}_j$, $\widehat{Y}_j$, $\widehat{Z}_j$ are replaced by the generators of the Clifford algebra (5.159) through Eq. (5.162). The generators of the Clifford algebra (5.159) are called Majorana operators. The Jordan-Wigner transformation is nonlocal. It maps the spin-1/2 cluster c chain $\widehat{H}_c^{(|c|)}$ defined in Eq. (5.25) with open boundary conditions ($b_c = |c|$) to the following quadratic form in the Majorana operators. If $c < 0$,

$$\widehat{H}_{c<0}^{(|c|)} = -\frac{\hbar\omega}{2} \sum_{j=1}^{N-|c|} \mathrm{i}\hat{\xi}_{j+|c|}\, \hat{\eta}_j. \tag{5.164a}$$

If $c = 0$,

$$\widehat{H}_{c=0}^{(c=0)} = +\frac{\hbar\omega}{2} \sum_{j=1}^{N} \mathrm{i}\hat{\xi}_j\, \hat{\eta}_j. \tag{5.164b}$$

If $c > 0$,

$$\widehat{H}_{c>0}^{(|c|)} = -\frac{\hbar\omega}{2} \sum_{j=1}^{N-|c|} \mathrm{i}\hat{\xi}_j\, \hat{\eta}_{j+|c|}. \tag{5.164c}$$

We infer the spectral degeneracies of the Majorana representation of $\widehat{H}_c^{(|c|)}$ from reading the c entry corresponding to "Any" or "OBC" boundary conditions in **Table** 5.2, which we report in **Table** 5.4.

The Majorana representation of the spin-1/2 cluster c chain $\widehat{H}_c^{(|c|)}$ defined in Eq. (5.25) with open boundary conditions ($b_c = |c|$) can have a hidden global $\mathrm{O}(|c|)$ symmetry. For example, when $c > 0$ and N is an integer multiple of c, this hidden symmetry is made explicit if one writes

$$\widehat{H}_{c>0}^{(c)} = -\frac{\hbar\omega}{2} \sum_{j=1}^{(N-b_c)/c} \mathrm{i}\hat{\boldsymbol{\xi}}_j \cdot \hat{\boldsymbol{\eta}}_{j+1}, \tag{5.165}$$

where $\hat{\boldsymbol{\xi}}_j \equiv (\hat{\xi}_{cj-(c-1)}, \cdots, \hat{\xi}_{cj})^{\mathsf{T}}$ and $\hat{\boldsymbol{\eta}}_j \equiv (\hat{\eta}_{cj-(c-1)}, \cdots, \hat{\eta}_{cj})^{\mathsf{T}}$.

The case of periodic boundary conditions, $b_c = 0$, must be treated separately since

$$\hat{\xi}_j := \widehat{X}_1 \cdots \widehat{X}_{j-1} \widehat{Y}_j = \hat{\xi}_{j+N}, \tag{5.166a}$$

$$\hat{\eta}_j := \widehat{X}_1 \cdots \widehat{X}_{j-1} \widehat{Z}_j = \hat{\eta}_{j+N}, \tag{5.166b}$$

for any $j = 1, \cdots, N$ if and only if we impose the global constraint

$$\widehat{X} \equiv \widehat{X}_1 \cdots \widehat{X}_N = 1, \quad \widehat{X}_j = -\mathrm{i}\hat{\xi}_j\, \hat{\eta}_j \iff 1 = \prod_{j=1}^{N} (-\mathrm{i})\hat{\xi}_j\, \hat{\eta}_j, \quad \widehat{X}_j = -\mathrm{i}\hat{\xi}_j\, \hat{\eta}_j. \tag{5.166c}$$

Table 5.4 *Spectral degeneracies of the spin-1/2 cluster c chains with open boundary conditions*

Spin-1/2 cluster c chain	GSD	SD E_n
$\widehat{H}_0^{(0)}$	1	$\frac{N!}{n!(N-n)!}$
$\widehat{H}_{2m}^{(\lvert 2m\rvert)}$, $m\neq 0$	$2^{\lvert 2m\rvert}$	$2^{\lvert 2m\rvert}\times\frac{(N-\lvert 2m\rvert)!}{n!(N-\lvert 2m\rvert-n)!}$
$\widehat{H}_{2m+1}^{(\lvert 2m+1\rvert)}$	$2^{\lvert 2m+1\rvert}$	$2^{\lvert 2m+1\rvert}\times\frac{(N-\lvert 2m+1\rvert)!}{n!(N-\lvert 2m+1\rvert-n)!}$

The spectral degeneracies of any spin-1/2 cluster $c\in\mathbb{Z}$ chain $\widehat{H}_c^{(|c|)}$ of length $N\geq|c|+1$ defined in Eq. (5.25) obeying open boundary conditions. We abbreviate ground-state degeneracy by GSD and spectral degeneracy by SD.

Imposing the constraint (5.166c) on the Hilbert space (5.3a) delivers the physical subspace

$$\mathfrak{H}^{\mathrm{PBC}}_{\widehat{X}=1}\cong\mathbb{C}^{2^{N-1}}. \tag{5.167a}$$

The constraint (5.166c) that defines the Hilbert space (5.167a) implies that Eq. (5.38) must be replaced with

$$\prod_{j=1}^{N}\widehat{\mathrm{C}}_j=\begin{cases}(-1)^{N\,(1-\delta_{c,0})}, & \text{if } c \text{ is even},\\ 1, & \text{if } c \text{ is odd}.\end{cases} \tag{5.167b}$$

In turn, Eq. (5.167b) implies that the eigenvalue of $\widehat{\mathrm{C}}_N=\widehat{\mathrm{C}}_N^{-1}$ is some function of the product of all $(N-1)$ eigenvalues of $\widehat{\mathrm{C}}_1=\widehat{\mathrm{C}}_1^{-1},\cdots,\widehat{\mathrm{C}}_{N-1}=\widehat{\mathrm{C}}_{N-1}^{-1}$ for both odd and even c. Constraint (5.166c) also implies, in view of Eq. (5.32f), that the spin-1/2 cluster $c\neq 0$ chains with odd c and obeying periodic boundary conditions have lost their intrinsic twofold degeneracy (that is the degeneracy that is independent of the choice between open and periodic boundary conditions) after projection onto the eigenspace of $\widehat{X}$ with eigenvalue $+1$. With constraint (5.166c), the degeneracies for the spin-1/2 cluster $c\neq 0$ chains obeying periodic boundary conditions in **Table** 5.2 become those obeying periodic boundary conditions from **Table** 5.5.

On the Hilbert space (5.167a), we may write the identities

Table 5.5 *Spectral degeneracies of the spin-1/2 cluster c chains with periodic boundary conditions*

Spin-1/2 cluster c chain	GSD	SD E_n
$\widehat{H}^{(0)}_{0}$	1	$\frac{(N-1)!}{n!(N-1-n)!}$
$\widehat{H}^{(0)}_{2m}$, $m \neq 0$	1	$\frac{(N-1)!}{n!(N-1-n)!}$
$\widehat{H}^{(0)}_{2m+1}$	1	$\frac{(N-1)!}{n!(N-1-n)!}$

The spectral degeneracies of any spin-1/2 cluster $c \in \mathbb{Z}$ chain $\widehat{H}^{(0)}_{c}$ with length $N \geq |c|+1$ defined in Eq. (5.25) with the constraint $\widehat{X} = 1$ [Eq. (5.166c)] imposed on the Hilbert space $\mathbb{C}^{2^N}$ when periodic boundary conditions are imposed.

$$\widehat{H}^{(0)}_{c<0} = -\frac{\hbar\omega}{2}\sum_{j=1}^{N} \mathrm{i}\hat{\xi}_{j+|c|}\,\hat{\eta}_{j}, \tag{5.167c}$$

$$\widehat{H}^{(0)}_{c=0} = +\frac{\hbar\omega}{2}\sum_{j=1}^{N} \mathrm{i}\hat{\xi}_{j}\,\hat{\eta}_{j}, \tag{5.167d}$$

$$\widehat{H}^{(0)}_{c>0} = -\frac{\hbar\omega}{2}\sum_{j=1}^{N} \mathrm{i}\hat{\xi}_{j}\,\hat{\eta}_{j+|c|}, \tag{5.167e}$$

and infer the spectral degeneracies of the Majorana representation of $\widehat{H}^{(0)}_{c}$ from **Table** 5.5.

We are going to explain why the constraint (5.166c) that defines the Hilbert space (5.167a) is to be interpreted in the Majorana representation as the conserved global fermion-parity operator corresponding to a (not necessarily conserved) global fermion-number operator whose eigenvalues are necessarily even integers. This conservation law in the Majorana representation is needed to relate the Majorana to the spin-1/2 representations of the cluster c chains with both open and periodic boundary conditions.

5.5 Majorana c Chains

5.5.1 Definition, Symmetries, and Energy Spectra

For any $c \in \mathbb{Z}$, for any integer $N \geq |c|+1$, and for any characteristic frequency $\omega > 0$, define the Majorana c chain to be

$$\widehat{H}_{c,b_c} := \begin{cases} -\frac{\hbar\omega}{2} \sum_{j=1}^{N-b_c} \mathrm{i}\hat{\xi}_{j+|c|}\, \hat{\eta}_j, & c<0, \\ +\frac{\hbar\omega}{2} \sum_{j=1}^{N-b_c} \mathrm{i}\hat{\xi}_{j}\, \hat{\eta}_j, & c=0, \\ -\frac{\hbar\omega}{2} \sum_{j=1}^{N-b_c} \mathrm{i}\hat{\xi}_{j}\, \hat{\eta}_{j+|c|}, & c>0, \end{cases} \tag{5.168a}$$

where

$$\hat{\xi}_{j+N} = \hat{\xi}_j = \hat{\xi}_j^{\dagger}, \qquad \hat{\eta}_{j+N} = \hat{\eta}_j = \hat{\eta}_j^{\dagger}, \qquad j=1,\cdots,N, \tag{5.168b}$$

and[19]

$$\left\{\hat{\xi}_j, \hat{\xi}_{j'}\right\} = \left\{\hat{\eta}_j, \hat{\eta}_{j'}\right\} = 2\delta_{j,j'}, \qquad \left\{\hat{\xi}_j, \hat{\eta}_{j'}\right\} = 0, \qquad j,j'=1,\cdots,N. \tag{5.168c}$$

The choice $b_c = 0$ selects periodic boundary conditions. The choice $b_c = |c|$ selects open boundary conditions. Hamiltonian $\widehat{H}_{c,b_c}$ acts on the 2^N-dimensional Hilbert space

$$\begin{aligned} \mathbb{C}^{2^N} \cong \operatorname{span}\Bigg\{ \left[\prod_{j=1}^{N}\left(\hat{c}_j^{\dagger}\right)^{n_j}\right]|0\rangle \,\Bigg|\, n_j = 0,1, \\ \hat{c}_j^{\dagger} := \frac{\hat{\eta}_j - \mathrm{i}\hat{\xi}_j}{2}, \qquad \hat{c}_j := \frac{\hat{\eta}_j + \mathrm{i}\hat{\xi}_j}{2}, \qquad \hat{c}_j|0\rangle = 0 \Bigg\}. \end{aligned} \tag{5.168d}$$

The operator $\hat{c}_j^{\dagger}$ ($\hat{c}_j$) creates (annihilates) a fermion with the quantum label $j = 1,\cdots,N$. Hence, the Hilbert space (5.168d) has the structure of a fermionic Fock space. Hamiltonian (5.168a) together with the Majorana operators (5.168c) and the fermionic operators (5.168d) are interpreted in Figure 5.3 for $c \geq 1$ as a bonded lattice with j labeling a repeat unit cell containing two distinct Majorana operators and the bonds labeling the hopping amplitude with the magnitude $\hbar\omega/2 > 0$. The convention for the change of sign with which the energy scale $\hbar\omega/2$ enters $\widehat{H}_{c,b_c}$ when $c = 0$ compared to when $c \neq 0$ is chosen so as to maintain the relation with the spin-1/2 cluster c chains defined in Eq. (5.25) that follows from the Jordan-Wigner transformation (5.157).

According to Eq. (5.164), an open Majorana c chain is nothing but an open spin-1/2 cluster c chain. According to Eq. (5.167), a closed Majorana c chain obeying periodic boundary conditions with definite parity (5.169) is nothing but a closed spin-1/2 cluster c chain obeying periodic boundary conditions and the constraint (5.166c). Thus, there follows the degeneracies from **Tables** 5.4 and 5.5. However, it

[19] When equating an operator to a mere $\mathbb{C}$ number, it is implied that the latter is multiplying the identity operator acting on the relevant Hilbert space.

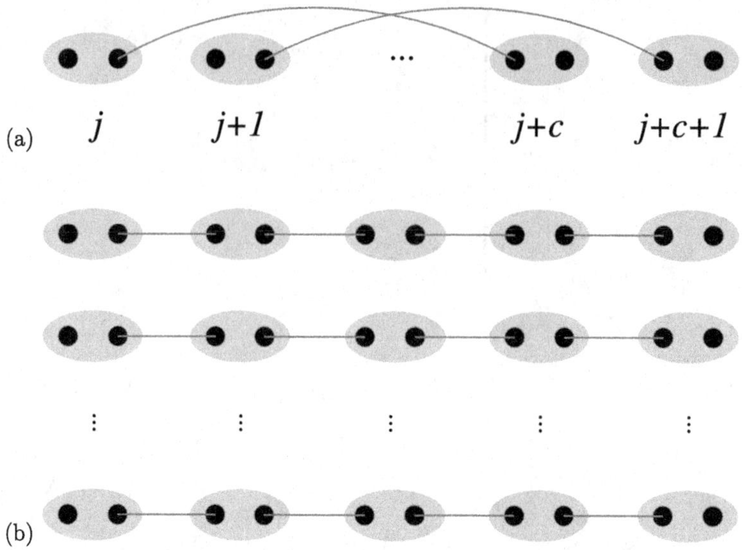

Figure 5.3 (a) A linear chain Λ labeled by $j = 1, \cdots, N$ with N an integer. Each site of the chain is represented by the ellipse colored in blue. To each site j of this linear chain, one associates the fermion annihilation operator $\hat{c}_j$. The additive decomposition $\hat{c}_j = (\hat{\eta}_j + \mathrm{i}\hat{\xi}_j)/2$ into the sum of two Majorana operators $\hat{\eta}_j$ and $\hat{\xi}_j$ is represented by associating $\hat{\xi}_j$ ($\hat{\eta}_j$) to the right (left) black disk inside the ellipse labeled by j. For any integer $c \geq 1$, the red bond connecting the right disk in ellipse j to the left disk in ellipse $j+c$ represents the term $\mathrm{i}\hat{\xi}_j\,\hat{\eta}_{j+c}$ in the Hamiltonian (5.164c) that defines a Majorana $c > 0$ chain. (b) A Majorana $c > 0$ chain can be thought of as a stacking of c Majorana 1 chains.

is instructive to revisit these degeneracies from the Majorana point of view. Without loss of generality, we do this for $c = 0, 1, 2, \cdots$.

First, we observe that Hamiltonian (5.168) is the sum of pairwise commuting operators.

Proof Without loss of generality, $c \geq 0$. The case $c = 0$ is obvious. Choose any $1 \leq c < N$. We choose $j = 1, \cdots, N$ and $j + c$ is defined modulo N. This fixes the pair j and $j+c$. We choose $k = 1, \cdots, N$ and $k+c$ is defined modulo N. This fixes the pair k and $k+c$, say (i) $k, k+c < j$, (ii) $k < j$ and $k+c = j$, (iii) $k = j$, (iv) $k > j$ and $k < j + c$, (v) $k = j + c$, and (vi) $k > j$ and $k + c < j$, for example. One verifies that $\hat{\xi}_j\,\hat{\eta}_{j+c}$ always commutes with $\hat{\xi}_k\,\hat{\eta}_{k+c}$ since passing $\hat{\xi}_j\,\hat{\eta}_{j+c}$ across $\hat{\xi}_k\,\hat{\eta}_{k+c}$ always produces an even number of multiplicative factors of (-1). □

Second, for $c \geq 0$ without loss of generality, we observe that $\mathrm{i}\hat{\eta}_k\,\hat{\xi}_k$ commutes with $\mathrm{i}\hat{\eta}_j\,\hat{\xi}_{j+c}$ when $k \neq j, j+c$. Otherwise, $\mathrm{i}\hat{\eta}_k\,\hat{\xi}_k$ anticommutes with $\mathrm{i}\hat{\eta}_j\,\hat{\xi}_{j+c}$. For $c = 0$, $\mathrm{i}\hat{\eta}_k\,\hat{\xi}_k$ always commutes with $\mathrm{i}\hat{\eta}_j\,\hat{\xi}_j$. It then follows that Hamiltonian $\widehat{H}_{c,b_c}$

commutes with the fermion parity operator

$$\widehat{P}^{\mathrm{f}} := \prod_{k=1}^{N} \mathrm{i}\hat{\eta}_k\,\hat{\xi}_k = \prod_{k=1}^{N} \left(2\hat{c}^{\dagger}_k\,\hat{c}_k - 1\right) = \exp\left(\mathrm{i}\sum_{k=1}^{N} \left(\hat{c}^{\dagger}_k\,\hat{c}_k - 1\right)\pi\right), \tag{5.169}$$

where we observe that $\hat{c}^{\dagger}_k\,\hat{c}_k$ counts the number of fermions with the quantum number k. When $N \geq c+1$ is even, the fermion parity operator carries the eigenvalue 1 (-1) if there is an even (odd) number of fermions. When $N \geq c+1$ is odd, the fermion parity operator carries the eigenvalue -1 (1) if there is an even (odd) number of fermions.

Third, Hamiltonian $\widehat{H}_{c,b_c}$ commutes with the antiunitary transformation

$$\widehat{T}' = \widehat{T}'^{-1} \tag{5.170a}$$

defined by the rules

$$\widehat{T}'\,\hat{c}^{\dagger}_j\,\widehat{T}' := \hat{c}^{\dagger}_j, \quad \widehat{T}'\,\hat{c}_j\,\widehat{T}' := \hat{c}_j \iff \widehat{T}'\,\hat{\eta}_j\,\widehat{T}' := +\hat{\eta}_j, \quad \widehat{T}'\,\hat{\xi}_j\,\widehat{T}' := -\hat{\xi}_j, \tag{5.170b}$$

that are extended in an antilinear fashion to the algebra over the complex numbers generated by the pair $\hat{c}^{\dagger}_j$ and $\hat{c}_j$ or, equivalently, the pair $\hat{\eta}_j$ and $\hat{\xi}_j$.

The energy spectrum of the Majorana c chain is derived as follows. Without loss of generality, c is chosen to be non negative, that is $c = 0, 1, 2, \cdots$. The identity

$$\left(\hat{\xi}_j\,\hat{\eta}_{j+c}\right)^{\dagger} = -\hat{\xi}_j\,\hat{\eta}_{j+c} \tag{5.171}$$

implies that $\hat{\xi}_j\,\hat{\eta}_{j+c}$ can be represented by the 2×2 antisymmetric matrix

$$B_{j,j+c} \equiv \begin{pmatrix} 0 & -1 \\ +1 & 0 \end{pmatrix}, \tag{5.172}$$

whose eigenvalues are $\pm\mathrm{i}$.

When $c = 0$, the spectrum of Hamiltonian (5.168a) is independent of the choice of boundary conditions and the ground state is selected by choosing the eigenvalue $+\mathrm{i}$ of $B_{j,j+c}$ for $j = 1, \cdots, N-c$.

When open boundary conditions ($b_c = c$ with $c > 0$) are chosen, the ground state of Hamiltonian (5.168a) is achieved by choosing the eigenvalue $-\mathrm{i}$ of $B_{j,j+c}$ for $j = 1, \cdots, N-c$. This leaves the eigenvalues $\pm\mathrm{i}$ of the c matrices $B_{N+1-c,N+1}, \cdots, B_{N,N+c}$ arbitrary. Hence, the ground states are 2^c-fold degenerate. Excited states are obtained by flipping one by one the eigenvalue $-\mathrm{i}$ of $B_{j,j+c}$ for $j = 1, \cdots, N-c$ to the value $+\mathrm{i}$. There are $(N-c)!/n!(N-c-n)!$ ways to flip n out of $N-c$ directed bonds $\langle j, j+c\rangle$, $j = 1, \cdots, N-c$. The excited state with n flipped bonds is thus $2^c \times (N-c)!/n!(N-c-n)!$-fold degenerate. We have recovered the degeneracies from **Table** 5.4.

When periodic boundary conditions ($b_c = 0$ with $c > 0$) are chosen, the ground state of Hamiltonian (5.168a) is achieved by choosing the eigenvalue $-\mathrm{i}$ of $B_{j,j+c}$ for

$j=1,\cdots,N$. This ground state is thus nondegenerate. Excited states are obtained by flipping one by one the eigenvalue $-\mathrm{i}$ of $B_{j,j+c}$ for $j=1,\cdots,N$ to the value $+\mathrm{i}$. There are $N!/n!(N-n)!$ ways to flip n out of N directed bonds $\langle j,j+c\rangle$, $j=1,\cdots,N$. The excited state with n flipped bonds is thus $N!/n!(N-n)!$-fold degenerate. We have not yet recovered the degeneracies from **Table** 5.5!

The fermion operator (5.169) has two eigenvalues ±1. Imposing the constraint

$$\widehat{P}^{\mathrm{f}}=\pm1 \tag{5.173}$$

is equivalent to imposing the contraint (5.166c), up to the multiplicative factor ±1, in view of Eq. (5.162a). Because of the constraint (5.173), the matrix $B_{N,N+c}$ is not independent of the matrices $B_{j,j+c}$ with $j=1,\cdots,N-1$, that is the physical subspace (5.167a) is spanned by the basis consisting of all possible direct products between the eigenstates of $B_{j,j+c}$ with $j=1,\cdots,N-1$. Excited states are obtained by flipping one by one the eigenvalue $-\mathrm{i}$ $(+\mathrm{i})$ of $B_{j,j+c}$ for $j=1,\cdots,N-1$ to the value $+\mathrm{i}$ $(-\mathrm{i})$ when $c>0$ $(c=0)$. There are $(N-1)!/n!(N-1-n)!$ ways to achieve this. Hence, the spectra and degeneracies of the Majorana c chains obeying periodic boundary conditions and subject to the constraint (5.173) are the same as the spectra and degeneracies of the spin-1/2 cluster c chains obeying periodic boundary conditions and subject to the constraint (5.166c). We have recovered the degeneracies from **Table** 5.5.

5.5.2 Fermionization

The Majorana 0 chain depicted in Figure 5.4(a) has the fermionic representation

$$\begin{aligned}\widehat{H}_{0,0}:=&+\frac{\hbar\omega}{2}\sum_{j=1}^{N}\mathrm{i}\hat{\xi}_j\,\hat{\eta}_j\\ =&+\frac{\hbar\omega}{2}\sum_{j=1}^{N}\left(\hat{c}_j-\hat{c}^{\dagger}_j\right)\left(\hat{c}_j+\hat{c}^{\dagger}_j\right)\\ =&-\hbar\omega\sum_{j=1}^{N}\left(\hat{c}^{\dagger}_j\,\hat{c}_j-\frac{1}{2}\right).\end{aligned} \tag{5.174}$$

This is the atomic limit of a chain of N "atoms," each of which can be thought of as a two-level system corresponding to one orbital with the eigenvalue $+\hbar\omega/2$ when it is empty and the eigenvalue $-\hbar\omega/2$ when it is occupied.

The Majorana 1 chain depicted in Figure 5.4(b) has the fermionic representation ($N\geq2$ and $b_1=0,1$)

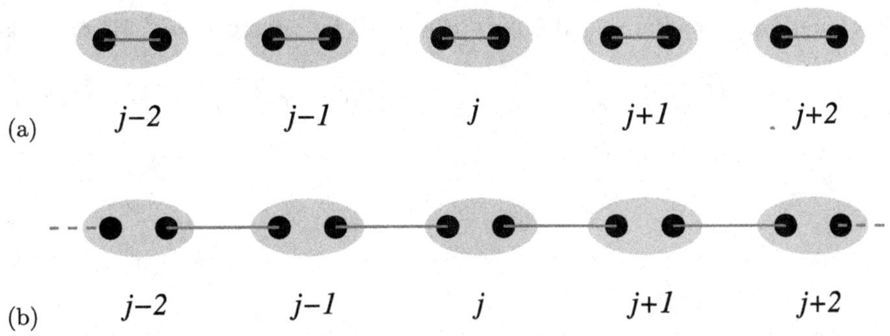

Figure 5.4 (a) The Majorana 0 chain. (b) The Majorana 1 chain. The red dashed bond is only present for periodic boundary conditions. For open boundary conditions, the pair consisting of the left- and right-most black disks are not connected to another black disk by a red bond.

$$\begin{aligned}\widehat{H}_{1,b_1} :=& -\frac{\hbar\omega}{2}\sum_{j=1}^{N-b_1} \mathrm{i}\hat{\xi}_j\,\hat{\eta}_{j+1}\\ =& -\frac{\hbar\omega}{2}\sum_{j=1}^{N-b_1}\left(\hat{c}_j-\hat{c}^\dagger_j\right)\left(\hat{c}_{j+1}+\hat{c}^\dagger_{j+1}\right)\\ =& +\frac{\hbar\omega}{2}\sum_{j=1}^{N-b_1}\left[\left(\hat{c}^\dagger_j\,\hat{c}_{j+1}+\text{H.c.}\right)+\left(\hat{c}^\dagger_j\,\hat{c}^\dagger_{j+1}+\text{H.c.}\right)\right]. \end{aligned} \tag{5.175}$$

This tight-binding Hamiltonian does not commute with the fermion number operator

$$\widehat{N}^{\mathrm{f}} := \sum_{j=1}^{N}\hat{c}^\dagger_j\,\hat{c}_j = \sum_{j=1}^{N}\frac{1}{2}\left(1+\mathrm{i}\hat{\eta}_j\,\hat{\xi}_j\right), \tag{5.176}$$

because of the presence of the local term $\hat{c}^\dagger_j\,\hat{c}^\dagger_{j+1}+\text{H.c}$ from the point of view of the fermion representation, or because of the fact that $\mathrm{i}\hat{\eta}_j\,\hat{\xi}_j$ and $\mathrm{i}\hat{\eta}_k\,\hat{\xi}_{k+1}$ anticommute when $k=j$ or $k+1=j$ from the point of view of the Majorana representation. Hamiltonian $\widehat{H}_{1,b_1}$ is the special case

$$t=-\frac{\hbar\omega}{2},\qquad \Delta=-\frac{\hbar\omega}{2},\qquad \mu=0 \tag{5.177}$$

of the superconducting Kitaev chain defined by the Hamiltonian

$$\widehat{H}_{\mathrm{Kit},b} := -\sum_{j=1}^{N-b}\left[\left(t\,\hat{c}^\dagger_j\,\hat{c}_{j+1}+\text{H.c.}\right)+\left(\Delta\,\hat{c}^\dagger_j\,\hat{c}^\dagger_{j+1}+\text{H.c.}\right)+\mu\left(\hat{c}^\dagger_j\,\hat{c}_j-\frac{1}{2}\right)\right], \tag{5.178}$$

where $b=0,1$ and $t,\Delta\in\mathbb{C}$, while $\mu\in\mathbb{R}$.

When periodic boundary conditions are selected by setting $b_1 = 0$, Hamiltonian

$$\widehat{H}_{1,0} = -\frac{\hbar\omega}{2}\left(\mathrm{i}\hat{\xi}_1\,\hat{\eta}_2 + \cdots + \mathrm{i}\hat{\xi}_{N-1}\,\hat{\eta}_N + \mathrm{i}\hat{\xi}_N\,\hat{\eta}_1\right) \tag{5.179}$$

is transformed into Hamiltonian

$$-\widehat{H}_{0,0} = -\frac{\hbar\omega}{2}\left(\mathrm{i}\hat{\xi}_1\,\hat{\eta}_1 + \cdots + \mathrm{i}\hat{\xi}_{N-1}\,\hat{\eta}_{N-1} + \mathrm{i}\hat{\xi}_N\,\hat{\eta}_N\right) \tag{5.180}$$

under

$$\hat{\xi}_j \mapsto \hat{\xi}_j, \qquad \hat{\eta}_j \mapsto \hat{\eta}_{j-1}. \tag{5.181}$$

The unitary transformation (5.181) is illegal when open boundary conditions are selected by setting $b_1 = 1$. All eigenenergies are at least twofold degenerate as follows from observing that

$$\widehat{H}_{1,1} := -\frac{\hbar\omega}{2}\left(\mathrm{i}\hat{\xi}_1\,\hat{\eta}_2 + \cdots + \mathrm{i}\hat{\xi}_{N-1}\,\hat{\eta}_N\right) \tag{5.182}$$

does not contain the operators $\hat{\eta}_1$ and $\hat{\xi}_N$. Hence, we have the algebra

$$\left[\widehat{H}_{1,1}, \hat{\eta}_1\right] = \left[\widehat{H}_{1,1}, \hat{\xi}_N\right] = 0, \qquad \left\{\hat{\eta}_1, \hat{\xi}_N\right\} = 0, \tag{5.183}$$

which implies that any eigenstate $|\Psi\rangle$ of $\widehat{H}_{1,1}$ is at least twofold degenerate.

Proof One verifies that

$$\left(\mathrm{i}\hat{\eta}_1\,\hat{\xi}_N\right)^2 = 1, \quad \left(\frac{1 \pm \mathrm{i}\hat{\eta}_1\,\hat{\xi}_N}{2}\right)^2 = \frac{1 \pm \mathrm{i}\hat{\eta}_1\,\hat{\xi}_N}{2}, \quad \frac{1 + \mathrm{i}\hat{\eta}_1\,\hat{\xi}_N}{2}\,\frac{1 - \mathrm{i}\hat{\eta}_1\,\hat{\xi}_N}{2} = 0. \tag{5.184}$$

The first identity is required to derive the next two identities. The first identity follows from the fact that $\hat{\eta}_1$ anticommutes with $\hat{\xi}_N$. Because $\hat{\eta}_1$ and $\hat{\xi}_N$ both commute with $\widehat{H}_{1,1}$,

$$|\Psi_\pm\rangle := \frac{1 \pm \mathrm{i}\hat{\eta}_1\,\hat{\xi}_N}{2}\,|\Psi\rangle \tag{5.185}$$

share the same eigenenergy with $|\Psi\rangle$. Moreover, they are orthogonal owing to the last identity in Eq. (5.184). Hence, the spectrum of $\widehat{H}_{1,1}$ is at least twofold degenerate. □

The twofold degeneracy of the spectrum of $\widehat{H}_{1,1}$ when $N \geq 2$ is even is a consequence of the three facts that (i) the fermion parity operator (5.169) factorizes for any N into

$$\widehat{P}^{\mathrm{f}} = \widehat{P}^{\mathrm{f}}_A\,\widehat{P}^{\mathrm{f}}_B, \tag{5.186a}$$

where the symbols A and B refer to the sublattices Λ_A and Λ_B made of odd- and even-numbered sites, respectively, (ii)

$$\widehat{P}^{\mathrm{f}}_S := \prod_{j\in\Lambda_S} \mathrm{i}\hat{\eta}_j\,\hat{\xi}_j, \qquad S = A, B, \qquad \left[\widehat{P}^{\mathrm{f}}_A, \widehat{P}^{\mathrm{f}}_B\right] = 0, \tag{5.186b}$$

and (iii) that the nontrivial action of $\widehat{P}^{\mathrm{f}}_A$ on the boundary $j = 1$ and the nontrivial action of $\widehat{P}^{\mathrm{f}}_B$ on the boundary $j = N$ with N even are represented by

$$\widehat{P}^{\mathrm{f}}_{\mathrm{L}} := \hat{\eta}_1, \qquad \widehat{P}^{\mathrm{f}}_{\mathrm{R}} := \hat{\xi}_N, \tag{5.187a}$$

respectively, so that the projective algebra

$$\left\{\widehat{P}^{\mathrm{f}}_{\mathrm{L}}, \widehat{P}^{\mathrm{f}}_{\mathrm{R}}\right\} = 0 \tag{5.187b}$$

is now a vanishing anticommutator on the boundaries instead of a vanishing commutator in the bulk. This twofold degeneracy is spread out across the two boundaries; it is nonlocal in the Majorana representation (recall that it was merely due to a global Ising symmetry in the spin-1/2 cluster $c = 1$ chain representation).

The Majorana $c > 1$ chains ($N \geq c + 1$ and $b_c = 0, c$)

$$\begin{aligned}
\widehat{H}_{c,b_c} :=& -\frac{\hbar\omega}{2}\sum_{j=1}^{N-b_c} \mathrm{i}\hat{\xi}_j\,\hat{\eta}_{j+c} \\
=& -\frac{\hbar\omega}{2}\sum_{j=1}^{N-b_c} \left(\hat{c}_j - \hat{c}^{\dagger}_j\right)\left(\hat{c}_{j+c} + \hat{c}^{\dagger}_{j+c}\right) \\
=& +\frac{\hbar\omega}{2}\sum_{j=1}^{N-b_c} \left[\left(\hat{c}^{\dagger}_j\,\hat{c}_{j+c} + \mathrm{H.c.}\right) + \left(\hat{c}^{\dagger}_j\,\hat{c}^{\dagger}_{j+c} + \mathrm{H.c.}\right)\right]
\end{aligned} \tag{5.188}$$

are nothing but c copies of the Majorana 1 chain (5.175), that is, nothing but c copies of the Kitaev superconducting chain (5.178) in the limit (5.177). All eigenenergies of $\widehat{H}_{c,c}$ are thus at least 2^c-fold degenerate when open boundary conditions are selected, as inspection of Figure 5.3 makes it evident. Of particular interest are the Majorana 2 chain and Majorana 4 chain owing to the equivalence of the former (latter) to the spinless (spin-1/2) SSH model defined in Section 2.7.1.

5.5.3 The Majorana 2 Chain Realizes the Spinless SSH Chain

We consider the Majorana 2 chain ($N \geq 3$ and $b_2 = 0, 2$) depicted in Figure 5.5

$$\widehat{H}_{2,b_2} := -\frac{\hbar\omega}{2}\sum_{j=1}^{N-b_2} \mathrm{i}\hat{\xi}_j\,\hat{\eta}_{j+2}. \tag{5.189}$$

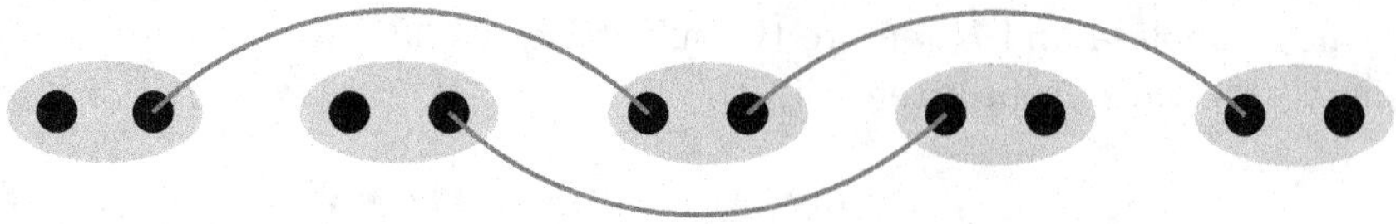

Figure 5.5 The Majorana 2 chain with open boundary conditions has two black discs belonging to the ellipses $j=1$ and $j=2$ and two black discs belonging to the ellipses $j=N-1$ and $j=N$ that are not the end points of a red bond. Correspondingly, the Majorana operators $\hat{\eta}_1$ and $\hat{\eta}_2$ on the left boundary and the Majorana operators $\hat{\xi}_{N-1}$ and $\hat{\xi}_N$ on the right boundary commute with Hamiltonian $\widehat{H}_{2,2}$ defined in Eq. (5.189).

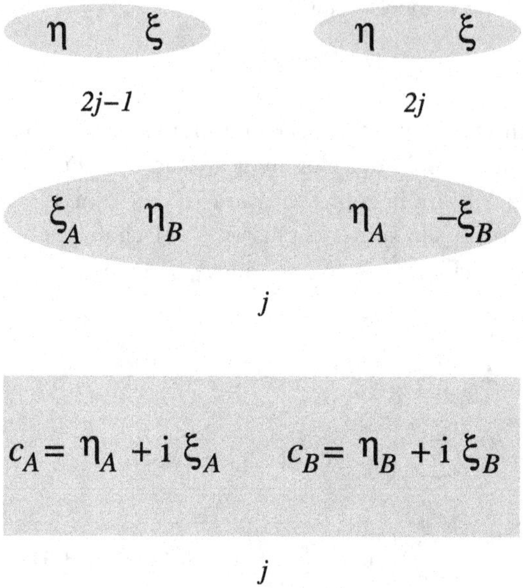

Figure 5.6 Pictorial representation of Eqs. (5.190) and (5.192).

When $N \geq 3$ is even, $\widehat{H}_{2,b_2}$ has an $\mathrm{O}(2) = \mathbb{Z}_2 \times \mathrm{SO}(2)$ symmetry according to Eq. (5.165). We are going to identify the connected component $\mathrm{SO}(2) \subset \mathrm{O}(2)$ with U(1) in three steps.

First, the Majorana operators are relabeled according to the rules

$$\hat{\eta}_{2j-1} =: +\hat{\xi}_{A,j}, \qquad \hat{\eta}_{2j} =: +\hat{\eta}_{A,j}, \tag{5.190a}$$

$$\hat{\xi}_{2j-1} =: +\hat{\eta}_{B,j}, \qquad \hat{\xi}_{2j} =: -\hat{\xi}_{B,j}. \tag{5.190b}$$

What we have done, according to Figure 5.6, is to replace two consecutive ellipses, each of which is assigned a pair of Majorana operators, by a single larger ellipse to which four Majorana operators are assigned. Correspondingly, we have the new Majorana bilinears

$$\mathrm{i}\hat{\xi}_{2j-1}\,\hat{\eta}_{2j-1+c} = \mathrm{i}\hat{\eta}_{B,j}\,\hat{\xi}_{A,j+\frac{c}{2}}, \qquad \mathrm{i}\hat{\xi}_{2j}\,\hat{\eta}_{2j+c} = \mathrm{i}\hat{\eta}_{A,j+\frac{c}{2}}\,\hat{\xi}_{B,j}, \tag{5.191}$$

for $j = 1, \cdots, N/2$ and $c = 0, 2$.

Second, we introduce the spinless fermionic operators

$$\hat{c}^{\dagger}_{S,j} := \frac{1}{2}\left(\hat{\eta}_{S,j} - \mathrm{i}\hat{\xi}_{S,j}\right), \qquad \hat{c}_{S,j} := \frac{1}{2}\left(\hat{\eta}_{S,j} + \mathrm{i}\hat{\xi}_{S,j}\right), \tag{5.192a}$$

for $S = A, B$ and $j = 1, \cdots, N/2$. By design,

$$\hat{c}_{A,j} = \frac{1}{2}\left(\hat{\eta}_{2j} + \mathrm{i}\hat{\eta}_{2j-1}\right), \qquad \hat{c}_{B,j} = \frac{1}{2}\left(\hat{\xi}_{2j-1} - \mathrm{i}\hat{\xi}_{2j}\right) \tag{5.192b}$$

so that

$$\begin{aligned}
\mathrm{i}\hat{\xi}_{2j-1}\,\hat{\eta}_{2j-1+c} + \mathrm{i}\hat{\xi}_{2j}\,\hat{\eta}_{2j+c} =& \,\mathrm{i}\hat{\eta}_{B,j}\,\hat{\xi}_{A,j+\frac{c}{2}} + \mathrm{i}\hat{\eta}_{A,j+\frac{c}{2}}\,\hat{\xi}_{B,j} \\
=& \left(\hat{c}_{B,j} + \hat{c}^{\dagger}_{B,j}\right)\left(\hat{c}_{A,j+\frac{c}{2}} - \hat{c}^{\dagger}_{A,j+\frac{c}{2}}\right) \\
&+ \left(B, j \leftrightarrow A, j + \frac{c}{2}\right) \\
=& \,\hat{c}^{\dagger}_{B,j}\,\hat{c}_{A,j+\frac{c}{2}} + \hat{c}^{\dagger}_{A,j+\frac{c}{2}}\,\hat{c}_{B,j} + \hat{c}^{\dagger}_{A,j+\frac{c}{2}}\,\hat{c}^{\dagger}_{B,j} + \hat{c}_{B,j}\hat{c}_{A,j+\frac{c}{2}} \\
&+ \left(B, j \leftrightarrow A, j + \frac{c}{2}\right) \\
=& \,2\left(\hat{c}^{\dagger}_{B,j}\,\hat{c}_{A,j+\frac{c}{2}} + \hat{c}^{\dagger}_{A,j+\frac{c}{2}}\,\hat{c}_{B,j}\right).
\end{aligned} \tag{5.192c}$$

Third, we define the family of Hamiltonians ($N \geq 3$ even)

$$\widehat{H}_{\mathrm{SSH},b_2}(\lambda) := -(1-\lambda)\,\frac{\hbar\omega}{2}\sum_{j=1}^{N}\mathrm{i}\hat{\xi}_j\,\hat{\eta}_j - \lambda\,\frac{\hbar\omega}{2}\sum_{j=1}^{N-b_2}\mathrm{i}\hat{\xi}_j\,\hat{\eta}_{j+2} \tag{5.193}$$

that interpolates between the Majorana 0 chain (5.174) for $\lambda = 0$ with the substitution $\omega \to -\omega$ and the Majorana 2 chain (5.189) for $\lambda = 1$ as a function of $0 \leq \lambda \leq 1$. By making use of the identities (5.192), we recover the fermionic Su-Schrieffer-Heeger (SSH) model

$$\widehat{H}_{\mathrm{SSH},b_2}(\lambda) := -\hbar\omega\left[(1-\lambda)\sum_{j=1}^{N/2}\hat{c}^{\dagger}_{A,j}\,\hat{c}_{B,j} + \lambda\sum_{j=1}^{(N-b_2)/2}\hat{c}^{\dagger}_{B,j}\,\hat{c}_{A,j+1} + \mathrm{H.c.}\right]. \tag{5.194}$$

As promised, the SO(2) symmetry of the Majorana 2 chain combines with the U(1) symmetry of the 0 chain into the U(1) symmetry of the SSH representation (5.194) under the transformation

$$\hat{c}^{\dagger}_{S,j} \mapsto \hat{c}^{\dagger}_{S,j}\,e^{-\mathrm{i}\phi}, \qquad \hat{c}_{S,j} \mapsto e^{+\mathrm{i}\phi}\,\hat{c}_{S,j}, \tag{5.195}$$

for any $S = A, B$, $j = 1, \cdots, N/2$, and $0 \leq \phi < 2\pi$. This U(1) global symmetry is equivalent to the conservation of the global fermionic charge.

As one could have guessed, the symmetries generated by $\widehat{P}^{\mathrm{f}}$ in Eq. (5.169) and by $\widehat{T}'$ in Eq. (5.170) can be interpreted as the Ising transformation

$$\widehat{P}^{\mathrm{f}}\,\hat{c}_{A,j}\,\widehat{P}^{\mathrm{f}}=-\hat{c}_{A,j},\qquad \widehat{P}^{\mathrm{f}}\,\hat{c}_{B,j}\,\widehat{P}^{\mathrm{f}}=-\hat{c}_{B,j},\tag{5.196}$$

and the particle-hole transformation

$$\widehat{T}'\,\hat{c}_{A,j}\,\widehat{T}'=+\hat{c}^{\dagger}_{A,j},\qquad \widehat{T}'\,\hat{c}_{B,j}\,\widehat{T}'=-\hat{c}^{\dagger}_{B,j},\tag{5.197}$$

for $j=1,\cdots,N/2$, respectively. As the representation (5.196) of $\widehat{P}^{\mathrm{f}}$ in Eq. (5.169) is the special case $\phi=\pi$ of Eq. (5.195), the SSH representation (5.194) of the Majorana 2 chain only inherits the particle-hole symmetry (5.197) as a symmetry distinct from its U(1) symmetry, out of the two symmetries $\widehat{X}$ and $\widehat{T}'$ obeyed by the spin-1/2 cluster $c=2$ chain (see **Table** 5.1).

When $0\leq\lambda\leq 1/2$, $\widehat{H}_{0,0}$ dominates over $\widehat{H}_{2,2}$ on the right-hand side of Eq. (5.193). The ground states are nondegenerate and separated from all excited states by a gap. This gap closes smoothly at $\lambda=1/2$. A quantum phase transition occurs. When $1/2\leq\lambda\leq 1$, $\widehat{H}_{2,2}$ dominates over $\widehat{H}_{0,0}$. The ground states are fourfold degenerate in the thermodynamic limit and is separated from all excited states by a gap. This degeneracy is the consequence of there being two fermionic zero modes that are localized on the left- and right-ends of the open chain, respectively. We say that the SSH Hamiltonian (5.194) is a trivial band insulator when $0\leq\lambda<1/2$, while it is a topological band insulator when $1/2<\lambda\leq 1$.

5.5.4 The Majorana 4 Chain Realizes the Spinfull SSH Chain

We consider the Majorana 4 chain ($N\geq 5$ and $b_4=0,4$)

$$\widehat{H}_{4,b_4}:=-\frac{\hbar\omega}{2}\sum_{k=1}^{N-b_4}\mathrm{i}\hat{\xi}_k\,\hat{\eta}_{k+4}.\tag{5.198}$$

When $N\geq 5$ is an integer multiple of the number 4, $\widehat{H}_{4,b_4}$ has an $\mathrm{O}(4)=\mathbb{Z}_2\times\mathrm{SO}(4)$ symmetry according to Eq. (5.165), for it can be rewritten as

$$\widehat{H}_{4,b_4}=-\frac{\hbar\omega}{2}\sum_{j=1}^{(N-b_4)/4}\mathrm{i}\hat{\boldsymbol{\xi}}_j\cdot\hat{\boldsymbol{\eta}}_{j+1},\tag{5.199a}$$

where we have introduced the column vectors [see Figure 5.7(a)]

$$\hat{\boldsymbol{\xi}}_j:=\begin{pmatrix}\hat{\xi}_{4j-3}&\hat{\xi}_{4j-2}&\hat{\xi}_{4j-1}&\hat{\xi}_{4j}\end{pmatrix}^{\mathsf{T}},\quad \hat{\boldsymbol{\eta}}_j:=\begin{pmatrix}\hat{\eta}_{4j-3}&\hat{\eta}_{4j-2}&\hat{\eta}_{4j-1}&\hat{\eta}_{4j}\end{pmatrix}^{\mathsf{T}}.\tag{5.199b}$$

A Majorana 4 chain can be thought of as four identical copies of a Majorana 1 chain. Alternatively, we can interpret the Majorana 4 chain as two identical copies of the SSH Hamiltonian (5.194) with $\lambda=1$. This interpretation is not unique, however, since there are six possible pairings out of four Majorana operators that enable the construction of fermionic operators. We will choose the following pairing rules (assuming that $N\geq 5$ is an integer multiple of the number 4)

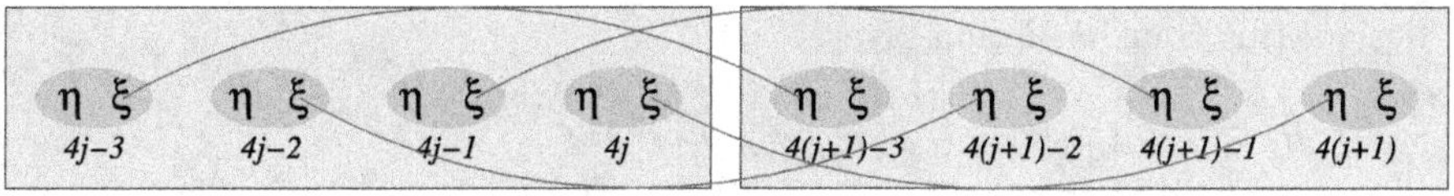

(a)

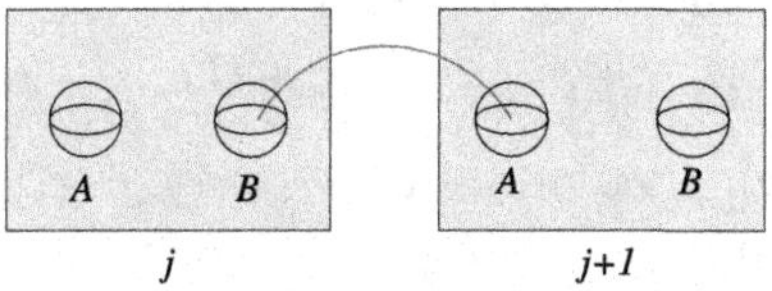

(b)

Figure 5.7 Pictorial representation of Eq. (5.200) and (5.202a). (a) A rectangle labeled by the index $j = 1, \cdots, N/4$ hosts four ellipses. These ellipses are labeled by the site index $k = 1, \cdots, N$. Each ellipse hosts one pair of Majorana fermions $\hat{\eta}_k$ and $\hat{\xi}_k$. The Majorana bilinear form $\mathrm{i}\hat{\xi}_k\, \hat{\eta}_{k+4}$ is represented by the red line. With open boundary conditions, there are four Majorana fermions $\hat{\eta}_1$, $\hat{\eta}_2$, $\hat{\eta}_3$, $\hat{\eta}_4$ and four Majorana fermions $\hat{\xi}_{N-3}\, \hat{\xi}_{N-2}\, \hat{\xi}_{N-1}\, \hat{\xi}_N$ that are unpaired by a red line. This implies a $(2^4 = 16)$-fold degeneracy of the Majorana 4 chain obeying open boundary conditions. (b) The fermionization of the four Majorana pairs per rectangle into two pairs of spin-1/2 fermions is represented by two Bloch spheres labeled A and B, each of which can be unoccupied, occupied by one spin-1/2 fermion with the North (South) poles corresponding to the $\sigma =\uparrow$ ($\sigma =\downarrow$) quantum numbers, or occupied by two spin-1/2 fermions with opposite quantum numbers σ. The red line now represents the hopping by which an electron with spin σ tunnels without spin flip back and forth between the Bloch sphere B at site j and the Bloch sphere A at site $j+1$ with $j = 1, \cdots, N/4$. The repulsive Hubbard interaction encoded by Eqs. (5.202d) and (5.202e) decreases the 16-fold degeneracy of the ground state to a fourfold degeneracy of the Majorana 4 chain obeying open boundary conditions.

$$\hat{c}_{A,j,\uparrow} := \frac{1}{2}\left(\hat{\eta}_{4j-1} + \mathrm{i}\hat{\eta}_{4j-3}\right), \qquad \hat{c}_{A,j,\downarrow} := \frac{1}{2}\left(\hat{\eta}_{4j} + \mathrm{i}\hat{\eta}_{4j-2}\right), \tag{5.200a}$$

$$\hat{c}_{B,j,\uparrow} := \frac{1}{2}\left(\hat{\xi}_{4j-3} - \mathrm{i}\hat{\xi}_{4j-1}\right), \qquad \hat{c}_{B,j,\downarrow} := \frac{1}{2}\left(\hat{\xi}_{4j-2} - \mathrm{i}\hat{\xi}_{4j}\right), \tag{5.200b}$$

for $j = 1, \cdots, N/4$. Inversion of Eq. (5.200) yields

$$\hat{\xi}_{4j-3} = \hat{c}_{B,j,\uparrow} + \hat{c}^{\dagger}_{B,j,\uparrow}, \qquad \hat{\eta}_{4j-3} = \frac{\hat{c}_{A,j,\uparrow} - \hat{c}^{\dagger}_{A,j,\uparrow}}{(+\mathrm{i})}, \tag{5.201a}$$

$$\hat{\xi}_{4j-2} = \hat{c}_{B,j,\downarrow} + \hat{c}^{\dagger}_{B,j,\downarrow}, \qquad \hat{\eta}_{4j-2} = \frac{\hat{c}_{A,j,\downarrow} - \hat{c}^{\dagger}_{A,j,\downarrow}}{(+\mathrm{i})}, \tag{5.201b}$$

$$\hat{\xi}_{4j-1} = \frac{\hat{c}_{B,j,\uparrow} - \hat{c}^{\dagger}_{B,j,\uparrow}}{(-\mathrm{i})}, \qquad \hat{\eta}_{4j-1} = \hat{c}_{A,j,\uparrow} + \hat{c}^{\dagger}_{A,j,\uparrow}, \tag{5.201c}$$

$$\hat{\xi}_{4j} = \frac{\hat{c}_{B,j,\downarrow} - \hat{c}^{\dagger}_{B,j,\downarrow}}{(-\mathrm{i})}, \qquad \hat{\eta}_{4j} = \hat{c}_{A,j,\downarrow} + \hat{c}^{\dagger}_{A,j,\downarrow}. \tag{5.201d}$$

By design, this transformation gives

$$
\begin{aligned}
\mathrm{i}\hat{\boldsymbol{\xi}}_j \cdot \hat{\boldsymbol{\eta}}_{j'} =& \quad \hat{c}^{\dagger}_{B,j,\uparrow}\,\hat{c}_{A,j',\uparrow} - \hat{c}^{\dagger}_{B,j,\uparrow}\,\hat{c}^{\dagger}_{A,j',\uparrow} + \text{H.c.} \\
&+ \hat{c}^{\dagger}_{B,j,\downarrow}\,\hat{c}_{A,j',\downarrow} - \hat{c}^{\dagger}_{B,j,\downarrow}\,\hat{c}^{\dagger}_{A,j',\downarrow} + \text{H.c.} \\
&+ \hat{c}^{\dagger}_{B,j,\uparrow}\,\hat{c}_{A,j',\uparrow} + \hat{c}^{\dagger}_{B,j,\uparrow}\,\hat{c}^{\dagger}_{A,j',\uparrow} + \text{H.c.} \\
&+ \hat{c}^{\dagger}_{B,j,\downarrow}\,\hat{c}_{A,j',\downarrow} + \hat{c}^{\dagger}_{B,j,\downarrow}\,\hat{c}^{\dagger}_{A,j',\downarrow} + \text{H.c.} \\
=& \, 2 \sum_{\sigma=\uparrow,\downarrow} \left(\hat{c}^{\dagger}_{B,j,\sigma}\,\hat{c}_{A,j',\sigma} + \text{H.c.} \right), \qquad j,j' = 1,\cdots,N/4. \qquad (5.202\text{a})
\end{aligned}
$$

Hence,

$$
\sum_{j'=j,j+1} \frac{t_{j,j'}}{2}\, \mathrm{i}\hat{\boldsymbol{\xi}}_j \cdot \hat{\boldsymbol{\eta}}_{j'} = \sum_{j'=j,j+1} t_{j,j'} \sum_{\sigma=\uparrow,\downarrow} \left(\hat{c}^{\dagger}_{B,j,\sigma}\,\hat{c}_{A,j',\sigma} + \text{H.c.} \right) \qquad (5.202\text{b})
$$

delivers the spinfull SSH Hamiltonian if

$$
t_{j,j'} = t + (-1)^{j-j'}\,\delta t, \qquad t, \delta t \in \mathbb{R}. \qquad (5.202\text{c})
$$

Whereas the left-hand side of Eq. (5.202b) is manifestly $\mathrm{O}(4) = \mathbb{Z}_2 \times \mathrm{SO}(4)$ symmetric, the right-hand side is only manifestly $\mathrm{U}(2) = \mathrm{U}(1) \times \mathrm{SU}(2)$ symmetric. Moreover, we have the identities

$$
\begin{aligned}
\hat{\eta}_{4j-3}\,\hat{\eta}_{4j-2}\,\hat{\eta}_{4j-1}\,\hat{\eta}_{4j} &= -\left(\hat{\eta}_{4j-1}\,\hat{\eta}_{4j-3}\right)\left(\hat{\eta}_{4j}\,\hat{\eta}_{4j-2}\right) \\
&= \left(2\hat{c}^{\dagger}_{A,j,\uparrow}\,\hat{c}_{A,j,\uparrow} - 1\right)\left(2\hat{c}^{\dagger}_{A,j,\downarrow}\,\hat{c}_{A,j,\downarrow} - 1\right), \qquad (5.202\text{d}) \\
\hat{\xi}_{4j-3}\,\hat{\xi}_{4j-2}\,\hat{\xi}_{4j-1}\,\hat{\xi}_{4j} &= -\left(\hat{\xi}_{4j-3}\,\hat{\xi}_{4j-1}\right)\left(\hat{\xi}_{4j-2}\,\hat{\xi}_{4j}\right) \\
&= \left(2\hat{c}^{\dagger}_{B,j,\uparrow}\,\hat{c}_{B,j,\uparrow} - 1\right)\left(2\hat{c}^{\dagger}_{B,j,\downarrow}\,\hat{c}_{B,j,\downarrow} - 1\right). \qquad (5.202\text{e})
\end{aligned}
$$

Let $U > 0$ carry the units of energy and define the interacting Hamiltonian

$$
\begin{aligned}
\widehat{H}_{AB-\text{Hub},b_4} :=& -\frac{\hbar\omega}{2} \sum_{j=1}^{(N-b_4)/4} \mathrm{i}\hat{\boldsymbol{\xi}}_j \cdot \hat{\boldsymbol{\eta}}_{j+1} \\
&+ \frac{U}{4} \sum_{j=1}^{N/4} \left(\hat{\eta}_{4j-3}\,\hat{\eta}_{4j-2}\,\hat{\eta}_{4j-1}\,\hat{\eta}_{4j} + \hat{\xi}_{4j-3}\,\hat{\xi}_{4j-2}\,\hat{\xi}_{4j-1}\,\hat{\xi}_{4j} \right).
\end{aligned} \qquad (5.203)
$$

The $\mathrm{O}(4) = \mathbb{Z}_2 \times \mathrm{SO}(4)$ symmetry of the bilinear contribution is broken down to $\mathrm{SO}(4) \subset \mathrm{O}(4)$ by the Hubbard interaction in view of the identity

$$
\hat{\eta}_1\,\hat{\eta}_2\,\hat{\eta}_3\,\hat{\eta}_4 = \frac{1}{4!} \sum_{a,b,c,d=1}^{4} \epsilon_{abcd}\,\hat{\eta}_a\,\hat{\eta}_b\,\hat{\eta}_c\,\hat{\eta}_d. \qquad (5.204)
$$

Define the elements $R_x, R_y \in \mathrm{SO}(4)$ acting globally (independently of j) on the column vectors $\hat{\boldsymbol{\xi}}$ and $\hat{\boldsymbol{\eta}}$ by

$$R_x := \begin{pmatrix} 0 & 1 & 0 & 0 \\ 1 & 0 & 0 & 0 \\ 0 & 0 & 0 & 1 \\ 0 & 0 & 1 & 0 \end{pmatrix} \equiv \Sigma_{10},$$

$$R_y := \begin{pmatrix} +1 & 0 & 0 & 0 \\ 0 & +1 & 0 & 0 \\ 0 & 0 & -1 & 0 \\ 0 & 0 & 0 & -1 \end{pmatrix} \equiv +\Sigma_{03}, \qquad (5.205a)$$

$$R_z := R_x\,R_y = \begin{pmatrix} 0 & +1 & 0 & 0 \\ +1 & 0 & 0 & 0 \\ 0 & 0 & 0 & -1 \\ 0 & 0 & -1 & 0 \end{pmatrix} \equiv +\Sigma_{13},$$

where we are using the shorthand notation $\Sigma_{\mu\nu} \equiv \tau_\mu \otimes \rho_\nu$ with $\mu,\nu = 0,1,2,3$ labeling the 2×2 identity matrix τ_0 and ρ_0 together with the Pauli matrices $\boldsymbol{\tau}$ and $\boldsymbol{\rho}$ and we have chosen the ordered basis of $\mathbb{R}^4$ given by

$$\{e_1\otimes e_1, e_2\otimes e_1, e_1\otimes e_2, e_2\otimes e_2\}. \qquad (5.205b)$$

One verifies that the roots S_x, S_y, and S_z to

$$R_a = e^{\mathrm{i}\pi S_a}, \qquad a = x,y,z, \qquad (5.206a)$$

obey the $\mathfrak{so}(3)$ algebra

$$[S_a, S_b] = \mathrm{i}\epsilon_{abc}\,S_c \qquad (5.206b)$$

(the summation convention over repeated indices is assumed) and are given by

$$S_x = \frac{1}{2}\left(\Sigma_{02} - \Sigma_{12}\right), \quad S_y = \frac{1}{2}\left(\Sigma_{20} - \Sigma_{23}\right), \quad S_z = -\frac{1}{2}\left(\Sigma_{21} + \Sigma_{32}\right). \quad (5.206c)$$

[To this end, one verifies the identities $(S_x)^{2n+1} = S_x$ and $(S_y)^{2n+1} = S_y$ for $n = 0,1,2,3,\cdots$.] The pair R_x and R_y generates the subgroup $\mathbb{Z}_2\times\mathbb{Z}_2$ of the subgroup SO(3) with the generators S_x, S_y, and S_z, which is itself a subgroup of the symmetry group SO(4). It is the chain of symmetry groups

$$\mathbb{Z}_2\times\mathbb{Z}_2 \subset \mathrm{SO}(3) \subset \mathrm{SO}(4) \qquad (5.207)$$

that is protecting the boundary states in this interacting Majorana 4 chain.

The fermionic representation of Hamiltonian (5.203) follows from the identities (5.202). It is

$$\begin{aligned} \widehat{H}_{AB-\mathrm{Hub},b_4} =& -\hbar\omega \sum_{j=1}^{(N-b_4)/4} \sum_{\sigma=\uparrow,\downarrow} \left(\hat{c}^\dagger_{B,j,\sigma}\,\hat{c}_{A,j+1,\sigma} + \mathrm{H.c.}\right) \\ &+ U \sum_{j=1}^{N/4} \sum_{S=A,B} \prod_{\sigma=\uparrow,\downarrow} \left(\hat{c}^\dagger_{S,j,\sigma}\,\hat{c}_{S,j,\sigma} - \frac{1}{2}\right). \end{aligned} \qquad (5.208)$$

The fermionic hopping is off-diagonal with respect to both the site index $j = 1,\cdots,N/4$ and the "orbital index" $S = A,B$, as is implied in Figure 5.7. The interaction is the usual repulsive Hubbard interaction. This is why we call this Hamiltonian the AB-Hubbard Hamiltonian. When $U/(\hbar\omega)=0$, $\widehat{H}_{AB-\mathrm{Hub},b_4}$ realizes four copies of the Majorana 1 chain or, alternatively, two copies of the $\lambda=1$ limit of the SSH Hamiltonian (5.194). Each eigenstate is at least 16-fold degenerate owing to the presence of two free spin-1/2 fermions at each ends of the chain when open boundary conditions are selected ($b_4=4$). When $U/(\hbar\omega)=\infty$, the Hubbard repulsive interaction prevents the two spheres in any rectangle from Figure 5.7(b) from both being either unoccupied or doubly occupied by spin-1/2 fermions. The ground state at half-filling has exactly one spin-1/2 fermion occupying each Bloch sphere in Figure 5.7(b). The degeneracy of this ground state is $2^{N/2}$ as there are $N/2$ block spheres. Second-order degenerate perturbation theory in powers of $(\hbar\omega)/U$ around the $(\hbar\omega)/U=0$ limit gives the nearest-neighbor antiferromagnetic quantum spin-1/2 Heisenberg Hamiltonian[20]

$$\widehat{H}_{AB-\mathrm{Heis},b_4} = \frac{4(\hbar\omega)^2}{U}\sum_{j=1}^{(N-b_4)/4} \hat{\boldsymbol{S}}_{B,j}\cdot\hat{\boldsymbol{S}}_{A,j+1},$$
$$\left[\hat{S}^{\alpha}_{S,j},\hat{S}^{\beta}_{S',j'}\right] = \mathrm{i}\delta_{S,S'}\,\delta_{j,j'}\,\hbar\,\epsilon^{\alpha\beta\gamma}\,\hat{S}^{\gamma}_{S,j}, \qquad \left(\hat{\boldsymbol{S}}_{S,j}\right)^2 = \frac{3}{4}\hbar^2, \tag{5.209}$$

on a chain with length $N/4$ and two spin-1/2 per unit cell. The ground state of $\widehat{H}_{AB-\mathrm{Heis},b_4}$ is a nondegenerate tensor product state of singlets when periodic boundary conditions are selected by $b_4=0$, while it is a tensor product state of singlets with one free spin-1/2 degree of freedom attached to each ends of the chain when open boundary conditions are selected by $b_4=4$ [see Figure 5.8(a)]. Hence, these ground states are fourfold degenerate, not 16-fold degenerate as it is in the noninteracting limit $U/(\hbar\omega)=0$. The same conclusion is reached using the Majorana representation as follows. The interaction on the left boundary is given by

$$\begin{aligned}
+\frac{U}{4}\hat{\eta}_1\hat{\eta}_2\hat{\eta}_3\hat{\eta}_4 \equiv& -\frac{U}{4}(\hat{\eta}_1\hat{\eta}_3)(\hat{\eta}_2\hat{\eta}_4)\\
=& +\frac{U}{4}\left(\hat{c}_{13}+\hat{c}^{\dagger}_{13}\right)\left(\hat{c}_{13}-\hat{c}^{\dagger}_{13}\right)\left(\hat{c}_{24}+\hat{c}^{\dagger}_{24}\right)\left(\hat{c}_{24}-\hat{c}^{\dagger}_{24}\right)\\
=& +U\left(\hat{c}^{\dagger}_{13}\hat{c}_{13}-\frac{1}{2}\right)\left(\hat{c}^{\dagger}_{24}\hat{c}_{24}-\frac{1}{2}\right)\\
\equiv& +\frac{U}{4}\frac{1}{\sqrt{2}}\left(|0,0\rangle\langle 0,0|\pm|1,1\rangle\langle 1,1|\right)\\
&-\frac{U}{4}\frac{1}{\sqrt{2}}\left(|1,0\rangle\langle 1,0|\pm|0,1\rangle\langle 0,1|\right)
\end{aligned} \tag{5.210}$$

[20] E. Fradkin, *Field Theories of Condensed Matter Physics,* Cambridge University Press, New York 2013. [Fradkin (2013)].

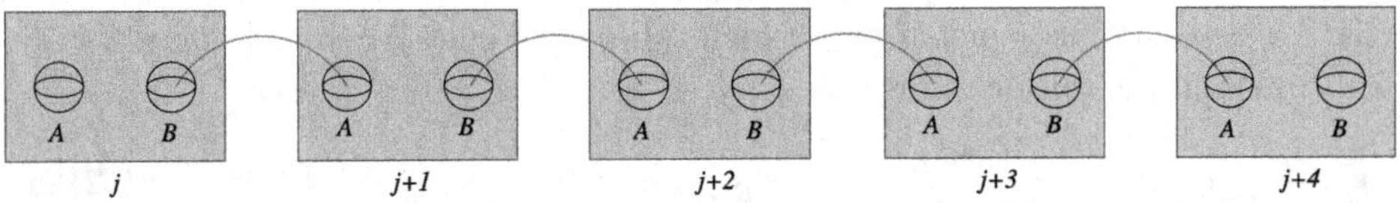

(a)

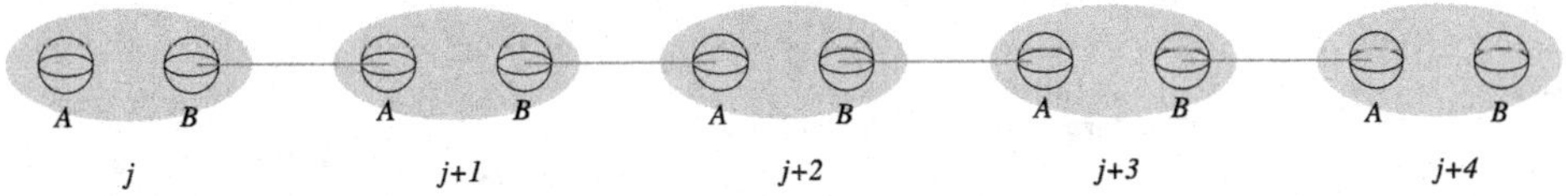

(b)

Figure 5.8 (a) A one-dimensional lattice with two inequivalent sites A and B per unit cell labeled by the integer $j = 1, \cdots, N/4$. Each site is represented by the Bloch sphere of a single spin-1/2 degree of freedom. The red bonds realize a dimer covering. Each dimer is the singlet state $(|\uparrow\downarrow\rangle - |\downarrow\uparrow\rangle)/\sqrt{2}$ out of two spin-1/2 degrees of freedom. The many-body ground state is a tensor product of singlet states, one per red bond. It is explicitly given in Eq. (5.215a). If periodic boundary conditions are imposed, every Bloch sphere is the end point of one and only one red dimer. This tensor product state is nondegenerate. If open boundary conditions are imposed, the spin-1/2 degrees of freedom on the first and last Bloch spheres are free. This implies a fourfold degeneracy of the dimer state tensored into two free spin-1/2 degrees of freedom. (b) A chain of ellipses such that each ellipse is associated with the spin-1 Hilbert space spanned by the states (5.211). The Hilbert space associated with an ellipse in (b) is a three-dimensional subspace of the four-dimensional Hilbert space associated with a rectangle in (a). A red bond is the singlet $|0,0\rangle = (|\uparrow\downarrow\rangle - |\downarrow\uparrow\rangle)/\sqrt{2}$ between two neighboring ellipses. The dimer covering of red bonds represents the AKLT state (5.221) when periodic boundary conditions are imposed.

if $\hat{\eta}_1$ ($\hat{\eta}_2$) is twice the real part of $\hat{c}_{13}$ ($\hat{c}_{24}$) and $\hat{\eta}_3$ ($\hat{\eta}_4$) is twice the imaginary part of $\hat{c}_{13}$ ($\hat{c}_{24}$). The ground states on the left boundary are thus spanned by the orthonormal states $|1,0\rangle$ and $|0,1\rangle$ in the basis defined by tensoring the eigenstates of the fermionic number operators $\hat{c}^{\dagger}_{13}\,\hat{c}_{13}$ and $\hat{c}^{\dagger}_{24}\,\hat{c}_{24}$. The same argument applies to the right boundary for which $\hat{\eta}_1 \mapsto \hat{\xi}_N$, $\hat{\eta}_2 \mapsto \hat{\xi}_{N-1}$, $\hat{\eta}_3 \mapsto \hat{\xi}_{N-2}$, and $\hat{\eta}_4 \mapsto \hat{\xi}_{N-3}$.

We are going to relate the ground state of the Majorana 4 chain to that of a quantum spin-1 chain christened after AKLT, who introduced this Hamiltonian.

First, to each ellipse shown in Figure 5.8(b), we associate the three-dimensional Hilbert space spanned by the orthonormal basis

$$|1_j, +1_j\rangle \equiv |\uparrow_{A,j}\rangle \otimes |\uparrow_{B,j}\rangle, \tag{5.211a}$$

$$|1_j, 0_j\rangle \equiv \frac{1}{\sqrt{2}}\left(|\uparrow_{A,j}\rangle \otimes |\downarrow_{B,j}\rangle + |\downarrow_{A,j}\rangle \otimes |\uparrow_{B,j}\rangle\right), \tag{5.211b}$$

$$|1_j, -1_j\rangle \equiv |\downarrow_{A,j}\rangle \otimes |\downarrow_{B,j}\rangle, \tag{5.211c}$$

where $|\sigma_{S,j}\rangle$ with $S = A, B$, $j = 1, \cdots, N/4$, and $\sigma_{S,j} = \uparrow, \downarrow$ is the eigenvalue of a spin-1/2 operator on sublattice S of the repeat unit cell j along the quantization axis. This spin-1/2 is depicted by a Bloch sphere in Figure 5.8(b). We may rewrite Eq. (5.211) in the tensor form

$$|1_j, m_j\rangle = \mathsf{T}^{(1_j,m_j)}_{\sigma_{A,j}\,\sigma_{B,j}}\,|\sigma_{A,j}, \sigma_{B,j}\rangle, \qquad m_j = -1, 0, +1, \tag{5.212a}$$

where we have introduced the triplet of tensors

$$\mathsf{T}^{(1,-1)}_{\sigma\sigma'} := \delta_{\sigma,\downarrow}\,\delta_{\sigma',\downarrow}, \tag{5.212b}$$

$$\mathsf{T}^{(1,0)}_{\sigma\sigma'} := \frac{1}{\sqrt{2}}\left(\delta_{\sigma,\uparrow}\,\delta_{\sigma',\downarrow} + \delta_{\sigma,\downarrow}\,\delta_{\sigma',\uparrow}\right), \tag{5.212c}$$

$$\mathsf{T}^{(1,+1)}_{\sigma\sigma'} := \delta_{\sigma,\uparrow}\,\delta_{\sigma',\uparrow}, \tag{5.212d}$$

respectively, and the summation convention over repeated indices is implied on the right-hand side of Eq. (5.212a). We shall also use the notation

$$\begin{aligned}|0_j, m_j\rangle &\equiv \frac{1}{\sqrt{2}}\left(|\uparrow_{A,j}\rangle \otimes |\downarrow_{B,j}\rangle - |\downarrow_{A,j}\rangle \otimes |\uparrow_{B,j}\rangle\right)\\ &= \mathsf{T}^{(0_j,0_j)}_{\sigma_{A,j}\,\sigma_{B,j}}\,|\sigma_{A,j}, \sigma_{B,j}\rangle, \qquad m_j = 0,\end{aligned} \tag{5.213a}$$

where the summation convention over repeated indices is implied on the right-hand side of Eq. (5.213a) and

$$\mathsf{T}^{(0,0)}_{\sigma\sigma'} := \frac{1}{\sqrt{2}}\left(\delta_{\sigma,\uparrow}\,\delta_{\sigma',\downarrow} - \delta_{\sigma,\downarrow}\,\delta_{\sigma',\uparrow}\right), \tag{5.213b}$$

for the orthogonal complement to the triplet basis (5.211) in the four-dimensional Hilbert space spanned by the basis

$$|\sigma_{A,j}, \sigma_{B,j}\rangle \equiv |\sigma_{A,j}\rangle \otimes |\sigma_{B,j}\rangle, \qquad \sigma_{A,j}, \sigma_{B,j} = \uparrow, \downarrow. \tag{5.214}$$

Second, for any choice of $\sigma_{A,1}, \sigma_{B,N/4} = \uparrow, \downarrow$, we define the product state

$$\begin{aligned}&\left|\sigma_{A,1}, \sigma_{B,N/4}\right\rangle := \\ &|\sigma_{A,1}\rangle \left[\bigotimes_{j=1}^{(N-4)/4} \frac{1}{\sqrt{2}}\left(|\uparrow_{B,j}\rangle \otimes |\downarrow_{A,j+1}\rangle - |\downarrow_{B,j}\rangle \otimes |\uparrow_{A,j+1}\rangle\right)\right] \otimes |\sigma_{B,N/4}\rangle \equiv \\ &\mathsf{S}_{\sigma_{B,1}\sigma_{A,2}} \cdots \mathsf{S}_{\sigma_{B,(N/4)-1}\sigma_{A,N/4}} \left|\sigma_{A,1}\,\sigma_{B,1}; \cdots; \sigma_{A,N/4}\,\sigma_{B,N/4}\right\rangle,\end{aligned} \tag{5.215a}$$

where we reserve the tensor S with the components

$$\mathsf{S}_{\sigma\sigma'} = \frac{1}{\sqrt{2}}\left(\delta_{\sigma,\uparrow}\,\delta_{\sigma',\downarrow} - \delta_{\sigma,\downarrow}\,\delta_{\sigma',\uparrow}\right), \tag{5.215b}$$

for the formation of a spin singlet across two consecutive repeat unit cell j and $j+1$ [we reserve the notation $\mathsf{T}^{(0,0)}_{\sigma\sigma'}$ as defined in Eq. (5.213b) for the formation of a spin singlet within a repeat unit cell j] and the summation convention over the repeated indices

$$\sigma_{B,1},\sigma_{A,2},\sigma_{B,2},\cdots,\sigma_{A,(N/4)-1},\sigma_{B,(N/4)-1},\sigma_{A,N/4}=\uparrow,\downarrow \tag{5.215c}$$

is implied on the right-hand side of the last equality in Eq. (5.215a). This product state is depicted in Figure 5.8(b). If we impose periodic boundary conditions, we obtain the nondegenerate ground state

$$\begin{aligned}|\mathrm{Dimer}\rangle \equiv\, & \mathsf{S}_{\sigma_{B,1}\sigma_{A,2}}\cdots\mathsf{S}_{\sigma_{B,(N/4)-1}\sigma_{A,N/4}}\,\mathsf{S}_{\sigma_{B,N/4}\sigma_{A,1}}\\ & \times\left|\sigma_{A,1}\,\sigma_{B,1};\cdots;\sigma_{A,(N/4)-1}\,\sigma_{B,(N/4)-1};\sigma_{A,N/4}\,\sigma_{B,N/4}\right\rangle\end{aligned} \tag{5.216a}$$

with the summation convention over the repeated indices

$$\sigma_{A,1},\sigma_{B,1},\sigma_{A,2},\sigma_{B,2},\cdots,\sigma_{A,(N/4)-1},\sigma_{B,(N/4)-1},\sigma_{A,N/4},\sigma_{B,N/4}=\uparrow,\downarrow, \tag{5.216b}$$

implied on the right-hand side of Eq. (5.216a).

Third, the AKLT state obeying periodic boundary conditions is obtained from the ground state (5.216) after projecting the local four-dimensional Hilbert spaces

$$\frac{\mathbf{1}}{\mathbf{2}}_{A,j}\otimes\frac{\mathbf{1}}{\mathbf{2}}_{B,j}:=\mathrm{span}\left\{\left|\sigma_{A,j},\sigma_{B,j}\right\rangle\right\}\equiv\mathbf{0}_j\oplus\mathbf{1}_j \tag{5.217a}$$

onto the three-dimensional Hilbert space

$$\mathbf{1}_j:=\mathrm{span}\left\{\left|1_j,m_j\right\rangle,\, m_j=-1,0,+1\right\} \tag{5.217b}$$

that is orthogonal to the singlet Hilbert space

$$\mathbf{0}_j:=\mathrm{span}\left\{\left|0_j,0_j\right\rangle\right\}. \tag{5.217c}$$

To this end, we use the identity

$$\begin{aligned}\left|\sigma_{A,j},\sigma_{B,j}\right\rangle &=\left(\left|0_j,0_j\right\rangle\left\langle 0_j,0_j\right|+\sum_{m_j=-1,0,+1}\left|1_j,m_j\right\rangle\left\langle 1_j,m_j\right|\right)\left|\sigma_{A,j},\sigma_{B,j}\right\rangle\\ &\overset{\text{Eqs. (5.212)-(5.213)}}{=}\mathsf{T}^{(0_j,0_j)}_{\sigma_{A,j}\,\sigma_{B,j}}\left|0_j,0_j\right\rangle+\sum_{m_j=-1,0,+1}\mathsf{T}^{(1_j,m_j)}_{\sigma_{A,j}\,\sigma_{B,j}}\left|1_j,m_j\right\rangle\end{aligned} \tag{5.218}$$

in terms of which we can rewrite the ground state (5.216) as

$$\begin{aligned}|\mathrm{Dimer}\rangle =& \sum_{J_1=0,1}\;\sum_{-J_1\le m_1\le +J_1}\cdots\sum_{J_{N/4}=0,1}\;\sum_{-J_{N/4}\le m_{N/4}\le +J_{N/4}}\\ &\times\mathrm{Tr}\left(\mathsf{T}^{(J_1,m_1)}\,\mathsf{S}\cdots\mathsf{T}^{(J_{N/4},m_{N/4})}\,\mathsf{S}\right)\left|J_1,m_1;\cdots;J_{N/4},m_{N/4}\right\rangle.\end{aligned} \tag{5.219}$$

The AKLT product state follows from the dimer state (5.219) by doing the substitution

$$\mathsf{T}^{(0_j,m_j)}\to\lambda\,\mathsf{T}^{(0_j,m_j)},\qquad m_j=0,\qquad j=1,\cdots,N/4, \tag{5.220}$$

with $0 \leq \lambda \leq 1$ and taking the limit $\lambda \to 0$. The AKLT product state for a ring (that is with periodic boundary conditions) is thus defined to be

$$\begin{aligned}|\Psi^{\mathrm{AKLT}}_{\mathrm{PBC}}\rangle := &\sum_{m_1=-1,0,+1} \cdots \sum_{m_{N/4}=-1,0,+1} \mathrm{Tr}\left(\mathsf{T}^{(1_1,m_1)}\,\mathsf{S}\cdots\mathsf{T}^{(1_{N/4},m_{N/4})}\,\mathsf{S}\right)\\ &\times|1_1,m_1;\cdots;1_{N/4},m_{N/4}\rangle.\end{aligned} \tag{5.221}$$

The AKLT product state (5.221) is an example of a matrix product state. (Matrix product states are themselves examples of tensor product states.)

We seek a Hamiltonian for which the AKLT product state (5.221) is the nondegenerate ground state. Affleck, Kennedy, Lieb, and Tasaki achieved this goal by defining the Hamiltonian

$$\widehat{H}^{\mathrm{AKLT}}_{b} := \sum_{j=1}^{(N/4)-b} P^{(2)}_{j,j+1}, \tag{5.222a}$$

where $b = 0, 1$ selects between periodic ($b = 0$) and open ($b = 1$) boundary conditions and $P^{(2)}_{j,j+1}$ is the projector onto the spin-2 subspace $\mathbf{2}_{j,j+1}$ in the irreducible decomposition

$$\mathbf{1}_j \otimes \mathbf{1}_{j+1} = \mathbf{0}_{j,j+1} \oplus \mathbf{1}_{j,j+1} \oplus \mathbf{2}_{j,j+1} \tag{5.222b}$$

of the tensor product of two spin-1 Hilbert spaces, one tensor product for each pair of neighboring ellipses in Figure 5.8(b). Because

$$\left(P^{(2)}_{j,j+1}\right)^2 = P^{(2)}_{j,j+1} = \left(P^{(2)}_{j,j+1}\right)^{\dagger}, \tag{5.223}$$

the eigenvalues of $P^{(2)}_{j,j+1}$ are nonnegative. Hence, any state annihilated by $\widehat{H}^{\mathrm{AKLT}}_{b}$ is a ground state.

Affleck, Kennedy, Lieb, and Tasaki showed that the ground states of $\widehat{H}^{\mathrm{AKLT}}_{b}$ are nondegenerate and separated from all excitations by a gap when periodic boundary conditions are selected by choosing $b = 0$. By design, the AKLT state defined by Eq. (5.221) is annihilated by $\widehat{H}^{\mathrm{AKLT}}_{b=0}$ and as such is thus the nondegenerate ground state. When open boundary conditions are selected by choosing $b = 1$, each end of the chain hosts one free spin-1/2. This claim is made plausible from inspection of Eq. (5.215a). The ground states are then fourfold degenerate.

An explicit representation of $P^{(2)}_{j,j+1}$ in terms of spin-1 operators is the polynomial (**Exercise** 5.4)

$$P^{(2)}_{j,j+1} = \frac{1}{3} + \frac{1}{2}\left(\widehat{\boldsymbol{S}}_j \cdot \widehat{\boldsymbol{S}}_{j+1}\right) + \frac{1}{6}\left(\widehat{\boldsymbol{S}}_j \cdot \widehat{\boldsymbol{S}}_{j+1}\right)^2, \qquad \widehat{\boldsymbol{S}}^2_j = 2\ (\hbar = 1). \tag{5.224}$$

What makes the AKLT Hamiltonian (5.222) remarkable is that it can be continuously deformed into the nearest-neighbor antiferromagnetic quantum spin-1 Heisenberg Hamiltonian

$$\widehat{H}_b^{\mathrm{AF}} := \sum_{j=1}^{(N/4)-b} \widehat{\boldsymbol{S}}_j \cdot \widehat{\boldsymbol{S}}_{j+1}, \tag{5.225}$$

where $b = 0, 1$, without closing the gap separating the ground state from all excited states when periodic boundary conditions ($b = 0$) are selected.[21] Hence, $\widehat{H}_1^{\mathrm{AF}}$ must support one free spin-1/2 mid-gap state at either ends of the chain. A signature consistent with these spin-1/2 mid-gap states has been observed in low-temperature electron-spin-resonance (ESR) spectra on $\mathrm{Ni(C_2H_8N_2)_2NO_2ClO_4}$ (NENP) containing selected impurities.[22] The substance NENP is a model material that realizes weakly coupled nearest-neighbor antiferromagnetic quantum spin-1 chains through the Ni^{2+} ions, in spite of the presence of substantial uniaxial anisotropy.[23] The construction of two spin-1/2 mid-gap end states supported by Hamiltonian $\widehat{H}_1^{\mathrm{AF}}$ is done in **Exercise** 5.5.

5.6 Quantum Phase Transitions between Two Majorana c Chains

5.6.1 The Case of the Majorana $c = 0$ and $c = 1$ Chains

Let $N \geq 2$ be an even integer, let $b = 0, 1$ select between periodic and open boundary conditions, respectively, let $0 \leq \lambda \leq 1$, and define the Hamiltonian

$$\widehat{H}_b(\lambda) := +\frac{\hbar\omega}{2}\left[\sum_{j=1}^{N}(1-\lambda)\,\mathrm{i}\hat{\xi}_j\,\hat{\eta}_j - \sum_{j=1}^{N-b}\lambda\,\mathrm{i}\hat{\xi}_j\,\hat{\eta}_{j+1}\right] \tag{5.226}$$

that interpolates between the Majorana $c = 0$ Hamiltonian (5.174) when $\lambda = 0$ and the Majorana $c = 1$ Hamiltonian (5.175) when $\lambda = 1$. The Majorana algebra was defined in Eq. (5.168c). We anticipate at least one quantum phase transition for some value $0 < \lambda_{\mathrm{c}} < 1$, since the limit $\lambda = 1$ supports a pair of Majorana boundary zero modes when $b = 1$, while the limit $\lambda = 0$ does not support boundary states when $b = 1$.

That this educated guess is plausible follows from taking the continuum limit of the equations of motion

$$\mathrm{i}\hbar\,\partial_t\,\hat{\eta}_j(t) := \left[\hat{\eta}_j(t), \widehat{H}_0(\lambda)\right] = -\hbar\omega\left[(1-\lambda)\,\mathrm{i}\hat{\xi}_j(t) - \lambda\,\mathrm{i}\hat{\xi}_{j-1}(t)\right], \tag{5.227a}$$

$$\mathrm{i}\hbar\,\partial_t\,\hat{\xi}_j(t) := \left[\hat{\xi}_j(t), \widehat{H}_0(\lambda)\right] = +\hbar\omega\left[(1-\lambda)\,\mathrm{i}\hat{\eta}_j(t) - \lambda\,\mathrm{i}\hat{\eta}_{j+1}(t)\right], \tag{5.227b}$$

with $j = 1, \cdots, N$, as periodic boundary conditions ($b = 0$) have been imposed.

We denote with $\mathfrak{a}$ the lattice spacing. We set

$$v_{\mathrm{F}}(\lambda) := \omega\,\mathfrak{a}\,\lambda, \qquad m(\lambda)\,v_{\mathrm{F}}^2(\lambda) := \hbar\omega\,(1-2\lambda). \tag{5.228}$$

[21] T. Kennedy, J. Phys. Condens. Matter **2**(26), 5737–5745 (1990). [Kennedy (1990)].

[22] M. Hagiwara, K. Katsumata, I. Affleck, B. I. Halperin, and J. P. Renard, Phys. Rev. Lett. **65**(25), 3181–3184 (1990). [Hagiwara et al. (1990)].

[23] J. P. Renard, M. Verdaguer, L. P. Regnault, et al., Europhys. Lett. **3**(8), 945–952 (1987). [Renard et al. (1987)].

The Fermi velocity $v_{\mathrm{F}}(\lambda)$ is a linearly increasing function of $0 \le \lambda \le 1$. The "relativistic rest energy" $m(\lambda)\, v_{\mathrm{F}}^{2}(\lambda)$ is a linearly decreasing function of $0 \le \lambda \le 1$ that is maximal when $\lambda = 0$, vanishes when $\lambda = 1/2$, and becomes negative for $1/2 < \lambda \le 1$. We define the quantum fields

$$\widehat{\chi}(x,t) := \begin{pmatrix} \widehat{\chi}_1(x,t) \\ \widehat{\chi}_2(x,t) \end{pmatrix}, \qquad \widehat{\chi}_1(x,t) \sim \frac{\hat{\eta}_j(t)}{\sqrt{2\mathfrak{a}}}, \qquad \widehat{\chi}_2(x,t) \sim \frac{\hat{\xi}_j(t)}{\sqrt{2\mathfrak{a}}}. \tag{5.229}$$

Linearization of the equations of motion (5.227) delivers the Dirac equation

$$\sigma_0\, \mathrm{i}\hbar\, \partial_t\, \widehat{\chi}(x,t) = \left[-\sigma_1\, v_{\mathrm{F}}(\lambda)\, \mathrm{i}\hbar\, \partial_x + \sigma_2\, m(\lambda)\, v_{\mathrm{F}}^{2}(\lambda)\right] \widehat{\chi}(x,t), \tag{5.230a}$$

where σ_0 denotes the 2×2 unit matrix, $\boldsymbol{\sigma}$ denotes the vector of Pauli matrices, the reality condition

$$\widehat{\chi}^{\dagger}(x,t) = \widehat{\chi}^{\mathsf{T}}(x,t), \tag{5.230b}$$

holds, and the equal-time algebra of the quantum fields is given by the Majorana algebra

$$\{\widehat{\chi}_a(x,t), \widehat{\chi}_{a'}(x',t)\} = \delta_{a,a'}\, \delta(x-x'), \qquad a,a' = 1,2. \tag{5.230c}$$

The mode expansion of Eq. (5.230) is

$$\widehat{\chi}\,(x,t) = \int \frac{\mathrm{d}p}{\sqrt{2\pi}}\, e^{\mathrm{i}[p\,x - \varepsilon(p)t]/\hbar}\, \widehat{\chi}\,(p), \tag{5.231a}$$

$$\left[v_{\mathrm{F}}(\lambda)\, p\, \sigma_1 + m(\lambda)\, v_{\mathrm{F}}^{2}(\lambda)\, \sigma_2\right] \widehat{\chi}\,(p) = \varepsilon(p)\, \widehat{\chi}\,(p), \tag{5.231b}$$

$$\varepsilon(p) \equiv \hbar\omega(p) = \pm v_{\mathrm{F}}(\lambda)\, \sqrt{p^2 + m^2(\lambda)\, v_{\mathrm{F}}^{2}(\lambda)}, \tag{5.231c}$$

$$\widehat{\chi}^{\dagger}\,(p) = \widehat{\chi}^{\mathsf{T}}(-p). \tag{5.231d}$$

Observe that the single-particle excitation spectrum derived in Eq. (2.82b) for the pair of spin-resolved quantum fields $\hat{\chi}^{\dagger}_{\mathrm{F},\sigma}(p)$ and $\hat{\chi}_{\mathrm{F},\sigma}(p)$ takes the same relativistic form as that in Eq. (5.231c). The single-particle spectrum (5.231c) is thus gapfull except when $\lambda = 1/2$. The point $\lambda_{\mathrm{c}} = 1/2$ realizes a quantum critical point. This quantum critical point is, however, not the same as the one corresponding to $\delta t = 0$ (vanishing dimerization) in Eq. (2.82). This difference arises from the fact that the pair of quantum fields $\hat{\chi}^{\dagger}_{\mathrm{F},\sigma}(p)$ and $\hat{\chi}_{\mathrm{F},\sigma}(p)$ entering Eq. (2.82) are not constrained by a reality condition such as (5.231d). The quantum critical point $\lambda_{\mathrm{c}} = 1/2$ realizes a conformal field theory in $(1+1)$-dimensional spacetime belonging to the Ising universality class, for it is equivalent to the critical point of the classical ferromagnetic Ising model on a two-dimensional lattice at its critical temperature, as was shown by Schultz, Mattis, and Lieb.[24] It differs from the quantum critical point corresponding to $\delta t = 0$ in Eq. (2.82b) through the behavior of the specific heat[25]

$$\frac{C(T)}{L} = \frac{\pi\, \mathsf{c}\, k_{\mathrm{B}}^2\, T}{3\hbar\, v_{\mathrm{F}}} + \mathcal{O}(T^2), \tag{5.232}$$

as the temperature T approaches $T = 0$ in a chain of fixed linear length $L \equiv \mathfrak{a}\, N$. The linear dependence on T of the right-hand side is a consequence of the conformal symmetry at criticality. The number c is called the central charge. It is given by

$$\mathsf{c} = \begin{cases} \frac{1}{2}, & \text{when } \lambda = 1/2 \text{ in Eq. (5.231)}, \\ \\ 1, & \text{when } \delta t = 0 \text{ in Eq. (2.82) for a given spin } \sigma = \uparrow, \downarrow. \end{cases} \tag{5.233}$$

The halving of the central charge when passing from Eq. (2.82) resolved in spin with $\delta t = 0$ to Eq. (5.231) with $\lambda = 1/2$ can be computed as follows. The partition function at the inverse temperature β for the noninteracting many-body Hamiltonian density

$$\widehat{\mathcal{H}}(x) := \widehat{\Psi}^\dagger(x)\, \mathcal{H}_{\mathrm{D}}(x)\, \widehat{\Psi}(x), \qquad \mathcal{H}_{\mathrm{D}}(x) := -\sigma_1\, v_{\mathrm{F}}(\lambda)\, \mathrm{i}\hbar\, \partial_x + \sigma_2\, m(\lambda)\, v_{\mathrm{F}}^2(\lambda), \tag{5.234}$$

where all the operator-valued components of the Dirac spinor $\widehat{\Psi}(x)$ are independent and obey the fermionic anticommuting algebra, is nothing but the thermal regularization of the functional determinant

$$\operatorname{Det} \mathcal{H}_{\mathrm{D}} = e^{\log \operatorname{Det} \mathcal{H}_{\mathrm{D}}} \tag{5.235a}$$

of the Dirac kernel $\mathcal{H}_{\mathrm{D}}$. If we do the substitutions

$$\widehat{\Psi}^\dagger(x) \to \widehat{\chi}^{\mathsf{T}}(x), \qquad \widehat{\Psi}(x) \to \widehat{\chi}(x), \tag{5.235b}$$

so that the reality (Majorana) condition (5.231d) holds, together with the global single-particle spectral rescaling[26]

$$\mathcal{H}_{\mathrm{D}}(x) \to \frac{1}{2} \mathcal{H}_{\mathrm{D}}(x), \tag{5.235c}$$

the resulting partition function is then the thermal regularization of

$$\sqrt{\operatorname{Det} \mathcal{H}_{\mathrm{D}}} = e^{\frac{1}{2} \log \operatorname{Det} \mathcal{H}_{\mathrm{D}}}. \tag{5.235d}$$

It is this square root that is responsible for $\mathsf{c} = 1 \to \mathsf{c} = 1/2$ in Eq. (5.233) upon imposing the reality condition (**Exercises** 5.6 and 5.7).

Alternatively, the halving of the central charge in Eq. (5.233) reflects the fact that the ground-state degeneracy of the open Majorana $c = 1$ chain is half that of the

24 T. D. Schultz, D. C. Mattis, and E. H. Lieb, Rev. Mod. Phys. **36**(3), 856–871 (1964). [Schultz et al. (1964)].

25 I. Affleck, Phys. Rev. Lett. **56**(7), 746–748 (1986); H. W. J. Blöte, John L. Cardy, and M. P. Nightingale, Phys. Rev. Lett. **56**(7), 742–745 (1986). [Affleck (1986); Blöte et al. (1986)].

26 The necessity to rescale the single-particle Dirac Hamiltonian is explained in **Exercise** 5.6.

open Majorana $c = 2$ chain, the latter chain being equivalent to the spinless SSH model as was shown in Section 5.5.3. This interpretation presumes that the quantum critical modes at $T = 0$ are related to the symmetry-protected boundary modes as follows. Indeed, the decay length $\xi \sim 1/m$ of the symmetry-protected boundary modes that is inherited from the single-particle gap $\sim m$ in the bulk diverges by tuning λ or δt to their critical values $\lambda = 1/2$ and $\delta t = 0$, respectively, and is also the bulk diverging length scale that controls the approach to the quantum critical point. Upon approaching the quantum critical points using open boundary conditions, the distinction between boundary and bulk modes disappears as the hybridization of the boundary modes with the bulk modes increases with increasing ξ. The quantum mechanics of boundary modes confined to a box of size ξ is turned into a quantum field theory in $(1+1)$-dimensional spacetime for extended and gapless modes in the limit $\xi \to \infty$.

We close this discussion by observing that performing the nonlocal Jordan-Wigner transformation on Hamiltonian (5.226) gives the Ising chain in a transverse magnetic field,

$$\widehat{H}_b(\lambda) = -\frac{\hbar\omega}{2}\left[\sum_{j=1}^{N}(1-\lambda)\,\widehat{X}_j + \sum_{j=1}^{N-b}\lambda\,\widehat{Z}_j\,\widehat{Z}_{j+1}\right]. \tag{5.236}$$

We have thus shown that the Ising chain in a transverse magnetic field supports two gapped ground states: the paramagnetic ground state when $0 \le \lambda < 1/2$, and the Ising long-range-ordered ground state when $1/2 < \lambda \le 1$. The quantum transition at $\lambda_{\mathrm{c}} = 1/2$ between these two phases is continuous and belongs to the Ising universality class.

5.6.2 The Case of the Majorana $c \ge 0$ and $c' \ge 0$ Chains

For c (c') any positive integer let $b_{c} = 0, c$ ($b_{c'} = 0, c'$). Let $N \ge \max\{c, c'\} + 1$ be an even integer. Select periodic ($b_{c} = b_{c'} = 0$) or open boundary conditions ($b_{c} = c, b_{c'} = c'$), respectively. Let $0 \le \lambda \le 1$ and define the Hamiltonian

$$\widehat{H}_{b_{c},b_{c'}}(\lambda) := -\frac{\hbar\omega}{2}\left[\sum_{j=1}^{N-b_{c}}(1-\lambda)\,\mathrm{i}\hat{\xi}_j\,\hat{\eta}_{j+c} - \sum_{j=1}^{N-b_{c'}}\lambda\,\mathrm{i}\hat{\xi}_j\,\hat{\eta}_{j+c'}\right] \tag{5.237}$$

that interpolates between a Majorana c chain when $\lambda = 0$ and a Majorana c' chain when $\lambda = 1$. We are going to show that the spectrum for periodic boundary conditions is gapless for $\lambda = 1/2$ and gapfull otherwise. We will also show that the quantum critical point

$$\lambda = 1/2 \equiv \lambda_{\mathrm{c}} \tag{5.238}$$

realizes the conformal field theory with the central charge

$$\mathsf{c} = \frac{|c - c'|}{2}, \tag{5.239}$$

consisting of $|c-c'|$ independent copies of the Ising conformal field theory in (1+1)-dimensional spacetime.

Proof *Step 1:* Choose periodic boundary conditions, that is, $b_c = b_{c'} = 0$. Define the Brillouin zone to be the set

$$\Lambda^{\star}_{\mathrm{BZ}} := \left\{ k = \frac{2\pi\, n}{N} - \pi \,\middle|\, n = 1, \cdots, N \right\}. \tag{5.240a}$$

Do the mode expansion

$$\hat{\eta}_j =: \frac{1}{\sqrt{N}} \sum_{k\in\Lambda^{\star}_{\mathrm{BZ}}} \hat{\eta}_k\, e^{\mathrm{i}k\,j}, \qquad \hat{\eta}^{\dagger}_k = \hat{\eta}_{-k}, \tag{5.240b}$$

$$\hat{\xi}_j =: \frac{1}{\sqrt{N}} \sum_{k\in\Lambda^{\star}_{\mathrm{BZ}}} \hat{\xi}_k\, e^{\mathrm{i}k\,j}, \qquad \hat{\xi}^{\dagger}_k = \hat{\xi}_{-k}, \tag{5.240c}$$

with the only nonvanishing anticommutators

$$\left\{\hat{\eta}_k, \hat{\eta}^{\dagger}_{k'}\right\} = 2\,\delta_{k,-k'}, \qquad \left\{\hat{\xi}_k, \hat{\xi}^{\dagger}_{k'}\right\} = 2\,\delta_{k,-k'}. \tag{5.240d}$$

In terms of these Fourier modes, we have found that

$$\widehat{H}_{b_c=0,b_{c'}=0}(\lambda) = -\frac{\hbar\omega}{2} \sum_{k\in\Lambda^{\star}_{\mathrm{BZ}}} \left[\mathrm{i}(1-\lambda)\, e^{-\mathrm{i}k\,c} - \mathrm{i}\lambda\, e^{-\mathrm{i}k\,c'}\right] \hat{\xi}_{+k}\, \hat{\eta}_{-k}. \tag{5.241}$$

The equations of motions for $\hat{\eta}_{+k} = \hat{\eta}^{\dagger}_{-k}$ and $\hat{\xi}_{+k} = \hat{\xi}^{\dagger}_{-k}$ are

$$\mathrm{i}\hbar\,\partial_t\, \hat{\eta}_{+k} := \left[\hat{\eta}_{+k}, \widehat{H}_{0,0}(\lambda)\right] = +\hbar\omega \left[\mathrm{i}(1-\lambda)\, e^{-\mathrm{i}k\,c} - \mathrm{i}\lambda\, e^{-\mathrm{i}k\,c'}\right] \hat{\xi}_{+k}, \tag{5.242a}$$

$$\mathrm{i}\hbar\,\partial_t\, \hat{\xi}_{-k} := \left[\hat{\xi}_{-k}, \widehat{H}_{0,0}(\lambda)\right] = -\hbar\omega \left[\mathrm{i}(1-\lambda)\, e^{+\mathrm{i}k\,c} - \mathrm{i}\lambda\, e^{+\mathrm{i}k\,c'}\right] \hat{\eta}_{+k}, \tag{5.242b}$$

for any $k \in \Lambda^{\star}_{\mathrm{BZ}}$.

Step 2: We split the summation over the wave vectors in the Brillouin zone into the strictly negative wave vectors, the vanishing wave vector, and the strictly positive wave vectors. Up to an error of order $1/N$, we may then write

$$\widehat{H}_{b_c=0,b_{c'}=0}(\lambda) = -\frac{\hbar\omega}{2} \sum_{k\in\Lambda^{\star}_{\mathrm{BZ}}}^{0<k\leq\pi} \begin{pmatrix} \hat{\eta}_{+k} & \hat{\xi}_{+k} \end{pmatrix} \begin{pmatrix} 0 & f_k(\lambda) \\ f^{*}_k(\lambda) & 0 \end{pmatrix} \begin{pmatrix} \hat{\eta}_{-k} \\ \hat{\xi}_{-k} \end{pmatrix} + \mathcal{O}\left(\frac{1}{N}\right), \tag{5.243a}$$

where

$$\begin{aligned} f_k(\lambda) &:= -\mathrm{i}\,(1-\lambda)\; e^{+\mathrm{i}k\,c} + \mathrm{i}\lambda\, e^{+\mathrm{i}k\,c'}, \\ \operatorname{Re} f_k(\lambda) &= (1-\lambda)\, \sin(k\,c) - \lambda \sin(k\,c'), \\ \operatorname{Im} f_k(\lambda) &= -\left[(1-\lambda)\, \cos(k\,c) - \lambda \cos(k\,c')\right]. \end{aligned} \tag{5.243b}$$

Step 3: Inspection of the complex-valued function

$$f(k,\lambda) \equiv f_k(\lambda) = -\mathrm{i}e^{+\mathrm{i}k\,c}\left[(1-\lambda) - \lambda\, e^{+\mathrm{i}k\,(c'-c)}\right] \tag{5.244}$$

of k over the reduced Brillouin zone $[0,\pi]$, holding the coupling $0 \leq \lambda \leq 1$ fixed, immediately tells us that it may vanish if and only if $\lambda = 1/2$, the only value of $0 \leq \lambda \leq 1$ for which $1-\lambda$ and λ are equal in magnitudes. Hence, the spectrum of Hamiltonian (5.243) is necessarily gapped over the entire reduced Brillouin zone $[0,\pi]$ given any $0 \leq \lambda \neq 1/2 \leq 1$.

Step 4: Define the Majorana spinor

$$\widehat{\Xi}^{\dagger}_{k} := \begin{pmatrix} \hat{\eta}_{+k} & \hat{\xi}_{+k} \end{pmatrix} \iff \widehat{\Xi}_{k} := \begin{pmatrix} \hat{\eta}_{-k} \\ \hat{\xi}_{-k} \end{pmatrix}, \tag{5.245a}$$

together with the unit 2×2 matrix σ_0 and the vector $\boldsymbol{\sigma}$ of Pauli matrices. We may then write

$$\begin{aligned} \widehat{H}_{b_c=0,b_{c'}=0}(\lambda) = &-\frac{\hbar\omega}{2}\sum_{k\in\Lambda^{\star}_{\mathrm{BZ}}}^{0<k\leq\pi} \widehat{\Xi}^{\dagger}_{k}\Big\{\left[(1-\lambda)\cos(k\,c) - \lambda\cos(k\,c')\right]\sigma_2 \\ &+ \left[(1-\lambda)\sin(k\,c) - \lambda\sin(k\,c')\right]\sigma_1\Big\}\widehat{\Xi}_{k} + \mathcal{O}\left(\frac{1}{N}\right). \end{aligned} \tag{5.245b}$$

The dispersion of Hamiltonian (5.245b) over the folded Brillouin zone $k\in[0,\pi]$

$$\varepsilon_{k,\pm}(\lambda) = \pm\frac{\hbar\omega}{2}\sqrt{[\mathrm{Im}\, f_k(\lambda)]^2 + [\mathrm{Re}\, f_k(\lambda)]^2} \tag{5.246}$$

is gapped for any $\lambda \neq 1/2$ in the interval $[0,1]$ according to *Step 3.*

Step 5: Observe that

$$\begin{aligned} \widehat{H}_{b_c=0,b_{c'}=0}(1/2) = &-\frac{\hbar\omega}{2}\sum_{k\in\Lambda^{\star}_{\mathrm{BZ}}}^{0<k\leq\pi} \widehat{\Xi}^{\dagger}_{k}\left[-\sin\left(\frac{k\,(c+c')}{2}\right)\sin\left(\frac{k\,(c-c')}{2}\right)\sigma_2\right. \\ &\left.+\cos\left(\frac{k\,(c+c')}{2}\right)\sin\left(\frac{k\,(c-c')}{2}\right)\sigma_1\right]\widehat{\Xi}_{k} + \mathcal{O}\left(\frac{1}{N}\right) \end{aligned} \tag{5.247}$$

has the single-particle eigenenergies

$$\varepsilon_{k,\pm}(1/2) = \pm\frac{\hbar\omega}{2}\left|\sin\left(\frac{k\,|c-c'|}{2}\right)\right|. \tag{5.248}$$

Linearization of Hamiltonian (5.247) around any one of the roots

$$\varepsilon_{k_n,\pm}(1/2) = \pm\frac{\hbar\omega}{2}\left|\sin\left(\frac{k_n\,|c-c'|}{2}\right)\right| = 0 \iff k_n = \frac{2\,\pi\,n}{|c-c'|}, \tag{5.249}$$

with n a nonnegative integer such that k_n belongs to the folded Brillouin zone $[0,\pi]$, delivers $|c-c'|$ independent Majorana fields, each of which obeys the Dirac equation (5.230).

When $|c-c'|=1$, the upper and lower branches of the single-particle dispersion (5.248) touch at the boundary $k=0$ of the folded Brillouin zone $[0,\pi]$, as is shown in Figure 5.9(a). The dispersion is linear in the vicinity of $k=0$ and is made of two Majorana branches with opposite group velocities. We call this pair of Majorana branches a Majorana half-cone.

When $|c-c'|=2$, the upper and lower branches of the single-particle dispersion (5.248) touch at the boundaries $k=0$ and $k=\pi$ of the folded Brillouin zone $[0,\pi]$, as is shown in Figure 5.9(b). The dispersion is linear in the vicinity of any one of $k=0$ or $k=\pi$ where it is made of two Majorana branches. There are thus two Majorana half-cones in the reduced Brillouin zone.

When $|c-c'|=3$, the upper and lower branches of the single-particle dispersion (5.248) touch at $k=0$ and at $k=2\pi/3$ in the folded Brillouin zone $[0,\pi]$, as is shown in Figure 5.9(c). The dispersion is linear in the vicinity of any one of $k=0$ or $k=2\pi/3$. It is comprised of one Majorana half-cone at $k=0$ and two Majorana half-cones at $k=2\pi/3$. The pair of Majorana half-cones at $k=2\pi/3$ make up a Dirac cone. There are thus three Majorana half-cones in the reduced Brillouin zone.

When $|c-c'|=4$, the upper and lower branches of the single-particle dispersion (5.248) touch at $k=0$, $k=\pi/2$, and $k=\pi$ in the folded Brillouin zone $[0,\pi]$, as is shown in Figure 5.9(d). The dispersion is linear in the vicinity of any one of $k=0$, $k=\pi/2$, and $k=\pi$. There are thus four Majorana half-cones in the reduced Brillouin zone.

In general, if we associate the central charge 1/2 to each Majorana half-cone, the central charge is $|c-c'|/2$ when $\lambda=1/2$. □

5.6.3 Application

Consider a linear combination with real-valued coefficients $\lambda_{c_1},\cdots,\lambda_{c_n}$ of Majorana c chains, say

$$\widehat{H}:=\sum_{c=0}^{n-1}\lambda_c\,\widehat{H}_{c,b_c=0}\equiv-\frac{\hbar\omega}{2}\sum_{c=0}^{n-1}\sum_{j=1}^{N}\lambda_c\,(-1)^{\delta_{c,0}}\mathrm{i}\hat{\xi}_j\,\hat{\eta}_{j+c}. \tag{5.250}$$

Because of the translation symmetry, this Hamiltonian can be diagonalized in the Brillouin zone and the free energy can be obtained in closed form. In practice, the phase diagram in the n-dimensional parameter space spanned by the real-valued couplings $\lambda_1,\cdots,\lambda_n$ is obtained for not too large n numerically. The universality class of any direct quantum phase transition between two fermionic invertible topological phases (FITPs) smoothly connected to the FITPs selected by λ_c and $\lambda_{c'}$ being much larger than any other couplings, respectively, has central charge $\mathsf{c}=|c-c'|/2$ provided the closing of the single-particle gap in the Brillouin zone occurs through a Dirac cone. However, the closing of the single-particle gap can be of higher than linear order in the momentum. This is known to happen at some

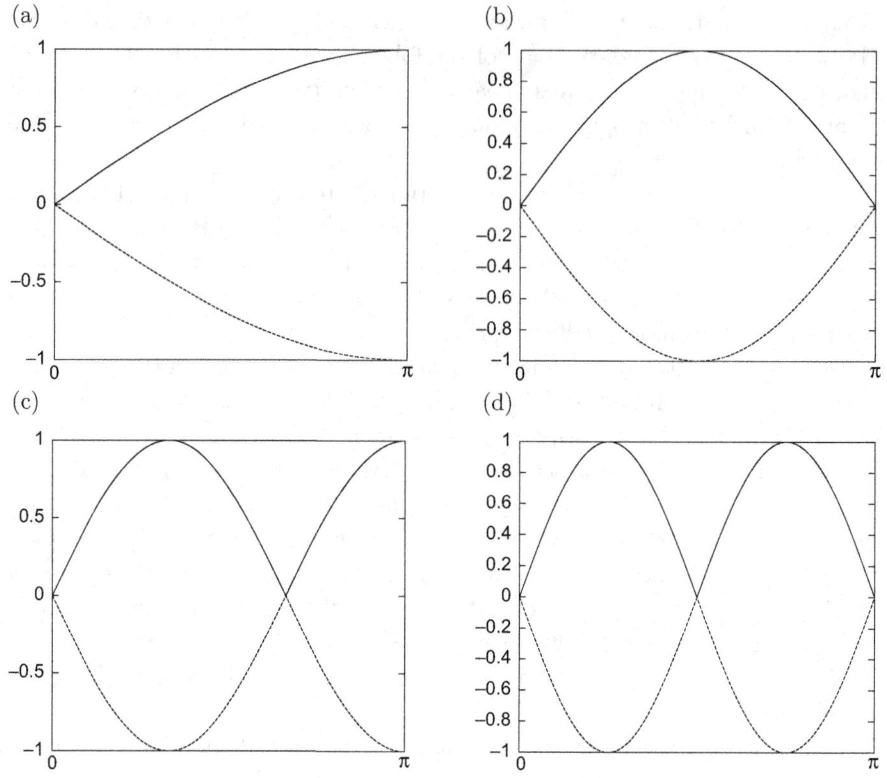

Figure 5.9 Upper (full line) and lower (dashed line) branches of the single-particle dispersion relation (5.248) for $|c-c'|=1,2,3,4$ in panels (a), (b), (c), and (d), respectively. The energy is measured in units of $\hbar\,\omega/2$.

multicritical points, that is points in coupling space at which more than two FITPs meet.[27] With the help of the nonlocal Jordan-Wigner transformation, we infer that this phase diagram is shared by the linear combination of spin-1/2 cluster c models

$$\widehat{H} = -\frac{\hbar\omega}{2}\sum_{j=1}^{N}\left[-\lambda_0\,\widehat{X}_j + \lambda_1\,\widehat{Z}_j\,\widehat{Z}_{j+1} + \cdots + \lambda_{n-1}\,\widehat{Z}_j\,\widehat{X}_{j+1}\cdots\widehat{X}_{j+n-2}\,\widehat{Z}_{j+n-1}\right]. \tag{5.251}$$

Of course, the terminology for the different gapped phases of matter in the phase diagram must be changed since the Jordan-Wigner transformation is nonlocal. For example, the FITP selected by λ_1 being the largest coupling in the Majorana representation (5.250) becomes a long-ranged ordered phase with spontaneous symmetry breaking of an Ising symmetry in the spin-1/2 representation (5.251).

27 R. Verresen, R. Moessner, and F. Pollmann, Phys. Rev. B **96**(16), 165124 (2017). [Verresen et al. (2017)].

5.6.4 A Conjecture

Verresen, Moessner, and Pollmann [27] have made the following conjecture. Let $\mathsf{d}^2 \in \mathbb{Z}$ be the ground-state degeneracy of a one-dimensional FITP when open boundary conditions are selected. The number d is called the quantum dimension. It can be an irrational real-valued number. Suppose that it is possible to find a Hamiltonian that interpolates between this FITP and a trivial FITP and suppose that the two gapped phases are separated in parameter space by a quantum critical point with conformal symmetry. This conformal field theory has then the central charge

$$\mathsf{c} \geq \frac{\ln \mathsf{d}}{\ln 2}. \tag{5.252}$$

The lower bound to the central charge is here deduced from the contributions arising from the symmetry-protected topological gapless boundary modes merging with the bulk states at the transition. The use of the logarithm in this conjecture is motivated by the fact that the ground-state degeneracy obtained by "stacking" FITPs with quantum dimensions d_1, d_2, $\cdots$ from tensoring the corresponding Hilbert spaces is $\mathsf{d}^2_{\mathrm{sta}} = \prod_{i=1,2,\cdots} \mathsf{d}^2_i$. The quantum dimension of this stacking of FITPs is then $\mathsf{d}_{\mathrm{sta}} = \prod_{i=1,2,\cdots} \mathsf{d}_i$. Taking the logarithm of $\mathsf{d}_{\mathrm{sta}}$ ensures that the lower bound on $\mathsf{c}_{\mathrm{sta}}$ is consistent with the additive law $\mathsf{c}_{\mathrm{sta}} = \sum_{i=1,2,,\cdots} \mathsf{c}_i$, as is expected from "stacking" conformal field theories (that is tensoring the Hilbert spaces on which they are defined).

5.7 Symmetry Fractionalization of Majorana *c* Chains

5.7.1 Overview

Because Majorana c chains (spin-1/2 cluster c chains) are the sums over pairwise commuting local operators indexed by the lattice sites, they are examples of integrable models.[28] This integrability can be broken by considering linear combinations of Majorana c chains (spin-1/2 cluster c chains). Nevertheless, the jump in the ground-state degeneracy of a Majorana c chain that arises from changing the boundary conditions from periodic to open ones is robust to perturbing weakly a Majorana c chain with a Majorana $c' \neq c$ chain, as we saw in Section 5.6.

Another way to lift the integrability of the Majorana c chains is to consider local interactions as perturbations. However, we showed in Section 5.5.4 that the jump in the degeneracy of the ground state when changing from periodic to open boundary conditions is not always stable when adding many-body interactions to a Majorana c chain that preserve the symmetries (5.169) and (5.170). Counting the number of Majorana zero modes per boundary is thus not a suitable topological index if many-body interactions that preserve the symmetries (5.169) and (5.170) of any Majorana c chain are considered.

[28] For integrable models, the number of locally defined operators that commute with the Hamiltonian scales with the system size.

This observation raises the following question. Are there indices associated with the symmetries (5.169) and (5.170) of any Majorana c chain that are robust to interactions allowed by these symmetries?

A positive answer to this question is the main result of Section 5.7 in the form of **Table** 5.8. This positive answer leads to the concept of symmetry fractionalization in physics terminology. The precise mathematical content of symmetry fractionalization consists in enumerating the projective representations of the symmetries (5.169) and (5.170) on any one of the left or right boundaries of an open Majorana c chain. To this end, one uses the tools of group cohomology to classify the projective representations of the symmetries (5.169) and (5.170) on Clifford algebras spanned by c generators that are represented by matrices of dimensionality $D(c)$ with

$$D(c) := 2^{[1-(-1)^c]/2} \times 2^{\lfloor c/2 \rfloor}, \qquad c = 0, 1, 2, 3, \cdots . \tag{5.253}$$

Here, the lower-floor function $\lfloor x \rfloor$ returns the largest integer that does not exceed $0 \leq x < \infty$. When c is even, we already encountred the dimensionality $D(c) = 1 \times 2^{c/2}$ when studying the irreducible representation with the smallest dimensionality of the Clifford algebra generated by c generators in **Exercise** 3.5. The origin of the multiplicative factor of 2 in the dimensionality $D(c) = 2 \times 2^{\lfloor c/2 \rfloor}$ for odd c will be explained in what follows.

This classification for the Majorana c chains (symmetry class BDI) is done in Section 5.7 without explicit reference to group cohomology. The underlying notions of group cohomology will be introduced in Section 6.3.2 together with the concept of the central extension of the symmetry group G by the group $\mathbb{Z}_2^{\mathrm{F}}$, whereby the latter group encodes the conservation of the fermion parity.

5.7.2 Periodic Boundary Conditions Revisited

We return to the Majorana $c = 0, 1, 2, \cdots$ chains defined by Eq. (5.168) with periodic boundary conditions ($b_{\mathrm{c}} = 0$), and choose the unit $\hbar\omega = 2$ of energy in Eq. (5.168a). We write

$$\widehat{H}_{c,0} := -(-1)^{\delta_{c,0}} \sum_{j=1}^{N} \mathrm{i}\hat{\xi}_j\, \hat{\eta}_{j+c}, \qquad \hat{\xi}_j \equiv \hat{\xi}_{j+N}, \qquad \hat{\eta}_j = \hat{\eta}_{j+N}, \tag{5.254}$$

where $\hat{\eta}_j := \hat{c}_j + \hat{c}_j^{\dagger}$ and $\mathrm{i}\hat{\xi}_j := \hat{c}_j - \hat{c}_j^{\dagger}$ are the Majorana operators stemming from the conjugate pair of creation and annihilation fermion operators $\hat{c}_j^{\dagger}$ and $\hat{c}_j$, respectively, that span the fermionic Fock space (5.168d).

Property 5.1 The (global) symmetry group (see Figure 5.10)

$$\mathrm{G}_{\mathrm{tot}} := \mathrm{G}_{\mathrm{trsl}} \times \mathrm{G}_f, \quad \mathrm{G}_{\mathrm{trsl}} := \mathbb{Z}_N, \quad \mathrm{G}_f := \mathrm{G} \times \mathbb{Z}_2^{\mathrm{F}}, \quad \mathrm{G} := \mathbb{Z}_2^{T}, \tag{5.255a}$$

obeyed by $\widehat{H}_{c,0}$ is represented by taking powers of the operators $\widehat{T}_1$, $\widehat{U}(t)$, and $\widehat{U}(p)$, whereby translation by one lattice spacing is realized by the unitary operator defined by ($j = 1, \cdots, N$ denotes the lattice sites)

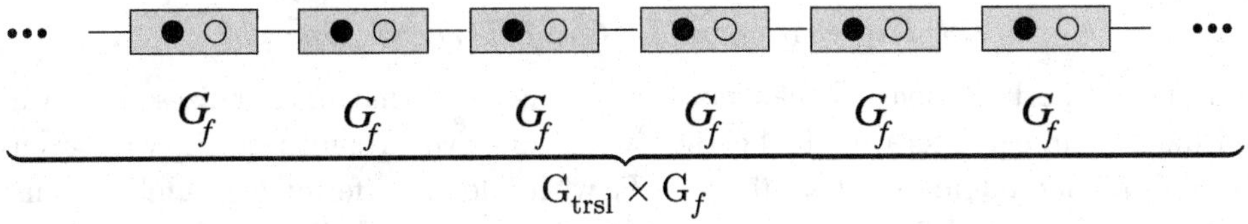

Figure 5.10 Majorana $c = 1$ chain obeying periodic boundary conditions. A repeat unit cell labeled by $j = 1, \cdots, N$ is a rectangle shaded in blue. It is assigned the two-dimensional fermionic Fock space $\mathfrak{F}_j$ that is generated by the fermionic creation operator $\hat{c}_j^\dagger$ out of the state $|0\rangle_j$ annihilated by $\hat{c}_j$. The real part of $\hat{c}_j$ is proportional to the Majorana operator $\hat{\eta}_j$ represented by a filled circle. The imaginary part of $\hat{c}_j$ is proportional to the Majorana operator $\hat{\xi}_j$ represented by an empty circle. The global symmetry is the direct product of the translation symmetry group $\mathrm{G}_{\mathrm{trsl}} := \mathbb{Z}_N$ with the global fermionic group $\mathrm{G}_f := \mathbb{Z}_2^T \times \mathbb{Z}_2^{\mathrm{F}}$, where $\mathbb{Z}_2^T$ denotes reversal of time for spinless fermions and $\mathbb{Z}_2^{\mathrm{F}}$ denotes the fermion parity. The action of the global fermionic group G_f is the same on every repeat unit cell, that is, G_f is an on-site symmetry. Each bond connecting two consecutive repeat unit cell represents a bilinear term in the Hamiltonian (5.254) with $c = 1$ labeled by a directed bond $\langle j, j+1 \rangle$.

$$\widehat{T}_1\, \hat{\eta}_j\, \widehat{T}_1^\dagger := \hat{\eta}_{j+1}, \qquad \widehat{T}_1\, \hat{\xi}_j\, \widehat{T}_1^\dagger := \hat{\xi}_{j+1}, \tag{5.255b}$$

reversal of time t is realized by the antiunitary operator defined by (K denotes complex conjugation)

$$\widehat{U}(t) := \mathbb{1}_{2^N}\, \mathsf{K}, \qquad \mathsf{K}\, \hat{\eta}_j\, \mathsf{K} := +\hat{\eta}_j, \qquad \mathsf{K}\, \hat{\xi}_j\, \mathsf{K} := -\hat{\xi}_j, \tag{5.255c}$$

and fermion parity p is realized by the unitary operator defined by

$$\widehat{U}(p) := \prod_{j=1}^{N} \mathrm{i}\hat{\eta}_j\, \hat{\xi}_j, \tag{5.255d}$$

respectively. One verifies that

$$\begin{aligned}
&\widehat{T}_1^N = \mathbb{1}_{2^N}, \qquad \widehat{U}(t)\,\widehat{U}(t) = \mathbb{1}_{2^N}, \qquad \widehat{U}(p)\,\widehat{U}(p) = \mathbb{1}_{2^N}, \\
&\left[\widehat{T}_1, \widehat{U}(t)\right] = \left[\widehat{U}(t), \widehat{U}(p)\right] = \left[\widehat{U}(p), \widehat{T}_1\right] = 0, \\
&\left[\widehat{H}_{c,0}, \widehat{T}_1\right] = \left[\widehat{H}_{c,0}, \widehat{U}(t)\right] = \left[\widehat{H}_{c,0}, \widehat{U}(p)\right] = 0.
\end{aligned} \tag{5.255e}$$

Property 5.2 A Majorana $c = 2, 3, 4, \cdots$ chain obeying periodic boundary conditions can be thought of as the stacking of $c = 2, 3, 4, \cdots$ Majorana 1 chains obeying periodic boundary conditions (see Figure 5.3).

Property 5.3 Any Majorana $c = 0, 1, 2, \cdots$ chain obeying periodic boundary conditions has a nondegenerate ground state separated from all excited states by a gap.

5.7.3 Digression on $\mathbb{Z}_2$-Graded Vector Spaces

Any fermionic Fock space $\mathcal{F}$ can be seen, in the basis that diagonalizes the total fermionic number operator, to be the direct sum over a subspace $\mathcal{F}_0$ with even total fermionic number and a subspace $\mathcal{F}_1$ with odd total fermionic number. This property endows fermionic Fock space with a natural $\mathbb{Z}_2$-grading.

Definition 5.4 ($\mathbb{Z}_2$-graded vector space) A $\mathbb{Z}_2$-graded vector space V admits the direct sum decomposition

$$V = V_0 \oplus V_1. \tag{5.256}$$

We shall identify the subscripts 0 and 1 as the elements of the additive group $\mathbb{Z}_2$. We say that V_0 (V_1) has parity 0 (1). Any vector space is $\mathbb{Z}_2$-graded since the choice $V_0 = V$ and $V_1 = \{0\}$ (a subspace of V contains at least the 0 vector) is always possible. Any subspace of V_0 shares its parity 0. Any subspace of V_1 shares its parity 1. A nonzero vector $|v\rangle \in V$ is called *homogeneous* if it entirely resides in either one of the subspaces V_0 and V_1. The parity $|v|$ of the homogeneous state $|v\rangle$ is either 0 if $|v\rangle \in V_0$ or 1 if $|v\rangle \in V_1$. These observations on the $\mathbb{Z}_2$-grading of a vector space V only become useful when one demands that any operation acting on V preserves the $\mathbb{Z}_2$-grading.

For example, certain operations need to be defined carefully between two $\mathbb{Z}_2$-graded vector spaces V and W that preserve their $\mathbb{Z}_2$ structure. One such operation is the $\mathbb{Z}_2$-graded tensor product.

Definition 5.5 (graded tensor product) Let $V = V_0 \oplus V_1$ and $W = W_0 \oplus W_1$ be two $\mathbb{Z}_2$-graded vector spaces. We define their graded tensor product as the map

$$\otimes_{\mathfrak{g}} : V \times W \to V \otimes W, \tag{5.257a}$$

such that

$$V_i \otimes_{\mathfrak{g}} W_j \subseteq (V \otimes W)_{(i+j) \bmod 2}, \qquad i,j = 0,1. \tag{5.257b}$$

By design, the operation $\otimes_{\mathfrak{g}}$ carries the $\mathbb{Z}_2$-grading of V and W to their $\mathbb{Z}_2$-graded tensor product. In particular, for any homogeneous vectors $|v\rangle \in V$ with parity $|v| = 0,1$ and $|w\rangle \in W$ with parity $|w| = 0,1$, the graded tensor product $|v\rangle \otimes_{\mathfrak{g}} |w\rangle$ of two homogeneous vectors has the parity

$$\big| |v\rangle \otimes_{\mathfrak{g}} |w\rangle \big| := (|v| + |w|) \bmod 2. \tag{5.257c}$$

The connection between the $\mathbb{Z}_2$-graded vector space $V = V_0 \oplus V_1$ and fermionic Fock spaces $\mathcal{F} = \mathcal{F}_0 \oplus \mathcal{F}_1$, is established through the identifications $\mathcal{F}_0 \to V_0$ and $\mathcal{F}_1 \to V_1$. However, a fermionic Fock space has more structure than a mere $\mathbb{Z}_2$-graded vector space. Wave functions in a fermionic Fock space are fully antisymmetric under the permutation of two fermions. This requirement can be implemented as follows on a $\mathbb{Z}_2$-graded vector space.

Definition 5.6 (reordering operation) The exchange of two fermions can be represented by the isomorphism

$$R\colon V\otimes_{\mathfrak{g}} W\to W\otimes_{\mathfrak{g}} V, \tag{5.258a}$$

by which the graded tensor product of the homogeneous vectors $|v\rangle\in V$ and $|w\rangle\in W$ obeys

$$|v\rangle\otimes_{\mathfrak{g}}|w\rangle\mapsto(-1)^{|v|\,|w|}\,|w\rangle\otimes_{\mathfrak{g}}|v\rangle. \tag{5.258b}$$

The map R is called the reordering operation. It is invertible with itself as inverse since R^2 is the identity map.

Definition 5.7 (dual $\mathbb{Z}_2$-graded vector space) For every $\mathbb{Z}_2$-graded vector space V, we define the dual $\mathbb{Z}_2$-graded vector space V^*. We denote an element of the dual $\mathbb{Z}_2$-graded vector space V^* by $\langle v|$, the dual to the vector $|v\rangle\in V$. The dual $\mathbb{Z}_2$-graded vector space V^* inherits a $\mathbb{Z}_2$ grading from assigning the parity $|v|$ to the vector $\langle v|\in V^*$ if $|v\rangle\in V$ is homogeneous with parity $|v|$.

Definition 5.8 The contraction $\mathcal{C}$ is the map

$$\begin{aligned}\mathcal{C}\colon V^*\otimes_{\mathfrak{g}} V&\to\mathbb{C},\\ \langle\psi|\otimes_{\mathfrak{g}}|\phi\rangle&\mapsto\langle\psi|\phi\rangle,\end{aligned} \tag{5.259a}$$

where $\langle\psi|\phi\rangle$ denotes the scalar product between the pair $|\psi\rangle,|\phi\rangle\in V$. Hence,

$$\mathcal{C}\left(\langle i|\otimes_{\mathfrak{g}}|j\rangle\right)=\delta_{ij} \tag{5.259b}$$

holds for any pair of orthonormal and homogeneous basis vectors $|i\rangle,|j\rangle\in V$.

Definition 5.9 The contraction $\mathcal{C}^*$ is the map $\mathcal{C}^*\colon V\otimes_{\mathfrak{g}} V^*\to\mathbb{C}$ defined by its action

$$\begin{aligned}\mathcal{C}^*\left(|i\rangle\otimes_{\mathfrak{g}}\langle j|\right)&:=\mathcal{C}\Big(R\left(|i\rangle\otimes_{\mathfrak{g}}\langle j|\right)\Big)\\ &=\mathcal{C}\Big((-1)^{|i||j|}\langle j|\otimes_{\mathfrak{g}}|i\rangle\Big)\\ &=(-1)^{|i||j|}\langle j|i\rangle=(-1)^{|i||j|}\delta_{ij}\end{aligned} \tag{5.260a}$$

for any pair of orthonormal basis vectors $|i\rangle,|j\rangle\in V$. It is common practice to use the same symbol $\mathcal{C}$ for both $\mathcal{C}$ and $\mathcal{C}^*$.

Any linear operator

$$M\colon V\to V \tag{5.261a}$$

can be represented in the orthonormal and homogeneous basis $\{|i\rangle\}$ of V by the matrix

$$M_{ij}=(-1)^{|i||j|}\,M_{ji} \tag{5.261b}$$

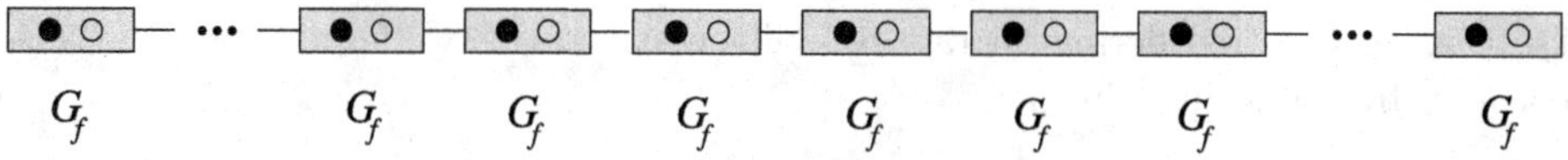

Figure 5.11 Majorana $c = 1$ chain obeying open boundary conditions. A repeat unit cell labeled by $j = 1, \cdots, N$ is a rectangle shaded in blue. It is assigned the two-dimensional fermionic Fock space $\mathfrak{F}_j$ that is generated by the fermionic creation operator $\hat{c}_j^\dagger$ out of the state $|0\rangle_j$ annihilated by $\hat{c}_j$. The real part of $\hat{c}_j$ is proportional to the Majorana operator $\hat{\eta}_j$ represented by a filled circle. The imaginary part of $\hat{c}_j$ is proportional to the Majorana operator $\hat{\xi}_j$ represented by an empty circle. The global symmetry is the global fermionic group $\mathrm{G}_f := \mathbb{Z}_2^T \times \mathbb{Z}_2^{\mathrm{F}}$, where $\mathbb{Z}_2^T$ denotes reversal of time for spinless fermions and $\mathbb{Z}_2^{\mathrm{F}}$ denotes the fermion parity. The action of the global fermionic group G_f is the same on every repeat unit cell, that is, G_f is an on-site symmetry. Each bond connecting two consecutive repeat unit cell represents a bilinear term in the Hamiltonian (5.263) with c=1 labeled by a directed bond $\langle j, j+1\rangle$.

through the expansion

$$M := \sum_{i,j} M_{ij}\, |i\rangle \otimes_{\mathfrak{g}} \langle j| \in V \otimes_{\mathfrak{g}} V^{*}. \tag{5.261c}$$

The linear operator M has a well-defined parity if and only if each term $|i\rangle \otimes_{\mathfrak{g}} \langle j|$ in the summation has the same parity, in which case

$$|M| := (|i| + |j|) \bmod 2. \tag{5.261d}$$

Definition 5.10 (parity of a tensor) More generally, if we define

$$T := \sum_{i_i,\cdots,i_n} T_{i_i,\cdots,i_n}\, |i_1\rangle \otimes_{\mathfrak{g}} \cdots \otimes_{\mathfrak{g}} |i_n\rangle \in \underbrace{V \otimes_{\mathfrak{g}} \cdots \otimes_{\mathfrak{g}} V}_{n\text{ times}}, \tag{5.262a}$$

we can assign the parity

$$|T| := (|i_1| + \cdots + |i_n|) \bmod 2 \tag{5.262b}$$

when all $|i_1\rangle \otimes_{\mathfrak{g}} \cdots \otimes_{\mathfrak{g}} |i_n\rangle$ share the same parity.

5.7.4 Open Boundary Conditions Revisited

For open boundary conditions and $c = 0, 1, 2, \cdots$, we write

$$\widehat{H}_{c,c} := -(-1)^{\delta_{c,0}} \sum_{j=1}^{N-c} \mathrm{i}\hat{\xi}_j\, \hat{\eta}_{j+c}, \tag{5.263}$$

where $\hat{\eta}_j := \hat{c}_j + \hat{c}_j^\dagger$ and $\mathrm{i}\hat{\xi}_j := \hat{c}_j - \hat{c}_j^\dagger$ are the Majorana operators stemming from the conjugate pair of creation and annihilation fermion operators $\hat{c}_j^\dagger$ and $\hat{c}_j$, respectively, that span the fermionic Fock space (5.168d).

Property 5.11 The (global) symmetry group is (see Figure 5.11)

$$\mathrm{G}_{\mathrm{tot}} := \mathrm{G}_f, \qquad \mathrm{G}_f := \mathrm{G} \times \mathbb{Z}_2^{\mathrm{F}}, \quad \mathrm{G} := \mathbb{Z}_2^{T}, \tag{5.264a}$$

where reversal of time t is realized by the antiunitary operator defined by ($j = 1, \cdots, N$ denotes the lattice sites and K denotes complex conjugation)

$$\widehat{U}(t) := \mathbb{1}_{2^N}\,\mathsf{K}, \qquad \mathsf{K}\,\hat{\eta}_j\,\mathsf{K} := +\hat{\eta}_j, \qquad \mathsf{K}\,\hat{\xi}_j\,\mathsf{K} := -\hat{\xi}_j, \tag{5.264b}$$

and fermion parity p is realized by the unitary operator defined by

$$\widehat{U}(p) := \prod_{j=1}^{N} \mathrm{i}\hat{\eta}_j\,\hat{\xi}_j, \tag{5.264c}$$

respectively. One verifies that

$$\begin{aligned} &\widehat{U}(t)\,\widehat{U}(t) = \mathbb{1}_{2^N}, \qquad \widehat{U}(p)\,\widehat{U}(p) = \mathbb{1}_{2^N}, \\ &\left[\widehat{U}(t), \widehat{U}(p)\right] = 0, \\ &\left[\widehat{H}_{c,c}, \widehat{U}(t)\right] = \left[\widehat{H}_{c,c}, \widehat{U}(p)\right] = 0. \end{aligned} \tag{5.264d}$$

Property 5.12 A Majorana $c = 2, 3, 4, \cdots$ chain obeying open boundary conditions can be thought of as the stacking of $c = 2, 3, 4, \cdots$ Majorana 1 chains obeying open boundary conditions (see Figure 5.3).

Property 5.13 The ground-state degeneracy of the Majorana $c = 0, 1, 2, \cdots$ chain obeying open boundary conditions is 2^c (see **Table** 5.4). When $c = 1, 2, 3, \cdots$, this degeneracy arises because $\hat{\eta}_1, \cdots, \hat{\eta}_c$ and $\hat{\xi}_{N-c+1}, \cdots, \hat{\xi}_N$ drop out from

$$\widehat{H}_{c>0,c>0} := -\sum_{j=1}^{N-c} \mathrm{i}\hat{\xi}_j\,\hat{\eta}_{j+c}. \tag{5.265a}$$

In this case, the Hilbert space spanned by the 2^c degenerate ground states is the fermionic Fock space

$$\begin{aligned} \mathfrak{F}_{\mathrm{bd}}^{(c)} := \mathrm{span}\Bigg\{ &\left[\prod_{j=1}^{c}\left(\hat{b}_j^{\dagger}\right)^{n_j}\right]|0\rangle \,\Bigg|\, n_j = 0, 1, \\ &\hat{b}_j^{\dagger} := \frac{\hat{\eta}_j - \mathrm{i}\hat{\xi}_{N+1-j}}{2}, \qquad \hat{b}_j := \frac{\hat{\eta}_j + \mathrm{i}\hat{\xi}_{N+1-j}}{2}, \qquad \hat{b}_j|0\rangle = 0 \Bigg\}. \end{aligned} \tag{5.265b}$$

The abbreviation "bd" refers to the boundary

$$\Lambda_{\mathrm{bd}} := \Lambda_{\mathrm{L}} \cup \Lambda_{\mathrm{R}}, \qquad \Lambda_{\mathrm{L}} := \{1, \cdots, c\}, \qquad \Lambda_{\mathrm{R}} := \{N - c + 1, \cdots, N\}. \tag{5.265c}$$

One observes that the creation $\hat{b}^{\dagger}_{j}$ and annihilation $\hat{b}_{j}$ fermionic operators are non-local in the original basis defined by Eq. (5.168d). The restriction of time reversal and fermion parity on $\mathfrak{F}^{(c)}_{\mathrm{bd}}$ can be chosen to be realized by

$$\widehat{U}_{\mathrm{bd}}(t) := \mathbb{1}_{2^c}\,\mathsf{K} \tag{5.265d}$$

and

$$\widehat{U}_{\mathrm{bd}}(p) := \prod_{j=1}^{c} \mathrm{i}\hat{\eta}_j\,\hat{\xi}_{N+1-j}, \tag{5.265e}$$

respectively. As was the case with Eq. (5.264d),

$$\widehat{U}_{\mathrm{bd}}(t)\,\widehat{U}_{\mathrm{bd}}(t) = \widehat{U}_{\mathrm{bd}}(p)\,\widehat{U}_{\mathrm{bd}}(p) = \mathbb{1}_{2^c}, \qquad \left[\widehat{U}_{\mathrm{bd}}(t), \widehat{U}_{\mathrm{bd}}(p)\right] = 0. \tag{5.265f}$$

For any positive integer $c = 1, 2, 3, \cdots$, the identity

$$\dim \mathfrak{F}^{(c)}_{\mathrm{bd}} = 2^{\frac{1-(-1)^c}{2}} \times 2^{\lfloor c/2 \rfloor} \times 2^{\lfloor c/2 \rfloor} \tag{5.266}$$

has the following interpretation (the lower-floor function $\lfloor x \rfloor$ returns the largest integer that does not exceed $0 \le x < \infty$).

Property 5.14 When c is even, the pair of sets $\hat{\eta}_1, \cdots, \hat{\eta}_c$ and $\hat{\xi}_{N-c+1}, \cdots, \hat{\xi}_N$ generates a local fermionic Fock space on the left

$$\Lambda_{\mathrm{L}} = \{1, \cdots, c\} \tag{5.267a}$$

and right

$$\Lambda_{\mathrm{R}} = \{N - c + 1, \cdots, N\} \tag{5.267b}$$

boundaries, respectively. It is thus possible to factorize $\mathfrak{F}^{(c)}_{\mathrm{bd}}$ into the antisymmetric product of two local fermionic Fock spaces $\mathfrak{F}_{\mathrm{L}}$ and $\mathfrak{F}_{\mathrm{R}}$. One may choose the "factorization"

$$\begin{aligned}
&\mathfrak{F}^{(c)}_{\mathrm{bd}} = \mathfrak{F}_{\mathrm{L}} \otimes_{\mathfrak{g}} \mathfrak{F}_{\mathrm{R}}, \\
&\mathfrak{F}_{\mathrm{L}} := \mathrm{span}\left\{ \prod_{a=0}^{\frac{c}{2}-1} \left(\hat{d}^{\dagger}_{\mathrm{L},a}\right)^{n_a} |0\rangle \;\middle|\; n_a = 0, 1,\ \hat{d}_{\mathrm{L},a}\,|0\rangle = 0,\ \hat{d}_{\mathrm{L},a} := \frac{\hat{\eta}_{1+2a} + \mathrm{i}\hat{\eta}_{2+2a}}{2} \right\}, \\
&\mathfrak{F}_{\mathrm{R}} := \mathrm{span}\left\{ \prod_{a=0}^{\frac{c}{2}-1} \left(\hat{d}^{\dagger}_{\mathrm{R},a}\right)^{n_a} |0\rangle \;\middle|\; n_a = 0, 1,\ \hat{d}_{\mathrm{R},a}\,|0\rangle = 0,\ \hat{d}_{\mathrm{R},a} := \frac{\hat{\xi}_{N-1-2a} + \mathrm{i}\hat{\xi}_{N-2a}}{2} \right\}.
\end{aligned} \tag{5.267c}$$

The notation $\mathfrak{F}_{\mathrm{L}} \otimes_{\mathfrak{g}} \mathfrak{F}_{\mathrm{R}}$ ($\otimes_{\mathfrak{g}}$ denotes the $\mathbb{Z}_2$-graded tensor product from Section 5.7.3) is used to emphasize that the fermionic operators that span $\mathfrak{F}_{\mathrm{L}}$ anti-commute with the fermionic operators that span $\mathfrak{F}_{\mathrm{R}}$ so that $\mathfrak{F}^{(c)}_{\mathrm{bd}}$ inherits from $\mathfrak{F}_{\mathrm{L}}$ and $\mathfrak{F}_{\mathrm{R}}$ their fermionic Fock structure.

Property 5.15 When c is odd, it is not possible to factorize $\mathfrak{F}^{(c)}_{\mathrm{bd}}$ into the antisymmetric product of two fermionic Fock spaces, one spanned by Majorana operators indexed in Λ_{L} and one spanned by Majorana operators indexed in Λ_{R}. Instead, one may choose the "factorization"

$$\begin{aligned}
&\mathfrak{F}^{(c)}_{\mathrm{bd}} = \mathfrak{F}_{\mathrm{LR}} \otimes_{\mathfrak{g}} \mathfrak{F}_{\mathrm{L}} \otimes_{\mathfrak{g}} \mathfrak{F}_{\mathrm{R}},\\
&\mathfrak{F}_{\mathrm{LR}} := \operatorname{span}\left\{\left(\hat{b}^{\dagger}_{\mathrm{LR}}\right)^{n} |0\rangle \;\middle|\; n=0,1,\quad \hat{b}_{\mathrm{LR}}|0\rangle = 0,\quad \hat{b}_{\mathrm{LR}} := \tfrac{\hat{\eta}_1 + \mathrm{i}\hat{\xi}_N}{2}\right\},\\
&\mathfrak{F}_{\mathrm{L}} := \operatorname{span}\left\{\prod_{a=0}^{\frac{c-1}{2}-1}\left(\hat{d}^{\dagger}_{\mathrm{L},a}\right)^{n_a}|0\rangle \;\middle|\; n_a = 0,1,\ \hat{d}_{\mathrm{L},a}|0\rangle = 0,\ \hat{d}_{\mathrm{L},a} := \frac{\hat{\eta}_{2+2a} + \mathrm{i}\hat{\eta}_{3+2a}}{2}\right\},\\
&\mathfrak{F}_{\mathrm{R}} := \operatorname{span}\left\{\prod_{a=0}^{\frac{c-1}{2}-1}\left(\hat{d}^{\dagger}_{\mathrm{R},a}\right)^{n_a}|0\rangle \;\middle|\; n_a = 0,1,\ \hat{d}_{\mathrm{R},a}|0\rangle = 0,\ \hat{d}_{\mathrm{R},a} := \frac{\hat{\xi}_{N-2-2a} + \mathrm{i}\hat{\xi}_{N-1-2a}}{2}\right\}.
\end{aligned} \tag{5.268a}$$

Hereto, the notation $\mathfrak{F}_{\mathrm{LR}} \otimes_{\mathfrak{g}} \mathfrak{F}_{\mathrm{L}} \otimes_{\mathfrak{g}} \mathfrak{F}_{\mathrm{R}}$ ($\otimes_{\mathfrak{g}}$ denotes the $\mathbb{Z}_2$-graded tensor product from Section 5.7.3) is used to emphasize that the fermionic operators that span $\mathfrak{F}_{\mathrm{B}}$ anticommute with the fermionic operators that span $\mathfrak{F}_{\mathrm{B}'}$ for any pair $\mathrm{B},\mathrm{B}' \in \mathrm{LR},\mathrm{L},\mathrm{R}$ so that $\mathfrak{F}^{(c)}_{\mathrm{bd}}$ inherits from $\mathfrak{F}_{\mathrm{L}}$, $\mathfrak{F}_{\mathrm{R}}$, and $\mathfrak{F}_{\mathrm{LR}}$ their fermionic Fock structure.

Property 5.16 We consider any Majorana c chains with $c = 2,4,6,8$. We choose to represent complex conjugation on the left boundary

$$\Lambda_{\mathrm{L}} := \{1,\cdots,c\} \tag{5.269a}$$

and on the right boundary

$$\Lambda_{\mathrm{R}} := \{N-c+1,\cdots,N\} \tag{5.269b}$$

by the pairs of antilinear operators K_{L} and K_{R} defined by the rules

$$\mathsf{K}_{\mathrm{L}}\,\hat{\eta}_{c-j+1}\,\mathsf{K}_{\mathrm{L}} = (-1)^{j+1}\,\hat{\eta}_{c-j+1},\qquad j=1,\cdots,c, \tag{5.269c}$$

and

$$\mathsf{K}_{\mathrm{R}}\,\hat{\xi}_{N-c+j}\,\mathsf{K}_{\mathrm{R}} = (-1)^{j+1}\,\hat{\xi}_{N-c+j},\qquad j=1,\cdots,c, \tag{5.269d}$$

respectively. Furthermore, we choose to represent the actions of time reversal t and fermion parity p on the left and right boundaries by the pair of operators $\widehat{U}_{\mathrm{L}}(t)$, $\widehat{U}_{\mathrm{L}}(p)$ and $\widehat{U}_{\mathrm{R}}(t)$, $\widehat{U}_{\mathrm{R}}(p)$, respectively, as defined in **Table** 5.6. The consistency condition[29]

$$\widehat{U}_{\mathrm{L}}(g)\,\hat{\eta}_{c-j+1}\,\widehat{U}^{\dagger}_{\mathrm{L}}(g) = \widehat{U}_{\mathrm{bd}}(g)\,\hat{\eta}_{c-j+1}\,\widehat{U}^{\dagger}_{\mathrm{bd}}(g),\qquad g=t,p, \tag{5.269e}$$

$$\widehat{U}_{\mathrm{R}}(g)\,\hat{\xi}_{N-c+j}\,\widehat{U}^{\dagger}_{\mathrm{R}}(g) = \widehat{U}_{\mathrm{bd}}(g)\,\hat{\xi}_{N-c+j}\,\widehat{U}^{\dagger}_{\mathrm{bd}}(g),\qquad g=t,p, \tag{5.269f}$$

[29] Given $\widehat{U}_{\mathrm{bd}}(g)$, the choice of $\widehat{U}_{\mathrm{B}}(g)$ with $\mathrm{B}=\mathrm{L},\mathrm{R}$ need not be unique since the left-hand side is invariant under multiplication from the right of $\widehat{U}_{\mathrm{B}}(g)$ with any norm preserving element from the one-dimensional center of the Clifford algebra generated by the even number c of generators.

Table 5.6 *Boundary projective representations of the protecting symmetries for the Majorana* $c = 2, 4, 6, 8$ *chains*

c	$\widehat{V}_{\mathrm{L}}(t)$	$\widehat{U}_{\mathrm{L}}(p)$	$\widehat{V}_{\mathrm{R}}(t)$	$\widehat{U}_{\mathrm{R}}(p)$
2	$\hat{\eta}_2$	$\mathrm{i}\hat{\eta}_2\,\hat{\eta}_1$	$\hat{\xi}_N$	$\mathrm{i}\hat{\xi}_{N-1}\,\hat{\xi}_N$
4	$\prod_{j=1}^{2}\hat{\eta}_{4-(2j-1)}$	$\prod_{j=1}^{4}\hat{\eta}_{4-j+1}$	$\prod_{j=1}^{2}\hat{\xi}_{N-4+(2j-1)}$	$\prod_{j=1}^{4}\hat{\xi}_{N-4+j}$
6	$\prod_{j=1}^{3}\hat{\eta}_{8-2j}$	$\mathrm{i}\prod_{j=1}^{6}\hat{\eta}_{6-j+1}$	$\prod_{j=1}^{3}\hat{\xi}_{N-6+2j}$	$\mathrm{i}\prod_{j=1}^{6}\hat{\xi}_{N-6+j}$
8	$\prod_{j=1}^{4}\hat{\eta}_{8-(2j-1)}$	$\prod_{j=1}^{8}\hat{\eta}_{8-j+1}$	$\prod_{j=1}^{4}\hat{\xi}_{N-8+(2j-1)}$	$\prod_{j=1}^{8}\hat{\xi}_{N-8+j}$

Representations of $\widehat{U}_{\mathrm{L}}(t) \equiv \widehat{V}_{\mathrm{L}}(t)\,\mathsf{K}_{\mathrm{L}}$ and $\widehat{U}_{\mathrm{L}}(p)$ on the left boundary and $\widehat{U}_{\mathrm{R}}(t) = \widehat{V}_{\mathrm{R}}(t)\,\mathsf{K}_{\mathrm{R}}$ and $\widehat{U}_{\mathrm{R}}(p)$ on the right boundary for the Majorana $c = 2, 4, 6, 8$ chains, respectively.

is then fullfilled for any $j = 1, \cdots, c$. Moreover, the pair of indices $\{[\nu_{\mathrm{L}}], [\rho_{\mathrm{L}}]\}$ that are defined by the (possibly) projective algebra

$$\widehat{U}_{\mathrm{L}}(t)\,\widehat{U}_{\mathrm{L}}(t) = (-1)^{[\nu_{\mathrm{L}}]}\,\mathbb{1}_{2^c}, \tag{5.269g}$$

$$\widehat{U}_{\mathrm{L}}(t)\,\widehat{U}_{\mathrm{L}}(p)\,\widehat{U}^{\dagger}_{\mathrm{L}}(t)\,\widehat{U}_{\mathrm{L}}(p) = (-1)^{[\rho_{\mathrm{L}}]}\,\mathbb{1}_{2^c} \tag{5.269h}$$

on the left boundary and the pair of indices $\{[\nu_{\mathrm{R}}], [\rho_{\mathrm{R}}]\}$ that are defined by the (possibly) projective algebra

$$\widehat{U}_{\mathrm{R}}(t)\,\widehat{U}_{\mathrm{R}}(t) = (-1)^{[\nu_{\mathrm{R}}]}\,\mathbb{1}_{2^c}, \tag{5.269i}$$

$$\widehat{U}_{\mathrm{R}}(t)\,\widehat{U}_{\mathrm{R}}(p)\,\widehat{U}^{\dagger}_{\mathrm{R}}(t)\,\widehat{U}_{\mathrm{R}}(p) = (-1)^{[\rho_{\mathrm{R}}]}\,\mathbb{1}_{2^c} \tag{5.269j}$$

on the right boundary are given in **Table** 5.8.[30] Evidently, it is only when $c = 8$ that the algebra obeyed by $\widehat{U}_{\mathrm{bd}}(t)$ and $\widehat{U}_{\mathrm{bd}}(p)$ is recovered. For $c = 2, 4, 6$, the algebra (5.269g)–(5.269j) are truely projective (symmetry fractionalization). Finally, we observe that the algebra (5.269g)–(5.269j) is left invariant by the gauge transformations

$$\widehat{U}_{\mathrm{B}}(t) \mapsto e^{\mathrm{i}\theta_{\mathrm{B}}(t)}\,\widehat{U}_{\mathrm{B}}(t), \qquad \widehat{U}_{\mathrm{B}}(p) \mapsto e^{\mathrm{i}\theta_{\mathrm{B}}(p)}\,\widehat{U}_{\mathrm{B}}(p), \tag{5.270}$$

for any $\theta_{\mathrm{B}}(t), \theta_{\mathrm{B}}(p) \in [0, 2\pi[$ and $\mathrm{B} = \mathrm{L}, \mathrm{R}$.

[30] The rational for bracketing the indices ν_{L} and ρ_{L} (ν_{R} and ρ_{R}) will be explained in Section 6.5 or can be found in Ö. M. Aksoy and C. Mudry, Phys. Rev. B **106**(3), 035117 (2022). [Aksoy and Mudry (2022)].

Property 5.17 We consider any Majorana c chains with $c = 1, 3, 5, 7$. We choose to represent complex conjugation on the left boundary

$$\Lambda_{\mathrm{L}} := \{1, \cdots, c\} \tag{5.271a}$$

and on the right boundary

$$\Lambda_{\mathrm{R}} := \{N - c + 1, \cdots, N\} \tag{5.271b}$$

by the pairs of antilinear operators K_{L} and K_{R} defined by the rules

$$\mathsf{K}_{\mathrm{L}}\, \hat{\eta}_{c-j+1}\, \mathsf{K}_{\mathrm{L}} = \begin{cases} (-1)^{j+1}\, \hat{\eta}_{c-j+1}, & j = 1, \cdots, c-1, \\ \\ +\hat{\eta}_1, & j = c, \end{cases} \tag{5.271c}$$

and

$$\mathsf{K}_{\mathrm{R}}\, \hat{\xi}_{N-c+j}\, \mathsf{K}_{\mathrm{R}} = \begin{cases} (-1)^{j+1}\, \hat{\xi}_{N-c+j}, & j = 1, \cdots, c-1, \\ \\ -\hat{\xi}_N, & j = c, \end{cases} \tag{5.271d}$$

respectively. Furthermore, we choose to represent the actions of time reversal t and fermion parity p on the left and right boundaries by the triplet of operators $\widehat{U}_{\mathrm{LR}}(t)$, $\widehat{U}_{\mathrm{LR}}(p)$, $\widehat{Y}_{\mathrm{L}}$, and $\widehat{U}_{\mathrm{RL}}(t)$, $\widehat{U}_{\mathrm{RL}}(p)$, $\widehat{Y}_{\mathrm{R}}$, respectively, as defined in **Table** 5.7.[31] The first consistency condition

$$\widehat{U}_{\mathrm{LR}}(g)\, \hat{\eta}_{c-j+1}\, \widehat{U}^{\dagger}_{\mathrm{LR}}(g) = \widehat{U}_{\mathrm{bd}}(g)\, \hat{\eta}_{c-j+1}\, \widehat{U}^{\dagger}_{\mathrm{bd}}(g), \qquad g = t, p, \tag{5.271f}$$

$$\widehat{U}_{\mathrm{RL}}(g)\, \hat{\xi}_{N-c+j}\, \widehat{U}^{\dagger}_{\mathrm{RL}}(g) = \widehat{U}_{\mathrm{bd}}(g)\, \hat{\xi}_{N-c+j}\, \widehat{U}^{\dagger}_{\mathrm{bd}}(t), \qquad g = t, p, \tag{5.271g}$$

is then fullfilled for any $j = 1, \cdots, c$. A second consistency condition

$$\left[\widehat{U}_{\mathrm{LR}}(t), \widehat{U}_{\mathrm{LR}}(p)\right] = 0, \qquad \left[\widehat{U}_{\mathrm{RL}}(t), \widehat{U}_{\mathrm{RL}}(p)\right] = 0 \tag{5.271h}$$

is also fulfilled. Moreover, the pair of indices $\{[\nu_{\mathrm{LR}}], [\rho_{\mathrm{LR}}]\}$ that are defined by the

[31] When c is odd, the product of all Majorana operators $\hat{\eta}_1 \cdots \hat{\eta}_c$ ($\hat{\xi}_1 \cdots \hat{\xi}_c$) on the left (right) boundary commutes with all the elements of the Clifford algebra generated by these c boundary Majorana operators. This operator is, by definition, proportional to $\widehat{Y}_{\mathrm{L}}$ ($\widehat{Y}_{\mathrm{R}}$). It, together with the identity operator in the Clifford algebra generated by c Majorana operators, span a vector space called the center of this Clifford algebra. As the left (right) boundary parity operator is defined to be

$$\widehat{U}_{\mathrm{LR}}(p) := \mathrm{i}\hat{\eta}_1\, \hat{\xi}_N \prod_{\alpha=1}^{(c-1)/2} \mathrm{i}\hat{\eta}_{2\alpha}\, \hat{\eta}_{2\alpha+1} \qquad \left(\widehat{U}_{\mathrm{RL}}(p) := \mathrm{i}\hat{\eta}_1\, \hat{\xi}_N \prod_{\alpha=1}^{(c-1)/2} \mathrm{i}\hat{\xi}_{N-2\alpha}\, \hat{\xi}_{N-2\alpha+1}\right), \tag{5.271e}$$

it follows that $\widehat{U}_{\mathrm{LR}}(p) \propto \widehat{Y}_{\mathrm{L}}\, \hat{\xi}_N$ ($\widehat{U}_{\mathrm{RL}}(p) \propto \widehat{Y}_{\mathrm{R}}\, \hat{\eta}_1$). The label LR (RL) refers to the fact that the representation of the parity operator on the left (right) boundary is necessarily nonlocal when c is odd. More details can be found in Ö. M. Aksoy and C. Mudry, Phys. Rev. B **106**(3), 035117 (2022). [Aksoy and Mudry (2022)].

Table 5.7 *Boundary projective representations of the protecting symmetries for the Majorana $c = 1, 3, 5, 7$ chains*

c	$\widehat{V}_{\mathrm{LR}}(t)$	$\widehat{U}_{\mathrm{LR}}(p)$	$\widehat{Y}_{\mathrm{L}}$	$\widehat{V}_{\mathrm{RL}}(t)$	$\widehat{U}_{\mathrm{RL}}(p)$	$\widehat{Y}_{\mathrm{R}}$
1	$\mathbb{1}_2$	$\mathbb{1}_2$	$\hat{\eta}_1$	$\mathbb{1}_2$	$\mathbb{1}_2$	$\hat{\xi}_N$
3	$\hat{\eta}_2\,\hat{\xi}_N$	$\downarrow$	$\mathrm{i}\prod_{j=1}^{3}\hat{\eta}_{3-j+1}$	$\hat{\xi}_{N-2}\,\hat{\eta}_1$	$\downarrow$	$\mathrm{i}\prod_{j=1}^{3}\hat{\xi}_{N-3+j}$
5	$\prod_{j=1}^{2}\hat{\eta}_{6-2j}$	$\downarrow$	$\prod_{j=1}^{5}\hat{\eta}_{5-j+1}$	$\prod_{j=1}^{2}\hat{\xi}_{N-6+2j}$	$\downarrow$	$\prod_{j=1}^{5}\hat{\xi}_{N-5+j}$
7	$\prod_{j=1}^{3}\hat{\eta}_{8-2j}\,\hat{\xi}_N$	$\downarrow$	$\mathrm{i}\prod_{j=1}^{7}\hat{\eta}_{7-j+1}$	$\prod_{j=1}^{3}\hat{\xi}_{N-8+2j}\,\hat{\eta}_1$	$\downarrow$	$\mathrm{i}\prod_{j=1}^{7}\hat{\xi}_{N-7+j}$

Representations of $\widehat{U}_{\mathrm{LR}}(t) \equiv \widehat{V}_{\mathrm{LR}}(t)\,\mathsf{K}_{\mathrm{L}}$, $\widehat{U}_{\mathrm{LR}}(p)$, and $\widehat{Y}_{\mathrm{L}}$ on the left boundary and $\widehat{U}_{\mathrm{RL}}(t) \equiv \widehat{V}_{\mathrm{RL}}(t)\,\mathsf{K}_{\mathrm{R}}$, $\widehat{U}_{\mathrm{RL}}(p)$, and $\widehat{Y}_{\mathrm{R}}$ on the right boundary for the Majorana $c = 1, 3, 5, 7$ chains, respectively. The entry $\downarrow$ for a Majorana $2m+1$ chain means "take the same entry as the one on the same column but for the Majorana m' chain with $m' = 2m$ and multiply by $\mathrm{i}\hat{\eta}_1\,\hat{\xi}_N$" as can be found explicitly in Eq. (5.271e).

strictly projective algebra

$$\widehat{U}_{\mathrm{LR}}(t)\,\widehat{U}_{\mathrm{LR}}(t) = (-1)^{[\nu_{\mathrm{LR}}]}\,\mathbb{1}_{2^c}, \tag{5.271i}$$

$$\widehat{U}_{\mathrm{LR}}(t)\,\widehat{Y}_{\mathrm{L}}\,\widehat{U}^{\dagger}_{\mathrm{LR}}(t)\,\widehat{Y}_{\mathrm{L}} = (-1)^{[\rho_{\mathrm{LR}}]}\,\mathbb{1}_{2^c} \tag{5.271j}$$

on the left boundary and the pair of indices $\{[\nu_{\mathrm{RL}}], [\rho_{\mathrm{RL}}]\}$ that are defined by the stricly projective algebra

$$\widehat{U}_{\mathrm{RL}}(t)\,\widehat{U}_{\mathrm{RL}}(t) = (-1)^{[\nu_{\mathrm{RL}}]}\,\mathbb{1}_{2^c}, \tag{5.271k}$$

$$\widehat{U}_{\mathrm{RL}}(t)\,\widehat{Y}_{\mathrm{R}}\,\widehat{U}^{\dagger}_{\mathrm{RL}}(t)\,\widehat{Y}_{\mathrm{R}} = (-1)^{[\rho_{\mathrm{RL}}]}\,\mathbb{1}_{2^c} \tag{5.271l}$$

on the right boundary are given in **Table** 5.8. [32] For $c = 1, 3, 5, 7$, the algebra (5.271i)–(5.271l) are truely projective (symmetry fractionalization). Finally, we observe that the algebra (5.271i)–(5.271l) is left invariant by the gauge transformations

$$\widehat{U}_{\mathrm{B}}(t) \mapsto e^{\mathrm{i}\theta_{\mathrm{B}}(t)}\,\widehat{U}_{\mathrm{B}}(t), \qquad \widehat{Y}_{\mathrm{B}} \mapsto e^{\mathrm{i}\theta_{\mathrm{B}}}\,\widehat{Y}_{\mathrm{B}}, \tag{5.272}$$

for any $\theta_{\mathrm{B}}(t), \theta_{\mathrm{B}} \in [0, 2\pi[$ and $\mathrm{B} = \mathrm{L}, \mathrm{R}, \mathrm{LR}, \mathrm{RL}$.

[32] The rationale for bracketing the indices ν_{L} and ρ_{L} (ν_{R} and ρ_{R}) will be explained in Section 6.5.

Table 5.8 *Cohomological indices for the Majorana c chains*

c	Triplet on left boundary	Triplet on right boundary	$\dim \mathfrak{F}^{(c)}_{\mathrm{bd,min}}$
1	$(0,0,1)$	$(0,1,1)$	2
2	$(0,1,0)$	$(1,1,0)$	4
3	$(1,1,1)$	$(1,0,1)$	8
4	$(1,0,0)$	$(1,0,0)$	4
5	$(1,0,1)$	$(1,1,1)$	8
6	$(1,1,0)$	$(0,1,0)$	4
7	$(0,1,1)$	$(0,0,1)$	2
8	$(0,0,0)$	$(0,0,0)$	1

The triplets $([\nu],[\rho],[\mu])$ on the left and right boundaries of a Majorana c chain with $c=1,2,3,4,5,6,7,8$. The last column $\mathfrak{F}^{(c)}_{\mathrm{bd,min}} \subset \mathfrak{F}^{(c)}_{\mathrm{bd}}$ is the smallest Fock space such that (i) it is a subset of $\mathfrak{F}^{(c)}_{\mathrm{bd}}$ defined in Eq. (5.265b) and (ii) it realizes both the projective algebra on the left and right boundaries and the bulk algebra for the reversal of time and fermion parity.

Table 5.8 is the main result of Chapter 5. It encodes the phenomenon of symmetry fractionalization, here that of the symmetry group

$$\mathrm{G}_f = \mathbb{Z}_2^T \times \mathbb{Z}_2^{\mathrm{F}}. \tag{5.273}$$

It suggests the breakdown of the $\mathbb{Z}$ classification of Majorana c chains down to the cyclic group

$$\mathbb{Z}_8 \equiv \mathbb{Z}/8\mathbb{Z} \tag{5.274}$$

solely on the basis of symmetry considerations. How to achieve this breakdown dynamically, that is by choosing interactions that gap all the boundary modes without explicit or spontaneous-symmetry breaking of G_f, will be derived explicitly on general grounds in Chapter 8.

Table 5.8 should be compared with **Table** 5.3. We see that the entries for $[\nu_{\mathrm{L}}]$, $[\rho_{\mathrm{L}}]$, and $\dim \mathfrak{F}^{(c)}_{\mathrm{bd,min}}$ in **Table** 5.8 agree with the entries for $[\nu_{\mathrm{L}}]$, $[\rho_{\mathrm{L}}]$, and $D(c)$ in **Table** 5.3 for $c=2,4,6$, and $c=0$ modulo 8, respectively. Unlike in **Table**

5.3, the left (B=L) and right (B=R) boundaries are distinct (except for $c = 4, 8$) in **Table** 5.8 as measured by the pair $[\nu_{\rm B}]$ and $[\rho_{\rm B}]$. This asymmetry has two interpretations. First, it is a consequence of the Jordan-Wigner transformation, for which we chose the convention that it is local on the left boundary but nonlocal on the right boundary. Second, it is a consequence of the fact that all elements of the group $\mathbb{Z}_8$ under the stacking rules (6.250) are not always their own inverse. The entries $c = 1, 3, 5, 7$ in **Table** 5.8 do not match the corresponding entries in **Table** 5.3. This is because the entries $c = 1, 3, 5, 7$ in **Table** 5.8 have no bosonic counterparts, that is, they realize fermionic invertible phases of matter that are not protected by symmetries.

Table 5.8 also suggests that the physical operation of stacking c Majorana 1 chains (an antisymmetric tensor product of fermionic Fock spaces) induces a composition law for the triplets encoding symmetry fractionalization. One deduces from inspection of any one of the columns left or right boundary in **Table** 5.8 that the triplet associated with a Majorana c chain must be the inverse triplet associated with the Majorana c' chain obeying the condition $c + c' = 0 \bmod 8$. A consistency check for this rule is that, after identifying those pair of triplets in a given column from **Table** 5.8 that are the inverse of each other, one verifies that the triplet on the left boundary is indeed the inverse of the triplet of the right boundary on each line from **Table** 5.8, as one would anticipate if one is to obtain a nondegenerate ground state upon imposing periodic boundary conditions, an operation that can also be interpreted as a stacking. Deriving the composition rule obeyed by the triplets in **Table** 5.8 under the physical operation of stacking will be done at the end of Chapter 6 for any arbitrary group G_f.

The parity of c defines the index $[\mu]$ in **Table** 5.8. This parity effect originates from the fact that the Hilbert space on the left or right boundary is not a fermionic Fock space when c is odd. For odd c, the number of Majorana zero modes per boundary is odd. The Clifford algebra spanned by an odd number of generators (the Majorana operators on one boundary) supports one element in addition to the identity that commutes with all elements of the Clifford algebra. This is to say that the Clifford algebra has a center generated by two elements. This property is the counterpart to the fact that spin-1/2 cluster c chains with c odd are invariant under reversal of time, while they break time-reversal symmetry explicitly when c is even. The existence of a nontrivial center when c is odd stems from the symmetry group $\mathbb{Z}_2^{\rm F}$. Even if the symmetry group G is broken explicitly in Eq. (5.264a), the center must be present when the number of Majorana zero modes on a boundary is odd, that is, all but one Majorana zero modes can be gapped by the breaking of the symmetry group G. The Majorana c chains with odd c can only be trivialized by stacking, not by reducing G_f down to $\mathbb{Z}_2^{\rm F}$. The Majorana c chains with odd c are examples of fermionic invertible topological phases of matter that are not protected by symmetries.

At last, we give in the last column of **Table** 5.8 the dimension of the smallest fermionic Fock subspace $\mathfrak{F}^{(c)}_{\mathrm{bd,min}}$ contained in $\mathfrak{F}^{(c)}_{\mathrm{bd}}$ [recall Eq. (5.265b)] such that $\mathfrak{F}^{(c)}_{\mathrm{bd,min}}$ realizes the pair of projective algebra with the left and right triplets corresponding to line c.

For $c=1$, one finds that

$$\mathfrak{F}^{(1)}_{\mathrm{bd,min}} = \mathfrak{F}^{(1)}_{\mathrm{bd}}, \tag{5.275}$$

from which follows

$$\dim \mathfrak{F}^{(1)}_{\mathrm{bd,min}} = 2 = (\sqrt{2}) \times (\sqrt{2}). \tag{5.276}$$

In the second equality, the first factor of $\sqrt{2}$ is the so-called quantum dimension of the Majorana operator $\hat{\eta}_1$, while the second factor of $\sqrt{2}$ is the so-called quantum dimension of the Majorana operator $\hat{\xi}_N$.

For $c=2$, one finds that

$$\mathfrak{F}^{(2)}_{\mathrm{bd,min}} = \mathfrak{F}^{(2)}_{\mathrm{bd}}, \tag{5.277}$$

from which follows

$$\dim \mathfrak{F}^{(2)}_{\mathrm{bd,min}} = 4 = 2 \times 2 = (\sqrt{2})^2 \times (\sqrt{2})^2. \tag{5.278}$$

The second equality can be interpreted as the presence of two projected generators of symmetries per boundary such that (i) they anticommute on the left boundary and (ii) one generator is antiunitary and squares to minus the unit operator on the right boundary (**Exercise** 5.8).

For $c=3$, one finds that

$$\mathfrak{F}^{(3)}_{\mathrm{bd,min}} = \mathfrak{F}^{(3)}_{\mathrm{bd}}, \tag{5.279}$$

from which follows

$$\dim \mathfrak{F}^{(3)}_{\mathrm{bd,min}} = 8 = (2 \times \sqrt{2}) \times (2 \times \sqrt{2}) = (\sqrt{2})^3 \times (\sqrt{2})^3. \tag{5.280}$$

The second equality can be interpreted as the presence of two generators of symmetries such that (i) one generator is antiunitary and squares to minus the unit operator on the left boundary, (ii) they anticommute on the left boundary, and (iii) $[\mu]=1$.

The case of $c=4$ is the first one for which

$$\mathfrak{F}^{(4)}_{\mathrm{bd,min}} \subset \mathfrak{F}^{(4)}_{\mathrm{bd}}, \qquad \mathfrak{F}^{(4)}_{\mathrm{bd}} \not\subset \mathfrak{F}^{(4)}_{\mathrm{bd,min}}. \tag{5.281}$$

One deduces from inspection of the triplets for the left and right boundaries, respectively, that they are identical and given by $(1,0,0)$. The first equality in

$$\dim \mathfrak{F}^{(4)}_{\mathrm{bd,min}} = 4 < (\sqrt{2})^4 \times (\sqrt{2})^4 \tag{5.282}$$

can be interpreted as the presence of two generators of symmetries per boundary such that (i) one generator is antiunitary and squares to minus the unit operator on the left boundary and (ii) one generator is antiunitary and squares to minus the unit operator on the right boundary.

The cases of $c = 5, 6, 7$ follow from the cases $c = 1, 2, 3$ for it suffices to interchange the meaning of left and right boundaries to get $c = 5$ from $c = 3$, $c = 6$ from $c = 2$, and $c = 7$ from $c = 1$, respectively. Hence,

$$\dim \mathfrak{F}^{(5)}_{\mathrm{bd,min}} = 8 < (\sqrt{2})^5 \times (\sqrt{2})^5, \tag{5.283}$$

$$\dim \mathfrak{F}^{(6)}_{\mathrm{bd,min}} = 4 < (\sqrt{2})^6 \times (\sqrt{2})^6, \tag{5.284}$$

$$\dim \mathfrak{F}^{(7)}_{\mathrm{bd,min}} = 2 < (\sqrt{2})^7 \times (\sqrt{2})^7. \tag{5.285}$$

The case of $c = 8$ is the last instance in **Table** 5.8 for which

$$\mathfrak{F}^{(8)}_{\mathrm{bd,min}} \subset \mathfrak{F}^{(8)}_{\mathrm{bd}}, \qquad \mathfrak{F}^{(8)}_{\mathrm{bd}} \not\subset \mathfrak{F}^{(8)}_{\mathrm{bd,min}}. \tag{5.286}$$

Because the generators of the projected symmetries on the left and right boundaries obey a faithful representation of the bulk symmetries,

$$\dim \mathfrak{F}^{(8)}_{\mathrm{bd,min}} = 1 < (\sqrt{2})^8 \times (\sqrt{2})^8. \tag{5.287}$$

According to the last column of **Table** 5.8, it is always possible to lift the ground-state degeneracy 2^c of the Majorana c chain obeying open boundary conditions by adding suitable G_f-symmetric interactions if and only if $c = 4, 5, 6, 7, 8$. We have recovered by purely algebraic means the result by Fidkowski and Kitaev (see footnote 6) that the topological classification by the Abelian group $\mathbb{Z}$ of noninteracting Majorana c chains breaks down to that by the Abelian cyclic group $\mathbb{Z}/8\mathbb{Z} \cong \mathbb{Z}_8$.

5.8 Exercises

5.1 We consider the $c = 2$ spin-1/2 cluster c chain with the ring Λ made of an even number $N \geq 3$ of sites. (The case $c = -2$ is unitarily equivalent to the case $c = 2$.) Because the sublattices $\Lambda_A = \{1, 3, \cdots, N-1\}$ and $\Lambda_B = \{2, 4, \cdots, N\}$ share the same cardinality $N/2$, we trade the one-site unit cell of Λ for a two-sites unit cell. This is to say that we write

$$\widehat{H}^{(b)}_2 := -\frac{\hbar\omega}{2} \sum_{j=1}^{(N-b)/2} \left(\widehat{C}_{2j-1} + \widehat{C}_{2j}\right), \tag{5.288a}$$

where, as before,

$$\widehat{C}_j := \widehat{Z}_j \, \widehat{X}_{j+1} \, \widehat{Z}_{j+2}, \tag{5.288b}$$

$b = 0$ selects periodic boundary conditions, and $b = 2$ selects open boundary conditions.

We are going to show that the unitary transformation $\widehat{U}_{\star\star}$, whose action on the "global" Hilbert space (5.3a) is defined in Eq. (5.294), delivers the transformation law

$$\widehat{H}^{(b)}_{2,\mathrm{U}(1)} := \widehat{U}_{\star\star}\,\widehat{H}^{(b)}_{2}\,\widehat{U}^{\dagger}_{\star\star} \tag{5.289a}$$

with

$$\widehat{H}^{(b)}_{2,\mathrm{U}(1)} := +\frac{\hbar\omega}{2}\sum_{j=1}^{(N-b)/2}\left(\widehat{Y}_{2j}\,\widehat{Y}_{2j+1}+\widehat{Z}_{2j}\,\widehat{Z}_{2j+1}\right). \tag{5.289b}$$

In other words, the spin-1/2 cluster $c=2$ chain (5.288) is unitarily equivalent to a quadratic form in the operators $\widehat{Y}$'s and $\widehat{Z}$'s that, together with the operators $\widehat{X}$'s, obey locally the Pauli algebra and commute otherwise.

Before proving Eq. (5.289), we stress how remarkable this identity is. Indeed, in the basis of the "global" Hilbert space (5.3a) that is specified by the unitary transformation $\widehat{U}_{\star\star}$, $\widehat{H}^{(b)}_{2,\mathrm{U}(1)}$ displays a continuous global symmetry that was hidden in the representation (5.288). More precisely, Hamiltonian $\widehat{H}^{(b)}_{2,\mathrm{U}(1)}$ is invariant under conjugation by any clockwise rotation with the angle $0\le\alpha<2\pi$ about the X axis,

$$\widehat{H}^{(b)}_{2,\mathrm{U}(1)} = \widehat{R}^{X}_{\alpha}\,\widehat{H}^{(b)}_{2,\mathrm{U}(1)}\,\widehat{R}^{X\dagger}_{\alpha}, \qquad \widehat{R}^{X}_{\alpha} := \prod_{j=1}^{N} e^{\mathrm{i}\alpha\,\widehat{X}_j/2}. \tag{5.290}$$

This symmetry is the justification for the subscript U(1) in $\widehat{H}^{(b)}_{2,\mathrm{U}(1)}$. The subscript $\star\star$ in $\widehat{U}_{\star\star}$ refers to choosing a repeat unit cell with two sites. Another consequence of enlarging the repeat unit cell from that of Λ to that of Λ_A is that

$$\begin{aligned}\left[\widehat{H}^{(b)}_{2,\mathrm{U}(1)},\widehat{X}\right] = \left[\widehat{H}^{(b)}_{2,\mathrm{U}(1)},\widehat{Y}\right] = \left[\widehat{H}^{(b)}_{2,\mathrm{U}(1)},\widehat{Z}\right] = 0,\\ \left[\widehat{H}^{(b)}_{2,\mathrm{U}(1)},\widehat{T}\right] = \left[\widehat{H}^{(b)}_{2,\mathrm{U}(1)},\widehat{T}'\right] = 0,\end{aligned} \tag{5.291}$$

that is the π-rotations $\widehat{Y}$ and $\widehat{Z}$ together with reversal of time $\widehat{T}$ are now symmetries.

Proof The proof proceeds in the following steps. In Step 1, we introduce a notation that has been optimized for an enlarged unit cell of Λ made of two consecutive sites of Λ. We also state the main result of the proof. In Step 2, we fix a non-standard basis for two spin-1/2. In Step 3, we choose a new basis for two spin-1/2, thereby defining a unitary transformation in U(4). In Step 4, this local transformation is turned into a global unitary transformation on the "global" Hilbert space (5.3a) by tensoring over all sites in $\Lambda_A=\{1,3,\cdots,N-1\}$. Equipped with this global unitary transformation, we can establish Eq. (5.289).

5.1.1 Step 1: For any site $j = \{1, 3, \cdots, N-1\} = \Lambda_A$, we adopt the notation

$$\begin{aligned}
&\widehat{\Sigma}^{00}_j := 1, &&\widehat{\Sigma}^{01}_j := \widehat{X}_{j+1}, &&\widehat{\Sigma}^{02}_j := \widehat{Y}_{j+1}, &&\widehat{\Sigma}^{03}_j := \widehat{Z}_{j+1},\\
&\widehat{\Sigma}^{10}_j := \widehat{X}_j, &&\widehat{\Sigma}^{11}_j := \widehat{X}_j\,\widehat{X}_{j+1}, &&\widehat{\Sigma}^{12}_j := \widehat{X}_j\,\widehat{Y}_{j+1}, &&\widehat{\Sigma}^{13}_j := \widehat{X}_j\,\widehat{Z}_{j+1},\\
&\widehat{\Sigma}^{20}_j := \widehat{Y}_j, &&\widehat{\Sigma}^{21}_j := \widehat{Y}_j\,\widehat{X}_{j+1}, &&\widehat{\Sigma}^{22}_j := \widehat{Y}_j\,\widehat{Y}_{j+1}, &&\widehat{\Sigma}^{23}_j := \widehat{Y}_j\,\widehat{Z}_{j+1},\\
&\widehat{\Sigma}^{30}_j := \widehat{Z}_j, &&\widehat{\Sigma}^{31}_j := \widehat{Z}_j\,\widehat{X}_{j+1}, &&\widehat{\Sigma}^{32}_j := \widehat{Z}_j\,\widehat{Y}_{j+1}, &&\widehat{\Sigma}^{33}_j := \widehat{Z}_j\,\widehat{Z}_{j+1}.
\end{aligned} \tag{5.292}$$

If we combine Eqs. (5.288) and (5.292), we obtain

$$\widehat{H}^{(b)}_2 = -\frac{\hbar\omega}{2}\sum_{j=1}^{(N-b)/2}\left(\widehat{\Sigma}^{31}_{2j-1}\,\widehat{\Sigma}^{30}_{2j+1} + \widehat{\Sigma}^{03}_{2j-1}\,\widehat{\Sigma}^{13}_{2j+1}\right). \tag{5.293}$$

We are going to construct explicitly a unitary transformation $\widehat{U}_j$ whose nontrivial action is restricted to the pair of sites $j \in \Lambda_A$ and $j+1 \in \Lambda_B$ and given by

$$\begin{aligned}
&\widehat{U}_j\,\widehat{\Sigma}^{00}_j\,\widehat{U}^\dagger_j = +\widehat{\Sigma}^{00}_j, &&\widehat{U}_j\,\widehat{\Sigma}^{01}_j\,\widehat{U}^\dagger_j = -\widehat{\Sigma}^{22}_j, &&\widehat{U}_j\,\widehat{\Sigma}^{02}_j\,\widehat{U}^\dagger_j = +\widehat{\Sigma}^{21}_j, &&\widehat{U}_j\,\widehat{\Sigma}^{03}_j\,\widehat{U}^\dagger_j = +\widehat{\Sigma}^{03}_j,\\
&\widehat{U}_j\,\widehat{\Sigma}^{10}_j\,\widehat{U}^\dagger_j = -\widehat{\Sigma}^{33}_j, &&\widehat{U}_j\,\widehat{\Sigma}^{11}_j\,\widehat{U}^\dagger_j = -\widehat{\Sigma}^{11}_j, &&\widehat{U}_j\,\widehat{\Sigma}^{12}_j\,\widehat{U}^\dagger_j = -\widehat{\Sigma}^{12}_j, &&\widehat{U}_j\,\widehat{\Sigma}^{13}_j\,\widehat{U}^\dagger_j = -\widehat{\Sigma}^{30}_j,\\
&\widehat{U}_j\,\widehat{\Sigma}^{20}_j\,\widehat{U}^\dagger_j = +\widehat{\Sigma}^{13}_j, &&\widehat{U}_j\,\widehat{\Sigma}^{21}_j\,\widehat{U}^\dagger_j = -\widehat{\Sigma}^{31}_j, &&\widehat{U}_j\,\widehat{\Sigma}^{22}_j\,\widehat{U}^\dagger_j = -\widehat{\Sigma}^{32}_j, &&\widehat{U}_j\,\widehat{\Sigma}^{23}_j\,\widehat{U}^\dagger_j = -\widehat{\Sigma}^{10}_j,\\
&\widehat{U}_j\,\widehat{\Sigma}^{30}_j\,\widehat{U}^\dagger_j = -\widehat{\Sigma}^{20}_j, &&\widehat{U}_j\,\widehat{\Sigma}^{31}_j\,\widehat{U}^\dagger_j = +\widehat{\Sigma}^{02}_j, &&\widehat{U}_j\,\widehat{\Sigma}^{32}_j\,\widehat{U}^\dagger_j = -\widehat{\Sigma}^{01}_j, &&\widehat{U}_j\,\widehat{\Sigma}^{33}_j\,\widehat{U}^\dagger_j = -\widehat{\Sigma}^{23}_j.
\end{aligned} \tag{5.294}$$

Show that

$$\begin{aligned}
\left(\prod_{j\in\Lambda_A}\widehat{U}_j\right)\widehat{H}^{(b)}_2\left(\prod_{j\in\Lambda_A}\widehat{U}_j\right)^\dagger &=\\
&+\frac{\hbar\omega}{2}\sum_{j=1}^{(N-b)/2}\left(\widehat{\Sigma}^{02}_{2j-1}\,\widehat{\Sigma}^{20}_{2j+1} + \widehat{\Sigma}^{03}_{2j-1}\,\widehat{\Sigma}^{30}_{2j+1}\right).
\end{aligned} \tag{5.295}$$

5.1.2 Step 2: For any odd integer $j \in \Lambda_A$, we define the four-dimensional Hilbert space $\mathbb{C}^4$ spanned by the basis

$$|1\rangle_j \equiv |\rightarrow\rightarrow\rangle_j := |\rightarrow\rangle_j \otimes |\rightarrow\rangle_{j+1}, \tag{5.296a}$$

$$|2\rangle_j \equiv |\rightarrow\leftarrow\rangle_j := |\rightarrow\rangle_j \otimes |\leftarrow\rangle_{j+1}, \tag{5.296b}$$

$$|3\rangle_j \equiv |\leftarrow\rightarrow\rangle_j := |\leftarrow\rangle_j \otimes |\rightarrow\rangle_{j+1}, \tag{5.296c}$$

$$|4\rangle_j \equiv |\leftarrow\leftarrow\rangle_j := |\leftarrow\rangle_j \otimes |\leftarrow\rangle_{j+1}, \tag{5.296d}$$

where [in contrast to Eq. (5.2)]

$$X_j\,|\rightarrow\rangle_j := +|\rightarrow\rangle_j,\quad X_j\,|\leftarrow\rangle_j := -|\leftarrow\rangle_j \implies X_j = \begin{pmatrix} +1 & 0 \\ 0 & -1 \end{pmatrix}, \tag{5.297a}$$

$$Y_j\,|\rightarrow\rangle_j := -\mathrm{i}|\leftarrow\rangle_j,\quad Y_j\,|\leftarrow\rangle_j := +\mathrm{i}|\rightarrow\rangle_j \implies Y_j = \begin{pmatrix} 0 & +\mathrm{i} \\ -\mathrm{i} & 0 \end{pmatrix}, \tag{5.297b}$$

$$Z_j\,|\rightarrow\rangle_j := +|\leftarrow\rangle_j,\quad Z_j\,|\leftarrow\rangle_j := +|\rightarrow\rangle_j \implies Z_j = \begin{pmatrix} 0 & +1 \\ +1 & 0 \end{pmatrix}. \tag{5.297c}$$

Starting from this basis and for any $j \in \Lambda_A$, construct the tensor-product representations

$$\Sigma^{31}_j = \begin{pmatrix} 0 & 0 & +1 & 0 \\ 0 & 0 & 0 & -1 \\ +1 & 0 & 0 & 0 \\ 0 & -1 & 0 & 0 \end{pmatrix}, \qquad \Sigma^{30}_{j+2} = \begin{pmatrix} 0 & 0 & +1 & 0 \\ 0 & 0 & 0 & +1 \\ +1 & 0 & 0 & 0 \\ 0 & +1 & 0 & 0 \end{pmatrix},$$
$$\Sigma^{03}_j = \begin{pmatrix} 0 & +1 & 0 & 0 \\ +1 & 0 & 0 & 0 \\ 0 & 0 & 0 & +1 \\ 0 & 0 & +1 & 0 \end{pmatrix}, \qquad \Sigma^{13}_{j+2} = \begin{pmatrix} 0 & +1 & 0 & 0 \\ +1 & 0 & 0 & 0 \\ 0 & 0 & 0 & -1 \\ 0 & 0 & -1 & 0 \end{pmatrix}, \tag{5.298a}$$

in the ordered basis

$$\{e_1 \otimes e_1, e_1 \otimes e_2, e_2 \otimes e_1, e_2 \otimes e_2\} \tag{5.298b}$$

of $\mathbb{R}^4$. These four traceless 4×4 Hermitean matrices represent the nontrivial action of the operators $\widehat{\Sigma}^{31}_j$, $\widehat{\Sigma}^{30}_{j+2}$, $\widehat{\Sigma}^{03}_j$, and $\widehat{\Sigma}^{13}_{j+2}$, respectively, on the "global" Hilbert space (5.3a). These four operators realize the local action of Hamiltonian (5.293).

5.1.3 Step 3: By the law of the addition of angular momentum, we decompose the four-dimensional Hilbert space spanned by the basis (5.296) into the direct sum of the one-dimensional Hilbert space $\mathbb{C}$ spanned by the single basis element

$$|0,0\rangle_j := \frac{1}{\sqrt{2}}\left(|\rightarrow\leftarrow\rangle_j - |\leftarrow\rightarrow\rangle_j\right) \tag{5.299a}$$

and the three-dimensional Hilbert space $\mathbb{C}^3$ spanned by the three basis elements

$$|1,+1\rangle_j := |\rightarrow\rightarrow\rangle_j, \tag{5.299b}$$

$$|1,0\rangle_j := \frac{1}{\sqrt{2}}\left(|\rightarrow\leftarrow\rangle_j + |\leftarrow\rightarrow\rangle_j\right), \tag{5.299c}$$

$$|1,-1\rangle_j := |\leftarrow\leftarrow\rangle_j. \tag{5.299d}$$

Show that the linear map that expresses the tensor-product basis (5.296) of two spin-1/2 in terms of the singlet-triplet basis (5.299) of two spin-1/2 is achieved by the unitary transformation

$$V_j := \begin{pmatrix} 0 & 1 & 0 & 0 \\ +\frac{1}{\sqrt{2}} & 0 & +\frac{1}{\sqrt{2}} & 0 \\ -\frac{1}{\sqrt{2}} & 0 & +\frac{1}{\sqrt{2}} & 0 \\ 0 & 0 & 0 & 1 \end{pmatrix}. \tag{5.300}$$

Show that the linear map that expresses the singlet-triplet basis (5.299) of two spin-1/2 in terms of the basis

$$|\mathrm{s}\rangle_j := |0,0\rangle_j, \tag{5.301a}$$

$$|x\rangle_j := +\frac{1}{\sqrt{2}}\left(|1,+1\rangle_j - |1,-1\rangle_j\right), \tag{5.301b}$$

$$|y\rangle_j := +\frac{\mathrm{i}}{\sqrt{2}}\left(|1,+1\rangle_j + |1,-1\rangle_j\right), \tag{5.301c}$$

$$|z\rangle_j := +\mathrm{i}\,|1,0\rangle_j, \tag{5.301d}$$

is achieved by the unitary transformation

$$W_j := \begin{pmatrix} 1 & 0 & 0 & 0 \\ 0 & \frac{+1}{\sqrt{2}} & \frac{-\mathrm{i}}{\sqrt{2}} & 0 \\ 0 & 0 & 0 & -\mathrm{i} \\ 0 & \frac{-1}{\sqrt{2}} & \frac{-\mathrm{i}}{\sqrt{2}} & 0 \end{pmatrix}. \tag{5.302}$$

We now define the local unitary transformation U_j by its action

$$U_j^\dagger\,|1\rangle_j = \frac{1}{\sqrt{2}}\left(|\rightarrow\leftarrow\rangle_j - |\leftarrow\rightarrow\rangle_j\right) \equiv |1'\rangle_j \equiv |\mathrm{s}\rangle_j, \tag{5.303a}$$

$$U_j^\dagger\,|2\rangle_j = \frac{1}{\sqrt{2}}\left(|\rightarrow\rightarrow\rangle_j - |\leftarrow\leftarrow\rangle_j\right) \equiv |2'\rangle_j \equiv |x\rangle_j, \tag{5.303b}$$

$$U_j^\dagger\,|3\rangle_j = \frac{\mathrm{i}}{\sqrt{2}}\left(|\rightarrow\rightarrow\rangle_j + |\leftarrow\leftarrow\rangle_j\right) \equiv |3'\rangle_j \equiv |y\rangle_j, \tag{5.303c}$$

$$U_j^\dagger\,|4\rangle_j = \frac{\mathrm{i}}{\sqrt{2}}\left(|\rightarrow\leftarrow\rangle_j + |\leftarrow\rightarrow\rangle_j\right) \equiv |4'\rangle_j \equiv |z\rangle_j \tag{5.303d}$$

on the tensor-product basis (5.296). Show that it is represented by the 4×4 matrix

$$U_j = \frac{1}{\sqrt{2}} \begin{pmatrix} 0 & +1 & -\mathrm{i} & 0 \\ +1 & 0 & 0 & -\mathrm{i} \\ -1 & 0 & 0 & -\mathrm{i} \\ 0 & -1 & -\mathrm{i} & 0 \end{pmatrix}. \tag{5.304}$$

5.1.4 Step 4: Define on the "global" Hilbert space (5.3a) the unitary operator

$$\widehat{U} := \prod_{j \in \Lambda_A} \widehat{U}_j \tag{5.305}$$

by its action through the 4×4 unitary matrix (5.304) on the local two-site Hilbert space $\mathfrak{h}_j \otimes \mathfrak{h}_{j+1}$ for any odd integer $j \in \Lambda_A$. We seek

$$\begin{aligned} \widehat{U}\, \widehat{H}_2^{(b)}\, \widehat{U}^\dagger = -\frac{\hbar\omega}{2} \sum_{j=1}^{(N-b)/2} & \left[\left(\widehat{U}\, \widehat{\Sigma}^{31}_{2j-1}\, \widehat{U}^\dagger \right) \left(\widehat{U}\, \widehat{\Sigma}^{30}_{2j+1}\, \widehat{U}^\dagger \right) \right. \\ & \left. + \left(\widehat{U}\, \widehat{\Sigma}^{03}_{2j-1}\, \widehat{U}^\dagger \right) \left(\widehat{U}\, \widehat{\Sigma}^{13}_{2j+1}\, \widehat{U}^\dagger \right) \right]. \end{aligned} \tag{5.306}$$

The action of the operators

$$\widehat{U}\, \widehat{\Sigma}^{31}_{2j-1}\, \widehat{U}^\dagger, \tag{5.307a}$$
$$\widehat{U}\, \widehat{\Sigma}^{30}_{2j+1}\, \widehat{U}^\dagger, \tag{5.307b}$$
$$\widehat{U}\, \widehat{\Sigma}^{03}_{2j-1}\, \widehat{U}^\dagger, \tag{5.307c}$$
$$\widehat{U}\, \widehat{\Sigma}^{13}_{2j+1}\, \widehat{U}^\dagger \tag{5.307d}$$

is nontrivial locally only and represented by the 4×4 matrices

$$U_{2j-1}\, \Sigma^{31}_{2j-1}\, U^\dagger_{2j-1}, \tag{5.308a}$$
$$U_{2j+1}\, \Sigma^{30}_{2j+1}\, U^\dagger_{2j+1}, \tag{5.308b}$$
$$U_{2j-1}\, \Sigma^{03}_{2j-1}\, U^\dagger_{2j-1}, \tag{5.308c}$$
$$U_{2j+1}\, \Sigma^{13}_{2j+1}\, U^\dagger_{2j+1}, \tag{5.308d}$$

respectively. Show that an exercise in matrix multiplication delivers Eq. (5.294) and, in particular,

$$U_{2j-1}\, \Sigma^{31}_{2j-1}\, U^\dagger_{2j-1} = +\Sigma^{02}_{2j-1}, \tag{5.309a}$$
$$U_{2j+1}\, \Sigma^{30}_{2j+1}\, U^\dagger_{2j+1} = -\Sigma^{20}_{2j+1}, \tag{5.309b}$$
$$U_{2j-1}\, \Sigma^{03}_{2j-1}\, U^\dagger_{2j-1} = +\Sigma^{03}_{2j-1}, \tag{5.309c}$$
$$U_{2j+1}\, \Sigma^{13}_{2j+1}\, U^\dagger_{2j+1} = -\Sigma^{30}_{2j+1}, \tag{5.309d}$$

respectively. Correspondingly,

$$\widehat{U}_{2j-1}\,\widehat{\Sigma}^{31}_{2j-1}\,\widehat{U}^{\dagger}_{2j-1} = +\widehat{\Sigma}^{02}_{2j-1}, \tag{5.310a}$$

$$\widehat{U}_{2j+1}\,\widehat{\Sigma}^{30}_{2j+1}\,\widehat{U}^{\dagger}_{2j+1} = -\widehat{\Sigma}^{20}_{2j+1}, \tag{5.310b}$$

$$\widehat{U}_{2j-1}\,\widehat{\Sigma}^{03}_{2j-1}\,\widehat{U}^{\dagger}_{2j-1} = +\widehat{\Sigma}^{03}_{2j-1}, \tag{5.310c}$$

$$\widehat{U}_{2j+1}\,\widehat{\Sigma}^{13}_{2j+1}\,\widehat{U}^{\dagger}_{2j+1} = -\widehat{\Sigma}^{30}_{2j+1}. \tag{5.310d}$$

We thus arrive at the representation

$$\begin{aligned}\widehat{U}\,\widehat{H}^{(b)}_{2}\,\widehat{U}^{\dagger} &= +\frac{\hbar\omega}{2}\sum_{j=1}^{(N-b)/2}\left(\widehat{\Sigma}^{02}_{2j-1}\,\widehat{\Sigma}^{20}_{2j+1} + \widehat{\Sigma}^{03}_{2j-1}\,\widehat{\Sigma}^{30}_{2j+1}\right)\\ &= +\frac{\hbar\omega}{2}\sum_{j=1}^{(N-b)/2}\left(\widehat{Y}_{2j}\,\widehat{Y}_{2j+1} + \widehat{Z}_{2j}\,\widehat{Z}_{2j+1}\right).\end{aligned} \tag{5.311}$$

We identify $\widehat{U}$ with $\widehat{U}_{\star\star}$ in Eq. (5.289).

5.1.5 Because we assume that $N \geq 3$ is even, Hamiltonian (5.288) commutes with the three unitary transformations

$$\widehat{X}_{A} := \prod_{j\in\Lambda_{A}} \widehat{X}_{j} = \prod_{j\in\Lambda_{A}} \widehat{\Sigma}^{10}_{j}, \tag{5.312a}$$

$$\widehat{X}_{B} := \prod_{j\in\Lambda_{A}} \widehat{X}_{j+1} = \prod_{j\in\Lambda_{A}} \widehat{\Sigma}^{01}_{j}, \tag{5.312b}$$

$$\widehat{X} := \prod_{j\in\Lambda_{A}} \widehat{X}_{j}\,\widehat{X}_{j+1} = \prod_{j\in\Lambda_{A}} \widehat{\Sigma}^{11}_{j}. \tag{5.312c}$$

Correspondingly, the transformed Hamiltonian (5.289) must commute with the three unitary transformations

$$\begin{aligned}\widehat{U}_{\star\star}\,\widehat{X}_{A}\,\widehat{U}^{\dagger}_{\star\star} &= \widehat{U}_{\star\star}\left(\prod_{j\in\Lambda_{A}}\widehat{\Sigma}^{10}_{j}\right)\widehat{U}^{\dagger}_{\star\star}\\ \text{Eq. (5.294)} &= \prod_{j\in\Lambda_{A}}(-1)\,\widehat{\Sigma}^{33}_{j} = e^{\mathrm{i}\frac{\pi}{2}\sum_{j=1}^{N}\widehat{Z}_{j}} = \left(\widehat{R}^{Z}_{\frac{\pi}{2}}\right)^{2},\end{aligned} \tag{5.313a}$$

$$\begin{aligned}\widehat{U}_{\star\star}\,\widehat{X}_{B}\,\widehat{U}^{\dagger}_{\star\star} &= \widehat{U}_{\star\star}\left(\prod_{j\in\Lambda_{A}}\widehat{\Sigma}^{01}_{j}\right)\widehat{U}^{\dagger}_{\star\star}\\ \text{Eq. (5.294)} &= \prod_{j\in\Lambda_{A}}(-1)\,\widehat{\Sigma}^{22}_{j} = e^{\mathrm{i}\frac{\pi}{2}\sum_{j=1}^{N}\widehat{Y}_{j}} = \left(\widehat{R}^{Y}_{\frac{\pi}{2}}\right)^{2},\end{aligned} \tag{5.313b}$$

$$\widehat{U}_{\star\star}\,\widehat{X}\;\widehat{U}^{\dagger}_{\star\star}=\widehat{U}_{\star\star}\left(\prod_{j\in\Lambda_A}\widehat{\Sigma}^{11}_{j}\right)\widehat{U}^{\dagger}_{\star\star}$$

$$\text{Eq. (5.294)}=\prod_{j\in\Lambda_A}(-1)\,\widehat{\Sigma}^{11}_{j}=e^{\mathrm{i}\frac{\pi}{2}\sum_{j=1}^{N}\widehat{X}_j}=\left(\widehat{R}^{X}_{\frac{\pi}{2}}\right)^{2}. \tag{5.313c}$$

One verifies Eq. (5.313) that this is indeed the case.

5.1.6 One also verifies that

$$\widehat{U}_{\star\star}\,\widehat{Y}\,\widehat{U}^{\dagger}_{\star\star}=\widehat{U}_{\star\star}\left(\prod_{j\in\Lambda_A}\widehat{\Sigma}^{22}_{j}\right)\widehat{U}^{\dagger}_{\star\star}=\prod_{j\in\Lambda_A}\left(-\widehat{\Sigma}^{32}_{j}\right), \tag{5.314a}$$

$$\widehat{U}_{\star\star}\,\widehat{Z}\,\widehat{U}^{\dagger}_{\star\star}=\widehat{U}_{\star\star}\left(\prod_{j\in\Lambda_A}\widehat{\Sigma}^{33}_{j}\right)\widehat{U}^{\dagger}_{\star\star}=\prod_{j\in\Lambda_A}\left(-\widehat{\Sigma}^{23}_{j}\right), \tag{5.314b}$$

$$\widehat{U}_{\star\star}\,\widehat{T}\,\widehat{U}^{\dagger}_{\star\star}=\widehat{U}_{\star\star}\left(\prod_{j\in\Lambda_A}\widehat{\Sigma}^{22}_{j}\right)\widehat{U}^{\mathsf{T}}_{\star\star}\,\mathsf{K}=\left[\prod_{j\in\Lambda_A}\left(-\mathrm{i}\,\widehat{\Sigma}^{01}_{j}\right)\right]\mathsf{K}, \tag{5.314c}$$

$$\widehat{U}_{\star\star}\,\widehat{T}'\,\widehat{U}^{\dagger}_{\star\star}=\widehat{U}_{\star\star}\left(\prod_{j\in\Lambda_A}\widehat{\Sigma}^{00}_{j}\right)\widehat{U}^{\mathsf{T}}_{\star\star}\,\mathsf{K}=\left[\prod_{j\in\Lambda_A}\left(-\widehat{\Sigma}^{33}_{j}\right)\right]\mathsf{K}. \tag{5.314d}$$

5.1.7 Finally, one observes that the transformed Hamiltonian (5.289) commutes with reversal of time $\widehat{T}$ defined in Eq. (5.14), since it is a quadratic form in the $\widehat{Y}$'s and $\widehat{Z}$'s.

5.1.8 The energy spectrum of the transformed Hamiltonian (5.289) with periodic ($b=0$) or open ($b=2$) boundary conditions follows from the identity [recall Eq. (5.299)]

$$\widehat{H}^{(b)}_{2,\mathrm{U}(1)}=-\hbar\omega\sum_{j=1}^{(N-b)/2}\left(|\mathrm{s}\rangle\langle\mathrm{s}|-|z\rangle\langle z|\right)\Big|_{\mathfrak{h}_{2j}\otimes\mathfrak{h}_{2j+1}}. \tag{5.315}$$

Proof Indeed, the nontrivial action of $\widehat{Y}_{2j}\,\widehat{Y}_{2j+1}+\widehat{Z}_{2j}\,\widehat{Z}_{2j+1}$ on the "global" Hilbert space (5.3a) is limited to the four-dimensional local Hilbert space

$$\mathfrak{h}_{2j}\otimes\mathfrak{h}_{2j+1}=\mathrm{span}\{|\mathrm{s}\rangle,|x\rangle,|y\rangle,|z\rangle\}. \tag{5.316}$$

With the choice (5.301) for the basis of $\mathfrak{h}_{2j}\otimes\mathfrak{h}_{2j+1}$, the action of $\widehat{Y}_{2j}\,\widehat{Y}_{2j+1}+\widehat{Z}_{2j}\,\widehat{Z}_{2j+1}$ is represented by $-2\left(|\mathrm{s}\rangle\langle\mathrm{s}|-|z\rangle\langle z|\right)\big|_{\mathfrak{h}_{2j}\otimes\mathfrak{h}_{2j+1}}$. □

5.1.9 With periodic boundary conditions ($b=0$), the ground state is proportional to the normalized state

$$|0\rangle=\bigotimes_{j=1}^{N/2}|\mathrm{s}\rangle\Big|_{\mathfrak{h}_{2j}\otimes\mathfrak{h}_{2j+1}}. \tag{5.317}$$

Verify that this ground state is a spin-singlet.

5.1.10 With open boundary conditions ($b=2$), the ground states are fourfold degenerate and given by

$$\left(\bigotimes_{j=1}^{(N-2)/2}|\mathrm{s}\rangle\Big|_{\mathfrak{h}_{2j}\otimes\mathfrak{h}_{2j+1}}\right)\bigotimes\mathrm{span}\bigotimes_{j=1,N}\left\{|+\rangle_j,|-\rangle_j\;\Big|\;\widehat{Z}_j\,|\pm\rangle_j=\pm|\pm\rangle_j\right\}. \tag{5.318}$$

Verify that the two-level degrees of freedom on the left and right boundaries each realize the spin-1/2 representation of the SU(2) group.

□

5.2 Without loss of generality, we consider the case of $c=4$. We also assume that $N\geq 5$, the number of sites from the chain Λ, is even. Show that the spin-1/2 cluster $c=4$ chain is given by

$$\begin{aligned}\widehat{H}^{(b)}_4 &:= -\frac{\hbar\omega}{2}\sum_{j=1}^{(N-b)/2}\Big(\widehat{Z}_{2j-1}\,\widehat{X}_{2j}\,\widehat{X}_{2j+1}\,\widehat{X}_{2j+2}\,\widehat{Z}_{2j+3}\\ &\qquad\qquad + \widehat{Z}_{2j}\,\widehat{X}_{2j+1}\,\widehat{X}_{2j+2}\,\widehat{X}_{2j+3}\,\widehat{Z}_{2j+4}\Big)\\ &\overset{\text{Eq. (5.292)}}{=} -\frac{\hbar\omega}{2}\sum_{j=1}^{(N-b)/2}\left(\widehat{\Sigma}^{31}_{2j-1}\,\widehat{\Sigma}^{11}_{2j+1}\,\widehat{\Sigma}^{30}_{2j+3}+\widehat{\Sigma}^{03}_{2j-1}\,\widehat{\Sigma}^{11}_{2j+1}\,\widehat{\Sigma}^{13}_{2j+3}\right).\end{aligned} \tag{5.319}$$

Verify that it commutes with the three unitary transformations (5.312).

Verify that if one performs the unitary transformation (5.289), then one obtains

$$\begin{aligned}\widehat{H}^{(b)}_4 &\mapsto \widehat{H}^{(b)\prime}_4\\ &\equiv \widehat{U}_{\star\star}\,\widehat{H}^{(b)}_4\,\widehat{U}^{\dagger}_{\star\star}\\ &= -\frac{\hbar\omega}{2}\sum_{j=1}^{(N-b)/2}\left(\widehat{\Sigma}^{02}_{2j-1}\,\widehat{\Sigma}^{11}_{2j+1}\,\widehat{\Sigma}^{20}_{2j+3}+\widehat{\Sigma}^{03}_{2j-1}\,\widehat{\Sigma}^{11}_{2j+1}\,\widehat{\Sigma}^{30}_{2j+3}\right)\\ &= -\frac{\hbar\omega}{2}\sum_{j=1}^{(N-b)/2}\left(\widehat{Y}_{2j}\,\widehat{X}_{2j+1}\,\widehat{X}_{2j+2}\,\widehat{Y}_{2j+3}+\widehat{Z}_{2j}\,\widehat{X}_{2j+1}\,\widehat{X}_{2j+2}\,\widehat{Z}_{2j+3}\right).\end{aligned} \tag{5.320}$$

Verify that the tranformed Hamiltonian (5.320) commutes with the unitary generators of π rotations $\widehat{X}$, $\widehat{Y}$, and $\widehat{Z}=\mathrm{i}\widehat{Y}\,\widehat{X}$.

Verify that the transformed Hamiltonian (5.320) also commutes with reversal of time $\widehat{T}$ defined in Eq. (5.14), since it is a quartic form in the $\widehat{X}$'s, $\widehat{Y}$'s, and $\widehat{Z}$'s, as well as with complex conjugation $\widehat{T}'$.

It follows that enlarging the repeat unit cell from that of Λ to that of Λ_A [compare with Eq. (5.291)] hereto results in an enlarged set of symmetries

$$\left[\widehat{H}_4^{(b)\prime},\widehat{X}\right]=\left[\widehat{H}_4^{(b)\prime},\widehat{Y}\right]=\left[\widehat{H}_4^{(b)\prime},\widehat{Z}\right]=\left[\widehat{H}_4^{(b)\prime},\widehat{T}\right]=\left[\widehat{H}_4^{(b)\prime},\widehat{T}'\right]=0. \quad (5.321)$$

5.3 For each site $j=1,\cdots,N$, we are given the three Hermitean operators $\widehat{X}_j$, $\widehat{Y}_j$, and $\widehat{Z}_j$ that obey the Pauli algebra

$$\left[\widehat{X}_j,\widehat{Y}_{j'}\right]=2\delta_{j,j'}\,\mathrm{i}\widehat{Z}_j,\quad \widehat{X}_j^2=\widehat{Y}_j^2=\widehat{Z}_j^2=1, \quad (5.322\text{a})$$

(with cyclic permutations for the letters X, Y, and Z) for any $j,j'=1,\cdots,N$. We define the Hermitean operators

$$\hat{\xi}_1:=\widehat{Y}_1,\quad \hat{\xi}_2:=\widehat{X}_1\widehat{Y}_2,\quad \cdots\quad \hat{\xi}_N:=\widehat{X}_1\cdots\widehat{X}_{N-1}\widehat{Y}_N, \quad (5.322\text{b})$$

and

$$\hat{\eta}_1:=\widehat{Z}_1,\quad \hat{\eta}_2:=\widehat{X}_1\widehat{Z}_2,\quad \cdots\quad \hat{\eta}_N:=\widehat{X}_1\cdots\widehat{X}_{N-1}\widehat{Z}_N. \quad (5.322\text{c})$$

One verifies that

$$\hat{\xi}_j^2=\hat{\eta}_j^2=1 \quad (5.323\text{a})$$

and that

$$\left\{\hat{\xi}_j,\hat{\xi}_{j'}\right\}=\left\{\hat{\eta}_j,\hat{\eta}_{j'}\right\}=\left\{\hat{\xi}_j,\hat{\eta}_{j'}\right\}=0 \quad (5.323\text{b})$$

for any pair $1\leq j\neq j'\leq N$. This proves Eq. (5.159).

5.4 Verify Eq. (5.224).

5.5 The goals of this exercise are twofold. We seek arguments for (i) the existence of a gap separating the ground state from all excited states of the nearest-neighbor antiferromagnetic quantum spin-1 Heisenberg ring defined in Eq. (5.225) and (ii) the existence of two spin-1/2 degrees of freedom with energy levels of order $\exp(-L/\xi)$ above that of the ground state, each of which are bound within a distance ξ to the opposite ends of an open chain of length L.

To this end, there are several approximate methods that are available. The Hamiltonian (5.225) defined on an open chain of length L can be studied by exact diagonalization methods.[21] The drawback of this approach is that the thermodynamic limit $L\to\infty$ is not accessible. The following two methods give access to the thermodynamic limit $L\to\infty$, but they involve uncontrolled approximations. One may represent a quantum spin-1 degree of freedom in terms of auxiliary bosons called Schwinger bosons and treat this representation within a mean-field approximation.[33] For an open chain, this approach was used by Ng.[34] The method that we will follow consists in treating the nearest-neighbor antiferromagnetic quantum spin-1 Heisenberg chain

[33] D. P. Arovas and A. Auerbach, Phys. Rev. B **38**(1), 316–332 (1988); A. Auerbach and D. P. Arovas, Phys. Rev. Lett. **61**(5), 617–620 (1988). [Arovas and Auerbach (1988); Auerbach and Arovas (1988)].

[34] T. K. Ng, Phys. Rev. B **45**(14), 8181–8184 (1992). [Ng (1992)].

in a semiclassical approximation that is valid in the large S limit, as orginally proposed by Haldane.[2] This was done for an open chain by Ng.[35]

We proceed with the following steps.

5.5.1 *Step 1: Representation of the $\mathfrak{su}(2)$ algebra in terms of coherent states*

The $\mathfrak{su}(2)$ algebra is defined by the commutators

$$\left[\widehat{S}^a,\widehat{S}^b,\right]=\mathrm{i}\epsilon^{abc}\,\widehat{S}^c,\qquad a,b=1,2,3, \tag{5.324}$$

where the summation over repeated indices is implied and ϵ^{abc} is the rank 3 fully antisymmetric Levi-Civita tensor with the convention that $\epsilon^{123}=1$. The irreducible representations of this algebra are fixed by the value $S=1/2,1,3/2,2,\cdots$ entering the eigenvalue of the quadratic Casimir operator

$$\widehat{\boldsymbol{S}}^2=S(S+1). \tag{5.325}$$

The irreducible representation S is $(2S+1)$-dimensional. An orthonormal basis $\mathfrak{B}_S$ of the associated Hilbert space $\mathfrak{H}_S$ is defined by the set

$$\left\{|S,m\rangle\;\middle|\;\widehat{S}^3\,|S,m\rangle=m\,|S,m\rangle,\ m=-S,-S+1,\cdots,S-1,S\right\}. \tag{5.326}$$

The states $|S,\pm S\rangle$ are called the highest and lowest states, respectively, as

$$\widehat{S}^\pm\,|S,\pm S\rangle=0\qquad \widehat{S}^\pm:=\widehat{S}^1\pm\mathrm{i}\widehat{S}^2. \tag{5.327}$$

The basis (5.326) can be interpreted as a finite ladder of states because of

$$\widehat{S}^\pm\,|S,m\rangle=\sqrt{S(S+1)-m(m\pm1)}\,|S,m\pm1\rangle. \tag{5.328}$$

The resolution of the identity is

$$1=\sum_{m=-S}^{+S}|S,m\rangle\langle S,m|. \tag{5.329}$$

Let S^2 denote the surface of the unit sphere in three-dimensional Euclidean space. We parametrize a point $\boldsymbol{\Omega}\in\mathsf{S}^2$ by the row vector

$$\boldsymbol{\Omega}^{\mathsf{T}}:=\begin{pmatrix}\sin\theta\,\cos\phi & \sin\theta\,\sin\phi & \cos\theta\end{pmatrix},\quad 0\leq\theta\leq\pi,\quad 0\leq\phi<2\pi. \tag{5.330}$$

We denote with

$$\boldsymbol{\Omega}_0^{\mathsf{T}}:=\begin{pmatrix}0 & 0 & 1\end{pmatrix} \tag{5.331}$$

the north pole of S^2. We denote with

$$\mathrm{d}^2\Omega:=\mathrm{d}\theta\,\mathrm{d}\phi\,\sin\theta \tag{5.332}$$

an infinitesimal surface element of S^2. Define the coherent state

$$|\boldsymbol{\Omega}\rangle:=e^{\zeta\,\widehat{S}^+-\zeta^*\,\widehat{S}^-}\,|S,S\rangle,\qquad \zeta:=-\frac{\theta}{2}\,e^{-\mathrm{i}\phi}. \tag{5.333}$$

[35] T. K. Ng, Phys. Rev. B **47**(17), 11575–11578 (1993); Phys. Rev. B **50**(1), 555–558 (1994). [Ng (1993, 1994)].

5.5.1.1 Show that

$$\langle \boldsymbol{\Omega}_1 \,|\, \boldsymbol{\Omega}_2 \rangle = e^{\mathrm{i}S\,\Phi(\boldsymbol{\Omega}_0,\boldsymbol{\Omega}_1,\boldsymbol{\Omega}_2)} \left(\frac{1+\boldsymbol{\Omega}_1\cdot\boldsymbol{\Omega}_2}{2} \right)^{S}, \tag{5.334a}$$

$$\langle \boldsymbol{\Omega} \,|\, \widehat{\boldsymbol{S}} \,|\, \boldsymbol{\Omega} \rangle = S\,\boldsymbol{\Omega}, \tag{5.334b}$$

$$1 = \frac{2S+1}{4\pi} \int\limits_{\mathsf{S}^2} \mathrm{d}^2\Omega \,|\, \boldsymbol{\Omega} \rangle \langle \boldsymbol{\Omega} \,|, \tag{5.334c}$$

where $\Phi(\boldsymbol{\Omega}_0,\boldsymbol{\Omega}_1,\boldsymbol{\Omega}_2)$ is the oriented area of the spherical triangle with vertices $\boldsymbol{\Omega}_0,\boldsymbol{\Omega}_1,\boldsymbol{\Omega}_2$. This area is defined modulo 4π, the area of S^2.

5.5.1.2 If $\widehat{H}[\widehat{\boldsymbol{S}}]$ is any quantum spin-S Hamiltonian, show that its partition function

$$Z_S(\beta) := \mathrm{tr}|_{\mathfrak{H}_S}\, e^{-\beta \widehat{H}[\widehat{\boldsymbol{S}}]} \tag{5.335a}$$

can be represented by the path integral

$$Z_S(\beta) = \int\limits_{\mathsf{S}^2} \mathcal{D}[\boldsymbol{\Omega}] \exp\left(-\mathrm{i}S\,\omega[\boldsymbol{\Omega}] - \int\limits_0^\beta \mathrm{d}\tau\, H[\boldsymbol{\Omega}] \right), \tag{5.335b}$$

where

$$\boldsymbol{\Omega}(\tau) = \boldsymbol{\Omega}(\tau+\beta), \tag{5.335c}$$

the so-called Berry phase $\omega[\boldsymbol{\Omega}]$ is the solid angle subtended by the closed trajectory $\boldsymbol{\Omega}(\tau)$ on the two sphere S^2, and $H[\boldsymbol{\Omega}]$ is obtained from $\widehat{H}[\widehat{\boldsymbol{S}}]$ by replacing all operators $\widehat{\boldsymbol{S}}$ with $S\,\boldsymbol{\Omega}$.

5.5.1.3 Explain why, had we defined the partition function starting from Eq. (5.335b) with S a positive real number, we would have to restrict $2S$ to be a positive integer for consistency.

5.5.1.4 Show that the solid angle $\omega[\boldsymbol{\Omega}]$ has the nonlocal representation

$$\omega[\boldsymbol{\Omega}] = \int\limits_0^\beta \mathrm{d}\tau \int\limits_0^1 \mathrm{d}u\, \bar{\boldsymbol{\Omega}} \cdot \left(\partial_u \bar{\boldsymbol{\Omega}} \wedge \partial_\tau \bar{\boldsymbol{\Omega}} \right), \tag{5.336a}$$

where it is understood that

$$\bar{\boldsymbol{\Omega}}^{\mathsf{T}}(\tau,u) = \left(\sin\left(u\,\theta(\tau)\right)\,\cos\left(\phi(\tau)\right), \sin\left(u\,\theta(\tau)\right)\,\sin\left(\phi(\tau)\right), \cos\left(u\,\theta(\tau)\right) \right), \tag{5.336b}$$

$$\bar{\boldsymbol{\Omega}}(\tau,u)\big|_{u=1} = \boldsymbol{\Omega}(\tau), \tag{5.336c}$$

and

$$\bar{\boldsymbol{\Omega}}(\tau,u)\big|_{u=0} = \boldsymbol{\Omega}_0. \tag{5.336d}$$

Observe that

$$\omega[-\boldsymbol{\Omega}] = -\omega[\boldsymbol{\Omega}] \bmod 4\pi. \tag{5.337}$$

The representation (5.336) is an example of a Wess-Zumino (WZ) term.[36]

5.5.1.5 Explain what is meant by "The existence of $\bar{\boldsymbol{\Omega}}$ is guaranteed because of $\pi_1(S^2) = \{\mathrm{id}\}$."

5.5.1.6 Let $\delta\boldsymbol{\Omega}$ be a variation of $\boldsymbol{\Omega}$ such that $(\boldsymbol{\Omega}+\delta\boldsymbol{\Omega})^2 = 1$ holds to linear order in $\delta\boldsymbol{\Omega}$. Expand $\omega[\boldsymbol{\Omega}+\delta\boldsymbol{\Omega}]$ to linear order in $\delta\boldsymbol{\Omega}$ and show that

$$\frac{\delta\,\omega[\boldsymbol{\Omega}]}{\delta\,\boldsymbol{\Omega}} = \partial_\tau \boldsymbol{\Omega} \wedge \boldsymbol{\Omega}. \tag{5.338}$$

To this end, use the fact that $\delta\bar{\boldsymbol{\Omega}}$, $\partial_u\bar{\boldsymbol{\Omega}}$, and $\partial_\tau\bar{\boldsymbol{\Omega}}$ are all orthogonal to $\bar{\boldsymbol{\Omega}}$ and thus lie in a plane.

5.5.1.7 To describe a many-spin system on a d-dimensional lattice whose lattice sites are denoted with the letters $i, j, \cdots$, one writes

$$\operatorname{Tr} e^{-\beta\hat{H}[\{\hat{\boldsymbol{S}}_i\}]} = \left[\prod_i \int\limits_{S^2} \mathcal{D}[\boldsymbol{\Omega}_i]\right] e^{-\mathrm{i}S\sum_i \omega[\boldsymbol{\Omega}_i] - \int_0^\beta \mathrm{d}\tau\, H[\{\boldsymbol{\Omega}_i\}]}. \tag{5.339}$$

The Berry phase

$$S\sum_i \omega[\boldsymbol{\Omega}_i] \tag{5.340}$$

is "simply" the sum of the Berry phases $S\,\omega[\boldsymbol{\Omega}_i]$, each one of which is labeled by the site index i. As the Berry phase $S\,\omega[\boldsymbol{\Omega}_i]$ is defined modulo $4\pi\,S$, so is the Berry phase $S\sum_i \omega[\boldsymbol{\Omega}_i]$. We shall use the notation

$$S_{\mathrm{B}} := S\sum_i \omega[\boldsymbol{\Omega}_i]. \tag{5.341}$$

5.5.2 *Step 2: Coherent-state representation assuming antiferromagnetism*

5.5.2.1 Verify that, for any bipartite lattice, the nearest-neighbor antiferromagnetic classical Heisenberg model has a ground state that supports Néel long-range order.

5.5.2.2 Explain why the limit $S \to \infty$ is classical.

5.5.2.3 For a bipartite lattice with lattice spacings measured in units of $\mathfrak{a}$ and on which a nearest-neighbor antiferromagnetic quantum spin-S Heisenberg Hamiltonian is defined, one makes the Ansatz

$$\boldsymbol{\Omega}_i(\tau) \approx \eta_i\, \boldsymbol{n}(\tau, \boldsymbol{r}) + \frac{\mathfrak{a}^d}{S}\,\boldsymbol{L}(\tau, \boldsymbol{r}), \tag{5.342}$$

where $\eta_i = +1$ on one sublattice and $\eta_i = -1$ on the other sublattice, $\boldsymbol{r}$ is a continuous coordinate in d-dimensional space that is substituted for the lattice site indexed by i, and

$$\boldsymbol{n}(\tau, \boldsymbol{r}) \in \mathrm{S}^2, \qquad \boldsymbol{n}(\tau, \boldsymbol{r}) \cdot \boldsymbol{L}(\tau, \boldsymbol{r}) = 0, \tag{5.343}$$

[36] J. Wess and B. Zumino, Phys. Lett. B **37**(1), 95–97 (1971). [Wess and Zumino (1971)].

with the classical Néel state $\boldsymbol{n}_{\mathrm{N}}$ any unit vector defined by

$$\begin{aligned}&\boldsymbol{L}_{\mathrm{N}}(\tau,\boldsymbol{r})=0,\\&(\partial_\tau\boldsymbol{n}_{\mathrm{N}})(\tau,\boldsymbol{r})=(\partial_{r_1}\boldsymbol{n}_{\mathrm{N}})(\tau,\boldsymbol{r})=\cdots=(\partial_{r_d}\boldsymbol{n}_{\mathrm{N}})(\tau,\boldsymbol{r})=0.\end{aligned}\tag{5.344}$$

Show that an expansion in powers of the deviation $\boldsymbol{L}$ and $\partial_\mu\boldsymbol{n}$ about the Néel state $\eta_i\,\boldsymbol{n}_{\mathrm{N}}$ gives the Berry phase

$$\mathrm{i}S\sum_i\eta_i\,\omega[\boldsymbol{n}]-\mathrm{i}\int\limits_0^\beta\mathrm{d}\tau\int\mathrm{d}^d r\,\boldsymbol{n}\cdot(\partial_\tau\boldsymbol{n}\wedge\boldsymbol{L})+\mathcal{O}(S^{-1}).\tag{5.345}$$

The first term is the only place where the discrete sum over lattice sites remains together with an explicit dependence on the microsopic value of the quantum spin number S. The factorization of the sign function η_i follows from inserting the first term on the right-hand side of Eq. (5.342). The second term fixes the classical dynamics of the smooth fluctuations about the Néel state $\eta_i\,\boldsymbol{n}_{\mathrm{N}}$ upon extremal variation of $\boldsymbol{L}$.

5.5.2.4 Show that the linearized form of the Hamiltonian for the antiferromagnet is given by

$$\frac{1}{2}\int\mathrm{d}^d r\,\left[\frac{1}{\gamma}\,(c\,\boldsymbol{L})^2+\gamma\,(\boldsymbol{\nabla}\cdot\boldsymbol{n})^2\right],\tag{5.346}$$

where γ is the stiffness of the Néel ground state and c is a characteristic velocity. Express γ and c in terms of J, S, and $\mathfrak{a}$ (use dimensional analysis as a guide).

5.5.2.5 Show that elimination of the fast fluctuations $\boldsymbol{L}$ in favor of the slow time-derivative $\partial_\tau\boldsymbol{n}$ gives the standard Euclidean action

$$S_{\mathrm{QNLSM}}:=\frac{1}{2g^2}\int\limits_0^\beta\mathrm{d}\tau\int\mathrm{d}^d r\,\left[(\partial_{c\,\tau}\boldsymbol{n})^2+(\boldsymbol{\nabla}\cdot\boldsymbol{n})^2\right],\quad g^2:=\frac{1}{\gamma},\tag{5.347a}$$

of the O(3) quantum nonlinear sigma model (QNLSM). The associated partition function is

$$Z_{\mathrm{QNLSM}}:=\int\mathcal{D}[\boldsymbol{n}]\,\delta(\boldsymbol{n}^2-1)\,e^{-S_{\mathrm{QNLSM}}[\boldsymbol{n}]}.\tag{5.347b}$$

Euclidean spacetime is $(1+d)$-dimensional at zero temperature. This is the base space. The space in which the image $\boldsymbol{n}(\tau,\boldsymbol{r})$ of a point $(\tau,\boldsymbol{r})$ from base space, here the two sphere S^2, is called the target space. Periodic boundary conditions always hold along the imaginary-time direction. The boundary conditions in space are not specified.

5.5.3 *Step 3: Application to a linear chain ($d=1$)* When $d=1$, the quantum field theory defined by Eq. (5.347) with periodic boundary conditions in space was shown by Wiegmann to be integrable and massive.[37] The generation of a gap had been anticipated by Polyakov 10 years earlier,[38] who had interpreted his computation of the one-loop infrared flow

$$\frac{\mathrm{d}\, g^2}{\mathrm{d}\ \ln \mathfrak{a}} = \frac{g^2}{4\pi} \tag{5.348}$$

to strong coupling as the signature of the instability of the ferromagnetic ground state of a classical ferromagnetic Heisenberg model on the square lattice with the lattice spacing $\mathfrak{a}$ to the paramagnetic phase for any nonvanishing temperature $\sim g^2$ due to the strongly interacting spin waves (Goldstone modes). From this perspective, it is the gaplessness of nearest-neighbor antiferromagnetic half-integer Heisenberg spin chains that is a remarkable result. It is a topological feature that distinguishes nearest-neighbor antiferromagnetic integer from half integer Heisenberg spin chains.

5.5.3.1 For the case of a chain of length $L = N\,\mathfrak{a}$ with N an even integer, argue that the Berry phase (5.341) is given by the Euclidean action

$$\begin{aligned} S_{\mathrm{B}} =& \eta_1\, \Theta(S) \int\limits_0^\beta \int\limits_0^L \frac{\mathrm{d}\tau\, \mathrm{d}x}{4\pi}\, \boldsymbol{n} \cdot (\partial_\tau \boldsymbol{n} \wedge \partial_x \boldsymbol{n}) \\ &+ \eta_1\, \frac{S}{2}\, \{\omega[\boldsymbol{n}_1] - \omega[\boldsymbol{n}_N]\} \\ &+ \mathcal{O}(S^{-1}) \end{aligned} \tag{5.349a}$$

with the boundary conditions $\boldsymbol{n}_1 = \boldsymbol{\Omega}_1$ and $\boldsymbol{n}_N = \boldsymbol{\Omega}_N$ and the so-called Θ angle given by

$$\Theta(S) \equiv 2\pi\, S \tag{5.349b}$$

and we recall that $\eta_1 = \pm 1$. *Hint:*

- Write

$$\sum_{i=1}^{N} \omega[\boldsymbol{\Omega}_i] = \sum_{i=1}^{N-1} \frac{\omega[\boldsymbol{\Omega}_i] + \omega[\boldsymbol{\Omega}_{i+1}]}{2} + \frac{\omega[\boldsymbol{\Omega}_1] + \omega[\boldsymbol{\Omega}_N]}{2}. \tag{5.350}$$

  The ratio $\Theta(S)/4\pi = S/2$ in Eq. (5.349) originates from the factor 1/2 in Eq. (5.350).
- Assume smoothness in that

$$\sum_{i=1}^{N-1} \frac{\omega[\boldsymbol{\Omega}_i] + \omega[\boldsymbol{\Omega}_{i+1}]}{2} \to \frac{\eta_1}{2} \int\limits_0^\beta \int\limits_0^L \mathrm{d}\tau\, \mathrm{d}x\, \frac{\delta\, \omega[\boldsymbol{n}]}{\delta \boldsymbol{n}} \cdot \partial_x \boldsymbol{n} \tag{5.351}$$

  when performing a gradient expansion of the sum in Eq. (5.350).

[37] P. B. Wiegmann, Phys. Lett. B **152**(3), 209–214 (1985). [Wiegmann (1985)].
[38] A. M. Polyakov, Phys. Lett. B **59**(1), 79–81 (1975). [Polyakov, A. M. (1975)].

5.5.3.2 Show by elementary geometry that (i) if the periodic boundary conditions

$$\boldsymbol{n}(\tau,x)=\boldsymbol{n}(\tau,x+L) \tag{5.352a}$$

hold in addition to the periodic boundary conditions

$$\boldsymbol{n}(\tau,x)=\boldsymbol{n}(\tau+\beta,x), \tag{5.352b}$$

(ii) if $\boldsymbol{n}(\tau,x)$ is a smooth function of τ and x, and (iii) if

$$\boldsymbol{n}(\tau,x)=\boldsymbol{n}_0, \qquad (\tau,x)\in\partial\left\{[0,\beta]\times[0,L]\right\}, \tag{5.352c}$$

then

$$Q:=\int\limits_0^\beta\int\limits_0^L\frac{\mathrm{d}\tau\,\mathrm{d}x}{4\pi}\,\boldsymbol{n}\cdot\left(\partial_\tau\boldsymbol{n}\wedge\partial_x\boldsymbol{n}\right)\in\mathbb{Z} \tag{5.352d}$$

and

$$S_{\mathrm{B}}=\pm2\pi\,S\,Q+\mathcal{O}(S^{-1}), \tag{5.352e}$$

where the sign $\pm1=\eta_1$ originates from the breaking of translation symmetry by one lattice spacing of the Néel state. Observe here that the condition

$$\boldsymbol{\Omega}_1(\tau)=\boldsymbol{\Omega}_{N+1}(\tau) \tag{5.352f}$$

implies the periodic boundary conditions (5.352a). However, if we do not impose these microscopic periodic boundary conditions but still demand that conditions (i)–(iii) hold, then

$$S_{\mathrm{B}}=\pm2\pi\,S\,Q\pm\frac{S}{2}\left\{\omega[\boldsymbol{n}_1]-\omega[\boldsymbol{n}_N]\right\}+\mathcal{O}(S^{-1}) \tag{5.352g}$$

can be interpreted as signaling the presence of two weakly hybridized spin-S/2 degrees of freedom provided the bulk is gapped.

5.5.3.3 When S is an integer, the bulk contribution $\pm2\pi\,S\,Q$ to the Berry phase is invisible in the partition function, since

$$e^{\pm2\pi\mathrm{i}\,S\,Q}=1. \tag{5.353}$$

The partition function when periodic boundary conditions are imposed is given by Eq. (5.347). According to Polyakov and Wiegmann, a gap Δ separates the ground state from all excited states and, in particular, the spin waves (strongly interacting Goldstone modes). When open boundary conditions are used, we may interpret the two Berry phases with the *fractionalized* spin $S/2$ in Eq. (5.352g) as the signature of free mid-gap states whose wave functions decay exponentially fast away from the ends of the chain with the characteristic decay length given by $\xi\sim1/\Delta$.

5.5.3.4 Show that the gap Δ can be estimated with the help of the one-loop infrared beta function (5.348) to be

$$\Delta \sim J\, e^{-2\pi/g^2}, \tag{5.354}$$

while the energy of the mid-gap end states is of order $J\, e^{-L/\xi}$.

5.5.3.5 When S is a half-integer, the bulk contribution $\pm 2\pi\, S\, Q$ to the Berry phase is visible in the partition function, since

$$e^{\pm 2\pi \mathrm{i}\, S\, Q} = (-1)^{Q}. \tag{5.355}$$

The partition function, when periodic boundary conditions that are compatible with base space being topologically equivalent to a sphere are imposed, is given by

$$Z^{(\Theta=\pi)}_{\mathrm{QNLSM}} := \int \mathcal{D}[\boldsymbol{n}]\, \delta(\boldsymbol{n}^2 - 1)\, (-1)^{Q[\boldsymbol{n}]}\, e^{-S_{\mathrm{QNLSM}}[\boldsymbol{n}]}. \tag{5.356}$$

Since $Q[\boldsymbol{n}]$ is quantized, it cannot be changed by smooth deformations of $\boldsymbol{n}$. Hence, the integer values taken by $Q[\boldsymbol{n}]$, the number of times compactified base space is wrapped around the target space S^2, defines a topological sector in the path integral. The path integral is an alternating sum over topological sectors differing by the parity of $Q[\boldsymbol{n}]$. Due to this alternating sum, Haldane conjectured that the one-loop infrared flow (5.348) is, *for any half-integer* S, to a quantum critical point at which all spin-spin correlation functions decay algebraically fast, as was believed to occur for $S = 1/2$ since the exact solution by Lieb, Schultz, and Mattis of the quantum XY chain.[39] This was a bold conjecture since the partition function (5.356) is derived in the semiclassical limit $S \to \infty$. Witten argued[40] that this quantum critical point is the so-called SU(2) level 1 ($\mathrm{SU}(2)_1$) conformal-field theory with central charge $\mathsf{c} = 1$ in $(1+1)$-dimensional spacetime. The conjecture of Haldane and the reasoning of Witten have been verified numerically.

5.5.3.6 What happens to the boundary degrees of freedom in an open chain when S is a half integer?

5.6 The goal of this exercise is to get familiarized with Majorana operators. We shall proceed in three steps. We start with a single-particle Hamiltonian and compute its partition function. We then define a fermionic Hamiltonian with the single-particle Hamiltonian as kernel and compute its partition function. Finally, we introduce a bilinear form for Majorana operators with the single-particle Hamiltonian as kernel, compute its partition function, thereby showing explicitly that it is the square root of the partition function in step 2.

[39] E. Lieb, T. Schultz, and D. Mattis, Ann. Phys. (N.Y.) **16**(3), 407–466 (1961). [Lieb et al. (1961)].
[40] E. Witten, Comm. Math. Phys. **92**(4), 455–472 (1984). [Witten (1984)].

5.6.1 Denote with $\boldsymbol{\sigma}$ the vector of Pauli matrices σ_1, σ_2, and σ_3, and let σ_0 denote the unit 2×2 matrix. Define the single-particle Hamiltonian

$$\mathcal{H} := \hbar\omega\,\sigma_2. \tag{5.357a}$$

Verify that this Hamiltonian has the eigenvalues

$$\varepsilon_{\pm} = \pm\hbar\omega \tag{5.357b}$$

and the spectral decomposition

$$\mathcal{H} = \begin{pmatrix} \varepsilon_+ & 0 \\ 0 & \varepsilon_- \end{pmatrix}. \tag{5.357c}$$

Verify that the partition function

$$Z(\beta) := \operatorname{tr}|_{\mathbb{C}^2}\, e^{-\beta\,\mathcal{H}} \tag{5.357d}$$

is given by

$$Z(\beta) = 2\cosh\left(\beta\,\hbar\omega\right). \tag{5.357e}$$

5.6.2 Define the Hamiltonian

$$\widehat{H}_{\mathrm{f}} := \widehat{\Psi}^{\dagger}\,\mathcal{H}\,\widehat{\Psi} \equiv \hbar\omega \begin{pmatrix} \hat{c}_1^{\dagger} & \hat{c}_2^{\dagger} \end{pmatrix} \begin{pmatrix} 0 & -\mathrm{i} \\ +\mathrm{i} & 0 \end{pmatrix} \begin{pmatrix} \hat{c}_1 \\ \hat{c}_2 \end{pmatrix}, \tag{5.358a}$$

and the total fermion-number operator

$$\widehat{N}_{\mathrm{f}} := \widehat{\Psi}^{\dagger}\,\mathbb{1}\,\widehat{\Psi} \equiv \begin{pmatrix} \hat{c}_1^{\dagger} & \hat{c}_2^{\dagger} \end{pmatrix} \begin{pmatrix} 1 & 0 \\ 0 & 1 \end{pmatrix} \begin{pmatrix} \hat{c}_1 \\ \hat{c}_2 \end{pmatrix}, \tag{5.358b}$$

where

$$\left\{\hat{c}_i, \hat{c}_j^{\dagger}\right\} = \delta_{ij}, \qquad \left\{\hat{c}_i, \hat{c}_j\right\} = \left\{\hat{c}_i^{\dagger}, \hat{c}_j^{\dagger}\right\} = 0, \qquad i,j = 1,2. \tag{5.358c}$$

The Hilbert space on which $\widehat{H}_{\mathrm{f}}$ is defined is the Fock space

$$\mathfrak{F}_{\mathrm{f}} := \operatorname{span}\left\{|0,0\rangle,\ |1,0\rangle,\ |0,1\rangle,\ |1,1\rangle\right\} \tag{5.358d}$$

with

$$|1,1\rangle := \hat{c}_-^{\dagger}\,\hat{c}_+^{\dagger}\,|0\rangle, \qquad |0,1\rangle := \hat{c}_+^{\dagger}\,|0\rangle, \qquad |1,0\rangle := \hat{c}_-^{\dagger}\,|0\rangle, \tag{5.358e}$$

where the (highest weight or vacuum) state $|0\rangle$ is defined by

$$\hat{c}_-\,|0\rangle = \hat{c}_+\,|0\rangle = 0, \tag{5.358f}$$

and

$$\hat{c}_{\mp}^{\dagger} := \frac{1}{\sqrt{2}}\left(\hat{c}_1^{\dagger} \mp \mathrm{i}\hat{c}_2^{\dagger}\right), \qquad \hat{c}_{\mp} := \frac{1}{\sqrt{2}}\left(\hat{c}_1 \pm \mathrm{i}\hat{c}_2\right). \tag{5.358g}$$

Verify that $\widehat{H}_{\mathrm{f}}$ and $\widehat{N}_{\mathrm{f}}$ have the spectral decompositions

$$\widehat{H}_{\mathrm{f}} = \varepsilon_{-}\,|1,0\rangle\langle 1,0| + \varepsilon_{+}\,|0,1\rangle\langle 0,1| \tag{5.358h}$$

and

$$\widehat{N}_{\mathrm{f}} = |1,0\rangle\langle 1,0| + |0,1\rangle\langle 0,1| + 2|1,1\rangle\langle 1,1| \tag{5.358i}$$

respectively. Verify that the partition function

$$Z_{\mathrm{f}}(\beta) := \operatorname{tr}|_{\mathfrak{F}_{\mathrm{f}}}\, e^{-\beta\,\widehat{H}_{\mathrm{f}}} \tag{5.358j}$$

is given by

$$Z_{\mathrm{f}}(\beta) = 2\left[1+\cosh\left(\beta\,\hbar\,\omega\right)\right]. \tag{5.358k}$$

5.6.3 We consider two objects $\hat{\eta}_1$ and $\hat{\eta}_2$ whose linear combination over the field of complex numbers we denote by

$$a\,\hat{\eta}_1 + b\,\hat{\eta}_2 \tag{5.359a}$$

for any $a, b \in \mathbb{C}$ and whose adjoint

$$(a\,\hat{\eta}_1 + b\,\hat{\eta}_2)^{\dagger} \tag{5.359b}$$

is defined by

$$(a\,\hat{\eta}_1 + b\,\hat{\eta}_2)^{\dagger} = a^{*}\,\hat{\eta}_1 + b^{*}\,\hat{\eta}_2. \tag{5.359c}$$

We also define the multiplication of $a\,\hat{\eta}_1 + b\,\hat{\eta}_2$ with $c\,\hat{\eta}_1 + d\,\hat{\eta}_2$ from the right through the Clifford algebra (Majorana algebra)

$$\left\{\hat{\eta}_i, \hat{\eta}_j\right\} = 2\delta_{ij}, \qquad i, j = 1, 2. \tag{5.359d}$$

Accordingly,

$$(a\,\hat{\eta}_1 + b\,\hat{\eta}_2)\,(c\,\hat{\eta}_1 + d\,\hat{\eta}_2) = a\,c + b\,d + (a\,d - b\,c)\,\hat{\eta}_1\,\hat{\eta}_2. \tag{5.359e}$$

Motivated by Eqs. (5.357a), (5.358a), and (5.358k), we define

$$\widehat{H}_{\mathrm{M}} := \frac{\hbar\,\omega}{2}\,\mathrm{i}\,\hat{\eta}_2\,\hat{\eta}_1 = \frac{\hbar\,\omega}{4}\begin{pmatrix}\hat{\eta}_1 & \hat{\eta}_2\end{pmatrix}\begin{pmatrix}0 & -\mathrm{i}\\ +\mathrm{i} & 0\end{pmatrix}\begin{pmatrix}\hat{\eta}_1\\ \hat{\eta}_2\end{pmatrix} = \widehat{H}^{\dagger}_{\mathrm{M}}. \tag{5.359f}$$

We would like to be able to interpret $\widehat{H}_{\mathrm{M}}$ as a Hamiltonian, that is, we "only" need to construct a vector space on which $\widehat{H}_{\mathrm{M}}$ would act as a linear operator on itself since $\widehat{H}_{\mathrm{M}}$ is formally Hermitean. This is achieved by defining the pair of ajoint operators

$$\hat{c}^{\dagger} := \frac{\hat{\eta}_2 - \mathrm{i}\hat{\eta}_1}{2}, \qquad \hat{c} := \frac{\hat{\eta}_2 + \mathrm{i}\hat{\eta}_1}{2}. \tag{5.359g}$$

5.6.3.1 Verify that $\hat{c}^{\dagger}$ and $\hat{c}$ realize the fermion algebra

$$\left\{\hat{c},\hat{c}^{\dagger}\right\}=1,\qquad \left\{\hat{c},\hat{c}\right\}=\left\{\hat{c}^{\dagger},\hat{c}^{\dagger}\right\}=0, \tag{5.360a}$$

together with the identity

$$\hat{c}^{\dagger}\,\hat{c}=\frac{1+\mathrm{i}\hat{\eta}_2\,\hat{\eta}_1}{2}. \tag{5.360b}$$

Hence, we may write

$$\widehat{H}_{\mathrm{M}}:=\hbar\omega\left(\hat{c}^{\dagger}\,\hat{c}-\frac{1}{2}\right). \tag{5.360c}$$

The Hilbert space on which $\widehat{H}_{\mathrm{M}}$ is defined is the Fock space

$$\mathfrak{F}_{\mathrm{M}}:=\text{span}\,\{|0\rangle,\ |1\rangle\} \tag{5.360d}$$

with

$$|1\rangle:=\hat{c}^{\dagger}\,|0\rangle, \tag{5.360e}$$

where the (highest weight or vacuum) state $|0\rangle$ is defined by

$$\hat{c}\,|0\rangle=0. \tag{5.360f}$$

5.6.3.2 Verify that $\widehat{H}_{\mathrm{M}}$ has the spectral decomposition

$$\widehat{H}_{\mathrm{M}}=\varepsilon_{-}\,|0\rangle\langle 0|+\varepsilon_{+}\,|1\rangle\langle 1|,\qquad \varepsilon_{+}=-\varepsilon_{-}=\frac{\hbar\omega}{2}. \tag{5.361}$$

5.6.3.3 Verify that the partition function

$$Z_{\mathrm{M}}(\beta):=\ \mathrm{tr}\big|_{\mathfrak{F}_{\mathrm{M}}}\ e^{-\beta\,\widehat{H}_{\mathrm{M}}} \tag{5.362a}$$

is given by

$$Z_{\mathrm{M}}(\beta)=2\,\cosh\left(\frac{\beta\,\hbar\omega}{2}\right). \tag{5.362b}$$

5.6.3.4 Verify that

$$\left[Z_{\mathrm{M}}(\beta)\right]^2=Z_{\mathrm{f}}(\beta). \tag{5.363}$$

We can thus interpret the Majorana partition function $Z_{\mathrm{M}}(\beta)$ as the square root of the fermionic partition function $Z_{\mathrm{f}}(\beta)$ provided the energy scale entering the Majorana Hamiltonian (5.359f) with the normalization convention (5.359d) for the Majorana algebra is half that entering the fermionic Hamiltonian (5.358a).

5.6.3.5 Do the transformation

$$\hat{\eta}_i=:\sqrt{\kappa}\,\hat{\xi}_i,\qquad i=1,2, \tag{5.364}$$

for some strictly positive real-valued number $\kappa>0$ on Eqs. (5.359d), (5.359f), and (5.359g). Show that this transformation leaves Eq. (5.363) unchanged. *Hint:* Trace the effects of this transformation on Eqs. (5.360a), (5.360b), and (5.360c).

5.6.4 Let $\mathcal{H}$ be an arbitrary $2n \times 2n$ real-valued antisymmetric matrix. Let

$$\widehat{H}_{\mathrm{M}} := \frac{\mathrm{i}}{4}\,\hat{\eta}^{\mathsf{T}}\,\mathcal{H}\,\hat{\eta} \tag{5.365}$$

be a bilinear form for $2n$ Hermitean operators arranged in the column vector $\hat{\eta} = (\hat{\eta}_i) = (\hat{\eta}_i^{\dagger})$ and obeying the Majorana algebra

$$\left\{\hat{\eta}_i, \hat{\eta}_j\right\} = 2\delta_{ij}, \qquad i,j = 1,2,\cdots,2n. \tag{5.366}$$

Construct the fermionic bilinear form such that its partition function is the square of the Majorana partition function generated by $\widehat{H}_{\mathrm{M}}$. [*Hint:* Use the normal form of real-valued symmetric matrices so that you may then use Eq. (5.363).]

5.7 We are going to adapt **Exercise** 3.14 in Chapter 3 to reproduce the result (5.363) from the point of view of Grassmann path integrals. For any integer $n = 1, 2, \cdots$, let $\eta_1, \cdots, \eta_{2n}$ denote $2n$ Grassmann numbers, that is they obey the algebra

$$\left\{\eta_i, \eta_j\right\} = 0, \qquad i,j = 1,2,\cdots,2n. \tag{5.367}$$

The difference with the Majorana algebra (5.366) is that we are not dealing with Hermitean operators that square to the identity and anticommute if different. Instead, we are dealing with numbers that square to zero and anticommute if different. We recall that all Grassmann integrals except

$$\int \mathrm{d}\eta_{2n} \cdots \mathrm{d}\eta_1\, \eta_1 \cdots \eta_{2n} = 1 \tag{5.368}$$

are vanishing. Let Q denote a $2n \times 2n$ complex-valued matrix that we assume to be antisymmetric, without loss of generality,

$$Q_{i,j} = -Q_{j,i}, \qquad i,j = 1,\cdots,2n. \tag{5.369}$$

The pfaffian of this antisymmetric matrix is defined to be

$$\mathrm{pf}(Q) := \frac{1}{2^n\, n!} \sum_{\sigma \in \mathcal{P}_{2n}} \mathrm{sgn}(\sigma)\, Q_{\sigma(1),\sigma(2)} \cdots Q_{\sigma(2n-1),\sigma(2n)}, \tag{5.370}$$

where σ is an element of the permutation group $\mathcal{P}_{2n}$ of $2n$ elements, its signature $\mathrm{sgn}(\sigma)$ is $+1$ (-1) if σ is the product of an even (odd) number of transpositions, and $\sigma(i)$ is the image of $i = 1, \cdots, 2n$ under the permutation σ. The pfaffian of a $(2n-1) \times (2n-1)$ antisymmetric matrix vanishes by definition. We define the quadratic Grassmann form

$$S := \frac{1}{2} \sum_{i,j=1}^{2n} \eta_i\, Q_{i,j}\, \eta_j. \tag{5.371a}$$

We want a closed expression for the Gaussian Grassmann integral

$$Z := \int \mathrm{d}\eta_{2n} \cdots \mathrm{d}\eta_1 \, e^S. \tag{5.371b}$$

5.7.1 Show that

$$\begin{aligned} e^S &= \prod_{1\le i<j\le 2n} \left(1 + Q_{i,j}\,\eta_i\,\eta_j\right) \\ &= \left(1+Q_{1,2}\,\eta_1\,\eta_2\right)\left(1+Q_{1,3}\,\eta_1\,\eta_3\right)\cdots\left(1+Q_{1,2n}\,\eta_1\,\eta_{2n}\right) \\ &\qquad\qquad\left(1+Q_{2,3}\,\eta_2\,\eta_3\right)\cdots\left(1+Q_{2,2n}\,\eta_2\,\eta_{2n}\right) \\ &\qquad\qquad\qquad\qquad\ddots \\ &\qquad\qquad\qquad\qquad\quad\left(1+Q_{2n-1,2n}\,\eta_{2n-1}\,\eta_{2n}\right). \end{aligned} \tag{5.372a}$$

The number of pairs $1 \le i < j \le 2n$ is here

$$1 + 2 + \cdots + 2n - 1 = (2n-1)\,n. \tag{5.372b}$$

5.7.2 Verify that

$$Z = \mathrm{pf}(Q). \tag{5.373}$$

5.7.3 Use Eq. (3.720) to show that if we arrange the $2n$ Grassmann numbers into the column vector $\boldsymbol{\eta}$ and if we perform the linear transformation

$$\boldsymbol{\eta} = J\,\boldsymbol{\xi}, \tag{5.374a}$$

then

$$\mathrm{pf}(J^{\mathsf{T}}\,Q\,J) = \det(J)\,\mathrm{pf}(Q). \tag{5.374b}$$

5.7.4 Show that if

$$Q = \begin{pmatrix} 0 & -K \\ K^{\mathsf{T}} & 0 \end{pmatrix}, \tag{5.375a}$$

where K is an arbitrary $n \times n$ matrix, then

$$(-1)^{n(n+1)/2}\,\mathrm{pf}(Q) = \det(K). \tag{5.375b}$$

5.7.5 Deduce from the previous identity that

$$[\mathrm{pf}(Q)]^2 = \det(Q). \tag{5.376}$$

5.8 Prove the degeneracy (5.278) by proving the following two independent results.

5.8.1 Show that if a Hamiltonian commutes with two Hermitean operators that anticommute pairwise, then each energy eigenstate is at least twofold degenerate.

5.8.2 Show that if a Hamiltonian commutes with an antiunitary operator that squares to minus the identity, then each energy eigenstate is at least twofold degenerate. If the symmetry is that of time-reversal symmetry, the ensuing twofold degeneracy is called Kramers' degeneracy.

6
The Lieb-Schultz-Mattis Theorem

6.1 Introduction

The fractionalization of the charge in polyacetylene was reinterpreted in Chapter 5 as the first historical example of a fermionic invertible topological phase of matter. More precisely, it was shown in Section 5.5.3 (Section 5.5.4) how the spinless (spinfull) SSH fermionic Hamiltonian (5.194) is related to the Majorana $c = 2$ ($c = 4$) chains. Moreover, it was shown in Chapter 5 that there is a one-to-one correspondence between the Majorana c chains and the spin-1/2 cluster c chains. It is the Jordan-Wigner transformation, a unitary but nonlocal transformation, that establishes this one-to-one correspondence. What is remarkable with this one-to-one correspondence is that, whereas the ground-state degeneracies of the Majorana c chains have a topological origin, this need not be so for the spin-1/2 cluster c chains, for which some of the ground-state degeneracies can be attributed to global symmetries and may be broken spontaneously. Undoing the topological character of a degeneracy by associating it to a global symmetry is possible because of the nonlocality of the Jordan-Wigner transformation.

The spin-1/2 cluster c chains or their Majorana c chains counterparts are Hamiltonians whose eigenstates and eigenenergies can be obtained in closed form. This is not the case for the following two celebrated Hamiltonians:

$$\widehat{H}_{b} := J \sum_{j=1}^{N-b} \left[\widehat{S}^{x}_{j}\, \widehat{S}^{x}_{j+1} + \Delta_{y}\, \widehat{S}^{y}_{j}\, \widehat{S}^{y}_{j+1} + \Delta_{z}\, \widehat{S}^{z}_{j}\, \widehat{S}^{z}_{j+1} \right], \qquad J > 0, \tag{6.1a}$$

where N is a positive integer, the three components $a = x, y, z = 1, 2, 3$ of the spin operator $\widehat{\boldsymbol{S}}_{j}$ obey the $\mathfrak{su}(2)$ algebra

$$\left[\widehat{S}^{\alpha}_{j}, \widehat{S}^{\beta}_{j'} \right] = \delta_{j,j'}\, \mathrm{i}\epsilon^{\alpha\beta\gamma}\, \widehat{S}^{\gamma}_{j}, \qquad \alpha, \beta = x, y, z = 1, 2, 3, \qquad \hbar = 1 \tag{6.1b}$$

(with the summation convention over repeated indices and $\epsilon^{\alpha\beta\gamma}$ the fully antisymmetric Levi-Civita tensor), while

$$1 = \Delta_{y} = \Delta_{z}, \qquad \widehat{S}^{2}_{j} = S(S+1) = \frac{3}{4}, \tag{6.1c}$$

for the nearest-neighbor quantum spin-1/2 antiferromagnetic Heisenberg chain, and

$$1 = \Delta_y = \Delta_z, \qquad \widehat{S}_j^2 = S(S+1) = 2, \tag{6.1d}$$

for the nearest-neighbor quantum spin-1 antiferromagnetic Heisenberg chain. The number $b = 0, 1$ selects the boundary conditions (twisted if $b = 0$ and open if $b = 1$), which we leave momentarily unspecified.

The antiferromagnetic quantum spin-1/2 Heisenberg chain is an integrable model, as was anticipated by Bethe in 1931[1] when he introduced what is now called the Bethe Ansatz. Given the representation of the eigenenergies and eigenfunctions, derived from the Bethe Ansatz, it is possible to deduce that the spectrum is gapless in the thermodynamic limit $N \to \infty$. However, computing spin correlation functions with the Bethe Ansatz has only become possible by combining 60 years of progress in understanding integrable models combined with the computing power made available in the 1990s. Knowledge of spin correlation functions is needed to decide if long-range order is present in the ground state.

One motivation by Lieb, Schultz, and Mattis in 1961[2] was to find an analytical argument that could decide if the nearest-neighbor antiferromagnetic quantum spin-1/2 Heisenberg chain supports antiferromagnetic long-range order at zero temperature. Although they could not answer this question rigorously,[3,4] they could show rigorously that the antiferromagnetic quantum spin-1/2 XY Hamiltonian defined by choosing $\Delta_y = 1$ and $\Delta_z = 0$ instead of $1 = \Delta_y = \Delta_z$ in Eq. (6.1c) has a gapless spectrum with all correlation functions of spins decaying algebraically in space. In modern parlance, the antiferromagnetic quantum spin-1/2 XY chain realizes the Gaussian conformal field theory with central charge $\mathsf{c} = 1$. During the 1970s, it was realized that the effect of perturbing the quantum XY limit with $\Delta_z\, J \sum_{j=1}^{N} \widehat{S}_j^z\, \widehat{S}_{j+1}^z$ amounts to adding an irrelevant perturbation for $0 < \Delta_z < 1$, which becomes exactly marginal at the Heisenberg point $\Delta_z = 1$, and relevant when $\Delta_z > 1$. Correspondingly, the ground state of the antiferromagnetic quantum spin-1/2 Heisenberg chain does not support long-range order. Instead, it realizes a quantum critical point, the SU(2) level 1 [$\mathrm{SU}(2)_1$] Wess-Zumino-Witten (WZW) conformal field theory,[5] as was shown during the 1980's.[6] This conformal field theory has the central charge $\mathsf{c} = 1$.

In their appendices, Lieb, Schultz, and Mattis also established two theorems, the second of which is now called the Lieb-Schultz-Mattis theorem (**Exercises** 6.1, 6.2, 6.3, and 6.4).

[1] H. Bethe, Z. Phys. **71**(3), 205–226 (1931). [Bethe (1931)].

[2] E. Lieb, T. Schultz, and D. Mattis, Ann. Phys. (N.Y.) **16**(3), 407–466 (1961). [Lieb et al. (1961)].

[3] Mermin and Wagner proved rigorously that the ground state does not support antiferromagnetic long-range order in 1966.[4]

[4] N. D. Mermin and H. Wagner, Phys. Rev. Lett. **17**(22), 1133–1136 (1966). [Mermin and Wagner (1966)].

[5] E. Witten, Commun. Math. Phys. **92**(4), 455–472 (1984). [Witten (1984)].

[6] I. Affleck and F. D. M. Haldane, Phys. Rev. B **36**(10), 5291–5300 (1987). [Affleck and Haldane (1987)].

Theorem 6.1 *The ground state of the antiferromagnetic quantum spin-1/2 Heisenberg chain defined by Eqs. (6.1) and (6.1c) is annihilated by the total spin operator*

$$\widehat{\boldsymbol{S}} := \sum_{j=1}^{N} \widehat{\boldsymbol{S}}_j \tag{6.2}$$

for any even integer N.

Theorem 6.2 (Lieb-Schultz-Mattis) *The antiferromagnetic quantum spin-1/2 Heisenberg chain defined by Eqs. (6.1) and (6.1c) and obeying periodic boundary conditions supports an excited eigenstate with an energy of order $1/N$ above the nondegenerate ground state for any even integer N.*

Remark (1) The proof of **Theorem** 6.2 does not apply to the antiferromagnetic quantum spin-1 Heisenberg chain. It is still possible to show for $S = 1$ that there exists a state with an energy of order $1/N$ above the ground state for any N; however, it is not possible to show that this state is orthogonal to the ground state.

Remark (2) **Theorem** 6.2 only proves the existence of at least one excited state with an energy that collapses like $1/N$ to that of the ground state in the thermodynamic limit $N \to \infty$. This state might or might not be isolated so that there is no guarantee that the thermodynamic limit of this excited state is distinct from that of the ground state. Hence, neither has the degeneracy of the ground states separated by a gap from all excited states been calculated, nor has the existence of a gapless continuum of states above a possibly nondegenerate ground state in the thermodynamic limit been shown (one would need the Bethe Ansatz).

Remark (3) The proof of **Theorem** 6.2 makes use of the global SU(2) symmetry, time-reversal symmetry, and of **Theorem** 6.1 (that is the fact that the ground state is nondegenerate).

Remark (4) The same constructive proof applied to a dimension of space d larger than one would imply the bound N^{d-2} between the ground state and the excited states. This bound is thus useless when $d > 1$.

The Lieb-Schultz-Mattis theorem 6.2 is not predictive for the antiferromagnetic quantum spin-1 Heisenberg chain. In the condensed matter community, it was believed until 1983 that the low-energy physics of antiferromagnetic quantum spin-S Heisenberg chains should be qualitatively the same for any integer values of $2S$. However, Haldane argued in 1983 that the parity of $2S$ matters greatly for the spectrum of the antiferromagnetic quantum spin-S Heisenberg chain.[7] Haldane identified the low-energy physics of antiferromagnetic quantum spin-S Heisenberg chains in the large S (semiclassical) limit with that of a conventional quantum nonlinear sigma model (QNLSM) in $(1+1)$-dimensional spacetime, provided the limit $S \gg 1$

[7] F. D. M. Haldane, Phys. Lett. A **93**(9), 464–468 (1983); Phys. Rev. Lett. **50**(15), 1153–1156 (1983). [Haldane (1983a,b)].

is taken by choosing S to be integer only, while a topological term must be added to the QNLSM when the limit $S \gg 1$ is taken by choosing S to be half-integer only. Haldane then relied on the renormalization-group analysis of the nonlinear sigma model (NLSM) in two-dimensional Euclidean space by Polyakov in 1975[8] to conjecture that any antiferromagnetic quantum spin-S Heisenberg chain with S integer should have a gap between its ground state and all excitations above it, while any antiferromagnetic quantum spin-S Heisenberg chain with S half-integer should be gapless as this was known to be the case for the $S = 1/2$ case. This conjecture is now called the Haldane's conjecture. Although there is no rigorous proof of this conjecture when periodic boundary conditions are imposed, it is supported by both numerical studies of antiferromagnetic quantum spin-S Heisenberg chains and neutron scattering studies of quasi-one-dimensional antiferromagnetic quantum spin-S magnets.

The qualitative difference between half-integer and integer antiferromagnetic quantum spin-S Heisenberg chains motivated the following important refinements of the Lieb-Schultz-Mattis theorem 6.2.

Affleck and Lieb in 1986[9] relaxed the assumption that $S = 1/2$ and that the global internal symmetry is SU(2) in the Lieb-Schultz-Mattis theorem 6.2. They showed that any nearest-neighbor antiferromagnetic quantum spin-S Heisenberg chain made of an even number N of sites with translation and global internal spin U(1) symmetries has a nondegenerate ground state, whereby the gap is of order $1/N$ for S any half-integer.

Oshikawa, Yamanaka, and Affleck[10] extended this result of Affleck and Lieb as follows. Without loss of generality, assume that (N an even integer)

$$\widehat{S}^z := \sum_{j=1}^{N} \widehat{S}^z_j \tag{6.3}$$

is the generator of the global U(1) internal spin symmetry, as would happen if a uniform magnetic field pointing along the quantization (z) axis was applied. Let M^z be the eigenvalue of $\widehat{S}^z$, that is the uniform magnetization along the quantization axis in spin space. Define the dimensionless rational number

$$\nu := \frac{M^z}{N} + S \tag{6.4}$$

for the antiferromagnetic quantum spin-S Heisenberg ring in the presence of a Zeeman term and in its ground state. If ν is not an integer, then there exists an excited eigenstate with an energy of order $1/N$ above the nondegenerate ground state.

8 A. M. Polyakov, Phys. Lett. B **59**(1), 79–81 (1975). [Polyakov, A. M. (1975)].
9 I. Affleck and E. H. Lieb, Lett. Math. Phys. **12**(1), 57–69 (1986). [Affleck and Lieb (1986)].
10 M. Oshikawa, M. Yamanaka, and I. Affleck, Phys. Rev. Lett. **78**(10), 1984–1987 (1997). [Oshikawa et al. (1997)].

An analogous theorem was proven by Yamanaka, Oshikawa, and Affleck[11] for any local lattice models of interacting electrons for which the electronic charge is conserved, translation symmetry holds, and

$$\nu := \frac{N_{\mathrm{f}}}{N} \tag{6.5}$$

is the ratio between the (conserved) total number of electrons N_{f} and the even number of sites N on the ring.

All four papers so far had always chosen Hamiltonians for which either reversal of time or inversion in space were symmetries. This assumption was shown by Koma to be superfluous.[12]

We shall present a more recent version of the Lieb-Schultz-Mattis theorem 6.2 that was proven by Tasaki,[13,14] in Section 6.2. Sections 6.3 and 6.4 are devoted to studying the consequences of replacing the continuous internal symmetry group by a discrete one in the assumptions of the Lieb-Schultz-Mattis theorem 6.2.

6.2 The Lieb-Schultz-Mattis Theorem with a Local Twist

6.2.1 Definitions and Assumptions

Define the chain

$$\Lambda := \{j \mid j = 1, \cdots, p\,N\}, \tag{6.6a}$$

where the integer $p = 1, 2, \cdots$ defines the length of the repeat unit cell, and $N = 2, 4, \cdots$ defines the large even number of repeat unit cells. The lattice spacing is denoted $\mathfrak{a}$ so that this chain has the length $L = p\,N\,\mathfrak{a}$. We endow Λ with the topology of a ring by demanding that

$$j + p\,N \equiv j, \qquad j \in \Lambda. \tag{6.6b}$$

We assign to Λ either the fermionic Fock space

$$\mathfrak{F}^{(\mathrm{f})}_{\Lambda} := \mathrm{span}\left\{ \left[\prod_{j\in\Lambda} \left(\hat{c}^{\dagger}_{j}\right)^{n_j}\right] |0\rangle \;\middle|\; n_j = 0, 1, \qquad \hat{c}_j\,|0\rangle = 0, \right.$$
$$\left. \left\{\hat{c}_j, \hat{c}^{\dagger}_{j'}\right\} = \delta_{j,j'}, \quad \left\{\hat{c}^{\dagger}_{j}, \hat{c}^{\dagger}_{j'}\right\} = \left\{\hat{c}_j, \hat{c}_{j'}\right\} = 0 \right\} \tag{6.7a}$$

[11] M. Yamanaka, M. Oshikawa, and I. Affleck, Phys. Rev. Lett. **79**(6), 1110–1113 (1997). [Yamanaka et al. (1997)].

[12] T. Koma, J. Stat. Phys. **99**(1), 313–381 (2000). [Koma (2000)].

[13] H. Tasaki, J. Stat. Phys. **170**(4), 653–671 (2018). [Tasaki (2018)].

[14] H. Tasaki, *Physics and Mathematics of Quantum Many-Body Systems,* Springer, Switzerland, 2020. [Tasaki (2020)].

or the bosonic Fock space

$$\mathfrak{F}^{(\mathrm{b})}_{\Lambda} := \mathrm{span}\left\{ \left[\prod_{j\in\Lambda}\left(\hat{c}^{\dagger}_{j}\right)^{n_j}\right]|0\rangle \,\middle|\, n_j = 0,1,2,\cdots, \quad \hat{c}_j\,|0\rangle = 0, \quad \left[\hat{c}_j, \hat{c}^{\dagger}_{j'}\right] = \delta_{j,j'}, \quad \left[\hat{c}^{\dagger}_j, \hat{c}^{\dagger}_{j'}\right] = \left[\hat{c}_j, \hat{c}_{j'}\right] = 0 \right\}. \tag{6.7b}$$

The local number operator is defined by

$$\hat{n}_j := \hat{c}^{\dagger}_j\,\hat{c}_j, \qquad j \in \Lambda. \tag{6.8}$$

The unitary operator $\widehat{T}_p$ that generates translations along the chain by a repeat unit cell is defined by its action

$$\widehat{T}_p\,\hat{c}_j\,\widehat{T}^{\dagger}_p = \hat{c}_{j+p}, \tag{6.9a}$$

where we demand that the chain has the topology of a ring through the constraint

$$\widehat{T}^{N}_p\,\hat{c}_j\,(\widehat{T}^{\dagger}_p)^{N} = \hat{c}_{j+p\,N} \equiv \hat{c}_j \tag{6.9b}$$

for any $j \in \Lambda$. Let $r = 0,1,2,\cdots$ define the range of the hopping amplitudes

$$t_{j,j'} = t^{*}_{j',j} \in \mathbb{C}, \qquad t_{j,j'} = t_{j+p,j'+p}, \tag{6.10a}$$

through the condition

$$|j - j'| > r \Longrightarrow t_{j,j'} = 0 \tag{6.10b}$$

for any $j, j' \in \Lambda$. Let

$$\widehat{V}^{\dagger}_j = \widehat{V}_j \text{ such that } \widehat{T}_p\left(\sum_{j=1}^{p\,N}\widehat{V}_j\right)\widehat{T}^{\dagger}_p = \sum_{j=1}^{p\,N}\widehat{V}_j \tag{6.11a}$$

be any p-periodic Hermitean operator that depends only on the number operators $\hat{n}_j, \cdots, \hat{n}_{j+r}$. For example, the 1-periodic choice

$$\widehat{V}_j := \mu\,\hat{n}_j + U\,\hat{n}_j\left(\hat{n}_j - 1\right), \qquad U > 0, \tag{6.11b}$$

is used for hard core bosons with the chemical potential μ, while the 1-periodic choice

$$\widehat{V}_j := U\left(\hat{n}_j - \frac{1}{2}\right)\left(\hat{n}_{j+1} - \frac{1}{2}\right), \qquad U > 0, \tag{6.11c}$$

is used for a short-range particle-hole-symmetric repulsive interaction between spinless fermions at vanishing chemical potential. By construction, the Hamiltonian

$$\widehat{H} := -\sum_{j=1}^{p\,N}\sum_{j'=j-r}^{j+r} t_{j,j'}\,\hat{c}^{\dagger}_j\,\hat{c}_{j'} + \sum_{j=1}^{p\,N}\widehat{V}_j \tag{6.12}$$

is invariant under the global U(1) gauge transformation

$$\hat{c}_j^\dagger \mapsto \hat{c}_j^\dagger \, e^{-\mathrm{i}\theta}, \qquad \hat{c}_j \mapsto e^{+\mathrm{i}\theta} \, \hat{c}_j, \tag{6.13}$$

for any $0 \leq \theta < 2\pi$ and for any $j \in \Lambda$ as it is under the translation by p sites

$$\begin{aligned}
\widehat{T}_p \, \widehat{H} \, \widehat{T}_p^\dagger &= -\sum_{j=1}^{pN} \sum_{j'=j-r}^{j+r} t_{j,j'} \, \hat{c}_{j+p}^\dagger \, \hat{c}_{j'+p} + \sum_{j=1}^{pN} \widehat{V}_{j+p} \\
&= -\sum_{j=1}^{pN} \sum_{j'=j-r}^{j+r} t_{j-p,j'-p} \, \hat{c}_j^\dagger \, \hat{c}_{j'} + \sum_{j=1}^{pN} \widehat{V}_j \\
&= \widehat{H}.
\end{aligned} \tag{6.14}$$

Because we allow the possibilities that

$$\arg t_{j,j'} \neq 0, \pi \tag{6.15a}$$

(that would break time-reversal symmetry on a ring),

$$t_{j+n,j'+n} \neq t_{j,j'}, \tag{6.15b}$$

or

$$\widehat{T}_n \left(\sum_{j=1}^{pN} \widehat{V}_j \right) \widehat{T}_n^\dagger \neq \sum_{j=1}^{pN} \widehat{V}_j \tag{6.15c}$$

for $n = 1, \cdots, p-1$ (that would break inversion symmetry within the repeat unit cell), neither time-reversal symmetry nor inversion symmetry are presumed. Let

$$\widehat{N}_{\mathrm{f/b}} := \sum_{j=1}^{pN} \hat{n}_j \tag{6.16}$$

denote the global number operator. Because of the global symmetry (6.13) and the translation symmetry (6.14),

$$\left[\widehat{N}_{\mathrm{f/b}}, \widehat{H}\right] = \left[\widehat{H}, \widehat{T}_p\right] = \left[\widehat{T}_p, \widehat{N}_{\mathrm{f/b}}\right] = 0. \tag{6.17}$$

Hence, $\widehat{H}$, $\widehat{N}_{\mathrm{f/b}}$, and $\widehat{T}_p$ can be diagonalized simultaneously. We shall restrict the Fock space defined in Eq. (6.7) to be the Hilbert space

$$\mathfrak{H}^\Lambda := \mathrm{span} \left\{ |n_1, \cdots, n_{pN}\rangle \, \middle| \, \sum_{j \in \Lambda} n_j = N_{\mathrm{f/b}} \right\} \tag{6.18}$$

with exactly $N_{\mathrm{f/b}} = 0, 1, 2, \cdots$ Fermi-like (f) or boson-like (b) particles. The filling fraction per repeat unit cell is then defined to be the rational number

$$\nu := \frac{N_{\mathrm{f/b}}}{N}, \qquad |\Lambda| := p\,N. \tag{6.19}$$

The thermodynamic limit $N \to \infty$ shall then always be taken holding ν fixed. The filling fraction per repeat unit cell ν is restricted to the interval $0 \leq \nu \leq p$ for spinless fermions and $0 \leq \nu \in \mathbb{R}$ for bosons. The ground state $|\Psi^{\Lambda}_{\mathrm{GS}}\rangle$ with the ground-state energy $E^{\Lambda}_{\mathrm{GS}}$ can always be chosen such that

$$\widehat{H}\,\left|\Psi^{\Lambda}_{\mathrm{GS}}\right\rangle = E^{\Lambda}_{\mathrm{GS}}\,\left|\Psi^{\Lambda}_{\mathrm{GS}}\right\rangle, \qquad \frac{\widehat{N}_{\mathrm{f/b}}}{N}\,\left|\Psi^{\Lambda}_{\mathrm{GS}}\right\rangle = \nu\,\left|\Psi^{\Lambda}_{\mathrm{GS}}\right\rangle, \qquad \widehat{T}_p\,\left|\Psi^{\Lambda}_{\mathrm{GS}}\right\rangle = e^{\mathrm{i}\kappa}\,\left|\Psi^{\Lambda}_{\mathrm{GS}}\right\rangle, \tag{6.20}$$

for some $\kappa \in [0, 2\pi[$.

The theorem that we seek makes use of the following operators. First, we need the coarse-grained density operator

$$\hat{\rho}_\ell := \frac{1}{\ell}\sum_{j=1}^{p\,\ell}\hat{n}_j, \qquad \ell = 1,\cdots,N, \tag{6.21}$$

from which we shall subtract the filling fraction per repeat unit cell ν:

$$\Delta\hat{\rho}_\ell := \hat{\rho}_\ell - \nu. \tag{6.22}$$

Second, we need the unitary ℓ-resolved twist operator

$$\widehat{U}_\ell := e^{+\mathrm{i}\sum\limits_{j=1}^{p\,\ell}\theta_j\,\hat{n}_j}, \qquad \theta_j := \begin{cases} \frac{2\pi}{\ell}\left\lfloor\frac{j-1}{p}\right\rfloor, & j = 1,\cdots,p\,\ell, \\[2ex] 2\pi, & j = p\,\ell+1,\cdots,p\,N, \end{cases} \tag{6.23}$$

where the lower-floor function $\lfloor x \rfloor$ returns the largest integer that does not exceed $0 \leq x < \infty$. If ℓ is held fixed in the thermodynamic limit $|\Lambda| \to \infty$, then this twist operator is local as it only acts in a nontrivial way on a bounded interval of the ring Λ. Third, we need the operator

$$\widehat{W}_\ell := \frac{\widehat{U}_\ell - \alpha_\ell}{\sqrt{1-|\alpha_\ell|^2}}, \qquad \alpha_\ell := \left\langle\Psi^{\Lambda}_{\mathrm{GS}}\right|\widehat{U}_\ell\left|\Psi^{\Lambda}_{\mathrm{GS}}\right\rangle \in \mathbb{C}. \tag{6.24}$$

The global twist operator obtained by choosing $\ell = N$ in Eq. (6.23) was used in the original Lieb-Schultz-Mattis theorem (see footnote 2), and by Oshikawa, Yamanaka, and Affleck (see footnote 10). The local twist operator (6.23) was used by Affleck and Lieb in 1986 (see footnote 9), Yamanaka, Oshikawa, and Affleck (see footnote 11), and Tasaki (see footnote 13). The advantage of using a local over a global twist operator is that one has a much better control of the thermodynamic limit $|\Lambda| \to \infty$ when using a local twist operator. Finally, we will need the positive kinetic energy scale C defined by

$$C := \frac{8\,\pi^2\,(p+1)(r+1)^2}{p^2}\,\nu\,\bar{t}, \qquad \bar{t} := \max_{j\in\Lambda}\sum_{j'\in\Lambda}|t_{j,j'}|. \tag{6.25}$$

6.2.2 Examples

Before stating and proving Tasaki's generalization of the Lieb-Schultz-Mattis theorem 6.2, we give some exactly soluble examples.

1 *Free spinless fermions:* Set

$$\widehat{V}_j = 0 \tag{6.26a}$$

for all $j \in \Lambda$ in Hamiltonian (6.12). There are p bands in the Brillouin zone $\Lambda^\star_{\mathrm{BZ}}$, whose dispersions are uniquely fixed by the hopping amplitudes $t_{j,j'}$ with $j, j' \in \Lambda$. We assume that all p bands have nonvanishing dispersions $\varepsilon^{(n)}_k$ with $k \in \Lambda^\star_{\mathrm{BZ}}$ and $n = 1, \cdots, p$. We also assume that band crossing never occurs, that is, the bands can be ordered by their minima on $\Lambda^\star_{\mathrm{BZ}}$ and any two consecutive bands are separated by a single-particle energy gap.

When ν is an integer, the lowest ν bands are fully filled while the remaining $p-\nu$ bands are empty in the nondegenerate ground state (6.20). All many-body excitations are separated from the ground-state energy

$$E^{\Lambda}_{\mathrm{GS}} = \sum_{n=1}^{\nu} \sum_{k\in\Lambda^\star_{\mathrm{BZ}}} \varepsilon^{(n)}_k \tag{6.26b}$$

by a many-body energy gap. The ground state is thus nondegenerate and gapfull to *all* excitations (see Figure 6.1). The latter property, as it survives the thermodynamic limit $|\Lambda| \to \infty$, defines the band insulating phase of matter when the fermionic charge is conserved. This scenario with $p = 2$ is the one realized by the spinless version of the single-particle Hamiltonian for polyaceylene (2.57c).

When ν is not an integer, the metallic ground state (6.20) consists of the first $\lfloor \nu \rfloor$ bands being fully filled, band $\lfloor \nu \rfloor + 1$ being partially filled up to the Fermi energy, and the remaining $p - \lfloor \nu \rfloor - 1$ bands being empty. In the thermodynamic limit $|\Lambda| \to \infty$, there is a continuum of gapless excitations. The ground state is nondegenerate and gapless (see Figure 6.2). If the filling fraction per repeat unit cell ν is not commensurate with the lattice,[15] the ground state remains nondegenerate and gapless when a weak $\widehat{V}_j$ is switched on adiabatically. The low-energy spectrum realizes a so-called Tomonaga-Luttinger liquid, the one-dimensional incarnation of a Fermi liquid in higher than one dimensions.

2 *Spinless fermions with nearest-neighbor repulsive interaction at $\nu = 1/2$:* Set

$$t_{j,j'} = 0 \tag{6.27a}$$

for all $j, j' \in \Lambda$ in Hamiltonian (6.12). Set

$$\widehat{V}_j = V\, \hat{n}_j\, \hat{n}_{j+1} \tag{6.27b}$$

[15] When $p = 1$, the commensurate filling fractions are $\nu = 1/n$ with $n = 2, 3, 4, \cdots$.

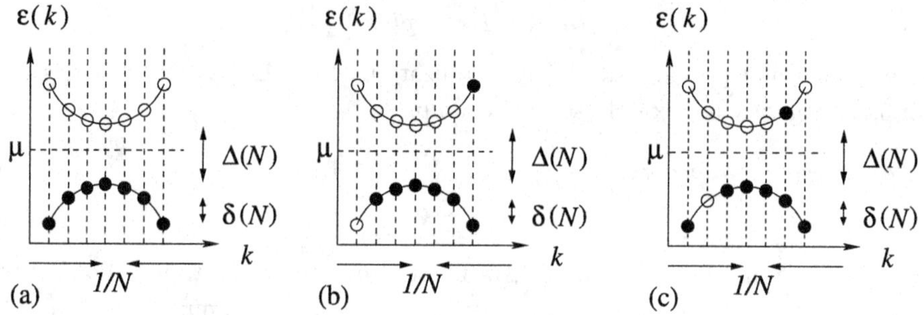

Figure 6.1 Many-body energy eigenstates of a (Bloch) two-band insulator in one-dimensional space. Single-particle energies are denoted $\varepsilon(k)$ in the thermodynamic limit $N \to \infty$ with N denoting the number of repeat unit cells. The chemical potential is denoted by μ. With periodic boundary conditions imposed, the Bloch momenta k are quantized with a spacing proportional to $1/N$. There follows the single-particle level spacing $\delta(N) \propto 1/N$. The single-particle gap $\Delta(N)$ as a function of N is $\Delta(N) = \Delta(\infty) + \mathcal{O}(1/N)$. (a) The many-body ground state for $\nu = 1$ (the insulating Fermi sea) is nondegenerate with all single-particle states occupied (filled circle) in the valence band and all single-particle states empty (open circle) in the conduction band. (b) A many-body excited state with energy of order $\Delta(N) + \mathcal{O}(1/N)$ obtained by creating a particle-hole pair with a momentum of order $1/N$ out of the insulating Fermi sea if measured relative to that of the insulating Fermi sea. (c) Another many-body excited state with energy $\Delta(N) + \mathcal{O}(1/N)$ obtained by creating a particle-hole pair with another momentum of order $1/N$ out of the insulating Fermi sea if measured relative to that of the insulating Fermi sea.

with $V > 0$ for all $j \in \Lambda$ in Hamiltonian (6.12). With these choices,

$$p = 1. \tag{6.27c}$$

Assume half-filling, that is, $\nu = 1/2$, for which the dimensionality of the Hilbert space is the binomial coefficient "choose $N/2$ out of N" (we recall that $|\Lambda| = N$ is assumed to be even). There are two orthonormal ground states with vanishing eigenenergy,

$$E^{\Lambda}_{\mathrm{GS}} = 0, \tag{6.27d}$$

namely,

$$\left|\Psi^{\Lambda}_{\mathrm{GS,\,odd}}\right\rangle := \prod_{j=1}^{N/2} \hat{c}^{\dagger}_{2j-1}|0\rangle \tag{6.27e}$$

and

$$\left|\Psi^{\Lambda}_{\mathrm{GS,\,even}}\right\rangle := \prod_{j=1}^{N/2} \hat{c}^{\dagger}_{2j}|0\rangle. \tag{6.27f}$$

Each one of these two ground states breaks spontaneously the translation symmetry. Having two spinless fermions occupying two consecutive sites costs the

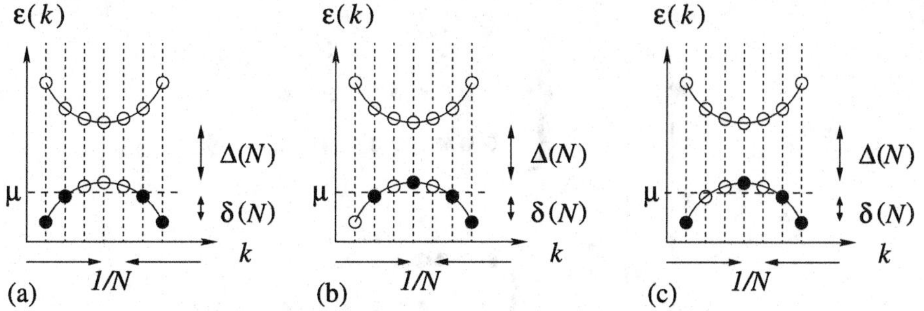

Figure 6.2 Many-body energy eigenstates of a (Bloch) two-band metal in one-dimensional space. Single-particle energies are denoted $\varepsilon(k)$ in the thermodynamic limit $N \to \infty$ with N denoting the number of repeat unit cells. The chemical potential is denoted by μ. With periodic boundary conditions imposed, the Bloch momenta k are quantized with a spacing proportional to $1/N$. There follows the single-particle level spacing $\delta(N) \propto 1/N$. (a) The many-body ground state for $\nu < 1$ (the metallic Fermi sea) is nondegenerate with all single-particle states occupied (filled circle) in the partially filled lower band and all single-particle states empty (open circle) in the upper band. (b) A many-body excited state with energy of order $1/N$ is obtained by creating a particle-hole pair with a momentum of order $1/N$ out of the metallic Fermi sea if measured relative to that of the metallic Fermi sea. (c) Another many-body excited state with energy of order $1/N$ is obtained by creating a particle-hole pair with another momentum of order $1/N$ out of the metallic Fermi sea if measured relative to that of the metallic Fermi sea.

energy V. Hence, an energy gap separates these two degenerate ground states from *all* excitations (see Figure 6.3). Define the linear combinations

$$\left|\Psi^{\Lambda}_{\mathrm{GS},\pm}\right\rangle := \frac{1}{\sqrt{2}}\left[\left|\Psi^{\Lambda}_{\mathrm{GS,\,odd}}\right\rangle \pm \left|\Psi^{\Lambda}_{\mathrm{GS,\,even}}\right\rangle\right]. \tag{6.27g}$$

These linear combinations are eigenstates of the generator of translations $\widehat{T}_1$ defined by Eq. (6.9) with the eigenvalues $e^{\mathrm{i}\kappa}$ whereby $\kappa = 0, \pi$, respectively, since

$$\widehat{T}_1 \left|\Psi^{\Lambda}_{\mathrm{GS},\pm}\right\rangle = \pm\left|\Psi^{\Lambda}_{\mathrm{GS},\pm}\right\rangle. \tag{6.27h}$$

Let $\widehat{O}$ denote any operator acting on the fermionic Fock space (6.7). We have the expectation values

$$\begin{aligned}\left\langle\Psi^{\Lambda}_{\mathrm{GS},\pm}\right|\widehat{O}\left|\Psi^{\Lambda}_{\mathrm{GS},\pm}\right\rangle =& \frac{1}{2}\left[\left\langle\Psi^{\Lambda}_{\mathrm{GS,\,odd}}\right|\widehat{O}\left|\Psi^{\Lambda}_{\mathrm{GS,\,odd}}\right\rangle\right.\\ &\left.+\left\langle\Psi^{\Lambda}_{\mathrm{GS,\,even}}\right|\widehat{O}\left|\Psi^{\Lambda}_{\mathrm{GS,\,even}}\right\rangle\right]\\ &\pm\frac{1}{2}\left[\left\langle\Psi^{\Lambda}_{\mathrm{GS,\,odd}}\right|\widehat{O}\left|\Psi^{\Lambda}_{\mathrm{GS,\,even}}\right\rangle\right.\\ &\left.+\left\langle\Psi^{\Lambda}_{\mathrm{GS,\,even}}\right|\widehat{O}\left|\Psi^{\Lambda}_{\mathrm{GS,\,odd}}\right\rangle\right].\end{aligned} \tag{6.27i}$$

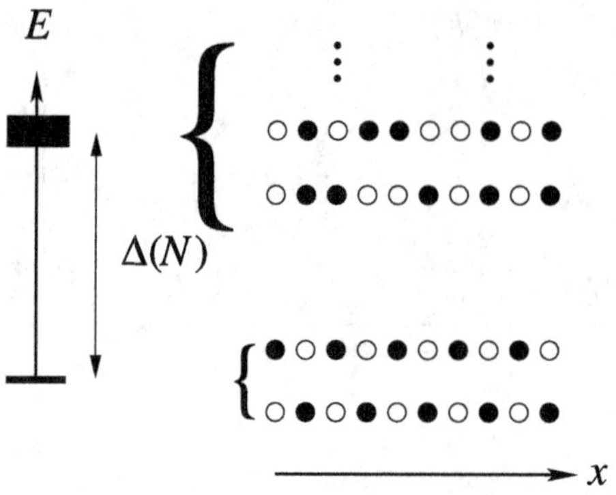

Figure 6.3 The basis of the Hilbert space for $\nu = 1/2$ is obtained by all the ways there are to color (occupy by a spinless fermion) half of the lattice sites (represented by circles) in black. There is an energy penality whenever two consecutive lattice sites are colored in black. In the continuum limit $N \to \infty$ with N the number of lattice sites, position in one-dimensional space is denoted by x. Many-body eigenenergies are denoted by E. The ground states are twofold degenerate (thick line at the bottom of the energy axis). The basis elements chosen to represent the ground states are interchanged by the translation by one lattice spacing. All excited states (lines along the energy axis whose thickness is indicative of the degeneracy) follow from creating defects in any one of the two ground states. Defects arise when two consecutive sites are colored in black. Hence, the degeneracy of all energy eigenvalues above the energy gap $\Delta(N) = \Delta(\infty) + \mathcal{O}(1/N)$ measured relative to the ground-state energy grows with N (in addition to a minimal degeneracy of two).

The choice $\widehat{O} = \hat{n}_j$ gives

$$\langle \Psi^{\Lambda}_{\mathrm{GS},\pm} | \, \hat{n}_j \, | \Psi^{\Lambda}_{\mathrm{GS},\pm} \rangle = \frac{1}{2} \tag{6.27j}$$

for any $j \in \Lambda$. The choice $\widehat{O} = \hat{n}_j \, \hat{n}_{j'}$ gives

$$\langle \Psi^{\Lambda}_{\mathrm{GS},\pm} | \, \hat{n}_j \, \hat{n}_{j'} \, | \Psi^{\Lambda}_{\mathrm{GS},\pm} \rangle = \begin{cases} 0, & \text{if } |j - j'| \text{ is odd,} \\ \\ \frac{1}{2}, & \text{if } |j - j'| \text{ is even,} \end{cases} \tag{6.27k}$$

for any $j, j' \in \Lambda$. There follows the truncated correlation function

$$\begin{aligned} &\langle \Psi^{\Lambda}_{\mathrm{GS},\pm} | \, \hat{n}_j \, \hat{n}_{j'} \, | \Psi^{\Lambda}_{\mathrm{GS},\pm} \rangle - \langle \Psi^{\Lambda}_{\mathrm{GS},\pm} | \, \hat{n}_j \, | \Psi^{\Lambda}_{\mathrm{GS},\pm} \rangle \langle \Psi^{\Lambda}_{\mathrm{GS},\pm} | \, \hat{n}_{j'} \, | \Psi^{\Lambda}_{\mathrm{GS},\pm} \rangle \\ &\qquad = \begin{cases} 0 - \frac{1}{4} = -\frac{1}{4}, & \text{if } |j - j'| \text{ is odd,} \\ \\ \frac{1}{2} - \frac{1}{4} = +\frac{1}{4}, & \text{if } |j - j'| \text{ is even,} \end{cases} \end{aligned} \tag{6.27l}$$

for any $j, j' \in \Lambda$. This is an oscillating function that never decays regardless of the value $|j - j'| = 0, 1, 2, \cdots, |\Lambda| - 1$. The absence of decay in the truncated correlation function (6.27l) is the finite-size signature of the spontaneous symmetry

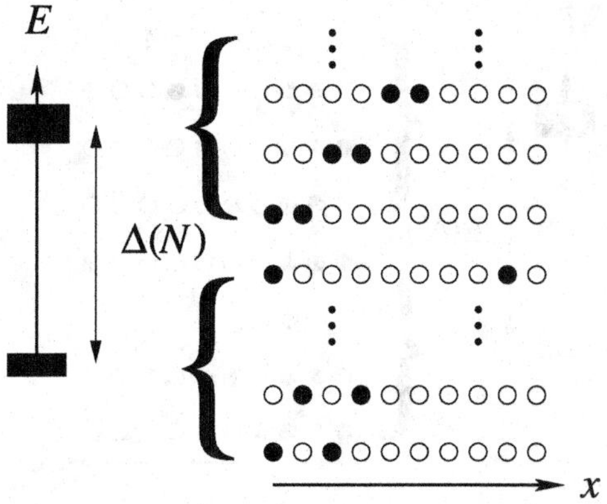

Figure 6.4 Same as Figure 6.3 except for the basis of the Hilbert space for $\nu < 1/2$ being obtained by all the ways there are to color (occupy by a spinless fermion) some number less than half that the lattice sites (represented by circles) in black.

breaking of translation symmetry by one lattice spacing in the thermodynamic limit $|\Lambda| \to \infty$ through the selection of a charge-density wave as the many-body gapped ground state.

3 *Spinless fermions with nearest-neighbor repulsive interaction at $\nu < 1/2$:* Assume that Eqs. (6.27a) and (6.27b) hold but that $\nu < 1/2$. As in example (6.27), the basis of the Hilbert space defined by Eq. (6.18) diagonalizes the Hamiltonian for spinless fermions. Because the Hamiltonian is nonnegative, any eigenstate with vanishing energy is a ground state, that is,

$$E^{\Lambda}_{\mathrm{GS}} = 0. \tag{6.28}$$

Any basis state from Eq. (6.18) with no consecutive occupied sites is a ground state. All other basis states have an eigenenergy bounded by V from below. The number of orthonormal states with vanishing eigenenergy grows exponentially fast with $|\Lambda|$.[16] A gap bounded from below by V separates these degenerate ground states from *all* excited eigenstates (see Figure 6.4).

4 *Spinless fermions with nearest-neighbor attractive interaction at $\nu < 1$:* Set

$$t_{j,j'} = 0 \tag{6.29a}$$

for all $j, j' \in \Lambda$ in Hamiltonian (6.12). Set

$$\widehat{V}_j = -V\, \hat{n}_j\, \hat{n}_{j+1} \tag{6.29b}$$

with $V > 0$ for all $j \in \Lambda$ in Hamiltonian (6.12). With these choices,

$$p = 1. \tag{6.29c}$$

[16] To estimate this number, one may approximate the growth with N of the central binomial coefficient "choose $N/4$ out of $N/2$" assuming that N is an integer multiple of 4. In turn, this can be done with the help of the Stirling approximation $n! \approx \sqrt{n}\, n^{n}$ for any large integer n.

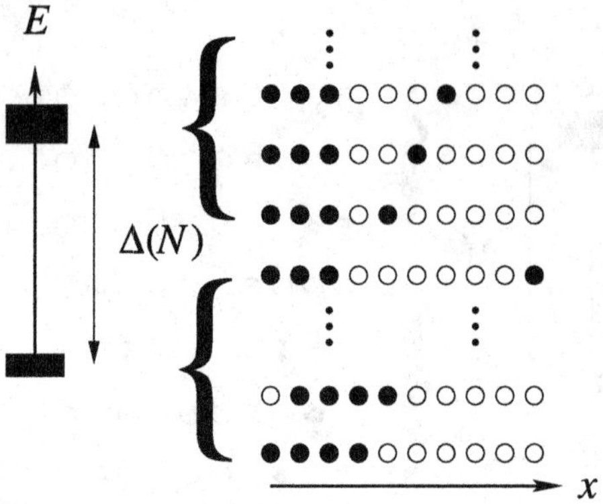

Figure 6.5 Same as Figure 6.3 except for two differences. First, the basis of the Hilbert space for $\nu < 1$ is obtained by all the ways there are to color (occupy by a spinless fermion) some number less than the number N of lattice sites (represented by circles) in black. Second, there is an energy gain whenever two consecutive lattice sites are colored in black.

Assume that

$$\nu \equiv N_{\mathrm{f}}/N < 1. \tag{6.29d}$$

We recall that $N = |\Lambda|$ is assumed to be even. One verifies that the $|\Lambda|$ orthonormal states

$$\begin{aligned} \left|\Psi^{\Lambda}_{\mathrm{GS},\,1}\right\rangle &:= \hat{c}^{\dagger}_{1} \cdots \hat{c}^{\dagger}_{N_{\mathrm{f}}} \,|0\rangle, \\ &\vdots \\ \left|\Psi^{\Lambda}_{\mathrm{GS},\,|\Lambda|}\right\rangle &:= \hat{c}^{\dagger}_{|\Lambda|} \cdots \hat{c}^{\dagger}_{|\Lambda|+N_{\mathrm{f}}-1} \,|0\rangle, \end{aligned} \tag{6.29e}$$

are ground states of Hamiltonian (6.12) with the eigenenergy (see Figure 6.5)

$$E^{\Lambda}_{\mathrm{GS}} = -\left(N_{\mathrm{f}} - 1\right) V = -\left(\nu\,|\Lambda| - 1\right) V. \tag{6.29f}$$

Here, we have used the identity

$$\sum_{j'=1}^{|\Lambda|} \hat{n}_{j'}\, \hat{n}_{j'+n} \left|\Psi^{\Lambda}_{\mathrm{GS},\,j}\right\rangle = \begin{cases} \left(N_{\mathrm{f}} - n\right) \left|\Psi^{\Lambda}_{\mathrm{GS},\,j}\right\rangle, & \text{if } 0 \leq n \leq N_{\mathrm{f}} - 1, \\ 0, & \text{if } N_{\mathrm{f}} - 1 < n, \end{cases} \tag{6.29g}$$

for any $n = 0, 1, \cdots, |\Lambda| - 1$. Each of these ground states is a single cluster made of N_{f} spinless fermions occupying consecutive sites of the ring. From Eq. (6.9), we deduce the relation

$$\widehat{T}_1 \left|\Psi^{\Lambda}_{\mathrm{GS},\,j}\right\rangle = \left|\Psi^{\Lambda}_{\mathrm{GS},\,j+1}\right\rangle \tag{6.29h}$$

for any $j \in \Lambda$, that is, any $|\Psi^{\Lambda}_{\mathrm{GS},\,j}\rangle$ breaks spontaneously the translation symmetry. On the other hand, translation invariance as defined by

$$\widehat{T}_1\, |\Psi^{\Lambda}_{\mathrm{GS},\,\kappa}\rangle = e^{+\mathrm{i}\kappa}\, |\Psi^{\Lambda}_{\mathrm{GS},\,\kappa}\rangle, \qquad |\Psi^{\Lambda}_{\mathrm{GS},\,\kappa}\rangle := \frac{1}{\sqrt{|\Lambda|}} \sum_{j=1}^{|\Lambda|} e^{-\mathrm{i}\kappa j}\, |\Psi^{\Lambda}_{\mathrm{GS},\,j}\rangle, \tag{6.29i}$$

holds for any $\kappa \in [0, 2\pi[$, and, in particular, for any $\kappa = (2\pi/|\Lambda|)\, n$ with $n = 0, 1, \cdots, |\Lambda| - 1$. Because $\hat{n}_i$ is diagonal in the basis (6.18), it is diagonal in the $|\Lambda|$-dimensional subspace spanned by the orthonormal states (6.29e) with the eigenvalues 0 or 1. However, $|\Psi^{\Lambda}_{\mathrm{GS},\,\kappa}\rangle$ is not an eigenstate of $\hat{n}_i$. Instead, we have the weaker identity

$$\langle\Psi^{\Lambda}_{\mathrm{GS},\,\kappa}|\, F[\hat{n}_i]\, |\Psi^{\Lambda}_{\mathrm{GS},\,\kappa}\rangle = \frac{1}{|\Lambda|} \sum_{j=1}^{|\Lambda|} \langle\Psi^{\Lambda}_{\mathrm{GS},\,j}|\, F[\hat{n}_i]\, |\Psi^{\Lambda}_{\mathrm{GS},\,j}\rangle \tag{6.29j}$$

for any functional $F[\hat{n}_i]$. Because of the transformation law (6.29i), we have the identities

$$\langle\Psi^{\Lambda}_{\mathrm{GS},\,\kappa}|\, \hat{n}_i\, |\Psi^{\Lambda}_{\mathrm{GS},\,\kappa}\rangle = \langle\Psi^{\Lambda}_{\mathrm{GS},\,\kappa}|\, \hat{n}_1\, |\Psi^{\Lambda}_{\mathrm{GS},\,\kappa}\rangle, \tag{6.29k}$$

$$\langle\Psi^{\Lambda}_{\mathrm{GS},\,\kappa}|\, \hat{n}_i\, \hat{n}_{i'}\, |\Psi^{\Lambda}_{\mathrm{GS},\,\kappa}\rangle = \langle\Psi^{\Lambda}_{\mathrm{GS},\,\kappa}|\, \hat{n}_1\, \hat{n}_{1+|i-i'|}\, |\Psi^{\Lambda}_{\mathrm{GS},\,\kappa}\rangle, \tag{6.29l}$$

for any $i, i' = 1, \cdots, |\Lambda|$. With the help of Eqs. (6.29c), (6.29j), and (6.29g), we conclude that

$$\langle\Psi^{\Lambda}_{\mathrm{GS},\,\kappa}|\, \hat{n}_i\, |\Psi^{\Lambda}_{\mathrm{GS},\,\kappa}\rangle = \nu, \tag{6.29m}$$

$$\langle\Psi^{\Lambda}_{\mathrm{GS},\,\kappa}|\, \hat{n}_i\, \hat{n}_{i'}\, |\Psi^{\Lambda}_{\mathrm{GS},\,\kappa}\rangle = \begin{cases} \nu - \frac{|i-i'|}{|\Lambda|}, & \text{if } 0 \le |i-i'| \le N_{\mathrm{f}} - 1, \\ 0, & \text{if } N_{\mathrm{f}} - 1 < |i-i'|, \end{cases} \tag{6.29n}$$

for any $i, i' = 1, \cdots, |\Lambda|$. The first punch line is that

$$\langle\Psi^{\Lambda}_{\mathrm{GS},\,\kappa}|\, \hat{\rho}_\ell\, |\Psi^{\Lambda}_{\mathrm{GS},\,\kappa}\rangle = \langle\Psi^{\Lambda}_{\mathrm{GS},\,\kappa}|\, \frac{1}{\ell} \sum_{i=1}^{\ell} \hat{n}_i\, |\Psi^{\Lambda}_{\mathrm{GS},\,\kappa}\rangle = \nu, \tag{6.29o}$$

that is

$$\langle\Psi^{\Lambda}_{\mathrm{GS},\,\kappa}|\, \Delta\hat{\rho}_\ell\, |\Psi^{\Lambda}_{\mathrm{GS},\,\kappa}\rangle = 0. \tag{6.29p}$$

The second punch line is that the expectation value in any eigenstate $|\Psi^{\Lambda}_{\mathrm{GS},\,\kappa}\rangle$ of

$$(\Delta\hat{\rho}_\ell)^2 = \frac{1}{\ell^2} \sum_{i,i'=1}^{\ell} \hat{n}_i\, \hat{n}_{i'} + \nu^2 - \frac{2\,\nu}{\ell} \sum_{i=1}^{\ell} \hat{n}_i \tag{6.29q}$$

is given by

$$\langle\Psi^{\Lambda}_{\mathrm{GS},\,\kappa}|\, (\Delta\hat{\rho}_\ell)^2\, |\Psi^{\Lambda}_{\mathrm{GS},\,\kappa}\rangle = \nu + \mathcal{O}(\nu^2) + \mathcal{O}(\ell/|\Lambda|). \tag{6.29r}$$

In other words, a simultaneous ground state and eigenstate of the translation operator $\widehat{T}_1$ is characterized by the nonvanishing variance

$$\begin{aligned}\operatorname{var}\Delta\hat{\rho}_\ell|_{\Psi^\Lambda_{\mathrm{GS},\,\kappa}} :=& \left\langle\Psi^\Lambda_{\mathrm{GS},\,\kappa}\right| \left(\Delta\hat{\rho}_\ell\right)^2 \left|\Psi^\Lambda_{\mathrm{GS},\,\kappa}\right\rangle - \left(\left\langle\Psi^\Lambda_{\mathrm{GS},\,\kappa}\right| \Delta\hat{\rho}_\ell \left|\Psi^\Lambda_{\mathrm{GS},\,\kappa}\right\rangle\right)^2 \\ =& \,\nu + \mathcal{O}(\nu^2) + \mathcal{O}(\ell/|\Lambda|) \end{aligned} \tag{6.29s}$$

of the coarse-grained density operator $\Delta\hat{\rho}_\ell$ for any choice of the ratio $0 < \ell/|\Lambda| \ll 1$ (**Exercise** 6.5).

6.2.3 Two Extensions of the Lieb-Schultz-Mattis Theorem

Theorem 6.3 (Tasaki) *With all symbols defined in Section 6.2.1:*
(i) Assume the ring Λ with the repeat unit cell of length p and a positive integer number N of repeat unit cells, that is, its cardinality is $|\Lambda| = p\,N$.
(ii) Assume the charge conserving and translation invariant (by integer multiples of p) lattice Hamiltonian $\widehat{H}$ for particles obeying either the fermion (f) or boson (b) algebra with hopping amplitudes and interactions restricted to the range r and whose domain of definition is the Hilbert space with $N_{\mathrm{f/b}}$ identical f/b particles.
(iii) Assume that the filling fraction per repeat unit cell

$$\nu \equiv \frac{N_{\mathrm{f/b}}}{N} \tag{6.30a}$$

is not an integer.
(iv) Assume that, for some sufficiently large $1 \leq \ell \leq N$, the bound

$$\left\langle\Psi^\Lambda_{\mathrm{GS},\,\kappa}\right| \left(\Delta\hat{\rho}_\ell\right)^2 \left|\Psi^\Lambda_{\mathrm{GS},\,\kappa}\right\rangle \leq \frac{\sin^2(\pi\,\nu)}{2\pi^2} \tag{6.30b}$$

holds on the expectation value of the particle density operator $\Delta\hat{\rho}_\ell$ that is coarse-grained (averaged) over $p\,\ell$ consecutive sites of Λ and measured relative to its expectation value ν in any normalizable ground state $|\Psi^\Lambda_{\mathrm{GS},\,\kappa}\rangle$ with the energy eigenvalue E^Λ_{GS} that is also an eigenstate of the translation by one repeat unit cell with the eigenvalue $\exp(\mathrm{i}\kappa)$.
Then, there exists an operator $\widehat{W}_\ell$ that depends only on the local particle number operators $\hat{n}_j$ with $j = 1, \cdots, p\,\ell$ and a strictly positive real-valued number C independent of ℓ and N such that

$$\left\langle\Psi^\Lambda_{\mathrm{GS},\,\kappa}\right| \widehat{W}_\ell \left|\Psi^\Lambda_{\mathrm{GS},\,\kappa}\right\rangle = 0, \tag{6.30c}$$

$$\left\langle\Psi^\Lambda_{\mathrm{GS},\,\kappa}\right| \widehat{W}^\dagger_\ell\, \widehat{W}_\ell \left|\Psi^\Lambda_{\mathrm{GS},\,\kappa}\right\rangle = 1, \tag{6.30d}$$

$$\left\langle\Psi^\Lambda_{\mathrm{GS},\,\kappa}\right| \widehat{W}^\dagger_\ell\, \widehat{H}\, \widehat{W}_\ell \left|\Psi^\Lambda_{\mathrm{GS},\,\kappa}\right\rangle - E^\Lambda_{\mathrm{GS}} \leq \frac{C}{\ell}. \tag{6.30e}$$

Remark Equation (6.30b) is a bound on the variance of the particle density coarse-grained over $p\,\ell$ consecutive sites of Λ.

Definition 6.4 (clustering property) In general, a quantum many-body state $|\Phi\rangle$ is said to satisfy the cluster decomposition property or clustering property if, for any pair of local operators $\widehat{O}_1(\boldsymbol{r})$ and $\widehat{O}_2(\boldsymbol{r})$, the identity

$$\lim_{|\boldsymbol{r}-\boldsymbol{r}'|\to\infty}\left\langle\Phi\middle|\widehat{O}_1(\boldsymbol{r})\,\widehat{O}_2(\boldsymbol{r}')\middle|\Phi\right\rangle=\lim_{|\boldsymbol{r}-\boldsymbol{r}'|\to\infty}\left\langle\Phi\middle|\widehat{O}_1(\boldsymbol{r})\middle|\Phi\right\rangle\left\langle\Phi\middle|\widehat{O}_2(\boldsymbol{r}')\middle|\Phi\right\rangle \tag{6.31}$$

holds.

If $\left|\Psi^{\Lambda}_{\mathrm{GS},\,\kappa}\right\rangle$ satisfies the cluster decomposition property, then

$$\begin{aligned}&\lim_{|j-j'|\to\infty}\lim_{|\Lambda|\to\infty}\langle\Psi^{\Lambda}_{\mathrm{GS},\,\kappa}\,|\,\hat{n}_j\,\hat{n}_{j'}\,|\,\Psi^{\Lambda}_{\mathrm{GS},\,\kappa}\rangle\\&\quad=\lim_{|j-j'|\to\infty}\lim_{|\Lambda|\to\infty}\langle\Psi^{\Lambda}_{\mathrm{GS},\,\kappa}\,|\,\hat{n}_j\,|\,\Psi^{\Lambda}_{\mathrm{GS},\,\kappa}\rangle\,\langle\Psi^{\Lambda}_{\mathrm{GS},\,\kappa}\,|\,\hat{n}_{j'}\,|\,\Psi^{\Lambda}_{\mathrm{GS},\,\kappa}\rangle\end{aligned} \tag{6.32}$$

for all $j,j'\in\Lambda$ so that Eq. (6.30b) will hold for ℓ sufficiently large. On the other hand, if

$$\lim_{|\Lambda|\to\infty}\langle\Psi^{\Lambda}_{\mathrm{GS},\,\kappa}|\,(\Delta\hat{\rho}_\ell)^2\,|\Psi^{\Lambda}_{\mathrm{GS},\,\kappa}\rangle=C'>0 \tag{6.33}$$

with $C'>0$ a constant independent of ℓ holds, then $\left|\Psi^{\Lambda}_{\mathrm{GS},\,\kappa}\right\rangle$ does not satisfy the cluster decomposition property. Instead, Eq. (6.33) is equivalent to the existence of long-range order and the spontaneous symmetry breaking of a symmetry in the thermodynamic limit $N\to\infty$, the symmetry under translations by integer multiples of p, since the U(1) continuous symmetry cannot be broken in one-dimensional space. Assumption (iv) is thus always satisfied if $\left|\Psi^{\Lambda}_{\mathrm{GS},\,\kappa}\right\rangle$ does not support any long-range order. We have arrived to the following corollary.

Corollary *Under the hypothesis (i)-(iii) of* ***Theorem*** *6.3 and (iv') if the variance*

$$\begin{aligned}\operatorname{var}\Delta\hat{\rho}_\ell|_{\Psi^{\Lambda}_{\mathrm{GS},\,\kappa}}&:=\langle\Psi^{\Lambda}_{\mathrm{GS},\,\kappa}|\,(\Delta\hat{\rho}_\ell)^2\,|\Psi^{\Lambda}_{\mathrm{GS},\,\kappa}\rangle-\left(\langle\Psi^{\Lambda}_{\mathrm{GS},\,\kappa}|\,\Delta\hat{\rho}_\ell\,|\Psi^{\Lambda}_{\mathrm{GS},\,\kappa}\rangle\right)^2\\&=\langle\Psi^{\Lambda}_{\mathrm{GS},\,\kappa}|\,(\Delta\hat{\rho}_\ell)^2\,|\Psi^{\Lambda}_{\mathrm{GS},\,\kappa}\rangle\end{aligned} \tag{6.34}$$

can be made as small as needed by letting ℓ be sufficiently large ($\ell\ll N$), then Eqs. (6.30c)-(6.30e) follow. Hypothesis (iv') is stronger than hypothesis (iv) in Eq. (6.30b). It is always met when the ground state $\left|\Psi^{\Lambda}_{\mathrm{GS},\,\kappa}\right\rangle$ obeys the clustering property.

Remark (1) If the constant C' on the right-hand side of Eq. (6.33) is larger than the right-hand side of Eq. (6.30b), Watanabe[17] has shown that we can replace $\widehat{W}_\ell$ by $\Delta\hat{\rho}_\ell$ and C/ℓ by C''/ℓ^2 with some $C''>0$ in Eqs. (6.30c), (6.30d), and (6.30e).

Remark (2) Equation (6.30c) asserts that $\widehat{W}_\ell\left|\Psi^{\Lambda}_{\mathrm{GS},\,\kappa}\right\rangle$ is orthogonal to the translation invariant (up to a multiplicative phase factor) ground state $\left|\Psi^{\Lambda}_{\mathrm{GS},\,\kappa}\right\rangle$.

Remark (3) Equation (6.30d) asserts that $\widehat{W}_\ell\left|\Psi^{\Lambda}_{\mathrm{GS},\,\kappa}\right\rangle$ is normalizable.

[17] H. Watanabe, Phys. Rev. Lett. **118**(11), 117205 (2017). [Watanabe (2017)].

Remark (4) By combining Eq. (6.30e) with the variational principle, there must exist at least one normalizable energy eigenstate that is orthogonal to $\left|\Psi^{\Lambda}_{\mathrm{GS},\,\kappa}\right\rangle$ and whose energy eigenvalue E^{Λ} satisfies

$$E^{\Lambda} - E^{\Lambda}_{\mathrm{GS}} < \frac{C}{\ell}. \tag{6.35}$$

In the thermodynamic limit $N \to \infty$, the energy of this state can be chosen to be arbitrarily close to $E^{\Lambda}_{\mathrm{GS}}$ by choosing ℓ sufficiently large but finite.

Remark (5) It was not assumed that the ground state $\left|\Psi^{\Lambda}_{\mathrm{GS},\,\kappa}\right\rangle$ is nondegenerate.

Remark (6) If ℓ is chosen to be N, then

$$\hat{\rho}_N = \frac{1}{N}\sum_{j=1}^{pN} \hat{n}_j \equiv \frac{\widehat{N}_{\mathrm{f/b}}}{N}, \qquad \Delta\hat{\rho}_N = \frac{\widehat{N}_{\mathrm{f/b}}}{N} - \nu. \tag{6.36}$$

In other words, the coarse-grained density operator with $\ell = N$ is nothing but the total number operator divided by the number of repeat unit cells N. On the Hilbert space (6.18), the operator identity

$$\hat{\rho}_N = \nu, \qquad \Delta\hat{\rho}_N = 0, \tag{6.37}$$

then follows. Hence, the bound (6.30b) in assumption (iv) is automatically satisfied. Define

$$\widehat{U}_N := \lim_{\ell\to N} \widehat{W}_{\ell}. \tag{6.38}$$

We may then state the following theorem.

Theorem 6.5 (Yamanaka, Oshikawa, Affleck, and Koma) *Under assumptions (i)-(ii) of* **Theorem** *6.3, there is a unitary operator $\widehat{U}_N$ such that*

$$\left\langle\Psi^{\Lambda}_{\mathrm{GS},\,\kappa}\right| \widehat{U}^{\dagger}_N\, \widehat{H}\, \widehat{U}_N \left|\Psi^{\Lambda}_{\mathrm{GS},\,\kappa}\right\rangle - E^{\Lambda}_{\mathrm{GS}} \leq \frac{C}{N}. \tag{6.39a}$$

Moreover, (iii) when the filling factor per unit repeat cell ν is not an integer, then

$$\left\langle\Psi^{\Lambda}_{\mathrm{GS},\,\kappa}\right| \widehat{U}_N \left|\Psi^{\Lambda}_{\mathrm{GS},\,\kappa}\right\rangle = 0 \tag{6.39b}$$

so that $\widehat{U}_N \left|\Psi^{\Lambda}_{\mathrm{GS},\,\kappa}\right\rangle$ is normalizable and orthogonal to $\left|\Psi^{\Lambda}_{\mathrm{GS},\,\kappa}\right\rangle$.

Remark According to **Theorem** 6.5 there exists at least one normalizable energy eigenstate that is orthogonal to $\left|\Psi^{\Lambda}_{\mathrm{GS},\,\kappa}\right\rangle$ and whose energy eigenvalue E^{Λ} satisfies

$$E^{\Lambda} - E^{\Lambda}_{\mathrm{GS}} < \frac{C}{N}. \tag{6.40}$$

However, as the thermodynamic limit $N \to \infty$ with ν held fixed is taken, one cannot decide between the possibility that $\left|\Psi^{\Lambda}_{\mathrm{GS},\,\kappa}\right\rangle$ and the variational state $\widehat{U}_N \left|\Psi^{\Lambda}_{\mathrm{GS},\,\kappa}\right\rangle$ converge to a unique state from the possibility that both states remain orthogonal. **Theorem** 6.3 is stronger in that, as the thermodynamic limit $N \to \infty$ with ν held

fixed is taken, $\left|\Psi^{\Lambda}_{\mathrm{GS},\,\kappa}\right\rangle$ and the variational state $\widehat{W}_{\ell}\left|\Psi^{\Lambda}_{\mathrm{GS},\,\kappa}\right\rangle$ remain orthogonal. Hence, if the assumptions of **Theorem** 6.3 are met as the thermodynamic limit $N\to\infty$ with ν held fixed is taken, it is then plausible to surmise that
(a) either the ground states are degenerate in the thermodynamic limit,
(b) or the ground states are nondegenerate with gapless excitations in the thermodynamic limit.

That this is indeed so was proven by Affleck and Lieb (see footnote 9).

Proof The proof of **Theorem** 6.3 is organized into three steps, the first two being formulated as lemmas.

Lemma 6.6 (Tasaki) *For any filling fraction per repeat unit cell ν, for any Hamiltonian of the form (6.12) with hopping amplitudes of range r, for any normalizable ground state $|\Psi^{\Lambda}_{\mathrm{GS},\,\kappa}\rangle$ with the energy eigenvalue $E^{\Lambda}_{\mathrm{GS}}$, and for any ℓ and N such that*

$$\max\left\{\frac{r+1}{p},2r-p\right\}\leq\ell\leq N,\qquad N=\frac{|\Lambda|}{p},\tag{6.41a}$$

there exists a unitary operator $\widehat{U}_{\ell}$ such that

$$\Delta E:=\left\langle\Psi^{\Lambda}_{\mathrm{GS},\,\kappa}\right|\widehat{U}^{\dagger}_{\ell}\,\widehat{H}\,\widehat{U}_{\ell}\left|\Psi^{\Lambda}_{\mathrm{GS},\,\kappa}\right\rangle-E^{\Lambda}_{\mathrm{GS}}\leq\frac{C}{2\,\ell},\tag{6.41b}$$

where the constant C is independent of ℓ and N and given in Eq. (6.25).

Proof For any $j,j'\in\Lambda$ and any $\theta\in[0,2\pi[$, we will make use of the identities

$$e^{-\mathrm{i}\theta\,\hat{n}_{j}}\,\hat{c}^{\dagger}_{j'}\,e^{+\mathrm{i}\theta\,\hat{n}_{j}}=e^{-\mathrm{i}\theta\,\delta_{j,j'}}\,\hat{c}^{\dagger}_{j'},\qquad e^{-\mathrm{i}\theta\,\hat{n}_{j}}\,\hat{c}_{j'}\,e^{+\mathrm{i}\theta\,\hat{n}_{j}}=e^{+\mathrm{i}\theta\,\delta_{j,j'}}\,\hat{c}_{j'},\tag{6.42}$$

that assert that particle (fermionic or bosonic) creation and annihilation operators carry the U(1) gauge charges $\mp$, respectively. It follows that the number operator is charge neutral:

$$e^{-\mathrm{i}\theta\,\hat{n}_{j}}\,\hat{n}_{j'}\,e^{+\mathrm{i}\theta\,\hat{n}_{j}}=\hat{n}_{j'}.\tag{6.43}$$

We recall that we had defined

$$\widehat{H}:=-\sum_{j,j'\in\Lambda}^{|j'-j|\leq r}t_{j,j'}\,\hat{c}^{\dagger}_{j}\,\hat{c}_{j'}+\sum_{j=1}^{pN}\widehat{V}_{j},\qquad t_{j,j'}=t^{*}_{j',j},\tag{6.44}$$

in Eq. (6.12) and

$$\widehat{U}_{\ell}:=e^{+\mathrm{i}\sum_{j=1}^{p\ell}\theta_{j}\,\hat{n}_{j}},\qquad\theta_{j}:=\begin{cases}\frac{2\pi}{\ell}\left\lfloor\frac{j-1}{p}\right\rfloor, & j=1,\cdots,p\ell,\\[2ex] 2\pi, & j=p\ell+1,\cdots,pN,\end{cases}\tag{6.45}$$

in Eq. (6.23). We thus have the pair of transformation laws

$$\widehat{U}_\ell^\dagger\,\widehat{H}\,\widehat{U}_\ell-\widehat{H}=\sum_{j,j'\in\Lambda}^{|j'-j|\le r}\left(1-e^{-\mathrm{i}(\theta_j-\theta_{j'})}\right)t_{j,j'}\,\hat{c}^\dagger_j\,\hat{c}_{j'},\tag{6.46}$$

$$\widehat{U}_\ell\,\widehat{H}\,\widehat{U}_\ell^\dagger-\widehat{H}=\sum_{j,j'\in\Lambda}^{|j'-j|\le r}\left(1-e^{+\mathrm{i}(\theta_j-\theta_{j'})}\right)t_{j,j'}\,\hat{c}^\dagger_j\,\hat{c}_{j'}.\tag{6.47}$$

We use the shorthand notation

$$\left\langle\Psi^\Lambda_{\mathrm{GS},\,\kappa}\right|\cdots\left|\Psi^\Lambda_{\mathrm{GS},\,\kappa}\right\rangle=\left\langle\cdots\right\rangle\tag{6.48}$$

for any ground-state expectation value. By the variational principle, $\Delta E\geq 0$. Hence, we have the first upper bound

$$\begin{aligned}
\Delta E&=\left\langle\left(\widehat{U}_\ell^\dagger\,\widehat{H}\,\widehat{U}_\ell-\widehat{H}\right)\right\rangle\\
\scriptstyle{\langle(\widehat{U}_\ell\widehat{H}\widehat{U}_\ell^\dagger-\widehat{H})\rangle\ge 0}\quad&\le\left\langle\left(\widehat{U}_\ell^\dagger\,\widehat{H}\,\widehat{U}_\ell-\widehat{H}\right)\right\rangle+\left\langle\left(\widehat{U}_\ell\,\widehat{H}\,\widehat{U}_\ell^\dagger-\widehat{H}\right)\right\rangle\\
&=\sum_{j,j'\in\Lambda}^{|j'-j|\le r}2\left[1-\cos\left(\theta_j-\theta_{j'}\right)\right]t_{j,j'}\left\langle\hat{c}^\dagger_j\,\hat{c}_{j'}\right\rangle\\
&\le\sum_{j,j'\in\Lambda}^{|j'-j|\le r}2\left[1-\cos\left(\theta_j-\theta_{j'}\right)\right]\left|t_{j,j'}\right|\left|\left\langle\hat{c}^\dagger_j\,\hat{c}_{j'}\right\rangle\right|.
\end{aligned}\tag{6.49}$$

For any pair $\widehat{A}$ and $\widehat{B}$ of operators, the Cauchy-Schwarz inequality states that

$$\left|\left\langle\widehat{A}^\dagger\,\widehat{B}\right\rangle\right|^2\le\left\langle\widehat{A}^\dagger\,\widehat{A}\right\rangle\left\langle\widehat{B}^\dagger\,\widehat{B}\right\rangle\tag{6.50}$$

(a generalization of the Cauchy inequality that states that the scalar product of two finite-dimensional vectors is smaller or equal in magnitude to the product of their norms). If we do the substitutions $\widehat{A}^\dagger\to\hat{c}^\dagger_j$ and $\widehat{B}\to\hat{c}_{j'}$, we have the second upper bound

$$\begin{aligned}
\left|\left\langle\hat{c}^\dagger_j\,\hat{c}_{j'}\right\rangle\right|&\le\sqrt{\left\langle\hat{n}_j\right\rangle\left\langle\hat{n}_{j'}\right\rangle}\\
&\le\max_{j\in\Lambda}\left\langle\hat{n}_j\right\rangle\\
\text{By translation symmetry}\quad&\le\sum_{j=1}^{p}\left\langle\hat{n}_j\right\rangle\\
\text{By definition of }\nu\quad&=\nu.
\end{aligned}\tag{6.51}$$

Because $\cos x\geq 1-(x^2/2)$, we have the third upper bound

$$2\left[1-\cos\left(\theta_j-\theta_{j'}\right)\right]\le\left(\Delta\theta_{j,j'}\right)^2,\quad\Delta\theta_{j,j'}:=\left(\theta_j-\theta_{j'}\right)\bmod 2\pi\in[0,2\pi[.\tag{6.52}$$

Combining all three upper bounds gives the fourth upper bound

$$\Delta E \leq \nu \sum_{j,j'\in\Lambda}^{|j'-j|\leq r} \left|\Delta\theta_{j,j'}\right|^2 \left|t_{j,j'}\right|, \quad \theta_j := \begin{cases} 0, \ j=1,\cdots,p, \\ \frac{2\pi}{\ell}\left\lfloor\frac{j-1}{p}\right\rfloor, \ j=p+1,\cdots,p\ell, \\ 2\pi, \ j=p\ell+1,\cdots,pN. \end{cases} \tag{6.53}$$

We need to find an upper bound on

$$\left|\Delta\theta_{j,j'}\right| = \begin{cases} \left|\frac{2\pi}{\ell}\left(\left\lfloor\frac{j-1}{p}\right\rfloor - \left\lfloor\frac{j'-1}{p}\right\rfloor\right) \bmod 2\pi\right|, \ j,j'=p+1,\cdots,p\ell, \\ \left|\frac{2\pi}{\ell}\left\lfloor\frac{j-1}{p}\right\rfloor \bmod 2\pi\right|, \ j=p+1,\cdots,p\ell, \ j'\neq p+1,\cdots,p\ell, \\ \left|\frac{2\pi}{\ell}\left\lfloor\frac{j'-1}{p}\right\rfloor \bmod 2\pi\right|, \ j\neq p+1,\cdots,p\ell, \ j'=p+1,\cdots,p\ell, \\ 0, \ j,j'\neq p+1,\cdots,p\ell, \end{cases} \tag{6.54}$$

given the finite range of the hopping amplitudes,

$$|j-j'|>r \implies t_{j,j'}=0. \tag{6.55}$$

Given the inequalities

$$0<x\in\mathbb{R} \implies \lfloor x\rfloor \leq x, \qquad -x\leq -\lfloor x\rfloor < -x+1, \tag{6.56}$$

$$0<x\leq y\in\mathbb{R} \implies \left|\lfloor x\rfloor-\lfloor y\rfloor\right| = \lfloor y\rfloor-\lfloor x\rfloor \leq y-\lfloor x\rfloor < y-x+1, \tag{6.57}$$

the desired fifth upper bound is

$$\left|\Delta\theta_{j,j'}\right| \leq \frac{2\pi}{p\ell}(r+1), \qquad \forall j,j'\in\Lambda, \qquad |j-j'|\leq r, \tag{6.58}$$

under the assumption that

$$\ell \geq \frac{r+1}{p}. \tag{6.59}$$

By combining the fourth and fifth upper bounds and taking into account that $\Delta\theta_{j,j'}=0$ if $j,j'\neq p+1,\cdots,p\ell$ while $t_{j,j'}=0$ if $|j-j'|>r$, we find the sixth upper bound

$$\begin{aligned} \Delta E &\leq \left[\frac{2\pi}{p\ell}(r+1)\right]^2 \nu \sum_{j,j'=p+1-r}^{p\ell+r} \left|t_{j,j'}\right| \\ &\leq \left[\frac{2\pi}{p\ell}(r+1)\right]^2 \nu \sum_{j=p+1-r}^{p\ell+r}\sum_{j'\in\Lambda} \left|t_{j,j'}\right| \end{aligned}$$

$$
\begin{aligned}
&\leq \left[\frac{2\pi}{p\,\ell}\,(r+1)\right]^2 \nu \sum_{j=p+1-r}^{p\,\ell+r} \max_{j\in\Lambda} \sum_{j'\in\Lambda} |t_{j,j'}| \\
\text{By Eq. (6.25)} &\equiv \left[\frac{2\pi}{p\,\ell}\,(r+1)\right]^2 \nu \sum_{j=p+1-r}^{p\,\ell+r} \bar{t} \\
&= \left[\frac{2\pi}{p\,\ell}\,(r+1)\right]^2 \nu\,[p\,(\ell-1)+2\,r]\,\bar{t}.
\end{aligned}
\tag{6.60}
$$

If ℓ is sufficiently large in the sense that

$$
\ell \geq 2\,r - p \iff 2\,r \leq \ell + p, \tag{6.61}
$$

it follows that

$$
p\,(\ell-1) + 2\,r \leq (p+1)\,\ell. \tag{6.62}
$$

Our desired final upper bound is then

$$
\begin{aligned}
\Delta E &\leq \left[\frac{2\pi}{p\,\ell}\,(r+1)\right]^2 \nu\,(p+1)\,\ell\,\bar{t} \\
&= \frac{4\pi^2\,(p+1)\,(r+1)^2\,\nu\,\bar{t}}{p^2} \times \frac{1}{\ell} \\
&\equiv \frac{C}{2\,\ell}.
\end{aligned}
\tag{6.63}
$$

□

So far, we have not made use of the assumptions that translation symmetry holds and that the filling fraction per repeat unit cell is not an integer. These assumptions are needed to establish a useful upper bound on the overlap between the ground state $|\Psi^{\Lambda}_{\mathrm{GS},\,\kappa}\rangle$ and the trial state $\widehat{U}_{\ell}\,|\Psi^{\Lambda}_{\mathrm{GS},\,\kappa}\rangle$. According to **Lemma** 6.7 of this proof, the overlap $\left|\langle\Psi^{\Lambda}_{\mathrm{GS},\,\kappa}|\,\widehat{U}_{\ell}\,|\Psi^{\Lambda}_{\mathrm{GS},\,\kappa}\rangle\right|^2$ is bounded by the variance of the coarse-grained density on $p\,\ell$ consecutive sites with a proportionality constant inversely proportional to $\sin^2(\pi\,\nu)$, the latter being nonvanishing if and only if ν is not an integer.

Lemma 6.7 (Tasaki) *Under assumptions (i)–(ii) of **Theorem** 6.3, for any filling fraction per repeat unit cell ν not an integer [that is assumption (iii) from **Theorem** 6.3], for any normalizable ground state $|\Psi^{\Lambda}_{\mathrm{GS},\,\kappa}\rangle$ with the energy eigenvalue $E^{\Lambda}_{\mathrm{GS}}$ that is also an eigenstate of the translation by one repeat unit cell with the eigenvalue $\exp(\mathrm{i}\kappa)$, the inequality*

$$
\left|\langle\Psi^{\Lambda}_{\mathrm{GS},\,\kappa}|\,\widehat{U}_{\ell}\,|\Psi^{\Lambda}_{\mathrm{GS},\,\kappa}\rangle\right|^2 \leq \frac{\pi^2\,\langle\Psi^{\Lambda}_{\mathrm{GS},\,\kappa}|\,(\Delta\hat{\rho}_{\ell})^2\,|\Psi^{\Lambda}_{\mathrm{GS},\,\kappa}\rangle}{\sin^2(\pi\,\nu)} \tag{6.64}
$$

holds.

Proof Because the phase

$$\theta_j = \begin{cases} 0, \ j=1,\cdots,p, \\ \frac{2\pi}{\ell}\left\lfloor\frac{j-1}{p}\right\rfloor, \ j=p+1,\cdots,p\,\ell, \\ 2\pi, \ j=p\,\ell+1,\cdots,p\,N, \end{cases} \tag{6.65}$$

that enters the definition (6.45) of the unitary operator $\widehat{U}_\ell$ is only distinct from the values 0 or 2π for the sites $p+1$, $p+2$, $\cdots$, $p\,\ell$, we may write

$$\widehat{U}_\ell = \exp\left(+\mathrm{i}\sum_{j=p+1}^{p\,(\ell+1)} \theta_j\,\hat{n}_j\right). \tag{6.66}$$

Conjugation of $\widehat{U}_\ell$ by the generator defined in Eq. (6.9) of a translation by one repeat unit cell gives

$$\begin{aligned}
\widehat{T}_p^\dagger\,\widehat{U}_\ell\,\widehat{T}_p &= \exp\left(+\mathrm{i}\sum_{j=p+1}^{p\,(\ell+1)} \theta_j\,\hat{n}_{j-p}\right) \\
&= \exp\left(+\mathrm{i}\sum_{j'=1}^{p\,\ell} \theta_{j'+p}\,\hat{n}_{j'}\right) \\
\text{Eq. (6.65)} \quad &= \exp\left(+\mathrm{i}\sum_{j'=1}^{p\,\ell}\left(\frac{2\pi}{\ell}+\theta_{j'}\right)\hat{n}_{j'}\right) \\
\text{Eq. (6.45)} \quad &= \exp\left(+\frac{2\pi\mathrm{i}}{\ell}\sum_{j'=1}^{p\,\ell}\hat{n}_{j'}\right)\widehat{U}_\ell \\
&\equiv e^{+2\pi\mathrm{i}\,\hat{\rho}_\ell}\,\widehat{U}_\ell.
\end{aligned} \tag{6.67}$$

Because the ground state that we chose is also an eigenstate of $\widehat{T}_p$ (we momentarily reinstate the full notation for ground-state expectation values to stress that we make use of the ground state being an eigenstate of $\widehat{T}_p$),

$$\begin{aligned}
\left\langle\Psi^\Lambda_{\mathrm{GS},\,\kappa}\right|\widehat{U}_\ell\left|\Psi^\Lambda_{\mathrm{GS},\,\kappa}\right\rangle &= \left\langle\Psi^\Lambda_{\mathrm{GS},\,\kappa}\right| e^{-\mathrm{i}\kappa}\,\widehat{U}_\ell\,e^{+\mathrm{i}\kappa}\left|\Psi^\Lambda_{\mathrm{GS},\,\kappa}\right\rangle \\
&= \left\langle\Psi^\Lambda_{\mathrm{GS},\,\kappa}\right|\widehat{T}_p^\dagger\,\widehat{U}_\ell\,\widehat{T}_p\left|\Psi^\Lambda_{\mathrm{GS},\,\kappa}\right\rangle \\
&= \left\langle\Psi^\Lambda_{\mathrm{GS},\,\kappa}\right| e^{+2\pi\mathrm{i}\,\hat{\rho}_\ell}\,\widehat{U}_\ell\left|\Psi^\Lambda_{\mathrm{GS},\,\kappa}\right\rangle.
\end{aligned} \tag{6.68}$$

If we subtract $e^{2\pi\mathrm{i}\nu}\left\langle\Psi^\Lambda_{\mathrm{GS},\,\kappa}\right|\widehat{U}_\ell\left|\Psi^\Lambda_{\mathrm{GS},\,\kappa}\right\rangle$ from both sides, we obtain the desired identity

$$\left(1-e^{2\pi\mathrm{i}\nu}\right)\left\langle\widehat{U}_\ell\right\rangle = \left\langle\left(e^{+2\pi\mathrm{i}\,\hat{\rho}_\ell}-e^{+2\pi\mathrm{i}\nu}\right)\widehat{U}_\ell\right\rangle. \tag{6.69}$$

We may then apply the Cauchy-Schwarz inequality (6.50) to the right-hand side of

$$\left|\left(1-e^{2\pi\mathrm{i}\nu}\right)\left\langle\widehat{U}_\ell\right\rangle\right|^2=\left|\left\langle\left(e^{+2\pi\mathrm{i}\,\hat{\rho}_\ell}-e^{+2\pi\mathrm{i}\nu}\right)\widehat{U}_\ell\right\rangle\right|^2, \tag{6.70}$$

that is

$$\left|\left(1-e^{2\pi\mathrm{i}\nu}\right)\left\langle\widehat{U}_\ell\right\rangle\right|^2\leq\left\langle\left(e^{+2\pi\mathrm{i}\,\hat{\rho}_\ell}-e^{+2\pi\mathrm{i}\nu}\right)\left(e^{-2\pi\mathrm{i}\,\hat{\rho}_\ell}-e^{-2\pi\mathrm{i}\nu}\right)\right\rangle\left\langle\widehat{U}^\dagger_\ell\widehat{U}_\ell\right\rangle. \tag{6.71}$$

The unitarity of $\widehat{U}_\ell$ implies that $\left\langle\widehat{U}^\dagger_\ell\widehat{U}_\ell\right\rangle=1$. The identities

$$\left|1-e^{-2\mathrm{i}a}\right|^2=\left|e^{+\mathrm{i}a}-e^{-\mathrm{i}a}\right|^2, \tag{6.72a}$$

$$e^{\mathrm{i}a}-e^{\mathrm{i}b}=e^{+\mathrm{i}(a+b)/2}\left[e^{+\mathrm{i}(a-b)/2}-e^{-\mathrm{i}(a-b)/2}\right], \tag{6.72b}$$

then imply that

$$\sin^2(\pi\,\nu)\left|\left\langle\widehat{U}_\ell\right\rangle\right|^2\leq\left\langle\sin^2\left(\pi\,(\hat{\rho}_\ell-\nu)\right)\right\rangle. \tag{6.73}$$

The left-hand side would vanish if ν was an integer. However, as ν is assumed not to be an integer, we can divide both sides by $\sin^2(\pi\,\nu)$,

$$\left|\left\langle\widehat{U}_\ell\right\rangle\right|^2\leq\frac{\left\langle\sin^2\left(\pi\,(\hat{\rho}_\ell-\nu)\right)\right\rangle}{\sin^2(\pi\,\nu)}. \tag{6.74}$$

Making use of the inequality $\sin^2 x\leq x^2$, we arrive at the proof of the inequality

$$\left|\left\langle\widehat{U}_\ell\right\rangle\right|^2\leq\frac{\pi^2\left\langle(\hat{\rho}_\ell-\nu)^2\right\rangle}{\sin^2(\pi\,\nu)}. \tag{6.75}$$

□

The proof of **Theorem** 6.3 can be completed as follows. First, we use the notation

$$\Delta\hat{\rho}_\ell\equiv\hat{\rho}_\ell-\nu,\qquad \alpha_\ell\equiv\left\langle\widehat{U}_\ell\right\rangle. \tag{6.76}$$

According to **Lemma** 6.7,

$$|\alpha_\ell|^2\leq\frac{\pi^2\left\langle(\Delta\hat{\rho}_\ell)^2\right\rangle}{\sin^2(\pi\,\nu)}. \tag{6.77}$$

Because of assumption (6.30b),

$$|\alpha_\ell|^2\leq\frac{\pi^2\sin^2(\pi\,\nu)}{2\pi^2\sin^2(\pi\,\nu)}\leq\frac{1}{2}. \tag{6.78}$$

Second, define the non-unitary operator

$$\widehat{W}_\ell:=\frac{\widehat{U}_\ell-\alpha_\ell}{\sqrt{1-|\alpha_\ell|^2}}. \tag{6.79}$$

We automatically have the orthogonality

$$\left\langle \widehat{W}_\ell \right\rangle = \frac{\left\langle \widehat{U}_\ell \right\rangle - \alpha_\ell}{\sqrt{1 - |\alpha_\ell|^2}} = \frac{\alpha_\ell - \alpha_\ell}{\sqrt{1 - |\alpha_\ell|^2}} = 0. \tag{6.80}$$

We have proven Eq. (6.30c). The state $\widehat{W}_\ell \left|\Psi^{\Lambda}_{\mathrm{GS},\,\kappa}\right\rangle$ is normalized to the number 1 since

$$\begin{aligned}\left\langle \widehat{W}^\dagger_\ell \widehat{W}_\ell - 1 \right\rangle &= \left\langle \frac{\left(\widehat{U}^\dagger_\ell - \alpha^*_\ell\right)\left(\widehat{U}_\ell - \alpha_\ell\right) - \left(1 - |\alpha_\ell|^2\right)}{1 - |\alpha_\ell|^2} \right\rangle \\ &= \frac{1 - |\alpha_\ell|^2 - |\alpha_\ell|^2 + |\alpha_\ell|^2 + |\alpha_\ell|^2 - 1}{1 - |\alpha_\ell|^2} \\ &= 0. \end{aligned} \tag{6.81}$$

We have proven Eq. (6.30d). Finally, the expectation value of $\widehat{H}$ in the trial state $\widehat{W}_\ell \left|\Psi^{\Lambda}_{\mathrm{GS},\,\kappa}\right\rangle$ relative to the ground-state eigenenergy $E^{\Lambda}_{\mathrm{GS}}$ can be bounded with the help of **Lemmas** 6.6 and 6.7 by

$$\begin{aligned}\left\langle \widehat{W}^\dagger_\ell \widehat{H} \widehat{W}_\ell \right\rangle - E^{\Lambda}_{\mathrm{GS}} &= \left\langle \frac{\left(\widehat{U}^\dagger_\ell - \alpha^*_\ell\right) \widehat{H} \left(\widehat{U}_\ell - \alpha_\ell\right) - \left(1 - |\alpha_\ell|^2\right) E^{\Lambda}_{\mathrm{GS}}}{1 - |\alpha_\ell|^2} \right\rangle \\ &= \left\langle \frac{\widehat{U}^\dagger_\ell \widehat{H} \widehat{U}_\ell - E^{\Lambda}_{\mathrm{GS}}}{1 - |\alpha_\ell|^2} \right\rangle \\ &\overset{\text{Lemma 6.6}}{\leq} \frac{C}{2\left(1 - |\alpha_\ell|^2\right)\ell} \\ &\overset{\text{Lemma 6.7}}{\leq} \frac{C}{\ell}. \end{aligned} \tag{6.82}$$

We have proven Eq. (6.30e). □

Theorem 6.3 identifies the sufficient conditions for which one excited eigenstate with eigenenergy of order $1/\ell$ (relative to the ground-state energy) can be constructed out of the unitary operator $\widehat{U}_\ell$ (the local twist operator whose action is only acting non trivially on an interval of size $p\,\ell$). The following theorem by Tasaki establishes what are sufficient conditions to have more than one such excited eigenstates.

Theorem 6.8 (Tasaki) *With all symbols defined in Section 6.2.1:*
(i) Assume the ring Λ with the repeat unit cell of length p and a positive integer number N of repeat unit cells, that is, its cardinality is $|\Lambda| = p\,N$.
(ii) Assume the charge conserving and translation invariant (by integer multiples of p) lattice Hamiltonian $\widehat{H}$ for particles obeying either the fermion (f) or boson

(b) algebra with hopping amplitudes and interactions restricted to the range r and whose domain of definition is the Hilbert space with $N_{\mathrm{f/b}}$ identical f/b particles.
(iii) Assume that the filling fraction per repeat unit cell

$$\nu \equiv \frac{N_{\mathrm{f/b}}}{N} \tag{6.83a}$$

is not an integer.
(iv) Assume that Eq. (6.30b) holds so that there exists the operator $\widehat{W}_\ell$ with the properties (6.30c)–(6.30e).
(v) Assume that, for any positive integer n larger than one, it is possible to choose N and $1 \leq \ell < N$ together with the set $s_1 \neq s_2 \neq \cdots \neq s_n$ of distinct positive integers such that

$$\left|\left\langle \Psi^{\Lambda}_{\mathrm{GS},\,\kappa} \right| \left(\widehat{T}^{s_\mu}_{p}\,\widehat{W}^{\dagger}_{\ell}\,\widehat{T}^{\dagger\,s_\mu}_{p}\right)\left(\widehat{T}^{s_\xi}_{p}\,\widehat{W}_{\ell}\,\widehat{T}^{\dagger\,s_\xi}_{p}\right)\left|\,\Psi^{\Lambda}_{\mathrm{GS},\,\kappa}\right\rangle\right| \leq \frac{1}{2n} \tag{6.83b}$$

for any choice of $\mu \neq \xi = 1, \cdots, n$. Here, $|\Psi^{\Lambda}_{\mathrm{GS},\,\kappa}\rangle$ is any normalizable ground state with the energy eigenvalue $E^{\Lambda}_{\mathrm{GS}}$ that is also an eigenstate of the translation by one repeat unit cell with the eigenvalue $\exp(\mathrm{i}\kappa)$.
Then, there exist n excited eigenstates $|\Psi^{\Lambda}_{\mu}\rangle$ of $\widehat{H}$ with the eigenenergies E^{Λ}_{μ} with $\mu, \xi = 1, \cdots, n$ such that

$$\langle \Psi^{\Lambda}_{\mu} |\, \Psi^{\Lambda}_{\xi}\rangle = \delta_{\mu,\xi}, \tag{6.83c}$$

$$\langle \Psi^{\Lambda}_{\mu} |\, \Psi^{\Lambda}_{\mathrm{GS},\,\kappa}\rangle = 0, \tag{6.83d}$$

$$E^{\Lambda}_{\mu} - E^{\Lambda}_{\mathrm{GS}} \leq \frac{2\,n\,C}{\ell}. \tag{6.83e}$$

The positive constant C is here independent of N and ℓ and given by Eq. (6.25).

Remark (1) The assumed upper bound (6.83b) is a bound on the amount by which $\left|\Psi^{\Lambda}_{\mathrm{GS},\,\kappa}\right\rangle$ could fail to satisfy the cluster decomposition property for the pair of operators

$$\widehat{W}^{\dagger}_{\ell,\,\mu} := \widehat{T}^{s_\mu}_{p}\,\widehat{W}^{\dagger}_{\ell}\,\widehat{T}^{\dagger\,s_\mu}_{p}, \qquad \widehat{W}_{\ell,\,\xi} := \widehat{T}^{s_\xi}_{p}\,\widehat{W}_{\ell}\,\widehat{T}^{\dagger\,s_\xi}_{p}, \qquad \mu,\xi = 1,\cdots,n, \tag{6.84}$$

since

$$\begin{aligned}
\left\langle \Psi^{\Lambda}_{\mathrm{GS},\,\kappa} \right| \widehat{W}^{\dagger}_{\ell,\,\mu} \left| \Psi^{\Lambda}_{\mathrm{GS},\,\kappa} \right\rangle &\equiv \left\langle \Psi^{\Lambda}_{\mathrm{GS},\,\kappa} \right| \left(\widehat{T}^{s_\mu}_{p}\,\widehat{W}^{\dagger}_{\ell}\,\widehat{T}^{\dagger\,s_\mu}_{p}\right)\left| \Psi^{\Lambda}_{\mathrm{GS},\,\kappa} \right\rangle \\
&= \left(\left\langle \Psi^{\Lambda}_{\mathrm{GS},\,\kappa} \right| \widehat{T}^{s_\mu}_{p}\right) \widehat{W}^{\dagger}_{\ell} \left(\widehat{T}^{\dagger\,s_\mu}_{p} \left| \Psi^{\Lambda}_{\mathrm{GS},\,\kappa} \right\rangle\right) \\
&= \left\langle \Psi^{\Lambda}_{\mathrm{GS},\,\kappa} \right| e^{+\mathrm{i}\kappa\, s_\mu}\,\widehat{W}^{\dagger}_{\ell}\,e^{-\mathrm{i}\kappa\, s_\mu} \left| \Psi^{\Lambda}_{\mathrm{GS},\,\kappa} \right\rangle \\
&= \left\langle \Psi^{\Lambda}_{\mathrm{GS},\,\kappa} \right| \widehat{W}^{\dagger}_{\ell} \left| \Psi^{\Lambda}_{\mathrm{GS},\,\kappa} \right\rangle \\
&= \left\langle \Psi^{\Lambda}_{\mathrm{GS},\,\kappa} \right| \widehat{W}_{\ell} \left| \Psi^{\Lambda}_{\mathrm{GS},\,\kappa} \right\rangle^{*} \\
&= 0
\end{aligned} \tag{6.85}$$

for any $\mu = 1, \cdots, n$. Here, we used the fact that $\widehat{T}_p \left|\Psi^{\Lambda}_{\mathrm{GS},\kappa}\right\rangle = e^{+\mathrm{i}\kappa} \left|\Psi^{\Lambda}_{\mathrm{GS},\kappa}\right\rangle$ to reach the third equality and Eq. (6.30c) to reach the last equality. Assumption (v) is always met if $\left|\Psi^{\Lambda}_{\mathrm{GS},\kappa}\right\rangle$ does not support *any long-range order.* Suppose that $\epsilon > 0$ very small and n a positive large integer are given. Select any integer $\ell \geq 2\,n\,C/\epsilon$. If $N = |\Lambda|/p$ and $s_1 \neq s_2 \neq \cdots \neq s_n$ can be chosen such that hypothesis (6.83b) holds, then the right-hand side of Eq. (6.83d) is replaced by ϵ. In other words, we have the following corollary.

Corollary *Under the hypothesis (i)-(iv) of* ***Theorem*** *6.8, and (v') if the co-variance*

$$\begin{aligned}
\textit{co-var}\left(\widehat{W}^{\dagger}_{\ell,\mu}\widehat{W}_{\ell,\xi}\right)\Big|_{\Psi^{\Lambda}_{\mathrm{GS},\kappa}} &:= \left\langle\Psi^{\Lambda}_{\mathrm{GS},\kappa}\right|\widehat{W}^{\dagger}_{\ell,\mu}\widehat{W}_{\ell,\xi}\left|\Psi^{\Lambda}_{\mathrm{GS},\kappa}\right\rangle\\
&\quad-\left\langle\Psi^{\Lambda}_{\mathrm{GS},\kappa}\right|\widehat{W}^{\dagger}_{\ell,\mu}\left|\Psi^{\Lambda}_{\mathrm{GS},\kappa}\right\rangle\left\langle\Psi^{\Lambda}_{\mathrm{GS},\kappa}\right|\widehat{W}_{\ell,\xi}\left|\Psi^{\Lambda}_{\mathrm{GS},\kappa}\right\rangle\\
&=\left\langle\Psi^{\Lambda}_{\mathrm{GS},\kappa}\right|\widehat{W}^{\dagger}_{\ell,\mu}\widehat{W}_{\ell,\xi}\left|\Psi^{\Lambda}_{\mathrm{GS},\kappa}\right\rangle
\end{aligned}\tag{6.86a}$$

can be made as small as one wishes by letting $|\xi-\mu|$ be sufficiently large ($|\xi-\mu| \ll N$), then for any $\epsilon > 0$ and any $n = 2, 3, \cdots$, one may choose $N = |\Lambda|/p$ large enough so that there are n excited eigenstates $|\Psi^{\Lambda}_{\mu}\rangle$ of $\widehat{H}$ with the eigenenergies E^{Λ}_{μ} ($\mu = 1, \cdots, n$) such that Eqs. (6.83c) and (6.83d) together with

$$E^{\Lambda}_{\mu} - E^{\Lambda}_{\mathrm{GS}} \leq \epsilon. \tag{6.86b}$$

hold. Hypothesis (v') is stronger than hypothesis (v) in Eq. (6.83b). It is always met when the ground state $\left|\Psi^{\Lambda}_{\mathrm{GS},\kappa}\right\rangle$ obeys the clustering property.

Remark (2) Equation (6.83c) is a consequence of the fact that the n trial states

$$\left|\Xi^{\Lambda}_{\ell,\mu}\right\rangle := \widehat{T}^{s_\mu}_{p}\,\widehat{W}_{\ell}\,\widehat{T}^{\dagger\,s_\mu}_{p}\left|\Psi^{\Lambda}_{\mathrm{GS},\kappa}\right\rangle, \qquad \mu = 1, \cdots, n, \tag{6.87}$$

will be shown to be linearly independent.

Remark (3) Equation (6.83d) is a consequence of the fact that the n trial states $\left|\Xi^{\Lambda}_{\ell,\mu}\right\rangle$ with $\mu = 1, \cdots, n$ will be shown to be orthogonal to $\left|\Psi^{\Lambda}_{\mathrm{GS},\kappa}\right\rangle$.

Proof Let $n = 2, 3, \cdots$ be any arbitrarily chosen positive integer larger than one. As all assumptions of **Theorem** 6.3 hold, there exists the local twist operator $\widehat{W}_{\ell}$ obeying Eqs. (6.30c)–(6.30e). Let $\left|\Xi^{\Lambda}_{\ell,\mu}\right\rangle$ with $\mu = 1, \cdots, n$ be defined as in Eq. (6.87). This state has the squared norm

$$\begin{aligned}
\left\langle\Xi^{\Lambda}_{\ell,\mu}\,\middle|\,\Xi^{\Lambda}_{\ell,\mu}\right\rangle &= \left\langle\Psi^{\Lambda}_{\mathrm{GS},\kappa}\right|\left(\widehat{T}^{s_\mu}_{p}\,\widehat{W}^{\dagger}_{\ell}\,\widehat{T}^{\dagger\,s_\mu}_{p}\right)\left(\widehat{T}^{s_\mu}_{p}\,\widehat{W}_{\ell}\,\widehat{T}^{\dagger\,s_\mu}_{p}\right)\left|\Psi^{\Lambda}_{\mathrm{GS},\kappa}\right\rangle\\
&= \left(\left\langle\Psi^{\Lambda}_{\mathrm{GS},\kappa}\right|\widehat{T}^{s_\mu}_{p}\right)\widehat{W}^{\dagger}_{\ell}\,\widehat{W}_{\ell}\left(\widehat{T}^{\dagger\,s_\mu}_{p}\left|\Psi^{\Lambda}_{\mathrm{GS},\kappa}\right\rangle\right)
\end{aligned}$$

$$\begin{aligned}
{\scriptstyle\text{Translation symmetry}} &= \left\langle \Psi^{\Lambda}_{\mathrm{GS},\,\kappa} \right| e^{+\mathrm{i}\kappa\, s_\mu}\, \widehat{W}^{\dagger}_{\ell}\, \widehat{W}_{\ell}\, e^{-\mathrm{i}\kappa\, s_\mu} \left| \Psi^{\Lambda}_{\mathrm{GS},\,\kappa} \right\rangle \\
&= \left\langle \Psi^{\Lambda}_{\mathrm{GS},\,\kappa} \right| \widehat{W}^{\dagger}_{\ell}\, \widehat{W}_{\ell} \left| \Psi^{\Lambda}_{\mathrm{GS},\,\kappa} \right\rangle \\
{\scriptstyle\text{Eq. (6.30d)}} &= 1
\end{aligned} \tag{6.88}$$

and it is orthogonal to $\left|\Psi^{\Lambda}_{\mathrm{GS},\,\kappa}\right\rangle$ as

$$\begin{aligned}
\left\langle \Xi^{\Lambda}_{\ell,\,\mu} \,\middle|\, \Psi^{\Lambda}_{\mathrm{GS},\,\kappa} \right\rangle &= \left\langle \Psi^{\Lambda}_{\mathrm{GS},\,\kappa} \right| \left(\widehat{T}^{\,s_\mu}_{p}\, \widehat{W}^{\dagger}_{\ell}\, \widehat{T}^{\dagger\, s_\mu}_{p} \right) \left| \Psi^{\Lambda}_{\mathrm{GS},\,\kappa} \right\rangle \\
&= \left(\left\langle \Psi^{\Lambda}_{\mathrm{GS},\,\kappa} \right| \widehat{T}^{\,s_\mu}_{p} \right) \widehat{W}^{\dagger}_{\ell} \left(\widehat{T}^{\dagger\, s_\mu}_{p} \left| \Psi^{\Lambda}_{\mathrm{GS},\,\kappa} \right\rangle \right) \\
{\scriptstyle\text{Translation symmetry}} &= \left\langle \Psi^{\Lambda}_{\mathrm{GS},\,\kappa} \right| e^{+\mathrm{i}\kappa\, s_\mu}\, \widehat{W}^{\dagger}_{\ell}\, e^{-\mathrm{i}\kappa\, s_\mu} \left| \Psi^{\Lambda}_{\mathrm{GS},\,\kappa} \right\rangle \\
&= \left\langle \Psi^{\Lambda}_{\mathrm{GS},\,\kappa} \right| \widehat{W}^{\dagger}_{\ell} \left| \Psi^{\Lambda}_{\mathrm{GS},\,\kappa} \right\rangle \\
{\scriptstyle\text{Eq. (6.30c)}} &= 0.
\end{aligned} \tag{6.89}$$

Next, we are going to show that the n (with $1 < n \in \mathbb{Z}$) states $\left|\Xi^{\Lambda}_{\ell,\,1}\right\rangle, \cdots, \left|\Xi^{\Lambda}_{\ell,\,n}\right\rangle$ are linearly independent. To this end, define the $n \times n$ complex-valued matrix G with the matrix elements

$$G_{\mu\,\xi} := \left\langle \Xi^{\Lambda}_{\ell,\,\mu} \,\middle|\, \Xi^{\Lambda}_{\ell,\,\xi} \right\rangle = G^{*}_{\xi\,\mu}, \qquad \mu, \xi = 1, \cdots, n. \tag{6.90}$$

This matrix is called the Gram matrix of the n states $\left|\Xi^{\Lambda}_{\ell,\,1}\right\rangle, \cdots, \left|\Xi^{\Lambda}_{\ell,\,n}\right\rangle$. Choose any n complex-valued numbers $c_1, \cdots, c_n \in \mathbb{C}$ such that

$$\sum_{\mu=1}^{n} |c_\mu|^2 = 1. \tag{6.91}$$

We seek a lower bound on

$$\sum_{\mu,\xi=1}^{n} c^{*}_{\mu}\, G_{\mu\,\xi}\, c_{\xi} = \sum_{\mu=1}^{n} c^{*}_{\mu}\, G_{\mu\mu}\, c_{\mu} + \sum_{\mu\neq\xi=1}^{n} c^{*}_{\mu}\, G_{\mu\,\xi}\, c_{\xi}. \tag{6.92}$$

By Eq. (6.88), $G_{\mu\mu} = 1$ for $\mu = 1, \cdots, n$. Hence,

$$\sum_{\mu,\xi=1}^{n} c^{*}_{\mu}\, G_{\mu\,\xi}\, c_{\xi} = \sum_{\mu=1}^{n} c^{*}_{\mu}\, c_{\mu} + \sum_{\mu\neq\xi=1}^{n} c^{*}_{\mu}\, G_{\mu\,\xi}\, c_{\xi}. \tag{6.93}$$

By Eq. (6.91),

$$\sum_{\mu,\xi=1}^{n} c^{*}_{\mu}\, G_{\mu\,\xi}\, c_{\xi} = 1 + \sum_{\mu\neq\xi=1}^{n} c^{*}_{\mu}\, G_{\mu\,\xi}\, c_{\xi}. \tag{6.94}$$

Our first lower bound is then

$$\sum_{\mu,\xi=1}^{n} c^{*}_{\mu}\, G_{\mu\,\xi}\, c_{\xi} \geq 1 - \sum_{\mu\neq\xi=1}^{n} |c^{*}_{\mu}|\, |G_{\mu\,\xi}|\, |c_{\xi}|. \tag{6.95}$$

By assumption (6.83b),

$$|G_{\mu\,\xi}| \le \frac{1}{2n} \tag{6.96}$$

for any $\mu \neq \xi = 1, \cdots, n$. We then have the second lower bound

$$\sum_{\mu,\xi=1}^{n} c^{*}_{\mu}\, G_{\mu\,\xi}\, c_{\xi} \ge 1 - \frac{1}{2n} \sum_{\mu\neq\xi=1}^{n} |c^{*}_{\mu}|\,|c_{\xi}|. \tag{6.97}$$

With the help of

$$(a-b)^2 = a^2 + b^2 - 2\,a\,b \ge 0 \quad \Longleftrightarrow \quad a\,b \le \frac{1}{2}(a^2+b^2) \tag{6.98}$$

for any positive real numbers a and b, we may insert the lower bound

$$\begin{aligned}\sum_{\mu\neq\xi=1}^{n} |c^{*}_{\mu}|\,|c_{\xi}| &\le \frac{1}{2} \sum_{\mu\neq\xi=1}^{n} \left(|c^{*}_{\mu}|^2 + |c_{\xi}|^2\right)\\ &= \sum_{\mu\neq\xi=1}^{n} |c^{*}_{\mu}|^2\\ &= (n-1) \sum_{\mu=1}^{n} |c^{*}_{\mu}|^2\\ &= (n-1)\end{aligned} \tag{6.99}$$

into Eq. (6.97). The final lower bound

$$\sum_{\mu,\xi=1}^{n} c^{*}_{\mu}\, G_{\mu\,\xi}\, c_{\xi} \ge 1 - \frac{n-1}{2\,n} = \frac{1}{2}\,\frac{n+1}{n} \ge \frac{1}{2} \tag{6.100}$$

ensues for all unit vectors $\boldsymbol{c} = (c_{\mu}) \in \mathbb{C}^{n}$. This bound is nothing but a bound on the norm of the Gram matrix G. In particular, because the Gram matrix is Hermitean, we may choose $\boldsymbol{c} = (c_{\mu}) \in \mathbb{C}^{n}$ to be a normalized eigenvector of G. We then infer that all eigenvalues of G are bounded from below by $1/2$. If so, this Gram matrix has a nonvanishing determinant, a necessary and sufficient condition for the linear independence of the n states $\left|\Xi^{\Lambda}_{\ell,\,1}\right\rangle, \cdots, \left|\Xi^{\Lambda}_{\ell,\,n}\right\rangle$.

The n-dimensional complex vector space

$$\operatorname{span}\left\{\left|\Xi^{\Lambda}_{\ell,\,1}\right\rangle, \cdots, \left|\Xi^{\Lambda}_{\ell,\,n}\right\rangle\right\} \tag{6.101}$$

is orthogonal to $|\Psi^{\Lambda}_{\mathrm{GS},\,\kappa}\rangle$, any ground state of $\widehat{H}$ that is an eigenstate of $\widehat{T}_{p}$. We are going to show that the expectation value of $\widehat{H}$ in any state from the vector space (6.101) is within a window of energy of order $1/\ell$ above the ground-state energy $E^{\Lambda}_{\mathrm{GS}}$. To this end, let

$$|\Gamma\rangle := \sum_{\mu=1}^{n} c_{\mu}\, \left|\Xi^{\Lambda}_{\ell,\,\mu}\right\rangle, \qquad \sum_{\mu=1}^{n} |c_{\mu}|^2 = 1. \tag{6.102}$$

The lower bound (6.100) implies that

$$\langle\Gamma|\Gamma\rangle \geq \frac{1}{2}. \tag{6.103}$$

Introduce the shorthand notation for the nonnegative Hamiltonian

$$\Delta\widehat{H} := \widehat{H} - E^{\Lambda}_{\mathrm{GS}}. \tag{6.104}$$

We seek an upper bound on the expectation value

$$\langle\Gamma|\,\Delta\widehat{H}\,|\Gamma\rangle = \sum_{\mu=1}^{n} c^{*}_{\mu}\,c_{\mu}\,\langle\Xi^{\Lambda}_{\ell,\mu}|\,\Delta\widehat{H}\,|\Xi^{\Lambda}_{\ell,\mu}\rangle + \sum_{\mu\neq\xi=1}^{n} c^{*}_{\mu}\,c_{\xi}\,\langle\Xi^{\Lambda}_{\ell,\mu}|\,\Delta\widehat{H}\,|\Xi^{\Lambda}_{\ell,\xi}\rangle. \tag{6.105}$$

For any term $\mu = 1,\cdots,n$ in the diagonal sum, we have the upper bound

$$\langle\Xi^{\Lambda}_{\ell,\mu}|\,\Delta\widehat{H}\,|\Xi^{\Lambda}_{\ell,\mu}\rangle = \langle\Psi^{\Lambda}_{\mathrm{GS},\kappa}|\,\widehat{W}^{\dagger}_{\ell}\,\Delta\widehat{H}\,\widehat{W}_{\ell}\,|\Psi^{\Lambda}_{\mathrm{GS},\kappa}\rangle \leq \frac{C}{\ell} \tag{6.106}$$

as a consequence of translation invariance and **Theorem** 6.3. For any term $\mu \neq \xi = 1,\cdots,n$ in the off-diagonal sum, we need to use the Cauchy-Schwarz inequality before being able to take advantage of translation symmetry and **Theorem** 6.3,

$$\begin{aligned}
\langle\Xi^{\Lambda}_{\ell,\mu}|\,\Delta\widehat{H}\,|\Xi^{\Lambda}_{\ell,\xi}\rangle &= \left(\langle\Xi^{\Lambda}_{\ell,\mu}|\,\sqrt{\Delta\widehat{H}}\right)\left(\sqrt{\Delta\widehat{H}}\,|\Xi^{\Lambda}_{\ell,\xi}\rangle\right)\\
&\leq \sqrt{\left\langle\Xi^{\Lambda}_{\ell,\mu}\right|\Delta\widehat{H}\left|\Xi^{\Lambda}_{\ell,\mu}\right\rangle\left\langle\Xi^{\Lambda}_{\ell,\xi}\right|\Delta\widehat{H}\left|\Xi^{\Lambda}_{\ell,\xi}\right\rangle}\\
&\leq \langle\Psi^{\Lambda}_{\mathrm{GS},\kappa}|\,\widehat{W}^{\dagger}_{\ell}\,\Delta\widehat{H}\,\widehat{W}_{\ell}\,|\Psi^{\Lambda}_{\mathrm{GS},\kappa}\rangle\\
&\leq \frac{C}{\ell}.
\end{aligned} \tag{6.107}$$

We thus have the upper bound

$$\begin{aligned}
\langle\Gamma|\,\Delta\widehat{H}\,|\Gamma\rangle &\leq \left(\sum_{\mu=1}^{n} c^{*}_{\mu}\,c_{\mu} + \sum_{\mu\neq\xi=1}^{n} c^{*}_{\mu}\,c_{\xi}\right)\frac{C}{\ell}\\
&\leq \left(1 + \sum_{\mu\neq\xi=1}^{n} |c^{*}_{\mu}|\,|c_{\xi}|\right)\frac{C}{\ell}\\
\text{Eq. (6.99)}\quad &\leq [1 + (n-1)]\,\frac{C}{\ell}\\
&\leq \frac{n\,C}{\ell}.
\end{aligned} \tag{6.108}$$

If we divide both sides by the norm $\langle\Gamma|\Gamma\rangle$ and replace it by its lower bound (6.103), we arrive at the final upper bound

$$\frac{\langle\Gamma|\,\widehat{H}\,|\Gamma\rangle}{\langle\Gamma|\Gamma\rangle} - E^{\Lambda}_{\mathrm{GS}} \leq \frac{2\,n\,C}{\ell}. \tag{6.109}$$

Equation (6.83e) then follows from the variational principle. □

6.2.4 Discussion

6.2.4.1 Sufficient and Necessary Conditions for not Having Finitely Degenerate Gapped Ground States

The consequence of the extension of the Lieb-Schultz-Mattis theorem by Tasaki that is perhaps the most remarkable occurs when the filling fraction per repeat unit cell ν is not an integer and when any ground state that is also an eigenstate of the generator for the translation-symmetry group obeys the clustering property, that is, it does not support any type of long-range order for observables that are local in the basis (6.18). If so, any of the local Hamiltonians defined in Section 6.2.1 supports low-energy eigenstates whose numbers increase indefinitely as the number of repeat unit cells $N = |\Lambda|/p$ grows. A gapless spectrum in the thermodynamic limit, that is a spectrum possessing a continuum of excitations directly above the ground state falls into this situation. More surprisingly perhaps, a spectrum in the thermodynamic limit with a gap separating the ground state from all excited states must necessarily support infinitely many pairwise orthogonal ground states.

The logical contraposition of the Lieb-Schultz-Mattis theorem by Tasaki is that the presence of finitely degenerate ground states that are separated by a gap from all excited states in the thermodynamic limit for the local Hamiltonians defined in Section 6.2.1 implies that either translation symmetry is spontaneously broken or the filling fraction per repeat unit cell ν is an integer.

We enumerate the following scenarios in the thermodynamic limit $N = |\Lambda|/p \to \infty$:

(A) The ground states are degenerate and violate the clustering property.

(B) The ground states are nondegenerate and the number of low-energy excited eigenstates collapsing to the ground states as the thermodynamic limit is taken grows indefinitely.

(C) The ground states are infinitely degenerate and separated from all excited states by a gap.

(D) The ground states are finitely degenerate and separated from all excited states by a gap.

- Example (6.26) of free spinless fermions with ν an integer realizes scenario **(D)**.
- Example (6.26) of free spinless fermions with ν not an integer realizes scenario **(B)**.
- Example (6.27) of strongly repulsively interacting spinless fermions at $\nu = 1/2$ realizes scenario **(A)**, for the commensurate charge-density wave encoded in the ground state of example (6.27) breaks spontaneously the translation symmetry by one lattice spacing.
- Example (6.28) of strongly repulsively interacting spinless fermions at $\nu < 1/2$ realizes scenario **(C)**, for a gap separates all excited states from the ground states that are infinitely degenerate.

- Example (6.29) of strongly attractively interacting spinless fermions at $\nu < 1$ realizes scenario **(A)**, for a gap separates all excited states from the ground states that are infinitely degenerate and violate the clustering property.

6.2.4.2 Accounting Explicitly for On-Site Continuous Symmetry Groups

The only changes to the notation in Section 6.2.1 stems from the substitution

$$\hat{c}^{\dagger}_{j} \to \hat{c}^{\dagger}_{j,\sigma}, \qquad \hat{c}_{j} \to \hat{c}_{j,\sigma}, \qquad \hat{n}_{j,\sigma} := \hat{c}^{\dagger}_{j,\sigma}\,\hat{c}_{j,\sigma}, \qquad \hat{n}_{j} := \sum_{\sigma=1}^{N_{\mathrm{c}}} \hat{n}_{j,\sigma}, \tag{6.110a}$$

with $j = 1, \cdots, p\,N$, as before, and $\sigma = 1, \cdots, N_{\mathrm{c}}$ runs over N_{c} values associated with the on-site symmetry group

$$\mathrm{U}(N_{\mathrm{c}}) = \mathrm{U}(1) \times \mathrm{SU}(N_{\mathrm{c}}). \tag{6.110b}$$

The index "c" stands here for color. We work in the canonical ensemble with the global number operator

$$\widehat{N} := \sum_{j=1}^{p\,N} \hat{n}_{j} \equiv \sum_{j=1}^{p\,N}\sum_{\sigma=1}^{N_{\mathrm{c}}} \hat{n}_{j,\sigma} \tag{6.110c}$$

fixed to be the integer

$$\widehat{N} = N_{\mathrm{c}} \times N_{\mathrm{f/b}}. \tag{6.110d}$$

The filling fraction per repeat unit cell (color unresolved) is the rational number[18]

$$\nu := \frac{N_{\mathrm{f/b}}}{N} \qquad |\Lambda| := p\,N. \tag{6.110e}$$

The Hamiltonian that we shall consider is given by

$$\widehat{H} := -\sum_{j,j'\in\Lambda}^{|j'-j|\leq r}\;\sum_{\sigma,\sigma'=1}^{N_{\mathrm{c}}} t_{j,\sigma;j',\sigma'}\,\hat{c}^{\dagger}_{j,\sigma}\,\hat{c}_{j',\sigma'} + \sum_{j=1}^{p\,N} \widehat{V}_{j}, \tag{6.110f}$$

where $t_{j,\sigma;j',\sigma'} \in \mathbb{C}$ is of finite range, that is, it vanishes if $|j' - j| > r$, and $\widehat{V}_{j}$ depends only on $\hat{n}_{j,\sigma}, \cdots, \hat{n}_{j+r,\sigma}$ for $\sigma = 1, \cdots, N_{\mathrm{c}}$. Whereas we allow the hopping amplitudes and the interactions to break explicitly the on-site $\mathrm{SU}(N_{\mathrm{c}})$ symmetry, we impose on $\widehat{H}$ the global U(1) symmetry associated with the conservation of the total fermion number (6.110d).

Hermiticity demands that

$$t_{j,\sigma;j',\sigma'} = t^{*}_{j',\sigma';j,\sigma}. \tag{6.110g}$$

[18] Alternatively, if we wanted the filling fraction per repeat unit cell (color unresolved) to be an arbitrary real-valued number, we could have defined $\tilde{\nu}$ implicitly from $N_{\mathrm{f/b}} = \lfloor \tilde{\nu}\,N \rfloor$. We would then have $\tilde{\nu} = \nu + \mathcal{O}(1/N)$.

Translation symmetry demands the p-periodicity

$$t_{j+p,\sigma;j'+p,\sigma'} = t_{j,\sigma;j',\sigma'}, \qquad \widehat{T}_p \left(\sum_{j=1}^{pN} \widehat{V}_j \right) \widehat{T}_p^{\dagger} = \sum_{j=1}^{pN} \widehat{V}_j. \tag{6.110h}$$

The local twist operator

$$\widehat{U}_\ell := \exp\left(\mathrm{i} \sum_{j=1}^{p\ell} \theta_j \, \hat{n}_j \right) \tag{6.110i}$$

can then be used to establish versions of **Theorems** 6.3 and 6.8 for the only change in the proof of **Theorems** 6.3 and 6.8 consists in replacing the upper bound on (6.49) by an upper bound on

$$\Delta E \leq \sum_{j,j'\in\Lambda}^{|j'-j|\leq r} 2\left[1-\cos\left(\theta_j - \theta_{j'}\right)\right] \sum_{\sigma,\sigma'=1}^{N_{\mathrm{c}}} \left|t_{j,\sigma;j',\sigma'}\right| \left|\left\langle \hat{c}_{j,\sigma}\, \hat{c}_{j',\sigma'} \right\rangle\right|. \tag{6.110j}$$

6.2.4.3 Coupling to Phonons

The notation of Section 6.2.1 holds up to the changes made in Eq. (6.110). **Theorems** 6.3 and 6.8 also hold if we couple the f/b Hamiltonian (6.110f) to the phonons associated with lattice Λ and assume that the quantum fluctuations of these phonons are tamed by some mechanism (there can be no breaking of a continuous symmetry in two- and one-dimensional space). Let $\hat{r}_j$ denote the position operator for an ion of mass M_j with $j \in \Lambda$. Let $\hat{p}_j$ denote the momentum operator for an ion of mass M_j with $j \in \Lambda$. Assume that these ions interact through the many-body potential $V'(\hat{r}_1, \cdots, \hat{r}_{pN})$ whose classical minima correspond to the sites of Λ. If we define the Hamiltonian

$$\begin{aligned} \widehat{H}_{\mathrm{f/b-phonons}} :=& \sum_{j\in\Lambda} \frac{\hat{p}_j^2}{2M_j} + V'(\hat{r}_1, \cdots, \hat{r}_{pN}) \\ &- \sum_{j,j'\in\Lambda}^{|j'-j|\leq r} \sum_{\sigma,\sigma'=1}^{N_{\mathrm{c}}} t_{j,\sigma;j',\sigma'}(\hat{r}_j, \hat{r}_{j'})\, \hat{c}_{j,\sigma}^{\dagger}\, \hat{c}_{j',\sigma'} + \sum_{j=1}^{pN} \widehat{V}_j, \end{aligned} \tag{6.111a}$$

where

$$t_{j,\sigma;j',\sigma'}(\hat{r}_j, \hat{r}_{j'}) = t^{*}_{j',\sigma';j,\sigma}(\hat{r}_{j'}, \hat{r}_j) = t_{j+p,\sigma;j'+p,\sigma'}(\hat{r}_{j+p}, \hat{r}_{j'+p}), \tag{6.111b}$$

the only change in the proof of **Theorems** 6.3 and 6.8 consists in replacing the upper bound (6.49) by the upper bound

$$\begin{aligned} \Delta E \leq& \sum_{j,j'\in\Lambda}^{|j'-j|\leq r} 2\left[1-\cos\left(\theta_j - \theta_{j'}\right)\right] \\ &\times \sum_{\sigma,\sigma'=1}^{N_{\mathrm{c}}} \sqrt{\left\langle t_{j,\sigma;j',\sigma'}(\hat{r}_j, \hat{r}_{j'})\, t^{*}_{j,\sigma;j',\sigma'}(\hat{r}_j, \hat{r}_{j'}) \right\rangle} \sqrt{\left\langle \hat{c}_{j',\sigma'}^{\dagger}\, \hat{c}_{j,\sigma}\, \hat{c}_{j,\sigma}^{\dagger}\, \hat{c}_{j',\sigma'} \right\rangle}. \end{aligned} \tag{6.111c}$$

Theorems 6.3 and 6.8 holds if an upper bound exists on the positive number

$$\left\langle t_{j,\sigma;j',\sigma'}(\hat{r}_j,\hat{r}_{j'})\, t^*_{j,\sigma;j',\sigma'}(\hat{r}_j,\hat{r}_{j'})\right\rangle . \tag{6.111d}$$

6.2.4.4 Quantum Spin Chains

The f/b creation and annihilation operators of Section 6.2.1 are replaced by the generators $\widehat{S}^a_j$ with $a = x, y, z \equiv 1, 2, 3$ and $j \in \Lambda$ obeying the $\mathfrak{su}(2)$ algebra

$$\left[\widehat{S}^a_j, \widehat{S}^{a'}_{j'}\right] = \delta_{j,j'} \sum_{a''=1}^{3} \mathrm{i}\epsilon^{aa'a''}\, \widehat{S}^{a''}_j, \qquad a, a' = 1, 2, 3, \tag{6.112a}$$

in the representation with the Casimir operator

$$\widehat{\boldsymbol{S}}^2_j \equiv \sum_{a=1,2,3} \left(\widehat{S}^a_j\right)^2 = S(S+1) \tag{6.112b}$$

fixed to the value corresponding to the spin quantum number

$$S = \frac{1}{2}, 1, \frac{3}{2}, 2, \frac{5}{2}, \cdots . \tag{6.112c}$$

The Hilbert space is $(2S+1)^{|\Lambda|}$-dimensional and given by

$$\mathfrak{H}^\Lambda := \mathrm{span}\left\{ \bigotimes_{j\in\Lambda} |S, S-m_j\rangle \,\middle|\, |S, S-m_j\rangle := \left(\widehat{S}^-_j\right)^{m_j} |S, S\rangle \right.$$
$$\left. \widehat{S}^\pm_j := \widehat{S}^1_j \pm \mathrm{i}\widehat{S}^2_j, \quad m_j = 0, \cdots, 2S, \qquad \widehat{S}^+_j\, |S, S\rangle = 0 \right\}. \tag{6.112d}$$

We shall demand that the Hamiltonian $\widehat{H}$ commutes with

$$\widehat{S}^z := \sum_{j\in\Lambda} \widehat{S}^z_j, \tag{6.112e}$$

commutes with the operator $\widehat{T}_p$ that generates the translation by a repeat unit cell

$$\widehat{T}^\dagger_p\, \widehat{\boldsymbol{S}}_j\, \widehat{T}_p = \widehat{\boldsymbol{S}}_{j-p}, \tag{6.112f}$$

and is local when expressed in terms of

$$\widehat{\boldsymbol{S}}_j = \widehat{\boldsymbol{S}}_{j+pN}, \qquad |\Lambda| \equiv p\,N. \tag{6.112g}$$

For example, we may choose

$$\widehat{H} := \frac{1}{2} \sum_{j,j'\in\Lambda}^{|j'-j|\leq r} \left[J^{(\mathrm{s})}_{j,j'} \left(\widehat{S}^x_j\, \widehat{S}^x_{j'} + \widehat{S}^y_j\, \widehat{S}^y_{j'}\right) + J^{(\mathrm{a})}_{j,j'} \left(\widehat{S}^x_j\, \widehat{S}^x_{j'} - \widehat{S}^y_j\, \widehat{S}^y_{j'}\right) + \delta_{j,j'} \widehat{V}_j \right], \tag{6.112h}$$

where

$$J^{(\mathrm{s})}_{j,j'} = +J^{(\mathrm{s})}_{j',j} = J^{(\mathrm{s})}_{j+p,j'+p} \in \mathbb{R}, \qquad J^{(\mathrm{a})}_{j,j'} = -J^{(\mathrm{a})}_{j',j} = J^{(\mathrm{a})}_{j+p,j'+p} \in \mathbb{R} \tag{6.112i}$$

are vanishing if $|j-j'|>r$ and the p-periodic Hermitean operator

$$\sum_{j=1}^{pN} \widehat{V}_j = \widehat{T}_p \left(\sum_{j=1}^{pN} \widehat{V}_j\right) \widehat{T}^{\dagger}_p \tag{6.112j}$$

with $\widehat{V}_j$ a polynomial of $\widehat{S}^z_j \cdots, \widehat{S}^z_{j+r}$. Let $|\Psi^{\Lambda}_{\mathrm{GS},\,\kappa}\rangle$ be simultaneously (i) a ground state of $\widehat{H}$, (ii) an eigenstate of $\widehat{T}_p$, and (iii) an eigenstate of $\widehat{S}^z$ with the eigenvalues (i) $E^{\Lambda}_{\mathrm{GS}} \in \mathbb{R}$, (ii) $\exp(\mathrm{i}\kappa) \in \mathrm{U}(1)$, and (iii) $M \in \mathbb{Z}$, respectively. Define the filling fraction per repeat unit cell to be the rational number

$$\nu := \frac{M}{N} + p\,S, \qquad 0 \le \nu \le 2p\,S. \tag{6.112k}$$

We can prove that **Theorems** 6.3 and 6.8 hold for $\widehat{H}$ in the sector of the Hilbert space with the magnetization M by defining the unitary local twist operator

$$\widehat{U}_{\ell} := \exp\left(\mathrm{i}\sum_{j=1}^{p\ell} \theta_j\, \hat{n}_j\right), \qquad \hat{n}_j \equiv \widehat{S}^z_j + S, \tag{6.112l}$$

and the coarse-grained magnetization density operator

$$\hat{\rho}_{\ell} := \frac{1}{\ell}\sum_{j=1}^{p\ell} \hat{n}_j, \qquad \hat{n}_j \equiv \widehat{S}^z_j + S. \tag{6.112m}$$

Hypothesis (6.30b) thus becomes a condition on how large the fluctuations of the coarse-grained magnetization along the quantization axis can be for **Theorem** 6.3 to hold.

6.3 A First Extension of the Lieb-Schultz-Mattis Theorem

6.3.1 Overview

The logical contraposition of Tasaki's extension of the Lieb-Schultz-Mattis theorem applies to a local lattice Hamiltonian that is invariant under translation by one repeat unit cell and invariant under a continuous on-site symmetry group.

It also presumes the existence of a positive real-valued number ν, the filling fraction of the repeat unit cell. It states that, if the ground states are finitely degenerate and separated by a gap from all excited states in the thermodynamic limit, then either translation symmetry is spontaneously broken or ν is an integer.

The proof of Tasaki's extension of the Lieb-Schultz-Mattis theorem relies on four steps. First, variational states are constructed. Second, the energy expectation values for these variational states are shown to collapse to the ground-state energy in the thermodynamic limit. Third, the variational states are shown to be orthogonal with each other and with the ground state for any fixed number of degrees of freedom. Finally, this orthogonality is shown to survive in the thermodynamic limit.

The variational states in the first step are constructed by deformations in position space of the ground state that are local and smooth. Here, the existence of an on-site global continuous symmetry of the Hamiltonian is crucial.

The degree of smoothness of these local deformations is controlled by the length of the one-dimensional lattice hosting the quantum degrees of freedom and the continuity of the on-site global symmetry of the Hamiltonian. The larger the length of the one-dimensional lattice, the smoother the local deformations in position space of the ground state are and the closer the energy expectation value of the variational state relative to the ground-state energy is. The locality of the Hamiltonian is needed to control the separation in energy between the variational and ground states. The proof that there are excited states that collapse to the ground state follows.

The proof of orthogonality in the third step hinges on the filling fraction ν not being integer-valued and the existence of translation symmetry in addition to the presence of an on-site continuous symmetry. No more information from the Hamiltonian is needed to complete this step of the proof.

Can the condition that the Hamiltonian is invariant under an on-site continuous symmetry be weaken by demanding instead that the on-site symmetry group is no more than a discrete group?

By assumption, we consider only one-dimensional lattice Hamiltonians that are local and invariant under all lattice translations of some repeat unit cell and an on-site discrete symmetry group. We are going to prove the following no-go theorem. If the on-site discrete symmetry group is realized by a nontrivial projective representation on the local Fock space, then the ground state cannot be a matrix product state [an example thereof being the representation (5.221) of the AKLT state] that is simultaneously invariant, up to a multiplicative phase factor, under all lattice translations and the on-site symmetries.

The importance of this no-go theorem is that it is believed that all nondegenerate gapped ground states of one-dimensional lattice Hamiltonians that are local are adiabatically related to matrix product states, that is, it is possible to deform adiabatically any one-dimensional local lattice Hamiltonian with a nondegenerate gapped ground state into a Hamiltonian with a nondegenerate matrix product state as ground state without ever closing the spectral gap separating the nondegenerate ground state from all excited states. Proving the latter claim corresponds to completing the second step in the proof of Tasaki's extension of the Lieb-Schultz-Mattis theorem for any local one-dimensional lattice Hamiltonian. This program has been completed rigorously when the local lattice degrees of freedom are bosonic but not when they are fermionic.[19] Proving the no-go theorem corresponds to performing the third step in the proof of Tasaki's extension of the Lieb-Schultz-Mattis theorem, a task that we now complete for the case of one-dimensional lattice Hamiltonians built out of Majorana degrees of freedom.

6.3.2 Projective Representations of Symmetries

Two Majorana operators can always be interpreted as the real and imaginary parts of the creation (annihilation) operator of one fermion. We shall only consider lattice Hamiltonians such that the total number of Majorana operators is even. If so, it is then always possible to construct a total fermion-number operator. Although this fermion-number operator does not need to commute with the lattice Hamiltonian, it is postulated that the total fermion-number parity operator must necessarily commute with the lattice Hamiltonian. In other words, the smallest symmetry group arising from an even number of Majorana operators originates from the conservation of the parity (evenness or oddness) of the total number of fermions. This symmetry is associated with a cyclic group of order 2 that we shall denote with $\mathbb{Z}_2^{\mathrm{F}}$. If it is postulated that the number of Majorana operators per site is also even, it is then possible to interpret the total fermion-number operator as the sum of site-resolved fermion-number operators, in which case $\mathbb{Z}_2^{\mathrm{F}}$ can be thought as being an on-site symmetry. Other on-site symmetries are possible, say time-reversal symmetry[20] or spin-rotation symmetry. All such additional on-site symmetries define a second group G (this group is arbitrary, that is it can be finite, countably infinite, or uncountably infinite). The first question to be answered is how many different ways are there to marry into a group G_f the intrinsic symmetry group $\mathbb{Z}_2^{\mathrm{F}}$ with the model-dependent symmetry group G.

Definition 6.9 This problem is known in group theory as the central extension of G by $\mathbb{Z}_2^{\mathrm{F}}$.

The solution to the central extension of G by $\mathbb{Z}_2^{\mathrm{F}}$ delivers a family of distinct equivalence classes with each equivalence class $[\gamma]$ in one-to-one correspondence with the cohomology group $H^2\left(\mathrm{G},\mathbb{Z}_2^{\mathrm{F}}\right)$, as we shall explain shortly.

Once a representative symmetry group G_f has been selected from the class $[\gamma]\in H^2\left(\mathrm{G},\mathbb{Z}_2^{\mathrm{F}}\right)$, its representation on the Fock space spanned by all the local quantum degrees of freedom, a set that includes Majorana operators, must be constructed. Hereto, there are many possibilities. Their enumeration amounts

19 Y. Ogata and H. Tasaki, Comm. Math. Phys. **372**(3), 951–962 (2019); Y. Ogata, Y. Tachikawa, and H. Tasaki, Comm. Math. Phys. **385**(1), 79–99 (2021); Y. Ogata, Comm. Math. Phys. **374**(2), 705–734 (2020). [Ogata and Tasaki (2019); Ogata et al. (2021); Ogata (2020)].

20 Inversion of a spatial coordinate is not an on-site operation in space. It is thus ruled out as an on-site symmetry. Reversal of time leaves all space coordinates invariant. Thus, reversal of time cannot be an on-site symmetry in spacetime. In particular, reversal of time is not considered an on-site symmetry in any relativistic theory, classical or quantum. With this caveat in mind, the terminology of an on-site symmetry includes the possibility of time-reversal symmetry in Chapters 5 and 6.

to classifying the inequivalent projective representations of the symmetry group G_f. All the inequivalent projective representations of any group G_f are in one-to-one correspondence with the cohomology group $H^2(G_f, U(1)_c)$ as we shall also explain.[21,22]

6.3.2.1 A Brief Review of Group Cohomology

Enumerating all distinct central extensions of a group G by an Abelian group A is solved by computing the second cohomology group $H^2(G, A)$. This mathematical milestone can be traced to the works of Schur, Schreier, and Brauer during the first quarter of the nineteenth century. However, it is only during the chaotic times of the Second World War that the abstract notion of group cohomology was independently formulated by Eilenberg and Mac Lane in the United States; Hopf and Eckmann in Switzerland; Hans Freudenthal in the Netherlands; and Dmitry Faddeev in the Soviet Union.[23,24]

In physics, Dijkgraaf and Witten have emphasized the interplay between topological gauge theories and group cohomology in 1990.[25] Whereas certain second cohomology groups enumerate the projective representations of symmetries in quantum mechanics, certain cohomology groups of order $n = 3, 4, \cdots$ are related to anomalies in quantum field theory in $(n-1)$-dimensional spacetime, say the chiral gauge anomaly encountered in Section 3.9. This point of view was emphasized by Chen, Gu, Liu, and Wen in their construction of symmetry-protected topological phases of matter.[26] The crash course on group cohomology that follows is inspired by the treatment thereof from Chen, Gu, Liu, and Wen.

Definition 6.10 (n-cochain) Given two groups G and M with M an Abelian group, an n-cochain is the map

$$\begin{aligned} \phi\colon \underbrace{G \times \cdots \times G}_{n\,\text{times}} &\to M, \\ (g_1, g_2, \cdots, g_n) &\mapsto \phi(g_1, g_2, \cdots, g_n) \end{aligned} \tag{6.113}$$

that maps an n-tuple $(g_1, g_2, \cdots, g_n) \in G^n$ to an element $\phi(g_1, g_2, \ldots, g_n) \in M$. The set of all n-cochains from G^n to M is denoted by $C^n(G, M)$. By convention, $C^0(G, M) \equiv M$. Henceforth, we will denote the group composition rule in G by $\cdot$ and the group composition rule in M additively by $+$ ($-$ denoting the inverse element).

[21] The label c in the notation $U(1)_c$ takes two values: one indicating that $g \in G_f$ is represented linearly, the other indicating that $g \in G_f$ is represented antilinearly.

[22] Observe that the cohomology group $H^2(G_f, U(1))$ can be thought as enumerating all the distinct central extensions of the group G_f by $U(1)$.

[23] J. J. Rotman, *An Introduction to the Theory of Groups*, Springer, New York, 1999. [Rotman (1999)].

[24] K. S. Brown, *Cohomology of Groups*, Springer, New York, 1982. [Brown (1982)].

[25] R. Dijkgraaf and E. Witten, Commun. Math. Phys. **129**(2), 393–429 (1990). [Dijkgraaf and Witten (1990)].

[26] X. Chen, Z.-C. Gu, Z.-X. Liu, and X.-G. Wen, Phys. Rev. B **87(15)**, 155114 (2013). [Chen et al. (2013)].

Definition 6.11 (addition of two n-cochains) For any pair $\phi_1, \phi_2 \in C^n(\mathrm{G}, \mathrm{M})$, we define $\phi_1 + \phi_2$ by

$$\begin{aligned}\phi_1 + \phi_2 \colon & \underbrace{\mathrm{G} \times \cdots \times \mathrm{G}}_{n \text{ times}} \to \mathrm{M}, \\ (g_1, g_2, \cdots, g_n) \mapsto & (\phi_1 + \phi_2)(g_1, g_2, \cdots, g_n) := \phi_1(g_1, g_2, \cdots, g_n) \\ & + \phi_2(g_1, g_2, \cdots, g_n).\end{aligned} \tag{6.114}$$

Property 6.12 (addition of two n-cochains) The set $C^n(\mathrm{G}, \mathrm{M})$ of n-cochains inherits from M an Abelian group structure through the additive composition law defined in Eq. (6.114).

Definition 6.13 (action of group homomorphism $\mathfrak{c}$) Given the group homomorphism $\mathfrak{c} \colon \mathrm{G} \to \{-1, 1\}$, for any $g \in \mathrm{G}$, we define the group action

$$\begin{aligned}\mathfrak{C}_g \colon & \mathrm{M} \to \mathrm{M}, \\ m \mapsto & \mathfrak{C}_g(m) := \begin{cases} m, & \text{if } \mathfrak{c}(g) = +1, \\ m^{-1} \equiv -m, & \text{if } \mathfrak{c}(g) = -1. \end{cases}\end{aligned} \tag{6.115}$$

The homomorphism $\mathfrak{c}$ indicates whether an element $g \in \mathrm{G}$ is to be represented by a unitary $[\mathfrak{c}(g) = +1]$ or antiunitary $[\mathfrak{c}(g) = -1]$ operator in quantum mechanics. If this distinction is irrelevant, $\mathfrak{c}(g) = 1$ for all $g \in \mathrm{G}$. In this case, both $\mathfrak{c}$ as a subscript or $\mathfrak{c}(g)$ as a multiplicative factor are to be omitted in what follows.

Definition 6.14 (coboundary operator) We define the map $\delta^n_{\mathfrak{c}}$

$$\begin{aligned}\delta^n_{\mathfrak{c}} \colon C^n(\mathrm{G}, \mathrm{M}) & \to C^{n+1}(\mathrm{G}, \mathrm{M}), \\ \phi & \mapsto (\delta^n_{\mathfrak{c}} \phi)\end{aligned} \tag{6.116a}$$

from n-cochains to $(n+1)$-cochains such that

$$\begin{aligned}(\delta^n_{\mathfrak{c}} \phi)(g_1, \cdots, g_{n+1}) := & \mathfrak{C}_{g_1}\left(\phi(g_2, \cdots, g_n, g_{n+1})\right) \\ & + \sum_{i=1}^{n} (-1)^i \phi(g_1, \cdots, g_i \cdot g_{i+1}, \cdots, g_{n+1}) \\ & - (-1)^n \phi(g_1, \cdots, g_n).\end{aligned} \tag{6.116b}$$

The map $\delta^n_{\mathfrak{c}}$ is called a coboundary operator.

Property 6.15 (coboundary operator is a group homomorphism) The coboundary operator (6.116) is a group homomorphism between the set $C^n(\mathrm{G}, \mathrm{M})$ of n-cochains and the set $C^{n+1}(\mathrm{G}, \mathrm{M})$ of $(n+1)$-cochains.

Example 6.16 ($n = 2$) The coboundary operator $\delta^2_{\mathfrak{c}}$ is defined by

$$\begin{aligned}\left(\delta^2_{\mathfrak{c}}\phi\right)(g_1, g_2, g_3) =& \mathfrak{C}_{g_1}(\phi(g_2, g_3)) + (-1)^1\phi(g_1 \cdot g_2, g_3) + (-1)^2\phi(g_1, g_2 \cdot g_3) \\ &- (-1)^2\,\phi(g_1, g_2) \\ =& \mathfrak{c}(g_1)\,\phi(g_2, g_3) - \phi(g_1 \cdot g_2, g_3) + \phi(g_1, g_2 \cdot g_3) - \phi(g_1, g_2).\end{aligned} \tag{6.117}$$

We observe that

$$\begin{aligned}&\left(\delta^2_{\mathfrak{c}}\phi\right)(g_1, g_2, g_3) = 0 \\ \Longleftrightarrow\ & \phi(g_1, g_2) + \phi(g_1 \cdot g_2, g_3) = \phi(g_1, g_2 \cdot g_3) + \mathfrak{c}(g_1)\,\phi(g_2, g_3).\end{aligned} \tag{6.118}$$

Example 6.17 ($n = 1$) The coboundary operator $\delta^1_{\mathfrak{c}}$ is defined by

$$\begin{aligned}\left(\delta^1_{\mathfrak{c}}\phi\right)(g_1, g_2) =& \mathfrak{C}_{g_1}(\phi(g_2)) + (-1)^1\phi(g_1 \cdot g_2) - (-1)^1\phi(g_1) \\ =& \mathfrak{c}(g_1)\,\phi(g_2) - \phi(g_1 \cdot g_2) + \phi(g_1).\end{aligned} \tag{6.119}$$

Property 6.18 One verifies the important identity

$$\Phi(g_1, g_2) := \left(\delta^1_{\mathfrak{c}}\phi\right)(g_1, g_2) \implies (\delta^2_{\mathfrak{c}}\Phi)(g_1, g_2, g_3) = 0. \tag{6.120}$$

Definition 6.19 (n-cocycles) We define the set of n-cocycles

$$Z^n(\mathrm{G}, \mathrm{M}_{\mathfrak{c}}) := \ker(\delta^n_{\mathfrak{c}}) = \{\phi \in C^n(\mathrm{G}, \mathrm{M}) \mid \delta^n_{\mathfrak{c}}\phi = 0\}. \tag{6.121}$$

The set of n-cocycles is made of those n-cochains ϕ that are mapped to the 0-map in $C^{n+1}(\mathrm{G}, \mathrm{M})$ by the coboundary operator, that is $\delta^n_{\mathfrak{c}}\phi$ maps each choice of inputs $g_1, \cdots, g_{n+1}$ to the neutral element 0 of M. The action of the coboundary operator on the elements of the group M is sensitive to the homomorphism $\mathfrak{c}$. For this reason, we label M by $\mathfrak{c}$ in $Z^n(\mathrm{G}, \mathrm{M}_{\mathfrak{c}})$.

Definition 6.20 (n-coboundaries) We define the set of n-coboundaries for $n \geq 1$

$$B^n(\mathrm{G}, \mathrm{M}_{\mathfrak{c}}) := \mathrm{im}(\delta^{n-1}_{\mathfrak{c}}) = \left\{\phi \in C^n(\mathrm{G}, \mathrm{M}) \mid \phi = \delta^{n-1}_{\mathfrak{c}}\phi',\ \phi' \in C^{n-1}(\mathrm{G}, \mathrm{M})\right\} \tag{6.122a}$$

with the convention

$$B^0(\mathrm{G}, \mathrm{M}_{\mathfrak{c}}) := \{0\}. \tag{6.122b}$$

The set of n-coboundaries is made of those n-cochains that are the image of an $(n-1)$-cochain by the coboundary operator.

Property 6.21 The set (6.122) of n-coboundaries is an Abelian subgroup of the set (6.113) of n-cochains.

Property 6.22 The importance of the coboundaries is that the identity (6.120) generalizes to

$$\phi = \delta_{\mathfrak{c}}^{n-1}\phi' \implies \delta_{\mathfrak{c}}^{n}\phi = 0. \tag{6.123}$$

In other words, the set (6.122) of n-coboundaries is an Abelian subgroup of the Abelian group (6.121) of n-cocycles.

Definition 6.23 (n-th cohomology group) The n-th cohomology group is defined as the quotient of the n-cocycles by the n-coboundaries, that is,

$$H^{n}(\mathrm{G},\mathrm{M}_{\mathfrak{c}}) := Z^{n}(\mathrm{G},\mathrm{M}_{\mathfrak{c}})/B^{n}(\mathrm{G},\mathrm{M}_{\mathfrak{c}}). \tag{6.124}$$

From now on, we omit the superscript n in $\delta_{\mathfrak{c}}^{n}$ for convenience. It should be understood that the map $\delta_{\mathfrak{c}}$ acting on a cochain ϕ maps it to a cochain of one higher degree.

Property 6.24 The n-th cohomology group $H^{n}(\mathrm{G},\mathrm{M}_{\mathfrak{c}})$ is an Abelian group. We denote its elements by $[\phi] \in H^{n}(\mathrm{G},\mathrm{M}_{\mathfrak{c}})$, that is the equivalence class of the n-cocycle ϕ.

Suppose that we are given the symmetry group G together with the three Abelian groups M, N, and O. Is it possible to relate the three corresponding cochains $\phi \in C^{m}(\mathrm{G},\mathrm{M})$, $\theta \in C^{n}(\mathrm{G},\mathrm{N})$, and $\upsilon \in C^{q}(\mathrm{G},\mathrm{O})$ for any choice of $m,n,q = 0,1,2,\cdots$? We answer this question affirmatively in the following steps, if we assume that the three Abelian groups M, N, and O all realize unital R-modules over the unital commutative ring R (see **Appendix** A).

Definition 6.25 (direct product of cochains) We denote with M and N two Abelian groups. Given two cochains $\phi \in C^{m}(\mathrm{G},\mathrm{M})$ and $\theta \in C^{n}(\mathrm{G},\mathrm{N})$, we produce the cochain $(\phi \cup \theta) \in C^{m+n}(\mathrm{G},\mathrm{M}\times\mathrm{N})$ through the action

$$\begin{aligned}(\phi \cup \theta)(g_1,\cdots,g_m,g_{m+1},\cdots,g_{m+n})& \\ := \Big(\phi(g_1,\cdots,g_m),\mathfrak{C}_{g_1\cdot g_2\cdots g_m}&\left(\theta(g_{m+1},\cdots,g_{m+n})\right)\Big)\end{aligned} \tag{6.125}$$

for any $g_1,\cdots,g_{m+n} \in \mathrm{G}^{m+n}$.

Definition 6.26 (pairing map) We denote with M, N, and O three R-modules over the commutative ring R. We denote their Abelian group-composition rule by the addition symbol. We ca ll the map

$$\begin{aligned} f\colon \mathrm{M}\times\mathrm{N} &\to \mathrm{O},\\ (m,n) &\mapsto f(m,n),\end{aligned} \tag{6.126a}$$

a pairing map if it is bilinear,

$$f(r\,m,n) = f(m,r\,n) = r\,f(m,n), \tag{6.126b}$$

$$f(m_1 + m_2, n) = f(m_1, n) + f(m_2, n), \tag{6.126c}$$

$$f(m, n_1 + n_2) = f(m, n_1) + f(m, n_2), \tag{6.126d}$$

for any $r \in R$, $m, m_1, m_2 \in \mathrm{M}$, and $n, n_1, n_2 \in \mathrm{N}$.

Definition 6.27 (cup product with the pairing map f) We denote with M, N, and O three R-modules over the commutative ring R that are related by a pairing map f as defined in Eq. (6.126). Given two cochains $\phi \in C^m(\mathrm{G}, \mathrm{M})$ and $\theta \in C^n(\mathrm{G}, \mathrm{N})$, we obtain the cup product $(\phi \smile \theta) \in C^{m+n}(\mathrm{G}, \mathrm{O})$ through the action

$$\begin{aligned}(\phi \smile \theta)(g_1, \cdots, g_m, g_{m+1}, \cdots, g_{m+n})& \\ := f\Big(\big(\phi(g_1, \cdots, g_m), \mathfrak{C}_{g_1 \cdot g_2 \cdots g_m}\left(\theta(g_{m+1}, \cdots, g_{m+n})\right)\big)\Big).&\end{aligned} \tag{6.127a}$$

In words, the cup product with the pairing map f is a binary operation that turns two cochains in $C^m(\mathrm{G}, \mathrm{M})$ and $C^n(\mathrm{G}, \mathrm{N})$, respectively, into a cochain in $C^{m+n}(\mathrm{G}, \mathrm{O})$, by composing operation (6.125) with the pairing map (6.126). The cup product between ϕ and θ with the pairing map f maps any element $(g_1, \cdots, g_{m+n}) \in \mathrm{G}^{m+n}$ to an element of the Abelian group O. For our purposes, both N and M are subsets of the integer numbers, while $\mathrm{O} = \mathbb{Z}_2$. The pairing map f is chosen to be

$$\begin{aligned}f\Big(\big(\phi(g_1, \cdots, g_m), \mathfrak{C}_{g_1 \cdot g_2 \cdots g_m}\left(\theta(g_{m+1}, \cdots, g_{m+n})\right)\big)\Big)& \\ := \phi(g_1, \cdots, g_m)\, \mathfrak{C}_{g_1 \cdot g_2 \cdots g_m}\left(\theta(g_{m+1}, \cdots, g_{m+n})\right) \bmod 2,&\end{aligned} \tag{6.127b}$$

for any $(g_1, \cdots, g_{m+n}) \in \mathrm{G}^{m+n}$, where multiplication of cochains ϕ and θ is treated as multiplication of integers numbers modulo 2.

Example 6.28 The cup product between a 1-cochain $\alpha \in C^1(\mathrm{G}, \mathbb{Z}_2)$ and a 2-cochain $\beta \in C^2(\mathrm{G}, \mathbb{Z}_2)$ is

$$(\alpha \smile \beta)(g_1, g_2, g_3) = \alpha(g_1)\, \mathfrak{C}_{g_1}(\beta(g_2, g_3)) = \alpha(g_1)\, \beta(g_2, g_3), \tag{6.128}$$

for any $g_1, g_2, g_3 \in \mathrm{G}$. Here, the cup product takes values in $\mathbb{Z}_2 = \{0, 1\}$ and multiplication of α and β is the multiplication of integers. In reaching the last equality, we have used the fact that the 2-cochain $\beta(g_2, g_3)$ takes values in $\mathbb{Z}_2$ for which $\mathfrak{C}_{g_1}(\beta(g_2, g_3)) = \beta(g_2, g_3)$ for any g_1.

Property 6.29 The cup product defined in Eqs. (6.127a) and (6.127b) satisfies

$$\delta^{m+n}_{\mathfrak{c}}(\phi \smile \theta) = (\delta^{m}_{\mathfrak{c}}\phi \smile \theta) + (-1)^m\, (\phi \smile \delta^{n}_{\mathfrak{c}}\theta)\,, \tag{6.129}$$

given two cochains $\phi \in C^m(\mathrm{G}, \mathrm{N})$ and $\theta \in C^n(\mathrm{G}, \mathrm{M})$. Hence, the cup product of two cocycles is again a cocycle as the right-hand side of Eq. (6.129) vanishes.

6.3.2.2 Marrying the Fermion Parity with the Symmetry Group G

For quantum systems built out of an even number of local Majorana operators, it is always possible to express all local Majorana operators as the real and imaginary parts of an integer-valued number of local fermionic creation or annihilation operators. The parity (evenness or oddness) of the total number of the local fermions is then necessarily a constant of the motion. If $\widehat{F}$ denotes the operator whose eigenvalues counts the total number of local fermions in the Fock space, then the parity operator $(-1)^{\widehat{F}}$ necessarily commutes with the Hamiltonian that dictates the quantum dynamics, even though $\widehat{F}$ might not, as is the case in any mean-field treatment of superconductivity.

We denote the group of two elements e and p

$$\mathbb{Z}_2^{\mathrm{F}} := \{e, p \,|\, e\,p = p\,e = p, \quad e = e\,e = p\,p\}, \tag{6.130}$$

whereby e is the identity element and we shall interpret the quantum representation of p as the fermion parity operator.[27] It is because of this interpretation of the group element p that we attach the upper index F to the cyclic group $\mathbb{Z}_2$. In addition to the symmetry group $\mathbb{Z}_2^{\mathrm{F}}$, we assume the existence of a second symmetry group G with the composition law $\cdot$ and the identity element id. We would like to construct a new symmetry group G_f out of the two groups G and $\mathbb{Z}_2^{\mathrm{F}}$. Here, the symmetry group G_f inherits the "fermionic" label f from its center $\mathbb{Z}_2^{\mathrm{F}}$.

One possibility is to consider the Cartesian product

$$\mathrm{G} \times \mathbb{Z}_2^{\mathrm{F}} := \left\{(g, h) \mid g \in \mathrm{G}, \quad h \in \mathbb{Z}_2^{\mathrm{F}}\right\} \tag{6.131a}$$

with the composition rule

$$(g_1, h_1) \circ (g_2, h_2) := (g_1 \cdot g_2, h_1\, h_2). \tag{6.131b}$$

The resulting group G_f is the direct product of G and $\mathbb{Z}_2^{\mathrm{F}}$.

However, the composition rule (6.131b) is not the only one compatible with the existence of a neutral element, inverse, and associativity. To see this, we assume first the existence of the map

$$\begin{aligned} \gamma\colon \mathrm{G} \times \mathrm{G} &\to \mathbb{Z}_2^{\mathrm{F}}, \\ (g_1, g_2) &\mapsto \gamma(g_1, g_2), \end{aligned} \tag{6.132a}$$

whereby we impose the conditions

$$\gamma(\mathrm{id}, g) = \gamma(g, \mathrm{id}) = e, \qquad \gamma(g^{-1}, g) = \gamma(g, g^{-1}), \tag{6.132b}$$

27 As we explained below Eq. (6.113), we reserve the symbol $\cdot$ for the group composition law in G. However, we use the Abelian multiplication table (6.130) instead of an additive symbol as below Eq. (6.113) to denote the composition rule in $\mathbb{Z}_2^{\mathrm{F}}$. We will move freely between an additive or a multiplicative representation of the group composition rule of M in the definition (6.113) of n-cochains.

for all $g \in \mathrm{G}$ and[28]

$$\gamma(g_1, g_2)\,\gamma(g_1 \cdot g_2, g_3) = \gamma(g_1, g_2 \cdot g_3)\,\gamma(g_2, g_3), \tag{6.132c}$$

for all $g_1, g_2, g_3 \in \mathrm{G}$. Second, we define G_f to be the set of all pairs (g, h) with $g \in \mathrm{G}$ and $h \in \mathbb{Z}_2^{\mathrm{F}}$ obeying the composition rule

$$\begin{aligned} \underset{\gamma}{\circ}\colon \left(\mathrm{G} \times \mathbb{Z}_2^{\mathrm{F}}\right) \times \left(\mathrm{G} \times \mathbb{Z}_2^{\mathrm{F}}\right) &\to \mathrm{G} \times \mathbb{Z}_2^{\mathrm{F}}, \\ \Big((g_1, h_1), (g_2, h_2)\Big) &\mapsto (g_1, h_1) \underset{\gamma}{\circ} (g_2, h_2), \end{aligned} \tag{6.132d}$$

where

$$(g_1, h_1) \underset{\gamma}{\circ} (g_2, h_2) := \Big(g_1 \cdot g_2, h_1\, h_2\, \gamma(g_1, g_2)\Big). \tag{6.132e}$$

Definition 6.30 The central extension G_f of G by $\mathbb{Z}_2^{\mathrm{F}}$ through the map γ is the set of all pairs $(g, h) \in \mathrm{G} \times \mathbb{Z}_2^{\mathrm{F}}$ together with the composition rule $\underset{\gamma}{\circ}$ defined by Eq. (6.132).

One may verify the following properties (**Exercise** 6.6).

Property 6.31 First, the ordering of the group elements in $h_1\, h_2\, \gamma(g_1, g_2)$ is arbitrary since $\mathbb{Z}_2^{\mathrm{F}}$ is Abelian.

Property 6.32 Second, conditions (6.132b) and (6.132c) ensure that G_f is a group such that

$$(\mathrm{id}, e) \tag{6.133a}$$

is the neutral element,

$$\left(g^{-1}, [\gamma(g, g^{-1})]^{-1}\, h^{-1}\right) \tag{6.133b}$$

is the inverse of (g, h), and

$$(\mathrm{id}, \mathbb{Z}_2^{\mathrm{F}}) \tag{6.133c}$$

is the group center,[29] that is the group G_f is a central extension of G by $\mathbb{Z}_2^{\mathrm{F}}$.

Property 6.33 Third, an equivalence relation between the pair of maps γ, γ' of the form (6.132) holds if there exists the one-to-one map

$$\begin{aligned} \widetilde{\kappa}\colon \mathrm{G} \times \mathbb{Z}_2^{\mathrm{F}} &\to \mathrm{G} \times \mathbb{Z}_2^{\mathrm{F}}, \\ (g, h) &\mapsto (g, \kappa(g)\, h), \end{aligned} \tag{6.134a}$$

[28] Equation (6.132c) is nothing but Eq. (6.118) with the identifications $\exp(\mathrm{i}\phi) \to \gamma$ and $\mathfrak{c}(g) = 1$ for all $g \in \mathrm{G}$ in Eq. (6.115).

[29] The center of a group is the set of elements that commute with every element of the group. A group is Abelian if and only if its center is the group itself. A group whose center is made of one element, the identity, is called centerless.

induced by the map

$$\begin{aligned} \kappa \colon \mathrm{G} &\to \mathbb{Z}_2^{\mathrm{F}}, \\ g &\mapsto \kappa(g), \end{aligned} \tag{6.134b}$$

such that the condition

$$\widetilde{\kappa}\Big((g_1, h_1) \underset{\gamma}{\circ} (g_2, h_2)\Big) = \widetilde{\kappa}\Big((g_1, h_1)\Big) \underset{\gamma'}{\circ} \widetilde{\kappa}\Big((g_2, h_2)\Big) \tag{6.135}$$

holds for all $(g_1, h_1), (g_2, h_2) \in \mathrm{G} \times \mathbb{Z}_2^{\mathrm{F}}$. In other words, γ and γ' generate two isomorphic groups if the identity

$$\kappa(g_1 \cdot g_2)\, \gamma(g_1, g_2) = \kappa(g_1)\, \kappa(g_2)\, \gamma'(g_1, g_2) \tag{6.136}$$

holds for all $g_1, g_2 \in \mathrm{G}$. The group isomorphism $\widetilde{\kappa}$ defines an equivalence relation.

Definition 6.34 We say that the group G_f obtained by extending the group G with the group $\mathbb{Z}_2^{\mathrm{F}}$ through the map γ splits when a map (6.134b) exists such that

$$\kappa(g_1 \cdot g_2)\, \gamma(g_1, g_2) = \kappa(g_1)\, \kappa(g_2) \tag{6.137}$$

holds for all $g_1, g_2 \in \mathrm{G}$, that is, G_f splits when it is isomorphic to the direct product (6.131).

Property 6.35 The task of classifying all the non-equivalent central extensions of G by $\mathbb{Z}_2^{\mathrm{F}}$ through γ is achieved by enumerating all the elements of the second cohomology group $H^2\left(\mathrm{G}, \mathbb{Z}_2^{\mathrm{F}}\right)$.[30] We define an index $[\gamma] \in H^2\left(\mathrm{G}, \mathbb{Z}_2^{\mathrm{F}}\right)$ to represent such an equivalence class, whereby the index $[\gamma] = 0$ is assigned to the case when G_f splits (we are using implicitly the convention that the group composition law of $H^2\left(\mathrm{G}, \mathbb{Z}_2^{\mathrm{F}}\right)$ is denoted by the addition with 0 denoting the neutral element).

6.3.2.3 Projective Representations of the Group G_f

We denote with Λ a d-dimensional lattice with $\boldsymbol{j} \in \mathbb{Z}^d$ labeling the repeat unit cells. We are going to attach to Λ a Fock space on which projective representations of the group G_f constructed so far are realized. To this end, we make four assumptions.

Assumption 6.36 We attach to each repeat unit cell $\boldsymbol{j} \in \Lambda$ the local Fock space $\mathfrak{F}_{\boldsymbol{j}}$. This step requires that the number of Majorana degrees of freedom in each repeat unit cell is even. It is then possible to define the local fermion number operator $\hat{f}_{\boldsymbol{j}}$ and the local fermion-parity operator

$$\hat{p}_{\boldsymbol{j}} := (-1)^{\hat{f}_{\boldsymbol{j}}}. \tag{6.138}$$

[30] This one-to-one correspondence arises from the fact that condition (6.132c) corresponds to Eq. (6.118), while the identity (6.136), which defines the equivalence relation, can be written using Eq. (6.119) as

$$(\delta^1 \kappa)(g_1, g_2)\, \gamma'(g_1, g_2) = \gamma(g_1, g_2).$$

One verifies that condition (6.132b) gives no further constraint.

We assume that all local Fock spaces $\mathfrak{F}_{\boldsymbol{j}}$ with $\boldsymbol{j} \in \Lambda$ are "identical," in particular they share the same dimensionality $\mathcal{D}$. This assumption is a prerequisite to imposing translation symmetry.

Assumption 6.37 Each repeat unit cell $\boldsymbol{j} \in \Lambda$ is equipped with a representation $\hat{u}_{\boldsymbol{j}}(g)$ of G_f through the conjugation

$$\hat{o}_{\boldsymbol{j}} \mapsto \hat{u}_{\boldsymbol{j}}(g)\, \hat{o}_{\boldsymbol{j}}\, \hat{u}^{\dagger}_{\boldsymbol{j}}(g), \qquad [\hat{u}_{\boldsymbol{j}}(g)]^{-1} = \hat{u}^{\dagger}_{\boldsymbol{j}}(g) \tag{6.139a}$$

of any operator $\hat{o}_{\boldsymbol{j}}$ acting on the local Fock space $\mathfrak{F}_{\boldsymbol{j}}$. The representation (6.139a) of $g \in \mathrm{G}_f$ can either be unitary or antiunitary. More precisely, let

$$\begin{aligned} \mathfrak{c}\colon \mathrm{G}_f &\to \{1,-1\}, \\ g &\mapsto \mathfrak{c}(g) \end{aligned} \tag{6.139b}$$

be a homomorphism. We then have the decomposition

$$\hat{u}_{\boldsymbol{j}}(g) := \begin{cases} \hat{v}_{\boldsymbol{j}}(g), & \text{if } \mathfrak{c}(g) = +1, \\ \\ \hat{v}_{\boldsymbol{j}}(g)\, \mathsf{K}, & \text{if } \mathfrak{c}(g) = -1, \end{cases} \tag{6.139c}$$

where

$$[\hat{v}_{\boldsymbol{j}}(g)]^{-1} = \hat{v}^{\dagger}_{\boldsymbol{j}}(g), \qquad \hat{p}_{\boldsymbol{j}}\, \hat{v}_{\boldsymbol{j}}(g)\, \hat{p}_{\boldsymbol{j}} = (-1)^{\rho(g)}\, \hat{v}_{\boldsymbol{j}}(g) \tag{6.139d}$$

is a unitary operator with the fermion parity $\rho(g) \in \{0,1\} \equiv \mathbb{Z}_2$ acting linearly on $\mathfrak{F}_{\boldsymbol{j}}$ and K denotes complex conjugation on the local Fock space $\mathfrak{F}_{\boldsymbol{j}}$. Accordingly, the homomorphism $\mathfrak{c}(g)$ dictates if the representation of the element $g \in \mathrm{G}_f$ is implemented through a unitary operator $[\mathfrak{c}(g) = 1]$ or an antiunitary operator $[\mathfrak{c}(g) = -1]$. Finally, we always choose to represent locally the fermion parity $p \in \mathbb{Z}_2^{\mathrm{F}}$ by the Hermitean operator $\hat{p}_{\boldsymbol{j}}$,

$$\hat{u}_{\boldsymbol{j}}(p) := \hat{p}_{\boldsymbol{j}} \equiv (-1)^{\hat{f}_{\boldsymbol{j}}}. \tag{6.139e}$$

Assumption 6.38 For any two elements $g, h \in \mathrm{G}_f$ [to simplify notation, $g \underset{\gamma}{\circ} h \equiv g\, h$ for all $g, h \in \mathrm{G}_f$], whereby $e = g\, g^{-1} = g^{-1}\, g$ denotes the neutral element and $g^{-1} \in \mathrm{G}_f$ the inverse of $g \in \mathrm{G}_f$, we postulate the projective representation

$$\hat{u}_{\boldsymbol{j}}(e) = \widehat{\mathbb{1}}_{\mathcal{D}}, \tag{6.140a}$$

$$\hat{u}_{\boldsymbol{j}}(g)\, \hat{u}_{\boldsymbol{j}}(h) = e^{\mathrm{i}\phi(g,h)}\, \hat{u}_{\boldsymbol{j}}(g\, h), \tag{6.140b}$$

$$\big[\hat{u}_{\boldsymbol{j}}(g)\, \hat{u}_{\boldsymbol{j}}(h)\big]\hat{u}_{\boldsymbol{j}}(f) = \hat{u}_{\boldsymbol{j}}(g)\big[\hat{u}_{\boldsymbol{j}}(h)\, \hat{u}_{\boldsymbol{j}}(f)\big], \tag{6.140c}$$

whereby the identity operator acting on $\mathfrak{F}_{\boldsymbol{j}}$ is denoted $\widehat{\mathbb{1}}_{\mathcal{D}}$ and the function

$$\begin{aligned} \phi\colon \mathrm{G}_f \times \mathrm{G}_f &\to [0, 2\pi[, \\ (g,h) &\mapsto \phi(g,h), \end{aligned} \tag{6.141a}$$

must be compatible with the associativity in G_f [Eq. (6.140c)], that is, [compare with Eqs. (6.132c) and (6.118)]

$$\phi(g,h)+\phi(g\,h,f)=\phi(g,h\,f)+\mathfrak{c}(g)\,\phi(h,f), \tag{6.141b}$$

for all $g,h,f\in G_f$ (**Exercise** 6.7). The map ϕ taking values in $[0,2\pi[$ and satisfying (6.141b) is an example of a 2-cocycle with the group action specified by the $\mathbb{Z}_2$-valued homomorphism $\mathfrak{c}$. In the vicinity of the value 0, ϕ generates the Lie algebra $\mathfrak{u}(1)$. The associated Lie group is denoted U(1). Given the neutral element $e\in G_f$, a normalized 2-cocycle obeys the additional constraint

$$\phi(e,g)=\phi(g,e)=0 \tag{6.141c}$$

for all $g\in G_f$. Two 2-cocycles $\phi(g,h)$ and $\phi'(g,h)$ are said to be equivalent if they can be consistently related through a map [compare with Eq. (6.134)]

$$\begin{aligned}\xi\colon G_f &\to [0,2\pi[,\\ g &\mapsto \xi(g),\end{aligned} \tag{6.142}$$

as follows. The equivalence relation $\phi\sim\phi'$ holds if the transformation

$$\hat{u}(g)=e^{\mathrm{i}\xi(g)}\,\hat{u}'(g) \tag{6.143a}$$

implies the relation [compare with Eq. (6.136)]

$$\phi(g,h)-\phi'(g,h)=\xi(g)+\mathfrak{c}(g)\,\xi(h)-\xi(g\,h), \tag{6.143b}$$

between the 2-cocycle $\phi(g,h)$ associated with the projective representation $\hat{u}(g)$ and the 2-cocycle $\phi'(g,h)$ associated with the projective representation $\hat{u}'(g)$ (**Exercise** 6.8). In particular, $\hat{u}$ is equivalent to an ordinary representation (a trivial projective representation) if $\phi'(g,h)=0$ for all $g,h\in G_f$. Any $\phi\sim 0$ is called a 2-coboundary. For any 2-coboundary ϕ there must exist a ξ such that [compare with Eqs. (6.119) and (6.120)]

$$\phi(g,h)=\xi(g)+\mathfrak{c}(g)\,\xi(h)-\xi(g\,h). \tag{6.144}$$

The space of equivalence classes of projective representations is obtained by taking the quotient of 2-cocycles (6.141b) by 2-coboundaries (6.144). The resulting set is the second cohomology group $H^2(G_f,U(1)_{\mathfrak{c}})$, which inherits from the Abelian group

$$U(1):=\{e^{\mathrm{i}\phi}\mid\phi\in[0,2\pi[\} \tag{6.145a}$$

the Abelian group structure

$$[\phi_1]+[\phi_2]=[\phi_1+\phi_2] \tag{6.145b}$$

for any pair of 2-cocycles $e^{\mathrm{i}\phi_1}$ and $e^{\mathrm{i}\phi_2}$. Here, given any pair of 2-cochains $e^{\mathrm{i}\phi_1}$ and $e^{\mathrm{i}\phi_2}$, we have defined the 2-cochain $e^{\mathrm{i}(\phi_1+\phi_2)}$ through the point-wise addition

$$\begin{aligned}\phi_1+\phi_2\colon \mathrm{G}_f\times\mathrm{G}_f &\to [0,2\pi[,\\ (g,h) &\mapsto (\phi_1+\phi_2)(g,h) := \phi_1(g,h)+\phi_2(g,h) \bmod 2\pi.\end{aligned} \tag{6.145c}$$

Assumption 6.39 We attach to Λ the global Fock space $\mathfrak{F}_\Lambda$ by taking the appropriate product over $\boldsymbol{j}$ of the local Fock spaces $\mathfrak{F}_{\boldsymbol{j}}$. This means that we impose some algebra on all local operators differing by their repeat unit cell labels.

Example 6.40 Any two local fermion number operators $\hat{f}_{\boldsymbol{j}}$ and $\hat{f}_{\boldsymbol{j}'}$ must commute

$$\left[\hat{f}_{\boldsymbol{j}},\hat{f}_{\boldsymbol{j}'}\right]=0 \tag{6.146}$$

for any two distinct repeat unit cells $\boldsymbol{j}\neq\boldsymbol{j}'\in\Lambda$. The total fermion and fermion-parity numbers are

$$\widehat{F}_\Lambda := \sum_{\boldsymbol{j}\in\Lambda}\hat{f}_{\boldsymbol{j}}, \qquad \widehat{P}_\Lambda := (-1)^{\widehat{F}_\Lambda}, \tag{6.147}$$

respectively.

Example 6.41 Any two Majorana operators labeled by $\boldsymbol{j}\neq\boldsymbol{j}'\in\Lambda$ must anti-commute.

Example 6.42 The algebra

$$\hat{u}_{\boldsymbol{j}}(g)\,\hat{u}_{\boldsymbol{j}'}(g') = (-1)^{\rho(g)\,\rho(g')}\,\hat{u}_{\boldsymbol{j}'}(g')\,\hat{u}_{\boldsymbol{j}}(g) \tag{6.148}$$

holds for any distinct $\boldsymbol{j}\neq\boldsymbol{j}'\in\Lambda$ and any $g,g'\in\mathrm{G}_f$ because of Eqs. (6.139d) and (6.139e). We then define the operator

$$\widehat{U}(g) := \begin{cases}\prod\limits_{\boldsymbol{j}\in\Lambda}\hat{v}_{\boldsymbol{j}}(g), & \text{if } \mathfrak{c}(g)=+1,\\[2ex] \left[\prod\limits_{\boldsymbol{j}\in\Lambda}\hat{v}_{\boldsymbol{j}}(g)\right]K, & \text{if } \mathfrak{c}(g)=-1,\end{cases} \tag{6.149}$$

that implements globally on the Fock space $\mathfrak{F}_\Lambda$ the operation corresponding to the group element $g\in\mathrm{G}_f$.

6.3.3 Fermionic Matrix Product States

Consider a one-dimensional lattice $\Lambda\cong\mathbb{Z}_N$. At the repeat unit cell $j=1,\cdots,N$, the local fermion number operator is denoted $\hat{f}_j$ and the local Fock space of dimension $\mathcal{D}_j$ is denoted $\mathfrak{F}_j\cong\mathbb{C}^{\mathcal{D}_j}$. We define with

$$|\psi_{j,\sigma_j}\rangle, \qquad \sigma_j=1,\cdots,\mathcal{D}_j, \tag{6.150a}$$

an orthonormal basis of $\mathfrak{F}_j$ such that

$$(-1)^{\hat{f}_j}\,|\psi_{j,\sigma_j}\rangle = (-1)^{|\sigma_j|}|\psi_{j,\sigma_j}\rangle. \tag{6.150b}$$

The fermion parity eigenvalue of the basis element $|\psi_{j,\sigma_j}\rangle$ is thus denoted $(-1)^{|\sigma_j|}$ with $|\sigma_j| \equiv 0, 1$. Thus, $|\sigma_j| \equiv 0\,(1)$ implies that the state $|\psi_{j,\sigma_j}\rangle$ contains an even (odd) number of fermions. The local Fock space $\mathfrak{F}_j$ admits the direct sum decomposition

$$\mathfrak{F}_j = \mathfrak{F}_j^{(0)} \oplus \mathfrak{F}_j^{(1)}, \tag{6.151a}$$

where, given $p = 0, 1$,

$$\mathfrak{F}_j^{(p)} := \operatorname{span}\left\{ |\psi_{j,\sigma_j}\rangle,\ \sigma_j = 1, \cdots, \mathcal{D}_j \;\middle|\; |\sigma_j| = p \right\}. \tag{6.151b}$$

One verifies that (**Exercise** 6.9)

$$\dim \mathfrak{F}_j^{(0)} = \dim \mathfrak{F}_j^{(1)} = \frac{\mathcal{D}_j}{2}. \tag{6.152}$$

To construct the Fock space $\mathfrak{F}_\Lambda$ for the lattice Λ, we demand that the direct sum (6.151) also holds for $\mathfrak{F}_\Lambda$. This is achieved with the help of the $\mathbb{Z}_2$ tensor product $\otimes_{\mathfrak{g}}$ defined in Section 5.7.3. This tensor product preserves the $\mathbb{Z}_2$-grading structure. We define the reordering rule

$$|\psi_{j,\sigma_j}\rangle \otimes_{\mathfrak{g}} |\psi_{j',\sigma_{j'}}\rangle \equiv (-1)^{|\sigma_j|\,|\sigma_{j'}|} |\psi_{j',\sigma_{j'}}\rangle \otimes_{\mathfrak{g}} |\psi_{j,\sigma_j}\rangle \tag{6.153}$$

on any two basis elements $|\psi_{j,\sigma_j}\rangle$ and $|\psi_{j',\sigma_{j'}}\rangle$ of $\mathfrak{F}_j$ and $\mathfrak{F}_{j'}$ for any two distinct sites $j \in \Lambda$ and $j' \in \Lambda$, respectively. The rule (6.153) guarantees that states are antisymmetric under the exchange of an odd number of fermions on site j with an odd number of fermions on site j', while symmetric otherwise. We then define the fermionic Fock space $\mathfrak{F}_\Lambda$ for the lattice Λ to be

$$\begin{aligned}\mathfrak{F}_\Lambda := \operatorname{span}\Bigg\{ |\Psi_{\boldsymbol{\sigma}}\rangle \;\Bigg|\; & \boldsymbol{\sigma} \equiv (\sigma_1, \cdots, \sigma_N) \in \{1, \cdots, \mathcal{D}_1\} \times \cdots \times \{1, \cdots, \mathcal{D}_N\}, \\ & |\Psi_{\boldsymbol{\sigma}}\rangle \equiv |\psi_{1,\sigma_1}\rangle \otimes_{\mathfrak{g}} |\psi_{2,\sigma_2}\rangle \otimes_{\mathfrak{g}} \cdots \otimes_{\mathfrak{g}} |\psi_{N-1,\sigma_{N-1}}\rangle \otimes_{\mathfrak{g}} |\psi_{N,\sigma_N}\rangle \Bigg\}.\end{aligned} \tag{6.154}$$

As the parity $|\sigma_j|$ of the state $|\psi_{j,\sigma_j}\rangle$ can be generalized to the parity $|\boldsymbol{\sigma}|$ of the state $|\Psi_{\boldsymbol{\sigma}}\rangle$ through the action of the global fermion number operator

$$\widehat{F}_\Lambda := \sum_{j=1}^{N} \hat{f}_j, \qquad |\boldsymbol{\sigma}| \equiv \sum_{j=1}^{N} |\sigma_j| \bmod 2, \tag{6.155}$$

the Fock space (6.154) inherits the direct sum decomposition (6.151a),

$$\mathfrak{F}_\Lambda = \mathfrak{F}_\Lambda^{(0)} \oplus \mathfrak{F}_\Lambda^{(1)}. \tag{6.156}$$

Any state $|\Psi\rangle \in \mathfrak{F}_\Lambda$ has the expansion

$$|\Psi\rangle = \sum_{\boldsymbol{\sigma}} c_{\boldsymbol{\sigma}} |\Psi_{\boldsymbol{\sigma}}\rangle \tag{6.157a}$$

with the expansion coefficient $c_{\boldsymbol{\sigma}} \in \mathbb{C}$. Such a state is homogeneous if it belongs to either $\mathfrak{F}_{\Lambda}^{(0)}$ or $\mathfrak{F}_{\Lambda}^{(1)}$, in which case it has a definite parity $|\Psi| \equiv 0, 1$.

From now on, we assume that all local Fock spaces are pairwise isomorphic, that is,

$$\mathcal{D}_j = \mathcal{D}, \qquad \mathfrak{F}_j \cong \mathfrak{F}_{j'} \qquad 1 \leq j < j' \leq N. \tag{6.158}$$

This assumption is needed to impose translation symmetry below.

We describe the construction of two families of states that lie in $\mathfrak{F}_{\Lambda}^{(0)}$ and $\mathfrak{F}_{\Lambda}^{(1)}$, respectively. To this end, we choose the positive integer M, denote with $\mathbb{1}_M$ the unit $M \times M$ matrix and define the following pair of $2M \times 2M$ matrices

$$P := \begin{pmatrix} \mathbb{1}_M & 0 \\ 0 & -\mathbb{1}_M \end{pmatrix}, \qquad Y := \begin{pmatrix} 0 & \mathbb{1}_M \\ -\mathbb{1}_M & 0 \end{pmatrix}. \tag{6.159}$$

The 2×2 grading that is displayed is needed to represent the $\mathbb{Z}_2$ grading in Eq. (6.156), as will soon become apparent. The anticommuting matrices P and Y belong to the set $\mathrm{Mat}(2M, \mathbb{C})$ of all $2M \times 2M$ matrices. This set is a $4M^2$-dimensional vector space over the complex numbers. For any $\sigma_j = 1, \cdots, \mathcal{D}$ with $j \in \Lambda$, we choose the same five families of $\mathcal{D}$ matrices

$$\begin{aligned} B_1, \cdots, B_{\mathcal{D}} &\in \mathrm{Mat}(M, \mathbb{C}), \\ C_1, \cdots, C_{\mathcal{D}} &\in \mathrm{Mat}(M, \mathbb{C}), \\ D_1, \cdots, D_{\mathcal{D}} &\in \mathrm{Mat}(M, \mathbb{C}), \\ E_1, \cdots, E_{\mathcal{D}} &\in \mathrm{Mat}(M, \mathbb{C}), \\ G_1, \cdots, G_{\mathcal{D}} &\in \mathrm{Mat}(M, \mathbb{C}), \end{aligned} \tag{6.160a}$$

with the help of which we define the matrices

$$A^{(0)}_{\sigma_j} := \begin{cases} \begin{pmatrix} B_{\sigma_j} & 0 \\ 0 & C_{\sigma_j} \end{pmatrix}, & \text{if } |\sigma_j| = 0, \\ \\ \begin{pmatrix} 0 & D_{\sigma_j} \\ E_{\sigma_j} & 0 \end{pmatrix}, & \text{if } |\sigma_j| = 1, \end{cases} \tag{6.160b}$$

and

$$A^{(1)}_{\sigma_j} := \begin{cases} \begin{pmatrix} G_{\sigma_j} & 0 \\ 0 & G_{\sigma_j} \end{pmatrix}, & \text{if } |\sigma_j| = 0, \\ \\ \begin{pmatrix} 0 & G_{\sigma_j} \\ -G_{\sigma_j} & 0 \end{pmatrix}, & \text{if } |\sigma_j| = 1, \end{cases} \tag{6.160c}$$

from Mat$(2M,\mathbb{C})$. Observe that Eq. (6.160c) is a special case of Eq. (6.160b). For any $\sigma_j = 1, \cdots, \mathcal{D}$ with $j \in \Lambda$, the matrix P commutes (anticommutes) with $A^{(p)}_{\sigma_j}$ when $|\sigma_j| = 0$ ($|\sigma_j| = 1$),

$$P\,A^{(p)}_{\sigma_j} = (-1)^{|\sigma_j|}\,A^{(p)}_{\sigma_j}\,P \tag{6.161}$$

for both $p = 0, 1$. In contrast, the matrix Y commutes with $A^{(1)}_{\sigma_j}$

$$Y\,A^{(1)}_{\sigma_j} = A^{(1)}_{\sigma_j}\,Y \tag{6.162}$$

for all $\sigma_j = 1, \cdots, \mathcal{D}$ with $j \in \Lambda$.

We are ready to define even ($p = 0$) and odd ($p = 1$) fermionic matrix product states (FMPS) that obey periodic boundary conditions.

Definition 6.43 An even parity FMPS obeying periodic boundary conditions is defined by

$$|\{A^{(0)}_{\sigma_j}\}\rangle := \sum_{\boldsymbol{\sigma}} \mathrm{tr}\left(P\,A^{(0)}_{\sigma_1}\cdots A^{(0)}_{\sigma_N}\right)|\Psi_{\boldsymbol{\sigma}}\rangle \tag{6.163}$$

for any choice of the matrices (6.160b) and with the basis (6.154) of the Fock space $\mathfrak{F}_\Lambda$.

Definition 6.44 An odd parity FMPS obeying periodic boundary conditions is defined by

$$|\{A^{(1)}_{\sigma_j}\}\rangle := \sum_{\boldsymbol{\sigma}} \mathrm{tr}\left(Y\,A^{(1)}_{\sigma_1}\cdots A^{(1)}_{\sigma_N}\right)|\Psi_{\boldsymbol{\sigma}}\rangle \tag{6.164}$$

for any choice of the matrices (6.160c) and with the basis (6.154) of the Fock space $\mathfrak{F}_\Lambda$.

The following properties follow from the cyclicity of the trace and from the fact that Y is traceless.

Property 6.45 The FMPS $|\{A^{(p)}_{\sigma_j}\}\rangle$ is homogeneous and belongs to $\mathfrak{F}^{(p)}_\Lambda$ for $p = 0, 1$.

This claim is a consequence of the identities

$$\sum_{j=1}^{N}|\sigma_j| = 1 \bmod 2 \implies \mathrm{tr}\left(P\,A^{(0)}_{\sigma_1}\cdots A^{(0)}_{\sigma_N}\right) = 0, \tag{6.165a}$$

$$\sum_{j=1}^{N}|\sigma_j| = 0 \bmod 2 \implies \mathrm{tr}\left(Y\,A^{(1)}_{\sigma_1}\cdots A^{(1)}_{\sigma_N}\right) = 0. \tag{6.165b}$$

Property 6.46 The FMPS $|\{A^{(p)}_{\sigma_j}\}\rangle$ is unchanged under a translation by one repeat unit cell (**Exercise** 6.10).

Property 6.47 The FMPS (6.163) and (6.164) are not uniquely specified by the choices $\{A^{(p)}_{\sigma_j}\}$ for $p = 0, 1$, respectively.

For example, the similarity transformation

$$A^{(0)}_{\sigma_j} \mapsto V\, A^{(0)}_{\sigma_j}\, V^{-1}, \qquad \sigma_j = 1, \cdots, \mathcal{D}, \tag{6.166}$$

with V any matrix that commutes with P leaves the trace unchanged. Another example occurs if there exists a nonvanishing matrix $Q = Q^2 \in \mathrm{Mat}(2M, \mathbb{C})$ such that

$$Q\, A^{(0)}_{\sigma_j} = Q\, A^{(0)}_{\sigma_j}\, Q, \qquad \sigma_j = 1, \cdots, \mathcal{D}. \tag{6.167}$$

Indeed, Eq. (6.167) implies the identity (**Exercise** 6.11)

$$\mathrm{tr}\left(P\, A^{(0)}_{\sigma_1} \cdots A^{(0)}_{\sigma_N}\right) = \mathrm{tr}\left(P\, \tilde{A}^{(0)}_{\sigma_1} \cdots \tilde{A}^{(0)}_{\sigma_N}\right) \tag{6.168a}$$

with $\tilde{A}^{(0)}_{\sigma_j}$ the matrix

$$\tilde{A}^{(0)}_{\sigma_j} := Q\, A^{(0)}_{\sigma_j}\, Q + (\mathbb{1}_{2M} - Q)\, A^{(0)}_{\sigma_j}\, (\mathbb{1}_{2M} - Q). \tag{6.168b}$$

While the $\mathcal{D}$ conditions in Eq. (6.167) imply that all matrices $A^{(0)}_1, \cdots, A^{(0)}_{\mathcal{D}}$ are reducible, the $\mathcal{D}$ definitions in Eq. (6.168b) imply that all matrices $\tilde{A}^{(0)}_1, \cdots, \tilde{A}^{(0)}_{\mathcal{D}}$ are decomposable into the same block diagonal form (**Exercise** 6.11). A necessary and sufficient condition on the $\mathcal{D}$ matrices $A^{(0)}_1, \cdots, A^{(0)}_{\mathcal{D}}$ to prevent that Eq. (6.167) holds for some $Q \in \mathrm{Mat}(2M, \mathbb{C})$ is to demand that there exists an integer $1 \le \ell^\star \le N$ such that the vector space spanned by the $\mathcal{D}^{\ell^\star}$ matrix products

$$A^{(0)}_{\sigma_1} \cdots A^{(0)}_{\sigma_{\ell^\star}}, \qquad \sigma_1, \cdots, \sigma_{\ell^\star} = 1, \cdots, \mathcal{D}, \tag{6.169a}$$

is $\mathrm{Mat}(2M, \mathbb{C})$.[31] More precisely, for any $A \in \mathrm{Mat}(2M, \mathbb{C})$, it is possible to find $\mathcal{D}^{\ell^\star}$ coefficients $a^{(0)}_{\sigma_1, \cdots, \sigma_{\ell^\star}} \in \mathbb{C}$ such that[32]

$$A = \sum_{\sigma_1, \cdots, \sigma_{\ell^\star} = 1}^{\mathcal{D}} a^{(0)}_{\sigma_1, \cdots, \sigma_{\ell^\star}}\, A^{(0)}_{\sigma_1} \cdots A^{(0)}_{\sigma_{\ell^\star}}. \tag{6.169b}$$

In order to restrict the redundancy in the choice of the matrices (6.160) that enter the FMPS (6.163) and (6.164), we make the following definitions.

Definition 6.48 The even-parity FMPS (6.163) is *injective* if there exists an integer $\ell^\star \ge 1$ such that the $\mathcal{D}^{\ell^\star}$ products $A^{(0)}_{\sigma_1} \cdots A^{(0)}_{\sigma_{\ell^\star}}$ of $2M \times 2M$ matrices span $\mathrm{Mat}(2M, \mathbb{C})$.

[31] On the one hand, if $\mathcal{D} = 2$, $M = 1$, $A^{(0)}_1$ is the second Pauli matrix τ_2, and $A^{(0)}_2$ is the third Pauli matrix τ_3, then $\ell^\star$ does not exist since the span of an odd (even) product of $A^{(0)}$'s is the two-dimensional vector space spanned by τ_2 and τ_3 (identity matrix τ_0 and τ_1). On the other hand, if $\mathcal{D} = 3$, $M = 1$, and the three $A^{(0)}$'s are the three Pauli matrices, then $\ell^\star = 2$.

[32] The basis (6.169a) is in general overcomplete owing to the condition $\mathcal{D}^{\ell^\star} \ge 4M^2$.

Definition 6.49 The odd-parity FMPS (6.164) is *injective* if there exists an integer $\ell^\star \geq 1$ such that the $(\mathcal{D}/2)^{\ell^\star}$ products $G_{\sigma_1} \cdots G_{\sigma_{\ell^\star}}$ with

$$|\sigma_1| = \cdots = |\sigma_{\ell^\star}| = 0 \tag{6.170}$$

of $M \times M$ matrices span $\mathrm{Mat}(M, \mathbb{C})$.

Translation-invariant bosonic matrix product states (BMPS) are defined in **Exercise** 6.12, together with an equivalent definition of injective matrix product states for which the use of the adjective injective is self explanatory.

The need to distinguish the definitions of injectivity for even- and odd-parity FMPS stems from the fact that for an odd-parity FMPS the matrix Y commutes with $A^{(1)}_1, \cdots, A^{(1)}_{\mathcal{D}}$. In other words, Y is in the center of the algebra closed by products of $A^{(1)}_1, \cdots, A^{(1)}_{\mathcal{D}}$. Injectivity requires this center to be generated by $\mathbb{1}_{2M}$ and Y, that is, the algebra closed by products of matrices $A^{(1)}_1, \cdots, A^{(1)}_{\mathcal{D}}$ is a $\mathbb{Z}_2$-graded simple algebra.[33] For the center to be generated by no more than $\mathbb{1}_{2M}$ and Y, the products of the $\mathcal{D}/2$ matrices

$$\left\{ G_{\sigma_1}, \cdots, G_{\sigma_{\mathcal{D}/2}} \,\middle|\, |\sigma_1| = \cdots = |\sigma_{\mathcal{D}/2}| = 0 \right\} \tag{6.171}$$

must close a simple algebra of $M \times M$ matrices, which is precisely the **Definition** 6.49. The following properties of FMPS are essential to the forthcoming proof of **Theorem** 6.52.

Property 6.50 Let $\ell \geq \ell^\star$. The $\mathcal{D}^\ell$ products $A^{(0)}_{\sigma_1} \cdots A^{(0)}_{\sigma_\ell}$ of $2M \times 2M$ matrices span $\mathrm{Mat}(2M, \mathbb{C})$ for any injective even-parity FMPS. The $(\mathcal{D}/2)^\ell$ products $G_{\sigma_1} \cdots G_{\sigma_\ell}$ with

$$|\sigma_1| = \cdots = |\sigma_\ell| = 0 \tag{6.172}$$

of $M \times M$ matrices span $\mathrm{Mat}(M, \mathbb{C})$ for any $\ell \geq \ell^\star$ injective odd-parity FMPS.

Property 6.51 If two sets of matrices $\{A^{(p)}_{\sigma_j}\}$ and $\{\widetilde{A}^{(p)}_{\sigma_j}\}$ generate the same injective FMPS, there then exists an invertible matrix V and a phase $\varphi_V \in [0, 2\pi[$ such that

$$\widetilde{A}^{(p)}_{\sigma_j} = e^{\mathrm{i}\varphi_V}\, V\, A^{(p)}_{\sigma_j}\, V^{-1}, \tag{6.173a}$$

for any $\sigma_j = 1, \cdots, \mathcal{D}$, and

$$P = \pm V\, P\, V^{-1}, \tag{6.173b}$$

for $p = 0$, while

$$P = V\, P\, V^{-1}, \qquad Y = \pm V\, Y\, V^{-1}, \tag{6.173c}$$

for $p = 1$. Here, the phase φ_V is needed to compensate for the possibility that the matrix V anticommutes with P or Y. We also observe that the index σ_j that

[33] See **Appendix** A for the definition of a $\mathbb{Z}_2$-graded simple algebra.

labels the local fermion number is preserved under the conjugation by V. The transformation (6.173) that leaves an injective FMPS invariant is called a gauge transformation.

We refer the reader for the proofs of **Properties** 6.50 and 6.51 to the study of BMPS made by Perez-Garcia et al.,[34] on the one hand, and to the study of FMPS made by Bultinck et al.,[35] on the other hand.

Theorem 6.52 *If an even- or odd-parity injective FMPS is invariant (up to a multiplicative phase) under the direct product of a representation of the translation symmetry group* $\mathrm{G}_{\mathrm{trsl}}$ *and a projective representation of the on-site symmetry group* G_f *defined in Section 6.3.2, then the projective representation* ϕ *of* G_f *must belong to the class* $[\phi]=0$ *from the second cohomology group.*[36]

Remark (1) The direct product structure of the symmetry group $\mathrm{G}_{\mathrm{trsl}}\times\mathrm{G}_f$ is crucial in **Theorem** 6.52. For example, **Theorem** 6.52 is not applicable for magnetic translation symmetries.

Remark (2) The contraposition of **Theorem** 6.52 of relevance to condensed matter physics is that any local, $\mathrm{G}_{\mathrm{trsl}}\times\mathrm{G}_f$-symmetric, and one-dimensional Hamiltonian cannot have as a ground state a $\mathrm{G}_{\mathrm{trsl}}\times\mathrm{G}_f$-symmetric injective FMPS if the on-site symmetry group G_f of the repeat unit cell realizes a nontrivial projective representation of G_f.

Remark (3) It is believed that, in the thermodynamic limit, any local, $\mathrm{G}_{\mathrm{trsl}}\times\mathrm{G}_f$-symmetric, and one-dimensional lattice Hamiltonian with a nondegenerate, $\mathrm{G}_{\mathrm{trsl}}\times\mathrm{G}_f$-symmetric, and gapped ground state can be deformed adiabatically into a local, $\mathrm{G}_{\mathrm{trsl}}\times\mathrm{G}_f$-symmetric, and one-dimensional lattice Hamiltonian with a $\mathrm{G}_{\mathrm{trsl}}\times\mathrm{G}_f$-symmetric, injective FMPS without closing the gap between the ground state and all the excited states. This belief combined with the contraposition to **Theorem** 6.52 implies that, in the thermodynamic limit, the $\mathrm{G}_{\mathrm{trsl}}\times\mathrm{G}_f$-symmetric ground state of any local, $\mathrm{G}_{\mathrm{trsl}}\times\mathrm{G}_f$-symmetric, and one-dimensional lattice Hamiltonian must be degenerate or gapless if the on-site symmetry group G_f of the repeat unit cell realizes a nontrivial projective representation of G_f.

Proof The proof[37] of **Theorem** 6.52 follows closely that for the bosonic case assuming the $\mathrm{G}_{\mathrm{trsl}}\times\mathrm{G}$ symmetry with the on-site symmetry group G (instead of the larger group G_f) as reviewed by Tasaki (see footnote 14).

34 D. Perez-Garcia, F. Verstraete, M. M. Wolf, and J. I. Cirac, Quantum Info. Comput. **7**(5), 401–430 (2007). [Perez-Garcia et al. (2007)].

35 N. Bultinck, D. J. Williamson, J. Haegeman, and F. Verstraete, Phys. Rev. B **95**(7), 075108 (2017). [Bultinck et al. (2017)].

36 When referring to Abelian groups, it is customary to choose the addition as the composition law. For the group U(1), this means representing a group element by the phase factor $e^{\mathrm{i}\phi}$ with $\phi\in[0,2\pi[$ the group label obeying an additive composition law (modulo 2π).

37 Ö. M. Aksoy, A. Tiwari, and C. Mudry, Phys. Rev. B **104(7)**, 075146 (2021). [Aksoy et al. (2021)].

We will show that a parity-even or parity-odd injective FMPS necessarily requires the local projective representation $\hat{u}_j$ of the symmetry group G_f to have trivial second cohomology class $[\phi] \cong [e^{i\phi}] \in H^2(G_f, U(1)_{\mathfrak{c}})$. In other words, when this cohomology class is nontrivial there is no compatible injective FMPS with even or odd parity. The general forms (6.163) and (6.164) as well as the definitions of injectivity 6.48 and 6.49 are distinct for even and odd parity FMPS, respectively. The proofs for the even- and the odd-parity cases are thus treated successively. For conciseness, we are going to supress the symbol $\otimes_{\mathfrak{g}}$ when working with the orthonormal and homogeneous basis

$$\left\{|\Psi_{\boldsymbol{\sigma}}\rangle \equiv |\psi_{1,\sigma_1}\rangle \otimes_{\mathfrak{g}} |\psi_{2,\sigma_2}\rangle \otimes_{\mathfrak{g}} \cdots \otimes_{\mathfrak{g}} |\psi_{N,\sigma_N}\rangle \,\Big|\, \sigma_j = 1, \cdots, \mathcal{D}\right\} \tag{6.174a}$$

of the Fock space

$$\mathfrak{F}_\Lambda \equiv \mathfrak{F}_1 \otimes_{\mathfrak{g}} \mathfrak{F}_2 \otimes_{\mathfrak{g}} \cdots \otimes_{\mathfrak{g}} \mathfrak{F}_N. \tag{6.174b}$$

We also need to introduce a notation that accomodates the possibility that an element $g \in G_f$ can be represented antiunitarily. For any $g \in G_f$,

$$\mathsf{K}_g := \begin{cases} 1, & \text{if } \mathfrak{c}(g) = +1, \\ \\ \mathsf{K}, & \text{if } \mathfrak{c}(g) = -1, \end{cases} \tag{6.175a}$$

where, as usual, K denotes complex conjugation. In words, K_g acts on the Fock space as the identity (complex conjugation) when g is represented by a unitary (antiunitary) operator acting on the Fock space. It is assumed that K acts on all symbols to its right. For this reason, we introduce the notation

$$\mathsf{K}_g\,[\![a\,b\,\cdots\,z]\!] := \begin{cases} a\,b\,\cdots\,z, & \text{if } \mathfrak{c}(g) = +1, \\ \\ \mathsf{K}\,a\,b\,\cdots\,z\,\mathsf{K}, & \text{if } \mathfrak{c}(g) = -1, \end{cases} \tag{6.175b}$$

to delimit the range over which K_g is acting. The meaning of

$$\mathsf{K}\,a\,b\,\cdots\,z\,\mathsf{K} \equiv (\mathsf{K}\,a\,\mathsf{K})\,(\mathsf{K}\,b\,\mathsf{K})\cdots(\mathsf{K}\,z\,\mathsf{K}) \tag{6.175c}$$

requires defining the action by conjugation of complex conjugation on the symbols a, b, $\cdots$, z. If these symbols stand for complex numbers, then

$$\mathsf{K}\,a\,b\,\cdots\,z\,\mathsf{K} = a^*\,b^*\,\cdots\,z^*. \tag{6.175d}$$

Property 6.53 For any $g, h \in G_f$ and for any finite multiplicative list of symbols $(a\,b\,\cdots\,z)$ and $(a'\,b'\,\cdots\,z')$, whereby conjugation by complex conjugation of each letter is well defined, the identities

$$\mathsf{K}_g [\![(a\,b \cdots z)]\!]\, \mathsf{K}_g [\![(a'\,b' \cdots z')]\!] = \mathsf{K}_g [\![(a\,b \cdots z)(a'\,b' \cdots z')]\!], \tag{6.176a}$$

$$\begin{aligned} \mathsf{K}_g [\![\mathsf{K}_h [\![(a\,b \cdots z)]\!]]\!] &= \mathsf{K}_{g\,h} [\![(a\,b \cdots z)]\!] \\ &= \mathsf{K}_{h\,g} [\![(a\,b \cdots z)]\!] \\ &= \mathsf{K}_h [\![\mathsf{K}_g [\![(a\,b \cdots z)]\!]]\!], \end{aligned} \tag{6.176b}$$

$$\mathsf{K}_g [\![\mathsf{K}_g [\![(a\,b \cdots z)]\!]]\!] = \mathsf{K}_{g\,g} [\![(a\,b \cdots z)]\!] = (a\,b \cdots z), \tag{6.176c}$$

hold (**Exercise** 6.13).

6.3.3.1 Proof for Even-Parity FMPS

Let

$$|\{A^{(0)}_{\sigma_j}\}\rangle \equiv \sum_{\boldsymbol{\sigma}} \operatorname{tr}\left(P\, A^{(0)}_{\sigma_1}\, A^{(0)}_{\sigma_2} \cdots A^{(0)}_{\sigma_N}\right) |\psi_{1,\sigma_1}\rangle\, |\psi_{2,\sigma_2}\rangle \cdots |\psi_{N,\sigma_N}\rangle \tag{6.177a}$$

be any translation-invariant (obeying periodic boundary conditions), G_f-symmetric, even-parity, and injective FMPS. For any $g \in \mathrm{G}_f$, the global representation $\widehat{U}(g)$ of g is defined in Eq. (6.149). For any $g \in \mathrm{G}_f$, there exists a phase $\eta(g) \in [0, 2\pi[$ such that

$$\widehat{U}(g)\, |\{A^{(0)}_{\sigma_j}\}\rangle = e^{\mathrm{i}\eta(g)} |\{A^{(0)}_{\sigma_j}\}\rangle. \tag{6.177b}$$

The action of the transformation $\widehat{U}(g)$ on the right-hand side of Eq. (6.177a) gives[38]

$$\begin{aligned} \widehat{U}(g)\, |\{A^{(0)}_{\sigma_j}\}\rangle =& \sum_{\boldsymbol{\sigma}'} \mathsf{K}_g \left[\!\!\left[\operatorname{tr}\left(P\, A^{(0)}_{\sigma'_1}\, A^{(0)}_{\sigma'_2} \cdots A^{(0)}_{\sigma'_N}\right)\right]\!\!\right] \\ & \times \left(\hat{u}_1(g)\, |\psi_{1,\sigma'_1}\rangle\right) \left(\hat{u}_2(g) |\psi_{2,\sigma'_2}\rangle\right) \cdots \left(\hat{u}_N(g)\, |\psi_{N,\sigma'_N}\rangle\right). \end{aligned} \tag{6.178}$$

Inserting N times the resolution of the identity, one for each local Fock space $\mathfrak{F}_j$, gives

$$\begin{aligned} \widehat{U}(g)\, |\{A^{(0)}_{\sigma_j}\}\rangle =& \sum_{\boldsymbol{\sigma}} \Bigg\{ \sum_{\boldsymbol{\sigma}'} \operatorname{tr}\left(P \mathsf{K}_g \left[\!\!\left[A^{(0)}_{\sigma'_1}\, A^{(0)}_{\sigma'_2} \cdots A^{(0)}_{\sigma'_N}\right]\!\!\right]\right) \\ & \times \prod_{j=1}^{N} \langle \psi_{j,\sigma_j} | \left(\hat{u}_j(g) |\psi_{j,\sigma'_j}\rangle\right) \Bigg\} |\psi_{1,\sigma_1}\rangle\, |\psi_{2,\sigma_2}\rangle \cdots |\psi_{N,\sigma_N}\rangle. \end{aligned} \tag{6.179}$$

The right-hand side can be written more elegantly with the definition of the g-dependent $2M \times 2M$ matrix

[38] Let X be an operator with some Hilbert space as domain of definition. By definition, X is linear if and only if $X(\alpha\, u + \beta\, v) = \alpha\,(X\, u) + \beta\,(X\, v)$ for any pair u and v from the Hilbert space and for any pair α and β from $\mathbb{C}$. In contrast, X is antilinear if and only if $X(\alpha\, u + \beta\, v) = \alpha^*\,(X\, u) + \beta^*\,(X\, v)$ for any pair u and v from the Hilbert space and for any pair α and β from $\mathbb{C}$.

$$A^{(0)}_{\sigma_j}(g) := \sum_{\sigma'_j=1}^{\mathcal{D}} \langle\psi_{j,\sigma_j}| \left(\hat{u}_j(g)\,|\psi_{j,\sigma'_j}\rangle\right) \mathsf{K}_g \left[\!\left[A^{(0)}_{\sigma'_j}\right]\!\right], \tag{6.180a}$$

where $\sigma_j = 1, \cdots, \mathcal{D}$ with $j = 1, \cdots, N$. Equation (6.179) becomes

$$\begin{aligned}\widehat{U}(g)\,|\{A^{(0)}_{\sigma_j}\}\rangle = \sum_{\boldsymbol{\sigma}} \mathrm{tr}\left(P\,A^{(0)}_{\sigma_1}(g)\,A^{(0)}_{\sigma_2}(g)\cdots A^{(0)}_{\sigma_N}(g)\right)\\ \times|\psi_{1,\sigma_1}\rangle\,|\psi_{2,\sigma_2}\rangle\cdots|\psi_{N,\sigma_N}\rangle,\end{aligned} \tag{6.180b}$$

which is nothing but the FMPS (6.177a) with $A^{(0)}_{\sigma_j}$ substituted for $A^{(0)}_{\sigma_j}(g)$.

Equating the right-hand sides of Eqs. (6.177b) and (6.180b) implies

$$\mathrm{tr}\left(P\,A^{(0)}_{\sigma_1}(g)\,A^{(0)}_{\sigma_2}(g)\cdots A^{(0)}_{\sigma_N}(g)\right) = e^{\mathrm{i}\eta(g)}\,\mathrm{tr}\left(P\,A^{(0)}_{\sigma_1}\,A^{(0)}_{\sigma_2}\cdots A^{(0)}_{\sigma_N}\right). \tag{6.181}$$

This equation is satisfied by the Ansatz

$$A^{(0)}_{\sigma_j}(g) = e^{\mathrm{i}\theta(g)}\,V^{-1}(g)\,A^{(0)}_{\sigma_j}\,V(g), \tag{6.182a}$$

$$P\,V(g)\,P = (-1)^{\kappa(g)}\,V(g), \tag{6.182b}$$

$$\theta(g) := \frac{1}{N}\left[\eta(g) - \pi\,\kappa(g)\right], \tag{6.182c}$$

where $\kappa(g) = 0, 1$ dictates if the $2M \times 2M$ invertible matrix $V(g)$ commutes or anticommutes with the $2M \times 2M$ parity matrix P defined in Eq. (6.159). Moreover,

$$V(p) = P, \qquad \kappa(p) = 0, \qquad \theta(p)\,N = \eta(p) \bmod 2\pi. \tag{6.182d}$$

Proof

Case $g = p \in \mathrm{G}_f$:

Equation (6.181) reads

$$\mathrm{tr}\left(P\,A^{(0)}_{\sigma_1}(p)\,A^{(0)}_{\sigma_2}(p)\cdots A^{(0)}_{\sigma_N}(p)\right) = e^{\mathrm{i}\eta(p)}\,\mathrm{tr}\left(P\,A^{(0)}_{\sigma_1}\,A^{(0)}_{\sigma_2}\cdots A^{(0)}_{\sigma_N}\right). \tag{6.183}$$

We try the Ansatz

$$A^{(0)}_{\sigma_j}(p) = e^{\mathrm{i}\theta(p)}\,V^{-1}(p)\,A^{(0)}_{\sigma_j}\,V(p), \tag{6.184a}$$

$$P\,V(p)\,P = (-1)^{\kappa(p)}\,V(p), \tag{6.184b}$$

$$\theta(p) := \frac{1}{N}\left[\eta(p) - \pi\,\kappa(p)\right]. \tag{6.184c}$$

Insertion of Eq. (6.184a) into the left-hand side of Eq. (6.183) for $\sigma_j = \sigma_1, \cdots, \sigma_N$ gives

$$\begin{aligned}&\mathrm{tr}\left(P\,A^{(0)}_{\sigma_1}(p)\cdots A^{(0)}_{\sigma_N}(p)\right)\\ &= e^{\mathrm{i}\theta(p)N}\,\mathrm{tr}\left[P\left(V^{-1}(p)\,A^{(0)}_{\sigma_1}\,V(p)\right)\cdots\left(V^{-1}(p)\,A^{(0)}_{\sigma_N}\,V(p)\right)\right]\\ &= e^{\mathrm{i}\theta(p)N}\,\mathrm{tr}\left(P\,V^{-1}(p)\,A^{(0)}_{\sigma_1}\cdots A^{(0)}_{\sigma_N}\,V(p)\right).\end{aligned} \tag{6.185}$$

Motivated by Eq. (6.139e), we make the Ansatz

$$V(p) = P, \tag{6.186}$$

in which case insertion of Eq. (6.159) into Eq. (6.184b) implies that

$$\kappa(p) = 0 \bmod 2\pi, \tag{6.187}$$

while Eq. (6.185) becomes

$$\operatorname{tr}\left(P\,A^{(0)}_{\sigma_1}(p)\cdots A^{(0)}_{\sigma_N}(p)\right) = e^{\mathrm{i}\theta(p)N}\operatorname{tr}\left(A^{(0)}_{\sigma_1}\cdots A^{(0)}_{\sigma_N}\,P\right). \tag{6.188}$$

On the one hand, if we use the cyclicity of the trace, we can bring P to the left of the product $A^{(0)}_{\sigma_1}\,A^{(0)}_{\sigma_2}\cdots A^{(0)}_{\sigma_N}$ in the trace on the right-hand side of Eq. (6.188). Comparison with the right-hand side of Eq. (6.183) then delivers the relation

$$\theta(p)N = \eta(p) \bmod 2\pi. \tag{6.189}$$

On the other hand, with the help of the identity (6.161), we can move $V(p) = P$ to the immediate left of $A^{(0)}_{\sigma_N}$ and iterate this exercise until $V(p) = P$ has been moved to the immediate left of $A^{(0)}_{\sigma_1}$. In doing so, the left-hand side of Eq. (6.183) becomes

$$\operatorname{tr}\left(P\,A^{(0)}_{\sigma_1}(p)\cdots A^{(0)}_{\sigma_N}(p)\right) = e^{\mathrm{i}\theta(p)N+\mathrm{i}\pi\sum_{j=1}^{N}|\sigma_j|}\operatorname{tr}\left(P\,A^{(0)}_{\sigma_1}\,A^{(0)}_{\sigma_2}\cdots A^{(0)}_{\sigma_N}\right). \tag{6.190}$$

Comparison with the right-hand side of Eq. (6.183) then delivers the relation

$$\theta(p)N + \pi\sum_{j=1}^{N}|\sigma_j| = \eta(p) \bmod 2\pi, \tag{6.191}$$

where

$$\pi\sum_{j=1}^{N}|\sigma_j| = 0 \bmod 2\pi \tag{6.192}$$

as $|\{A^{(0)}_{\sigma_j}\}\rangle$ is an even-parity state. In either cases, we have shown that

$$\kappa(p) = 0, \qquad \theta(p)N = \eta(p) \bmod 2\pi. \tag{6.193}$$

Case $g \neq p \in \mathrm{G}_f$:
With the help of Eq. (6.182a),

$$\begin{aligned}\operatorname{tr}&\left[P\,A^{(0)}_{\sigma_1}(g)\,A^{(0)}_{\sigma_2}(g)\cdots A^{(0)}_{\sigma_N}(g)\right]\\ &= e^{\mathrm{i}\theta(g)N}\operatorname{tr}\left(P\,V^{-1}(g)\,A^{(0)}_{\sigma_1}\,A^{(0)}_{\sigma_2}\cdots A^{(0)}_{\sigma_N}\,V(g)\right).\end{aligned} \tag{6.194}$$

By the cyclicity of the trace,

$$\begin{aligned}\operatorname{tr}&\left(P\,A^{(0)}_{\sigma_1}(g)\,A^{(0)}_{\sigma_2}(g)\cdots A^{(0)}_{\sigma_N}(g)\right]\\ &= e^{\mathrm{i}\theta(g)N}\operatorname{tr}\left(V(g)\,P\,V^{-1}(g)\,A^{(0)}_{\sigma_1}\,A^{(0)}_{\sigma_2}\cdots A^{(0)}_{\sigma_N}\right).\end{aligned} \tag{6.195}$$

With the help of Eq. (6.182b),

$$\begin{aligned}\operatorname{tr}&\left(P\,A^{(0)}_{\sigma_1}(g)\,A^{(0)}_{\sigma_2}(g)\,\cdots A^{(0)}_{\sigma_N}(g)\right)\\&=e^{\mathrm{i}\theta(g)N}\,(-1)^{\kappa(g)}\operatorname{tr}\left(P\,A^{(0)}_{\sigma_1}\,A^{(0)}_{\sigma_2}\,\cdots A^{(0)}_{\sigma_N}\right).\end{aligned}\tag{6.196}$$

With the help of Eq. (6.182c),

$$e^{\mathrm{i}\theta(g)N}\,(-1)^{\kappa(g)}=e^{\mathrm{i}\theta(g)N+\mathrm{i}\pi\,\kappa(g)}\equiv e^{\mathrm{i}\eta(g)}.\tag{6.197}$$

Insertion of Eq. (6.197) into Eq. (6.196) gives

$$\begin{aligned}\operatorname{tr}&\left(P\,A^{(0)}_{\sigma_1}(g)\,A^{(0)}_{\sigma_2}(g)\,\cdots A^{(0)}_{\sigma_N}(g)\right)\\&=e^{\mathrm{i}\eta(g)}\operatorname{tr}\left(P\,A^{(0)}_{\sigma_1}\,A^{(0)}_{\sigma_2}\,\cdots A^{(0)}_{\sigma_N}\right).\end{aligned}\tag{6.198}$$

□

The existence of $V(g)$ in the Ansatz (6.182) is guaranteed because of the injectivity of the FMPS. In an injective even-parity FMPS, the matrices $A^{(0)}_{\sigma_1},\cdots,A^{(0)}_{\sigma_\ell}$ span the simple algebra of all $2M\times 2M$ matrices for any $\ell>\ell^{\star}$ for some nonvanishing integer $\ell^{\star}$. Hence, provided N is sufficiently large, the family of matrices

$$\{A^{(0)}_{\sigma_1}(g),\cdots,A^{(0)}_{\sigma_N}(g)\}\tag{6.199}$$

is related to the family of matrices

$$\{e^{\mathrm{i}\eta(g)/N}\,A^{(0)}_{\sigma_1},\cdots,e^{\mathrm{i}\eta(g)/N}A^{(0)}_{\sigma_N}\}\tag{6.200}$$

that give the same FMPS (6.177a) by the similarity transformation [see Eq. (6.173)]

$$A^{(0)}_{\sigma_j}(g)=e^{\mathrm{i}\varphi_{V(g)}}\,V^{-1}(g)\,\left[e^{\mathrm{i}\eta(g)/N}\,A^{(0)}_{\sigma_j}\right]V(g),\tag{6.201a}$$

for some phase $\varphi_{V(g)}=[0,2\pi[$ and some invertible $2M\times 2M$ matrix $V(g)$ that must also obey

$$P\,V(g)\,P=(-1)^{\kappa(g)}\,V(g).\tag{6.201b}$$

Here, the map $\kappa\colon \mathrm{G}_f\to\{0,1\}$ specifies the algebra between the similarity transformation $V(g)$ corresponding to element $g\in\mathrm{G}_f$ and the fermion parity P. The effect of the factor $(-1)^{\kappa(g)}$ is nothing but the phase

$$\varphi_{V(g)}=-\frac{1}{N}\,\pi\,\kappa(g),\tag{6.202}$$

as follows from Eq. (6.182).

For any $\sigma_j=1,\cdots,\mathcal{D}$ with $j=1,\cdots,N$, equating the right-hand sides of Eqs. (6.201a) and (6.180a) implies

$$e^{\mathrm{i}\theta(g)}\,V^{-1}(g)\,A^{(0)}_{\sigma_j}\,V(g)=\sum_{\sigma'_j=1}^{\mathcal{D}}\left[\mathcal{U}(g)\right]_{\sigma_j\,\sigma'_j}\,\mathsf{K}_g\left[\!\left[A^{(0)}_{\sigma'_j}\right]\!\right],\tag{6.203a}$$

where we have introduced the shorthand notation

$$
\begin{aligned}
[\mathcal{U}(g)]_{\sigma_j\,\sigma'_j} &:= \langle\psi_{j,\sigma_j}|\left(\hat{u}_j(g)\,|\psi_{j,\sigma'_j}\rangle\right)\\
&= \begin{cases} \left[\langle\psi_{j,\sigma'_j}|\left(\hat{u}^\dagger_j(g)\,|\psi_{j,\sigma_j}\rangle\right)\right]^*, & \text{if } \mathfrak{c}(g)=+1,\\[2ex] \left[\langle\psi_{j,\sigma'_j}|\left(\hat{u}^\dagger_j(g)\,|\psi_{j,\sigma_j}\rangle\right)\right], & \text{if } \mathfrak{c}(g)=-1, \end{cases}
\end{aligned}
\tag{6.203b}
$$

for the $\mathcal{D}$-dimensional local representation of $\hat{u}_j(g)$ in the basis $\{|\psi_{j,\sigma_j}\rangle,\,\sigma_j = 1,\cdots,\mathcal{D}\}$ of $\mathfrak{F}_j$.[39] We would like to isolate $A^{(0)}_{\sigma'_j}$ on the right-hand side of Eq. (6.203a). To this end, we need the inverse $\mathcal{U}^{-1}(g) = \mathcal{U}^\dagger(g)$ of $\mathcal{U}(g)$. It is given by

$$
[\mathcal{U}^\dagger(g)]_{\sigma_j\,\sigma'_j} = \mathsf{K}_g\left[\!\left[\langle\psi_{j,\sigma_j}|\left(\hat{u}^\dagger_j(g)\,|\psi_{j,\sigma'_j}\rangle\right)\right]\!\right]. \tag{6.204}
$$

Proof The matrix product between $\mathcal{U}^\dagger(g)$ and $\mathcal{U}(g)$ is

$$
[\mathcal{U}^\dagger(g)\mathcal{U}(g)]_{\sigma''_j\,\sigma'_j} = \sum_{\sigma_j=1}^{\mathcal{D}} \mathsf{K}_g\left[\!\left[\langle\psi_{j,\sigma''_j}|\left(\hat{u}^\dagger_j(g)\,|\psi_{j,\sigma_j}\rangle\right)\right]\!\right]\langle\psi_{j,\sigma_j}|\left(\hat{u}_j(g)|\psi_{j,\sigma'_j}\rangle\right). \tag{6.205}
$$

Here, because $\hat{u}_j(g)$ can be antiunitary, it is necessary to use a pair of parentheses with the rule

$$
\langle\psi_{j,\sigma''_j}|\left(\hat{u}^\dagger_j(g)\,|\psi_{j,\sigma_j}\rangle\right) = \mathsf{K}_g\left[\!\left[\left(\langle\psi_{j,\sigma''_j}|\hat{u}_j(g)\right)|\psi_{j,\sigma_j}\rangle\right]\!\right] \tag{6.206}
$$

to move a pair of parentheses that surrounds a ket to one that surrounds a bra. Because K_g squares to the identity,

$$
\begin{aligned}
[\mathcal{U}^\dagger(g)\mathcal{U}(g)]_{\sigma''_j\,\sigma'_j} &= \sum_{\sigma_j=1}^{\mathcal{D}} (\mathsf{K}_g)^2\left[\!\left[\left(\langle\psi_{j,\sigma''_j}|\hat{u}_j(g)\right)|\psi_{j,\sigma_j}\rangle\right]\!\right]\langle\psi_{j,\sigma_j}|\left(\hat{u}_j(g)|\psi_{j,\sigma'_j}\rangle\right)\\
&= \sum_{\sigma_j=1}^{\mathcal{D}} \left[\left(\langle\psi_{j,\sigma''_j}|\hat{u}_j(g)\right)|\psi_{j,\sigma_j}\rangle\right]\langle\psi_{j,\sigma_j}|\left(\hat{u}_j(g)|\psi_{j,\sigma'_j}\rangle\right),
\end{aligned}
\tag{6.207}
$$

we may drop the standard square bracket on the right-hand side of the last equality and use the resolution of the identity to write

$$
\begin{aligned}
[\mathcal{U}^\dagger(g)\mathcal{U}(g)]_{\sigma''_j\,\sigma'_j} &= \left(\langle\psi_{j,\sigma''_j}|\hat{u}_j(g)\right)\left(\sum_{\sigma_j=1}^{\mathcal{D}}|\psi_{j,\sigma_j}\rangle\langle\psi_{j,\sigma_j}|\right)\left(\hat{u}_j(g)|\psi_{j,\sigma'_j}\rangle\right)\\
&= \left(\langle\psi_{j,\sigma''_j}|\hat{u}_j(g)\right)\left(\hat{u}_j(g)|\psi_{j,\sigma'_j}\rangle\right).
\end{aligned}
\tag{6.208}
$$

[39] We recall that if we use the notation $\langle u, v\rangle$ for the scalar product between any pair u and v from some given Hilbert space, then the adjoint of any linear operator X acting on the Hilbert space is defined by $\langle u, X\,v\rangle = \langle X^\dagger\,u, v\rangle$, while the adjoint of any antilinear operator X acting on the Hilbert space is defined by $\langle u, X\,v\rangle = \langle X^\dagger\,u, v\rangle^*$, for all pairs u, v from the Hilbert space.

We may now move the pair of parentheses that surrounds the bra to one that surrounds the ket:

$$\left[\mathcal{U}^{\dagger}(g)\mathcal{U}(g)\right]_{\sigma_j''\,\sigma_j'} = \mathsf{K}_g\left[\!\left[\langle\psi_{j,\sigma_j''}|\left(\hat{u}_j^{\dagger}(g)\,\hat{u}_j(g)\,|\psi_{j,\sigma_j'}\rangle\right)\right]\!\right]. \tag{6.209}$$

At last, we can safely use the identity $\hat{u}_j^{\dagger}(g)\,\hat{u}_j(g) = \mathbb{1}_{\mathcal{D}}$ to prove that

$$\left[\mathcal{U}^{\dagger}(g)\mathcal{U}(g)\right]_{\sigma_j''\,\sigma_j'} = \delta_{\sigma_j'',\sigma_j'}, \tag{6.210}$$

owing to the orthonormality of the basis $\{|\psi_{j,\sigma_j}\rangle,\ \sigma_j = 1,\cdots,\mathcal{D}\}$ of $\mathfrak{F}_j$. □

With this inverse, we obtain the self-consistency condition

$$A^{(0)}_{\sigma_j} = \mathsf{K}_g\left[\!\left[e^{\mathrm{i}\theta(g)}\sum_{\sigma_j'=1}^{\mathcal{D}}\left[\mathcal{U}^{\dagger}(g)\right]_{\sigma_j\sigma_j'}\,V^{-1}(g)\,A^{(0)}_{\sigma_j'}\,V(g)\right]\!\right]. \tag{6.211}$$

To proceed we need to act with K_g defined by Eq. (6.175) on the multiplicative factors $e^{\mathrm{i}\theta(g)}$ and $\left[\mathcal{U}^{\dagger}(g)\right]_{\sigma_j\sigma_j'}$. First, we shall use the identity

$$\mathsf{K}_g\left[\!\left[e^{\mathrm{i}\theta(g)}\right]\!\right] = e^{\mathrm{i}\mathfrak{c}(g)\,\theta(g)}. \tag{6.212}$$

Second, we shall make use of Eqs. (6.204) and (6.176c) to establish that

$$\begin{aligned}\mathsf{K}_g\left[\!\left[\left[\mathcal{U}^{\dagger}(g)\right]_{\sigma_j\sigma_j'}\right]\!\right] &= \mathsf{K}_g\left[\!\left[\mathsf{K}_g\left[\!\left[\langle\psi_{j,\sigma_j}|\left(\hat{u}_j^{\dagger}(g)\,|\psi_{j,\sigma_j'}\rangle\right)\right]\!\right]\right]\!\right]\\ &= \langle\psi_{j,\sigma_j}|\left(\hat{u}_j^{\dagger}(g)\,|\psi_{j,\sigma_j'}\rangle\right).\end{aligned} \tag{6.213}$$

By combining all three identities we arrive at the desired self-consistency condition

$$A^{(0)}_{\sigma_j} = e^{\mathrm{i}\mathfrak{c}(g)\,\theta(g)}\sum_{\sigma_j'=1}^{\mathcal{D}}\langle\psi_{j,\sigma_j}|\left(\hat{u}_j^{\dagger}(g)|\psi_{j,\sigma_j'}\rangle\right)\mathsf{K}_g\left[\!\left[V^{-1}(g)\,A^{(0)}_{\sigma_j'}\,V(g)\right]\!\right]. \tag{6.214a}$$

Had we chosen the elements $h \in G_f$ and $g\,h \in G_f$, Eq. (6.214a) would give the self-consistency conditions

$$A^{(0)}_{\sigma_j''} = e^{\mathrm{i}\mathfrak{c}(h)\,\theta(h)}\sum_{\sigma_j=1}^{\mathcal{D}}\langle\psi_{j,\sigma_j''}|\left(\hat{u}_j^{\dagger}(h)|\psi_{j,\sigma_j}\rangle\right)\mathsf{K}_h\left[\!\left[V^{-1}(h)\,A^{(0)}_{\sigma_j}\,V(h)\right]\!\right], \tag{6.214b}$$

for $\sigma_j'' = 1,\cdots,\mathcal{D}$ with $j = 1,\cdots,N$, and

$$A^{(0)}_{\sigma_j} = e^{\mathrm{i}\mathfrak{c}(g\,h)\,\theta(g\,h)}\sum_{\sigma_j'=1}^{\mathcal{D}}\langle\psi_{j,\sigma_j}|\left(\hat{u}_j^{\dagger}(g\,h)|\psi_{j,\sigma_j'}\rangle\right)\mathsf{K}_{g\,h}\left[\!\left[V^{-1}(g\,h)\,V^{(0)}_{\sigma_j'}\,V(g\,h)\right]\!\right], \tag{6.214c}$$

for $\sigma_j = 1,\cdots,\mathcal{D}$ with $j = 1,\cdots,N$, respectively.

Inserting the right-hand side of the self-consistency condition (6.214a) into the right-hand side of the self-consistency condition (6.214b) gives

$$
\begin{aligned}
A^{(0)}_{\sigma''_j} = e^{\mathrm{i}\mathfrak{c}(h)\,\theta(h)} \sum_{\sigma_j=1}^{\mathcal{D}} \langle\psi_{j,\sigma''_j}| \left(\hat{u}^{\dagger}_j(h)|\psi_{j,\sigma_j}\rangle\right) \mathsf{K}_h \Bigg[\!\!\Bigg[V^{-1}(h) \\
\times \left(e^{\mathrm{i}\mathfrak{c}(g)\,\theta(g)} \sum_{\sigma'_j=1}^{\mathcal{D}} \langle\psi_{j,\sigma_j}| \left(\hat{u}^{\dagger}_j(g)|\psi_{j,\sigma'_j}\rangle\right) \mathsf{K}_g \left[\!\left[V^{-1}(g)\, A^{(0)}_{\sigma'_j}\, V(g) \right]\!\right] \right) V(h) \Bigg]\!\!\Bigg].
\end{aligned}
\tag{6.215}
$$

If we bring the $\mathbb{C}$-numbers $e^{\mathrm{i}\mathfrak{c}(g)\,\theta(g)}$ and $\langle\psi_{j,\sigma_j}| \left(\hat{u}^{\dagger}_j(g)|\psi_{j,\sigma'_j}\rangle\right)$ to the left of K_h, we may recast the self-consistency condition as

$$
\begin{aligned}
A^{(0)}_{\sigma''_j} =& e^{\mathrm{i}\mathfrak{c}(h)\,\theta(h)+\mathrm{i}\mathfrak{c}(h)\,\mathfrak{c}(g)\,\theta(g)} \\
&\times \sum_{\sigma_j,\sigma'_j=1}^{\mathcal{D}} \langle\psi_{j,\sigma''_j}| \left(\hat{u}^{\dagger}_j(h)|\psi_{j,\sigma_j}\rangle\right) \mathsf{K}_h \left[\!\left[\langle\psi_{j,\sigma_j}| \left(\hat{u}^{\dagger}_j(g)|\psi_{j,\sigma'_j}\rangle\right) \right]\!\right] \\
&\times \mathsf{K}_h \left[\!\left[V^{-1}(h)\, \mathsf{K}_g \left[\!\left[V^{-1}(g)\, A^{(0)}_{\sigma'_j}\, V(g) \right]\!\right] V(h) \right]\!\right].
\end{aligned}
\tag{6.216}
$$

To proceed, we observe that

$$
\begin{aligned}
&\sum_{\sigma_j=1}^{\mathcal{D}} \langle\psi_{j,\sigma''_j}| \left(\hat{u}^{\dagger}_j(h)\,|\psi_{j,\sigma_j}\rangle\right) \mathsf{K}_h \left[\!\left[\langle\psi_{j,\sigma_j}| \left(\hat{u}^{\dagger}_j(g)\,|\psi_{j,\sigma'_j}\rangle\right) \right]\!\right] \\
&\quad= \sum_{\sigma_j=1}^{\mathcal{D}} \mathsf{K}_h \left[\!\left[\left(\langle\psi_{j,\sigma''_j}|\, \hat{u}_j(h)\right)\, |\psi_{j,\sigma_j}\rangle \right]\!\right] \mathsf{K}_h \left[\!\left[\langle\psi_{j,\sigma_j}| \left(\hat{u}^{\dagger}_j(g)\,|\psi_{j,\sigma'_j}\rangle\right) \right]\!\right] \\
&\quad= \mathsf{K}_h \left[\!\!\left[\left(\langle\psi_{j,\sigma''_j}|\, \hat{u}_j(h)\right) \left(\sum_{\sigma_j=1}^{\mathcal{D}} |\psi_{j,\sigma_j}\rangle\,\langle\psi_{j,\sigma_j}| \right) \left(\hat{u}^{\dagger}_j(g)\,|\psi_{j,\sigma'_j}\rangle\right) \right]\!\!\right] \\
&\quad= \langle\psi_{j,\sigma''_j}| \left(\hat{u}^{\dagger}_j(h)\,\hat{u}^{\dagger}_j(g)\,|\psi_{\sigma'_j}\rangle\right).
\end{aligned}
\tag{6.217}
$$

If we use the projective representation (6.140), we obtain

$$
\begin{aligned}
\hat{u}^{\dagger}_j(h)\,\hat{u}^{\dagger}_j(g) &= \left[\hat{u}_j(g)\,\hat{u}_j(h)\right]^{\dagger} \\
&= \left[e^{+\mathrm{i}\phi(g,h)}\,\hat{u}_j(g\,h)\right]^{\dagger} \\
&= \hat{u}^{\dagger}_j(g\,h)\,e^{-\mathrm{i}\phi(g,h)} \\
&= e^{-\mathrm{i}\mathfrak{c}(g\,h)\,\phi(g,h)}\,\hat{u}^{\dagger}_j(g\,h).
\end{aligned}
\tag{6.218}
$$

Hence, the self-consistency condition (6.216) takes the simpler form

$$A^{(0)}_{\sigma''_j} = e^{\mathrm{i}\mathfrak{c}(h)\,\theta(h)+\mathrm{i}\mathfrak{c}(h)\,\mathfrak{c}(g)\,\theta(g)-\mathrm{i}\mathfrak{c}(g\,h)\phi(g,h)} \sum_{\sigma'_j=1}^{\mathcal{D}} \langle\psi_{j,\sigma''_j}| \left(\hat{u}^{\dagger}_j(g\,h)\,|\psi_{j,\sigma'_j}\rangle\right) \times \mathsf{K}_h \left[\!\left[V^{-1}(h)\,\mathsf{K}_g \left[\!\left[V^{-1}(g)\,A^{(0)}_{\sigma'_j}\,V(g)\right]\!\right] V(h)\right]\!\right]. \tag{6.219}$$

Using the fact that $\mathfrak{c}$ is a homomorphism taking the values ±1 so that $\mathfrak{c}(g\,h) = \mathfrak{c}(g)\,\mathfrak{c}(h)$ and $\mathfrak{c}(g)\,\mathfrak{c}(g\,h) = \mathfrak{c}(h)$ hold,

$$A^{(0)}_{\sigma''_j} = e^{\mathrm{i}\mathfrak{c}(g\,h)[\mathfrak{c}(g)\,\theta(h)+\theta(g)-\phi(g,h)]} \sum_{\sigma'_j=1}^{\mathcal{D}} \langle\psi_{j,\sigma''_j}| \left(\hat{u}^{\dagger}_j(g\,h)\,|\psi_{j,\sigma'_j}\rangle\right) \times \mathsf{K}_h \left[\!\left[V^{-1}(h)\,\mathsf{K}_g \left[\!\left[V^{-1}(g)\,A^{(0)}_{\sigma'_j}\,V(g)\right]\!\right] V(h)\right]\!\right]. \tag{6.220}$$

Equating the right-hand sides of Eqs. (6.220) and (6.214c) ties the three self-consistency conditions (6.214a), (6.214b), and (6.214c) by

$$\begin{aligned} &e^{\mathrm{i}\mathfrak{c}(g\,h)[\mathfrak{c}(g)\,\theta(h)+\theta(g)-\phi(g,h)]}\,\mathsf{K}_h \left[\!\left[V^{-1}(h)\,\mathsf{K}_g \left[\!\left[V^{-1}(g)\,A^{(0)}_{\sigma'_j}\,V(g)\right]\!\right] V(h)\right]\!\right] \\ &\quad= e^{\mathrm{i}\mathfrak{c}(g\,h)\,\theta(g\,h)}\,\mathsf{K}_{g\,h} \left[\!\left[V^{-1}(g\,h)\,A^{(0)}_{\sigma'_j}\,V(g\,h)\right]\!\right] \\ &\quad= \mathsf{K}_{g\,h} \left[\!\left[e^{\mathrm{i}\theta(g\,h)}\,V^{-1}(g\,h)\,A^{(0)}_{\sigma'_j}\,V(g\,h)\right]\!\right]. \end{aligned} \tag{6.221}$$

We arrive at

$$V(g\,h)\,\mathsf{K}_{g\,h} \left[\!\left[\mathsf{K}_h \left[\!\left[V^{-1}(h)\,\mathsf{K}_g \left[\!\left[V^{-1}(g)\,A^{(0)}_{\sigma_j}\,V(g)\right]\!\right] V(h)\right]\!\right]\right]\!\right] V^{-1}(g\,h) = e^{-\mathrm{i}\delta(g,h)}\,A^{(0)}_{\sigma_j}, \tag{6.222a}$$

for $\sigma_j = 1,\cdots,\mathcal{D}$ with $j = 1,\cdots,N$, where

$$\delta(g,h) := \mathfrak{c}(g)\,\theta(h) + \theta(g) - \phi(g,h) - \theta(g\,h). \tag{6.222b}$$

We close this sequence of algebraic manipulations by simplifying the left-hand side of Eq. (6.222a) so that we may recast Eq. (6.222) as the special case of

$$W^{-1}(g,h)\,A^{(0)}_{\sigma_1}\,A^{(0)}_{\sigma_2}\cdots A^{(0)}_{\sigma_\ell}\,W(g,h) = e^{-\mathrm{i}\ell\,\delta(g,h)}\,A^{(0)}_{\sigma_1}\,A^{(0)}_{\sigma_2}\cdots A^{(0)}_{\sigma_\ell}, \tag{6.223a}$$

$$W(g,h) := V(g)\,\mathsf{K}_g[\![V(h)]\!]\,V^{-1}(g\,h), \tag{6.223b}$$

$$\delta(g,h) := \mathfrak{c}(g)\,\theta(h) + \theta(g) - \phi(g,h) - \theta(g\,h), \tag{6.223c}$$

for any positive integer ℓ.

Proof The left-hand side of Eq. (6.222a) can be simplified with the help of properties (6.176) as follows:

$$\begin{aligned}
e^{-\mathrm{i}\delta(g,h)}\,A^{(0)}_{\sigma_j} &= V(g\,h)\mathsf{K}_{gh}\left[\!\left[\mathsf{K}_h\left[\!\left[V^{-1}(h)\mathsf{K}_g\left[\!\left[V^{-1}(g)A^{(0)}_{\sigma_j}V(g)\right]\!\right]V(h)\right]\!\right]\right]\!\right]V^{-1}(g\,h)\\
&= V(g\,h)\,\mathsf{K}_g\left[\!\left[V^{-1}(h)\,\mathsf{K}_g\left[\!\left[V^{-1}(g)\,A^{(0)}_{\sigma_j}\,V(g)\right]\!\right]\,V(h)\right]\!\right]\,V^{-1}(g\,h)\\
&= V(g\,h)\,\mathsf{K}_g\,[\![V^{-1}(h)]\!]\,\left[V^{-1}(g)\,A^{(0)}_{\sigma_j}\,V(g)\right]\,\mathsf{K}_g\,[\![V(h)]\!]\,V^{-1}(g\,h)\\
\overset{\text{Eq. (6.223b)}}{} &= W^{-1}(g,h)\,A^{(0)}_{\sigma_j}\,W(g,h).
\end{aligned}\tag{6.224}$$

Equation (6.223a) follows from the multiplication of ℓ copies of Eq. (6.224) with indices $\sigma_1,\cdots,\sigma_\ell$. □

Injectivity of a FMPS implies that for some integer $\ell^\star > 1$ and any $\ell \geq \ell^\star$ all the products of the form $A^{(0)}_{\sigma_1}\,A^{(0)}_{\sigma_2}\cdots A^{(0)}_{\sigma_\ell}$ span the space of all $2M \times 2M$ matrices. Therefore, Eq. (6.223) combined with injectivity implies that the $2M \times 2M$ matrix $W(g,h)$ is an element from the center of the algebra defined by the vector space of all $2M \times 2M$ matrices, that is span$\{\mathbb{1}_{2M}\}$. Condition (6.223) thus simplifies to

$$A^{(0)}_{\sigma_1}\,A^{(0)}_{\sigma_2}\cdots A^{(0)}_{\sigma_\ell} = e^{-\mathrm{i}\ell\,\delta(g,h)}\,A^{(0)}_{\sigma_1}\,A^{(0)}_{\sigma_2}\cdots A^{(0)}_{\sigma_\ell}\tag{6.225}$$

for any $\ell \geq \ell^\star$. Choosing a linear combination of $A^{(0)}_{\sigma_1}\,A^{(0)}_{\sigma_2}\cdots A^{(0)}_{\sigma_\ell}$ equating the identity matrix $\mathbb{1}_{2M}$, delivers the constraint

$$\ell\,\delta(g,h) = 0, \quad \forall \ell > \ell^\star \implies \delta(g,h) = 0.\tag{6.226a}$$

Inserting the value of $\delta(g,h)$ given in Eq. (6.222) implies the final constraint

$$\phi(g,h) = \mathfrak{c}(g)\theta(h) + \theta(g) - \theta(g\,h).\tag{6.226b}$$

This is the coboundary condition (6.143) when $\phi' = 0$. In other words, the local representation $\hat{u}_j$ is equivalent to the trivial projective representation.

6.3.3.2 Proof for Odd-Parity FMPS

Let

$$|\{A^{(1)}_{\sigma_j}\}\rangle \equiv \sum_{\sigma} \mathrm{tr}\left(Y\,A^{(1)}_{\sigma_1}\,A^{(1)}_{\sigma_2}\cdots A^{(1)}_{\sigma_N}\right)|\psi_{1,\sigma_1}\rangle\,|\psi_{2,\sigma_2}\rangle\cdots|\psi_{N,\sigma_N}\rangle\tag{6.227}$$

be any translation-invariant (obeying periodic boundary conditions), G_f-symmetric, odd-parity (each matrix $A^{(1)}_{\sigma_j}$ commutes with the matrix Y), and injective FMPS. For any $g \in \mathrm{G}_f$, the global representation $\widehat{U}(g)$ of g is defined in Eq. (6.149). Hence, for any $g \in \mathrm{G}_f$, there exists a phase $\eta(g) \in [0, 2\pi[$ such that

$$\widehat{U}(g)\,|\{A^{(1)}_{\sigma_j}\}\rangle = e^{\mathrm{i}\eta(g)}|\{A^{(1)}_{\sigma_j}\}\rangle.\tag{6.228}$$

The counterpart to Eq. (6.180) is

$$\begin{aligned}\widehat{U}(g)\,|\{A^{(1)}_{\sigma_j}\}\rangle &= \sum_{\sigma} \mathrm{tr}\left(Y\,A^{(1)}_{\sigma_1}(g)\,A^{(1)}_{\sigma_2}(g)\cdots A^{(1)}_{\sigma_N}(g)\right)\\ &\qquad\times|\psi_{1,\sigma_1}\rangle\,|\psi_{2,\sigma_2}\rangle\cdots|\psi_{N,\sigma_N}\rangle, \end{aligned} \tag{6.229a}$$

$$\begin{aligned}A^{(1)}_{\sigma_j}(g) &:= \sum_{\sigma'_j}\langle\psi_{j,\sigma_j}|\left(\hat{u}_j(g)\,|\psi_{j,\sigma'_j}\rangle\right)\mathsf{K}_g\left[\!\left[A^{(1)}_{\sigma'_j}\right]\!\right]\\ &= \sum_{\sigma'_j}\mathcal{U}(g)_{\sigma_j,\sigma'_j}\,\mathsf{K}_g\left[\!\left[A^{(1)}_{\sigma'_j}\right]\!\right],\end{aligned} \tag{6.229b}$$

$$\mathsf{K}_g\left[\!\left[A^{(1)}_{\sigma_j}\right]\!\right] := \begin{cases} A^{(1)}_{\sigma_j}, & \text{if } \mathfrak{c}(g)=+1,\\ \\ \mathsf{K}\,A^{(1)}_{\sigma_j}\,\mathsf{K}, & \text{if } \mathfrak{c}(g)=-1.\end{cases} \tag{6.229c}$$

[Equation (6.203b) was used to reach the second equality in Eq. (6.229b).]

The same steps that lead to the solution (6.182) to Eq. (6.181) then give

$$\mathrm{tr}\left(Y\,A^{(1)}_{\sigma_1}(g)\,A^{(1)}_{\sigma_2}(g)\cdots A^{(1)}_{\sigma_N}(g)\right) = e^{\mathrm{i}\eta(g)}\,\mathrm{tr}\left(Y\,A^{(1)}_{\sigma_1}\,A^{(1)}_{\sigma_2}\cdots A^{(1)}_{\sigma_N}\right) \tag{6.230}$$

with the solution

$$A^{(1)}_{\sigma_j}(g) = e^{\mathrm{i}\theta(g)}\,V^{-1}(g)\,A^{(1)}_{\sigma_j}\,V(g), \tag{6.231a}$$

$$Y\,V(g)\,Y = (-1)^{\varsigma(g)}\,V(g), \tag{6.231b}$$

$$\theta(g) := \frac{1}{N}\left[\eta(g) - \pi\,\varsigma(g)\right]. \tag{6.231c}$$

The difference between Eq. (6.182b) and Eq. (6.231b) can be understood as follows. Odd-parity injective FMPS differ from the even ones in one crucial way. There exists a positive integer $\ell^\star \geq 1$ such that for any $\ell \geq \ell^\star$ the products of the form $A^{(1)}_{\sigma_1}\,A^{(1)}_{\sigma_2}\cdots A^{(1)}_{\sigma_\ell}$ span the $\mathbb{Z}_2$-graded algebra of $2M\times 2M$ matrices with the center span $\{\mathbb{1}_{2M}, Y\}$. Consequently, there exists an invertible $2M\times 2M$ $V(g)$ and a phase $\theta(g)\in[0,2\pi[$ such that [recall Eq. (6.173)]

$$V(g) = P\,V(g)\,P, \qquad V(g) = (-1)^{\varsigma(g)}\,Y\,V(g)\,Y, \qquad \varsigma(g) = 0,1, \tag{6.232a}$$

with $\varsigma\colon \mathrm{G}_f \to \{-1,+1\}$ a group homomorphism and

$$A^{(1)}_{\sigma_j}(g) = e^{\mathrm{i}\theta(g)}\,V^{-1}(g)\,A^{(1)}_{\sigma_j}\,V(g), \tag{6.232b}$$

for $\sigma_j = 1,\cdots,\mathcal{D}$ with $j = 1,\cdots,N$.

All the steps leading to Eq. (6.222) deliver

$$W^{-1}(g,h)\,A^{(1)}_{\sigma_j}\,W(g,h) = e^{-\mathrm{i}\delta(g,h)}\,A^{(1)}_{\sigma_j}, \tag{6.233a}$$

for $\sigma_j = 1,\cdots,\mathcal{D}$ with $j = 1,\cdots,N$, where

$$\delta(g,h) := \mathfrak{c}(g)\,\theta(h) + \theta(g) - \phi(g,h) - \theta(g\,h), \tag{6.233b}$$

and

$$W(g,h) := V(g)\,\mathsf{K}_g[\![V(h)]\!]\,V^{-1}(g\,h). \tag{6.233c}$$

Because $V(g)$ commutes with P so does $W(g,h)$. Because all possible products of the form $A^{(1)}_{\sigma_1}A^{(1)}_{\sigma_2}\cdots A^{(1)}_{\sigma_\ell}$ span the $\mathbb{Z}_2$-graded algebra of $2M\times 2M$ matrices with the center span $\{\mathbb{1}_{2M}, Y\}$, $W(g,h)$ is, up to a phase factor, proportional to $\mathbb{1}_{2M}$. The counterpart to the even-parity coboundary condition (6.226) then follows, thereby completing the proof of **Theorem** 6.52 for the parity-odd FMPS. □

6.4 A Second Extension of the Lieb-Schultz-Mattis Theorem

Theorem 6.52 presumes the existence of a local fermionic Fock space, that is of an even number of Majorana degrees of freedom per repeat unit cell. This hypothesis precludes translation invariant lattice Hamiltonians with an odd number of Majorana operators per repeat unit cell such as

$$\widehat{H}_{\mathrm{Kit},0} := \sum_{j=1}^{2M} \mathrm{i}\hat{\gamma}_j\,\hat{\gamma}_{j+1}. \tag{6.234a}$$

Here, the Hermitean operators $\{\hat{\gamma}_j = \hat{\gamma}^\dagger_j,\ j = 1,\cdots,2M\}$ obey the Majorana algebra

$$\{\hat{\gamma}_j, \hat{\gamma}_{j'}\} = 2\delta_{jj'}, \qquad j,j' = 1,\cdots,2M, \tag{6.234b}$$

and the total number $2M$ of repeat unit cells is an even integer. Hamiltonian $\widehat{H}_{\mathrm{Kit},0}$ realizes the quantum critical point between the two topologically distinct phases of the Kitaev chain (5.178). In the continuum limit, it describes a helical pair of Majorana fields and has a gapless spectrum.

Motivated by this example, we now prove a separate LSM constraint on Majorana lattice models with an odd number of Majorana flavors per repeat unit site. We use **Theorem** 6.52 for the proof.

Theorem 6.54 *A local Majorana Hamiltonian with an odd number of Majorana degrees of freedom per repeat unit cell that is invariant under the symmetry group* $\mathrm{G}_{\mathrm{trsl}}\times\mathrm{G}_f$ *cannot have a nondegenerate,* $\mathrm{G}_{\mathrm{trsl}}\times\mathrm{G}_f$*-invariant, and gapped ground state.*

Proof Let $m \geq 0$ be an integer and

$$\hat{\chi}_j := \left(\hat{\chi}_{j,1}, \hat{\chi}_{j,2}, \cdots, \hat{\chi}_{j,2m+1}\right)^{\mathsf{T}}, \qquad j = 1,\cdots,2M, \tag{6.235}$$

be the spinor made of $2m+1$ Majorana operators. Let the Hamiltonian $\widehat{H}$ be local and translationally invariant. We write

$$\widehat{H} \equiv \sum_{j=1}^{2M} \hat{h}\big(\hat{\chi}_{j-q+1},\cdots,\hat{\chi}_j,\cdots,\hat{\chi}_{j+q}\big), \tag{6.236}$$

where $\hat{h}$ is a Hermitean polynomial of $2q$ Majorana spinors $\left\{\hat{\chi}_{j-q+1},\cdots,\hat{\chi}_{j+q}\right\}$ with q a positive integer. The finiteness of q renders $\widehat{H}$ local. Hamiltonian (6.236) is defined over $2M$ sites, since an even number of Majorana operators are needed to have a well-defined fermionic Fock space. We assume that $\widehat{H}$ has a nondegenerate, $\mathrm{G}_{\mathrm{trsl}}\times\mathrm{G}_{f}$-invariant, and gapped ground state $|\Psi_0\rangle$. We are going to deliver a contradiction by making use of **Theorem** 6.52, thereby proving **Theorem** 6.54.

Define the Hamiltonian,

$$\widehat{H}':=\sum_{j=1}^{2M}\sum_{\alpha=1}^{2}\hat{h}\left(\hat{\chi}^{(\alpha)}_{j-q+1},\cdots,\hat{\chi}^{(\alpha)}_{j+q}\right), \tag{6.237a}$$

which is the sum of two copies of Hamiltonian (6.236). The repeat unit cell labeled by $j=1,\cdots,2M$ now contains two Majorana spinors labeled by $\alpha=1,2$. Hamiltonian (6.237a) thus acts on a fermionic Fock space which is locally spanned by an even number of Majorana flavors. At each site $j=1,\cdots,2M$ one can define a local fermionic Fock space. Since there is no coupling between the two copies $\alpha=1,2$ of Majorara spinors, $\widehat{H}'$ inherits from $\widehat{H}$ the nondegenerate, $\mathrm{G}_{\mathrm{trsl}}\times\mathrm{G}_{f}$-invariant, and gapped ground state

$$|\Psi'_0\rangle:=|\Psi_0\rangle\otimes_{\mathfrak{g}}|\Psi_0\rangle. \tag{6.237b}$$

Since at each site j, there is no term coupling the two copies $\hat{\chi}^{(1)}_j$ and $\hat{\chi}^{(2)}_j$, $\widehat{H}'$ is invariant under any local permutation

$$\begin{pmatrix}\hat{\chi}^{(1)}_j\\ \hat{\chi}^{(2)}_j\end{pmatrix}\mapsto\begin{pmatrix}\hat{\chi}^{(2)}_j\\ \hat{\chi}^{(1)}_j\end{pmatrix}, \tag{6.238a}$$

that is the on-site symmetry group G'_f of $\widehat{H}'$ must contain the transformation (6.238a). The local representation of the fermion parity operator is [compare with Eq. (5.169)]

$$\widehat{P}_j:=\prod_{l=1}^{2m+1}\left[\mathrm{i}\,\hat{\chi}^{(1)}_{j,l}\,\hat{\chi}^{(2)}_{j,l}\right]. \tag{6.238b}$$

Under the transformation (6.238a), the local fermion parity operator $\widehat{P}_j$ acquires the phase $(-1)^{2m+1}=-1$. Therefore, the symmetry transformation (6.238a) anticommutes with $\widehat{P}_j$. This anticommutation relation implies a nontrivial second cohomology group class $[\phi]\neq 0$ of G'_f, independently of the group of on-site symmetries of Hamiltonian (6.236). Therefore, by **Theorem** 6.52 Hamiltonian $\widehat{H}'$ cannot have a nondegenerate, $\mathrm{G}_{\mathrm{trsl}}\times\mathrm{G}'_f$-invariant, and gapped ground state. This is in contradiction with the initial assumption that Hamiltonian (6.236) has the nondegenerate, $\mathrm{G}_{\mathrm{trsl}}\times\mathrm{G}_{f}$-invariant, and gapped ground state $|\Psi_0\rangle$. □

Remark (1) One can interpret **Theorem** 6.54 as the inability to write down a $\mathrm{G}_{\mathrm{trsl}}\times\mathrm{G}_{f}$-symmetric injective FMPS for the ground state of translationally invariant local Hamiltonians with an odd number of Majorana flavors per repeat unit cell. This is so because one cannot define the matrices A_{σ_j} as there is no well-defined fermionic Fock space at site j to begin with.

Remark (2) The importance of this no-go theorem is that any $\mathrm{G}_{\mathrm{trsl}} \times \mathrm{G}_f$-symmetric ground state, when the local number of Majorana flavors per repeat unit cell is odd, must then be degenerate or gapless in the thermodynamic limit. This deduction follows from combining **Theorem** 6.54 with the belief that, in the thermodynamic limit, any nondegenerate, $\mathrm{G}_{\mathrm{trsl}} \times \mathrm{G}_f$-symmetric, and gapped ground state of a local, $\mathrm{G}_{\mathrm{trsl}} \times \mathrm{G}_f$-symmetric, and one-dimensional lattice Hamiltonian can be deformed adiabatically to a $\mathrm{G}_{\mathrm{trsl}} \times \mathrm{G}_f$-symmetric and injective FMPS without ever closing the spectral gap separating the nondegenerate and $\mathrm{G}_{\mathrm{trsl}} \times \mathrm{G}_f$-symmetric ground state from all excited states.

Remark (3) **Theorem** 6.54 is closely related to the fact that when the number of Majorana flavors per repeat unit cell is odd, any translation-invariant one-dimensional lattice Hamiltonian is supersymmetric in the sense of Section 3.6.3, as shown by Hsieh et al.[40]

Remark (4) The dimensionality d of space played no role in the proof of **Theorem** 6.54 until **Theorem** 6.52 was used. In fact, it can be shown that **Theorem** 6.54 holds for any d-dimensional lattice, as the proof can be made independent of **Theorem** 6.52.

6.5 Fermionic Invertible Topological Phases

6.5.1 Definitions

Definition 6.55 Space will refer to some d-dimensional manifold embedded in $\mathbb{R}^n$ with $n \geq d$ that, prior to taking the thermodynamic limit, is discretized by some lattice. Each repeat unit cell of the lattice hosts an even number of local Majorana degrees of freedom that generates a local fermionic Fock space.

Definition 6.56 A family of local Hamiltonians governing the quantum dynamics of locally defined Majorana degrees of freedom, an even number per repeat unit cell of a lattice, is said to realize fermionic invertible topological (FIT) phases of matter if their ground states in the thermodynamic limit obey the following properties.

1 Whenever space is boundaryless, the ground state of any of these Hamiltonians is nondegenerate and separated from all excitations by a nonvanishing energy gap.
2 This family of local Hamiltonians contains Hamiltonians whose eigenstates are obtained from occupying decoupled local fermionic two-level systems. We shall

40 T. H. Hsieh, G. B. Halász, and T. Grover, Phys. Rev. Lett. **117**(16), 166802 (2016). [Hsieh et al. (2016b)].

call such Hamiltonians together with their local Majorana degrees of freedom trivial.[41]

3 Whenever space is boundaryless, the property that any two of these Hamiltonians can be deformed into each other without closing the spectral gap through the combined operations of adding trivial local Majorana degrees of freedom or the adiabatic tuning of local interactions defines an equivalence relation on this family of local Hamiltonians.

4 There exist stacking rules between any two of these Hamiltonians such that
   (1) the new repeat unit cell under stacking accomodates all local Majorana degrees of freedom stemming from the two Hamiltonians, and
   (2) these rules induce the composition rule of an Abelian group between their equivalence classes.

The equivalence class corresponding to the neutral element of this Abelian group is called the trivial fermionic invertible topological phase. Its existence is guaranteed by condition 2.

Definition 6.57 Fermionic symmetry-protected topological (FSPT) phases of matter are trivial fermionic invertible topological phases of matter that acquire an Abelian group structure distinct from the Abelian group made of one element if and only if an on-site symmetry is imposed that is compatible with conditions 1, 2, 3, and 4 of **Definition** 6.56.

Nontrivial invertible topological phases of matter support gapless boundary states when space is not boundaryless. Examples of fermionic invertible topological phases of matter that are not protected by symmetries are the Majorana c-odd chains. Examples of fermionic invertible topological phases of matter that are protected by symmetries are the Majorana c-even chains with time-reversal and fermion parity as the protecting symmetries.[42] Majorana c chains in the symmetry class BDI can be assigned eight distinct patterns of symmetry fractionalization as is done in **Table** 5.8. We can deduce stacking rules for the triplets $([\nu],[\rho],\mu)$ from **Table** 5.8 from the rule that stacking a Majorana c chain to a Majorana c' chain delivers a Majorana c'' chain with $c''\equiv c+c' \bmod 8$. We are going to generalize these stacking rules to an arbitrary fermionic symmetry group G_f from **Definition** 6.30 when space is one-dimensional.

6.5.2 Stacking Rules

The classification of the fermionic invertible topological phases in one-dimensional space is intimately related to the classification of the projective representations

[41] They are the Majorana counterparts to the atomic limit of (topologically trivial) band insulators in band theory.

[42] Without the protecting symmetry t for time-reversal in the symmetry class BDI, the values taken by the indices $[\nu]$ and $[\rho]$ are $[\nu]=[\rho]=0$, while the values taken by the index $[\mu]$ (the index of relevance to the symmetry class D as we shall see in Chapter 7) remain unchanged in **Table** 5.8.

of the fermionic symmetry group G_f, an on-site symmetry acting globally on the fermionic Fock space.

Given is the fermionic symmetry group G_f (see Section 6.3.2) and the *local* fermionic Fock space $\mathfrak{F}$.[43] The map

$$\begin{aligned}\hat{u}\colon G_f &\to \mathrm{Aut}\,(\mathfrak{F}),\\ g &\mapsto \hat{u}(g),\end{aligned} \tag{6.239a}$$

from the fermionic symmetry group G_f to the set of automorphisms $\mathrm{Aut}\,(\mathfrak{F})$ of $\mathfrak{F}$[44] is a norm-preserving projective representation of G_f if it satisfies

$$\hat{u}(\mathrm{id}) = \widehat{\mathbb{1}}, \tag{6.239b}$$

$$\hat{u}(g)\,\hat{u}(h) = e^{\mathrm{i}\phi(g,h)}\,\hat{u}(g\,h), \qquad e^{\mathrm{i}\phi(g,h)} \in C^2\big(G_f, \mathrm{U}(1)_{\mathfrak{c}}\big), \tag{6.239c}$$

$$[\hat{u}(g)\,\hat{u}(h)]\,\hat{u}(f) = \hat{u}(g)\,[\hat{u}(h)\,\hat{u}(f)], \tag{6.239d}$$

and

$$\big|\big(\langle\psi|\hat{u}(g)\big)\big(\hat{u}(g)|\varphi\rangle\big)\big| = |\langle\psi|\varphi\rangle|, \qquad \forall\psi,\varphi\in\mathfrak{F}, \tag{6.239e}$$

where $\mathrm{id}\in G_f$ is the identity element of G_f, $\widehat{\mathbb{1}}\in\mathrm{Aut}(\mathfrak{F})$ is the identity map, $C^2\big(G_f,\mathrm{U}(1)_{\mathfrak{c}}\big)$ is the set of 2-cochains from G_f to U(1), and

$$\mathfrak{c}\colon G_f \to \{-1,+1\} \tag{6.239f}$$

is the group-homomorphic map that distinguishes the case when $\hat{u}(g)\in\mathrm{Aut}(\mathfrak{F})$ is antilinear as is indicated by $\mathfrak{c}(g)=-1$ from the case when $\hat{u}(g)\in\mathrm{Aut}(\mathfrak{F})$ is linear as is indicated by $\mathfrak{c}(g)=+1$.

It is always possible to do the decomposition

$$\hat{u}(g) = \hat{v}(g)\,\mathsf{K}^{\mathfrak{d}(g)}, \qquad \mathsf{K}\,\mathrm{i}\,\mathsf{K} = -\mathrm{i}, \qquad \mathfrak{d}(g) := \frac{1-\mathfrak{c}(g)}{2}, \tag{6.240}$$

with $\hat{v}(g)\in\mathrm{Aut}(\mathfrak{F})$ a unitary linear operator and K denoting complex conjugation for any $g\in G_f$.

The phase factor $\phi(g,h)\in[0,2\pi[$ on the right-hand side of Eq. (6.239) can be identified with a 2-cochain (see Section 6.3.2). It is not uniquely determined due to the gauge ambiguity that is inherent to defining norm-preserving operators in $\mathrm{Aut}(\mathfrak{F})$. The pair of functions ϕ and ϕ' are equivalent to each other if they are related by the gauge transformation

$$\hat{u}(g) \mapsto e^{\mathrm{i}\xi(g)}\,\hat{u}(g) \tag{6.241a}$$

through

$$\phi(g,h) = \phi'(g,h) + \xi(g\,h) - \xi(g) - \mathfrak{c}(g)\,\xi(h) \tag{6.241b}$$

[43] The *local* fermionic Fock space $\mathfrak{F}$ is the one spanned by the Majorana degrees of freedom hosted by a repeat unit cell that discretizes one-dimensional space in **Definition** 6.55.

[44] The set $\mathrm{Aut}\,(\mathfrak{F})$ is the set of all one-to-one maps from $\mathfrak{F}$ to $\mathfrak{F}$.

for some map

$$\begin{aligned}\xi\colon \mathrm{G}_f &\to [0,2\pi[,\\ g &\mapsto \xi(g).\end{aligned} \tag{6.241c}$$

The equivalence classes $[\phi]$ of a projective representation of G_f are enumerated by the elements $[e^{\mathrm{i}\phi}]$ of the second cohomology group $H^2(\mathrm{G}_f, \mathrm{U}(1)_{\mathfrak{c}})$ (recall Section 6.3.2).

By **Definition** 6.56, fermionic invertible topological phases of matter are gapped phases of matter with nondegenerate ground states under any closed boundary conditions. Local Hamiltonians with the symmetry group G_f that realize fermionic invertible topological phases of matter must then necessarily have a nondegenerate ground state that transforms by the multiplication of a phase under the symmetry group G_f under any closed boundary conditions.

We restrict our attention to one-dimensional space and to fermionic invertible topological phases of matter with translation symmetry $\mathrm{G}_{\mathrm{trsl}}$ in addition to the on-site fermionic symmetry group G_f. In other words, the total symmetry group $\mathrm{G}_{\mathrm{tot}}$ is by hypothesis the direct product

$$\mathrm{G}_{\mathrm{tot}} \equiv \mathrm{G}_{\mathrm{trsl}} \times \mathrm{G}_f. \tag{6.242}$$

If so, **Theorem** 6.52 applies, that is, the local representation of the on-site fermionic symmetry group G_f for each repeat unit cell[45] must be a trivial projective representation, that is, it is associated with the neutral element of the second cohomology group $H^2(\mathrm{G}_f, \mathrm{U}(1)_{\mathfrak{c}})$.

Open boundary conditions break the hypothesis of translation symmetry in **Theorem** 6.52. The breaking of translation symmetry can happen in two ways as is shown in Figure 6.6. This binary choice for the breaking of translation symmetry dictates the fate of the change in the degeneracy of the ground state of an fermionic invertible topological phase when replacing periodic with open boundary conditions.

Without loss of generality, we assume that the range of couplings for the $2n = 2, 4, 6, 8, \cdots$ local Majorana degrees of freedom hosted by a repeat unit cell in Figure 6.6 is one lattice spacing. If the number of Majorana degrees of freedom hosted in the upmost left or right cell is unchanged from that of a repeat unit cell in a ring geometry, as is illustrated with Figure 6.6(b), the action of the projective representation of G_f is unchanged relative to that for a single repeat unit cell in the ring geometry. For an fermionic invertible topological phase of matter, the action of the projective representation on the left (right) boundary is then the trivial one. Hence, the ground state remains nondegenerate and symmetric with respect to G_f when open boundary conditions are selected such that they do not change the number of Majorana degrees of freedom localized at one of the two boundaries of

[45] **Definition** 6.56 presumes that there exists a local fermionic Fock space attached to each repeat unit cell.

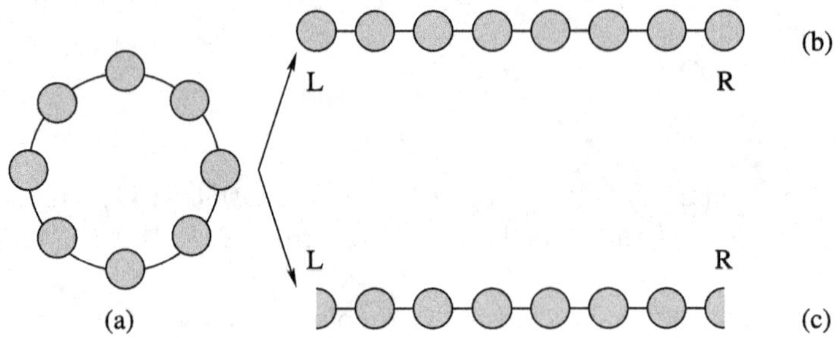

Figure 6.6 Repeat unit cells of a one-dimensional lattice are pictured by colored discs. Each repeat unit cell hosts an even number $2n$ of Majorana degrees of freedom. Without loss of generality, the range of the couplings between Majorana degrees of freedom is one lattice spacing (the thick line between the repeat unit cells). (a) Translation symmetry is imposed by choosing periodic boundary conditions, in which case the one-dimensional lattice is the discretization of a ring. (b-c) Open boundary conditions break the translation symmetry. This can be achieved by cutting a thick line connecting two repeat unit cell (b) or by cutting open a repeat unit cell (c). In the former case, the number of Majorana degrees of freedom on any one of the upmost left or right cells is the same even number $2n$ of Majorana degrees of freedom as in a single repeat unit cell. In the latter case, the number of Majorana degrees of freedom on the upmost left cell is any integer $1 < n_{\mathrm{L}} < 2n$, while that on the upmost cell is $n_{\mathrm{R}} = 2n - n_{\mathrm{L}}$. Figure 6.6 should be compared with Figure 5.1.

the lattice compared to the number $2n$ of local Majorana degrees of freedom hosted by a repeat unit cell.

If the number $1 < n_{\mathrm{L}} < 2n$ of Majorana degrees of freedom in the upmost left cell is reduced from the integer number $2n$ of Majorana degrees of freedom per repeat unit cell in a ring geometry, as is illustrated with Figure 6.6(c), in which case the number $n_{\mathrm{R}} < 2n$ of Majorana degrees of freedom in the upmost right cell is necessarily given by $n_{\mathrm{R}} = 2n - n_{\mathrm{L}}$, one must distinguish the cases of $1 < n_{\mathrm{L}} < 2n$ an even integer from the case of $1 < n_{\mathrm{L}} < 2n$ an odd integer.

In the former case, it is possible to define the projective representation of G_f on either one of the left or right boundaries in a local way. However, the action of the projective representation of G_f on either one of the left or right boundaries can change relative to that of a single repeat unit cell in the ring geometry. If so, for an fermionic invertible topological phase of matter, symmetry fractionalization takes place in a complementary fashion on the left and right boundaries. The ground state is then no longer nondegenerate, with a ground-state degeneracy governed by the minimal dimensionality of the nontrivial projective representation of the symmetry group G_f on one of the boundaries.

In the latter case, it is not possible to define the projective representation of G_f on either one of the left or right boundaries in a local way. When $2n = 2$, the two-dimensional fermionic Fock space associated with the union of the left and right boundary is nonlocal by construction. When $2n = 4, 6, 8, \cdots$ and even if the action of the projective representation of G_f on the left (right) boundary is closed on the local fermionic Fock space spanned by $1 < n_{\mathrm{L}} - 1 < 2n - 1$ ($1 < n_{\mathrm{R}} - 1 < 2n - 1$) Majorana degrees of freedom, the operators representing G_f on the left (right) boundary must still be constructed by making use of the remaining n_{R}-th (n_{L}-th) Majorana degree of freedom from the right (left) boundary. For an fermionic invertible topological phase of matter, the ground state is then no longer nondegenerate, with a ground-state degeneracy bounded from below by the number 2, the dimensionality of the nonlocal fermionic Fock space spanned by the n_{R}-th and n_{L}-th Majorana degrees of freedom localized on the right and left boundaries, respectively. This minimal degeneracy of two reflects the fact that the center of the Clifford algebra spanned by an odd number of Majorana operators is two-dimensional.

Any fermionic invertible topological phase of matter, when space is one-dimensional and open boundary conditions are selected, is thus characterized by the following data.

1 There is a $\mathbb{Z}_2$-valued indicator $[\mu] \in \{0,1\}$ that measures the parity of the number of Majorana degrees of freedom that are localized on either one of the left or right boundaries of an open chain $\Lambda_{\mathrm{bd}} = \Lambda_{\mathrm{L}} \cup \Lambda_{\mathrm{R}}$. The index $[\mu]$ can be viewed as element of the zero-th cohomology group $H^0(\mathrm{G}_f, \mathbb{Z}_2) = \mathbb{Z}_2$.
2 There is the equivalence class $[\phi] \cong [e^{\mathrm{i}\phi}] \in H^2(\mathrm{G}_f, \mathrm{U}(1)_{\mathfrak{c}})$ that characterizes the projective representation of the on-site symmetry group G_f at either one of the left or right boundaries of an open chain $\Lambda_{\mathrm{bd}} = \Lambda_{\mathrm{L}} \cup \Lambda_{\mathrm{R}}$.

There is no need to specify to what boundary the data $[\mu] \in H^0(\mathrm{G}_f, \mathbb{Z}_2)$ or $[\phi] \cong [e^{\mathrm{i}\phi}] \in H^2(\mathrm{G}_f, \mathrm{U}(1)_{\mathfrak{c}})$ are assigned since specifying $[\mu]$ and $[\phi]$ on the left boundary Λ_{L} fixes their counterparts on the right boundary Λ_{R}, owing to the condition that the ground state of an fermionic invertible topological phase of matter must be nondegenerate and G_f-symmetric when periodic boundary conditions are selected. The pair of data

$$([\phi],[\mu]) \in H^2(\mathrm{G}_f, \mathrm{U}(1)_{\mathfrak{c}}) \times H^0(\mathrm{G}_f, \mathbb{Z}_2) \tag{6.243a}$$

fixes the minimal dimension

$$D^{\mathrm{min}}_{([\phi],[\mu])} = 1, 2, 3, \cdots \tag{6.243b}$$

of the projective representation of G_f that is realized by the Majorana zero modes localized on the disconnected components Λ_{L} and Λ_{R} of the boundary $\Lambda_{\mathrm{bd}} = \Lambda_{\mathrm{L}} \cup \Lambda_{\mathrm{R}}$, respectively. The physical interpretation of this dimensionality is that of the dimensionality of a multiplet of pairwise orthogonal many-body states that includes the ground state for any finite chain of repeat unit cells whose energies converge to that of the ground states while retaining their pairwise orthogonality in the thermodynamic limit. As the thermodynamic limit is required to establish the lower bound (6.243b) on the dimensionality of the eigenspace for the ground states, the pair of data (6.243a) that quantifies the notion of symmetry fractionalization is an emergent property.[46]

Remark (1) The pair of data (6.243a) does not fix the dimensionality of the Fock space associated with the boundary $\Lambda_{\rm bd}$ of one-dimensional space on which the projective representation of G_f is realized other than through the lower bound (6.243b).[47]

Remark (2) The pair of data (6.243a) does not resolve the number of Majorana zero modes that are localized on the disconnected components $\Lambda_{\rm L}$ and $\Lambda_{\rm R}$ of the boundary $\Lambda_{\rm bd} = \Lambda_{\rm L} \cup \Lambda_{\rm R}$, respectively.

Remark (3) The pair of data (6.243a) is stable to any G_f-symmetric perturbation that is localized on $\Lambda_{\rm L}$ or $\Lambda_{\rm R}$ except for the possibility that the G_f symmetry is spontaneously broken.

This latter stability of the pair of data (6.243a) suggests that it is also a bulk property. Indeed, Bourne and Ogata[48] have shown rigorously that the pair of data (6.243a) that characterizes some given Hamiltonian realizing an fermionic invertible topological phase of matter in the thermodynamic limit is invariant under any perturbation that is G_f-symmetric and local as long as this perturbation does not close the gap or break spontaneously the G_f symmetry when periodic boundary conditions are imposed.[49] In short, Bourne and Ogata have shown that the pair of data (6.243a) is a pair of topological indices that can be used to classify fermionic invertible topological phases of matter in one-dimensional space.

The index $[\mu] = 0, 1$ in the pair of data (6.243a) is independent from the fermionic symmetry group G_f. It measures the parity of the number of Majorana zero modes that are localized on any one of the disconnected components $\Lambda_{\rm L}$ and $\Lambda_{\rm R}$ of the boundary $\Lambda_{\rm bd} = \Lambda_{\rm L} \cup \Lambda_{\rm R}$ for any fermionic invertible topological phase of matter in one-dimensional space. When $[\mu] = 1$, the minimal degeneracy of the eigenspace for the ground states when open boundary conditions are selected is two for any fermionic symmetry group G_f, including the smallest possible fermionic symmetry group $\mathrm{G}_f = \mathbb{Z}_2^{\rm F}$. Hence, one-dimensional fermionic invertible topological phases of matter with $[\mu] = 1$ cannot be deformed adiabatically to a trivial fermionic invertible topological phase of matter at the expense of breaking explicitly any of the protecting symmetries in G_f other than $\mathbb{Z}_2^{\rm F}$. A fortiori, these phases of matter cannot realize fermionic symmetry-protected phases of matter in one-dimensional

[46] The Majorana c chains from Chapter 5 are atypical examples of Hamiltonians realizing fermionic invertible topological phases of matter in that the support of their boundary states is on a discrete subset of the lattice as opposed to all sites, but with exponentially decaying amplitudes. They realize zero modes before taking the thermodynamic limit, unlike for polyacetylene.

[47] Open Majorana c chains from Chapter 5 with $c > 3$ have a ground-state degeneracy larger than $D^{\rm min}_{([\phi],[\mu])}$.

[48] C. Bourne and Y. Ogata, Forum of Mathematics, Sigma, **9**:e25 1–45 (2021). [Bourne and Ogata (2021)].

[49] For a mathematically rigorous proof, one cannot rely only on the stability of the boundaries as the very definition of the boundaries might not be preserved by a bulk G_f-symmetric and local perturbation.

space. In one-dimensional space, fermionic symmetry-protected phases of matter are only possible when $[\mu]=0$.

The index $[\phi]\cong[e^{i\phi}]\in H^2\big(\mathrm{G}_f,\mathrm{U}(1)_{\mathfrak{c}}\big)$ in the pair of data (6.243a) depends on both $[\mu]=0,1$ and the fermionic symmetry group G_f. This is so because G_f is the central extension of the on-site symmetry group G by the fermion-parity symmetry group $\mathbb{Z}_2^{\mathrm{F}}$ through γ, that is G is isomorph to the quotient group

$$\mathrm{G}\cong\mathrm{G}_f/\mathbb{Z}_2^{\mathrm{F}}. \tag{6.244}$$

As the center of the fermionic symmetry group G_f is the fermion-parity subgroup $\mathbb{Z}_2^{\mathrm{F}}$, its projective representations are sensitive to the values of μ. This sensitivity can be made explicit if, following Turzillo and You,[50] one trades the equivalence classes $[\phi]\cong[e^{i\phi}]\in H^2\big(\mathrm{G}_f,\mathrm{U}(1)_{\mathfrak{c}}\big)$ for the equivalence classes $[(\nu,\rho)]\in\ker\mathcal{D}^2_{\gamma,\mathfrak{c}}/\mathrm{im}\,\mathcal{D}^1_{\gamma,\mathfrak{c}}$ owing to the set isomorphism

$$H^2\big(\mathrm{G}_f,\mathrm{U}(1)_{\mathfrak{c}}\big)\cong\ker\mathcal{D}^2_{\gamma,\mathfrak{c}}\Big/\mathrm{im}\,\mathcal{D}^1_{\gamma,\mathfrak{c}}. \tag{6.245a}$$

Here, we have defined the generalized co-boundary operators [the cup product was defined in Eqs. (6.127a) and (6.127b)]

$$\begin{aligned}\mathcal{D}^2_{\gamma,\mathfrak{c}}\colon C^2[\mathrm{G},\mathrm{U}(1)]\times C^1[\mathrm{G},\mathbb{Z}_2]&\to C^3[\mathrm{G},\mathrm{U}(1)]\times C^2[\mathrm{G},\mathbb{Z}_2],\\ (\nu,\rho)\mapsto\mathcal{D}^2_{\gamma,\mathfrak{c}}(\nu,\rho)&:=\big(\delta^2_{\mathfrak{c}}\nu-\pi\,\rho\smile\gamma,\delta^1_{\mathfrak{c}}\rho\big)\,,\end{aligned} \tag{6.245b}$$

and

$$\begin{aligned}\mathcal{D}^1_{\gamma,\mathfrak{c}}\colon C^1[\mathrm{G},\mathrm{U}(1)]\times C^0[\mathrm{G},\mathbb{Z}_2]&\to C^2[\mathrm{G},\mathrm{U}(1)]\times C^1[\mathrm{G},\mathbb{Z}_2],\\ (\alpha,\beta)\mapsto\mathcal{D}^1_{\gamma,\mathfrak{c}}(\alpha,\beta)&:=\big(\delta^1_{\mathfrak{c}}\alpha+\pi\,\beta\smile\gamma,\delta^0_{\mathfrak{c}}\beta\big)\,.\end{aligned} \tag{6.245c}$$

The 2-cochain $\nu\cong e^{i\nu}\in C^2[\mathrm{G},\mathrm{U}(1)]$ encodes the projective representations of the quotient group G of G_f. The interpretation of the 1-cochain $\rho\in C^1[\mathrm{G},\mathbb{Z}_2]$ depends on the value of $[\mu]=0,1$. When $[\mu]=0$, the 1-cochain $\rho\in C^1[\mathrm{G},\mathbb{Z}_2]$ encodes the relation between the representations of the elements of the quotient group G and the representation of the fermion-parity from $\mathbb{Z}_2^{\mathrm{F}}$. When $[\mu]=1$, the 1-cochain $\rho\in C^1[\mathrm{G},\mathbb{Z}_2]$ encodes the relation between the representations of the elements of the quotient group G and the representation of the nontrivial central element of a Clifford algebra $\mathrm{C}\ell_{2k+1}$ with an odd number $2k+1=1,3,5,\cdots$ of generators.

Once equivalence classes of fermionic invertible topological phases in one-dimension are characterized by the triplet $([(\nu,\rho)],[\mu])$, it is imperative to derive the stacking rules, that is, group composition rules of the triplets $([(\nu,\rho)],[\mu])$ that are compatible with the $\mathbb{Z}_2$-graded tensor product between fermionic Fock spaces (in physics terminology, antisymmetrization). Stacking rules can be derived by considering the boundary representations of an fermionic invertible topological phase of matter that is constructed by combining the boundary degrees of freedom of two other fermionic invertible topological phases with topological indices $([(\nu_1,\rho_1)],[\mu_1])$

[50] A. Turzillo and M. You, Phys. Rev. B **99**(3), 035103 (2019). [Turzillo and You (2019)].

and $([(\nu_2, \rho_2)], [\mu_2])$, respectively. These rules can be derived by elementary but lengthy algebraic methods and are given by[51]

$$\begin{aligned}
&([(\nu_1, \rho_1)], 0) \wedge ([(\nu_2, \rho_2)], 0) \\
&\qquad = ([(\nu_1 + \nu_2 + \pi\,(\rho_1 \smile \rho_2)\,,\, \rho_1 + \rho_2)], 0)\,, && \text{(6.246a)}\\
&([(\nu_1, \rho_1)], 0) \wedge ([(\nu_2, \rho_2)], 1) \\
&\qquad = ([(\nu_1 + \nu_2 + \pi\,(\rho_1 \smile \rho_2 + \rho_1 \smile \mathfrak{d})\,,\, \rho_1 + \rho_2)], 1)\,, && \text{(6.246b)}\\
&([(\nu_1, \rho_1)], 1) \wedge ([(\nu_2, \rho_2)], 0) \\
&\qquad = ([(\nu_1 + \nu_2 + \pi\,(\rho_1 \smile \rho_2 + \rho_2 \smile \mathfrak{d})\,,\, \rho_1 + \rho_2)], 1)\,, && \text{(6.246c)}\\
&([(\nu_1, \rho_1)], 1) \wedge ([(\nu_2, \rho_2)], 1) \\
&\qquad = ([(\nu_1 + \nu_2 + \pi\,(\rho_1 \smile \rho_2)\,,\, \rho_1 + \rho_2 + \mathfrak{d})], 0)\,, && \text{(6.246d)}
\end{aligned}$$

where $\mathfrak{d}$ was defined in Eq. (6.240) and the addition is defined modulo 2π (2) for the index ν (ρ). Here, we denote the stacking operation with the symbol $\wedge$. We also made use of the cup product $\smile$ defined in Eqs. (6.127a) and (6.127b) to construct 2-cochains out of two 1-cochains.

These stacking rules are essential properties of fermionic invertible topological phases of matter. They enforce a group composition law between fermionic invertible topological phases of matter sharing the same fermionic symmetry group G_f. This group composition law can be interpreted as the physical operation by which two blocks of matter, each realizing fermionic invertible topological phases of matter sharing the same fermionic symmetry group G_f, are brought into contact so as to form a single larger block of matter. This group composition law is also needed to implement a consistency condition corresponding to changing from open to closed boundary conditions, a process by which the trivial fermionic invertible topological phase of matter is obtained. Topological data associated with the left and the right disconnected components of the one-dimensional boundary must be the inverse of each other with respect to the stacking rules (6.246), that is, one should obtain the trivial data $([0,0],0)$ if the change from open to periodic boundary conditions is interpreted as the stacking of opposite boundaries.

Example 6.58 We now specialize to the case of the Majorana c chains (recall Section 5.7), for which

$$G_f = \mathbb{Z}_2^T \times \mathbb{Z}_2^F = \{(e,e), (e,p), (t,e), (t,p)\}. \qquad (6.247)$$

The group G_f is thus split according to the definition (6.137). If so, the modified coboundary operators (6.245b) and (6.245c) reduce to the coboundary operator (6.116) with $n = 2$ and $n = 1$, respectively. In turn, the cochains ν and ρ are both cocycles, that is

$$(\nu, \rho) \in Z^2(G, U(1)_{\mathfrak{c}}) \times Z^1(G, \mathbb{Z}_2), \qquad G = \mathbb{Z}_2^T. \qquad (6.248)$$

[51] Ö. M. Aksoy and C. Mudry, Phys. Rev. B **106**(3), 035117 (2022). [Aksoy and Mudry (2022)].

The set of equivalence classes $[(\nu,\rho)]$ is then given by

$$[(\nu,\rho)] = ([\nu],[\rho]) \in H^2(\mathrm{G},\mathrm{U}(1)_{\mathfrak{c}}) \times H^1(\mathrm{G},\mathbb{Z}_2), \qquad \mathrm{G}=\mathbb{Z}_2^T, \tag{6.249a}$$

with the assignments

$$\mathfrak{d}(t)=1, \qquad \mathfrak{d}(p)=0. \tag{6.249b}$$

The indices $[\nu]$ and $[\rho]$ from **Table** 5.8 are obtained from computing $\nu(t,t)/\pi$ and $\rho(t)$ in Eq. (6.249), respectively.

Given two projective representations $\widehat{U}_1$ and $\widehat{U}_2$ of the group (6.247) acting on the Fock spaces $\mathfrak{F}_1$ and $\mathfrak{F}_2$, respectively, the indices associated with the projective representation $\widehat{U}_\wedge$ acting on the Fock space $\mathfrak{F}_\wedge$ constructed from the graded tensor product of the Fock spaces $\mathfrak{F}_1$ and $\mathfrak{F}_2$ follow from Eq. (6.246). They can be shown to be given by the triplet $([\nu_\wedge],[\rho_\wedge],[\mu_\wedge])$,[52] where

$$[\nu_\wedge] = \begin{cases} [\nu_1]+[\nu_2]+[\rho_1]\,[\rho_2], & \text{if } [\mu_\wedge]\equiv[\mu_1]+[\mu_2]=0 \bmod 2,\\ [\nu_1]+[\nu_2]+[\rho_1]\,[\rho_2]+[\rho_1], & \text{if } [\mu_1]=0,\ [\mu_2]=1,\\ [\nu_1]+[\nu_2]+[\rho_1]\,[\rho_2]+[\rho_2], & \text{if } [\mu_1]=1,\ [\mu_2]=0, \end{cases} \tag{6.250a}$$

and

$$[\rho_\wedge] = \begin{cases} [\rho_1]+[\rho_2]+1, & \text{if } [\mu_1]=1,\ [\mu_2]=1,\\ [\rho_1]+[\rho_2], & \text{otherwise.} \end{cases} \tag{6.250b}$$

fermionic invertible topological phases with the symmetry group $\mathbb{Z}_2^T\times\mathbb{Z}_2^F$ form the cyclic group $\mathbb{Z}_8$ under the stacking rule (6.250). Without loss of generality, the generator of the group $\mathbb{Z}_8$ can be chosen as the fermionic invertible topological phase with indices $([\nu],[\rho],[\mu])=(0,0,1)$. In Table 6.1, the triplets $([\nu],[\rho],[\mu])$ for all elements of $\mathbb{Z}_8$ are computed using the stacking rules (6.250) (compare with **Table** 5.8).

The fermionic stacking rule (6.246a) differs from the composition law of the second cohomology group $H^2(\mathrm{G}_f,\mathrm{U}(1)_{\mathfrak{c}})$. This difference comes about because of the term $\pi\,\rho_1 \smile \rho_2$ on the right-hand side of Eq. (6.246a). This term arises because distinct Majorana degrees of freedom anticommute with each other. Accordingly, the fermion-parity symmetry $\mathbb{Z}_2^{\mathrm{F}}$ that is contained in the center of G_f is a distinguished symmetry, for it may neither be broken explicitly nor spontaneously. Fermion-parity symmetry acts as a superselection rule and provides a $\mathbb{Z}_2$ grading for any fermionic Fock space. Any operation done on fermionic Fock spaces, such as taking the tensor product of two Fock spaces, must respect the grading imposed by the fermion-parity symmetry. The fermionic stacking rule (6.246a) must be compatible with the Pauli principle.

[52] See footnote 51 and recall that the indices ν_1 and ν_2 in Eq. (6.250) and in **Table** 5.8 differ from those in Eq. (6.246) by the multiplicative factor π.

Table 6.1 *Stacking rules for the Majorana c chains*

c	$([\nu],[\rho],[\mu])$
1	(0,0,1)
2	(0,1,0)
3	(1,1,1)
4	(1,0,0)
5	(1,0,1)
6	(1,1,0)
7	(0,1,1)
8	(0,0,0)

The indices $([\nu],[\rho],[\mu])$ generated by stacking the indices $(0,0,1)$ $(c=1)$ with itself form the cyclic group $\mathbb{Z}_8$ under the stacking rules (6.250).

There is no Pauli principle at work when all local degrees of freedom are exclusively bosonic. In one-dimensional space, bosonic symmetry-protected topological (BSPT) phases of matter for G_b-symmetric and local Hamiltonians are classified by $H^2(G_b, U(1)_{\mathfrak{c}})$. The group composition law induced by stacking BSPT phases of G_b-symmetric and local Hamiltonians is that of the second cohomology group $H^2(G_b, U(1)_{\mathfrak{c}})$.

In particular, if G_b accomodates the center $\mathbb{Z}_2$ such that $G = G_b/\mathbb{Z}_2$ is isomorphic to $G_f/\mathbb{Z}_2^{\mathrm{F}}$, the bosonic stacking rules are fixed by the composition law of $H^2(G_b, U(1)_{\mathfrak{c}})$, whereas the fermionic stacking rules with $[\mu]=0$ are fixed by (6.246a). These stacking rules are different even though the number of topologically distinct bosonic symmetry-protected topological and fermionic symmetry-protected topological phases of matter are equal and given by the cardinality of $H^2(G_b, U(1)_{\mathfrak{c}})$.

In fact, we have shown with Eq. (5.155a) that the spin-1/2 cluster $c=2m$-even chains realize the bosonic invertible topological phases of matter with

$$H^2(G_b, U(1)_{\mathfrak{c}}) \equiv H^2(\mathbb{Z}_2^{T'} \times \mathbb{Z}_2, U(1)_{\mathfrak{c}}) = \mathbb{Z}_2 \times \mathbb{Z}_2. \tag{6.251}$$

We have also shown with Eq. (5.155b) that the spin-1/2 cluster $c=2m+1$-odd chains, if treated within the mean-field approximation that implements the spontaneous symmetry breaking of time-reversal symmetry, realize the bosonic invertible topological phases of matter with

$$H^2(G_b, U(1)_{\mathfrak{c}}) \equiv H^2(\widetilde{\mathbb{Z}}_2^{\widetilde{T}'} \times \widetilde{\mathbb{Z}}_2, U(1)_{\mathfrak{c}}) = \mathbb{Z}_2 \times \mathbb{Z}_2. \tag{6.252}$$

The fact that spin-1/2 cluster $c = 2m + 1$-odd chains support a twofold degenerate ground state when periodic boundary conditions hold does not contradict the bosonic version of **Theorem** 6.52 as the symmetry group (5.59) is represented projectively owing to the time-reversal symmetry.

6.5.3 Ground-State Degeneracies

Distinct fermionic invertible topological phases are characterized by the projective character of the boundary representation $\widehat{U}_{\mathrm{B}}$ of the symmetry group G_f, whereby $\mathrm{B} = \mathrm{L}, \mathrm{R}$ stands for either the left or right disconnected component of the boundary. In turn this projective character is captured by the triplet of indices $([(\nu_{\mathrm{B}}, \rho_{\mathrm{B}})], [\mu_{\mathrm{B}}])$. What are the implications of this triplet being nontrivial, that is,

$$([(\nu_{\mathrm{B}}, \rho_{\mathrm{B}})], [\mu_{\mathrm{B}}]) \neq ([(0,0)], [0]) \tag{6.253}$$

for the spectral degeneracy of the boundary states?

The foremost consequence of the nontrivial indices $([(\nu_{\mathrm{B}}, \rho_{\mathrm{B}})], [\mu_{\mathrm{B}}])$ is the robustness of the boundary degeneracy that is protected by a combination of the symmetry group G_f being represented projectively and the existence of a nonlocal boundary Fock space, denoted $\mathfrak{F}_{\mathrm{LR}}$, whenever the left and right disconnected components of the boundary host odd numbers of Majorana degrees of freedom.

We are going to deduce from the indices $([(\nu_{\mathrm{B}}, \rho_{\mathrm{B}})], [\mu_{\mathrm{B}}])$ the presence of a quantum mechanical supersymmetry at the left and right boundaries, respectively. To this end, we consider the two cases $[\mu_{\mathrm{B}}] = 0$ and $[\mu_{\mathrm{B}}] = 1$ separately under the assumption that $[(\nu_{\mathrm{B}}, \rho_{\mathrm{B}})]$ is not the neutral element.

6.5.3.1 The Case of $[\mu_{\mathrm{B}}] = 0$

When $[\mu_{\mathrm{B}}] = 0$, there always are even numbers of Majorana degrees of freedom localized on the left, Λ_{L}, and right, Λ_{R}, disconnected components of the boundary, Λ_{bd}. In this case, the boundary Fock space $\mathfrak{F}_{\Lambda_{\mathrm{bd}}}$ spanned by the Majorana degrees of freedom supported on Λ_{bd} decomposes as

$$\mathfrak{F}_{\Lambda_{\mathrm{bd}}} = \mathfrak{F}_{\Lambda_{\mathrm{L}}} \otimes_{\mathfrak{g}} \mathfrak{F}_{\Lambda_{\mathrm{R}}}, \tag{6.254}$$

where $\otimes_{\mathfrak{g}}$ denotes a $\mathbb{Z}_2$-graded tensor product, while $\mathfrak{F}_{\Lambda_{\mathrm{L}}}$ and $\mathfrak{F}_{\Lambda_{\mathrm{R}}}$ are the Fock spaces spanned by the Majorana degrees of freedom localized on the left, Λ_{L}, and right, Λ_{R}, disconnected components of the boundary, Λ_{bd}, respectively. We denote with $\widehat{H}_{\mathrm{L}}$ and $\widehat{H}_{\mathrm{R}}$ the Hamiltonians that act on Fock spaces $\mathfrak{F}_{\Lambda_{\mathrm{L}}}$ and $\mathfrak{F}_{\Lambda_{\mathrm{R}}}$ and govern the dynamics of the local Majorana degrees of freedom localized at boundaries Λ_{L} and Λ_{R}, respectively. By assumption, the Hamiltonians $\widehat{H}_{\mathrm{L}}$ and $\widehat{H}_{\mathrm{R}}$ are invariant under the representations $\widehat{U}_{\mathrm{L}}$ and $\widehat{U}_{\mathrm{R}}$ of the given symmetry group G_f, respectively.

Since $[\mu_{\mathrm{B}}] = 0$, the only nontrivial fermionic invertible topological phases are those with nontrivial equivalence classes $[(\nu_{\mathrm{B}}, \rho_{\mathrm{B}})] \neq [(0,0)]$, that is the fermionic invertible topological phases. According to **Definitions** 6.56 and 6.57, the indices

$([\nu_{\mathrm{L}},\rho_{\mathrm{L}}],0)$ and $([\nu_{\mathrm{R}},\rho_{\mathrm{R}}],0)$ associated with the representations $\widehat{U}_{\mathrm{L}}$ and $\widehat{U}_{\mathrm{R}}$, respectively, satisfy

$$([(\nu_{\mathrm{L}},\rho_{\mathrm{L}})],0)\wedge([(\nu_{\mathrm{R}},\rho_{\mathrm{R}})],0)=([(0,0)],0) \tag{6.255}$$

under the stacking rule (6.246).

For the disconnected component $\mathrm{B}=\mathrm{L},\mathrm{R}$ of the boundary, the equivalence class $[(\nu_{\mathrm{B}},\rho_{\mathrm{B}})]$ characterizes the nontrivial projective nature of the boundary representation $\widehat{U}_{\mathrm{B}}$. Whenever $[(\nu_{\mathrm{B}},\rho_{\mathrm{B}})]\neq[(0,0)]$, it is guaranteed that there is no state in the Fock space $\mathfrak{F}_{\Lambda_{\mathrm{B}}}$ that is invariant under the action of $\widehat{U}_{\mathrm{B}}(g)$ for all $g\in\mathrm{G}_f$.[53] In other words, there is no state in the Fock space $\mathfrak{F}_{\Lambda_{\mathrm{B}}}$ that transforms as a singlet under the representation $\widehat{U}_{\mathrm{B}}$. Any eigenenergy of a G_f-symmetric boundary Hamiltonian $\widehat{H}_{\mathrm{B}}$ must be degenerate. The degeneracy is protected by the particular representation $\widehat{U}_{\mathrm{B}}$ of the symmetry group G_f. This degeneracy cannot be lifted without breaking the G_f symmetry. The minimal degeneracy that is protected by the G_f symmetry depends on the explicit structure of the group G_f and the equivalence class $[(\nu_{\mathrm{B}},\rho_{\mathrm{B}})]$ of the boundary representation $\widehat{U}_{\mathrm{B}}$.

Since for $[\mu]=0$, the boundary representations $\widehat{U}_{\mathrm{L}}$ and $\widehat{U}_{\mathrm{R}}$ act on two independent Fock spaces $\mathfrak{F}_{\Lambda_{\mathrm{L}}}$ and $\mathfrak{F}_{\Lambda_{\mathrm{R}}}$, the total protected ground-state degeneracy $\mathrm{GSD}^{[\mu]=0}_{\mathrm{bd}}$ when open boundary conditions are imposed is nothing but the product of the protected ground-state degeneracies $\mathrm{GSD}^{[\mu]=0}_{\mathrm{L}}$ and $\mathrm{GSD}^{[\mu]=0}_{\mathrm{R}}$ of the Hamiltonians $\widehat{H}_{\mathrm{L}}$ and $\widehat{H}_{\mathrm{R}}$, respectively, that is,

$$\mathrm{GSD}^{[\mu]=0}_{\mathrm{bd}}=\mathrm{GSD}^{[\mu]=0}_{\mathrm{L}}\times\mathrm{GSD}^{[\mu]=0}_{\mathrm{R}}. \tag{6.256}$$

When $[\mu]=0$, the 1-cochain $\rho_{\mathrm{B}}(g)=0,1$ encodes the commutation relation between the representations $\widehat{U}_{\mathrm{B}}(g)$ of group element $g\in\mathrm{G}_f$ and $\widehat{U}_{\mathrm{B}}(p)$ of fermion parity $p\in\mathrm{G}_f$. A nonzero second entry in the equivalence class $[(\nu_{\mathrm{B}},\rho_{\mathrm{B}})]$ implies that there exists at least one group element $g\in\mathrm{G}_f$ with $\rho_{\mathrm{B}}(g)=1$, that is the operator $\widehat{U}_{\mathrm{B}}(g)$ is of odd fermion parity. If this is so, the boundary Hamiltonian $\widehat{H}_{\mathrm{B}}$ must possess an emergent quantum mechanical supersymmetry (SUSY). The supercharges associated with the boundary SUSY are constructed as follows.[54] Assume without loss of generality that all energy eigenvalues ε_α of a boundary Hamiltonian $\widehat{H}_{\mathrm{B}}$ are shifted to the positive energies, that is, $\varepsilon_\alpha>0$. Also assume that there exists a group element $g\in\mathrm{G}_f$ with $\rho_{\mathrm{B}}(g)=1$. For any orthonormal eigenstate $|\psi_\alpha\rangle$ of $\widehat{H}_{\mathrm{B}}$ with energy ε_α, the state

$$|\psi'_\alpha\rangle:=\widehat{U}_{\mathrm{B}}(g)\,|\psi_\alpha\rangle, \tag{6.257a}$$

[53] An intuition for this claim is gained from considering the example of time-reversal symmetry with reversal of time squaring to the identity in the bulk while squaring to minus the identity (projectively) on the boundary. The ground states are then at least twofold degenerate and one may choose the two degenerate and orthogonal ground states to be even and odd under reversal of time, respectively. Reversal of time then acts as a ladder operator on their bonding and anti-bonding linear combinations.

[54] J. Behrends and B. Béri, Phys. Rev. Lett. **124**(23), 236804 (2020). [Behrends and Béri (2020)].

is also an orthonormal eigenstate of $\widehat{H}_{\mathrm{B}}$ with the same energy but opposite fermion parity. Since the fermion parities of $|\psi'_\alpha\rangle$ and $|\psi_\alpha\rangle$ are different, they are orthogonal. Two supercharges can then be defined as

$$\widehat{Q}_1 := \sum_\alpha \sqrt{\varepsilon_\alpha}\left[\left(\widehat{U}_{\mathrm{B}}(g)\,|\psi_\alpha\rangle\right)\langle\psi_\alpha| + |\psi_\alpha\rangle\left(\langle\psi_\alpha|\,\widehat{U}^\dagger_{\mathrm{B}}(g)\right)\right], \tag{6.257b}$$

$$\widehat{Q}_2 := \sum_\alpha \mathrm{i}\sqrt{\varepsilon_\alpha}\left[\left(\widehat{U}_{\mathrm{B}}(g)\,|\psi_\alpha\rangle\right)\langle\psi_\alpha| - |\psi_\alpha\rangle\left(\langle\psi_\alpha|\,\widehat{U}^\dagger_{\mathrm{B}}(g)\right)\right]. \tag{6.257c}$$

Operators $\widehat{Q}_1$ and $\widehat{Q}_2$ are Hermitian, carry odd fermion parity, and satisfy the defining properties

$$\left\{\widehat{Q}_i,\widehat{Q}_j\right\} = 2\widehat{H}_{\mathrm{B}}\,\delta_{i,j}, \qquad \left[\widehat{Q}_i,\widehat{H}_{\mathrm{B}}\right] = 0, \qquad i,j = 1,2, \tag{6.257d}$$

of $N = 2$ SUSY supercharges (recall Section 3.6.3). The precise number of supercharges on the boundary Λ_{B} depends on the pair $[(\nu_{\mathrm{B}},\rho_{\mathrm{B}})]$ that characterizes the number of symmetry operators $\widehat{U}_{\mathrm{B}}(g)$ that carry odd fermion parity and their mutual algebra.

6.5.3.2 The Case of $[\mu_{\mathrm{B}}] = 1$

When $[\mu_{\mathrm{B}}] = 1$, there are odd numbers of Majorana degrees of freedom localized on each disconnected component Λ_{L} and Λ_{R} of the boundary Λ_{bd}. In this case, the boundary Fock space $\mathfrak{F}_{\Lambda_{\mathrm{bd}}}$ spanned by Majorana degrees of freedom supported on Λ_{bd} decomposes as

$$\mathfrak{F}_{\Lambda_{\mathrm{bd}}} = \mathfrak{F}_{\Lambda_{\mathrm{L}}} \otimes_{\mathfrak{g}} \mathfrak{F}_{\mathrm{LR}} \otimes_{\mathfrak{g}} \mathfrak{F}_{\Lambda_{\mathrm{R}}}, \tag{6.258}$$

where $\otimes_{\mathfrak{g}}$ denotes a $\mathbb{Z}_2$-graded tensor product. The Fock spaces $\mathfrak{F}_{\Lambda_{\mathrm{B}}}$ with $\mathrm{B} = \mathrm{L},\mathrm{R}$ are spanned by all the Majorana operators localized on the disconnected components Λ_{B} except one. The two-dimensional Fock space $\mathfrak{F}_{\mathrm{LR}}$ is spanned by the two remaining Majorana operators with one localized on the left boundary Λ_{L} and the other localized on the right boundary Λ_{R} of the open chain. Correspondingly, the pair of fermionic creation and annihilation operators that span $\mathfrak{F}_{\mathrm{LR}}$ are nonlocal in the sense that they are formed by Majorana operators supported on opposite boundaries. One can define Hamiltonians $\widehat{H}_{\mathrm{L}}$ and $\widehat{H}_{\mathrm{R}}$ that are constructed out of Majorana operators localized at the boundaries Λ_{L} and Λ_{R}. If so, the Hamiltonians $\widehat{H}_{\mathrm{L}}$ and $\widehat{H}_{\mathrm{R}}$ act on the Fock spaces

$$\mathfrak{F}_{\Lambda_{\mathrm{L}}} \otimes_{\mathfrak{g}} \mathfrak{F}_{\mathrm{LR}}, \tag{6.259a}$$

and

$$\mathfrak{F}_{\Lambda_{\mathrm{R}}} \otimes_{\mathfrak{g}} \mathfrak{F}_{\mathrm{LR}}, \tag{6.259b}$$

respectively. By assumption, the Hamiltonians $\widehat{H}_{\mathrm{L}}$ and $\widehat{H}_{\mathrm{R}}$ are invariant under the representations $\widehat{U}_{\mathrm{L}}$ and $\widehat{U}_{\mathrm{R}}$ of a given symmetry group G_f, respectively.

On each boundary $\Lambda_{\rm B}$, there exists a local Hermitean and unitary operator $\widehat{Y}_{\rm B}$ that commutes with any other local operator supported on $\Lambda_{\rm B}$. The operator $\widehat{Y}_{\rm B}$ is the representation of the nontrivial central element of a Clifford algebra $C\ell_n$ with n an odd number of generators (recall **Table** 5.7). It therefore carries an odd fermion parity and anticommutes with the representation $\widehat{U}_{\rm B}(p)$ of fermion parity. It follows that $\widehat{Y}_{\rm B}$ must commute with $\widehat{H}_{\rm B}$. We label the simultaneous eigenstates of $\widehat{H}_{\rm B}$ and $\widehat{Y}_{\rm B}$ by $|\psi_{{\rm B},\alpha,\pm}\rangle$, that is,

$$\widehat{Y}_{\rm B}\,|\psi_{{\rm B},\alpha,\pm}\rangle = \pm|\psi_{{\rm B},\alpha,\pm}\rangle, \qquad \widehat{H}_{\rm B}\,|\psi_{{\rm B},\alpha,\pm}\rangle = \varepsilon_\alpha\,|\psi_{{\rm B},\alpha,\pm}\rangle, \tag{6.260}$$

where ε_α is the corresponding energy eigenvalue which we assume without loss of generality to be strictly positive. Hence, all eigenstates of $\widehat{H}_{\rm B}$ are at least twofold degenerate. Since $\widehat{Y}_{\rm B}$ carries odd fermion parity, the eigenstates $|\psi_{{\rm B},\alpha,\pm}\rangle$ do not have definite fermion parities. The simultaneous eigenstates of $\widehat{H}_{\rm B}$ and $\widehat{U}_{\rm B}(p)$ must be the bonding and anti-bonding linear combinations of $|\psi_{{\rm B},\alpha,+}\rangle$ and $|\psi_{{\rm B},\alpha,-}\rangle$ that are exchanged under the action of $\widehat{Y}_{\rm B}$. A twofold degeneracy of $\widehat{H}_{\rm B}$ when $[\mu]=1$ is thus due to the presence of the two-dimensional Fock space $\mathfrak{F}_{\rm LR}$. This twofold degeneracy is of SUSY nature with the associated supercharges

$$\widehat{Q}_1 := \sum_\alpha \sqrt{\varepsilon_\alpha}\,\big(|\psi_{\alpha,+}\rangle\langle\psi_{\alpha,+}| - |\psi_{\alpha,-}\rangle\langle\psi_{\alpha,-}|\big)\,, \tag{6.261a}$$

$$\widehat{Q}_2 := \sum_\alpha {\rm i}\sqrt{\varepsilon_\alpha}\,\big(|\psi_{\alpha,+}\rangle\langle\psi_{\alpha,-}| - |\psi_{\alpha,-}\rangle\langle\psi_{\alpha,+}|\big)\,. \tag{6.261b}$$

Operators $\widehat{Q}_1$ and $\widehat{Q}_2$ are Hermitian. They carry odd fermion parity since the operator $\widehat{U}_{\rm B}(p)$ exchanges the states $|\psi_{\alpha,\pm}\rangle$ with $|\psi_{\alpha,\mp}\rangle$. They satisfy the defining properties

$$\left\{\widehat{Q}_i,\widehat{Q}_j\right\} = 2\widehat{H}_{\rm B}\,\delta_{i,j}, \qquad \left[\widehat{Q}_i,\widehat{H}_{\rm B}\right] = 0, \qquad i,j=1,2, \tag{6.261c}$$

of $N=2$ SUSY supercharges (recall Section 3.6.3).

There may be other supercharges in addition to the ones defined in Eq. (6.261) due to the representation $\widehat{U}_{\rm B}$ of the group G_f. The precise number of these additional supercharges on the boundary $\Lambda_{\rm B}$ depends on the pair $[(\nu_{\rm B},\rho_{\rm B})]$ that characterizes the number of symmetry operators $\widehat{U}_{\rm B}(g)$ that carry odd fermion parity and their mutual algebra. They can be constructed in the same fashion as in Eq. (6.257).

According to **Definitions** 6.56 and 6.57, the indices $([(\nu_{\rm L},\rho_{\rm L})],1)$ and $([(\nu_{\rm R},\rho_{\rm R})],1)$ associated with the representations $\widehat{U}_{\rm L}$ and $\widehat{U}_{\rm R}$, respectively, satisfy

$$([(\nu_{\rm L},\rho_{\rm L})],1)\wedge([(\nu_{\rm R},\rho_{\rm R})],1) = ([(0,0)],0) \tag{6.262}$$

under the stacking rule (6.246).

For the disconnected component B $=$ L, R of the boundary, the equivalence class $[(\nu_{\rm B},\rho_{\rm B})]$ characterizes the nontrivial projective nature of the boundary representation $\widehat{U}_{\mathrm{B}}$. Whenever $[(\nu_{\mathrm{B}},\rho_{\mathrm{B}})] \neq [(0,0)]$, it is guaranteed that there is no state in the Fock space $\mathfrak{F}_{\Lambda_{\mathrm{B}}}$ that is invariant under the action of $\widehat{U}_{\mathrm{B}}(g)$ for all $g \in \mathrm{G}_f$. In other words, there is no state in the Fock space $\mathfrak{F}_{\Lambda_{\mathrm{B}}}$ that transforms as a singlet under the representation $\widehat{U}_{\mathrm{B}}$. Each eigenstate of a symmetric boundary Hamiltonian $\widehat{H}_{\mathrm{B}}$ must carry degeneracies in addition to the twofold degeneracy due to $[\mu_{\mathrm{B}}] = 1$. The degeneracy is protected by the particular representation $\widehat{U}_{\mathrm{B}}$ of the symmetry group G_f and cannot be lifted without breaking the G_f symmetry. The exact degeneracy protected by the representation depends on the explicit form of the group G_f, and the boundary representation $\widehat{U}_{\mathrm{B}}$ with the equivalence class $[(\nu_{\mathrm{B}},\rho_{\mathrm{B}})]$.

Since for $[\mu_{\mathrm{B}}] = 1$, the boundary representations $\widehat{U}_{\mathrm{L}}$ and $\widehat{U}_{\mathrm{R}}$ do not act on two decoupled Fock spaces. The total protected ground-state degeneracy $\mathrm{GSD}^{[\mu]=1}_{\mathrm{bd}}$ when open boundary conditions are imposed cannot be computed by taking the products of degeneracies associated with the Hamiltonians $\widehat{H}_{\mathrm{L}}$ and $\widehat{H}_{\mathrm{R}}$ separately. However, $\mathrm{GSD}^{[\mu]=1}_{\mathrm{bd}}$ can be computed by multiplying the "naive" protected ground-state degeneracies of the Hamiltonians at the two boundaries and modding out the twofold degeneracy due to $\mathfrak{F}_{\mathrm{LR}}$ shared by the two Hamiltonians, that is,

$$\mathrm{GSD}^{[\mu]=1}_{\mathrm{bd}} = \frac{1}{2} \times \mathrm{GSD}^{[\mu]=1}_{\mathrm{L}} \times \mathrm{GSD}^{[\mu]=1}_{\mathrm{R}}, \tag{6.263}$$

where $\mathrm{GSD}^{[\mu]=1}_{\mathrm{L}}$ and $\mathrm{GSD}^{[\mu]=1}_{\mathrm{R}}$ are the protected ground-state degeneracies of $\widehat{H}_{\mathrm{L}}$ and $\widehat{H}_{\mathrm{R}}$, respectively.

6.6 Exercises

6.1 We are going to prove as an exercise[55]

Theorem 6.59 (Marshall) *The antiferromagnetic quantum spin-1/2 Heisenberg ring defined by Eqs. (6.1) and (6.1c) has a ground state that is annihilated by the total spin operator*

$$\widehat{\boldsymbol{S}} := \sum_{j=1}^{N} \widehat{\boldsymbol{S}}_j \tag{6.264}$$

for any even integer $N = 2, 4, \cdots$.

Proof The proof is organized into three steps.

6.1.1 *Step 1:*

6.1.1.1 For any even integer $N = 2, 4, 6, 8, \cdots$, we start from the quantum spin-1/2 antiferromagnetic Heisenberg ring, that is the SU(2)-symmetric limit

$$\widehat{H} := 2J \sum_{j=1}^{N} \left(\widehat{S}^x_j\,\widehat{S}^x_{j+1} + \widehat{S}^y_j\,\widehat{S}^y_{j+1} + \widehat{S}^z_j\,\widehat{S}^z_{j+1} - \frac{1}{4} \right), \qquad J > 0. \tag{6.265a}$$

[55] W. Marshall and R. E. Peirls, Proc. R. Soc. Lond. A **232**(1188), 48–68 (1955). [Marshall (1955)].

Up to a rescaling of J and an additive shift by an overall constant, this is the SU(2) symmetric limit of (6.1) for $S = 1/2$. Define

$$\widehat{S}_j^{\pm} := \widehat{S}_j^x \pm \mathrm{i}\widehat{S}_j^y, \qquad j = 1, \cdots, N. \tag{6.265b}$$

Show that

$$\widehat{H} := \sum_{j=1}^{N} \left[2J \left(\widehat{S}_j^z \, \widehat{S}_{j+1}^z - \frac{1}{4} \right) + J \left(\widehat{S}_j^+ \, \widehat{S}_{j+1}^- + \widehat{S}_j^- \, \widehat{S}_{j+1}^+ \right) \right]. \tag{6.265c}$$

6.1.1.2 This Hamiltonian acts on the Hilbert space $\mathfrak{H}_N$ spanned by the basis

$$\begin{aligned} &|\sigma_1, \cdots, \sigma_j, \cdots, \sigma_N\rangle, \\ &\widehat{S}_j^z \, |\sigma_1, \cdots, \sigma_j, \cdots, \sigma_N\rangle = \frac{\sigma_j}{2} \, |\sigma_1, \cdots, \sigma_j, \cdots, \sigma_N\rangle, \\ &\sigma_j := \pm 1, \qquad j = 1, \cdots, N. \end{aligned} \tag{6.266a}$$

Verify that the dimensionality of $\mathfrak{H}_N$ is 2^N.

6.1.1.3 We use the Greek letter $\mu = 1, \cdots, 2^N$ to label this basis,

$$\mathfrak{H}_N = \mathrm{span} \left\{ |\Phi^{(\mu)}\rangle := |\sigma_1^{(\mu)}, \cdots, \sigma_j^{(\mu)}, \cdots, \sigma_N^{(\mu)}\rangle \mid \mu = 1, \cdots, 2^N \right\}. \tag{6.266b}$$

Verify that, for any $|\Phi^{(\mu)}\rangle \equiv |\sigma_1^{(\mu)}, \cdots, \sigma_j^{(\mu)}, \cdots, \sigma_N^{(\mu)}\rangle$,

$$\widehat{S}_j^z \, \widehat{S}_{j+1}^z \, |\Phi^{(\mu)}\rangle = \frac{\sigma_j^{(\mu)} \, \sigma_{j+1}^{(\mu)}}{4} \, |\Phi^{(\mu)}\rangle \tag{6.267a}$$

and

$$\begin{aligned} &\left(\widehat{S}_j^+ \, \widehat{S}_{j+1}^- + \widehat{S}_j^- \, \widehat{S}_{j+1}^+ \right) |\sigma_1^{(\mu)}, \cdots, \sigma_j^{(\mu)}, \sigma_{j+1}^{(\mu)}, \cdots, \sigma_N^{(\mu)}\rangle \\ &\quad = \delta_{\sigma_j^{(\mu)}, -\sigma_{j+1}^{(\mu)}} \, |\sigma_1^{(\mu)}, \cdots, \sigma_{j+1}^{(\mu)}, \sigma_j^{(\mu)}, \cdots, \sigma_N^{(\mu)}\rangle. \end{aligned} \tag{6.267b}$$

6.1.1.4 For any $|\Phi^{(\mu)}\rangle \equiv |\sigma_1^{(\mu)}, \cdots, \sigma_j^{(\mu)}, \cdots, \sigma_N^{(\mu)}\rangle$, let

$$M^{(\mu)} := \sum_{j=1}^{N} \sigma_j^{(\mu)} \tag{6.268a}$$

denote its magnetization (in units of 2) along the quantization axis,

$$Q_{\pm}^{(\mu)} := \sum_{j=1}^{N} \delta_{\sigma_j^{(\mu)}, \pm\sigma_{j+1}^{(\mu)}}, \tag{6.268b}$$

denote its number of consecutive parallel spins (+) and antiparallel spins (−), and

$$P^{(\mu)} := \sum_{j=1}^{N/2} \delta_{\sigma_{2j-1}^{(\mu)}, +1}, \tag{6.268c}$$

denote its number of up spins on the sublattice made of odd sites. Assume that

$$|\Psi\rangle := \sum_{\mu=1}^{2^N} c_\Psi^{(\mu)}\,|\Phi^{(\mu)}\rangle, \qquad c_\Psi^{(\mu)} \in \mathbb{C} \tag{6.269}$$

is a normalized eigenstate of Hamiltonian (6.265c) with the energy eigenvalue E_Ψ. Show that one may always choose all the expansion coefficients $c_\Psi^{(\mu)}$ to be real-valued. Hence, from now on,

$$c_\Psi^{(\mu)} \in \mathbb{R}, \qquad \mu = 1,\cdots,2^N. \tag{6.270}$$

6.1.1.5 Since

$$\begin{aligned}\widehat{H}\,|\Phi^{(\mu)}\rangle =&\, \frac{J}{2}\sum_{j=1}^{N}\left(\sigma_j^{(\mu)}\,\sigma_{j+1}^{(\mu)} - 1\right)|\sigma_1^{(\mu)},\cdots,\sigma_j^{(\mu)},\sigma_{j+1}^{(\mu)},\cdots,\sigma_N^{(\mu)}\rangle \\ &+ J\sum_{j=1}^{N}\delta_{\sigma_j^{(\mu)},-\sigma_{j+1}^{(\mu)}}\,|\sigma_1^{(\mu)},\cdots,\sigma_{j+1}^{(\mu)},\sigma_j^{(\mu)},\cdots,\sigma_N^{(\mu)}\rangle,\end{aligned} \tag{6.271}$$

verify that

$$\widehat{H}\,|\Psi\rangle = J\sum_{\mu=1}^{2^N} c_\Psi^{(\mu)}\left(-Q_-^{(\mu)}\,|\Phi^{(\mu)}\rangle + \sum_{\mu'_\mu}|\Phi^{(\mu'_\mu)}\rangle\right) \tag{6.272}$$

follows. Here, the sum over the index μ'_μ contains $Q_-^{(\mu)}$ terms and $|\Phi^{(\mu'_\mu)}\rangle$ is the image of $|\Phi^{(\mu)}\rangle$ obtained by interchanging two consecutive antiparallel spins in $|\Phi^{(\mu)}\rangle$.

6.1.1.6 Justify the step

$$\widehat{H}\,|\Psi\rangle = J\sum_{\mu=1}^{2^N}\sum_{\mu'_\mu}\left(c_\Psi^{(\mu'_\mu)} - c_\Psi^{(\mu)}\right)|\Phi^{(\mu)}\rangle. \tag{6.273}$$

6.1.1.7 Show that the energy eigenvalue E_Ψ of $|\Psi\rangle$ can be written as

$$E_\Psi = J\sum_{\mu=1}^{2^N}\sum_{\mu'_\mu} c_\Psi^{(\mu)}\left(c_\Psi^{(\mu'_\mu)} - c_\Psi^{(\mu)}\right). \tag{6.274}$$

6.1.2 *Step 2:*

Lemma 6.60 (Peierls) *For any energy eigenstate $|\Psi\rangle$, the choice (the Marshall sign)*

$$c_\Psi^{(\mu)} = (-1)^{P^{(\mu)}}\,a_\Psi^{(\mu)}, \qquad a_\Psi^{(\mu)} \geq 0, \qquad \mu = 1,\cdots,2^N \tag{6.275}$$

insures that E_Ψ in Eq. (6.274) is minimized, that is, that $|\Psi\rangle$ is a ground state.

Proof We prove this lemma by contradiction. Assume that not all $a^{(\mu)}_{\Psi}$ are real-valued and positive even though $|\Psi\rangle$ is a ground state. If so, show that [*Hint:* Show that $P^{(\mu'_{\mu})} = P^{(\mu)} \pm 1$] Eq. (6.274) becomes

$$E_{\Psi} = -J\sum_{\mu=1}^{2^N}\sum_{\mu'_{\mu}} a^{(\mu)*}_{\Psi}\left(a^{(\mu'_{\mu})}_{\Psi} + a^{(\mu)}_{\Psi}\right) \tag{6.276}$$

from which follows the strict lower bound

$$E_{\Psi} > -J\sum_{\mu=1}^{2^N}\sum_{\mu'_{\mu}} |a^{(\mu)}_{\Psi}|\left(|a^{(\mu'_{\mu})}_{\Psi}| + |a^{(\mu)}_{\Psi}|\right). \tag{6.277}$$

Hence, the choice

$$c^{(\mu)}_{\Psi} = (-1)^{P^{(\mu)}}\,|a^{(\mu)}_{\Psi}|, \qquad \mu = 1,\cdots,2^N \tag{6.278}$$

lowers the ground-state energy. This contradiction proves **Lemma** 6.60. □

6.1.3 *Step 3:*

Lemma 6.61 (Marshall) *The normalized ground state (6.269) whose expansion coefficients obey Eq. (6.275) is annihilated by the total spin operator (6.264).*

Proof We prove this lemma by contradiction.

6.1.3.1 Show that

$$\left[\widehat{S}^z, \widehat{H}\right] = 0. \tag{6.279}$$

All eigenstates of $\widehat{H}$ can thus be simultaneously diagonalized with $\widehat{S}^z$. We demand that the normalized ground state (6.269) has $M^{(\mu)} = 0$. This means that $c^{(\mu)}_{\Psi} = 0$ whenever $M^{(\mu)} \neq 0$ in the expansion (6.269).

6.1.3.2 Define the state

$$|\Psi'\rangle := S^+\,|\Psi\rangle, \qquad S^+ := S^x + \mathrm{i}S^y \equiv \sum_{j=1}^{N}\left(S^x_j + \mathrm{i}S^y_j\right). \tag{6.280}$$

6.1.3.3 Show that either $|\Psi'\rangle = 0$ if $\widehat{\boldsymbol{S}}^2\,|\Psi\rangle = 0$ or it is a ground state that is degenerate with $|\Psi\,\rangle$ and has the eigenvalue 1 with respect to $\widehat{S}^z$. We are going to assume that $|\Psi'\rangle \neq 0$, in which case it is a ground state that is degenerate with $|\Psi\,\rangle$ with the eigenvalue 1 of $\widehat{S}^z$.

6.1.3.4 Show that one can write

$$|\Psi'\rangle = \sum_{\mu'} c^{(\mu')}_{\Psi'}\,|\Phi^{(\mu')}\rangle, \qquad c^{(\mu')}_{\Psi'} := \sum_{\mu_{\mu'}} c^{(\mu_{\mu'})}_{\Psi}, \tag{6.281}$$

where the sum over μ' runs over all states $|\Phi^{(\mu')}\rangle$ with $M^{(\mu')} = 2$ owing to the normalization chosen in the definition (6.268a) and the coefficient

$c_{\Psi}^{(\mu_{\mu'})}$ corresponds to the state $|\Phi^{(\mu_{\mu'})}\rangle$ obtained from the state $|\Phi^{(\mu')}\rangle$ by replacing $\sigma_j^{(\mu')}=+1$ with $\sigma_j^{(\mu_{\mu'})}=-1$ for some $j=1,\cdots,N$.

6.1.3.5 Show that the sum over μ' ranges over $\binom{N}{(N/2)+1}$ terms.

6.1.3.6 Show that the sum over $\mu_{\mu'}$ ranges over $(N/2)+1$ terms.

6.1.3.7 Define the translation operator $\widehat{T}$ by its action

$$\begin{aligned}\widehat{T}\,|\Phi^{(\mu)}\rangle &:= |\sigma_N^{(\mu)},\cdots,\sigma_{j-1}^{(\mu)},\cdots,\sigma_1^{(\mu)}\rangle,\\ |\Phi^{(\mu)}\rangle &:= |\sigma_1^{(\mu)},\cdots,\sigma_j^{(\mu)},\cdots,\sigma_N^{(\mu)}\rangle.\end{aligned} \tag{6.282}$$

Show that, if $|\Phi^{(\mu')}\rangle := |\sigma_1^{(\mu')},\cdots,\sigma_j^{(\mu')},\cdots,\sigma_N^{(\mu')}\rangle$ has all its up spins located at

$$\sigma_{j_1}^{(\mu')}=\sigma_{j_2}^{(\mu')}=\cdots=\sigma_{j_{(N/2)+1}}^{(\mu')}=+1, \tag{6.283}$$

then $\widehat{T}\,|\Phi^{(\mu')}\rangle := |\sigma_N^{(\mu')},\cdots,\sigma_{j-1}^{(\mu')},\cdots,\sigma_1^{(\mu')}\rangle$ has all its up spins located at

$$\sigma_{j_1+1}^{(\mu')}=\sigma_{j_2+1}^{(\mu')}=\cdots=\sigma_{j_{(N/2)+1}+1}^{(\mu')}=+1. \tag{6.284}$$

Instead of the label μ in $c_{\Psi}^{(\mu)}$ we shall use as label the positions of all the up spins. With this convention in mind and given Eqs. (6.283) and (6.284), we may write

$$\begin{aligned}c_{\Psi'}^{(\mu')} =& c_{\Psi}(j_2,j_3,\cdots,j_{N/2+1})\\ &+c_{\Psi}(j_1,j_3,\cdots,j_{N/2+1})+\cdots+c_{\Psi}(j_1,j_2,\cdots,j_{N/2})\end{aligned} \tag{6.285}$$

for the sum over $\mu_{\mu'}$ corresponding to the state $|\Phi^{(\mu')}\rangle$ and

$$\begin{aligned}c_{\Psi'}^{T\,(\mu')} =& c_{\Psi}(j_2+1,j_3+1,\cdots,j_{(N/2)+1}+1)\\ &+c_{\Psi}(j_1+1,j_3+1,\cdots,j_{(N/2)+1}+1)\\ &+\cdots+c_{\Psi}(j_1+1,j_2+1,\cdots,j_{N/2}+1)\end{aligned} \tag{6.286}$$

for the sum over $\mu_{\mu'}$ corresponding to the state $\widehat{T}\,|\Phi^{(\mu')}\rangle$.

6.1.3.8 Explain why

$$\left|c_{\Psi}(j_2,j_3,\cdots,j_{(N/2)+1})\right| = \left|c_{\Psi}(j_2+1,j_3+1,\cdots,j_{(N/2)+1}+1)\right|. \tag{6.287}$$

6.1.3.9 Prove that the number $P^{(\mu)}$ of up spins on the sublattice made of odd sites for $M^{(\mu)}$ fixed transforms under translation by one lattice spacing as

$$P^{(\mu)}\mapsto\frac{N}{2}-P^{(\mu)}\pm\frac{M^{(\mu)}}{2}. \tag{6.288}$$

6.1.3.10 Deduce using Eq. (6.275) that

$$c_{\Psi'}^{(\mu')}\equiv\sum_{\mu_{\mu'}}c_{\Psi}^{(\mu_{\mu'})}=+(-1)^{N/2}\sum_{\mu_{\mu'}}c_{\Psi}^{T(\mu_{\mu'})}\equiv+(-1)^{N/2}c_{\Psi'}^{T(\mu')} \tag{6.289}$$

must hold if we interpret these coefficients as originating from the ground state $|\Psi\rangle$, while

$$c_{\Psi'}^{(\mu')} \equiv \sum_{\mu_{\mu'}} c_{\Psi}^{(\mu_{\mu'})} = -(-1)^{N/2} \sum_{\mu_{\mu'}} c_{\Psi}^{T(\mu_{\mu'})} \equiv -(-1)^{N/2} c_{\Psi'}^{T(\mu')} \tag{6.290}$$

must hold if we interpret these coefficients as those of the ground state $|\Psi'\rangle$. It follows that

$$\sum_{\mu_{\mu'}} c_{\Psi}^{(\mu_{\mu'})} = \sum_{\mu_{\mu'}} c_{\Psi}^{T(\mu_{\mu'})} = 0 \tag{6.291}$$

so that $|\Psi'\rangle = 0$ in contradiction with our initial assumption. The **Lemma** 6.61 follows. □

A ground state was constructed explicitly with **Lemma** 6.60. This ground state was shown to be annihilated by the total spin operator (6.264) with **Lemma** 6.61. □

Convince yourself that the proof never used the dimensionality one of space. Only the symmetries under reversal of time, translation symmetry, and the bipartiteness of the lattice were used. Marshall also showed that the proof extends to any $S = 1/2, 1, 3/2, \cdots$.

6.2 We are going to prove as an exercise the following theorem:[2]

Theorem 6.62 (Lieb-Schultz-Mattis I) *The ground state of the antiferromagnetic quantum spin-1/2 Heisenberg ring defined by Eqs. (6.1) and (6.1c) is non-degenerate and annihilated by*

$$\widehat{\boldsymbol{S}}^2, \qquad \widehat{\boldsymbol{S}} := \sum_{j=1}^{N} \widehat{\boldsymbol{S}}_j \tag{6.292}$$

for any even integer N.

Proof The proof is organized into six steps.

6.2.1 *Step 1:* For any even integer $N = 2, 4, 6, 8, \cdots$, we start from the quantum spin-1/2 antiferromagnetic Heisenberg ring, that is the SU(2)-symmetric limit

$$\widehat{H} := J \sum_{j=1}^{N} \left(\widehat{S}_j^x \, \widehat{S}_{j+1}^x + \widehat{S}_j^y \, \widehat{S}_{j+1}^y + \widehat{S}_j^z \, \widehat{S}_{j+1}^z \right), \quad J > 0, \tag{6.293a}$$

of (6.1) for $S = 1/2$. Define

$$\widehat{S}_j^{\pm} := \widehat{S}_j^x \pm \mathrm{i} \widehat{S}_j^y, \qquad j = 1, \cdots, N. \tag{6.293b}$$

Show that

$$\widehat{H} := J \sum_{j=1}^{N} \left[\widehat{S}_j^z \, \widehat{S}_{j+1}^z + \frac{1}{2} \left(\widehat{S}_j^+ \, \widehat{S}_{j+1}^- + \widehat{S}_j^- \, \widehat{S}_{j+1}^+ \right) \right]. \tag{6.293c}$$

Construct explicitly the unitary transformation

$$\widehat{U}^z_\pi := \prod_{j=1}^{N} \widehat{U}^z_{\pi,j} \tag{6.294}$$

acting on the algebra spanned by the spin operators that rotates every other on-site spin operator by π around the z-axis in spin space. Show that

$$\begin{aligned}\widehat{H}' &:= \widehat{U}^z_\pi\,\widehat{H}\,\widehat{U}^{z\dagger}_\pi \\ &= J\sum_{j=1}^{N}\left[\widehat{S}^z_j\,\widehat{S}^z_{j+1} - \frac{1}{2}\left(\widehat{S}^+_j\,\widehat{S}^-_{j+1} + \widehat{S}^-_j\,\widehat{S}^+_{j+1}\right)\right].\end{aligned} \tag{6.295}$$

This Hamiltonian acts on the Hilbert space defined by the basis (6.266).

6.2.2 *Step 2:*

6.2.2.1 Show that

$$\left[\widehat{S}^z, \widehat{H}\right] = \left[\widehat{S}^z, \widehat{H}'\right] = 0. \tag{6.296}$$

All eigenstates of $\widehat{H}$ and $\widehat{H}'$ can thus be simultaneously diagonalized with $\widehat{S}^z$.

6.2.2.2 Give the condition on the $\left\{\sigma^{(\mu)}_j,\ \mu\in\mathcal{S}\right\}$ entering each orthonormal state of the subset

$$\left\{|\Phi^{(\mu)}\rangle \equiv |\sigma^{(\mu)}_1,\cdots,\sigma^{(\mu)}_j,\cdots,\sigma^{(\mu)}_N\rangle,\ \mu\in\mathcal{S}\right\}, \qquad \mathcal{S}\subset\{1,\cdots,2^N\} \tag{6.297}$$

of the basis (6.266) that is annihilated by $\widehat{S}^z$.

6.2.2.3 What is the dimensionality $|\mathcal{S}|$ of the vector space

$$\mathfrak{H}^{S^z=0}_N := \operatorname{span}\{|\Phi^{(\mu)}\rangle, \qquad \mu\in\mathcal{S}\} \tag{6.298}$$

spanned by this subset of the basis (6.266)?

6.2.2.4 Let $|\Psi\rangle$ be any normalizable eigenstate defined by

$$\widehat{H}'\,|\Psi\rangle = E_\Psi\,|\Psi\rangle, \qquad \widehat{S}^z\,|\Psi\rangle = 0. \tag{6.299}$$

Express the expansion coefficients $c^{(\mu)}_\Psi\in\mathbb{C}$ in

$$|\Psi\rangle = \sum_{\mu\in\mathcal{S}} c^{(\mu)}_\Psi\,|\Phi^{(\mu)}\rangle, \qquad \sum_{\mu\in\mathcal{S}}\left|c^{(\mu)}_\Psi\right|^2 = 1 \tag{6.300}$$

in terms of $|\Psi\rangle$ and $|\Phi^{(\mu)}\rangle$.

6.2.2.5 Verify that Eq. (6.267) implies that

$$J\sum_{j=1}^{N}\widehat{S}^z_j\,\widehat{S}^z_{j+1}\,|\Phi^{(\mu)}\rangle = E_\mu\,|\Phi^{(\mu)}\rangle, \qquad E_\mu := \frac{J}{4}\left(\sum_{j=1}^{N}\sigma^{(\mu)}_j\,\sigma^{(\mu)}_{j+1}\right), \tag{6.301}$$

and

$$\frac{J}{2}\sum_{j=1}^{N}\left(\widehat{S}_j^{+}\,\widehat{S}_{j+1}^{-}+\widehat{S}_j^{-}\,\widehat{S}_{j+1}^{+}\right)|\Phi^{(\mu)}\rangle=\frac{J}{2}\sum_{\mu'_\mu}|\Phi^{(\mu'_\mu)}\rangle, \tag{6.302}$$

where the sum over μ'_μ extends over all the basis states that span $\mathfrak{H}_N^{S^z=0}$ and are the image of $|\Phi^{(\mu)}\rangle$ by $\left(\widehat{S}_j^{+}\,\widehat{S}_{j+1}^{-}+\widehat{S}_j^{-}\,\widehat{S}_{j+1}^{+}\right)$ for some $j=1,\cdots,N$.

6.2.2.6 Verify that the eigenvalue problem (6.299) can be rewritten as the system of linear equations

$$\left(E_\Psi-E_\mu\right)c_\Psi^{(\mu)}=-\frac{J}{2}\sum_{\mu'_\mu}c_\Psi^{(\mu'_\mu)},\qquad \mu\in\mathcal{S}. \tag{6.303}$$

6.2.2.7 Explain why all the coefficients $c_\Psi^{(\mu)}$ in the expansion (6.300) can be chosen real, without loss of generality. We will make this choice from now on.

6.2.2.8 If so, verify that the expectation value of the Hamiltonian $\widehat{H}'$ in the normalized eigenstate $|\Psi\rangle$ with vanishing total spin along the quantization axis is

$$E_\Psi=\sum_{\mu\in\mathcal{S}}E_\mu\,c_\Psi^{(\mu)}\,c_\Psi^{(\mu)}-\frac{J}{2}\sum_{\mu\in\mathcal{S}}\sum_{\mu'_\mu}c_\Psi^{(\mu)}\,c_\Psi^{(\mu'_\mu)}. \tag{6.304}$$

6.2.3 *Step 3:* Show that, for any pair $\mu,\nu\in\mathcal{S}$, there exists a finite sequence $\{j_\iota,\ \iota=1,\cdots,n\}$ such that

$$|\Phi^{(\nu)}\rangle=\prod_{\iota=1}^{n}\left(\widehat{S}_{j_\iota}^{+}\,\widehat{S}_{j_\iota+1}^{-}+\widehat{S}_{j_\iota}^{-}\,\widehat{S}_{j_\iota+1}^{+}\right)|\Phi^{(\mu)}\rangle. \tag{6.305}$$

6.2.4 *Step 4:* We need to prove the following lemma.

Lemma 6.63 *If we select the eigenstate (6.299) to be a ground state of $\widehat{H}'$ with the ground-state energy E_0, then all the real-valued expansion coefficients $c_\Psi^{(\mu)}\in\mathbb{R}$ in Eq. (6.300) are nonvanishing.*

Proof We prove **Lemma** 6.63 by contradiction.

6.2.4.1 Assume that

$$\mu=\mu_1,\cdots,\mu_p,\cdots,\mu_r\Rightarrow c_\Psi^{(\mu)}=0. \tag{6.306}$$

6.2.4.2 Verify that

$$0=\sum_{\mu'_\mu}c_\Psi^{(\mu'_\mu)},\qquad \mu=\mu_1,\cdots,\mu_p,\cdots,\mu_r. \tag{6.307}$$

6.2.4.3 Explain with the help of Step 3 why not all $c_\Psi^{(\mu'_\mu)}$ can vanish when

$$\mu=\mu_1,\cdots,\mu_p,\cdots,\mu_r. \tag{6.308}$$

6.2.4.4 Deduce that there must be nonvanishing $c_{\Psi}^{(\mu'_{\mu})}$ with $\mu=\mu_1,\cdots,\mu_p,\cdots,\mu_r$ of both signs.

6.2.4.5 Define the trial wave function

$$|\Psi'\rangle := \sum_{\mu\in\mathcal{S}} \left|c_{\Psi}^{(\mu)}\right| |\Phi^{(\mu)}\rangle, \qquad \sum_{\mu\in\mathcal{S}} \left(c_{\Psi}^{(\mu)}\right)^2 = 1. \tag{6.309}$$

On the one hand, explain why

$$\left|c_{\Psi}^{(\mu_p)}\right| = 0, \qquad 0 \neq \sum_{\mu'_{\mu_p}} \left|c_{\Psi}^{(\mu'_{\mu_p})}\right| \tag{6.310}$$

prevent $|\Psi'\rangle$ from being an eigenstate of $\widehat{H}'$. Explain why the lower bound

$$E'_0 \equiv \langle\Psi'|\widehat{H}'|\Psi'\rangle > E_0 \tag{6.311}$$

then follows.

6.2.4.6 On the other hand,

$$\begin{aligned} E_0 &= \sum_{\mu\in\mathcal{S}} E_{\mu}\, c_{\Psi}^{(\mu)}\, c_{\Psi}^{(\mu)} - \frac{J}{2} \sum_{\mu\in\mathcal{S}} \sum_{\mu'_{\mu}} c_{\Psi}^{(\mu)}\, c_{\Psi}^{(\mu'_{\mu})} \\ &\geq \sum_{\mu\in\mathcal{S}} E_{\mu}\, c_{\Psi}^{(\mu)}\, c_{\Psi}^{(\mu)} - \frac{J}{2} \sum_{\mu\in\mathcal{S}} \sum_{\mu'_{\mu}} \left|c_{\Psi}^{(\mu)}\right| \left|c_{\Psi}^{(\mu'_{\mu})}\right| \\ &= E'_0. \end{aligned} \tag{6.312}$$

The contradiction between the last two bounds proves **Lemma** 6.63. □

6.2.5 *Step 5:* We need to prove the following lemma.

Lemma 6.64 *If we select the eigenstate (6.299) to be a ground state of $\widehat{H}'$ with the ground-state energy E_0, then all the real-valued expansion coefficients $c_{\Psi}^{(\mu)} \in \mathbb{R}$ in Eq. (6.300) have the same sign.*

Proof If the eigenstate (6.299) is to be a ground state, all the terms $c_{\Psi}^{(\mu)}\, c_{\Psi}^{(\mu'_{\mu})}$ entering the bound (6.312) are nonvanishing by **Lemma** 6.63 and the equality must hold in the bound (6.312). This equality requires that all the terms $c_{\Psi}^{(\mu)}\, c_{\Psi}^{(\mu'_{\mu})}$ entering the bound (6.312) are strictly positive. This is to say that all the coefficients $c_{\Psi}^{(\mu)}$ in the expansion (6.300) of the ground state (6.299) pairwise connected by

$$\sum_{j=1}^{N} \left(\widehat{S}_j^{+}\, \widehat{S}_{j+1}^{-} + \widehat{S}_j^{-}\, \widehat{S}_{j+1}^{+}\right) \tag{6.313}$$

have the same sign. By the ergodicity of step 3, this implies that $c_{\Psi}^{(\mu)}$ have the same sign for all $\mu \in \mathcal{S}$. □

6.2.6 *Step 6:* Two ground states with vanishing total spins must each have all their expansion coefficients $c^{(\mu)}_{\Psi}$ strictly positive. Hence, they cannot be orthogonal. A ground state with vanishing total spin must then be nondegenerate. To complete the proof of **Theorem** 6.62, it suffices to appeal to the **Theorem** 6.61 of Marshall, who showed the existence of a ground state annihilated by

$$\widehat{\boldsymbol{S}}^2, \qquad \widehat{\boldsymbol{S}} := \sum_{j=1}^{N} \widehat{\boldsymbol{S}}_j. \tag{6.314}$$

□

6.3 We are going to prove the following theorem:[2]

Theorem 6.65 (Lieb-Schultz-Mattis II) *The antiferromagnetic quantum spin-1/2 Heisenberg ring defined by Eqs. (6.1) and (6.1c) and obeying periodic boundary conditions supports an excited eigenstate with an energy of order $1/N$ above the nondegenerate ground state for any even integer N.*

Proof The proof is organized into four steps.

6.3.1 *Step 1:* According to **Theorem** 6.62, the ground state is a total spin singlet and is nondegenerate. The normalized ground state is denoted $|\Psi_0\rangle$. Its eigenenergy is denoted E_0. We choose the units of energy such that $J = 1$.

6.3.2 *Step 2:* Define the trial state

$$|\Psi_k\rangle := \exp\left(\mathrm{i}k\sum_{j=1}^{N} j\,\widehat{S}^z_j\right)|\Psi_0\rangle \equiv \widehat{O}^k\,|\Psi_0\rangle, \qquad k := \frac{2\pi}{N}\,m \tag{6.315}$$

for any $m = 0, 1, \cdots, N-1$. Verify that

$$\widehat{O} := \exp\left(\mathrm{i}\sum_{j=1}^{N} j\,\widehat{S}^z_j\right) \tag{6.316}$$

is a unitary operator.

6.3.3 *Step 3:* We need to prove

Lemma 6.66

$$\langle\Psi_k|\,\widehat{H}\,|\Psi_k\rangle \leq E_0 + \frac{2\pi^2}{N} + \mathcal{O}(N^{-3}). \tag{6.317}$$

Proof By definition

$$\langle\Psi_k|\,\widehat{H}\,|\Psi_k\rangle = \langle\Psi_0|\,\widehat{O}^{k\,\dagger}\,\widehat{H}\,\widehat{O}^k\,|\Psi_0\rangle. \tag{6.318}$$

6.3.3.1 Verify that

$$\widehat{O}^{k\,\dagger}\,\widehat{S}^x_j\,\widehat{O}^k = +\cos(k\,j)\,\widehat{S}^x_j + \sin(k\,j)\,\widehat{S}^y_j, \tag{6.319}$$

$$\widehat{O}^{k\,\dagger}\,\widehat{S}^y_j\,\widehat{O}^k = -\sin(k\,j)\,\widehat{S}^x_j + \cos(k\,j)\,\widehat{S}^y_j, \tag{6.320}$$

$$\widehat{O}^{k\,\dagger}\,\widehat{S}^z_j\,\widehat{O}^k = \widehat{S}^z_j \tag{6.321}$$

for any $j = 1, \cdots, N$.

6.3.3.2 Deduce that

$$\widehat{O}^{k\,\dagger}\,\widehat{H}\,\widehat{O}^{k} = \cos k \sum_{j=1}^{N}\left(\widehat{S}^{x}_{j}\,\widehat{S}^{x}_{j+1}+\widehat{S}^{y}_{j}\,\widehat{S}^{y}_{j+1}\right)+\sin k \sum_{j=1}^{N}\left(\widehat{S}^{x}_{j}\,\widehat{S}^{y}_{j+1}-\widehat{S}^{y}_{j}\,\widehat{S}^{x}_{j+1}\right) \tag{6.322}$$

from which it follows that

$$\widehat{O}^{k\,\dagger}\,\widehat{H}\,\widehat{O}^{k}+\widehat{O}^{(-k)\,\dagger}\,\widehat{H}\,\widehat{O}^{(-k)} = 2\cos k \sum_{j=1}^{N}\left(\widehat{S}^{x}_{j}\,\widehat{S}^{x}_{j+1}+\widehat{S}^{y}_{j}\,\widehat{S}^{y}_{j+1}\right). \tag{6.323}$$

6.3.3.3 Show that

$$\begin{aligned} 0 \leq & \langle\Psi_{k}|\left(\widehat{H}-E_{0}\right)|\Psi_{k}\rangle \\ = & \langle\Psi_{0}|\left(\widehat{O}^{k\,\dagger}\,\widehat{H}\,\widehat{O}^{k}-\widehat{H}\right)|\Psi_{0}\rangle \\ \leq & -2\,(1-\cos k)\sum_{j=1}^{N}\langle\Psi_{0}|\left(\widehat{S}^{x}_{j}\,\widehat{S}^{x}_{j+1}+\widehat{S}^{y}_{j}\,\widehat{S}^{y}_{j+1}\right)|\Psi_{0}\rangle. \end{aligned} \tag{6.324}$$

6.3.3.4 Choose $k=2\pi/N$. Prove that the bound

$$\langle\Psi_{k=2\pi/N}|\left(\widehat{H}-E_{0}\right)|\Psi_{k=2\pi/N}\rangle \leq \left[\left(\frac{2\pi}{N}\right)^{2}+\mathcal{O}(N^{-4})\right]\frac{N}{2} \tag{6.325}$$

holds. The Lemma follows.

6.3.3.5 We turn our attention to the second term on the right-hand side of Eq. (6.322). Verify the remarkable identity

$$\left[\sum_{j'=1}^{N} j'\,\widehat{S}^{z}_{j'},\widehat{H}\right] = \mathrm{i}\sum_{j=1}^{N}\left(\widehat{S}^{x}_{j}\,\widehat{S}^{y}_{j+1}-\widehat{S}^{y}_{j}\,\widehat{S}^{x}_{j+1}\right)-\mathrm{i}N\left(\widehat{S}^{x}_{N}\,\widehat{S}^{y}_{1}-\widehat{S}^{y}_{N}\,\widehat{S}^{x}_{1}\right). \tag{6.326}$$

On the right-hand side, the contribution $j=N$ in the sum over j together with the parenthesis multiplied by N arise from the periodic boundary conditions. To interpret the local term

$$\hat{j}^{z}_{j\to j+1} := \widehat{S}^{x}_{j}\,\widehat{S}^{y}_{j+1}-\widehat{S}^{y}_{j}\,\widehat{S}^{x}_{j+1} \tag{6.327}$$

on the right-hand side as a conserved current-density operator, show that it obeys the discretized continuity equation

$$\frac{\partial}{\partial t}\widehat{S}^{z}_{j} \equiv -\mathrm{i}\left[\widehat{S}^{z}_{j},\widehat{H}\right] = -\hat{j}^{z}_{j\to j+1}+\hat{j}^{z}_{j-1\to j}. \tag{6.328}$$

Define the current density in the ground state to be

$$j^{z}_{0} := \frac{1}{N}\sum_{j=1}^{N}\langle\Psi_{0}|\,\hat{j}^{z}_{j\to j+1}\,|\Psi_{0}\rangle. \tag{6.329}$$

With the help of the Lemma, show the inequality

$$|j_0^z| \leq \frac{\pi}{2N}\left[1+\mathcal{O}(N^{-2})\right], \tag{6.330}$$

that is the current density in the ground state vanishes in the limit $N \to \infty$. □

6.3.4 *Step 4:* We need to prove

Lemma 6.67 *The trial state $|\Psi_k\rangle$ with $k = 2\pi\, m/N$ and m any odd integer is orthogonal to $|\Psi_0\rangle$.*

Proof Define the translation operator $\widehat{T}$ by its action

$$\widehat{T}\,\widehat{\boldsymbol{S}}_j\,\widehat{T}^\dagger = \widehat{\boldsymbol{S}}_{j+1} \tag{6.331}$$

for any $j = 1, \cdots, N$. Explain why

$$\widehat{T}\,|\Psi_0\rangle = e^{\mathrm{i}\alpha}\,|\,\Psi_0\rangle \tag{6.332}$$

holds for some $\alpha \in \mathbb{R}$ and prove the identity

$$\widehat{T}\,\widehat{O}^k\,\widehat{T}^\dagger = \widehat{O}^k \times e^{\mathrm{i}k\,N\,\widehat{S}_1^z} \times e^{-\mathrm{i}k\,\widehat{S}^z} \tag{6.333}$$

valid for any real-valued k. Deduce that

$$\langle\Psi_0\,|\,\Psi_k\rangle = 0 \tag{6.334}$$

for any $k = 2\pi\, m/N$ with m an odd integer. □

The theorem follows by combining **Lemmas** 6.66 and 6.67. □

6.4 The Marshall and Lieb-Schultz-Mattis theorems all assume that the ring was made of an even number N of sites. What changes when N is odd?

6.5 Consider the state $\left|\Psi^{\Lambda}_{\mathrm{GS},\,\kappa}\right\rangle$ defined in Eq. (6.29i). Define the fermionic coarse-grained density operator

$$\hat{\rho}_\ell := \frac{1}{\ell}\sum_{i=1}^{\ell}\hat{n}_i \tag{6.335}$$

[recall Eqs. (6.21) and (6.29o)]. With ν the dimensionless filling fraction defined in Eq. (6.29d) define [recall Eqs. (6.22) and (6.29p)]

$$\Delta\hat{\rho}_\ell := \hat{\rho}_\ell - \nu. \tag{6.336}$$

Verify the identity (with $m=\min\{\ell-1, N_{\mathfrak{f}}-1\}$)

$$\begin{aligned}
\langle\Psi^{\Lambda}_{\mathrm{GS},\kappa}|\,(\Delta\hat{\rho}_\ell)^2\,|\Psi^{\Lambda}_{\mathrm{GS},\kappa}\rangle
&=\frac{\langle\Psi^{\Lambda}_{\mathrm{GS},\kappa}|\left(\nu+\sum\limits_{n=1}^{\ell-1}\hat{n}_1\,\hat{n}_{1+n}\right)|\Psi^{\Lambda}_{\mathrm{GS},\kappa}\rangle}{\ell}-\nu^2\\
&=\frac{\nu+\sum\limits_{n=1}^{m}\left(\nu-\frac{n}{|\Lambda|}\right)}{\ell}-\nu^2\\
&=\frac{\nu\,(m+1)-\frac{m(m+1)}{2\,|\Lambda|}}{\ell}-\nu^2\\
&=\frac{\nu(m+1)}{\ell}-\nu^2-\frac{m(m+1)}{2\,\ell\,|\Lambda|}.
\end{aligned}\tag{6.337}$$

6.6 Verify that the **Properties** 6.31, 6.32, and 6.33 hold.

6.7 Verify that Eq. (6.141b) is a generalization of Eq. (6.132c) if one identifies the exponential of ϕ in Eq. (6.141b) with γ in Eq. (6.132c) [up to the homomorphism (6.139b)].

6.8 Verify that Eq. (6.143b) is a generalization of Eq. (6.136) if one identifies the exponential of ξ in Eq. (6.143b) with κ in Eq. (6.136) [up to the homomorphism (6.139b)].

6.9 Verify Eq. (6.152).

6.10 To prove **Property** 6.46, use the action

$$\begin{aligned}
&|\psi_{1,\sigma_1}\rangle\otimes_{\mathfrak{g}}|\psi_{2,\sigma_2}\rangle\otimes_{\mathfrak{g}}\cdots\otimes_{\mathfrak{g}}|\psi_{N-1,\sigma_{N-1}}\rangle\otimes_{\mathfrak{g}}|\psi_{N,\sigma_N}\rangle\\
&\qquad\mapsto|\psi_{2,\sigma_1}\rangle\otimes_{\mathfrak{g}}|\psi_{3,\sigma_2}\rangle\otimes_{\mathfrak{g}}\cdots\otimes_{\mathfrak{g}}|\psi_{N,\sigma_{N-1}}\rangle\otimes_{\mathfrak{g}}|\psi_{1,\sigma_N}\rangle
\end{aligned}\tag{6.338}$$

of the translation by one lattice spacing to the right. Reorder the kets with the lattice site 1 to the left of the lattice site 2, and so on. (In doing so, multiplicative factors of -1 must be accounted for.) Change the name of the internal indices σ_j that are contracted with the matrices in the matrix product coefficients. Reorder the matrices in the matrix product coefficients with the internal index σ_1 to the left of the internal index σ_2, and so on. (In doing so, multiplicative factors of -1 must be accounted for.)

6.11 Let $\mathrm{Mat}(n,\mathbb{C})$ denote the set of $n\times n$ complex-valued matrices. Assume the additive decomposition

$$n=n_1+n_2,\qquad n_1=1,\cdots,n-1.\tag{6.339a}$$

For any $A\in\mathrm{Mat}(n,\mathbb{C})$, write

$$A\equiv\begin{pmatrix}A_{11}&A_{12}\\A_{21}&A_{22}\end{pmatrix}\tag{6.339b}$$

with A_{ij} the corresponding rectangular complex-valued block matrices. Define the projectors

$$Q \equiv \begin{pmatrix} \mathbb{1}_{n_1} & 0 \\ 0 & 0 \end{pmatrix}, \qquad \mathbb{1}_n - Q \equiv \begin{pmatrix} 0 & 0 \\ 0 & \mathbb{1}_{n_2} \end{pmatrix}. \tag{6.339c}$$

6.11.1 Show that (reducibility of a matrix)

$$Q\,A = Q\,A\,Q \implies A = \begin{pmatrix} A_{11} & 0 \\ A_{21} & A_{22} \end{pmatrix}, \tag{6.340a}$$

while (decomposability of a matrix)

$$Q\,A = A\,Q \implies A = \begin{pmatrix} A_{11} & 0 \\ 0 & A_{22} \end{pmatrix}. \tag{6.340b}$$

6.11.2 Show that

$$\widetilde{A} := Q\,A\,Q + (\mathbb{1}_n - Q)\,A\,(\mathbb{1}_n - Q) = \begin{pmatrix} A_{11} & 0 \\ 0 & A_{22} \end{pmatrix}. \tag{6.341}$$

6.11.3 Show that

$$Q\,A = Q\,A\,Q, \qquad Q\,B = Q\,B\,Q \implies Q\,(A\,B) = Q\,(A\,B)\,Q. \tag{6.342}$$

Hence, any matrix from the set of all matrices obtained by adding or multiplying the $\mathcal{D}$ matrices $A^{(0)}_1, \cdots, A^{(0)}_{\mathcal{D}}$, each one obeying condition (6.340a), takes the same reducible form (6.340a).

6.11.4 Show that Eq. (6.168) holds.

6.12 We define translation-invariant BMPS and give an alternate definition of injective BMPS.

Definition 6.68 Given are $\mathcal{D}$ matrices $A_\sigma \in \mathrm{Mat}(M,\mathbb{C})$ with $\sigma = 1, \cdots, \mathcal{D}$, each of which is of dimension M and acts on the vector space

$$\mathbb{C}^M := \mathrm{span}\{|m\rangle \mid m = 1, \cdots, M\}. \tag{6.343}$$

They define the translation-invariant BMPS

$$|\mathrm{MPS}\rangle = \sum_{\sigma_1, \cdots, \sigma_L = 1}^{\mathcal{D}} \mathrm{tr}\left(A_{\sigma_1} \cdots A_{\sigma_L}\right) |\sigma_1, \cdots, \sigma_L\rangle. \tag{6.344}$$

Definition 6.69 The BMPS $|\mathrm{MPS}\rangle$ is said to be injective if there exists a strictly positive integer I^* such that, for any integer I larger than I^*, the anti-linear map

$$\begin{aligned} \Gamma_I \colon \mathrm{Mat}(M,\mathbb{C}) &\to \mathrm{span}\left\{|\sigma_1, \cdots, \sigma_I\rangle \mid \sigma_i = 1, \cdots, \mathcal{D}\right\}, \\ Y &\mapsto \Gamma_I(Y), \end{aligned} \tag{6.345a}$$

with

$$\Gamma_I(Y) := \sum_{\sigma_1, \cdots, \sigma_I = 1}^{\mathcal{D}} \mathrm{tr}\left(Y^\dagger A_{\sigma_1} \cdots A_{\sigma_I}\right) |\sigma_1, \cdots, \sigma_I\rangle \tag{6.345b}$$

is injective.

6.12.1 Formulate as alternate definition of an injective translation-invariant BMPS the bosonic counterparts to definitions 6.48 and 6.49.

6.12.2 Show that this alternate definition of an injective translation-invariant BMPS is equivalent to the **Definition** 6.69.

Hint I: Show that if

$$\operatorname{span}\left\{A_{\sigma_1}\cdots A_{\sigma_I}\mid \sigma_1,\cdots,\sigma_I=1,\cdots,\mathcal{D}\right\}=\operatorname{Mat}(M,\mathbb{C}), \tag{6.346}$$

then Γ_I is injective. To this end, make use of the linearity of Γ_I.

Hint II: Show that if

$$\operatorname{span}\left\{A_{\sigma_1}\cdots A_{\sigma_I}\mid \sigma_1,\cdots,\sigma_I=1,\cdots,\mathcal{D}\right\}\neq\operatorname{Mat}(M,\mathbb{C}), \tag{6.347}$$

then Γ_I is not injective. To this end, show that

$$\begin{aligned}\langle\cdot|\cdot\rangle:\operatorname{Mat}(M,\mathbb{C})\times\operatorname{Mat}(M,\mathbb{C})&\to\mathbb{C},\\ (X,Y)&\mapsto\langle X|Y\rangle:=\operatorname{tr}\left(X^{\dagger}Y\right),\end{aligned} \tag{6.348}$$

defines a scalar product on the vector space $\operatorname{Mat}(M,\mathbb{C})$ such that the M^2 $M\times M$ matrices $L_{m,m'}$, whose only nonvanishing matrix element is the number 1 on line $m=1,\cdots,M$ and column $m'=1,\cdots,M$, form an orthonormal basis. Show that the orthogonal complement in $\operatorname{Mat}(M,\mathbb{C})$ to the vector space

$$\operatorname{span}\left\{A_{\sigma_1}\cdots A_{\sigma_I}\mid \sigma_1,\cdots,\sigma_I=1,\cdots,\mathcal{D}\right\}\subset\operatorname{Mat}(M,\mathbb{C}) \tag{6.349}$$

is the ker of Γ_I.

6.13 Prove with the help of definition (6.175) Eq. (6.176).

7
Fractionalization in Quantum Wires

7.1 Introduction

It is an experimental fact that the bond-density instability of polyacetylene is the dominant one at half-filling. As long as the characteristic energy scale of electron-electron interactions is smaller than the dimerization gap, we may treat electron-electron interactions perturbatively and work to leading order in this approximation (the single-particle approximation), as we have done so far. At half-filling and when the phonons are treated as a static background within the Born-Oppenheimer approximation, polyacetylene can thus be considered as an example of a band insulator. The dimerization gap takes the form (2.123) in the weak coupling limit (2.115) of the electron-phonon interaction.

In strictly one-dimensional space, the spontaneous symmetry breaking of a continuous symmetry at zero temperature is prevented by quantum fluctuations. This is why the nearest-neighbor antiferromagnetic quantum spin-1/2 Heisenberg chain realizes a quantum critical point with algebraic spin-spin correlation functions. For the same reason, the U(1) symmetry associated with the conservation of the electric charge of a one-dimensional electron gas is not broken spontaneously by local attractive interactions through the selection of superconducting ground states, as would be the case in two and higher dimensions. Superconductivity at nonvanishing temperature in a one-dimensional quantum wire is only possible by the proximity effect to a superconducting substrate that is three-dimensional. It is this scenario that we will have in mind when discussing superconductivity in one-dimensional quantum wires.

In a mean-field treatment of a superconducting instability, the superconducting order parameter is taken to be a static background in which electrons hop with the caveat that the global conservation of the electronic number operator is spontaneously broken down to a discrete subgroup $\mathbb{Z}_2$ of U(1). The only remnant of the global conservation of the electronic number operator is that the parity of the total number operator is a constant of the motion. As for polyacetylene, the quasi-particle spectrum at the mean-field level acquires a gap, a superconducting gap.

Assuming implicitly that superconductivity is established by the proximity effect to a higher-dimensional superconducting reservoir, we are going to determine when a one-dimensional superconductor treated within the so-called Bogoliubov-de-Gennes approximation can support point defects that bind zero modes and why this phenomenon can be thought of as a counterpart to the fractionalization of the charge in polyacetylene.

The following question then arises. Is there an organizing principle in which one may place polyacetylene and those examples of one-dimensional Bogoliubov-de-Gennes Hamiltonians that support zero modes at point defects? The answer will be positive for the so-called (one-dimensional) *strong topological insulators or superconductors.* This organizing principle applies to fermionic single-particle Hamiltonians with a combination of three global discrete symmetries acting locally in space. In particular, this organizing principle does not presume the existence of space-group symmetries of the underlying crystal, that is it applies to statistical ensembles of static single-particle fermionic Hamiltonians. This organizing principle is related to the ten symmetry classes of random matrix theory, the tenfold way. Strong topological insulators or superconductors are examples of fermionic invertible topological phases of matter.[1] The defining attribute of an invertible topological phase of matter is a ground state that is nondegenerate and gapped when space has no boundary, while it coexists with zero modes that are simultaneously extended along (when the dimensionality of space is larger than one) while decaying exponentially fast away from certain boundaries of space. These boundary states are anomalous in that they carry a fractional part of some conserved quantum number or through their transformation laws under some global symmetry.

We devote Section 7.2 to an elementary introduction of Bogoliubov-de-Gennes Hamiltonians. We then classify single-particle fermionic Hamiltonians in terms of ten symmetry classes in Section 7.3. This classification is known as the Altland-Zirnbauer tenfold way. If space is zero dimensional, a triplet of ensembles of random matrices can be constructed, namely, random Hermitean matrices in Section 7.4, random unitary matrices in Section 7.5, and random scattering (transfer) matrices in Section 7.6, for any 1 of the ten Altland-Zirnbauer symmetry classes. From these tools of random matrix theory, it is possible to compute transport properties of quasi-one-dimensional wires for any 1 of the ten Altland-Zirnbauer symmetry classes, as is explained in Section 7.7. It then becomes manifest that five out of the ten Altland-Zirnbauer symmetry classes are special, namely, the chiral symmetry classes AIII, CII, BDI and the Bogoliubov-de-Gennes symmetry classes D and DIII. Finally, starting from quasi-one-dimensional band insulators or band superconductors in Section 7.8, we identify these five Altland-Zirnbauer symmetry classes as the ones that realize the strong topological insulators or superconductors in one-dimensional space.

[1] Z.-C. Gu and X.-G. Wen, Phys. Rev. B **80**(15), 155131 (2009). [Gu and Wen (2009)].

7.2 Bogoliubov-de-Gennes (BdG) Hamiltonians

7.2.1 Motivation

We shall consider a quantum interacting model for two spinless fermions that we define with the help of a Grassmann path integral (**Exercise** 3.14). The partition function is a double integral over the two pairs (a^*, a) and (b^*, b) of Grassmann variables, that is,

$$Z := \int \mathcal{D}[a^*, a] \int \mathcal{D}[b^*, b] \, e^{-\int \mathrm{d}\tau (\mathcal{L}_0 + \mathcal{L}_1)(\tau)}. \tag{7.1a}$$

The Lagrangian in imaginary time τ is the sum over the quadratic contribution

$$\mathcal{L}_0(\tau) := \Big(a^* \left(\partial_\tau + \varepsilon_a\right) a + b^* \left(\partial_\tau + \varepsilon_b\right) b - t\, a^*\, b - t^*\, b^*\, a\Big)(\tau) \tag{7.1b}$$

and the quartic contribution

$$\mathcal{L}_1(\tau) := -(a^*\, b^*\, b\, a)(\tau). \tag{7.1c}$$

The parameters ε_a and ε_b are real-valued. The parameter t is complex-valued. The sign of the interaction is "attractive." This is to say that it "favors" the Hubbard-Stratonovich decoupling

$$e^{-\int \mathrm{d}\tau\, \mathcal{L}_1(\tau)} \propto \int \frac{\mathcal{D}\Delta^*\, \mathcal{D}\Delta}{2\pi \mathrm{i}} e^{-\int \mathrm{d}\tau \left[|\Delta(\tau)|^2 + (\Delta\, a^*\, b^*)(\tau) + (\Delta^*\, b\, a)(\tau)\right]} \tag{7.2}$$

with the auxiliary field $\Delta(\tau)$ complex-valued. Whereas the identity (7.2) is exact, the uncontrolled mean-field approximation of the right-hand side (7.2) that consists in truncating the integral over the real and imaginary parts of the integration variable Δ is expected to be a "good" approximation for an "attractive" interaction.

If one approximates the path integral by its saddle-point value for a time-independent Δ, one must then solve the spectrum of the Hamiltonian (7.6). This mean-field theory is typically encountered in the context of superconductivity.[2] The sign of the quartic interaction favors Cooper pairing. The field $\Delta(\tau)$ is to be interpreted as the expectation value of the product of two creation operators for fermions that are separated in imaginary time by τ. A nonvanishing value of Δ at the saddle point breaks spontaneously the global conservation of the fermion number, that is the U(1) symmetry of Eq. (7.1) under the continuous transformation

$$a^* \mapsto a^*\, e^{-\mathrm{i}\phi}, \quad a \mapsto e^{+\mathrm{i}\phi}\, a, \qquad b^* \mapsto b^*\, e^{-\mathrm{i}\phi}, \quad b \mapsto e^{+\mathrm{i}\phi}\, b, \tag{7.3a}$$

labeled by $0 \leq \phi < 2\pi$, leaving only the residual discrete symmetry $\mathbb{Z}_2$ obtained by restricting $\phi = 0, \pi$, that is,

$$a^* \mapsto \pm a^*, \quad a \mapsto \pm a, \qquad b^* \mapsto \pm b^*, \quad b \mapsto \pm b. \tag{7.3b}$$

[2] C. Mudry, *Lecture Notes on Field Theory in Condensed Matter Physics*, World Scientific, Singapore 2014. [Mudry (2014)].

7.2.2 A 4 × 4 Bogoliubov-de-Gennes Hamiltonian

As a prelude to the one-dimensional case, we consider the pair of labels a and b. We assign to the label a the pair of creation and annihilation fermion operators $\hat{a}^\dagger$ and $\hat{a}$ with the only nonvanishing anticommutator

$$\{\hat{a}, \hat{a}^\dagger\} = 1. \tag{7.4a}$$

We assign to the label b the pair of creation and annihilation fermion operators $\hat{b}^\dagger$ and $\hat{b}$ with the only nonvanishing anticommutator

$$\{\hat{b}, \hat{b}^\dagger\} = 1. \tag{7.4b}$$

Finally, we impose that

$$\{\hat{b}, \hat{a}^\dagger\} = \{\hat{b}, \hat{a}\} = 0. \tag{7.4c}$$

The fermionic Fock space $\mathfrak{F}$ associated with the two labels a and b for the fermions is the four-dimensional vector space over the complex numbers,

$$\mathfrak{F} \sim \mathbb{C}^4. \tag{7.4d}$$

Here, the power 4 should be interpreted as 2^n with $n = 2$. The fermionic Fock space $\mathfrak{F}$ is spanned by the basis

$$|0\rangle, \qquad \hat{a}^\dagger\,|0\rangle, \qquad \hat{b}^\dagger\,|0\rangle, \qquad \hat{a}^\dagger\,\hat{b}^\dagger\,|0\rangle, \qquad \hat{a}\,|0\rangle = \hat{b}\,|0\rangle = 0. \tag{7.4e}$$

Had we traded the fermion labels a and b for n labels $\iota = 1, \cdots, n$ and demanded that all fermion operators with different labels anticommute, the dimensionality of the fermionic Fock space would grow like 2^n, that is exponentially fast with n as it should be for the Hilbert space of a many-body Hamiltonian.

The total number operator is defined to be

$$\widehat{N} := \hat{a}^\dagger\,\hat{a} + \hat{b}^\dagger\,\hat{b} \tag{7.5a}$$

with its domain of definition taken to be the four-dimensional fermionic Fock space (7.4d). Its action on $\mathfrak{F}$ is represented by the 4 × 4 diagonal matrix

$$\mathcal{N}_{\mathfrak{F}} = \begin{pmatrix} 0 & 0 & 0 & 0 \\ 0 & 1 & 0 & 0 \\ 0 & 0 & 1 & 0 \\ 0 & 0 & 0 & 2 \end{pmatrix} \tag{7.5b}$$

in the basis (7.4e).

We also define the Hermitean operator

$$\widehat{H} := \varepsilon_a\,\hat{a}^\dagger\,\hat{a} + \varepsilon_b\,\hat{b}^\dagger\,\hat{b} - t\,\hat{a}^\dagger\,\hat{b} - t^*\,\hat{b}^\dagger\,\hat{a} + \Delta\,\hat{a}^\dagger\,\hat{b}^\dagger + \Delta^*\,\hat{b}\,\hat{a} \tag{7.6a}$$

with its domain of definition taken to be the four-dimensional fermionic Fock space (7.4d). The parameters ε_a and ε_b are real-valued. They assign the single-particle energy ε_a and ε_b to the single-particle state $\hat{a}^\dagger|0\rangle$ and $\hat{b}^\dagger|0\rangle$, respectively. The parameters t and Δ are complex-valued. The parameter t quantifies the hybridization between the single-particle states $\hat{a}^\dagger|0\rangle$ and $\hat{b}^\dagger|0\rangle$ in the many-body eigenstates. The parameter Δ quantifies the amount by which $\widehat{N}$ does not commute with $\widehat{H}$:

$$\left[\widehat{N},\widehat{H}\right] = 2\left(\Delta\,\hat{a}^\dagger\,\hat{b}^\dagger - \Delta^*\,\hat{b}\,\hat{a}\right). \tag{7.6b}$$

Up to an additive $\mathbb{C}$ number, Hamiltonian $\widehat{H}$ is the most general Hermitean bilinear form that can be built out of the pair of creation operators $\hat{a}^\dagger$ and $\hat{b}^\dagger$ and the pair of annihilation operators $\hat{a}$ and $\hat{b}$ with the rule that any creation operator is to be placed to the left of an annihilation operator. The action of $\widehat{H}$ on the four-dimensional fermionic Fock space (7.4d) is represented by the 4×4 non-diagonal matrix

$$\mathcal{H}_{\mathfrak{F}} := \begin{pmatrix} 0 & 0 & 0 & \Delta^* \\ 0 & \varepsilon_a & -t & 0 \\ 0 & -t^* & \varepsilon_b & 0 \\ \Delta & 0 & 0 & \varepsilon_a+\varepsilon_b \end{pmatrix} \tag{7.7}$$

in the basis (7.4e). This matrix decomposes into two 2×2 blocks, as becomes apparent after interchanging line 2 with line 4 and column 2 with column 4, in which case

$$\mathcal{N}_{\mathfrak{F}} \sim \begin{pmatrix} 0 & 0 & 0 & 0 \\ 0 & 2 & 0 & 0 \\ 0 & 0 & 1 & 0 \\ 0 & 0 & 0 & 1 \end{pmatrix}, \qquad \mathcal{H}_{\mathfrak{F}} \sim \begin{pmatrix} 0 & \Delta^* & 0 & 0 \\ \Delta & \varepsilon_a+\varepsilon_b & 0 & 0 \\ 0 & 0 & \varepsilon_b & -t^* \\ 0 & 0 & -t & \varepsilon_a \end{pmatrix}. \tag{7.8}$$

This is the consequence of the fact that the occupation number of any of the elements from the basis (7.4e) of the fermionic Fock space $\mathfrak{F}$ can only change by ± 2 or 0 under the action of $\widehat{H}$, that is $\widehat{H}$ conserves the fermion parity operator

$$\widehat{P}_{\mathrm{f}} := (-1)^{\widehat{N}}. \tag{7.9}$$

The matrix $\mathcal{H}_{\mathfrak{F}}$ has thus the four eigenvalues:

$$E_{|\Delta|,\pm} := \frac{\varepsilon_a+\varepsilon_b}{2} \pm \sqrt{\left(\frac{\varepsilon_a+\varepsilon_b}{2}\right)^2 + |\Delta|^2}, \tag{7.10a}$$

$$E_{|t|,\pm} := \frac{\varepsilon_a+\varepsilon_b}{2} \pm \sqrt{\left(\frac{\varepsilon_a-\varepsilon_b}{2}\right)^2 + |t|^2}. \tag{7.10b}$$

This many-body spectrum is symmetric about the energy $(\varepsilon_a+\varepsilon_b)/2$ along the energy axis, as is depicted in Figure 7.1.

There is an alternative way to understand this many-body spectral symmetry.

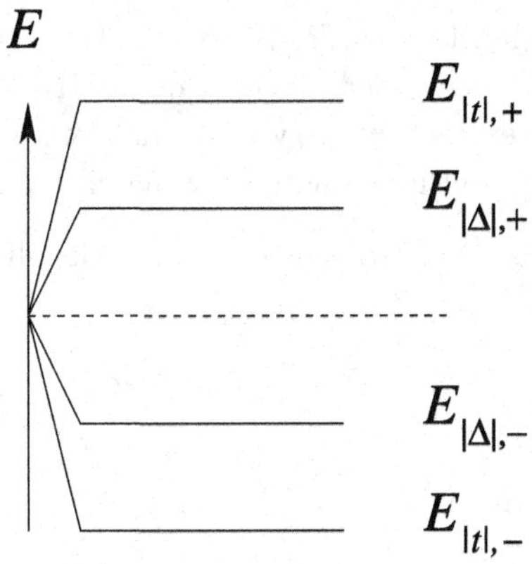

Figure 7.1 Spectrum (7.10) if $[(\varepsilon_a-\varepsilon_b)/2]^2+|t|^2>[(\varepsilon_a+\varepsilon_b)/2]^2+|\Delta|^2$. The dashed line is the energy $(\varepsilon_a+\varepsilon_b)/2$.

One introduces the Nambu spinors

$$\widehat{\Psi}^\dagger \equiv \begin{pmatrix}\hat{a}^\dagger & \hat{b}^\dagger & \hat{a} & \hat{b}\end{pmatrix}, \qquad \widehat{\Psi} \equiv \begin{pmatrix}\hat{a}\\ \hat{b}\\ \hat{a}^\dagger \\ \hat{b}^\dagger\end{pmatrix}, \tag{7.11a}$$

together with the single-particle Bogoliubov-de-Gennes Hamiltonian

$$\mathcal{H}_{\mathrm{BdG}} := \begin{pmatrix} +\varepsilon_a & -t & 0 & +\Delta \\ -t^* & +\varepsilon_b & -\Delta & 0 \\ 0 & -\Delta^* & -\varepsilon_a & +t^* \\ +\Delta^* & 0 & +t & -\varepsilon_b \end{pmatrix}. \tag{7.11b}$$

The domain of definition of $\mathcal{H}_{\mathrm{BdG}}$ is the four-dimensional Hilbert space

$$\mathfrak{H}_{\mathrm{BdG}} = \mathbb{C}^4. \tag{7.11c}$$

It is a coincidence that the dimensionality of the single-particle Hilbert space $\mathfrak{H}_{\mathrm{BdG}}$ matches that of the many-body Hilbert space $\mathfrak{F}$ defined in Eq. (7.4). Had we traded the two labels a and b for n labels $\iota=1,\cdots,n$ and demanded that all fermion operators with different labels ι anticommute, the dimensionality of the many-body fermionic Fock space would grow like 2^n, whereas the dimensionality of the single-particle Hilbert space $\mathfrak{H}_{\mathrm{BdG}}$ would grow as $2n$. Hence, $\mathfrak{H}_{\mathrm{BdG}}$ and $\mathfrak{F}$ are distinct. By design,

$$\operatorname{tr}|_{\mathfrak{H}_{\mathrm{BdG}}}\, \mathcal{H}_{\mathrm{BdG}} = 0 \tag{7.11d}$$

and

$$\widehat{H} = \frac{\varepsilon_a+\varepsilon_b}{2} + \frac{1}{2}\left(\widehat{\Psi}^\dagger\, \mathcal{H}_{\mathrm{BdG}}\, \widehat{\Psi}\right). \tag{7.11e}$$

The four single-particle eigenstates of $\mathcal{H}_{\mathrm{BdG}}$ come in two pairs with opposite eigenenergies. If $\Upsilon \in \mathbb{C}^4$ is a single-particle eigenstate of $\mathcal{H}_{\mathrm{BdG}}$ with the nonvanishing eigenenergy ξ, then there exists a unitary 4×4 matrix $\mathcal{C}$ such that $\mathcal{C}\,\Upsilon^*$ is a single-particle eigenstate of $\mathcal{H}_{\mathrm{BdG}}$ with the nonvanishing eigenenergy $-\xi$.

Proof This is a consequence of the single-particle Bogoliubov-de-Gennes spectral symmetry implied by the relation

$$\mathcal{C}\,\mathcal{H}^*_{\mathrm{BdG}}\,\mathcal{C}^{-1} = -\mathcal{H}_{\mathrm{BdG}}, \tag{7.12a}$$

where

$$\mathcal{C} = \begin{pmatrix} 0 & 0 & 1 & 0 \\ 0 & 0 & 0 & 1 \\ 1 & 0 & 0 & 0 \\ 0 & 1 & 0 & 0 \end{pmatrix} \equiv \begin{pmatrix} 0 & \sigma_0 \\ \sigma_0 & 0 \end{pmatrix} \equiv \sigma_0 \otimes \rho_1, \qquad \mathcal{C}^{-1} = \mathcal{C}, \tag{7.12b}$$

and

$$\mathcal{H}_{\mathrm{BdG}} \equiv \begin{pmatrix} \mathcal{H}_{\mathrm{pp}} & \mathcal{H}_{\mathrm{ph}} \\ \mathcal{H}^{\dagger}_{\mathrm{ph}} & -\mathcal{H}^{\mathsf{T}}_{\mathrm{pp}} \end{pmatrix}, \quad \mathcal{H}_{\mathrm{pp}} \equiv \begin{pmatrix} +\varepsilon_a & -t \\ -t^* & +\varepsilon_b \end{pmatrix}, \quad \mathcal{H}_{\mathrm{ph}} \equiv \begin{pmatrix} 0 & +\Delta \\ -\Delta & 0 \end{pmatrix}. \tag{7.12c}$$

The 2×2 block $\mathcal{H}_{\mathrm{pp}}$ is Hermitean. It depends on four real-valued parameters. The 2×2 block $\mathcal{H}_{\mathrm{ph}}$ is antisymmetric. It depends on two real-valued parameters. The claim then follows from

$$\mathcal{H}_{\mathrm{BdG}}\,\mathcal{C}\,\Upsilon^* = \left(\mathcal{H}^*_{\mathrm{BdG}}\,\mathcal{C}\,\Upsilon\right)^* = -\left(\mathcal{C}\,\mathcal{H}_{\mathrm{BdG}}\,\Upsilon\right)^* = -\xi\,\mathcal{C}\,\Upsilon^*. \tag{7.12d}$$

□

The single-particle Bogoliubov-de-Gennes spectral symmetry (7.12a) implies the following unitary symmetry of the second-quantized Hamiltonian (7.11e). Let $\mu = 1,\cdots,4$ run over the four components $\widehat{\Psi}_\mu$ of a Nambu spinor $\widehat{\Psi}$. Let α and β be a pair of complex numbers. Define the unitary transformation $\widehat{\mathrm{C}}$ on the fermionic Fock space $\mathfrak{F}$ by the transformation laws

$$\widehat{\mathrm{C}}\left(\alpha^*\,\widehat{\Psi}^\dagger_\mu + \beta\,\widehat{\Psi}_\mu\right)\widehat{\mathrm{C}}^{-1} := \sum_{\mu'=1}^{4}\left(\alpha^*\,\widehat{\Psi}_{\mu'}\,\mathcal{C}^*_{\mu'\mu} + \beta\,\mathcal{C}_{\mu'\mu}\widehat{\Psi}^\dagger_{\mu'}\right) \tag{7.13}$$

for $\mu = 1,\cdots 4$. The 4×4 matrix $\mathcal{C}$ was defined in Eq. (7.12b). The particle-hole symmetry (PHS)

$$\widehat{\mathrm{C}}\,\widehat{H}\,\widehat{\mathrm{C}}^{-1} = \widehat{H} \tag{7.14}$$

must then hold.

Proof By Eq. (7.11e),

$$\widehat{\mathrm{C}}\,\widehat{H}\,\widehat{\mathrm{C}}^{-1} = \widehat{\mathrm{C}}\left[\frac{\varepsilon_a + \varepsilon_b}{2} + \frac{1}{2}\left(\widehat{\Psi}^\dagger\,\mathcal{H}_{\mathrm{BdG}}\,\widehat{\Psi}\right)\right]\widehat{\mathrm{C}}^{-1}. \tag{7.15}$$

The linearity of Eq. (7.13) implies that

$$\widehat{\mathrm{C}}\,\widehat{H}\,\widehat{\mathrm{C}}^{-1}=\frac{\varepsilon_a+\varepsilon_b}{2}+\frac{1}{2}\sum_{\mu,\mu'=1}^{4}\left(\widehat{\mathrm{C}}\,\widehat{\Psi}^{\dagger}_{\mu}\,\widehat{\mathrm{C}}^{-1}\right)\mathcal{H}_{\mathrm{BdG}\,\mu\mu'}\left(\widehat{\mathrm{C}}\,\widehat{\Psi}_{\mu'}\widehat{\mathrm{C}}^{-1}\right). \tag{7.16}$$

From Eq. (7.13),

$$\widehat{\mathrm{C}}\,\widehat{H}\,\widehat{\mathrm{C}}^{-1}=\frac{\varepsilon_a+\varepsilon_b}{2}+\frac{1}{2}\sum_{\mu,\mu'=1}^{4}\left(\sum_{\nu=1}^{4}\widehat{\Psi}_{\nu}\,\mathcal{C}^{*}_{\nu\mu}\right)\mathcal{H}_{\mathrm{BdG}\,\mu\mu'}\left(\sum_{\nu'=1}^{4}\mathcal{C}_{\nu'\mu'}\widehat{\Psi}^{\dagger}_{\nu'}\right). \tag{7.17}$$

We do the manipulations

$$\begin{aligned}
\widehat{\mathrm{C}}\,\widehat{H}\,\widehat{\mathrm{C}}^{-1}&=\frac{\varepsilon_a+\varepsilon_b}{2}+\frac{1}{2}\sum_{\nu,\nu'=1}^{4}\widehat{\Psi}_{\nu}\left(\sum_{\mu,\mu'=1}^{4}\mathcal{C}^{*}_{\nu\mu}\,\mathcal{H}_{\mathrm{BdG}\,\mu\mu'}\,\mathcal{C}_{\nu'\mu'}\right)\widehat{\Psi}^{\dagger}_{\nu'}\\
&=\frac{\varepsilon_a+\varepsilon_b}{2}+\frac{1}{2}\sum_{\nu,\nu'=1}^{4}\widehat{\Psi}_{\nu}\left(\sum_{\mu,\mu'=1}^{4}\mathcal{C}_{\nu'\mu'}\,\mathcal{H}^{*}_{\mathrm{BdG}\,\mu'\mu}\,\mathcal{C}^{*}_{\nu\mu}\right)\widehat{\Psi}^{\dagger}_{\nu'}\\
&=\frac{\varepsilon_a+\varepsilon_b}{2}+\frac{1}{2}\sum_{\nu,\nu'=1}^{4}\widehat{\Psi}_{\nu}\left(\mathcal{C}\,\mathcal{H}^{*}_{\mathrm{BdG}}\,\mathcal{C}^{-1}\right)_{\nu'\nu}\widehat{\Psi}^{\dagger}_{\nu'},
\end{aligned} \tag{7.18}$$

where we used the Hermiticity of $\mathcal{H}_{\mathrm{BdG}}$ to reach the second equality and the unitarity of $\mathcal{C}$ to reach the last equality. According to Eq. (7.12a),

$$\widehat{\mathrm{C}}\,\widehat{H}\,\widehat{\mathrm{C}}^{-1}=\frac{\varepsilon_a+\varepsilon_b}{2}-\frac{1}{2}\sum_{\nu,\nu'=1}^{4}\widehat{\Psi}_{\nu}\,\mathcal{H}_{\mathrm{BdG}\,\nu'\nu}\,\widehat{\Psi}^{\dagger}_{\nu'}. \tag{7.19}$$

In turn, Eqs. (7.4) and (7.11d) dictate that

$$\widehat{\mathrm{C}}\,\widehat{H}\,\widehat{\mathrm{C}}^{-1}=\frac{\varepsilon_a+\varepsilon_b}{2}+\frac{1}{2}\sum_{\nu,\nu'=1}^{4}\widehat{\Psi}^{\dagger}_{\nu'}\,\mathcal{H}_{\mathrm{BdG}\,\nu'\nu}\,\widehat{\Psi}_{\nu}=\widehat{H}. \tag{7.20}$$

□

For any many-body Hamiltonian that is quadratic in the fermion creation and annihilation operators, the single-particle energy eigenvalues must be related linearly to the many-body energy eigenvalues. In particular, the $2\times 2=4$ single-particle energy eigenvalues of $\mathcal{H}_{\mathrm{BdG}}$ must be linearly related to the $2^2=4$ many-body energy eigenvalues given in Eq. (7.10) and displayed in Figure 7.1. For simplicity, we verify this assertion for two special cases.

First, we make the assumption that $t,\Delta\in\mathbb{R}$ with

$$\varepsilon_a=\varepsilon_b=\mu>0,\qquad t>\sqrt{\mu^2+\Delta^2}>0,\qquad \Delta\geq 0. \tag{7.21}$$

We must compare the many-body spectrum

$$E_{t,-}<E_{\Delta,-}<E_{\Delta,+}<E_{t,+} \tag{7.22a}$$

with

$$E_{\Delta,\pm} = \mu \pm \sqrt{\mu^2 + \Delta^2}, \qquad E_{t,\pm} = \mu \pm t, \tag{7.22b}$$

to the single-particle Bogoliubov-de-Gennes spectrum of [3]

$$\widetilde{\mathcal{H}}_{\mathrm{BdG}} = \mu\, X_{00} + \mathcal{H}_{\mathrm{BdG}} = \mu\, X_{00} + \mu\, X_{03} - t\, X_{13} - \Delta\, X_{22}, \tag{7.23}$$

where we are using the notation introduced in Eq. (3.507). The 4×4 matrix X_{13} has the twofold degenerate eigenvalues ± 1 and commutes with both X_{03} and X_{22}. The pair X_{03} and X_{22} are anticommuting so that $\mu\, X_{03} - \Delta\, X_{22}$ squares to $(\mu^2 + \Delta^2) X_{00}$ and it has the twofold degenerate eigenvalues $\pm\sqrt{\mu^2 + \Delta^2}$. We conclude that $\widetilde{\mathcal{H}}_{\mathrm{BdG}}$ has the four eigenvalues

$$\xi_1 < \xi_2 < \xi_3 < \xi_4, \tag{7.24a}$$

with

$$\xi_1 := \mu - t - \sqrt{\mu^2 + \Delta^2}, \qquad \xi_2 := \mu - t + \sqrt{\mu^2 + \Delta^2}, \tag{7.24b}$$

and

$$\xi_3 := \mu + t - \sqrt{\mu^2 + \Delta^2}, \qquad \xi_4 := \mu + t + \sqrt{\mu^2 + \Delta^2}. \tag{7.24c}$$

The linear relation that we seek is then

$$\begin{pmatrix} E_{t,-} \\ E_{\Delta,-} \\ E_{\Delta,+} \\ E_{t,+} \end{pmatrix} = \frac{1}{2} \begin{pmatrix} 1 & 1 & 0 & 0 \\ 1 & 0 & 1 & 0 \\ 0 & 1 & 0 & 1 \\ 0 & 0 & 1 & 1 \end{pmatrix} \begin{pmatrix} \xi_1 \\ \xi_2 \\ \xi_3 \\ \xi_4 \end{pmatrix}. \tag{7.25}$$

Second, we make the assumption that $t, \Delta \in \mathbb{R}$ with

$$\varepsilon_a = -\varepsilon_b = \varepsilon > 0, \qquad t > \Delta > 0. \tag{7.26}$$

We must compare the many-body spectrum

$$E_{t,-} < E_{\Delta,-} < E_{\Delta,+} < E_{t,+} \tag{7.27a}$$

with

$$E_{\Delta,\pm} = \pm\Delta, \qquad E_{t,\pm} = \pm\sqrt{\varepsilon^2 + t^2}, \tag{7.27b}$$

to the single-particle Bogoliubov-de-Gennes spectrum of

$$\mathcal{H}_{\mathrm{BdG}} = \varepsilon\, X_{33} - t\, X_{13} - \Delta\, X_{22}, \tag{7.28}$$

where we are using the notation introduced in Eq. (3.507). The 4×4 matrix X_{22} has the twofold degenerate eigenvalues ± 1 and commutes with both X_{13} and X_{33}. The pair X_{13} and X_{33} are anticommuting so that $\varepsilon\, X_{33} - t\, X_{13}$ squares to $(\varepsilon^2 + t^2) X_{00}$

[3] The need for the term $\mu\, X_{00}$ originates from the $\mathbb{C}$-number $(\varepsilon_a + \varepsilon_b)/2 = \mu$ on the right-hand side of Eq. (7.11e).

and it has the twofold degenerate eigenvalues $\pm\sqrt{\varepsilon^2+t^2}$. We conclude that $\mathcal{H}_{\mathrm{BdG}}$ has the four eigenvalues

$$\xi_1 < \xi_2 < \xi_3 < \xi_4, \tag{7.29a}$$

with

$$\xi_1 := -\sqrt{\varepsilon^2+t^2} - \Delta, \qquad \xi_2 := -\sqrt{\varepsilon^2+t^2} + \Delta, \tag{7.29b}$$

and

$$\xi_3 := +\sqrt{\varepsilon^2+t^2} - \Delta, \qquad \xi_4 := +\sqrt{\varepsilon^2+t^2} + \Delta. \tag{7.29c}$$

The linear relation that we seek is then

$$\begin{pmatrix} E_{t,-} \\ E_{\Delta,-} \\ E_{\Delta,+} \\ E_{t,+} \end{pmatrix} = \frac{1}{2} \begin{pmatrix} 1 & 1 & 0 & 0 \\ 1 & 0 & 1 & 0 \\ 0 & 1 & 0 & 1 \\ 0 & 0 & 1 & 1 \end{pmatrix} \begin{pmatrix} \xi_1 \\ \xi_2 \\ \xi_3 \\ \xi_4 \end{pmatrix}. \tag{7.30}$$

The single-particle Bogoliubov-de-Gennes Hamiltonian $\mathcal{H}_{\mathrm{BdG}}$ defined in Eq. (7.11b) gives an alternative construction to the many-body fermionic Fock space $\mathfrak{F}$ defined in Eq. (7.4). Because $\mathcal{H}_{\mathrm{BdG}}$ is Hermitean and because of Eq. (7.12a), it can be shown that there exists a unitary 4×4 matrix $\mathcal{U}_{\mathrm{BdG}}$ with the following two properties (**Exercise** 7.1). First,

$$\mathcal{H}_{\mathrm{BdG}} = \mathcal{U}^{\dagger}_{\mathrm{BdG}}\, \mathcal{E}_{\mathrm{BdG}}\, \mathcal{U}_{\mathrm{BdG}} \tag{7.31a}$$

with

$$\mathcal{E}_{\mathrm{BdG}} := \begin{pmatrix} +\xi_1 & 0 & 0 & 0 \\ 0 & +\xi_2 & 0 & 0 \\ 0 & 0 & -\xi_1 & 0 \\ 0 & 0 & 0 & -\xi_2 \end{pmatrix}, \qquad \xi_1, \xi_2 \geq 0. \tag{7.31b}$$

Define the transformation

$$\widehat{\Upsilon}^{\dagger} := \widehat{\Psi}^{\dagger}\, \mathcal{U}^{\dagger}_{\mathrm{BdG}}, \qquad \widehat{\Upsilon} := \mathcal{U}_{\mathrm{BdG}}\, \widehat{\Psi}, \tag{7.32}$$

on the Nambu spinors. Second, we demand that

$$\widehat{\Upsilon}^{\dagger} \equiv \begin{pmatrix} \hat{\gamma}^{\dagger}_1 & \hat{\gamma}^{\dagger}_2 & \hat{\gamma}_1 & \hat{\gamma}_2 \end{pmatrix}, \qquad \widehat{\Upsilon} \equiv \begin{pmatrix} \hat{\gamma}_1 \\ \hat{\gamma}_2 \\ \hat{\gamma}^{\dagger}_1 \\ \hat{\gamma}^{\dagger}_2 \end{pmatrix}, \tag{7.33a}$$

with

$$\{\hat{\gamma}_i, \hat{\gamma}^{\dagger}_j\} = \delta_{ij}, \qquad \{\hat{\gamma}_i, \hat{\gamma}_j\} = \{\hat{\gamma}^{\dagger}_i, \hat{\gamma}^{\dagger}_j\} = 0, \tag{7.33b}$$

for $i, j = 1, 2$. A unitary transformation (7.32) on the fermionic Fock space $\mathfrak{F}$ that meets conditions (7.31) and (7.33) is called a Bogoliubov transformation. If such a Bogoliubov transformation exists, then Eq. (7.11e) becomes

$$\begin{aligned}\widehat{H} &= \frac{\varepsilon_a + \varepsilon_b}{2} + \frac{1}{2} \sum_{i=1,2} \xi_i \left(\hat{\gamma}_i^\dagger \hat{\gamma}_i - \hat{\gamma}_i \hat{\gamma}_i^\dagger\right) \\ &= \frac{\varepsilon_a + \varepsilon_b}{2} - \frac{\xi_1 + \xi_2}{2} + \sum_{i=1,2} \xi_i \hat{\gamma}_i^\dagger \hat{\gamma}_i. \end{aligned} \tag{7.34}$$

The many-body ground state $|\mathrm{GS}\rangle$ with the many-body eigenenergy

$$E_{\mathrm{GS}} = \frac{\varepsilon_a + \varepsilon_b}{2} - \frac{\xi_1 + \xi_2}{2} \tag{7.35a}$$

of $\widehat{H}$ is defined by the pair of conditions

$$\hat{\gamma}_i \, |\mathrm{GS}\rangle = 0, \qquad i = 1, 2. \tag{7.35b}$$

The remaining three many-body excited states of $\widehat{H}$ are

$$\hat{\gamma}_1^\dagger \, |\mathrm{GS}\rangle, \qquad \hat{\gamma}_2^\dagger \, |\mathrm{GS}\rangle, \qquad \hat{\gamma}_1^\dagger \hat{\gamma}_2^\dagger \, |\mathrm{GS}\rangle. \tag{7.35c}$$

The fermionic Fock space

$$\mathfrak{F}_{\mathrm{BdG}} \sim \mathfrak{F} \tag{7.36a}$$

is spanned by the basis

$$|\mathrm{GS}\rangle, \qquad \hat{\gamma}_1^\dagger \, |\mathrm{GS}\rangle, \qquad \hat{\gamma}_2^\dagger \, |\mathrm{GS}\rangle, \qquad \hat{\gamma}_1^\dagger \hat{\gamma}_2^\dagger \, |\mathrm{GS}\rangle. \tag{7.36b}$$

In this basis of the fermionic Fock space, the single-particle states $\hat{\gamma}_1^\dagger \, |\mathrm{GS}\rangle$ and $\hat{\gamma}_2^\dagger \, |\mathrm{GS}\rangle$ are called Bogoliubov-de-Gennes (BdG) quasiparticles. They have the single-particle BdG eigenenergies ξ_1 and ξ_2 measured relative to the many-body ground-state energy E_{GS}. We recall that taking the commutator between the total number operator (7.5a) and the operators $\hat{a}^\dagger$ and $\hat{b}^\dagger$ renders $+\hat{a}^\dagger$ and $+\hat{b}^\dagger$, respectively. This is why one can assign the fermion charge $+1$ to the single-particle states $\hat{a}^\dagger \, |0\rangle$ and $\hat{b}^\dagger \, |0\rangle$. Because $\hat{\gamma}_1^\dagger$ and $\hat{\gamma}_2^\dagger$ are linear combinations that mix the creation and annihilation operators $\hat{a}^\dagger$, $\hat{b}^\dagger$, $\hat{a}$, and $\hat{b}$, the BdG quasiparticle operators $\hat{\gamma}_1^\dagger$ and $\hat{\gamma}_2^\dagger$ are not mere raising operators for the fermion occupation numbers. The BdG quasiparticle operators $\hat{\gamma}_1^\dagger$ and $\hat{\gamma}_2^\dagger$ mix the subspaces of the fermionic Fock space with one more and one less fermion number than the fermion number set by the chemical potential. The BdG quasiparticle states $\hat{\gamma}_1^\dagger \, |\mathrm{GS}\rangle$ and $\hat{\gamma}_2^\dagger \, |\mathrm{GS}\rangle$ cannot be assigned a sharp fermion number. This is also why the energy scales Δ, on the one hand, and t, on the other hand, are mixed in the eigenvalues ξ_1 and ξ_2 [recall Eqs. (7.24) and (7.29)].

Cooper pairing is only possible between pairs made of two distinct fermions (this is a consequence of the Pauli principle). Hence, single-particle BdG Hamiltonians that can be represented by finite dimensional matrices are necessarily represented by even-dimensional matrices. Owing to the spectral symmetry of single-particle BdG Hamiltonians, it follows that a finite-dimensional BdG matrix cannot support an odd number of zero modes. It can only support an even number of zero modes. A zero mode Υ_0 of $\mathcal{H}_{\mathrm{BdG}}$ is called a Majorana mode. It does not enter the many-body Hamiltonian. The goal of this chapter is to explain why Majorana modes in superconducting chains are the superconducting counterparts to the zero modes in polyacetylene. However, before we move to the one-dimensional superconducting chains, we need to introduce the concept of the Altland-Zirnbauer superconducting symmetry classes.

7.3 The Tenfold Way: Altland-Zirnbauer Symmetry Classes

Wigner introduced the idea that the statistical properties of the eigenvalues and eigenfunctions of certain (Gaussian) ensembles of Hermitean random matrices could explain the statistical behavior of the large number of resonances observed when a beam of slow nucleons is scattered inelastically from heavy nuclei such as thorium or uranium.[4] Slow means here that the phase relation between the incoming nucleon and the outgoing nucleon has been lost due to a large number of inelastic processes taking place in the heavy nuclei between the absorption of the incoming nucleon and the emission of the outgoing nucleon.

Dyson proposed another class of random matrices, namely, the (circular) ensembles of random unitary matrices.[5] In a physics setting, these unitary matrices can be thought of as the exponentials of Hamiltonians multipled by an imaginary number, as occurs with the unitary time evolution of closed quantum systems.

The Gaussian and circular ensembles come in three kinds, depending on whether the entries of the Hamiltonians are real numbers, complex numbers, or real quaternions, respectively. (See **Exercise** 7.2 for a definition of quaternions.) Dyson referred to the recurrence of the number 3 for the Gaussian or circular ensembles as the threefold way.[5] The intuition of Dyson was that the recurrence of the number 3 originates from the Frobenius theorem of algebra, that is, there exist precisely three associative division algebras, namely, the real numbers, the complex numbers, and the real quaternions, over the field of real numbers.

An elementary intuition for why there could be three ensembles is that the level repulsion for a Hermitean 2×2 matrix made of real, complex, or real quaternion

[4] E. P. Wigner, Ann. Math. **53**(1), 36–67 (1951); **62**(3), 548–564 (1955); **65**(2), 203–207 (1957); **67**(2), 325–327 (1958); Math. Proc. Cambridge Phil. Soc. **47**(4), 790–798 (1951). [Wigner (1951a, 1955, 1957, 1958, 1951b)].

[5] F. J. Dyson, J. Math. Phys. **3**(1), 140–156 (1962); **3**(1), 157–165 (1962); **3**(1), 166–175 (1962). [Dyson (1962a,b,c)].

numbers can be quantified by the number β of real parameters needed to switch off the level repulsion, namely,

$$\beta = 1, 2, 4, \tag{7.37}$$

respectively (**Exercises** 7.3 and 7.4).

It turns out that the probability density functions $p_\beta(s)$, with s the spacing between two consecutive energy eigenvalues divided by the mean level spacing, are the three universal functions of s that depend only on $\beta = 1, 2, 4$ and are given by (**Exercise** 7.5)

$$p_\beta(s) = N_\beta\, s^\beta\, e^{-c_\beta\, s^2}, \qquad (N_\beta, c_\beta) = \begin{cases} \left(\frac{\pi}{2}, \frac{\pi}{4}\right), & \text{for } \beta = 1, \\ \left(\frac{32}{\pi^2}, \frac{4}{\pi}\right), & \text{for } \beta = 2, \\ \left(\frac{2^{18}}{3^6\pi^3}, \frac{64}{9\pi}\right), & \text{for } \beta = 4, \end{cases} \tag{7.38}$$

for the Gaussian ensembles of random Hamiltonians.[6] The Wigner-Dyson ensembles with $\beta = 1, 2, 4$ are also called the orthogonal, unitary, and symplectic ensembles, respectively.

Quantum chromodynamics is a generalization of quantum electrodynamics obtained by replacing the relativistic electron with (a family of) quarks and the photon by gluons. If all members of the (family of) quarks are assumed to be massless, the Hamiltonian for the quarks, a Dirac Hamiltonian, has a symmetry called the chiral symmetry. In the approximation in which the gluons are not treated as quantum fields, but thought of as random static backgrounds to the quarks, the quark Dirac Hamiltonian can be thought of as being associated with a random single-particle Hamiltonian. This point of view motivated Shuryak and Verbaarschot to introduce three additional ensembles of random Hermitean matrices called the chiral ensembles.[7] The same ensemble had also been introduced independently in physics by Slevin and Nagao in the study of the statistical distribution of the matrix $\log(\mathcal{T}\mathcal{T}^\dagger)$ with $\mathcal{T}$ the transfer matrix of a mesoscopic disordered quantum wire.[8] In mathematics, the chiral ensemble is related to the Laguerre ensemble introduced by Bronk.[9] What distinguishes the chiral ensembles of Hamiltonians from the Wigner ensembles of Hamiltonians is that a chiral Hamiltonian can always be represented by a block off-diagonal Hermitean matrix owing to the chiral symmetry. Otherwise, their entries consist of real numbers for the chiral orthogonal ensemble, complex numbers for the chiral unitary ensemble, and real quaternions for the chiral symplectic ensemble, very much as was the case with the threefold way.

[6] The standard reference for a pedagogical derivation of this result is M. L. Mehta, *Random Matrices,* Elsevier Science, Amsterdam, 2004. [Mehta (2004)].

[7] E. V. Shuryak and J. J. M. Verbaarschot, Nucl. Phys. A **560**, 306–320 (1993); J. J. M. Verbaarschot, Phys. Rev. Lett. **72**(16), 2531–2533 (1994). [Shuryak and Verbaarschot (1993); Verbaarschot (1994)].

[8] K. Slevin and T. Nagao, Phys. Rev. Lett. **70**(5), 635–638 (1993); Phys. Rev. B **50**(4), 2380–2392 (1994). [Slevin and Nagao (1993, 1994)].

[9] B. V. Bronk, J. Math. Phys. **6**(2), 228–237 (1965). [Bronk (1965)].

In the context of condensed matter physics, the Dyson-Wigner ensembles of random matrices capture some universal effects of disorder on a metallic grain or a quantum dot such as the statistics of the level spacings or that of the conductance fluctuations.[10] The chiral symmetry for the chiral ensembles of random matrix theory originates from demanding that the single-particle electronic Hamiltonian obeys a sublattice symmetry,[11] as was the case in the approximation in which we have described polyacetylene so far. The existence of the three chiral ensembles of random matrices suggests that the threefold way is not exhaustive. Hereto, it was also the discipline of condensed matter physics that paved the way for a unifying principle in which to organize the threefold way and chiral ensembles of random matrices, namely, within the tenfold way.

The tenfold way comes into play if the threefold way and chiral ensembles are complemented by random matrices that encode the Bogoliubov-de-Gennes structure (7.12) obeyed by the quasiparticles of a superconductor within a mean-field approximation of superconductivity. Altland and Zirnbauer proposed that some universal properties of superconducting quantum dots are encoded by four, not three, additional ensembles of random matrices obeying the particle-hole (Bogoliubov-de-Gennes) spectral symmetry implied by Eq. (7.12a),[12] bringing the number of ensembles of random matrices to

$$3+3+4=10. \tag{7.39}$$

Altland and Zirnbauer also realized a deep connection between these ten ensembles and the geometry of the ten compact symmetric spaces associated with the classical simple Lie groups SU(N), SO(N), and Sp($2N$) and their coset spaces (see **Appendix** A for the formal definition of a coset space) as is summarized by **Table** 7.1.[13,14,15]

We are now going to explain how one derives **Table** 7.1 when considering unitary evolution operators. For this purpose, we review the definitions and some properties of the classical Lie groups in **Exercise** 7.6.

7.3.1 The Bogoliubov-de-Gennes Symmetry Class D

The symmetry class with the least symmetries is the Bogoliubov-de-Gennes symmetry class D. Let $\mathcal{H}$ be a Hermitean matrix of dimension $N=2,3,\cdots$. Its complex-valued matrix elements must obey the constraints

10 C. W. J. Beenakker, Rev. Mod. Phys. **69**(3), 731–808 (1997). [Beenakker (1997)].

11 R. Gade and F. Wegner, Nucl. Phys. B **360**, 213–218 (1991); R. Gade, Nucl. Phys. B **398**, 499–515 (1993). [Gade and Wegner (1991); Gade (1993)].

12 M. R. Zirnbauer, J. Math. Phys. **37**(10), 4986–5018 (1996); A. Altland and M. R. Zirnbauer, Phys. Rev. B **55**(2), 1142–1161 (1997). [Zirnbauer (1996); Altland and Zirnbauer (1997)].

13 From a purely algebraic point of view, the tenfold way arises from enumerating all the associative division algebras over the field of the real numbers that are compatible with a direct sum decomposition into even and odd elements (a $\mathbb{Z}_2$ grading or superalgebra) needed to encode the conservation of fermion parity from a physicist point of view.

14 C. T. C. Wall, J. Reine Angew. Math. **213**, 187–199 (1964). [Wall (1964)].

15 J. C. Baez, Notices Amer. Math. Soc. **67**, 1599–1601 (2020). [Baez (2020)].

Table 7.1 *The tenfold way: Cartan labels and Altland-Zirnbauer symmetry classes*

Cartan label	TRS	PHS	CHS	$\exp(\mathrm{i}\mathcal{H})$
A (unitary)	0	0	0	$\mathrm{SU}(N)$
AI (orthogonal)	+1	0	0	$\mathrm{SU}(N)/\mathrm{SO}(N)$
AII (symplectic)	−1	0	0	$\mathrm{SU}(2N)/\mathrm{Sp}(2N)$
AIII (chiral unitary)	0	0	1	$\mathrm{SU}(M+N)/\mathrm{S}\big(\mathrm{U}(M)\times\mathrm{U}(N)\big)$
BDI (chiral orthogonal)	+1	+1	1	$\mathrm{SO}(M+N)/[\mathrm{SO}(M)\times\mathrm{SO}(N)]$
CII (chiral symplectic)	−1	−1	1	$\mathrm{Sp}(2M+2N)/[\mathrm{Sp}(2M)\times\mathrm{Sp}(2N)]$
D (BdG)	0	+1	0	$\mathrm{SO}(M),\quad M=2N,2N+1$
C (BdG)	0	−1	0	$\mathrm{Sp}(2N)$
DIII (BdG)	−1	+1	1	$\mathrm{SO}(2M)/\mathrm{U}(M),\quad M=2N,2N+1$
CI (BdG)	+1	−1	1	$\mathrm{Sp}(2N)/\mathrm{U}(N)$

The tenfold way distinguishes the behavior of electronic Hamiltonian $\mathcal{H}$ under reversal of time ($\mathcal{T}$), charge conjugation ($\mathcal{C}$), and the chiral transformation ($\mathcal{S}$), with the gauge convention that $\mathcal{T}^2=\pm1$, $\mathcal{C}^2=\pm1$, $[\mathcal{T},\mathcal{C}]=0$, and $\mathcal{S}^2=1$. The entries under the columns "TRS," "PHS," and "CHS" are the values of $\mathcal{T}^2=\pm1$, $\mathcal{C}^2=\pm1$, $\mathcal{S}^2=1$ when these (spectral) symmetries are present, with the value 0 indicating the absence of these (spectral) symmetries. The column "Cartan label" is the name given by Altland and Zirnbauer to the value of the triplet $(\mathcal{T},\mathcal{C},\mathcal{S})$ that is deduced from the columns "TRS," "PHS," and "CHS" and that characterizes $\mathcal{H}$. These names are chosen, by convention, to match the Cartan notation for the irreducible Riemannian globally symmetric compact spaces associated with the simple Lie groups $\mathrm{SU}(N)$, $\mathrm{SO}(N)$, $\mathrm{Sp}(2N)$, and their coset spaces through the unitary evolution $\exp(\mathrm{i}\mathcal{H})$. The grouping in rows of 3, 3, and 4 follows the time line of applications of random matrix theory in physics.

$$\mathcal{H}_{ij}=\mathcal{H}^{*}_{ji},\qquad i,j=1,\cdots,N, \tag{7.40a}$$

that is $\mathcal{H}$ is parameterized by N^2 real-valued matrix elements. Let Δ be a complex-valued and antisymmetric matrix of dimension $N=2,3,\cdots$. Its complex-valued matrix elements must obey the constraints

$$\Delta_{ij}=-\Delta_{ji},\qquad i,j=1,\cdots,N, \tag{7.40b}$$

that is Δ is parameterized by $N(N-1)$ real-valued matrix elements. Finally, we define the single-particle Bogoliubov-de-Gennes Hamiltonian

$$\mathcal{H}_{\mathrm{BdG}}:=\begin{pmatrix}\mathcal{H} & \Delta\\ \Delta^{\dagger} & -\mathcal{H}^{\mathsf{T}}\end{pmatrix}=-\mathcal{C}\,\mathcal{H}^{*}_{\mathrm{BdG}}\,\mathcal{C}^{-1}=\mathcal{H}^{\dagger}_{\mathrm{BdG}},\qquad \mathcal{C}:=\begin{pmatrix}0 & \mathbb{1}\\ \mathbb{1} & 0\end{pmatrix}. \tag{7.40c}$$

Observe that the single-particle Hamiltonian $\mathcal{H}_{\mathrm{BdG}}$ is a square matrix of dimension $2N$ that is parameterized by

$$N^2+N(N-1)=(2N-1)N \tag{7.41}$$

real-valued parameters. It acts on the single-particle BdG Hilbert space

$$\mathfrak{H}_{\mathrm{BdG}} = \mathbb{C}^N \otimes \mathbb{C}^2, \tag{7.42}$$

that is $\mathfrak{H}_{\mathrm{BdG}}$ is constructed from tensoring the single-particle Hilbert space

$$\mathfrak{H} := \mathbb{C}^N \tag{7.43}$$

of $\mathcal{H}$ with the auxiliary Hilbert space $\mathbb{C}^2$. This auxiliary Hilbert space is introduced for convenience. The two auxiliary degrees of freedom that it encodes are customarily called particle and hole. They should not be confused with the physical interpretation of the single-particle states of $\mathcal{H}$ in terms of particle and holes of a degenerate Fermi gas. The dimension of the single-particle Bogoliubov-de-Gennes Hilbert space grows like N. We shall reserve the Pauli matrices ρ_1, ρ_2, and ρ_3, together with the unit 2×2 matrix ρ_0 for the auxiliary (Bogoliubov-de-Gennes) particle-hole grading. The antiunitary spectral symmetry

$$\mathcal{C}\,\mathcal{H}^{*}_{\mathrm{BdG}}\,\mathcal{C}^{-1} = -\mathcal{H}_{\mathrm{BdG}}, \qquad \mathcal{C} := \mathbb{1} \otimes \rho_1, \qquad \mathcal{C}^2 = \mathbb{1} \otimes \rho_0, \tag{7.44}$$

where $\mathbb{1}$ is the unit $N \times N$ matrix, is called the single-particle Bogoliubov-de-Gennes spectral symmetry.

We now explain how the single-particle Bogoliubov-de-Gennes spectral symmetry (7.44) imposes on the set of all unitary time evolutions $\exp(\mathrm{i}\mathcal{H}_{\mathrm{BdG}})$ matrices of dimension $2N$ the structure of a compact Lie group. To this end, the notation

$$\mathcal{X}_{\mathrm{BdG}} := \mathrm{i}\mathcal{H}_{\mathrm{BdG}} = +\mathcal{C}\,\mathcal{X}^{*}_{\mathrm{BdG}}\,\mathcal{C}^{-1} = -\mathcal{X}^{\dagger}_{\mathrm{BdG}} \tag{7.45}$$

is used. According to **Exercise** 7.7, if $\mathcal{X}_{\mathrm{BdG}}$ and $\mathcal{Y}_{\mathrm{BdG}}$ are both anti-Hermitean matrices of dimension $2N$ that satisfy Eq. (7.45), so is their commutator $[\mathcal{X}_{\mathrm{BdG}}, \mathcal{Y}_{\mathrm{BdG}}]$. In other words, the set of anti-Hermitean matrices of dimension $2N$ that satisfies Eq. (7.45) forms a Lie algebra. Upon exponentiation, the set with elements $\exp(\mathrm{i}\mathcal{H}_{\mathrm{BdG}})$ forms a Lie group.

To identify this Lie algebra and its Lie group, we seek a unitary transformation

$$\mathcal{X}_{\mathrm{BdG}} =: \mathcal{U}^{\dagger}_{\mathrm{M}}\,\mathcal{X}_{\mathrm{M}}\,\mathcal{U}_{\mathrm{M}} = -\mathcal{X}^{\dagger}_{\mathrm{BdG}} \Longleftrightarrow \mathcal{X}_{\mathrm{M}} =: \mathcal{U}_{\mathrm{M}}\,\mathcal{X}_{\mathrm{BdG}}\,\mathcal{U}^{\dagger}_{\mathrm{M}} = -\mathcal{X}^{\dagger}_{\mathrm{M}} \tag{7.46a}$$

such that

$$\mathcal{X}_{\mathrm{M}} = \mathcal{X}^{*}_{\mathrm{M}} = -\mathcal{X}^{\mathsf{T}}_{\mathrm{M}} \tag{7.46b}$$

is real-valued and thus an antisymmetric $2N \times 2N$ matrix. The label M in this putative new basis of $\mathfrak{H}_{\mathrm{BdG}}$ stands for the Majorana representation. If such a unitary transformation exists, we have shown that the $2N \times 2N$ real-valued antisymmetric matrix $\mathcal{X}_{\mathrm{M}} = \mathrm{i}\mathcal{H}_{\mathrm{M}}$ is a generator of the Lie algebra $\mathfrak{so}(2N)$, that is,

$$\begin{aligned}
&e^{t\,\mathcal{X}_{\mathrm{M}}} = e^{\mathrm{i}t\,\mathcal{H}_{\mathrm{M}}} \Longrightarrow \det\left(e^{t\,\mathcal{X}_{\mathrm{M}}}\right) = 1, \qquad \forall t \in \mathbb{R},\\
&I\,e^{t\,\mathcal{X}_{\mathrm{M}}} = I\,e^{\mathrm{i}t\,\mathcal{H}_{\mathrm{M}}},\ I \in \mathrm{O}(2N),\ \det I = -1 \Longrightarrow \det\left(I\,e^{t\,\mathcal{X}_{\mathrm{M}}}\right) = -1,\ \forall t \in \mathbb{R},
\end{aligned} \tag{7.46c}$$

are two elements of the orthogonal group O(2N) (the set of $2N \times 2N$ orthogonal matrices with determinant equal to ± 1) (**Exercise** 7.8). We are going to show that a solution to Eq. (7.46) is provided by the choice

$$\mathcal{U}_{\mathrm{M}} := \mathbb{1} \otimes \frac{1}{\sqrt{2}} \begin{pmatrix} 1 & 1 \\ +\mathrm{i} & -\mathrm{i} \end{pmatrix}. \tag{7.46d}$$

Proof We seek a solution to the reality condition

$$\begin{aligned} \mathcal{X}^{*}_{\mathrm{M}} =& \mathcal{U}^{*}_{\mathrm{M}} \, \mathcal{X}^{*}_{\mathrm{BdG}} \, \mathcal{U}^{\mathsf{T}}_{\mathrm{M}} \\ =& \mathcal{U}^{*}_{\mathrm{M}} \, \mathcal{C} \, \mathcal{X}_{\mathrm{BdG}} \, \mathcal{C}^{-1} \mathcal{U}^{\mathsf{T}}_{\mathrm{M}} \\ =& \mathcal{U}_{\mathrm{M}} \, \mathcal{X}_{\mathrm{BdG}} \, \mathcal{U}^{\dagger}_{\mathrm{M}} \\ \equiv& \mathcal{X}_{\mathrm{M}}. \end{aligned} \tag{7.47}$$

Hence, the unitary transformation $\mathcal{U}_{\mathrm{M}}$ must obey the condition

$$\mathcal{U}_{\mathrm{M}} = \mathcal{U}^{*}_{\mathrm{M}} \, \mathcal{C}. \tag{7.48}$$

This condition holds because of the identity

$$\begin{pmatrix} 1 & 1 \\ +\mathrm{i} & -\mathrm{i} \end{pmatrix}^{*} \begin{pmatrix} 0 & 1 \\ 1 & 0 \end{pmatrix} = \begin{pmatrix} 1 & 1 \\ -\mathrm{i} & +\mathrm{i} \end{pmatrix} \begin{pmatrix} 0 & 1 \\ 1 & 0 \end{pmatrix} = \begin{pmatrix} 1 & 1 \\ +\mathrm{i} & -\mathrm{i} \end{pmatrix}. \tag{7.49}$$

Observe that the transformation of the Bogoliubov-de-Gennes particle-hole transformation $\mathcal{C}\,\mathsf{K}$ (K denotes complex conjugation) under the Majorana unitary transformation is

$$\begin{aligned} \mathcal{C}\,\mathsf{K} \mapsto& \mathcal{U}_{\mathrm{M}} \, \mathcal{C} \, \mathsf{K} \, \mathcal{U}^{\dagger}_{\mathrm{M}} \\ =& \mathcal{U}_{\mathrm{M}} \, \mathcal{C} \, \mathcal{U}^{\mathsf{T}}_{\mathrm{M}} \, \mathsf{K} \\ =& \mathbb{1} \otimes \rho_0 \, \mathsf{K}. \end{aligned} \tag{7.50}$$

□

In mathematics,[16,17] the family of simple complex Lie algebra $\mathfrak{so}(2n, \mathbb{C})$ labeled by $n = 1, 2, 3, \cdots$ is referred to with the symbol $\mathfrak{d}_n$ (**Exercise** 7.6). This is the rational of Altland and Zirnbauer for using the terminology of the symmetry class D in **Table** 7.1. Here, it should be emphasized that the symbol D is attached to a symmetry attribute, namely, the conservation of the fermion parity. This is why in the last column of **Table** 7.1, the unitary time evolution on the line labeled D is associated with the group SO(M) with M allowed to be an odd integer, even though the family of simple complex Lie algebra $\mathfrak{so}(2n+1, \mathbb{C})$ labeled by $n = 0, 1, 2, \cdots$ is associated with the symbol $\mathfrak{b}_n$ in the mathematical literature. In physics, the reality condition (7.46b) leads to the Clifford (Majorana) algebra

[16] S. Helgason, *Differential Geometry, Lie Groups, and Symmetric Spaces,* American Mathematical Soc., 2001. [Helgason (2001)].

[17] D. H. Sattinger and O. L. Weaver, *Lie Groups and Algebras with Applications to Physics, Geometry, and Mechanics,* Springer, New York, 1986. [Sattinger and Weaver (1986)].

upon second quantization, as we shall verify at the end of this section. As shown in Chapter 5, an open (Kitaev) Majorana $c=1$ chain supports a pair of Majorana zero modes in the thermodynamic limit. The unitary time evolution for this pair of Majorana zero modes is that of a Hamiltonian that vanishes [the right-hand side of Eq. (7.62) with $N=1$ and $\xi_1=0$]. Correspondingly, the time evolution of this pair of Majorana zero modes is associated with the single element making up the group $\mathrm{SO}(M=1)$. A fortiori, the unitary time evolution with elements from the group $\mathrm{SO}(M=2N+1)$ necessarily entails a pair of Majorana zero modes, while the remaining N independent energy eigenvalues entering an expression of the form given by the right-hand side of Eq. (7.62) can but need not vanish.

Because $\mathcal{H}_{\mathrm{BdG}}$ is Hermitean, there exists a unitary $2N\times 2N$ matrix $\mathcal{U}_{\mathrm{BdG}}$ with

$$\mathcal{H}_{\mathrm{BdG}}=\mathcal{U}^{\dagger}_{\mathrm{BdG}}\,\mathcal{E}_{\mathrm{BdG}}\,\mathcal{U}_{\mathrm{BdG}},\qquad \mathcal{U}^{-1}_{\mathrm{BdG}}=\mathcal{U}^{\dagger}_{\mathrm{BdG}},\tag{7.51a}$$

where the $2N\times 2N$ Hermitean matrix $\mathcal{E}_{\mathrm{BdG}}$ with

$$\mathrm{tr}|_{\mathfrak{H}_{\mathrm{BdG}}}\,\mathcal{E}_{\mathrm{BdG}}=\mathrm{tr}|_{\mathfrak{H}_{\mathrm{BdG}}}\,\mathcal{H}_{\mathrm{BdG}}=0\tag{7.51b}$$

is diagonal and traceless, with the eigenvalues ξ_μ of $\mathcal{H}_{\mathrm{BdG}}$ along the diagonal. In other words, we have the spectral decomposition of $\mathcal{H}_{\mathrm{BdG}}$ given by

$$\mathcal{H}_{\mathrm{BdG}}:=\sum_{\mu=1}^{2N}\xi_\mu\,\Upsilon_\mu\,\Upsilon^{\dagger}_{\mu}.\tag{7.51c}$$

Here, the column vector $\Upsilon_\mu\in\mathbb{C}^{2N}$ is an eigenvector of $\mathcal{H}_{\mathrm{BdG}}$ with the eigenvalue ξ_μ and $\mu=1,\cdots,2N$. The choice of $\mathcal{U}_{\mathrm{BdG}}$ is not unique. For example, we may permute the eigenvalues so that there is no guarantee that $\mathcal{E}_{\mathrm{BdG}}=-\mathcal{C}\,\mathcal{E}_{\mathrm{BdG}}\,\mathcal{C}^{-1}$, even though the single-particle Bogoliubov-de-Gennes spectral symmetry (7.44) does imply the Bogoliubov-de-Gennes spectral condition

$$\Big\{(\Upsilon_\mu,\xi_\mu)\ \Big|\ \mu=1,\cdots,2N\Big\}=\Big\{(\Upsilon_i,\xi_i;\mathcal{C}\,\Upsilon^{*}_{i},-\xi_i)\ \Big|\ i=1,\cdots,N\Big\}.\tag{7.51d}$$

However, we may always choose $\mathcal{U}^{-1}_{\mathrm{BdG}}=\mathcal{U}^{\dagger}_{\mathrm{BdG}}\in\mathrm{U}(2N)$ such that it obeys the compatibility condition

$$\mathcal{C}\,\mathcal{U}^{*}_{\mathrm{BdG}}\,\mathcal{C}^{-1}=\mathcal{U}_{\mathrm{BdG}}\iff\mathcal{U}^{\mathsf{T}}_{\mathrm{BdG}}\,\mathcal{C}\,\mathcal{U}_{\mathrm{BdG}}=\mathcal{C}\tag{7.51e}$$

so that

$$\mathcal{E}_{\mathrm{BdG}}=-\mathcal{C}\,\mathcal{E}_{\mathrm{BdG}}\,\mathcal{C}\iff\mathcal{E}_{\mathrm{BdG}}=\begin{pmatrix}\mathcal{E}&0\\0&-\mathcal{E}\end{pmatrix},\qquad \mathcal{E}=\mathrm{diag}\begin{pmatrix}\xi_1&\cdots&\xi_N\end{pmatrix}.\tag{7.51f}$$

The generalization of the Bogoliubov-de-Gennes Hamiltonian (7.6) that we seek is

$$\widehat{H}:=\sum_{i,j=1}^{N}\left(\hat{c}^{\dagger}_{i}\,\mathcal{H}_{ij}\hat{c}_{j}+\Delta_{ij}\,\hat{c}^{\dagger}_{i}\,\hat{c}^{\dagger}_{j}+\Delta^{*}_{ij}\,\hat{c}_{j}\,\hat{c}_{i}\right),\tag{7.52a}$$

where

$$\{\hat{c}_i, \hat{c}_j^\dagger\} = \delta_{ij}, \qquad \{\hat{c}_i^\dagger, \hat{c}_j^\dagger\} = \{\hat{c}_j, \hat{c}_i\} = 0. \tag{7.52b}$$

The dimension of the fermionic Fock space $\mathfrak{F}$ on which $\widehat{H}$ acts grows like 2^N. If we introduce the Nambu spinors

$$\widehat{\Psi}^\dagger := \begin{pmatrix} \hat{c}_1^\dagger & \cdots & \hat{c}_N^\dagger & \hat{c}_1 & \cdots & \hat{c}_N \end{pmatrix}, \qquad \widehat{\Psi} := \begin{pmatrix} \hat{c}_1 \\ \vdots \\ \hat{c}_N \\ \hat{c}_1^\dagger \\ \vdots \\ \hat{c}_N^\dagger \end{pmatrix}, \tag{7.53a}$$

together with the single-particle Bogoliubov-de-Gennes Hamiltonian (7.40c), we may then write

$$\widehat{H} = \frac{1}{2}\mathrm{tr}|_{\mathfrak{H}}\, \mathcal{H} + \frac{1}{2}\widehat{\Psi}^\dagger\, \mathcal{H}_{\mathrm{BdG}}\, \widehat{\Psi}, \tag{7.53b}$$

whereby the $2N$ components of $\widehat{\Psi}^\dagger$ and $\widehat{\Psi}$ obey the algebra

$$\left\{\widehat{\Psi}_\mu, \widehat{\Psi}_{\mu'}^\dagger\right\} = \delta_{\mu\mu'}, \qquad \left\{\widehat{\Psi}_\mu, \widehat{\Psi}_{\mu'}\right\} = \left\{\widehat{\Psi}_\mu^\dagger, \widehat{\Psi}_{\mu'}^\dagger\right\} = \mathcal{C}_{\mu\mu'}, \tag{7.53c}$$

for $\mu, \mu' = 1, \cdots, 2N$.

The single-particle Bogoliubov-de-Gennes spectral symmetry (7.44) implies that the unitary transformation $\widehat{\mathrm{C}}$ on the fermionic Fock space $\mathfrak{F}$ defined by

$$\widehat{\mathrm{C}}\left(\alpha^*\, \widehat{\Psi}_\mu^\dagger + \beta\, \widehat{\Psi}_\mu\right)\widehat{\mathrm{C}}^{-1} := \sum_{\mu'=1}^{2N}\left(\alpha^*\, \widehat{\Psi}_{\mu'}\, \mathcal{C}_{\mu'\mu}^* + \beta\, \mathcal{C}_{\mu'\mu}\widehat{\Psi}_{\mu'}^\dagger\right) \tag{7.54a}$$

for any complex numbers α and β and for $\mu = 1, \cdots, 2N$, is a unitary symmetry,

$$\widehat{\mathrm{C}}\,\widehat{H}\,\widehat{\mathrm{C}}^{-1} = \widehat{H}. \tag{7.54b}$$

We shall call this unitary symmetry a Bogoliubov-de-Gennes particle-hole symmetry (PHS).

We choose a transformation $\mathcal{U}_{\mathrm{BdG}}$ of the form (7.51) on the single-particle Hilbert space $\mathfrak{H}_{\mathrm{BdG}}$ that implements a diagonalization compatible with the Bogoliubov-de-Gennes particle-hole symmetry. The many-body transformation

$$\widehat{\Upsilon}^\dagger := \widehat{\Psi}^\dagger\, \mathcal{U}_{\mathrm{BdG}}^\dagger, \qquad \widehat{\Upsilon} := \mathcal{U}_{\mathrm{BdG}}\, \widehat{\Psi}, \tag{7.55a}$$

on the Nambu spinors is called a Bogoliubov transformation. Under this transformation, the components

$$\widehat{\Upsilon}^{\dagger} \equiv \begin{pmatrix} \hat{\gamma}_1^{\dagger} & \cdots & \hat{\gamma}_N^{\dagger} & \hat{\gamma}_1 & \cdots & \hat{\gamma}_N \end{pmatrix}, \qquad \widehat{\Upsilon} \equiv \begin{pmatrix} \hat{\gamma}_1 \\ \vdots \\ \hat{\gamma}_N \\ \hat{\gamma}_1^{\dagger} \\ \vdots \\ \hat{\gamma}_N^{\dagger} \end{pmatrix}, \tag{7.55b}$$

obey the algebra

$$\left\{\widehat{\Upsilon}_{\mu}, \widehat{\Upsilon}^{\dagger}_{\mu'}\right\} = \delta_{\mu\mu'}, \qquad \left\{\widehat{\Upsilon}_{\mu}, \widehat{\Upsilon}_{\mu'}\right\} = \left\{\widehat{\Upsilon}^{\dagger}_{\mu}, \widehat{\Upsilon}^{\dagger}_{\mu'}\right\} = \mathcal{C}_{\mu\mu'}, \tag{7.55c}$$

for $\mu, \mu' = 1, \cdots, 2N$. The fermion algebra

$$\{\hat{\gamma}_i, \hat{\gamma}_j^{\dagger}\} = \delta_{ij}, \qquad \{\hat{\gamma}_i, \hat{\gamma}_j\} = \{\hat{\gamma}_i^{\dagger}, \hat{\gamma}_j^{\dagger}\} = 0, \tag{7.55d}$$

for $i, j = 1, \cdots, N$ follows. Moreover, Eq. (7.53b) becomes

$$\widehat{H} = \frac{1}{2}\mathrm{tr}|_{\mathfrak{H}}\, \mathcal{H} + \frac{1}{2}\sum_{i=1}^{N} \xi_i \left(\hat{\gamma}_i^{\dagger}\, \hat{\gamma}_i - \hat{\gamma}_i\, \hat{\gamma}_i^{\dagger}\right). \tag{7.55e}$$

The many-body ground state $|\mathrm{BCS}\rangle$, where BCS stands for Bardeen, Cooper, and Schrieffer, of $\widehat{H}$ is defined by the pair of conditions

$$\hat{\gamma}_i\, |\mathrm{BCS}\rangle = 0, \qquad \xi_i > 0, \tag{7.56a}$$

and

$$\hat{\gamma}_i^{\dagger}\, |\mathrm{BCS}\rangle = 0, \qquad \xi_i < 0, \tag{7.56b}$$

for any $i = 1, \cdots, N$. With the BCS ground state in mind, we do the manipulation

$$\begin{aligned}
\widehat{H} &= \frac{1}{2}\mathrm{tr}|_{\mathfrak{H}}\, \mathcal{H} + \frac{1}{2}\sum_{i=1,\cdots,N}^{\xi_i>0} \xi_i \left(\hat{\gamma}_i^{\dagger}\, \hat{\gamma}_i - \hat{\gamma}_i\, \hat{\gamma}_i^{\dagger}\right) + \frac{1}{2}\sum_{i=1,\cdots,N}^{\xi_i<0} \xi_i \left(\hat{\gamma}_i^{\dagger}\, \hat{\gamma}_i - \hat{\gamma}_i\, \hat{\gamma}_i^{\dagger}\right) \\
&= \frac{1}{2}\mathrm{tr}|_{\mathfrak{H}}\, \mathcal{H} + \frac{1}{2}\sum_{i=1,\cdots,N}^{\xi_i>0} \xi_i \left(2\hat{\gamma}_i^{\dagger}\, \hat{\gamma}_i - 1\right) + \frac{1}{2}\sum_{i=1,\cdots,N}^{\xi_i<0} \xi_i \left(1 - 2\hat{\gamma}_i\, \hat{\gamma}_i^{\dagger}\right) \\
&= \frac{1}{2}\mathrm{tr}|_{\mathfrak{H}}\, \mathcal{H} - \frac{1}{2}\sum_{i=1}^{N} |\xi_i| + \sum_{i=1,\cdots,N}^{\xi_i>0} |\xi_i|\, \hat{\gamma}_i^{\dagger}\, \hat{\gamma}_i + \sum_{i=1,\cdots,N}^{\xi_i<0} |\xi_i|\, \hat{\gamma}_i\, \hat{\gamma}_i^{\dagger},
\end{aligned} \tag{7.57a}$$

whereby we recognize that

$$\langle \mathrm{BCS}|\,\widehat{H}\,|\mathrm{BCS}\rangle = \frac{1}{2}\mathrm{tr}|_{\mathfrak{H}}\,\mathcal{H} - \frac{1}{2}\sum_{i=1}^{N}|\xi_i|\,. \tag{7.57b}$$

Normal ordering of $\widehat{H}$ with respect to the ground state $|\mathrm{BCS}\rangle$ is denoted by $:\widehat{H}:$ and defined by subtracting from $\widehat{H}$ the expectation value of $\widehat{H}$ in the ground state $|\mathrm{BCS}\rangle$,

$$\begin{aligned} :\widehat{H}: &:= \widehat{H} - \langle \mathrm{BCS}|\,\widehat{H}\,|\mathrm{BCS}\rangle \\ &= \sum_{i=1,\cdots,N}^{\xi_i>0} |\xi_i|\,\hat{\gamma}^\dagger_i\,\hat{\gamma}_i + \sum_{i=1,\cdots,N}^{\xi_i<0} |\xi_i|\,\hat{\gamma}_i\,\hat{\gamma}^\dagger_i. \end{aligned} \tag{7.58}$$

This normal ordered Hamiltonian has only positive single-particle energy eigenvalues as it measures all many-body energies relative to the ground-state energy.

We close this discussion of a superconducting Hamiltonian in the symmetry class D by presenting Hamiltonians (7.52) and (7.58) in their Majorana representations. As a prelude to this task, observe that if we represent the fermionic creation and annihilation operators $\hat{c}^\dagger$ and $\hat{c}$ by the matrices

$$\begin{pmatrix} 0 & 1 \\ 0 & 0 \end{pmatrix}, \qquad \begin{pmatrix} 0 & 0 \\ 1 & 0 \end{pmatrix}, \tag{7.59}$$

respectively, we deduce that the pair of Majorana operators $\hat{\eta} := \sqrt{2}\,\mathrm{Re}\,\hat{c}$ and $\hat{\zeta} := \sqrt{2}\,\mathrm{Im}\,\hat{c}$ are then represented by the pair of matrices

$$\frac{1}{\sqrt{2}}\begin{pmatrix} 0 & 1 \\ 1 & 0 \end{pmatrix}, \qquad \frac{1}{\sqrt{2}}\begin{pmatrix} 0 & +\mathrm{i} \\ -\mathrm{i} & 0 \end{pmatrix}. \tag{7.60}$$

In other words, creation and annihilation fermion operators are proportional to the raising and lowering ladder operators in the Pauli algebra, while the corresponding pair of Majorana operators are proportional to the two Pauli matrices entering the Pauli ladder operators.

One verifies that the transformation

$$\hat{c}^\dagger_i := \frac{1}{\sqrt{2}}\left(\hat{\eta}_i - \mathrm{i}\hat{\zeta}_i\right), \qquad \hat{c}_i := \frac{1}{\sqrt{2}}\left(\hat{\eta}_i + \mathrm{i}\hat{\zeta}_i\right), \tag{7.61a}$$

owing to the orthonormal normalization that we have chosen, gives the Majorana algebra

$$\hat{\eta}_i = \hat{\eta}^\dagger_i, \quad \left\{\hat{\eta}_i, \hat{\eta}_j\right\} = \delta_{ij}, \quad \hat{\zeta}_i = \hat{\zeta}^\dagger_i, \quad \left\{\hat{\zeta}_i, \hat{\zeta}_j\right\} = \delta_{ij}, \quad \left\{\hat{\zeta}_i, \hat{\eta}_j\right\} = 0, \tag{7.61b}$$

for any $i,j = 1,\cdots,N$, together with the Majorana representation

$$\begin{aligned} \widehat{H} = &+\frac{1}{2}\sum_{i,j=1}^{N}\left(\hat{\eta}_i\,\mathcal{H}_{ij}\,\hat{\eta}_j + \hat{\zeta}_i\,\mathcal{H}_{ij}\,\hat{\zeta}_j + 2\,\hat{\eta}_i\,\mathrm{iRe}\,\mathcal{H}_{ij}\,\hat{\zeta}_j\right) \\ &+\sum_{i,j=1}^{N}\left(\mathrm{i}\,\hat{\eta}_i\,\mathrm{Im}\,\Delta_{ij}\,\hat{\eta}_j - \mathrm{i}\,\hat{\zeta}_i\,\mathrm{Im}\,\Delta_{ij}\,\hat{\zeta}_j - 2\,\mathrm{i}\hat{\eta}_i\,\mathrm{Re}\,\Delta_{ij}\,\hat{\zeta}_j\right). \end{aligned} \tag{7.61c}$$

Alternatively, if we replace $\hat{c}$ by $\hat{\gamma}$ on the left-hand side of Eq. (7.61a), we get in place of the Majorana representation (7.61c) of Eq. (7.52), the Majorana representation of Eq. (7.58) given by

$$:\widehat{H}: = \frac{1}{2}\sum_{i=1}^{N}|\xi_i| + \sum_{i=1,\cdots,N}^{\xi_i>0} |\xi_i|\, \mathrm{i}\hat{\eta}_i\,\hat{\zeta}_i + \sum_{i=1,\cdots,N}^{\xi_i<0} |\xi_i|\, \mathrm{i}\hat{\zeta}_i\,\hat{\eta}_i. \tag{7.62}$$

The quadratic form in terms of the operator-valued vector $\widehat{\Upsilon}_{\mathrm{M}}$ with

$$\left(\widehat{\Upsilon}_{\mathrm{M}}\right)^{\mathsf{T}} \equiv \begin{pmatrix}\hat{\eta}_1 & \cdots & \hat{\eta}_N & \hat{\zeta}_1 & \cdots & \hat{\zeta}_N\end{pmatrix}^{\mathsf{T}} \tag{7.63}$$

that appears on the right-hand side of Eq. (7.62) defines a Majorana representation of the diagonal Hamiltonian (7.58). The Hermitean $2N \times 2N$ matrix associated with this quadratic form is purely imaginary. Hence, the reality condition (7.46b) is satisfied.

7.3.2 The Bogoliubov-de-Gennes Symmetry Class C

So far, for any $N = 1, 2, 3, \cdots$, we have defined the symmetry class D to be the set $\{\mathcal{H}_{\mathrm{BdG}}\}$ of $2N \times 2N$ Hermitean matrices obeying the Bogoliubov-de-Gennes spectral symmetry (7.44). No reference was made to the fact that electrons that participate to a superconducting ground state have spin-1/2 degrees of freedom. This omission implicitly assumes that the electronic spin-SU(2) symmetry is completely broken, say by spin-orbit coupling. What if the electronic spin-SU(2) symmetry was preserved explicitly by the Hamiltonian and unbroken in the superconducting ground state? If the electronic spin-1/2 symmetry was preserved, one would need to account for the spin-1/2 degrees of freedom by reserving a two-valued label. This means that we should start with $4N \times 4N$ Hermitean matrices obeying the Bogoliubov-de-Gennes spectral symmetry (7.44). Let

$$\mathcal{X}_{\mathrm{BdG}} := \mathrm{i}\mathcal{H}_{\mathrm{BdG}} \equiv \mathrm{i}\begin{pmatrix}\mathcal{H} & \Delta \\ \Delta^{\dagger} & -\mathcal{H}^{\mathsf{T}}\end{pmatrix} \equiv \begin{pmatrix}\mathcal{A} & \mathcal{B} \\ \mathcal{C} & \mathcal{D}\end{pmatrix}, \tag{7.64a}$$

where $\mathcal{H}$ and Δ are $2N \times 2N$ with $N = 1, 2, 3, \cdots$ complex-valued matrices obeying the conditions

$$\mathcal{H} = \mathcal{H}^{\dagger},\ \Delta = -\Delta^{\mathsf{T}} \iff \mathcal{A} = -\mathcal{A}^{\dagger},\ \mathcal{B} = -\mathcal{B}^{\mathsf{T}},\ \mathcal{C} = -\mathcal{B}^{\dagger},\ \mathcal{D} = -\mathcal{A}^{\mathsf{T}}. \tag{7.64b}$$

If we make explicit the tensor structure in the Bogoliubov-de-Gennes particle-hole grading, we have the representation

$$\mathcal{X}_{\mathrm{BdG}} = \mathcal{A}\otimes\rho_{\mathrm{pp}} + \mathcal{B}\otimes\rho_{\mathrm{ph}} + \mathcal{C}\otimes\rho_{\mathrm{hp}} + \mathcal{D}\otimes\rho_{\mathrm{hh}}, \tag{7.65a}$$

where

$$\rho_{\mathrm{pp}} := \frac{\rho_0+\rho_3}{2},\quad \rho_{\mathrm{ph}} := \frac{\rho_1+\mathrm{i}\rho_2}{2},\quad \rho_{\mathrm{hp}} := \frac{\rho_1-\mathrm{i}\rho_2}{2},\quad \rho_{\mathrm{hh}} := \frac{\rho_0-\rho_3}{2}, \tag{7.65b}$$

and

$$\mathcal{A} = -\mathcal{A}^{\dagger}, \qquad \mathcal{B} = -\mathcal{B}^{\mathsf{T}}, \qquad \mathcal{C} = -\mathcal{B}^{\dagger}, \qquad \mathcal{D} = -\mathcal{A}^{\mathsf{T}}. \tag{7.65c}$$

Moreover, if we do the same with the spin-1/2 degrees of freedom, whereby we reserve the Pauli matrices σ_1, σ_2, and σ_3, together with the unit 2×2 matrix σ_0 for this (tensorial) factor $\mathbb{C}^{2N}$ of the Hilbert space $\mathbb{C}^{4N}$, we have

$$\mathcal{A} \equiv \mathsf{A} \otimes (\sigma_0 + \boldsymbol{a} \cdot \boldsymbol{\sigma}), \qquad \mathsf{A} = -\mathsf{A}^{\dagger}, \tag{7.65d}$$

with $\boldsymbol{a} \in \mathbb{R}^3$ and A a complex-valued $N \times N$ matrix together with

$$\mathcal{B} \equiv \mathsf{B} \otimes \mathrm{i}\sigma_2, \qquad \mathsf{B} = \mathsf{B}^{\mathsf{T}}, \tag{7.65e}$$

with B a complex-valued $N \times N$ matrix. We must also supply the representation of the spin-1/2 rotation in this Bogoliubov-de-Gennes tensorial basis. These are the $\mathfrak{su}(2)$ generators

$$\mathcal{J}_{\mathrm{BdG}} := \mathcal{J} \otimes \rho_{\mathrm{pp}} - \mathcal{J}^{\mathsf{T}} \otimes \rho_{\mathrm{hh}}, \qquad \mathcal{J} := \mathbb{1} \otimes \boldsymbol{\sigma}, \qquad \mathcal{J}^{\mathsf{T}} = \mathbb{1} \otimes \boldsymbol{\sigma}^{\mathsf{T}}, \tag{7.66}$$

with $\mathbb{1}$ the $N \times N$ unit matrix.

The Bogoliubov-de-Gennes symmetry class C is defined to be isomorphic to the set $\{\mathcal{X}_{\mathrm{BdG}} \equiv \mathrm{i}\mathcal{H}_{\mathrm{BdG}}\}$ of $4N \times 4N$ matrices defined by Eq. (7.65) that commute with all three generators $\mathcal{J}$ of SU(2) rotations of the spin-1/2 degrees of freedom defined by Eq. (7.66). One verifies that

$$[\mathcal{X}_{\mathrm{BdG}}, \mathcal{J}_{\mathrm{BdG}}] = 0 \iff [\mathcal{A}, \mathcal{J}] = 0, \qquad \mathcal{B}\,\mathcal{J}^{\mathsf{T}} + \mathcal{J}\,\mathcal{B} = 0. \tag{7.67}$$

The condition

$$[\mathcal{A}, \mathcal{J}] = 0 \iff \boldsymbol{a} = \boldsymbol{0} \tag{7.68}$$

is constraining, whereas the condition

$$\mathcal{B}\,\mathcal{J}^{\mathsf{T}} + \mathcal{J}\,\mathcal{B} = 0 \tag{7.69}$$

is not since

$$\sigma_2\,\boldsymbol{\sigma}^{\mathsf{T}} + \boldsymbol{\sigma}\,\sigma_2 = 0 \tag{7.70}$$

implies that Eq. (7.68) holds for any B in the tensor product (7.65e). Hence, we may write

$$\mathcal{X}_{\mathrm{BdG}} = \begin{pmatrix} +\mathsf{A} & 0 & 0 & +\mathsf{B} \\ 0 & +\mathsf{A} & -\mathsf{B} & 0 \\ 0 & +\mathsf{B}^{\dagger} & -\mathsf{A}^{\mathsf{T}} & 0 \\ -\mathsf{B}^{\dagger} & 0 & 0 & -\mathsf{A}^{\mathsf{T}} \end{pmatrix}, \qquad \mathsf{A} = -\mathsf{A}^{\dagger}, \qquad \mathsf{B} = +\mathsf{B}^{\mathsf{T}}, \tag{7.71}$$

for any $4N \times 4N$ matrix belonging to the symmetry class C.

Equation (7.71) makes explicit the fact that any $4N \times 4N$ matrix belonging to the symmetry class C is decomposable. If the rows and column in the 4×4 grading encoding the Bogoliubov-de-Gennes particle-hole and spin gradings are permuted according to the rule $(1,2,3,4) \mapsto (1,4,2,3)$, we obtain the representation

$$\mathcal{X}_{\mathrm{BdG}} \mapsto \begin{pmatrix} +\mathsf{A} & +\mathsf{B} & 0 & 0 \\ -\mathsf{B}^{\dagger} & -\mathsf{A}^{\mathsf{T}} & 0 & 0 \\ 0 & 0 & +\mathsf{A} & -\mathsf{B} \\ 0 & 0 & +\mathsf{B}^{\dagger} & -\mathsf{A}^{\mathsf{T}} \end{pmatrix}, \qquad \mathsf{A} = -\mathsf{A}^{\dagger}, \qquad \mathsf{B} = +\mathsf{B}^{\mathsf{T}}. \tag{7.72}$$

The lower right block is related to the upper left block by $\mathsf{B} \to -\mathsf{B}$ (**Exercise** 7.9).

Without loss of generality, we only need to study the irreducible block

$$\mathcal{X}^{\mathrm{ire}}_{\mathrm{BdG}} := \begin{pmatrix} +\mathsf{A} & +\mathsf{B} \\ -\mathsf{B}^{\dagger} & -\mathsf{A}^{\mathsf{T}} \end{pmatrix}, \qquad \mathsf{A} = -\mathsf{A}^{\dagger}, \qquad \mathsf{B} = +\mathsf{B}^{\mathsf{T}}. \tag{7.73}$$

Observe that (**Exercise** 7.10) the condition $\mathsf{A} = -\mathsf{A}^{\dagger}$ follows from demanding that

$$\mathcal{X}^{\mathrm{ire}}_{\mathrm{BdG}} = -\left(\mathcal{X}^{\mathrm{ire}}_{\mathrm{BdG}}\right)^{\dagger}, \tag{7.74a}$$

while the condition $\mathsf{B} = +\mathsf{B}^{\mathsf{T}}$ follows from demanding that

$$\mathcal{C}^{\mathrm{ire}} \left(\mathcal{X}^{\mathrm{ire}}_{\mathrm{BdG}}\right)^{\mathsf{T}} \left(\mathcal{C}^{\mathrm{ire}}\right)^{-1} = -\mathcal{X}^{\mathrm{ire}}_{\mathrm{BdG}}, \tag{7.74b}$$

whereby

$$\mathcal{C}^{\mathrm{ire}} := \mathbb{1} \otimes \mathrm{i} \begin{pmatrix} 0 & -\mathrm{i} \\ +\mathrm{i} & 0 \end{pmatrix}, \qquad \left(\mathcal{C}^{\mathrm{ire}}\right)^{2} = -\mathbb{1} \otimes \begin{pmatrix} 1 & 0 \\ 0 & 1 \end{pmatrix}. \tag{7.74c}$$

The set of $2N \times 2N$ matrices $\{\mathcal{X}^{\mathrm{ire}}_{\mathrm{BdG}}\}$ that obeys conditions (7.74) is closed under the commutator and, by definition, is the Lie algebra $\mathfrak{sp}(2N)$.[18] We conclude that imposing the spin-1/2 rotation symmetry on any $4N \times 4N$ Hamiltonian from the symmetry class D, that is, a matrix unitarily isomorph to a matrix from $\mathfrak{so}(4N)$ multiplied by $-\mathrm{i}$, brings it to a decomposable $4N \times 4N$ matrix that is in one-to-one correspondence with matrices from $\mathfrak{sp}(2N)$. In mathematics, the family of simple complex Lie algebra $\mathfrak{sp}(2n, \mathbb{C})$ labeled by $n = 1, 2, 3, \cdots$ is referred to with the symbol $\mathfrak{c}_n$ [see Eq. (A.32) and **Exercise** 7.6]. This is the rational of Altland and Zirnbauer for using the terminology of the symmetry class C in **Table** 7.1.

The single-particle Bogoliubov-de-Gennes Hamiltonian $\mathcal{H}^{\mathrm{ire}}_{\mathrm{BdG}} = -\mathrm{i}\mathcal{X}^{\mathrm{ire}}_{\mathrm{BdG}}$ corresponding to the irreducible block (7.73) can be diagonalized to the traceless $2N \times 2N$ Hermitean diagonal matrix

$$\mathcal{E}^{\mathrm{ire}}_{\mathrm{BdG}} := \mathcal{U}^{\mathrm{ire}}_{\mathrm{BdG}}\, \mathcal{H}^{\mathrm{ire}}_{\mathrm{BdG}}\, \left(\mathcal{U}^{\mathrm{ire}}_{\mathrm{BdG}}\right)^{\dagger}, \qquad \mathrm{tr}|_{\mathbb{C}^{2N}}\, \mathcal{E}^{\mathrm{ire}}_{\mathrm{BdG}} = \mathrm{tr}|_{\mathbb{C}^{2N}}\, \mathcal{H}^{\mathrm{ire}}_{\mathrm{BdG}} = 0, \tag{7.75a}$$

with a unitary matrix

$$\left(\mathcal{U}^{\mathrm{ire}}_{\mathrm{BdG}}\right)^{-1} = \left(\mathcal{U}^{\mathrm{ire}}_{\mathrm{BdG}}\right)^{\dagger} \in \mathrm{U}(2N) \tag{7.75b}$$

[18] The matrix $\mathcal{C}^{\mathrm{ire}}$ corresponds to the matrix J_{2n} of **Exercise** 7.6.

that is also symplectic

$$\mathrm{Sp}(2N) \ni \mathcal{U}^{\mathrm{ire}}_{\mathrm{BdG}} = \mathcal{C}^{\mathrm{ire}}\, \left(\mathcal{U}^{\mathrm{ire}}_{\mathrm{BdG}}\right)^{*} \left(\mathcal{C}^{\mathrm{ire}}\right)^{-1} \iff \left(\mathcal{U}^{\mathrm{ire}}_{\mathrm{BdG}}\right)^{\mathsf{T}} \mathcal{C}^{\mathrm{ire}}\, \mathcal{U}^{\mathrm{ire}}_{\mathrm{BdG}} = \mathcal{C}^{\mathrm{ire}}. \qquad (7.75c)$$

Proof Given are the pair of matrices X and C obeying

$$X = -X^{\dagger}, \qquad C^{\mathsf{T}} = -C, \qquad C^2 = -1, \qquad X = -C\, X^{\mathsf{T}} C^{-1}. \qquad (7.76a)$$

The condition $X = -X^{\dagger}$ implies that there exists the pair of matrices U and D such that

$$U^{-1} = U^{\dagger}, \qquad D = -D^{*} \text{ is diagonal}, \qquad X = U\, D\, U^{\dagger}. \qquad (7.76b)$$

In terms of this pair of matrices,

$$\begin{aligned}
0 &= X + C\, X^{\mathsf{T}}\, C^{-1} \\
&= U\, D\, U^{\dagger} + C\, \left(U\, D\, U^{\dagger}\right)^{\mathsf{T}}\, C^{-1} \\
&= U\, D\, U^{\dagger} + C\, \left(U^{*}\, D^{\mathsf{T}}\, U^{\mathsf{T}}\right)\, C^{-1} \\
\overset{D \text{ is diagonal}}{} &= U\, D\, U^{\dagger} + C\, \left(U^{*}\, D\, U^{\mathsf{T}}\right)\, C^{-1} \\
&= \left(U\, C^{-1}\right) \left(C\, D\, C^{-1}\right) \left(C\, U^{\dagger}\right) + C\, \left(U^{*}\, D\, U^{\mathsf{T}}\right)\, C^{-1}.
\end{aligned} \qquad (7.76c)$$

If we demand that

$$D = -C\, D^{\mathsf{T}}\, C^{-1} = -C\, D\, C^{-1}, \qquad (7.76d)$$

we can then rewrite condition (7.76c) as

$$0 = -\left(U\, C^{-1}\right)\, D\, \left(C\, U^{\dagger}\right) + \left(C\, U^{*}\right)\, D\, \left(U^{\mathsf{T}}\, C^{-1}\right). \qquad (7.76e)$$

Condition (7.76e) is met if

$$\begin{aligned}
U &= +e^{+\mathrm{i}\phi}\, C\, U^{*}\, C = -e^{+\mathrm{i}\phi}\, C\, U^{*}\, C^{-1}, \\
U^{\dagger} &= +e^{-\mathrm{i}\phi}\, C^{-1}\, U^{\mathsf{T}}\, C^{-1} = -e^{-\mathrm{i}\phi}\, C\, U^{\mathsf{T}}\, C^{-1},
\end{aligned} \qquad (7.76f)$$

for some phase $\phi \in [0, 2\pi[$. As we may always write

$$U = e^{Y}, \qquad Y = -Y^{\dagger}, \qquad (7.76g)$$

condition (7.76f) becomes

$$\begin{aligned}
e^{Y} &= -\, e^{+\mathrm{i}\phi}\, C\, e^{Y^{*}}\, C^{-1} \\
&= -\, e^{+\mathrm{i}\phi}\, e^{C\, Y^{*} C^{-1}} \\
\overset{Y = -Y^{\dagger}}{} &= -\, e^{+\mathrm{i}\phi}\, e^{-C\, Y^{\mathsf{T}}\, C^{-1}}.
\end{aligned} \qquad (7.76h)$$

If we choose $\phi = \pi$ and demand that $Y = -Y^{\dagger}$ also obeys

$$Y = -C\, Y^{\mathsf{T}}\, C^{-1}, \qquad (7.76i)$$

condition (7.76h) then becomes the condition that Y belongs to the symplectic Lie algebra. □

The Nambu representation to the counterpart of Eq. (7.52) in the symmetry class C is

$$\begin{aligned}\widehat{H} :=& -\frac{\mathrm{i}}{2}\mathrm{tr}|_{\mathfrak{H}}\,\mathsf{A} - \frac{\mathrm{i}}{2}\sum_{m,n=1}^{N}\begin{pmatrix}\widehat{\Psi}^{\dagger}_{m,\uparrow} & \widehat{\Psi}_{m,\downarrow}\end{pmatrix}\begin{pmatrix}+\mathsf{A}_{mn} & +\mathsf{B}_{mn}\\ -\mathsf{B}^{\dagger}_{mn} & -\mathsf{A}^{\mathsf{T}}_{mn}\end{pmatrix}\begin{pmatrix}\widehat{\Psi}_{n,\uparrow}\\ \widehat{\Psi}^{\dagger}_{n,\downarrow}\end{pmatrix}\\ & -\frac{\mathrm{i}}{2}\mathrm{tr}|_{\mathfrak{H}}\,\mathsf{A} - \frac{\mathrm{i}}{2}\sum_{m,n=1}^{N}\begin{pmatrix}\widehat{\Psi}^{\dagger}_{m,\downarrow} & \widehat{\Psi}_{m,\uparrow}\end{pmatrix}\begin{pmatrix}+\mathsf{A}_{mn} & -\mathsf{B}_{mn}\\ +\mathsf{B}^{\dagger}_{mn} & -\mathsf{A}^{\mathsf{T}}_{mn}\end{pmatrix}\begin{pmatrix}\widehat{\Psi}_{n,\downarrow}\\ \widehat{\Psi}^{\dagger}_{n,\uparrow}\end{pmatrix}.\end{aligned} \tag{7.77}$$

The counterpart in the symmetry class C to Eq. (7.55) is then

$$\begin{aligned}\widehat{H} =& -\frac{\mathrm{i}}{2}\mathrm{tr}|_{\mathfrak{H}}\,\mathsf{A} + \frac{1}{2}\sum_{i=1}^{N}\xi_i\left(\hat{\gamma}^{\dagger}_{\mathrm{upper}\,i}\,\hat{\gamma}_{\mathrm{upper}\,i} - \hat{\gamma}_{\mathrm{upper}\,i}\,\hat{\gamma}^{\dagger}_{\mathrm{upper}\,i}\right)\\ & -\frac{\mathrm{i}}{2}\mathrm{tr}|_{\mathfrak{H}}\,\mathsf{A} + \frac{1}{2}\sum_{i=1}^{N}\xi_i\left(\hat{\gamma}^{\dagger}_{\mathrm{lower}\,i}\,\hat{\gamma}_{\mathrm{lower}\,i} - \hat{\gamma}_{\mathrm{lower}\,i}\,\hat{\gamma}^{\dagger}_{\mathrm{lower}\,i}\right),\end{aligned} \tag{7.78a}$$

where

$$\begin{pmatrix}\hat{\gamma}_{\mathrm{upper}}\\ \hat{\gamma}^{\dagger}_{\mathrm{upper}}\end{pmatrix} := \mathcal{U}^{\mathrm{ire}}_{\mathrm{BdG}}\begin{pmatrix}\widehat{\Psi}_{\uparrow}\\ \widehat{\Psi}^{\dagger}_{\downarrow}\end{pmatrix}, \qquad \begin{pmatrix}\hat{\gamma}_{\mathrm{lower}}\\ \hat{\gamma}^{\dagger}_{\mathrm{lower}}\end{pmatrix} := \mathcal{U}^{\mathrm{ire}}_{\mathrm{BdG}}\begin{pmatrix}+\widehat{\Psi}_{\downarrow}\\ -\widehat{\Psi}^{\dagger}_{\uparrow}\end{pmatrix}. \tag{7.78b}$$

The Bogoliubov-de-Gennes quasi-particle energies are twofold degenerate as a result of the SU(2) symmetry.

7.3.3 The Bogoliubov-de-Gennes Symmetry Class DIII

We fix the $\mathfrak{so}(4N)$ Lie algebra with $N = 1, 2, 3, \cdots$ by demanding that the $4N \times 4N$ anti-Hermitean matrix

$$\mathcal{X}_{\mathrm{BdG}} := \mathrm{i}\mathcal{H}_{\mathrm{BdG}} = -\mathcal{X}^{\dagger}_{\mathrm{BdG}} \tag{7.79a}$$

obeys the condition

$$\mathcal{C}\,\mathcal{X}^{\mathsf{T}}_{\mathrm{BdG}}\,\mathcal{C}^{-1} = -\mathcal{X}_{\mathrm{BdG}}, \tag{7.79b}$$

whereby

$$\mathcal{C} := \mathbb{1}\otimes\sigma_0\otimes\rho_1, \qquad \mathcal{C}^2 = \mathbb{1}\otimes\sigma_0\otimes\rho_0. \tag{7.79c}$$

We recall that the Pauli matrices ρ_1, ρ_2, and ρ_3, together with the unit 2×2 matrix ρ_0, act on the auxiliary Bogoliubov-de-Gennes particle-hole degrees of freedom, while Pauli matrices σ_1, σ_2, and σ_3, together with the unit 2×2 matrix σ_0, act on the spin-1/2 degrees of freedom. The Bogoliubov-de-Gennes symmetry class DIII is defined by imposing the symmetry

$$\mathcal{T}\,\mathcal{X}^{\mathsf{T}}_{\mathrm{BdG}}\,\mathcal{T}^{-1} = +\mathcal{X}_{\mathrm{BdG}}, \tag{7.79d}$$

whereby

$$\mathcal{T} := \mathbb{1} \otimes \mathrm{i}\sigma_2 \otimes \rho_0, \qquad \mathcal{T}^2 = -\mathbb{1} \otimes \sigma_0 \otimes \rho_0, \tag{7.79e}$$

composed with complex conjugation K represents reversal of time for spinfull electrons. We emphasize the very important difference with the signs on the right-hand sides of Eqs. (7.79b) and (7.79d). Notice that

$$\mathcal{S} := -\mathrm{i}\mathcal{T}\,\mathcal{C} = \mathbb{1} \otimes \sigma_2 \otimes \rho_1, \qquad \mathcal{S}^2 = \mathbb{1} \otimes \sigma_0 \otimes \rho_0, \tag{7.79f}$$

is the generator of a chiral symmetry.[19]

We define two sets. There is the set

$$\mathfrak{p} := \left\{ \mathcal{X}_{\mathrm{M}} := \mathcal{U}_{\mathrm{M}}\, \mathcal{X}_{\mathrm{BdG}}\, \mathcal{U}_{\mathrm{M}}^{\dagger} \in \mathfrak{so}(4N) \mid \mathcal{T}\, \mathcal{X}_{\mathrm{BdG}}^{\mathsf{T}}\, \mathcal{T}^{-1} = +\mathcal{X}_{\mathrm{BdG}} \right\} \tag{7.80}$$

of matrices that are even under TRS. There is the set

$$\mathfrak{k} := \left\{ \mathcal{Y}_{\mathrm{M}} := \mathcal{U}_{\mathrm{M}}\, \mathcal{Y}_{\mathrm{BdG}}\, \mathcal{U}_{\mathrm{M}}^{\dagger} \in \mathfrak{so}(4N) \mid \mathcal{T}\, \mathcal{Y}_{\mathrm{BdG}}^{\mathsf{T}}\, \mathcal{T}^{-1} = -\mathcal{Y}_{\mathrm{BdG}} \right\} \tag{7.81}$$

of matrices that are odd under TRS. The only matrix that belongs to both $\mathfrak{p}$ and $\mathfrak{k}$ is the matrix with all vanishing entries. The identity

$$[A, B]^{\mathsf{T}} = \left[B^{\mathsf{T}}, A^{\mathsf{T}}\right] = -\left[A^{\mathsf{T}}, B^{\mathsf{T}}\right] \tag{7.82}$$

for any pair A and B of square matrices sharing the same dimension implies that

$$[\mathfrak{p}, \mathfrak{p}] \subset \mathfrak{k}, \qquad [\mathfrak{p}, \mathfrak{k}] \subset \mathfrak{p}, \qquad [\mathfrak{k}, \mathfrak{k}] \subset \mathfrak{k}. \tag{7.83}$$

This means that $\mathfrak{k}$ is a subalgebra of $\mathfrak{so}(4N)$ and that $\mathfrak{p}$ is the complement of $\mathfrak{k}$ in $\mathfrak{so}(4N)$, a relation that is denoted by (**Exercise** 7.6)

$$\mathfrak{so}(4N) = \mathfrak{k} \oplus \mathfrak{p}. \tag{7.84}$$

What is the subalgebra of $\mathfrak{so}(4N)$ that is realized by $\mathfrak{k}$? To answer this question, we enumerate the identities obeyed by $\mathcal{Y}_{\mathrm{BdG}} = \mathcal{U}_{\mathrm{M}}^{\dagger}\, \mathcal{Y}_{\mathrm{M}}\, \mathcal{U}_{\mathrm{M}}$ with $\mathcal{Y}_{\mathrm{M}} \in \mathfrak{k} \subset \mathfrak{so}(4N)$. First,

$$\mathcal{Y}_{\mathrm{BdG}} = -\mathcal{Y}_{\mathrm{BdG}}^{\dagger}. \tag{7.85a}$$

Second,

$$\mathcal{Y}_{\mathrm{BdG}} = \mathcal{C}\, \mathcal{Y}_{\mathrm{BdG}}^{*}\, \mathcal{C}^{-1} = -\mathcal{C}\, \mathcal{Y}_{\mathrm{BdG}}^{\mathsf{T}}\, \mathcal{C}^{-1}, \qquad \mathcal{C}^{-1} = \mathcal{C}. \tag{7.85b}$$

Third,

$$\mathcal{Y}_{\mathrm{BdG}} = -\mathcal{T}\, \mathcal{Y}_{\mathrm{BdG}}^{\mathsf{T}}\, \mathcal{T}^{-1} = +\mathcal{T}\,\mathcal{C}\; \mathcal{Y}_{\mathrm{BdG}}\; \mathcal{C}^{-1}\mathcal{T}^{-1} = -\mathcal{T}\,\mathcal{C}\; \mathcal{Y}_{\mathrm{BdG}}\; \mathcal{T}\,\mathcal{C}. \tag{7.85c}$$

To reach the second equality, we used Eq. (7.85b). To reach the third equality, we used the facts that $\mathcal{C}^{-1} = +\mathcal{C}$, $\mathcal{T}^{-1} = -\mathcal{T}$, and $[\mathcal{C}, \mathcal{T}] = 0$.

[19] The multiplicative factor of $-\mathrm{i}$ in the definition of $\mathcal{S}$ is chosen so that $\mathcal{S}$ squares to the identity, as is done by convention in **Table** 7.1.

Suppose that there exists a unitary transformation $\mathcal{U}_{\text{DIII}} \in \text{U}(4N)$ such that

$$\mathcal{U}^{\dagger}_{\text{DIII}}\, \mathcal{C}\, \mathcal{U}^{*}_{\text{DIII}} = \mathrm{i}\mathcal{C}, \qquad \mathcal{C} = \mathbb{1} \otimes \sigma_0 \otimes \rho_1, \tag{7.86a}$$

$$\mathcal{U}^{\dagger}_{\text{DIII}}\, \mathcal{S}\, \mathcal{U}_{\text{DIII}} = \mathbb{1} \otimes \sigma_0 \otimes \rho_3 \equiv \widetilde{\mathcal{S}}, \qquad \widetilde{\mathcal{S}}^2 = \mathbb{1} \otimes \sigma_0 \otimes \rho_0. \tag{7.86b}$$

If so, define

$$\widetilde{\mathcal{Y}}_{\text{BdG}} := \mathcal{U}^{\dagger}_{\text{DIII}}\, \mathcal{Y}_{\text{BdG}}\, \mathcal{U}_{\text{DIII}}. \tag{7.87}$$

Multiplication of Eq. (7.85c) from the left by $\mathcal{U}^{\dagger}_{\text{DIII}}$ and from the right by $\mathcal{U}_{\text{DIII}}$ gives the condition

$$\widetilde{\mathcal{Y}}_{\text{BdG}} = \widetilde{\mathcal{S}}\, \widetilde{\mathcal{Y}}_{\text{BdG}}\, \widetilde{\mathcal{S}}^{-1}, \qquad \widetilde{\mathcal{S}} = \mathbb{1} \otimes \sigma_0 \otimes \rho_3, \tag{7.88a}$$

which supplements the two conditions

$$\widetilde{\mathcal{Y}}_{\text{BdG}} = \mathcal{C}\, \widetilde{\mathcal{Y}}^{\mathsf{T}}_{\text{BdG}}\, \mathcal{C} \tag{7.88b}$$

and

$$\left(\widetilde{\mathcal{Y}}_{\text{BdG}}\right)^{\dagger} = -\widetilde{\mathcal{Y}}_{\text{BdG}}. \tag{7.88c}$$

Condition (7.88a) implies that $\widetilde{\mathcal{Y}}_{\text{BdG}}$ is block diagonal in the BdG particle-hole grading. Conditions (7.88b) and (7.88c) imply that

$$\widetilde{\mathcal{Y}}_{\text{BdG}} = \begin{pmatrix} \mathcal{Z} & 0 \\ 0 & \mathcal{Z}^{\mathsf{T}} \end{pmatrix}, \qquad \mathcal{Z} = -\mathcal{Z}^{\dagger} \in \mathfrak{u}(2N). \tag{7.89}$$

We have thus shown that if the hypothesis (7.86) holds, then $\mathfrak{k} \subset \mathfrak{so}(4N)$ is isomorphic to the Lie algebra $\mathfrak{u}(2N)$. In turn, the set $\mathfrak{p}$ that is complementary to $\mathfrak{k}$ in $\mathfrak{so}(4N)$ is the difference of two Lie algebras, namely, $\mathfrak{so}(4N)$ and $\mathfrak{k} \cong \mathfrak{u}(2N)$. This difference has the geometric interpretation that it is the tangent space of the coset space $\text{SO}(4N)/\text{U}(2N)$ of the corresponding Lie groups.[20] This coset space is an example of a compact symmetric space. In mathematics, the family of compact symmetric spaces $\text{SO}(2n)/\text{U}(n)$ labeled by $n = 1, 2, 3, \cdots$ is referred to with the symbol DIII. This is the rational of Altland and Zirnbauer for using the terminology of the symmetry class DIII in **Table** 7.1 (**Exercise** 7.6). The intersection between the row DIII and the last column in **Table** 7.1 allows for the compact symmetric space DIII defined by the manifold $\text{SO}(2M)/\text{U}(M)$ with $M = 2N + 1$. As was already the case for the symmetry class D with the compact symmetric space defined by the manifold $\text{SO}(M)$ with $M = 2N + 1$, such entries describe the quantum time evolution of isolated topological boundary states of open one-dimensional superconducting chains realizing fermionic invertible phases of matter in the symmetry class DIII. Examples of one-dimensional superconducting Hamiltonians that realize such boundary states will be presented in detail in Section 7.8.

[20] The coset space $\text{SO}(4N)/\text{U}(2N)$ is the set of all equivalence classes with two elements $\exp(X)$ and $\exp(Y)$ from $\text{SO}(4N)$ equivalent if and only if $X - Y \in \mathfrak{k} \cong \mathfrak{u}(2N)$.

It is left to verify (**Exercise** 7.11) that the choice

$$\mathcal{U}_{\mathrm{DIII}} := \frac{1}{\sqrt{2}} \begin{pmatrix} \mathbb{1}\otimes\sigma_0 & +\mathbb{1}\otimes \mathrm{i}\sigma_2 \\ \mathbb{1}\otimes\sigma_2 & -\mathbb{1}\otimes \mathrm{i}\sigma_0 \end{pmatrix} \in \mathrm{U}(4N) \tag{7.90}$$

validates the hypothesis (7.86). For completeness, we also refer to **Exercise** 7.11 according to which

$$\widetilde{\mathcal{X}}_{\mathrm{BdG}} := \mathcal{U}^{\dagger}_{\mathrm{DIII}}\,\mathcal{X}_{\mathrm{BdG}}\,\mathcal{U}_{\mathrm{DIII}} = \begin{pmatrix} 0 & \mathcal{W} \\ -\mathcal{W}^{\dagger} & 0 \end{pmatrix}, \qquad \mathcal{W} = -\mathcal{W}^{\mathsf{T}}, \tag{7.91a}$$

so that

$$\widetilde{\mathcal{X}}_{\mathrm{BdG}} = -\widetilde{\mathcal{S}}\,\widetilde{\mathcal{X}}_{\mathrm{BdG}}\,\widetilde{\mathcal{S}}^{-1}, \qquad \widetilde{\mathcal{S}} = \mathbb{1}\otimes\sigma_0\otimes\rho_3. \tag{7.91b}$$

From the general theory of symmetric spaces (see footnote 16), we know that an element from $\mathfrak{p}$ can be diagonalized by a unitary transformation from the Lie group $\mathrm{U}(2N)$ with $\mathfrak{k}$ as its Lie algebra. Every eigenvalue of $\mathcal{H}_{\mathrm{BdG}}$ in the symmetry class DIII is doubly degenerate by Kramers' theorem.

7.3.4 The Bogoliubov-de-Gennes Symmetry Class CI

We start from the Bogoliubov-de-Gennes symmetry class C and consider the set of $4N\times 4N$ anti-Hermitean matrices that can be represented in the form (7.72) with a $2N\times 2N$ block obeying the conditions (7.74). The Bogoliubov-de-Gennes symmetry class CI is defined to be the set of $4N\times 4N$ anti-Hermitean matrices with $N=1,2,3,\cdots$ for which their $2N\times 2N$ block obeys the conditions

$$\mathcal{X}^{\mathrm{ire}}_{\mathrm{BdG}} = -\left(\mathcal{X}^{\mathrm{ire}}_{\mathrm{BdG}}\right)^{\dagger}, \tag{7.92a}$$

$$\mathcal{C}^{\mathrm{ire}}\left(\mathcal{X}^{\mathrm{ire}}_{\mathrm{BdG}}\right)^{\mathsf{T}}\left(\mathcal{C}^{\mathrm{ire}}\right)^{-1} = -\mathcal{X}^{\mathrm{ire}}_{\mathrm{BdG}}, \tag{7.92b}$$

$$\mathcal{T}^{\mathrm{ire}}\left(\mathcal{X}^{\mathrm{ire}}_{\mathrm{BdG}}\right)^{\mathsf{T}}\left(\mathcal{T}^{\mathrm{ire}}\right)^{-1} = +\mathcal{X}^{\mathrm{ire}}_{\mathrm{BdG}}, \tag{7.92c}$$

whereby the matrix

$$\mathcal{C}^{\mathrm{ire}} := \mathbb{1}\otimes \mathrm{i}\begin{pmatrix} 0 & -\mathrm{i} \\ +\mathrm{i} & 0 \end{pmatrix}, \qquad \left(\mathcal{C}^{\mathrm{ire}}\right)^2 = -\mathbb{1}\otimes\begin{pmatrix} 1 & 0 \\ 0 & 1 \end{pmatrix}, \tag{7.92d}$$

composed with complex conjugation K represents charge conjugation and the matrix

$$\mathcal{T}^{\mathrm{ire}} := \mathbb{1}\otimes\begin{pmatrix} 1 & 0 \\ 0 & 1 \end{pmatrix}, \qquad \left(\mathcal{T}^{\mathrm{ire}}\right)^2 = +\mathbb{1}\otimes\begin{pmatrix} 1 & 0 \\ 0 & 1 \end{pmatrix}, \tag{7.92e}$$

composed with complex conjugation K represents reversal of time. The choice (7.92e) is motivated by the fact that all spin-1/2 degrees of freedom are paired into spin singlets in Eq. (7.77). Reversal of time merely acts on the degrees of freedom labeled by the integers $m,n=1,\cdots,N$ in Eq. (7.77) through complex conjugation, since we have chosen to represent reversal of time by the matrix $\mathbb{1}$ for these degrees of freedom. Notice that

$$\mathcal{S} := -\mathrm{i}\mathcal{T}\,\mathcal{C} = \mathbb{1}\otimes\begin{pmatrix}0 & -\mathrm{i}\\ +\mathrm{i} & 0\end{pmatrix}, \qquad \mathcal{S}^2 = \mathbb{1}\otimes\begin{pmatrix}1 & 0\\ 0 & 1\end{pmatrix} \tag{7.92f}$$

is the generator of a chiral symmetry. The multiplicative factor of $-\mathrm{i}$ in the definiton of $\mathcal{S}$ is chosen so that $\mathcal{S}$ squares to the identity, as is done by convention in **Table** 7.1.

As was the case for the symmetry class DIII, we define two sets. There is the set

$$\mathfrak{p}^{\mathrm{ire}} := \left\{\mathcal{X}^{\mathrm{ire}}_{\mathrm{BdG}} \in \mathfrak{sp}(2N)\;\middle|\; \mathcal{T}^{\mathrm{ire}}\,\mathcal{X}^{\mathsf{T}}_{\mathrm{BdG}}\,\left(\mathcal{T}^{\mathrm{ire}}\right)^{-1} = +\mathcal{X}^{\mathrm{ire}}_{\mathrm{BdG}}\right\} \tag{7.93}$$

of anti-Hermitean matrices from $\mathfrak{sp}(2N)$ that are even under reversal of time. Because $\mathcal{T}^{\mathrm{ire}}$ is the identity matrix, $\mathfrak{p}^{\mathrm{ire}}$ is nothing but the subset of symmetric matrices in $\mathfrak{sp}(2N)$. There is the set

$$\mathfrak{k}^{\mathrm{ire}} := \left\{\mathcal{Y}^{\mathrm{ire}}_{\mathrm{BdG}} \in \mathfrak{sp}(2N)\;\middle|\; \mathcal{T}^{\mathrm{ire}}\,\mathcal{Y}^{\mathsf{T}}_{\mathrm{BdG}}\,\left(\mathcal{T}^{\mathrm{ire}}\right)^{-1} = -\mathcal{Y}^{\mathrm{ire}}_{\mathrm{BdG}}\right\} \tag{7.94}$$

of anti-Hermitean matrices from $\mathfrak{sp}(2N)$ that are odd under reversal of time. As for the symmetry class DIII, the identities

$$\left[\mathfrak{p}^{\mathrm{ire}},\mathfrak{p}^{\mathrm{ire}}\right]\subset\mathfrak{k}^{\mathrm{ire}}, \qquad \left[\mathfrak{p}^{\mathrm{ire}},\mathfrak{k}^{\mathrm{ire}}\right]\subset\mathfrak{p}^{\mathrm{ire}}, \qquad \left[\mathfrak{k}^{\mathrm{ire}},\mathfrak{k}^{\mathrm{ire}}\right]\subset\mathfrak{k}^{\mathrm{ire}} \tag{7.95}$$

imply that (**Exercise** 7.6)

$$\mathfrak{sp}(2N) = \mathfrak{k}^{\mathrm{ire}}\oplus\mathfrak{p}^{\mathrm{ire}} \tag{7.96}$$

with $\mathfrak{k}^{\mathrm{ire}}$ a subalgebra of $\mathfrak{sp}(2N)$. We are going to show that the set $\mathfrak{k}^{\mathrm{ire}}$ in $\mathfrak{sp}(2N)$ that complements the set of symmetric matrices $\mathfrak{p}^{\mathrm{ire}}$ in $\mathfrak{sp}(2N)$ is the Lie algebra $\mathfrak{u}(N)$.

Proof We parameterize any $\mathcal{Y}^{\mathrm{ire}}_{\mathrm{BdG}} \in \mathfrak{sp}(2N)$ by

$$\mathcal{Y}^{\mathrm{ire}}_{\mathrm{BdG}} := \begin{pmatrix}+\mathsf{A} & +\mathsf{B}\\ -\mathsf{B}^{\dagger} & -\mathsf{A}^{\mathsf{T}}\end{pmatrix}, \qquad \mathsf{A}=-\mathsf{A}^{\dagger}, \qquad \mathsf{B}=+\mathsf{B}^{\mathsf{T}} \tag{7.97}$$

(recall Eq. (7.73)). For $\mathcal{Y}^{\mathrm{ire}}_{\mathrm{BdG}} \in \mathfrak{k}^{\mathrm{ire}}$ to hold, we must solve the condition

$$\begin{aligned}\begin{pmatrix}+\mathsf{A} & +\mathsf{B}\\ -\mathsf{B}^{*} & +\mathsf{A}^{*}\end{pmatrix} &= -\begin{pmatrix}+\mathsf{A} & +\mathsf{B}\\ -\mathsf{B}^{*} & -\mathsf{A}^{\mathsf{T}}\end{pmatrix}^{\mathsf{T}}\\ &= -\begin{pmatrix}+\mathsf{A}^{\mathsf{T}} & -\mathsf{B}^{*}\\ +\mathsf{B} & -\mathsf{A}\end{pmatrix}\\ &= \begin{pmatrix}+\mathsf{A}^{*} & +\mathsf{B}^{*}\\ -\mathsf{B} & \mathsf{A}\end{pmatrix}.\end{aligned} \tag{7.98}$$

We conclude that $\mathcal{Y}^{\mathrm{ire}}_{\mathrm{BdG}} \in \mathfrak{k}^{\mathrm{ire}}$ if and only if

$$\mathcal{Y}^{\mathrm{ire}}_{\mathrm{BdG}} := \begin{pmatrix} +\mathsf{A} & +\mathsf{B} \\ -\mathsf{B} & +\mathsf{A} \end{pmatrix}, \qquad \mathsf{A} = -\mathsf{A}^{\dagger} = \mathsf{A}^{*} = -\mathsf{A}^{\mathsf{T}}, \qquad \mathsf{B} = +\mathsf{B}^{\mathsf{T}} = \mathsf{B}^{*} = +\mathsf{B}^{\dagger} \tag{7.99}$$

in the parameterization (7.97). In particular, if $\mathcal{Y}^{\mathrm{ire}}_{\mathrm{BdG}} \in \mathfrak{k}^{\mathrm{ire}}$ then $\mathcal{Y}^{\mathrm{ire}}_{\mathrm{BdG}}$ is a $2N \times 2N$ anti-Hermitean matrix (in fact, it is real-valued and antisymmetric). Now, any $N \times N$ matrix $\mathsf{U} \in \mathfrak{u}(N)$ obeys, by definition,

$$\mathsf{U} = -\mathsf{U}^{\dagger} \tag{7.100a}$$

so that

$$\mathrm{Re}\,\mathsf{U} := \frac{\mathsf{U} + \mathsf{U}^{*}}{2} = +\left(\mathrm{Re}\,\mathsf{U}\right)^{*} = -\left(\mathrm{Re}\,\mathsf{U}\right)^{\dagger}, \tag{7.100b}$$

$$\mathrm{Im}\,\mathsf{U} := \frac{\mathsf{U} - \mathsf{U}^{*}}{2\mathrm{i}} = +\left(\mathrm{Im}\,\mathsf{U}\right)^{*} = +\left(\mathrm{Im}\,\mathsf{U}\right)^{\dagger}. \tag{7.100c}$$

Consequently, the map

$$\begin{aligned} \mathfrak{u}(N) &\to \mathfrak{k}^{\mathrm{ire}}, \\ \mathsf{U} &\mapsto \begin{pmatrix} +\mathrm{Re}\,\mathsf{U} & +\mathrm{Im}\,\mathsf{U} \\ -\mathrm{Im}\,\mathsf{U} & +\mathrm{Re}\,\mathsf{U} \end{pmatrix}, \end{aligned} \tag{7.101}$$

is one to one and a Lie algebra homomorphism (**Exercise** 7.12). □

We have thus shown that $\mathfrak{k}^{\mathrm{ire}} \subset \mathfrak{sp}(2N)$ is isomorphic to the Lie algebra $\mathfrak{u}(N)$. In turn, the set $\mathfrak{p}^{\mathrm{ire}}$ that is complementary to $\mathfrak{k}^{\mathrm{ire}}$ in $\mathfrak{sp}(2N)$ is the difference of two Lie algebras, namely, $\mathfrak{sp}(2N)$ and $\mathfrak{k}^{\mathrm{ire}} \cong \mathfrak{u}(N)$. This difference has the geometric interpretation that it is the tangent space of the coset space $\mathrm{Sp}(2N)/\mathrm{U}(N)$ of the corresponding Lie groups. This coset space is an example of a compact symmetric space. In mathematics, the family of compact symmetric spaces $\mathrm{Sp}(2n)/\mathrm{U}(n)$ labeled by $n = 1, 2, 3, \cdots$ is referred to with the symbol CI. This is the rational of Altland and Zirnbauer for using the terminology of the symmetry class CI in **Table** 7.1 (**Exercise** 7.6).

From the general theory of symmetric spaces (see footnote 16), we know that an element from $\mathfrak{p}^{\mathrm{ire}}$ can be diagonalized by a unitary transformation from the Lie group $\mathrm{U}(N)$ with $\mathfrak{k}^{\mathrm{ire}}$ as its Lie algebra.

7.3.5 The Wigner-Dyson Symmetry Class A

We define the symmetry class A as the set of all $2N \times 2N$ matrices of the symmetry class D obeying one additional condition, namely,

$$\exp\left(-\mathrm{i}\mathbb{1} \otimes \rho_3\, \phi\right) \mathcal{X}_{\mathrm{BdG}} \exp\left(+\mathrm{i}\mathbb{1} \otimes \rho_3\, \phi\right) = \mathcal{X}_{\mathrm{BdG}} \tag{7.102}$$

for any $\mathcal{X}_{\mathrm{BdG}} = \mathcal{U}^{\dagger}_{\mathrm{M}}\, \mathcal{X}_{\mathrm{M}}\, \mathcal{U}_{\mathrm{M}}$ with $\mathcal{X}_{\mathrm{M}} \in \mathfrak{so}(2N)$ and any $0 \leq \phi < 2\pi$. Condition (7.102) prevents the superconductivity instability by demanding that the electronic charge be conserved by the single-particle Hamiltonian. Correspondingly, we must set $\Delta = 0$ in Eq. (7.40); this is to say that $\mathcal{X}_{\mathrm{BdG}}$ is decomposable down to the $N \times N$ anti-Hermitean block matrix $\mathrm{i}\mathcal{H}$ that generates the Lie algebra $\mathfrak{u}(N)$. As we do not demand that $\mathcal{H}$ is traceless, setting $\Delta = 0$ amounts to selecting the subgroup of $\mathfrak{so}(2N)$ that is isomorphic to $\mathfrak{u}(N)$. In mathematics, the family of simple Lie algebras $\mathfrak{su}(n)$ labeled by $n = 1, 2, 3, \cdots$, that would follow from demanding that $\mathcal{H}$ in Eq. (7.40a) is traceless, is referred to with the symbol A. This is the rational of Altland and Zirnbauer for using the terminology of the symmetry class A in **Table** 7.1 (**Exercise** 7.6).

If we set $\Delta = 0$ in Section 7.3.2, we immediately see that imposing spin-rotation symmetry simply renders the element $\mathcal{X}_{\mathrm{BdG}} = \mathcal{U}_{\mathrm{M}}^{\dagger}\,\mathcal{X}_{\mathrm{M}}\,\mathcal{U}_{\mathrm{M}}$ with $\mathcal{X}_{\mathrm{M}} \in \mathfrak{so}(4N)$ decomposable down to four $N \times N$ diagonal blocks that solely depend on the anti-Hermitean matrix $\mathsf{A} = -\mathsf{A}^{\dagger}$. Imposing spin-rotation symmetry alone does not produce a new symmetry class when charge conservation holds. New symmetry classes only appear if time-reversal symmetry (TRS) or chiral symmetry (CHS) are imposed.

7.3.6 The Wigner-Dyson Symmetry Class AI

We impose on the symmetry class A spin-rotation symmetry and time-reversal symmetry. This can be achieved in an economical way by starting from the symmetry class CI from Section 7.3.4, setting $\mathsf{B} = 0$ in the parameterization of the sets (7.93) and (7.94). By inspection of Eq. (7.99) with $\mathsf{B} = 0$, we infer that the symmetry class AI is the tangent space to the coset space $\mathrm{U}(N)/\mathrm{SO}(N)$. In mathematics, the family of compact symmetric spaces $\mathrm{SU}(n)/\mathrm{SO}(n)$ labeled by $n = 1, 2, 3, \cdots$, that would follow from demanding that $\mathcal{H}$ in Eq. (7.64) is traceless, is referred to with the symbol AI. This is the rational of Altland and Zirnbauer for using the terminology of the symmetry class AI in **Table** 7.1 (**Exercise** 7.6).

7.3.7 The Wigner-Dyson Symmetry Class AII

Let $N = 1, 2, 3, \cdots$. We start from the set of Hermitean $2N \times 2N$ matrices with the representative

$$\mathcal{H} = \mathcal{H}^{\dagger}. \tag{7.103}$$

The anti-Hermitean matrix

$$\mathcal{X} := \mathrm{i}\mathcal{H} = -\mathcal{X}^{\dagger} \tag{7.104}$$

generates the Lie algebra $\mathfrak{u}(2N)$. We make explicit the spin-1/2 grading by writing

$$\mathcal{X} = \begin{pmatrix} \mathsf{A} & \mathsf{B} \\ -\mathsf{B}^{\dagger} & \mathsf{C} \end{pmatrix}, \qquad \mathsf{A} = -\mathsf{A}^{\dagger}, \qquad \mathsf{C} = -\mathsf{C}^{\dagger}, \tag{7.105}$$

with A, B, and C $N \times N$ complex-valued matrices.

The symmetry class AII is obtained from the symmetry class A by demanding that

$$\mathcal{T}\mathcal{H}^{*}\mathcal{T}^{-1} = \mathcal{T}\mathcal{H}^{\mathsf{T}}\mathcal{T}^{-1} = \mathcal{H}, \qquad \mathcal{T} := \begin{pmatrix} 0 & +\mathbb{1} \\ -\mathbb{1} & 0 \end{pmatrix} = -\mathcal{T}^{-1}, \tag{7.106a}$$

with $\mathbb{1}$ the unit $N \times N$ matrix or, equivalently,

$$-\mathcal{T}\mathcal{X}^{*}\mathcal{T}^{-1} = \mathcal{T}\mathcal{X}^{\mathsf{T}}\mathcal{T}^{-1} = \mathcal{X}, \qquad \mathcal{T} := \begin{pmatrix} 0 & +\mathbb{1} \\ -\mathbb{1} & 0 \end{pmatrix} = -\mathcal{T}^{-1}. \tag{7.106b}$$

On the one hand, condition (7.106b), defines the set

$$\begin{aligned} \mathfrak{p} &:= \left\{\mathcal{X} \in \mathfrak{u}(2N) \mid \mathcal{T}\mathcal{X}^{*}\mathcal{T}^{-1} = -\mathcal{X}\right\} \\ &= \left\{\begin{pmatrix} \mathsf{A} & \mathsf{B} \\ \mathsf{B}^{*} & -\mathsf{A}^{*} \end{pmatrix} \middle| \ \mathsf{A} = -\mathsf{A}^{\dagger}, \qquad \mathsf{B} = -\mathsf{B}^{\mathsf{T}}\right\} \end{aligned} \tag{7.107}$$

that does not close under the commutator. On the other hand, the complementary set

$$\begin{aligned} \mathfrak{k} &:= \left\{\mathcal{X} \in \mathfrak{u}(2N) \mid \mathcal{T}\mathcal{X}^{*}\mathcal{T}^{-1} = +\mathcal{X}\right\} \\ &= \left\{\begin{pmatrix} \mathsf{A} & \mathsf{B} \\ -\mathsf{B}^{*} & \mathsf{A}^{*} \end{pmatrix} \middle| \ \mathsf{A} = -\mathsf{A}^{\dagger}, \qquad \mathsf{B} = +\mathsf{B}^{\mathsf{T}}\right\} \end{aligned} \tag{7.108}$$

is closed under the commutator. More precisely,

$$[\mathfrak{p}, \mathfrak{p}] \subset \mathfrak{k}, \qquad [\mathfrak{p}, \mathfrak{k}] \subset \mathfrak{p}, \qquad [\mathfrak{k}, \mathfrak{k}] \subset \mathfrak{k}. \tag{7.109}$$

The set $\mathfrak{k}$ is thus a Lie subalgebra of $\mathfrak{u}(2N)$, given by

$$\mathfrak{k} = \mathfrak{sp}(2N) \tag{7.110}$$

according to **Exercise** 7.6. The set $\mathfrak{p}$ that is complementary to $\mathfrak{k}$ in $\mathfrak{u}(2N)$ is the difference of two Lie algebras, namely, $\mathfrak{u}(2N)$ and $\mathfrak{sp}(2N)$. This difference has the geometric interpretation that it is the tangent space of the coset space $\mathrm{U}(2N)/\mathrm{Sp}(2N)$ of the corresponding Lie groups. This coset space is an example of a compact symmetric space. In mathematics, the family of compact symmetric spaces $\mathrm{SU}(2n)/\mathrm{Sp}(2n)$ labeled by $n = 1, 2, 3, \cdots$, that would follow from demanding that $\mathcal{H}$ in Eq. (7.103) is traceless, is referred to with the symbol AII. This is the rational of Altland and Zirnbauer for using the terminology of the symmetry class AII in **Table** 7.1 (**Exercise** 7.6).

7.3.8 The Chiral Symmetry Class AIII

Let $M, N = 1, 2, 3, \cdots$. We start from any $(M+N) \times (M+N)$ Hermitean matrix

$$\mathcal{H} = \mathcal{H}^{\dagger} \tag{7.111}$$

from the symmetry class A. The anti-Hermitean matrix

$$\mathcal{X} := \mathrm{i}\mathcal{H} = -\mathcal{X}^\dagger \tag{7.112}$$

generates the Lie algebra $\mathfrak{u}(M+N)$. We make explicit the chiral grading by writing

$$\mathcal{X} = \begin{pmatrix} \mathsf{A} & \mathsf{B} \\ -\mathsf{B}^\dagger & \mathsf{C} \end{pmatrix}, \qquad \mathsf{A} = -\mathsf{A}^\dagger, \qquad \mathsf{C} = -\mathsf{C}^\dagger, \tag{7.113}$$

with A, B, and C $M\times M$, $M\times N$, and $N\times N$ complex-valued matrices, respectively.

The symmetry class AIII is defined from the symmetry class A by demanding that

$$\Sigma\,\mathcal{H}\,\Sigma^{-1} = -\mathcal{H}, \qquad \Sigma := \begin{pmatrix} +\mathbb{1}_M & 0 \\ 0 & -\mathbb{1}_N \end{pmatrix} = \Sigma^{-1}, \tag{7.114a}$$

with $\mathbb{1}_n$ the unit $n\times n$ matrix or, equivalently,

$$\Sigma\,\mathcal{X}\,\Sigma^{-1} = -\mathcal{X}, \qquad \Sigma := \begin{pmatrix} +\mathbb{1}_M & 0 \\ 0 & -\mathbb{1}_N \end{pmatrix} = \Sigma^{-1}. \tag{7.114b}$$

On the one hand, condition (7.114b) defines the set

$$\begin{aligned} \mathfrak{p} &:= \left\{\mathcal{X} \in \mathfrak{u}(M+N) \mid \Sigma\,\mathcal{X}\,\Sigma^{-1} = -\mathcal{X}\right\} \\ &= \left\{\begin{pmatrix} 0 & \mathsf{B} \\ -\mathsf{B}^\dagger & 0 \end{pmatrix} \middle| \ \mathsf{B}\ \text{complex-valued}\right\} \end{aligned} \tag{7.115}$$

that does not close under the commutator. On the other hand, the complementary set

$$\begin{aligned} \mathfrak{k} &:= \left\{\mathcal{X} \in \mathfrak{u}(M+N) \mid \Sigma\,\mathcal{X}\,\Sigma^{-1} = +\mathcal{X}\right\} \\ &= \left\{\begin{pmatrix} \mathsf{A} & 0 \\ 0 & \mathsf{C} \end{pmatrix} \middle| \ \mathsf{A} = -\mathsf{A}^\dagger,\ \mathsf{C} = -\mathsf{C}^\dagger\ \text{complex-valued}\right\} \end{aligned} \tag{7.116}$$

is closed under the commutator. More precisely,

$$[\mathfrak{p},\mathfrak{p}] \subset \mathfrak{k}, \qquad [\mathfrak{p},\mathfrak{k}] \subset \mathfrak{p}, \qquad [\mathfrak{k},\mathfrak{k}] \subset \mathfrak{k}. \tag{7.117}$$

The set $\mathfrak{k}$ is thus a Lie subalgebra of $\mathfrak{u}(M+N)$, given by

$$\mathfrak{k} = \mathfrak{u}(M) \oplus \mathfrak{u}(N) \tag{7.118}$$

(**Exercise** 7.6). The set $\mathfrak{p}$ that is complementary to $\mathfrak{k}$ in $\mathfrak{u}(M+N)$ is the difference of two Lie algebras, namely, $\mathfrak{u}(M+N)$ and $\mathfrak{u}(M)\oplus\mathfrak{u}(N)$. This difference has the geometric interpretation that it is the tangent space of the coset space $\mathrm{U}(M+N)/\big(\mathrm{U}(M)\times\mathrm{U}(N)\big)$ of the corresponding Lie groups. This coset space is an example of a compact symmetric space. In mathematics, the family of compact symmetric spaces $\mathrm{SU}(m+n)/\mathrm{S}\big(\mathrm{U}(m)\times\mathrm{U}(n)\big)$ labeled by $m,n = 1,2,3,\cdots$, that would follow from demanding that $\mathcal{H}$ in Eq. (7.111) is traceless, is referred to with the symbol AIII. This is the rational of Altland and Zirnbauer for using the terminology of the symmetry class AIII in **Table** 7.1 (**Exercise** 7.6).

7.3.9 The Chiral Symmetry Class BDI

Let $M, N = 1, 2, 3, \cdots$. We start from any $(M+N)\times(M+N)$ BdG matrix in its Majorana representation (7.46) for which

$$\mathcal{H}_{\mathrm{M}} = \mathcal{H}_{\mathrm{M}}^{\dagger} = -\mathcal{H}_{\mathrm{M}}^{*}. \tag{7.119}$$

The real-valued antisymmetric matrix

$$\mathcal{X}_{\mathrm{M}} := \mathrm{i}\mathcal{H}_{\mathrm{M}} = -\mathcal{X}_{\mathrm{M}}^{\dagger} = \mathcal{X}_{\mathrm{M}}^{*} = -\mathcal{X}_{\mathrm{M}}^{\mathsf{T}} \tag{7.120}$$

generates the Lie algebra $\mathfrak{so}(M+N)$. We make explicit the chiral grading by writing

$$\mathcal{X}_{\mathrm{M}} = \begin{pmatrix} \mathsf{A} & \mathsf{B} \\ -\mathsf{B}^{\mathsf{T}} & \mathsf{C} \end{pmatrix}, \quad \mathsf{A} = \mathsf{A}^{*} = -\mathsf{A}^{\mathsf{T}},\ \mathsf{C} = \mathsf{C}^{*} = -\mathsf{C}^{\mathsf{T}},\ \mathsf{B} = \mathsf{B}^{*}, \tag{7.121}$$

with A, B, and C $M \times M$, $M \times N$, and $N \times N$ real-valued matrices, respectively.

The symmetry class BDI is defined from the symmetry class D by demanding that

$$\Sigma\,\mathcal{H}_{\mathrm{M}}\,\Sigma^{-1} = -\mathcal{H}_{\mathrm{M}}, \qquad \Sigma := \begin{pmatrix} +\mathbb{1}_M & 0 \\ 0 & -\mathbb{1}_N \end{pmatrix} = \Sigma^{-1}, \tag{7.122a}$$

with $\mathbb{1}_n$ the unit $n \times n$ matrix or, equivalently,

$$\Sigma\,\mathcal{X}_{\mathrm{M}}\,\Sigma^{-1} = -\mathcal{X}_{\mathrm{M}}, \qquad \Sigma := \begin{pmatrix} +\mathbb{1}_M & 0 \\ 0 & -\mathbb{1}_N \end{pmatrix} = \Sigma^{-1}. \tag{7.122b}$$

On the one hand, condition (7.122b) defines the set

$$\begin{aligned} \mathfrak{p} &:= \left\{\mathcal{X}_{\mathrm{M}} \in \mathfrak{so}(M+N) \mid \Sigma\,\mathcal{X}_{\mathrm{M}}\,\Sigma^{-1} = -\mathcal{X}_{\mathrm{M}}\right\} \\ &= \left\{ \begin{pmatrix} 0 & \mathsf{B} \\ -\mathsf{B}^{\mathsf{T}} & 0 \end{pmatrix} \,\middle|\, \mathsf{B} \text{ real-valued} \right\} \end{aligned} \tag{7.123}$$

that does not close under the commutator. On the other hand, the complementary set

$$\begin{aligned} \mathfrak{k} &:= \left\{\mathcal{X}_{\mathrm{M}} \in \mathfrak{so}(M+N) \mid \Sigma\,\mathcal{X}_{\mathrm{M}}\,\Sigma^{-1} = +\mathcal{X}_{\mathrm{M}}\right\} \\ &= \left\{ \begin{pmatrix} \mathsf{A} & 0 \\ 0 & \mathsf{C} \end{pmatrix} \,\middle|\, \mathsf{A} = \mathsf{A}^{*} = -\mathsf{A}^{\mathsf{T}},\ \mathsf{C} = \mathsf{C}^{*} = -\mathsf{C}^{\mathsf{T}} \right\} \end{aligned} \tag{7.124}$$

is closed under the commutator. More precisely,

$$[\mathfrak{p},\mathfrak{p}] \subset \mathfrak{k}, \qquad [\mathfrak{p},\mathfrak{k}] \subset \mathfrak{p}, \qquad [\mathfrak{k},\mathfrak{k}] \subset \mathfrak{k}. \tag{7.125}$$

The set $\mathfrak{k}$ is thus a Lie subalgebra of $\mathfrak{so}(M+N)$, given by

$$\mathfrak{k} = \mathfrak{so}(M) \oplus \mathfrak{so}(N) \tag{7.126}$$

(**Exercise** 7.6). The set $\mathfrak{p}$ that is complementary to $\mathfrak{k}$ in $\mathfrak{so}(M+N)$ is the difference of two Lie algebras, namely, $\mathfrak{so}(M+N)$ and $\mathfrak{so}(M) \oplus \mathfrak{so}(N)$. This difference has the geometric interpretation that it is the tangent space of the coset space

$\mathrm{SO}(M+N)/\big(\mathrm{SO}(M)\times\mathrm{SO}(N)\big)$ of the corresponding Lie groups. In mathematics, the family of compact symmetric spaces $\mathrm{SO}(m+n)/\big(\mathrm{SO}(m)\times\mathrm{SO}(n)\big)$ labeled by $m,n=1,2,3,\cdots$ is referred to with the symbol BDI. This is the rational of Altland and Zirnbauer for using the terminology of the symmetry class BDI in **Table** 7.1 (**Exercise** 7.6).

7.3.10 The Chiral Symmetry Class CII

We start from an arbitrary element of $\mathfrak{sp}(2M+2N)$ with $M,N=1,2,3,\cdots$ from the symmetry class C, which we parameterize as

$$\mathcal{X} := \begin{pmatrix} \mathsf{X}_{11} & \mathsf{X}_{12} & \mathsf{X}_{13} & \mathsf{X}_{14} \\ -\mathsf{X}_{12}^{\dagger} & \mathsf{X}_{22} & \mathsf{X}_{14}^{\mathsf{T}} & \mathsf{X}_{24} \\ & & & \\ -\mathsf{X}_{13}^{\dagger} & -\mathsf{X}_{14}^{*} & -\mathsf{X}_{11}^{\mathsf{T}} & \mathsf{X}_{12}^{*} \\ -\mathsf{X}_{14}^{\dagger} & -\mathsf{X}_{24}^{\dagger} & -\mathsf{X}_{12}^{\mathsf{T}} & -\mathsf{X}_{22}^{\mathsf{T}} \end{pmatrix}, \tag{7.127a}$$

where, according to Eq. (7.73),

$$\mathsf{X}_{11}=-\mathsf{X}_{11}^{\dagger}, \qquad \mathsf{X}_{22}=-\mathsf{X}_{22}^{\dagger}, \qquad \mathsf{X}_{13}=\mathsf{X}_{13}^{\mathsf{T}}, \qquad \mathsf{X}_{24}=\mathsf{X}_{24}^{\mathsf{T}}. \tag{7.127b}$$

The matrices X_{11} and X_{13} are $M\times M$ complex-valued. The matrices X_{22} and X_{24} are $N\times N$ complex-valued. The matrices X_{12} and X_{14} are $M\times N$ complex-valued. This representation of $\mathcal{X}$ emphasizes that

(1) the $(M+N)\times(M+N)$ upper-left block is anti-Hermitean,
(2) the $(M+N)\times(M+N)$ lower-right block is minus the transpose of the $(M+N)\times(M+N)$ upper-left block,
(3) the $(M+N)\times(M+N)$ upper-right block is symmetric, and
(4) the $(M+N)\times(M+N)$ lower-left block is minus the adjoint of the $(M+N)\times(M+N)$ upper-right block.

Hence, as it should be, $\mathcal{X}\in\mathfrak{u}(2M+2N)\cap\mathfrak{sp}(2M+2N,\mathbb{C})\equiv\mathfrak{sp}(2M+2N)$. The symmetry class CII is defined by demanding that all $\mathcal{X}\in\mathfrak{sp}(2M+2N)$ obey the condition

$$\Sigma\,\mathcal{X}\,\Sigma^{-1}=-\mathcal{X},\ \Sigma:=\mathrm{diag}\left(+\mathbb{1}_M \quad -\mathbb{1}_N \quad +\mathbb{1}_M \quad -\mathbb{1}_N\right)=\Sigma^{-1}, \tag{7.127c}$$

where $\mathbb{1}_n$ is the unit $n\times n$ matrix.

Because of

$$\Sigma\,\mathcal{X}\,\Sigma^{-1} = \begin{pmatrix} \mathsf{X}_{11} & -\mathsf{X}_{12} & \mathsf{X}_{13} & -\mathsf{X}_{14} \\ +\mathsf{X}_{12}^{\dagger} & \mathsf{X}_{22} & -\mathsf{X}_{14}^{\mathsf{T}} & \mathsf{X}_{24} \\ & & & \\ -\mathsf{X}_{13}^{\dagger} & +\mathsf{X}_{14}^{*} & -\mathsf{X}_{11}^{\mathsf{T}} & -\mathsf{X}_{12}^{*} \\ +\mathsf{X}_{14}^{\dagger} & -\mathsf{X}_{24}^{\dagger} & +\mathsf{X}_{12}^{\mathsf{T}} & -\mathsf{X}_{22}^{\mathsf{T}} \end{pmatrix}, \tag{7.128}$$

one finds, on the one hand, that condition (7.127) defines the set

$$\mathfrak{p} := \left\{\mathcal{X} \in \mathfrak{sp}(2M+2N) \mid \Sigma\,\mathcal{X}\,\Sigma^{-1} = -\mathcal{X}\right\} \tag{7.129}$$

with

$$\mathcal{X} := \begin{pmatrix} 0 & \mathsf{X}_{12} & 0 & \mathsf{X}_{14} \\ -\mathsf{X}_{12}^{\dagger} & 0 & \mathsf{X}_{14}^{\mathsf{T}} & 0 \\ 0 & -\mathsf{X}_{14}^{*} & 0 & \mathsf{X}_{12}^{*} \\ -\mathsf{X}_{14}^{\dagger} & 0 & -\mathsf{X}_{12}^{\mathsf{T}} & 0 \end{pmatrix} \tag{7.130}$$

that does not close under the commutator. On the other hand, the complementary set

$$\mathfrak{k} := \left\{\mathcal{X} \in \mathfrak{sp}(2M+2N) \mid \Sigma\,\mathcal{X}\,\Sigma^{-1} = +\mathcal{X}\right\} \tag{7.131}$$

with

$$\begin{pmatrix} \mathsf{X}_{11} & 0 & \mathsf{X}_{13} & 0 \\ 0 & \mathsf{X}_{22} & 0 & \mathsf{X}_{24} \\ -\mathsf{X}_{13}^{\dagger} & 0 & -\mathsf{X}_{11}^{\mathsf{T}} & 0 \\ 0 & -\mathsf{X}_{24}^{\dagger} & 0 & -\mathsf{X}_{22}^{\mathsf{T}} \end{pmatrix} \tag{7.132}$$

is closed under the commutator. More precisely,

$$[\mathfrak{p},\mathfrak{p}] \subset \mathfrak{k}, \qquad [\mathfrak{p},\mathfrak{k}] \subset \mathfrak{p}, \qquad [\mathfrak{k},\mathfrak{k}] \subset \mathfrak{k}. \tag{7.133}$$

The set $\mathfrak{k}$ is thus a Lie subalgebra of $\mathfrak{sp}(2M+2N)$, given by

$$\mathfrak{k} = \mathfrak{sp}(2M) \oplus \mathfrak{sp}(2N) \tag{7.134}$$

(**Exercise** 7.6). The set $\mathfrak{p}$ that is complementary to $\mathfrak{k}$ in $\mathfrak{sp}(2M+2N)$ is the difference of two Lie algebras, namely, $\mathfrak{sp}(2M+2N)$ and $\mathfrak{sp}(2M) \oplus \mathfrak{sp}(2N)$. This difference has the geometric interpretation that it is the tangent space of the coset space $\mathrm{Sp}(2M+2N)/\big(\mathrm{Sp}(2M) \times \mathrm{Sp}(2N)\big)$ of the corresponding Lie groups. This coset space is an example of a compact symmetric space. In mathematics, the family of compact symmetric spaces $\mathrm{Sp}(2m+2n)/\big(\mathrm{Sp}(2m) \times \mathrm{Sp}(2n)\big)$ labeled by $m,n = 1,2,3,\cdots$ is referred to with the symbol CII. This is the rational of Altland and Zirnbauer for using the terminology of the symmetry class CII in **Table** 7.1 (**Exercise** 7.6).

7.3.11 The Geometry of Compact Symmetric Spaces

The main result of Section 7.3 is **Table** 7.1. It was derived using the tools of linear algebra. The symmetry classes A, AI, AII, and AIII stand out in **Table** 7.1 in that the corresponding spaces defined by the unitary time evolution $\exp(\mathrm{i}t\mathcal{H})$ are all built starting from the group $\mathrm{U}(N) = \mathrm{U}(1) \times \mathrm{SU}(N)$. The group $\mathrm{U}(N) = \mathrm{U}(1) \times \mathrm{SU}(N)$ is an example of a semi-simple Lie group. The adjective semi-simple encodes the fact that $\mathrm{U}(N)$ factors into the product of two commuting subgroups, an Abelian subgroup U(1) and the subgroup SU(N). The reference to Lie stems from the fact that U(N) is made of an uncountable number of group elements and that these elements can be deformed into each other smoothly. The interpretation of the factor subgroup U(1) is that the global phase of all states in the Hilbert space on which $\mathcal{H}$ is defined is arbitrary and that this arbitrariness results in the conservation of the fermion number from the application of Noether's theorem. The complementary symmetry classes BDI, CII, D, C, DIII, and CI are built from either the group SO(N) or the group Sp($2N$). These groups cannot be factored out into two commuting subgroups, one of which is Abelian. The groups SU(N), SO(N), and Sp($2N$) are examples of simple Lie groups. The adjective simple means that these groups, unlike U(N) or SU(M) $\times$ SU(N), cannot be factored out into subgroups other than the identity and the group itself that commute.

The classical matrix groups SU(N), SO(N), and Sp($2N$) inherit from GL($N,\mathbb{K}$) and SL($N,\mathbb{K}$) the property that the operation of matrix multiplication and matrix inversion are both smooth (continuous and infinitely many times differentiable). In other words, if $\mathrm{G} = \mathrm{SU}(N), \mathrm{SO}(N), \mathrm{Sp}(2N)$, then the map

$$\begin{aligned} \mathrm{G}\times\mathrm{G} &\to \mathrm{G}, \\ (g,g') &\mapsto g^{-1}\,g' \end{aligned} \tag{7.135}$$

is a smooth map. This notion of smoothness is made possible because we can define a norm on the space of matrices G and the inverse element with respect to the matrix multiplication necessarily exists.

The (real) Lie algebra of the smooth group $\mathrm{G} = \mathrm{SU}(N), \mathrm{SO}(N), \mathrm{Sp}(2N)$ is the set

$$\mathrm{Lie}\,(\mathrm{G}) \equiv \mathfrak{g} := \left\{ X \;\middle|\; e^{tX} \equiv \sum_{n=0}^{\infty} \frac{t^n}{n!} X^n \in \mathrm{G} \qquad \forall t \in \mathbb{R} \right\}. \tag{7.136}$$

By expanding the exponential to linear order in t, one verifies that the Lie algebra $\mathfrak{g}$ of G is a vector space over the real numbers that is closed under commutation

$$X, Y \in \mathfrak{g} \implies [X,Y] := XY - YX \in \mathfrak{g} \tag{7.137}$$

and obeys the Jacobi identity

$$X,Y,Z \in \mathfrak{g} \implies \big[X,[Y,Z]\big] + \big[Y,[Z,X]\big] + \big[Z,[X,Y]\big] = 0. \tag{7.138}$$

The relation between matrix Lie groups and their Lie algebras has a geometric meaning that requires the notion of differentiable manifolds.

The simplest differentiable manifold is the real line $\mathbb{R}$. As a set, we denote an element of $\mathbb{R}$ to be the point x. The identity map

$$\begin{aligned} \mathbb{R} &\to \mathbb{R}, \\ x &\mapsto x \end{aligned} \tag{7.139}$$

is a smooth function that establishes that $\mathbb{R}$ is a smooth manifold of dimension one.

Next, we consider the unit circle

$$\mathsf{S}^1 := \left\{\boldsymbol{x}\in\mathbb{R}^2 \mid \boldsymbol{x}^2 = 1\right\} \equiv \{z\in\mathbb{C} \mid |z| = 1\}. \tag{7.140}$$

We want to establish that S^1 is a smooth manifold in that we can find for any open subset of S^1 a homeomorphism, that is a continuous one-to-one map with continuous inverse, between this open subset of S^1 and an open interval of $\mathbb{R}$, thus providing the manifold with local coordinates. In addition, the maps in this collection have to be compatible with each other, meaning that all changes of coordinates have to be smooth, in order for the manifold to be smooth.

One coordinate system is provided by the continuous map[21]

$$\begin{aligned}\psi_{\mathrm{N}}\colon\ \mathsf{S}^1\setminus\{-\mathrm{i}\} &\to]-\pi,+\pi[\\ z &\mapsto \psi_{\mathrm{N}}(z) := \arg\left(z\,e^{-\mathrm{i}\pi/2}\right),\end{aligned} \tag{7.141}$$

with the continuous inverse

$$\begin{aligned}\psi_{\mathrm{N}}^{-1}\colon\ ]-\pi,+\pi[&\to \mathsf{S}^1\setminus\{-\mathrm{i}\},\\ \varphi &\mapsto \psi_{\mathrm{N}}^{-1}(\varphi) = e^{+\mathrm{i}\left(\frac{\pi}{2}+\varphi\right)}.\end{aligned} \tag{7.142}$$

Another coordinate system is provided by the continuous map

$$\begin{aligned}\psi_{\mathrm{S}}\colon\ \mathsf{S}^1\setminus\{+\mathrm{i}\} &\to]-\pi,+\pi[\\ z &\mapsto \psi_{\mathrm{S}}(z) := -\arg\left(z\,e^{\mathrm{i}\pi/2}\right),\end{aligned} \tag{7.143}$$

with the continuous inverse

$$\begin{aligned}\psi_{\mathrm{S}}^{-1}\colon\ ]-\pi,+\pi[&\to \mathsf{S}^1\setminus\{+\mathrm{i}\},\\ \varphi &\mapsto \psi_{\mathrm{S}}^{-1}(\varphi) = e^{-\mathrm{i}\left(\frac{\pi}{2}+\varphi\right)}.\end{aligned} \tag{7.144}$$

Both coordinate systems are compatible in that the change of coordinates (for the region where both systems are defined)

$$\begin{aligned}\psi_{\mathrm{S}}\circ\psi_{\mathrm{N}}^{-1}\colon\ ]-\pi,+\pi[\setminus\{0\} &\to]-\pi,+\pi[\setminus\{0\},\\ \varphi &\mapsto \psi_{\mathrm{S}}\circ\psi_{\mathrm{N}}^{-1}(\varphi) = \psi_{\mathrm{S}}\left(\psi_{\mathrm{N}}^{-1}(\varphi)\right) = -(\pi+\varphi)\,\mathrm{mod}\,2\pi,\end{aligned} \tag{7.145}$$

is one-to-one and smooth. This example suggests the following definition of a smooth manifold.

Definition 7.1 (smooth manifold) A smooth manifold of dimension n is a topological manifold M (see **Appendix** A) on which it is possible to define a family of open sets $\{U_\alpha\}$ such that

1. for any element $p\in M$ there exists a U_α that contains p,

[21] We use the convention according to which the range of the argument function is $[-\pi,+\pi[$.

2. there exists a family of homeomorphisms $\{\psi_\alpha\}$ between the open sets U_α of the manifold M and open sets of $\mathbb{R}^n$,
3. the transitions functions $\psi_\alpha \circ \psi_\beta^{-1}$ are one-to-one and smooth from $\psi_\beta(U_\alpha \cap U_\beta)$ to $\psi_\alpha(U_\alpha \cap U_\beta)$ (both subsets of $\mathbb{R}^n$), and
4. the family (atlas) $\{(U_\alpha, \psi_\alpha)\}$ of pairs (charts) (U_α, ψ_α) is maximal.[22]

In other words, the homeomorphism ψ_α allows to identify each point $p \in U_\alpha$ on the n-dimensional manifold with local coordinates in $\mathbb{R}^n$,

$$\begin{aligned} \psi_\alpha : U_\alpha &\to \mathbb{R}^n, \\ p &\mapsto (x_1, \cdots, x_n). \end{aligned} \tag{7.146}$$

The compatibility of the charts guarantees that these local coordinates are well defined. The smoothness of all coordinates changes ensures the smoothness of the manifold. If the n-dimensional manifold is embedded in a higher dimensional $\mathbb{R}^k$ (as is the case for the Lie groups and coset spaces we are studying, since one can identify $N \times N$ real-valued matrices with elements of $\mathbb{R}^{N^2}$), then it is also possible to speak of differentiability for the charts ψ_α and their inverses. The smoothness of all coordinates changes ensures then that all these maps are smooth (as one can easily verify for the charts of S^1). At each point p of a smooth manifold M of dimension n we can define the n-dimensional tangent space as follows.

Definition 7.2 (tangent space) To each curve on the differentiable manifold M passing through p,

$$\gamma\colon\ ]-\varepsilon, \varepsilon[\to M \qquad \text{with } \gamma(0) = p, \tag{7.147a}$$

and with $\varepsilon > 0$ small enough so that $\gamma(]-\varepsilon, \varepsilon[) \subset U_\alpha$, there corresponds a curve

$$\begin{aligned} \widetilde{\gamma}\colon\ ]-\varepsilon, \varepsilon[&\to \mathbb{R}^n, \\ t &\mapsto \widetilde{\gamma}(t) = \psi_\alpha(\gamma(t)) = (x_1(t), \cdots, x_n(t)), \end{aligned} \tag{7.147b}$$

passing through $(x_0(0), \cdots, x_n(0)) = \psi_\alpha(p)$ in the local coordinates representation (U_α, ψ_α) around p. The tangent vector to γ in $p \in M$, denoted $\frac{\mathrm{d}\gamma}{\mathrm{d}t}\big|_{t=0}$, has the representation

$$\left(\frac{\mathrm{d}\widetilde{\gamma}}{\mathrm{d}t}\right)(t)\bigg|_{t=0} = (\dot{x}_1(0), \cdots, \dot{x}_n(0)) \in \mathbb{R}^n, \tag{7.147c}$$

in the local coordinates (U_α, ψ_α) around p. The tangent space T_pM is defined as the n-dimensional vector space of all tangent vectors in p, with a local basis represented by the standard basis $\boldsymbol{e}_i = (0, \cdots, 0, 1, 0, \cdots, 0)$, $i = 1, \cdots, n$, along the coordinates axis. If the n-dimensional manifold M is embedded in a higher dimensional $\mathbb{R}^k$,

[22] The union of two atlas $\{(U_\alpha, \psi_\alpha)\}$ and $\{(U'_{\alpha'}, \psi'_{\alpha'})\}$ is not necessarily an atlas. An atlas $\{(U_\alpha, \psi_\alpha)\}$ is called maximal if there does not exist any atlas $\{(U'_{\alpha'}, \psi'_{\alpha'})\}$ such that $\{(U_\alpha, \psi_\alpha)\} \subset \{(U'_{\alpha'}, \psi'_{\alpha'})\}$. If $\{(U_\alpha, \psi_\alpha)\}$ is maximal and if $\{(U'_{\alpha'}, \psi'_{\alpha'})\}$ is an atlas such that their union is an atlas, then $\{(U'_{\alpha'}, \psi'_{\alpha'})\} \subseteq \{(U_\alpha, \psi_\alpha)\}$.

then it is possible to directly differentiate the curves $\gamma\colon\ ]-\varepsilon,\varepsilon[\to M\subset\mathbb{R}^k$ and the tangent vectors $\frac{\mathrm{d}\gamma}{\mathrm{d}t}\big|_{t=0}$ span a n-dimensional vector space embedded in $\mathbb{R}^k$: a concrete realization of the tangent space T_pM.

If we interpret the classical matrix groups $\mathrm{G}=\mathrm{SU}(N),\mathrm{SO}(N),\mathrm{Sp}(2N)$ as manifolds, we may interpret their Lie algebras as the tangent space to the identity matrix in these manifolds since, for any curve of the form $\gamma(t)=e^{t\,X}\in\mathrm{G}$,

$$X=\lim_{t\to 0}\frac{\mathrm{d}\,e^{t\,X}}{\mathrm{d}\,t} \tag{7.148a}$$

and

$$\gamma(0)=\mathbb{1}. \tag{7.148b}$$

The geometric interpretation of the unitary time evolution operator $\exp(\mathrm{i}\mathcal{H})$ in the symmetry classes A, D, and C is that $\mathrm{i}\mathcal{H}$ is a tangent vector to the manifold at the origin. What about the symmetry classes AI, AII, AIII, CI, CII, DIII, and BDI? We will illustrate their geometric interpretation by studying the case of the symmetry class BDI when the unitary time evolution operator in **Table** 7.1 corresponds to the case of $M=1$ and $N=2$ given by the set $\mathrm{SO}(3)/\mathrm{SO}(2)$. We are going to show explicitly that $\mathrm{SO}(3)/\mathrm{SO}(2)$ is isomorphic to the surface of the unit sphere in three-dimensional space.

We start from the Lie algebra $\mathfrak{so}(3)$, namely, the three-dimensional vector space over the real numbers of 3×3 matrices that are real-valued and antisymmetric. Define

$$L^1:=\frac{1}{2}\begin{pmatrix}0&0&0\\0&0&+1\\0&-1&0\end{pmatrix},\qquad L^2:=\frac{1}{2}\begin{pmatrix}0&0&+1\\0&0&0\\-1&0&0\end{pmatrix}, \tag{7.149a}$$

on the one hand, and

$$L^3:=\frac{1}{2}\begin{pmatrix}0&+1&0\\-1&0&0\\0&0&0\end{pmatrix} \tag{7.149b}$$

on the other hand. One verifies that

$$\left[L^i,L^j\right]=\frac{1}{2}\,\epsilon^{ijk}\,L^k,\qquad i,j,k=1,2,3. \tag{7.149c}$$

The fully antisymmetric Levi-Civita tensor ϵ^{ijk} multiplied by $1/2$ defines the structure constants of $\mathfrak{so}(3)$ [in this basis of the Lie algebra $\mathfrak{so}(3)$]. Any element X of the Lie algebra $\mathfrak{so}(3)$ can be written as

$$X=\sum_{i=1}^{3}t_i\,L^i \tag{7.150}$$

for some choice of $t_i \in \mathbb{R}$. We define the involutive automorphism

$$\begin{aligned} \theta\colon\ & \mathfrak{so}(3) \to \mathfrak{so}(3), \\ & X \mapsto \theta(X) = I_{2,1}\, X\, I_{2,1}^{-1}, \qquad I_{2,1} := \operatorname{diag}(-1,-1,+1). \end{aligned} \tag{7.151}$$

One verifies that

$$\theta(L^1) = -L^1, \qquad \theta(L^2) = -L^2, \qquad \theta(L^3) = +L^3. \tag{7.152}$$

Hence, we have the decomposition

$$\mathfrak{so}(3) = \mathfrak{k} \oplus \mathfrak{p}, \tag{7.153a}$$

where

$$\mathfrak{k} := \operatorname{span}\{L^3\} = \left\{ \begin{pmatrix} 0 & +t_3 & 0 \\ -t_3 & 0 & 0 \\ 0 & 0 & 0 \end{pmatrix},\ t_3 \in \mathbb{R} \right\} \tag{7.153b}$$

is isomorphic to the closed Lie subalgebra $\mathfrak{so}(2)$, while

$$\mathfrak{p} := \operatorname{span}\{L^1, L^2\} = \left\{ \begin{pmatrix} 0 & 0 & +t_2 \\ 0 & 0 & +t_1 \\ -t_2 & -t_1 & 0 \end{pmatrix},\ t_1, t_2 \in \mathbb{R} \right\} \tag{7.153c}$$

is not closed under commutation.

The vector space on which the matrices of $\mathfrak{so}(3)$ act is

$$\mathbb{R}^3 := \left\{ \begin{pmatrix} x \\ y \\ z \end{pmatrix},\ x, y, z \in \mathbb{R} \right\} = \operatorname{span}\{\boldsymbol{e}_1, \boldsymbol{e}_2, \boldsymbol{e}_3\} \tag{7.154a}$$

with

$$\boldsymbol{e}_1 := \begin{pmatrix} 1 \\ 0 \\ 0 \end{pmatrix}, \qquad \boldsymbol{e}_2 := \begin{pmatrix} 0 \\ 1 \\ 0 \end{pmatrix}, \qquad \boldsymbol{e}_3 := \begin{pmatrix} 0 \\ 0 \\ 1 \end{pmatrix}. \tag{7.154b}$$

The two sphere S^2 is the surface of the unit ball in $\mathbb{R}^3$, that is

$$\mathrm{S}^2 := \left\{ \begin{pmatrix} x \\ y \\ z \end{pmatrix},\ x, y, z \in \mathbb{R},\ x^2 + y^2 + z^2 = 1 \right\}. \tag{7.155}$$

The north pole N of S^2 is represented by $\boldsymbol{e}_3$. It is the null state of L^3, since

$$L^1\, \boldsymbol{e}_3 = \frac{1}{2}\, \boldsymbol{e}_2, \qquad L^2\, \boldsymbol{e}_3 = \frac{1}{2}\, \boldsymbol{e}_1, \qquad L^3\, \boldsymbol{e}_3 = 0. \tag{7.156}$$

We may define the Lie subgroup

$$\mathrm{H} \equiv \mathrm{SO}(2) := \left\{ e^{t_3\, L^3},\ t_3 \in \mathbb{R} \right\}. \tag{7.157}$$

This is a subgroup of the Lie group

$$\mathrm{G} \equiv \mathrm{SO}(3) := \left\{ e^{t_1 L^1 + t_2 L^2 + t_3 L^3}, \; t_i \in \mathbb{R} \right\}. \tag{7.158}$$

Moreover, one verifies by representing the exponential map of matrices by its series that

$$X \in \mathrm{H} \equiv \mathrm{SO}(2) \implies X\, \boldsymbol{e}_3 = \boldsymbol{e}_3. \tag{7.159}$$

In fact, H is the largest subset of G that leaves $\boldsymbol{e}_3$ unchanged. One says that $\mathrm{H} \equiv \mathrm{SO}(2)$ is the isotropy subgroup (or stabilizer) of G at the point $\boldsymbol{e}_3 \in \mathbb{R}^3$, which coincides with the north pole of S^2. The orbit of $\mathrm{G} = \mathrm{SO}(3)$ at $\boldsymbol{e}_3 \in \mathbb{R}^3$ is defined to be the subset of $\mathbb{R}^3$ given by

$$G\, \boldsymbol{e}_3 := \left\{ \boldsymbol{v} \in \mathbb{R}^3 \mid \exists g \in \mathrm{G}, \qquad \boldsymbol{v} = g\, \boldsymbol{e}_3 \right\}. \tag{7.160}$$

It is nothing but the two sphere S^2,

$$\mathrm{G}\, \boldsymbol{e}_3 = \mathrm{S}^2. \tag{7.161}$$

Proof Let $\boldsymbol{x} \in \mathrm{S}^2$ and $\boldsymbol{x} \neq \boldsymbol{e}_3$. Let $\boldsymbol{r} := \boldsymbol{e}_3 \wedge \boldsymbol{x}$. There exists a proper rotation, a 3×3 orthogonal matrix with unit determinant, about the axis $\boldsymbol{r}$ that maps $\boldsymbol{e}_3$ to $\boldsymbol{x}$. □

More generally, $\mathrm{G} = \mathrm{SO}(3)$ acts transitively on S^2 since, for any two elements $\boldsymbol{x}$ and $\boldsymbol{y}$ from S^2, one may construct a proper rotation $g_{\boldsymbol{x},\boldsymbol{y}} \in \mathrm{G}$ such that $\boldsymbol{x} = g_{\boldsymbol{x},\boldsymbol{y}}\, \boldsymbol{y}$.[23] However, this element $g_{\boldsymbol{x},\boldsymbol{y}}$ is not unique since if $h_{\boldsymbol{y}}$ is a proper rotation about $\boldsymbol{y}$, we have

$$\boldsymbol{y} = h_{\boldsymbol{y}}\, \boldsymbol{y}, \qquad \boldsymbol{x} = g_{\boldsymbol{x},\boldsymbol{y}}\, h_{\boldsymbol{y}}\, \boldsymbol{y}. \tag{7.162}$$

The same is true of the orbit S^2 of $\mathrm{G} = \mathrm{SO}(3)$ at $\boldsymbol{e}_3 \in \mathbb{R}^3$. If $\boldsymbol{x} \in \mathrm{S}^2$ and $\boldsymbol{x} \neq \boldsymbol{e}_3$, one may always find a $g_{\boldsymbol{x}} \in \mathrm{G}$ such that $\boldsymbol{x} = g_{\boldsymbol{x}}\, \boldsymbol{e}_3$. However, $g_{\boldsymbol{x}}$ is not unique since any element from $g_{\boldsymbol{x}}\, \mathrm{H}$ will also map $\boldsymbol{e}_3$ to $\boldsymbol{x}$. On the other hand, because of the orbit-stabilizer theorem, there must exist a bijection between S^2 and $\mathrm{G}/\mathrm{H} \equiv \mathrm{SO}(3)/\mathrm{SO}(2)$.[24] Under this bijection, the subset $h\,\mathrm{H}$ with $h \in \mathrm{H}$ is the image of the north pole $\boldsymbol{e}_3$ of S^2, while any subset $g\,\mathrm{H}$ with $\mathrm{G} \ni g \notin \mathrm{H}$ is the image of one and only one point from $\mathrm{S}^2 \setminus \{\boldsymbol{e}_3\}$.

To establish this bijection explicitly, we have the definition

$$\begin{aligned} \mathrm{G}/\mathrm{H} &\equiv \mathrm{SO}(3)/\mathrm{SO}(2) \\ &= \left\{ e^{t_1 L^1 + t_2 L^2}\, \mathrm{SO}(2) \,\middle|\, t_1, t_2 \in \mathbb{R} \right\}. \end{aligned} \tag{7.163}$$

An explicit calculation for a representative of the coset $e^{t_1 L^1 + t_2 L^2}\, \mathrm{SO}(2)$ in G/H gives

[23] In contrast, $\mathrm{G} = \mathrm{SO}(3)$ does not act transitively on $\mathbb{R}^3$.

[24] An element of the coset space G/H is any set of the form $g\,\mathrm{H} := \{ g\, h \mid h \in H \}$ with $g \in \mathrm{G}$. We often identify G/H with a set of representatives: one representative g for each coset $g\,\mathrm{H}$.

$$e^{t_1 L^1 + t_2 L^2} = \begin{pmatrix} 1 + t_2\, t_2 \frac{\cos(|\boldsymbol{t}|/2)-1}{\boldsymbol{t}^2} & t_1\, t_2 \frac{\cos(|\boldsymbol{t}|/2)-1}{\boldsymbol{t}^2} & t_2 \frac{\sin(|\boldsymbol{t}|/2)}{|\boldsymbol{t}|} \\ t_1\, t_2 \frac{\cos(|\boldsymbol{t}|/2)-1}{\boldsymbol{t}^2} & 1 + t_1\, t_1 \frac{\cos(|\boldsymbol{t}|/2)-1}{\boldsymbol{t}^2} & t_1 \frac{\sin(|\boldsymbol{t}|/2)}{|\boldsymbol{t}|} \\ -t_2 \frac{\sin(|\boldsymbol{t}|/2)}{|\boldsymbol{t}|} & -t_1 \frac{\sin(|\boldsymbol{t}|/2)}{|\boldsymbol{t}|} & \cos(|\boldsymbol{t}|/2) \end{pmatrix}, \qquad (7.164a)$$

where

$$\boldsymbol{t} := \begin{pmatrix} t_1 \\ t_2 \end{pmatrix}, \qquad |\boldsymbol{t}| := \sqrt{t_1^2 + t_2^2}. \qquad (7.164b)$$

If we introduce the parametrization

$$x := t_2 \frac{\sin(|\boldsymbol{t}|/2)}{|\boldsymbol{t}|}, \qquad y := t_1 \frac{\sin(|\boldsymbol{t}|/2)}{|\boldsymbol{t}|}, \qquad z := \cos(|\boldsymbol{t}|/2), \qquad (7.165a)$$

together with the column vector

$$X := \begin{pmatrix} x \\ y \end{pmatrix}, \qquad (7.165b)$$

we may then write

$$x^2 + y^2 + z^2 = 1 \qquad (7.165c)$$

together with[25]

$$e^{t_1 L^1 + t_2 L^2} = \begin{pmatrix} \sqrt{\mathbb{1}_2 - X\, X^{\mathsf{T}}} & X \\ -X^{\mathsf{T}} & \sqrt{1 - X^{\mathsf{T}}\, X} \end{pmatrix} \qquad (7.165d)$$

for a representative of any element of SO(3)/SO(2), whereby the signs of the square roots are determined by the sign of z (otherwise the map would not be one-to-one).

We have thus constructed an explicit one-to-one map between G/H $\equiv$ SO(3)/SO(2) and S^2. By design,

1 the north pole

$$\mathrm{N} \equiv \boldsymbol{e}_3 = \begin{pmatrix} 0 \\ 0 \\ 1 \end{pmatrix} \qquad (7.166)$$

of S^2 corresponds to the identity matrix

$$\mathbb{1}_3 = e^{t_1 L^1 + t_2 L^2}\Big|_{t_1 = t_2 = 0} \qquad (7.167)$$

in G/H $\equiv$ SO(3)/SO(2),

2 the action of the involutive automorphism θ defined in Eq. (7.151) by conjugation with the matrix $I_{2,1}$ on elements of G/H $\equiv$ SO(3)/SO(2) leaves the north pole $p \equiv \mathrm{N}$ invariant but acts as minus the identity on the tangent space $T_{p=\mathrm{N}}M$ as it induces the transformation $\boldsymbol{t} \mapsto -\boldsymbol{t}$ in Eq. (7.164) or, equivalently, the transformation $(x, y, z) \mapsto (-x, -y, z)$ in Eq. (7.165a),

[25] A matrix B is the square root of a matrix A if $B^2 = A$.

3 the north pole has $\mathrm{H} \equiv \mathrm{SO}(2)$ as isotropy Lie subgroup of $\mathrm{G} \equiv \mathrm{SO}(3)$, and
4 the orbit of $\mathrm{G} \equiv \mathrm{SO}(3)$ at the north pole is S^2 since

$$e^{t_1 L^1 + t_2 L^2} e_3 = \begin{pmatrix} x \\ y \\ z \end{pmatrix}, \qquad x^2 + y^2 + z^2 = 1. \tag{7.168}$$

All ten spaces associated with the unitary time evolution in **Table** 7.1 are examples of compact and connected symmetric spaces. Compact symmetric spaces (but not necessarily connected) can be constructed out of elementary building blocks consisting of all the distinct pairs of simple and compact Lie groups G together with an involutive automorphism θ on their Lie algebras. This program was completed by Cartan in 1929.

7.4 The Tenfold Way: Gaussian Ensembles

Each compact and connected symmetric space M from **Table** 7.1 is a coset space of the form[26]

$$\mathrm{M} \equiv \mathrm{G}/\mathrm{H}, \tag{7.169a}$$

where H is a semi-simple compact and connected Lie group that generates the isotropy Lie subgroup of the semi-simple compact and connected Lie group G. Each symmetric space M from **Table** 7.1 may be thought of as being in one-to-one correspondence through the exponential map with a set $\{\mathrm{i}\mathcal{H}\}$ built out of random Hermitean matrices $\mathcal{H} = \mathcal{H}^\dagger$ such that

(i) the set $\{\mathrm{i}\mathcal{H}\}$ is closed under conjugation by the isotropy Lie subgroup H, that is,

$$\mathrm{i}\mathcal{H} \in \{\mathrm{i}\mathcal{H}\} \implies \mathcal{U}^{-1}\,\mathrm{i}\mathcal{H}\mathcal{U} \in \{\mathrm{i}\mathcal{H}\}, \qquad \forall \mathcal{U} \in \mathrm{H}. \tag{7.169b}$$

(ii) if the anti-Hermitean random matrix $\mathrm{i}\mathcal{H} \in \{\mathrm{i}\mathcal{H}\}$ is thought of as an element of a vector space of dimension n over the real numbers, then all its real-valued expansion coefficients are identically and independently distributed with a distribution that is invariant under the transformation

$$\mathrm{i}\mathcal{H} \mapsto \mathcal{U}^{-1}\,\mathrm{i}\mathcal{H}\mathcal{U} \in \{\mathrm{i}\mathcal{H}\}, \qquad \forall \mathcal{U} \in \mathrm{H}. \tag{7.169c}$$

Motivated by Eq. (7.169), we define the Gaussian ensemble (GE) to be an extension of the set (7.169) equipped with the measure

$$\mu_{\mathrm{GE}}(\mathrm{d}\mathcal{H}) \equiv \mu(\mathrm{d}\mathcal{H})\, P_{\mathrm{GE}}[\mathcal{H}] \tag{7.170a}$$

[26] For the symmetry classes D, C, and A, the semi-simple compact and connected group G is given by $\mathrm{H} \times \mathrm{H}$ with $\mathrm{H} = \mathrm{SO}(2N)$, $\mathrm{Sp}(2N)$, and $\mathrm{SU}(N)$, respectively.

such that it returns the number of random Hermitean matrices $\mathcal{H} = \mathcal{H}^\dagger$ within a volume element

$$\mu(\mathrm{d}\mathcal{H}) := \mathcal{V}^{-1}\sqrt{|\det(\mathfrak{g})|}\prod_{\iota}\mathrm{d}h_{\iota}, \qquad \mathcal{V} := \int \mu(\mathrm{d}\mathcal{H}). \tag{7.170b}$$

Here, the set $\{h_\iota\}$ is the set of independent real-valued variables that parametrize the Hermitean matrix $\mathcal{H} = \mathcal{H}^\dagger$, the metric $\mathfrak{g}$ is defined by the change of the invariant arclength

$$\mathrm{tr}\,(\mathrm{d}\mathcal{H})^2 \equiv \sum_{i,j}\left(\mathrm{d}\mathcal{H}_{ij}\,\mathrm{d}\mathcal{H}_{ji}\right) = \sum_{i,j}\left(\mathrm{d}\mathcal{H}_{ij}\,\mathrm{d}\mathcal{H}^{*}_{ij}\right) = \sum_{\iota,\iota'}\mathfrak{g}_{\iota\iota'}\,\mathrm{d}h_{\iota}\,\mathrm{d}h_{\iota'} \tag{7.170c}$$

resulting from the infinitesimal change $\mathrm{d}h_\iota$, and $P_{\mathrm{GE}}[\mathcal{H}]$ is given by the Gaussian weight

$$P_{\mathrm{GE}}[\mathcal{H}] \propto \exp\left(-\frac{\mathrm{tr}\,(\mathcal{H}^\dagger\,\mathcal{H})}{2\sigma^2}\right) = \exp\left(-\frac{\mathrm{tr}\,(\mathcal{H}^2)}{2\sigma^2}\right). \tag{7.170d}$$

The extension of the set $\{\mathrm{i}\mathcal{H}\}$ in Eq. (7.169) that we seek consists in enlarging G to (**Exercise** 7.13)

$$\mathrm{O}(2N) \equiv \mathrm{O}(1)\times\mathrm{SO}(2N) \text{ and } \mathrm{U}(N) \equiv \mathrm{U}(1)\times\mathrm{SU}(N) \tag{7.170e}$$

for the symmetry classes D and A, respectively, so that we may choose H to be the largest Lie subgroup of G for which the measure (7.170) is left invariant under conjugation by elements of H. The corresponding compact symmetric spaces are to be found in the second column of **Table** 7.2. One observes that these symmetric spaces are not always connected when they descend from the orthogonal group.

One aim of random matrix theory is the study of the statistics obeyed by the eigenvalues of random Hermitean matrices. For any random Hermitean matrix $\mathcal{H} = \mathcal{H}^\dagger$ from one of the ten symmetry classes associated with the symmetric spaces G/H in **Table** 7.2, we may always diagonalize $\mathcal{H}$ with a random unitary matrix from the isotropy Lie subgroup H,

$$\mathcal{H} = \mathcal{U}\,\mathrm{diag}(\varepsilon_1,\varepsilon_2,\cdots)\,\mathcal{U}^\dagger, \qquad \mathcal{U}\in\mathrm{H}. \tag{7.171}$$

It can be shown that the joint probability distribution $P(\varepsilon_1,\cdots,\varepsilon_{N_{\mathrm{AZ}}})$ for the N_{AZ} independent random energy eigenvalues $\varepsilon_1,\cdots,\varepsilon_{N_{\mathrm{AZ}}}$ of $\mathcal{H} = \mathcal{H}^\dagger$ to be found in the volume element $\mathrm{d}\varepsilon_1\cdots\mathrm{d}\varepsilon_{N_{\mathrm{AZ}}}$ is proportional to

$$\begin{aligned}&\left(\prod_{j=1}^{N_{\mathrm{AZ}}}\mathrm{d}\varepsilon_j\,e^{-\varepsilon_j^2/(2\,\sigma_{\mathrm{AZ}}^2)}\right)\\ &\times\left(\prod_{1\le j<k\le N_{\mathrm{AZ}}}|\varepsilon_k-\varepsilon_j|^{\beta_{\mathrm{AZ}}}\right)\end{aligned} \tag{7.172a}$$

Table 7.2 *The tenfold way: Gaussian random-matrix ensembles*

AZ	$\mathrm{M} \equiv \mathrm{G/H}$	N_{AZ}	d_{AZ}	β_{AZ}	α_{AZ}	ν_{AZ}
A	$\mathrm{U}(N)$	N	1	2	0	0
AI	$\mathrm{U}(N)/\mathrm{O}(N)$	N	1	1	0	0
AII	$\mathrm{U}(2N)/\mathrm{Sp}(2N)$	N	2	4	0	0
AIII	$\mathrm{U}(M+N)/[\mathrm{U}(M)\times\mathrm{U}(N)]$	N	1	2	1	$M-N\geq 0$
BDI	$\mathrm{O}(M+N)/[\mathrm{SO}(M)\times\mathrm{SO}(N)]$	N	1	1	0	$M-N\geq 0$
CII	$\mathrm{Sp}(2M+2N)/[\mathrm{Sp}(2M)\times\mathrm{Sp}(2N)]$	N	2	4	3	$M-N\geq 0$
D	$\mathrm{O}(2N)$	N	1	2	0	0
	$\mathrm{O}(2N+1)$	N	1	2	0	1
C	$\mathrm{Sp}(2N)$	N	1	2	2	0
DIII	$\mathrm{O}(4N)/\mathrm{U}(2N)$	N	2	4	1	0
	$\mathrm{O}(4N+2)/\mathrm{U}(2N+1)$	N	2	4	1	1
CI	$\mathrm{Sp}(2N)/\mathrm{U}(N)$	N	1	1	1	0

The tenfold way for the Gaussian random-matrix ensembles. The Altland-Zirnbauer (AZ) symmetry classes are labeled in the first column by the Cartan labels of the symmetric spaces $\mathrm{M} \equiv \mathrm{G/H}$ that descend from the semi-simple Lie group $\mathrm{O}(2N)$ for the symmetry class D, the simple Lie group $\mathrm{Sp}(2N)$ for the symmetry class C, and the semi-simple Lie group $\mathrm{U}(N)$ for the symmetry class A. The third column gives the number N_{AZ} of independent energy eigenvalues. The fourth column gives the (Kramer's) degeneracy d_{AZ} of the energy eigenvalues. The fifth column gives the exponent β_{AZ} controlling the level repulsion between two energy eigenvalues. The sixth column gives the exponent α_{AZ} controlling the level repulsion from the energy zero. The seventh column gives the number ν_{AZ} of zero modes. There are two entries for the symmetry classes D and DIII as these symmetry classes are sensitive to the parity of the dimensionality of their random Hamiltonians.

for the Wigner-Dyson symmetry classes A, AI, and AII, while it is proportional to

$$\left(\prod_{j=1}^{N_{\mathrm{AZ}}} \mathrm{d}\varepsilon_j\,\Theta(\varepsilon_j)\,|\varepsilon_j|^{\alpha_{\mathrm{AZ}}+\nu_{\mathrm{AZ}}\,\beta_{\mathrm{AZ}}}\,e^{-\varepsilon_j^2/(2\,\sigma_{\mathrm{AZ}}^2)}\right) \times \left(\prod_{1\leq j<k\leq N_{\mathrm{AZ}}} |\varepsilon_k^2-\varepsilon_j^2|^{\beta_{\mathrm{AZ}}}\right) \tag{7.172b}$$

for the chiral symmetry classes AIII, BDI, CII, and for the Bogoliubov-de-Gennes symmetry classes D, DIII, C, and CI. The number N_{AZ} of independent energy eigenvalues, their degeneracies d_{AZ}, the exponent β_{AZ} controlling the level repulsion between two energy eigenvalues, the exponent α_{AZ} controlling the level repulsion at the origin for the positive energy eigenvalues [$\Theta(\varepsilon_j)$ denotes the Heaviside function], and the number ν_{AZ} of zero modes are derived in **Exercises** 7.14 and 7.15 and tabulated in **Table** 7.2. The width σ_{AZ} of the Gaussian distribution is proportional to the mean level spacing δ,[27]

$$\sigma^2_{\mathrm{AZ}} = \begin{cases} \frac{2N_{\mathrm{AZ}}\,\delta^2}{\pi^2\,\beta_{\mathrm{AZ}}}, & \mathrm{A, AI, AII}, \\ \\ \frac{2(2N_{\mathrm{AZ}}+\nu_{\mathrm{AZ}})\,\delta^2}{\pi^2\,\beta_{\mathrm{AZ}}}, & \mathrm{AIII, BDI, CII, D, DIII, C, CI}. \end{cases} \tag{7.172c}$$

The variance σ^2_{AZ} of the Gaussian distribution scales linearly with the number of independent energy eigenvalues N_{AZ} in order for the limit $N_{\mathrm{AZ}} \to \infty$ to be well defined. Random matrix theory (7.170) thus delivers universal predictions in the scaling limit $N_{\mathrm{AZ}} \to \infty$.[6]

According to Eq. (7.172) and **Table** 7.2, all ten random-matrix ensembles are characterized by power laws for the probability to find two levels $\varepsilon_j \neq 0$ and $\varepsilon_k \neq 0$ close to each other. Level crossing at any nonvanishing energy is thus avoided in all ten symmetry classes. In the scaling limit $N_{\mathrm{AZ}} \to \infty$, the joint-probability distributions for the energy eigenvalues in the Wigner-Dyson Gaussian random-matrix ensembles (A, AI, and AII) are invariant under a constant shift of all energy eigenvalues, as well as under a global mirror symmetry about the energy zero. In the scaling limit $N_{\mathrm{AZ}} \to \infty$, there is no translation invariance of the joint-probability distributions for the chiral (AIII, BDI, CII) and Bogoliubov-de-Gennes (BdG) (D, DIII, C, CI) Gaussian random-matrix ensembles. However, the joint-probability distributions for the chiral and BdG Gaussian random-matrix ensembles are invariant under changing the sign of energy eigenvalues, one eigenvalue at a time. This is a consequence of the mirror symmetry of the spectrum about the zero energy for each Hamiltonian from the chiral or BdG Gaussian random-matrix ensembles. As nearby energy eigenvalues are expected to repel, a spectral mirror symmetry should cause a level repulsion from the energy zero. This repulsion is present whenever $\alpha_{\mathrm{AZ}} \neq 0$, as is the case for the chiral symmetry classes AIII and CII and the BdG symmetry classes DIII, C, and CI. Symmetry classes D and BDI have $\alpha_{\mathrm{AZ}} = 0$ even though their spectra are mirror symmetric about the energy zero. Level repulsion from the energy zero is guaranteed whenever a symmetry class admits one or more zero modes, as is the case for all three chiral symmetry classes and the BdG symmetry classes D and DIII. However, level crossing is permissible at zero energy in the symmetry classes D and BDI in the absence of zero modes. Finally, for all chiral and BdG symmetry classes, the positive energy eigenvalue $\varepsilon_j > 0$ repels not only from the positive energy eigenvalue $\varepsilon_k > 0$ but also from the negative energy eigenvalue $-\varepsilon_k < 0$ in view of the identity

$$\varepsilon_j^2 - \varepsilon_k^2 = \left(\varepsilon_j - \varepsilon_k\right)\left(\varepsilon_j + \varepsilon_k\right). \tag{7.173}$$

[27] C. W. J. Beenakker, Rev. Mod. Phys. **87**(3), 1037-1066 (2015). [Beenakker (2015)].

According to **Table** 7.2, zero modes are possible in five out of the ten symmetry classes. The condition for the presence of $M - N$ zero modes in the three chiral symmetry classes AIII, BDI, and CII is that $M > N$. One zero mode is present whenever the dimension of the random Hamiltonian is $2N+1$ in the Bogoliubov-de-Gennes symmetry class D or $2 \times (2N+1)$ in the Bogoliubov-de-Gennes symmetry class DIII. Such random matrices can only be realized as effective theories on any one of a pair of point-like defects embedded in a space of dimensions equal to or larger than one, provided a large distance separates this pair of point-like defects. A long open Majorana $c = 1$ chain (Kitaev chain) realizes on any one of its two boundaries the case $2N+1$ in the symmetry class D.

7.5 The Tenfold Way: Circular Ensembles

We are going to construct the generalization of the three circular ensembles of Dyson to the remaining seven symmetry classes. As was the case with Section 7.4, we start from any one of the ten symmetric spaces from **Table** 7.1. For each Altland-Zirnbauer symmetry class (abbreviated by AZ in Table 7.1), there is a coset of the form[28]

$$\mathrm{M} \equiv \mathrm{G}/\mathrm{H}, \tag{7.174}$$

where H is a semi-simple Lie group that generates an isotropy semi-simple Lie subgroup of the semi-simple Lie group G.

Let $S \in \mathrm{U}(N_{\mathrm{M}})$ be an arbitrary $N_{\mathrm{M}} \times N_{\mathrm{M}}$ unitary matrix from M. There are uncountably many pairs $U, V \in \mathrm{U}(N_{\mathrm{M}})$ such that

$$S = U\,V. \tag{7.175}$$

Proof This factorization is a consequence of the fact that, for any pair $U_1 \neq U_2 \in \mathrm{U}(N_{\mathrm{M}})$,

$$U_2 = \left(U_2\,U_1^{\dagger}\right)\,U_1 = U_1\,\left(U_1^{\dagger}\,U_2\right). \tag{7.176}$$

That this factorization is not unique is a consequence of the fact that it is invariant under

$$U \mapsto U\,W, \qquad V \mapsto W^{\dagger}\,V \tag{7.177}$$

for any $W \in \mathrm{U}(N_{\mathrm{M}})$. □

We would like to define a small change of S that is compatible with the factorization (7.175). We define the infinitesimal $N_{\mathrm{M}} \times N_{\mathrm{M}}$ matrix increment $\mathrm{d}S$ by

$$S + \mathrm{d}S := U\,(\mathbb{1} + \mathrm{i}\mathrm{d}H)\,V \tag{7.178a}$$

[28] For the symmetry classes D, C, and A, the semi-simple group G is given by $\mathrm{H} \times \mathrm{H}$ with $\mathrm{H} = \mathrm{SO}(N)$, $\mathrm{Sp}(2N)$, and $\mathrm{U}(N)$, respectively.

($\mathbb{1}$ the unit $N_{\mathrm{M}}\times N_{\mathrm{M}}$ matrix) for some infinitesimal Hermitean $N_{\mathrm{M}}\times N_{\mathrm{M}}$ matrix

$$\mathrm{d}H=(\mathrm{d}H)^{\dagger}\,. \tag{7.178b}$$

We assign to the infinitesimal matrix $\mathrm{d}H$ the infinitesimal volume element

$$\mu_{\mathrm{AZ}}(\mathrm{d}H):=\left(\prod_{j,k\in\Omega^{(\mathrm{Re})}_{\mathrm{AZ}}}\mathrm{d}\left(\operatorname{Re}H_{jk}\right)\right)\left(\prod_{j,k\in\Omega^{(\mathrm{Im})}_{\mathrm{AZ}}}\mathrm{d}\left(\operatorname{Im}H_{jk}\right)\right), \tag{7.178c}$$

where $\Omega^{(\mathrm{Re})}_{\mathrm{AZ}}$ and $\Omega^{(\mathrm{Im})}_{\mathrm{AZ}}$ are two subsets from the set made of all N^2_{M} pairs $j,k=1,\cdots,N_{\mathrm{M}}$. This pair of subsets is constructed explicitly in **Exercises** 7.14 and 7.15 for each symmetric space M from **Table** 7.1 corresponding to an Altland-Zirnbauer symmetry class. According to **Exercise** 7.16, this measure is invariant under the transformation

$$U\mapsto U',\qquad V\mapsto V', \tag{7.179a}$$

for any choice of $U',V'\in\mathrm{U}(N_{\mathrm{M}})$ such that

$$S=U\,V=U'\,V'. \tag{7.179b}$$

We have constructed a Haar measure for the symmetric space M corresponding to any one of the ten Altland-Zirnbauer symmetry classes. The corresponding circular ensemble is defined by the partition function

$$Z_{\mathrm{AZ}}:=\int\mu_{\mathrm{AZ}}(\mathrm{d}H)\,1. \tag{7.180}$$

Any unitary $N_{\mathrm{M}}\times N_{\mathrm{M}}$ matrix S can be diagonalized. This is to say that there exists the nonunique factorization

$$S=W\,E\,W^{\dagger}, \tag{7.181a}$$

with $W\in\mathrm{U}(N_{\mathrm{M}})$ and E a diagonal matrix of the form

$$E=e^{\mathrm{i}\Theta},\qquad \Theta=\operatorname{diag}\left(\theta_1,\cdots,\theta_{N_{\mathrm{M}}}\right),\qquad 0\le\theta_1,\cdots,\theta_{N_{\mathrm{M}}}<2\pi. \tag{7.181b}$$

It is shown in **Exercise** 7.17 that the partition function (7.180) corresponding to the symmetric space M (that is the Altland-Zirnbauer symmetry class AZ) in **Table** 7.2 is now represented by

$$Z_{\mathrm{AZ}}\propto\int_0^{2\pi}\mathrm{d}\theta_1\cdots\int_0^{2\pi}\mathrm{d}\theta_{N_{\mathrm{AZ}}}\,J^{(+)}_{\mathrm{AZ}}\left(\theta_1,\cdots,\theta_{N_{\mathrm{AZ}}}\right)\int\mu_{\mathrm{AZ}}(\mathrm{d}A)\,1, \tag{7.182a}$$

where a Haar measure $\mu_{\mathrm{AZ}}(\mathrm{d}A)$ for

$$\mathrm{d}A=-\left(\mathrm{d}A\right)^{\dagger}=W^{\dagger}\,\mathrm{d}W \tag{7.182b}$$

was derived in **Exercise** 7.14 for each symmetric space M (that is the Altland-Zirnbauer symmetry class AZ) and $J^{(+)}_{\mathrm{AZ}}\left(\theta_1,\cdots,\theta_{N_{\mathrm{AZ}}}\right)$ is the Jacobian associated with the spectral representation. For the standard symmetry classes A, AI, and AII, one finds

$$J^{(+)}_{\mathrm{AZ}}\left(\theta_1,\cdots,\theta_{N_{\mathrm{AZ}}}\right)=\prod_{1\leq j<k\leq N_{\mathrm{AZ}}}\left|2\sin\left(\frac{\theta_j-\theta_k}{2}\right)\right|^{\beta_{\mathrm{AZ}}}. \tag{7.182c}$$

For the chiral symmetry classes AIII, BDI, and CII or for the Bogoliubov-de-Gennes symmetry classes D, C, DIII, and CI, one finds

$$\begin{aligned}J^{(+)}_{\mathrm{AZ}}\left(\theta_1,\cdots,\theta_{N_{\mathrm{AZ}}}\right)=&\prod_{j=1}^{N_{\mathrm{AZ}}}\left|2\sin\theta_j\right|^{\alpha_{\mathrm{AZ}}}\left|2\sin\left(\frac{\theta_j}{2}\right)\right|^{\nu_{\mathrm{AZ}}\,\beta_{\mathrm{AZ}}}\\&\times\prod_{1\leq j<k\leq N_{\mathrm{AZ}}}\left|4\sin\left(\frac{\theta_k-\theta_j}{2}\right)\sin\left(\frac{\theta_k+\theta_j}{2}\right)\right|^{\beta_{\mathrm{AZ}}}.\end{aligned} \tag{7.182d}$$

The coset space M, the integer N_{AZ}, the exponents α_{AZ} and β_{AZ}, and the number ν_{AZ} of zero modes are to be found in **Table** 7.2.

Remark (1) Dyson chose the representation

$$J^{(+)}_{\mathrm{AZ}}\left(\theta_1,\cdots,\theta_{N_{\mathrm{AZ}}}\right)=\prod_{1\leq j<k\leq N_{\mathrm{AZ}}}\left|e^{+\mathrm{i}\theta_k}-e^{-\mathrm{i}\theta_j}\right|^{\beta_{\mathrm{AZ}}}, \tag{7.183a}$$

for the symmetry classes A, AI, and AII. In this representation, the chiral and Bogoliubov-de-Gennes symmetry classes have the Jacobian

$$\begin{aligned}J^{(+)}_{\mathrm{AZ}}\left(\theta_1,\cdots,\theta_{N_{\mathrm{AZ}}}\right)=&\prod_{j=1}^{N_{\mathrm{AZ}}}\left|e^{+\mathrm{i}2\,\theta_j}-1\right|^{\alpha_{\mathrm{AZ}}}\left|e^{+\mathrm{i}\theta_j}-1\right|^{\nu_{\mathrm{AZ}}\,\beta_{\mathrm{AZ}}}\\&\times\prod_{1\leq j<k\leq N_{\mathrm{AZ}}}\left|\left(e^{+\mathrm{i}\theta_k}-e^{-\mathrm{i}\theta_j}\right)\left(e^{+\mathrm{i}\theta_k}-e^{+\mathrm{i}\theta_j}\right)\right|^{\beta_{\mathrm{AZ}}}.\end{aligned} \tag{7.183b}$$

Remark (2) The limit $\sigma^2_{\mathrm{AZ}}\to\infty$ of the Jacobian (7.172) is recovered if we write

$$\theta_j=\eta\,\varepsilon_j,\qquad j=1,\cdots,N_{\mathrm{AZ}}, \tag{7.184}$$

and take the limit $\eta\to 0$ keeping ε_j fixed.

Remark (3) If $\{e_j\mid j=1,\cdots,N_{\mathrm{AZ}}\}$ is the Cartesian basis of $\mathbb{R}^{N_{\mathrm{AZ}}}$, $\boldsymbol{\varepsilon}\in\mathbb{R}^{N_{\mathrm{AZ}}}$ is the vector with the energy eigenvalues ε_j as components, $\boldsymbol{\theta}\in\mathbb{R}^{N_{\mathrm{AZ}}}$ is the vector with the phase eigenvalues θ_j as components, and we define the set of "positive" roots to be the union of the short (s), ordinary (o), and long (l) positive roots defined by

$$\begin{aligned}
\mathfrak{Roots}_+ &:= \mathfrak{Roots}_{\mathrm{s}} \cup \mathfrak{Roots}_{\mathrm{o}}^- \cup \mathfrak{Roots}_{\mathrm{o}}^+ \cup \mathfrak{Roots}_{\mathrm{l}},\\
\mathfrak{Roots}_{\mathrm{s}} &:= \{\boldsymbol{e}_j \mid 1 \leq j \leq N_{\mathrm{AZ}}\},\\
\mathfrak{Roots}_{\mathrm{o}}^{\pm} &:= \{\boldsymbol{e}_k \pm \boldsymbol{e}_j \mid 1 \leq j < k \leq N_{\mathrm{AZ}}\},\\
\mathfrak{Roots}_{\mathrm{l}} &:= \{2\boldsymbol{e}_j \mid 1 \leq j \leq N_{\mathrm{AZ}}\},
\end{aligned} \tag{7.185a}$$

we have then constructed the pair of Jacobians

$$\begin{aligned}
J_{\mathrm{AZ}}^{(0)}(\boldsymbol{\varepsilon}) := & \prod_{\boldsymbol{\alpha}\in\mathfrak{Roots}_{\mathrm{s}}} |\boldsymbol{\alpha}\cdot\boldsymbol{\varepsilon}|^{m_{\mathrm{s}}}\\
& \times \left(\prod_{\boldsymbol{\alpha}\in\mathfrak{Roots}_{\mathrm{o}}^-} |\boldsymbol{\alpha}\cdot\boldsymbol{\varepsilon}|^{m_{\mathrm{o}}^-}\right)\left(\prod_{\boldsymbol{\alpha}\in\mathfrak{Roots}_{\mathrm{o}}^+} |\boldsymbol{\alpha}\cdot\boldsymbol{\varepsilon}|^{m_{\mathrm{o}}^+}\right)\\
& \times \prod_{\boldsymbol{\alpha}\in\mathfrak{Roots}_{\mathrm{l}}} |\boldsymbol{\alpha}\cdot\boldsymbol{\varepsilon}|^{m_{\mathrm{l}}}
\end{aligned} \tag{7.185b}$$

and

$$\begin{aligned}
J_{\mathrm{AZ}}^{(+)}(\boldsymbol{\theta}) := & \left(\prod_{\boldsymbol{\alpha}\in\mathfrak{Roots}_{\mathrm{s}}} \left|2\sin\left(\frac{\boldsymbol{\alpha}\cdot\boldsymbol{\theta}}{2}\right)\right|^{m_{\mathrm{s}}}\right)\\
& \times \left(\prod_{\sigma=\pm}\prod_{\boldsymbol{\alpha}\in\mathfrak{Roots}_{\mathrm{o}}^{\sigma}} \left|2\sin\left(\frac{\boldsymbol{\alpha}\cdot\boldsymbol{\theta}}{2}\right)\right|^{m_{\mathrm{o}}^{\sigma}}\right)\\
& \times \left(\prod_{\boldsymbol{\alpha}\in\mathfrak{Roots}_{\mathrm{l}}} \left|2\sin\left(\frac{\boldsymbol{\alpha}\cdot\boldsymbol{\theta}}{2}\right)\right|^{m_{\mathrm{l}}}\right)
\end{aligned} \tag{7.185c}$$

for the Gaussian and circular ensembles of any one of the ten Altland-Zirnbauer symmetry classes. The exponent m_{s} is called the multiplicity of the short root $\boldsymbol{e}_j$. The exponent $m_{\mathrm{o}}^{\pm}$ is called the multiplicity of the ordinary root $\boldsymbol{e}_k \pm \boldsymbol{e}_j$. The exponent m_{l} is called the multiplicity of the long root $2\boldsymbol{e}_j$. Their values are tabulated in **Table** 7.3 for the ten Altland-Zirnbauer symmetry classes. In mathematics, a root lattice is a configuration of vectors in an Euclidean space satisfying certain geometrical properties. The concept or root lattice is fundamental to the classification and representation theory of simple Lie algebras over the field of complex numbers (see footnote 16 for a mathematical treatment of this connection or Georgi[29] for a a physicist perspective on this connection).

Definition 7.3 (reduced root lattice) A reduced root lattice of a finite-dimensional Euclidean space $\mathbb{E}$ is a finite set $\mathfrak{Roots}(\mathbb{E})$ of nonvanishing vectors (called roots and denoted $\boldsymbol{\alpha}$) that satisfy the following conditions.

[29] H. Georgi, *Lie Algebras in Particle Physics: From Isospin to Unified Theories,* CRS Press, 2000. [Georgi (2000)].

Table 7.3 *The tenfold way: Root lattices and their multiplicities for symmetric spaces with positive curvature*

AZ	M ≡ G/H	$\mathfrak{Roots}_+$	$m_{\rm s}$	$m_{\rm o}^-$	$m_{\rm o}^+$	$m_{\rm l}$
A	$\mathrm{U}(N)$	$\boldsymbol{e}_k - \boldsymbol{e}_j$	0	2	0	0
AI	$\mathrm{U}(N)/\mathrm{O}(N)$	$\boldsymbol{e}_k - \boldsymbol{e}_j$	0	1	0	0
AII	$\mathrm{U}(2N)/\mathrm{Sp}(2N)$	$\boldsymbol{e}_k - \boldsymbol{e}_j$	0	4	0	0
AIII	$\mathrm{U}(2N)/[\mathrm{U}(N)\times\mathrm{U}(N)]$	$\boldsymbol{e}_k \pm \boldsymbol{e}_j, 2\boldsymbol{e}_j$	0	2	2	1
	$\mathrm{U}(M+N)/[\mathrm{U}(M)\times\mathrm{U}(N)]$	$\boldsymbol{e}_j, \boldsymbol{e}_k \pm \boldsymbol{e}_j, 2\boldsymbol{e}_j$	2ν	2	2	1
BDI	$\mathrm{O}(2N)/[\mathrm{O}(N)\times\mathrm{O}(N)]$	$\boldsymbol{e}_k \pm \boldsymbol{e}_j$	0	1	1	0
	$\mathrm{O}(M+N)/[\mathrm{O}(M)\times\mathrm{O}(N)]$	$\boldsymbol{e}_j, \boldsymbol{e}_k \pm \boldsymbol{e}_j$	ν	1	1	0
CII	$\mathrm{Sp}(4N)/[\mathrm{Sp}(2N)\times\mathrm{Sp}(2N)]$	$\boldsymbol{e}_k \pm \boldsymbol{e}_j, 2\boldsymbol{e}_j$	0	4	4	3
	$\mathrm{Sp}(2M+2N)/[\mathrm{Sp}(2M)\times\mathrm{Sp}(2N)]$	$\boldsymbol{e}_j, \boldsymbol{e}_k \pm \boldsymbol{e}_j, 2\boldsymbol{e}_j$	4ν	4	4	3
D	$\mathrm{O}(2N)$	$\boldsymbol{e}_k \pm \boldsymbol{e}_j$	0	2	2	0
	$\mathrm{O}(2N+1)$	$\boldsymbol{e}_j, \boldsymbol{e}_k \pm \boldsymbol{e}_j$	2	2	2	0
C	$\mathrm{Sp}(2N)$	$\boldsymbol{e}_k \pm \boldsymbol{e}_j, 2\boldsymbol{e}_j$	0	2	2	2
DIII	$\mathrm{O}(4N)/\mathrm{U}(2N)$	$\boldsymbol{e}_k \pm \boldsymbol{e}_j, 2\boldsymbol{e}_j$	0	4	4	1
	$\mathrm{O}(4N+2)/\mathrm{U}(2N+1)$	$\boldsymbol{e}_j, \boldsymbol{e}_k \pm \boldsymbol{e}_j, 2\boldsymbol{e}_j$	4	4	4	1
CI	$\mathrm{Sp}(2N)/\mathrm{U}(N)$	$\boldsymbol{e}_k \pm \boldsymbol{e}_j, 2\boldsymbol{e}_j$	0	1	1	1

The Jacobian for the spectral decomposition of a random Hermitean matrix H or its exponential $\exp(\mathrm{i}H)$ is determined by the positive roots and their multiplicities associated with the symmetric space $\mathrm{M} \equiv \mathrm{G}/\mathrm{H} \equiv \mathrm{M}^{(+)}$ with positive curvature generated by the isotropy subgroup H, the maximal subgroup that leaves the measure for the random matrices invariant under left or right multiplication by any element of H. For the ten Altland-Zirnbauer (AZ) symmetry classes, the positive roots and their multiplicities (as derived in **Exercise** 7.14) are tabulated as follows ($1 \leq j < k \leq N \equiv N_{\mathrm{AZ}}$, $N \leq M = 1, 2, 3, \cdots$). For the chiral classes AIII, BDI, and CII, we introduced the shorthand notation $|M - N| = \nu = 0, 1, 2, \cdots$.

(1) The roots span Euclidean space $\mathbb{E}$.

(2) The only scalar multiples of a root $\boldsymbol{\alpha} \in \mathfrak{Roots}(\mathbb{E})$ that belong to $\mathfrak{Roots}(\mathbb{E})$ are $\boldsymbol{\alpha}$ and $-\boldsymbol{\alpha}$.

(3) For every root $\boldsymbol{\alpha} \in \mathfrak{Roots}(\mathbb{E})$, $\mathfrak{Roots}(\mathbb{E})$ is closed under reflection through the hyperplane perpendicular to $\boldsymbol{\alpha}$.

(4) For any pair of roots $\boldsymbol{\alpha}, \boldsymbol{\beta} \in \mathfrak{Roots}(\mathbb{E})$, the projection of $\boldsymbol{\beta}$ onto the unit vector parallel to $\boldsymbol{\alpha}$ is either a half integer or an integer multiple of $\boldsymbol{\alpha}$.

Definition 7.4 (positive roots) Given the reduced root lattice $\mathfrak{Roots}(\mathbb{E})$, a non-empty subset $\mathfrak{Roots}_+(\mathbb{E}) \subset \mathfrak{Roots}(\mathbb{E})$ must obey the following conditions.

(1) For any $\boldsymbol{\alpha} \in \mathfrak{Roots}(\mathbb{E})$, exactly one of $+\boldsymbol{\alpha}$ or $-\boldsymbol{\alpha}$ belongs to $\mathfrak{Roots}_+(\mathbb{E})$.

(2) For any pair $\boldsymbol{\alpha} \neq \boldsymbol{\beta} \in \mathfrak{Roots}_{+}(\mathbb{E})$ such that $\boldsymbol{\alpha}+\boldsymbol{\beta} \in \mathfrak{Roots}(\mathbb{E})$, then $\boldsymbol{\alpha}+\boldsymbol{\beta} \in \mathfrak{Roots}_{+}(\mathbb{E})$.

Elements of $\mathfrak{Roots}_{+}(\mathbb{E})$ are called positive roots. Given the set $\mathfrak{Roots}_{+}(\mathbb{E})$, the set $\mathfrak{Roots}_{-}(\mathbb{E}) := -\mathfrak{Roots}_{+}(\mathbb{E})$ has no common elements with the set $\mathfrak{Roots}_{+}(\mathbb{E})$ and their union gives back the set $\mathfrak{Roots}(\mathbb{E})$. Elements of $\mathfrak{Roots}_{+}(\mathbb{E})$ are called negative roots.

Definition 7.5 (simple roots) A positive root $\boldsymbol{\alpha} \in \mathfrak{Roots}_{+}(\mathbb{E})$ is a simple (fundamental) root if it cannot be written as the sum of two distinct positive roots.

Property 7.6 The set of simple roots gives a basis of $\mathbb{E}$ such that every root is a linear combination of simple roots with integer coefficients that are either all nonnegative or all nonpositive.

Property 7.7 In any simple Lie algebra $\mathfrak{g}$ over the field of complex numbers, there are two kinds of generators:

(1) There is a maximal Abelian Lie subalgebra

$$\mathfrak{h}_0 = \{H_1, \cdots, H_r \mid [H_i, H_j] = 0,\ i,j = 1, \cdots, r\} \tag{7.186a}$$

called the Cartan subalgebra. The integer r is called the rank of the simple Lie algebra.

(2) There is a reduced root lattice $\boldsymbol{\alpha} \in \mathfrak{Roots}_{\mathfrak{g}}(\mathbb{R}^r)$ and a set $\{E_{\boldsymbol{\alpha}}\}$ labeled by the elements $\boldsymbol{\alpha}$ of the reduced root lattice such that

$$[H_i, E_{\boldsymbol{\alpha}}] = \alpha_i\, E_{\boldsymbol{\alpha}} \tag{7.186b}$$

for any $i = 1, \cdots, r$ and any $\boldsymbol{\alpha} \in \mathfrak{Roots}_{\mathfrak{g}}(\mathbb{R}^r)$. Correspondingly, the generator $E_{\boldsymbol{\alpha}}$ is either a raising or a lowering operator.

Example 7.8 Let $\{\boldsymbol{e}_i \mid i = 1, \cdots, N\}$ be the Cartesian basis of $\mathbb{R}^N$. The reduced roots of the classical Lie algebras over the field of complex numbers $\mathfrak{su}(N,\mathbb{C})$, $\mathfrak{so}(2N+1,\mathbb{C})$, $\mathfrak{sp}(2N,\mathbb{C})$, and $\mathfrak{so}(2N,\mathbb{C})$ are

$$A_{N-1} \equiv \bigcup_{\sigma=\pm} \Big\{ \sigma(\boldsymbol{e}_i - \boldsymbol{e}_j) \ \Big|\ i<j=1,\cdots,N\Big\}, \tag{7.187a}$$

$$B_N \equiv \bigcup_{\sigma=\pm} \Big\{ \sigma\,\boldsymbol{e}_i, \sigma(\boldsymbol{e}_i - \boldsymbol{e}_j), \sigma(\boldsymbol{e}_i + \boldsymbol{e}_j) \ \Big|\ i<j=1,\cdots,N\Big\}, \tag{7.187b}$$

$$C_N \equiv \bigcup_{\sigma=\pm} \Big\{ \sigma(\boldsymbol{e}_i - \boldsymbol{e}_j), \sigma(\boldsymbol{e}_i + \boldsymbol{e}_j), \sigma\,2\boldsymbol{e}_i \ \Big|\ i<j=1,\cdots,N\Big\}, \tag{7.187c}$$

$$D_N \equiv \bigcup_{\sigma=\pm} \Big\{ \sigma(\boldsymbol{e}_i - \boldsymbol{e}_j), \sigma(\boldsymbol{e}_i + \boldsymbol{e}_j) \ \Big|\ i<j=1,\cdots,N\Big\}, \tag{7.187d}$$

respectively. A non reduced root system is the union of B_N and C_N:

$$BC_N \equiv \bigcup_{\sigma=\pm} \Big\{ \sigma\,\boldsymbol{e}_i, \sigma(\boldsymbol{e}_i - \boldsymbol{e}_j), \sigma(\boldsymbol{e}_i + \boldsymbol{e}_j), \sigma\,2\boldsymbol{e}_i \ \Big|\ i<j=1,\cdots,N\Big\}. \tag{7.187e}$$

Property 7.9 As any semi-simple Lie algebra $\mathfrak{g}$ is the direct sum

$$\mathfrak{g} = \bigoplus_{\iota} \mathfrak{g}_{\iota} \tag{7.188a}$$

over a finite set $\{\mathfrak{g}_{\iota}\}$ with the simple Lie algebras $\mathfrak{g}_{\iota}$ labeled by ι, the set $\mathfrak{Roots}_{\mathfrak{g}}(\mathbb{E}_{\mathfrak{g}})$ is the direct sum

$$\mathfrak{Roots}_{\mathfrak{g}}(\mathbb{E}_{\mathfrak{g}}) = \bigoplus_{\iota} \mathfrak{Roots}_{\mathfrak{g}_{\iota}}(\mathbb{E}_{\mathfrak{g}_{\iota}}), \qquad \dim \mathbb{E}_{\mathfrak{g}} = \sum_{\iota} \dim \mathbb{E}_{\mathfrak{g}_{\iota}}. \tag{7.188b}$$

Remark (4) If we do the analytical continuation

$$\boldsymbol{\theta} \mapsto \mathrm{i}\boldsymbol{x}, \qquad \boldsymbol{x} \in \begin{cases} \mathbb{R}^{N_{\mathrm{AZ}}}, & \text{A, AI, AII}, \\ x_j \geq 0, & \text{AIII, BDI, CII, D, C, DIII, CI}, \end{cases} \tag{7.189a}$$

we obtain the Jacobian (see **Table** 7.4)

$$\begin{aligned} J^{(-)}_{\mathrm{AZ}}(\boldsymbol{x}) := & \left(\prod_{\alpha \in \mathfrak{Roots}_{\mathrm{s}}} \left| 2\sinh\left(\frac{\boldsymbol{\alpha}\cdot\boldsymbol{x}}{2}\right) \right|^{m_{\mathrm{s}}} \right) \\ & \times \left(\prod_{\sigma=\pm} \prod_{\alpha \in \mathfrak{Roots}^{\sigma}_{\mathrm{o}}} \left| 2\sinh\left(\frac{\boldsymbol{\alpha}\cdot\boldsymbol{x}}{2}\right) \right|^{m^{\sigma}_{\mathrm{o}}} \right) \\ & \times \left(\prod_{\alpha \in \mathfrak{Roots}_{\mathrm{l}}} \left| 2\sinh\left(\frac{\boldsymbol{\alpha}\cdot\boldsymbol{x}}{2}\right) \right|^{m_{\mathrm{l}}} \right). \end{aligned} \tag{7.189b}$$

We have imposed the restriction $x_j \geq 0$ whenever the Jacobian (7.189b) is invariant under the transformation

$$x_j \mapsto \zeta_j\, x_j, \qquad \zeta_j = \pm 1. \tag{7.189c}$$

This analytical continuation is nothing but the consequence of the analytical continuation

$$\mathfrak{g} \equiv \mathfrak{k} \oplus \mathfrak{p} \mapsto \mathfrak{g}^{\star} \equiv \mathfrak{k} \oplus \mathrm{i}\mathfrak{p} \tag{7.190a}$$

by which

$$e^{\mathrm{i}H} \mapsto e^{\mathrm{i}H^{\star}}, \qquad H = H^{\dagger}. \tag{7.190b}$$

Here, $\mathrm{i}H^{\star}$ follows from $\mathrm{i}H$ by multiplication with i of all the generators from the Lie algebra of G that are not in the subalgebra of the isotropy subgroup H of G. As is

Table 7.4 *The tenfold way: Root lattices and their multiplicities for symmetric spaces with negative curvature*

AZ	$\mathrm{M}^\star \equiv \mathrm{G}^\star/\mathrm{H}$	$\mathfrak{Roots}_+$	m_s	m_o^-	m_o^+	m_l
A	$\mathrm{GL}(N,\mathbb{C})/\mathrm{U}(N)$	$e_k - e_j$	0	2	0	0
AI	$\mathrm{GL}(N,\mathbb{R})/\mathrm{O}(N)$	$e_k - e_j$	0	1	0	0
AII	$\mathrm{U}^\star(2N)/\mathrm{Sp}(2N)$	$e_k - e_j$	0	4	0	0
AIII	$\mathrm{U}(N,N)/[\mathrm{U}(N)\times\mathrm{U}(N)]$	$e_k \pm e_j, 2e_j$	0	2	2	1
	$\mathrm{U}(M,N)/[\mathrm{U}(M)\times\mathrm{U}(N)]$	$e_j, e_k \pm e_j, 2e_j$	2ν	2	2	1
BDI	$\mathrm{O}(N,N)/[\mathrm{O}(N)\times\mathrm{O}(N)]$	$e_k \pm e_j$	0	1	1	0
	$\mathrm{O}(M,N)/[\mathrm{O}(M)\times\mathrm{O}(N)]$	$e_j, e_k \pm e_j$	ν	1	1	0
CII	$\mathrm{Sp}(2N,2N)/[\mathrm{Sp}(2N)\times\mathrm{Sp}(2N)]$	$e_k \pm e_j, 2e_j$	0	4	4	3
	$\mathrm{Sp}(2M,2N)/[\mathrm{Sp}(2M)\times\mathrm{Sp}(2N)]$	$e_j, e_k \pm e_j, 2e_j$	4ν	4	4	3
D	$\mathrm{O}(2N,\mathbb{C})/\mathrm{O}(2N)$	$e_k \pm e_j$	0	2	2	0
	$\mathrm{O}(2N+1,\mathbb{C})/\mathrm{O}(2N+1)$	$e_j, e_k \pm e_j$	2	2	2	0
C	$\mathrm{Sp}(2N,\mathbb{C})/\mathrm{Sp}(2N)$	$e_k \pm e_j, 2e_j$	0	2	2	2
DIII	$\mathrm{O}^\star(4N)/\mathrm{U}(2N)$	$e_k \pm e_j, 2e_j$	0	4	4	1
	$\mathrm{O}^\star(4N+2)/\mathrm{U}(2N+1)$	$e_j, e_k \pm e_j, 2e_j$	4	4	4	1
CI	$\mathrm{Sp}(2N,\mathbb{R})/\mathrm{U}(N)$	$e_k \pm e_j, 2e_j$	0	1	1	1

The Jacobian for the spectral decomposition of a random Hermitean matrix H or its nonunitary exponential $\exp(\mathrm{i}H^\star)$ is determined by the positive roots and their multiplicities associated with the symmetric space $\mathrm{M}^\star \equiv \mathrm{G}^\star/\mathrm{H} \equiv \mathrm{M}^{(-)}$ with negative curvature generated by the isotropy subgroup H, the maximal subgroup that leaves the measure for the random matrices invariant under left or right multiplication by any element of H. Here, $\mathrm{i}H^\star$ follows from $\mathrm{i}H$ by multiplication with i of all the generators from the Lie algebra of G that are not in the subalgebra of the isotropy subgroup H of G. For the ten Altland-Zirnbauer (AZ) symmetry classes, the positive roots and their multiplicities are tabulated as follows ($1 \le j < k \le N \equiv N_{\mathrm{AZ}}$, $N \le M = 1,2,3,\cdots$). For the chiral classes AIII, BDI, and CII, we introduced the shorthand notation $|M-N| = \nu = 0,1,2,\cdots$.

verified in **Exercise** 7.6, the analytical continuation (7.190) implies the one-to-one maps[30]

[30] For the symmetry classes A, D, and C, one uses the identities

$$\mathrm{SU}(N) \cong (\mathrm{SU}(N)\times\mathrm{SU}(N))/\mathrm{SU}(N),$$
$$\mathrm{SO}(M) \cong (\mathrm{SO}(M)\times\mathrm{SO}(M))/\mathrm{SO}(M),$$
$$\mathrm{Sp}(2N) \cong (\mathrm{Sp}(2N)\times\mathrm{Sp}(2N))/\mathrm{Sp}(2N).$$

Analytical continuation (7.190a) is then equivalent to

$$(\mathrm{SU}(N)\times\mathrm{SU}(N))/\mathrm{SU}(N) \mapsto \mathrm{SU}(N,\mathbb{C})/\mathrm{SU}(N) \cong \mathrm{SL}(N,\mathbb{C})/\mathrm{SU}(N),$$
$$(\mathrm{SO}(M)\times\mathrm{SO}(M))/\mathrm{SO}(M) \mapsto \mathrm{SO}(M,\mathbb{C})/\mathrm{SO}(M),$$
$$(\mathrm{Sp}(2N)\times\mathrm{Sp}(2N))/\mathrm{Sp}(2N) \mapsto \mathrm{Sp}(2N,\mathbb{C})/\mathrm{Sp}(2N).$$

$$\mathrm{SU}(N) \mapsto \mathrm{SL}(N,\mathbb{C})/\mathrm{SU}(N), \tag{7.191a}$$
$$\mathrm{SU}(N)/\mathrm{SO}(N) \mapsto \mathrm{SL}(N,\mathbb{R})/\mathrm{SO}(N), \tag{7.191b}$$
$$\mathrm{SU}(2N)/\mathrm{Sp}(2N) \mapsto \mathrm{SU}^{\star}(2N)/\mathrm{Sp}(2N), \tag{7.191c}$$

for the standard symmetry classes A, AI, and AII, respectively,

$$\mathrm{SU}(M+N)/\mathrm{S}\big(\mathrm{U}(M)\times \mathrm{U}(N)\big) \mapsto \mathrm{SU}(M,N)/\mathrm{S}\big(\mathrm{U}(M)\times \mathrm{U}(N)\big), \tag{7.191d}$$
$$\mathrm{SO}(M+N)/\big(\mathrm{SO}(M)\times \mathrm{SO}(N)\big) \mapsto \mathrm{SO}_0(M,N)/\big(\mathrm{SO}(M)\times \mathrm{SO}(N)\big), \tag{7.191e}$$
$$\mathrm{Sp}(2M+2N)/\big(\mathrm{Sp}(2M)\times \mathrm{Sp}(2N)\big) \mapsto \mathrm{Sp}(2M,2N)/\big(\mathrm{Sp}(2M)\times \mathrm{Sp}(2N)\big), \tag{7.191f}$$

where $\mathrm{SO}_0(M,N)$ denotes the component of $\mathrm{SO}(M,N)$ connected to the unit matrix, for the chiral symmetry classes AIII, BDI, CII, respectively, and[31]

$$\mathrm{SO}(M) \mapsto \mathrm{SO}(M,\mathbb{C})/\mathrm{SO}(M), \qquad M = 2N, 2N+1, \tag{7.191g}$$
$$\mathrm{Sp}(2N) \mapsto \mathrm{Sp}(2N,\mathbb{C})/\mathrm{Sp}(2N), \tag{7.191h}$$
$$\mathrm{SO}(2M)/\mathrm{U}(M) \mapsto \mathrm{SO}^{\star}(2M)/\mathrm{U}(M), \qquad M = 2N, 2N+1, \tag{7.191i}$$
$$\mathrm{Sp}(2N)/\mathrm{U}(N) \mapsto \mathrm{SL}(2N,\mathbb{R})/\mathrm{U}(N), \tag{7.191j}$$

for the Bogoliubov-de-Gennes symmetry classes D, C, DIII, and CI, respectively. Upon this analytical continuation (see footnote 16 or Caselle[32]),

$$\mathrm{M} \equiv \mathrm{G}/\mathrm{H} \mapsto \mathrm{M}^{\star} \equiv \mathrm{G}^{\star}/\mathrm{H}. \tag{7.192}$$

Whereas M is a compact Riemannian manifold with positive curvature [say the two sphere $\mathrm{SO}(3)/\mathrm{SO}(2)$], $\mathrm{M}^{\star}$ is a non-compact Riemannian manifold with negative curvature [say the hyperboloid $\mathrm{SO}_0(2,1)/\mathrm{SO}(2)$]. Both M and $\mathrm{M}^{\star}$ share the same tangent space since both tangent spaces are generated by those Hamiltonians H belonging to an Altland-Zirnbauer symmetry class. It is thus meaningful to introduce the triplet of manifolds

$$\mathrm{M}^{(0)} := \mathrm{Tangent}\left(\mathrm{M}^{(\pm)}\right), \tag{7.193a}$$
$$\mathrm{M}^{(+)} := \mathrm{G}/\mathrm{H}, \tag{7.193b}$$
$$\mathrm{M}^{(-)} := \mathrm{G}^{\star}/\mathrm{H}, \tag{7.193c}$$

with the corresponding Jacobians

$$J^{(0)}_{\mathrm{AZ}}(\boldsymbol{\varepsilon}) := \prod_{\boldsymbol{\alpha}\in\mathfrak{Roots}_+} |\boldsymbol{\alpha}\cdot\boldsymbol{\varepsilon}|^{m_{\boldsymbol{\alpha}}}, \tag{7.194a}$$

$$J^{(+)}_{\mathrm{AZ}}(\boldsymbol{\theta}) := \prod_{\boldsymbol{\alpha}\in\mathfrak{Roots}_+} \left|2\sin\left(\frac{\boldsymbol{\alpha}\cdot\boldsymbol{\theta}}{2}\right)\right|^{m_{\boldsymbol{\alpha}}}, \tag{7.194b}$$

[31] According to Eq. (7.493g), the identification in $\mathrm{SL}(2N,\mathbb{R})/\mathrm{U}(N)$ of H with $\mathrm{U}(N)$ is to be interpreted as a group isomorphism.

[32] M. Caselle and U. Magnea, Physics Reports **394**(2), 41–156 (2004). [Caselle and Magnea (2004)].

$$J^{(-)}_{\mathrm{AZ}}(\boldsymbol{x}) := \prod_{\boldsymbol{\alpha}\in\mathfrak{Roots}_+} \left|2\sinh\left(\frac{\boldsymbol{\alpha}\cdot\boldsymbol{x}}{2}\right)\right|^{m_{\boldsymbol{\alpha}}}. \tag{7.194c}$$

The positive roots $\boldsymbol{\alpha} \in \mathfrak{Roots}_+$ and their multiplicities $m_{\boldsymbol{\alpha}}$ for each Altland-Zirnbauer symmetry class are found in either **Table** 7.3 or **Table** 7.4.

It remains to explain in what context of condensed matter physics the non-compact symmetric space (7.193c) is of relevance.

7.6 The Tenfold Way: Scattering or Transfer Matrix Ensembles

The Gaussian and circular ensembles of random matrix theory describe the statistical properties of closed quantum systems. We have shown that the components of the vector $\boldsymbol{\varepsilon}$, whose joint probability distribution is proportional to the Jacobian (7.194a), are random energy eigenvalues of a closed system that are distributed on the real line. We have shown that the components of the vector $\boldsymbol{\theta}$, whose joint probability distribution is proportional to the Jacobian (7.194b), are also random energy eigenvalues of a closed system, but they are distributed on the circle. We are going to show that the components of the vector $\boldsymbol{x}$, whose joint probability distribution is proportional to the Jacobian (7.194c), are related to the transmission eigenvalues of a scattering matrix or, equivalently, the Lyapunov exponents of a transfer matrix, describing an open quantum system connected to two reservoirs of delocalized quasiparticles (leads or terminals) that we shall denote Ω and Ω', respectively. We begin by introducing the transmission eigenvalues of a scattering matrix. We continue by introducing the Lyapunov exponents of a transfer matrix. We stress that choosing between the scattering-matrix or the transfer-matrix approaches is just a matter of preference.

7.6.1 Scattering Matrix

Assume that lead Ω support $M = 1, 2, 3, \cdots$ delocalized single-particle states, while lead Ω' support $M' = 1, 2, 3, \cdots$ delocalized single-particle states. The two leads are separated by a region Λ, a chaotic quantum dot. We assume that any incoming flux on Λ is scattered into an outgoing flux without dissipation, that is flux is conserved by the scattering. Furthermore, we assign to Λ a unitary matrix $\mathcal{S} \in \mathrm{U}(M + M')$ (because of flux conservation) drawn from the circular ensemble of unitary matrices. Incoming states may originate from either lead Ω or lead Ω', that is, given the incoming vector

$$\mathcal{I} \equiv \begin{pmatrix} \mathsf{i} \\ \mathsf{i}' \end{pmatrix} \in \mathbb{C}^{M+M'}, \qquad \mathsf{i} \in \mathbb{C}^{M}, \qquad \mathsf{i}' \in \mathbb{C}^{M'}, \tag{7.195a}$$

we may define the outgoing vector

$$\mathcal{O} \equiv \begin{pmatrix} \mathsf{o} \\ \mathsf{o}' \end{pmatrix} \in \mathbb{C}^{M+M'}, \qquad \mathsf{o} \in \mathbb{C}^{M}, \qquad \mathsf{o}' \in \mathbb{C}^{M'}, \tag{7.195b}$$

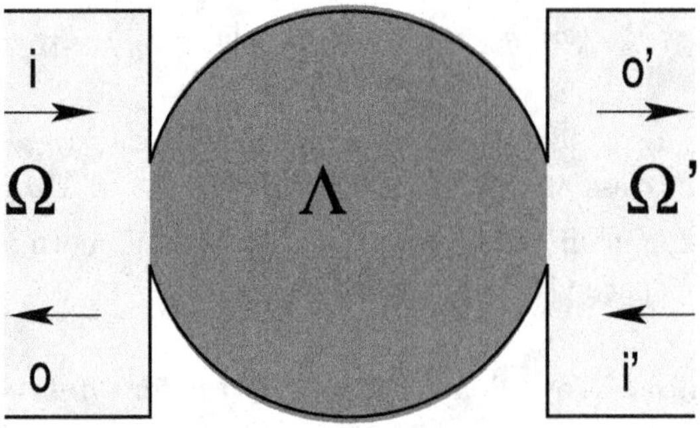

Figure 7.2 Chaotic quantum dot Λ weakly connected to two reservoirs Ω and Ω' that support M and M' delocalized single-particle states at the chemical potential μ. Incoming single-particle states $\mathcal{I}$ are mapped into outgoing single-particle states $\mathcal{O}$ by a unitary matrix $\mathcal{S}$. This description of transport across the region Λ applies at low frequencies, at low temperatures, and at low voltage differences between the two reservoirs provided electron-electron interactions can be neglected. The reservoirs are in thermodynamic equilibrium at zero temperature and the chemical potential (or Fermi energy) $\mu \equiv \varepsilon_{\mathrm{F}}$. They serve as ideal electron wave guides that define a basis for the elastic scattering matrix across Λ.

through the relation

$$\mathcal{O} = \mathcal{S}\mathcal{I}, \qquad \mathcal{S} = \begin{pmatrix} \mathsf{r} & \mathsf{t}' \\ \mathsf{t} & \mathsf{r}' \end{pmatrix}. \tag{7.195c}$$

Here, r and r' are $M \times M$ and $M' \times M'$ complex-valued square matrices, respectively, and the matrices t and t' are $M' \times M$ and $M \times M'$ complex-valued rectangular matrices, respectively. Figure 7.2 is a cartoon representation of this setup. We are going to show that the four matrices r, t', t, and r' are not independent owing to the unitarity of $\mathcal{S}$. Indeed, it is possible to find four unitary matrices $\mathsf{u}, \mathsf{v} \in \mathrm{U}(M)$ and $\mathsf{u}', \mathsf{v}' \in \mathrm{U}(M')$ together with the rectangular diagonal $M' \times M$ matrix d, whose nonvanishing matrix elements are necessarily positive, such that the singular-value decomposition

$$\mathcal{S} = \begin{pmatrix} \mathsf{u} & 0 \\ 0 & \mathsf{u}' \end{pmatrix} \begin{pmatrix} -\sqrt{\mathbb{1}_M - \mathsf{d}^\dagger \mathsf{d}} & \mathsf{d}^\dagger \\ \mathsf{d} & \sqrt{\mathbb{1}_{M'} - \mathsf{d}\,\mathsf{d}^\dagger} \end{pmatrix} \begin{pmatrix} \mathsf{v} & 0 \\ 0 & \mathsf{v}' \end{pmatrix} \tag{7.196}$$

holds.[33]

Proof We shall use the singular-value decomposition of an arbitrary rectangular matrix (**Exercise** 7.15).

The square matrix r, with the singular-value decomposition

[33] Th. Martin and R. Landauer, Phys. Rev. B **45**(4), 1742–1755 (1992). [Martin and Landauer (1992)].

$$\mathsf{r}=\mathsf{u}_{11}\,\mathsf{d}_{11}\,\mathsf{v}_{11},\qquad \begin{cases} \mathsf{r}^{\dagger}\,\mathsf{r}=\mathsf{v}_{11}^{\dagger}\,\mathsf{d}_{11}^{\dagger}\,\mathsf{d}_{11}\,\mathsf{v}_{11}, & \mathsf{v}_{11}\in \mathrm{U}(M),\\ \mathsf{r}\,\mathsf{r}^{\dagger}=\mathsf{u}_{11}\,\mathsf{d}_{11}\,\mathsf{d}_{11}^{\dagger}\,\mathsf{u}_{11}^{\dagger}, & \mathsf{u}_{11}\in \mathrm{U}(M), \end{cases} \tag{7.197a}$$

in terms of the diagonal *sign-definite* square $M\times M$ matrix d_{11},[34] describes the process by which incoming single-particle states from reservoir Ω are reflected into reservoir Ω. The square matrix r', with the singular-value decomposition

$$\mathsf{r}'=\mathsf{u}_{22}\,\mathsf{d}_{22}\,\mathsf{v}_{22},\qquad \begin{cases} \mathsf{r}'^{\dagger}\,\mathsf{r}'=\mathsf{v}_{22}^{\dagger}\,\mathsf{d}_{22}^{\dagger}\,\mathsf{d}_{22}\,\mathsf{v}_{22}, & \mathsf{v}_{22}\in \mathrm{U}(M'),\\ \mathsf{r}'\,\mathsf{r}'^{\dagger}=\mathsf{u}_{22}\,\mathsf{d}_{22}\,\mathsf{d}_{22}^{\dagger}\,\mathsf{u}_{22}^{\dagger}, & \mathsf{u}_{22}\in \mathrm{U}(M'), \end{cases} \tag{7.197b}$$

in terms of the diagonal *nonnegative* square $M'\times M'$ matrix d_{22}, describes the process by which incoming single-particle states from reservoir Ω' are reflected into reservoir Ω'. The rectangular matrix t, with the singular-value decomposition

$$\mathsf{t}=\mathsf{u}_{21}\,\mathsf{d}_{21}\,\mathsf{v}_{21},\qquad \begin{cases} \mathsf{t}^{\dagger}\,\mathsf{t}=\mathsf{v}_{21}^{\dagger}\,\mathsf{d}_{21}^{\dagger}\,\mathsf{d}_{21}\,\mathsf{v}_{21}, & \mathsf{v}_{21}\in \mathrm{U}(M),\\ \mathsf{t}\,\mathsf{t}^{\dagger}=\mathsf{u}_{21}\,\mathsf{d}_{21}\,\mathsf{d}_{21}^{\dagger}\,\mathsf{u}_{21}^{\dagger}, & \mathsf{u}_{21}\in \mathrm{U}(M'), \end{cases} \tag{7.197c}$$

in terms of the diagonal *nonnegative* rectangular $M'\times M$ matrix d_{21}, describes the process by which incoming single-particle states from reservoir Ω are transmitted into reservoir Ω'. The rectangular matrix t' with the singular-value decomposition

$$\mathsf{t}'=\mathsf{u}_{12}\,\mathsf{d}_{12}\,\mathsf{v}_{12},\qquad \begin{cases} \mathsf{t}'^{\dagger}\,\mathsf{t}'=\mathsf{v}_{12}^{\dagger}\,\mathsf{d}_{12}^{\dagger}\,\mathsf{d}_{12}\,\mathsf{v}_{12}, & \mathsf{v}_{12}\in \mathrm{U}(M'),\\ \mathsf{t}'\,\mathsf{t}'^{\dagger}=\mathsf{u}_{12}\,\mathsf{d}_{12}\,\mathsf{d}_{12}^{\dagger}\,\mathsf{u}_{12}^{\dagger}, & \mathsf{u}_{12}\in \mathrm{U}(M), \end{cases} \tag{7.197d}$$

in terms of the diagonal *nonnegative* rectangular $M\times M'$ matrix d_{12}, describes the process by which incoming single-particle states from reservoir Ω' are transmitted into reservoir Ω. The four blocks of $\mathcal{S}$ are parameterized by eight unitary matrices and four diagonal and real-valued matrices.

The four matrices (7.197) are not independent, for the conservation of flux is implemented by the condition that $\mathcal{S}$ is a unitary matrix. In other words, $\mathcal{S}$ obeys two conditions. First,

$$\mathcal{S}^{\dagger}\,\mathcal{S}=\mathbb{1}_{M+M'}\iff \begin{cases} \mathsf{r}^{\dagger}\,\mathsf{r}+\mathsf{t}^{\dagger}\,\mathsf{t}=\mathbb{1}_{M}, & \mathsf{r}^{\dagger}\,\mathsf{t}'+\mathsf{t}^{\dagger}\,\mathsf{r}'=0,\\ \mathsf{t}'^{\dagger}\,\mathsf{r}+\mathsf{r}'^{\dagger}\,\mathsf{t}=0, & \mathsf{t}'^{\dagger}\,\mathsf{t}'+\mathsf{r}'^{\dagger}\,\mathsf{r}'=\mathbb{1}_{M'}. \end{cases} \tag{7.198a}$$

Second,

[34] The matrix d_{11} is sign definite, say either nonnegative or nonpositive, because it is real-valued and all its nonvanishing diagonal matrix elements share the same sign. In the standard singular-value decomposition, the singular values are nonnegative. However, this condition on d_{11} would not be compatible with flux conservation, as will shortly become apparent.

$$\mathcal{S}\,\mathcal{S}^{\dagger}=\mathbb{1}_{M+M'} \iff \begin{cases} \mathsf{r}\,\mathsf{r}^{\dagger}+\mathsf{t}'\,\mathsf{t}'^{\dagger}=\mathbb{1}_{M}, & \mathsf{r}\,\mathsf{t}^{\dagger}+\mathsf{t}'\,\mathsf{r}'^{\dagger}=0, \\ \mathsf{t}\,\mathsf{r}^{\dagger}+\mathsf{r}'\,\mathsf{t}'^{\dagger}=0, & \mathsf{t}\,\mathsf{t}^{\dagger}+\mathsf{r}'\,\mathsf{r}'^{\dagger}=\mathbb{1}_{M'}. \end{cases} \tag{7.198b}$$

This gives the four independent inhomogeneous matrix equations

$$\mathbb{1}_{M}=\mathsf{v}^{\dagger}_{11}\,\mathsf{d}^{\dagger}_{11}\,\mathsf{d}_{11}\,\mathsf{v}_{11}+\mathsf{v}^{\dagger}_{21}\,\mathsf{d}^{\dagger}_{21}\,\mathsf{d}_{21}\,\mathsf{v}_{21}, \tag{7.199a}$$

$$\mathbb{1}_{M'}=\mathsf{v}^{\dagger}_{12}\,\mathsf{d}^{\dagger}_{12}\,\mathsf{d}_{12}\,\mathsf{v}_{12}+\mathsf{v}^{\dagger}_{22}\,\mathsf{d}^{\dagger}_{22}\,\mathsf{d}_{22}\,\mathsf{v}_{22}, \tag{7.199b}$$

$$\mathbb{1}_{M}=\mathsf{u}_{11}\mathsf{d}_{11}\,\mathsf{d}^{\dagger}_{11}\,\mathsf{u}^{\dagger}_{11}+\mathsf{u}_{12}\mathsf{d}_{12}\,\mathsf{d}^{\dagger}_{12}\,\mathsf{u}^{\dagger}_{12}, \tag{7.199c}$$

$$\mathbb{1}_{M'}=\mathsf{u}_{21}\mathsf{d}_{21}\,\mathsf{d}^{\dagger}_{21}\,\mathsf{u}^{\dagger}_{21}+\mathsf{u}_{22}\mathsf{d}_{22}\,\mathsf{d}^{\dagger}_{22}\,\mathsf{u}^{\dagger}_{22}, \tag{7.199d}$$

and the two independent homogeneous matrix equations

$$0=\mathsf{v}^{\dagger}_{11}\,\mathsf{d}^{\dagger}_{11}\,\mathsf{u}^{\dagger}_{11}\mathsf{u}_{12}\,\mathsf{d}_{12}\,\mathsf{v}_{12}+\mathsf{v}^{\dagger}_{21}\,\mathsf{d}^{\dagger}_{21}\,\mathsf{u}^{\dagger}_{21}\,\mathsf{u}_{22}\,\mathsf{d}_{22}\,\mathsf{v}_{22}, \tag{7.200a}$$

$$0=\mathsf{u}_{11}\,\mathsf{d}_{11}\,\mathsf{v}_{11}\,\mathsf{v}^{\dagger}_{21}\,\mathsf{d}^{\dagger}_{21}\,\mathsf{u}^{\dagger}_{21}+\mathsf{u}_{12}\,\mathsf{d}_{12}\,\mathsf{v}_{12}\,\mathsf{v}^{\dagger}_{22}\,\mathsf{d}^{\dagger}_{22}\,\mathsf{u}^{\dagger}_{22}. \tag{7.200b}$$

A particular solution to the four inhomogeneous equations (7.199) is

$$\mathsf{v}_{11}=\mathsf{v}_{21}\equiv\mathsf{v}\in\mathrm{U}(M), \qquad \mathsf{d}^{\dagger}_{11}\,\mathsf{d}_{11}=\mathbb{1}_{M}-\mathsf{d}^{\dagger}_{21}\,\mathsf{d}_{21}, \tag{7.201a}$$

$$\mathsf{v}_{22}=\mathsf{v}_{12}\equiv\mathsf{v}'\in\mathrm{U}(M'), \qquad \mathsf{d}^{\dagger}_{22}\,\mathsf{d}_{22}=\mathbb{1}_{M'}-\mathsf{d}^{\dagger}_{12}\,\mathsf{d}_{12}, \tag{7.201b}$$

$$\mathsf{u}_{11}=\mathsf{u}_{12}\equiv\mathsf{u}\in\mathrm{U}(M), \qquad \mathsf{d}_{11}\,\mathsf{d}^{\dagger}_{11}=\mathbb{1}_{M}-\mathsf{d}_{12}\,\mathsf{d}^{\dagger}_{12}, \tag{7.201c}$$

$$\mathsf{u}_{22}=\mathsf{u}_{21}\equiv\mathsf{u}'\in\mathrm{U}(M'), \qquad \mathsf{d}_{22}\,\mathsf{d}^{\dagger}_{22}=\mathbb{1}_{M'}-\mathsf{d}_{21}\,\mathsf{d}^{\dagger}_{21}. \tag{7.201d}$$

If so, a solution to the two homogeneous equations (7.200) is

$$0=\mathsf{d}^{\dagger}_{11}\,\mathsf{d}_{12}+\mathsf{d}^{\dagger}_{21}\,\mathsf{d}_{22}, \tag{7.202a}$$

$$0=\mathsf{d}_{11}\,\mathsf{d}^{\dagger}_{21}+\mathsf{d}_{12}\,\mathsf{d}^{\dagger}_{22}, \tag{7.202b}$$

that is

$$\mathsf{d}_{12}\equiv\mathsf{d}^{\dagger}, \qquad \mathsf{d}_{21}\equiv\mathsf{d}, \qquad \mathsf{d}^{\dagger}_{11}\,\mathsf{d}^{\dagger}=-\mathsf{d}^{\dagger}\,\mathsf{d}_{22}. \tag{7.203}$$

Hence, if d has at least one nonvanishing diagonal element, then d_{11} has at least one nonpositive diagonal element, since d_{22} is nonnegative definite. Furthermore, since d_{11} is sign-definite, all diagonal elements of d_{11} are nonpositive. By combining the singular-value decompositions (7.197) with unitarity (7.198), we have found the particular solution

$$\mathsf{r}=-\mathsf{u}\sqrt{(\mathbb{1}_{M}-\mathsf{d}^{\dagger}\,\mathsf{d})}\,\mathsf{v}, \qquad \begin{cases}\mathsf{r}^{\dagger}\,\mathsf{r}=\mathsf{v}^{\dagger}\left(\mathbb{1}_{M}-\mathsf{d}^{\dagger}\,\mathsf{d}\right)\mathsf{v}, & \mathsf{v}\in\mathrm{U}(M),\\ \mathsf{r}\,\mathsf{r}^{\dagger}=\mathsf{u}\left(\mathbb{1}_{M}-\mathsf{d}^{\dagger}\,\mathsf{d}\right)\mathsf{u}^{\dagger}, & \mathsf{u}\in\mathrm{U}(M),\end{cases} \tag{7.204a}$$

$$\mathsf{t}' = \mathsf{u}\,\mathsf{d}^{\dagger}\,\mathsf{v}', \qquad \begin{cases} \mathsf{t}'^{\dagger}\,\mathsf{t}' = \mathsf{v}'^{\dagger}\,\mathsf{d}\,\mathsf{d}^{\dagger}\,\mathsf{v}', & \mathsf{v}' \in \mathrm{U}(M'), \\ \mathsf{t}'\,\mathsf{t}'^{\dagger} = \mathsf{u}\,\mathsf{d}^{\dagger}\,\mathsf{d}\,\mathsf{u}^{\dagger}, & \mathsf{u} \in \mathrm{U}(M), \end{cases} \tag{7.204b}$$

$$\mathsf{t} = \mathsf{u}'\,\mathsf{d}\,\mathsf{v}, \qquad \begin{cases} \mathsf{t}^{\dagger}\,\mathsf{t} = \mathsf{v}^{\dagger}\,\mathsf{d}^{\dagger}\,\mathsf{d}\,\mathsf{v}, & \mathsf{v} \in \mathrm{U}(M), \\ \mathsf{t}\,\mathsf{t}^{\dagger} = \mathsf{u}'\,\mathsf{d}\,\mathsf{d}^{\dagger}\,\mathsf{u}'^{\dagger}, & \mathsf{u}' \in \mathrm{U}(M'), \end{cases} \tag{7.204c}$$

$$\mathsf{r}' = \mathsf{u}'\,\sqrt{(\mathbb{1}_{M'} - \mathsf{d}\,\mathsf{d}^{\dagger})}\,\mathsf{v}', \qquad \begin{cases} \mathsf{r}'^{\dagger}\,\mathsf{r}' = \mathsf{v}'^{\dagger}\left(\mathbb{1}_{M'} - \mathsf{d}\,\mathsf{d}^{\dagger}\right)\mathsf{v}', & \mathsf{v}' \in \mathrm{U}(M'), \\ \mathsf{r}'\,\mathsf{r}'^{\dagger} = \mathsf{u}'\left(\mathbb{1}_{M'} - \mathsf{d}\,\mathsf{d}^{\dagger}\right)\mathsf{u}'^{\dagger}, & \mathsf{u}' \in \mathrm{U}(M'), \end{cases} \tag{7.204d}$$

that is of the form (7.196). It can be shown that the most general solution of Eqs. (7.199) and (7.200) is also of the form (7.196).[35] Equation (7.196) is not unique since it is invariant under

$$\begin{aligned} &\mathsf{u} \mapsto \mathsf{u}\,\mathsf{w}, && \mathsf{v} \mapsto \mathsf{w}^{\dagger}\,\mathsf{v}, && \left[\mathsf{w}, \mathsf{d}^{\dagger}\,\mathsf{d}\right] = 0, && \mathsf{w} \in \mathrm{U}(M), \\ &\mathsf{u}' \mapsto \mathsf{u}'\,\mathsf{w}', && \mathsf{v}' \mapsto \mathsf{w}'^{\dagger}\,\mathsf{v}', && \left[\mathsf{w}', \mathsf{d}\,\mathsf{d}^{\dagger}\right] = 0, && \mathsf{w}' \in \mathrm{U}(M'), \\ &\mathsf{w}\,\mathsf{d}^{\dagger}\,\mathsf{w}'^{\dagger} = \mathsf{d}^{\dagger}, && \mathsf{w}'\,\mathsf{d}\,\mathsf{w}^{\dagger} = \mathsf{d}. \end{aligned} \tag{7.205}$$

Finally, Eq. (7.196) transforms under

$$\mathsf{u} \mapsto \mathrm{i}\mathsf{u}, \qquad \mathsf{v} \mapsto \mathrm{i}\mathsf{v}, \qquad \mathsf{u}' \mapsto \mathsf{u}', \qquad \mathsf{v}' \mapsto \mathsf{v}', \tag{7.206a}$$

into

$$\mathcal{S} \mapsto \begin{pmatrix} \mathsf{u} & 0 \\ 0 & \mathsf{u}' \end{pmatrix} \begin{pmatrix} \sqrt{\mathbb{1}_M - \mathsf{d}^{\dagger}\,\mathsf{d}} & \mathrm{i}\mathsf{d}^{\dagger} \\ \mathrm{i}\mathsf{d} & \sqrt{\mathbb{1}_{M'} - \mathsf{d}\,\mathsf{d}^{\dagger}} \end{pmatrix} \begin{pmatrix} \mathsf{v} & 0 \\ 0 & \mathsf{v}' \end{pmatrix}. \tag{7.206b}$$

□

According to the singular-value decomposition (7.196),

$$\mathrm{tr}\left(\mathsf{t}^{\dagger}\,\mathsf{t}\right) = \mathrm{tr}\left(\mathsf{d}^{\dagger}\,\mathsf{d}\right) = \mathrm{tr}\left(\mathsf{d}\,\mathsf{d}^{\dagger}\right) = \mathrm{tr}\left(\mathsf{t}'^{\dagger}\,\mathsf{t}'\right), \tag{7.207a}$$

$$\mathrm{tr}\left(\mathsf{r}^{\dagger}\,\mathsf{r}\right) = M - \mathrm{tr}\left(\mathsf{d}^{\dagger}\,\mathsf{d}\right) = \mathrm{tr}\left(\mathsf{r}'^{\dagger}\,\mathsf{r}'\right) + M - M'. \tag{7.207b}$$

The dimensionless Landauer conductance is defined to be[36]

$$g_{\mathrm{L}} := \mathrm{tr}\left(\mathsf{t}^{\dagger}\,\mathsf{t}\right) = \mathrm{tr}\left(\mathsf{t}'^{\dagger}\,\mathsf{t}'\right). \tag{7.208}$$

This definition is motivated by the following considerations.

First, if $\mathrm{rank}(\mathsf{t}^{\dagger}\,\mathsf{t}) = \min(M, M')$ and all the nonvanishing eigenvalues of $\mathsf{t}^{\dagger}\,\mathsf{t}$ equal one, then

$$g_{\mathrm{L}} = \min(M, M') \tag{7.209}$$

depends in an extensive way on $\min(M, M')$. This rules out interpreting g_{L} as a conductivity.

35 P. A. Mello and J. L. Pichard, J. Phys. I France **1**(4), 493–513 (1991). [Mello and Pichard (1991)].

36 R. Landauer, IBM J. Res. Dev. **1**(3), 223–231- (1957); Philos. Mag. **21**(172), 863–867 (1970). [Landauer (1957, 1970)].

Second, when $M = M' = 1$, we can describe the chaotic quantum dot Λ by the 2×2 unitary scattering matrix

$$\begin{aligned} \mathcal{S} &= \begin{pmatrix} r & t' \\ t & r' \end{pmatrix}, \qquad r, t', t, r' \in \mathbb{C}, \\ R &\equiv |r|^2 = |r'|^2 = 1 - T, \qquad T := |t|^2 = |t'|^2. \end{aligned} \tag{7.210}$$

Assume that the reservoir Ω consists of noninteracting spinless electrons at the Fermi energy $\mu + (e\,\delta V) > \mu$ (e the electron charge and δV an infinitesimal electric potential). Assume that the reservoir Ω' consists of noninteracting spinless electrons at the Fermi energy μ. Assume the excess density δn of spinless electrons at the Fermi energy $\mu + (e\,\delta V) > \mu$ in Ω relative to the density of electrons at the Fermi energy μ in Ω' to be related to the density n of spinless electrons at the Fermi energy μ by

$$\delta n = \left(\frac{\mathrm{d}n}{\mathrm{d}\varepsilon}\right)(e\,\delta V). \tag{7.211}$$

Assume that this excess equals the difference between the magnitude j of the particle current in Ω divided by the Fermi speed v_{F} in Ω and the magnitude j' of the particle current in Ω' divided by the Fermi speed v'_{F} in Ω', that is

$$\delta n = \frac{j}{v_{\mathrm{F}}} - \frac{j'}{v'_{\mathrm{F}}}. \tag{7.212}$$

We make the scattering Ansatz by doing the decomposition[37]

$$j = j_{\mathsf{i}} + j_{\mathsf{o}}, \qquad j' = j'_{\mathsf{i}'} + j'_{\mathsf{o}'}, \tag{7.213a}$$

where flux conservation is expressed by the condition

$$j_{\mathsf{i}} + j'_{\mathsf{i}'} = j_{\mathsf{o}} + j'_{\mathsf{o}'}. \tag{7.213b}$$

Accordingly, the net incoming flux into the region Λ $j_{\mathsf{i}} - j'_{\mathsf{i}'}$ is either transmitted as the net outgoing flux from Λ to Ω',

$$j'_{\mathsf{o}'} - j'_{\mathsf{i}'} = T\,(j_{\mathsf{i}} - j'_{\mathsf{i}'}), \tag{7.214a}$$

which is equal to the net incoming flux from Ω to Λ

$$j_{\mathsf{i}} - j_{\mathsf{o}} = T\,(j_{\mathsf{i}} - j'_{\mathsf{i}'}) \tag{7.214b}$$

by flux conservation, or, equivalently owing to flux conservation ($1 = R + T$), reflected as

$$j_{\mathsf{i}} - j'_{\mathsf{o}'} = R\,(j_{\mathsf{i}} - j'_{\mathsf{i}'}), \tag{7.215a}$$

which is equal to

$$j_{\mathsf{o}} - j'_{\mathsf{i}'} = R\,(j_{\mathsf{i}} - j'_{\mathsf{i}'}) \tag{7.215b}$$

[37] P. W. Anderson, D. J. Thouless, E. Abrahams, and D. S. Fisher, Phys. Rev. B **22**(8), 3519–3526 (1980). [Anderson et al. (1980)].

by flux conservation. Hence, we arrive at

$$j - j' = (j_{\mathrm{i}} - j'_{\mathrm{o}'}) + (j_{\mathrm{o}} - j'_{\mathrm{i}'}) = 2R\,(j_{\mathrm{i}} - j'_{\mathrm{i}'})\,. \tag{7.216}$$

If we do the approximation

$$v_{\mathrm{F}} = v'_{\mathrm{F}}, \tag{7.217a}$$

we find

$$\delta n = 2R\,(j_{\mathrm{i}} - j'_{\mathrm{i}'})\,(v_{\mathrm{F}})^{-1} \tag{7.217b}$$

and

$$(e\,\delta V)^{-1} = \left[2R\,(j_{\mathrm{i}} - j'_{\mathrm{i}'})\right]^{-1} v_{\mathrm{F}} \left(\frac{\mathrm{d}n}{\mathrm{d}\varepsilon}\right). \tag{7.217c}$$

The electrical current δI that flows from Ω to Ω' is

$$\delta I := e\,(j_{\mathrm{i}} - j_{\mathrm{o}}) = e\,T\,(j_{\mathrm{i}} - j'_{\mathrm{i}'})\,. \tag{7.218a}$$

The electrical conductance would then be the ratio

$$G := \frac{\delta I}{\delta V} = \frac{e\,\delta I}{(e\,\delta V)} = e^2\,\frac{T}{2R}\,v_{\mathrm{F}}\left(\frac{\mathrm{d}n}{\mathrm{d}\varepsilon}\right). \tag{7.218b}$$

If we assume that

$$v_{\mathrm{F}}\left(\frac{\mathrm{d}n}{\mathrm{d}\varepsilon}\right) = \frac{1}{\pi\,\hbar}, \tag{7.218c}$$

the electrical conductance

$$G = \frac{e^2}{2\pi\hbar}\,\frac{T}{R} \tag{7.219}$$

would follow from 2×2 scattering theory. This result is suggestive of the Landauer conductance (7.208) but not quite it, because of the denominator R and the nonuniversality of the step (7.218c). Equation (7.219) converges to the Landauer conductance (7.208) in the limit $T \to 0$. However, it predicts that the electrical conductance is infinity in the limit $T \to 1$, in complete disagreement with the Landauer conductance (7.208).

The electrical conductance would be given by the Landauer conductance (7.208) if the electrical current δI through region Λ in Figure 7.2 induced by imposing the infinitesimal potential difference δV between reservoir Ω and Ω' was given by (**Exercise** 7.18)

$$\delta I = \frac{e}{2\pi\hbar}\left[\int^{\mu+\delta\mu} \mathrm{d}\varepsilon\,\mathrm{tr}\left(\mathsf{t}(\varepsilon)\,\mathsf{t}^{\dagger}(\varepsilon)\right) - \int^{\mu} \mathrm{d}\varepsilon\,\mathrm{tr}\left(\mathsf{t}'^{\dagger}(\varepsilon)\,\mathsf{t}'(\varepsilon)\right)\right] \tag{7.220}$$

with the understanding that $\delta\mu = (e\,\delta V)$ is infinitesimal.

A derivation of the electrical conductance in terms of the Landauer conductance (7.208), that is of Eq. (7.220), was obtained by Fisher and Lee in 1981[38] by combining linear response theory and scattering theory. We refer the reader to the papers by Stone and Szafer, on the one hand, and Landauer, on the other hand, for a discussion of the conditions under which the Landauer conductance gives the electrical conductance.[39] These conditions essentially amount to an idealization of how the reservoirs Ω and Ω' turn into the scattering region Λ, an idealization that must be validated by comparison with experiments. One prediction of the Landauer formula is that if the quantum dot Λ is described by $M = M' = N$ (identical reservoirs) and $\mathsf{t} = \mathbb{1}_N$ (ballistic regime), then its electrical conductance should be quantized to the value $e^2\, N/(\pi\hbar)$,[40] up to a multiplicative degeneracy factor. Moreover, if the number N of channels at the Fermi energy could be changed, then the conductance of the quantum dot would jump by integer multiples of the quantum of conductance $2\times e^2/(2\pi\hbar)$ (the multiplicative factor of 2 arises from the spin degeneracy) at zero temperature (this jump would be smoothened at nonvanishing temperature). This prediction was verified in 1988 by patterning sufficiently small constrictions in the two-dimensional degenerate electron gas of a high-mobility GaAs-AlGaAs heterostructure as is shown in Figures 7.3 and 7.4 taken from the paper of van Wees et al. The Landauer formula and its extensions by Büttiker and Imry have been the workhorses for quantum transport since the 1980s.[41]

The scattering matrix $\mathcal{S}$ and the Landauer conductance g_{L} associated with the quantum dot Λ depend on the Fermi energy ε selected by the condition $\mu = \varepsilon$ in reservoirs Ω and Ω', that is the scattering matrix $\mathcal{S}(\varepsilon)$ depends parametrically on the chemical potential $\mu = \varepsilon$ from the reservoirs. In order to deduce the transformation law obeyed by the scattering matrix $\mathcal{S}(\varepsilon)$ for each one of the ten Altland-Zirnbauer symmetry classes, it is necessary to model how the reservoirs couple to the quantum dot, which in isolation is described by Hamiltonian $\mathcal{H}$. If the coupling between the quantum dot and the reservoirs is modeled by a Hermitean operator that does not break the combinations of time-reversal symmetry, spectral particle-hole symmetry, or spectral chiral symmetry that define each one of the Altland-Zirnbauer symmetry class, one may then deduce the transformation law obeyed by the scattering matrix $\mathcal{S}(\varepsilon)$ from that obeyed by $(\varepsilon\mathbb{1} - \mathcal{H})$ for each one of the ten Altland-Zirnbauer symmetry classes. This program has been carried through by Fulga, Hassler, and Akhmerov[42] (see footnote 27). For the symmetry classes D, DIII, C, CI, AIII, BDI, and CII, the transformation laws

[38] D. S. Fisher and P. A. Lee, Phys. Rev. B **23**(12), 6851–6854 (1981). [Fisher and Lee (1981)].

[39] A. D. Stone and A. Szafer, IBM J. Res. Dev. **32**(3), 384–413 (1988); R. Landauer, J. Phys. Cond. Matter **1**(43), 8099–8110 (1989). [Stone and Szafer (1988); Landauer (1989)].

[40] The quantum of conductance $2e^2/h = e^2/(\pi\,\hbar)$ is defined by the elementary electric charge e and Planck constant $\hbar = h/(2\,\pi)$.

[41] M. Büttiker, Phys. Rev. Lett. **57**(14), 1761–1764 (1986); M. Büttiker, IBM J. Res. Dev. **32(3)**, 317–334 (1988); Y. Imry, in: *Directions in Condensed Matter Physics,* pages 101–164, edited by G. Grinstein and G. Mazenko, World Scientific, Singapore, 1986. [Büttiker (1986); Buttiker (1988); Imry (1986)].

[42] I. C. Fulga, F. Hassler, and A. R. Akhmerov, 2012, Phys. Rev. B **85**(16), 165409 (2012). [Fulga et al. (2012)].

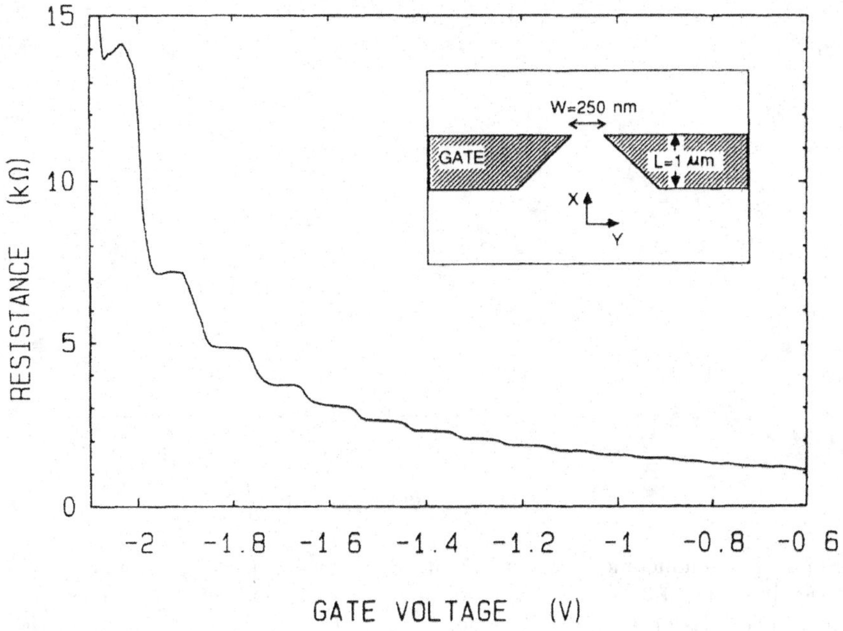

Figure 7.3 A constriction of width W and length L can be patterned into the two-dimensional degenerate electron gas (2DEG) of a high-mobility GaAs-AlGaAs heterostructure by electrostatic depletion of the 2DEG underneath a gate. Transport through such constrictions is ballistic if $W, L \ll \ell$ with ℓ the mean free path in the 2DEG. The resistance is then determined by the point-contact geometry only. At the gate voltage $V_{\mathrm{g}} = -0.6\,\mathrm{V}$, $W \approx 150\,\mathrm{nm} \ll \ell$ for the 2DEG mobility $85\,\mathrm{m}^2/\mathrm{Vs}$ (at the temperature 0.6 K). At the gate voltage $V_{\mathrm{g}} = -2.2\,\mathrm{V}$, $W \approx 0\,\mathrm{nm}$. The measured resistance consists of the resistance of the point contact, a non-monotonous function of V_{g}, and a constant series resistance that arises from the interface between the 2DEG lead and the point contact, here $400\,\Omega$. Steps in the dependence of the resistance as a function of V_{g} are seen, after Figure 1 from van Wees et al. (1988).

$$\varepsilon \mathbb{1} - \mathcal{H} \mapsto \begin{cases} \varepsilon \mathbb{1} + \mathcal{H}^* = -\left[(-\varepsilon)\mathbb{1} - \mathcal{H}^*\right], & \text{for D, DIII, C, CI,} \\ \\ \varepsilon \mathbb{1} + \mathcal{H} = -\left[(-\varepsilon)\mathbb{1} - \mathcal{H}\right], & \text{for AIII, BDI, CII,} \end{cases} \tag{7.221}$$

under charge conjugation or the chiral transformation, respectively, translate to the transformation law

$$\mathcal{S}(\varepsilon) = \mathcal{S}'(-\varepsilon) \tag{7.222}$$

for some $\mathcal{S}'(-\varepsilon)$ [say $\mathcal{S}'(-\varepsilon) = \mathcal{S}^*(-\varepsilon)$ for D] for each of the symmetry classes D, DIII, C, CI, AIII, BDI, and CII. This condition is only a restriction on the unitary matrix $\mathcal{S}(\varepsilon)$ if the chemical potential from the reservoirs vanishes, $\mu = \varepsilon = 0$. In contrast, one finds the condition

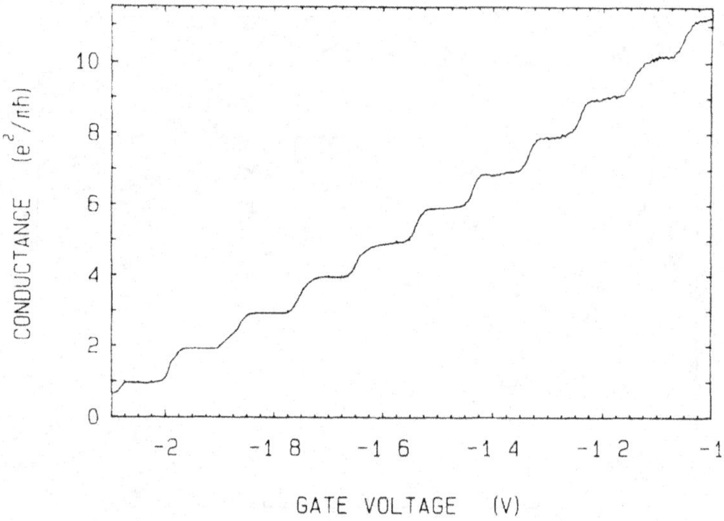

Figure 7.4 The conductance as a function of gate voltage V_{g} extracted from the measured resistance in Figure 7.3, after subtraction of a lead resistance of $400\,\Omega$, shows plateaus at integer multiples of $2 \times e^2/(2\pi\hbar)$ (the multiplicative factor of 2 arises from the spin degeneracy), after Figure 2 from van Wees et al. (1988).

$$\mathcal{S}(\varepsilon) = \mathcal{S}'(\varepsilon) \tag{7.223}$$

for some $\mathcal{S}'(\varepsilon)$ [say $\mathcal{S}'(\varepsilon) = \mathcal{S}^{\mathsf{T}}(\varepsilon)$ for AI] for each of the symmetry classes A, AI, and AII, which is a restriction on the unitary matrix $\mathcal{S}(\varepsilon)$ for any value of the chemical potential $\mu = \varepsilon$ from the reservoirs. By working out the consequences of these constraints on the singular-value decomposition (7.196), one may deduce the joint probability distribution of the diagonal matrix elements of the rectangular matrix $\mathsf{d}(\varepsilon = 0)$. This joint probability distribution is related to the joint probability distribution proportional to the Jacobian (7.194c) by a variable transformation. We are going to show this by trading the scattering matrix $\mathcal{S}(\varepsilon)$ for the transfer matrix $\mathcal{M}(\varepsilon)$.

7.6.2 Transfer Matrix

For notational simplicity, we treat the case when the numbers M and M' of propagating single-particle modes in reservoirs Ω and Ω', respectively, are equal. The generalization to $M \neq M'$ is left to the reader.

The transfer matrix $\mathcal{M}$ (the dependence on $\varepsilon = \mu$ is not made explicit) associated with the chaotic quantum dot separating reservoirs Ω and Ω' in Figure 7.2 is defined by the linear rule

$$\mathcal{R} = \mathcal{M}\,\mathcal{L}, \qquad \mathcal{L} := \begin{pmatrix} \mathsf{i} \\ \mathsf{o} \end{pmatrix} \in \mathbb{C}^{2M}, \qquad \mathcal{R} := \begin{pmatrix} \mathsf{o}' \\ \mathsf{i}' \end{pmatrix} \in \mathbb{C}^{2M}, \tag{7.224a}$$

where flux conservation implies that

$$|\mathsf{i}|^2-|\mathsf{o}|^2 = |\mathsf{o}'|^2 - |\mathsf{i}'|^2 \iff \mathcal{M}^\dagger\,(\mathbb{1}_M \otimes \tau_3)\,\mathcal{M} = \mathbb{1}_M \otimes \tau_3. \tag{7.224b}$$

The unit 2×2 matrix τ_0 and the vector $\boldsymbol{\tau}$ of Pauli matrices are defined such that the state $\begin{pmatrix}\mathsf{i} & 0\end{pmatrix}^{\mathsf{T}}$ is an eigenstate of $\mathbb{1}_M \otimes \tau_3$ with eigenvalue $+1$ that travels from Ω to Ω', while the state $\begin{pmatrix}0 & \mathsf{i}'\end{pmatrix}^{\mathsf{T}}$ is an eigenstate of $\mathbb{1}_M \otimes \tau_3$ with eigenvalue -1 that travels from Ω' to Ω. Conservation of flux implies that $\mathcal{M} \in \mathrm{U}(M, M)$.

Remark (1) We had already encountered a transfer matrix in **Exercise** 3.2. There, the transfer matrix was an alternative representation of the stationary Schrödinger equation for a one-dimensional tight-binding model. Here, the transfer matrix is defined without explicit reference to any Hamiltonian. A pair of transfer matrices was defined in Eq. (4.3) to construct continuum states in the same spirit as that in Eq. (7.224).

Remark (2) Observe that the pseudounitarity of $\mathcal{M}$ implies that

$$\begin{aligned}\mathbb{1}_M \otimes \tau_3 \left(\mathcal{M}\,\mathcal{M}^\dagger\right) \mathbb{1}_M \otimes \tau_3 \left(\mathcal{M}\,\mathcal{M}^\dagger\right) &= \mathbb{1}_M \otimes \tau_0 \\ \Longrightarrow \left(\mathcal{M}\,\mathcal{M}^\dagger\right)^{-1} &= \mathbb{1}_M \otimes \tau_3 \left(\mathcal{M}\,\mathcal{M}^\dagger\right) \mathbb{1}_M \otimes \tau_3,\end{aligned} \tag{7.225a}$$

$$\begin{aligned}\mathbb{1}_M \otimes \tau_3 \left(\mathcal{M}^\dagger\,\mathcal{M}\right) \mathbb{1}_M \otimes \tau_3 \left(\mathcal{M}^\dagger\,\mathcal{M}\right) &= \mathbb{1}_M \otimes \tau_0 \\ \Longrightarrow \left(\mathcal{M}^\dagger\,\mathcal{M}\right)^{-1} &= \mathbb{1}_M \otimes \tau_3 \left(\mathcal{M}^\dagger\,\mathcal{M}\right) \mathbb{1}_M \otimes \tau_3,\end{aligned} \tag{7.225b}$$

provided $\mathcal{M}\,\mathcal{M}^\dagger$ and $\mathcal{M}^\dagger\,\mathcal{M}$ are invertible. As the positive definite $2M \times 2M$ matrix product $\mathcal{M}\,\mathcal{M}^\dagger$ is unitarily equivalent to its inverse $(\mathcal{M}\,\mathcal{M}^\dagger)^{-1}$, they must share the same (positive) eigenvalues. Consequently, $\mathcal{M}\,\mathcal{M}^\dagger$ and $(\mathcal{M}\,\mathcal{M}^\dagger)^{-1}$, are isospectral[43] and their positive eigenvalues must come in inverse pairs. The same reasoning applies to $\mathcal{M}^\dagger\,\mathcal{M}$ and $(\mathcal{M}^\dagger\,\mathcal{M})^{-1}$. Moreover, both pairs must share the same eigenvalue spectrum since $\mathcal{M}\,\mathcal{M}^\dagger$ and $\mathcal{M}^\dagger\,\mathcal{M}$ are necessarily essential isospectral.[44]

Remark (3) The advantage of the transfer matrix $\mathcal{M}$ over the scattering matrix $\mathcal{S}$ is the following. If the two reservoirs Ω and Ω' are separated by two chaotic quantum dots Λ_1 and Λ_2 that are aligned sequentially as is pictured in Figure 7.5, one may not trade the two scattering matrices $\mathcal{S}_1$ and $\mathcal{S}_2$ associated with each chaotic quantum dot by their products $\mathcal{S}_1\,\mathcal{S}_2$ (**Exercise** 7.19). This composition law is, however, obeyed by the two transfer matrices $\mathcal{M}_1$ and $\mathcal{M}_2$ associated with each chaotic quantum dot, that is,

$$\left.\begin{aligned}&\Lambda_{12} := \Lambda_1 \cup \Lambda_2, \\ &\Lambda_1 \sim \mathcal{M}_1,\ \Lambda_2 \sim \mathcal{M}_2,\end{aligned}\right\} \Longrightarrow \left\{\begin{aligned}&\mathcal{M}_{12} = \mathcal{M}_1\,\mathcal{M}_2, \\ &\Lambda_{12} \sim \mathcal{M}_{12}.\end{aligned}\right. \tag{7.226}$$

43 In other words, $\mathcal{M}\,\mathcal{M}^\dagger$ and $(\mathcal{M}\,\mathcal{M}^\dagger)^{-1}$ must share the same eigenvalues with the same multiplicities.

44 In other words, $\mathcal{M}\,\mathcal{M}^\dagger$ and $\mathcal{M}^\dagger\,\mathcal{M}$ must share the same nonvanishing eigenvalues with the same multiplicities.

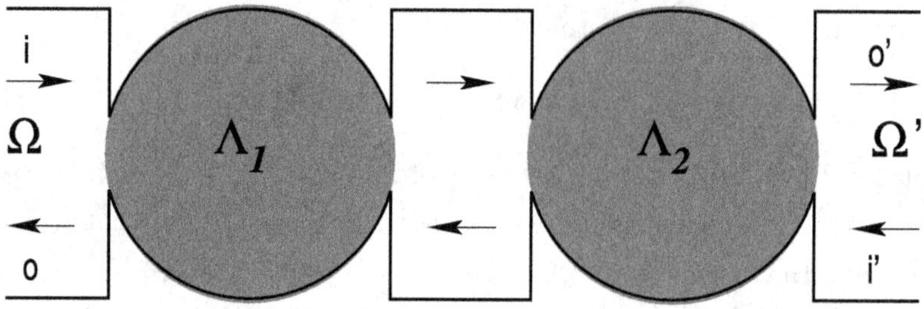

Figure 7.5 Two sequential chaotic quantum dots Λ_1 and Λ_2 that are weakly connected to two reservoirs Ω and Ω' that each support M delocalized single-particle states at the chemical potential μ. On the one hand, incoming single-particle states $\mathcal{I}$ are mapped into outgoing single-particle states $\mathcal{O}$ by a unitary matrix $\mathcal{S}_{12}$ that is not the product of the unitary scattering matrices $\mathcal{S}_1$ and $\mathcal{S}_2$ associated with Λ_1 and Λ_2, respectively. On the other hand, the pseudounitary transfer matrix $\mathcal{M}_{12} \equiv \mathcal{M}_1 \mathcal{M}_2$ maps $(\mathsf{i}, \mathsf{o})^{\mathsf{T}}$ into $(\mathsf{o}', \mathsf{i}')^{\mathsf{T}}$. Here, $\mathcal{M}_1$ and $\mathcal{M}_2$ are the pseudounitary transfer matrices associated with Λ_1 and Λ_2, respectively.

The relationship between the $2M \times 2M$ scattering and transfer matrix is one-to-one and given by (**Exercise** 7.20)

$$\begin{aligned} \mathcal{M} \equiv \begin{pmatrix} \mathsf{m}_{11} & \mathsf{m}_{12} \\ \mathsf{m}_{21} & \mathsf{m}_{22} \end{pmatrix} &= \begin{pmatrix} \mathsf{t}^{\dagger\,-1} & \mathsf{r}'\,\mathsf{t}'^{\,-1} \\ -\mathsf{t}'^{\,-1}\,\mathsf{r} & \mathsf{t}'^{\,-1} \end{pmatrix} \\ \Longleftrightarrow \qquad \mathcal{S} \equiv \begin{pmatrix} \mathsf{r} & \mathsf{t}' \\ \mathsf{t} & \mathsf{r}' \end{pmatrix} &= \begin{pmatrix} -\mathsf{m}_{22}^{-1}\,\mathsf{m}_{21} & \mathsf{m}_{22}^{-1} \\ \mathsf{m}_{11}^{\dagger\,-1} & \mathsf{m}_{12}\,\mathsf{m}_{22}^{-1} \end{pmatrix}, \end{aligned} \tag{7.227}$$

provided t and t' are invertible (this is why we made the assumption $M = M'$). Equipped with the relation (7.227) that relates the four blocks of the transfer matrix $\mathcal{M}$ to the reflection and transfer submatrices from the scattering matrix $\mathcal{S}$, we deduce the singular-value decomposition (**Exercise** 7.21)[45]

$$\mathcal{M} = \begin{pmatrix} \mathsf{u}' & 0 \\ 0 & \mathsf{v}'^{\dagger} \end{pmatrix} \begin{pmatrix} \sqrt{(\mathsf{d}^{\dagger}\,\mathsf{d})^{-1}} & \sqrt{(\mathsf{d}^{\dagger}\,\mathsf{d})^{-1} - \mathbb{1}_M} \\ \sqrt{(\mathsf{d}^{\dagger}\,\mathsf{d})^{-1} - \mathbb{1}_M} & \sqrt{(\mathsf{d}^{\dagger}\,\mathsf{d})^{-1}} \end{pmatrix} \begin{pmatrix} \mathsf{v} & 0 \\ 0 & \mathsf{u}^{\dagger} \end{pmatrix}, \tag{7.228}$$

provided the $M \times M$ diagonal and nonnegative matrix $\mathsf{d}^{\dagger}\,\mathsf{d} = \mathsf{d}\,\mathsf{d}^{\dagger} = \mathsf{d}^2$ is invertible (that is positive definite).

By combining the representation (7.225) of pseudounitarity with the singular-value decomposition (7.228), one obtains the identities (remember that $d = d^{\dagger}$ is a real-valued diagonal $M \times M$ matrix)

[45] P. A. Mello, P. Pereyra, and N. Kumar, Ann. Phys. (N.Y.) **181**(2), 290–317 (1988); P. A. Mello and J. L. Pichard, J. Phys. I France **1**(4), 493–513 (1991). [Mello et al. (1988); Mello and Pichard (1991)].

$$2\mathbb{1}_{2M} + \mathcal{M}\,\mathcal{M}^{\dagger} + \left(\mathcal{M}\,\mathcal{M}^{\dagger}\right)^{-1}$$
$$= \begin{pmatrix} \mathsf{u}' & 0 \\ 0 & \mathsf{v}'^{\dagger} \end{pmatrix} \begin{pmatrix} 4\left(\mathsf{d}\,\mathsf{d}^{\dagger}\right)^{-1} & 0 \\ 0 & 4\left(\mathsf{d}\,\mathsf{d}^{\dagger}\right)^{-1} \end{pmatrix} \begin{pmatrix} \mathsf{u}'^{\dagger} & 0 \\ 0 & \mathsf{v}' \end{pmatrix} \tag{7.229a}$$

and

$$2\mathbb{1}_{2M} + \mathcal{M}^{\dagger}\,\mathcal{M} + \left(\mathcal{M}^{\dagger}\,\mathcal{M}\right)^{-1}$$
$$= \begin{pmatrix} \mathsf{v}^{\dagger} & 0 \\ 0 & \mathsf{u} \end{pmatrix} \begin{pmatrix} 4\left(\mathsf{d}^{\dagger}\,\mathsf{d}\right)^{-1} & 0 \\ 0 & 4\left(\mathsf{d}^{\dagger}\,\mathsf{d}\right)^{-1} \end{pmatrix} \begin{pmatrix} \mathsf{v} & 0 \\ 0 & \mathsf{u}^{\dagger} \end{pmatrix}. \tag{7.229b}$$

After inverting the left- and right-hand sides of both equalities, the right-hand sides are seen to be related by Eqs. (7.204b) and (7.204c) to the transmission matrices. We conclude that

$$\left[2\mathbb{1}_{2M} + \mathcal{M}\,\mathcal{M}^{\dagger} + \left(\mathcal{M}\,\mathcal{M}^{\dagger}\right)^{-1}\right]^{-1} = \frac{1}{4} \begin{pmatrix} \mathsf{t}\,\mathsf{t}^{\dagger} & 0 \\ 0 & \mathsf{t}'^{\dagger}\,\mathsf{t}' \end{pmatrix} \tag{7.230a}$$

and

$$\left[2\mathbb{1}_{2M} + \mathcal{M}^{\dagger}\,\mathcal{M} + \left(\mathcal{M}^{\dagger}\,\mathcal{M}\right)^{-1}\right]^{-1} = \frac{1}{4} \begin{pmatrix} \mathsf{t}^{\dagger}\,\mathsf{t} & 0 \\ 0 & \mathsf{t}'\,\mathsf{t}'^{\dagger} \end{pmatrix}, \tag{7.230b}$$

respectively. As we have discussed with Eq. (7.225), the pseudounitarity of $\mathcal{M}$ together with the existence of $\left(\mathcal{M}\,\mathcal{M}^{\dagger}\right)^{-1}$ imply that the eigenvalues of the positive definite product $\mathcal{M}\,\mathcal{M}^{\dagger}$ may always be parameterized by $\exp(\pm 2x_j)$ with $x_j \in \mathbb{R}$ for $j = 1, \cdots, M$. The same applies to $\mathcal{M}^{\dagger}\,\mathcal{M}$ and $(\mathcal{M}^{\dagger}\,\mathcal{M})^{-1}$ with the same eigenvalues $\exp(\pm 2x_j)$ for $j = 1, \cdots, M$, since $\mathcal{M}^{\dagger}\,\mathcal{M}$ is essential isospectral to $\mathcal{M}\,\mathcal{M}^{\dagger}$. By tracing either one of Eqs. (7.230a) or (7.230b), one obtains the representation

$$g_{\mathrm{L}} = \sum_{j=1}^{M} \frac{1}{\cosh^2 x_j} \tag{7.231}$$

for the dimensionless Landauer conductance (7.208). The choice of the notation x_j in the representation (7.231) of the Landauer conductance is not accidental. The variable x_j, the so-called Lyapunov exponent of the transfer matrix, also enters the Jacobian (7.194c), as will become explicit once we have established the transformation laws obeyed by the transfer matrix under reversal of time or charge conjugation or chiral (sublattice interchange) transformation.

We seek the transformation laws of the linear equation (7.224) implementing reversal of time, chiral (sublattice) transformation, or charge conjugation. Complex conjugation of

$$\mathcal{R}(\varepsilon) = \mathcal{M}(\varepsilon)\,\mathcal{L}(\varepsilon) \tag{7.232}$$

gives

$$\mathcal{R}^{*}(\varepsilon) = \mathcal{M}^{*}(\varepsilon)\,\mathcal{L}^{*}(\varepsilon). \tag{7.233}$$

If the unitary matrix $\mathcal{T}$ represents the unitary component of reversal of time in the representation (7.224), then

$$\mathcal{T}\,\mathcal{R}^{*}(\varepsilon)=\left[\mathcal{T}\,\mathcal{M}^{*}(\varepsilon)\,\mathcal{T}^{-1}\right]\,\mathcal{T}\,\mathcal{L}^{*}(\varepsilon). \tag{7.234}$$

Similarly, the unitary chiral spectral symmetry is

$$\Gamma\,\mathcal{R}(\varepsilon)=\left[\Gamma\,\mathcal{M}(\varepsilon)\,\Gamma^{-1}\right]\,\Gamma\,\mathcal{L}(\varepsilon), \tag{7.235}$$

while the antiunitary spectral symmetry is

$$\mathcal{C}\,\mathcal{R}^{*}(\varepsilon)=\left[\mathcal{C}\,\mathcal{M}^{*}(\varepsilon)\,\mathcal{C}^{-1}\right]\,\mathcal{C}\,\mathcal{L}^{*}(\varepsilon). \tag{7.236}$$

To proceed, needed is a model for the left and right reservoirs and how they couple to the intervening region, here the quantum dot. This is done explicitly in Sections 7.7.2 and 7.7.3 for a quantum wire. We can deduce from Eq. (7.348) that the transfer matrix for a quantum dot (of size δL and in units in which $\hbar=v_{\mathrm{F}}=1$) is given by

$$\mathcal{M}(\varepsilon)=\exp\Big(\mathrm{i}\delta L\left[(\mathbb{1}\otimes\tau_{3})\,\mathcal{H}-\varepsilon\,(\mathbb{1}\otimes\tau_{3})\right]\Big), \tag{7.237}$$

where the rank of the unit matrix $\mathbb{1}$ depends on the choice made for the Altland-Zirnbauer symmetry class and the Hermitean matrix $\mathcal{H}$ is a random matrix from one of the Altland-Zirnbauer symmetry classes sharing the same rank as the matrix $\mathbb{1}\otimes\tau_{3}$. Knowledge of the transformation laws of $\mathcal{H}$ and $\mathbb{1}\otimes\tau_{3}$ under reversal of time, charge conjugation, and the chiral transformation fixes the transformation law obeyed by the transfer matrix $\mathcal{M}(\varepsilon)$ for each of the ten Altland-Zirnbauer symmetry classes. It then follows that the time-reversal symmetry of $\mathcal{H}$ implies

$$\mathcal{T}\,\mathcal{M}^{*}(\varepsilon)\,\mathcal{T}^{-1}=\mathcal{M}(\varepsilon), \tag{7.238a}$$

the antiunitary charge-conjugation (Bogoliubov-de-Gennes) spectral symmetry of $\mathcal{H}$ implies

$$\mathcal{C}\,\mathcal{M}^{*}(\varepsilon)\,\mathcal{C}^{-1}=\mathcal{M}(-\varepsilon), \tag{7.238b}$$

and the unitary chiral spectral symmetry of $\mathcal{H}$ implies

$$\Gamma\,\mathcal{M}(\varepsilon)\,\Gamma^{-1}=\mathcal{M}(-\varepsilon). \tag{7.238c}$$

For completeness, we present in **Table** 7.5 the representations of the unitary matrices $\mathcal{T}$, $\mathcal{C}$, and Γ in the basis defined by Eq. (7.348).

The next task consists in identifying which are the independent Lyapunov exponents, their range, and their joint probability distribution for each of the ten

Table 7.5 *The tenfold way: Symmetries of the transfer matrix*

Cartan label	TRS	PHS	CHS
A	0	0	0
AI	$\mathbb{1}\otimes\tau_1\,\mathsf{K}$	0	0
AII	$\mathbb{1}\otimes\mathrm{i}\sigma_2\otimes\tau_1\,\mathsf{K}$	0	0
AIII	0	0	$\mathbb{1}\otimes\tau_1$
BDI	$\mathbb{1}\otimes\tau_1\,\mathsf{K}$	$\mathbb{1}\otimes\tau_0\,\mathsf{K}$	$\mathbb{1}\otimes\tau_1$
CII	$\mathbb{1}\otimes\mathrm{i}\sigma_2\otimes\tau_1\,\mathsf{K}$	$\mathbb{1}\otimes\mathrm{i}\sigma_2\otimes\tau_0\,\mathsf{K}$	$\mathbb{1}\otimes\sigma_0\otimes\tau_1$
D	0	$\mathbb{1}\otimes\rho_1\otimes\tau_0\,\mathsf{K}$	0
C	0	$\mathbb{1}\otimes\mathrm{i}\kappa_2\otimes\tau_0\,\mathsf{K}$	0
DIII	$\mathbb{1}\otimes\rho_0\otimes\mathrm{i}\sigma_2\otimes\tau_1\,\mathsf{K}$	$\mathbb{1}\otimes\rho_1\otimes\sigma_0\otimes\tau_0\,\mathsf{K}$	$\mathbb{1}\otimes\rho_1\otimes\sigma_2\otimes\tau_1$
CI	$\mathbb{1}\otimes\kappa_0\otimes\tau_1\,\mathsf{K}$	$\mathbb{1}\otimes\mathrm{i}\kappa_2\otimes\tau_0\,\mathsf{K}$	$\mathbb{1}\otimes\kappa_2\otimes\tau_1$

Representations of time reversal, charge conjugation, and the chiral transformation in the basis defined by the transfer matrix (7.237) [or Eq. (7.348)]. The unit Pauli matrix τ_0 and the three Pauli matrices $\boldsymbol{\tau}$ denote the 2×2 block decomposition (left-right movers grading) in the representation (7.224). The unit Pauli matrix σ_0 and the three Pauli matrices $\boldsymbol{\sigma}$ denote the 2×2 block decomposition for the spin-1/2 degrees of freedom, if present. The unit Pauli matrix ρ_0 and the three Pauli matrices $\boldsymbol{\rho}$ denote the 2×2 block decomposition for the Bogoliubov-de-Gennes particle-hole grading, if present. The unit Pauli matrix κ_0 and the three Pauli matrices $\boldsymbol{\kappa}$ denote the effective 2×2 grading that mixes the Bogoliubov-de-Gennes and spin gradings when spin-rotation symmetry holds, if present. The dimensionality of the unit matrix $\mathbb{1}$ is left unspecified for each Altland-Zirnbauer symmetry class.

Altland-Zirnbauer symmetry classes. To this end, we start from the singular-value decomposition[46]

$$
\begin{aligned}
&\mathcal{M}=\begin{pmatrix}\mathsf{u}_1 & 0\\ 0 & \mathsf{u}_2\end{pmatrix}\begin{pmatrix}\cosh\mathsf{x} & \sinh\mathsf{x}\\ \sinh\mathsf{x} & \cosh\mathsf{x}\end{pmatrix}\begin{pmatrix}\mathsf{v}_1 & 0\\ 0 & \mathsf{v}_2\end{pmatrix},\\
&\mathcal{U}\equiv\operatorname{diag}\left(\mathsf{u}_1\quad \mathsf{u}_2\right),\mathcal{V}\equiv\operatorname{diag}\left(\mathsf{v}_1\quad \mathsf{v}_2\right)\in \mathrm{U}(M)\times\mathrm{U}(M),\\
&\mathcal{X}\equiv\begin{pmatrix}\cosh\mathsf{x} & \sinh\mathsf{x}\\ \sinh\mathsf{x} & \cosh\mathsf{x}\end{pmatrix}\in U(M,M),\\
&\mathsf{x}\equiv\operatorname{diag}\left(x_1\quad\cdots\quad x_M\right),\qquad x_1,\cdots,x_M\in\mathbb{R}
\end{aligned}
\tag{7.239a}
$$

of a generic transfer matrix $\mathcal{M}\in\mathrm{U}(M,M)$ with $M=1,2,3,\cdots$ that follows from imposing flux conservation

$$
\mathcal{M}\,(\mathbb{1}\otimes\tau_3)\,\mathcal{M}^{\dagger}=\mathcal{M}^{\dagger}\,(\mathbb{1}\otimes\tau_3)\,\mathcal{M}=\mathbb{1}\otimes\tau_3. \tag{7.239b}
$$

[46] The singular-value decomposition (7.239) is equivalent to Eq. (7.228).

Our goal is to identify the symmetric space with its radial coordinates associated with the Lie group spanned by transfer matrices of the form (7.239a) and to relate for each Altland-Zirnbauer symmetry class the radial coordinates to the Lyapunov exponents

$$\boldsymbol{x} \equiv \begin{pmatrix} x_1 & \cdots & x_M \end{pmatrix}^{\mathsf{T}}, \qquad x_1, \cdots, x_M \in \mathbb{R}. \tag{7.239c}$$

entering the singular-value decomposition (7.239a).

The singular-value decomposition (7.239a) is invariant under the transformation

$$\mathcal{U} \mapsto \mathcal{U} \operatorname{diag}\begin{pmatrix} e^{+\mathrm{i}\Theta} & e^{+\mathrm{i}\Theta} \end{pmatrix}, \qquad \mathcal{V} \mapsto \operatorname{diag}\begin{pmatrix} e^{-\mathrm{i}\Theta} & e^{-\mathrm{i}\Theta} \end{pmatrix} \mathcal{V}, \tag{7.240a}$$

where the $M \times M$ matrix Θ is diagonal and given by

$$\Theta = \operatorname{diag}\begin{pmatrix} \theta_1 & \cdots & \theta_M \end{pmatrix}, \qquad 0 \le \theta_1, \cdots, \theta_M < 2\pi. \tag{7.240b}$$

The singular-value decomposition (7.239a) is thus one-to-many. Moreover, this one-to-many correspondence is here not countable as the left-hand side is a matrix that depends on

$$(2M)^2 = 4M^2 \tag{7.241}$$

independent real-valued matrix elements, while the right-hand side depends on

$$M^2 + M^2 + M + M^2 + M^2 = 4M^2 + M \tag{7.242}$$

independent real-valued matrix elements. This uncountable redundancy of the singular-value decomposition (7.239a) is lifted by quotienting either one of the two groups of matrices of the form

$$\mathcal{U} \equiv \begin{pmatrix} \mathsf{u}_1 & 0 \\ 0 & \mathsf{u}_2 \end{pmatrix}, \qquad \mathcal{V} \equiv \begin{pmatrix} \mathsf{v}_1 & 0 \\ 0 & \mathsf{v}_2 \end{pmatrix}, \tag{7.243}$$

each of which is isomorphic to $\mathrm{U}(M) \times \mathrm{U}(M)$, by the group of matrices of the form

$$\begin{pmatrix} e^{\mathrm{i}\Theta} & 0 \\ 0 & e^{\mathrm{i}\Theta} \end{pmatrix}, \quad \Theta = \operatorname{diag}\begin{pmatrix} \theta_1 & \cdots & \theta_M \end{pmatrix}, \quad 0 \le \theta_1, \cdots, \theta_M < 2\pi, \tag{7.244}$$

that is isomorphic to $[\mathrm{U}(1)]^M$.

We seek a measure for the space of transfer matrices that is invariant under left or right multiplications

$$\mathcal{M} \mapsto \mathcal{M}_0\, \mathcal{M}, \qquad \mathcal{M} \mapsto \mathcal{M}\, \mathcal{M}_0 \tag{7.245}$$

by some arbitrarily chosen transfer matrix $\mathcal{M}_0 \in \mathcal{U}(M, M)$. Define the squared infinitesimal invariant arclength

$$\mathrm{d}s^2 := \operatorname{tr}\left(\Sigma_3\, \mathrm{d}\mathcal{M}^{\dagger}\, \Sigma_3\, \mathrm{d}\mathcal{M}\right), \qquad \Sigma_3 := \mathbb{1}_M \otimes \tau_3, \qquad \mathcal{M} \equiv \mathcal{U}\, \mathcal{X}\, \mathcal{V}, \tag{7.246}$$

where $\mathcal{U} \in U(M) \times U(M)$, $\mathcal{X} \in \mathrm{U}(M, M)$, and $\mathcal{V} \in U(M) \times U(M)$ are shorthand notations for the matrices entering on the right-hand side of the singular-value decomposition (7.239a). By design, $\mathrm{d}s^2$ is invariant under the left or right multiplications (7.245) as we will motivate in the following. The infinitesimal transfer matrix is

$$\mathrm{d}\mathcal{M} = \mathcal{U}\left[\mathrm{d}\mathcal{A}_{\mathcal{U}}\mathcal{X} + \mathrm{d}\mathcal{X} + \mathcal{X}\,\mathrm{d}\mathcal{A}_{\mathcal{V}}\right]\mathcal{V}, \tag{7.247a}$$

$$\mathrm{d}\mathcal{A}_{\mathcal{U}} := \mathcal{U}^{\dagger}\,\mathrm{d}\mathcal{U} = -\left(\mathrm{d}\mathcal{A}_{\mathcal{U}}\right)^{\dagger}, \qquad \mathrm{d}\mathcal{A}_{\mathcal{V}} := (\mathrm{d}\mathcal{V})\,\mathcal{V}^{\dagger} = -\left(\mathrm{d}\mathcal{A}_{\mathcal{V}}\right)^{\dagger}. \tag{7.247b}$$

Because of

$$\mathcal{U}\,\Sigma_3\,\mathcal{U}^{\dagger} = \mathcal{U}^{\dagger}\,\Sigma_3\,\mathcal{U} = \mathcal{V}\,\Sigma_3\,\mathcal{V}^{\dagger} = \mathcal{V}^{\dagger}\,\Sigma_3\,\mathcal{V} = \Sigma_3, \tag{7.248}$$

pseudounitarity, that is,[47]

$$\mathcal{M}^{-1} = \Sigma_3\,\mathcal{M}^{\dagger}\,\Sigma_3, \tag{7.249}$$

and

$$[\mathcal{U}, \Sigma_3] = [\mathrm{d}\mathcal{A}_{\mathcal{U}}, \Sigma_3] = [\mathcal{V}, \Sigma_3] = [\mathrm{d}\mathcal{A}_{\mathcal{V}}, \Sigma_3] = 0, \tag{7.250}$$

we arrive at

$$\begin{aligned}\mathrm{d}s^2 &= \mathrm{tr}\left((\mathrm{d}\mathcal{M})^{-1}\,\mathrm{d}\mathcal{M}\right)\\ &= \mathrm{tr}\Big\{\left[\mathcal{X}^{-1}\,\mathrm{d}\,(\mathcal{A}_{\mathcal{U}})^{\dagger} + \mathrm{d}\,(\mathcal{X})^{-1} + (\mathrm{d}\mathcal{A}_{\mathcal{V}})^{\dagger}\,\mathcal{X}^{-1}\right]\left[\mathrm{d}\mathcal{A}_{\mathcal{U}}\,\mathcal{X} + (\mathrm{d}\mathcal{X}) + \mathcal{X}\,\mathrm{d}\mathcal{A}_{\mathcal{V}}\right]\Big\}.\end{aligned} \tag{7.251}$$

The task of identifying the symmetric space with the radial coordinates (7.239c) is achieved once $4M^2$ independent variables $\mathrm{d}\kappa_{\mu} \in \mathbb{R}$ with $\mu = 1, \cdots, 4M^2$ can be extracted from the matrices $\mathrm{d}\mathcal{X}$, $\mathrm{d}\mathcal{A}_{\mathcal{U}}$, and $\mathrm{d}\mathcal{A}^{\dagger}_{\mathcal{V}}$ together with a Riemannian metric $\mathfrak{g}_{\mu\nu}(\kappa)$ such that

$$\mathrm{d}s^2 = \sum_{\mu,\nu=1}^{4M} \mathrm{d}\kappa_{\mu}\,\mathfrak{g}_{\mu\nu}(\kappa)\,\mathrm{d}\kappa_{\nu}, \tag{7.252}$$

since differential geometry then dictates that

$$\mu(\mathrm{d}\mathcal{M}) := \prod_{\mu=1}^{4M^2} \mathrm{d}\kappa_{\mu}\,|\mathfrak{g}(\kappa)|^{1/2}, \qquad \mathfrak{g}(\kappa) := \det\left(\mathfrak{g}_{\mu\nu}(\kappa)\right), \tag{7.253}$$

is an invariant measure under the left or right multiplications (7.245). This task is left as an exercise to the reader.

Instead, we will take a shortcut to identify the symmetric spaces upon imposing on the space of transfer matrices time-reversal symmetry or spectral symmetries

[47] Because of pseudounitarity, we have

$$\mathcal{M}\,\Sigma_3\,\mathcal{M}^{\dagger} = \Sigma_3.$$

Hence,

$$\mathcal{M}\left(\Sigma_3\,\mathcal{M}^{\dagger}\,\Sigma_3\right) = \left(\mathcal{M}\,\Sigma_3\,\mathcal{M}^{\dagger}\right)\Sigma_3 = \Sigma_3\,\Sigma_3 = \mathbb{1}_{2M}$$

implies that $\Sigma_3\,\mathcal{M}^{\dagger}\,\Sigma_3$ is the inverse of $\mathcal{M}$.

originating from either the chiral transformation or charge conjugation. This shortcut consists in identifying the non-compact Lie group $\mathrm{G}^{\star}$ in the second column of **Table** 7.4 from a symmetry analysis by which once selects M to be a multiple of the integer N that is appropriate for the Altland-Zirnbauer symmetry class corresponding to the combination of symmetries and spectral symmetries imposed on the space of transfer matrices. We will consider the transfer matrix and its singular-value decomposition for each Altland-Zirnbauer symmetry class by going down the ten rows of **Table** 7.5.

Symmetry class A We choose $M = N$ with $N = 1, 2, 3, \cdots$. Transfer matrices are elements of the non-compact Lie group $\mathrm{U}(N, N)$. Thus, we identify

$$\mathrm{G}^{\star} = \mathrm{U}(N, N) = \mathrm{U}(1) \times \mathrm{SU}(N, N) \tag{7.254a}$$

and

$$\mathrm{H} = \mathrm{U}(N) \times \mathrm{U}(N) \tag{7.254b}$$

in the second column of **Table** 7.4.[48] Accordingly, the non-compact symmetric space associated with such random transfer matrices is the coset space

$$\mathrm{G}^{\star}/\mathrm{H} = \mathrm{U}(N, N)/[\mathrm{U}(N) \times \mathrm{U}(N)]. \tag{7.254c}$$

One interpretation for quotienting by the subgroup $\mathrm{U}(N) \times \mathrm{U}(N)$ follows from the observation that

$$\mathcal{M}^{\dagger}\,\mathcal{M} = \mathcal{U}^{\dagger}\,\mathcal{X}^{2}\,\mathcal{U}, \qquad \mathcal{U} \in \mathrm{U}(N) \times \mathrm{U}(N), \tag{7.255}$$

is independent of the factor $\mathcal{V} \in \mathrm{U}(N) \times \mathrm{U}(N)$ that enters the singular-value decomposition (7.239a). But the $2N \times 2N$ positive definite matrix $\mathcal{M}^{\dagger}\,\mathcal{M}$ is all that is needed to study the statistics of the transmission eigenvalues according to Eq. (7.230b). The joint probability distribution

$$P(x_1, \cdots, x_N) \tag{7.256a}$$

for the N independent Lyapunov exponents entering the dimensionless charge conductance[49]

$$g_{\mathrm{L}} = \sum_{j=1}^{N} \frac{1}{\cosh^2 x_j}, \qquad 0 \le x_j < \infty, \qquad j = 1, \cdots, N \tag{7.256b}$$

for the symmetry class A is then proportional to the Jacobian (7.194c) with its radial coordinates $\boldsymbol{x}$ substituted by two times the Lyapunov exponents ($\boldsymbol{x} \to 2\boldsymbol{x}$) and the roots on the line AIII of **Table** 7.4, namely,

[48] A. Hüffmann, J. Phys. A: Math. and Gen. **23**(24), 5733–5744 (1990). [Hüffmann (1990)].

[49] Observe that it is always possible to choose v_1 and v_2 relative to u_1 and u_2 in the singular-value decomposition (7.239a) such that $0 \le x_1, \cdots, x_N < \infty$.

$$
\begin{aligned}
P(x_1,\cdots,x_N) \propto & \prod_{1\le j<k\le N} \left|4\sinh\left(x_j - x_k\right)\sinh\left(x_j + x_k\right)\right|^2 \\
& \times \prod_{j=1}^{N} \left|2\sinh(2\,x_j)\right|^1 \\
\propto & \left[\prod_{1\le j<k\le N} \left|\cosh(2\,x_j) - \cosh(2\,x_k)\right|^2\right] \left[\prod_{j=1}^{N} \left|\sinh(2\,x_j)\right|\right].
\end{aligned}
\tag{7.256c}
$$

The singular-value decomposition (7.239a) allows to choose all Lyapunov exponents entering the Landauer conductance (7.256b) to be positive. Correspondingly, $P(x_1,\cdots,x_N)$ is invariant under the transformation

$$
x_j \mapsto \zeta_j\, x_j, \qquad \zeta_j = \pm 1, \qquad j = 1,\cdots,N. \tag{7.257}
$$

Symmetry class AI We choose $M = N$ with $N = 1,2,3,\cdots$. Any transfer matrix belongs to the non-compact Lie group $\mathrm{U}(N,N)$. Time-reversal symmetry for spinless fermions is implemented by demanding that

$$
\mathcal{T}\mathcal{M}^*\mathcal{T}^{-1} = \mathcal{M}, \qquad \mathcal{T} := \mathbb{1}_N \otimes \tau_1, \tag{7.258}
$$

whatever the value of the chemical potential in the reservoirs Ω and Ω'. (Conjugation by $\mathcal{T}$ is needed since states from reservoir Ω moving toward reservoir Ω' are exchanged with states from reservoir Ω moving away from reservoir Ω' upon reversing time.) Imposing this representation of time-reversal symmetry on the singular-value decomposition (7.239a) gives the singular-value decomposition[50]

$$
\begin{aligned}
&\mathcal{M} = \begin{pmatrix} \mathsf{u} & 0 \\ 0 & \mathsf{u}^* \end{pmatrix} \begin{pmatrix} \cosh\mathsf{x} & \sinh\mathsf{x} \\ \sinh\mathsf{x} & \cosh\mathsf{x} \end{pmatrix} \begin{pmatrix} \mathsf{v} & 0 \\ 0 & \mathsf{v}^* \end{pmatrix}, \\
&\mathsf{u},\mathsf{v} \in \mathrm{U}(N), \\
&\mathsf{x} \equiv \mathrm{diag}\left(x_1 \quad \cdots \quad x_N\right), \qquad 0 \le x_1,\cdots,x_N < \infty.
\end{aligned}
\tag{7.259}
$$

The singular-value decomposition (7.259) is countably one-to-many since we must restrict $\theta_j = 0,\pi$ for $j = 1,\cdots,N$ in the transformation (7.240) for it to leave Eq. (7.259) invariant. It follows that the right-hand side of Eq. (7.259) depends on

$$
N^2 + N + N^2 = 2N^2 + N \tag{7.260a}
$$

independent real-valued parameters, a number that coincides with the dimensionality of the non-compact Lie group

$$
\mathrm{Sp}(2N,\mathbb{R}) \tag{7.260b}
$$

[50] Observe that it is always possible to choose v relative to u such that $0 \le x_1,\cdots,x_N < \infty$.

[see Eq. (7.487) from **Exercise** 7.6]. Thus, we make the identifications

$$\mathrm{G}^{\star} = \mathrm{Sp}(2N, \mathbb{R}) \tag{7.260c}$$

and

$$\mathrm{H} \cong \mathrm{U}(N) = \mathrm{U}(1) \times \mathrm{SU}(N) \tag{7.260d}$$

in the second column of **Table** 7.4 (see footnote 48). Accordingly, the non-compact symmetric space associated with such random transfer matrices is the coset space[51]

$$\mathrm{G}^{\star}/\mathrm{H} \cong \mathrm{Sp}(2N, \mathbb{R})/\mathrm{U}(N). \tag{7.260e}$$

One interpretation for quotienting by the subgroup $\mathrm{U}(N)$ follows from the observation that

$$\mathcal{M}^{\dagger}\,\mathcal{M} = \mathcal{U}^{\dagger}\,\mathcal{X}^{2}\,\mathcal{U} \tag{7.261}$$

is independent of the factor $\mathcal{V}$ that enters the counterpart to the singular-value decomposition (7.259). But the $2N \times 2N$ positive definite matrix $\mathcal{M}^{\dagger}\,\mathcal{M}$ is all that is needed to study the statistics of the transmission eigenvalues according to Eq. (7.230b). The joint probability distribution

$$P(x_1, \cdots, x_N) \tag{7.262a}$$

for the N independent Lyapunov exponents entering the dimensionless charge conductance

$$g_{\mathrm{L}} = \sum_{j=1}^{N} \frac{1}{\cosh^2 x_j}, \qquad 0 \leq x_j < \infty, \qquad j = 1, \cdots, N \tag{7.262b}$$

for the symmetry class AI is then proportional to the Jacobian (7.194c) with its radial coordinates $\boldsymbol{x}$ substituted by two times the Lyapunov exponents ($\boldsymbol{x} \to 2\boldsymbol{x}$) and the roots on the line CI of **Table** 7.4, namely,

$$\begin{aligned}
P(x_1, \cdots, x_N) \propto & \prod_{1 \leq j < k \leq N} \left|4 \sinh\left(x_j - x_k\right) \sinh\left(x_j + x_k\right)\right|^{1} \\
& \times \prod_{j=1}^{N} \left|2 \sinh(2\,x_j)\right|^{1} \\
\propto & \left[\prod_{1 \leq j < k \leq N} \left|\cosh(2\,x_j) - \cosh(2\,x_k)\right|\right] \left[\prod_{j=1}^{N} \left|\sinh(2\,x_j)\right|\right].
\end{aligned} \tag{7.262c}$$

[51] We have an isomorphism instead of an equality because neither is $\mathrm{Sp}(2N, \mathbb{R})$ a subgroup of $\mathrm{U}(N, N)$ nor is $\mathrm{U}(N)$ a subgroup of $\mathrm{Sp}(2N, \mathbb{R})$.

The singular-value decomposition (7.259) allows to choose all Lyapunov exponents entering the Landauer conductance (7.262b) to be positive. Correspondingly, $P(x_1, \cdots, x_N)$ is invariant under the transformation

$$x_j \mapsto \zeta_j\, x_j, \qquad \zeta_j = \pm 1, \qquad j = 1, \cdots, N. \tag{7.263}$$

Symmetry class AII We choose $M = 2N$ with $N = 1, 2, 3, \cdots$ to accommodate the spin-1/2 degrees of freedom of electrons. We reserve the unit 2×2 matrix σ_0 and the vector $\boldsymbol{\sigma}$ of Pauli matrices for the spin-1/2 degrees of freedom. Any transfer matrix belongs to the non-compact Lie group $\mathrm{U}(2N, 2N)$. Time-reversal symmetry for spin-1/2 fermions is implemented by demanding that

$$\mathcal{T} \mathcal{M}^{*} \mathcal{T}^{-1} = \mathcal{M}, \qquad \mathcal{T} := \mathbb{1}_N \otimes \mathrm{i}\sigma_2 \otimes \tau_1, \tag{7.264}$$

whatever the value of the chemical potential in the reservoirs Ω and Ω'. (Conjugation by $\mathcal{T}$ is needed since states from reservoir Ω moving toward reservoir Ω' with spin up are exchanged with states with spin down from reservoir Ω moving away from reservoir Ω' upon reversing time.) Imposing this representation of time-reversal symmetry on the singular-value decomposition (7.239a) gives the singular-value decomposition[52]

$$\begin{aligned}
&\mathcal{M} = \begin{pmatrix} \mathsf{u} & 0 \\ 0 & \mathsf{u}^{\mathsf{R}} \end{pmatrix} \begin{pmatrix} \cosh \mathsf{x} & \sinh \mathsf{x} \\ \sinh \mathsf{x} & \cosh \mathsf{x} \end{pmatrix} \begin{pmatrix} \mathsf{v} & 0 \\ 0 & \mathsf{v}^{\mathsf{R}} \end{pmatrix}, \qquad \mathsf{u}, \mathsf{v} \in \mathrm{U}(2N), \\
&\mathsf{u}^{\mathsf{R}} \equiv (\sigma_2 \otimes \mathbb{1}_N)\, \mathsf{u}^{*}\, (\sigma_2 \otimes \mathbb{1}_N), \quad \mathsf{v}^{\mathsf{R}} \equiv (\sigma_2 \otimes \mathbb{1}_N)\, \mathsf{v}^{*}\, (\sigma_2 \otimes \mathbb{1}_N), \\
&\mathsf{x} \equiv \operatorname{diag}\begin{pmatrix} x_1 & x_1 & \cdots & x_N & x_N \end{pmatrix}, \qquad 0 \leq x_1, \cdots, x_N < \infty.
\end{aligned} \tag{7.265}$$

The singular-value decomposition (7.265) is uncountably one-to-many since it is invariant under[53]

$$\mathsf{u} \mapsto \mathsf{u}\,\mathsf{w}, \qquad \mathsf{v} \mapsto \mathsf{w}^{\dagger}\, \mathsf{v}, \qquad \forall \mathsf{w} \in [\mathrm{SU}(2)]^{N}. \tag{7.266}$$

It follows that the right-hand side of Eq. (7.265) depends on[54]

$$8N^2 + N - 3N = 8N^2 - 2N \tag{7.267a}$$

independent real-valued parameters, a number that coincides with the dimensionality of the non-compact Lie group

$$\mathrm{O}^{\star}(4N) \tag{7.267b}$$

[52] The Kramers' degeneracy is made explicit in the diagonal matrix x. Observe that it is always possible to choose v relative to u such that $0 \leq x_1, \cdots, x_N < \infty$.

[53] One uses here the identity

$$\sigma_2\, \mathsf{w}^{*}\, \sigma_2 = \mathsf{w}, \qquad \forall \mathsf{w} \in \mathrm{SU}(2).$$

[54] The number of independent real parameters that parameterize $[\mathrm{SU}(2)]^{N}$ is $3N$.

[see Eq. (7.484) from **Exercise** 7.6]. Thus, we make the identifications

$$\mathrm{G}^{\star} \cong \mathrm{O}^{\star}(4N) \tag{7.267c}$$

and

$$\mathrm{H} \cong \mathrm{U}(2N) = \mathrm{U}(1) \times \mathrm{SU}(2N) \tag{7.267d}$$

in the second column of **Table** 7.4 (see footnote 48). Accordingly, the non-compact symmetric space associated with such random transfer matrices is the coset space

$$\mathrm{G}^{\star}/\mathrm{H} \cong \mathrm{O}^{\star}(4N)/\mathrm{U}(2N). \tag{7.267e}$$

One interpretation for quotienting by the subgroup $\mathrm{U}(2N)$ follows from the observation that

$$\mathcal{M}^{\dagger}\,\mathcal{M} = \mathcal{U}^{\dagger}\,\mathcal{X}^{2}\,\mathcal{U} \tag{7.268}$$

is independent of the factor $\mathcal{V}$ that enters the counterpart to the singular-value decomposition (7.265). But the $4N \times 4N$ positive definite matrix $\mathcal{M}^{\dagger}\,\mathcal{M}$ is all that is needed to study the statistics of the transmission eigenvalues according to Eq. (7.230b). The joint probability distribution

$$P(x_1,\cdots,x_N) \tag{7.269a}$$

for the N independent Lyapunov exponents entering the dimensionless charge conductance

$$g_{\mathrm{L}} = 2 \times \sum_{j=1}^{N} \frac{1}{\cosh^2 x_j}, \qquad 0 \leq x_j < \infty, \qquad j = 1,\cdots,N \tag{7.269b}$$

for the symmetry class AII is then proportional to the Jacobian (7.194c) with its radial coordinates $\boldsymbol{x}$ substituted by two times the Lyapunov exponents ($\boldsymbol{x} \to 2\boldsymbol{x}$) and the roots on the line DIII of **Table** 7.4, namely,

$$\begin{aligned}
P(x_1,\cdots,x_N) \propto & \prod_{1\leq j<k\leq N} \left|4\sinh\left(x_j - x_k\right)\sinh\left(x_j + x_k\right)\right|^4 \\
& \times \prod_{j=1}^{N} \left|2\sinh(2\,x_j)\right|^1 \\
\propto & \left[\prod_{1\leq j<k\leq N} \left|\cosh(2\,x_j) - \cosh(2\,x_k)\right|^4\right] \left[\prod_{j=1}^{N} \left|\sinh(2\,x_j)\right|\right].
\end{aligned} \tag{7.269c}$$

The multiplicative factor 2 on the right-hand side of the Landauer conductance in the symmetry class AII originates from the Kramers' degeneracy. The singular-value decomposition (7.265) allows to choose all Lyapunov exponents entering the

Landauer conductance (7.269b) to be positive. Correspondingly, $P(x_1,\cdots,x_N)$ is invariant under the transformation

$$x_j \mapsto \zeta_j\, x_j, \qquad \zeta_j = \pm 1, \qquad j = 1,\cdots,N. \tag{7.270}$$

Symmetry class AIII We choose $M = N$ with $N = 1,2,3,\cdots$. Any transfer matrix belongs to the non-compact Lie group $\mathrm{U}(N,N)$. A chiral transformation is implemented by demanding that

$$\Gamma\,\mathcal{M}(\varepsilon)\,\Gamma = \mathcal{M}(-\varepsilon), \qquad \Gamma := \mathbb{1}_N \otimes \tau_1, \tag{7.271}$$

for the value $\mu = \varepsilon$ of the chemical potential in the reservoirs Ω and Ω'. This transformation law is a condition on

$$\mathcal{M}(\varepsilon = 0) \equiv \mathcal{M} \tag{7.272}$$

at the single-particle energy $\varepsilon = 0$. Imposing this condition on the singular-value decomposition (7.239a) gives the singular-value decomposition[55]

$$\begin{aligned}
&\mathcal{M} = \begin{pmatrix} \mathsf{u} & 0 \\ 0 & \mathsf{u} \end{pmatrix} \begin{pmatrix} \cosh \mathsf{x} & \sinh \mathsf{x} \\ \sinh \mathsf{x} & \cosh \mathsf{x} \end{pmatrix} \begin{pmatrix} \mathsf{v} & 0 \\ 0 & \mathsf{v} \end{pmatrix}, \\
&\mathsf{u}, \mathsf{v} \in \mathrm{U}(N), \\
&\mathsf{x} \equiv \mathrm{diag}\begin{pmatrix} x_1 & \cdots & x_N \end{pmatrix}, \qquad x_1,\cdots,x_N \in \mathbb{R}.
\end{aligned} \tag{7.273}$$

The singular-value decomposition (7.273) is uncountably one-to-many since it is invariant under

$$\mathsf{u} \mapsto \mathsf{u}\,\mathsf{w}, \qquad \mathsf{v} \mapsto \mathsf{w}^{\dagger}\,\mathsf{v}, \qquad \forall \mathsf{w} \in [\mathrm{U}(1)]^N. \tag{7.274}$$

It follows that the right-hand side of Eq. (7.273) depends on

$$2N^2 + N - N = 2N^2 \tag{7.275a}$$

independent real-valued parameters, a number that coincides with the dimensionality of the non-compact Lie group

$$\mathrm{GL}(N,\mathbb{C}) \tag{7.275b}$$

[see the comment after Eq. (7.477) from **Exercise** 7.6]. Thus, we make the identifications

$$G^{\star} = \mathrm{GL}(N,\mathbb{C}) = \mathbb{C} \times \mathrm{SL}(N,\mathbb{C}) \tag{7.275c}$$

and

$$H = \mathrm{U}(N) = \mathrm{U}(1) \times \mathrm{SU}(N) \tag{7.275d}$$

[55] Observe that it is not possible anymore to choose $x_j \geq 0$, as it is the same $\mathbb{C}$ number originating from the matrix elements of u and v that multiplies $\cosh x_j$ and $\sinh x_j$.

in the second column of **Table** 7.4. Accordingly, the non-compact symmetric space associated with such random transfer matrices is the coset space

$$\mathrm{G}^{\star}/\mathrm{H} = \mathrm{GL}(N,\mathbb{C})/\mathrm{U}(N). \tag{7.275e}$$

One interpretation for quotienting by the subgroup $\mathrm{U}(N)$ follows from the observation that

$$\mathcal{M}^{\dagger}\,\mathcal{M} = \mathcal{U}^{\dagger}\,\mathcal{X}^{2}\,\mathcal{U} \tag{7.276}$$

is independent of the factor $\mathcal{V}$ that enters the counterpart to the singular-value decomposition (7.273). But the $2N\times 2N$ positive definite matrix $\mathcal{M}^{\dagger}\,\mathcal{M}$ is all that is needed to study the statistics of the transmission eigenvalues according to Eq. (7.230b). The joint probability distribution

$$P(x_{1},\cdots,x_{N}) \tag{7.277a}$$

for the N independent Lyapunov exponents entering the dimensionless charge conductance

$$g_{\mathrm{L}} = \sum_{j=1}^{N}\frac{1}{\cosh^{2}x_{j}}, \qquad x_{j}\in\mathbb{R}, \qquad j=1,\cdots,N \tag{7.277b}$$

for the symmetry class AIII is then proportional to the Jacobian (7.194c) with its radial coordinates $\boldsymbol{x}$ substituted by two times the Lyapunov exponents ($\boldsymbol{x}\to 2\boldsymbol{x}$) and the roots on the line A of **Table** 7.4, namely,

$$P(x_{1},\cdots,x_{N}) \propto \prod_{1\le j<k\le N} \left|2\sinh\left(x_{j}-x_{k}\right)\right|^{2}. \tag{7.277c}$$

The singular-value decomposition (7.273) does not allow to choose all Lyapunov exponents entering the Landauer conductance (7.277b) to be positive. Correspondingly, $P(x_{1},\cdots,x_{N})$ is not invariant under the transformation

$$x_{j}\mapsto\zeta_{j}\,x_{j}, \qquad \zeta_{j}=\pm 1, \qquad j=1,\cdots,N. \tag{7.278}$$

Symmetry class BDI We choose $M=N$ with $N=1,2,3,\cdots$. Any transfer matrix belongs to the non-compact Lie group $\mathrm{U}(N,N)$. A chiral transformation is implemented by demanding that

$$\Gamma\,\mathcal{M}(\varepsilon)\,\Gamma^{-1} = \mathcal{M}(-\varepsilon), \qquad \Gamma := \mathbb{1}_{N}\otimes\tau_{1}, \tag{7.279}$$

for the value $\mu=\varepsilon$ of the chemical potential in the reservoirs Ω and Ω'. This transformation law is a condition on

$$\mathcal{M}(\varepsilon=0)\equiv\mathcal{M} \tag{7.280}$$

at the single-particle energy $\varepsilon=0$. Imposing this condition together with the condition

$$\mathcal{T}\,\mathcal{M}^{*}\,\mathcal{T}^{-1} = \mathcal{M}, \qquad \mathcal{T} := \mathbb{1}_{N}\otimes\tau_{1} \tag{7.281}$$

for time-reversal symmetry (which holds for all ε, not only $\varepsilon = 0$) on the singular-value decomposition (7.259) gives the singular-value decomposition[56]

$$\begin{aligned} &\mathcal{M} = \begin{pmatrix} \mathsf{u} & 0 \\ 0 & \mathsf{u} \end{pmatrix} \begin{pmatrix} \cosh \mathsf{x} & \sinh \mathsf{x} \\ \sinh \mathsf{x} & \cosh \mathsf{x} \end{pmatrix} \begin{pmatrix} \mathsf{v} & 0 \\ 0 & \mathsf{v} \end{pmatrix}, \\ &\mathsf{u}, \mathsf{v} \in \mathrm{O}(N), \\ &\mathsf{x} \equiv \mathrm{diag}\left(x_1 \quad \cdots \quad x_N\right), \qquad x_1, \cdots, x_N \in \mathbb{R}. \end{aligned} \tag{7.282}$$

The singular-value decomposition (7.282) is countably one-to-many since we must restrict $\theta_j = 0, \pi$ for $j = 1, \cdots, N$ in the transformation (7.240) for it to leave Eq. (7.282) invariant. It follows that the right-hand side of Eq. (7.282) depends on

$$2 \times \frac{1}{2}(N-1)N + N = N^2 \tag{7.283a}$$

independent real-valued parameters, a number that coincides with the dimensionality of the non-compact Lie group

$$\mathrm{GL}(N, \mathbb{R}) \tag{7.283b}$$

[see the comment after Eq. (7.477) from **Exercise** 7.6]. Hence, we make the identifications

$$\mathrm{G}^\star = \mathrm{GL}(N, \mathbb{R}) = \mathbb{R} \times \mathrm{SL}(N, \mathbb{R}) \tag{7.283c}$$

and

$$\mathrm{H} = \mathrm{O}(N) \tag{7.283d}$$

in the second column of **Table** 7.4. Accordingly, the non-compact symmetric space associated with such random transfer matrices is the coset space

$$\mathrm{G}^\star/\mathrm{H} = \mathrm{GL}(N, \mathbb{R})/\mathrm{O}(N). \tag{7.283e}$$

One interpretation for quotienting by the subgroup $\mathrm{O}(N)$ follows from the observation that

$$\mathcal{M}^\dagger \mathcal{M} = \mathcal{U}^\dagger \mathcal{X}^2 \mathcal{U} \tag{7.284}$$

is independent of the factor $\mathcal{V}$ that enters the counterpart to the singular-value decomposition (7.282). But the $2N \times 2N$ positive definite matrix $\mathcal{M}^\dagger \mathcal{M}$ is all that is needed to study the statistics of the transmission eigenvalues according to Eq. (7.230b). The joint probability distribution

$$P(x_1, \cdots, x_N) \tag{7.285a}$$

[56] Observe that it is not possible anymore to choose $x_j \geq 0$, as it is the same $\mathbb{C}$ number originating from the matrix elements of u and v that multiplies $\cosh x_j$ and $\sinh x_j$.

for the N independent Lyapunov exponents entering the dimensionless charge conductance

$$g_{\mathrm{L}}=\sum_{j=1}^{N}\frac{1}{\cosh^{2}x_{j}}, \qquad x_{j}\in\mathbb{R}, \qquad j=1,\cdots,N \tag{7.285b}$$

for the symmetry class BDI is then proportional to the Jacobian (7.194c) with its radial coordinates $\boldsymbol{x}$ substituted by two times the Lyapunov exponents ($\boldsymbol{x}\to 2\boldsymbol{x}$) and the roots on the line AI of **Table** 7.4, namely,

$$P(x_{1},\cdots,x_{N})\propto\prod_{1\leq j<k\leq N}\left|2\sinh\left(x_{j}-x_{k}\right)\right|^{1}. \tag{7.285c}$$

The singular-value decomposition (7.282) does not allow to choose all Lyapunov exponents entering the Landauer conductance (7.285b) to be positive. Correspondingly, $P(x_{1},\cdots,x_{N})$ is not invariant under the transformation

$$x_{j}\mapsto\zeta_{j}\,x_{j}, \qquad \zeta_{j}=\pm1, \qquad j=1,\cdots,N. \tag{7.286}$$

Symmetry class CII We choose $M=2N$ with $N=1,2,3,\cdots$. Any transfer matrix belongs to the non-compact Lie group $\mathrm{U}(2N,2N)$. A chiral transformation is implemented by demanding that

$$\Gamma\,\mathcal{M}(\varepsilon)\,\Gamma=\mathcal{M}(-\varepsilon), \qquad \Gamma:=\mathbb{1}_{N}\otimes\sigma_{0}\otimes\tau_{1}, \tag{7.287}$$

for the value $\mu=\varepsilon$ of the chemical potential in the reservoirs Ω and Ω'. This transformation law is a condition on

$$\mathcal{M}(\varepsilon=0)\equiv\mathcal{M} \tag{7.288}$$

at the single-particle energy $\varepsilon=0$. Imposing this condition together with the condition

$$\mathcal{T}\mathcal{M}^{*}\mathcal{T}^{-1}=\mathcal{M}, \qquad \mathcal{T}:=\mathbb{1}_{N}\otimes\mathrm{i}\sigma_{2}\otimes\tau_{1} \tag{7.289}$$

for time-reversal symmetry (which holds for all ε, not only $\varepsilon=0$) on the singular-value decomposition (7.265) gives the singular-value decomposition[57]

$$\begin{aligned}&\mathcal{M}=\begin{pmatrix}\mathsf{u}&0\\0&\mathsf{u}\end{pmatrix}\begin{pmatrix}\cosh\mathsf{x}&\sinh\mathsf{x}\\\sinh\mathsf{x}&\cosh\mathsf{x}\end{pmatrix}\begin{pmatrix}\mathsf{v}&0\\0&\mathsf{v}\end{pmatrix},\\&\mathsf{u}=\mathsf{u}^{\mathsf{R}},\mathsf{v}=\mathsf{v}^{\mathsf{R}}\in\mathrm{Sp}(2N),\\&\mathsf{x}\equiv\mathrm{diag}\left(x_{1}\quad x_{1}\quad\cdots\quad x_{N}\quad x_{N}\right), \qquad x_{1},\cdots,x_{N}\in\mathbb{R}.\end{aligned} \tag{7.290}$$

[57] The Kramers' degeneracy is made explicit in the diagonal matrix x. Observe that it is not possible anymore to choose $x_{j}\geq0$, as it is the same $\mathbb{C}$ number originating from the matrix elements of u and v that multiplies $\cosh x_{j}$ and $\sinh x_{j}$.

The singular-value decomposition (7.290) is uncountably one-to-many since it is invariant under

$$\mathsf{u} \mapsto \mathsf{u}\,\mathsf{w}, \qquad \mathsf{v} \mapsto \mathsf{w}^{\dagger}\,\mathsf{v}, \qquad \forall \mathsf{w} \in [\mathrm{SU}(2)]^{N}. \tag{7.291}$$

It follows that the right-hand side of Eq. (7.290) depends on

$$2 \times N(2N+1) + N - 3N = 4N^2 \tag{7.292a}$$

independent real-valued parameters, a number that coincides with the dimensionality of the non-compact Lie group

$$\mathrm{U}^{\star}(2N) \tag{7.292b}$$

[see Eq. (7.481) from **Exercise** 7.6]. Hence, we make the identifications

$$\mathrm{G}^{\star} = \mathrm{U}^{\star}(2N) = \mathbb{C} \times \mathrm{SU}^{\star}(2N) \tag{7.292c}$$

and

$$\mathrm{H} \cong \mathrm{Sp}(2N) \tag{7.292d}$$

in the second column of **Table** 7.4. Accordingly, the non-compact symmetric space associated with such random transfer matrices is the coset space

$$\mathrm{G}^{\star}/\mathrm{H} \cong \mathrm{U}^{\star}(2N)/\mathrm{Sp}(2N). \tag{7.292e}$$

One interpretation for quotienting by the subgroup $\mathrm{Sp}(2N)$ follows from the observation that

$$\mathcal{M}^{\dagger}\,\mathcal{M} = \mathcal{U}^{\dagger}\,\mathcal{X}^{2}\,\mathcal{U} \tag{7.293}$$

is independent of the factor $\mathcal{V}$ that enters the counterpart to the singular-value decomposition (7.290). But the $4N \times 4N$ positive definite matrix $\mathcal{M}^{\dagger}\,\mathcal{M}$ is all that is needed to study the statistics of the transmission eigenvalues according to Eq. (7.230b). The joint probability distribution

$$P(x_1, \cdots, x_N) \tag{7.294a}$$

for the N independent Lyapunov exponents entering the dimensionless charge conductance

$$g_{\mathrm{L}} = 2 \times \sum_{j=1}^{N} \frac{1}{\cosh^2 x_j}, \qquad x_j \in \mathbb{R}, \qquad j = 1, \cdots, N \tag{7.294b}$$

for the symmetry class CII is then proportional to the Jacobian (7.194c) with its radial coordinates $\boldsymbol{x}$ substituted by two times the Lyapunov exponents ($\boldsymbol{x} \to 2\boldsymbol{x}$) and the roots on the line AII of **Table** 7.4, namely,

$$P(x_1, \cdots, x_N) \propto \prod_{1 \leq j < k \leq N} \left|2 \sinh\left(x_j - x_k\right)\right|^4. \tag{7.294c}$$

The multiplicative factor 2 on the right-hand side of the Landauer conductance in the symmetry class CII originates from the Kramers' degeneracy. The singular-value decomposition (7.290) does not allow to choose all Lyapunov exponents entering the Landauer conductance (7.294b) to be positive. Correspondingly, $P(x_1,\cdots,x_N)$ is not invariant under the transformation

$$x_j \mapsto \zeta_j\, x_j, \qquad \zeta_j = \pm 1, \qquad j = 1,\cdots,N. \tag{7.295}$$

Symmetry class D We choose $M = 2N$ with $N = 1,2,3,\cdots$. Any transfer matrix belongs to the non-compact Lie group $\mathrm{U}(2N,2N)$. Charge conjugation is implemented by demanding that, in the Majorana representation of the symmetry class D (**Exercise** 7.22),

$$\mathcal{M}^*(\varepsilon) = \mathcal{M}(-\varepsilon) \tag{7.296}$$

for the value $\mu = \varepsilon$ of the chemical potential in the reservoirs Ω and Ω'. This transformation law is a condition on

$$\mathcal{M}(\varepsilon = 0) \equiv \mathcal{M} \tag{7.297}$$

at the single-particle energy $\varepsilon = 0$. Imposing this condition on the singular-value decomposition (7.239a) gives the singular-value decomposition[58]

$$\begin{aligned}
&\mathcal{M} = \begin{pmatrix} \mathsf{u} & 0 \\ 0 & \mathsf{u}' \end{pmatrix} \begin{pmatrix} \cosh \mathsf{x} & \sinh \mathsf{x} \\ \sinh \mathsf{x} & \cosh \mathsf{x} \end{pmatrix} \begin{pmatrix} \mathsf{v} & 0 \\ 0 & \mathsf{v}' \end{pmatrix}, \\
&\mathsf{u} = \mathsf{u}^*, \mathsf{u}' = \mathsf{u}'^*, \mathsf{v} = \mathsf{v}^*, \mathsf{v}' = \mathsf{v}'^* \in \mathrm{O}(2N), \\
&\mathsf{x} \equiv \mathrm{diag}\begin{pmatrix} x_1 & \cdots & x_{2N} \end{pmatrix}, \qquad 0 \le x_1,\cdots,x_{2N} < \infty.
\end{aligned} \tag{7.298}$$

The singular-value decomposition (7.298) is countably one-to-many since it is invariant under

$$\begin{aligned}
&\mathsf{u} \mapsto \mathsf{u}\,\mathsf{w}, \qquad \mathsf{u}' \mapsto \mathsf{u}'\,\mathsf{w}, \qquad \mathsf{v} \mapsto \mathsf{w}^{\mathsf{T}}\,\mathsf{v}, \qquad \mathsf{v}' \mapsto \mathsf{w}^{\mathsf{T}}\,\mathsf{v}', \\
&\forall \mathsf{w} = \mathsf{w}^* \in [\mathrm{O}(1)]^{2N} \subset [\mathrm{U}(1)]^{2N}.
\end{aligned} \tag{7.299}$$

It follows that the right-hand side of Eq. (7.298) depends on

$$4 \times \frac{1}{2}(2N-1)2N + 2N = 8N^2 - 2N \tag{7.300a}$$

independent real-valued parameters, a number that coincides with the dimensionality of the non-compact Lie group

$$\mathrm{O}(2N,2N) \subset \mathrm{U}(2N,2N) \tag{7.300b}$$

[see Eq. (7.483) from **Exercise** 7.6]. Hence, we make the identifications

$$\mathrm{G}^\star = \mathrm{O}(2N,2N) \tag{7.300c}$$

[58] Observe that it is always possible to choose v and v' relative to u and u' such that $0 \le x_1,\cdots,x_N < \infty$.

and

$$\mathrm{H} = \mathrm{O}(2N) \times \mathrm{O}(2N) \tag{7.300d}$$

in the second column of **Table** 7.4. Accordingly, the symmetric space associated with such random transfer matrices is the coset space

$$\mathrm{G}^{\star}/\mathrm{H} = \mathrm{O}(2N, 2N)/\left[\mathrm{O}(2N) \times \mathrm{O}(2N)\right]. \tag{7.300e}$$

One interpretation for quotienting by the subgroup $\mathrm{O}(2N) \times \mathrm{O}(2N)$ follows from the observation that

$$\mathcal{M}^{\dagger}\,\mathcal{M} = \mathcal{U}^{\dagger}\,\mathcal{X}^{2}\,\mathcal{U} \tag{7.301}$$

is independent of the factor $\mathcal{V}$ that enters the counterpart to the singular-value decomposition (7.298). But the $4N \times 4N$ positive definite matrix $\mathcal{M}^{\dagger}\,\mathcal{M}$ is all that is needed to study the statistics of the transmission eigenvalues according to Eq. (7.230b). The joint probability distribution

$$P(x_1, \cdots, x_{2N}) \tag{7.302a}$$

for the $2N$ independent Lyapunov exponents entering the dimensionless *thermal* conductance[59]

$$g_{\mathrm{L}} = \sum_{j=1}^{2N} \frac{1}{\cosh^{2} x_j}, \qquad 0 \le x_j < \infty, \qquad j = 1, \cdots, 2N \tag{7.302b}$$

for the symmetry class D is then proportional to the Jacobian (7.194c) with its radial coordinates $\boldsymbol{x}$ substituted by two times the Lyapunov exponents ($\boldsymbol{x} \to 2\boldsymbol{x}$) and the roots on the line BDI of **Table** 7.4, namely,

$$\begin{aligned} P(x_1, \cdots, x_{2N}) &\propto \prod_{1 \le j < k \le 2N} \left|4 \sinh\left(x_j - x_k\right) \sinh\left(x_j + x_k\right)\right|^{1} \\ &\propto \prod_{1 \le j < k \le 2N} \left|\cosh(2\,x_j) - \cosh(2\,x_k)\right|. \end{aligned} \tag{7.302c}$$

The singular-value decomposition (7.298) allows to choose all Lyapunov exponents entering the Landauer conductance (7.302b) to be positive. Correspondingly, $P(x_1, \cdots, x_{2N})$ is invariant under the transformation

$$x_j \mapsto \zeta_j\, x_j, \qquad \zeta_j = \pm 1, \qquad j = 1, \cdots, 2N. \tag{7.303}$$

[59] The concept of Landauer conductance was defined for the transport of electrical charge obeying the continuity equation. In the mean-field treatment of a superconductor, total electric charge is not conserved, that is there is no continuity equation relating the local charge density to the local current density. However, the total energy is conserved. This conservation law is implemented locally by a continuity equation for the flow of energy. If so, the notion of a thermal conductance is meaningful. In turn, a scattering theory for thermal transport applied to Bogoliubov-de-Gennes quasiparticles can be developed in analogy to that for charge transport.

Symmetry class C We choose $M = 2N$ with $N = 1, 2, 3, \cdots$. Any transfer matrix belongs to the non-compact Lie group $\mathrm{U}(2N, 2N)$. Charge conjugation is implemented by demanding that

$$\mathcal{C}\,\mathcal{M}^{*}(\varepsilon)\,\mathcal{C}^{-1} = \mathcal{M}(-\varepsilon), \qquad \mathcal{C} := \mathbb{1}_{N} \otimes \mathrm{i}\kappa_{2} \otimes \tau_{0}, \tag{7.304}$$

for the value $\mu = \varepsilon$ of the chemical potential in the reservoirs Ω and Ω'. This transformation law is a condition on

$$\mathcal{M}(\varepsilon = 0) \equiv \mathcal{M} \tag{7.305}$$

at the single-particle energy $\varepsilon = 0$. Imposing this condition gives the singular-value decomposition[60]

$$\begin{aligned}
&\mathcal{M} = \begin{pmatrix} \mathsf{u} & 0 \\ 0 & \mathsf{u}' \end{pmatrix} \begin{pmatrix} \cosh \mathsf{x} & \sinh \mathsf{x} \\ \sinh \mathsf{x} & \cosh \mathsf{x} \end{pmatrix} \begin{pmatrix} \mathsf{v} & 0 \\ 0 & \mathsf{v}' \end{pmatrix}, \\
&\mathsf{u} = \mathsf{u}^{\mathsf{R}}, \mathsf{u}' = \mathsf{u}'^{\mathsf{R}}, \mathsf{v} = \mathsf{v}^{\mathsf{R}}, \mathsf{v}' = \mathsf{v}'^{\mathsf{R}} \in \mathrm{Sp}(2N), \\
&\mathsf{x} \equiv \mathrm{diag}\begin{pmatrix} x_{1} & x_{1} & \cdots & x_{N} & x_{N} \end{pmatrix}, \qquad 0 \leq x_{1}, \cdots, x_{N} < \infty.
\end{aligned} \tag{7.306}$$

The singular-value decomposition (7.306) is uncountably one-to-many since it is invariant under

$$\begin{aligned}
&\mathsf{u} \mapsto \mathsf{u}\,\mathsf{w}, \qquad \mathsf{u}' \mapsto \mathsf{u}'\,\mathsf{w}, \qquad \mathsf{v} \mapsto \mathsf{w}^{\dagger}\,\mathsf{v}, \qquad \mathsf{v}' \mapsto \mathsf{w}^{\dagger}\,\mathsf{v}', \\
&\forall \mathsf{w} \in [\mathrm{SU}(2)]^{N}.
\end{aligned} \tag{7.307}$$

It follows that the right-hand side of Eq. (7.306) depends on

$$4 \times N(2N+1) + N - 3N = 8N^{2} + 2N \tag{7.308a}$$

independent real-valued parameters, a number that coincides with the dimensionality of the non-compact Lie group

$$\mathrm{Sp}(2N, 2N) \tag{7.308b}$$

[see Eq. (7.488) from **Exercise** 7.6]. Hence, we make the identifications

$$\mathrm{G}^{\star} = \mathrm{Sp}(2N, 2N) \tag{7.308c}$$

and

$$\mathrm{H} = \mathrm{Sp}(2N) \times \mathrm{Sp}(2N) \tag{7.308d}$$

in the second column of **Table** 7.4. Accordingly, the symmetric space associated with such random transfer matrices is the coset space

$$\mathrm{G}^{\star}/\mathrm{H} = \mathrm{Sp}(2N, 2N)/\left[\mathrm{Sp}(2N) \times \mathrm{Sp}(2N)\right]. \tag{7.308e}$$

[60] The effective Kramers' degeneracy is made explicit in the diagonal matrix x. This degeneracy originates from spin-rotation symmetry. Observe that it is always possible to choose v and v' relative to u and u' such that $0 \leq x_{1}, \cdots, x_{N} < \infty$.

One interpretation for quotienting by the subgroup $\mathrm{Sp}(2N) \times \mathrm{Sp}(2N)$ follows from the observation that

$$\mathcal{M}^{\dagger}\,\mathcal{M} = \mathcal{U}^{\dagger}\,\mathcal{X}^{2}\,\mathcal{U} \tag{7.309}$$

is independent of the factor $\mathcal{V}$ that enters the counterpart to the singular-value decomposition (7.306). But the $4N \times 4N$ positive definite matrix $\mathcal{M}^{\dagger}\,\mathcal{M}$ is all that is needed to study the statistics of the transmission eigenvalues according to Eq. (7.230b). The joint probability distribution

$$P(x_1, \cdots, x_N) \tag{7.310a}$$

for the N independent Lyapunov exponents entering the dimensionless *thermal* conductance

$$g_{\mathrm{L}} = 2 \times \sum_{j=1}^{N} \frac{1}{\cosh^2 x_j}, \qquad 0 \leq x_j < \infty, \qquad j = 1, \cdots, N \tag{7.310b}$$

for the symmetry class C is then proportional to the Jacobian (7.194c) with its radial coordinates $\boldsymbol{x}$ substituted by two times the Lyapunov exponents ($\boldsymbol{x} \to 2\boldsymbol{x}$) and the roots on the line CII of **Table** 7.4, namely,

$$\begin{aligned} P(x_1, \cdots, x_N) &\propto \left[\prod_{1 \leq j < k \leq N} \left|4 \sinh\left(x_j - x_k\right) \sinh\left(x_j + x_k\right)\right|^4\right] \\ &\quad \times \left[\prod_{j=1}^{N} \left|2 \sinh(2\,x_j)\right|^3\right] \\ &\propto \left[\prod_{1 \leq j < k \leq N} \left|\cosh(2\,x_j) - \cosh(2\,x_k)\right|^4\right] \left[\prod_{j=1}^{N} \left|\sinh(2\,x_j)\right|^3\right]. \end{aligned} \tag{7.310c}$$

The multiplicative factor 2 on the right-hand side of the thermal Landauer conductance in the symmetry class C originates from the effective Kramers' degeneracy in the Bogoliubov-de-Gennes particle-hole grading. This degeneracy originates from the spin-rotation symmetry as was shown in Section 7.3.2. Contrary to the standard and chiral symmetry classes, having or not having spin-rotation symmetry when time-reversal symmetry is explicitly broken matters for the transfer matrices in the superconducting symmetry classes. The singular-value decomposition (7.306) allows to choose all Lyapunov exponents entering the Landauer conductance (7.310b) to be positive. Correspondingly, $P(x_1, \cdots, x_N)$ is invariant under the transformation

$$x_j \mapsto \zeta_j\, x_j, \qquad \zeta_j = \pm 1, \qquad j = 1, \cdots, N. \tag{7.311}$$

Symmetry class DIII We choose $M = 4N$ with $N = 1, 2, 3, \cdots$. We reserve the unit Pauli matrix ρ_0 (σ_0) and the three Pauli matrices $\boldsymbol{\rho}$ ($\boldsymbol{\sigma}$) for the

Bogoliubov-de-Gennes particle-hole grading (spin-1/2 grading). Charge conjugation is implemented by demanding that (**Exercise** 7.22)

$$\mathcal{M}^*(\varepsilon) = \mathcal{M}(-\varepsilon) \tag{7.312}$$

for the value $\mu = \varepsilon$ of the chemical potential in the reservoirs Ω and Ω'. This transformation law is a condition on

$$\mathcal{M}(\varepsilon = 0) \equiv \mathcal{M} \tag{7.313}$$

at the single-particle energy $\varepsilon = 0$. If one imposes the additional constraint

$$\mathcal{T}\mathcal{M}^*\mathcal{T}^{-1} = \mathcal{M}, \qquad \mathcal{T} := \mathbb{1}_N \otimes \rho_1 \otimes \mathrm{i}\sigma_2 \otimes \tau_1 \tag{7.314}$$

that represents time-reversal symmetry in the Majorana basis (7.312) [see Eq. (7.658)], it is shown in **Exercise** 7.22 that $\mathcal{M} \in \mathrm{U}(4N, 4N)$ is in one-to-one correspondence with a matrix $M_{\mathrm{c}} \in \mathrm{O}(4N, \mathbb{C})$. There follows the singular-value decomposition[61]

$$\begin{aligned} &M_{\mathrm{c}} = U \begin{pmatrix} \cosh \mathsf{x} & +\mathrm{i}\sinh \mathsf{x} \\ -\mathrm{i}\sinh \mathsf{x} & \cosh \mathsf{x} \end{pmatrix} V, \\ &U, V \in \mathrm{O}(4N), \\ &\mathsf{x} \equiv \mathrm{diag}\begin{pmatrix} x_1 & \cdots & x_{2N} \end{pmatrix}, \qquad 0 \leq x_1, \cdots, x_{2N} < \infty. \end{aligned} \tag{7.315}$$

The singular-value decomposition (7.315) is uncountably one-to-many since it is invariant under

$$U \mapsto U\,W, \qquad V \mapsto W^{\mathsf{T}}\,V, \qquad W := \begin{pmatrix} \cos \mathsf{z} & \sin \mathsf{z} \\ -\sin \mathsf{z} & \cos \mathsf{z} \end{pmatrix} \in \mathrm{O}(4N), \tag{7.316}$$

with z any $2N \times 2N$ real-valued and diagonal matrix. As it should be, the right-hand side of Eq. (7.315) depends on

$$2 \times \frac{1}{2}(4N - 1)4N + 2N - 2N = 16N^2 - 4N \tag{7.317a}$$

independent real-valued parameters, a number that coincides with the dimensionality of the non-compact Lie group

$$\mathrm{O}(4N, \mathbb{C}) \tag{7.317b}$$

[see Eq. (7.482) from **Exercise** 7.6]. Hence, we make the identifications

$$\mathrm{G}^\star = \mathrm{O}(4N, \mathbb{C}) \tag{7.317c}$$

and

$$\mathrm{H} = \mathrm{O}(4N) \tag{7.317d}$$

[61] Observe that it is always possible to choose V relative to U such that $0 \leq x_1, \cdots, x_{2N} < \infty$. Moreover, observe that $\mathcal{M}$ is a $8N \times 8N$ matrix, while M_{c} is a $4N \times 4N$ matrix. If follows that the $2N$ independent Lyapunov exponents of M_{c} appear as $2N$ twofold degenerate pairs of Lyapunov exponents of $\mathcal{M}$. This is how Kramers' degeneracy manifests itself for the transfer matrix in the symmetry class DIII.

in the second column of **Table** 7.4. Accordingly, the symmetric space associated with such random transfer matrices is the coset space

$$\mathrm{G}^{\star}/\mathrm{H} = \mathrm{O}(4N,\mathbb{C})/\mathrm{O}(4N). \tag{7.317e}$$

One interpretation for quotienting by the subgroup $\mathrm{O}(4N)$ follows from the observation that

$$\mathcal{M}^{\dagger}\,\mathcal{M} = \mathcal{U}^{\dagger}\,\mathcal{X}^{2}\,\mathcal{U} \tag{7.318}$$

is independent of the factor $\mathcal{V}$ that enters the counterpart to the singular-value decomposition (7.315). But the $8N \times 8N$ positive definite matrix $\mathcal{M}^{\dagger}\,\mathcal{M}$ is all that is needed to study the statistics of the transmission eigenvalues according to Eq. (7.230b). The joint probability distribution

$$P(x_1,\cdots,x_{2N}) \tag{7.319a}$$

for the $2N$ independent Lyapunov exponents entering the dimensionless *thermal* conductance

$$g_{\mathrm{L}} = \sum_{j=1}^{2N} \frac{1}{\cosh^2 x_j}, \qquad 0 \le x_j < \infty, \qquad j = 1,\cdots,2N \tag{7.319b}$$

for the symmetry class DIII is then proportional to the Jacobian (7.194c) with its radial coordinates $\boldsymbol{x}$ substituted by two times the Lyapunov exponents ($\boldsymbol{x} \to 2\boldsymbol{x}$) and the roots on the line D of **Table** 7.4, namely,

$$\begin{aligned} P(x_1,\cdots,x_{2N}) &\propto \prod_{1\le j<k\le 2N} \left|4\sinh\left(x_j - x_k\right)\sinh\left(x_j + x_k\right)\right|^2 \\ &\propto \prod_{1\le j<k\le 2N} \left|\cosh(2\,x_j) - \cosh(2\,x_k)\right|^2 . \end{aligned} \tag{7.319c}$$

The singular-value decomposition (7.315) allows to choose all Lyapunov exponents entering the Landauer conductance (7.319b) to be positive. Correspondingly, $P(x_1,\cdots,x_{2N})$ is invariant under the transformation

$$x_j \mapsto \zeta_j\,x_j, \qquad \zeta_j = \pm 1, \qquad j = 1,\cdots,2N. \tag{7.320}$$

Symmetry class CI We choose $M = 2N$ with $N = 1,2,3,\cdots$. Any transfer matrix belongs to the non-compact Lie group $\mathrm{U}(2N,2N)$. Charge conjugation is implemented by demanding that

$$\mathcal{C}\,\mathcal{M}^{*}(\varepsilon)\,\mathcal{C}^{-1} = \mathcal{M}(-\varepsilon), \qquad \mathcal{C} := \mathbb{1}_N \otimes \mathrm{i}\kappa_2 \otimes \tau_0, \tag{7.321}$$

for the value $\mu = \varepsilon$ of the chemical potential in the reservoirs Ω and Ω'. This transformation law is a condition on

$$\mathcal{M}(\varepsilon = 0) \equiv \mathcal{M} \tag{7.322}$$

at the single-particle energy $\varepsilon = 0$. Imposing this condition gives the singular-value decomposition (7.306). It is shown in **Exercise** 7.22.4 that if the condition

$$\mathcal{T}\mathcal{M}^{*}\mathcal{T}^{-1} = \mathcal{M} \qquad \mathcal{T} := \mathbb{1}_{N} \otimes \kappa_{0} \otimes \tau_{1}, \tag{7.323}$$

for time-reversal symmetry is also imposed, then $\mathcal{M} \in \mathrm{U}(2N, 2N)$ is in one-to-one correspondence with a matrix $M_{\mathrm{c}} \in \mathrm{Sp}(2N, \mathbb{C})$. There follows the singular-value decomposition[62]

$$\begin{aligned} &M_{\mathrm{c}} = U \begin{pmatrix} \cosh \mathsf{x} & +\mathrm{i}\sinh \mathsf{x} \\ -\mathrm{i}\sinh \mathsf{x} & \cosh \mathsf{x} \end{pmatrix} V, \\ &U, V \in \mathrm{Sp}(2N), \\ &\mathsf{x} \equiv \mathrm{diag}\left(x_{1} \quad \cdots \quad x_{N}\right), \qquad 0 \leq x_{1}, \cdots, x_{N} < \infty. \end{aligned} \tag{7.324}$$

The singular-value decomposition (7.324) is uncountably one-to-many since it is invariant under

$$U \mapsto U\,W, \qquad V \mapsto W^{\mathsf{T}}\,V, \qquad W := \begin{pmatrix} \cosh \mathsf{z} & +\mathrm{i}\sinh \mathsf{z} \\ -\mathrm{i}\sinh \mathsf{z} & \cosh \mathsf{z} \end{pmatrix} \in \mathrm{Sp}(2N), \tag{7.325}$$

with z any $N \times N$ real-valued and diagonal matrix. As it should be, the right-hand side of Eq. (7.324) depends on

$$2 \times N(2N+1) + N - N = 4N^{2} + 2N \tag{7.326a}$$

independent real-valued parameters, a number that coincides with the dimensionality of the non-compact Lie group

$$\mathrm{Sp}(2N, \mathbb{C}) \tag{7.326b}$$

[see Eq. (7.485) from **Exercise** 7.6]. Hence, we make the identifications

$$\mathrm{G}^{\star} = \mathrm{Sp}(2N, \mathbb{C}) \tag{7.326c}$$

and

$$\mathrm{H} = \mathrm{Sp}(2N) \tag{7.326d}$$

in the second column of **Table** 7.4. Accordingly, the symmetric space associated with such random transfer matrices is the coset space

$$\mathrm{G}^{\star}/\mathrm{H} = \mathrm{Sp}(2N, \mathbb{C})/\mathrm{Sp}(2N). \tag{7.326e}$$

One interpretation for quotienting by the subgroup $\mathrm{Sp}(2N)$ follows from the observation that

$$\mathcal{M}^{\dagger}\,\mathcal{M} = \mathcal{U}^{\dagger}\,\mathcal{X}^{2}\,\mathcal{U} \tag{7.327}$$

[62] Observe that it is always possible to choose V relative to U such that $0 \leq x_{1}, \cdots, x_{N} < \infty$. Moreover, observe that $\mathcal{M}$ is a $4N \times 4N$ matrix, while M_{c} is a $2N \times 2N$ matrix. If follows that the N independent Lyapunov exponents of M_{c} appear as N twofold degenerate pairs of Lyapunov exponents of $\mathcal{M}$. This is how spin-rotations symmetry manifests itself for the transfer matrix in the symmetry class CI.

is independent of the factor $\mathcal{V}$ that enters the counterpart to the singular-value decomposition (7.324). But the $4N \times 4N$ positive definite matrix $\mathcal{M}^{\dagger}\,\mathcal{M}$ is all that is needed to study the statistics of the transmission eigenvalues according to Eq. (7.230b). The joint probability distribution

$$P(x_1, \cdots, x_N) \tag{7.328a}$$

for the N independent Lyapunov exponents entering the dimensionless *thermal* conductance

$$g_{\mathrm{L}} = \sum_{j=1}^{N} \frac{1}{\cosh^2 x_j}, \qquad 0 \leq x_j < \infty, \qquad j = 1, \cdots, N \tag{7.328b}$$

for the symmetry class CI is then proportional to the Jacobian (7.194c) with its radial coordinates $\boldsymbol{x}$ substituted by two times the Lyapunov exponents ($\boldsymbol{x} \to 2\boldsymbol{x}$) and the roots on the line C of **Table** 7.4, namely,

$$\begin{aligned} P(x_1, \cdots, x_N) \propto & \prod_{1 \leq j < k \leq N} \left|4 \sinh\left(x_j - x_k\right) \sinh\left(x_j + x_k\right)\right|^2 \\ & \times \prod_{j=1}^{N} \left|2 \sinh(2\,x_j)\right|^2 \\ \propto & \left[\prod_{1 \leq j < k \leq N} \left|\cosh(2\,x_j) - \cosh(2\,x_k)\right|^2\right] \left[\prod_{j=1}^{N} \left|\sinh(2\,x_j)\right|^2\right]. \end{aligned} \tag{7.328c}$$

The singular-value decomposition (7.324) allows to choose all Lyapunov exponents entering the Landauer conductance (7.328b) to be positive. Correspondingly, $P(x_1, \cdots, x_N)$ is invariant under the transformation

$$x_j \mapsto \zeta_j\, x_j, \qquad \zeta_j = \pm 1, \qquad j = 1, \cdots, N. \tag{7.329}$$

Table 7.6 summarizes the correspondence between the Altland-Zirnbauer symmetry classes, the non-compact Lie group $\mathrm{G}^{\star}$ attached to the set $\{\mathcal{M}\}_{\mathrm{AZ}}$ of transfer matrices associated with an Altland-Zirnbauer symmetry class when the chemical potential of the reservoirs is fixed to be at the band center, the factor group H in the singular-value decomposition of a transfer matrix that falls out from the set $\{\mathcal{M}^{\dagger}\,\mathcal{M}\}_{\mathrm{AZ}}$ of Hermitean and positive definite matrices built from transfer matrices in an Altland-Zirnbauer symmetry class, the positive roots of the symmetric space $\mathrm{M}^{\star} \equiv \mathrm{G}^{\star}/\mathrm{H} \equiv \mathrm{M}^{-}$, the multiplicities of the positive roots, and the domain of definition of the Lyapunov exponents for each Altland-Zirnbauer symmetry class. **Table** 7.6 follows from **Table** 7.4 by a permutation of the ten lines in **Table** 7.4.

Table 7.6 *The tenfold way: Lyapunov exponents for the transfer matrices*

AZ	$M^\star \equiv G^\star/H$ for $\mathcal{M}$	$\mathfrak{Roots}_+$	m_s	m_o^-	m_o^+	m_l	x_j
A	$U(N,N)/[U(N)\times U(N)]$	$e_k \pm e_j, 2e_j$	0	2	2	1	$\mathbb{R}_+$
	$U(M,N)/[U(M)\times U(N)]$	$e_j, e_k \pm e_j, 2e_j$	2ν	2	2	1	$\mathbb{R}_+$
AI	$Sp(2N,\mathbb{R})/U(N)$	$e_k \pm e_j, 2e_j$	0	1	1	1	$\mathbb{R}_+$
AII	$O^\star(4N)/U(2N)$	$e_k \pm e_j, 2e_j$	0	4	4	1	$\mathbb{R}_+$
	$O^\star(4N+2)/U(2N+1)$	$e_j, e_k \pm e_j, 2e_j$	4	4	4	1	$\mathbb{R}_+$
AIII	$GL(N,\mathbb{C})/U(N)$	$e_k - e_j$	0	2	0	0	$\mathbb{R}$
BDI	$GL(N,\mathbb{R})/O(N)$	$e_k - e_j$	0	1	0	0	$\mathbb{R}$
CII	$U^\star(2N)/Sp(2N)$	$e_k - e_j$	0	4	0	0	$\mathbb{R}$
D	$O(2N,2N)/[O(2N)\times O(2N)]$	$e_k \pm e_j$	0	1	1	0	$\mathbb{R}_+$
	$O(2M,2N)/[O(2M)\times O(2N)]$	$e_j, e_k \pm e_j$	ν	1	1	0	$\mathbb{R}_+$
C	$Sp(2N,2N)/[Sp(2N)\times Sp(2N)]$	$e_k \pm e_j, 2e_j$	0	4	4	3	$\mathbb{R}_+$
	$Sp(2M,2N)/[Sp(2M)\times Sp(2N)]$	$e_j, e_k \pm e_j, 2e_j$	4ν	4	4	3	$\mathbb{R}_+$
DIII	$O(4N,\mathbb{C})/O(4N)$	$e_k \pm e_j$	0	2	2	0	$\mathbb{R}_+$
	$O(4N+1,\mathbb{C})/O(4N+1)$	$e_j, e_k \pm e_j$	2	2	2	0	$\mathbb{R}_+$
CI	$Sp(2N,\mathbb{C})/Sp(2N)$	$e_k \pm e_j, 2e_j$	0	2	2	2	$\mathbb{R}_+$

The joint probability distribution for the Lyapunov exponents $\{x_j\}$ of random transfer matrices $\{\mathcal{M}\}$ is proportional to the Jacobian for the radial coordinates of non-compact symmetric spaces with negative curvature. The first column defines the Altland-Zirnbauer (AZ) symmetry class through discrete symmetries (time-reversal symmetry or charge-conjugation or chiral spectral symmetries). The second column gives the symmetric space. This symmetric space follows from selecting the maximal isotropy subgroup H that leaves the measure for the random transfer matrices $\{\mathcal{M}\}$ invariant under left or right multiplication by any element of H. The third column gives the positive roots of the symmetric space. The next four columns give the multiplicities of the positive roots. The last column gives the domain of definition of a Lyapunov exponent x_j, whereby $\mathbb{R}_+ \subset \mathbb{R}$ denotes all nonnegative real numbers. The possibility that reservoirs Ω and Ω' differ by the number $\nu = |M - N|$ of channels is also allowed.

7.7 The Tenfold Way: Disordered Quantum Wires

7.7.1 Overview

Solving the stationary Schrödinger equation when space is one-dimensional allows to introduce notions from scattering theory such as reflection and transmission amplitudes, scattering phase shifts, resonances, and bound states. In this context, it is assumed that a static potential breaks translation symmetry in a bounded region Λ_L of size L of one-dimensional space (the static potential is a mere constant outside of this region). Scattering theory expresses the outgoing solutions of Schrödinger's equation in terms of two incoming plane waves impinging on the region Λ_L from the left and right, respectively.

One question that is not covered in the standard textbooks on quantum mechanics is the following. What is the dependence on L in the limit $L \to \infty$ of the squared magnitude of the transmission amplitude if the static potential belongs to some statistical ensemble of real-valued functions? Answering this question is fundamental to the phenomenon of *Anderson localization.* Here, *localization* refers to the situation when the conductance of a disordered region Λ_L of d-dimensional space decreases exponentially fast with its linear size L, that is the probability for a quantum particle to be transmitted across a disordered region Λ_L is exponentially suppressed with increasing L.

Anderson localization is necessarily non-perturbative if the static potential is to be treated as a perturbation to the kinetic energy in the stationary Schrödinger equation. To substantiate this claim, we begin with identifying the competing characteristic length scales. For electrons, the characteristic length scale is the Fermi wave length k_{F}^{-1} with k_{F} the Fermi wave vector. The characteristic length of the disorder is the mean free path ℓ, which for a static potential modelled by a set of point scatterers, is the average distance between the point scatterers as is shown in Figure 7.6. The static potential is weak if

$$\frac{\ell}{k_{\mathrm{F}}^{-1}} \gg 1 \iff (k_{\mathrm{F}}\,\ell)^{-1} \ll 1. \tag{7.330}$$

The conductivity σ, the proportionality constant that relates the electric current density[63] to an applied electric field, is an intrinsic quantity that is expected to become independent of L provided $\ell \leq L$. It is approximated by

$$\sigma_{\mathrm{D}} = n_e\,\frac{e^2}{\hbar}\,\frac{\ell}{k_{\mathrm{F}}} \tag{7.331a}$$

in Drude's theory of the conductivity of an electron gas with the electronic density n_e and e the electric charge of the electron. The conductance G, the proportionality constant that gives the electric current[64] that flows across Λ_L in Figure 7.6 upon multiplication of G with the applied potential difference across Λ_L, has the same units as $e^2/\hbar$. In the same approximation that gives (7.331a), the dependence of the dimensionless conductance

$$g_{\mathrm{D},d} := \frac{G_{\mathrm{D},d}}{e^2/\hbar} \tag{7.331b}$$

on L and d associated with the disordered region $\Lambda_L \subset \mathbb{R}^d$ is given by Ohm's law

$$g_{\mathrm{D},d}(L) = \sigma_{\mathrm{D}}\,L^{d-2}, \qquad L \geq \ell. \tag{7.331c}$$

[63] The local electric current density $\boldsymbol{j}$ is related to the local electric charge density ρ by the continuity equation $\partial_t\rho + \boldsymbol{\nabla}\cdot\boldsymbol{j} = 0$.

[64] The number of electric charges per unit time through a boundary ∂V that follows from integrating over the volume V the continuity equation $\partial_t\rho + \boldsymbol{\nabla}\cdot\boldsymbol{j} = 0$.

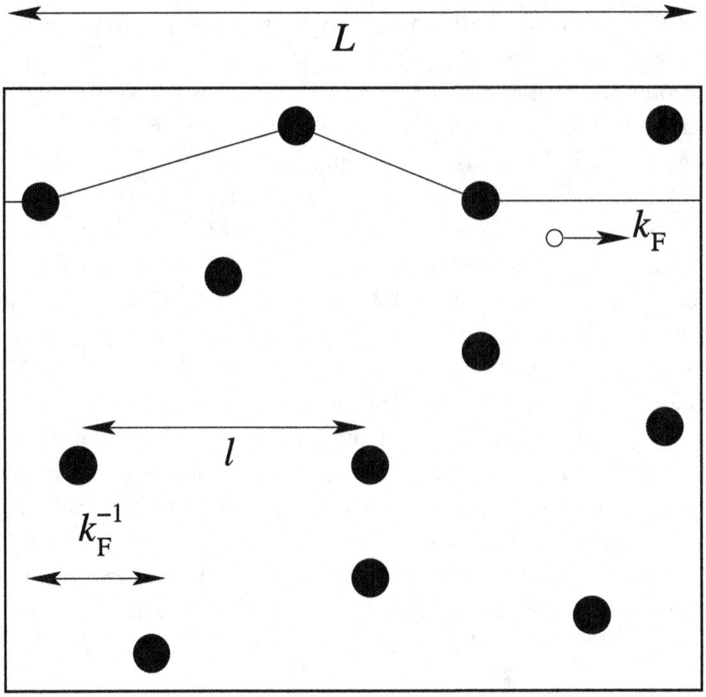

Figure 7.6 Region of linear size L in which a static random potential arises because of point scatterers (black circles). The average separation between these point scatterers defines the mean-free path ℓ. An electron (open circle) with characteristic momentum k_{F} (in units for which $\hbar = 1$) undergoes elastic scattering whenever it collides with a point scatterer. One possible classical trajectory of an electron that enters and leaves the disordered region is shown through a line. A necessary (but not sufficient) condition for scattering to be weak is that $k_{\mathrm{F}}\,\ell \gg 1$ holds.

The crossover from the semiclassical[65] functional scaling form (7.331c) to the exponential decay

$$g_{\mathrm{D},d}(L) = g_{\mathrm{UV}}\, e^{-L/\xi_d(\ell)}, \qquad L \geq \xi_d(\ell), \tag{7.332}$$

is necessarily non-perturbative in the expansion parameter $(k_{\mathrm{F}}\,\ell)^{-1} \ll 1$. The signature of this non-perturbative phenomenon is the emergence of the length scale $\xi_d(\ell)$, *the localization length.* When $d = 3$, it turns out that the asymptotic behavior (7.332) requires the disorder to be strong, that is the strong coupling regime

$$(k_{\mathrm{F}}\,\ell)^{-1} \gg 1 \tag{7.333}$$

[65] Quantum mechanics only enters the right-hand side of Eq. (7.331a) after expressing the mean-scattering time τ in terms of the mean free path ℓ and the electron momentum $\hbar\, k_{\mathrm{F}}$. A necessary but not always sufficient condition for this approximation to be self-consistent is that the weak coupling regime (7.330) holds.

must hold. When $d = 3$, quantum corrections to Ohm's law (7.331c) can be shown to be small in the weak coupling regime (7.330). The remarkable non-perturbative crossover to the exponential suppression (7.332) was deduced by Anderson in 1958[66] by treating the kinetic energy as a perturbation to the disorder potential. When $d = 1$, the Drude scaling law

$$g_{\mathrm{D},d=1}(L) = g_{\mathrm{UV}}\,\frac{\ell}{L}, \qquad L \geq \ell, \tag{7.334a}$$

with the constant g_{UV} set by the physics at short distances (ultra-violet momenta), is replaced by the exponential decay

$$g_{\mathrm{D},d=1}(L) = g_{\mathrm{UV}}\, e^{-L/\xi_{d=1}(\ell)} \qquad L \geq \xi_{d=1}(\ell), \qquad \xi_{d=1}(\ell) \propto \ell, \tag{7.334b}$$

for any nonvanishing ℓ in the standard universality classes A, AI, and AII.[67,68]

We are going to derive Eq. (7.334b) for the symmetry classes A, AI, AII, C, and CI. More importantly, we are going to show that the three chiral symmetry classes AIII, CII, and BDI, on the one hand, and the two Bogoliubov-de-Gennes symmetry classes D and DIII, on the other hand, evade exponential localization, that is Anderson localization, in one-dimensional space.

7.7.2 Hamiltonian

In order to treat all ten Altland-Zirnabuer symmetry classes in one go, we reserve the unit Pauli matrices τ_0, σ_0, ρ_0 and the triplets of Pauli matrices $\boldsymbol{\tau}$, $\boldsymbol{\sigma}$, $\boldsymbol{\rho}$ for the left- or right-moving, spin-1/2, and Bogoliubov-de-Gennes particle-hole grading, respectively. We shall also need a 2×2 grading generated by the unit 2×2 matrix κ_0 and the triplet of Pauli matrices $\boldsymbol{\kappa}$ that mixes two of the three 2×2 left- or right-moving, spin-1/2, and Bogoliubov-de-Gennes particle-hole gradings. Our model Hamiltonian is defined to be

$$\mathcal{H}(x) := \mathcal{K} + \mathcal{V}(x), \tag{7.335a}$$

where the operator-valued $8\overline{N} \times 8\overline{N}$ matrix

$$\mathcal{K} := -\mathbb{1}_{\overline{N}} \otimes \rho_0 \otimes \sigma_0 \otimes \tau_3\, v_{\mathrm{F}}\, \hat{p}, \qquad \hat{p} := -\mathrm{i}\hbar\,\partial_x \equiv \hbar\,\frac{\partial}{\mathrm{i}\partial x}, \tag{7.335b}$$

describes the propagation of right- or left-moving fermionic quasiparticles distributed among $\overline{N}$ channels at the Fermi level with the same Fermi velocity v_{F}, while the complex-valued Hermitean $8\overline{N} \times 8\overline{N}$ matrix

$$\mathcal{V}(x) := \begin{pmatrix} V(x) & \Delta(x) \\ \Delta^{\dagger}(x) & -V^{\mathsf{T}}(x) \end{pmatrix}, \qquad V^{\dagger}(x) = V(x), \qquad \Delta^{\mathsf{T}}(x) = -\Delta(x), \tag{7.335c}$$

[66] P. W. Anderson, Phys. Rev. **109**(5), 1492–1505 (1958). [Anderson (1958)].
[67] N. F. Mott and W. D. Twose, Adv. Phys. **10**(38), 107–163 (1961). [Mott and Twose (1961)].
[68] R. E. Borland, Proc. R. Soc. Lond. A **274**(1359), 529–545 (1963). [Borland (1963)].

(in the ρ grading) is the one-body potential. From now on, we set

$$\hbar = v_{\mathrm{F}} = 1 \tag{7.335d}$$

unless stated otherwise.

Any nonvanishing $4\overline{N} \times 4\overline{N}$ antisymmetric matrix $\Delta(x)$ for some x encodes superconducting correlations, in which case:

1 According to Section 7.3.1, symmetry class D is defined by demanding that

$$\mathcal{C}\,\mathcal{V}^{*}(x)\,\mathcal{C}^{-1} = -\mathcal{V}(x), \qquad \mathcal{C} := \mathbb{1}_{\overline{N}} \otimes \rho_1 \otimes \sigma_0 \otimes \tau_0. \tag{7.336}$$

2 According to Section 7.3.2, symmetry class C is defined by demanding that $\mathcal{V}(x)$ is decomposable down to an irreducible (ire) $4\overline{N} \times 4\overline{N}$ Hermitean block $\mathcal{V}^{\mathrm{ire}}(x)$ that satisfies

$$\mathcal{C}\,\mathcal{V}^{\mathrm{ire}\,*}(x)\,\mathcal{C}^{-1} = -\mathcal{V}^{\mathrm{ire}}(x), \qquad \mathcal{C} := \mathbb{1}_{\overline{N}} \otimes \mathrm{i}\kappa_2 \otimes \tau_0. \tag{7.337}$$

Spin-1/2 rotation symmetry implies a twofold degeneracy of the eigenvalues of the $8\overline{N} \times 8\overline{N}$ Hermitean matrix $\mathcal{V}(x)$.

3 According to Section 7.3.3, symmetry class DIII is defined by demanding that

$$\mathcal{C}\,\mathcal{V}^{*}(x)\,\mathcal{C}^{-1} = -\mathcal{V}(x), \qquad \mathcal{C} := \mathbb{1}_{\overline{N}} \otimes \rho_1 \otimes \sigma_0 \otimes \tau_0, \tag{7.338a}$$

$$\mathcal{T}\,\mathcal{V}^{*}(x)\,\mathcal{T}^{-1} = +\mathcal{V}(x), \qquad \mathcal{T} := \mathbb{1}_{\overline{N}} \otimes \rho_0 \otimes \mathrm{i}\sigma_2 \otimes \tau_1. \tag{7.338b}$$

Time-reversal symmetry implies a twofold Kramers' degeneracy of the eigenvalues of the $8\overline{N} \times 8\overline{N}$ Hermitean matrix $\mathcal{V}(x)$.

4 According to Section 7.3.4, symmetry class CI is defined by demanding that $\mathcal{V}(x)$ is decomposable down to an irreducible (ire) $4\overline{N} \times 4\overline{N}$ Hermitean block that satisfies

$$\mathcal{C}\,\mathcal{V}^{\mathrm{ire}*}(x)\,\mathcal{C}^{-1} = -\mathcal{V}^{\mathrm{ire}}(x), \qquad \mathcal{C} := \mathbb{1}_{\overline{N}} \otimes \mathrm{i}\kappa_2 \otimes \tau_0, \tag{7.339a}$$

$$\mathcal{T}\,\mathcal{V}^{\mathrm{ire}\,*}(x)\,\mathcal{T}^{-1} = +\mathcal{V}^{\mathrm{ire}}(x), \qquad \mathcal{T} := \mathbb{1}_{\overline{N}} \otimes \kappa_0 \otimes \tau_1. \tag{7.339b}$$

Spin-1/2 rotation symmetry implies a twofold degeneracy of the eigenvalues of the $8\overline{N} \times 8\overline{N}$ Hermitean matrix $\mathcal{V}(x)$.

In the absence of superconducting correlations, $\Delta(x) = 0$ for all x, a Fermi sea is perturbed by the one-body Hermitean $4\overline{N} \times 4\overline{N}$ matrix $V(x)$, in which case:

1 According to Section 7.3.5, symmetry class A is defined by demanding that the $4\overline{N} \times 4\overline{N}$ matrix $V(x)$ is no more than Hermitean, possibly decomposable down to irreducible (ire) $2\overline{N} \times 2\overline{N}$ Hermitean blocks if either spin U(1) or spin SU(2) are conserved.

2 According to Section 7.3.6, symmetry class AI is defined by demanding that the Hermitean $4\overline{N} \times 4\overline{N}$ matrix $V(x)$ is decomposable down to an irreducible (ire) $2\overline{N} \times 2\overline{N}$ Hermitean block $V^{\mathrm{ire}}(x)$ that satisfies

$$\mathcal{T}\,V^{\mathrm{ire}\,*}(x)\,\mathcal{T}^{-1} = V^{\mathrm{ire}}(x), \qquad \mathcal{T} := \mathbb{1}_{\overline{N}} \otimes \tau_1. \tag{7.340}$$

Spin-1/2 rotation symmetry implies a twofold degeneracy of the eigenvalues of the $4\overline{N} \times 4\overline{N}$ Hermitean matrix V(x).

3 According to Section 7.3.7, symmetry class AII is defined by demanding that the Hermitean $4\overline{N} \times 4\overline{N}$ matrix $V(x)$ satisfies

$$\mathcal{T}\, V^{*}(x)\, \mathcal{T}^{-1} = V(x), \qquad \mathcal{T} := \mathbb{1}_{\overline{N}} \otimes \mathrm{i}\sigma_2 \otimes \tau_1. \tag{7.341}$$

Time-reversal symmetry implies a twofold Kramers' degeneracy of the eigenvalues of the $4\overline{N} \times 4\overline{N}$ Hermitean matrix $V(x)$.

Finally, in the absence of superconducting correlations, $\Delta(x) = 0$ for all x but after imposing a unitary spectral symmetry, a Fermi sea is perturbed by the single-particle Hermitean $4\overline{N} \times 4\overline{N}$ matrix

$$V(x) = -\Gamma\, V(x)\, \Gamma^{-1} = V^{\dagger}(x), \qquad \Gamma := \mathbb{1}_{\overline{N}} \otimes \sigma_0 \otimes \tau_1, \tag{7.342}$$

in which case:

1 In the symmetry class AIII and with the choice of basis made to represent Γ in Eq. (7.342), the Hermitean $4\overline{N} \times 4\overline{N}$ matrix $V(x)$ takes the form

$$V(x) = \begin{pmatrix} \mathsf{A}(x) & \mathrm{i}\mathsf{B}(x) \\ -\mathrm{i}\mathsf{B}(x) & -\mathsf{A}(x) \end{pmatrix}, \quad \mathsf{A}(x) = \mathsf{A}^{\dagger}(x), \mathsf{B}(x) = \mathsf{B}^{\dagger}(x) \in \mathrm{Mat}(2\overline{N}, \mathbb{C}), \tag{7.343}$$

in the 2×2 grading of left- and right-movers. nonvanishing eigenvalues of $V(x)$ occur in pairs of opposite signs. If either spin U(1) or spin SU(2) are conserved, the blocks $\mathsf{A}(x)$ and $\mathsf{B}(x)$ are decomposable down to two irreducible $\overline{N} \times \overline{N}$ blocks so that all eigenvalues of $V(x)$ are twofold degenerate.

2 In the symmetry class BDI and with the choice of basis made to represent Γ in Eq. (7.342), time-reversal symmetry is represented by the constraint

$$\mathcal{T}\, V^{*}(x)\, \mathcal{T}^{-1} = V(x), \qquad \mathcal{T} := \mathbb{1}_{\overline{N}} \otimes \sigma_0 \otimes \tau_1, \tag{7.344}$$

on any $4\overline{N} \times 4\overline{N}$ Hermitean matrix $V(x)$ from the symmetry class AIII. This implies that

$$\mathsf{A}(x) = \mathsf{A}^{\dagger}(x) = -\mathsf{A}^{*}(x), \qquad \mathsf{B}(x) = \mathsf{B}^{\dagger}(x) = +\mathsf{B}^{*}(x), \tag{7.345}$$

in Eq. (7.343). Furthermore, spin-1/2 rotation symmetry implies that the blocks $\mathsf{A}(x)$ and $\mathsf{B}(x)$ are decomposable down to two irreducible $\overline{N} \times \overline{N}$ blocks. Accordingly, the eigenvalues of the $4\overline{N} \times 4\overline{N}$ Hermitean matrix V(x) are twofold degenerate.

3 In the symmetry class CII and with the choice of basis made to represent Γ in Eq. (7.342), time-reversal symmetry is represented by the constraint

$$\mathcal{T}\, V^{*}(x)\, \mathcal{T}^{-1} = V(x), \qquad \mathcal{T} := \mathbb{1}_{\overline{N}} \otimes \mathrm{i}\sigma_2 \otimes \tau_1, \tag{7.346}$$

on any $4\overline{N} \times 4\overline{N}$ Hermitean matrix $V(x)$ from the symmetry class AIII. This implies that

$$\mathsf{A}(x) = \mathsf{A}^{\dagger}(x) = -\mathsf{A}^{\mathrm{R}}(x), \qquad \mathsf{B}(x) = \mathsf{B}^{\dagger}(x) = +\mathsf{B}^{\mathrm{R}}(x), \tag{7.347}$$

in Eq. (7.343). Accordingly, time-reversal symmetry without spin-rotation symmetry implies a twofold Kramers' degeneracy of the eigenvalues of the $4\overline{N} \times 4\overline{N}$ Hermitean matrix $V(x)$.

7.7.3 Transfer Matrix

The solution to the stationary Schrödinger equation

$$\mathcal{H}(x)\,\Psi(x) = \varepsilon\,\Psi(x) \tag{7.348a}$$

for some given seed value $\Psi(x_0) \in \mathbb{C}^{8\overline{N}}$ and the single-particle Hamiltonian $\mathcal{H}(x)$ defined in Eq. (7.335) is

$$\Psi(x) = \mathcal{M}_{\varepsilon}(x, x_0)\,\Psi(x_0) \tag{7.348b}$$

with the transfer matrix[69]

$$\mathcal{M}_{\varepsilon}(x, x_0) := \mathsf{P}_{x'} \exp\left(\mathrm{i}\int\limits_{x_0}^{x} \mathrm{d}x' \, \left[\left(-\mathbb{1}_{4\overline{N}} \otimes \tau_3\right)\varepsilon + \left(\mathbb{1}_{4\overline{N}} \otimes \tau_3\right)\,\mathcal{V}(x')\right]\right). \tag{7.348c}$$

The symbol $\mathsf{P}_{x'}$ stands for path ordering.

Remark (1) The quantum wire described by the transfer matrix (7.348c) applies to spin-1/2 electrons confined to propagating in a one-dimensional geometry. One-dimensional tight-binding Hamiltonians with an underlying Bravais lattice on which spin-1/2 electrons hop in the background of a static, random, but not too strong one-body potential have a low-energy and long wavelength-effective description in terms of the transfer matrix (7.348c), provided the clean limit accomodates a Brillouin zone with two Fermi points.

Remark (2) The composition law (7.226) and the transformation laws (7.238) can be deduced explicitly from Eq. (7.348c) and the transformation laws of $\mathcal{V}(x)$ defined in Section 7.7.2.

Remark (3) The choice of basis for the transfer matrix (7.348c) is different from the choice of basis made in **Exercise** 3.2.

Remark (4) Deducing the tenfold way from the symmetry class D with its transfer matrices from the non-compact symmetric space $O(4\overline{N}, 4\overline{N})/\big(O(4\overline{N}) \times O(4\overline{N})\big)$ removes the short positive roots that are associated with zero modes of Hamiltonian (7.335). We shall see that the option $\overline{M} \neq \overline{N}$ for the symmetry classes A, D,

[69] Although the matrix $\mathcal{V}(x)$ is Hermitean, the product $\left(\mathbb{1}_{4\overline{N}} \otimes \tau_3\right)\mathcal{V}(x)$ is not Hermitean for a generic $\mathcal{V}(x)$.

and C from **Table** 7.6 or the options $O^{*}(4\overline{N}+2)$ for symmetry class AII and $O(4\overline{N}+1,\mathbb{C})$ for symmetry class DIII can be realized on the boundaries of two-dimensional topological insulators.

7.7.4 Anderson Localization

We assume that reservoirs Ω and Ω' from Figure 7.2 are the semi-infinite intervals $]-\infty,0]$ and $[L,+\infty[$, respectively. The disordered region Λ_{L} is the open interval $]0,L[$. The single-particle potential $\mathcal{V}(x)$, that is, the complex-valued $8\overline{N}\times 8\overline{N}$ Hermitean matrix (7.335c), vanishes for all $x\in\Omega\cup\Omega'$. In the scattering region Λ_{L}, the matrix-valued function $\mathcal{V}$ is random, whereby $\mathcal{V}$ is assumed to be of vanishing mean and white-noise (WN) distributed with the functional measure (the notation is adapted from that in Sections 7.4 and 7.5)

$$\mu_{\mathrm{WNGE}}[\mathrm{d}\mathcal{V}]\propto\mu_{\mathrm{AZ}}[\mathrm{d}\mathcal{V}]\,\exp\left(-\frac{\gamma\,\ell}{4\,c}\int\limits_{0}^{L}\mathrm{d}x\,\mathrm{tr}\left[\mathcal{V}^{\dagger}(x)\,\mathcal{V}(x)\right]\right). \tag{7.349a}$$

The functional measure $\mu_{\mathrm{AZ}}[\mathrm{d}\mathcal{V}]$ is defined point wise to be any one of the ten Haar measures from Section 7.5. The index Altland-Zirnbauer refers to a specific Altland-Zirnbauer symmetry class. The presence of the length scale ℓ in Eq. (7.349a) is dictated by dimensional analysis [recall Eq. (7.335d)]. It will be interpreted as the mean free path. The numerical constants c and γ, a linear function of $\overline{N}$, are read from **Table** 7.7 and

$$\begin{aligned}\gamma &:= \frac{1}{2}\left(m_{\mathrm{o}}^{-}+m_{\mathrm{o}}^{+}\right)\left(N^{\star}-1\right)+m_{\mathrm{l}}+1,\\ N^{\star} &\equiv \begin{cases}\frac{4\overline{N}}{D}, & \text{for D, C, DIII, CI,}\\[2ex] \frac{2\overline{N}}{D}, & \text{for A, AI, AII, AIII, BDI, CII,}\end{cases}\end{aligned} \tag{7.349b}$$

respectively, for each Altland-Zirnbauer symmetry class. The dimensionless parameter γ and the integer c are chosen for each Altland-Zirnbauer symmetry class such that the diffusion constant in the Fokker-Planck equation (7.359a) that will be derived from Eqs. (7.348) and (7.349) is given by

$$\frac{1}{2\,\gamma\,\ell}. \tag{7.350}$$

The integer c that appears in the definition of the probability distribution (7.349a) provides the desired variance for the independent real-valued random matrix elements of the $8\overline{N}\times 8\overline{N}$ Hermitean matrix $\mathcal{V}(x)$ for each Altland-Zirnbauer symmetry class.

The transfer matrix $\mathcal{M}_{\varepsilon}(L,0)$ defined by Eq. (7.348c) for a given realization of the random disorder potential $\mathcal{V}(x)$ accounts for the quantum transport of heat for all ten Altland-Zirnbauer symmetry classes, of spin for the symmetry classes C, CI, AI, and BDI, and electric charge for the symmetry classes A, AI, AII, AIII, BDI, and CII.

Table 7.7 *The tenfold way: Geometric data and transport for a disordered quantum wire*

AZ	$\mathrm{M}^- \equiv \mathrm{G}^\star/\mathrm{H}$ for $\mathcal{M}_{\varepsilon=0}(L,0)$	$\mathfrak{Roots}_+$	m_{o}^-	m_{o}^+	m_{l}	x_j	D	c
A	$\mathrm{U}(2\overline{N},2\overline{N})/[\mathrm{U}(2\overline{N})\times\mathrm{U}(2\overline{N})]$	$e_k \pm e_j, 2e_j$	2	2	1	$\mathbb{R}_+$	1(2)	2
AI	$\mathrm{Sp}(4\overline{N},\mathbb{R})/\mathrm{U}(2\overline{N})$	$e_k \pm e_j, 2e_j$	1	1	1	$\mathbb{R}_+$	2	4
AII	$\mathrm{O}^\star(4\overline{N})/\mathrm{U}(2\overline{N})$	$e_k \pm e_j, 2e_j$	4	4	1	$\mathbb{R}_+$	2	4
AIII	$\mathrm{GL}(2\overline{N},\mathbb{C})/\mathrm{U}(2\overline{N})$	$e_k - e_j$	2	0	0	$\mathbb{R}$	1(2)	2
BDI	$\mathrm{GL}(2\overline{N},\mathbb{R})/\mathrm{O}(2\overline{N})$	$e_k - e_j$	1	0	0	$\mathbb{R}$	2	4
CII	$\mathrm{U}^\star(2\overline{N})/\mathrm{Sp}(2\overline{N})$	$e_k - e_j$	4	0	0	$\mathbb{R}$	2	4
D	$\mathrm{O}(4\overline{N},4\overline{N})/[\mathrm{O}(4\overline{N})\times\mathrm{O}(4\overline{N})]$	$e_k \pm e_j$	1	1	0	$\mathbb{R}_+$	1	1
C	$\mathrm{Sp}(2\overline{N},2\overline{N})/[\mathrm{Sp}(2\overline{N})\times\mathrm{Sp}(2\overline{N})]$	$e_k \pm e_j, 2e_j$	4	4	3	$\mathbb{R}_+$	4	1
DIII	$\mathrm{O}(4\overline{N},\mathbb{C})/\mathrm{O}(4\overline{N})$	$e_k \pm e_j$	2	2	0	$\mathbb{R}_+$	2	2
CI	$\mathrm{Sp}(2\overline{N},\mathbb{C})/\mathrm{Sp}(2\overline{N})$	$e_k \pm e_j, 2e_j$	2	2	2	$\mathbb{R}_+$	4	2

A disordered quantum wire of length L is modeled by Eq. (7.348). Transport across a disordered quantum wire connected to ideal reservoirs at the chemical potential $\varepsilon = 0$ is encoded by the random $8\overline{N}\times 8\overline{N}$ transfer matrix $\mathcal{M}_{\varepsilon=0}(L,0)$ given by Eq. (7.348c). The differences compared to **Table** 7.6 are the following. First, the number N of channels in **Table** 7.6 is $N = 2\overline{N}$ for the symmetry classes A, AI, AIII, BDI, and D in order to accommodate the spin-1/2 of electrons; while $N = \overline{N}$ for the symmetry classes AII, CII, DIII, C, and CI. Second, there are two new columns labeled by D and c. The label D for the symmetry classes D, C, DIII, and CI is the degeneracy of the $N^\star$ independent Lyapunov exponents x_j defined through the eigenvalues $\exp(\pm 2x_j)$ of $\mathcal{M}^\dagger_{\varepsilon=0}(L,0)\,\mathcal{M}_{\varepsilon=0}(L,0)$. The label D for the symmetry classes A, AI, AII AIII, BDI, and CII is the degeneracy of the $N^\star$ independent Lyapunov exponents if the Bogoliubov-de-Gennes grading is removed. [The entry (2) for lines A and AIII presumes spin-rotation symmetry.] The integer c is related to the multiplicity of the independent random matrix elements present in the trace $\mathrm{tr}\left[\mathcal{V}^\dagger(x)\,\mathcal{V}(x)\right]$ that enters the measure (7.349a). Third, deducing the tenfold way from $\mathrm{O}(4\overline{N},4\overline{N})$ removes the short positive roots.

The transfer matrix (7.348c) is a $8\overline{N}\times 8\overline{N}$ representation of an element of a non-compact Lie group $\mathrm{G}^\star$, one for each of the ten Altland-Zirnbauer symmetry classes. This $8\overline{N}\times 8\overline{N}$ representation, for a typical realization of the random disorder potential $\mathcal{V}(x)$, is only irreducible for the symmetry class D. It is decomposable otherwise.

We have shown in Section 7.6.2 that the $8\overline{N}\times 8\overline{N}$ transfer matrix (7.348c) admits a singular-value decomposition of the form

$$
\mathcal{U}\,\mathcal{X}\,\mathcal{V},\qquad \mathcal{X} = \begin{cases} \begin{pmatrix} \cosh X & \sinh X \\ \sinh X & \cosh X \end{pmatrix}, & \text{A, AI, AII, AIII, BDI, CII, D, C,} \\[2ex] \begin{pmatrix} \cosh X & \mathrm{i}\sinh X \\ -\mathrm{i}\sinh X & \cosh X \end{pmatrix}, & \text{DIII, CI,} \end{cases} \tag{7.351a}
$$

with the $4\overline{N} \times 4\overline{N}$ matrix

$$X = \operatorname{diag}\begin{pmatrix} x_1 & \cdots & x_{4\overline{N}} \end{pmatrix} \tag{7.351b}$$

diagonal and real-valued. The $8\overline{N} \times 8\overline{N}$ "angular" matrices $\mathcal{U}$ and $\mathcal{V}$ are representations of elements of the factor group H in the second column of **Table** 7.7 for all ten Altland-Zirnbauer symmetry classes. Hereto, the angular matrices $\mathcal{U}$ and $\mathcal{V}$ for the symmetry class D are irreducible representations of an element of $\mathrm{H} = \mathrm{O}(4\overline{N}, 4\overline{N}) \times \mathrm{O}(4\overline{N}, 4\overline{N})$ for a typical realization of the random disorder potential $\mathcal{V}(x)$. They are decomposable otherwise. The dimensionality of the largest irreducible block of this representation can be deduced from the isotropy subgroup H in the second column of **Table** 7.7.

For the four Bogoliubov-de-Gennes symmetry classes D, C, DIII, and CI the eigenvalues of the $8\overline{N} \times 8\overline{N}$ positive definite matrix $\mathcal{M}_\varepsilon(L,0)\,\mathcal{M}^\dagger_\varepsilon(L,0)$ are the D-fold degenerate and independent inverse pairs

$$e^{\pm 2\,x_j}, \qquad j = 1, \cdots, N^\star, \qquad N^\star := \frac{4\overline{N}}{D}, \tag{7.352}$$

with the value of D listed for the Bogoliubov-de-Gennes lines of **Table** 7.7. The degeneracy $D = 4$ for the symmetry class C can be understood as follows. For a typical realization of the random disorder potential $\mathcal{V}(x)$, the $8\overline{N} \times 8\overline{N}$ transfer matrix $\mathcal{M}_\varepsilon(L,0)$ is the direct sum of two irreducible $4\overline{N} \times 4\overline{N}$ transfer matrices from the Lie group $\mathrm{Sp}(2\overline{N}, 2\overline{N})$ that share the same Lyapunov exponents. This brings about a twofold degeneracy for the Lyapunov exponents of $\mathcal{M}_\varepsilon(L,0)$. According to the singular-value decomposition (7.306), the Lyapunov exponents of any one of these two irreducible $4\overline{N} \times 4\overline{N}$ transfer matrices are also twofold degenerate due to an effective Kramers' degeneracy. Hence, the Lyapunov exponents of $\mathcal{M}_\varepsilon(L,0)$ are at least fourfold degenerate in the symmetry class C. The degeneracy $D = 4$ for the symmetry class CI can be interpreted as a consequence of the degeneracy $D = 4$ for the symmetry class C. The (Kramers) degeneracy $D = 2$ for the symmetry class DIII is a consequence of the singular-value decomposition (7.315).

For the three Wigner-Dyson symmetry classes A, AI, AII and the chiral symmetry classes AIII, BDI, and CII, we trade the $8\overline{N} \times 8\overline{N}$ matrix $\mathcal{V}(x)$ for the $4\overline{N} \times 4\overline{N}$ block $V(x)$ in Eq. (7.335c). The eigenvalues of the resulting $4\overline{N} \times 4\overline{N}$ positive definite matrix $\mathcal{M}_\varepsilon(L,0)\,\mathcal{M}^\dagger_\varepsilon(L,0)$ are the D-fold degenerate and independent inverse pairs

$$e^{\pm 2\,x_j}, \qquad j = 1, \cdots, N^\star, \qquad N^\star := \frac{2\overline{N}}{D}, \tag{7.353}$$

with the value of D listed for the Wigner-Dyson and chiral lines of **Table** 7.7. The degeneracy $D = 2$ for the symmetry classes AI and BDI is a trivial consequence of spin-rotation symmetry. The degeneracy $D = 2$ for the symmetry classes AII and CII is a consequence of the Kramers' degeneracy in the singular-value decompositions (7.265) and (7.290), respectively.

Knowledge of the dependence on L of the joint probability distribution

$$P_{\varepsilon}(x_1,\cdots,x_{N^\star};L), \tag{7.354a}$$

where $N^\star$ is the number of independent Lyapunov exponents

$$x_1,\cdots,x_{N^\star} \tag{7.354b}$$

in the transfer matrix, enables to compute the dependence on L of the probability distribution of the dimensionless Landauer conductance

$$g_{\mathrm{L},\varepsilon}(L) := D\sum_{j=1}^{N^\star}\cosh^{-2}x_j. \tag{7.354c}$$

However, the full probability distribution of $g_{\mathrm{L},\varepsilon}(L)$ can be calculated in closed form for $N^\star=1$ only. For large $N^\star$, *the thick quantum wire limit*, cumulants[70] of $g_{\mathrm{L},\varepsilon}(L)$ can be computed in three regimes, the *ballistic* regime

$$L\lesssim\ell, \tag{7.355}$$

in which the transport is hardly affected by the random potential $\mathcal{V}$, the *diffusive* regime

$$\ell\ll L\ll N^\star\,\ell, \tag{7.356}$$

in which the random potential $\mathcal{V}$ can be treated perturbatively, and the asymptotic regime

$$N^\star\,\ell\ll L, \tag{7.357}$$

in which the effects of the random potential $\mathcal{V}$ are non-perturbative. Anderson localization takes place if the first cumulant of $g_{\mathrm{L},\varepsilon}(L)$ decays like

$$\exp\left(-\frac{L}{\xi_{\mathrm{ave}}(\varepsilon)}\right) \tag{7.358}$$

when $N^\star\,\ell\ll L$, in which case the length scale $\xi_{\mathrm{ave}}(\varepsilon)$ is called the average *localization length.*

[70] If x is a random variable with probability distribution $p(x)$, then its first moment (mean), second moment, and corresponding cumulants are denoted by

$$\langle x\rangle\equiv\int\mathrm{d}x\,p(x)\,x\equiv\bar{x},\qquad \langle x^2\rangle\equiv\int\mathrm{d}x\,p(x)\,x^2,$$
$$\langle x\rangle_{\mathrm{c}}\equiv\int\mathrm{d}x\,p(x)\,(x-\bar{x})=0,\qquad \langle x^2\rangle_{\mathrm{c}}\equiv\int\mathrm{d}x\,p(x)\,(x-\bar{x})^2=\langle x^2\rangle-(\bar{x})^2,$$

respectively.

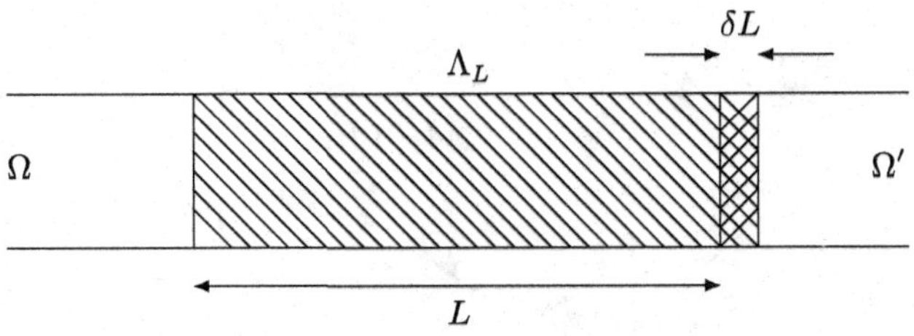

Figure 7.7 A thin slice of length δL with $\mathfrak{a} \ll \delta L \ll \ell \ll L$ is added to the disordered region Λ_L of length L connected to reservoirs Ω and Ω'. The short-distance cutoff $\mathfrak{a}$ is the lattice spacing.

7.7.5 One-Parameter Scaling of Anderson Localization

We seek the parametric dependence on L of the joint probability distribution (7.354a) for all ten Altland-Zirnauber symmetry classes from **Table** 7.7. The answer for symmetry classes A, AI, and AII is captured for any Fermi energy ε by the so-called Dorokhov-Mello-Pereyra-Kumar (DMPK) scaling equation.[71] The counterpart to the DMPK scaling equation for the chiral and Bogoliubov-de-Gennes symmetry classes, provided the Fermi energy is at the band center $\varepsilon = 0$ of the fixed point under $\varepsilon \mapsto -\varepsilon$ (of the single-particle spectra in the reservoirs Ω and Ω' and in the region Λ_L when each region is closed), was derived by Brouwer et al.[72,73] The DMPK scaling equation is an example of one-parameter scaling. According to the one-parameter scaling hypothesis of Anderson localization, the statistical distributions of the conductance, energy levels, or wave functions, are entirely determined by the fundamental symmetries and the dimensionality of the sample.[74] Once symmetry and dimensionality are taken into account, all microscopic details of Λ_L can be represented by a single length scale ℓ, the "mean free path," such that on length scales L larger than ℓ, the sample is completely characterized by the ratio L/ℓ. The concept of universality that underlies this one-parameter scaling hypothesis is the cornerstone of various field-theoretic, diagrammatic, and random-matrix approaches to Anderson localization.

Increasing the length L of the quantum wire by a small increment δL (see Figure 7.7) amounts to multiplication of its transfer matrix $\mathcal{M}_\varepsilon(L,0)$ by a transfer matrix $\mathcal{M}_\varepsilon(L+\delta L, L)$. Since $\mathcal{M}_\varepsilon(L+\delta L, L)$ is close to the unit matrix for small

[71] O. N. Dorokhov, Pis'ma Zh. Eksp. Teor. Fiz. **36**, 259 (1982); JETP Letters **36**(7), 318–321 (1982); P. A. Mello, P. Pereyra, and N. Kumar, Ann. Phys. (NY) **181**(2), 290–317 (1988). [Dorokhov (1982); Mello et al. (1988)].

[72] P. W. Brouwer, C. Mudry, B. D. Simons, and A. Altland, Phys. Rev. Lett. **81**(4), 862–865 (1998). [Brouwer et al. (1998)].

[73] P. W. Brouwer, A. Furusaki, I. A. Gruzberg, and C. Mudry, Phys. Rev. Lett. **85**(5), 1064–1067 (2000). [Brouwer et al. (2000b)].

[74] J. T. Edwards and D. J. Thouless, J. Phys. C: Solid St. Phys. **5**(8), 807–820 (1972); F. J. Wegner, Z. Phys. B **25**(4), 327–337 (1976); D. C. Licciardello and D. J. Thouless, J. Phys. C: Solid St. Phys. **8**(24), 4157–4170 (1975); **11**(5), 925–936 (1978); E. Abrahams, P. W. Anderson, D. C. Licciardello, and T. V. Ramakrishnan, Phys. Rev. Lett. **42**(10), 673–676 (1979). [Edwards and Thouless (1972); Wegner (1976); Licciardello and Thouless (1975, 1978); Abrahams et al. (1979)].

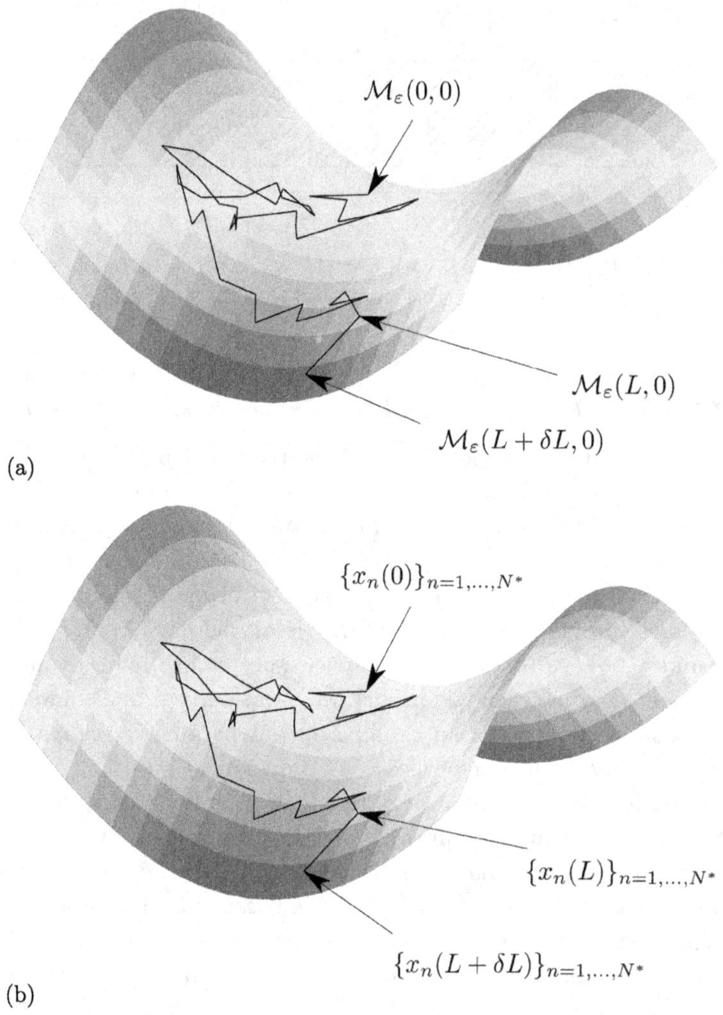

Figure 7.8 (a) Brownian motion of a transfer matrix (7.348) in the non-compact Lie group $G^\star$ upon a small increase of the length of the disordered region from L to $L+\delta L$. (b) Brownian motion of the Lyapunov exponents of a transfer matrix in the symmetric space $G^\star/H$ upon a small increase of the length of the disordered region from L to $L+\delta L$. The Lyapunov exponents of the transfer matrix are twice the values of the radial coordinates of the associated symmetric space $G^\star/H$.

δL, random, and statistically independent from $\mathcal{M}_\varepsilon(L,0)$ for white-noise correlated disorder, one finds that, as a function of L, $\mathcal{M}_\varepsilon(L,0)$ performs a random trajectory on its non-compact Lie group $G^\star$ as shown in Figure 7.8(a). Actually, the full trajectory on $G^\star$ is not needed if we are only interested in the Landauer conductance (7.354c). It is sufficient to know the trajectories of the Lyapunov exponents $\{x_j(L)\}$

of $\mathcal{M}_\varepsilon(L,0)$ as a function of L. This Brownian motion is depicted in Figure 7.8(b). One can show that the trajectory obeyed by the Lyapunov exponents $\{x_j(L)\}$ is a Brownian motion on the symmetric space M* encoded by the joint probability distribution (7.354a). This Brownian motion is described by a Fokker-Planck equation, which follows either from a direct calculation starting from Eqs. (7.335), (7.348), and (7.349) or from the general theory of symmetric spaces (see footnote 16 or 32) supplemented by the rule that the radial coordinates of the non-compact symmetric spaces are identified with the Lyapunov exponents rescaled by the multiplicative factor of 2. In both cases, one finds

$$\frac{\partial P_\varepsilon}{\partial L} = \frac{1}{2\gamma\ell}\sum_{j=1}^{N^\star}\frac{\partial}{\partial x_j}\left[J_\varepsilon\frac{\partial}{\partial x_j}\left(J_\varepsilon^{-1}\,P_\varepsilon\right)\right], \tag{7.359a}$$

$$J_\varepsilon(\boldsymbol{x}) := \left[\prod_{j=1}^{N^\star}|\sinh 2x_j|^{m_\mathrm{l}}\right]\left[\prod_{1\leq j<k\leq N^\star}|\sinh(x_j-x_k)|^{m_\mathrm{o}^-}\,|\sinh(x_j+x_k)|^{m_\mathrm{o}^+}\right], \tag{7.359b}$$

with $\boldsymbol{x}=(x_1,\cdots,x_{N^\star})$ the independent Lyapunov exponents of the transfer matrix and the constant γ given in Eq. (7.349b). The number of independent Lyapunov exponents $N^\star$, the choice of the positive roots and their multiplicities (m_l for long and $m_\mathrm{o}^\pm$ for ordinary ones) are tabulated in **Table** 7.7 for each Altland-Zirnbauer symmetry class when $\varepsilon=0$, while one must only consider the three symmetry classes A, AI, and AII when $\varepsilon\neq 0$. The Fokker-Planck equation (7.359a) has a unique solution if it is supplemented with the boundary conditions

$$\frac{\partial P_\varepsilon}{\partial x_j} = \frac{P_\varepsilon}{J_\varepsilon}\frac{\partial J_\varepsilon}{\partial x_j}, \qquad \text{for } x_1=\cdots=x_{N^\star}=0, \tag{7.359c}$$

$$P_\varepsilon = 0, \qquad \text{for } x_1=\cdots=x_{N^\star}=\infty, \tag{7.359d}$$

that enforce the condition that the joint probability distribution P_ε is normalized for all L and ε (**Exercise** 7.23), and the initial condition

$$\mathcal{M}_\varepsilon(L=0,0)=\mathbb{1} \iff P_\varepsilon(x_1,\cdots,x_{N^\star};L=0)=\prod_{j=1}^{N^\star}\delta(x_j). \tag{7.359e}$$

Remark No short positive roots are allowed in the Jacobian (7.359b) because neither an unequal number of left- and right-moving channels[75] nor an odd number

[75] A necessary and sufficient condition for the presence of the non-compact symmetric spaces

$$\mathrm{U}(\overline{M},\overline{N})/[\mathrm{U}(\overline{M})\times\mathrm{U}(\overline{N})] \text{ for the symmetry class A,}$$
$$\mathrm{O}(\overline{M},\overline{N})/[\mathrm{O}(\overline{M})\times\mathrm{O}(\overline{N})] \text{ for the symmetry class D,}$$
$$\mathrm{Sp}(2\overline{M},2\overline{N})/[\mathrm{Sp}(2\overline{M})\times\mathrm{Sp}(2\overline{N})] \text{ for the symmetry class C.}$$

of propagating Kramers' degenerate pairs[76] are permissible when deriving the continuum Hamiltonian from a one-dimensional lattice model for which reversal of time, charge conjugation, and the chiral (sublattice) transformations are all represented locally in space. This is a consequence of the Nielsen-Ninomiya theorem.[77] In our context, we use this theorem to fix the kinetic energy starting from a one-dimensional Bravais lattice such that the kinetic energy is time-reversal and inversion symmetric. The number of Fermi points (we are in one-dimensional space) is then necessarily even since (i) the set of the Fermi wave vectors in the Brillouin zone must be closed under reversal of time or inversion and (ii) locality in position space implies smoothness of the single-particle dispersions over the Brillouin zone. Linearization of the single-particle dispersions at the Fermi points delivers generically a kinetic energy of the form $\mathcal{K}$ given in Eq. (7.335b).

7.7.5.1 *Ballistic Regime*

In the ballistic regime $L \lesssim \ell$, all Lyapunov exponents x_j are close to zero, all transmission eigenvalues $\cosh^{-2} x_j$ are close to unity, and the dimensionless Landauer conductance is close to its initial value $D\,N^\star$ when $L=0$. The fluctuations of the dimensionless Landauer conductance about its mean value are small for $N^\star \gg 1$ large, that is, (see footnote 10)

$$\operatorname{var} g_{\mathrm{L},\varepsilon} := \left\langle \left(g_{\mathrm{L},\varepsilon} - \langle g_{\mathrm{L},\varepsilon}\rangle\right)^2 \right\rangle \simeq \left(\frac{\ell}{L}\right)^2 . \tag{7.360}$$

7.7.5.2 *Diffusive Regime*

The effect of disorder in the diffusive regime $\ell \ll L \ll N^\star\,\ell$ can be treated perturbatively in powers of the dimensionless parameter

$$\frac{L}{N^\star\,\ell} \ll 1, \tag{7.361}$$

since the Landauer conductance (7.354c) has only small fluctuations around its mean $\langle g_{\mathrm{L},\varepsilon}\rangle$ in this regime. The first step to compute $\langle g_{\mathrm{L},\varepsilon}\rangle$ from the Fokker-Planck equation (7.359) in the diffusive regime is exact and consists in deriving a countably infinite set of coupled evolution equations for the expectation values of the quantities

$$g^{(a)}_{\mathrm{L},\varepsilon} := D \sum_{j=1}^{N^\star} \cosh^{-2a} x_j, \qquad a = 1,2,3,\cdots . \tag{7.362}$$

[76] A necessary and sufficient condition for the presence of the non-compact symmetric spaces

$$\mathrm{O}^\star(4\overline{N}+2)/\mathrm{U}(2\overline{N}+1) \text{ for the symmetry class AII,}$$
$$\mathrm{O}(4\overline{N}+1,\mathbb{C})/\mathrm{O}(4\overline{N}+1) \text{ for the symmetry class DIII.}$$

[77] H. B. Nielsen and M. Ninomiya, Nucl. Phys. B **185**, 20–40 (1981). [Nielsen and Ninomiya (1981)].

For example, for all the Wigner-Dyson symmetry classes A, AI, and AII irrespective of ε and for the four Bogoliubov-de-Gennes symmetry classes D, C, DIII, CI at the band center $\varepsilon = 0$ (these seven symmetry classes share the property $m_{\rm o} \equiv m_{\rm o}^{+} = m_{\rm o}^{-}$), one finds (**Exercise** 7.24)

$$\begin{aligned}\frac{\gamma\,\ell}{a}\frac{\partial\left\langle g^{(a)}_{\mathrm{L},\varepsilon=0}\right\rangle}{\partial L} =& \frac{m_{\rm o}}{D}\sum_{n=1}^{a-1}\left\langle g^{(a-n)}_{\mathrm{L},\varepsilon=0}\, g^{(n)}_{\mathrm{L},\varepsilon=0}\right\rangle - \frac{m_{\rm o}}{D}\sum_{n=1}^{a}\left\langle g^{(a-n+1)}_{\mathrm{L},\varepsilon=0}\, g^{(n)}_{\mathrm{L},\varepsilon=0}\right\rangle\\ &+ (a\,m_{\rm o} - 2\,a - 1 + m_{\rm l})\left\langle g^{(a+1)}_{\mathrm{L},\varepsilon=0}\right\rangle\\ &+ (2a - a\,m_{\rm o} + m_{\rm o} - 2m_{\rm l})\left\langle g^{(a)}_{\mathrm{L},\varepsilon=0}\right\rangle, \qquad a = 1,2,3,\cdots.\end{aligned} \tag{7.363}$$

In the diffusive regime, one may do the approximation by which the average of a product is replaced by the product of the averages, thereby decoupling these first-order ordinary differential equations. The s dependence of $\left\langle g^{(a)}_{\mathrm{L},\varepsilon=0}\right\rangle$ in the diffusive regime is then obtained recursively starting from $a = 1$. Repeating this strategy for all ten Altland-Zirnbauer symmetry classes delivers

$$\left\langle g_{\mathrm{L},\varepsilon=0}\right\rangle = \begin{cases} \frac{D\,N^{\star}\,\ell}{L+\ell} + \frac{D\,(m_{\rm o}-2\,m_{\rm l})}{3\,m_{\rm o}} + \mathcal{O}\left(\frac{\ell}{L}, \frac{L}{N^{\star}\,\ell}\right), & \text{A, AI, AII, D, C, DIII, CI,}\\ \\ \frac{D\,N^{\star}\,\ell}{L+\ell} + 0 + \mathcal{O}\left(\frac{\ell}{L}, \frac{L}{N^{\star}\,\ell}\right), & \text{AIII, BDI, CII.}\end{cases} \tag{7.364a}$$

The first term

$$g_{\rm D} \equiv \frac{D\,N^{\star}\ell}{L+\ell} \tag{7.364b}$$

on the right-hand side is the Drude (D) conductance. The second term on the right-hand side is the first quantum interference correction to the average conductance. It is referred to as the *weak localization* (WL) correction. For the standard symmetry classes A, AI, and AII, it takes the value

$$\delta g^{\rm WL}_{\mathrm{L},\varepsilon=0} \equiv 0, -2/3, +1/3, \tag{7.364c}$$

respectively. For the Bogoliubov-de-Gennes symmetry classes D, C,[78] DIII, and CI, it takes the values

$$\delta g^{\rm WL}_{\mathrm{L},\varepsilon=0} \equiv +1/3, -2/3, +2/3, -4/3, \tag{7.364d}$$

respectively. For the symmetry classes AII, D, and DIII, the correction is positive, that is, quantum interferences enhance the conductance relative to the classical Drude-like leading behavior, a phenomenon called *antilocalization*. For the

[78] R. Bundschuh, C. Cassanello, D. Serban, and M. R. Zirnbauer, Nucl. Phys. B **532**, 689–732 (1998); Phys. Rev. B **59**(6), 4382–4389 (1999). [Bundschuh et al. (1998, 1999)].

symmetry class A and the chiral symmetry classes, the first quantum interference correction to the average conductance is

$$\delta g^{\mathrm{WL}}_{\mathrm{L},\varepsilon=0} = 0. \tag{7.364e}$$

The correction $\delta g^{\mathrm{WL}}_{\mathrm{L},\varepsilon=0}$ is universal in the thick quantum wire limit $N^\star \gg 1$. It quantifies the phenonemon of *weak localization.* With more efforts, one finds the *universal conductance fluctuations* (see footnotes 19 and 27)

$$\operatorname{var} g^{\mathrm{WL}}_{\mathrm{L},\varepsilon=0} \equiv \frac{2}{15}\left(\frac{m^-_{\mathrm{o}} + m^+_{\mathrm{o}}}{2}\right)^{-1} + \mathcal{O}\left(\frac{\ell}{L}, \frac{L}{N^\star\,\ell}\right) \tag{7.364f}$$

in the thick quantum wire limit $N^\star \gg 1$ at the band center $\varepsilon = 0$ for each symmetry class.

For any nonvanishing ε, neither the Bogoliubov-de-Gennes nor the chiral spectral symmetries have constraining effects on the transfer matrix. The only symmetry that matters is the presence or absence of time-reversal symmetry. The weak localization corrections in the chiral and Bogoliubov-de-Gennes symmetry classes should smoothly change as a function of ε to those of the standard symmetry classes for sufficienlty large ε. The crossover energy scale at which the weak localization corrections of the standard symmetry classes are recovered should be of the order of

$$\frac{\hbar}{\tau_{\mathrm{cross}}}, \qquad \tau_{\mathrm{cross}} \sim \frac{(N^\star\,\ell)^2}{v_{\mathrm{F}}\,\ell} = \frac{(N^\star)^2\,\ell}{v_{\mathrm{F}}}. \tag{7.365}$$

Here, τ_{cross} is the characteristic time needed for a classical particle with the characteristic speed v_{F} to diffuse with the diffusion constant $\sim (v_{\mathrm{F}}\,\ell)^{-1}$ through a region of linear size $\sim N^\star\,\ell$. This expectation has been verified numerically.[79]

7.7.5.3 Localized Regime

In the localized regime $L \gg N^\star\,\ell$, all Lyapunov exponents x_j and their spacings are typically much larger than unity. If so, the conductance is governed by the smallest radial coordinate

$$x_{\mathrm{min}} := \min\{x_1, \cdots, x_{N^\star}\}. \tag{7.366}$$

If the multiplicity m_{l} of the long positive roots is nonvanishing, then x_{min} has a Gaussian distribution, with mean (**Exercise** 7.25)

$$\langle x_{\mathrm{min}} \rangle \approx \frac{m_{\mathrm{l}}\,L}{\gamma\,\ell} \tag{7.367a}$$

and variance

$$\operatorname{var} x_{\mathrm{min}} := \left\langle (x_{\mathrm{min}} - \langle x_{\mathrm{min}} \rangle)^2 \right\rangle \approx \frac{L}{\gamma\,\ell}. \tag{7.367b}$$

[79] C. Mudry, P. W. Brouwer, and A. Furusaki, Phys. Rev. B **62**(12), 8249–8268 (2000); Erratum Phys. Rev. B **63**(12), 129901 (2001). [Mudry et al. (2000, 2001)].

For the standard symmetry classes (A, AI, and AII) and the Bogoliubov-de-Gennes symmetry classes C and CI this implies that the Landauer conductance at the band center is exponentially small, with (**Exercise** 7.25)

$$\langle -\ln g_{\mathrm{L},\varepsilon=0} \rangle \approx \frac{2\,m_{\mathrm{l}}\,L}{\gamma\,\ell}, \qquad \operatorname{var} \ln g_{\mathrm{L},\varepsilon=0} \approx \frac{4\,L}{\gamma\,\ell}. \tag{7.367c}$$

For those symmetry classes with $m_{\mathrm{l}} \neq 0$, Anderson localization rules, the Landauer conductance at the band center $g_{\mathrm{L},\varepsilon=0}$ has a log-normal distribution in the localized regime, that is its logarithm is a Gaussian distribution in the localized regime. However, $m_{\mathrm{l}} = 0$ for the chiral symmetry classes, AIII, BDI, and CII and the Bogoliubov-de-Gennes symmetry classes D and DIII. For the chiral symmetry classes with $N^{\star}$ an odd integer and for the Bogoliubov-de-Gennes symmetry classes D and DIII, the Landauer conductance at the band center $g_{\mathrm{L},\varepsilon=0}$ has a distribution much broader than log-normal, with an algebraic decay of the mean and the variance and an $L^{1/2}$-dependence of $\ln g_{\mathrm{L},\varepsilon=0}$. For example, one finds

$$\langle g_{\mathrm{L},\varepsilon=0} \rangle \approx D\sqrt{\frac{2\,\gamma\,\ell}{\pi\,L}}, \qquad \operatorname{var} g_{\mathrm{L},\varepsilon=0} \approx \frac{2\,D}{3} \langle g_{\mathrm{L},\varepsilon=0} \rangle, \tag{7.368a}$$

$$\langle \ln g_{\mathrm{L},\varepsilon=0} \rangle \approx -4\sqrt{\frac{L}{2\pi\,\gamma\,\ell}}, \qquad \operatorname{var} \ln g_{\mathrm{L},\varepsilon=0} \approx \frac{4(\pi-2)L}{\pi\,\gamma\,\ell}, \tag{7.368b}$$

for the symmetry classes D and DIII. Hence, for the chiral symmetry classes with $N^{\star}$ an odd integer and for the Bogoliubov-de-Gennes symmetry classes D and DIII, quasiparticle states are not localized at the fixed point $\varepsilon = 0$ under $\varepsilon \mapsto -\varepsilon$ of their dispersions. Since they are neither truly extended (typically $g_{\mathrm{L},\varepsilon=0} \ll 1$) nor (exponentially) localized, we christen them critical. Finally, Eq. (7.367) with the substitution $m_{\mathrm{l}} \to m_{\mathrm{o}}^{-}$ applies to the chiral symmetry classes with $N^{\star}$ an even integer. We are going to derive explicitly the Fokker-Planck equation (7.359) for the chiral symmetry classes in Section 7.7.6 to explain in more detail this remarkable even-odd effect. Hereto, the crossover energy scale at which Anderson localization in the standard symmetry classes is recovered is given by Eq. (7.365).

7.7.5.4 Four Exact Solutions to the Fokker-Planck Equations

For any one of the symmetry classes A,[80] AIII,[72] CI, and DIII,[73] the corresponding DMPK equation (7.359) is solvable. This solution can be used to represent the moments (see footnote 70) of the Landauer conductance as integrals over special functions associated with the symmetric space corresponding to the symmetry classes A, AIII, CI, and DIII, respectively, for any L and any $N^{\star}$.[81]

In the large $N^{\star} \gg 1$ limit, the moments of the Landauer conductance for the symmetry classes AIII, CI, and DIII admit a series respresentation in terms of

[80] C. W. J. Beenakker and B. Rejaei, Phys. Rev. Lett. **71**(22), 3689–3692 (1993); Phys. Rev. B **49**(11), 7499–7510 (1994). [Beenakker and Rejaei (1993, 1994)].
[81] K. Frahm, Phys. Rev. Lett. **74**(23), 4706–4709 (1995). [Frahm (1995)].

elementary functions of the dimensionless ratio L/ℓ.[82] For example, the dependence on

$$0 \leq s := \frac{L}{4\,N^{\star}\,\ell} < \infty \tag{7.369a}$$

of the mean Landauer conductance at the band center is (for the symmetry class AIII, we are here considering the spin-rotation symmetric case so that the degeneracy of the Lyapunov exponents is twofold)

$$\left\langle g_{\mathrm{L},\varepsilon=0}\right\rangle = \begin{cases} \frac{1}{2\,s} + 0 + \frac{1}{s}\sum\limits_{n=1}^{\infty}(-1)^{n(N^{\star}+1)}e^{-\pi^2\,n^2/(8\,s)}, & \text{AIII for } N^{\star} \gg 1, \\ \frac{1}{s} - \frac{4}{3} + 4\sum\limits_{n=1}^{\infty} e^{-\pi^2\,n^2/(4\,s)}\left(\frac{1}{s} + \frac{2}{\pi^2\,n^2}\right), & \text{CI for } N^{\star} \gg 1, \\ \frac{1}{s} + \frac{2}{3} - 4\sum\limits_{n=1}^{\infty} e^{-\pi^2\,n^2/(2s)}\frac{1}{\pi^2\,n^2}, & \text{DIII for } N^{\star} \gg 1, \end{cases} \tag{7.369b}$$

in the thick quantum wire limit.[83] The even-odd effect in $N^{\star}$ for the symmetry class AIII stems from the factor

$$(-1)^{n(N^{\star}+1)} = \begin{cases} +1, & N^{\star} \text{ odd}, \\ (-1)^{n}, & N^{\star} \text{ even}. \end{cases} \tag{7.370}$$

It is a non-perturbative effect in powers of $0 < s \ll 1$, for the difference between the mean conductance $\left\langle g_{\mathrm{L},\varepsilon=0}\right\rangle$ in the symmetry class AIII for even and odd $N^{\star} \gg 1$ is a sum over functions that have an essential singularity at $s = 0$.[84] The weak-localization corrections are recovered in the diffusive limit $0 < s \ll 1$. It is shown in **Exercise** 7.26 how to recover the asymptotic behavior of the mean Landauer conductance $\left\langle g_{\mathrm{L},\varepsilon=0}\right\rangle$ in the symmetry classes AIII, CI, and DIII as $s \to \infty$ from Eq. (7.369).

7.7.5.5 The Density of States

The following variant of the Fokker-Planck equation (7.359) can also be used to compute the dependence on the energy $\varepsilon > 0$ of the mean number of states

$$\mathcal{N}(\varepsilon) \equiv \int\limits_{-\varepsilon}^{+\varepsilon} \mathrm{d}\varepsilon'\,\nu(\varepsilon') \tag{7.371a}$$

[82] The case of the symmetry class A had been solved earlier by representing the mean conductance using a supersymmetric nonlinear sigma model by M. R. Zirnbauer, Phys. Rev. Lett. **69**(10), 1584–1587 (1992). [Zirnbauer (1992)].

[83] The case AIII was solved in C. Mudry, P. W. Brouwer, and A. Furusaki, Phys. Rev. B **59**(20), 13221–13234 (1999) [Mudry et al. (1999)]; while the cases CI and DIII were solved by Brouwer et al. in footnote 73. The reader is referred to these references for the detailed derivation.

[84] The functions $\exp(\pm 1/z)$ with $z \in \mathbb{C}$ are the paradigmatic examples of functions that are analytic in $\mathbb{C}\setminus\{0\}$ but their singularities at the origin $z = 0$ cannot be removed by multiplication with some function that is analytic in an open neighborhood of $z = 0$ where it possesses a zero of finite order.

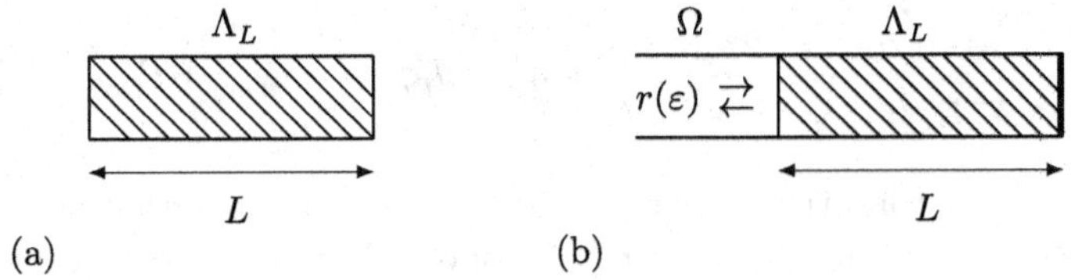

Figure 7.9 (a) A closed quantum wire on the interval $\Lambda_L \equiv [0, L]$. (b) A reservoir $\Omega \equiv]-\infty, 0[$ connected to a a quantum wire on the interval $\Lambda_L \equiv]0, L]$ that is closed by a perfectly reflecting wall at $x = L$.

in the energy interval $[-\varepsilon, +\varepsilon]$ and per unit length in the thermodynamic limit $L \to \infty$ when the disordered region Λ_L is closed on one side by a perfectly reflecting wall as shown in Figure 7.9(a). In the chiral and Bogoliubov-de-Gennes symmetry classes, the density of states per unit length is an even function of ε for each realization of the disorder. For the Wigner-Dyson symmetry classes, we expect that the average density of states per unit length is also an even function of ε. As it is anticipated that the density of states per unit length is self-averaging, that is, the variance of the density of states per unit length converges to zero as $L \to \infty$, we can safely assume that $\nu(\varepsilon)$ is also an even function of ε in the thermodynamic limit for the standard symmetry classes A, AI, and AII. Hence,

$$\mathcal{N}(\varepsilon) \equiv 2 \int\limits_0^{\varepsilon} \mathrm{d}\varepsilon' \, \nu(\varepsilon') \tag{7.371b}$$

and

$$\nu(\varepsilon) \equiv \frac{1}{2} \left(\frac{\mathrm{d}\mathcal{N}}{\mathrm{d}\varepsilon} \right) (\varepsilon) \tag{7.371c}$$

holds in the thermodynamic limit for all ten Altland-Zirnbauer symmetry classes.

Consider the geometry shown in Figure 7.9(b). Accordingly, the disordered region Λ_L is closed on its right side by a perfectly reflecting wall. Any incoming states from the reservoir Ω must necessarily be reflected from Λ_L. Let the chemical potential in the reservoir Ω be set at the single-particle energy ε. Denote with $r(\varepsilon)$ the (random) reflection matrix associated with the region $\Lambda_L =]0, L[$. It is unitary owing to the choice of perfectly reflecting boundary conditions when $x = L$. The matrix product

$$r^{\dagger}(-\varepsilon)\, r(+\varepsilon) = \mathcal{U}\, e^{2\mathrm{i}\Theta}\, \mathcal{U}^{\dagger}, \quad \Theta := \operatorname{diag}\left(\cdots \quad \phi_j \quad \cdots\right), \quad 0 \leq \phi_j < 2\pi, \tag{7.372}$$

is unitary because $r^{\dagger}(-\varepsilon)$ and $r(+\varepsilon)$ are ($\mathcal{U}$ is also unitary). As was the case with the Lyapunov exponents of the transfer matrix, the eigenvalues $e^{2\mathrm{i}\phi_j}$ perform a Brownian motion upon changing L to $L + \delta L$. It can be shown that the joint probability distribution $P_{\mathrm{DoS},\varepsilon}$ of the $N^{\star}$ independent angular variables ϕ_j obeys the Fokker-Planck equation

$$\frac{\partial P_{\mathrm{DoS},\varepsilon}}{\partial L} = \sum_{j=1}^{N^\star} \frac{\partial}{\partial \phi_j} \left(-\frac{2\,\varepsilon}{\hbar v_{\mathrm{F}}} + \frac{2}{\gamma \ell} \sin^2 \phi_j \, J_{\mathrm{DoS}} \frac{\partial}{\partial \phi_j} J_{\mathrm{DoS}}^{-1} \right) P_{\mathrm{DoS},\varepsilon}, \tag{7.373a}$$

with the boundary conditions corresponding to the normalization of $P_{\mathrm{DoS},\varepsilon}$ being independent of L, the initial conditions corresponding to the reflection matrix being the unit matrix when $L = 0$, and where the Jacobian J_{DoS} is given by[85]

$$J_{\mathrm{DoS}} := \left[\prod_{j=1}^{N^\star} \frac{1}{\sin^\gamma \phi_j} \right] \left[\prod_{1 \le j < k \le N^\star} \sin^{m_{\mathrm{o}}} (\phi_j - \phi_k) \right], \qquad m_{\mathrm{o}} := \frac{m_{\mathrm{o}}^- + m_{\mathrm{o}}^+}{2}, \tag{7.373b}$$

for the standard and Bogoliubov-de-Gennes symmetry classes and[86]

$$J_{\mathrm{DoS}} := \left[\prod_{j=1}^{N^\star} \frac{1}{\sin^\gamma \phi_j} \right] \left[\prod_{1 \le j < k \le N^\star} \sin^{m_{\mathrm{o}}^-} \left(\frac{\phi_j - \phi_k}{2} \right) \right] \tag{7.373c}$$

for the chiral symmetry classes.

The density of states $\nu(\varepsilon)$ (DoS) is to be computed from the "node-counting theorem."[87] We thus assume that the mean number of states $\mathcal{N}(\varepsilon)$ in the energy interval $]0, \varepsilon]$ and per unit length is proportional to the mean of the sum over j of the phase derivatives $\partial \phi_j / \partial L$.[88,89] We also assume that the counting function $-\mathrm{Im}\,\ln \sin(\phi_j + \mathrm{i}0^+)$ has the same average slope as ϕ_j. If so, we write

$$\mathcal{N}(\varepsilon) := \int_{-\varepsilon}^{+\varepsilon} \mathrm{d}\varepsilon' \, \nu(\varepsilon') = -\frac{D}{\pi} \sum_{j=1}^{N^\star} \frac{\partial}{\partial L} \left\langle \mathrm{Im}\,\ln \sin(\phi_j + \mathrm{i}0^+) \right\rangle_{\varepsilon, L}, \tag{7.374}$$

where $\langle \cdot \rangle_{\varepsilon,L}$ denotes the expectation value with the joint probability distribution $P_{\mathrm{DoS},\varepsilon}$.

The techniques used to evaluate Eq. (7.374) make full use of random matrix theory. We will only quote the results. In the window of energy

$$0 < \left| \frac{\varepsilon\, \tau_{\mathrm{cross}}}{\hbar} \right| \ll 1, \tag{7.375}$$

where τ_{cross} was defined in Eq. (7.365), the density of states $\nu(\varepsilon)$ in the thermodynamic limit $L \to \infty$ for the chiral and Bogoliubov-de-Gennes universality classes differ from the constant

85 M. Titov, P. W. Brouwer, A. Furusaki, and C. Mudry, Phys. Rev. B **63**(23), 235318 (2001). [Titov et al. (2001)].

86 P. W. Brouwer, C. Mudry, and A. Furusaki, Phys. Rev. Lett. **84**(13), 2913–2916 (2000). [Brouwer et al. (2000a)].

87 The node-counting theorem is the generalization to an arbitrary potential well of the fact that the energy eigenvalues of the one-dimensional Schrödinger equation with a potential square well such that their eigenfunctions are bound to the potential well are discretely spaced and strictly increasing with the number of nodes of the eigenfunctions.

88 H. Schmidt, Phys. Rev. **105**(2), 425–441 (1957). [Schmidt (1957)].

89 M. Büttiker, J. Phys.: Condens. Matter **5**(50), 9361–9378 (1993). [Büttiker (1993)].

$$\nu_0 = \frac{D\,N^\star}{\pi\,\hbar\,v_{\mathrm{F}}} \tag{7.376}$$

found for the standard symmetry classes because of the mirror symmetry in the spectrum. This difference is tabulated in **Table** 7.8. It originates from the fact that the pairing of nonvanishing eigenvalues into mirror symmetric pairs in the chiral and Bogoliubov-de-Gennes symmetry classes implies enhanced level repulsion at the band center $\varepsilon = 0$.

To appreciate the results in **Table** 7.8, we shall make a comparison with the average density of states of a chaotic quantum dot for each of the ten Altland-Zirnbauer symmetry classes, that is the average density of states of the large random $8\overline{N} \times 8\overline{N}$ Hermitean matrix $\mathcal{V}(x)$ in Eq. (7.335a) for some arbitrarily chosen x. The average density of states of a chaotic quantum dot for the chiral and Bogoliubov-de-Gennes symmetry classes is suppressed upon approaching the band center $\varepsilon = 0$ as the power law with the exponents (see **Table** 7.2)

$$\alpha_{\mathrm{AZ}} = \begin{cases} 1, & \text{symmetry class AIII}, \\ 0, & \text{symmetry class BDI}, \\ 3, & \text{symmetry class CII}, \\ & \\ 0, & \text{symmetry class D}, \\ 2, & \text{symmetry class C}, \\ 1, & \text{symmetry class DIII}, \\ 1, & \text{symmetry class CI}. \end{cases} \tag{7.377}$$

This suppression is a manifestation of the fact that eigenenergies come in pairs of opposite signs in the chiral and Bogoliubov-de-Gennes symmetry classes and of level repulsion. For a quantum wire in the thermodynamic limit $L \to \infty$, this behavior is only present for the symmetry classes C and CI among the chiral and Bogoliubov-de-Gennes symmetry classes according to **Table** 7.8.

The condition for the applicability of random matrix theory, if understood as the theory describing a chaotic quantum dot, to a quantum wire in the thermodynamic limit $L \to \infty$ is the existence of a length scale ξ, the localization length, together with the following properties. Consider two eigenfunctions ψ_1 and ψ_2 with eigenenergies ε_1 and ε_2, respectively, for a given realization of the disordered quantum wire. These eigenfunctions and their eigenvalues should be interpreted as random functions and random eigenvalues, respectively. If the location of the maximum of ψ_1 is a distance larger than ξ away from the location of the maximum of ψ_2, then both ψ_1 and ψ_2 (ε_1 and ε_1) are uncorrelated random functions (eigenvalues). Conversely, if the location of the maximum of ψ_1 is a distance less than ξ away from the location of the maximum of ψ_2, then the statistical correlations between ψ_1 and ψ_2 (ε_1 and ε_1) are captured by random matrix theory.

Table 7.8 *The tenfold way: Density of states for a disordered quantum wire in the thermodynamic limit*

AZ	$\mathrm{M}^- \equiv \mathrm{G}^\star/\mathrm{H}$ for $\mathcal{M}_{\varepsilon=0}(L,0)$	$\nu(\varepsilon)$ for $0<\varepsilon\,\tau_{\mathrm{cross}}\ll\hbar$
Standard		$\nu_0=\frac{D\,N^\star}{\pi\,\hbar\,v_{\mathrm{F}}}$
AIII, N even	$\mathrm{GL}(2N,\mathbb{C})/\mathrm{U}(2N)$	$\pi\,\nu_0\left\lvert\frac{\varepsilon\,\tau_{\mathrm{cross}}}{\hbar}\ln\left(\frac{\varepsilon\,\tau_{\mathrm{cross}}}{\hbar}\right)\right\rvert$
BDI, N even	$\mathrm{GL}(2N,\mathbb{R})/\mathrm{O}(2N)$	$\nu_0\left\lvert\ln\left(\frac{\varepsilon\,\tau_{\mathrm{cross}}}{\hbar}\right)\right\rvert$
CII, N even	$\mathrm{U}^\star(2N)/\mathrm{Sp}(2N)$	$\frac{\pi\,\nu_0}{3}\left\lvert\left(\frac{\varepsilon\,\tau_{\mathrm{cross}}}{\hbar}\right)^3\ln\left(\frac{\varepsilon\,\tau_{\mathrm{cross}}}{\hbar}\right)\right\rvert$
Chiral, N odd		$\frac{\pi\nu_0}{\left\lvert\frac{\varepsilon\,\tau_{\mathrm{cross}}}{\hbar}\ln^3\left(\frac{\varepsilon\,\tau_{\mathrm{cross}}}{\hbar}\right)\right\rvert}$
D	$\mathrm{O}(4N,4N)/[\mathrm{O}(4N)\times\mathrm{O}(4N)]$	$\frac{\pi\nu_0}{\left\lvert\frac{\varepsilon\,\tau_{\mathrm{cross}}}{\hbar}\ln^3\left(\frac{\varepsilon\,\tau_{\mathrm{cross}}}{\hbar}\right)\right\rvert}$
C	$\mathrm{Sp}(2N,2N)/[\mathrm{Sp}(2N)\times\mathrm{Sp}(2N)]$	$\nu_0\left\lvert\frac{\varepsilon\,\tau_{\mathrm{cross}}}{\hbar}\right\rvert^2$
DIII	$\mathrm{O}(4N,\mathbb{C})/\mathrm{O}(4N)$	$\frac{\pi\nu_0}{\left\lvert\frac{\varepsilon\,\tau_{\mathrm{cross}}}{\hbar}\ln^3\left(\frac{\varepsilon\,\tau_{\mathrm{cross}}}{\hbar}\right)\right\rvert}$
CI	$\mathrm{Sp}(2N,\mathbb{C})/\mathrm{Sp}(2N)$	$\frac{\pi\,\nu_0}{2}\left\lvert\frac{\varepsilon\,\tau_{\mathrm{cross}}}{\hbar}\right\rvert$

Asymptotes of the density of states $\nu(\varepsilon)$ of a quantum wire in the thermodynamic limit for small energies $0<\varepsilon\,\tau_{\mathrm{cross}}\ll\hbar$ and for $N^\star\gg1$, where τ_{cross} is given in Eq. (7.365). The chiral symmetry classes with $N^\star$ odd and the Bogoliubov-de-Gennes symmetry classes D, DIII show the same singularity at the band center.

For a quantum wire in the thermodynamic limit $L\to\infty$ and in one of the chiral symmetry classes with an even number $N^\star$ of independent Lyapunov exponents, the contribution to the average density of states from **Table** 7.8 that could have been ascribed to the density of states of a chaotic quantum dot is modified by the presence of multiplicative logarithmic corrections. These corrections are marginal since the power laws overcome the weaker logarithmic divergence at the band center. Although not exact, random matrix theory is here nevertheless "morally correct."

For a quantum wire in the thermodynamic limit $L\to\infty$ and in one of the chiral symmetry classes with an odd number $N^\star$ of independent Lyapunov exponents or for the Bogoliubov-de-Gennes symmetry classes D and DIII, the density of states displays a singularity, *the Dyson singularity*,[90]

$$\nu(\varepsilon)=\frac{\pi\nu_0}{\left\lvert\left(\frac{\varepsilon\,\tau_{\mathrm{cross}}}{\hbar}\right)\ln^3\left(\frac{\varepsilon\,\tau_{\mathrm{cross}}}{\hbar}\right)\right\rvert}. \tag{7.378}$$

The breakdown of random matrix theory upon approaching the band center $\varepsilon=0$ as measured by the Dyson singularity (7.378) is yet another signature (aside from the algebraic decay with L of the mean conductance) for a quantum phase

[90] F. J. Dyson, Phys. Rev. **92**(6), 1331–1338 (1953). [Dyson (1953)]. Dyson found this singularity in his study of the density of states of phonons in a disordered chain of coupled harmonic oscillators, a Hamiltonian which is equivalent, upon bosonization, to the continuum limit (3.243a) of the Hamiltonian used to model polyacetylene, whereby the mass $\phi(x)\equiv m(x)$ is a random function of space that is white-noise correlated and of vanishing mean and the staggered chemical potential $\mu_{\mathrm{s}}=0$ vanishes identically.

transition accompanied by diverging length scales that can be tuned by changing the sign of ε. Such a quantum phase transition is induced by disorder. At the quantum critical point, the statistical distributions of observables are different from the fixed-point distributions sufficiently far away from the quantum critical point. These critical distributions can be much broader than the latter. For this reason, many (possibly infinitely many) independent characteristic length scales diverge at such quantum critical points. By this measure, criticality induced by disorder can be richer than criticality in clean systems, for which the number of independent diverging characteristic length scales is usually finite.

7.7.6 Examples: The Symmetry Classes AIII, BDI, and CII

The Fokker-Planck equation (7.359) for the three chiral symmetry classes AIII, BDI, and CII is going to be derived starting from a specific lattice (tight-binding) Hamiltonian. Deriving the Fokker-Planck equation (7.359) for any one of the remaining seven Altland-Zirnbauer symmetry classes proceeds similarly. One important lesson from this exercise is that the Lyapunov exponents for each one of the ten Altland-Zirnbauer symmetry classes in the first column of **Table** 7.6 are two times the radial coordinates of the non-compact symmetric spaces in the second column of **Table** 7.6.

The choice of the chiral symmetry classes is made for the following reasons:

1 First, our modeling of polyacetylene with a fractional charge of 1/2 bound to a soliton in Chapters 2–4 puts polyacetylene in the chiral symmetry BDI. The inclusion of spin-orbit coupling as the electrons hop between nearest-neighbor sites puts polyacetylene in the chiral symmetry CII. Imposing the geometry of a ring in the presence of a magnetic field normal to the plane containing the ring put polyacetylene in the chiral symmetry AIII (when $N^{\star} > 1$).
2 Second, the chiral symmetry classes are the only ones for which the Jacobian (7.359b) is invariant under translation of all the Lyapunov exponents by the same real-valued number. This property facilitates the treatment of a one-body random potential (7.335c) of nonvanishing mean, that is, the treatment of the effect of disorder in the presence of a spectral gap in the clean limit is simpler for the chiral classes than it is for the standard and Bogoliubov-de-Gennes symmetry classes. In doing so, it will be possible to study the effects of disorder on (i) the chiral invertible topological phases of fermionic matter and (ii) on the quantum phase transitions separating them.
3 Third, the non-compact Lie groups for the chiral symmetry classes are semisimple with the non-compact factor subgroup $\mathbb{R}^{+}$. This brings about a breakdown of one-parameter scaling, as will be shown by explicit calculation.
4 Finally, tuning the chemical potential away from the value $\varepsilon = 0$ results in a functional change of the transfer matrix $\mathcal{M}(\varepsilon)$ that describes the crossover of

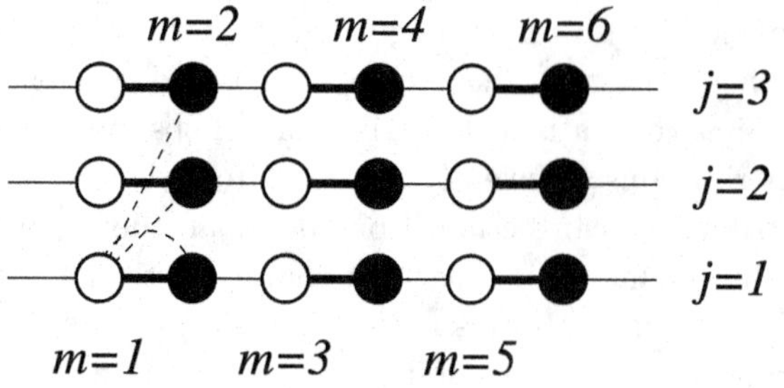

Figure 7.10 Quasi-one-dimensional lattice consisting of three chains (labeled by $j = 1, 2, 3$), each of which is made of six sites (labeled by $m = 1, \cdots, 6$). A single chain is made of open and filled circles. Deterministic hopping amplitudes only occur along a single chain. They are $t \pm t'$ with $t - t'$ $(t + t')$ being represented by a thick (thin) line. Random hopping amplitudes are represented by dashed lines [only the random hopping amplitudes between the site $(m, j) = (1, 1)$ and the three sites $(2, 1)$, $(2, 2)$, and $(2, 3)$ are shown]. Only random hopping amplitudes between open and filled circles belonging to two consecutive columns are allowed. The random hopping amplitudes are assumed to be identically and independently distributed even though they connect sites on different rows. Random hopping amplitudes are complex-valued in the symmetry class AIII, while they are real-valued in the symmetry class BDI. For the symmetry class CII, we replace $t \pm t'$ by $(t \pm t')\,\sigma_0$ with σ_0 the unit 2×2 matrix for the spin-1/2 degrees of freedom, while the random hopping amplitudes are real-quaternion-valued (see **Exercise** 7.2 for a definition of quaternions).

the statistical properties of the transfer matrix $\mathcal{M}(\varepsilon)$ from the chiral symmetry classes AIII, BDI, and CII to the Wigner-Dyson or standard symmetry classes A, AI, and AII, respectively. This crossover can be studied numerically.

We consider a quasi-one-dimensional lattice as is exemplified in Figure 7.10. Identical chains are labeled with the integer-valued indices i or j. Equilibrium atomic positions $m\,\mathfrak{a}$ along the chains are labeled with the integer-valued index m, whereby $\mathfrak{a}$ denotes the lattice spacing along any chain. A thick (thin) line represents a strong (weak) hopping amplitude that connects exclusively a pair of consecutive sites along a chain labeled by the integer j, whereby the left site from a pair is a white (black) circle and the right site from a pair is a black (white) circle. White (black) circles are all stacked along columns labeled by the integer-valued index m that are odd (even). Without disorder, all the chains are decoupled, each one realizing a deterministic tight-binding model for polyacetylene in the symmetry classes AIII, BDI, or CII. Chains are only coupled by disorder through random hopping amplitudes between any pair of sites labeled by two consecutive column indices. These random hopping amplitudes of vanishing means are independently and identically distributed, irrespective of the pair (i, j) of chains they connect, up to the symmetry constraints associated with the symmetry classes AIII, BDI, or CII. Because the hopping, whether deterministic or random, is exclusively between white and black circles, the single-particle Hamiltonian belongs to one of the chiral symmetry classes. All the directed hoppings between the site $m+1$ and m are collected into the matrix T_m. All the directed hoppings between the site m and $m+1$ are collected into the matrix $T_m^{\dagger}$. The dimensionality of the matrix T_m is the number of chains $N^{\star}$ for the symmetry classes AIII and BDI, as is appropriate for the bi-partite hopping of spinless fermions. It is $2N^{\star}$ for the symmetry classes CII, as is appropriate for the bi-partite hopping of spinfull fermions.

We assign to each site m, i of the lattice creation $\hat{c}^{\dagger}_{m,i}$ and annihilation $\hat{c}_{m,i}$ operators obeying the fermionic algebra. For the symmetry classes AIII and BDI, the fermions are spinless and their only nonvanishing anticommutators are

$$\left\{\hat{c}_{m,i}, \hat{c}^{\dagger}_{m',i'}\right\} = \delta_{mm'}\,\delta_{ii'}. \tag{7.379a}$$

For the symmetry class CII, the fermions also carry a spin-1/2 degree of freedom that is two-valued (spin up or down) that we have supressed, that is,

$$\left\{\hat{c}_{m,i}, \hat{c}^{\dagger}_{m',i'}\right\} = \delta_{mm'}\,\delta_{ii'}\,\sigma_0 \tag{7.379b}$$

(σ_0 denotes the unit 2×2 matrix in spin-1/2 space) are the only nonvanishing anticommutators. The one-body Hamiltonian is

$$\widehat{H} := \sum_{m=1}^{M-b} \sum_{i,j=1}^{N^{\star}} \left(\hat{c}^{\dagger}_{m,i}\, T_{m,ij}\, \hat{c}_{m+1,j} + \text{H.c.}\right). \tag{7.379c}$$

The choice $b=0$ ($b=1$) selects periodic (open) boundary conditions along the chain direction and it is assumed that the number M of sites per chain is even. Open boundary conditions are always selected along the direction orthogonal to the chain direction.[91] The single-particle matrix

$$T_m = \left(T_{m,ij}\right), \qquad T_{m,ij} = [t + (-1)^m\, t']\,\delta_{ij} + \delta T_{m,ij} \in \mathbb{C}, \tag{7.379d}$$

for each $m = 1, \cdots, M$ is a $N^{\star} \times N^{\star}$ complex-valued matrix in the symmetry class AIII. The single-particle matrix

$$T_m = \left(T_{m,ij}\right), \qquad T_{m,ij} = [t + (-1)^m\, t']\,\delta_{ij} + \delta T_{m,ij} \in \mathbb{R}, \tag{7.379e}$$

for each $m = 1, \cdots, M$ is a $N^{\star} \times N^{\star}$ real-valued matrix in the symmetry class BDI. The single-particle matrix

$$T_m = \left(T_{m,ij}\right), \qquad (T^{\mathsf{R}}_m)_{ij} = T_{m,ij} = [t - (-1)^m\, t']\,\delta_{ij}\,\sigma_0 + \delta T_{m,ij}, \tag{7.379f}$$

for each $m = 1, \cdots, M$ is a $N^{\star} \times N^{\star}$ real-quaternion-valued matrix in the symmetry

[91] Imposing periodic boundary conditions in the direction transversal to the chains breaks the chiral symmetry when $N^{\star}$ is odd. This sensitivity is not a mere finite-size effect. We shall see that the conductance and the density of states remain sensitive to the parity of $N^{\star}$ in the thermodynamic limit $M \to \infty$, holding either $N^{\star}$ (strict quasi-one-dimensional geometry) or $N^{\star}/M$ (thick quantum wire limit) fixed.

class CII.[92] The uniform t and staggered t' hopping amplitudes are deterministic parameters that are complex-valued in the symmetry class AIII and real-valued in the symmetry classes BDI and CII. The magnitude $|t|$ is large compared to the typical magnitude of the random variables $\delta T_{m,ij}$, that we choose to be white-noise and Gaussian correlated with the vanishing mean

$$\left\langle \delta T_{m,ij} \right\rangle_{v^2,\eta,m_{\rm o}^-} := 0 \tag{7.379g}$$

and the variance

$$\left\langle \delta T_{m,ij} \left(\delta T_{m',i'j'}\right)^{\mathsf{C}} \right\rangle_{v^2,\eta,m_{\rm o}^-} := \frac{m_{\rm o}^-}{2\gamma}\, v^2\, \delta_{mm'} \left(\delta_{ii'}\, \delta_{jj'} - \frac{1-\eta}{N^\star}\, \delta_{ij}\, \delta_{i'j'} \right), \tag{7.379h}$$

whereby the additional symmetry constraint[93]

$$\left\langle \delta T_{m,ij} \quad \delta T_{m',i'j'} \quad \right\rangle_{v^2,\eta,m_{\rm o}^-} = \frac{2-m_{\rm o}^-}{m_{\rm o}^-} \left\langle \delta T_{m,ij} \left(\delta T_{m',i'j'}\right)^{\mathsf{C}} \right\rangle_{v^2,\eta,m_{\rm o}^-} \tag{7.379i}$$

must hold for the symmetry classes BDI and CII. This distribution depends on two continuous parameters: (i) the dimensionfull continuous parameter $v^2 \geq 0$ with $v > 0$ carrying the units of energy (the same units as the hopping matrix elements) and (ii) on the dimensionless continuous parameter $\eta \geq 0$. The superscript C stands for complex conjugation in the symmetry classes AIII and BDI, while it stands for the operation of Hermitean conjugation for quaternions (7.451d) with the identification $q \to \delta T_{m,ij}$ for the symmetry class CII. Each symmetry class is assigned

$$m_{\rm o}^- := \begin{cases} 2, & \text{for the symmetry class AIII (A)}, \\ 1, & \text{for the symmetry class BDI (AI)}, \\ 4, & \text{for the symmetry class CII (AII)}, \end{cases} \tag{7.379j}$$

and[94]

$$\gamma(\eta) := \frac{1}{2} \left[m_{\rm o}^- \left(N^\star - 1\right) + 2 - \frac{2(1-\eta)}{N^\star} \right]. \tag{7.379k}$$

The effect of η in $\gamma(\eta)$ becomes vanishingly small in the limit $N^\star \to \infty$. Moreover,

$$|t| \gg |t'|, v, \qquad v \equiv \sqrt{v^2} \tag{7.379l}$$

implements the condition that the one-dimensional band width $|t|$ is much larger

[92] According to the definition (7.456b),

$$T^{\mathsf{R}}_{m,ij} \equiv \left(T^{\mathsf{R}}_{m}\right)_{ij} := \sigma_2 \left(T^*_m\right)_{ij} \sigma_2 = (\mathrm{i}\sigma_2) \left(T^*_m\right)_{ij} (\mathrm{i}\sigma_2)^{-1} \equiv (\mathrm{i}\sigma_2)\, T^*_{m,ij}\, (\mathrm{i}\sigma_2)^{-1}.$$

The condition (7.379f) is thus that of R-self-duality. The matrix T_m, if expressed in terms of complex numbers, is a $2N^\star \times 2N^\star$ matrix.

[93] For example, if the matrix A is real-valued and symmetric, then the matrix element A_{ij} is correlated with the matrix element A_{ji}.

[94] When $\eta = 1$, Eq. (7.379k) is nothing but Eq. (7.349b) for the chiral symmetry classes AIII, BDI, and CII.

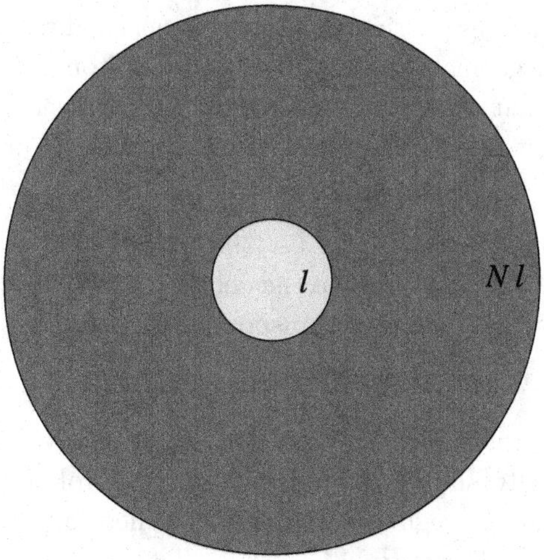

Figure 7.11 When the localization length ξ is much larger than the mean free path ℓ, one may identify three distinct regimes in Anderson localization. The ballistic regime when the linear system size $L \lesssim \ell$, the diffusive regime $\ell \ll L \ll \xi$, and the localized regime $\xi \ll L$. In thick quantum wires, the role of ξ is played by $N\,\ell$ with $N \gg 1$ the number of quasi-one-dimensional channels.

than either the dimerization gap $4\,|t'|$ or the disorder strength v. It is convenient to characterize the disorder strength v by a length scale, the mean free path ℓ

$$\ell := \frac{(\hbar\, v_{\mathrm{F}})^2}{2\, v^2\, \mathfrak{a}}. \tag{7.379m}$$

As ℓ is inversely proportional to v^2, the power ℓ^{-1} must appear to second order in perturbation theory of the disorder.[95] The dimensionfull constants v_{F} and $\mathfrak{a}$ are the Fermi speed and the lattice spacing along a chain, respectively. The unitary (gauge) transformation

$$\hat{c}_{m,i} \mapsto (-1)^m\, \hat{c}_{m,i} \tag{7.379n}$$

is the origin of a spectral symmetry. We call this unitary transformation a sublattice transformation and the associated unitary spectral symmetry is identified with the attribute that defines the chiral symmetry classes.

Definition 7.10 (Strict quasi-one-dimensional geometry) The limit

$$M \to \infty \text{ holding } N^{\star} \text{ fixed} \tag{7.380}$$

[95] Perturbation theory is defined by two steps. First, an observable is expanded in powers of the random matrix elements of the Hamiltonian. Second, this expansion is averaged with the Gaussian distribution defined by Eqs. (7.379g), (7.379h), and (7.379i). Because of Eq. (7.379g), the first nonvanishing contribution cannot be lower than the order $v^2 \sim \ell^{-1}$.

of the model (7.379) defines a strict quasi-one-dimensional wire. This definition of a quasi-one-dimensional geometry is strict because Hamiltonian (7.379c) is necessarily local in this limit in that all matrix elements of T_m connect pair of sites separated by a finite distance no greater than a distance of order $N^\star$ in units of the lattice spacing.[96]

The choice (7.379g), (7.379h), and (7.379i) amounts to presuming a diffusive transport in the direction transverse to the chains for any choice of $N^\star$. This choice precludes two-dimensional quantum transport in the limit

$$M, N^\star \to \infty \text{ holding } N^\star/M \text{ fixed}, \tag{7.381}$$

for the probability to hop between chains that are many lattice spacings apart should be exponentially suppressed relative to the probability to hop between nearby chains in any lattice regularization of two-dimensional Euclidean space compatible with the notion of locality.

Definition 7.11 (thick quantum wire limit) The thick quantum wire limit is defined by the limit (7.381) in combination with the choice (7.379g), (7.379h), and (7.379i) for the hopping matrix elements.

The thick quantum wire limit allows to explore the crossover from diffusive to localized quantum transport in the standard symmetry classes A, AI, and AII (see Figure 7.11). Whereas, one-parameter scaling is not expected to hold in the strict quasi-one-dimensional limit (7.380) for the standard symmetry classes A, AI, and AII, one-parameter scaling as embodied by the corresponding Fokker-Planck equation (7.359) is believed[97] to become exact in the thick quantum wire limit for the standard universality classes. This is not so for the chiral symmetry classes if η is allowed to scale with $N^\star$. To substantiate this claim, the continuum limit is taken in order to derive explicitly the Fokker-Planck equation for any one of the chiral symmetry classes AIII, BDI, and CII. The resulting Fokker-Planck equation (7.387) will be seen to depart, if $\eta \neq 1$, from the Fokker-Planck equation (7.359) selected by the chiral symmetry classes.

The hierarchy of energy scales (7.379l) suggests to take the continuum limit along each chain with the lattice spacing $\mathfrak{a}$.[98] This gives equations of motion of the Dirac type, which we choose to represent by the stationary eigenvalue equation (**Exercise** 7.27)

$$\varepsilon\,\Psi(x) = \mathcal{H}(x)\,\Psi(x) \tag{7.382a}$$

[96] Locality in space is defined by the condition that the elementary building blocks of matter, say electrons, entering a Hamiltonian in the position basis are coupled by terms with a magnitude that decreases no less than exponentially fast with the distance separating the elementary building blocks of matter.

[97] This belief is grounded on comparing predictions for quantum transport based on the Fokker-Planck equation in the thick quantum wire limit (or, equivalently, on a nonlinear sigma model) against numerical simulations of distinct random lattice Hamiltonians within one of the standard symmetry classes A, AI, or AII.

[98] Numerical simulations are required otherwise.

with the single-particle Dirac Hamiltonian (as usual, the 2×2 grading for the left and right movers is encoded by the unit 2×2 matrix τ_0 and the vector $\boldsymbol{\tau}$ of Pauli matrices)

$$\begin{aligned}\mathcal{H}(x) &= \mathbb{1}_N \otimes \tau_3\, \hbar v_{\mathrm{F}}\, \mathrm{i}\partial_x + \mathbb{1}_N \otimes \tau_2\, m\, v_{\mathrm{F}}^2 + \hbar v_{\mathrm{F}}\, \mathsf{v}(x) \otimes \tau_3 + \hbar v_{\mathrm{F}}\, \mathsf{w}(x) \otimes \tau_2 \\ &= -\left(\mathbb{1}_N \otimes \tau_1\right) \mathcal{H}(x) \left(\mathbb{1}_N \otimes \tau_1\right),\end{aligned} \tag{7.382b}$$

where $(t>0)$

$$v_{\mathrm{F}} = \frac{2\,|t|\,\mathfrak{a}}{\hbar} = \frac{2\,t\,\mathfrak{a}}{\hbar}, \qquad m\, v_{\mathrm{F}}^2 = 2\,t', \tag{7.382c}$$

and

$$N = N^{\star}, \quad \mathsf{v}^{\dagger}(x) = \mathsf{v}(x), \qquad \mathsf{w}^{\dagger}(x) = \mathsf{w}(x), \tag{7.382d}$$

$$N = N^{\star}, \quad \mathsf{v}^{\dagger}(x) = \mathsf{v}^{*}(x) = \mathsf{v}(x), \quad \mathsf{w}^{\dagger}(x) = \mathsf{w}^{*}(x) = \mathsf{w}(x), \tag{7.382e}$$

$$N = 2N^{\star}, \quad \mathsf{v}^{\mathsf{R}}(x) = \mathsf{v}(x) = \mathsf{v}^{\dagger}(x), \quad \mathsf{w}^{\mathsf{R}}(x) = \mathsf{w}(x) = \mathsf{w}^{\dagger}(x), \tag{7.382f}$$

for the chiral symmetry classes AIII, BDI, and CII, respectively. It is convenient to measure the ratio between the dimerization $t' \in \mathbb{R}$ and the variance $v^2 \geq 0$ of the disorder with the dimensionless parameter

$$f := -\frac{2\,\gamma(\eta)\, t'\, \hbar v_{\mathrm{F}}}{m_{\mathrm{o}}^{-}\, v^2\, \mathfrak{a}} = -\frac{4\,\gamma(\eta)\, t'\, \ell}{m_{\mathrm{o}}^{-}\, \hbar v_{\mathrm{F}}}. \tag{7.382g}$$

We choose units such that $\hbar = v_{\mathrm{F}} = 1$. The matrix-valued random potentials $\mathsf{v} = (v_{ij})$ and $\mathsf{w} = (w_{ij})$ have then the dimensions of inverse length. Their distributions are fixed by Eqs. (7.379g), (7.379h), and (7.379i). Their means are

$$\left\langle v_{ij}(x) \right\rangle_{\ell,\eta,m_{\mathrm{o}}^{-}} = \left\langle w_{ij}(x) \right\rangle_{\ell,\eta,m_{\mathrm{o}}^{-}} = 0. \tag{7.382h}$$

Their variances are

$$\begin{aligned}\left\langle v_{ij}(x)\, \left[v_{kl}(y)\right]^{\mathsf{C}} \right\rangle_{\ell,\eta,m_{\mathrm{o}}^{-}} = \frac{m_{\mathrm{o}}^{-}}{2\,\gamma(\eta)\,\ell}\, \delta(x-y) \Bigg[&\delta_{ik}\, \delta_{jl} - \frac{2-m_{\mathrm{o}}^{-}}{m_{\mathrm{o}}^{-}}\, \delta_{il}\, \delta_{jk} \\ &- \frac{2(m_{\mathrm{o}}^{-}-1)(1-\eta)}{m_{\mathrm{o}}^{-}\, N^{\star}}\, \delta_{ij}\, \delta_{kl} \Bigg],\end{aligned} \tag{7.382i}$$

$$\begin{aligned}\left\langle w_{ij}(x)\, \left[w_{kl}(y)\right]^{\mathsf{C}} \right\rangle_{\ell,\eta,m_{\mathrm{o}}^{-}} = \frac{m_{\mathrm{o}}^{-}}{2\,\gamma(\eta)\,\ell}\, \delta(x-y) \Bigg[&\delta_{ik}\, \delta_{jl} + \frac{2-m_{\mathrm{o}}^{-}}{m_{\mathrm{o}}^{-}}\, \delta_{il}\, \delta_{jk} \\ &- \frac{2(1-\eta)}{m_{\mathrm{o}}^{-}\, N^{\star}}\, \delta_{ij}\, \delta_{kl} \Bigg],\end{aligned} \tag{7.382j}$$

$$\left\langle v_{ij}(x)\, \left[w_{kl}(y)\right]^{\mathsf{C}} \right\rangle_{\ell,\eta,m_{\mathrm{o}}^{-}} = 0. \tag{7.382k}$$

Hereto, the superscript C stands for complex conjugation in the symmetry classes AIII and BDI, while it stands for the operation of Hermitean conjugation for quaternions (7.451d) with the identification $q \to v_{ij}, w_{ij}$ for the symmetry class CII. Moreover, we are using the mean-free path ℓ defined by Eq. (7.379m) as a measure of the disorder strength (the disorder is weak if the ratio $\ell/\mathfrak{a} \gg 1$ with $\mathfrak{a}$ the lattice spacing).

The stationary eigenvalue equation (7.382a) is solved by [as was the case with Eq. (7.348c)]

$$\Psi(x) = \mathcal{M}_{\varepsilon}(x, x_0)\, \Psi(x_0), \tag{7.383a}$$

$$\mathcal{M}_{\varepsilon}(x, x_0) := \mathsf{P}_y\, e^{\int_{x_0}^{x} \mathrm{d}y \left[-\mathrm{i}\varepsilon\, \mathbb{1}_N \otimes \tau_0 + m\, \mathbb{1}_N \otimes \tau_1 + \mathrm{i}\mathsf{v}(y) \otimes \tau_0 + \mathsf{w}(y) \otimes \tau_1\right]}, \tag{7.383b}$$

where

$$N = \begin{cases} N^\star, & \text{in the symmetry classes A or AIII,} \\ N^\star, & \text{in the symmetry classes AI or BDI,} \\ 2N^\star, & \text{in the symmetry classes AII or CII.} \end{cases} \tag{7.383c}$$

If the disordered region in the scattering geometry corresponding to Figure (7.7) is the interval $]0, L[$, we define $\mathcal{M}_{\varepsilon}(L)$ to be the transfer matrix (7.383b) with $x_0 = 0$ and $x = L$.

On the one hand, the singular-value decomposition of $\mathcal{M}_{\varepsilon \neq 0}(L)$ with the left- and right-moving grading made explicit is

$$\mathcal{M}_{\varepsilon \neq 0}(L) = \begin{pmatrix} \mathsf{m}_{11} & \mathsf{m}_{12} \\ \mathsf{m}_{21} & \mathsf{m}_{22} \end{pmatrix} = \begin{pmatrix} \mathsf{u}_1 & 0 \\ 0 & \mathsf{u}_2 \end{pmatrix} \begin{pmatrix} \cosh \mathsf{x} & \sinh \mathsf{x} \\ \sinh \mathsf{x} & \cosh \mathsf{x} \end{pmatrix} \begin{pmatrix} \mathsf{v}_1 & 0 \\ 0 & \mathsf{v}_2 \end{pmatrix}, \tag{7.384a}$$

where

$$\begin{aligned} &\mathsf{u}_1, \mathsf{u}_2 \in \mathrm{U}(N), \qquad \mathsf{v}_1, \mathsf{v}_2 \in \mathrm{U}(N), \qquad N = N^\star, \\ &\mathsf{x} := \mathrm{diag}\left(x_1 \quad \cdots \quad x_{N^\star}\right), \qquad x_1, \cdots, x_{N^\star} \geq 0 \end{aligned} \tag{7.384b}$$

for the symmetry class A,

$$\begin{aligned} &\mathsf{u}_2 = \mathsf{u}_1^* \in \mathrm{U}(N), \qquad \mathsf{v}_2 = \mathsf{v}_1^* \in \mathrm{U}(N), \qquad N = N^\star, \\ &\mathsf{x} := \mathrm{diag}\left(x_1 \quad \cdots \quad x_{N^\star}\right), \qquad x_1, \cdots, x_{N^\star} \geq 0 \end{aligned} \tag{7.384c}$$

for the symmetry class AI, and

$$\begin{aligned} &\mathsf{u}_2 = \mathsf{u}_1^{\mathsf{R}} \in \mathrm{U}(N), \qquad \mathsf{v}_2 = \mathsf{v}_1^{\mathsf{R}} \in \mathrm{U}(N), \qquad N = 2N^\star, \\ &\mathsf{x} := \mathrm{diag}\left(x_1 \quad x_1 \quad \cdots \quad x_{N^\star} \quad x_{N^\star}\right), \qquad x_1, \cdots, x_{N^\star} \geq 0 \end{aligned} \tag{7.384d}$$

for the symmetry class AII.

On the other hand, the singular-value decomposition of $\mathcal{M}_{\varepsilon=0}(L)$ is

$$\mathcal{M}_{\varepsilon=0}(L) = \begin{pmatrix} \mathsf{m}_{11} & \mathsf{m}_{12} \\ \mathsf{m}_{21} & \mathsf{m}_{22} \end{pmatrix} = \begin{pmatrix} \mathsf{u} & 0 \\ 0 & \mathsf{u} \end{pmatrix} \begin{pmatrix} \cosh \mathsf{x} & \sinh \mathsf{x} \\ \sinh \mathsf{x} & \cosh \mathsf{x} \end{pmatrix} \begin{pmatrix} \mathsf{v} & 0 \\ 0 & \mathsf{v} \end{pmatrix}, \tag{7.385a}$$

where

$$\begin{aligned}&\mathsf{u}\in \mathrm{U}(N), \qquad \mathsf{v}\in \mathrm{U}(N), \qquad N=N^{\star},\\ &\mathsf{x}:=\mathrm{diag}\left(x_1 \quad \cdots \quad x_{N^\star}\right), \qquad x_1,\cdots,x_{N^\star}\in\mathbb{R}\end{aligned} \tag{7.385b}$$

for the symmetry class AIII,

$$\begin{aligned}&\mathsf{u}=\mathsf{u}^{*}\in \mathrm{O}(N)\subset \mathrm{U}(N), \quad \mathsf{v}=\mathsf{v}^{*}\in \mathrm{O}(N)\subset \mathrm{U}(N), \qquad N=N^{\star},\\ &\mathsf{x}:=\mathrm{diag}\left(x_1 \quad \cdots \quad x_{N^\star}\right), \qquad x_1,\cdots,x_{N^\star}\in\mathbb{R}\end{aligned} \tag{7.385c}$$

for the symmetry class BDI, and

$$\begin{aligned}&\mathsf{u}=\mathsf{u}^{\mathsf{R}}\in \mathrm{Sp}(2N^{\star})\subset \mathrm{U}(2N^{\star}), \quad \mathsf{v}=\mathsf{v}^{\mathsf{R}}\in \mathrm{Sp}(2N^{\star})\subset \mathrm{U}(2N^{\star}),\\ &\mathsf{x}:=\mathrm{diag}\left(x_1 \quad x_1 \quad \cdots \quad x_{N^\star} \quad x_{N^\star}\right),\ x_1,\cdots,x_{N^\star}\in\mathbb{R},\ N=2N^{\star}\end{aligned} \tag{7.385d}$$

for the symmetry class CII.

At the fixed point $\varepsilon=0$ of the single-particle spectrum of the Dirac Hamiltonian (7.382b) under the unitary implementation of $\varepsilon\to-\varepsilon$, the chiral spectral symmetry implies the singular-value decomposition (7.385). Given the Lyapunov exponents $x_1, \cdots, x_{N^\star}\in\mathbb{R}$ entering the Hermitean matrix

$$2\,\mathsf{m}_{11}\,\mathsf{m}_{12}^{\dagger}=2\,\mathsf{u}\,\cosh(\mathsf{x})\,\mathsf{v}\,\mathsf{v}^{\dagger}\,\sinh(\mathsf{x})\,\mathsf{u}^{\dagger}=\mathsf{u}\,\sinh(2\,\mathsf{x})\,\mathsf{u}^{\dagger}, \tag{7.386}$$

we seek their changes if the length of the disordered region is increased by the infinitesimal amount δL. As is shown in **Exercise** 7.28, computing the changes of the eigenvalues of the Hermitean matrix $2\,\mathsf{m}_{11}\,\mathsf{m}_{12}^{\dagger}$ induced by $L\to L+\delta L$ is an exercise in nondegenerate perturbation theory. The resulting random walk obeyed by the Lyapunov exponents $x_1, \cdots, x_{N^\star}\in\mathbb{R}$ is encoded by the Fokker-Planck equation[99]

$$\frac{\partial P_{\varepsilon=0}}{\partial L}=\frac{1}{2\,\gamma(\eta)\,\ell}\sum_{i,j=1}^{N^\star}\frac{\partial}{\partial x_i}\left[m_{\mathrm{o}}^{-}\,f\,\delta_{ij}+\left(\delta_{ij}-\frac{1-\eta}{N^\star}\right)J_{\varepsilon=0}\,\frac{\partial}{\partial x_j}\,J_{\varepsilon=0}^{-1}\right]P_{\varepsilon=0}, \tag{7.387a}$$

$$J_{\varepsilon=0}(x_1,\cdots,x_{N^\star}):=\prod_{1\le j<k\le N^\star}|\sinh(x_j-x_k)|^{m_{\mathrm{o}}^{-}}. \tag{7.387b}$$

The Jacobian (7.194c) for the chiral symmetry classes follow if one rescales the Lyapunov exponents by the multiplicative factor 1/2.

For any chiral symmetry class, the Fokker-Planck equation (7.387) depends on three continuous parameters. There is the mean free path $\ell\ge 0$. There is the dimensionless dimerization $f\in\mathbb{R}$. There is the ratio $\eta\ge 0$ of two independent diffusion constants, as we now explain. Introduce the "center of mass"

$$\bar{x}:=\frac{x_1+\cdots+x_{N^\star}}{N^\star}\in\mathbb{R}^{+} \tag{7.388a}$$

[99] P. W. Brouwer, C. Mudry, and A. Furusaki, Nucl. Phys. B **565**, 653–663 (2000). [Brouwer et al. (2000)].

that we may always choose to be positive owing to the global $\mathbb{Z}_2$ symmetry of the Fokker-Planck equation (7.387) under $x_j \to -x_j$ for all independent channels. Introduce the $N^\star$ "relative coordinates"

$$y_j := x_j - \bar{x} \in \mathbb{R}, \qquad j = 1, \cdots, N^\star, \tag{7.388b}$$

of which only $N^\star - 1$ are independent since

$$\bar{y} := \frac{1}{N^\star} \sum_{j=1}^{N^\star} y_j = 0. \tag{7.388c}$$

One then verifies (**Exercise** 7.29) that the Fokker-Planck equation separates into two independent Fokker-Planck equations given by

$$\frac{\partial \bar{P}_{\varepsilon=0}}{\partial L} = \frac{\partial}{\partial \bar{x}} \left(D_{\text{rel}}\, m_{\text{o}}^{-}\, f + \frac{\overline{D}}{N^\star} \frac{\partial}{\partial \bar{x}} \right) \bar{P}_{\varepsilon=0} \tag{7.389a}$$

for the center of mass and

$$\left. \frac{\partial P^{\text{rel}}_{\varepsilon=0}}{\partial L} \right|_{\bar{y}=0} = D_{\text{rel}} \sum_{j=1}^{N^\star} \frac{\partial}{\partial y_j} \left[J_{\varepsilon=0} \frac{\partial}{\partial y_j} \left(J^{-1}_{\varepsilon=0}\, P^{\text{rel}}_{\varepsilon=0} \right) \right] \Bigg|_{\bar{y}=0} \tag{7.389b}$$

for the relative coordinates, where

$$\overline{D} = D_{\text{rel}}\, \eta, \qquad D_{\text{rel}} = \frac{1}{2\,\gamma(\eta)\, \ell}. \tag{7.389c}$$

The choice $\eta = 0$ corresponds to a purely driven motion of the center of mass. The choice $\eta = 1$ corresponds to the case when the motion of the center of mass and the motion of the relative coordinates share the same diffusion constant. It is only for $\eta = 1$ and $f = 0$ that one recovers the Fokker-Planck equation (7.359) selected by the chiral symmetry classes. The impact of η is the strongest the smaller $N^\star$. Changing η will only affect transport properties that are dominated by the Lyapunov exponent that remains close to the value zero by terms of order $1/N^\star$. Hence, when $f = 0$, a universal conductance distribution is still expected in the thick quantum wire limit $N^\star \to \infty$, an expectation that has been confirmed numerically (see footnote 83).

If the chemical potential ε in the reservoirs Ω and Ω' from Figure (7.7) is much larger in magnitude than the dimerization gap $2\,\Delta_{\text{dimer}}$, we may set $t' = 0$ and $f = 0$ in Eq. (7.382g). As the chemical potential in the reservoirs selects the nonvanishing single-particle energy ε of the left- and right-movers in the reservoirs, the chiral spectral symmetry is inoperative as a constraint on the functional form of the transfer matrix and the only symmetries that may affect the functional form of the transfer matrix are the absence of reversal of time symmetry or its presence with or without spin-rotation symmetry. The resulting symmetry constraints are those associated with the Wigner-Dyson or standard symmetry classes, for which the Lyapunov exponents $x_1, \cdots, x_{N^\star}$ can always be chosen to be nonnegative. Given the Lyapunov exponents $x_1, \cdots, x_{N^\star} \geq 0$ entering the Hermitean matrix $\frac{1}{2}\log \mathcal{M}_\varepsilon \mathcal{M}_\varepsilon^\dagger$, we seek their changes if the length of the disordered region is increased by the infinitesimal amount δL. As was shown by Dorokhov, Mello, Pereyra, and Kumar,[100]

$$\frac{\partial P_\varepsilon}{\partial L} = \frac{1}{2\gamma\ell}\sum_{j=1}^{N^\star}\frac{\partial}{\partial x_j}\left[J_\varepsilon\frac{\partial}{\partial x_j}\left(J_\varepsilon^{-1}P_\varepsilon\right)\right], \tag{7.390a}$$

$$J_\varepsilon(\boldsymbol{x}) := \left[\prod_{j=1}^{N^\star}|\sinh 2x_j|^{m_1}\right]\left[\prod_{1\leq j<k\leq N^\star}|\sinh(x_j-x_k)|^{m_o^-}\,|\sinh(x_j+x_k)|^{m_o^+}\right], \tag{7.390b}$$

where $\boldsymbol{x} = (x_1, \cdots, x_{N^\star})$, $x_j \geq 0$, and with the long positive root $m_l = 1$ and the ordinary positive roots $m_0^- = 2, 1, 4$ for the symmetry classes A, AI, and AII, respectively. The Fokker-Planck equation (7.390) applies for sufficiently large $|\varepsilon|$, that is, if

$$1 \ll \left|\frac{\varepsilon}{4\,t'}\right|, \qquad 1 \lesssim \left|\frac{\varepsilon\,\tau_{\text{cross}}}{\hbar}\right|, \tag{7.391}$$

where $2\,\Delta_{\text{dimer}} = 4\,|t'|$ was computed in Eq. (2.59) for polyacetylene and τ_{cross} was defined in Eq. (7.365).

The most important differences between the Fokker-Planck equations (7.387) and (7.390) are the symmetries of the Jacobians. In Eq. (7.387), the Jacobian is invariant under a simultaneous translation

$$x_j \to x_j + \delta x, \qquad x_j, \delta x \in \mathbb{R}, \tag{7.392}$$

and under a simultaneous reflection

$$x_j \to -x_j \tag{7.393}$$

for all independent channels $j = 1, \cdots, N^\star$. The translation invariance decouples the (Brownian) motion of the "center of mass" (7.388a) from that of the "relative coordinates" (7.388b). This decoupling is a consequence of the fact that the Lyapunov exponents perform a (Brownian) motion on the non-compact and semi-simple symmetric spaces

$$\mathrm{G}^\star/\mathrm{H} = \begin{cases} \mathrm{GL}(N^\star,\mathbb{C})/\mathrm{U}(N^\star) = \mathbb{R}^+\times\mathrm{U}(1)\times\mathrm{SL}(N^\star,\mathbb{C})/\mathrm{U}(N^\star), & \text{for AIII},\\ \mathrm{GL}(N^\star,\mathbb{R})/\mathrm{O}(N^\star) = \mathbb{R}^+\times\mathbb{Z}_2\times\mathrm{SL}(N^\star,\mathbb{R})/\mathrm{O}(N^\star), & \text{for BDI},\\ \mathrm{U}^\star(2N^\star)/\mathrm{Sp}(2N^\star) = \mathbb{R}^+\times\mathrm{U}(1)\times\mathrm{SU}^\star(2N^\star)/\mathrm{Sp}(2N^\star), & \text{for CII}. \end{cases} \tag{7.394}$$

[100] Dorokhov, Mello, Pereyra, and Kumar express Eq. (7.390) in terms of the transmission eigenvalues $\lambda_j := (1-\tanh^2 x_j)^{1/2}$, not the Lyapunov exponents x_j. They undergo a random walk encoded by the Fokker-Planck equation.

Each simple and non-compact factor in the Cartesian product

$$\begin{aligned} \mathrm{G}^{\star} &= \bar{\mathrm{G}}^{\star} \times \widetilde{\mathrm{G}}^{*}, \\ \bar{\mathrm{G}}^{\star} &\equiv \mathbb{R}^{+} \times \mathrm{U}(1),\ \mathbb{R}^{+} \times \mathbb{Z}_2,\ \mathbb{R}^{+} \times \mathrm{U}(1), \\ \widetilde{\mathrm{G}}^{*} &= \mathrm{SL}(N^{\star},\mathbb{C}),\ \mathrm{SL}(N^{\star},\mathbb{R}),\ \mathrm{SU}^{\star}(2N^{\star}), \end{aligned} \tag{7.395}$$

can acquire independent diffusion constants that are parameterized by η and ℓ, respectively. The diffusion of the center of mass with the diffusion constant

$$\frac{\overline{D}}{N^{\star}} = \frac{D_{\mathrm{rel}}\,\eta}{N^{\star}}, \qquad D_{\mathrm{rel}} = \frac{1}{2\,\gamma(\eta)\,\ell}, \qquad \eta \neq 1 \tag{7.396}$$

in the Fokker-Planck equation (7.387) induces corrections of order $1/N^{\star}$ to the one-parameter scaling encoded by the Fokker-Planck equations (7.359) for the chiral symmetry classes that apply when $\varepsilon = 0$ and there is no dimerization gap, that is, $f = 0$. In the standard Fokker-Planck equation (7.390) that applies for energies ε sufficiently far away from the band center (that is $\varepsilon = 0$), the Jacobian is invariant under any reflection

$$x_j \to \zeta_j\, x_j, \qquad \zeta_j = \pm 1. \tag{7.397}$$

It is the reduction of this "local" reflection symmetry $(\mathbb{Z}_2)^{N^{\star}}$ sufficiently far away from $\varepsilon = 0$ to the "global" reflection symmetry $\mathbb{Z}_2$ that is responsible for anomalies in transport properties at $\varepsilon = 0$. Moreover, the absence of translation invariance (7.392) in the standard Fokker-Planck equation (7.390) can be interpreted as the fact that they describe Brownian motions on non-compact symmetric spaces that, although not simple, have no non-compact factor groups. This is why η drops out of the standard Fokker-Planck equation (7.390) and one-parameter scaling is recovered sufficiently far away from $\varepsilon = 0$.

As opposed to the diffusive regime $\ell \ll L \ll N^{\star}\,\ell$, [79] no analytical approach is known that describes the full crossover for any given L as a function of ε from the band center, where Eq. (7.387) rules, and sufficiently far away from the band center, where Eq. (7.390) rules. More generally, this is also true of symmetry crossovers such as between the symmetry classes AI and A upon switching a magnetic field.

We are going to solve the chiral Fokker-Planck equation (7.387) in the limit of very large L and contrast this solution to that of the standard Fokker-Planck equation (7.390).

Our starting point is the Fokker-Planck equation (7.387). For notational simplicity, we drop the subscript $\varepsilon = 0$ refering to the band center and make the Ansatz (**Exercise** 7.29)

$$P(x_1,\cdots,x_{N^{\star}},L) =: \bar{P}(\bar{x},L)\,P_{\mathrm{rel}}(y_1,\cdots,y_{N^{\star}},L)\big|_{\bar{y}=0} \tag{7.398}$$

in terms of the center of mass coordinate (7.388a) and the relative coordinates (7.388b), the latter obeying the condition (7.388c). With this change of variables, the Fokker-Planck equation (7.387) separates into the Fokker-Planck equation for the center of mass given by

$$\frac{\partial \bar{P}}{\partial L} = \frac{1}{2\,\gamma(\eta)\,\ell}\frac{\partial}{\partial \bar{x}}\left(m_{\mathrm{o}}^{-}\,f + \frac{\eta}{N^{\star}}\frac{\partial}{\partial \bar{x}}\right)\bar{P} \tag{7.399}$$

and the Fokker-Planck equation for the relative coordinates given by

$$\left.\frac{\partial P_{\mathrm{rel}}}{\partial L}\right|_{\bar{y}=0} = \frac{1}{2\,\gamma(\eta)\,\ell}\sum_{j=1}^{N^{\star}}\frac{\partial}{\partial y_j}\left.\left[\left(\frac{\partial}{\partial y_j} + \frac{\partial \Omega}{\partial y_j}\right)P_{\mathrm{rel}}\right]\right|_{\bar{y}=0}, \tag{7.400a}$$

where we have introduced the "potential"

$$\begin{aligned}\Omega(y_1,\cdots,y_{N^{\star}}) :=&\ \ln\left(J^{-1}(y_1,\cdots,y_{N^{\star}})\right)\\ =&-m_{\mathrm{o}}^{-}\sum_{k=1}^{N^{\star}}\sum_{l=1}^{k-1}\ln|\sinh(y_l-y_k)|.\end{aligned} \tag{7.400b}$$

According to Eq. (7.399), the center of mass $\bar{x}$ of the fictitious particles with coordinates $x_j \in \mathbb{R}$ performs a Brownian motion on the real line $\mathbb{R}$ subject to a "constant force" $m_{\mathrm{o}}^{-}\,f$ as L increases. According to Eq. (7.400), the fictitious particles with the relative coordinates $y_j \in \mathbb{R}$ perform a Brownian motion on the real line $\mathbb{R}$ subject to the repulsive, hard-core, translation-invariant, two-body potential Ω as L increases. Because Ω has a hard core, we may assume without loss of generality that $y_1 < \cdots < y_{N^{\star}}$ for all L. Moreover, as a result of their long-range repulsive interactions, the distances between the y_j's will grow with increasing L, until the inequalities

$$y_1 \ll \cdots \ll y_{N^{\star}} \tag{7.401a}$$

hold for sufficiently large L. As $|y_l - y_k|$ for $l \neq k$ is a large positive number, we may then do the approximations

$$\ln|\sinh(y_l - y_k)| \approx |y_l - y_k| \tag{7.401b}$$

and

$$\frac{\partial \Omega}{\partial y_j} \approx -m_{\mathrm{o}}^{-}\,(N^{\star}+1-2j) \tag{7.401c}$$

in Eq. (7.400a). These approximations give the asymptotic limit

$$\left.\frac{\partial P_{\mathrm{rel}}}{\partial L}\right|_{\bar{y}=0} \approx \frac{1}{2\,\gamma(\eta)\,\ell}\sum_{j=1}^{N^{\star}}\frac{\partial}{\partial y_j}\left.\left[\left(\frac{\partial}{\partial y_j} - m_{\mathrm{o}}^{-}\,(N^{\star}+1-2j)\right)\right]P_{\mathrm{rel}}\right|_{\bar{y}=0} \tag{7.401d}$$

for $L \gg N^{\star}\,\ell$. In this limit, the solution of the Fokker-Planck equation with the initial conditions

$$P(x_1,\cdots,x_{N^{\star}};L=0) = \delta(x_j)\cdots\delta(x_{N^{\star}}) \tag{7.402a}$$

is given by (**Exercise** 7.30)

$$
\begin{aligned}
P(x_1,\cdots,x_{N^\star};L) &= \bar{P}(\bar{x},L)\times P_{\mathrm{rel}}(y_1,\cdots,y_{N^\star};L)\big|_{\bar{y}=0} \\
&\sim \left[\frac{\gamma(\eta)\,N^\star\,\ell}{2\,\pi\,\eta\,L}\right]^{\frac{1}{2}} e^{-\frac{\gamma(\eta)\,N^\star\,\ell}{2\,\eta\,L}\left(\bar{x}+\frac{m_{\mathrm{o}}^-\,f}{2\,\gamma(\eta)\,\ell}\,L\right)^2} \\
&\quad\times \left[\frac{\gamma(\eta)\,\ell}{2\,\pi\,L}\right]^{\frac{N^\star}{2}} \prod_{j=1}^{N^\star} e^{-\frac{\gamma(\eta)\,\ell}{2\,L}\left(y_i-\frac{L}{\xi_j}\right)^2}\Bigg|_{\bar{y}=0},
\end{aligned}
\tag{7.402b}
$$

where $y_1,\cdots,y_{N^\star},\bar{x}\in\mathbb{R}$ and with

$$
\xi_j=\frac{2\,\gamma(\eta)\,\ell}{m_{\mathrm{o}}^-\,(N^\star+1-2j)}\in\mathbb{R} \tag{7.402c}
$$

and

$$
\gamma(\eta)=\frac{1}{2}\left[m_{\mathrm{o}}^-\,(N^\star-1)+2-\frac{2(1-\eta)}{N^\star}\right]. \tag{7.402d}
$$

In the thermodynamic limit $L\to\infty$, the probability distribution for the Lyapunov exponents thus factorizes into the product of $N^\star$ independent Gaussian distributions with the means and variances

$$
\langle\bar{x}\rangle=-\frac{m_{\mathrm{o}}^-\,f}{2\,\gamma(\eta)\,\ell},\qquad \operatorname{var}\bar{x}=\frac{\eta\,L}{\gamma(\eta)\,N^\star\,\ell}>0, \tag{7.403}
$$

for the center of mass coordinate and

$$
\langle y_j\rangle=\frac{L}{\xi_j}\in\mathbb{R},\qquad \operatorname{var}y_j=\frac{L}{\gamma(\eta)\,\ell}>0,\qquad j=1,\cdots,N^\star \tag{7.404}
$$

for the relative coordinates.

For comparison, the same treatment of the Fokker-Planck equation (7.390) delivers[101]

$$
P(x_1,\cdots,x_{N^\star};L)\sim 2^{N^\star}\times\left(\frac{\gamma^{\mathrm{std}}\,\ell}{2\,\pi\,L}\right)^{\frac{N^\star}{2}}\prod_{j=1}^{N^\star}e^{-\frac{\gamma^{\mathrm{std}}\,\ell}{2\,L}\left(x_i-\frac{L}{\xi_j}\right)^2}, \tag{7.405a}
$$

where $x_1,\cdots,x_{N^\star}\geq 0$ and with

$$
\xi_j:=\frac{\gamma^{\mathrm{std}}\,\ell}{1+\frac{m_{\mathrm{o}}^-+m_{\mathrm{o}}^+}{2}(j-1)}>0 \tag{7.405b}
$$

and

$$
\gamma^{\mathrm{std}}:=\frac{1}{2}\left(m_{\mathrm{o}}^-+m_{\mathrm{o}}^+\right)(N^\star-1)+m_{\mathrm{l}}+1. \tag{7.405c}
$$

The standard symmetry classes A, AI, and AII have the same long root $m_{\mathrm{l}}=1$ but the different ordinary roots $m_{\mathrm{o}}^{\mp}=1,2,4$, respectively.

[101] The normalization is reduced by the factor 1/2 per independent Lyapunov exponent compared to the chiral symmetry classes as the Lyapunov exponents are strictly positive for the standard classes.

7.7.6.1 Case without Dimerization, $f = 0$

We treat first the case without dimerization, $f = 0$. In the localized regime $L \gg N^\star \ell$ only the x_j that are closest to 0 contribute to the conductance.

For even $N^\star$, they are $x_{N^\star/2}$ and $x_{(N^\star/2)+1}$, both of which are an average distance

$$-\langle x_{N^\star/2}\rangle = +\langle x_{(N^\star/2)+1}\rangle = \frac{L}{\xi}, \qquad \xi := \frac{2\,\gamma(\eta)\,\ell}{m_{\text{o}}^{-}} \tag{7.406}$$

away from zero. The length scale ξ, which is a length of order $N^\star \ell$, serves as the "localization length" for even $N^\star$.

For odd $N^\star$, the conductance is determined by that Lyapunov exponent $x_{(N^\star+1)/2}$ for which its localization length (7.402c) is diverging:

$$N^\star + 1 - 2j = 0 \iff j = \frac{N^\star + 1}{2}, \tag{7.407a}$$

so that its average is vanishing

$$\langle x_{(N^\star+1)/2}\rangle = 0. \tag{7.407b}$$

The presence of the eigenvalue $x_{(N^\star+1)/2}$ with zero average is responsible for the absence of exponential localization in this case.

The average and variance of the conductance and the average and variance of its logarithm follow from the probability distribution (7.402).

For even $N^\star$,

$$\ln\langle g\rangle = -\frac{m_{\text{o}}^{-}}{4}\left(1 - \frac{1-\eta}{N^\star}\right)^{-1/2}\frac{L}{\xi} - \frac{1}{2}\ln\left(\frac{L}{\xi}\right) + \mathcal{O}\left(\frac{L^0}{\xi^0}\right), \tag{7.408a}$$

$$\ln \operatorname{var} g = \ln\langle g\rangle + \mathcal{O}\left(\frac{L^0}{\xi^0}\right), \tag{7.408b}$$

$$\xi = \frac{2\,\gamma(\eta)\,\ell}{m_{\text{o}}^{-}}. \tag{7.408c}$$

The regime $L \gg \xi$ for $N^\star$ even can be identified with a regime of localization since $\langle g\rangle$ decays exponentially fast with $L/\xi \gg 1$ according to Eq. (7.408a). However, the conductance g fluctuates strongly around its average $\langle g\rangle$ in that the logarithm of the conductance g converges in the limit $\xi/L \to 0$ to a random variable that is Gaussian distributed with the mean and the variance given by

$$\langle \ln g\rangle = -\frac{2L}{\xi} + 2\sqrt{\frac{2}{m_{\text{o}}^{-}\,\pi}\left(1 - 2\frac{1-\eta}{N^\star}\right)\frac{L}{\xi}} + \mathcal{O}\left(\frac{L^0}{\xi^0}\right), \tag{7.408d}$$

$$\operatorname{var}\ln g = \frac{4}{m_{\text{o}}^{-}}\left[1 + \left(1 - \frac{2}{\pi}\right)\left(1 - 2\frac{1-\eta}{N^\star}\right)\right]\frac{L}{\xi} + \mathcal{O}\left(\frac{L^0}{\xi^0}\right). \tag{7.408e}$$

In the localized regime, the conductance distribution is thus well approximated by a log-normal distribution. Unlike for the statistical ensemble of g for which all cumulants higher than $\langle g\rangle$ and $\operatorname{var} g$ are needed to characterize the statistical ensemble of g in the localized regime, $\langle \ln g\rangle$ and var $\ln g$ provide good characteristic of the statistical ensemble of $\ln g$ in the localized regime.

For odd $N^\star$, there is no exponential localization when $L \gg \xi$ with the crossover length scale ξ given by Eq. (7.406). The conductance has a broad distribution, which is neither characterized by the (average of the) conductance nor its logarithm, as

$$P(g) \propto \frac{\exp\left(-\frac{\gamma\ell}{2L}\left(1-\frac{1-\eta}{N^\star}\right)^{-1}\operatorname{arccosh}^2(\sqrt{g})\right)}{g\sqrt{1-g}}. \tag{7.409}$$

With this distribution, the average conductance decays algebraically,

$$\langle g\rangle = \left(\frac{m_{\rm o}^-}{\pi}\right)^{\frac{1}{2}}\left(1-\frac{1-\eta}{N^\star}\right)^{-\frac{1}{2}}\left(\frac{\xi}{L}\right)^{\frac{1}{2}} + \mathcal{O}\left(\frac{L^0}{\xi^0}\right), \tag{7.410a}$$

$$\langle g^2\rangle = \frac{2}{3}\langle g\rangle + \mathcal{O}\left(\frac{L^0}{\xi^0}\right), \tag{7.410b}$$

while the average of its logarithm grows proportional to $L^{1/2}$ rather than L,

$$\langle \ln g\rangle = -4\sqrt{\frac{1}{m_{\rm o}^-\pi}\left(1-\frac{1-\eta}{N^\star}\right)\frac{L}{\xi}} + \mathcal{O}\left(\frac{L^0}{\xi^0}\right), \tag{7.410c}$$

$$\operatorname{var}\ln g = \frac{8}{m_{\rm o}^-}\left(1-\frac{2}{\pi}\right)\left(1-\frac{1-\eta}{N^\star}\right)\frac{L}{\xi} + \mathcal{O}\left(\frac{L^0}{\xi^0}\right). \tag{7.410d}$$

Away from the band center $\varepsilon = 0$, the conductance distribution for both $N^\star$ even and odd follows from the standard Fokker-Planck equation (7.390). It is close to log-normal, with

$$\ln\langle g\rangle = -\frac{L}{2\xi_{\rm st}} - \frac{3}{2}\ln\left(\frac{L}{\xi_{\rm st}}\right), \qquad \ln\operatorname{var} g = \ln\langle g\rangle, \tag{7.411a}$$

$$\langle \ln g\rangle = -\frac{2L}{\xi_{\rm st}}, \qquad \operatorname{var}\ln g = \frac{4L}{\xi_{\rm st}}, \tag{7.411b}$$

up to corrections of order $\mathcal{O}(L^0/\xi^0)$. Here, the localization length for the standard symmetry classes is given by

$$\xi_{\rm st} = \left[m_{\rm o}^-(N^\star - 1) + 2\right]\ell. \tag{7.411c}$$

The most striking difference in the conductance distribution appears for odd $N^\star$, for which the absence of exponential localization at $\varepsilon = 0$ is to be contrasted with the exponential decay of the conductance for $\varepsilon \neq 0$. For even $N^\star$, there also is an important difference. At $\varepsilon = 0$, the localization length $\xi \approx N^\star\ell$ is $m_{\rm o}^-$-independent for large $N^\star$, while the localization length $\xi_{\rm st} \approx m_{\rm o}^- N^\star\,\ell$ away from the band center is proportional to $m_{\rm o}^-$ for $\varepsilon \neq 0$. Hence, upon moving away from the band center, the localization length increases by a factor $m_{\rm o}^-$ (the mean free path is independent of ε).

The average and variance of the conductance in the localized regime are dominated by rare events, for which the smallest x_j is close to zero (corresponding to a transmission eigenvalue close to unity). For wires without chiral symmetry, approximation of $P(x_1, \cdots, x_{N^\star}; L)$ by a Gaussian fails for x_j close to zero because it does not account for the repulsion between x_j and its mirror image $-x_j$ that is encoded by the nonvanishing long root in Eq. (7.390).[102] While this failure does not affect the leading $\mathcal{O}(L)$ behavior of $\ln\langle g\rangle$ and $\ln \operatorname{var} g$, this failure shows up as the subleading logarithmic term present in Eq. (7.411a), a difference from what one would have obtained from a Gaussian distribution for x_j close to zero. In the presence of the chiral symmetry, there is no repulsion between x_j and $-x_j$, so that the approximation (7.402) remains valid for x_j close to zero.

7.7.6.2 Case with Dimerization, $f \neq 0$

The existence of a disorder-induced critical point in the absence of dimerization, $f = 0$, depends on the parity of the number $N^\star$ of chains. This parity effect can be understood from the "level repulsion" between the Lyapunov exponents $\{x_j\}$. Indeed, in the large-L limit, $x_1 \ll \cdots \ll x_{N^\star}$ holds and the Lyapunov exponents repel by constant forces. For an even number of channels, there is a net force on all Lyapunov exponents driving them away from their initial value 0 and resulting in an exponential suppression of the conductance as is shown in Figure 7.12(a). However, as is depicted in Figure 7.12(b), when the number of channels is odd, there is no force on the middle Lyapunov exponent. Therefore, this random variable will remain close to zero and give rise to a diverging localization length and a critical state. For comparison, in the case of a wire with on-site disorder,[103] the repulsion between the Lyapunov exponent x_j and its mirror image $-x_j$ results in a nonvanishing force for all radial coordinates [see Figure 7.12(c)].

By fine-tuning the dimensionless dimerization f, the number $N^\star$ of disorder-induced critical points can be reached, for both even and odd numbers of chains $N^\star$. Indeed, as the staggering parameter f approaches the critical value

$$f_{c,j} := N^\star + 1 - 2j, \qquad j = 1, \cdots, N^\star, \tag{7.412}$$

the localization length (7.402c) diverges with the critical exponent 1. The statistical properties of the conductance for large L at this critical point are captured by Eq. (7.409) for both $N^\star$ odd and even.

102 The long root contributes to the Jacobian for the Lyapunov exponents the term

$$\prod_j \sinh(2x_j) = \prod_j \sinh\left(x_j - (-x_j)\right),$$

which is intrepreted as a repulsion of x_j from its mirror image $(-x_j)$ about 0.

103 On-site disorder breaks the sublattice symmetry of the lattice model that is responsible for the chiral spectral symmetry as the chemical potential

$$\varepsilon\,\mathbb{1} \otimes \tau_0$$

does in Eq. (7.237).

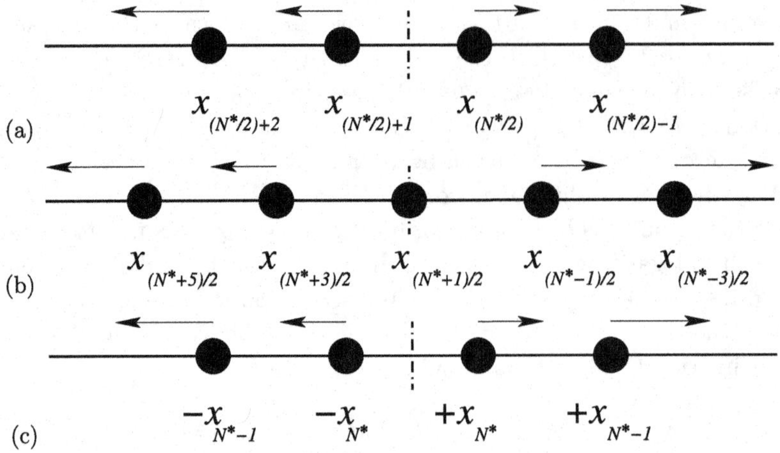

Figure 7.12 The parity effect results for chiral quantum wires from level repulsion between the transmission eigenvalues. (a) For an even number of chains, all radial coordinates x_j (represented by a black disc) are repelled away from 0 (the vertical dashed line), while (b) $x_{(N^\star+1)/2}$ remains close to 0 for an odd number of chains. (c) Repulsion from mirror images for a quantum wire with on-site disorder results in a positive driving force for all x_j.

7.8 The Tenfold Way: Gapped Phases of Quantum Wires

The theory of one-dimensional quantum transport defined by Eqs. (7.348) and (7.349) for all ten Altland-Zirbauer symmetry classes assumes a vanishing expectation value for the static white-noise correlated random potential $\mathcal{V}(x)$, as it aims at quantifying how the ballistic dimensionless Landauer conductance

$$g_{\mathrm{ballistic}}(L) := D \sum_{j=1}^{N^\star} \frac{1}{\cosh^2(x_j = 0)} = D\, N^\star \tag{7.413}$$

is changed by disorder as a function of the length L of the disordered region. Had we assumed instead the probability distribution

$$P[\mathcal{V}(x)] = \delta\big(\mathcal{V}(x) - \mathcal{V}_{\mathrm{gap}}\big) \tag{7.414a}$$

for the random potential $\mathcal{V}(x)$ in Eq. (7.348a) such that the x-independent matrix $\mathcal{V}_{\mathrm{gap}}$ opens a spectral gap Δ at the Fermi energy $\varepsilon = 0$, we would have found the exponentially decreasing dependence

$$g_{\mathrm{gap}}(L) := D \sum_{j=1}^{N^\star} \frac{1}{\cosh^2(L/\ell_j)} \tag{7.414b}$$

with each channel-resolved length scale ℓ_j of the order of $v_{\mathrm{F}}\, \hbar/\Delta$.

Our purpose will now be to study those channels of the random potential that open a gap in the spectrum of translation-invariant quantum wires and to identify which of the gapped symmetry classes, in addition to the chiral symmetry classes AIII, BDI, and CII, support zero modes that are robust to (not too strong) static and short-ranged correlated disorder allowed by the Altland-Zirnbauer symmetry classes.

To this end, we shall start with a continuum description that presumes a zero-dimensional Fermi surface consisting of two Fermi points with an internal symmetry that encodes the number of channels in a quantum wire. We shall depart from Section 7.7 in that we do not demand that this continuum description is necessarily the low-energy limit of a one-dimensional tight-binding model. A criterion will then be introduced to establish when this continuum description is the low-energy limit of a one-dimensional tight-binding model. This exercise will be performed for each symmetry class from the tenfold way. The following terminology will be used.

Definition 7.12 (rank of a Dirac Hamiltonian) The rank of a Dirac Hamiltonian is the dimension of the Dirac matrices used to represent the Dirac Hamiltonian.

7.8.1 Dirac Hamiltonians of Rank 2

A generic translation-invariant Dirac Hamiltonian of rank $r = 2$ can be represented by

$$\mathcal{H}(k) := \tau_3\, k + \tau_3\, A_1 + \tau_0\, A_0 + \tau_2\, M_2 + \tau_1\, M_1. \tag{7.415}$$

The Fermi velocity and the Planck constant have been set to unity. The unit 2×2 matrix is denoted by τ_0 and $\boldsymbol{\tau} = (\tau_1, \tau_2, \tau_3)$ are three Pauli matrices. The real number A_1 couples as a vector potential induced by a magnetic flux at the center of a ring would do. The real number A_0 couples as a scalar potential would do, that is as the chemical potential. The real parameters M_1 and M_2 couple to the two available Pauli matrices that anticommute with the kinetic energy $\tau_3\, k$. The four real-valued parameters A_1, M_1, M_2, and A_0 furnish an exhaustive parameterization of a 2×2 Hermitean matrix.

When $A_1 = A_0 = 0$, the Dirac Hamiltonian (7.415) has the eigenvalues

$$\varepsilon(k) = \pm\sqrt{k^2 + M^2}, \qquad M^2 := M_1^2 + M_2^2. \tag{7.416}$$

The real parameters M_1 and M_2 thus parameterize a Dirac mass. More precisely, we may write

$$\tau_2\, M_2 + \tau_1\, M_1 = M\, \beta(\theta), \tag{7.417a}$$

where

$$M := \sqrt{M_1^2 + M_2^2}, \qquad \theta := \arctan \frac{M_2}{M_1}, \qquad \beta(\theta) := \tau_1 \cos\theta + \tau_2 \sin\theta, \quad (7.417b)$$

in which case ($\alpha \equiv \tau_3$)

$$\beta^2(\theta) = 1, \qquad \{\beta(\theta), \alpha\} = 0. \quad (7.417c)$$

Thus, the normalized Dirac mass $\beta(\theta)$ is parameterized by the angle $0 \leq \theta < 2\pi$. As a topological set, the normalized Dirac masses $\{\beta(\theta) \,|\, 0 \leq \theta < 2\pi\}$ are homeomorphic to the circle S^1. They are also homeomorphic to U(1) through the map

$$\begin{aligned} \beta_{12} \colon [0, 2\pi[&\to \mathrm{U}(1), \\ \theta &\mapsto [\beta(\theta)]_{12}, \end{aligned} \quad (7.417d)$$

where $[\beta(\theta)]_{12}$ is the upper-right matrix element of the 2×2 matrix $\beta(\theta)$.

Symmetry class A: The representation of the Dirac Hamiltonian (7.415) in the position basis with A_1, M_1, M_2, A_0 smooth and nonvanishing functions of the position x along the real line that break translation symmetry explicitly is said to belong to the Altland-Zirnbauer symmetry class A. As we have seen, the set

$$V^{\mathrm{A}}_{d=1,r=2} := \{\beta(\theta) \,|\, 0 \leq \theta < 2\pi\} \cong \mathrm{S}^1, \quad (7.418a)$$

a circle, is the topological space of normalized Dirac masses associated with the Dirac Hamiltonian by adding to the kinetic contribution with the Dirac matrix α the normalized mass matrix $\beta(\theta)$ for $0 \leq \theta < 2\pi$. We note that $V^{\mathrm{A}}_{d=1,r=2}$ and U(1) are homeomorphic as topological spaces:

$$V^{\mathrm{A}}_{d=1,r=2} \cong \mathrm{U}(1). \quad (7.418b)$$

Thus, they share the same homotopy groups (see **Appendix** A for the formal definition of a homotopy group). On the other hand, $V^{\mathrm{A}}_{d=1,r=2}$ is not a group under matrix multiplication, while U(1) is. The massive Dirac Hamiltonian (7.415) is the low-energy and long-wavelength limit of a one-dimensional electronic tight-binding noninteracting Hamiltonian in the Altland-Zirnbauer symmetry class A.

In addition to the conservation of the fermion number, we may impose TRS on the Dirac Hamiltonian (7.415). There are two possibilities to do so.

Symmetry class AIII: If charge conservation holds together with the CHS and TRS is imposed through

$$\mathcal{H}(k) = +\tau_2\, \mathcal{H}^*(-k)\, \tau_2, \quad (7.419a)$$

then

$$\mathcal{H}(k) = \tau_3\, k + \tau_0\, A_0. \quad (7.419b)$$

No mass matrix is permissible if TRS squares to minus the identity.[104] The topological space of normalized Dirac masses in the symmetry class AII is the empty set

$$V^{\mathrm{AII}}_{d=1,r=2} = \emptyset. \tag{7.419c}$$

Because of the fermion-doubling obstruction (see footnote 77), the only way to realize (7.419b) as the low-energy and long-wavelength limit of an electronic lattice model is on the boundary of a two-dimensional topological insulator in the symmetry class AII.

Symmetry class AI: If charge conservation holds and TRS is imposed through

$$\mathcal{H}(k) = +\tau_1\,\mathcal{H}^{*}(-k)\,\tau_1, \tag{7.420a}$$

then

$$\mathcal{H}(k) = \tau_3\,k + \tau_2\,M_2 + \tau_1\,M_1 + \tau_0\,A_0. \tag{7.420b}$$

The same mass matrix as in the symmetry class A is permissible if TRS squares to the identity.[105] The homeomorphy between the allowed masses in the symmetry classes A and AI is accidental. It does not hold for representations with larger Dirac matrices of the Dirac Hamiltonian as we shall verify explicitly when considering a rank 4 Dirac Hamiltonian below. The topological space of normalized Dirac masses obtained by augmenting the Dirac kinetic contribution by adding a mass matrix squaring to unity and obeying this TRS is

$$V^{\mathrm{AI}}_{d=1,r=2} := \{\beta(\theta)\,|\,0\le\theta<2\pi\} \cong \{e^{\mathrm{i}\theta}\,|\,0\le\theta<2\pi\} =: \mathrm{S}^1 \equiv \mathrm{U}(1). \tag{7.420c}$$

The topological spaces $V^{\mathrm{AI}}_{d=1,r=2}$ and $\mathrm{U}(1)$ are homeomorphic. (This homeomorphism is not a group homomorphism, for $V^{\mathrm{A}}_{d=1,r=2}$ is not a group with respect to matrix multiplication, while $\mathrm{U}(1)$ is.) Consequently, they share the same homotopy groups. This massive Dirac Hamiltonian is the low-energy and long-wavelength limit of a one-dimensional electronic tight-binding noninteracting Hamiltonian in the Altland-Zirnbauer symmetry class AI.

The standard symmetry classes A, AII, and AI can be further constrained by imposing chiral spectral symmetry (CHS). This gives the following three possibilities.

Symmetry class AIII: If charge conservation holds together with the CHS

$$\mathcal{H}(k) = -\tau_1\,\mathcal{H}(k)\,\tau_1, \tag{7.421a}$$

[104] The transformation law of the spinor $\psi(k)$ in the momentum basis under reversal of time is

$$(\mathcal{T}\,\psi)(-k) := \mathrm{i}\tau_2\,\mathsf{K}\,\psi(-k) = \mathrm{i}\tau_2\,\psi^{*}(-k).$$

It squares to minus the identity since

$$(\mathcal{T}^2\,\psi)(k) = \mathrm{i}\tau_2\,\mathsf{K}\,\mathrm{i}\tau_2\,\psi^{*}(+k) = (\mathrm{i}\tau_2)^2\,\mathsf{K}\,\psi^{*}(+k) = -\psi\,(+k).$$

[105] As can be seen by replacing $\mathrm{i}\tau_2$ by τ_1 in footnote 104.

then

$$\mathcal{H}(k) = \tau_3\, k + \tau_3\, A_1 + \tau_2\, M_2. \tag{7.421b}$$

There is a unique normalized mass matrix up to a sign. The topological space of normalized Dirac masses obtained by adding to the Dirac kinetic contribution a mass matrix squaring to unity and obeying the CHS is

$$V^{\text{AIII}}_{d=1,r=2} = \{\pm\tau_2\}. \tag{7.421c}$$

This Dirac Hamiltonian is the low-energy and long-wavelength limit of a one-dimensional electronic tight-binding noninteracting Hamiltonian (that of polyacetylene, say) in the Altland-Zirnbauer symmetry class AIII.

Symmetry class CII: It is not possible to write down a Dirac equation of rank 2 in the symmetry class CII. For example, imposing

$$\mathcal{H}(k) = -\tau_1\, \mathcal{H}(k)\, \tau_1, \qquad \mathcal{H}(k) = +\tau_2\, \mathcal{H}^{*}(-k)\, \tau_2, \tag{7.422}$$

enforces the symmetry class DIII, for composing the CHS with the TRS delivers a particle-hole spectral symmetry that squares to the unity and not minus the unity (replace $\mathrm{i}\tau_2\,\mathsf{K}$ by $\mathrm{i}\tau_1\,\tau_2\,\mathsf{K}$ in footnote 104). In order to implement the symmetry constraints of the symmetry class CII, we need to consider a Dirac equation of rank 4 (see Section 7.8.2). This obstruction to constructing a Dirac Hamiltonian of rank $r = 2$ for the Altland-Zirnbauer symmetry class CII can also be interpreted as arising from the fermion-doubling obstruction.

Symmetry class BDI: If charge conservation holds together with

$$\mathcal{H}(k) = -\tau_1\, \mathcal{H}(k)\, \tau_1, \qquad \mathcal{H}(k) = +\tau_1\, \mathcal{H}^{*}(-k)\, \tau_1, \tag{7.423a}$$

then

$$\mathcal{H}(k) = \tau_3\, k + \tau_2\, M_2. \tag{7.423b}$$

There is a unique normalized mass matrix up to a sign. The topological space of normalized Dirac masses obtained by adding to the Dirac kinetic contribution a mass matrix squaring to unity while preserving TRS and PHS (a product of TRS and CHS), both of which square to unity, is

$$V^{\text{BDI}}_{d=1,r=2} = \{\pm\tau_2\}. \tag{7.423c}$$

This massive Dirac Hamiltonian is the low-energy and long-wavelength limit of a one-dimensional electronic tight-binding noninteracting Hamiltonian (that of polyacetylene, say) in the Altland-Zirnbauer symmetry class BDI.

Now, we move on to the four Bogoliubov-de-Gennes symmetry classes with particle-hole spectral symmetry.

Symmetry class D: If we impose PHS through

$$\mathcal{H}(k) = -\mathcal{H}^{*}(-k), \tag{7.424a}$$

then

$$\mathcal{H}(k) = \tau_3\, k + \tau_2\, M_2. \tag{7.424b}$$

There is a unique normalized mass matrix up to a sign. The topological space of normalized Dirac masses obtained by adding to the Dirac kinetic contribution a mass matrix squaring to unity and preserving the PHS squaring to unity is

$$V^{\mathrm{D}}_{d=1,r=2} = \{\pm\tau_2\}. \tag{7.424c}$$

Because of the fermion-doubling obstruction,[106] this massive Dirac Hamiltonian is not the low-energy and long-wavelength limit of a one-dimensional electronic tight-binding noninteracting Hamiltonian (the Kitaev chain, say) in the Altland-Zirnbauer symmetry class D. However, it is the low-energy and long-wavelength limit of the one-dimensional quantum Ising model in a transverse field.[107]

Symmetry class DIII: If we impose PHS and TRS through

$$\mathcal{H}(k) = -\mathcal{H}^{*}(-k), \qquad \mathcal{H}(k) = +\tau_2\, \mathcal{H}^{*}(-k)\, \tau_2, \tag{7.425a}$$

respectively, then

$$\mathcal{H}(k) = \tau_3\, k. \tag{7.425b}$$

No mass matrix is permissible if TRS squares to minus the identity. The topological space of normalized Dirac masses in the symmetry class DIII is the empty set

$$V^{\mathrm{DIII}}_{d=1,r=2} = \emptyset. \tag{7.425c}$$

Because of the fermion-doubling obstruction, the only way to realize (7.425b) as the low-energy and long-wavelength limit of an electronic lattice model is on the boundary of a two-dimensional topological superconductor in the symmetry class DIII.

Symmetry class C: If we impose PHS through

$$\mathcal{H}(k) = -\tau_2\, \mathcal{H}^{*}(-k)\, \tau_2, \tag{7.426a}$$

then

$$\mathcal{H}(k) = \tau_3\, A_1 + \tau_2\, M_2 + \tau_1\, M_1. \tag{7.426b}$$

PHS squaring to minus unity prohibits a kinetic energy in any Dirac Hamiltonian of rank 2 in the symmetry class C.

[106] Assume that we start with a one-dimensional lattice hosting one electron with its quantum spin-1/2 per site and allow for a mean-field treatment of superconductivity, that is, we must add a particle-hole grading. Linearization at the Fermi energy delivers an additional left- and right-moving grading. The minimal rank of the Dirac Hamiltonian is thus $2 \times 2 \times 2 = 8$, the minimal rank in Eq. (7.335). If the electronic spin is polarized by an external magnetic field, we can ignore the spin-1/2 degrees of freedom. The minimal rank of the Dirac Hamiltonian reduces to $2 \times 2 = 4$.

[107] The quantum Ising model in a transverse field can be realized by adding the Hamiltonians describing the spin-1/2 cluster $c = 1$ and $c = 0$ chains, respectively, that were defined in Eq. (5.25b). After the Jordan-Wigner transformation, such a linear combination becomes the Hamiltonian (5.226) with the continuum limit (5.230a).

Symmetry class CI: If we impose PHS and TRS through

$$\mathcal{H}(k) = -\tau_2\,\mathcal{H}^*(-k)\,\tau_2, \qquad \mathcal{H}(k) = +\tau_1\,\mathcal{H}^*(-k)\,\tau_1, \tag{7.427a}$$

respectively, then

$$\mathcal{H}(k) = \tau_2\,M_2 + \tau_1\,M_1. \tag{7.427b}$$

PHS squaring to minus unity prohibits a kinetic energy in the symmetry class CI.

7.8.2 Dirac Hamiltonian of Rank 4

A generic translation-invariant Dirac Hamiltonian of rank $r = 4$ can be represented by

$$\begin{aligned}\mathcal{H}(k) :=& \tau_3 \otimes \sigma_0\,k + \tau_3 \otimes \sigma_\nu\,A_{1,\nu} + \tau_0 \otimes \sigma_\nu\,A_{0,\nu} \\ &+ \tau_2 \otimes \sigma_\nu\,M_{2,\nu} + \tau_1 \otimes \sigma_\nu\,M_{1,\nu}.\end{aligned} \tag{7.428}$$

The Fermi velocity and the Planck constant have been set to unity. The matrices τ_0 and $\boldsymbol{\tau}$ were defined in Eq. (7.415). A second unit 2×2 matrix is denoted by σ_0 and $\boldsymbol{\sigma} = (\sigma_1, \sigma_2, \sigma_3)$ are another set of three Pauli matrices. The summation convention over the repeated index $\nu = 0,1,2,3$ is implied. There are four real-valued parameters for the components $A_{1,\nu}$ with $\nu = 0,1,2,3$ of an U(2) vector potential, eight for the components $M_{1,\nu}$ and $M_{2,\nu}$ with $\nu = 0,1,2,3$ of two independent U(2) masses, and four for the components $A_{0,\nu}$ with $\nu = 0,1,2,3$ of an U(2) scalar potential. As it should be, there are 16 real-valued free parameters (functions if we opt to break translation invariance).

Symmetry class A: The representation of the Dirac Hamiltonian (7.428) in the position basis with $A_{1,\nu}$, $M_{1,\nu}$, $M_{2,\nu}$, $A_{0,\nu}$ smooth and nonvanishing functions of the position x along the real line is said to belong to the symmetry class A. All eight mass matrices of rank $r = 4$ in the 1D symmetry class A can be arranged into the four pairs $(\tau_1 \otimes \sigma_\nu\,M_{1,\nu}, \tau_2 \otimes \sigma_\nu\,M_{2,\nu})$ with $\nu = 0,1,2,3$ of anticommuting masses. Alternatively, all eight mass matrices can be arranged into the pair

$$\boldsymbol{X} := (+\tau_2 \otimes \sigma_0, \quad \tau_1 \otimes \sigma_1, \quad \tau_1 \otimes \sigma_2, \quad \tau_1 \otimes \sigma_3), \tag{7.429a}$$

$$\boldsymbol{Y} := (-\tau_1 \otimes \sigma_0, \quad \tau_2 \otimes \sigma_1, \quad \tau_2 \otimes \sigma_2, \quad \tau_2 \otimes \sigma_3) \tag{7.429b}$$

of quadruplets. All four mass matrices making up the quadruplet $\boldsymbol{X}$ anticommute pairwise. The same is true for all four mass matrices making up the quadruplet $\boldsymbol{Y}$. Hence, if we define the pair

$$\boldsymbol{M} := (+M_{2,0}, \quad M_{1,1}, \quad M_{1,2}, \quad M_{1,3}), \tag{7.430a}$$

$$\boldsymbol{N} := (-M_{1,0}, \quad M_{2,1}, \quad M_{2,2}, \quad M_{2,3}) \tag{7.430b}$$

of vectors from $\mathbb{R}^4$, it follows that

$$(\boldsymbol{M}\cdot\boldsymbol{X})^2 = |\boldsymbol{M}|^2\,\tau_0\otimes\sigma_0, \tag{7.431a}$$

$$(\boldsymbol{N}\cdot\boldsymbol{Y})^2 = |\boldsymbol{N}|^2\,\tau_0\otimes\sigma_0, \tag{7.431b}$$

respectively. We demand that

$$\begin{aligned}\tau_0\otimes\sigma_0 &= (\boldsymbol{M}\cdot\boldsymbol{X}+\boldsymbol{N}\cdot\boldsymbol{Y})^2\\ &= (\boldsymbol{M}\cdot\boldsymbol{X})^2+(\boldsymbol{N}\cdot\boldsymbol{Y})^2+\sum_{\mu,\nu=1}^{4} M_\mu\,N_\nu\{X_\mu,Y_\nu\}\\ &= \left(|\boldsymbol{M}|^2+|\boldsymbol{N}|^2\right)\,\tau_0\otimes\sigma_0+2\sum_{1\leq\mu\neq\nu\leq 4} M_\mu\,N_\nu\,X_\mu\,Y_\nu,\end{aligned} \tag{7.432a}$$

where the identities

$$\{X_\mu,Y_\mu\}=0,\qquad [X_\mu,Y_\nu]=0,\qquad 1\leq\mu<\nu\leq 4 \tag{7.432b}$$

were used to reach the last equality. With the help of

$$X_1\,(Y_2+Y_3+Y_4) = +\tau_0\otimes\sigma_1+\tau_0\otimes\sigma_2+\tau_0\otimes\sigma_3, \tag{7.433a}$$

$$X_2\,(Y_1+Y_3+Y_4) = -\tau_0\otimes\sigma_1-\tau_3\otimes\sigma_3+\tau_3\otimes\sigma_2, \tag{7.433b}$$

$$X_3\,(Y_1+Y_2+Y_4) = -\tau_0\otimes\sigma_2+\tau_3\otimes\sigma_3-\tau_3\otimes\sigma_1, \tag{7.433c}$$

$$X_4\,(Y_1+Y_2+Y_3) = -\tau_0\otimes\sigma_3-\tau_3\otimes\sigma_2+\tau_3\otimes\sigma_1, \tag{7.433d}$$

we conclude that the cross terms on the right-hand side of Eq. (7.432a) vanish if and only if

$$\begin{aligned}M_1\,N_2=M_2\,N_1,\quad M_1\,N_3&=M_3\,N_1,\quad M_1\,N_4=M_4\,N_1,\\ M_2\,N_3&=M_3\,N_2,\quad M_2\,N_4=M_4\,N_2,\\ &\qquad\quad\; M_3\,N_4=M_4\,N_3\end{aligned} \tag{7.433e}$$

holds for all allowed $\boldsymbol{M},\boldsymbol{N}$ in $\mathbb{R}^4$. Cancelling the cross terms on the right-hand side of Eq. (7.432a) is achieved by choosing

$$\boldsymbol{N}=\alpha_{\parallel}\,\boldsymbol{M},\qquad \alpha_{\parallel}\in\mathbb{R}, \tag{7.434a}$$

while normalizing the mass matrix is achieved by choosing

$$|\boldsymbol{M}|^2=\alpha^2,\qquad |\boldsymbol{N}|^2=\alpha_{\parallel}^2\,\alpha^2,\qquad \alpha^2(1+\alpha_{\parallel}^2)=1,\qquad \alpha\in\mathbb{R}. \tag{7.434b}$$

Two solutions of Eq. (7.434b) are

$$\alpha=\cos\theta,\qquad \alpha_{\parallel}=\tan\theta,\qquad -\pi\leq\theta<\pi,\qquad \theta\neq-\frac{\pi}{2},+\frac{\pi}{2}, \tag{7.434c}$$

$$\alpha=\sin\theta,\qquad \alpha_{\parallel}=\cot\theta,\qquad -\pi\leq\theta<\pi,\qquad \theta\neq-\pi,0, \tag{7.434d}$$

if four overlaping intervals on the unit circle centered at 0, $\pm\pi/2$, $\pm\pi$ are used to define an atlas of the unit circle. For conciseness, we will only refer to the solution (7.434c). We have thus shown that the space of mass matrices for the Dirac Hamiltonian of rank 4 in the symmetry class A is given by

$$\begin{aligned} V^{\mathrm{A}}_{d=1,r=4} := \Big\{ \boldsymbol{M}\cdot\boldsymbol{X} + \boldsymbol{N}\cdot\boldsymbol{Y} \,\Big|\, |\boldsymbol{M}|^2 = \cos^2\theta, \boldsymbol{N} = \tan\theta\,\boldsymbol{M}, \quad \boldsymbol{M}\in\mathbb{R}^4, \\ \boldsymbol{X} := (+\tau_2\otimes\sigma_0,\ \tau_1\otimes\sigma_1,\ \tau_1\otimes\sigma_2,\ \tau_1\otimes\sigma_3), \\ \boldsymbol{Y} := (-\tau_1\otimes\sigma_0,\ \tau_2\otimes\sigma_1,\ \tau_2\otimes\sigma_2,\ \tau_2\otimes\sigma_3) \Big\}. \end{aligned} \tag{7.435}$$

By inspection, $V^{\mathrm{A}}_{d=1,r=4}$ is parameterized by the set of real-valued vectors $\boldsymbol{M}\in\mathbb{R}^4$ of fixed length, a set homeomorphic to the unit sphere S^3, together with the angle θ, a set homeomorphic to the unit circle S^1. This gives the homeomorphism

$$V^{\mathrm{A}}_{d=1,r=4} \cong \mathrm{S}^1\times\mathrm{S}^3. \tag{7.436}$$

For any fixed $\theta\in]-\pi/2,+\pi/2[$, assign to any $\boldsymbol{M}\in\mathbb{R}^4$ of length $|\cos\theta|$ the 2×2 matrix

$$\boldsymbol{M} \mapsto e^{\mathrm{i}\theta/2}\left(u_0\,\sigma_0 + \mathrm{i}u_1\,\sigma_1 + \mathrm{i}u_2\,\sigma_2 + \mathrm{i}u_3\,\sigma_3\right) =: U, \tag{7.437a}$$

where

$$u_0 := \frac{M_1}{\cos\theta}, \qquad u_1 := \frac{M_2}{\cos\theta}, \qquad u_2 := \frac{M_3}{\cos\theta}, \qquad u_3 := \frac{M_4}{\cos\theta}. \tag{7.437b}$$

In words, any $\boldsymbol{M}\in\mathbb{R}^4$ of fixed length $|\cos\theta|$ is assigned an element of SU(2) multiplied by the phase factor $e^{\mathrm{i}\theta/2}$. This map is a bijection to the set $e^{\mathrm{i}\theta/2}\,\mathrm{SU}(2)$ of 2×2 unitary matrices with determinant $e^{\mathrm{i}\theta}$. Upon varying θ, we have the homeomorphism [108]

$$V^{\mathrm{A}}_{d=1,r=4} \cong \left\{ \begin{pmatrix} 0 & U \\ U^{\dagger} & 0 \end{pmatrix} \,\middle|\, U\in\mathrm{U}(2) \right\}. \tag{7.438}$$

Because the Lie group U(2) is semi-simple with the decomposition $\mathrm{U}(2)=\mathrm{U}(1)\times\mathrm{SU}(2)$, we have the homeomorphism

$$\mathrm{U}(2)\cong\mathrm{S}^3\times\mathrm{S}^1, \tag{7.439}$$

in agreement with Eq. (7.436). A Dirac Hamiltonian with the Dirac mass of the form (7.435) is the low-energy and long-wavelength limit of a one-dimensional electronic tight-binding noninteracting Hamiltonian in the symmetry class A.

In addition to the conservation of the fermion number, we may impose TRS on the Dirac Hamiltonian (7.428). There are two possibilities to do so.

Symmetry class AII: If charge conservation holds together with TRS through

$$\mathcal{H}(k) = +\tau_1\otimes\sigma_2\,\mathcal{H}^{*}(-k)\,\tau_1\otimes\sigma_2, \tag{7.440a}$$

[108] We note that the right-hand side of Eq. (7.438) is not closed under matrix multiplication so that it does not carry the group structure of U(2).

then

$$\begin{aligned}\mathcal{H}(k) =& \tau_3 \otimes \sigma_0\, k + \sum_{\nu=1,2,3} \tau_3 \otimes \sigma_\nu\, A_{1,\nu} + \tau_2 \otimes \sigma_0\, M_{2,0} \\ &+ \tau_1 \otimes \sigma_0\, M_{1,0} + \tau_0 \otimes \sigma_0\, A_{0,0}.\end{aligned} \tag{7.440b}$$

Observe that by doubling the rank of the Dirac Hamiltonian (7.419b), we went from no mass matrix to two anticommuting mass matrices. The topological space of normalized Dirac masses obtained by adding to the Dirac kinetic contribution a mass matrix squaring to unity and obeying the TRS (7.440a) is

$$\begin{aligned}V^{\mathrm{AII}}_{d=1,r=4} :=& \left\{ \boldsymbol{M}\cdot\boldsymbol{X} \,\middle|\, |\boldsymbol{M}|^2 = 1, \quad \boldsymbol{M} \in \mathbb{R}^2, \quad \boldsymbol{X} := (\tau_2 \otimes \sigma_0, \tau_1 \otimes \sigma_0) \right\} \\ \cong& \, \mathsf{S}^1.\end{aligned} \tag{7.440c}$$

Alternatively, we could have started from the homeomorphism (7.438) and imposed the condition of TRS to obtain

$$V^{\mathrm{AII}}_{d=1,r=4} \cong \left\{ \begin{pmatrix} 0 & U \\ U^\dagger & 0 \end{pmatrix} \,\middle|\, U = +\sigma_2\, U^* \sigma_2 \in \mathrm{U}(2) \right\}, \tag{7.440d}$$

where the conjugacy classes induced by the TRS define the quotient space

$$\mathrm{U}(2)/\mathrm{Sp}(2) \cong [\mathrm{U}(1) \times \mathrm{SU}(2)]/\mathrm{SU}(2) \cong \mathrm{U}(1) \cong \mathsf{S}^1. \tag{7.440e}$$

A Dirac Hamiltonian with a Dirac mass of the form (7.440c) is the low-energy and long-wavelength limit of a one-dimensional electronic tight-binding noninteracting Hamiltonian in the symmetry class AII.

Symmetry class AI: If charge conservation holds together with TRS through

$$\mathcal{H}(k) = +\tau_1 \otimes \sigma_0\, \mathcal{H}^*(-k)\, \tau_1 \otimes \sigma_0, \tag{7.441a}$$

then

$$\begin{aligned}\mathcal{H}(k) =& \tau_3 \otimes \sigma_0\, k + \tau_3 \otimes \sigma_2\, A_{1,2} + \sum_{\nu=0,1,3} \big(\tau_2 \otimes \sigma_\nu\, M_{2,\nu} \\ &+ \tau_1 \otimes \sigma_\nu\, M_{1,\nu} + \tau_0 \otimes \sigma_\nu\, A_{0,\nu}\big).\end{aligned} \tag{7.441b}$$

There are six mass matrices of rank $r = 4$ in the 1D symmetry class AI that can be arranged into the three pairs $(\tau_1 \otimes \sigma_\nu\, M_{1,\nu}, \tau_2 \otimes \sigma_\nu\, M_{2,\nu})$ with $\nu = 0,1,3$ of anticommuting masses. The topological space of normalized Dirac masses obtained by adding to the Dirac kinetic contribution a mass matrix squaring to unity and obeying the TRS (7.441a) is

$$\begin{aligned}V^{\mathrm{AI}}_{d=1,r=4} :=& \Big\{ \boldsymbol{M}\cdot\boldsymbol{X} + \boldsymbol{N}\cdot\boldsymbol{Y} \,\Big|\, |\boldsymbol{M}|^2 = \cos^2\theta,\ \boldsymbol{N} = \tan\theta\, \boldsymbol{M},\ \boldsymbol{M} \in \mathbb{R}^3, \\ & \boldsymbol{X} := (+\tau_2 \otimes \sigma_0, \tau_1 \otimes \sigma_1, \tau_1 \otimes \sigma_3), \\ & \boldsymbol{Y} := (-\tau_1 \otimes \sigma_0, \tau_2 \otimes \sigma_1, \tau_2 \otimes \sigma_3) \Big\} \\ \cong& \, \mathsf{S}^1 \times \mathsf{S}^2.\end{aligned} \tag{7.441c}$$

Alternatively, we could have started from the homeomorphism (7.438) and imposed the condition of TRS to obtain

$$V^{\mathrm{AI}}_{d=1,r=4} \cong \left\{ \begin{pmatrix} 0 & U \\ U^{\dagger} & 0 \end{pmatrix} \middle| \, U = +U^{\mathsf{T}} \in \mathrm{U}(2) \right\}, \tag{7.441d}$$

where the conjugacy classes induced by the TRS define the quotient space

$$\mathrm{U}(2)/\mathrm{O}(2) \cong [\mathrm{U}(1) \times \mathrm{SU}(2)]/[\mathrm{O}(1) \times \mathrm{SO}(2)] \cong \mathsf{S}^1 \times \mathsf{S}^2. \tag{7.441e}$$

A Dirac Hamiltonian with a Dirac mass of the form (7.441c) is the low-energy and long-wavelength limit of a one-dimensional electronic tight-binding noninteracting Hamiltonian in the symmetry class AI.

The standard symmetry classes A, AII, and AI can be further constrained by imposing CHS. This gives the following three possibilities.

Symmetry class AIII If charge conservation holds together with the CHS

$$\mathcal{H}(k) = -\tau_1 \otimes \sigma_0 \, \mathcal{H}(k) \, \tau_1 \otimes \sigma_0, \tag{7.442a}$$

then

$$\mathcal{H}(k) = \tau_3 \otimes \sigma_0 \, k + \sum_{\nu=0,1,2,3} \left(\tau_3 \otimes \sigma_\nu \, A_{1,\nu} + \tau_2 \otimes \sigma_\nu \, M_{2,\nu} \right). \tag{7.442b}$$

The Dirac mass matrix $\tau_2 \otimes \sigma_0 \, M_{2,0}$ that descends from Eq. (7.421b) commutes with the triplet of anticommuting mass matrices $\tau_2 \otimes \sigma_1 \, M_{2,1}$, $\tau_2 \otimes \sigma_2 \, M_{2,2}$, and $\tau_2 \otimes \sigma_3 \, M_{2,3}$. The topological space of normalized Dirac masses obtained by adding to the Dirac kinetic contribution a mass matrix squaring to the unit matrix and obeying the CHS (7.442) is

$$\begin{aligned} V^{\mathrm{AIII}}_{d=1,r=4} := & \{-\tau_2 \otimes \sigma_0\} \\ & \cup \left\{ \boldsymbol{M} \cdot \boldsymbol{X} \,\middle|\, |\boldsymbol{M}|^2 = 1, \ \boldsymbol{M} \in \mathbb{R}^3, \ \boldsymbol{X} := (\tau_2 \otimes \sigma_1, \tau_2 \otimes \sigma_2, \tau_2 \otimes \sigma_3) \right\} \\ & \cup \{+\tau_2 \otimes \sigma_0\} \\ \cong & \{-\tau_2 \otimes \sigma_0\} \cup \mathsf{S}^2 \cup \{+\tau_2 \otimes \sigma_0\}. \end{aligned} \tag{7.442c}$$

Alternatively, all 4×4 Dirac mass matrices β must anticommute with $\tau_3 \otimes \sigma_0$ (the Dirac kinetic energy) and with $\tau_1 \otimes \sigma_0$ (CHS). Hence, they must be of the form

$$\beta = \tau_2 \otimes A, \tag{7.442d}$$

where Hermiticity demands that

$$A = A^{\dagger} \tag{7.442e}$$

and the normalization of the Dirac mass matrices demands that

$$A^2 = \sigma_0. \tag{7.442f}$$

This means that all eigenvalues of A are either -1 or $+1$. We arrive at the parameterization

$$V^{\mathrm{AIII}}_{d=1,r=4} \cong \Bigg\{\beta = \tau_2 \otimes A \,\Bigg|\, A := U\, I_{m,n}\, U^{\dagger},$$
$$m,n = 0,1,2, \quad m+n=2, \quad U \in \mathrm{U}(2), \tag{7.442g}$$
$$I_{m,n} := \mathrm{diag}(\overbrace{-1,\cdots,-1}^{m-\text{times}},\overbrace{+1,\cdots,+1}^{n-\text{times}})\Bigg\}.$$

As the set of unitary matrices that leaves the diagonal quadratic matrix $I_{m,n}$ invariant upon conjugation defines the Grassmannian manifold

$$\mathrm{Gr}(m,n;\mathbb{C}) := \mathrm{U}(m+n)/[\mathrm{U}(m) \times U(n)], \tag{7.442h}$$

we have found the finite union of complex Grassmannians

$$V^{\mathrm{AIII}}_{d=1,r=4} \cong \mathrm{U}(2)/[\mathrm{U}(2)\times U(0)] \cup \mathrm{U}(2)/[\mathrm{U}(1)\times \mathrm{U}(1)] \cup \mathrm{U}(2)/[U(0)\times \mathrm{U}(2)] \tag{7.442i}$$

(recall that $\mathrm{S}^2 \cong \mathrm{SU}(2)/\mathrm{U}(1)$ so that $\mathrm{U}(2)/[\mathrm{U}(1)\times\mathrm{U}(1)] \cong \mathrm{S}^2$). A Dirac Hamiltonian with a Dirac mass of the form (7.442c) is the low-energy and long-wavelength limit of a one-dimensional electronic tight-binding noninteracting Hamiltonian in the symmetry class AIII.

Symmetry class CII: If charge conservation holds together with

$$\mathcal{H}(k) = -\tau_1 \otimes \sigma_0\, \mathcal{H}(k)\, \tau_1 \otimes \sigma_0, \tag{7.443a}$$
$$\mathcal{H}(k) = +\tau_1 \otimes \sigma_2\, \mathcal{H}^{*}(-k)\, \tau_1 \otimes \sigma_2, \tag{7.443b}$$

representing CHS and TRS, respectively, then

$$\mathcal{H}(k) = \tau_3 \otimes \sigma_0\, k + \sum_{\nu=1,2,3} \tau_3 \otimes \sigma_\nu\, A_{1,\nu} + \tau_2 \otimes \sigma_0\, M_{2,0}. \tag{7.443c}$$

There is a unique normalized mass matrix up to a sign, as was the case for rank 2 Hamiltonians in Eqs. (7.421b) and (7.423b). The topological space of normalized Dirac masses obtained by adding to the Dirac kinetic contribution a mass matrix squaring to the unit matrix and obeying CHS (7.443a) and TRS (7.443b) for the symmetry class CII is

$$V^{\mathrm{CII}}_{d=1,r=4} := \{-\tau_2 \otimes \sigma_0\} \cup \{+\tau_2 \otimes \sigma_0\}. \tag{7.443d}$$

Alternatively, we could have started from the parameterization (7.442g) and imposed the condition of TRS to obtain

$$V^{\mathrm{CII}}_{d=1,r=4} \cong \left\{\beta \in V^{\mathrm{AIII}}_{d=1,r=4} \,\middle|\, \beta = (\tau_1 \otimes \sigma_2)\, \beta^{*}\, (\tau_1 \otimes \sigma_2)\right\}, \tag{7.443e}$$

where the conjugacy classes induced by the TRS define the finite union of quaternionic Grassmannians

$$V^{\mathrm{CII}}_{d=1,r=4} \cong \mathrm{Sp}(2)/[\mathrm{Sp}(2) \times \mathrm{Sp}(0)] \cup \mathrm{Sp}(2)/[\mathrm{Sp}(0) \times \mathrm{Sp}(2)]. \tag{7.443f}$$

A Dirac Hamiltonian with a Dirac mass of the form (7.443d) is the low-energy and long-wavelength limit of a one-dimensional electronic tight-binding noninteracting Hamiltonian in the symmetry class CII.

Symmetry class BDI: If charge conservation holds together with

$$\mathcal{H}(k) = -\tau_1 \otimes \sigma_0\, \mathcal{H}(k)\, \tau_1 \otimes \sigma_0, \tag{7.444a}$$

$$\mathcal{H}(k) = +\tau_1 \otimes \sigma_0\, \mathcal{H}^*(-k)\, \tau_1 \otimes \sigma_0, \tag{7.444b}$$

representing CHS and TRS, respectively, then

$$\mathcal{H}(k) = \tau_3 \otimes \sigma_0\, k + \tau_3 \otimes \sigma_2\, A_{1,2} + \sum_{\nu=0,1,3} \tau_2 \otimes \sigma_\nu\, M_{2,\nu}. \tag{7.444c}$$

The Dirac mass matrix $\tau_2 \otimes \sigma_0\, M_{2,0}$ that descends from Eq. (7.423b) commutes with the pair of anticommuting mass matrices $\tau_2 \otimes \sigma_1\, M_{2,1}$ and $\tau_2 \otimes \sigma_3\, M_{2,3}$. The topological space of normalized Dirac masses obtained by adding to the Dirac kinetic contribution a mass matrix squaring to the unit matrix and obeying CHS (7.444a) and TRS (7.444b) is

$$\begin{aligned} V^{\mathrm{BDI}}_{d=1,r=4} =& \{-\tau_2 \otimes \sigma_0\} \\ & \cup \left\{ \boldsymbol{M} \cdot \boldsymbol{X} \,|\, |\boldsymbol{M}|^2 = 1, \quad \boldsymbol{M} \in \mathbb{R}^2, \quad \boldsymbol{X} := (\tau_2 \otimes \sigma_1, \tau_2 \otimes \sigma_3) \right\} \\ & \cup \{+\tau_2 \otimes \sigma_0\} \\ \cong& \{-\tau_2 \otimes \sigma_0\} \cup \mathsf{S}^1 \cup \{+\tau_2 \otimes \sigma_0\}. \end{aligned} \tag{7.444d}$$

Alternatively, we could have started from the parameterization (7.442g) and imposed the condition of TRS to obtain

$$V^{\mathrm{BDI}}_{d=1,r=4} \cong \left\{ \beta \in V^{\mathrm{AIII}}_{d=1,r=4} \,\middle|\, \beta = (\tau_1 \otimes \sigma_0)\, \beta^*\, (\tau_1 \otimes \sigma_0) \right\}, \tag{7.444e}$$

where the conjugacy classes induced by the TRS define the finite union of real Grassmannians

$$V^{\mathrm{BDI}}_{d=1,r=4} \cong \mathrm{O}(2)/[\mathrm{O}(2)\times\mathrm{O}(0)] \cup \mathrm{O}(2)/[\mathrm{O}(1)\times\mathrm{O}(1)] \cup \mathrm{O}(2)/[\mathrm{O}(0)\times\mathrm{O}(2)] \tag{7.444f}$$

(recall that $\mathsf{S}^1 \cong \mathrm{O}(2)/[\mathrm{O}(1) \times \mathrm{O}(1)]$). A Dirac Hamiltonian with a Dirac mass of the form (7.444d) is the low-energy and long-wavelength limit of a one-dimensional electronic tight-binding noninteracting Hamiltonian in the symmetry class BDI.

Next, we move on to the four Bogoliubov-de-Gennes symmetry classes.

Symmetry class D: If we impose PHS through

$$\mathcal{H}(k) = -\mathcal{H}^*(-k), \tag{7.445a}$$

then

$$\begin{aligned} \mathcal{H}(k) =& \tau_3 \otimes \sigma_0\, k + \tau_3 \otimes \sigma_2\, A_{1,2} + \sum_{\nu=0,1,3} \tau_2 \otimes \sigma_\nu\, M_{2,\nu} \\ & + \tau_1 \otimes \sigma_2\, M_{1,2} + \tau_0 \otimes \sigma_2\, A_{0,2}. \end{aligned} \tag{7.445b}$$

There are four Dirac linearly independent mass matrices. None commutes with all remaining ones. However, each of them is antisymmetric and so is their sum. They can be arranged into two pairs of anticommuting mass matrices

$$\boldsymbol{X} := (\tau_2 \otimes \sigma_1, \tau_2 \otimes \sigma_3)\,, \tag{7.445c}$$

$$\boldsymbol{Y} := (\tau_2 \otimes \sigma_0, \tau_1 \otimes \sigma_2)\,. \tag{7.445d}$$

The elements of one pair commute with the elements of the second pair. Multiplying out all the elements of one pair against all elements of the other pair delivers four linearly independent and traceless matrices. Consequently, the generic linear combination of mass matrices $\boldsymbol{M}\cdot\boldsymbol{X}+\boldsymbol{N}\cdot\boldsymbol{Y}$ with $\boldsymbol{M},\boldsymbol{N}\in\mathbb{R}^2$ can only square to the unit 4×4 matrix $\tau_0\otimes\sigma_0$ if either one of $\boldsymbol{M}$ or $\boldsymbol{N}$ is vanishing. The topological space of normalized Dirac masses obtained by adding to the Dirac kinetic contribution a mass matrix squaring to the unit matrix and obeying PHS squaring to unity is thus

$$\begin{aligned} V^{\mathrm{D}}_{d=1,r=4} := &\left\{\boldsymbol{M}\cdot\boldsymbol{X} \mid |\boldsymbol{M}|^2=1, \quad \boldsymbol{M}\in\mathbb{R}^2, \quad \boldsymbol{X}:=(\tau_2\otimes\sigma_1,\tau_2\otimes\sigma_3)\right\} \\ &\cup\left\{\boldsymbol{N}\cdot\boldsymbol{Y} \mid |\boldsymbol{N}|^2=1, \quad \boldsymbol{N}\in\mathbb{R}^2, \quad \boldsymbol{Y}:=(\tau_2\otimes\sigma_0,\tau_1\otimes\sigma_2)\right\}. \end{aligned} \tag{7.445e}$$

As a topological space, $V^{\mathrm{D}}_{d=1,r=4}$ is homeomorphic to $\mathrm{O}(2)$, as

$$V^{\mathrm{D}}_{d=1,r=4} \cong \mathrm{S}^1\cup\mathrm{S}^1 \cong \mathrm{U}(1)\times\mathbb{Z}_2 \cong \mathrm{O}(2). \tag{7.445f}$$

Observe that $V^{\mathrm{D}}_{d=1,r=4}$ is not closed under matrix multiplication, that is, it does not carry the group structure of $\mathrm{O}(2)$. A Dirac Hamiltonian with a Dirac mass of the form (7.445e) is the low-energy and long-wavelength limit of a one-dimensional electronic tight-binding noninteracting Hamiltonian in the symmetry class D.

Symmetry class DIII: If we impose PHS and TRS through

$$\mathcal{H}(k) = -\mathcal{H}^*(-k), \qquad \mathcal{H}(k) = +\tau_2\otimes\sigma_0\,\mathcal{H}^*(-k)\,\tau_2\otimes\sigma_0, \tag{7.446a}$$

then

$$\mathcal{H}(k) = \tau_3\otimes\sigma_0\,k + \tau_3\otimes\sigma_2\,A_{1,2} + \tau_1\otimes\sigma_2\,M_{1,2}. \tag{7.446b}$$

Observe that there is only one Dirac mass matrix [there was none in Eq. (7.425b)]. Moreover, this Dirac mass matrix is Hermitean and antisymmetric. The topological space of normalized Dirac masses obtained by adding to the Dirac kinetic contribution a mass matrix squaring to the unit matrix and obeying this PHS and this TRS is

$$V^{\mathrm{DIII}}_{d=1,r=4} := \{-\tau_1\otimes\sigma_2\}\cup\{+\tau_1\otimes\sigma_2\}. \tag{7.446c}$$

As a topological space, $V^{\mathrm{DIII}}_{d=1,r=4}$ is homeomorphic to $\mathrm{O}(2)/\mathrm{U}(1)$,

$$V^{\mathrm{DIII}}_{d=1,r=4} \cong \mathrm{O}(2)/\mathrm{U}(1). \tag{7.446d}$$

Because of the fermion-doubling obstruction,[109] a Dirac Hamiltonian with a Dirac mass of the form (7.446c) is not the low-energy and long-wavelength limit of a one-dimensional electronic tight-binding noninteracting Hamiltonian in the symmetry class DIII. However, it is the low-energy and long-wavelength limit of the Majorana lattice Hamiltonian of the form (5.237) with $|c-c'|=2$.

Symmetry class C: If we impose PHS through

$$\mathcal{H}(k) = -\tau_0 \otimes \sigma_2\, \mathcal{H}^*(-k)\, \tau_0 \otimes \sigma_2, \tag{7.447a}$$

then

$$\begin{aligned}\mathcal{H}(k) =& \tau_3 \otimes \sigma_0\, k + \sum_{\nu=1,2,3} \tau_3 \otimes \sigma_\nu\, A_{1,\nu} + \tau_2 \otimes \sigma_0\, M_{2,0} \\ &+ \sum_{\nu=1,2,3} \tau_1 \otimes \sigma_\nu\, M_{1,\nu} + \sum_{\nu=1,2,3} \tau_0 \otimes \sigma_\nu\, A_{0,\nu}.\end{aligned} \tag{7.447b}$$

There are four mass matrices that anticommute pairwise. The topological space of normalized Dirac masses obtained by adding to the Dirac kinetic contribution a mass matrix squaring to the unit matrix and obeying this PHS is

$$\begin{aligned}V^{\mathrm{C}}_{d=1,r=4} =& \left\{\boldsymbol{M}\cdot\boldsymbol{X}\,\middle|\,|\boldsymbol{M}|^2=1,\ \boldsymbol{M}\in\mathbb{R}^4,\ \boldsymbol{X} := (\tau_2\otimes\sigma_0, \tau_1\otimes\sigma_1, \tau_1\otimes\sigma_2, \tau_1\otimes\sigma_3)\right\} \\ \cong& \, S^3.\end{aligned} \tag{7.447c}$$

As a topological space, $V^{\mathrm{C}}_{d=1,r=4}$ is homeomorphic to Sp(2) since we have

$$\mathrm{Sp}(2) \cong \mathrm{SU}(2) \cong S^3. \tag{7.447d}$$

Observe that $V^{\mathrm{C}}_{d=1,r=4}$ is not closed under matrix multiplication, that is, it does not carry the group structure of Sp(2). A Dirac Hamiltonian with a Dirac mass of the form (7.447c) is the low-energy and long-wavelength limit of a one-dimensional electronic tight-binding noninteracting Hamiltonian in the symmetry class C.

Symmetry class CI: If we impose PHS and TRS through

$$\mathcal{H}(k) = -\tau_0 \otimes \sigma_2\, \mathcal{H}^*(-k)\, \tau_0 \otimes \sigma_2, \tag{7.448a}$$

$$\mathcal{H}(k) = +\tau_1 \otimes \sigma_0\, \mathcal{H}^*(-k)\, \tau_1 \otimes \sigma_0, \tag{7.448b}$$

respectively, then

$$\begin{aligned}\mathcal{H}(k) =& \tau_3 \otimes \sigma_0\, k + \tau_3 \otimes \sigma_2\, A_{1,2} + \tau_2 \otimes \sigma_0\, M_{2,0} \\ &+ \sum_{\nu=1,3} \left(\tau_1 \otimes \sigma_\nu\, M_{1,\nu} + \tau_0 \otimes \sigma_\nu\, A_{0,\nu}\right).\end{aligned} \tag{7.448c}$$

[109] We must start with a one-dimensional lattice hosting one electron with its quantum spin-1/2 per site. To allow for a mean-field treatment of superconductivity, we must add a particle-hole grading. Linearization at the Fermi energy delivers an additional left- and right-moving grading. The minimal rank of the Dirac Hamiltonian is thus $2\times 2\times 2 = 8$, the minimal rank in Eq. (7.335).

There are three mass matrices that anticommute pairwise. The topological space of normalized Dirac masses obtained by adding to the Dirac kinetic contribution a mass matrix squaring to the unit matrix and obeying this PHS and this TRS is

$$\begin{aligned} V^{\mathrm{CI}}_{d=1,r=4} &:= \left\{ \boldsymbol{M}\cdot\boldsymbol{X} \mid |\boldsymbol{M}|^2=1, \quad \boldsymbol{M}\in\mathbb{R}^3, \quad \boldsymbol{X}:=(\tau_2\otimes\sigma_0,\tau_1\otimes\sigma_1,\tau_1\otimes\sigma_3)\right\} \\ &\cong S^2. \end{aligned} \tag{7.448d}$$

As a topological space, $V^{\mathrm{CI}}_{d=1,r=4}$ is homeomorphic to $\mathrm{Sp}(2)/\mathrm{U}(1)$ since we have the homeomorphism

$$\mathrm{Sp}(2)/\mathrm{U}(1) \cong \mathrm{SU}(2)/\mathrm{U}(1) \cong S^2. \tag{7.448e}$$

A Dirac Hamiltonian with a Dirac mass of the form (7.448d) is the low-energy and long-wavelength limit of a one-dimensional electronic tight-binding noninteracting Hamiltonian in the symmetry class CI.

7.8.3 Existence and Uniqueness of Normalized Dirac Masses

We summarize the properties that can be deduced from repeating the exercises from Sections 7.8.1 and 7.8.2 when the rank r of the Dirac matrices entering the Dirac Hamiltonian in one-dimensional space for any one of the ten Altland-Zirnbauer symmetry classes is given by

$$r = r^{\mathrm{AZ}}_{\min}\, N \tag{7.449}$$

with $N = 1, 2, 3, \cdots$. We shall denote with $V^{\mathrm{AZ}}_{1,r}$ the space of $r\times r$ mass matrices that square to unity in a given Altland-Zirnbauer symmetry class when space is one-dimensional.[110,111]

1 For each Altland-Zirnbauer symmetry class, there exists a minimum rank $r^{\mathrm{AZ}}_{\min}$, an even integer equal to or larger than the integer 2, for which the Dirac Hamiltonian in one-dimensional space supports a Dirac mass matrix and below which either no Dirac mass matrix or no Dirac kinetic contributions are allowed. Lattice regularizations within an Altland-Zirnbauer symmetry class of a Dirac Hamiltonian within the same Altland-Zirnbauer symmetry class are only possible when the rank of the Dirac matrices is equal to or larger than $r^{\mathrm{AZ}}_{\min}$.

2 When there exists a Dirac mass matrix that

 2.1 is unique up to a sign,

 2.2 squares to unity,

 2.3 and that commutes with all other symmetry-allowed Dirac mass matrices;

 we call it the unique Dirac mass matrix.

[110] T. Morimoto and A. Furusaki, Phys. Rev. B **88**(12), 125129 (2013). [Morimoto and Furusaki (2013)].

[111] T. Morimoto, A. Furusaki, and C. Mudry, Phys. Rev. B **91**(23), 235111 (2015). [Morimoto et al. (2015a)].

3 There are the following three cases depending on the Altland-Zirnbauer symmetry class.

 3.1 There is always a unique Dirac mass matrix for $N \geq 1$ in the symmetry classes AIII, BDI, and CII.
 3.2 There is a unique Dirac mass matrix β_{min} only when $N = 1$ in the symmetry classes D and DIII. When N is an even integer, for any given Dirac mass matrix, there exists another matrix that anticommutes with it. When N is an odd integer larger than one, the Dirac matrix $\beta_{\text{min}} \otimes \mathbb{1}_N$ plays a role similar to that of the unique Dirac mass matrix in that the two matrices $\pm\beta_{\text{min}} \otimes \mathbb{1}_N$ belong to different path-connected components of $V^{\text{AZ}}_{1,r}$, even though there exists a normalized mass matrix that neither commutes nor anticommutes with $\beta_{\text{min}} \otimes \mathbb{1}_N$.
 3.3 For any given Dirac mass matrix, there exists another Dirac mass matrix that anticommutes with it in the symmetry classes A, AI, AII, C, and CI. That is, there is no unique mass matrix for $N \geq 1$ in those Altland-Zirnbauer symmetry classes.

4 Under the assumptions that (i) $N = 1$, (ii) the Dirac Hamiltonian in one-dimensional space has a unique mass matrix that realizes topologically distinct ground states for different signs of its mass, and (iii) the mass is varied smoothly in space, domain boundaries along which the mass vanishes are accompanied by localized zero modes (up to an energy split that is exponentially small with the separation between nearest-neighbor domain walls).

In summary, out of the ten Altland-Zirnbauer symmetry classes, there are five symmetry classes that can realize a strong topological insulator (superconductor) in a quantum wire, namely, the three chiral symmetry classes AIII, CII, and BDI and the two Bogoliubov-de-Gennes symmetry classes D and DIII. Tight-binding lattice realizations of the chiral symmetry classes AIII, CII, and BDI are variants of poly-acetylene. Tight-binding lattice realizations of the Bogoliubov-de-Gennes symmetry classes D and DIII are variants of the Kitaev chain. The disorder-induced quantum critical points with diverging density of states of quantum wires in the symmetry classes AIII, CII, BDI, D, and DIII that were derived in Section 7.7 are quantum phase transitions between topologically distinct phases of strong topological insulators (superconductors).

7.9 Exercises

7.1 Prove Eqs. (7.31) and (7.33).

7.2 William Rowan Hamilton introduced in 1843 the set $\mathbb{H}$ of real quaternions by constructing a non-commutative division algebra over the field of the real numbers. The vector space is defined to be the set of all linear combinations with real-valued coefficients of the four objects 1, $\boldsymbol{i}$, $\boldsymbol{j}$, and $\boldsymbol{k}$ where 1 is the neutral element of the multiplication and where the multiplicative rules are uniquely fixed by

$$\boldsymbol{i}^2 = \boldsymbol{j}^2 = \boldsymbol{k}^2 = \boldsymbol{i}\,\boldsymbol{j}\,\boldsymbol{k} = -1. \tag{7.450}$$

There is nothing sacred about choosing the real numbers as the field. We are going to choose the field to be that of the complex numbers. We shall also choose a two-dimensional representation of Eq. (7.450) with $\boldsymbol{i}$ represented by $\mathrm{i}\sigma_3$, $\boldsymbol{j}$ represented by $\mathrm{i}\sigma_2$, and $\boldsymbol{k}$ represented by $\mathrm{i}\sigma_1$, whereby $\boldsymbol{\sigma}$ is the vector of Pauli matrices.

Let $q_0, q_1, q_2, q_3 \in \mathbb{C}$. Let σ_0 denote the 2×2 unit matrix and $\boldsymbol{\sigma}$ the standard Pauli matrices. A quaternion q, its complex-conjugate q^*, its transpose q^{T}, its adjoint $q^{\dagger}$, and its quaternion-conjugate $\overline{q}$ are defined by the linear combinations

$$q := q_0\,\sigma_0 + \mathrm{i}\,q_1\,\sigma_1 + \mathrm{i}\,q_2\,\sigma_2 + \mathrm{i}\,q_3\,\sigma_3, \tag{7.451a}$$
$$q^* := q_0^*\,\sigma_0 - \mathrm{i}\,q_1^*\,\sigma_1 + \mathrm{i}\,q_2^*\,\sigma_2 - \mathrm{i}\,q_3^*\,\sigma_3, \tag{7.451b}$$
$$q^{\mathsf{T}} := q_0\,\sigma_0 + \mathrm{i}\,q_1\,\sigma_1 - \mathrm{i}\,q_2\,\sigma_2 + \mathrm{i}\,q_3\,\sigma_3, \tag{7.451c}$$
$$q^{\dagger} := q_0^*\,\sigma_0 - \mathrm{i}\,q_1^*\,\sigma_1 - \mathrm{i}\,q_2^*\,\sigma_2 - \mathrm{i}\,q_3^*\,\sigma_3, \tag{7.451d}$$
$$\overline{q} := q_0\,\sigma_0 - \mathrm{i}\,q_1\,\sigma_1 - \mathrm{i}\,q_2\,\sigma_2 - \mathrm{i}\,q_3\,\sigma_3, \tag{7.451e}$$

respectively. A quaternion $q = q_0\,\sigma_0 + \mathrm{i}\,q_1\,\sigma_1 + \mathrm{i}\,q_2\,\sigma_2 + \mathrm{i}\,q_3\,\sigma_3$ is defined to be real if and only if $q_0, q_1, q_2, q_3 \in \mathbb{R}$. For the product $q_1\,q_2 \cdots q_n$ of n quaternions, we have the definitions

$$(q_1\,q_2 \cdots q_n)^* := q_1^*\,q_2^* \cdots q_n^*, \tag{7.451f}$$
$$(q_1\,q_2 \cdots q_n)^{\mathsf{T}} := q_n^{\mathsf{T}}\,q_{n-1}^{\mathsf{T}} \cdots q_1^{\mathsf{T}}, \tag{7.451g}$$
$$(q_1\,q_2 \cdots q_n)^{\dagger} := q_n^{\dagger}\,q_{n-1}^{\dagger} \cdots q_1^{\dagger}, \tag{7.451h}$$
$$\overline{(q_1\,q_2 \cdots q_n)} := \bar{q}_n\,\bar{q}_{n-1} \cdots \bar{q}_1, \tag{7.451i}$$

for its complex-conjugate, transpose, adjoint, and quaternion-conjugate, respectively.

7.2.1 Verify that a quaternion q is real if

$$q = (\mathrm{i}\sigma_2)\,q^*\,(\mathrm{i}\sigma_2)^{-1}, \text{that is if } q^{\dagger} = \overline{q}. \tag{7.452}$$

7.2.2 Verify that transposition of a quaternion is related to conjugation of a quaternion by

$$q^{\mathsf{T}} = (\mathrm{i}\sigma_2)\,\overline{q}\,(\mathrm{i}\sigma_2)^{-1}. \tag{7.453}$$

7.2.3 Any $2N \times 2N$ matrix with complex-valued matrix elements can be written as a $N \times N$ matrix with quaternion-valued matrix elements. A $N \times N$ quaternion-valued matrix Q with the matrix elements

$$Q_{ij} := q_{ij\,0}\,\sigma_0 + \mathrm{i}\,q_{ij\,1}\,\sigma_1 + \mathrm{i}\,q_{ij\,2}\,\sigma_2 + \mathrm{i}\,q_{ij\,3}\,\sigma_3 \tag{7.454}$$

for $i,j=1,\cdots,N$ can be operated on as follows. Its complex-conjugate Q^{*}, transpose Q^{T}, adjoint $Q^{\dagger}$, and quaternion-conjugate $\overline{Q}$ are defined by the matrix elements

$$\left(Q^{*}\right)_{ij} := \left(Q_{ij}\right)^{*}, \tag{7.455a}$$

$$\left(Q^{\mathsf{T}}\right)_{ij} := \left(Q_{ji}\right)^{\mathsf{T}}, \tag{7.455b}$$

$$\left(Q^{\dagger}\right)_{ij} := \left(Q_{ji}\right)^{\dagger}, \tag{7.455c}$$

$$\left(\overline{Q}\right)_{ij} := \overline{\left(Q_{ij}\right)}, \tag{7.455d}$$

respectively, for $i,j=1,\cdots,N$.

7.2.3.1 Any $2N\times 2N$ matrix X is quaternion-real if and only if

$$X := J\,X^{*}\,J^{-1}, \qquad J := \mathbb{1}_{N}\otimes \mathrm{i}\sigma_{2}, \tag{7.456a}$$

with $\mathbb{1}_{N}$ the unit $N\times N$ matrix. The R-dual matrix Q^{R} to the quaternion-valued matrix Q is defined by the matrix elements

$$\left(Q^{\mathsf{R}}\right)_{ij} := (\mathrm{i}\sigma_{2})\,\left(Q^{*}\right)_{ij}\,(\mathrm{i}\sigma_{2})^{-1} = (\mathrm{i}\sigma_{2})\,\left(Q_{ij}\right)^{*}\,(\mathrm{i}\sigma_{2})^{-1} \tag{7.456b}$$

for $i,j=1,\cdots,N$. A quaternion-valued matrix Q is R-self-dual if and only if

$$Q^{\mathsf{R}} = Q. \tag{7.456c}$$

Show that if one reverts to representing the matrix element Q_{ij} as the 2×2 complex-valued matrix

$$Q_{ij} = \begin{pmatrix} a_{ij} & b_{ij} \\ c_{ij} & d_{ij} \end{pmatrix}, \qquad a_{ij}, b_{ij}, c_{ij}, d_{ij} \in \mathbb{C}, \tag{7.456d}$$

then Q is R-self-dual if and only if

$$a_{ij} = +d^{*}_{ij}, \qquad b_{ij} = -c^{*}_{ij}, \tag{7.456e}$$

for $i,j=1,\cdots,N$. Verify that the condition

$$Q^{\mathsf{R}} = Q \tag{7.456f}$$

also implies that Q is quaternion-real. Finally, verify that if

$$Q^{-1} = Q^{\dagger} = \left(Q^{\mathsf{R}}\right)^{\dagger}, \tag{7.456g}$$

then $Q\in \mathrm{Sp}(2N)$ with $\mathrm{Sp}(2N)$ defined in Eq. (7.474).[112]

[112] Whereas we chose here to represent an element of $\mathrm{Sp}(2N)$ by a matrix consisting of N^{2} 2×2 blocks, an element of $\mathrm{Sp}(2N)$ is represented by a matrix with $4=2^{2}$ $N\times N$ blocks in Eq. (7.474).

7.2.3.2 The RT-dual matrix Q^{RT} to the quaternion-valued matrix Q is defined by the matrix elements

$$\left(Q^{\mathsf{RT}}\right)_{ij} := (\mathrm{i}\sigma_2)\,\left(Q^{\mathsf{T}}\right)_{ij}\,(\mathrm{i}\sigma_2)^{-1} = (\mathrm{i}\sigma_2)\,\left(Q_{ji}\right)^{\mathsf{T}}\,(\mathrm{i}\sigma_2)^{-1} = \overline{\left(Q_{ji}\right)} \tag{7.457a}$$

for $i,j = 1,\cdots,N$. A quaternion-valued matrix Q is RT-self-dual if and only if

$$Q^{\mathsf{RT}} = Q. \tag{7.457b}$$

Show that if one reverts to representing the matrix element Q_{ij} as the 2×2 complex-valued matrix

$$Q_{ij} = \begin{pmatrix} a_{ij} & b_{ij} \\ c_{ij} & d_{ij} \end{pmatrix}, \qquad a_{ij}, b_{ij}, c_{ij}, d_{ij} \in \mathbb{C}, \tag{7.457c}$$

then Q is RT-self-dual if and only if

$$a_{ij} = +d_{ji}, \qquad b_{ij} = -b_{ji}, \qquad c_{ij} = -c_{ji} \tag{7.457d}$$

for $i,j = 1,\cdots,N$. Verify that the condition

$$Q^{\mathsf{RT}} = Q^{\dagger} \tag{7.457e}$$

also implies that Q is quaternion-real. Finally, verify that if

$$Q^{-1} = Q^{\dagger} = Q^{\mathsf{RT}}, \tag{7.457f}$$

then $Q \in \mathrm{Sp}(2N)$ with $\mathrm{Sp}(2N)$ defined in Eq. (7.474).

7.3 Let r be the (integer) number of rows and columns of a square matrix R.

7.3.1 What is the smallest $r > 1$ such that R is a real-valued and Hermitean matrix? What happens to R under complex conjugation? Enumerate all the real parameters parameterizing R. Which and how many parameters in R must be set to 0 so that there is no level repulsion (define what is meant by level repulsion).

7.3.2 Repeat the same exercise if the $r \times r$ matrix R is complex-valued.

7.3.3 Let $T := \sigma_2 \otimes \mathbb{1} \otimes \mathsf{K}$ where σ_2 is the antisymmetric Pauli matrix, $\mathbb{1}$ is a $n \times n$ unit matrix, and K denotes complex conjugation. Let R be any $r \times r$ Hermitean matrix such that $T\,R\,T^{-1} = R$. What is the smallest $r > 1$? How many real-valued parameters are needed to parameterize R? How many of these need to be set to 0 to prevent level repulsion? Is the spectrum of R degenerate?

7.4 Repeat **Exercise** 7.3 after demanding that the matrix R is block off-diagonal. Be careful with the size of the off-diagonal blocks, you may choose them to be rectangular.

7.5 At a conference on “Neutron Physics by Time-of-Flight,” held at the Oak Ridge National Laboratory in 1957, E. P. Wigner went from the audience to the blackboard during a discussion and surmised that

$$P(s)=\frac{\pi}{2}\,s\,e^{-\frac{\pi}{4}\,s^{2}} \tag{7.458}$$

governs the spacing of energy levels in units of the average spacing in a nucleus. An intuition for this surmise can be obtained as follows.

Consider a 2×2 symmetric matrix

$$\mathcal{H}:=\begin{pmatrix}h_{11} & h_{12}\\ h_{12} & h_{22}\end{pmatrix} \tag{7.459}$$

with h_{11}, h_{12}, and h_{22} arbitrary real-valued numbers. Define the probability to find Hamiltonian $\mathcal{H}$ in the infinitesimal volume element

$$\mathcal{D}[\mathcal{H}]:=\mathrm{d}h_{11}\,\mathrm{d}h_{12}\,\mathrm{d}h_{22} \tag{7.460a}$$

to be

$$P[\mathcal{H}]\,\mathcal{D}[\mathcal{H}]:=N_{\sigma^2}\,e^{-\frac{1}{2\sigma^2}\operatorname{tr}(\mathcal{H}^2)}\,\mathcal{D}[\mathcal{H}]. \tag{7.460b}$$

Here, $N_{\sigma^2}>0$ is a normalization constant and $\sigma^2>0$ is any positive number.

7.5.1 Show that there exists an angle $0\le\theta<2\pi$ and two real numbers ε_+ and ε_- such that

$$\begin{pmatrix}h_{11} & h_{12}\\ h_{12} & h_{22}\end{pmatrix}=\begin{pmatrix}\cos\theta & +\sin\theta\\ -\sin\theta & \cos\theta\end{pmatrix}\begin{pmatrix}\varepsilon_+ & 0\\ 0 & \varepsilon_-\end{pmatrix}\begin{pmatrix}\cos\theta & -\sin\theta\\ +\sin\theta & \cos\theta\end{pmatrix}, \tag{7.461a}$$

where

$$\varepsilon_\pm=\frac{h_{11}+h_{22}}{2}\pm\sqrt{(h_{12})^2+\left(\frac{h_{11}-h_{22}}{2}\right)^2} \tag{7.461b}$$

and

$$\tan\theta=\frac{\frac{h_{11}-h_{22}}{2}-\sqrt{(h_{12})^2+\left(\frac{h_{11}-h_{22}}{2}\right)^2}}{h_{12}}. \tag{7.461c}$$

7.5.2 Show that the probability to find Hamiltonian $\mathcal{H}$ in the infinitesimal volume element

$$\mathcal{D}[\mathcal{H}]:=\mathrm{d}\varepsilon_+\,\mathrm{d}\varepsilon_-\,\mathrm{d}\theta \tag{7.462a}$$

is

$$P[\mathcal{H}]\,\mathcal{D}[\mathcal{H}]=N_{\sigma^2}\,e^{-\frac{1}{2\sigma^2}(\varepsilon_+^2+\varepsilon_-^2)}\left|\frac{\partial(h_{11},h_{22},h_{12})}{\partial(\varepsilon_+,\varepsilon_-,\theta)}\right|\mathrm{d}\varepsilon_+\,\mathrm{d}\varepsilon_-\,\mathrm{d}\theta \tag{7.462b}$$

and give the explicit dependence of the Jacobian

$$J(\varepsilon_+,\varepsilon_-,\theta) := \left|\frac{\partial\,(h_{11},h_{22},h_{12})}{\partial\,(\varepsilon_+,\varepsilon_-,\theta)}\right| \tag{7.462c}$$

on ε_+, ε_-, and θ.

7.5.3 Do the change of variables

$$s := \varepsilon_+ - \varepsilon_-, \qquad \Delta := \frac{\varepsilon_+ + \varepsilon_-}{2}, \tag{7.463}$$

integrate over $\Delta \in \mathbb{R}$ and $0 \leq \theta < 2\pi$ the probability distribution (7.462) and show that Eq. (7.458) follows.

7.6 Let $n = 1, 2, 3, \cdots$ denote an integer. An element of $\mathbb{R}^n$ is denoted by the column vector $\boldsymbol{x}$ with the components $x_i \in \mathbb{R}$, $i = 1, \cdots, n$. An element of $\mathbb{C}^n$ is denoted by the column vector $\boldsymbol{z}$ with the components $z_i \in \mathbb{C}$, $i = 1, \cdots, n$. The corresponding row and adjoint vectors are $\boldsymbol{x}^{\mathsf{T}}$, $\boldsymbol{z}^{\dagger} \equiv (\boldsymbol{z}^*)^{\mathsf{T}}$, respectively. We reserve capital letters for linear maps of $\mathbb{K}^n$ to itself where the field is $\mathbb{K} = \mathbb{R}, \mathbb{C}$. If A denotes an $n \times n$ matrix over the field $\mathbb{C}$, then A^{T}, A^*, and $A^{\dagger}$ denote its transpose, complex-conjugate, and adjoint, respectively. The $n \times n$ matrix is symmetric, antisymmetric, Hermitean, and anti-Hermitean if

$$A = +A^{\mathsf{T}}, \quad A = -A^{\mathsf{T}}, \quad A = +A^{\dagger}, \quad A = -A^{\dagger}, \tag{7.464}$$

respectively. The number of linearly independent columns of A is called the rank of A and denoted rank$\,(A)$. The determinant of the matrix A is denoted det$\,(A)$. If det$\,(A) \neq 0$, then rank$\,(A) = n$, the dimension of the square matrix A. The set of all $n \times n$ matrices with nonvanishing determinant is denoted by $\mathrm{GL}(n, \mathbb{K})$. The set of all $n \times n$ matrices with determinant equal to one is denoted by $\mathrm{SL}(n, \mathbb{K})$. It is a subset of $\mathrm{GL}(n, \mathbb{K})$. Both sets realize a group under matrix multiplication because of the identity

$$\det(A\,B) = \det(A)\det(B) \tag{7.465}$$

for any two matrices A and B from these sets. Hence, $\mathrm{SL}(n, \mathbb{K})$ is a subgroup of $\mathrm{GL}(n, \mathbb{K})$.

We are going to define several subgroups of $\mathrm{GL}(n, \mathbb{K})$ and $\mathrm{SL}(n, \mathbb{K})$. To this end, we define the following matrices

$$\begin{aligned} I_{p,q} &:= \begin{pmatrix} -I_p & 0 \\ 0 & +I_q \end{pmatrix}, \\ K_{2p,2q} &:= \begin{pmatrix} -I_p & 0 & 0 & 0 \\ 0 & +I_q & 0 & 0 \\ 0 & 0 & -I_p & 0 \\ 0 & 0 & 0 & +I_q \end{pmatrix}, \\ J_{2n} &:= \begin{pmatrix} 0 & I_n \\ -I_n & 0 \end{pmatrix}, \end{aligned} \tag{7.466}$$

where $p, q, n = 1, 2, 3, \cdots$ and I_n is the $n \times n$ unit matrix. Observe that $I_{p,q}$, $K_{2p,2q}$, and I_n square to the unit matrix, but J_{2n} squares to minus the unit matrix, that is $J_{2n}^{-1} = -J_{2n}$. The classical reference for what follows can be found in Chapter X, paragraph 2 of Helgason from footnote 16.

7.6.1 Denote with $\mathrm{U}(p,q) \subset \mathrm{GL}(p+q,\mathbb{C})$ those matrices V such that

$$V^{\dagger}\, I_{p,q}\, V = I_{p,q}. \tag{7.467a}$$

Show that these matrices are the representations of the linear coordinate transformations of $\mathbb{C}^{p+q}$ that leave the sesquilinear form

$$\langle \boldsymbol{z} | \boldsymbol{w} \rangle_{I_{p,q}} := \boldsymbol{z}^{\dagger}\, I_{p,q}\, \boldsymbol{w}, \qquad \boldsymbol{z}, \boldsymbol{w} \in \mathbb{C}^{p+q} \tag{7.467b}$$

invariant. Show that $\mathrm{U}(p,q)$ is a group under matrix multiplication.
If we formally set $p = 0$ and $q = n$, we obtain the group $\mathrm{U}(n) \subset \mathrm{GL}(n,\mathbb{C})$ of matrices V such that

$$V^{\dagger}\, I_n\, V = I_n. \tag{7.467c}$$

7.6.2 The intersections of $\mathrm{U}(p,q)$ and $\mathrm{U}(n)$ with $\mathrm{SL}(n,\mathbb{C})$ are denoted $\mathrm{SU}(p,q)$ and $\mathrm{SU}(n)$, respectively. The group $\mathrm{U}(p) \times \mathrm{U}(q)$ denotes the set of matrices V of the form

$$V := \begin{pmatrix} V_p & 0 \\ 0 & V_q \end{pmatrix}, \qquad V_p \in \mathrm{U}(p), \qquad V_q \in \mathrm{U}(q) \tag{7.468a}$$

for any $p, q = 1, 2, 3, \cdots$. Its subgroup $\mathrm{S}\big(\mathrm{U}(p) \times \mathrm{U}(q)\big)$ is defined by demanding that

$$\det(V_p)\det(V_q) = 1. \tag{7.468b}$$

If we formally set $p = 0$ and $q = n$, we obtain the group $\mathrm{SU}(n) \subset \mathrm{GL}(n,\mathbb{C})$ of matrices V such that

$$V^{\dagger}\, I_n\, V = I_n, \qquad \det(V) = 1. \tag{7.468c}$$

7.6.3 Denote with $\mathrm{SU}^{\star}(2n) \subset \mathrm{SL}(2n,\mathbb{C})$ those matrices V such that

$$[V, J_{2n}\mathsf{K}] = 0, \tag{7.469}$$

where K denotes complex conjugation. Show that $\mathrm{SU}^{\star}(2n)$ is a group under matrix multiplication.

7.6.4 Denote with $\mathrm{SO}(n,\mathbb{C}) \subset \mathrm{SL}(n,\mathbb{C})$ those matrices V such that

$$V^{\mathsf{T}}\, I_n\, V = I_n. \tag{7.470a}$$

Show that these matrices are the representations of the linear coordinate transformations of $\mathbb{C}^n$ that leave the bilinear form

$$\langle \boldsymbol{z}^{*} | \boldsymbol{w} \rangle_{I_n} := \boldsymbol{z}^{\mathsf{T}}\, I_n\, \boldsymbol{w}, \quad \boldsymbol{z}, \boldsymbol{w} \in \mathbb{C}^n \tag{7.470b}$$

invariant. Show that $\mathrm{SO}(n,\mathbb{C})$ is a group under matrix multiplication.

7.6.5 Denote with $\mathrm{SO}(p,q) \subset \mathrm{SL}(p+q,\mathbb{R})$ those matrices V such that

$$V^{\mathsf{T}}\, I_{p,q}\, V = I_{p,q}. \tag{7.471a}$$

Show that these matrices are the representations of the linear coordinate transformations of $\mathbb{R}^{p+q}$ that leave the bilinear form

$$\langle \boldsymbol{x}|\boldsymbol{y}\rangle_{I_{p,q}} := \boldsymbol{x}^{\mathsf{T}}\, I_{p,q}\, \boldsymbol{y}, \quad \boldsymbol{x},\boldsymbol{y} \in \mathbb{R}^{p+q} \tag{7.471b}$$

invariant. Show that $\mathrm{SO}(p,q)$ is a group under matrix multiplication. If we formally set $p=0$ and $q=n$, we obtain the group $\mathrm{SO}(n) \subset \mathrm{SL}(n,\mathbb{R})$ of matrices V such that

$$V^{\mathsf{T}}\, I_n\, V = I_n. \tag{7.471c}$$

7.6.6 Denote with $\mathrm{SO}^{\star}(2n) \subset \mathrm{SO}(2n,\mathbb{C})$ those matrices V such that

$$V^{\dagger}\, J_{2n}\, V = J_{2n}. \tag{7.472a}$$

Show that these matrices are the representations of the linear coordinate transformations of $\mathbb{C}^{2n}$ that leave the bilinear and sesquilinear forms

$$\langle \boldsymbol{z}^*|\boldsymbol{w}\rangle_{I_{2n}} := \boldsymbol{z}^{\mathsf{T}}\, I_{2n}\, \boldsymbol{w}, \qquad \langle \boldsymbol{z}|\boldsymbol{w}\rangle_{J_{2n}} := \boldsymbol{z}^{\dagger}\, J_{2n}\, \boldsymbol{w}, \quad \boldsymbol{z},\boldsymbol{w} \in \mathbb{C}^{2n}, \tag{7.472b}$$

respectively, invariant. Show that $\mathrm{SO}^{\star}(2n)$ is a group under matrix multiplication.

7.6.7 Denote with $\mathrm{Sp}(2n,\mathbb{C}) \subset \mathrm{GL}(2n,\mathbb{C})$ those matrices V such that

$$V^{\mathsf{T}}\, J_{2n}\, V = J_{2n}. \tag{7.473a}$$

Show that these matrices are the representations of the linear coordinate transformations of $\mathbb{C}^{2n}$ that leave the bilinear form

$$\langle \boldsymbol{z}^*|\boldsymbol{w}\rangle_{J_{2n}} := \boldsymbol{z}^{\mathsf{T}}\, J_{2n}\, \boldsymbol{w}, \quad \boldsymbol{z},\boldsymbol{w} \in \mathbb{C}^{2n} \tag{7.473b}$$

invariant. Show that $\mathrm{Sp}(2n,\mathbb{C})$ is a group under matrix multiplication.

7.6.8 Define

$$\mathrm{Sp}(2n) := \mathrm{Sp}(2n,\mathbb{C}) \cap \mathrm{U}(2n) \tag{7.474}$$

and show that it is a group under matrix multiplication.

7.6.9 Denote with $\mathrm{Sp}(2n,\mathbb{R}) \subset \mathrm{GL}(2n,\mathbb{R})$ those matrices V such that

$$V^{\mathsf{T}}\, J_{2n}\, V = J_{2n}. \tag{7.475a}$$

Show that these matrices are the representations of the linear coordinate transformations of $\mathbb{R}^{2n}$ that leave the bilinear form

$$\langle \boldsymbol{x}|\boldsymbol{y}\rangle_{J_{2n}} := \boldsymbol{x}^{\mathsf{T}}\, J_{2n}\, \boldsymbol{y}, \quad \boldsymbol{x},\boldsymbol{y} \in \mathbb{R}^{2n} \tag{7.475b}$$

invariant. Show that $\mathrm{Sp}(2n,\mathbb{R})$ is a group under matrix multiplication.

7.6.10 Denote with $\mathrm{Sp}(2p,2q) \subset \mathrm{Sp}(2p+2q,\mathbb{C})$ those matrices V such that

$$V^{\dagger}\, K_{2p,2q}\, V = K_{2p,2q}. \tag{7.476a}$$

Show that these matrices are the representations of the linear coordinate transformations of $\mathbb{C}^{2p+2q}$ that leave the bilinear form (7.473b) and the sesquilinear form

$$(\boldsymbol{z}|\boldsymbol{w})_{K_{2p,2q}} := \boldsymbol{z}^{\dagger}\, K_{2p,2q}\, \boldsymbol{w}, \qquad \boldsymbol{z}, \boldsymbol{w} \in \mathbb{C}^{2p+2q} \tag{7.476b}$$

invariant. Show that $\mathrm{Sp}(2p,2q)$ is a group under matrix multiplication. Observe that setting formally $p=0$ and $q=n$ or conversely $p=n$ and $q=0$ in $\mathrm{Sp}(2p,2q)$ defines $\mathrm{Sp}(2n)$.
We treat the corresponding Lie algebras.
Denote with $\mathrm{gl}(n,\mathbb{K})$ all $\mathbb{K}$-valued $n\times n$ matrices. Argue that for any $A \in \mathrm{GL}(n,\mathbb{K})$, there exist a $t\in\mathbb{R}$ and a $X \in \mathrm{gl}(n,\mathbb{K})$ such that

$$A = e^{t\,X}. \tag{7.477}$$

The set $\mathrm{gl}(n,\mathbb{K})$ is the Lie algebra of the Lie group $\mathrm{GL}(n,\mathbb{K})$. Show that $\mathrm{gl}(n,\mathbb{K})$ is a vector space over the field of real numbers of dimension $2n^2$ if $\mathbb{K}=\mathbb{C}$ and n^2 if $\mathbb{K}=\mathbb{R}$.
Denote with $\mathfrak{sl}(n,\mathbb{K})$ all $\mathbb{K}$-valued $n\times n$ matrices that are traceless. Argue that

$$\mathfrak{sl}(n,\mathbb{K}) = \left\{X \in \mathrm{gl}(n,\mathbb{K}) \;\middle|\; e^{t\,X} \in \mathrm{SL}(n,\mathbb{K}) \quad \forall t\in\mathbb{R}\right\}. \tag{7.478}$$

The set $\mathfrak{sl}(n,\mathbb{K})$ is a Lie subalgebra of $\mathrm{gl}(n,\mathbb{K})$. Show that $\mathfrak{sl}(n,\mathbb{K})$ is a vector space over the field of real numbers of dimension $2(n^2-1)$ if $\mathbb{K}=\mathbb{C}$ and n^2-1 if $\mathbb{K}=\mathbb{R}$.

7.6.11 Argue that the Lie algebra $\mathfrak{u}(p,q)$ of $\mathrm{U}(p,q)$ is the set of matrices of the form

$$\begin{pmatrix} A & B \\ B^{\dagger} & C \end{pmatrix}, \qquad A=-A^{\dagger}, \qquad C=-C^{\dagger}, \tag{7.479a}$$

with A, B, and C complex-valued $p\times p$, $p\times q$, and $q\times q$ matrices, respectively. Show that $\mathfrak{u}(p,q)$ is a vector space over the field of real numbers of dimension

$$p^2+q^2+2p\,q = (p+q)^2. \tag{7.479b}$$

If we formally set $p=0$ and $q=n$, we obtain the Lie algebra $\mathfrak{u}(n) \subset \mathrm{gl}(n,\mathbb{C})$, a vector space over the field of real numbers of dimension n^2.

7.6.12 Argue that the Lie algebra $\mathfrak{su}(p,q)$ of $\mathrm{SU}(p,q)$ is the set of matrices of the form

$$\begin{pmatrix} A & B \\ B^{\dagger} & C \end{pmatrix}, \; A=-A^{\dagger}, \; C=-C^{\dagger}, \qquad \mathrm{tr}\,(A)+\mathrm{tr}\,(C)=0, \tag{7.480a}$$

with A, B, and C complex-valued $p \times p$, $p \times q$, and $q \times q$ matrices, respectively. Show that $\mathfrak{su}(p,q)$ is a vector space over the field of real numbers of dimension

$$p^2 + q^2 + 2pq - 1 = (p+q)^2 - 1. \tag{7.480b}$$

If we formally set $p = 0$ and $q = n$, we obtain the Lie algebra $\mathfrak{su}(n) \subset \mathfrak{sl}(n,\mathbb{C})$, a vector space over the field of real numbers of dimension $n^2 - 1$.

7.6.13 Argue that the Lie algebra $\mathfrak{su}^\star(2n)$ of $\mathrm{SU}^\star(2n)$ is the set of $2n \times 2n$ complex-valued matrices of the form

$$\begin{pmatrix} A & B \\ -B^* & A^* \end{pmatrix}, \qquad \mathrm{tr}\,(A + A^*) = 0, \tag{7.481a}$$

with A and B complex-valued $n \times n$ matrices. Show that $\mathfrak{su}^\star(2n)$ is a vector space over the field of real numbers of dimension

$$2n^2 + 2n^2 - 1 = 4n^2 - 1. \tag{7.481b}$$

7.6.14 Argue that the Lie algebra $\mathfrak{so}(n,\mathbb{C})$ of $\mathrm{SO}(n,\mathbb{C})$ is the set of $n \times n$ complex-valued matrices of the form

$$A, \qquad A = -A^\mathsf{T}. \tag{7.482a}$$

Show that $\mathfrak{so}(n,\mathbb{C})$ is a vector space over the field of real numbers of dimension

$$2 \times \frac{(n-1)\,n}{2} = (n-1)\,n. \tag{7.482b}$$

7.6.15 Argue that the Lie algebra $\mathfrak{so}(p,q)$ of $\mathrm{SO}(p,q)$ is the set of matrices of the form

$$\begin{pmatrix} A & B \\ B^\mathsf{T} & C \end{pmatrix}, \qquad A = -A^\mathsf{T}, \qquad C = -C^\mathsf{T}, \tag{7.483a}$$

with A, B, and C real-valued $p \times p$, $p \times q$, and $q \times q$ matrices, respectively. Show that $\mathfrak{so}(p,q)$ is a vector space over the field of real numbers of dimension

$$\frac{(p-1)\,p}{2} + \frac{(q-1)\,q}{2} + p\,q = \frac{[(p+q)-1](p+q)}{2}. \tag{7.483b}$$

If we formally set $p = 0$ and $q = n$, we obtain the Lie algebra $\mathfrak{so}(n) \subset \mathfrak{sl}(n,\mathbb{R})$, a vector space over the field of real numbers of dimension $(n-1)n/2$.

7.6.16 Argue that the Lie algebra $\mathfrak{so}^\star(2n)$ of $\mathrm{SO}^\star(2n)$ is the set of matrices of the form

$$\begin{pmatrix} A & B \\ -B^* & A^* \end{pmatrix}, \qquad A = -A^\mathsf{T}, \qquad B = B^\dagger, \tag{7.484a}$$

with A and B complex-valued $n\times n$ matrices. Show that $\mathfrak{so}^\star(2n)$ is a vector space over the field of real numbers of dimension

$$(n-1)n+n^2=2n^2-n. \tag{7.484b}$$

7.6.17 Argue that the Lie algebra $\mathfrak{sp}(2n,\mathbb{C})$ of $\mathrm{Sp}(2n,\mathbb{C})$ is the set of matrices of the form

$$\begin{pmatrix} A & B \\ C & -A^{\mathsf{T}} \end{pmatrix}, \qquad B=B^{\mathsf{T}}, \qquad C=C^{\mathsf{T}}, \tag{7.485a}$$

with A, B, and C $n\times n$ complex-valued matrices. Show that $\mathfrak{sp}(2n,\mathbb{C})$ is a vector space over the field of real numbers of dimension

$$2n^2+n\,(n+1)+n\,(n+1)=2n\,(2n+1). \tag{7.485b}$$

7.6.18 Argue that the Lie algebra $\mathfrak{sp}(2n)$ of $\mathrm{Sp}(2n)$ is the set of matrices of the form

$$\begin{pmatrix} A & B \\ -B^{\dagger} & -A^{\mathsf{T}} \end{pmatrix}, \qquad A=-A^{\dagger}, \quad B=B^{\mathsf{T}}, \tag{7.486a}$$

with A and B $n\times n$ complex-valued matrices. Show that $\mathfrak{sp}(2n)$ is a vector space over the field of real numbers of dimension

$$n^2+n(n+1)=n\,(2n+1)\,. \tag{7.486b}$$

7.6.19 Argue that the Lie algebra $\mathfrak{sp}(2n,\mathbb{R})$ of $\mathrm{Sp}(2n,\mathbb{R})$ is the set of matrices of the form

$$\begin{pmatrix} A & B \\ C & -A^{\mathsf{T}} \end{pmatrix}, \qquad B=B^{\mathsf{T}}, \qquad C=C^{\mathsf{T}}, \tag{7.487a}$$

with A, B, and C $n\times n$ real-valued matrices. Show that $\mathfrak{sp}(2n,\mathbb{R})$ is a vector space over the field of real numbers of dimension

$$n^2+n\,(n+1)=n\,(2n+1). \tag{7.487b}$$

7.6.20 Argue that the Lie algebra $\mathfrak{sp}(2p,2q)$ of $\mathrm{Sp}(2p,2q)$ is the set of matrices of the form

$$\begin{pmatrix} A_{11} & A_{12} & A_{13} & A_{14} \\ A^{\dagger}_{12} & A_{22} & A^{\mathsf{T}}_{14} & A_{24} \\ -A^{*}_{13} & A^{*}_{14} & A^{*}_{11} & -A^{*}_{12} \\ A^{\dagger}_{14} & -A^{*}_{24} & -A^{\mathsf{T}}_{12} & A^{*}_{22} \end{pmatrix}, \tag{7.488a}$$

where

$$A_{11}=-A^{\dagger}_{11}, \quad A_{22}=-A^{\dagger}_{22}, \quad A_{13}=A^{\mathsf{T}}_{13}, \quad A_{24}=A^{\mathsf{T}}_{24}, \tag{7.488b}$$

with all six A_{11} $(p \times p)$, A_{12} $(p \times q)$, A_{13} $(p \times p)$, A_{14} $(p \times q)$, A_{22} $(q \times q)$, A_{24} $(q \times q)$, complex-valued matrices. Show that $\mathfrak{sp}(2p, 2q)$ is a vector space over the field of real numbers of dimension

$$p^2 + 2p\,q + p(p+1) + (p \leftrightarrow q) = p(2p+1) + q(2q+1) + 4p\,q. \tag{7.488c}$$

We are going to map one-to-one and onto themselves each Lie algebra $\mathfrak{su}(n)$, $\mathfrak{su}(2n)$, $\mathfrak{su}(p+q)$, $\mathfrak{so}(p+q)$, $\mathfrak{so}(2n)$, $\mathfrak{sp}(2n)$, $\mathfrak{sp}(2p+2q)$ in such a way that each map is either linear or antilinear and involutive (an involutive automorphism), that is its composition with itself is the identity. This will allow an additive decomposition of each Lie algebra when interpreted as a vector space on which the involutive automorphism acts.

7.6.21 Cartan notation AI Choose the Lie algebra $\mathfrak{su}(n)$ with $n = 1, 2, 3, \cdots$. Define the map

$$\begin{aligned} \theta\colon \mathfrak{su}(n) &\to \mathfrak{su}(n), \\ X &\mapsto X^*. \end{aligned} \tag{7.489a}$$

Show that this map is antilinear and that it is involutive. Hence, the eigenvalues of this antilinear map are ± 1. Define the sets

$$\mathfrak{k} := \{X \in \mathfrak{su}(n) \mid X^* = +X\}, \tag{7.489b}$$

$$\mathfrak{p} := \{X \in \mathfrak{su}(n) \mid X^* = -X\}. \tag{7.489c}$$

Show that $\mathfrak{p}$ consists of all symmetric purely imaginary $n \times n$ matrices of vanishing traces,

$$\mathfrak{p} := \left\{ \mathrm{i}X \;\middle|\; X = X^{\mathsf{T}} \in \mathrm{gl}(n, \mathbb{R}), \qquad \mathrm{tr}\,(X) = 0 \right\}. \tag{7.489d}$$

Show that $\mathfrak{p}$ is not closed under the commutator. Show that

$$\mathfrak{k} = \mathfrak{so}(n) \tag{7.489e}$$

and that $\mathfrak{k}$ is a Lie subalgebra of $\mathfrak{su}(n)$. Give the proper interpretation to the notations

$$\begin{aligned} \mathfrak{su}(n) &= \mathfrak{k} \oplus \mathfrak{p}, \qquad && \mathrm{SU}(n)/\mathrm{SO}(n), \\ \mathfrak{sl}(n, \mathbb{R}) &= \mathfrak{k} \oplus \mathrm{i}\mathfrak{p}, \qquad && \mathrm{SL}(n, \mathbb{R})/\mathrm{SO}(n). \end{aligned} \tag{7.489f}$$

Hint: The coset space $\mathrm{SU}(n)/\mathrm{SO}(n)$ is the set of all equivalence classes with two elements e^X and e^Y from $\mathrm{SU}(n)$ equivalent if and only if $X - Y \in \mathfrak{so}(n)$. What happens to the generators of the subgroup $\mathrm{SO}(n)$ of $\mathrm{SU}(n)$ in $\mathrm{SU}(n)/\mathrm{SO}(n)$? How are the generators from $\mathrm{SU}(n)$ present in $\mathrm{SU}(n)/\mathrm{SO}(n)$ related to those present in $\mathrm{SL}(n, \mathbb{R})/\mathrm{SO}(n)$?

7.6.22 Cartan notation AII Choose the Lie algebra $\mathfrak{su}(2n)$ with $n = 1, 2, 3, \cdots$. Define the map

$$\begin{aligned} \theta\colon \mathfrak{su}(2n) &\to \mathfrak{su}(2n), \\ X &\mapsto J_{2n}\, X^*\, J_{2n}^{-1}. \end{aligned} \tag{7.490a}$$

Show that this map is antilinear and that it is involutive. Hence, the eigenvalues of this antilinear map are ± 1. Define the sets

$$\mathfrak{k} := \left\{ X \in \mathfrak{su}(2n) \mid J_{2n}\, X^{*}\, J_{2n}^{-1} = +X \right\}, \tag{7.490b}$$

$$\mathfrak{p} := \left\{ X \in \mathfrak{su}(2n) \mid J_{2n}\, X^{*}\, J_{2n}^{-1} = -X \right\}. \tag{7.490c}$$

Show that $\mathfrak{p}$ consists of all matrices of the form

$$\begin{pmatrix} A & B \\ B^{*} & -A^{*} \end{pmatrix}, \qquad A \in \mathfrak{su}(n), \qquad B \in \mathfrak{so}(n,\mathbb{C}). \tag{7.490d}$$

Show that $\mathfrak{p}$ is not closed under the commutator. Show that

$$\mathfrak{k} = \mathfrak{sp}(2n) \tag{7.490e}$$

and that $\mathfrak{k}$ is a Lie subalgebra of $\mathfrak{su}(2n)$. Give the proper interpretation to the notations

$$\begin{aligned} \mathfrak{su}(2n) &= \mathfrak{k} \oplus \mathfrak{p}, \qquad \mathrm{SU}(2n)/\mathrm{Sp}(2n), \\ \mathfrak{su}^{\star}(2n) &= \mathfrak{k} \oplus \mathrm{i}\mathfrak{p}, \qquad \mathrm{SU}^{\star}(2n)/\mathrm{Sp}(2n). \end{aligned} \tag{7.490f}$$

Hint: The coset space $\mathrm{SU}(2n)/\mathrm{Sp}(2n)$ is the set of all equivalence classes with two elements e^{X} and e^{Y} from $\mathrm{SU}(2n)$ equivalent if and only if $X - Y \in \mathfrak{sp}(2n)$.

7.6.23 Cartan notation AIII Choose the Lie algebra $\mathfrak{su}(p+q)$ with $p, q = 1, 2, 3, \cdots$. Define the map

$$\begin{aligned} \theta\colon \mathfrak{su}(p+q) &\to \mathfrak{su}(p+q), \\ X &\mapsto I_{p,q}\, X\, I_{p,q}^{-1}. \end{aligned} \tag{7.491a}$$

Show that this map is linear and that it is involutive. Hence, the eigenvalues of this linear map are ± 1. Define the sets

$$\mathfrak{k} := \left\{ X \in \mathfrak{su}(p+q) \mid I_{p,q}\, X\, I_{p,q}^{-1} = +X \right\}, \tag{7.491b}$$

$$\mathfrak{p} := \left\{ X \in \mathfrak{su}(p+q) \mid I_{p,q}\, X\, I_{p,q}^{-1} = -X \right\}. \tag{7.491c}$$

Show that $\mathfrak{p}$ consists of all matrices of the form

$$\begin{pmatrix} 0 & A \\ -A^{\dagger} & 0 \end{pmatrix}, \tag{7.491d}$$

with A any arbitrary complex-valued $p \times q$ matrix. Show that $\mathfrak{p}$ is not closed under the commutator. Show that $\mathfrak{k}$ consists of all matrices of the form

$$\begin{pmatrix} A & 0 \\ 0 & B \end{pmatrix}, \quad A \in \mathfrak{u}(p), \; B \in \mathfrak{u}(q), \; \mathrm{tr}\,(A) + \mathrm{tr}\,(B) = 0, \tag{7.491e}$$

that is

$$\mathfrak{k} = \mathfrak{s}\big(\mathfrak{u}(p) \oplus \mathfrak{u}(q)\big), \tag{7.491f}$$

and that $\mathfrak{k}$ is a Lie subalgebra of $\mathfrak{su}(p+q)$. Give the proper interpretation to the notations

$$\begin{aligned} \mathfrak{su}(p+q) &= \mathfrak{k} \oplus \mathfrak{p}, \qquad \mathrm{SU}(p+q)/\mathrm{S}\big(\mathrm{U}(p) \times \mathrm{U}(q)\big), \\ \mathfrak{su}(p,q) &= \mathfrak{k} \oplus \mathrm{i}\mathfrak{p}, \qquad \mathrm{SU}(p,q)/\mathrm{S}\big(\mathrm{U}(p) \times \mathrm{U}(q)\big). \end{aligned} \tag{7.491g}$$

Hint: The coset space $\mathrm{SU}(p+q)/\mathrm{S}\big(\mathrm{U}(p) \times \mathrm{U}(q)\big)$ is the set of all equivalence classes with two elements e^X and e^Y from $\mathrm{SU}(p+q)$ equivalent if and only if $X - Y \in \mathfrak{s}\big(\mathfrak{u}(p) \oplus \mathfrak{u}(q)\big)$.

7.6.24 Cartan notation BDI Choose the Lie algebra $\mathfrak{so}(p+q)$ with $p \geq q = 1, 2, 3, \cdots$. Define the map[113]

$$\begin{aligned} \theta \colon \mathfrak{so}(p+q) &\to \mathfrak{so}(p+q), \\ X &\mapsto I_{p,q}\, X\, I_{p,q}^{-1}. \end{aligned} \tag{7.492a}$$

Show that this map is linear and that it is involutive. Hence, the eigenvalues of this linear map are ± 1. Define the sets

$$\mathfrak{k} := \left\{ X \in \mathfrak{so}(p+q) \mid I_{p,q}\, X\, I_{p,q}^{-1} = +X \right\}, \tag{7.492b}$$

$$\mathfrak{p} := \left\{ X \in \mathfrak{so}(p+q) \mid I_{p,q}\, X\, I_{p,q}^{-1} = -X \right\}. \tag{7.492c}$$

Show that $\mathfrak{p}$ consists of all matrices of the form

$$\begin{pmatrix} 0 & X \\ -X^{\mathsf{T}} & 0 \end{pmatrix}, \tag{7.492d}$$

with X an arbitrary $p \times q$ real-valued matrix. Show that $\mathfrak{p}$ is not closed under the commutator. Show that $\mathfrak{k}$ consists of all matrices of the form

$$\begin{pmatrix} A & 0 \\ 0 & B \end{pmatrix}, \qquad A \in \mathfrak{so}(p), \qquad B \in \mathfrak{so}(q), \tag{7.492e}$$

that is

$$\mathfrak{k} = \mathfrak{so}(p) \oplus \mathfrak{so}(q), \tag{7.492f}$$

and that $\mathfrak{k}$ is a Lie subalgebra of $\mathfrak{so}(p+q)$. Convince yourself that the map

$$\begin{pmatrix} A & \mathrm{i}X \\ -\mathrm{i}X^{\mathsf{T}} & B \end{pmatrix} \mapsto \begin{pmatrix} A & X \\ X^{\mathsf{T}} & B \end{pmatrix} \tag{7.492g}$$

is an isomorphism onto $\mathfrak{so}(p,q)$. Give the proper interpretation to the notations

$$\begin{aligned} \mathfrak{so}(p+q) &= \mathfrak{k} \oplus \mathfrak{p}, \qquad \mathrm{SO}(p+q)/\mathrm{SO}(p) \times \mathrm{SO}(q), \\ \mathfrak{so}(p,q) &\cong \mathfrak{k} \oplus \mathrm{i}\mathfrak{p}, \qquad \mathrm{SO}_0(p,q)/\mathrm{SO}(p) \times \mathrm{SO}(q). \end{aligned} \tag{7.492h}$$

Here, $\mathrm{SO}_0(p,q)$ denotes the component of $\mathrm{SO}(p,q)$ connected to the unit matrix. *Hint:* The coset space $\mathrm{SO}(p+q)/[\mathrm{SO}(p) \times \mathrm{SO}(q)]$ is the set of all

[113] The assumption $p \geq q$ fixes the choice of $I_{p,q}$ as opposed to $I_{q,p}$.

equivalence classes with two elements e^X and e^Y from $\mathrm{SO}(p+q)$ equivalent if and only if $X - Y \in \mathfrak{so}(p) \oplus \mathfrak{so}(q)$.

7.6.25 Cartan notation DIII Choose the Lie algebra $\mathfrak{so}(2n)$ with $n = 1, 2, 3, \cdots$. Define the map

$$\begin{aligned}\theta\colon \mathfrak{so}(2n) &\to \mathfrak{so}(2n),\\ X &\mapsto J_{2n}\, X\, J_{2n}^{-1}.\end{aligned} \tag{7.493a}$$

Show that this map is linear and that it is involutive. Hence, the eigenvalues of this linear map are ± 1. Define the sets

$$\mathfrak{k} := \left\{X \in \mathfrak{so}(2n) \mid J_{2n}\, X\, J_{2n}^{-1} = +X\right\}, \tag{7.493b}$$

$$\mathfrak{p} := \left\{X \in \mathfrak{so}(2n) \mid J_{2n}\, X\, J_{2n}^{-1} = -X\right\}. \tag{7.493c}$$

Show that $\mathfrak{p}$ consists of all matrices of the form

$$\begin{pmatrix} A & B \\ B & -A \end{pmatrix}, \qquad A, B \in \mathfrak{so}(n). \tag{7.493d}$$

Show that $\mathfrak{p}$ is not closed under the commutator. Show that $\mathfrak{k}$ consists of all matrices of the form

$$\begin{pmatrix} A & B \\ -B & A \end{pmatrix}, \qquad A = A^* = -A^{\mathsf{T}}, \qquad B = B^* = B^{\mathsf{T}}. \tag{7.493e}$$

Show that the map defined

$$A + \mathrm{i}B \mapsto \begin{pmatrix} A & B \\ -B & A \end{pmatrix} \tag{7.493f}$$

for all $A = -A^{\mathsf{T}} \in \mathrm{gl}(n, \mathbb{R})$ and all $B = +B^{\mathsf{T}} \in \mathrm{gl}(n, \mathbb{R})$ is a Lie algebra isomorphism between $\mathfrak{u}(n)$ and $\mathfrak{k}$, that is it is linear, bijective, and compatible with the commutator. Thus, we have the Lie algebra isomorphism

$$\mathfrak{k} \cong \mathfrak{u}(n). \tag{7.493g}$$

Give the proper interpretation to the notations

$$\begin{aligned}\mathfrak{so}(2n) &= \mathfrak{k} \oplus \mathfrak{p}, \qquad \mathrm{SO}(2n)/\mathrm{U}(n),\\ \mathfrak{so}^\star(2n) &= \mathfrak{k} \oplus \mathrm{i}\mathfrak{p}, \qquad \mathrm{SO}^\star(2n)/\mathrm{U}(n).\end{aligned} \tag{7.493h}$$

Hint: The coset space $\mathrm{SO}(2n)/\mathrm{U}(n)$ is the set of all equivalence classes with two elements e^X and e^Y from $\mathrm{SO}(2n)$ equivalent if and only if $X - Y \in \mathfrak{k} \cong \mathfrak{u}(n)$.

7.6.26 Cartan notation CI Choose the Lie algebra $\mathfrak{sp}(2n)$ with $n = 1, 2, 3, \cdots$. Define the map

$$\begin{aligned}\theta\colon \mathfrak{sp}(2n) &\to \mathfrak{sp}(2n),\\ X &\mapsto J_{2n}\, X\, J_{2n}^{-1} \equiv X^*.\end{aligned} \tag{7.494a}$$

Show that this map is antilinear and that it is involutive. Hence, the eigenvalues of this antilinear map are ± 1. Define the sets

$$\mathfrak{k} := \left\{X \in \mathfrak{sp}(2n) \mid X^* = J_{2n}\, X\, J_{2n}^{-1} = +X\right\}, \tag{7.494b}$$

$$\mathfrak{p} := \left\{X \in \mathfrak{sp}(2n) \mid X^* = J_{2n}\, X\, J_{2n}^{-1} = -X\right\}. \tag{7.494c}$$

Show that $\mathfrak{p}$ consists of all matrices of the form

$$\begin{pmatrix} A & B \\ B & -A \end{pmatrix}, \quad A = -A^* = -A^\dagger, \quad B = -B^* = +B^\mathsf{T}. \tag{7.494d}$$

Show that $\mathfrak{p}$ is not closed under the commutator. Show that $\mathfrak{k}$ consists of all matrices of the form

$$\begin{pmatrix} A & B \\ -B & A \end{pmatrix}, \quad A = -A^\dagger = -A^\mathsf{T} = A^*, \quad B = B^\mathsf{T} = B^*, \tag{7.494e}$$

that is we have the Lie algebra isomorphism

$$\mathfrak{k} \cong \mathfrak{u}(n). \tag{7.494f}$$

Give the proper interpretation to the notations

$$\begin{aligned} \mathfrak{sp}(2n) &= \mathfrak{k} \oplus \mathfrak{p}, \qquad \mathrm{Sp}(2n)/\mathrm{U}(n), \\ \mathfrak{sp}(2n,\mathbb{R}) &= \mathfrak{k} \oplus \mathrm{i}\mathfrak{p}, \qquad \mathrm{Sp}(2n,\mathbb{R})/\mathrm{U}(n). \end{aligned} \tag{7.494g}$$

Hint: The coset space $\mathrm{Sp}(2n)/\mathrm{U}(n)$ is the set of all equivalence classes with two elements e^X and e^Y from $\mathrm{Sp}(2n)$ equivalent if and only if $X - Y \in \mathfrak{k} \cong \mathfrak{u}(n)$.

7.6.27 Cartan notation CII Choose the Lie algebra $\mathfrak{sp}(2p+2q)$ with $p, q = 1, 2, 3, \cdots$. Define the map

$$\begin{aligned} \theta\colon \mathfrak{sp}(2p+2q) &\to \mathfrak{sp}(2p+2q), \\ X &\mapsto K_{2p,2q}\, X\, K_{2p,2q}^{-1}. \end{aligned} \tag{7.495a}$$

Show that this map is linear and that it is involutive. Hence, the eigenvalues of this linear map are ± 1. Define the sets

$$\mathfrak{k} := \left\{X \in \mathfrak{sp}(2p+2q) \mid K_{2p,2q}\, X\, K_{2p,2q}^{-1} = +X\right\}, \tag{7.495b}$$

$$\mathfrak{p} := \left\{X \in \mathfrak{sp}(2p+2q) \mid K_{2p,2q}\, X\, K_{2p,2q}^{-1} = -X\right\}. \tag{7.495c}$$

Show that $\mathfrak{p}$ consists of all matrices of the form

$$\begin{pmatrix} 0 & A_{12} & 0 & A_{14} \\ -A_{12}^\dagger & 0 & A_{14}^\mathsf{T} & 0 \\ 0 & -A_{14}^* & 0 & A_{12}^* \\ -A_{14}^\dagger & 0 & -A_{12}^\mathsf{T} & 0 \end{pmatrix}, \tag{7.495d}$$

with A_{12} and A_{14} any $p \times q$ complex-valued matrices. Show that $\mathfrak{p}$ is not closed under the commutator. Show that $\mathfrak{k}$ consists of all matrices of the form

$$\begin{pmatrix} A_{11} & 0 & A_{13} & 0 \\ 0 & A_{22} & 0 & A_{24} \\ -A_{13}^{*} & 0 & A_{11}^{*} & 0 \\ 0 & -A_{24}^{*} & 0 & A_{22}^{*} \end{pmatrix}, \tag{7.495e}$$

where

$$A_{11} = -A_{11}^{\dagger}, \quad A_{22} = -A_{22}^{\dagger}, \quad A_{13} = A_{13}^{\mathsf{T}}, \quad A_{24} = A_{24}^{\mathsf{T}}, \tag{7.495f}$$

with all four A_{11} $(p \times p)$, A_{13} $(p \times p)$, A_{22} $(q \times q)$, A_{24} $(q \times q)$, complex-valued matrices. Deduce that

$$\mathfrak{k} = \mathfrak{sp}(2p) \oplus \mathfrak{sp}(2q). \tag{7.495g}$$

Give the proper interpretation to the notations

$$\begin{aligned} &\mathfrak{sp}(2p+2q) = \mathfrak{k} \oplus \mathfrak{p}, \qquad && \mathrm{Sp}(2p+2q)/\mathrm{Sp}(2p) \times \mathrm{Sp}(2q), \\ &\mathfrak{sp}(2p,2q) = \mathfrak{k} \oplus i\mathfrak{p}, \qquad && \mathrm{Sp}(2p,2q)/\mathrm{Sp}(2p) \times \mathrm{Sp}(2q). \end{aligned} \tag{7.495h}$$

Hint: The coset space $\mathrm{Sp}(2p+2q)/[\mathrm{Sp}(2p) \times \mathrm{Sp}(2q)]$ is the set of all equivalence classes with two elements e^{X} and e^{Y} from $\mathrm{Sp}(2p+2q)$ equivalent if and only if $X - Y \in \mathfrak{sp}(2p) \oplus \mathfrak{sp}(2q)$.

7.7 Show that if $\mathcal{X}_{\mathrm{BdG}}$ and $\mathcal{Y}_{\mathrm{BdG}}$ are both anti-Hermitean matrices of dimension $2N$, so is their commutator $[\mathcal{X}_{\mathrm{BdG}}, \mathcal{Y}_{\mathrm{BdG}}]$.

7.8 How many independent real parameters are needed to parameterize a real-valued antisymmetric $2N \times 2N$ matrix? Does the counting of real parameters entering the single-particle BdG Hamiltonian (7.40) agree with the number of generators of $\mathfrak{so}(2N)$?

7.9 The matrix

$$\begin{pmatrix} +\mathsf{A} & +\mathsf{B} & 0 & 0 \\ -\mathsf{B}^{\dagger} & -\mathsf{A}^{\mathsf{T}} & 0 & 0 \\ 0 & 0 & +\mathsf{A} & -\mathsf{B} \\ 0 & 0 & +\mathsf{B}^{\dagger} & -\mathsf{A}^{\mathsf{T}} \end{pmatrix}, \qquad \mathsf{A} = -\mathsf{A}^{\dagger}, \qquad \mathsf{B} = +\mathsf{B}^{\mathsf{T}} \tag{7.496}$$

from Eq. (7.72) has two diagonal blocks. Show that each block is invariant under conjugation by the second Pauli matrix followed by complex conjugation. Show that the lower block is obtained from the upper block by conjugation with the first Pauli matrix followed by complex conjugation. Verify Eq. (7.74).

7.10 Show that any $2N \times 2N$ matrix $\mathcal{X}^{\mathrm{ire}}$ with the block structure

$$\mathcal{X}^{\mathrm{ire}} = \begin{pmatrix} \mathsf{A} & \mathsf{B} \\ \mathsf{C} & \mathsf{D} \end{pmatrix},$$

where A, B, C, and D are four arbitrary $N \times N$ matrices, is of the form (7.73) if and only if

$$\mathcal{X}^{\mathrm{ire}} = -\left(\mathcal{X}^{\mathrm{ire}}\right)^{\dagger}, \qquad \mathcal{C}^{\mathrm{ire}}\left(\mathcal{X}^{\mathrm{ire}}\right)^{\mathsf{T}} + \mathcal{X}^{\mathrm{ire}}\,\mathcal{C}^{\mathrm{ire}} = 0 \tag{7.497}$$

hold.

7.11 We are going to verify that the choice (7.90) validates the hypothesis (7.86).

7.11.1 Let

$$\mathcal{H} = \mathcal{H}^{\dagger} = (\mathbb{1} \otimes \mathrm{i}\sigma_2)\, \mathcal{H}^{*}\, (\mathbb{1} \otimes \mathrm{i}\sigma_2)^{-1} \tag{7.498a}$$

be a $2N \times 2N$ Hermitean matrix obeying spin-1/2 time-reversal symmetry. Let

$$\Delta = -\Delta^{\mathsf{T}} = (\mathbb{1} \otimes \mathrm{i}\sigma_2)\, \Delta^{*}\, (\mathbb{1} \otimes \mathrm{i}\sigma_2)^{-1} \tag{7.498b}$$

be a $2N \times 2N$ antisymmetric matrix obeying spin-1/2 time-reversal symmetry. Define the Bogoliubov-de-Gennes $4N \times 4N$ Hamiltonian

$$\mathcal{H}_{\mathrm{BdG}} := \begin{pmatrix} \mathcal{H} & \Delta \\ \Delta^{\dagger} & -\mathcal{H}^{\mathsf{T}} \end{pmatrix}. \tag{7.498c}$$

Show that

$$\mathcal{C}_{\mathrm{BdG}}\, \mathcal{H}^{\mathsf{T}}_{\mathrm{BdG}}\, \mathcal{C}^{-1}_{\mathrm{BdG}} = -\mathcal{H}_{\mathrm{BdG}}, \quad \mathcal{C}_{\mathrm{BdG}} := \mathbb{1} \otimes \sigma_0 \otimes \rho_1, \quad \mathcal{C}^{2}_{\mathrm{BdG}} = +\mathbb{1}_{4N}, \tag{7.498d}$$

and

$$\mathcal{T}_{\mathrm{BdG}}\, \mathcal{H}^{\mathsf{T}}_{\mathrm{BdG}}\, \mathcal{T}^{-1}_{\mathrm{BdG}} = +\mathcal{H}_{\mathrm{BdG}}, \quad \mathcal{T}_{\mathrm{BdG}} := \mathbb{1} \otimes \mathrm{i}\sigma_2 \otimes \rho_0, \quad \mathcal{T}^{2}_{\mathrm{BdG}} = -\mathbb{1}_{4N}, \tag{7.498e}$$

respectively.

7.11.2 Show that the 2×2 blocks

$$\mathcal{H}_{jk} := \mathcal{H}^{(0)}_{jk}\, \sigma_0 + \mathcal{H}^{(1)}_{jk}\, \sigma_1 + \mathcal{H}^{(2)}_{jk}\, \sigma_2 + \mathcal{H}^{(3)}_{jk}\, \sigma_3, \qquad \mathcal{H}^{(a)}_{jk} \in \mathbb{C} \tag{7.499a}$$

with $j, k = 1, \cdots, N$ and $a = 0, 1, 2, 3$ of $\mathcal{H} = \mathcal{H}^{\dagger}$ are parameterized by the

$$N + 4 \times \frac{1}{2}(N-1)N = 2N^2 - N \tag{7.499b}$$

independent real-valued matrix elements

$$\mathrm{Re}\left(\mathcal{H}^{(0)}_{jj}\right), \quad \mathrm{Re}\left(\mathcal{H}^{(0)}_{jk}\right), \quad \mathrm{Im}\left(\mathcal{H}^{(1)}_{jk}\right), \quad \mathrm{Im}\left(\mathcal{H}^{(2)}_{jk}\right), \quad \mathrm{Im}\left(\mathcal{H}^{(3)}_{jk}\right) \tag{7.499c}$$

for $1 \le j < k \le N$.

7.11.3 Show that the 2×2 blocks

$$\Delta_{jk} := \Delta^{(0)}_{jk}\, \sigma_0 + \Delta^{(1)}_{jk}\, \sigma_1 + \Delta^{(2)}_{jk}\, \sigma_2 + \Delta^{(3)}_{jk}\, \sigma_3, \qquad \Delta^{(a)}_{jk} \in \mathbb{C} \tag{7.500a}$$

with $j,k=1,\cdots,N$ and $a=0,1,2,3$ of $\Delta=-\Delta^{\mathsf{T}}$ are parameterized by the

$$N+4\times\frac{1}{2}(N-1)N=2N^2-N \tag{7.500b}$$

independent real-valued matrix elements

$$\mathrm{Im}\left(\Delta^{(2)}_{jj}\right),\quad \mathrm{Re}\left(\Delta^{(0)}_{jk}\right),\quad \mathrm{Im}\left(\Delta^{(1)}_{jk}\right),\quad \mathrm{Im}\left(\Delta^{(2)}_{jk}\right),\quad \mathrm{Im}\left(\Delta^{(3)}_{jk}\right) \tag{7.500c}$$

for $1\le j<k\le N$. The Bogoliubov-de-Gennes Hamiltonian defined by Eq. (7.498) is thus parameterized by

$$\left(2N^2-N\right)+\left(2N^2-N\right)=4N^2-2N \tag{7.501}$$

independent real-valued matrix elements.

7.11.4 Define the $4N\times 4N$ matrix

$$\mathcal{U}^{\mathrm{DIII}}_{\mathrm{BdG}}:=\frac{1}{\sqrt{2}}\begin{pmatrix}\mathbb{1}\otimes\sigma_0 & +\mathbb{1}\otimes\mathrm{i}\sigma_2\\ \mathbb{1}\otimes\sigma_2 & -\mathbb{1}\otimes\mathrm{i}\sigma_0\end{pmatrix}. \tag{7.502a}$$

Verify that (K denotes complex conjugation)

$$\left(\mathcal{U}^{\mathrm{DIII}}_{\mathrm{BdG}}\right)^{\dagger}\mathcal{U}^{\mathrm{DIII}}_{\mathrm{BdG}}=\mathcal{U}^{\mathrm{DIII}}_{\mathrm{BdG}}\left(\mathcal{U}^{\mathrm{DIII}}_{\mathrm{BdG}}\right)^{\dagger}=\mathbb{1}_{4N}, \tag{7.502b}$$

$$\left(\mathcal{U}^{\mathrm{DIII}}_{\mathrm{BdG}}\right)^{\dagger}\mathcal{C}_{\mathrm{BdG}}\,\mathcal{U}^{\mathrm{DIII}}_{\mathrm{BdG}}\equiv\mathcal{C}^{\mathrm{DIII}}_{\mathrm{BdG}}=\mathbb{1}\otimes\sigma_2\otimes\rho_3, \tag{7.502c}$$

$$\left(\mathcal{U}^{\mathrm{DIII}}_{\mathrm{BdG}}\right)^{\dagger}\mathcal{C}_{\mathrm{BdG}}\,\mathsf{K}\,\mathcal{U}^{\mathrm{DIII}}_{\mathrm{BdG}}=\mathbb{1}\otimes\sigma_0\otimes\mathrm{i}\rho_1\,\mathsf{K}, \tag{7.502d}$$

$$\left(\mathcal{U}^{\mathrm{DIII}}_{\mathrm{BdG}}\right)^{\dagger}\mathcal{T}_{\mathrm{BdG}}\,\mathcal{U}^{\mathrm{DIII}}_{\mathrm{BdG}}\equiv\mathcal{T}^{\mathrm{DIII}}_{\mathrm{BdG}}=\mathbb{1}\otimes\mathrm{i}\sigma_2\otimes\rho_0, \tag{7.502e}$$

$$\left(\mathcal{U}^{\mathrm{DIII}}_{\mathrm{BdG}}\right)^{\dagger}\mathcal{T}_{\mathrm{BdG}}\,\mathsf{K}\,\mathcal{U}^{\mathrm{DIII}}_{\mathrm{BdG}}=-\mathbb{1}\otimes\sigma_0\otimes\mathrm{i}\rho_2\,\mathsf{K}, \tag{7.502f}$$

$$\left(\mathcal{U}^{\mathrm{DIII}}_{\mathrm{BdG}}\right)^{\dagger}\mathcal{C}_{\mathrm{BdG}}\,\mathcal{T}_{\mathrm{BdG}}\,\mathcal{U}^{\mathrm{DIII}}_{\mathrm{BdG}}\equiv\mathrm{i}\mathcal{S}^{\mathrm{DIII}}_{\mathrm{BdG}}=\mathbb{1}\otimes\mathrm{i}\sigma_0\otimes\rho_3. \tag{7.502g}$$

7.11.5 Verify that

$$\left(\mathcal{U}^{\mathrm{DIII}}_{\mathrm{BdG}}\right)^{\dagger}\mathcal{H}_{\mathrm{BdG}}\,\mathcal{U}^{\mathrm{DIII}}_{\mathrm{BdG}}\equiv\mathcal{H}^{\mathrm{DIII}}_{\mathrm{BdG}}=\begin{pmatrix}\mathcal{H}^{\mathrm{DIII}} & \Delta^{\mathrm{DIII}}\\ \left(\Delta^{\mathrm{DIII}}\right)^{\dagger} & -\left(\mathcal{H}^{\mathrm{DIII}}\right)^{\mathsf{T}}\end{pmatrix}, \tag{7.503a}$$

where (we are abreviating $\mathbb{1}\otimes\sigma_2$ by σ_2)

$$\begin{aligned}2\mathcal{H}^{\mathrm{DIII}}&:=\mathcal{H}-\sigma_2\,\mathcal{H}^{\mathsf{T}}\,\sigma_2+\Delta\,\sigma_2+\sigma_2\,\Delta^{\dagger}=\left(2\mathcal{H}^{\mathrm{DIII}}\right)^{\dagger},\\ 2\Delta^{\mathrm{DIII}}&:=\mathrm{i}\mathcal{H}\,\sigma_2+\mathrm{i}\sigma_2\,\mathcal{H}^{\mathsf{T}}-\mathrm{i}\Delta+\mathrm{i}\sigma_2\,\Delta^{\dagger}\,\sigma_2=-\left(2\Delta^{\mathrm{DIII}}\right)^{\mathsf{T}}\end{aligned} \tag{7.503b}$$

for any $\mathcal{H}_{\mathrm{BdG}}$ of the form (7.498c), while (we are abreviating $\mathbb{1}\otimes\sigma_2$ by σ_2)

$$\begin{aligned}\mathcal{H}^{\mathrm{DIII}}&=0,\\ \Delta^{\mathrm{DIII}}&:=\mathrm{i}\mathcal{H}\,\sigma_2-\mathrm{i}\Delta=-\left(\Delta^{\mathrm{DIII}}\right)^{\mathsf{T}},\end{aligned} \tag{7.503c}$$

that is $\mathcal{H}_{\mathrm{BdG}}$ of the form (7.498c) is block off-diagonal if time-reversal symmetry is imposed through Eq. (7.498e). Equation (7.91) has been proved.

7.11.6 Verify that the $4N \times 4N$ complex-valued Hermitean matrix

$$\mathcal{H}^{\mathrm{DIII}}_{\mathrm{BdG}} = \begin{pmatrix} 0 & \Delta^{\mathrm{DIII}} \\ \left(\Delta^{\mathrm{DIII}}\right)^{\dagger} & 0 \end{pmatrix}, \qquad \Delta^{\mathrm{DIII}} = -\left(\Delta^{\mathrm{DIII}}\right)^{\mathsf{T}} \tag{7.504a}$$

has the

$$(2N-1)\,2N = 4N^2 - 2N \tag{7.504b}$$

independent real-valued matrix elements

$$\operatorname{Re}\left(\Delta^{\mathrm{DIII}}_{jk}\right), \qquad \operatorname{Im}\left(\Delta^{\mathrm{DIII}}_{jk}\right), \qquad 1 \leq j < k \leq 2N. \tag{7.504c}$$

7.11.7 Verify that the symmetric space $\mathrm{SO}(4N)/\mathrm{U}(2N)$ is of dimensionality

$$\frac{1}{2}(4N-1)\,4N - (2N)^2 = 4N^2 - 2N. \tag{7.505}$$

7.12 Verify that the map (7.101) is a Lie algebra homomorphism.[114]

7.13 The compact Lie groups $\mathrm{SU}(N)$ and $\mathrm{SO}(M)$ with $M = 2N$ or $M = 2N + 1$ that enter **Table** 7.1 in the symmetry classes A and D, respectively, are subgroups of the special groups $\mathrm{SL}(N, \mathbb{C})$ and $\mathrm{SL}(M, \mathbb{R})$, respectively, as their determinants is the number 1. Thus, the compact Lie groups $\mathrm{SU}(N)$ and $\mathrm{SO}(M)$ are simple Lie groups from an algebraic point of view that are also connected from a topological point of view.

If we interpret an element $\mathcal{X}$ of the Lie algebra $\mathfrak{su}(N)$ as being the imaginary unit multiplying the Hamiltonian $\mathcal{H}$, that is $\mathcal{X} = \mathrm{i}\mathcal{H}$, it follows that the complex-valued Hermitean matrix $\mathcal{H} = \mathcal{H}^{\dagger}$ is traceless. This condition amounts to choosing in a special way the origin of the energy axis along which the eigenvalues of $\mathcal{H}$ are measured. Any other choice of the origin would imply a nonvanishing value of the trace of $\mathcal{H}$, an arbitrariness that is encoded by the semi-simple Lie algebra

$$\mathfrak{u}(N) = \mathfrak{u}(1) \oplus \mathfrak{su}(N) \tag{7.506a}$$

of the semi-simple, compact, and connected Lie group

$$\mathrm{U}(N) = \mathrm{U}(1) \times \mathrm{SU}(N). \tag{7.506b}$$

This is one rational to replace $\mathrm{SU}(N)$ by $\mathrm{U}(N)$ in the symmetry class A from **Table** 7.1, a change that affects the symmetry classes AI, AII, AIII, BDI, and CII.

[114] A Lie algebra homomorphism Φ is a linear map between Lie algebras, which is compatible with the commutator, that is, $\Phi([U,V]) = [\Phi(U), \Phi(V)]$ for any pair U, V from the Lie algebra.

7.13.1 Another rational to replace SU(N) by U(N) in the row of **Table** 7.1 corresponding to the symmetry class A is to recall that we are dealing with N fermions, that is, we seek to study the many-body spectrum of

$$\widehat{H} := \sum_{i,j=1}^{N} \hat{c}_i^\dagger \, \mathcal{H}_{ij} \, \hat{c}_j = \widehat{H}^\dagger, \tag{7.507a}$$

where $\mathcal{H} = \mathcal{H}^\dagger$ and

$$\{\hat{c}_i, \hat{c}_j^\dagger\} = \delta_{ij}, \qquad \{\hat{c}_i^\dagger, \hat{c}_j^\dagger\} = \{\hat{c}_i, \hat{c}_j\} = 0. \tag{7.507b}$$

Here, the set of states $\{\hat{c}_i^\dagger \,|0\rangle\}$ with $\hat{c}_i \,|0\rangle = 0$ labeled by the index $i = 1, \cdots, N$ define an orthonormal basis of the Hilbert space $\mathbb{C}^N$. Any other (not necessarily orthonormal) basis is obtained by doing a linear transformation from the group GL($N, \mathbb{C}$). What singles out the subgroup U(N) $\subset$ GL($N, \mathbb{C}$) is the following. Show that the largest group that leaves the fermion algebra (7.507b) invariant is U(N). This rational to replace SU(N) by U(N) in the symmetry class A from **Table** 7.1 can be extended to the case of the symmetry class D.

7.13.2 An element $\mathcal{X}$ of the Lie algebra $\mathfrak{so}(M)$ with $M = 2N$ or $M = 2N+1$ must be a real-valued and antisymmetric matrix. The rational to replace SO(M) by O(M) in the symmetry class D is to recall that we are dealing with M Majorana operators, that is, we seek to study the many-body spectrum of[115]

$$\widehat{H} := \sum_{i,j=1}^{M} \hat{\chi}_i \, \mathrm{i}\mathcal{X}_{ij} \, \hat{\chi}_j = \widehat{H}^\dagger, \tag{7.508a}$$

where $\mathcal{X} = \mathcal{X}^* = -\mathcal{X}^\mathsf{T}$ and

$$\hat{\chi}_i = \hat{\chi}_i^\dagger, \qquad \{\hat{\chi}_i, \hat{\chi}_j\} = \delta_{ij}. \tag{7.508b}$$

Show that the largest group that leaves the Majorana algebra (7.508b) invariant is O(M). This is one rational to replace SO(M) by O(M) in the symmetry class D from **Table** 7.1, a change that affects the symmetry classes DIII, C, and CI. The group O(M) is not connected as a topological space. The group O(M) has two connected components. The one that contains the identity element is the normal subgroup SO(M). The other component consists of all orthogonal matrices of determinant equal to the number -1. This component does not form a group, as the determinant of the product of any two of its elements is the number 1. With the definition of the group

$$\begin{aligned} \mathbb{Z}_2 &:= \{I, C_2\}, \\ I &:= \operatorname{diag}\begin{pmatrix}1 & 1 & \cdots & 1 & 1\end{pmatrix}, \\ C_2 &:= \operatorname{diag}\begin{pmatrix}-1 & 1 & \cdots & 1 & 1\end{pmatrix}, \end{aligned} \tag{7.508c}$$

[115] For instance, see Eq. (7.61c).

with I the $M \times M$ unit matrix and C_2 a reflection, we have the semi-direct product decomposition (see **Appendix** A)

$$\mathrm{O}(M) = \mathbb{Z}_2 \ltimes \mathrm{SO}(M). \tag{7.508d}$$

Observe that when $M = 2N + 1$ is odd, we also have the direct product decomposition

$$\mathrm{O}(2N+1) = \{I, -I\} \times \mathrm{SO}(2N+1) \tag{7.508e}$$

with I the $(2N+1) \times (2N+1)$ unit matrix.

7.14 This exercise is a guide for a proof of Eq. (7.172) that is elementary in that it relies on little more than linear algebra. To this end, we will need to use the singular-value decomposition of a matrix that is proven in **Exercise** 7.15 when dealing with chiral (block off-diagonal) representations of Hamiltonians. We will also refer to the invariance properties of Haar measures discussed in **Exercise** 7.16.

7.14.1 Common to all ten symmetry classes from **Table** 7.2 is that any Hermitean matrix $\mathcal{H}$ can be diagonalized by a unitary matrix $\mathcal{U}$ to the real-valued diagonal matrix $\mathcal{E}$ through the relation

$$\mathcal{H} = \mathcal{U}\,\mathcal{E}\,\mathcal{U}^{\dagger}. \tag{7.509a}$$

It must be kept in mind that the choice of the unitary matrix $\mathcal{U}$ need not be unique.

One verifies that

$$\mathrm{d}\mathcal{H} = \mathcal{U}\left(\mathrm{d}\mathcal{A}\,\mathcal{E} + \mathrm{d}\mathcal{E} - \mathcal{E}\,\mathrm{d}\mathcal{A}\right)\mathcal{U}^{\dagger}, \tag{7.509b}$$

where

$$\mathrm{d}\mathcal{A} := \mathcal{U}^{\dagger}\,\mathrm{d}\mathcal{U} = -\mathrm{d}\mathcal{U}^{\dagger}\,\mathcal{U} =: -\mathrm{d}\mathcal{A}^{\dagger} \tag{7.509c}$$

relates the infinitesimal $N \times N$ complex-valued Hermitean matrix $\mathrm{d}\mathcal{H}$ to the infinitesimal $N \times N$ complex-valued matrix $\mathrm{d}\mathcal{A} \in \mathfrak{u}(N)$ and to the infinitesimal $N \times N$ real-valued diagonal matrix $\mathrm{d}\mathcal{E}$.

There are two steps to be completed. One must first define the Haar measure $\mu(\mathrm{d}\mathcal{H})$ of $\mathcal{H}$ (**Exercise** 7.16). One must then express this Haar measure in terms of the measures entering the spectral decomposition $\mathcal{U}\,\mathcal{E}\,\mathcal{U}^{\dagger}$ of $\mathcal{H}$. In doing so, we shall use the notation

$$(\mathrm{d}\mathcal{H})_{jk} \equiv \mathrm{d}\mathcal{H}_{jk} \equiv \mathrm{d}\left(\mathrm{Re}\,\mathcal{H}_{jk}\right) + \mathrm{i}\,\mathrm{d}\left(\mathrm{Im}\,\mathcal{H}_{jk}\right), \tag{7.510a}$$

$$(\mathrm{d}\mathcal{A})_{jk} \equiv \mathrm{d}\mathcal{A}_{jk} \equiv \mathrm{d}\left(\mathrm{Re}\,\mathcal{A}_{jk}\right) + \mathrm{i}\,\mathrm{d}\left(\mathrm{Im}\,\mathcal{A}_{jk}\right), \tag{7.510b}$$

for $j, k = 1, \cdots, N$.

7.14.2 Symmetry class A We start with the symmetry class A from the three standard symmetry classes A, AI, and AII. Let

$$\mathcal{H}=\mathcal{H}^{\dagger} \tag{7.511}$$

be any complex-valued $N\times N$ matrix with $N=1,2,3,\cdots$. Verify that $\mathcal{H}=\mathcal{H}^{\dagger}$ can be parameterized by the N^2 real-valued independent matrix elements

$$\mathcal{H}_{jj}=\mathcal{H}^{*}_{jj},\qquad j=1,\cdots,N, \tag{7.512a}$$

and

$$\operatorname{Re}\mathcal{H}_{jk},\qquad \operatorname{Im}\mathcal{H}_{jk},\qquad 1\leq j<k\leq N. \tag{7.512b}$$

Define the measure

$$\mu(\mathrm{d}\mathcal{H}):=\left(\prod_{j=1}^{N}\mathrm{d}\mathcal{H}_{jj}\right)\left(\prod_{1\leq j<k\leq N}\mathrm{d}\left(\operatorname{Re}\mathcal{H}_{jk}\right)\right)\left(\prod_{1\leq j<k\leq N}\mathrm{d}\left(\operatorname{Im}\mathcal{H}_{jk}\right)\right) \tag{7.512c}$$

and verify that it is a Haar measure (**Exercise** 7.16).

There exists a unitary matrix $\mathcal{U}\in\mathrm{U}(N)$ and a diagonal real-valued matrix $\mathcal{E}=\operatorname{diag}\left(\varepsilon_1\ \cdots\ \varepsilon_N\right)$ for which Eq. (7.509) holds. Verify that the relation (7.509) is invariant under the transformation

$$\mathcal{U}\mapsto\mathcal{U}\,e^{\mathrm{i}\Theta},\quad \Theta=\operatorname{diag}\left(\theta_1\ \cdots\ \theta_N\right),\quad 0\leq\theta_j<2\pi,\quad j=1,\cdots,N. \tag{7.513}$$

Deduce from this ambiguity when choosing $\mathcal{U}$ in Eq. (7.509) that we may always choose $\mathrm{d}\mathcal{A}$ so that

$$\mathrm{d}\mathcal{A}_{jj}=0,\qquad j=1,\cdots,N. \tag{7.514}$$

We are left with the N^2-N independent matrix elements

$$\mathrm{d}\left(\operatorname{Re}\mathcal{A}_{jk}\right),\qquad \mathrm{d}\left(\operatorname{Im}\mathcal{A}_{jk}\right),\qquad 1\leq j<k\leq N. \tag{7.515}$$

Verify that, for any $1\leq j<k\leq N$,

$$\left(\mathcal{U}^{\dagger}\,\mathrm{d}\mathcal{H}\mathcal{U}\right)_{jk}=\left(\varepsilon_k-\varepsilon_j\right)\mathrm{d}\mathcal{A}_{jk}, \tag{7.516a}$$

$$\left(\mathcal{U}^{\dagger}\,\mathrm{d}\mathcal{H}\mathcal{U}\right)_{jj}=\mathrm{d}\varepsilon_j. \tag{7.516b}$$

Deduce that

$$\begin{aligned}\mu\left(\mathrm{d}\left(\mathcal{U}\mathcal{E}\mathcal{U}^{\dagger}\right)\right)\propto&\left(\prod_{j=1}^{N}\mathrm{d}\varepsilon_j\right)\\&\times\left[\prod_{1\leq j<k\leq N}\left(\varepsilon_k-\varepsilon_j\right)\mathrm{d}\left(\operatorname{Re}\mathcal{A}_{jk}\right)\right]\\&\times\left[\prod_{1\leq j<k\leq N}\left(\varepsilon_k-\varepsilon_j\right)\mathrm{d}\left(\operatorname{Im}\mathcal{A}_{jk}\right)\right].\end{aligned} \tag{7.517}$$

By counting the number of real-valued parameters needed to parameterize the manifold $\mathrm{U}(N)/\left[\mathrm{U}(1)\right]^N$, convince yourself that (i)

$$\mu(\mathrm{d}\,(\mathcal{U}\,\mathcal{E}\,\mathcal{U}^{\dagger})\,) \propto \left(\prod_{j=1}^{N} \mathrm{d}\varepsilon_j\right)\left(\prod_{1\leq j<k\leq N} \left|\varepsilon_k - \varepsilon_j\right|^2\right) \mu(\mathrm{d}\mathcal{A}) \tag{7.518a}$$

and (ii)

$$\mu(\mathrm{d}\mathcal{H}) \propto \mu(\mathrm{d}\,(\mathcal{U}\,\mathcal{E}\,\mathcal{U}^{\dagger})\,), \tag{7.518b}$$

whereby (iii) the measure

$$\mu(\mathrm{d}\mathcal{A}) := \prod_{1\leq j<k\leq N} \mathrm{d}\left(\mathrm{Re}\,\mathcal{A}_{jk}\right)\mathrm{d}\left(\mathrm{Im}\,\mathcal{A}_{jk}\right), \qquad \mathrm{d}\mathcal{A} := \mathcal{U}^{\dagger}\,\mathrm{d}\mathcal{U} \tag{7.518c}$$

is a Haar measure of $\mathrm{U}(N)/\left[\mathrm{U}(1)\right]^N$. We have found that the line A in **Table** 7.2 corresponds to the degeneracy $d_{\mathrm{A}} = 1$, the level repulsion exponent $\beta_{\mathrm{A}} = 2$, the repulsion exponent from zero energy $\alpha_{\mathrm{A}} = 0$, and the number $\nu_{\mathrm{A}} = 0$ of zero modes. Moreover, if $\boldsymbol{e}_j$ is the Cartesian basis of $\mathbb{R}^N$, $\boldsymbol{\varepsilon} \in \mathbb{R}^N$ is the vector with the eigenvalues ε_j as components, and we define the set of "positive" roots to be

$$\mathfrak{Roots}_{+} := \{\boldsymbol{e}_k - \boldsymbol{e}_j \mid 1 \leq j < k \leq N\}, \tag{7.519a}$$

we have then constructed the Jacobian

$$J(\boldsymbol{\varepsilon}) := \prod_{\boldsymbol{\alpha}\in\mathfrak{Roots}_{+}} \left|\boldsymbol{\alpha}\cdot\boldsymbol{\varepsilon}\right|^{m_{\mathrm{o}}}, \qquad m_{\mathrm{o}} = 2. \tag{7.519b}$$

The exponent m_{o} is called the multiplicity of the ordinary root $\boldsymbol{e}_k - \boldsymbol{e}_j$.

7.14.3 Symmetry class AI For the symmetry class AI, we must modify Eq. (7.511) by demanding that the $N \times N$ matrix $\mathcal{H}$ obeys

$$\mathcal{H} = \mathcal{H}^{\dagger} = \mathcal{H}^{*} = \mathcal{H}^{\mathsf{T}}. \tag{7.520a}$$

A symmetric and real-valued $N \times N$ matrix depends on

$$N + \frac{1}{2}(N-1)N = \frac{1}{2}(N+1)N \tag{7.520b}$$

real-valued parameters with the Haar measure

$$\mu(\mathrm{d}\mathcal{H}) := \prod_{1\leq j\leq k\leq N} \mathrm{d}\mathcal{H}_{jk}. \tag{7.520c}$$

It can be diagonalized by an orthogonal matrix $\mathcal{O} \in \mathrm{SO}(N)$,

$$\mathcal{H} = \mathcal{O}\,\mathcal{E}\,\mathcal{O}^{\mathsf{T}} \tag{7.521a}$$

to the diagonal matrix

$$\mathcal{E} = \mathrm{diag}\left(\varepsilon_1 \quad \cdots \quad \varepsilon_N\right). \tag{7.521b}$$

The relation (7.521a) is invariant under the transformation

$$\mathcal{O}\mapsto\mathcal{O}\,e^{\mathrm{i}\Theta},\qquad \Theta=\mathrm{diag}\left(\theta_1\quad\cdots\quad\theta_N\right),\quad \theta_j=0,\pi,\quad j=1,\cdots,N. \tag{7.521c}$$

Verify that the reality condition (7.520a) turns Eq. (7.518) into

$$\mu(\mathrm{d}\,(\mathcal{O}\,\mathcal{E}\,\mathcal{O}^{\mathsf{T}})\,)\propto\left(\prod_{j=1}^{N}\mathrm{d}\varepsilon_j\right)\left(\prod_{1\le j<k\le N}\left|\varepsilon_j-\varepsilon_k\right|\right)\mu(\mathrm{d}\mathcal{A}) \tag{7.522a}$$

and (ii)

$$\mu(\mathrm{d}\mathcal{H})\propto\mu(\mathrm{d}\,(\mathcal{O}\,\mathcal{E}\,\mathcal{O}^{\mathsf{T}})\,), \tag{7.522b}$$

whereby (iii) the measure

$$\mu(\mathrm{d}\mathcal{A}):=\prod_{1\le j<k\le N}\mathrm{d}\mathcal{A}_{jk},\qquad \mathrm{d}\mathcal{A}\equiv\mathcal{O}^{\mathsf{T}}\mathrm{d}\mathcal{O} \tag{7.522c}$$

is a Haar measure of SO(N). We have found that the line AI in **Table** 7.2 corresponds to the degeneracy $d_{\mathrm{AI}}=1$, the level repulsion exponent $\beta_{\mathrm{AI}}=1$, the repulsion exponent from zero energy $\alpha_{\mathrm{AI}}=0$, and the number $\nu_{\mathrm{AI}}=0$ of zero modes. Moreover, if $\boldsymbol{e}_j$ is the Cartesian basis of $\mathbb{R}^N$, $\boldsymbol{\varepsilon}\in\mathbb{R}^N$ is the vector with the eigenvalues ε_j as components, and we define the set of "positive" roots to be

$$\mathfrak{Roots}_{+}:=\{\boldsymbol{e}_k-\boldsymbol{e}_j\mid 1\le j<k\le N\}, \tag{7.523a}$$

we have then constructed the Jacobian

$$J(\boldsymbol{\varepsilon}):=\prod_{\boldsymbol{\alpha}\in\mathfrak{Roots}_{+}}|\boldsymbol{\alpha}\cdot\boldsymbol{\varepsilon}|^{m_{\mathrm{o}}},\qquad m_{\mathrm{o}}=1. \tag{7.523b}$$

The exponent m_{o} is called the multiplicity of the ordinary root $\boldsymbol{e}_k-\boldsymbol{e}_j$.

7.14.4 Symmetry class AII For the symmetry class AII, we must modify Eq. (7.511) by demanding that the $2N\times2N$ matrix $\mathcal{H}$ obeys

$$\mathcal{H}=\mathcal{H}^{\dagger}=(\mathrm{i}\sigma_2\otimes\mathbb{1})\,\mathcal{H}^{*}\,(\mathrm{i}\sigma_2\otimes\mathbb{1})^{-1}=(\mathrm{i}\sigma_2\otimes\mathbb{1})\,\mathcal{H}^{\mathsf{T}}\,(\mathrm{i}\sigma_2\otimes\mathbb{1})^{-1}\equiv\mathcal{H}^{\mathsf{R}}, \tag{7.524a}$$

where the R-self-duality coincides with the RT-self-duality [recall Eqs. (7.456b) and (7.457a).]

Verify that the parameterization

$$\mathcal{H}_{jk}=\mathcal{H}^{(0)}_{jk}\,\sigma_0+\mathrm{i}\sum_{a=1}^{3}\mathcal{H}^{(a)}_{jk}\,\sigma_a,\qquad j,k=1,\cdots,N \tag{7.524b}$$

depends on

$$N+\frac{1}{2}(N-1)N+3\times\frac{1}{2}(N-1)N=N+2(N-1)N=N(2N-1) \tag{7.524c}$$

real-valued parameters with the Haar measure

$$\mu(\mathrm{d}\mathcal{H}) := \left(\prod_{1\le j\le k\le N} \mathrm{d}\mathcal{H}^{(0)}_{jk}\right)\prod_{a=1}^{3}\left(\prod_{1\le j<k\le N} \mathrm{d}\mathcal{H}^{(a)}_{jk}\right). \tag{7.524d}$$

It can be diagonalized by a symplectic and unitary matrix $\mathcal{S}\in \mathrm{Sp}(2N)$,

$$\mathcal{H} = \mathcal{S}\,\mathcal{E}\,\mathcal{S}^{\dagger}, \qquad \mathcal{S}^{\dagger} = \mathcal{S}^{\mathsf{RT}} \tag{7.525a}$$

[the RT-dual is defined in Eq. (7.457a)] to the diagonal matrix

$$\mathcal{E} = \sigma_0 \otimes \operatorname{diag}\begin{pmatrix}\varepsilon_1 & \cdots & \varepsilon_N\end{pmatrix}. \tag{7.525b}$$

The relation (7.525a) is invariant under the transformation

$$\begin{aligned} &\mathcal{S}\mapsto \mathcal{S}\mathcal{U}, \qquad \mathcal{U} = \operatorname{diag}\begin{pmatrix}U_1 & \cdots & U_N\end{pmatrix},\\ &U_j \in \mathrm{SU}(2), \qquad j=1,\cdots,N. \end{aligned} \tag{7.525c}$$

Verify that the quaternion-real condition (7.524a) turns Eq. (7.518) into

$$\mu(\mathrm{d}\,(\mathcal{S}\,\mathcal{E}\,\mathcal{S}^{\mathsf{RT}})\,) \propto \left(\prod_{j=1}^{N}\mathrm{d}\varepsilon_j\right)\left(\prod_{1\le j<k\le N}\left|\varepsilon_j-\varepsilon_k\right|^4\right)\mu(\mathrm{d}\mathcal{A}) \tag{7.526a}$$

and (ii)

$$\mu(\mathrm{d}\mathcal{H}) \propto \mu(\mathrm{d}\,(\mathcal{S}\,\mathcal{E}\,\mathcal{S}^{\mathsf{RT}})\,), \tag{7.526b}$$

whereby (iii) the measure

$$\mu(\mathrm{d}\mathcal{A}) := \prod_{a=0}^{3}\left(\prod_{1\le j<k\le N}\mathrm{d}\mathcal{A}^{(a)}_{jk}\right), \quad \mathrm{d}\mathcal{A}^{(a)} \equiv \frac{1}{2}\operatorname{tr}\left(\sigma_a\,\mathcal{S}^{\mathsf{RT}}\mathrm{d}\mathcal{S}\right), \tag{7.526c}$$

(the symbol tr is reserved for the partial trace over the Hilbert space spanned by the 2×2 matrices σ_0 and $\boldsymbol{\sigma}$) is a Haar measure of $\mathrm{Sp}(2N)/[\mathrm{SU}(2)]^N$. We have found that the line AII in **Table** 7.2 corresponds to the degeneracy $d_{\mathrm{AII}}=2$, the level repulsion exponent $\beta_{\mathrm{AII}}=4$, the repulsion exponent from zero energy $\alpha_{\mathrm{AII}}=0$, and the number $\nu_{\mathrm{AII}}=0$ of zero modes. Moreover, if $\boldsymbol{e}_j$ is the Cartesian basis of $\mathbb{R}^N$, $\boldsymbol{\varepsilon}\in\mathbb{R}^N$ is the vector with the eigenvalues ε_j as components, and we define the set of "positive" roots

$$\mathfrak{Roots}_{+} := \{\boldsymbol{e}_k - \boldsymbol{e}_j \mid 1\le j<k\le N\}, \tag{7.527a}$$

we have then constructed the Jacobian

$$J(\boldsymbol{\varepsilon}) := \prod_{\boldsymbol{\alpha}\in\mathfrak{Roots}_{+}} |\boldsymbol{\alpha}\cdot\boldsymbol{\varepsilon}|^{m_{\mathrm{o}}}, \qquad m_{\mathrm{o}} = 4. \tag{7.527b}$$

The exponent m_{o} is called the multiplicity of the ordinary root $\boldsymbol{e}_k - \boldsymbol{e}_j$.

7.14.5 Symmetry class AIII We start with the symmetry class AIII from the three chiral symmetry classes AIII, BDI, and CII. Let

$$\mathcal{H}=\mathcal{H}^{\dagger}=\begin{pmatrix}0 & B\\ B^{\dagger} & 0\end{pmatrix} \tag{7.528}$$

be any $(M+N)\times(M+N)$ block off-diagonal Hermitean matrix that is parameterized by the rectangular $M\times N$ complex-valued matrix B. Without loss of generality $M\geq N$ is assumed, that is,

$$M=N+\nu, \qquad \nu=0,1,2,3,\cdots. \tag{7.529}$$

The Hermitean matrix thus depends on the

$$2\times M\times N=2(N+\nu)N=2N^{2}+2N\,\nu \tag{7.530}$$

real-valued parameters

$$\operatorname{Re}B_{jk}, \qquad \operatorname{Im}B_{jk}, \qquad j=1,\cdots,N+\nu, \qquad k=1,\cdots,N. \tag{7.531}$$

Define the measure

$$\mu(\mathrm{d}\mathcal{H}):=\prod_{j=1}^{N+\nu}\prod_{k=1}^{N}\mathrm{d}\left(\operatorname{Re}B_{jk}\right)\mathrm{d}\left(\operatorname{Im}B_{jk}\right) \tag{7.532}$$

and verify that it is a Haar measure on the symmetric space $\mathrm{U}(2N+\nu)/[\mathrm{U}(N+\nu)\times\mathrm{U}(N)]$ (**Exercise** 7.16).
The singular-value decomposition of a matrix (7.607) (**Exercise** 7.15) implies the existence of $U\in\mathrm{U}(N+\nu)$, $V\in\mathrm{U}(N)$, the singular values $\varepsilon_1\geq 0,\cdots,\varepsilon_N\geq 0$, and D a $(N+\nu)\times N$ rectangular matrix such that

$$B=U\,D\,V^{\dagger}, \quad D_{jk}=\varepsilon_k\delta_{jk}, \quad j=1,\cdots,N+\nu, \quad k=1,\cdots,N. \tag{7.533}$$

Verify with the help of the block decomposition

$$U\equiv\begin{pmatrix}U_{\theta,\theta} & U_{\theta,\nu}\\ U_{\nu,\theta} & U_{\nu,\nu}\end{pmatrix}, \qquad D\,V^{\dagger}=\begin{pmatrix}E\,V^{\dagger}\\ 0_{\nu,N}\end{pmatrix}, \tag{7.534}$$

where $U_{\theta,\theta}$ is a $N\times N$ matrix, $U_{\theta,\nu}$ is a $N\times\nu$ matrix, $U_{\nu,\theta}$ is a $\nu\times N$ matrix, $U_{\nu,\nu}$ is a $\nu\times\nu$ matrix, $E=\operatorname{diag}\left(\varepsilon_1 \ \cdots \ \varepsilon_N\right)$, and $0_{\nu,N}$ is the $\nu\times N$ matrix with vanishing entries, that the singular-value decomposition (7.533) is unchanged under

$$\begin{pmatrix}U_{\theta,\theta} & U_{\theta,\nu}\\ U_{\nu,\theta} & U_{\nu,\nu}\end{pmatrix}\mapsto\begin{pmatrix}U_{\theta,\theta} & U_{\theta,\nu}\\ U_{\nu,\theta} & U_{\nu,\nu}\end{pmatrix}\begin{pmatrix}e^{\mathrm{i}\Theta} & 0\\ 0 & Q_{\nu,\nu}\end{pmatrix}, \qquad V\mapsto V\,e^{\mathrm{i}\Theta}, \tag{7.535a}$$

where

$$\Theta=\operatorname{diag}\left(\theta_1 \ \cdots \ \theta_N\right), \qquad 0\leq\theta_j<2\pi, \qquad j=1,\cdots,N, \tag{7.535b}$$

and

$$Q_{\nu,\nu} \in \mathrm{U}(\nu). \tag{7.535c}$$

Deduce that out of the

$$(N+\nu)^2 + N + N^2 = 2N^2 + 2N\,\nu + N + \nu^2 \tag{7.536a}$$

real-valued generators on the right-hand side of the singular-value decomposition (7.533), only

$$(N+\nu)^2 + N + N^2 - N - \nu^2 = 2N^2 + 2N\,\nu \tag{7.536b}$$

are independent. Verify that

$$\mathrm{d}B = U\,(\mathrm{d}A_U\,D + \mathrm{d}D - D\,\mathrm{d}A_V\,)\,V^\dagger, \tag{7.537a}$$

where

$$\mathrm{d}A_U := U^\dagger\,\mathrm{d}U \in \mathfrak{u}(N+\nu), \qquad \mathrm{d}A_V := V^\dagger\,\mathrm{d}V \in \mathfrak{u}(N), \tag{7.537b}$$

with the conditions

$$(\mathrm{d}A_V)_{jj} = 0, \qquad j=1,\cdots,N, \tag{7.537c}$$

and

$$(\mathrm{d}A_U)_{(N+j)(N+k)} = 0, \qquad j,k=1,\cdots,\nu. \tag{7.537d}$$

Verify that

$$\begin{aligned}
\left(U^\dagger\,\mathrm{d}B\,V\right)_{jj} &= \mathrm{d}\varepsilon_j + (\mathrm{d}A_U)_{jj}\,\varepsilon_j, \qquad j=1,\cdots,N, &&(7.538\mathrm{a})\\
\left(U^\dagger\,\mathrm{d}B\,V\right)_{jk} &= +\,(\mathrm{d}A_U)_{jk}\,\varepsilon_k - \varepsilon_j\,(\mathrm{d}A_V)_{jk}, \qquad 1\le j\neq k\le N, &&(7.538\mathrm{b})\\
\left(U^\dagger\,\mathrm{d}B\,V\right)_{kj} &= -\,(\mathrm{d}A_U)^*_{jk}\,\varepsilon_j + \varepsilon_k\,(\mathrm{d}A_V)^*_{jk}, \qquad 1\le j\neq k\le N, &&(7.538\mathrm{c})\\
\left(U^\dagger\,\mathrm{d}B\,V\right)_{(N+j)k} &= (\mathrm{d}A_U)_{jk}\,\varepsilon_k \qquad j=1,\cdots,\nu, \qquad k=1,\cdots,N. &&(7.538\mathrm{d})
\end{aligned}$$

Infer from Eq. (7.538) that (i)

$$\mu(\mathrm{d}\,(U\,D\,V^\dagger)\,) \propto \left(\prod_{j=1}^{N}\mathrm{d}\varepsilon_j\,|\varepsilon_j|^{1+2\nu}\right)\left(\prod_{1\le j<k\le N}\left|\varepsilon_j^2-\varepsilon_k^2\right|^2\right) \times \mu(\mathrm{d}A_U)\,\mu(\mathrm{d}A_V) \tag{7.539a}$$

and (ii)

$$\mu(\mathrm{d}\mathcal{H}) \propto \mu(\mathrm{d}\,(U\,D\,V^\dagger)\,), \tag{7.539b}$$

whereby (iii) the measures

$$\mu(\mathrm{d}A_U) := \left[\prod_{j=1}^{N} \mathrm{d}\,(\mathrm{Im}\,A_U)_{jj}\right] \times \left[\prod_{1\leq j<k\leq N} \mathrm{d}\,(\mathrm{Re}\,A_U)_{jk}\,\mathrm{d}\,(\mathrm{Im}\,A_U)_{jk}\right] \times \left[\prod_{j=1}^{\nu}\prod_{k=1}^{N} \mathrm{d}\,(\mathrm{Re}\,A_U)_{(N+j)k}\,\mathrm{d}\,(\mathrm{Im}\,A_U)_{(N+j)k}\right], \tag{7.539c}$$

$$\mu(\mathrm{d}A_V) := \prod_{1\leq j<k\leq N} \mathrm{d}\,(\mathrm{Re}\,A_V)_{jk}\,\mathrm{d}\,(\mathrm{Im}\,A_V)_{jk}, \tag{7.539d}$$

are the Haar measures of $\mathrm{U}(N+\nu)/\mathrm{U}(\nu)$ and $\mathrm{U}(N)/[\mathrm{U}(1)]^N$, respectively [recall conditions (7.537c) and (7.537d)]. We have found that the line AIII in **Table** 7.2 corresponds to the degeneracy $d_{\mathrm{AIII}} = 1$, the level repulsion exponent $\beta_{\mathrm{AIII}} = 2$, the repulsion exponent from zero energy $\alpha_{\mathrm{AIII}} = 1$, and the number $\nu_{\mathrm{AIII}} = \nu$ of zero modes. Moreover, if $\boldsymbol{e}_j$ is the Cartesian basis of $\mathbb{R}^N$, $\boldsymbol{\varepsilon} \in \mathbb{R}^N$ is the vector with the eigenvalues $\varepsilon_j \geq 0$ as components, and we define the set of "positive" roots to be

$$\begin{aligned}
\mathfrak{Roots}_+ &:= \mathfrak{Roots}_{\mathrm{s}} \cup \mathfrak{Roots}_{\mathrm{o}} \cup \mathfrak{Roots}_{\mathrm{l}},\\
\mathfrak{Roots}_{\mathrm{s}} &:= \{\boldsymbol{e}_j \mid 1 \leq j \leq N\},\\
\mathfrak{Roots}_{\mathrm{o}} &:= \{\boldsymbol{e}_k - \boldsymbol{e}_j, \boldsymbol{e}_k + \boldsymbol{e}_j \mid 1 \leq j < k \leq N\},\\
\mathfrak{Roots}_{\mathrm{l}} &:= \{2\boldsymbol{e}_j \mid 1 \leq j \leq N\},
\end{aligned} \tag{7.540a}$$

we have then constructed the Jacobian

$$\begin{aligned}
J(\boldsymbol{\varepsilon}) &:= \prod_{\boldsymbol{\alpha}\in\mathfrak{Roots}_{\mathrm{s}}} |\boldsymbol{\alpha}\cdot\boldsymbol{\varepsilon}|^{m_{\mathrm{s}}} \prod_{\boldsymbol{\alpha}\in\mathfrak{Roots}_{\mathrm{o}}} |\boldsymbol{\alpha}\cdot\boldsymbol{\varepsilon}|^{m_{\mathrm{o}}} \prod_{\boldsymbol{\alpha}\in\mathfrak{Roots}_{\mathrm{l}}} |\boldsymbol{\alpha}\cdot\boldsymbol{\varepsilon}|^{m_{\mathrm{l}}},\\
m_{\mathrm{s}} &= 2\,\nu, \qquad m_{\mathrm{o}} = 2, \qquad m_{\mathrm{l}} = 1.
\end{aligned} \tag{7.540b}$$

The exponent m_{s} is called the multiplicity of the short root $\boldsymbol{e}_j$. The exponent m_{o} is called the multiplicity of the ordinary root $\boldsymbol{e}_k \pm \boldsymbol{e}_j$. The exponent m_{l} is called the multiplicity of the long root $2\boldsymbol{e}_j$.

7.14.6 Symmetry class BDI For the symmetry class BDI, verify that Eq. (7.528) must be modified by demanding that the rectangular $(N+\nu)\times N$ matrix B is real-valued:

$$B = B^{*}. \tag{7.541a}$$

Define the measure

$$\mu(\mathrm{d}\mathcal{H}) := \prod_{j=1}^{N+\nu}\prod_{k=1}^{N} \mathrm{d}B_{jk} \tag{7.541b}$$

and verify that it is a Haar measure on the symmetric space $\mathrm{SO}(2N+\nu)/[\mathrm{SO}(N+\nu)\times\mathrm{SO}(N)]$ (**Exercise** 7.16). Verify that the counterpart to Eq. (7.539) is then given by (i)

$$\mu(\mathrm{d}\,(U\,D\,V^{\mathsf{T}})\,)\propto\left(\prod_{j=1}^{N}\mathrm{d}\varepsilon_j\,|\varepsilon_j|^{\nu}\right)\left(\prod_{1\le j<k\le N}|\varepsilon_j^2-\varepsilon_k^2|\right)\times\mu(\mathrm{d}A_U)\,\mu(\mathrm{d}A_V) \tag{7.542a}$$

and (ii)

$$\mu(\mathrm{d}\mathcal{H})\propto\mu(\mathrm{d}\,(U\,D\,V^{\mathsf{T}})\,), \tag{7.542b}$$

whereby (iii) the measures

$$\mu(\mathrm{d}A_U):=\left[\prod_{1\le j<k\le N}\mathrm{d}\,(A_U)_{jk}\right]\times\left[\prod_{j=1}^{\nu}\prod_{k=1}^{N}\mathrm{d}\,(A_U)_{(N+j)k}\right],\qquad \mathrm{d}A_U:=U^{\mathsf{T}}\,\mathrm{d}U, \tag{7.542c}$$

$$\mu(\mathrm{d}A_V):=\prod_{1\le j<k\le N}\mathrm{d}\,(A_V)_{jk},\qquad \mathrm{d}A_V:=V^{\mathsf{T}}\,\mathrm{d}V, \tag{7.542d}$$

are the Haar measures of $\mathrm{SO}(N+\nu)/\mathrm{SO}(\nu)$ and $\mathrm{SO}(N)$, respectively. We have found that the line BDI in **Table** 7.2 corresponds to the degeneracy $d_{\mathrm{BDI}}=1$, the level repulsion exponent $\beta_{\mathrm{BDI}}=1$, the repulsion exponent from zero energy $\alpha_{\mathrm{BDI}}=0$, and the number $\nu_{\mathrm{BDI}}=\nu$ of zero modes. Moreover, if $\boldsymbol{e}_j$ is the Cartesian basis of $\mathbb{R}^N$, $\boldsymbol{\varepsilon}\in\mathbb{R}^N$ is the vector with the eigenvalues $\varepsilon_j\ge 0$ as components, and we define the set of "positive" roots to be

$$\begin{aligned}
&\mathfrak{Roots}_{+}:=\mathfrak{Roots}_{\mathrm{s}}\cup\mathfrak{Roots}_{\mathrm{o}}\cup\mathfrak{Roots}_{\mathrm{l}},\\
&\mathfrak{Roots}_{\mathrm{s}}:=\{\boldsymbol{e}_j\mid 1\le j\le N\},\\
&\mathfrak{Roots}_{\mathrm{o}}:=\{\boldsymbol{e}_k-\boldsymbol{e}_j,\boldsymbol{e}_k+\boldsymbol{e}_j\mid 1\le j<k\le N\},\\
&\mathfrak{Roots}_{\mathrm{l}}:=\{2\boldsymbol{e}_j\mid 1\le j\le N\},
\end{aligned} \tag{7.543a}$$

we have then constructed the Jacobian

$$J(\boldsymbol{\varepsilon}):=\prod_{\boldsymbol{\alpha}\in\mathfrak{Roots}_{\mathrm{s}}}|\boldsymbol{\alpha}\cdot\boldsymbol{\varepsilon}|^{m_{\mathrm{s}}}\prod_{\boldsymbol{\alpha}\in\mathfrak{Roots}_{\mathrm{o}}}|\boldsymbol{\alpha}\cdot\boldsymbol{\varepsilon}|^{m_{\mathrm{o}}}\prod_{\boldsymbol{\alpha}\in\mathfrak{Roots}_{\mathrm{l}}}|\boldsymbol{\alpha}\cdot\boldsymbol{\varepsilon}|^{m_{\mathrm{l}}}, \tag{7.543b}$$
$$m_{\mathrm{s}}=\nu,\qquad m_{\mathrm{o}}=1,\qquad m_{\mathrm{l}}=0.$$

The exponent m_{s} is called the multiplicity of the short root $\boldsymbol{e}_j$. The exponent m_{o} is called the multiplicity of the ordinary root $\boldsymbol{e}_k\pm\boldsymbol{e}_j$. The exponent m_{l} is called the multiplicity of the long root $2\boldsymbol{e}_j$.

7.14.7 Symmetry class CII For the symmetry class CII, verify that Eq. (7.528) must be modified by demanding that the Hermitean $(2M+2N)\times(2M+2N)$ matrix with $M=N+\nu$ satisfies

$$\mathcal{H}=\begin{pmatrix}0 & B\\ B^{\dagger} & 0\end{pmatrix},\qquad B=\mathsf{B}^{(0)}\otimes\sigma_0+\mathrm{i}\sum_{a=1}^{3}\mathsf{B}^{(a)}\otimes\sigma_a,\tag{7.544a}$$

where

$$\mathsf{B}^{(0)}=+\mathsf{B}^{(0)\,*},\qquad \mathsf{B}^{(a)}=+\mathsf{B}^{(a)\,*},\qquad a=1,2,3,\tag{7.544b}$$

are rectangular and real-valued $(N+\nu)\times N$ matrices with $N=1,2,\cdots$ and $\nu=0,1,2,\cdots$. Define the measure

$$\mu(\mathrm{d}\mathcal{H}):=\left(\prod_{j=1}^{N+\nu}\prod_{k=1}^{N}\mathrm{d}\mathsf{B}^{(0)}_{jk}\right)\left(\prod_{a=1}^{3}\prod_{j=1}^{N+\nu}\prod_{k=1}^{N}\mathrm{d}\mathsf{B}^{(a)}_{jk}\right)\tag{7.544c}$$

and verify that it is a Haar measure on the symmetric space $\mathrm{Sp}(4N+2\nu)/[\mathrm{Sp}(2N+2\nu)\times\mathrm{Sp}(2N)]$ (**Exercise** 7.16). Verify that the counterpart to Eq. (7.539) is then given by (i)

$$\begin{aligned}\mu(\mathrm{d}\,(U\,D\,V^{\mathsf{RT}}))\propto&\left(\prod_{j=1}^{N}\mathrm{d}\varepsilon_j\,|\varepsilon_j|^{3+4\nu}\right)\left(\prod_{1\le j<k\le N}\left|\varepsilon_j^2-\varepsilon_k^2\right|^4\right)\\&\times\prod_{a=0}^{3}\mu(\mathrm{d}A^{(a)}_U)\,\mu(\mathrm{d}A^{(a)}_V)\end{aligned}\tag{7.545a}$$

[the RT-dual is defined in Eq. (7.457a)] and (ii)

$$\mu(\mathrm{d}\mathcal{H})\propto\mu(\mathrm{d}\,(U\,D\,V^{\mathsf{RT}})),\tag{7.545b}$$

whereby (iii) the measures $\prod_{a=0}^{3}\mu(\mathrm{d}A^{(a)}_U)$ with

$$\mathrm{d}A^{(a)}_U:=\frac{1}{2}\,\mathrm{tr}\,\left(\sigma_a\,U^{\mathsf{RT}}\,\mathrm{d}U\right)\tag{7.545c}$$

(the symbol tr is reserved for the partial trace over the Hilbert space spanned by the 2×2 matrices σ_0 and $\boldsymbol{\sigma}$) and $\prod_{a=0}^{3}\mu(\mathrm{d}A^{(a)}_V)$ with

$$\mathrm{d}A^{(a)}_V:=\frac{1}{2}\,\mathrm{tr}\,\left(\sigma_a\,V^{\mathsf{RT}}\,\mathrm{d}V\right)\tag{7.545d}$$

are the Haar measures of $\mathrm{Sp}(2N+2\nu)/\mathrm{Sp}(2\nu)$ and $\mathrm{Sp}(2N)/[\mathrm{SU}(2)]^N$, respectively. We have found that the line CII in **Table** 7.2 corresponds to the degeneracy $d_{\mathrm{CII}}=2$, the level repulsion exponent $\beta_{\mathrm{CII}}=4$, the repulsion exponent from zero energy $\alpha_{\mathrm{CII}}=3$, and the number $\nu_{\mathrm{CII}}=\nu$ of Kramer's degenerate pairs of zero modes. Moreover, if $\boldsymbol{e}_j$ is the Cartesian basis of $\mathbb{R}^N$, $\boldsymbol{\varepsilon} \in \mathbb{R}^N$ is the vector with the eigenvalues $\varepsilon_j \geq 0$ as components, and we define the set of "positive" roots to be

$$\begin{aligned}
\mathfrak{Roots}_+ &:= \mathfrak{Roots}_{\mathrm{s}} \cup \mathfrak{Roots}_{\mathrm{o}} \cup \mathfrak{Roots}_{\mathrm{l}}, \\
\mathfrak{Roots}_{\mathrm{s}} &:= \{\boldsymbol{e}_j \mid 1 \leq j \leq N\}, \\
\mathfrak{Roots}_{\mathrm{o}} &:= \{\boldsymbol{e}_k - \boldsymbol{e}_j, \boldsymbol{e}_k + \boldsymbol{e}_j \mid 1 \leq j < k \leq N\}, \\
\mathfrak{Roots}_{\mathrm{l}} &:= \{2\boldsymbol{e}_j \mid 1 \leq j \leq N\},
\end{aligned} \tag{7.546a}$$

we have then constructed the Jacobian

$$\begin{aligned}
J(\boldsymbol{\varepsilon}) &:= \prod_{\boldsymbol{\alpha}\in\mathfrak{Roots}_{\mathrm{s}}} |\boldsymbol{\alpha}\cdot\boldsymbol{\varepsilon}|^{m_{\mathrm{s}}} \prod_{\boldsymbol{\alpha}\in\mathfrak{Roots}_{\mathrm{o}}} |\boldsymbol{\alpha}\cdot\boldsymbol{\varepsilon}|^{m_{\mathrm{o}}} \prod_{\boldsymbol{\alpha}\in\mathfrak{Roots}_{\mathrm{l}}} |\boldsymbol{\alpha}\cdot\boldsymbol{\varepsilon}|^{m_{\mathrm{l}}}, \\
m_{\mathrm{s}} &= 4\,\nu, \qquad m_{\mathrm{o}} = 4, \qquad m_{\mathrm{l}} = 3.
\end{aligned} \tag{7.546b}$$

The exponent m_{s} is called the multiplicity of the short root $\boldsymbol{e}_j$. The exponent m_{o} is called the multiplicity of the ordinary root $\boldsymbol{e}_k \pm \boldsymbol{e}_j$. The exponent m_{l} is called the multiplicity of the long root $2\boldsymbol{e}_j$.

7.14.8 Symmetry class D Let

$$\mathcal{H} = \mathcal{H}^{\dagger} = -\mathcal{H}^{\mathsf{T}} \tag{7.547}$$

be any complex-valued $M \times M$ matrix with $M = 1, 2, 3, 4, 5, \cdots$. Verify that $\mathcal{H}$ is a purely imaginary valued and antisymmetric matrix

$$\operatorname{Re}\mathcal{H}_{jk} = 0, \qquad \operatorname{Im}\mathcal{H}_{jk} = -\operatorname{Im}\mathcal{H}_{kj}, \qquad j,k = 1,\cdots,N, \tag{7.548a}$$

and as such depends on

$$\frac{1}{2}(M-1)M = \frac{1}{2}M^2 - \frac{1}{2}M = 0, 1, 3, 6, 10, \cdots \tag{7.548b}$$

real-valued matrix elements. Verify that if there exists $\psi \in \mathbb{C}^M$ and some nonvanishing energy eigenvalue ε such that

$$\mathcal{H}\,\psi = +\varepsilon\,\psi, \tag{7.549a}$$

then ψ^* is an eigenstate of $\mathcal{H}$ with the nonvanishing energy eigenvalue $-\varepsilon$,

$$\mathcal{H}\,\psi^* = -\varepsilon\,\psi^*. \tag{7.549b}$$

Infer that the maximum number of nonvanishing energy eigenvalues of $\mathcal{H}$ is

$$2\left\lfloor\frac{M}{2}\right\rfloor = \begin{cases} M, & \text{if } M \text{ is even}, \\ M-1, & \text{if } M \text{ is odd}, \end{cases} \tag{7.550}$$

that is there must always be a vanishing energy eigenvalue when M is an odd integer. We are going to treat the cases of $M \equiv 2N$ even and $M \equiv 2N+1$ odd successively.

Symmetry class D with $\mathcal{H}$ acting on the Hilbert space $\mathbb{C}^{2N}$
Verify that $\mathcal{H}$ is parameterized by

$$\frac{1}{2}\,(2N-1)\,2N = 2N^2 - N = 4\times\frac{1}{2}(N-1)N + N \tag{7.551}$$

real-valued matrix elements. Define the measure

$$\begin{aligned}\mu(\mathrm{d}\mathcal{H}) :=& \left[\prod_{1\le j<k\le N} \mathrm{d}\left(\mathrm{Im}\,\mathcal{H}_{jk}\right)\right]\left[\prod_{1\le j<k\le N} \mathrm{d}\left(\mathrm{Im}\,\mathcal{H}_{(N+j)(N+k)}\right)\right]\\ &\times\left[\prod_{1\le j<k\le N} \mathrm{d}\left(\mathrm{Im}\,\mathcal{H}_{j(N+k)}\right)\right]\left[\prod_{1\le j<k\le N} \mathrm{d}\left(\mathrm{Im}\,\mathcal{H}_{k(N+j)}\right)\right]\\ &\times\prod_{1\le j\le N} \mathrm{d}\left(\mathrm{Im}\,\mathcal{H}_{j(N+j)}\right)\end{aligned} \tag{7.552}$$

and verify that it is a Haar measure on the group SO($2N$) (**Exercise** 7.16). Verify that $\mathcal{H}$ can be diagonalized, by a unitary matrix $\mathcal{U}\in \mathrm{U}(2N)$ that is in one-to-one correspondence with the group SO($2N$). (*Hint:* What is the normal form of a real-valued antisymmetric matrix?) More precisely, there exists $\mathcal{O}\in \mathrm{SO}(2N)$ such that

$$\mathcal{H} = \mathcal{U}\,\mathcal{E}\,\mathcal{U}^{\dagger} \tag{7.553a}$$

with

$$\mathcal{U} = \mathcal{O}\,\exp\left(\mathbb{1}\otimes \mathrm{i}\frac{\pi}{4}\,\rho_1\right), \tag{7.553b}$$

where ρ_0 and $\boldsymbol{\rho}$ are the unit 2×2 and Pauli matrices in the Majorana particle-hole grading, and

$$\mathcal{E} = \mathrm{diag}\,\begin{pmatrix}\varepsilon_1 & \cdots & \varepsilon_N\end{pmatrix}\otimes\rho_3, \qquad \varepsilon_1,\cdots,\varepsilon_N \ge 0. \tag{7.553c}$$

We shall loosely say that a unitary matrix $\mathcal{U}$ of the form given in Eq. (7.553b) belongs to SO($2N$). Verify that the relation (7.553) is invariant under the $\mathrm{U}(1)\cong \mathrm{SO}(2)$ transformations

$$\begin{aligned}&\mathcal{U}\mapsto \mathcal{U}\,\mathcal{V}_j, \qquad j=1,\cdots,N,\\ &\mathcal{V}_j := \mathrm{diag}\,\begin{pmatrix}1 & \cdots & 1 & e^{\mathrm{i}\theta_j} & 1 & \cdots & 1 & e^{\mathrm{i}\theta_j} & 1 & \cdots & 1\end{pmatrix},\\ &0\le\theta_j<2\pi,\end{aligned} \tag{7.554}$$

where the phase $\exp(\mathrm{i}\theta_j)$ enters the j-th and $(N+j)$-th diagonal elements. Verify that the right-hand side of Eq. (7.553a) with

$$\mathcal{U} \in \mathrm{SO}(2N)/[\mathrm{SO}(2)]^N \tag{7.555}$$

depends on

$$\frac{1}{2}(2N-1)\,2N - N = 2N^2 - 2N \tag{7.556}$$

independent real-valued matrix elements. Verify that if we define

$$\mathrm{d}\mathcal{A} := \mathcal{U}^\dagger \mathrm{d}\mathcal{U} = -\mathrm{d}\mathcal{A}^\dagger, \qquad \mathcal{U} \in \mathrm{SO}(2N)/[\mathrm{SO}(2)]^N, \tag{7.557a}$$

we may then parameterize $\mathrm{d}\mathcal{A}$ in terms of the

$$4 \times \frac{1}{2}(N-1)N = 2N^2 - 2N \tag{7.557b}$$

independent real-valued matrix elements given by

$$\mathrm{d}\left(\mathrm{Im}\,\mathcal{A}_{jk}\right),\ \mathrm{d}\left(\mathrm{Im}\,\mathcal{A}_{(N+j)(N+k)}\right),\ \mathrm{d}\left(\mathrm{Im}\,\mathcal{A}_{j(N+k)}\right),\ \mathrm{d}\left(\mathrm{Im}\,\mathcal{A}_{k(N+j)}\right) \tag{7.557c}$$

for $1 \leq j < k \leq N$. Verify that

$$\mathrm{Im}\left(\mathrm{d}\mathcal{A}\,\mathcal{E} + \mathrm{d}\mathcal{E} - \mathcal{E}\,\mathrm{d}\mathcal{A}\right)_{jk} = +\left(\varepsilon_k - \varepsilon_j\right)\mathrm{d}\left(\mathrm{Im}\,\mathcal{A}_{jk}\right), \tag{7.558a}$$

$$\mathrm{Im}\left(\mathrm{d}\mathcal{A}\,\mathcal{E} + \mathrm{d}\mathcal{E} - \mathcal{E}\,\mathrm{d}\mathcal{A}\right)_{(N+j)(N+k)} = -\left(\varepsilon_k - \varepsilon_j\right)\mathrm{d}\left(\mathrm{Im}\,\mathcal{A}_{(N+j)(N+k)}\right), \tag{7.558b}$$

$$\mathrm{Im}\left(\mathrm{d}\mathcal{A}\,\mathcal{E} + \mathrm{d}\mathcal{E} - \mathcal{E}\,\mathrm{d}\mathcal{A}\right)_{j(N+k)} = -\left(\varepsilon_k + \varepsilon_j\right)\mathrm{d}\left(\mathrm{Im}\,\mathcal{A}_{j(N+k)}\right), \tag{7.558c}$$

$$\mathrm{Im}\left(\mathrm{d}\mathcal{A}\,\mathcal{E} + \mathrm{d}\mathcal{E} - \mathcal{E}\,\mathrm{d}\mathcal{A}\right)_{k(N+j)} = -\left(\varepsilon_j + \varepsilon_k\right)\mathrm{d}\left(\mathrm{Im}\,\mathcal{A}_{k(N+j)}\right) \tag{7.558d}$$

for $1 \leq j < k \leq N$, while

$$\mathrm{Re}\left(\mathrm{d}\mathcal{A}\,\mathcal{E} + \mathrm{d}\mathcal{E} - \mathcal{E}\,\mathrm{d}\mathcal{A}\right)_{jj} = \mathrm{d}\varepsilon_j \tag{7.558e}$$

for $1 \leq j \leq N$. Verify that (i)

$$\mu(\mathrm{d}\,(\mathcal{U}\,\mathcal{E}\,\mathcal{U}^\dagger)\,) \propto \left(\prod_{j=1}^{N} \mathrm{d}\varepsilon_j\right)\left(\prod_{1\leq j<k\leq N} \left|\varepsilon_j^2 - \varepsilon_k^2\right|^2\right)\mu(\mathrm{d}\mathcal{A}) \tag{7.559a}$$

and (ii)

$$\mu(\mathrm{d}\mathcal{H}) \propto \mu(\mathrm{d}\,(\mathcal{U}\,\mathcal{E}\,\mathcal{U}^\dagger)\,), \tag{7.559b}$$

whereby (iii) the measure

$$\mu(\mathrm{d}\mathcal{A}) := \prod_{1\le j<k\le N} \mathrm{d}\left(\mathrm{Im}\,\mathcal{A}_{jk}\right)\mathrm{d}\left(\mathrm{Im}\,\mathcal{A}_{(N+j)(N+k)}\right) \times \mathrm{d}\left(\mathrm{Im}\,\mathcal{A}_{j(N+k)}\right)\mathrm{d}\left(\mathrm{Im}\,\mathcal{A}_{k(N+j)}\right), \qquad \mathrm{d}\mathcal{A} := \mathcal{U}^\dagger\,\mathrm{d}\mathcal{U} \tag{7.559c}$$

is a Haar measure of $\mathrm{SO}(2N)/[\mathrm{SO}(2)]^N$. We have found that the line D with $\mathrm{SO}(2N)$ in **Table** 7.2 corresponds to the degeneracy $d_{\mathrm{D}}=1$, the level repulsion exponent $\beta_{\mathrm{D}}=2$, the repulsion exponent from zero energy $\alpha_{\mathrm{D}}=0$, and the number $\nu_{\mathrm{D}}=0$ of zero modes. Moreover, if $\boldsymbol{e}_j$ is the Cartesian basis of $\mathbb{R}^N$, $\boldsymbol{\varepsilon}\in\mathbb{R}^N$ is the vector with the eigenvalues $\varepsilon_j\ge 0$ as components, and we define the set of "positive" roots to be

$$\begin{aligned}
\mathfrak{Roots}_{+} &:= \mathfrak{Roots}_{\mathrm{s}} \cup \mathfrak{Roots}_{\mathrm{o}} \cup \mathfrak{Roots}_{\mathrm{l}},\\
\mathfrak{Roots}_{\mathrm{s}} &:= \{\boldsymbol{e}_j \mid 1\le j\le N\},\\
\mathfrak{Roots}_{\mathrm{o}} &:= \{\boldsymbol{e}_k-\boldsymbol{e}_j, \boldsymbol{e}_k+\boldsymbol{e}_j \mid 1\le j<k\le N\},\\
\mathfrak{Roots}_{\mathrm{l}} &:= \{2\boldsymbol{e}_j \mid 1\le j\le N\},
\end{aligned} \tag{7.560a}$$

we have then constructed the Jacobian

$$J(\boldsymbol{\varepsilon}) := \prod_{\boldsymbol{\alpha}\in\mathfrak{Roots}_{\mathrm{s}}} |\boldsymbol{\alpha}\cdot\boldsymbol{\varepsilon}|^{m_{\mathrm{s}}} \prod_{\boldsymbol{\alpha}\in\mathfrak{Roots}_{\mathrm{o}}} |\boldsymbol{\alpha}\cdot\boldsymbol{\varepsilon}|^{m_{\mathrm{o}}} \prod_{\boldsymbol{\alpha}\in\mathfrak{Roots}_{\mathrm{l}}} |\boldsymbol{\alpha}\cdot\boldsymbol{\varepsilon}|^{m_{\mathrm{l}}}, \qquad m_{\mathrm{s}}=0, \quad m_{\mathrm{o}}=2, \quad m_{\mathrm{l}}=0. \tag{7.560b}$$

The exponent m_{s} is called the multiplicity of the short root $\boldsymbol{e}_j$. The exponent m_{o} is called the multiplicity of the ordinary root $\boldsymbol{e}_k\pm\boldsymbol{e}_j$. The exponent m_{l} is called the multiplicity of the long root $2\boldsymbol{e}_j$.

Symmetry class D with $\mathcal{H}$ acting on the Hilbert space $\mathbb{C}^{2N+1}$
Verify that $\mathcal{H}$ is parameterized by

$$\frac{1}{2}\,2N\,(2N+1) = 2N^2+N = 4\times\frac{1}{2}(N-1)N+N+2N \tag{7.561}$$

real-valued matrix elements. Define the measure

$$\begin{aligned}
\mu(\mathrm{d}\mathcal{H}) := &\left[\prod_{1\le j<k\le N}\mathrm{d}\left(\mathrm{Im}\,\mathcal{H}_{jk}\right)\right]\left[\prod_{1\le j<k\le N}\mathrm{d}\left(\mathrm{Im}\,\mathcal{H}_{(N+j)(N+k)}\right)\right]\\
&\times\left[\prod_{1\le j<k\le N}\mathrm{d}\left(\mathrm{Im}\,\mathcal{H}_{j(N+k)}\right)\right]\left[\prod_{1\le j<k\le N}\mathrm{d}\left(\mathrm{Im}\,\mathcal{H}_{k(N+j)}\right)\right]\\
&\times\left[\prod_{1\le j\le N}\mathrm{d}\left(\mathrm{Im}\,\mathcal{H}_{j(N+j)}\right)\right]\left[\prod_{1\le j\le 2N}\mathrm{d}\left(\mathrm{Im}\,\mathcal{H}_{j(2N+1)}\right)\right]
\end{aligned} \tag{7.562}$$

and verify that it is a Haar measure on the group $\mathrm{SO}(2N+1)$ (**Exercise** 7.16). Verify that $\mathcal{H}$ can be diagonalized, by a unitary matrix $\mathcal{U} \in \mathrm{U}(2N+1)$ that is in one-to-one correspondence with the group $\mathrm{SO}(2N+1)$. (*Hint:* What is the normal form of a real-valued antisymmetric matrix?) More precisely, there exists $\mathcal{O} \in \mathrm{SO}(2N+1)$ such that

$$\mathcal{H} = \mathcal{U}\,\mathcal{E}\,\mathcal{U}^{\dagger} \tag{7.563a}$$

with

$$\mathcal{U} = \mathcal{O}\left(\exp\left(\mathbb{1} \otimes \mathrm{i}\frac{\pi}{4}\rho_1 \right) \oplus \mathrm{diag}(1) \right), \tag{7.563b}$$

where ρ_0 and $\boldsymbol{\rho}$ are the unit 2×2 and Pauli matrices in the Majorana particle-hole grading, and

$$\mathcal{E} = \left(\mathrm{diag}\,(\varepsilon_1 \quad \cdots \quad \varepsilon_N) \otimes \rho_3 \right) \oplus \mathrm{diag}(0), \qquad \varepsilon_1, \cdots, \varepsilon_N \geq 0. \tag{7.563c}$$

We shall loosely say that a unitary matrix $\mathcal{U}$ of the form given in Eq. (7.563b) belongs to $\mathrm{SO}(2N+1)$. Verify that the relation (7.563) is invariant under the $\mathrm{U}(1) \cong \mathrm{SO}(2)$ transformations

$$\begin{aligned} &\mathcal{U} \mapsto \mathcal{U}\,\mathcal{V}_j, \qquad j = 1, \cdots, N, \\ &\mathcal{V}_j := \mathrm{diag}\,(1 \quad \cdots \quad 1 \quad e^{\mathrm{i}\theta_j} \quad 1 \quad \cdots \quad 1 \quad e^{\mathrm{i}\theta_j} \quad 1 \quad \cdots \; 1\; 1), \quad 0 \leq \theta_j < 2\pi, \end{aligned} \tag{7.564}$$

where the phase $\exp(\mathrm{i}\theta_j)$ enters the j-th and $(N+j)$-th diagonal elements. Verify that the right-hand side of (7.563a) with

$$\mathcal{U} \in \mathrm{SO}(2N+1)/[\mathrm{SO}(2)]^N \tag{7.565}$$

depends on

$$\frac{1}{2}\,2N\,(2N+1) - N = 2N^2 \tag{7.566}$$

independent real-valued matrix elements. Verify that if we define

$$\mathrm{d}\mathcal{A} := \mathcal{U}^{\dagger}\,\mathrm{d}\mathcal{U} = -\,\mathrm{d}\mathcal{A}^{\dagger}, \qquad \mathcal{U} \in \mathrm{SO}(2N+1)/[\mathrm{SO}(2)]^N, \tag{7.567a}$$

we may then parameterize $\mathrm{d}\mathcal{A}$ in terms of the

$$4 \times \frac{1}{2}(N-1)N + 2N = 2N^2 \tag{7.567b}$$

independent real-valued matrix elements given by

$$\mathrm{d}\left(\mathrm{Im}\,\mathcal{A}_{jk}\right), \; \mathrm{d}\left(\mathrm{Im}\,\mathcal{A}_{(N+j)(N+k)}\right), \; \mathrm{d}\left(\mathrm{Im}\,\mathcal{A}_{j(N+k)}\right), \; \mathrm{d}\left(\mathrm{Im}\,\mathcal{A}_{k(N+j)}\right) \tag{7.567c}$$

for $1 \leq j < k \leq N$ and

$$\mathrm{d}\left(\mathrm{Im}\,\mathcal{A}_{j(2N+1)}\right) \tag{7.567d}$$

for $j=1,\cdots,2N$. Verify that

$$\mathrm{Im}\left(\mathrm{d}\mathcal{A}\,\mathcal{E}+\mathrm{d}\mathcal{E}-\mathcal{E}\,\mathrm{d}\mathcal{A}\right)_{jk}=+\left(\varepsilon_k-\varepsilon_j\right)\mathrm{d}\left(\mathrm{Im}\,\mathcal{A}_{jk}\right), \tag{7.568a}$$

$$\mathrm{Im}\left(\mathrm{d}\mathcal{A}\,\mathcal{E}+\mathrm{d}\mathcal{E}-\mathcal{E}\,\mathrm{d}\mathcal{A}\right)_{(N+j)(N+k)}=-\left(\varepsilon_k-\varepsilon_j\right)\mathrm{d}\left(\mathrm{Im}\,\mathcal{A}_{(N+j)(N+k)}\right), \tag{7.568b}$$

$$\mathrm{Im}\left(\mathrm{d}\mathcal{A}\,\mathcal{E}+\mathrm{d}\mathcal{E}-\mathcal{E}\,\mathrm{d}\mathcal{A}\right)_{j(N+k)}=-\left(\varepsilon_k+\varepsilon_j\right)\mathrm{d}\left(\mathrm{Im}\,\mathcal{A}_{j(N+k)}\right), \tag{7.568c}$$

$$\mathrm{Im}\left(\mathrm{d}\mathcal{A}\,\mathcal{E}+\mathrm{d}\mathcal{E}-\mathcal{E}\,\mathrm{d}\mathcal{A}\right)_{k(N+j)}=-\left(\varepsilon_j+\varepsilon_k\right)\mathrm{d}\left(\mathrm{Im}\,\mathcal{A}_{k(N+j)}\right) \tag{7.568d}$$

for $1\leq j<k\leq N$, while

$$\mathrm{Im}\left(\mathrm{d}\mathcal{A}\,\mathcal{E}+\mathrm{d}\mathcal{E}-\mathcal{E}\,\mathrm{d}\mathcal{A}\right)_{j(2N+1)}=\left(0-\varepsilon_j\right)\mathrm{d}\left(\mathrm{Im}\,\mathcal{A}_{j(2N+1)}\right), \tag{7.568e}$$

$$\mathrm{Im}\left(\mathrm{d}\mathcal{A}\,\mathcal{E}+\mathrm{d}\mathcal{E}-\mathcal{E}\,\mathrm{d}\mathcal{A}\right)_{(N+j)(2N+1)}=\left(0+\varepsilon_j\right)\mathrm{d}\left(\mathrm{Im}\,\mathcal{A}_{(N+j)(2N+1)}\right), \tag{7.568f}$$

$$\mathrm{Re}\left(\mathrm{d}\mathcal{A}\,\mathcal{E}+\mathrm{d}\mathcal{E}-\mathcal{E}\,\mathrm{d}\mathcal{A}\right)_{jj}=\mathrm{d}\varepsilon_j \tag{7.568g}$$

for $1\leq j\leq N$. Verify that (i)

$$\mu(\mathrm{d}\,(\mathcal{U}\,\mathcal{E}\,\mathcal{U}^{\dagger})\,)\propto\left(\prod_{j=1}^{N}\mathrm{d}\varepsilon_j\,|\varepsilon_j|^2\right)\left(\prod_{1\leq j<k\leq N}\left|\varepsilon_j^2-\varepsilon_k^2\right|^2\right)\mu(\mathrm{d}\mathcal{A}) \tag{7.569a}$$

and (ii)

$$\mu(\mathrm{d}\mathcal{H})\propto\mu(\mathrm{d}\,(\mathcal{U}\,\mathcal{E}\,\mathcal{U}^{\dagger})\,), \tag{7.569b}$$

whereby (iii) the measure

$$\begin{aligned}\mu(\mathrm{d}\mathcal{A}):=&\left[\prod_{1\leq j<k\leq N}\mathrm{d}\left(\mathrm{Im}\,\mathcal{A}_{jk}\right)\mathrm{d}\left(\mathrm{Im}\,\mathcal{A}_{(N+j)(N+k)}\right)\right.\\&\left.\times\,\mathrm{d}\left(\mathrm{Im}\,\mathcal{A}_{j(N+k)}\right)\mathrm{d}\left(\mathrm{Im}\,\mathcal{A}_{k(N+j)}\right)\right]\\&\times\left[\prod_{1\leq j\leq 2N}\mathrm{d}\left(\mathrm{Im}\,\mathcal{A}_{j(2N+1)}\right)\right],\qquad \mathrm{d}\mathcal{A}:=\mathcal{U}^{\dagger}\,\mathrm{d}\mathcal{U}\end{aligned} \tag{7.569c}$$

is a Haar measure of $\mathrm{SO}(2N+1)/[\mathrm{SO}(2)]^N$. We have found that the line D with $\mathrm{SO}(2N+1)$ in **Table** 7.2 corresponds to the degeneracy $d_{\mathrm{D}}=1$, the level repulsion exponent $\beta_{\mathrm{D}} = 2$, the repulsion exponent from zero energy $\alpha_{\mathrm{D}} = 0$, and the number $\nu_{\mathrm{D}} = 1$ of zero modes. Moreover, if $\boldsymbol{e}_j$ is the Cartesian basis of $\mathbb{R}^N$, $\boldsymbol{\varepsilon} \in \mathbb{R}^N$ is the vector with the eigenvalues $\varepsilon_j \geq 0$ as components, and we define the set of "positive" roots to be

$$\begin{aligned} \mathfrak{Roots}_+ &:= \mathfrak{Roots}_{\mathrm{s}} \cup \mathfrak{Roots}_{\mathrm{o}} \cup \mathfrak{Roots}_{\mathrm{l}}, \\ \mathfrak{Roots}_{\mathrm{s}} &:= \{\boldsymbol{e}_j \mid 1 \leq j \leq N\}, \\ \mathfrak{Roots}_{\mathrm{o}} &:= \{\boldsymbol{e}_k - \boldsymbol{e}_j, \boldsymbol{e}_k + \boldsymbol{e}_j \mid 1 \leq j < k \leq N\}, \\ \mathfrak{Roots}_{\mathrm{l}} &:= \{2\boldsymbol{e}_j \mid 1 \leq j \leq N\}, \end{aligned} \tag{7.570a}$$

we have then constructed the Jacobian

$$\begin{aligned} J(\boldsymbol{\varepsilon}) &:= \prod_{\boldsymbol{\alpha}\in\mathfrak{Roots}_{\mathrm{s}}} |\boldsymbol{\alpha}\cdot\boldsymbol{\varepsilon}|^{m_{\mathrm{s}}} \prod_{\boldsymbol{\alpha}\in\mathfrak{Roots}_{\mathrm{o}}} |\boldsymbol{\alpha}\cdot\boldsymbol{\varepsilon}|^{m_{\mathrm{o}}} \prod_{\boldsymbol{\alpha}\in\mathfrak{Roots}_{\mathrm{l}}} |\boldsymbol{\alpha}\cdot\boldsymbol{\varepsilon}|^{m_{\mathrm{l}}}, \\ m_{\mathrm{s}} &= 2, \qquad m_{\mathrm{o}} = 2, \qquad m_{\mathrm{l}} = 0. \end{aligned} \tag{7.570b}$$

The exponent m_{s} is called the multiplicity of the short root $\boldsymbol{e}_j$. The exponent m_{o} is called the multiplicity of the ordinary root $\boldsymbol{e}_k \pm \boldsymbol{e}_j$. The exponent m_{l} is called the multiplicity of the long root $2\boldsymbol{e}_j$.

7.14.9 Symmetry class C Consider the $2N \times 2N$ complex-valued matrix

$$\mathcal{H} = \mathcal{H}^\dagger = -\mathcal{S}\,\mathcal{H}^{\mathsf{T}}\,\mathcal{S}^{-1}, \tag{7.571a}$$

where

$$\mathcal{S} := \mathbb{1} \otimes \mathrm{i}\kappa_2 \tag{7.571b}$$

with κ_0 the 2×2 unit matrix and $\boldsymbol{\kappa}$ the Pauli matrices in the irreducible 2×2 grading introduced in Eqs. (7.72) and (7.73). Verify that Eq. (7.571) implies that

$$\mathcal{H} := \begin{pmatrix} \mathsf{A} & \mathsf{B} \\ \mathsf{B}^\dagger & -\mathsf{A}^{\mathsf{T}} \end{pmatrix}, \qquad \mathsf{A} = \mathsf{A}^\dagger, \qquad \mathsf{B} = \mathsf{B}^{\mathsf{T}}, \tag{7.572}$$

with A and B complex-valued matrices. Verify that $\mathcal{H}$ depends on

$$N^2 + 2 \times \left[N + \frac{1}{2}(N-1)N\right] = 2N^2 + N \tag{7.573}$$

independent real-valued matrix elements that we may choose to be

$$\begin{aligned} &\operatorname{Re}\mathsf{A}_{jj}, \quad \operatorname{Re}\mathsf{B}_{jj}, \quad \operatorname{Im}\mathsf{B}_{jj}, \quad j = 1, \cdots, N, \\ &\operatorname{Re}\mathsf{A}_{jk}, \quad \operatorname{Im}\mathsf{A}_{jk}, \quad 1 \leq j < k \leq N, \\ &\operatorname{Re}\mathsf{B}_{jk}, \quad \operatorname{Im}\mathsf{B}_{jk}, \quad 1 \leq j < k \leq N. \end{aligned} \tag{7.574}$$

Verify that the measure

$$\mu(\mathrm{d}\mathcal{H}) := \prod_{j=1}^{N} \mathrm{d}\mathsf{A}_{jj}\, \mathrm{d}\left(\mathrm{Re}\,\mathsf{B}_{jj}\right)\, \mathrm{d}\left(\mathrm{Im}\,\mathsf{B}_{jj}\right) \times \prod_{1\le j<k\le N} \mathrm{d}\left(\mathrm{Re}\,\mathsf{A}_{jk}\right)\, \mathrm{d}\left(\mathrm{Im}\,\mathsf{A}_{jk}\right)\, \mathrm{d}\left(\mathrm{Re}\,\mathsf{B}_{jk}\right)\, \mathrm{d}\left(\mathrm{Im}\,\mathsf{B}_{jk}\right) \tag{7.575}$$

is a Haar measure on the group Sp($2N$) (**Exercise** 7.16). Verify that the unitary transformation

$$\mathcal{H}' := \mathcal{U}^{\dagger}\,\mathcal{H}\,\mathcal{U}, \qquad \mathcal{U}\in \mathrm{U}(2N) \tag{7.576}$$

leaves Eq. (7.571) invariant if and only if

$$\mathcal{U}\,\mathcal{S}\,\mathcal{U}^{\mathsf{T}} = \mathcal{S}, \tag{7.577}$$

that is $\mathcal{U}\in \mathrm{Sp}(2N)$. Verify that the diagonal matrix

$$\mathcal{E} := \begin{pmatrix}\varepsilon_1 & \cdots & \varepsilon_N\end{pmatrix}\otimes \kappa_3, \qquad \varepsilon_1 \ge 0, \cdots, \varepsilon_N \ge 0 \tag{7.578a}$$

obeys Eq. (7.571). It is thus consistent to assume that $\mathcal{H}$ can always be diagonalized by a matrix from Sp($2N$). We make this assumption from now on, that is, there always exist a $\mathcal{U}\in \mathrm{Sp}(2N)\subset \mathrm{U}(2N)$ and a diagonal matrix $\mathcal{E}$ of the form (7.578a) such that

$$\mathcal{H} = \mathcal{U}\,\mathcal{E}\,\mathcal{U}^{\dagger}. \tag{7.578b}$$

This spectral decomposition is one-to-many. For example, it is invariant under the transformation

$$\begin{aligned}&\mathcal{U}\mapsto \mathcal{U}\,e^{\mathrm{i}\Theta},\\ &\Theta := \mathrm{diag}\begin{pmatrix}\theta_1 & \cdots & \theta_N\end{pmatrix}\otimes\kappa_3, \qquad 0\le \theta_1,\cdots,\theta_N < 2\pi,\end{aligned} \tag{7.579}$$

for which $\mathcal{U}\,e^{\mathrm{i}\Theta}\in \mathrm{Sp}(2N)$. Verify that this ambiguity allows to perform the diagonalization (7.578) with an element $\mathcal{U}\in \mathrm{Sp}(2N)/[\mathrm{U}(1)]^N$ such that

$$\mathrm{d}\mathcal{A} := \mathcal{U}^{\dagger}\,\mathrm{d}\mathcal{U} = -\left(\mathrm{d}\mathcal{A}\right)^{\dagger} \in \mathfrak{sp}(2N) \tag{7.580a}$$

obeys

$$\mathrm{d}\mathcal{A}_{jj} = \mathrm{d}\mathcal{A}_{(N+j)(N+j)} = 0, \qquad j=1,\cdots,N. \tag{7.580b}$$

We may then parameterize such a constrained $\mathrm{d}\mathcal{A}\in\mathfrak{sp}(2N)$ by the

$$4\times\frac{1}{2}(N-1)N + 2\times N = 2N^2 \tag{7.580c}$$

independent real-valued matrix elements

$$
\begin{aligned}
&\mathrm{d}\,(\mathrm{Re}\,\mathcal{A})_{jk}\,, && \mathrm{d}\,(\mathrm{Im}\,\mathcal{A})_{jk}\,, && 1\le j<k\le N,\\
&\mathrm{d}\,(\mathrm{Re}\,\mathcal{A})_{j(N+k)}\,, && \mathrm{d}\,(\mathrm{Im}\,\mathcal{A})_{j(N+k)} && 1\le j<k\le N,\\
&\mathrm{d}\,(\mathrm{Re}\,\mathcal{A})_{j(N+j)}\,, && \mathrm{d}\,(\mathrm{Im}\,\mathcal{A})_{j(N+j)}\,, && j=1,\cdots,N.
\end{aligned}
\tag{7.580d}
$$

Verify that for $\mathrm{d}\mathcal{A}$ constrained as in Eq. (7.580),

$$(\mathrm{d}\mathcal{A}\,\mathcal{E}+\mathrm{d}\mathcal{E}-\mathcal{E}\,\mathrm{d}\mathcal{A})_{jk}=+\left(\varepsilon_k-\varepsilon_j\right)\mathrm{d}\mathcal{A}_{jk}, \tag{7.581a}$$

$$(\mathrm{d}\mathcal{A}\,\mathcal{E}+\mathrm{d}\mathcal{E}-\mathcal{E}\,\mathrm{d}\mathcal{A})_{j(N+k)}=-\left(\varepsilon_k+\varepsilon_j\right)\mathrm{d}\mathcal{A}_{j(N+k)}, \tag{7.581b}$$

$$(\mathrm{d}\mathcal{A}\,\mathcal{E}+\mathrm{d}\mathcal{E}-\mathcal{E}\,\mathrm{d}\mathcal{A})_{jj}=\mathrm{d}\varepsilon_j, \tag{7.581c}$$

$$(\mathrm{d}\mathcal{A}\,\mathcal{E}+\mathrm{d}\mathcal{E}-\mathcal{E}\,\mathrm{d}\mathcal{A})_{j(N+j)}=-2\varepsilon_j\,\mathrm{d}\mathcal{A}_{j(N+j)} \tag{7.581d}$$

for $1\le j<k\le N$. Verify that the counterpart to Eq. (7.559) is (i)

$$\mu(\mathrm{d}\,(\mathcal{U}\,\mathcal{E}\,\mathcal{U}^\dagger)\,)\propto\left(\prod_{j=1}^{N}\mathrm{d}\varepsilon_j\,|2\varepsilon_j|^2\right)\left(\prod_{1\le j<k\le N}\left|\varepsilon_j^2-\varepsilon_k^2\right|^2\right)\mu(\mathrm{d}\mathcal{A}) \tag{7.582a}$$

and (ii)

$$\mu(\mathrm{d}\mathcal{H})\propto\mu(\mathrm{d}\,(\mathcal{U}\,\mathcal{E}\,\mathcal{U}^\dagger)\,), \tag{7.582b}$$

whereby (iii) the measure

$$
\begin{aligned}
\mu(\mathrm{d}\mathcal{A}):=&\left[\prod_{1\le j<k\le N}\mathrm{d}\left(\mathrm{Re}\,\mathcal{A}_{jk}\right)\mathrm{d}\left(\mathrm{Im}\,\mathcal{A}_{jk}\right)\right.\\
&\left.\times\,\mathrm{d}\left(\mathrm{Re}\,\mathcal{A}_{j(N+k)}\right)\mathrm{d}\left(\mathrm{Im}\,\mathcal{A}_{j(N+k)}\right)\right]\\
&\times\left[\prod_{1\le j\le N}\mathrm{d}\left(\mathrm{Re}\,\mathcal{A}_{j(N+j)}\right)\mathrm{d}\left(\mathrm{Im}\,\mathcal{A}_{j(N+j)}\right)\right],\qquad \mathrm{d}\mathcal{A}:=\mathcal{U}^\dagger\,\mathrm{d}\mathcal{U}
\end{aligned}
\tag{7.582c}
$$

is a Haar measure of $\mathrm{Sp}(2N)/[\mathrm{U}(1)]^N$. We have found that the line C in **Table** 7.2 corresponds to the degeneracy $d_{\mathrm{C}}=1$, the level repulsion exponent $\beta_{\mathrm{C}}=2$, the repulsion exponent from zero energy $\alpha_{\mathrm{C}}=2$, and the number $\nu_{\mathrm{C}}=0$ of zero modes. Moreover, if $\boldsymbol{e}_j$ is the Cartesian basis of $\mathbb{R}^N$, $\boldsymbol{\varepsilon}\in\mathbb{R}^N$ is the vector with the eigenvalues $\varepsilon_j\ge 0$ as components, and we define the set of "positive" roots to be

$$
\begin{aligned}
&\mathfrak{Roots}_+:=\mathfrak{Roots}_{\mathrm{s}}\cup\mathfrak{Roots}_{\mathrm{o}}\cup\mathfrak{Roots}_{\mathrm{l}},\\
&\mathfrak{Roots}_{\mathrm{s}}:=\{\boldsymbol{e}_j\mid 1\le j\le N\},\\
&\mathfrak{Roots}_{\mathrm{o}}:=\{\boldsymbol{e}_k-\boldsymbol{e}_j,\boldsymbol{e}_k+\boldsymbol{e}_j\mid 1\le j<k\le N\},\\
&\mathfrak{Roots}_{\mathrm{l}}:=\{2\boldsymbol{e}_j\mid 1\le j\le N\},
\end{aligned}
\tag{7.583a}
$$

we have then constructed the Jacobian

$$J(\varepsilon) := \prod_{\alpha\in\mathfrak{Roots}_{\rm s}} |\alpha\cdot\varepsilon|^{m_{\rm s}} \prod_{\alpha\in\mathfrak{Roots}_{\rm o}} |\alpha\cdot\varepsilon|^{m_{\rm o}} \prod_{\alpha\in\mathfrak{Roots}_{\rm l}} |\alpha\cdot\varepsilon|^{m_{\rm l}}, \tag{7.583b}$$

$$m_{\rm s}=0, \qquad m_{\rm o}=2, \qquad m_{\rm l}=2.$$

The exponent $m_{\rm s}$ is called the multiplicity of the short root $\boldsymbol{e}_j$. The exponent $m_{\rm o}$ is called the multiplicity of the ordinary root $\boldsymbol{e}_k\pm\boldsymbol{e}_j$. The exponent $m_{\rm l}$ is called the multiplicity of the long root $2\boldsymbol{e}_j$.

7.14.10 Symmetry class CI Consider the $2N\times 2N$ complex-valued matrix

$$\mathcal{H}=\mathcal{H}^{\dagger}=-\mathcal{S}\,\mathcal{H}^{\mathsf{T}}\,\mathcal{S}^{-1}=\mathcal{H}^{\mathsf{T}}, \tag{7.584a}$$

where

$$\mathcal{S}:=\mathbb{1}\otimes\mathrm{i}\kappa_2 \tag{7.584b}$$

with κ_0 the 2×2 unit matrix and $\boldsymbol{\kappa}$ the Pauli matrices in the ireducible 2×2 grading introduced in Eqs. (7.72) and (7.73). Verify that Eq. (7.584) implies that

$$\mathcal{H}:=\begin{pmatrix}\mathsf{A} & \mathsf{B}\\ \mathsf{B}^{\dagger} & -\mathsf{A}^{\mathsf{T}}\end{pmatrix}, \qquad \mathsf{A}=\mathsf{A}^{\dagger}=\mathsf{A}^{\mathsf{T}}, \qquad \mathsf{B}=\mathsf{B}^{\mathsf{T}}=\mathsf{B}^{*}. \tag{7.585}$$

Thus, A and B are now $N\times N$ real-valued matrices. Verify that $\mathcal{H}$ depends on

$$2\times\left[N+\frac{1}{2}(N-1)N\right]=N^2+N \tag{7.586}$$

independent real-valued matrix elements that we may choose to be

$$\mathsf{A}_{jj}, \qquad \mathsf{B}_{jj}, \qquad \mathsf{A}_{jk}, \qquad \mathsf{B}_{jk}, \qquad 1\le j<k\le N. \tag{7.587}$$

Verify that the measure

$$\mu(\mathrm{d}\mathcal{H}) := \left(\prod_{j=1}^{N}\mathrm{d}\mathsf{A}_{jj}\,\mathrm{d}\mathsf{B}_{jj}\right)\left(\prod_{1\le j<k\le N}\mathrm{d}\mathsf{A}_{jk}\,\mathrm{d}\mathsf{B}_{jk}\right) \tag{7.588}$$

is a Haar measure on the symmetric space $\mathrm{Sp}(2N)/\mathrm{U}(N)$ (**Exercise** 7.16). Verify that the counterpart to Eq. (7.582) is (i)

$$\mu(\mathrm{d}\,(\mathcal{U}\,\mathcal{E}\,\mathcal{U}^{\dagger})\,)\propto\left(\prod_{j=1}^{N}\mathrm{d}\varepsilon_j\,|2\varepsilon_j|\right)\left(\prod_{1\le j<k\le N}|\varepsilon_j^2-\varepsilon_k^2|\right)\mu(\mathrm{d}\mathcal{A}) \tag{7.589a}$$

and (ii)

$$\mu(\mathrm{d}\mathcal{H})\propto\mu(\mathrm{d}\,(\mathcal{U}\,\mathcal{E}\,\mathcal{U}^{\dagger})\,), \tag{7.589b}$$

whereby (iii) the measure

$$\mu(\mathrm{d}\mathcal{A}) := \left(\prod_{1\le j<k\le N} \mathrm{d}\mathcal{A}_{jk}\,\mathrm{d}\mathcal{A}_{j(N+k)}\right)\left(\prod_{j=1}^{N}\mathrm{d}\mathcal{A}_{j(N+j)}\right), \qquad \mathrm{d}\mathcal{A} := \mathcal{U}^{\dagger}\,\mathrm{d}\mathcal{U}, \qquad \mathcal{U}=\mathcal{U}^{*} \tag{7.589c}$$

is a Haar measure of $\mathrm{U}(N) \cong \mathrm{Sp}(2N)/\{\mathrm{U}(N)\otimes[\mathrm{U}(1)]^{N}\}$. We have found that the line CI in **Table** 7.2 corresponds to the degeneracy $d_{\mathrm{CI}}=1$, the level repulsion exponent $\beta_{\mathrm{CI}}=1$, the repulsion exponent from zero energy $\alpha_{\mathrm{CI}}=1$, and the number $\nu_{\mathrm{CI}}=0$ of zero modes. Moreover, if $\boldsymbol{e}_j$ is the Cartesian basis of $\mathbb{R}^N$, $\boldsymbol{\varepsilon}\in\mathbb{R}^N$ is the vector with the eigenvalues $\varepsilon_j\ge 0$ as components, and we define the set of "positive" roots to be

$$\begin{aligned}
&\mathfrak{Roots}_{+} := \mathfrak{Roots}_{\mathrm{s}} \cup \mathfrak{Roots}_{\mathrm{o}} \cup \mathfrak{Roots}_{\mathrm{l}},\\
&\mathfrak{Roots}_{\mathrm{s}} := \{\boldsymbol{e}_j \mid 1\le j\le N\},\\
&\mathfrak{Roots}_{\mathrm{o}} := \{\boldsymbol{e}_k-\boldsymbol{e}_j, \boldsymbol{e}_k+\boldsymbol{e}_j \mid 1\le j<k\le N\},\\
&\mathfrak{Roots}_{\mathrm{l}} := \{2\boldsymbol{e}_j \mid 1\le j\le N\},
\end{aligned} \tag{7.590a}$$

we have then constructed the Jacobian

$$J(\boldsymbol{\varepsilon}) := \prod_{\boldsymbol{\alpha}\in\mathfrak{Roots}_{\mathrm{s}}}|\boldsymbol{\alpha}\cdot\boldsymbol{\varepsilon}|^{m_{\mathrm{s}}}\prod_{\boldsymbol{\alpha}\in\mathfrak{Roots}_{\mathrm{o}}}|\boldsymbol{\alpha}\cdot\boldsymbol{\varepsilon}|^{m_{\mathrm{o}}}\prod_{\boldsymbol{\alpha}\in\mathfrak{Roots}_{\mathrm{l}}}|\boldsymbol{\alpha}\cdot\boldsymbol{\varepsilon}|^{m_{\mathrm{l}}}, \qquad m_{\mathrm{s}}=0, \qquad m_{\mathrm{o}}=1, \qquad m_{\mathrm{l}}=1. \tag{7.590b}$$

The exponent m_{s} is called the multiplicity of the short root $\boldsymbol{e}_j$. The exponent m_{o} is called the multiplicity of the ordinary root $\boldsymbol{e}_k\pm\boldsymbol{e}_j$. The exponent m_{l} is called the multiplicity of the long root $2\boldsymbol{e}_j$.

7.14.11 **Symmetry class DIII** We choose the chiral representation (7.91). Let the integer M be either even, $M=2N$, or odd, $M=2N+1$, for $N=1,2,3,\cdots$. Consider the $2M\times 2M$ complex-valued matrix

$$\mathcal{H}_{\mathrm{DIII}}=\mathcal{H}_{\mathrm{DIII}}^{\dagger}=\begin{pmatrix}0 & \mathcal{B}\\ \mathcal{B}^{\dagger} & 0\end{pmatrix}, \qquad \mathcal{B}=-\mathcal{B}^{\mathsf{T}}. \tag{7.591}$$

Verify that $\mathcal{H}_{\mathrm{DIII}}$ can be parameterized by the

$$(M-1)M = M^2-M = \begin{cases}4N^2-2N, & \text{if } M=2N,\\ 4N^2+2N, & \text{if } M=2N+1,\end{cases} \tag{7.592a}$$

independent real-valued matrix elements

$$\mathrm{Re}\,\mathcal{B}_{jk}, \qquad \mathrm{Im}\,\mathcal{B}_{jk}, \qquad 1\le j<k\le M. \tag{7.592b}$$

Verify that the measure

$$\mu(\mathrm{d}\mathcal{H}_{\mathrm{DIII}}) := \prod_{1\le j<k\le M}\mathrm{d}\left(\mathrm{Re}\,\mathcal{B}_{jk}\right)\mathrm{d}\left(\mathrm{Im}\,\mathcal{B}_{jk}\right) \tag{7.593}$$

is a Haar measure on the symmetric space $\mathrm{SO}(2M)/\mathrm{U}(M)$ (**Exercise** 7.16). **Case** $M = 2N$**:** We shall use the 2×2 unit matrix κ_0 and the triplet $\boldsymbol{\kappa}$ of Pauli matrices. Verify that the spectral decomposition

$$\mathcal{B} = \mathcal{U}\,\mathcal{E}\,\mathcal{V}, \qquad \mathcal{E} := (\mathsf{E} \otimes \kappa_0)\,, \tag{7.594a}$$

whereby

$$\mathcal{U} \in \mathrm{U}(2N), \qquad \mathcal{V} = [\mathcal{U}\,(-\mathbb{1} \otimes \mathrm{i}\kappa_2)]^{\mathsf{T}} \in \mathrm{U}(2N), \tag{7.594b}$$

and the diagonal matrix

$$\mathsf{E} = \mathrm{diag}\,\begin{pmatrix}\varepsilon_1 & \cdots & \varepsilon_N\end{pmatrix} \tag{7.594c}$$

has no more than the N nonvanishing real-valued matrix elements $\varepsilon_1 \geq 0, \cdots, \varepsilon_N \geq 0$, is compatible with the singular-value decomposition of the antisymmetric matrix $\mathcal{B} = -\mathcal{B}^{\mathsf{T}}$ (**Exercise** 7.15).[116] Verify that the spectral representation (7.594) is not unique. For example, show that it is invariant under the transformation

$$\mathcal{U} \mapsto \mathcal{U}\,\left(\kappa_0 \oplus \cdots \oplus \kappa_0 \oplus e^{\mathrm{i}\boldsymbol{\alpha}_j \cdot \boldsymbol{\kappa}} \oplus \kappa_0 \oplus \cdots \oplus \kappa_0\right), \qquad \boldsymbol{\alpha}_j \in \mathbb{R}^3, \tag{7.595}$$

for $j = 1, \cdots, N$. Verify that the dimensionality of the symmetric space

$$\mathrm{U}(2N)/[\mathrm{SU}(2)]^N \tag{7.596a}$$

is

$$4N^2 - 3N = \left(4N^2 - 2N\right) - N. \tag{7.596b}$$

Show that if one defines the pair of matrices

$$\begin{aligned} \mathrm{d}\mathcal{A}_{\mathcal{U}} &:= \mathcal{U}^{\dagger}\,\mathrm{d}\mathcal{U} = -\left(\mathrm{d}\mathcal{A}_{\mathcal{U}}\right)^{\dagger} \in \mathfrak{u}(2N), \\ \mathrm{d}\mathcal{A}_{\mathcal{V}} &:= \mathrm{d}\mathcal{V}\,\mathcal{V}^{\dagger} = -\left(\mathrm{d}\mathcal{A}_{\mathcal{V}}\right)^{\dagger} \in \mathfrak{u}(2N), \end{aligned} \tag{7.597a}$$

then

$$\mathcal{U}^{\dagger}\,\mathrm{d}\mathcal{B}\,\mathcal{V}^{\dagger} = \mathrm{d}\mathcal{A}_{\mathcal{U}}\,(\mathsf{E} \otimes \kappa_0) + (\mathrm{d}\mathsf{E} \otimes \kappa_0) + (\mathsf{E} \otimes \kappa_0)\,\mathrm{d}\mathcal{A}_{\mathcal{V}}, \tag{7.597b}$$

where

$$\mathrm{d}\mathcal{A}_{\mathcal{V}} = (\mathbb{1} \otimes \kappa_2)\,(\mathrm{d}\mathcal{A}_{\mathcal{U}})^{\mathsf{T}}\,(\mathbb{1} \otimes \kappa_2)\,. \tag{7.597c}$$

We shall use the parameterizations

$$(\mathrm{d}\mathcal{A}_{\mathcal{U}})_{jk} \equiv \left(\mathrm{d}\mathcal{A}^{(0)}_{\mathcal{U}}\right)_{jk} \kappa_0 + \left(\mathrm{d}\mathcal{A}^{(1)}_{\mathcal{U}}\right)_{jk} \kappa_1 + \left(\mathrm{d}\mathcal{A}^{(2)}_{\mathcal{U}}\right)_{jk} \kappa_2 + \left(\mathrm{d}\mathcal{A}^{(3)}_{\mathcal{U}}\right)_{jk} \kappa_3, \tag{7.598a}$$

$$\left[(\mathrm{d}\mathcal{A}_{\mathcal{U}})_{jk}\right]^{\dagger} = \left(\mathrm{d}\mathcal{A}^{(0)}_{\mathcal{U}}\right)^{*}_{jk} \kappa_0 + \left(\mathrm{d}\mathcal{A}^{(1)}_{\mathcal{U}}\right)^{*}_{jk} \kappa_1 + \left(\mathrm{d}\mathcal{A}^{(2)}_{\mathcal{U}}\right)^{*}_{jk} \kappa_2 + \left(\mathrm{d}\mathcal{A}^{(3)}_{\mathcal{U}}\right)^{*}_{jk} \kappa_3, \tag{7.598b}$$

[116] D. C. Youla, Can. J. Math. **13**, 694–704 (1961). [Youla (1961)].

$$\left[(\mathrm{d}\mathcal{A}_{\mathcal{U}})_{jk}\right]^{\mathsf{T}}=\left(\mathrm{d}\mathcal{A}^{(0)}_{\mathcal{U}}\right)_{jk}\kappa_0+\left(\mathrm{d}\mathcal{A}^{(1)}_{\mathcal{U}}\right)_{jk}\kappa_1-\left(\mathrm{d}\mathcal{A}^{(2)}_{\mathcal{U}}\right)_{jk}\kappa_2+\left(\mathrm{d}\mathcal{A}^{(3)}_{\mathcal{U}}\right)_{jk}\kappa_3, \tag{7.598c}$$

$$(\mathrm{d}\mathcal{A}_{\mathcal{V}})_{jk}=\left(\mathrm{d}\mathcal{A}^{(0)}_{\mathcal{U}}\right)_{kj}\kappa_0-\left(\mathrm{d}\mathcal{A}^{(1)}_{\mathcal{U}}\right)_{kj}\kappa_1-\left(\mathrm{d}\mathcal{A}^{(2)}_{\mathcal{U}}\right)_{kj}\kappa_2-\left(\mathrm{d}\mathcal{A}^{(3)}_{\mathcal{U}}\right)_{kj}\kappa_3 \tag{7.598d}$$

for any 2×2 block of $\mathrm{d}\mathcal{A}_{\mathcal{U}}$ indexed by $j,k=1,\cdots,N$ in terms of the four complex-valued numbers

$$\left(\mathrm{d}\mathcal{A}^{(0)}_{\mathcal{U}}\right)_{jk},\left(\mathrm{d}\mathcal{A}^{(1)}_{\mathcal{U}}\right)_{jk},\left(\mathrm{d}\mathcal{A}^{(2)}_{\mathcal{U}}\right)_{jk},\left(\mathrm{d}\mathcal{A}^{(3)}_{\mathcal{U}}\right)_{jk}\in\mathbb{C}. \tag{7.598e}$$

Show that the fact that $\mathrm{d}\mathcal{A}_{\mathcal{U}}$ is anti-Hermitean implies the relations

$$\begin{aligned}
&\left(\mathrm{d}\mathcal{A}^{(0)}_{\mathcal{U}}\right)_{jk}=-\left(\mathrm{d}\mathcal{A}^{(0)}_{\mathcal{U}}\right)^{*}_{kj}\in\mathbb{C}, &&\left(\mathrm{d}\mathcal{A}^{(1)}_{\mathcal{U}}\right)_{jk}=-\left(\mathrm{d}\mathcal{A}^{(1)}_{\mathcal{U}}\right)^{*}_{kj}\in\mathbb{C},\\
&\left(\mathrm{d}\mathcal{A}^{(2)}_{\mathcal{U}}\right)_{jk}=-\left(\mathrm{d}\mathcal{A}^{(2)}_{\mathcal{U}}\right)^{*}_{kj}\in\mathbb{C}, &&\left(\mathrm{d}\mathcal{A}^{(3)}_{\mathcal{U}}\right)_{jk}=-\left(\mathrm{d}\mathcal{A}^{(3)}_{\mathcal{U}}\right)^{*}_{kj}\in\mathbb{C}
\end{aligned} \tag{7.598f}$$

for any $j,k=1,\cdots,N$. In particular, the N constraints

$$\begin{aligned}
&\mathrm{Re}\left(\mathrm{d}\mathcal{A}^{(0)}_{\mathcal{U}}\right)_{jj}=0, &&\mathrm{Re}\left(\mathrm{d}\mathcal{A}^{(1)}_{\mathcal{U}}\right)_{jj}=0,\\
&\mathrm{Re}\left(\mathrm{d}\mathcal{A}^{(2)}_{\mathcal{U}}\right)_{jj}=0, &&\mathrm{Re}\left(\mathrm{d}\mathcal{A}^{(3)}_{\mathcal{U}}\right)_{jj}=0
\end{aligned} \tag{7.598g}$$

follow for the N diagonal 2×2 blocks entering $(\mathrm{d}\mathcal{A}_{\mathcal{U}})_{jj}$ with $j=1,\cdots,N$. Moreover,

$$(\mathrm{d}\mathcal{A}_{\mathcal{V}})_{jk}=-\left(\mathrm{d}\mathcal{A}^{(0)}_{\mathcal{U}}\right)^{*}_{jk}\kappa_0+\left(\mathrm{d}\mathcal{A}^{(1)}_{\mathcal{U}}\right)^{*}_{jk}\kappa_1+\left(\mathrm{d}\mathcal{A}^{(2)}_{\mathcal{U}}\right)^{*}_{jk}\kappa_2+\left(\mathrm{d}\mathcal{A}^{(3)}_{\mathcal{U}}\right)^{*}_{jk}\kappa_3 \tag{7.598h}$$

must hold for $j,k=1,\cdots,N$. Verify that, for any $1\le j<k\le N$,

$$\left(\mathcal{U}^{\dagger}\,\mathrm{d}\mathcal{B}\,\mathcal{V}^{\dagger}\right)_{jj}=\left[\mathrm{d}\varepsilon_j+\mathrm{i}(2\varepsilon_j)\left(\mathrm{Im}\,\mathrm{d}\mathcal{A}^{(0)}_{\mathcal{U}}\right)_{jj}\right]\kappa_0, \tag{7.598i}$$

$$\begin{aligned}
\left(\mathcal{U}^{\dagger}\,\mathrm{d}\mathcal{B}\,\mathcal{V}^{\dagger}\right)_{jk}=&\left(\mathrm{Re}\,\mathrm{d}\mathcal{A}^{(0)}_{\mathcal{U}}\right)_{jk}(\varepsilon_k-\varepsilon_j)\,\kappa_0+\mathrm{i}\left[\left(\mathrm{Im}\,\mathrm{d}\mathcal{A}^{(0)}_{\mathcal{U}}\right)_{jk}(\varepsilon_k+\varepsilon_j)\right]\kappa_0\\
&+\left(\mathrm{Re}\,\mathrm{d}\mathcal{A}^{(1)}_{\mathcal{U}}\right)_{jk}(\varepsilon_k+\varepsilon_j)\,\kappa_1+\mathrm{i}\left[\left(\mathrm{Im}\,\mathrm{d}\mathcal{A}^{(1)}_{\mathcal{U}}\right)_{jk}(\varepsilon_k-\varepsilon_j)\right]\kappa_1\\
&+\left(\mathrm{Re}\,\mathrm{d}\mathcal{A}^{(2)}_{\mathcal{U}}\right)_{jk}(\varepsilon_k+\varepsilon_j)\,\kappa_2+\mathrm{i}\left[\left(\mathrm{Im}\,\mathrm{d}\mathcal{A}^{(2)}_{\mathcal{U}}\right)_{jk}(\varepsilon_k-\varepsilon_j)\right]\kappa_2\\
&+\left(\mathrm{Re}\,\mathrm{d}\mathcal{A}^{(3)}_{\mathcal{U}}\right)_{jk}(\varepsilon_k+\varepsilon_j)\,\kappa_3+\mathrm{i}\left[\left(\mathrm{Im}\,\mathrm{d}\mathcal{A}^{(3)}_{\mathcal{U}}\right)_{jk}(\varepsilon_k-\varepsilon_j)\right]\kappa_3.
\end{aligned} \tag{7.598j}$$

Verify that [we are using the notation (7.510)]

$$\begin{aligned}
&\prod_{j=1}^{N} \mathrm{d}\left(\mathrm{Im}\,\mathcal{A}^{(0)}_{\mathcal{U}}\right)_{jj} \mathrm{d}\left(\mathrm{Im}\,\mathcal{A}^{(1)}_{\mathcal{U}}\right)_{jj} \mathrm{d}\left(\mathrm{Im}\,\mathcal{A}^{(2)}_{\mathcal{U}}\right)_{jj} \mathrm{d}\left(\mathrm{Im}\,\mathcal{A}^{(3)}_{\mathcal{U}}\right)_{jj} \\
&\times \prod_{1\leq j<k\leq N} \mathrm{d}\left(\mathrm{Re}\,\mathcal{A}^{(0)}_{\mathcal{U}}\right)_{jk} \mathrm{d}\left(\mathrm{Re}\,\mathcal{A}^{(1)}_{\mathcal{U}}\right)_{jk} \mathrm{d}\left(\mathrm{Re}\,\mathcal{A}^{(2)}_{\mathcal{U}}\right)_{jk} \mathrm{d}\left(\mathrm{Re}\,\mathcal{A}^{(3)}_{\mathcal{U}}\right)_{jk} \\
&\times \prod_{1\leq j<k\leq N} \mathrm{d}\left(\mathrm{Im}\,\mathcal{A}^{(0)}_{\mathcal{U}}\right)_{jk} \mathrm{d}\left(\mathrm{Im}\,\mathcal{A}^{(1)}_{\mathcal{U}}\right)_{jk} \mathrm{d}\left(\mathrm{Im}\,\mathcal{A}^{(2)}_{\mathcal{U}}\right)_{jk} \mathrm{d}\left(\mathrm{Im}\,\mathcal{A}^{(3)}_{\mathcal{U}}\right)_{jk}
\end{aligned} \tag{7.599}$$

is a Haar measure for U(2N). Verify that the counterpart to Eq. (7.559) is (i)

$$\mu(\mathrm{d}\,(\mathcal{U}\,\mathcal{E}\,\mathcal{V})\,) \propto \left(\prod_{j=1}^{N} \mathrm{d}\varepsilon_j\, |2\varepsilon_j|\right) \left(\prod_{1\leq j<k\leq N} \left|\varepsilon^2_j - \varepsilon^2_k\right|^4\right) \mu(\mathrm{d}\mathcal{A}) \tag{7.600a}$$

and (ii)

$$\mu(\mathrm{d}\mathcal{H}_{\mathrm{DIII}}) \propto \mu(\mathrm{d}\,(\mathcal{U}\,\mathcal{E}\,\mathcal{V})\,), \tag{7.600b}$$

whereby (iii) the measure

$$\begin{aligned}
&\mu(\mathrm{d}\mathcal{A}) := \left[\prod_{j=1}^{N} \mathrm{d}\left(\mathrm{Im}\,\mathcal{A}^{(0)}\right)_{jj}\right] \left[\prod_{1\leq j<k\leq N} \prod_{a=0}^{3} \mathrm{d}\left(\mathrm{Re}\,\mathcal{A}^{(a)}\right)_{jk} \mathrm{d}\left(\mathrm{Im}\,\mathcal{A}^{(a)}\right)_{jk}\right], \\
&\mathrm{d}\mathcal{A} := \mathcal{U}^{\dagger}\,\mathrm{d}\mathcal{U}
\end{aligned} \tag{7.600c}$$

is a Haar measure of U(2N)/[SU(2)]^N. We have found that the line DIII with SO(4N)/U(2N) in **Table** 7.2 corresponds to the degeneracy $d_{\mathrm{DIII}} = 2$, the level repulsion exponent $\beta_{\mathrm{DIII}} = 4$, the repulsion exponent from zero energy $\alpha_{\mathrm{DIII}} = 1$, and the number $\nu_{\mathrm{DIII}} = 0$ of zero modes. Moreover, if $\boldsymbol{e}_j$ is the Cartesian basis of $\mathbb{R}^N$, $\boldsymbol{\varepsilon} \in \mathbb{R}^N$ is the vector with the eigenvalues $\varepsilon_j \geq 0$ as components, and we define the set of "positive" roots to be

$$\begin{aligned}
&\mathfrak{Roots}_{+} := \mathfrak{Roots}_{\mathrm{s}} \cup \mathfrak{Roots}_{\mathrm{o}} \cup \mathfrak{Roots}_{\mathrm{l}}, \\
&\mathfrak{Roots}_{\mathrm{s}} := \{\boldsymbol{e}_j \mid 1 \leq j \leq N\}, \\
&\mathfrak{Roots}_{\mathrm{o}} := \{\boldsymbol{e}_k - \boldsymbol{e}_j, \boldsymbol{e}_k + \boldsymbol{e}_j \mid 1 \leq j < k \leq N\}, \\
&\mathfrak{Roots}_{\mathrm{l}} := \{2\boldsymbol{e}_j \mid 1 \leq j \leq N\},
\end{aligned} \tag{7.601a}$$

we have then constructed the Jacobian

$$\begin{aligned}
&J(\boldsymbol{\varepsilon}) := \prod_{\boldsymbol{\alpha}\in\mathfrak{Roots}_{\mathrm{s}}} |\boldsymbol{\alpha}\cdot\boldsymbol{\varepsilon}|^{m_{\mathrm{s}}} \prod_{\boldsymbol{\alpha}\in\mathfrak{Roots}_{\mathrm{o}}} |\boldsymbol{\alpha}\cdot\boldsymbol{\varepsilon}|^{m_{\mathrm{o}}} \prod_{\boldsymbol{\alpha}\in\mathfrak{Roots}_{\mathrm{l}}} |\boldsymbol{\alpha}\cdot\boldsymbol{\varepsilon}|^{m_{\mathrm{l}}}, \\
&m_{\mathrm{s}} = 0, \qquad m_{\mathrm{o}} = 4, \qquad m_{\mathrm{l}} = 1.
\end{aligned} \tag{7.601b}$$

The exponent m_{s} is called the multiplicity of the short root $\boldsymbol{e}_j$. The exponent m_{o} is called the multiplicity of the ordinary root $\boldsymbol{e}_k \pm \boldsymbol{e}_j$. The exponent m_{l} is called the multiplicity of the long root $2\boldsymbol{e}_j$.

Case $M = 2N+1$: The $(2N+1)\times(2N+1)$ antisymmetric matrix $\mathcal{B}$ in Eq. (7.591) must necessarily have one eigenvalue that vanishes. Verify that the spectral decomposition (7.594) becomes

$$\mathcal{B} = \mathcal{U}\,\mathcal{E}\,\mathcal{V}, \qquad \mathcal{E} := (\mathsf{E}\otimes\kappa_0)\oplus\mathrm{diag}(0), \tag{7.602a}$$

whereby

$$\mathcal{U}\in\mathrm{U}(2N+1), \qquad \mathcal{V} = \left\{\mathcal{U}\,\left[(-\mathbb{1}\otimes\mathrm{i}\kappa_2)\oplus\mathrm{diag}(1)\right]\right\}^{\mathsf{T}}\in\mathrm{U}(2N+1), \tag{7.602b}$$

and the diagonal matrix

$$\mathsf{E} = \mathrm{diag}\,\begin{pmatrix}\varepsilon_1 & \cdots & \varepsilon_N\end{pmatrix} \tag{7.602c}$$

is built out of the nonnegative square roots $\varepsilon_1,\cdots,\varepsilon_N\geq 0$ of the nonnegative eigenvalues of the positive semidefinite $(2N+1)\times(2N+1)$ Hermitean matrix $\mathcal{B}^\dagger\,\mathcal{B}$.

The Haar measure of $\mathrm{U}(2N+1)$ is defined out of the

$$(2N+1)^2 = 4N^2+4N+1 \tag{7.603a}$$

independent real-valued matrix elements of

$$\mathrm{d}\mathcal{A}_{\mathcal{U}} := \mathcal{U}^\dagger\,\mathrm{d}\mathcal{U} = -\left(\mathrm{d}\mathcal{A}_{\mathcal{U}}\right)^\dagger \tag{7.603b}$$

to be [we are using the notation (7.510)]

$$\begin{aligned}
&\prod_{j=1}^{N}\mathrm{d}\left(\mathrm{Im}\,\mathcal{A}^{(0)}_{\mathcal{U}}\right)_{jj}\mathrm{d}\left(\mathrm{Im}\,\mathcal{A}^{(1)}_{\mathcal{U}}\right)_{jj}\mathrm{d}\left(\mathrm{Im}\,\mathcal{A}^{(2)}_{\mathcal{U}}\right)_{jj}\mathrm{d}\left(\mathrm{Im}\,\mathcal{A}^{(3)}_{\mathcal{U}}\right)_{jj}\\
&\times\prod_{1\leq j<k\leq N}\mathrm{d}\left(\mathrm{Re}\,\mathcal{A}^{(0)}_{\mathcal{U}}\right)_{jk}\mathrm{d}\left(\mathrm{Re}\,\mathcal{A}^{(1)}_{\mathcal{U}}\right)_{jk}\mathrm{d}\left(\mathrm{Re}\,\mathcal{A}^{(2)}_{\mathcal{U}}\right)_{jk}\mathrm{d}\left(\mathrm{Re}\,\mathcal{A}^{(3)}_{\mathcal{U}}\right)_{jk}\\
&\times\prod_{1\leq j<k\leq N}\mathrm{d}\left(\mathrm{Im}\,\mathcal{A}^{(0)}_{\mathcal{U}}\right)_{jk}\mathrm{d}\left(\mathrm{Im}\,\mathcal{A}^{(1)}_{\mathcal{U}}\right)_{jk}\mathrm{d}\left(\mathrm{Im}\,\mathcal{A}^{(2)}_{\mathcal{U}}\right)_{jk}\mathrm{d}\left(\mathrm{Im}\,\mathcal{A}^{(3)}_{\mathcal{U}}\right)_{jk}\\
&\times\mathrm{d}\left(\mathrm{Im}\,\mathcal{A}_{\mathcal{U}}\right)_{(2N+1)(2N+1)}\prod_{\mathrm{j}=1}^{2N}\mathrm{d}\left(\mathrm{Re}\,\mathcal{A}_{\mathcal{U}}\right)_{\mathrm{j}(2N+1)}\,\mathrm{d}\left(\mathrm{Im}\,\mathcal{A}_{\mathcal{U}}\right)_{\mathrm{j}(2N+1)}
\end{aligned} \tag{7.603c}$$

in the block representation

$$\mathrm{d}\mathcal{A}_{\mathcal{U}} = \begin{pmatrix}\left(\sum\limits_{a=0}^{3}\left(\mathrm{d}\mathcal{A}^{(a)}_{\mathcal{U}}\right)_{jk}\kappa_a\right) & \left(\mathrm{d}\,\mathcal{A}_{\mathcal{U}}\right)_{\mathrm{j}(2N+1)}\\ -\left(\mathrm{d}\,\mathcal{A}_{\mathcal{U}}\right)^{*}_{\mathrm{j}(2N+1)} & \mathrm{Im}\,\left(\mathrm{d}\,\mathcal{A}_{\mathcal{U}}\right)_{(2N+1)(2N+1)}\end{pmatrix}. \tag{7.603d}$$

The symmetric space

$$\mathrm{U}(2N+1)/\{[\mathrm{SU}(2)]^N \otimes \mathrm{U}(1)\} \tag{7.604a}$$

is of dimensionality

$$(2N+1)^2 - 3N - 1 = 4N^2 + N \tag{7.604b}$$

and has the Haar measure

$$\begin{aligned}
&\prod_{j=1}^{N} \mathrm{d}\left(\mathrm{Im}\,\mathcal{A}^{(0)}_{\mathcal{U}}\right)_{jj} \\
&\times \prod_{1\le j<k\le N} \mathrm{d}\left(\mathrm{Re}\,\mathcal{A}^{(0)}_{\mathcal{U}}\right)_{jk} \mathrm{d}\left(\mathrm{Re}\,\mathcal{A}^{(1)}_{\mathcal{U}}\right)_{jk} \mathrm{d}\left(\mathrm{Re}\,\mathcal{A}^{(2)}_{\mathcal{U}}\right)_{jk} \mathrm{d}\left(\mathrm{Re}\,\mathcal{A}^{(3)}_{\mathcal{U}}\right)_{jk} \\
&\times \prod_{1\le j<k\le N} \mathrm{d}\left(\mathrm{Im}\,\mathcal{A}^{(0)}_{\mathcal{U}}\right)_{jk} \mathrm{d}\left(\mathrm{Im}\,\mathcal{A}^{(1)}_{\mathcal{U}}\right)_{jk} \mathrm{d}\left(\mathrm{Im}\,\mathcal{A}^{(2)}_{\mathcal{U}}\right)_{jk} \mathrm{d}\left(\mathrm{Im}\,\mathcal{A}^{(3)}_{\mathcal{U}}\right)_{jk} \\
&\times \prod_{\mathrm{j}=1}^{2N} \mathrm{d}\left(\mathrm{Re}\,\mathcal{A}_{\mathcal{U}}\right)_{\mathrm{j}(2N+1)} \left(\mathrm{Im}\,\mathcal{A}_{\mathcal{U}}\right)_{\mathrm{j}(2N+1)} .
\end{aligned} \tag{7.604c}$$

Verify that the counterpart to Eq. (7.569) is (i)

$$\mu(\mathrm{d}\,(\mathcal{U}\,\mathcal{E}\,\mathcal{V})) \propto \left(\prod_{j=1}^{N} \mathrm{d}\varepsilon_j\, |2\varepsilon_j||\varepsilon_j|^4\right) \left(\prod_{1\le j<k\le N} |\varepsilon_j^2 - \varepsilon_k^2|^4\right) \mu(\mathrm{d}\mathcal{A}) \tag{7.605a}$$

and (ii)

$$\mu(\mathrm{d}\mathcal{H}_{\mathrm{DIII}}) \propto \mu(\mathrm{d}\,(\mathcal{U}\,\mathcal{E}\,\mathcal{V})), \tag{7.605b}$$

whereby (iii) the measure

$$\begin{aligned}
\mu(\mathrm{d}\mathcal{A}) :=& \left[\prod_{j=1}^{N} \mathrm{d}\left(\mathrm{Im}\,\mathcal{A}^{(0)}\right)_{jj}\right] \\
&\times \left[\prod_{1\le j<k\le N} \prod_{a=0}^{3} \mathrm{d}\left(\mathrm{Re}\,\mathcal{A}^{(a)}\right)_{jk} \mathrm{d}\left(\mathrm{Im}\,\mathcal{A}^{(a)}\right)_{jk}\right] \\
&\times \prod_{\mathrm{j}=1}^{2N} \mathrm{d}\left(\mathrm{Re}\,\mathcal{A}\right)_{\mathrm{j}(2N+1)} \left(\mathrm{Im}\,\mathcal{A}\right)_{\mathrm{j}(2N+1)}, \qquad \mathrm{d}\mathcal{A} := \mathcal{U}^{\dagger}\, \mathrm{d}\mathcal{U}
\end{aligned} \tag{7.605c}$$

is a Haar measure of $\mathrm{U}(2N+1)/\{[\mathrm{SU}(2)]^N \otimes \mathrm{U}(1)\}$. We have found that the line DIII with $\mathrm{SO}(4N+2)/\mathrm{U}(2N+1)$ in **Table** 7.2 corresponds to the degeneracy $d_{\mathrm{DIII}} = 2$, the level repulsion exponent $\beta_{\mathrm{DIII}} = 4$, the repulsion exponent from zero energy $\alpha_{\mathrm{DIII}} = 1$, and the number $\nu_{\mathrm{DIII}} = 1$

with $d_{\mathrm{DIII}}\,\nu_{\mathrm{DIII}} = 2$ the number of zero modes. Moreover, if $\boldsymbol{e}_j$ is the Cartesian basis of $\mathbb{R}^N$, $\boldsymbol{\varepsilon} \in \mathbb{R}^N$ is the vector with the eigenvalues $\varepsilon_j \geq 0$ as components, and we define the set of "positive" roots to be

$$\begin{aligned}
\mathfrak{Roots}_+ &:= \mathfrak{Roots}_{\mathrm{s}} \cup \mathfrak{Roots}_{\mathrm{o}} \cup \mathfrak{Roots}_{\mathrm{l}},\\
\mathfrak{Roots}_{\mathrm{s}} &:= \{\boldsymbol{e}_j \mid 1 \leq j \leq N\},\\
\mathfrak{Roots}_{\mathrm{o}} &:= \{\boldsymbol{e}_k - \boldsymbol{e}_j, \boldsymbol{e}_k + \boldsymbol{e}_j \mid 1 \leq j < k \leq N\},\\
\mathfrak{Roots}_{\mathrm{l}} &:= \{2\boldsymbol{e}_j \mid 1 \leq j \leq N\},
\end{aligned} \tag{7.606a}$$

we have then constructed the Jacobian

$$\begin{aligned}
J(\boldsymbol{\varepsilon}) &:= \prod_{\boldsymbol{\alpha}\in\mathfrak{Roots}_{\mathrm{s}}} |\boldsymbol{\alpha}\cdot\boldsymbol{\varepsilon}|^{m_{\mathrm{s}}} \prod_{\boldsymbol{\alpha}\in\mathfrak{Roots}_{\mathrm{o}}} |\boldsymbol{\alpha}\cdot\boldsymbol{\varepsilon}|^{m_{\mathrm{o}}} \prod_{\boldsymbol{\alpha}\in\mathfrak{Roots}_{\mathrm{l}}} |\boldsymbol{\alpha}\cdot\boldsymbol{\varepsilon}|^{m_{\mathrm{l}}},\\
m_{\mathrm{s}} &= 4, \qquad m_{\mathrm{o}} = 4, \qquad m_{\mathrm{l}} = 1.
\end{aligned} \tag{7.606b}$$

The exponent m_{s} is called the multiplicity of the short root $\boldsymbol{e}_j$. The exponent m_{o} is called the multiplicity of the ordinary root $\boldsymbol{e}_k \pm \boldsymbol{e}_j$. The exponent m_{l} is called the multiplicity of the long root $2\boldsymbol{e}_j$.

7.15 We are going to prove the following theorem for matrices.

Definition 7.13 (singular values of a matrix) Let M, N be any pair of non-negative integers. Let A be any $M \times N$ complex-valued matrix with at least one nonvanishing matrix element. As we have shown with Eq. (3.139), the Hermitean matrices $A^{\dagger}\,A$ and $A\,A^{\dagger}$ share the same real-valued, nonvanishing, and positive eigenvalues $\sigma_1^2 > 0$, $\sigma_2^2 > 0$, $\cdots$, $\sigma_r^2 > 0$, with $1 \leq r \leq \min(M, N)$. We define their positive roots to be $\sigma_1 > 0$, $\sigma_2 > 0$, $\cdots$, $\sigma_r > 0$, respectively. The *singular values of the matrix* A are the r real-valued numbers $\sigma_1 > 0$, $\sigma_2 > 0$, $\cdots$, $\sigma_r > 0$. The integer r defines the *rank of the matrix* A, $\mathrm{rank}(A) = r$.

7.15.1 Prove the following theorem.

Theorem 7.14 (Singular-value decomposition of a matrix) *Let A be any $M \times N$ complex-valued matrix with the singular values $\sigma_1, \sigma_2, \cdots, \sigma_r$, where $1 \leq r \leq \min(M, N)$. Then, there exists the pair of unitary matrices $U \in \mathrm{U}(M)$ and $U \in \mathrm{U}(N)$ such that*

$$A = U\,D\,V, \tag{7.607a}$$

where

$$D := \mathrm{diag}\begin{pmatrix}\sigma_1 & \cdots & \sigma_r & 0 & \cdots & 0\end{pmatrix} \tag{7.607b}$$

is a diagonal $M \times N$ matrix of rank r.

Proof

7.15.1.1 *Step 1:* Let $M = N = 1$ and $A = (a)$ with $a \in \mathbb{C}$. Verify that $U = (1)$, $D = (|a|)$, and $V = e^{\mathrm{i}\arg(a)}$.

7.15.1.2 *Step 2:* Let $M=1$, $N>1$, and $A=(\boldsymbol{a})$ with $\boldsymbol{a}:=\begin{pmatrix}a_1 & \cdots & a_N\end{pmatrix}\in\mathbb{C}^N$. Verify that $U=(1)$, $D=\begin{pmatrix}\sigma_1 & 0 & \cdots & 0\end{pmatrix}$ with $\sigma_1:=+\sqrt{\boldsymbol{a}^*\cdot\boldsymbol{a}}$, and $V\equiv(V_{ij}):=(\boldsymbol{e}^i_j)$ with $\boldsymbol{e}^1\equiv\frac{1}{\sigma_1}\boldsymbol{a}$ the first element of an orthonormal basis of $\mathbb{C}^N$, that is, $\boldsymbol{e}^{k*}\cdot\boldsymbol{e}^l\equiv\sum_{j=1}^{N}\boldsymbol{e}^{k*}_j\,\boldsymbol{e}^l_j=\delta^{kl}$, $k,l=1,\cdots,N$.

7.15.1.3 *Step 3:* Let $M>1$, $N=1$, and $A=(\boldsymbol{a}^{\mathsf{T}})$ with $\boldsymbol{a}:=\begin{pmatrix}a_1 & \cdots & a_M\end{pmatrix}\in\mathbb{C}^M$. Verify that $V=(1)$, $D^{\mathsf{T}}=\begin{pmatrix}\sigma_1 & 0 & \cdots & 0\end{pmatrix}$ with $\sigma_1:=+\sqrt{\boldsymbol{a}^*\cdot\boldsymbol{a}}$, and $U\equiv(U_{ij}):=(\boldsymbol{e}^j_i)$ with $\boldsymbol{e}^1\equiv\frac{1}{\sigma_1}\boldsymbol{a}$ the first element of an orthonormal basis of $\mathbb{C}^M$, that is, $\boldsymbol{e}^{k*}\cdot\boldsymbol{e}^l\equiv\sum_{j=1}^{M}\boldsymbol{e}^{k*}_j\,\boldsymbol{e}^l_j=\delta^{kl}$, $k,l=1,\cdots,M$.

7.15.1.4 *Step 4:* Let $M>1$, $N>1$, and $A=(A_{ij})\neq 0$ with $i=1,\cdots,M$ and $j=1,\cdots,N$. There exists at least one unit column vector $\boldsymbol{u}_1\in\mathbb{C}^N$ and one positive number $\sigma_1^2>0$ such that

$$A^\dagger\,A\,\boldsymbol{u}_1=\sigma_1^2\,\boldsymbol{u}_1,\qquad \boldsymbol{u}_1^\dagger\,A^\dagger\,A=\sigma_1^2\,\boldsymbol{u}_1^\dagger\qquad \boldsymbol{u}_1^\dagger\,\boldsymbol{u}_1=1. \tag{7.608}$$

Verify that

$$\boldsymbol{v}_1:=\frac{1}{\sigma_1}A\,\boldsymbol{u}_1,\iff \boldsymbol{u}_1^\dagger\,A^\dagger=:\sigma_1\,\boldsymbol{v}_1^\dagger \tag{7.609}$$

is a unit column vector in $\mathbb{C}^M$ obeying

$$A^\dagger\,\boldsymbol{v}_1=\sigma_1\,\boldsymbol{u}_1,\qquad \boldsymbol{u}_1^\dagger\,A^\dagger\,\boldsymbol{v}_1=\sigma_1. \tag{7.610}$$

Verify that for any $P\in\mathrm{U}(N)$ with $\boldsymbol{u}_1$ as the first column and for any $Q\in\mathrm{U}(M)$ with $\boldsymbol{v}_1$ as the first column there exists a $(N-1)\times(M-1)$ complex-valued matrix $B^\dagger$ such that

$$P^\dagger\,A^\dagger\,Q=\begin{pmatrix}\sigma_1 & 0\\ 0 & B^\dagger\end{pmatrix}\iff A=Q\begin{pmatrix}\sigma_1 & 0\\ 0 & B\end{pmatrix}P^\dagger. \tag{7.611}$$

7.15.1.5 *Step 5:* Complete the proof by induction.

□

7.15.2 Show that the singular-value decomposition (7.607) is not unique and prove the following theorem.

Theorem 7.15 (Polar decomposition) *For any complex-valued $N\times N$ matrix A, there exist a positive semidefinite Hermitean complex-valued $N\times N$ matrix P and a unitary complex-valued $N\times N$ matrix W such that*

$$A=P\,W. \tag{7.612}$$

Hint: The $N\times N$ matrix P is positive semidefinite if and only if

$$\boldsymbol{v}^\dagger\,P\,\boldsymbol{v}\geq 0 \tag{7.613}$$

for any column vector $\boldsymbol{v} \in \mathbb{C}^N$. By choosing $\boldsymbol{v}$ to be the column vector whose i-th component is 1 with all other components vanishing, show that $P_{ii} \geq 0$. By choosing $\boldsymbol{v}$ to be the column vector whose i-th component is 1, whose j-th component is 1, with all other components vanishing, show that $\mathrm{Im}\, P_{ij} = -\mathrm{Im}\, P_{ji}$. By choosing $\boldsymbol{v}$ to be the column vector whose i-th component is 1, whose j-th component is i, with all other components vanishing, show that $\mathrm{Re}\, P_{ij} = +\mathrm{Re}\, P_{ji}$. Deduce that P must be Hermitean and that any eigenvalue of P must be nonnegative. Insert $V = U^\dagger U V$ in the singular-value decomposition (7.607).

7.16 First, show that Eqs. (7.179) and

$$U\, \mathrm{d}H\, V = U'\, \mathrm{d}H'\, V' \tag{7.614}$$

imply

$$\mathrm{d}H \mapsto \mathrm{d}H', \qquad \mathrm{d}H' = U'^\dagger\, U\, \mathrm{d}H\, V\, V'^\dagger. \tag{7.615}$$

Second, let X be any $N \times N$ matrix with matrix elements X_{ij}. Let U and V be any two $N \times N$ matrices with $|\det U| = |\det V| = 1$. Define the transformation

$$X' := U\, X\, V. \tag{7.616}$$

We are going to show that the Jacobian

$$J := \left| \det \left(\frac{\partial X'}{\partial X} \right) \right| = 1. \tag{7.617}$$

Proof We start from

$$X'_{ij} = \sum_{k,l=1}^{N} U_{ik}\, X_{kl}\, V_{lj}, \qquad i,j = 1,\cdots,N, \tag{7.618}$$

which we rewrite as

$$X'_{ij} = \sum_{k,l=1}^{N} U_{ik}\, \left(V^{\mathsf{T}}\right)_{jl}\, X_{kl}, \qquad i,j = 1,\cdots,N. \tag{7.619}$$

Definition 7.16 (Kronecker product of two matrices) Define the Kronecker product of the $m \times m$ matrix A with the $n \times n$ matrix B to be

$$(A \times B)_{ij,kl} := A_{ik} B_{jl}, \quad i,k = 1,\cdots,m, \quad j,l = 1,\cdots,n. \tag{7.620a}$$

It can be represented by a $mn \times mn$ block matrix of the form

$$A \times B = \begin{pmatrix} A_{11}B & \cdots & A_{1m}B \\ \vdots & & \vdots \\ A_{m1}B & \cdots & A_{mm}B \end{pmatrix}, \tag{7.620b}$$

with $i,k=1,\cdots,m$ indexing the blocks and $j,l=1,\cdots,n$ indexing the components in the block.

Example 7.17 Equation (7.619) becomes

$$X'=\left(U\times V^{\mathsf{T}}\right)\,X \tag{7.621}$$

by making use of the Kronecker product between U and V^{T}.

Verify the matrix identities

$$(A\times B)\,(A'\times B')=(A\,A')\times(B\,B'), \tag{7.622a}$$

$$\det(A\times B)\,\det(A'\times B')=\det\Big((A\,A')\times(B\,B')\Big) \tag{7.622b}$$

for any pair A and A' of $m\times m$ matrices and for any pair B and B' of $n\times n$ matrices.

Verify that the substitutions

$$A\to U,\qquad A'\to\mathbb{1}_N,\qquad B\to\mathbb{1}_N,\qquad B'\to V^{\mathsf{T}}, \tag{7.623}$$

imply that

$$\det\left(U\times V^{\mathsf{T}}\right)=[\det(U)]^N\left[\det\left(V^{\mathsf{T}}\right)\right]^N. \tag{7.624}$$

Convince yourself with the help of Eq. (7.619) that

$$\det\left(\frac{\partial X'}{\partial X}\right)=\det\left(U\times V^{\mathsf{T}}\right). \tag{7.625}$$

Deduce that

$$\begin{aligned}J:=&\left|\det\left(\frac{\partial X'}{\partial X}\right)\right|\\=&\left|[\det(U)]^N\left[\det\left(V^{\mathsf{T}}\right)\right]^N\right|\\=&\,1.\end{aligned} \tag{7.626}$$

□

7.17 We are going to derive the representation (7.182) of the partition function (7.180) for the circular ensemble in any one of the ten Altland-Zirnbauer symmetry classes.

We start from the symmetric space M, an element S of which is a certain unitary $N_{\mathrm{M}}\times N_{\mathrm{M}}$ matrix. As a unitary matrix, S can be diagonalized by a unitary matrix W into a diagonal unitary matrix E:

$$\begin{aligned}&S=W\,E\,W^{\dagger},\qquad S,W\in\mathrm{U}(N_{\mathrm{M}}),\\&E=e^{\mathrm{i}\Theta},\qquad \Theta=\Theta^{\dagger}=\mathrm{diag}\left(\theta_1\quad\cdots\quad\theta_{N_{\mathrm{M}}}\right),\\&0\leq\theta_1,\cdots,\theta_{N_{\mathrm{M}}}<2\pi.\end{aligned} \tag{7.627}$$

This is a spectral representation of the matrix S. It is not unique.

7.17.1 *Step 1:* Apply the product rule for differentiation to evaluate $\mathrm{d}\left(W\,E\,W^{\dagger}\right)$ and $\mathrm{d}\left(W\,W^{\dagger}\right)$. Show that

$$\mathrm{d}S = W\left(\mathrm{i}E\,\mathrm{d}\Theta + [\mathrm{d}A, E]\right)W^{\dagger}, \tag{7.628a}$$

where

$$\mathrm{d}A := W^{\dagger}\,\mathrm{d}W = -\left(\mathrm{d}A\right)^{\dagger} \tag{7.628b}$$

is an infinitesimal anti-Hermitean matrix.

7.17.2 *Step 2:* Define the unitary and diagonal matrix F by

$$F^2 := E. \tag{7.629a}$$

Show that

$$F = \mathrm{diag}\left(\sigma_1\,e^{\mathrm{i}\theta_1/2} \quad \cdots \quad \sigma_{N_{\mathrm{M}}}\,e^{\mathrm{i}\theta_{N_{\mathrm{M}}}/2}\right) \tag{7.629b}$$

and verify that the sign ambiguity $\sigma_i = \pm 1$ does not matter in all the following manipulations. We will choose $\sigma_i = 1$ for all $i = 1, \cdots, N_{\mathrm{M}}$. Show that one can choose

$$U = W\,F, \qquad V = F\,W^{\dagger}, \tag{7.630a}$$

in the factorization

$$S = U\,V, \qquad U^{-1} = U^{\dagger}, \qquad V^{-1} = V^{\dagger}. \tag{7.630b}$$

With this choice for U and V and with the definition

$$\mathrm{d}S := U\,\mathrm{i}\mathrm{d}H\,V, \qquad \mathrm{d}H = \left(\mathrm{d}H\right)^{\dagger}, \tag{7.631a}$$

for some infinitesimal Hermitean matrix $\mathrm{d}H$, show that

$$W^{\dagger}\,\mathrm{d}S\,W = F\,\mathrm{i}\mathrm{d}H\,F \tag{7.631b}$$

with

$$F\,\mathrm{i}\mathrm{d}H\,F = F^2\,\mathrm{i}\mathrm{d}\Theta + \left[\mathrm{d}A, F^2\right], \qquad \mathrm{d}A = W^{\dagger}\,\mathrm{d}W. \tag{7.631c}$$

Finally, show that

$$\mathrm{d}H = \mathrm{d}\Theta + F\,\mathrm{i}\mathrm{d}A\,F^{\dagger} - F^{\dagger}\,\mathrm{i}\mathrm{d}A\,F, \qquad \mathrm{d}A = W^{\dagger}\,\mathrm{d}W. \tag{7.632}$$

From now on, the manifold M will be any one of the ten manifolds in **Table** 7.2, while the integer N in what follows is the corresponding N_{AZ} from **Table** 7.2 for the selected Altland-Zirnbauer symmetry class.

7.17.3 *Step 3:* We consider first the standard symmetry classes A, AI, and AII. Verify that the components of the matrix equation (7.632) read

$$\mathrm{d}H_{jk} = \mathrm{d}\theta_j\,\delta_{jk} + 2\sin\left(\frac{\theta_k - \theta_j}{2}\right)\mathrm{d}A_{jk}, \tag{7.633}$$

with

$$E = F^2 = \operatorname{diag}\left(e^{+\mathrm{i}\theta_1} \quad \cdots \quad e^{+\mathrm{i}\theta_N}\right) \tag{7.634}$$

for symmetry class A and AI, while

$$E = F^2 = \operatorname{diag}\left(e^{+\mathrm{i}\theta_1} \quad \cdots \quad e^{+\mathrm{i}\theta_N}\right) \otimes \sigma_0 \tag{7.635}$$

for symmetry class AII. One is then left with enumerating the independent and real-valued matrix elements of the $N_{\mathrm{M}} \times N_{\mathrm{M}}$ anti-Hermitean matrix $\mathrm{d}A = -(\mathrm{d}A)^{\dagger}$. This number β_{AZ} is one, two, or four for any $1 \leq j < k \leq N$ as H_{jk} is either real-valued, complex-valued, or quaternion-real-valued, respectively.

7.17.4 *Step 4:* For the chiral symmetry classes AIII, BDI, and CII or for the Bogoliubov-de-Gennes symmetry classes D, C, DIII, CI,

$$E = \operatorname{diag}\left(e^{+\mathrm{i}\theta_1} \quad \cdots \quad e^{+\mathrm{i}\theta_N} \quad \middle| \quad e^{-\mathrm{i}\theta_1} \quad \cdots \quad e^{-\mathrm{i}\theta_N} \quad \middle| \quad 1 \quad \cdots \quad 1\right), \tag{7.636}$$

where the entries $1 \cdots 1$ in E are only present if there are ν zero modes (up to a Kramers' degeneracy). Hence, for any $1 \leq j < k \leq N$,

$$\mathrm{d}H_{jk} = 2\sin\left(\frac{\theta_k - \theta_j}{2}\right)\mathrm{d}A_{jk} \tag{7.637}$$

must be supplemented with

$$\mathrm{d}H_{(N+j)k} = 2\sin\left(\frac{\theta_k + \theta_j}{2}\right)\mathrm{d}A_{(N+j)k}. \tag{7.638}$$

Moreover, for any $j = 1, \cdots, N$,

$$\mathrm{d}H_{jj} = \mathrm{d}\theta_{jj} \tag{7.639}$$

must be supplemented with

$$\mathrm{d}H_{(N+j)j} = 2\sin\left(\frac{\theta_j + \theta_j}{2}\right)\mathrm{d}A_{(N+j)j} \tag{7.640}$$

and

$$\mathrm{d}H_{(N+j+\nu)j} = 2\sin\left(\frac{\theta_j + 0}{2}\right)\mathrm{d}A_{(N+j+\nu)j}. \tag{7.641}$$

One is then left with enumerating the independent and real-valued matrix elements of the $N_{\mathrm{M}} \times N_{\mathrm{M}}$ anti-Hermitean matrix $\mathrm{d}A = -(\mathrm{d}A)^{\dagger}$. This number β_{AZ} is one, two, or four for any $1 \leq j < k \leq N$ as H_{jk} is either real-valued, complex-valued, or quaternion-real-valued, respectively. This number α_{AZ} for any diagonal matrix element $\mathrm{d}A_{(N+j)j}$ is tabulated in **Table** 7.2.

7.18 Assume that the fermionic reservoirs Ω and Ω' in Figure 7.2 are held at the chemical potentials $\mu + e\,\delta V$ and μ, respectively. Fermionic single-particle states with incoming wave boundary conditions for lead Ω are fed from reservoir Ω. Hence, they are only occupied for the single-particle energies $\varepsilon < \mu + e\,\delta V$. Fermionic single-particle states with incoming wave boundary conditions for lead Ω' are fed from reservoir Ω'. Hence, they are only occupied for the single-particle energies $\varepsilon < \mu$. Assume that the current that is injected from a reservoir into the leads is $e/(2\pi\,\hbar)$ per mode and per unit of energy of the leads. Assume that for the current injected from reservoir Ω through the mode $n = 1, \cdots, M$ at the single-particle energy ε of lead Ω, the fraction $\sum_m |\mathtt{t}_{mn}(\varepsilon)|^2$ is transmitted into lead Ω', while the rest is reflected back into lead Ω. Conversely, assume that for the current injected from reservoir Ω' through the mode $n' = 1, \cdots, M' \equiv M$ at the single-particle energy ε of lead Ω', the fraction $\sum_{m'} |\mathtt{t}'_{m'n'}(\varepsilon)|^2$ is transmitted into lead Ω, while the rest is reflected back into lead Ω'.

7.18.1 Show that the current δI flowing through the region Λ in Figure 7.2 is given by

$$\delta I = \frac{e}{2\pi\hbar}\left[\int^{\mu+e\,\delta V} \mathrm{d}\varepsilon\,\mathrm{tr}\left(\mathtt{t}^{\dagger}(\varepsilon)\,\mathtt{t}(\varepsilon)\right) - \int^{\mu} \mathrm{d}\varepsilon\,\mathrm{tr}\left(\mathtt{t}'^{\dagger}(\varepsilon)\,\mathtt{t}'(\varepsilon)\right)\right] \tag{7.642a}$$

$$= \frac{e}{2\pi\hbar}\left[\int^{\mu+e\,\delta V} \mathrm{d}\varepsilon\,\mathrm{tr}\left(\mathbb{1} - \mathtt{r}^{\dagger}(\varepsilon)\,\mathtt{r}(\varepsilon)\right) - \int^{\mu} \mathrm{d}\varepsilon\,\mathrm{tr}\left(\mathtt{t}'^{\dagger}(\varepsilon)\,\mathtt{t}'(\varepsilon)\right)\right] \tag{7.642b}$$

$$= \frac{e}{2\pi\hbar}\left[\int^{\mu+e\,\delta V} \mathrm{d}\varepsilon\,\mathrm{tr}\left(\mathtt{t}^{\dagger}(\varepsilon)\,\mathtt{t}(\varepsilon)\right) - \int^{\mu} \mathrm{d}\varepsilon\,\mathrm{tr}\left(\mathbb{1} - \mathtt{r}'^{\dagger}(\varepsilon)\,\mathtt{r}'(\varepsilon)\right)\right]. \tag{7.642c}$$

7.18.2 Show that this current vanishes for $\delta V = 0$.

7.18.3 If the coefficients g_n with $n = 1, 2, 3, \cdots$ are defined by

$$\delta I =: \sum_{n=1}^{\infty} g_n\,\delta V^n, \tag{7.643}$$

show that

$$g_n = \frac{e}{2\pi\hbar}\,\frac{e^n}{n!}\,\left.\frac{\mathrm{d}^{n-1}\,\mathrm{tr}\left(\mathtt{t}^{\dagger}(\varepsilon)\,\mathtt{t}(\varepsilon)\right)}{\mathrm{d}\varepsilon^{n-1}}\right|_{\varepsilon=\mu}. \tag{7.644}$$

7.18.4 Denote with $f_{\mathrm{FD}}(\varepsilon) := \left[e^{\beta(\varepsilon-\mu)} + 1\right]^{-1}$ the Fermi-Dirac distribution function at the inverse reduced temperature β and chemical potential μ. Show that

$$\delta I = \frac{e}{2\pi\hbar}\int_{\mathbb{R}} \mathrm{d}\varepsilon\,\left[f_{\mathrm{FD}}(\varepsilon - e\,\delta V) - f_{\mathrm{FD}}(\varepsilon)\right]\,\mathrm{tr}\left(\mathtt{t}^{\dagger}(\varepsilon)\,\mathtt{t}(\varepsilon)\right) \tag{7.645}$$

generalizes Eq. (7.642) to any nonvanishing temperature.

7.19 Derive the composition law obeyed by the two $2M \times 2M$ scattering matrices depicted in Figure 7.5. *Hint:* It is given by the scattering matrix

$$\mathcal{S}_{12} = \begin{pmatrix} \mathsf{r}_{12} & \mathsf{t}'_{12} \\ \mathsf{t}_{12} & \mathsf{r}'_{12} \end{pmatrix}, \tag{7.646a}$$

where

$$\mathsf{r}_{12} = \mathsf{r}_1 + \mathsf{t}'_1 \left(\mathbb{1}_M - \mathsf{r}_2 \, \mathsf{r}'_1\right)^{-1} \mathsf{r}_2 \, \mathsf{t}_1, \tag{7.646b}$$

$$\mathsf{t}'_{12} = \mathsf{t}'_1 \left(\mathbb{1}_M - \mathsf{r}_2 \, \mathsf{r}'_1\right)^{-1} \mathsf{t}'_2, \tag{7.646c}$$

$$\mathsf{t}_{12} = \mathsf{t}_2 \left(\mathbb{1}_M - \mathsf{r}'_1 \, \mathsf{r}_2\right)^{-1} \mathsf{t}_1, \tag{7.646d}$$

$$\mathsf{r}'_{12} = \mathsf{r}'_2 + \mathsf{t}_2 \left(\mathbb{1}_M - \mathsf{r}'_1 \, \mathsf{r}_2\right)^{-1} \mathsf{r}'_1 \, \mathsf{t}'_2, \tag{7.646e}$$

in terms of the pair

$$\mathcal{S}_1 = \begin{pmatrix} \mathsf{r}_1 & \mathsf{t}'_1 \\ \mathsf{t}_1 & \mathsf{r}'_1 \end{pmatrix}, \qquad \mathcal{S}_2 = \begin{pmatrix} \mathsf{r}_2 & \mathsf{t}'_2 \\ \mathsf{t}_2 & \mathsf{r}'_2 \end{pmatrix} \tag{7.646f}$$

of scattering matrices.

7.20 Verify Eq. (7.227).

7.21 Verify the singular-value decomposition (7.228) of the transfer matrix.

7.22 We use the notation

$$\Sigma_{0;\mu_1 \cdots \mu_i \cdots \mu_n} := \mathbb{1}_M \otimes \sigma^{(1)}_{\mu_1} \otimes \cdots \sigma^{(i)}_{\mu_i} \otimes \cdots \sigma^{(n)}_{\mu_n} = \left(\Sigma_{0;\mu_1 \cdots \mu_i \cdots \mu_n}\right)^{\dagger}, \tag{7.647a}$$

where $\sigma^{(i)}_{\mu_i}$ denotes the 2×2 unit matrix when $\mu_i = 0$ and the Pauli matrices when $\mu_i = 1, 2, 3$, while $\mathbb{1}_M$ is the $M \times M$ unit matrix for some positive integer M. Observe that

$$\left(\Sigma_{0;\mu_1 \cdots \mu_i \cdots \mu_n}\right)^{-1} = \Sigma_{0;\mu_1 \cdots \mu_i \cdots \mu_n} \tag{7.647b}$$

as $\Sigma_{0;\mu_1 \cdots \mu_i \cdots \mu_n}$ squares to the identity.

7.22.1 Symmetry class D Choose $M = N$ and $n = 2$ in Eq. (7.647a). We denote $\sigma^{(1)}$ with ρ (Bogoliubov-de-Gennes particle-hole grading), and $\sigma^{(2)}$ with τ (left-right movers grading). We begin with the element $\mathcal{M}(\varepsilon)$ from $\mathrm{U}(2N, 2N)$. In addition to the condition

$$\mathcal{M}^{\dagger}(\varepsilon)\, \Sigma_{0;03}\, \mathcal{M}(\varepsilon) = \mathcal{M}(\varepsilon)\, \Sigma_{0;03}\, \mathcal{M}^{\dagger}(\varepsilon) = \Sigma_{0;03} \tag{7.648a}$$

for (thermal) current conservation [so that $\mathcal{M}(\varepsilon) \in \mathrm{U}(2N, 2N)$], we impose the transformation law (for charge conjugation)

$$\Sigma_{0;10}\, \mathcal{M}^{*}(\varepsilon)\, \Sigma_{0;10} = \mathcal{M}(-\varepsilon) \tag{7.648b}$$

for the value $\mu = \varepsilon$ of the chemical potential in the reservoirs Ω and Ω'. This transformation law is a condition at the band center $\varepsilon = 0$ on the form taken by $\mathcal{M}(\varepsilon = 0)$. Observe that

$$\Sigma_{0;10} = +\mathrm{i}\mathcal{A}\,\mathcal{A}^{*} = -\mathrm{i}\mathcal{A}^{*}\,\mathcal{A}, \tag{7.649a}$$

where

$$\mathcal{A} := \mathbb{1}_{N} \otimes \frac{1}{\sqrt{2}} \begin{pmatrix} +1 & +\mathrm{i} \\ -\mathrm{i} & -1 \end{pmatrix} \otimes \tau_{0} \in \mathrm{U}(4N). \tag{7.649b}$$

With the help of the identities

$$\mathcal{A}^{-1} = \mathcal{A}^{\dagger} = \mathcal{A}, \qquad \mathcal{A}^{\mathsf{T}} = \mathcal{A}^{*}, \tag{7.650}$$

verify that the one-to-one map

$$\mathcal{M}(\varepsilon) \mapsto \mathcal{A}\,\mathcal{M}(\varepsilon)\,\mathcal{A} \tag{7.651}$$

turns Eq. (7.648) into the Majorana representation

$$\mathcal{M}^{\dagger}(\varepsilon)\,\Sigma_{0;03}\,\mathcal{M}\,(\varepsilon) = \mathcal{M}\,(\varepsilon)\,\Sigma_{0;03}\,\mathcal{M}^{\dagger}(\varepsilon) = \Sigma_{0;03}, \tag{7.652a}$$

$$\mathcal{M}^{*}(\varepsilon) = \mathcal{M}(-\varepsilon). \tag{7.652b}$$

We have constructed the (Majorana) basis for which Eq. (7.296) holds. At the band center $\varepsilon = 0$, the Majorana representation (7.652) for

$$\mathcal{M}(\varepsilon = 0) \equiv \mathcal{M} \tag{7.653a}$$

simplifies to

$$\mathcal{M}^{\mathsf{T}}\,\Sigma_{0;03}\,\mathcal{M} = \mathcal{M}\,\Sigma_{0;03}\,\mathcal{M}^{\mathsf{T}} = \Sigma_{0;03}, \tag{7.653b}$$

$$\mathcal{M}^{*} = \mathcal{M}. \tag{7.653c}$$

Hence, $\mathcal{M}$ is an element of $\mathrm{O}(2N, 2N)$ according to Eq. (7.471) from **Exercise** 7.6.

7.22.2 Symmetry class DIII Choose $M = N$ and $n = 3$ in Eq. (7.647a). We denote $\sigma^{(1)}$ with ρ (BdG particle-hole grading), $\sigma^{(2)}$ with σ (spin-1/2 grading), and $\sigma^{(3)}$ with τ (left-right movers grading). We begin with the element $\mathcal{M}(\varepsilon)$ from $\mathrm{U}(4N, 4N)$. In addition to the conditions

$$\mathcal{M}^{\dagger}(\varepsilon)\,\Sigma_{0;003}\,\mathcal{M}\,(\varepsilon) = \mathcal{M}\,(\varepsilon)\,\Sigma_{0;003}\,\mathcal{M}^{\dagger}(\varepsilon) = \Sigma_{0;003} \tag{7.654a}$$

for (thermal) current conservation [so that $\mathcal{M}\,(\varepsilon) \in \mathrm{U}(4N, 4N)$] and

$$\Sigma_{0;021}\,\mathcal{M}^{*}(\varepsilon)\,\Sigma_{0;021} = \mathcal{M}(+\varepsilon) \tag{7.654b}$$

for time-reversal symmetry, we impose the transformation law (for charge conjugation)

$$\Sigma_{0;100}\,\mathcal{M}^{*}(\varepsilon)\,\Sigma_{0;100} = \mathcal{M}(-\varepsilon) \tag{7.654c}$$

for the value $\mu = \varepsilon$ of the chemical potential in the reservoirs Ω and Ω'. This transformation law is a condition at the band center $\varepsilon = 0$ on the form taken by $\mathcal{M}(\varepsilon = 0)$. Observe that

$$\Sigma_{0;100} = +\mathrm{i}\mathcal{A}\,\mathcal{A}^{*} = -\mathrm{i}\mathcal{A}^{*}\,\mathcal{A}, \tag{7.655a}$$

where

$$\mathcal{A} := \mathbb{1}_N \otimes \frac{1}{\sqrt{2}}\begin{pmatrix} +1 & +\mathrm{i} \\ -\mathrm{i} & -1 \end{pmatrix} \otimes \sigma_0 \otimes \tau_0 \in \mathrm{U}(8N). \tag{7.655b}$$

7.22.2.1 With the help of the identities

$$\mathcal{A}^{-1} = \mathcal{A}^{\dagger} = \mathcal{A}, \qquad \mathcal{A}^{\mathsf{T}} = \mathcal{A}^{*}, \qquad \mathcal{A}\,\Sigma_{0;021}\,\mathcal{A}^{*} = -\mathrm{i}\Sigma_{0;121}, \tag{7.656}$$

verify that the one-to-one map

$$\mathcal{M}(\varepsilon) \mapsto \mathcal{A}\,\mathcal{M}(\varepsilon)\,\mathcal{A} \tag{7.657}$$

turns Eq. (7.654) into the Majorana representation

$$\mathcal{M}^{\dagger}(\varepsilon)\,\Sigma_{0;003}\,\mathcal{M}(\varepsilon) = \mathcal{M}(\varepsilon)\,\Sigma_{0;003}\,\mathcal{M}^{\dagger}(\varepsilon) = \Sigma_{0;003}, \tag{7.658a}$$

$$\Sigma_{0;121}\,\mathcal{M}^{*}(\varepsilon)\,\Sigma_{0,121} = \mathcal{M}(+\varepsilon), \tag{7.658b}$$

$$\mathcal{M}^{*}(\varepsilon) = \mathcal{M}(-\varepsilon). \tag{7.658c}$$

7.22.2.2 Define the orthogonal $8N \times 8N$ matrix

$$\mathcal{B} := \begin{pmatrix} \mathbb{1}_N \otimes \rho_0 \otimes \tau_0 & 0 \\ 0 & \mathbb{1}_N \otimes \rho_1 \otimes \tau_1 \end{pmatrix} = \mathcal{B}^{*} = \mathcal{B}^{\mathsf{T}}, \qquad \mathcal{B}^{2} = \mathbb{1}_{8N}. \tag{7.659}$$

Here, the 2×2 blocks belong to the σ (spin-1/2) grading. Show that the one-to-one map

$$\mathcal{M}(\varepsilon) \mapsto \mathcal{B}\,\mathcal{M}(\varepsilon)\,\mathcal{B} \tag{7.660}$$

turns the Majorana representation (7.658) into

$$\mathcal{M}^{\dagger}(\varepsilon)\,\Sigma_{0;033}\,\mathcal{M}(\varepsilon) = \mathcal{M}(\varepsilon)\,\Sigma_{0;033}\,\mathcal{M}^{\dagger}(\varepsilon) = \Sigma_{0;033}, \tag{7.661a}$$

$$\Sigma_{0;020}\,\mathcal{M}^{*}(\varepsilon)\,\Sigma_{0;020} = \mathcal{M}(+\varepsilon), \tag{7.661b}$$

$$\mathcal{M}^{*}(\varepsilon) = \mathcal{M}(-\varepsilon). \tag{7.661c}$$

7.22.2.3 At the band center $\varepsilon = 0$, the Majorana representation (7.661) for

$$\mathcal{M}(\varepsilon = 0) \equiv \mathcal{M} \tag{7.662a}$$

simplifies to

$$\mathcal{M}^{\mathsf{T}}\,\Sigma_{0;033}\,\mathcal{M} = \mathcal{M}\,\Sigma_{0;033}\,\mathcal{M}^{\mathsf{T}} = \Sigma_{0;033}, \tag{7.662b}$$

$$\Sigma_{0;020}\,\mathcal{M}\,\Sigma_{0;020} = \mathcal{M}, \tag{7.662c}$$

$$\mathcal{M}^{*} = \mathcal{M}. \tag{7.662d}$$

7.22.2.4 Show that conditions (7.662c) and (7.662d) are equivalent to the block decomposition

$$\mathcal{M} = \begin{pmatrix} M_1 & M_2 \\ -M_2 & M_1 \end{pmatrix}, \qquad M_1 = M_1^{*}, \qquad M_2 = M_2^{*}, \tag{7.663}$$

in the σ (spin-1/2) grading.

7.22.2.5 Show that inserting this block decomposition into condition (7.662b) is equivalent to the condition

$$(M_1 + \mathrm{i}M_2)\,(\mathbb{1}_N \otimes \rho_0 \otimes \tau_3)\,(M_1 + \mathrm{i}M_2)^{\mathsf{T}} = (\mathbb{1}_N \otimes \rho_0 \otimes \tau_3) \tag{7.664}$$

in the τ, ρ, and channel gradings.

7.22.2.6 Take advantage of the identity

$$C^2 = \left(C^{\dagger}\right)^2 = (\mathbb{1}_N \otimes \rho_0 \otimes \tau_3)\,, \tag{7.665a}$$

where the block decomposition of C in the τ grading is given by

$$C := \begin{pmatrix} \mathbb{1}_N \otimes \rho_0 & 0 \\ 0 & \mathbb{1}_N \otimes \mathrm{i}\rho_0 \end{pmatrix} = C^{\mathsf{T}}, \tag{7.665b}$$

to deduce that

$$M_{\mathrm{c}} := C\,(M_1 + \mathrm{i}M_2)\,C^{\dagger} \tag{7.666a}$$

obeys the condition

$$M_{\mathrm{c}}\,\mathbb{1}_{4N}\,M_{\mathrm{c}}^{\mathsf{T}} = M_{\mathrm{c}}^{\mathsf{T}}\,\mathbb{1}_{4N}\,M_{\mathrm{c}} = \mathbb{1}_{4N}. \tag{7.666b}$$

Hence, M_{c} is an element of $\mathrm{O}(4N,\mathbb{C})$ according to Eq. (7.470) from **Exercise** 7.6.

7.22.2.7 Argue that the singular-value decomposition of any $M_{\mathrm{c}} \in \mathrm{O}(4N,\mathbb{C})$ is

$$M_{\mathrm{c}} = U \begin{pmatrix} \cosh \mathsf{x} & +\mathrm{i}\sinh \mathsf{x} \\ -\mathrm{i}\sinh \mathsf{x} & \cosh \mathsf{x} \end{pmatrix} V, \qquad U, V \in \mathrm{O}(4N), \tag{7.667}$$

where x is a $2N \times 2N$ diagonal matrix with the eigenvalues $0 \leq x_j < \infty$ for $j = 1, \cdots, 2N$.

7.22.2.8 Verify that the right-hand side of Eq. (7.667) is invariant under the transformation

$$U \mapsto U\,W, \qquad V \mapsto W^{\mathsf{T}}\,V, \qquad W := \begin{pmatrix} \cos \mathsf{z} & \sin \mathsf{z} \\ -\sin \mathsf{z} & \cos \mathsf{z} \end{pmatrix} \in \mathrm{O}(4N), \tag{7.668}$$

with z any real-valued and diagonal $2N \times 2N$ matrix. *Hint:* Here, one may find it useful to verify that

$$\begin{pmatrix} \cosh \mathrm{z}' & +\mathrm{i}\sinh \mathrm{z}' \\ -\mathrm{i}\sinh \mathrm{z}' & \cosh \mathrm{z}' \end{pmatrix} \begin{pmatrix} \cosh \mathrm{z}'' & +\mathrm{i}\sinh \mathrm{z}'' \\ -\mathrm{i}\sinh \mathrm{z}'' & \cosh \mathrm{z}'' \end{pmatrix} = \begin{pmatrix} \cosh(\mathrm{z}'+\mathrm{z}'') & +\mathrm{i}\sinh(\mathrm{z}'+\mathrm{z}'') \\ -\mathrm{i}\sinh(\mathrm{z}'+\mathrm{z}'') & \cosh(\mathrm{z}'+\mathrm{z}'') \end{pmatrix} \tag{7.669}$$

for any commuting pair of $2N \times 2N$ matrices z' and z''.

7.22.3 Symmetry class C Choose $M = N$ and $n = 2$ in Eq. (7.647a). We denote $\sigma^{(1)}$ with κ [irreducible-block grading for the symmetry class C introduced in Eqs. (7.72) and (7.73)] and $\sigma^{(2)}$ with τ (left-right movers grading). We begin with the element $\mathcal{M}(\varepsilon)$ from $\mathrm{U}(2N, 2N)$. In addition to the conditions

$$\mathcal{M}^{\dagger}(\varepsilon)\,\Sigma_{0;03}\,\mathcal{M}(\varepsilon) = \mathcal{M}(\varepsilon)\,\Sigma_{0;03}\,\mathcal{M}^{\dagger}(\varepsilon) = \Sigma_{0;03} \tag{7.670a}$$

for (thermal) current conservation [so that $\mathcal{M}(\varepsilon) \in \mathrm{U}(2N, 2N)$], we impose the transformation law (that combines PHS with spin-rotation symmetry)

$$\Sigma_{0;20}\,\mathcal{M}^{*}(\varepsilon)\,\Sigma_{0;20} = \mathcal{M}(-\varepsilon) \tag{7.670b}$$

for the value $\mu = \varepsilon$ of the chemical potential in the reservoirs Ω and Ω'. This transformation law is a condition at the band center $\varepsilon = 0$ on the form taken by $\mathcal{M}(\varepsilon = 0)$. At the band center $\varepsilon = 0$, Eq. (7.670) for

$$\mathcal{M}(\varepsilon = 0) \equiv \mathcal{M} \tag{7.671a}$$

simplifies to

$$\mathcal{M}^{\dagger}\,\Sigma_{0;03}\,\mathcal{M} = \mathcal{M}\,\Sigma_{0;03}\,\mathcal{M}^{\dagger} = \Sigma_{0;03}, \tag{7.671b}$$

$$\Sigma_{0;20}\,\mathcal{M}^{*}\,\Sigma_{0;20} = \mathcal{M}. \tag{7.671c}$$

Verify that condition (7.671b) dictates that $\log \mathcal{M}$ belongs to the Lie algebra $\mathfrak{u}(2N, 2N)$. Verify that condition (7.671c) dictates that $\log \mathcal{M}$ obeys the transformation law (7.74). Convince yourself that, hence, $\mathcal{M}$ is an element of $\mathrm{Sp}(2N, 2N)$ according to Eq. (7.476a) from **Exercise** 7.6.

7.22.4 Symmetry class CI Choose $M = N$ and $n = 2$ in Eq. (7.647a). We denote $\sigma^{(1)}$ with κ [irreducible-block grading for the symmetry class C introduced in Eqs. (7.72) and (7.73)] and $\sigma^{(2)}$ with τ (left-right movers grading). We begin with the element $\mathcal{M}(\varepsilon)$ from $\mathrm{U}(2N, 2N)$. In addition to the conditions

$$\mathcal{M}^{\dagger}(\varepsilon)\,\Sigma_{0;03}\,\mathcal{M}(\varepsilon) = \mathcal{M}(\varepsilon)\,\Sigma_{0;03}\,\mathcal{M}^{\dagger}(\varepsilon) = \Sigma_{0;03} \tag{7.672a}$$

for (thermal) current conservation [so that $\mathcal{M}(\varepsilon) \in \mathrm{U}(2N, 2N)$] and

$$\Sigma_{0;01}\,\mathcal{M}^{*}(\varepsilon)\,\Sigma_{0;01} = \mathcal{M}(+\varepsilon) \tag{7.672b}$$

for time-reversal symmetry, we impose the transformation law (for charge conjugation)

$$\Sigma_{0;20}\,\mathcal{M}^{*}(\varepsilon)\,\Sigma_{0;20} = \mathcal{M}(-\varepsilon) \tag{7.672c}$$

for the value $\mu = \varepsilon$ of the chemical potential in the reservoirs Ω and Ω'. This transformation law is a condition at the band center $\varepsilon = 0$ on the form taken by $\mathcal{M}(\varepsilon = 0)$. Observe that

$$\Sigma_{0;01} = +\mathrm{i}\mathcal{A}\,\mathcal{A}^{*} = -\mathrm{i}\mathcal{A}^{*}\,\mathcal{A}, \tag{7.673a}$$

where

$$\mathcal{A} := \mathbb{1}_N \otimes \kappa_0 \otimes \frac{1}{\sqrt{2}}\begin{pmatrix} +1 & +\mathrm{i} \\ -\mathrm{i} & -1 \end{pmatrix} \in \mathrm{U}(4N). \tag{7.673b}$$

7.22.4.1 With the help of the identities

$$\mathcal{A}^{-1} = \mathcal{A}^{\dagger} = \mathcal{A}, \qquad \mathcal{A}^{\mathsf{T}} = \mathcal{A}^{*}, \qquad \mathcal{A}\,\Sigma_{0;03}\,\mathcal{A} = -\Sigma_{0;02}, \tag{7.674}$$

verify that the one-to-one map

$$\mathcal{M}(\varepsilon) \mapsto \mathcal{A}\,\mathcal{M}(\varepsilon)\,\mathcal{A} \tag{7.675}$$

turns Eq. (7.672) into the Majorana representation

$$\mathcal{M}^{\dagger}(\varepsilon)\,\Sigma_{0;02}\,\mathcal{M}(\varepsilon) = \mathcal{M}(\varepsilon)\,\Sigma_{0;02}\,\mathcal{M}^{\dagger}(\varepsilon) = \Sigma_{0;02}, \tag{7.676a}$$

$$\mathcal{M}^{*}(\varepsilon) = \mathcal{M}(\varepsilon), \tag{7.676b}$$

$$\Sigma_{0;21}\,\mathcal{M}(\varepsilon)\,\Sigma_{0;21} = \mathcal{M}^{*}(-\varepsilon). \tag{7.676c}$$

7.22.4.2 Define the orthogonal $4N \times 4N$ matrix

$$\mathcal{B} := \begin{pmatrix} \mathbb{1}_N \otimes \tau_0 & 0 \\ 0 & \mathbb{1}_N \otimes \tau_1 \end{pmatrix} = \mathcal{B}^{*} = \mathcal{B}^{\mathsf{T}}, \qquad \mathcal{B}^2 = \mathbb{1}_{4N}. \tag{7.677}$$

Here, the 2×2 blocks belong to the κ (irreducible-block) grading. Show that the one-to-one map

$$\mathcal{M}(\varepsilon) \mapsto \mathcal{B}\,\mathcal{M}(\varepsilon)\,\mathcal{B} \tag{7.678}$$

turns the Majorana representation (7.676) into

$$\mathcal{M}^{\dagger}(\varepsilon)\,\Sigma_{0;32}\,\mathcal{M}(\varepsilon) = \mathcal{M}(\varepsilon)\,\Sigma_{0;32}\,\mathcal{M}^{\dagger}(\varepsilon) = \Sigma_{0;32}, \tag{7.679a}$$

$$\mathcal{M}^{*}(\varepsilon) = \mathcal{M}(\varepsilon), \tag{7.679b}$$

$$\Sigma_{0;20}\,\mathcal{M}^{*}(\varepsilon)\,\Sigma_{0;20} = \mathcal{M}(-\varepsilon). \tag{7.679c}$$

7.22.4.3 At the band center $\varepsilon = 0$, the Majorana representation (7.679) for

$$\mathcal{M}(\varepsilon = 0) \equiv \mathcal{M} \tag{7.680a}$$

simplifies to

$$\mathcal{M}^{\mathsf{T}}\,\Sigma_{0;32}\,\mathcal{M} = \mathcal{M}\,\Sigma_{0;32}\,\mathcal{M}^{\mathsf{T}} = \Sigma_{0;32}, \tag{7.680b}$$

$$\mathcal{M}^{*} = \mathcal{M}, \tag{7.680c}$$

$$\Sigma_{0;20}\,\mathcal{M}\,\Sigma_{0;20} = \mathcal{M}. \tag{7.680d}$$

Show that conditions (7.680c) and (7.680d) are equivalent to the block decomposition

$$\mathcal{M} = \begin{pmatrix} M_1 & M_2 \\ -M_2 & M_1 \end{pmatrix}, \qquad M_1 = M_1^*, \qquad M_2 = M_2^* \tag{7.681}$$

in the κ (irreducible-block) grading.

7.22.4.4 Show that inserting this block decomposition into condition (7.680b) is equivalent to the condition

$$(M_1 + \mathrm{i}M_2)\,(\mathbb{1}_N \otimes \tau_2)\,(M_1 + \mathrm{i}M_2)^{\mathsf{T}} = (\mathbb{1}_N \otimes \tau_2) \tag{7.682}$$

in the τ and channel gradings. Hence,

$$M_{\mathrm{c}} := M_1 + \mathrm{i}M_2 \tag{7.683}$$

is an element of $\mathrm{Sp}(2N,\mathbb{C})$ according to Eq. (7.473) from **Exercise** 7.6.

7.22.4.5 Argue that the singular-value decomposition of any $M_{\mathrm{c}} \in \mathrm{Sp}(2N,\mathbb{C})$ is

$$M_{\mathrm{c}} = U \begin{pmatrix} \cosh \mathsf{x} & +\mathrm{i}\sinh \mathsf{x} \\ -\mathrm{i}\sinh \mathsf{x} & \cosh \mathsf{x} \end{pmatrix} V, \qquad U, V \in \mathrm{Sp}(2N), \tag{7.684}$$

where x is a $N \times N$ diagonal matrix with the eigenvalues $0 \le x_j < \infty$ for $j = 1, \cdots, N$.

7.22.4.6 Verify that the right-hand side of Eq. (7.684) is invariant under the transformation

$$U \mapsto U\,W, \quad V \mapsto W^{\mathsf{T}}\,V, \quad W := \begin{pmatrix} \cosh \mathsf{z} & +\mathrm{i}\sinh \mathsf{z} \\ -\mathrm{i}\sinh \mathsf{z} & \cosh \mathsf{z} \end{pmatrix} \in \mathrm{Sp}(2N), \tag{7.685}$$

with z any real-valued and diagonal $N \times N$ matrix. *Hint:* Verify that

$$\begin{pmatrix} \cosh \mathsf{z}' & +\mathrm{i}\sinh \mathsf{z}' \\ -\mathrm{i}\sinh \mathsf{z}' & \cosh \mathsf{z}' \end{pmatrix} \begin{pmatrix} \cosh \mathsf{z}'' & +\mathrm{i}\sinh \mathsf{z}'' \\ -\mathrm{i}\sinh \mathsf{z}'' & \cosh \mathsf{z}'' \end{pmatrix} = \begin{pmatrix} \cosh(\mathsf{z}'+\mathsf{z}'') & +\mathrm{i}\sinh(\mathsf{z}'+\mathsf{z}'') \\ -\mathrm{i}\sinh(\mathsf{z}'+\mathsf{z}'') & \cosh(\mathsf{z}'+\mathsf{z}'') \end{pmatrix} \tag{7.686}$$

for any commuting pair of $N \times N$ matrices z' and z''.

7.23 Prove that Eqs. (7.359c) and (7.359d) imply that the solution to the Fokker-Planck equation (7.359a) is normalized for all L and ε. *Hint:* Define

$$N_\varepsilon(L) := \int \mathrm{d}x_1 \cdots \int \mathrm{d}x_{N^\star}\, P_\varepsilon(x_1, \cdots, x_{N^\star}). \tag{7.687}$$

Demand that $N_\varepsilon(L) \equiv N_\varepsilon$ is independent of L. Take the derivative of N_ε with respect to L and assume that the differentiation with respect to L commutes with the integration over the $N^\star$ Lyapunov exponents. Combine Eqs. (7.359c) and (7.359d) with partial integrations to complete the proof.

7.24 With the help of Eqs. (7.354a) and (7.362), define

$$\left\langle g^{(a)}_{\mathrm{L},\varepsilon=0}\right\rangle := \int\limits_0^\infty \mathrm{d}x_1 \cdots \int\limits_0^\infty \mathrm{d}x_{N^\star}\, P_{\varepsilon=0}(x_1,\cdots,x_{N^\star};L)\, D \sum_{j=1}^{N^\star} \cosh^{-2a}(x_j), \tag{7.688}$$

where $P_{\varepsilon=0}(x_1,\cdots,x_{N^\star};L)$ is the solution to the Fokker-Planck equation (7.359).

7.24.1 Show that, when $\varepsilon=0$, Eq. (7.363) holds for the symmetry classes A, AI, AII, D, C, DIII, and CI, for which

$$m_{\mathrm{o}} \equiv m^{+}_{\mathrm{o}} = m^{-}_{\mathrm{o}}. \tag{7.689}$$

Hint: Assume that the integration over the Lyapunov exponents commutes with the differentiation with respect to L. Use the Fokker-Planck equation (7.359) and partial integration.

7.24.2 Derive the counterpart to Eq. (7.363) for the remaining three chiral symmetry classes AIII, BDI, and CII.

7.25 Derive the asymptotic behavior of the typical and mean conductance in the localized regime for all ten symmetry classes.

Definition 7.18 (typical and mean) If $x>0$ is a random variable, then its mean is $\langle x\rangle$, while its typical value is $\exp(\langle \ln x\rangle)$.

Remark The distinction between the typical and mean values of x is particularly important if $x=\prod_\iota y_\iota$ is the product of random variables $y_\iota>0$. For example, even though the random variable

$$\ln x = \sum_\iota \ln y_\iota \tag{7.690}$$

has a narrow distribution if the random variables y_ι are independently and identically distributed and the range of ι is large, this might not be so for x.

7.25.1 *Step 1:* Show that the Fokker-Planck equation

$$\frac{\partial P}{\partial L} = \frac{1}{2\gamma\ell}\sum_{j=1}^{N^\star}\frac{\partial}{\partial x_j}\left[J\frac{\partial}{\partial x_j}\left(J^{-1}P\right)\right] \tag{7.691a}$$

can be manipulated into the form

$$\frac{\partial P}{\partial L} = \frac{1}{2\gamma\ell}\sum_{j=1}^{N^\star}\frac{\partial}{\partial x_j}\left(\frac{\partial P}{\partial x_j}+P\frac{\partial\Omega}{\partial x_j}\right), \qquad \Omega := -\ln J. \tag{7.691b}$$

7.25.2 *Step 2:* For which of the ten Altland-Zirnbauer symmetry classes is the Ansatz

$$0 \leq x_1 \ll x_2 \ll \cdots \ll x_{N^\star-1} \ll x_{N^\star}, \tag{7.692}$$

called the crystallization of the transmission eigenvalues, justified?

7.25.3 *Step 3:* Show that if Eq. (7.692) holds, then the corresponding Fokker-Planck equations separates. *Hint:* Show that

$$\begin{aligned}\Omega(x_1,\cdots,x_{N^\star}) :=& -\ln J(x_1,\cdots,x_{N^\star})\\ \sim& -\sum_{j=1}^{N^\star} 2\left[(j-1)m_{\mathrm{o}} + m_{\mathrm{l}}\right] x_j\\ =:& \, \Omega_{\mathrm{loc}}(x_1,\cdots,x_{N^\star}).\end{aligned} \tag{7.693}$$

7.25.4 *Step 4:* Solve the Fokker-Planck equation after replacing $\Omega(x_1,\cdots,x_{N^\star})$ with the potential $\Omega_{\mathrm{loc}}(x_1,\cdots,x_{N^\star})$ and compute the mean values of and the covariances between the Lyapunov exponents.

7.25.5 *Step 5:* Use the solution to the Fokker-Planck equation with $\Omega(x_1,\cdots,x_{N^\star})$ substituted by $\Omega_{\mathrm{loc}}(x_1,\cdots,x_{N^\star})$ to compute the mean and variance of the dimensionless Landauer conductance. *Hint:* Explain why you need only compute the average of $\cosh^{-2} x_1$ if Eq. (7.693) holds.

7.25.6 *Step 6:* Comment on how the vanishing of m_{l} for the symmetry classes D and DIII affect the average value of the conductance as a function of L.

7.25.7 *Step 7:* Repeat and adapt *Steps* 1 – 6 for the three chiral classes.

7.26 Derive the asymptotic behavior as $s \to \infty$ of $\langle g_{\mathrm{L},\varepsilon=0}\rangle$ for the symmetry classes AIII, CI, and DIII from Eq. (7.369). *Hint:* Use the Poisson summation formula

$$\sum_{m\in\mathbb{Z}} \delta(x-2m-1) = \frac{1}{2}\sum_{n\in\mathbb{Z}} e^{\mathrm{i}\pi n\,(x-1)}, \tag{7.694a}$$

$$\sum_{m\in\mathbb{Z}} \delta(x-m) = \sum_{n\in\mathbb{Z}} e^{2\pi\mathrm{i}n\,x}. \tag{7.694b}$$

7.27 We start from the stationary eigenvalue problem

$$\varepsilon\,\Psi_m = T_m\,\Psi_{m+1} + T^\dagger_{m-1}\,\Psi_{m-1}, \quad m = 1,2,3,\cdots,M-2,M-1,M. \tag{7.695}$$

7.27.1 Explain how this stationary eigenvalue problem is related to the second quantized Hamiltonian (7.379c) when $b=0$. Do the gauge transformation

$$\Psi_m \mapsto \mathrm{i}^m\,\Psi_m \tag{7.696a}$$

and rewrite Eq. (7.695) as

$$\varepsilon\,\Psi_m = \mathrm{i}T_m\,\Psi_{m+1} - \mathrm{i}T^\dagger_{m-1}\,\Psi_{m-1}, \qquad m = 1,3,\cdots,M-1, \tag{7.696b}$$

$$\varepsilon\,\Psi_{m+1} = \mathrm{i}T_{m+1}\,\Psi_{m+2} - \mathrm{i}T^\dagger_{m}\,\Psi_{m}, \qquad m = 1,3,\cdots,M-1, \tag{7.696c}$$

under the assumption that M is an even integer.

7.27.2 Deduce from the identifications

$$\begin{aligned}
&\Psi_{m} \to \sqrt{2\mathfrak{a}}\,\psi^{(\mathrm{o})}(x), && (7.697\mathrm{a})\\
&\Psi_{m+1} \to \sqrt{2\mathfrak{a}}\,\psi^{(\mathrm{e})}(x), && (7.697\mathrm{b})\\
&T_{m} \to t - t' + (2\,t\,\mathfrak{a})\,\delta T^{(\mathrm{o})}(x), && (7.697\mathrm{c})\\
&T_{m+1} \to t + t' + (2\,t\,\mathfrak{a})\,\delta T^{(\mathrm{e})}(x), && (7.697\mathrm{d})\\
&m = 1,3,\cdots,M-1 \to -L/2 \leq x < L/2, && (7.697\mathrm{e})
\end{aligned}$$

the stationary eigenvalue problem

$$\varepsilon\,\psi^{(\mathrm{o})}(x) = \Big\{(2\,t\,\mathfrak{a})\,\mathrm{i}\partial_{x} - (2\,t')\,\mathrm{i} + (2\,t\,\mathfrak{a})\,\mathrm{i}\left[\delta T^{(\mathrm{o})}(x) - \delta T^{(\mathrm{e})\dagger}(x)\right]\Big\}\psi^{(\mathrm{e})}(x), \tag{7.697f}$$

$$\varepsilon\,\psi^{(\mathrm{e})}(x) = \Big\{(2\,t\,\mathfrak{a})\,\mathrm{i}\partial_{x} + (2\,t')\,\mathrm{i} + (2\,t\,\mathfrak{a})\,\mathrm{i}\left[\delta T^{(\mathrm{e})}(x) - \delta T^{(\mathrm{o})\dagger}(x)\right]\Big\}\psi^{(\mathrm{o})}(x). \tag{7.697g}$$

7.27.3 Choose units in which

$$\hbar = 1, \qquad v_{\mathrm{F}} := 2\,|t|\,\mathfrak{a} = 1, \qquad m := 2\,t', \tag{7.698a}$$

define

$$\begin{aligned}
&\Psi(x) := \begin{pmatrix} \psi^{(\mathrm{o})}(x) \\ \psi^{(\mathrm{e})}(x) \end{pmatrix}, && (7.698\mathrm{b})\\
&v(x) := +\frac{\mathrm{i}}{2}\left[\delta T^{(\mathrm{o})}(x) - \delta T^{(\mathrm{e})\dagger}(x) - \mathrm{H.c.}\right], && (7.698\mathrm{c})\\
&w(x) := -\frac{1}{2}\left[\delta T^{(\mathrm{o})}(x) - \delta T^{(\mathrm{e})\dagger}(x) + \mathrm{H.c.}\right], && (7.698\mathrm{d})
\end{aligned}$$

and show that we have obtained the representation

$$\varepsilon\,\Psi(x) = \mathcal{H}(x)\,\Psi(x), \tag{7.698e}$$

where

$$\mathcal{H}(x) = \mathrm{sgn}(t)\,\mathbb{1}_{N} \otimes \tau_{1}\,\mathrm{i}\partial_{x} + \mathbb{1}_{N} \otimes \tau_{2}\,m + \mathrm{sgn}(t)\left[v(x) \otimes \tau_{1} + w(x) \otimes \tau_{2}\right], \tag{7.698f}$$

of the stationary Dirac equation.

7.27.4 What unitary transformation brings $\mathcal{H}(x)$ to the representation (7.382b)?

7.28 Equation (7.387) will be derived in a self-contained and explicit way for the chiral symmetry classes AIII and BDI.

Definitions Start from the Schrödinger equation[117]

$$\begin{aligned}\mathcal{H}\,\psi_j(x) &:= \sum_{k=1}^{N}\left[\mathrm{i}\tau_3\,\delta_{jk}\,\partial_x + \tau_3\,v_{jk}(x) + \tau_2\,w_{jk}(x)\right]\psi_k(x)\\ &= \varepsilon\,\psi_j(x), \qquad x\in\mathbb{R}.\end{aligned} \tag{7.699a}$$

Here, the wave function $\psi(x)$ is represented by a vector whose N components $\psi_j(x)$, $j=1,\cdots,N$, are themselves two-component spinors. We reserve the 2×2 unit matrix τ_0 and the three Pauli matrices τ_1,τ_2,τ_3 when acting on the indices of these spinors. The matrix elements $v_{jk}(x)$ and $w_{jk}(x)$ of the Hermitean $N\times N$ matrices $\mathsf{v}(x)=\mathsf{v}^{\dagger}(x)$ and $\mathsf{w}(x)=\mathsf{w}^{\dagger}(x)$, respectively, obey

$$v_{jk}(x) = v^{*}_{kj}(x), \qquad w_{jk}(x) = w^{*}_{kj}(x). \tag{7.699b}$$

Moreover, in the presence of time-reversal symmetry,[118]

$$v_{jk}(x) = -v^{*}_{jk}(x), \qquad w_{jk}(x) = w^{*}_{jk}(x). \tag{7.699c}$$

7.28.1 Generalize this definition for the chiral symmetry class CII.

7.28.2 In matrix notation, Hamiltonian (7.699) is represented by

$$\mathcal{H} = \mathrm{i}\mathbb{1}_N\otimes\tau_3\,\partial_x + \mathsf{v}\otimes\tau_3 + \mathsf{w}\otimes\tau_2, \tag{7.700}$$

where the unit $N\times N$ matrix is denoted by $\mathbb{1}_N$. Hamiltonian (7.699) possesses a unitary spectral symmetry, called chiral symmetry, which is encoded by

$$(\mathbb{1}_N\otimes\tau_1)\,\mathcal{H}\,(\mathbb{1}_N\otimes\tau_1) = -\mathcal{H} \tag{7.701}$$

in the representation of Eq. (7.700). In addition, TRS (for BDI) holds when

$$\mathsf{v}(x) = -\mathsf{v}^{*}(x), \qquad \mathsf{w}(x) = \mathsf{w}^{*}(x), \tag{7.702a}$$

in which case

$$(\mathbb{1}_N\otimes\tau_1)\,\mathcal{H}^{*}\,(\mathbb{1}_N\otimes\tau_1) = (-1)^2\,\mathcal{H} = \mathcal{H}. \tag{7.702b}$$

Adapt Eq. (7.702) to the chiral symmetry class CII.

7.28.3 We want to solve Schrödinger equation at zero energy,

$$\mathcal{H}\,\psi(x) = 0, \qquad \psi(0)\text{ given}. \tag{7.703}$$

[117] We have set $\hbar = v_{\mathrm{F}} = 1$ and absorbed the deterministic mass m by demanding that the mean value of the matrix-valued random potential $\mathsf{w}(x)$ is m times the unit $N\times N$ matrix in Eq. (7.382b).

[118] Reversal of time implies that the interchange of left and right movers (represented by conjugation with τ_1) is to be composed with complex conjugation.

Verify that, in analogy to the solution of the time-dependent Schrödinger equation, the solution of Eq. (7.703) is given by the *space*-ordered exponential

$$\psi(x) = \mathsf{P}_y \left[e^{+\mathrm{i}\int\limits_0^x \mathrm{d}y\,(\mathsf{v}\otimes\tau_0 - \mathrm{i}\mathsf{w}\otimes\tau_1)(y)} \right] \psi(0). \tag{7.704}$$

We are going to make use of Eq. (7.704) to construct the transfer matrix describing the transmission of an incoming plane wave through a disordered region of length L [the disordered region is the support of the matrix-valued random potentials $\mathsf{v}(x)$ and $\mathsf{w}(x)$].

Transfer matrix Assume that the disorder is confined between $0 < x < L$. We shall use the roman font L (R) to denote the left (right) reservoir $x < 0$ ($x > L$). Given that $\psi_j(x) = \psi_j(\mathrm{L})$ for $x = 0$ and $\psi_j(x) = \psi_j(\mathrm{R})$ for $x = L$ obey Eq. (7.703), the $2N \times 2N$ transfer matrix $\mathcal{M}$ is defined by

$$\psi_j(\mathrm{R}) = \mathcal{M}_{jk}\,\psi_k(\mathrm{L}), \qquad j = 1, \cdots, N. \tag{7.705a}$$

According to Eq. (7.704), the transfer matrix $\mathcal{M}$ and its adjoint $\mathcal{M}^\dagger$ are given by

$$\mathcal{M} = \mathsf{P}_y \left[e^{+\int\limits_0^L \mathrm{d}y\,(\mathrm{i}\mathsf{v}\otimes\tau_0 + \mathsf{w}\otimes\tau_1)(y)} \right], \qquad \mathcal{M}^\dagger = \mathsf{P}_y \left[e^{-\int\limits_0^L \mathrm{d}y\,(\mathrm{i}\mathsf{v}\otimes\tau_0 - \mathsf{w}\otimes\tau_1)(y)} \right], \tag{7.705b}$$

respectively. By construction, the transfer matrix is a pseudounitary $2N \times 2N$ matrix,

$$\mathcal{M}\,(\mathbb{1}_N \otimes \tau_3)\,\mathcal{M}^\dagger = \mathcal{M}^\dagger\,(\mathbb{1}_N \otimes \tau_3)\,\mathcal{M} = (\mathbb{1}_N \otimes \tau_3). \tag{7.706}$$

The physical interpretation of Eq. (7.706) is that the probability density $(\psi^\dagger\psi)(x)$ obeys a continuity equation, that is, probability density (or flux in short) is conserved. In addition, the transfer matrix has the chiral symmetry

$$(\mathbb{1}_N \otimes \tau_1)\,\mathcal{M}\,(\mathbb{1}_N \otimes \tau_1) = \mathcal{M}, \tag{7.707}$$

and, provided $\mathsf{v}(x) = -\mathsf{v}^*(x)$ and $\mathsf{w}(x) = \mathsf{w}^*(x)$ hold, the TRS (for BDI)

$$\mathcal{M}^* = \mathcal{M}. \tag{7.708}$$

7.28.1 Adapt Eq. (7.708) to the chiral symmetry class CII.

We will make use of the singular-value decomposition

$$\begin{aligned}\mathcal{M} &= \begin{pmatrix}\mathsf{u} & 0\\ 0 & \mathsf{u}'\end{pmatrix}\begin{pmatrix}\cosh(\mathsf{x}) & \sinh(\mathsf{x})\\ \sinh(\mathsf{x}) & \cosh(\mathsf{x})\end{pmatrix}\begin{pmatrix}\mathsf{v} & 0\\ 0 & \mathsf{v}'\end{pmatrix}\\ &= \begin{pmatrix}\mathsf{u}\cosh(\mathsf{x})\,\mathsf{v} & \mathsf{u}\sinh(\mathsf{x})\,\mathsf{v}'\\ \mathsf{u}'\sinh(\mathsf{x})\,\mathsf{v} & \mathsf{u}'\cosh(\mathsf{x})\,\mathsf{v}'\end{pmatrix}\\ &\equiv \begin{pmatrix}\mathsf{m}_{11} & \mathsf{m}_{12}\\ \mathsf{m}_{21} & \mathsf{m}_{22}\end{pmatrix}.\end{aligned} \tag{7.709}$$

Equation (7.709) expresses $\mathcal{M}$ in terms of the $N\times N$ random matrices $\mathsf{u},\mathsf{u}',\mathsf{v},\mathsf{v}'$ that are unitary (orthogonal for BDI), and in terms of the $N\times N$ diagonal and real-valued random matrix x with eigenvalues $x_j\in\mathbb{R}$, $j=1,\cdots,N$.

7.28.2 Verify that imposing chiral symmetry implies that

$$\mathsf{u}'=\mathsf{u},\qquad \mathsf{v}'=\mathsf{v}. \tag{7.710}$$

7.28.3 What changes for the chiral symmetry class CII?

Probability distribution obeyed by the transfer matrix We seek the probability distribution of the N eigenvalues x_j, $j=1,\cdots,N$, of x given that the random Hermitean potentials $\mathsf{v}(x)=\big(v_{ij}(x)\big)$ and $\mathsf{w}(x)=\big(w_{ij}(x)\big)$ are independent, white-noise, and Gaussian-correlated with the means [the mean of the random backscattering matrix $\mathsf{w}(x)$ is no more vanishing since the mass m was absorbed in it]

$$\langle v_{ij}(x)\rangle = 0,\qquad \langle w_{ij}(x)\rangle = m\,\delta_{ij}, \tag{7.711a}$$

and the (only nonvanishing) second cumulants (see footnote 70),

$$\begin{aligned}\langle v_{ij}(x)\,v^{*}_{kl}(y)\rangle_{\mathrm{c}} = \frac{m_{\mathrm{o}}^{-}}{2\,\gamma(\eta)\,\ell}\,\delta(x-y)\Bigg[\delta_{ik}\,\delta_{jl} - \left(\frac{2}{m_{\mathrm{o}}^{-}}-1\right)\delta_{il}\,\delta_{jk}\\ - \frac{2(m_{\mathrm{o}}^{-}-1)(1-\eta)}{m_{\mathrm{o}}^{-}\,N}\,\delta_{ij}\,\delta_{kl}\Bigg],\end{aligned} \tag{7.711b}$$

$$\begin{aligned}\langle w_{ij}(x)\,w^{*}_{kl}(y)\rangle_{\mathrm{c}} = \frac{m_{\mathrm{o}}^{-}}{2\,\gamma(\eta)\,\ell}\,\delta(x-y)\Bigg[\delta_{ik}\,\delta_{jl} + \left(\frac{2}{m_{\mathrm{o}}^{-}}-1\right)\delta_{il}\,\delta_{jk}\\ - \frac{2(1-\eta)}{m_{\mathrm{o}}^{-}\,N}\,\delta_{ij}\,\delta_{kl}\Bigg].\end{aligned} \tag{7.711c}$$

Here, ℓ is a length scale (the mean free path) that must be determined from a microscopic calculation,

$$\gamma(\eta) := \frac{1}{2}\left[m_{\mathrm{o}}^{-}\,(N-1) + 2 - \frac{2(1-\eta)}{N}\right], \tag{7.711d}$$

and

$$m_{\mathrm{o}}^{-} := \begin{cases} 2, & \text{in the chiral symmetry class AIII,} \\ 1, & \text{in the chiral symmetry class BDI.} \end{cases} \tag{7.711e}$$

The rational for the definition (7.711d) of $\gamma(\eta)$ will be established with Eqs. (7.739c) and (7.740). Parameter η is needed because the Brownian motion that will be derived takes place on a non-compact symmetric space M that is associated with a semi-simple Lie algebra, that is the direct sum over two simple Lie algebras. One Lie algebra is associated with the Brownian motion of the center of mass of the radial coordinates of M, the other Lie algebra is associated with the Brownian motion of their relative coordinates, as is shown in **Exercise** 7.29. As these two random processes are independent, we parameterize the cumulants of the random potentials by the dimensionless parameter η and by the dimensionfull mean-free path ℓ, respectively.

7.28.1 Construct the measure $\mathcal{D}[\mathsf{v}(x),\mathsf{w}(x)]\,P[\mathsf{v}(x),\mathsf{w}(x)]$ with those means and second cumulants (with all higher-order cumulants vanishing).

7.28.2 Show that this measure is invariant under the simultaneous transformation

$$\mathsf{v}(x) \to \mathsf{u}\,\mathsf{v}(x)\,\mathsf{u}^{\dagger}, \qquad \mathsf{w}(x) \to \mathsf{u}\,\mathsf{w}(x)\,\mathsf{u}^{\dagger}, \tag{7.712}$$

where u is any $N \times N$ unitary (orthogonal for BDI) matrix that is independent of x.

7.28.3 What changes for the chiral symmetry class CII?

7.28.4 Under these assumptions, it will be shown that if the support L of the disordered region is increased by the infinitesimal amount δL, the statistical change δx_j for given x_j is captured by the moments

$$\frac{\ell}{\delta L}\left\langle \delta x_{j_1}\right\rangle_{\delta L} = m\,\ell + \frac{m_{\mathrm{o}}^{-}}{2\,\gamma(\eta)} \sum_{1\leq k\leq N}^{k\neq j_1} \coth(x_{j_1} - x_k), \tag{7.713a}$$

$$\frac{\ell}{\delta L}\left\langle \delta x_{j_1}\,\delta x_{j_2}\right\rangle_{\delta L} = \frac{m_{\mathrm{o}}^{-}}{2\,\gamma(\eta)}\left[\frac{2}{m_{\mathrm{o}}^{-}}\,\delta_{j_1 j_2} - \frac{2(1-\eta)}{m_{\mathrm{o}}^{-}\,N}\right], \tag{7.713b}$$

$$\frac{\ell}{\delta L}\left\langle \delta x_{j_1}\,\delta x_{j_2}\cdots\delta x_{j_n}\right\rangle_{\delta L} = 0, \tag{7.713c}$$

for $j_1, j_2, \cdots, j_n = 1, \cdots, N$ that are calculated by performing a disorder average over the infinitesimal slice of length δL up to first order in δL.

It will also be shown how this Brownian process can be recast as the following Fokker-Planck equation for the probability distribution $P(x_1, \cdots, x_N; L)$

$$\frac{\partial P}{\partial L}=\frac{1}{2\,\gamma(\eta)\,\ell}\sum_{i,j=1}^{N}\frac{\partial}{\partial x_i}\left[m_{\rm o}^{-}\,f\,\delta_{ij}+\left(\delta_{ij}-\frac{1-\eta}{N}\right)J\,\frac{\partial}{\partial x_j}\,J^{-1}\right]P, \tag{7.714a}$$

$$J(x_1,\cdots,x_N):=\prod_{1\leq j<k\leq N}|\sinh(x_j-x_k)|^{m_{\rm o}^{-}}, \tag{7.714b}$$

$$f:=-\frac{2\,\gamma(\eta)\,\ell}{m_{\rm o}^{-}}\,m. \tag{7.714c}$$

Verify that the solution of (7.714) for $f=0$, $m_{\rm o}^{-}=2$, and $\eta=1$ is

$$P(x_1,\cdots,x_N;L)\propto\prod_{j=1}^{N}e^{-\frac{N\ell}{2L}x_j^2}\prod_{1\leq j<k\leq N}\frac{(x_k-x_j)\,\sinh\left(x_k-x_j\right)}{k-j}, \tag{7.715a}$$

with the initial conditions

$$P(x_1,\cdots,x_N;0)=\prod_{j=1}^{N}\delta(x_j). \tag{7.715b}$$

Moments of the radial eigenvalues We will use the parameterizations[119]

$$\widetilde{\mathcal{M}}=\begin{pmatrix}\tilde{\mathtt{u}}&0\\0&\tilde{\mathtt{u}}\end{pmatrix}\begin{pmatrix}\cosh(\tilde{\mathtt{x}})&\sinh(\tilde{\mathtt{x}})\\\sinh(\tilde{\mathtt{x}})&\cosh(\tilde{\mathtt{x}})\end{pmatrix}\begin{pmatrix}\tilde{\mathtt{v}}&0\\0&\tilde{\mathtt{v}}\end{pmatrix} \tag{7.716}$$

$$=\begin{pmatrix}\tilde{\mathtt{u}}\cosh(\tilde{\mathtt{x}})\,\tilde{\mathtt{v}}&\tilde{\mathtt{u}}\sinh(\tilde{\mathtt{x}})\,\tilde{\mathtt{v}}\\\tilde{\mathtt{u}}\sinh(\tilde{\mathtt{x}})\,\tilde{\mathtt{v}}&\tilde{\mathtt{u}}\cosh(\tilde{\mathtt{x}})\,\tilde{\mathtt{v}}\end{pmatrix} \tag{7.717}$$

$$=\begin{pmatrix}\widetilde{\mathsf{m}}_{11}&\widetilde{\mathsf{m}}_{12}\\\widetilde{\mathsf{m}}_{21}&\widetilde{\mathsf{m}}_{22}\end{pmatrix} \tag{7.718}$$

$$\begin{aligned}&=\mathbb{1}_N\otimes\tau_0\\&\quad+\int_{L}^{L+\delta L}dx\,\left(\mathrm{i}\mathsf{v}\otimes\tau_0+\mathsf{w}\otimes\tau_1\right)(x)\\&\quad+\int_{L}^{L+\delta L}dx\,\left(\mathrm{i}\mathsf{v}\otimes\tau_0+\mathsf{w}\otimes\tau_1\right)(x)\int_{L}^{x}dy\,\left(\mathrm{i}\mathsf{v}\otimes\tau_0+\mathsf{w}\otimes\tau_1\right)(y)\\&\quad+\cdots\end{aligned} \tag{7.719}$$

of the chiral transfer matrix $\widetilde{\mathcal{M}}$ for a slab of infinitesimal size δL. We will need

[119] Type-writer fonts are used for the pair of matrices $\tilde{\mathtt{u}}$ and $\tilde{\mathtt{v}}$ entering the polar decomposition of the transfer matrix in order to avoid any confusions with the random forward scattering matrix $\mathsf{v}(x)$.

$$\widetilde{\mathsf{m}}_{11} \approx \mathbb{1}_N + \mathrm{i} \int\limits_{L}^{L+\delta L} \mathrm{d}x \, \mathsf{v}(x) - \int\limits_{L}^{L+\delta L} \mathrm{d}x \int\limits_{L}^{x} \mathrm{d}y \left[\mathsf{v}(x)\,\mathsf{v}(y) - \mathsf{w}(x)\,\mathsf{w}(y)\right], \quad (7.720a)$$

$$\widetilde{\mathsf{m}}_{12} \approx \int\limits_{L}^{L+\delta L} \mathrm{d}x \, \mathsf{w}(x) + \mathrm{i} \int\limits_{L}^{L+\delta L} \mathrm{d}x \int\limits_{L}^{x} \mathrm{d}y \left[\mathsf{v}(x)\,\mathsf{w}(y) + \mathsf{w}(x)\,\mathsf{v}(y)\right]. \quad (7.720b)$$

Also needed will be the product $\widehat{\mathcal{M}}$ of the two transfer matrices $\mathcal{M}$ and $\widetilde{\mathcal{M}}$:

$$\begin{aligned}
\widehat{\mathcal{M}} :=& \, \mathcal{M}\,\widetilde{\mathcal{M}} \\
=& \begin{pmatrix} \mathsf{m}_{11} & \mathsf{m}_{12} \\ \mathsf{m}_{21} & \mathsf{m}_{22} \end{pmatrix} \begin{pmatrix} \widetilde{\mathsf{m}}_{11} & \widetilde{\mathsf{m}}_{12} \\ \widetilde{\mathsf{m}}_{21} & \widetilde{\mathsf{m}}_{22} \end{pmatrix} \\
=& \begin{pmatrix} \mathsf{m}_{11}\,\widetilde{\mathsf{m}}_{11} + \mathsf{m}_{12}\,\widetilde{\mathsf{m}}_{21} & \mathsf{m}_{11}\,\widetilde{\mathsf{m}}_{12} + \mathsf{m}_{12}\,\widetilde{\mathsf{m}}_{22} \\ \mathsf{m}_{21}\,\widetilde{\mathsf{m}}_{11} + \mathsf{m}_{22}\,\widetilde{\mathsf{m}}_{21} & \mathsf{m}_{21}\,\widetilde{\mathsf{m}}_{12} + \mathsf{m}_{22}\,\widetilde{\mathsf{m}}_{22} \end{pmatrix} \\
\equiv& \begin{pmatrix} \widehat{\mathsf{m}}_{11} & \widehat{\mathsf{m}}_{12} \\ \widehat{\mathsf{m}}_{21} & \widehat{\mathsf{m}}_{22} \end{pmatrix}.
\end{aligned} \quad (7.721)$$

Our strategy will be to infer the probability distribution of x_j, $j = 1, \cdots, N$ from the transformation law of the Hermitean matrix

$$\mathsf{u}^\dagger\,\mathsf{m}_{11}\,\mathsf{m}_{12}^\dagger\,\mathsf{u} = \cosh(\mathsf{x})\,\sinh(\mathsf{x}) = \frac{1}{2}\,\sinh(2\mathsf{x}) \quad (7.722)$$

under multiplication of $\mathcal{M}$ by $\widetilde{\mathcal{M}}$ from the right.

We begin with

$$\begin{aligned}
\widehat{\mathsf{m}}_{11}\widehat{\mathsf{m}}_{12}^\dagger =& \left(\mathsf{m}_{11}\,\widetilde{\mathsf{m}}_{11} + \mathsf{m}_{12}\,\widetilde{\mathsf{m}}_{21}\right)\left(\widetilde{\mathsf{m}}_{12}^\dagger\,\mathsf{m}_{11}^\dagger + \widetilde{\mathsf{m}}_{22}^\dagger\,\mathsf{m}_{12}^\dagger\right) \\
=& \quad \mathsf{m}_{11}\,\widetilde{\mathsf{m}}_{11}\widetilde{\mathsf{m}}_{22}^\dagger\,\mathsf{m}_{12}^\dagger \\
& + \mathsf{m}_{12}\,\widetilde{\mathsf{m}}_{21}\widetilde{\mathsf{m}}_{12}^\dagger\,\mathsf{m}_{11}^\dagger \\
& + \mathsf{m}_{11}\,\widetilde{\mathsf{m}}_{11}\widetilde{\mathsf{m}}_{12}^\dagger\,\mathsf{m}_{11}^\dagger \\
& + \mathsf{m}_{12}\,\widetilde{\mathsf{m}}_{21}\widetilde{\mathsf{m}}_{22}^\dagger\,\mathsf{m}_{12}^\dagger.
\end{aligned} \quad (7.723)$$

Because $\widetilde{\mathcal{M}}$ is pseudounitary and chiral, the parameterization (7.717) holds so that, with the help of $\cosh^2(\tilde{\mathsf{x}}) - \sinh^2(\tilde{\mathsf{x}}) = \mathbb{1}_N$,

$$\widetilde{\mathsf{m}}_{11}\widetilde{\mathsf{m}}_{22}^\dagger = \widetilde{\mathsf{m}}_{11}\widetilde{\mathsf{m}}_{11}^\dagger = \mathbb{1}_N + \widetilde{\mathsf{m}}_{12}\widetilde{\mathsf{m}}_{12}^\dagger. \quad (7.724)$$

This allows us to express the change in $\mathsf{m}_{11}\mathsf{m}_{12}^\dagger$ due to the addition of a very small disordered slab solely in terms of m_{11}, m_{12}, $\widetilde{\mathsf{m}}_{11}$, $\widetilde{\mathsf{m}}_{12}$ and their adjoints,

$$\begin{aligned}
\widehat{\mathsf{m}}_{11}\,\widehat{\mathsf{m}}_{12}^{\dagger}-\mathsf{m}_{11}\,\mathsf{m}_{12}^{\dagger}=\;&\mathsf{m}_{11}\,\widetilde{\mathsf{m}}_{12}\,\widetilde{\mathsf{m}}_{12}^{\dagger}\,\mathsf{m}_{12}^{\dagger}\\
&+\mathsf{m}_{12}\,\widetilde{\mathsf{m}}_{12}\,\widetilde{\mathsf{m}}_{12}^{\dagger}\,\mathsf{m}_{11}^{\dagger}\\
&+\mathsf{m}_{11}\,\widetilde{\mathsf{m}}_{11}\,\widetilde{\mathsf{m}}_{12}^{\dagger}\,\mathsf{m}_{11}^{\dagger}\\
&+\mathsf{m}_{12}\,\widetilde{\mathsf{m}}_{12}\,\widetilde{\mathsf{m}}_{11}^{\dagger}\,\mathsf{m}_{12}^{\dagger}\\
=\;&\mathrm{u}\cosh(\mathsf{x})\left(\mathrm{v}\,\widetilde{\mathsf{m}}_{12}\,\widetilde{\mathsf{m}}_{12}^{\dagger}\,\mathrm{v}^{\dagger}\right)\sinh(\mathsf{x})\,\mathrm{u}^{\dagger}\\
&+\mathrm{u}\sinh(\mathsf{x})\left(\mathrm{v}\,\widetilde{\mathsf{m}}_{12}\,\widetilde{\mathsf{m}}_{12}^{\dagger}\,\mathrm{v}^{\dagger}\right)\cosh(\mathsf{x})\,\mathrm{u}^{\dagger}\\
&+\mathrm{u}\cosh(\mathsf{x})\left(\mathrm{v}\,\widetilde{\mathsf{m}}_{11}\,\widetilde{\mathsf{m}}_{12}^{\dagger}\,\mathrm{v}^{\dagger}\right)\cosh(\mathsf{x})\,\mathrm{u}^{\dagger}\\
&+\mathrm{u}\sinh(\mathsf{x})\left(\mathrm{v}\,\widetilde{\mathsf{m}}_{12}\,\widetilde{\mathsf{m}}_{11}^{\dagger}\,\mathrm{v}^{\dagger}\right)\sinh(\mathsf{x})\,\mathrm{u}^{\dagger}.
\end{aligned}\tag{7.725}$$

The advantage of Eq. (7.725) is that it shows that we do not need to worry about v and its adjoint. Indeed, the four blocks $\widetilde{\mathsf{m}}_{ab}$ making up $\widetilde{\mathcal{M}}$ are polynomials in the random potentials $\mathsf{v}(x)=\mathsf{v}^{\dagger}(x)$ and $\mathsf{w}(x)=\mathsf{w}^{\dagger}(x)$ to any finite order in perturbation theory in $\mathsf{v}(x)$ or $\mathsf{w}(x)$. Hence, v implements the unitary transformation

$$\mathsf{v}(x)\to\mathrm{v}\,\mathsf{v}(x)\,\mathrm{v}^{\dagger},\qquad \mathsf{w}(x)\to\mathrm{v}\,\mathsf{w}(x)\,\mathrm{v}^{\dagger},\tag{7.726}$$

which leaves the probability measure $\mathcal{D}[\mathsf{v}(x),\mathsf{w}(x)]\,P[\mathsf{v}(x),\mathsf{w}(x)]$ invariant. Consequently, we can absorb v and its adjoint in a redefinition of $\mathsf{v}(x)$ and $\mathsf{w}(x)$ without loss of generality.

Next, we use again the fact that the pseudounitarity and chirality of $\widehat{\mathcal{M}}$ imply that there exists a unitary $\hat{\mathrm{u}}$ and diagonal $\hat{\mathsf{x}}$ such that

$$\widehat{\mathsf{m}}_{11}\widehat{\mathsf{m}}_{12}^{\dagger}=\hat{\mathrm{u}}\,\cosh(\hat{\mathsf{x}})\,\sinh(\hat{\mathsf{x}})\,\hat{\mathrm{u}}^{\dagger}\tag{7.727a}$$

has N real eigenvalues

$$\cosh(\hat{x}_j)\,\sinh(\hat{x}_j)=\frac{1}{2}\,\sinh(2\hat{x}_j),\qquad j=1,\cdots,N.\tag{7.727b}$$

We want to express the set $\{\sinh(2\hat{x}_j)\}$ in terms of the set $\{\sinh(2x_j)\}$ up to second order in $\mathsf{v}(x)$ or $\mathsf{w}(x)$ with $L<x<L+\delta L$. This goal motivates the definition of the $N\times N$ Hermitean random matrix $\Delta=(\Delta_{ij})$,

$$\begin{aligned}
\Delta:=\;&\cosh(\mathsf{x})\,\widetilde{\mathsf{m}}_{12}\,\widetilde{\mathsf{m}}_{12}^{\dagger}\,\sinh(\mathsf{x})+\sinh(\mathsf{x})\,\widetilde{\mathsf{m}}_{12}\,\widetilde{\mathsf{m}}_{12}^{\dagger}\,\cosh(\mathsf{x})\\
&+\cosh(\mathsf{x})\,\widetilde{\mathsf{m}}_{11}\,\widetilde{\mathsf{m}}_{12}^{\dagger}\,\cosh(\mathsf{x})+\sinh(\mathsf{x})\,\widetilde{\mathsf{m}}_{12}\,\widetilde{\mathsf{m}}_{11}^{\dagger}\,\sinh(\mathsf{x}).
\end{aligned}\tag{7.728}$$

7.28.1 Show that, upon multiplication of $\mathcal{M}$ by $\widetilde{\mathcal{M}}$ from the right, the random eigenvalue $\cosh(\hat{x}_j)\,\sinh(\hat{x}_j)$ with $j=1,\cdots,N$ is related to the random eigenvalues from the set $\{\cosh(x_k)\,\sinh(x_k)\}$ by

$$\begin{aligned}\delta_j &:= \cosh(\hat{x}_j)\sinh(\hat{x}_j) - \cosh(x_j)\sinh(x_j)\\ &\approx \Delta_{jj} + \sum_{1\le k\le N}^{k\neq j} \frac{\Delta_{jk}\,\Delta_{kj}}{\cosh(x_j)\sinh(x_j) - \cosh(x_k)\sinh(x_k)}\end{aligned} \tag{7.729}$$

up to second-order nondegenerate perturbation theory.

7.28.2 Justify the use of the nondegenerate perturbation theory.

We can now calculate all moments of the random shift δ_j in the eigenvalue $\cosh(x_j)\sinh(x_j)$ with $j = 1, \cdots, N$ upon multiplication of $\mathcal{M}$ by $\widetilde{\mathcal{M}}$ from the right by averaging over $\mathsf{v}(x)$ and $\mathsf{w}(x)$ with $L < x < L + \delta L$ owing from the fact that $\mathsf{v}(x)$ and $\mathsf{w}(x)$ are white-noise correlated in space. This averaging will be denoted $\langle\cdot\rangle_{\delta L}$.

First moment Before calculating $\langle\delta_j\rangle_{\delta L}$, $j = 1, \cdots, N$, notice that

$$\begin{aligned}\langle\Delta\rangle_{\delta L} \approx\ & \cosh(\mathsf{x})\left\langle\widetilde{\mathsf{m}}_{12}\,\widetilde{\mathsf{m}}_{12}^{\dagger}\right\rangle_{\delta L}\sinh(\mathsf{x}) + \sinh(\mathsf{x})\left\langle\widetilde{\mathsf{m}}_{12}\,\widetilde{\mathsf{m}}_{12}^{\dagger}\right\rangle_{\delta L}\cosh(\mathsf{x})\\ &+ \cosh(\mathsf{x})\left\langle\widetilde{\mathsf{m}}_{11}\,\widetilde{\mathsf{m}}_{12}^{\dagger}\right\rangle_{\delta L}\cosh(\mathsf{x}) + \sinh(\mathsf{x})\left\langle\widetilde{\mathsf{m}}_{12}\,\widetilde{\mathsf{m}}_{11}^{\dagger}\right\rangle_{\delta L}\sinh(\mathsf{x}).\end{aligned} \tag{7.730}$$

Needed is

$$\left\langle\widetilde{\mathsf{m}}_{ab}\,\widetilde{\mathsf{m}}_{cd}^{\dagger}\right\rangle_{\delta L}, \qquad a,b,c,d = 1,2. \tag{7.731}$$

7.28.3 With the help of Eqs. (7.720a) and (7.720b), show that, up to second order in the random potentials $\mathsf{v}(x)$ or $\mathsf{w}(x)$,

$$\left\langle\widetilde{\mathsf{m}}_{12}\widetilde{\mathsf{m}}_{12}^{\dagger}\right\rangle_{\delta L} \approx \int\limits_{L}^{L+\delta L} dx \int\limits_{L}^{L+\delta L} dx' \left\langle\mathsf{w}(x)\,\mathsf{w}(x')\right\rangle_{\delta L}, \tag{7.732a}$$

$$\begin{aligned}\left\langle\widetilde{\mathsf{m}}_{11}\,\widetilde{\mathsf{m}}_{12}^{\dagger}\right\rangle_{\delta L} \approx & \int\limits_{L}^{L+\delta L} dx' \left[\left\langle\mathsf{w}(x')\right\rangle_{\delta L} - \mathrm{i}\int\limits_{L}^{x'} dy' \left\langle\mathsf{w}(y')\,\mathsf{v}(x') + \mathsf{v}(y')\,\mathsf{w}(x')\right\rangle_{\delta L}\right]\\ &+\mathrm{i}\int\limits_{L}^{L+\delta L} dx \int\limits_{L}^{x} dx' \left\langle\mathsf{v}(x)\,\mathsf{w}(x') + \mathsf{v}(x')\,\mathsf{w}(x)\right\rangle_{\delta L},\end{aligned} \tag{7.732b}$$

$$\begin{aligned}\left\langle\widetilde{\mathsf{m}}_{12}\,\widetilde{\mathsf{m}}_{11}^{\dagger}\right\rangle_{\delta L} \approx & \int\limits_{L}^{L+\delta L} dx \left[\left\langle\mathsf{w}(x)\right\rangle_{\delta L} + \mathrm{i}\int\limits_{L}^{x} dy \left\langle\mathsf{v}(x)\,\mathsf{w}(y) + \mathsf{w}(x)\,\mathsf{v}(y)\right\rangle_{\delta L}\right]\\ &-\mathrm{i}\int\limits_{L}^{L+\delta L} dx \int\limits_{L}^{x'} dx' \left\langle\mathsf{w}(x)\,\mathsf{v}(x') + \mathsf{w}(x')\,\mathsf{v}(x)\right\rangle_{\delta L}.\end{aligned} \tag{7.732c}$$

7.28.4 Show that insertion of Eq. (7.732) into Eq. (7.730) gives[120]

$$\begin{aligned}\langle\Delta\rangle_{\delta L} \approx\ & \cosh(\mathsf{x}) \int\limits_{L}^{L+\delta L} \mathrm{d}x \int\limits_{L}^{L+\delta L} \mathrm{d}x' \left\langle \mathsf{w}(x)\,\mathsf{w}(x')\right\rangle_{\delta L} \sinh(\mathsf{x}) \\ &+ \sinh(\mathsf{x}) \int\limits_{L}^{L+\delta L} \mathrm{d}x \int\limits_{L}^{L+\delta L} \mathrm{d}x' \left\langle \mathsf{w}(x)\,\mathsf{w}(x')\right\rangle_{\delta L} \cosh(\mathsf{x}) \\ &+ \cosh(\mathsf{x}) \int\limits_{L}^{L+\delta L} \mathrm{d}x \left\langle \mathsf{w}(x)\right\rangle_{\delta L} \cosh(\mathsf{x}) \\ &+ \sinh(\mathsf{x}) \int\limits_{L}^{L+\delta L} \mathrm{d}x \left\langle \mathsf{w}(x)\right\rangle_{\delta L} \sinh(\mathsf{x}) \end{aligned} \tag{7.733}$$

and

$$\langle\Delta^2\rangle_{\delta L} \approx \left\langle \left(\cosh(\mathsf{x}) \int\limits_{L}^{L+\delta L} \mathrm{d}x\, \mathsf{w}(x)\, \cosh(\mathsf{x}) + \sinh(\mathsf{x}) \int\limits_{L}^{L+\delta L} \mathrm{d}x\, \mathsf{w}(x)\, \sinh(\mathsf{x}) \right)^2 \right\rangle_{\delta L} \tag{7.734}$$

up to second order in the random potentials $\mathsf{v}(x)$ or $\mathsf{w}(x)$.

7.28.5 Show that, on average, the matrix elements of Eqs. (7.733) and (7.734) are given by (summation convention over repeated indices)

$$\begin{aligned}\langle\Delta_{jl}\rangle_{\delta L} \approx\ & \int\limits_{L}^{L+\delta L} \mathrm{d}x \int\limits_{L}^{L+\delta L} \mathrm{d}x' \left\langle w_{jk}(x)\, w_{kl}(x')\right\rangle_{\delta L} \\ &\times \left[\cosh(x_j)\, \sinh(x_l) + \sinh(x_j)\, \cosh(x_l\right] \\ &+ \int\limits_{L}^{L+\delta L} \mathrm{d}x \left\langle w_{jl}(x)\right\rangle_{\delta L} \\ &\times \left[\cosh(x_j)\, \cosh(x_l) + \sinh(x_j)\, \sinh(x_l)\right] \end{aligned} \tag{7.735}$$

and (no summation convention over repeated indices)

$$\begin{aligned}\langle\Delta_{jk}\,\Delta_{kl}\rangle_{\delta L} \approx\ & \left[\cosh(x_j)\, \cosh(x_k) + \sinh(x_j)\, \sinh(x_k)\right] \\ &\times \left[\cosh(x_k)\, \cosh(x_l) + \sinh(x_k)\, \sinh(x_l)\right] \\ &\times \int\limits_{L}^{L+\delta L} \mathrm{d}x \int\limits_{L}^{L+\delta L} \mathrm{d}x' \,\langle w_{jk}(x)\, w_{kl}(x')\rangle_{\delta L}\,, \end{aligned} \tag{7.736}$$

respectively.

[120] *Hint:* Use

$$\langle \mathsf{v}(x)\,\mathsf{w}(y)\rangle_{\delta L} = \langle \mathsf{v}(x)\rangle_{\delta L} \langle \mathsf{w}(y)\rangle_{\delta L}.$$

7.28.6 Use Eq. (7.711) to show that (summation convention over repeated indices)

$$\begin{aligned}\langle\Delta_{jl}\rangle_{\delta L} \approx& \frac{m_{\rm o}^{-}\,\delta L}{2\,\gamma(\eta)\,\ell}\left[\delta_{jl}\,\delta_{kk} + \left(\frac{2}{m_{\rm o}^{-}} - 1\right)\delta_{jk}\,\delta_{kl} - \frac{2(1-\eta)}{m_{\rm o}^{-}\,N}\,\delta_{jk}\,\delta_{kl}\right]\\ &\times\left[\cosh(x_j)\,\sinh(x_l) + \sinh(x_j)\,\cosh(x_l)\right]\\ &+ m\,\delta L\,\delta_{jl}\left[\cosh(x_j)\,\cosh(x_l) + \sinh(x_j)\,\sinh(x_l)\right]\end{aligned} \tag{7.737}$$

and (no summation convention over repeated indices)

$$\begin{aligned}\langle\Delta_{jk}\,\Delta_{kl}\rangle_{\delta L} \approx& \frac{m_{\rm o}^{-}\,\delta L}{2\,\gamma(\eta)\,\ell}\left[\delta_{jl}\,\delta_{kk} + \left(\frac{2}{m_{\rm o}^{-}} - 1\right)\delta_{jk}\,\delta_{kl} - \frac{2(1-\eta)}{m_{\rm o}^{-}\,N}\,\delta_{jk}\,\delta_{kl}\right]\\ &\times\left[\cosh(x_j)\,\cosh(x_k) + \sinh(x_j)\,\sinh(x_k)\right]\\ &\times\left[\cosh(x_k)\,\cosh(x_l) + \sinh(x_k)\,\sinh(x_l)\right].\end{aligned} \tag{7.738}$$

Carrying the sum over repeated indices explicitly in Eq. (7.737) gives

$$\begin{aligned}\langle\Delta_{jl}\rangle_{\delta L} \approx& \,\delta_{jl}\,\frac{m_{\rm o}^{-}\,\delta L}{2\,\gamma(\eta)\,\ell}\left[N + \left(\frac{2}{m_{\rm o}^{-}} - 1\right) - \frac{2(1-\eta)}{m_{\rm o}^{-}\,N}\right]\sinh(2x_l)\\ &+ \delta_{jl}\,m\,\delta L\,\cosh(2x_j)\end{aligned} \tag{7.739a}$$

whereas, for any choice $j,k,l = 1,\cdots,N$ with $j \neq k$,

$$\langle\Delta_{jk}\,\Delta_{kl}\rangle_{\delta L} \approx \delta_{jl}\,\frac{m_{\rm o}^{-}\,\delta L}{2\,\gamma(\eta)\,\ell}\,\cosh^2(x_j + x_k), \tag{7.739b}$$

where the square bracket on the right-hand side of Eq. (7.739a) is none but

$$\frac{2\,\gamma(\eta)}{m_{\rm o}^{-}} = N + \left(\frac{2}{m_{\rm o}^{-}} - 1\right) - \frac{2(1-\eta)}{m_{\rm o}^{-}\,N}. \tag{7.739c}$$

We now return to the calculation of $\langle\delta_j\rangle_{\delta L}$ using Eqs. (7.729), (7.739a), and (7.739b).

7.28.7 Show that

$$\begin{aligned}\langle\delta_j\rangle_{\delta L} \approx& \,m\,\delta L\,\cosh(2x_j) + \frac{\delta L}{\ell}\,\sinh(2x_j)\\ &+ \frac{m_{\rm o}^{-}\,\delta L}{2\,\gamma(\eta)\,\ell}\sum_{1\leq k\leq N}^{k\neq j}\frac{2\,\cosh^2(x_j + x_k)}{\sinh(2x_j) - \sinh(2x_k)}\end{aligned} \tag{7.740}$$

for $j = 1,\cdots,N$.

Second moment Needed are $\langle\delta_j\,\delta_k\rangle_{\delta L}$ up to second order in $\mathsf{v}(x)$ or $\mathsf{w}(x)$. From the fact that the random Hermitean matrix Δ is of first order in $\mathsf{v}(x)$ or $\mathsf{w}(x)$, follows that

$$\Big\langle\delta_j\,\delta_k\Big\rangle_{\delta L} \approx \Big\langle\Delta_{jj}\,\Delta_{kk}\Big\rangle_{\delta L}, \qquad j,k = 1,\cdots,N, \tag{7.741}$$

where only the contributions proportional to $\mathsf{w}(x)$ should be included in Δ, that is,

$$\begin{aligned}\Delta_{jj} \approx&\left[\cosh^2(x_j)+\sinh^2(x_j)\right]\int\limits_{L}^{L+\delta L}\mathrm{d}x\, w_{jj}(x)\\ =&\cosh(2x_j)\int\limits_{L}^{L+\delta L}\mathrm{d}x\, w_{jj}(x).\end{aligned} \tag{7.742}$$

7.28.8 Show that

$$\begin{aligned}\left\langle\delta_j\,\delta_k\right\rangle_{\delta L}\approx&\cosh(2x_j)\cosh(2x_k)\int\limits_{L}^{L^{+}\delta L}\mathrm{d}x\int\limits_{L}^{L^{+}\delta L}\mathrm{d}x'\left\langle w_{jj}(x)\,w_{kk}(x')\right\rangle_{\delta L}\\ =&\frac{m_{\mathrm{o}}^{-}\,\delta L}{2\,\gamma(\eta)\,\ell}\left[\frac{2}{m_{\mathrm{o}}^{-}}\delta_{jk}-\frac{2(1-\eta)}{m_{\mathrm{o}}^{-}\,N}\right]\cosh(2x_j)\cosh(2x_k)\end{aligned} \tag{7.743}$$

for $j,k=1,\cdots,N$.

7.28.9 **Higher moments** Explain why all higher moments of δ_j vanish within the present approximation.

7.28.10 **Some algebra** We have calculated all integer moments of the random shift δ_j in the eigenvalues $\cosh(x_j)\sinh(x_j)$, $j=1,\cdots,N$ caused by the multiplication from the right of the transfer matrix $\mathcal{M}$ for a slab of length L by the transfer matrix $\widetilde{\mathcal{M}}$ for a slab of infinitesimal length δL (up to second order in the disorder in the region $L<x<L+\delta L$). We now calculate all integer moments of the corresponding change in x_j.

Do the Taylor expansion

$$\begin{aligned}2\delta_j\equiv&\sinh\left(2(x_j+\delta x_j)\right)-\sinh(2x_j)\\ =&2\cosh(2x_j)\,\delta x_j+\frac{4}{2}\sinh(2x_j)\,(\delta x_j)^2+\cdots\end{aligned} \tag{7.744a}$$

for any $j=1,\cdots,N$. Verify that inversion of this Taylor expansion yields

$$\delta x_j=\frac{1}{2\cosh(2x_j)}2\delta_j-\frac{\sinh(2x_j)}{4\cosh^3(2x_j)}(2\delta_j)^2+\cdots, \tag{7.744b}$$

$$\delta x_j\,\delta x_k=\frac{1}{2\cosh(2x_j)}2\delta_j\,\frac{1}{2\cosh(2x_k)}2\delta_k+\cdots, \tag{7.744c}$$

for $j,k=1,\cdots,N$.

7.28.11 Verify that insertion of Eqs. (7.740) and (7.743) on the right-hand side of Eq. (7.744b) gives

$$
\begin{aligned}
\left\langle\delta x_j\right\rangle_{\delta L} \approx\, & m\,\delta L + \frac{\delta L}{\ell}\,\underline{\frac{\gamma(\eta)}{\gamma(\eta)}}\,\frac{\sinh(2x_j)}{\cosh(2x_j)} \\
& + \frac{m_{\rm o}^{-}\,\delta L}{2\,\gamma(\eta)\,\ell}\sum_{1\leq k\leq N}^{k\neq j}\frac{2\cosh^2(x_j+x_k)}{\cosh(2x_j)\,[\sinh(2x_j)-\sinh(2x_k)]} \\
& - \frac{\delta L}{\ell}\,\underline{\left[\frac{1}{\gamma(\eta)}-\frac{(1-\eta)}{\gamma(\eta)\,N}\right]}\,\frac{\sinh(2x_j)}{\cosh(2x_j)}.
\end{aligned}
\tag{7.745}
$$

As the underlined multiplicative terms combine into

$$
\frac{1}{\gamma(\eta)}\left[\gamma(\eta)-1+\frac{(1-\eta)}{N}\right]=\frac{m_{\rm o}^{-}\,(N-1)}{2\,\gamma(\eta)},
\tag{7.746}
$$

one verifies that

$$
\left\langle\delta x_j\right\rangle_{\delta L} \approx \left[m\,\ell + \frac{m_{\rm o}^{-}}{2\,\gamma(\eta)}\sum_{1\leq k\leq N}^{k\neq j}\coth(x_j-x_k)\right]\frac{\delta L}{\ell}
\tag{7.747}
$$

must hold for any $j=1,\cdots,N$. *Hint:* Use the hyperbolic identity

$$
\tanh(2x_j)+\frac{2\cosh^2(x_j+x_k)}{\cosh(2x_j)\,[\sinh(2x_j)-\sinh(2x_k)]}=\coth(x_j-x_k).
\tag{7.748}
$$

Equation (7.713a) has been proven.

7.28.12 Verify that insertion of Eq. (7.743) on the right-hand side of Eq. (7.744c) gives

$$
\left\langle\delta x_j\,\delta x_k\right\rangle_{\delta L} \approx \frac{m_{\rm o}^{-}}{2\,\gamma(\eta)}\left[\frac{2}{m_{\rm o}^{-}}\,\delta_{jk}-\frac{2(1-\eta)}{m_{\rm o}^{-}\,N}\right]\frac{\delta L}{\ell}
\tag{7.749}
$$

for any $j,k=1,\cdots,N$. Equation (7.713b) has been proven.

7.28.13 Prove Eq. (7.713c).

Brownian motion The random process

$$
\left\langle\delta x_j\right\rangle_{\delta L} = \left[m\,\ell + \frac{m_{\rm o}^{-}}{2\,\gamma(\eta)}\sum_{1\leq k\leq N}^{k\neq j}\coth(x_j-x_k)\right]\frac{\delta L}{\ell}+\mathcal{O}\left[\left(\frac{\delta L}{\ell}\right)^2\right],
\tag{7.750a}
$$

$$
\left\langle\delta x_j\,\delta x_k\right\rangle_{\delta L} = \frac{m_{\rm o}^{-}}{2\,\gamma(\eta)}\left[\frac{2}{m_{\rm o}^{-}}\,\delta_{jk}-\frac{2(1-\eta)}{m_{\rm o}^{-}\,N}\right]\frac{\delta L}{\ell}+\mathcal{O}\left[\left(\frac{\delta L}{\ell}\right)^2\right],
\tag{7.750b}
$$

$$
\left\langle\delta x_j\,\delta x_k\,\cdots\right\rangle_{\delta L} = 0+\mathcal{O}\left[\left(\frac{\delta L}{\ell}\right)^2\right],
\tag{7.750c}
$$

describes the Brownian motion obeyed by the Lyapunov exponents of the chiral transfer matrices. Equation (7.750) is a discrete version of a first-order differential equation as $L \to L+\delta L$. In Eq. (7.750a), all Lyapunov exponents are subjected to the same constant "force" $m\,\ell$. They also exert a translation invariant two-body "force" that is repulsive if $x_j - x_k > 0$ with an infinite hardcore as $x_j - x_k \to 0^+$ and nonvanishing but finite limiting value as $x_j - x_k \to \infty$.

7.28.1 Explain why, when $x_1 \ll \cdots \ll x_N$, the scaling equations (7.750) obeyed by the mean values of the Lyapunov exponents separate and show small Gaussian fluctuations with the variance

$$\operatorname{var} x_j = \frac{m_{\mathrm{o}}^-}{2\,\gamma(\eta)}\left[\frac{2}{m_{\mathrm{o}}^-}1 - \frac{2(1-\eta)}{m_{\mathrm{o}}^-\,N}\right]\frac{L}{\ell} \tag{7.751a}$$

around the equilibrium positions

$$\langle x_j\rangle = (N+1-2j-f)\,\frac{m_{\mathrm{o}}^-}{2\,\gamma(\eta)}\,\frac{L}{\ell}, \qquad f := -\frac{2\,\gamma(\eta)}{m_{\mathrm{o}}^-}\,m\,\ell. \tag{7.751b}$$

This is the so-called *"crystallization of the transmission eigenvalues."*

7.28.2 Explain under what conditions this "crystallization of transmission eigenvalues" is a signature of exponential localization for the chiral symmetry classes. To this end, define the channel-dependent length scale

$$\xi_j := \frac{L}{\langle x_j\rangle}, \qquad j=1,\cdots,N, \tag{7.752}$$

and give an interpretation for this length scale.

7.28.3 Explain why the condition

$$(N+1-2j-f)=0, \qquad j=1,\cdots,N \tag{7.753}$$

is a signature for the absence of exponential localization that can be interpreted as a disorder-induced quantum critical point. Observe that for any given N, there are N such quantum critical points upon varying the dimensionless dimerization gap f.

7.28.4 When the dimensionless dimerization gap f is vanishing, explain the difference between chiral quantum wires with an even and odd number N of channels.

According to the general theory of Brownian motion,[121] the joint probability distribution of the Lyapunov exponents obeys the Fokker-Planck equation

$$\frac{\partial P}{\partial L} := \sum_{j,k=1}^{N}\frac{\partial}{\partial x_j}\left(\frac{1}{2}\frac{\langle\delta x_j\,\delta x_k\rangle_{\delta L}}{\delta L}\frac{\partial}{\partial x_k} - \delta_{jk}\,\frac{\langle\delta x_k\rangle_{\delta L}}{\delta L}\right)P, \tag{7.754a}$$

[121] N. G. Van Kampen, *Stochastic Processes in Physics and Chemistry*, Elsevier Science, Amsterdam, 2011. [Van Kampen (2011)].

where

$$\left\langle \delta x_j \right\rangle_{\delta L} = \left[m\,\ell + \frac{m_{\rm o}^-}{2\,\gamma(\eta)} \sum_{1\leq k\leq N}^{k\neq j} \coth(x_j - x_k) \right] \frac{\delta L}{\ell}, \tag{7.754b}$$

$$\left\langle \delta x_j\, \delta x_k \right\rangle_{\delta L} = \frac{m_{\rm o}^-}{2\,\gamma(\eta)} \left[\frac{2}{m_{\rm o}^-}\,\delta_{jk} - \frac{2(1-\eta)}{m_{\rm o}^-\,N} \right] \frac{\delta L}{\ell}. \tag{7.754c}$$

7.28.5 Show that Eq. (7.754) takes the form (7.714).

7.29 We are going to prove Eqs. (7.389a) and (7.389b). Let

$$\bar{x} := \frac{1}{N} \sum_{j=1}^{N} x_j \tag{7.755a}$$

define the center of mass coordinate and let

$$y_j := x_j - \bar{x}, \qquad j = 1, \cdots, N \tag{7.755b}$$

define the N relative coordinates, of which only $N-1$ are independent in view of

$$\sum_{j=1}^{N} y_j = N\,\bar{x} - N\bar{x} = 0. \tag{7.755c}$$

7.29.1 Show that, for $j = 1, \cdots, N-1$,

$$\frac{\partial}{\partial x_j} = \frac{1}{N}\left(\frac{\partial}{\partial \bar{x}} - \bar{\partial}\right) + \frac{\partial}{\partial y_j}, \qquad \bar{\partial} := \sum_{k=1}^{N-1} \frac{\partial}{\partial y_k}, \tag{7.756a}$$

while

$$\frac{\partial}{\partial x_N} = \frac{1}{N}\left(\frac{\partial}{\partial \bar{x}} - \bar{\partial}\right), \qquad \bar{\partial} := \sum_{k=1}^{N-1} \frac{\partial}{\partial y_k}. \tag{7.756b}$$

7.29.2 Verify that

$$\nabla := \sum_{j=1}^{N} \frac{\partial}{\partial x_j} = \left(\frac{\partial}{\partial \bar{x}} - \bar{\partial}\right) + \bar{\partial} = \frac{\partial}{\partial \bar{x}}. \tag{7.757}$$

Accordingly, ∇ only depends on the derivative with respect to the center of mass.

7.29.3 Define the Laplacian

$$\Delta := \sum_{j=1}^{N} \frac{\partial^2}{\partial x_j^2}. \tag{7.758}$$

Show that

$$\Delta = \frac{1}{N}\frac{\partial^2}{\partial \bar{x}^2} + \sum_{j,k=1}^{N-1} \frac{\partial}{\partial y_j}\left(\delta_{jk} - \frac{1}{N}\right)\frac{\partial}{\partial y_k}. \tag{7.759}$$

Accordingly, the center of mass and the relative coordinates separate in the Laplacian.

7.29.4 Finally, show that

$$\nabla \otimes \nabla \equiv \sum_{j,k=1}^{N} \frac{\partial}{\partial x_j}\frac{\partial}{\partial x_k} = \frac{\partial^2}{\partial \bar{x}^2} \tag{7.760}$$

when represented in terms of the center of mass and relative coordinates.

7.29.5 We can now also manipulate Eq. (7.759) as follows. We write

$$\begin{aligned}\Delta =& \frac{1}{N}\frac{\partial^2}{\partial \bar{x}^2} + \sum_{j,k=1}^{N-1} \frac{\partial}{\partial y_j}\left(\delta_{jk} - \frac{1}{N}\right)\frac{\partial}{\partial y_k}\\ =& \frac{1}{N}\frac{\partial^2}{\partial \bar{x}^2} + \delta(y_1 + \cdots + y_N)\sum_{j,k=1}^{N} \frac{\partial}{\partial y_j}\left(\delta_{jk} - \frac{1}{N}\right)\frac{\partial}{\partial y_k}\\ =& \frac{1}{N}\frac{\partial^2}{\partial \bar{x}^2} + \delta(N\,\bar{y})\sum_{j=1}^{N} \frac{\partial}{\partial y_j}\frac{\partial}{\partial y_j} - \frac{1}{N}\,\delta(N\,\bar{y})\,\frac{\partial^2}{\partial \bar{y}^2},\end{aligned} \tag{7.761a}$$

where we have introduced the notation

$$\bar{y} := \frac{y_1 + \cdots + y_N}{N}. \tag{7.761b}$$

Verify that the Fokker Planck equation [see Eq. (7.714a)]

$$\frac{\partial P}{\partial L} = \frac{1}{2\,\gamma(\eta)\,\ell}\sum_{i,j=1}^{N}\frac{\partial}{\partial x_i}\left[m_{\mathrm{o}}^{-}\,f\,\delta_{ij} + \left(\delta_{ij} - \frac{1-\eta}{N}\right) J\,\frac{\partial}{\partial x_j}\,J^{-1}\right] P, \tag{7.762a}$$

with

$$P(x_1,\cdots,x_N,L) = \bar{P}(\bar{x},L) \times P_{\mathrm{rel}}(y_1,\cdots,y_N,L)|_{\bar{y}=0} \tag{7.762b}$$

and

$$J(x_1,\cdots,x_N) = J_{\mathrm{rel}}(y_1,\cdots,y_N)|_{\bar{y}=0} \tag{7.762c}$$

separates into the pair of independent Fokker-Planck equations

$$\frac{\partial \bar{P}}{\partial L} = \frac{\partial}{\partial \bar{x}}\left(D_{\mathrm{rel}}\,m_{\mathrm{o}}^{-}\,f + \frac{\overline{D}}{N}\frac{\partial}{\partial \bar{x}}\right)\bar{P}, \quad D_{\mathrm{rel}} := \frac{1}{2\,\gamma(\eta)\,\ell}, \quad \overline{D} := \frac{\eta}{2\,\gamma(\eta)\,\ell}, \tag{7.763a}$$

and

$$\left.\frac{\partial P_{\mathrm{rel}}}{\partial L}\right|_{\bar{y}=0} = D_{\mathrm{rel}} \sum_{j=1}^{N} \frac{\partial}{\partial y_j} \left[J_{\mathrm{rel}} \frac{\partial}{\partial y_j} \left(J_{\mathrm{rel}}^{-1} P_{\mathrm{rel}} \right) \right]\Bigg|_{\bar{y}=0} . \tag{7.763b}$$

Here, we chose to write the diffusion coefficient for the center of mass $\bar{x}$ as $\overline{D}/N$, that is, the diffusion coefficient that one would obtain for $\bar{x}$ if the N variables x_j were diffusing independently with the diffusion coefficient D_{rel}. We may thus interpret the choice $\eta = 1$ as corresponding to the "diffusion constants" $\overline{D}$ and D_{rel} for the independent Fokker-Planck equations of the center of mass and relative coordinates, respectively, being equal and given by $(2\,\gamma\,\ell)^{-1}$.

7.30 Equation (7.401d) is an example of $N^{\star}$ independent diffusion-convection equations in one-dimensional space. Let $u(x,t)$ be a real-valued function that satisfies the partial second-order differential equation

$$u_t = D\,u_{xx} + C\,u_x, \qquad D \geq 0, \qquad C \in \mathbb{R}, \tag{7.764a}$$

with the boundary equations

$$\lim_{x \to \pm\infty} u(x,t) = 0 \tag{7.764b}$$

and the initial condition

$$u(x, t=0) = \delta(x). \tag{7.764c}$$

Here, u_t denotes the first-order partial derivative with respect to $t \in \mathbb{R}$ of u, u_x denotes the first-order partial derivative with respect to $x \in \mathbb{R}$ of u, and u_{xx} denotes the second-order partial derivative with respect to $x \in \mathbb{R}$ of u. In physics and engineering, x is interpreted as a spatial coordinate, while t is interpreted as time. In Eq. (7.401d), the role of time is played by the length L of the quantum wire and one-dimensional space is the domain of definition of the Lyapunov exponents from the chiral symmetry classes. We are going to show that the normalized solution of Eq. (7.764) is given by

$$u(x,t) = \sqrt{\frac{1}{4\,\pi\,D\,t}}\, e^{-\frac{1}{4\,D\,t}(x+C\,t)^2}. \tag{7.765}$$

7.30.1 *Step 1:* Show that a solution to the diffusion-convection equation (7.764) is of the form

$$u(x,t) = e^{-\frac{C^2}{4\,D}\,t - \frac{C}{2\,D}\,x}\, w(x,t), \tag{7.766a}$$

where the function $w(x,t)$ obeys the pure diffusion equation

$$w_t = D\,w_{xx}, \qquad D > 0, \tag{7.766b}$$

with the boundary conditions

$$\lim_{x\to\pm\infty} e^{-\frac{C^2}{4D}t-\frac{C}{2D}x}\,w(x,t)=0 \tag{7.766c}$$

and the initial condition

$$e^{-\frac{C}{2D}x}\,w(x,t=0)=\delta(x) \tag{7.766d}$$

in one-dimensional space.

7.30.2 *Step 2:* Show that the normalized solution to Eqs. (7.766b)–(7.766d) is

$$w(x,t)=\sqrt{\frac{1}{2\pi\,(2\,D\,t)}}\,e^{-\frac{x^2}{4\,D\,t}}=\sqrt{\frac{1}{4\,\pi\,D\,t}}\,e^{-\frac{x^2}{4\,D\,t}}. \tag{7.767}$$

7.30.3 *Step 3:* Insert Eq. (7.767) into Eq. (7.766a) to complete the proof of Eq. (7.765).

8

The Tenfold Way: Gapped Phases in Any Dimensions

8.1 Introduction

We have seen in Chapter 7 (Section 7.8) that the set of allowed normalized Dirac masses of the Dirac equation with Dirac matrices of rank $r \geq r_{\text{min}}$ is a topological space for each one of the ten Altland-Zirnbauer symmetry classes in one-dimensional space. The same exercise can be repeated for any given dimension d of space and for any rank $r = r_{\text{min}} N$ ($N = 1, 2, 3, \cdots$) of the Dirac matrices. Here, r_{min} is the minimal rank of the Dirac matrices that is compatible with the existence of a Dirac mass matrix given the Altland-Zirnbauer symmetry class and the dimensionality d of space. This chapter carries out this program.

It is explained in Section 8.2 that the set of allowed normalized Dirac masses is a classifying space associated with the extension problem of a Clifford algebra. (A mathematical reference to classifying spaces was written by Karoubi.[1]) The solution of this extension problem is given in the sixth column of **Table** 8.1. The homotopy groups of these classifying spaces are known from the mathematical literature. They are given in **Table** 8.2. A zeroth homotopy group in **Table** 8.2 dictates if the classifying space is path-connected or not. When it is not, the cardinality of the zeroth homotopy group is the number of distinct connected topological subspaces making up the classifying space. All zeroth homotopy groups are reported in **Table** 8.3. This table defines what are strong topological insulators or superconductors, as will be explained. The entries of this table are derived in a certain limit ($N \to \infty$). It is only when this limit is taken ($N \to \infty$) that **Table** 8.3 is "exhaustive."

Strong topological insulators or superconductors support boundary states with the remarkable property that they are delocalized along the boundary.[2,3,4] There is a sense in which these boundary states evade the phenomenon of Anderson localization. The existence of these delocalized boundary states has consequences for the

[1] M. Karoubi, *K-Theory: An Introduction,* Springer, Berlin Heidelberg, 1978. [Karoubi (2008)].

[2] A. P. Schnyder, S. Ryu, A. Furusaki, and A. W. W. Ludwig, Phys. Rev. B **78**(19), 195125 (2008). [Schnyder et al. (2008)].

[3] S. Ryu, A. P. Schnyder, A. Furusaki, and A. W. W. Ludwig, New J. Phys. **12**(6), 065010 (2010). [Ryu et al. (2010)].

[4] A. Kitaev, AIP Conf. Proc. **1134**, 22 (2009). [Kitaev (2009)].

Table 8.1 *The tenfold way: Classifying spaces*

Class	TRS	PHS	CHS	Extension	V_d	$V_{d=0,r_{\min}N}$
A	0	0	0	$C\ell_d \to C\ell_{d+1}$	C_{0+d}	$\cup_{k=0}^{N}\mathrm{Gr}(k,N;\mathbb{C})$
AIII	0	0	1	$C\ell_{d+1} \to C\ell_{d+2}$	C_{1+d}	$\mathrm{U}(N)$
AI	+1	0	0	$C\ell_{0,d+2} \to C\ell_{1,d+2}$	R_{0-d}	$\cup_{k=0}^{N}\mathrm{Gr}(k,N;\mathbb{R})$
BDI	+1	+1	1	$C\ell_{d+1,2} \to C\ell_{d+1,3}$	R_{1-d}	$\mathrm{O}(N)$
D	0	+1	0	$C\ell_{d,2} \to C\ell_{d,3}$	R_{2-d}	$\mathrm{O}(2N)/\mathrm{U}(N)$
DIII	−1	+1	1	$C\ell_{d,3} \to C\ell_{d,4}$	R_{3-d}	$\mathrm{U}(2N)/\mathrm{Sp}(2N)$
AII	−1	0	0	$C\ell_{2,d} \to C\ell_{3,d}$	R_{4-d}	$\cup_{k=0}^{N}\mathrm{Gr}(k,N;\mathbb{H})$
CII	−1	−1	1	$C\ell_{d+3,0} \to C\ell_{d+3,1}$	R_{5-d}	$\mathrm{Sp}(2N)$
C	0	−1	0	$C\ell_{d+2,0} \to C\ell_{d+2,1}$	R_{6-d}	$\mathrm{Sp}(2N)/\mathrm{U}(N)$
CI	+1	−1	1	$C\ell_{d+2,1} \to C\ell_{d+2,2}$	R_{7-d}	$\mathrm{U}(N)/\mathrm{O}(N)$

The tenfold way distinguishes the behavior of massive Dirac Hamiltonians $\mathcal{H}$ under reversal of time (T), charge conjugation (C), and the chiral transformation (Γ), with the gauge convention that $T^2 = \pm 1$, $C^2 = \pm 1$, $[T,C]=0$, and $\Gamma^2 = 1$. The entries under the columns "TRS," "PHS," and "CHS" are the values of $T^2 = \pm 1$, $C^2 = \pm 1$, $\Gamma^2 = 1$ when these (spectral) symmetries are present, with the value 0 indicating the absence of these (spectral) symmetries. The entries under the column "class" are the names of the symmetry classes assigned by Altland and Zirnbauer to the values of the triplet (T, C, Γ) in the columns "TRS," "PHS," and "CHS" that characterizes $\mathcal{H}$. The Clifford extension in column five delivers in column six the classifying space V_d in d-dimensional space associated with the family of manifolds labeled by N in column seven. Column seven gives $V_{d,r_{\min}N}$ when $d = 0$. We have introduced the shorthand notation $\mathrm{Gr}(k,N;\mathbb{C}) \equiv \mathrm{U}(N)/\big[\mathrm{U}(k)\times\mathrm{U}(N-k)\big]$, $\mathrm{Gr}(k,N;\mathbb{R}) \equiv \mathrm{O}(N)/\big[\mathrm{O}(k)\times\mathrm{O}(N-k)\big]$, and $\mathrm{Gr}(k,N;\mathbb{H}) \equiv \mathrm{Sp}(2N)/\big[\mathrm{Sp}(2k)\times\mathrm{Sp}(2N-2k)\big]$, where $\mathbb{C}$, $\mathbb{R}$, and $\mathbb{H}$ refer to the complex, real, and quaternion-real numbers.

zero-temperature phase diagram of strong topological insulators or superconductors when, as is explained in Section 8.3, the characteristic energy scale responsible for opening a band gap competes with the characteristic energy scale encoding disorder.

Table 8.3 is derived in Section 8.2 under the assumption that the electrons under consideration are the noninteracting mean-field excitations of some underlying tight-binding Hamiltonian belonging to one of the Altland-Zirnbauer symmetry classes. Interactions among electrons were shown by Fidkowski and Kitaev[5,6] to modify some of the entries of **Table** 8.3 when $d = 1$. It is shown in Section 8.4 in a systematic way how interactions modify **Table** 8.3 and, in particular, how

[5] L. Fidkowski and A. Kitaev, Phys. Rev. B **81**(13), 134509 (2010). [Fidkowski and Kitaev (2010)].
[6] L. Fidkowski and A. Kitaev, Phys. Rev. B **83**(7), 075103 (2011). [Fidkowski and Kitaev (2011)].

Table 8.2 *The tenfold way: Homotopy groups of the classifying spaces*

Label	$\pi_0(V)$	$\pi_1(V)$	$\pi_2(V)$	$\pi_3(V)$	$\pi_4(V)$	$\pi_5(V)$	$\pi_6(V)$	$\pi_7(V)$
C_0	$\mathbb{Z}$	0	$\mathbb{Z}$	0	$\mathbb{Z}$	0	$\mathbb{Z}$	0
C_1	0	$\mathbb{Z}$	0	$\mathbb{Z}$	0	$\mathbb{Z}$	0	$\mathbb{Z}$
R_0	$\mathbb{Z}$	$\mathbb{Z}_2$	$\mathbb{Z}_2$	0	$\mathbb{Z}$	0	0	0
R_1	$\mathbb{Z}_2$	$\mathbb{Z}_2$	0	$\mathbb{Z}$	0	0	0	$\mathbb{Z}$
R_2	$\mathbb{Z}_2$	0	$\mathbb{Z}$	0	0	0	$\mathbb{Z}$	$\mathbb{Z}_2$
R_3	0	$\mathbb{Z}$	0	0	0	$\mathbb{Z}$	$\mathbb{Z}_2$	$\mathbb{Z}_2$
R_4	$\mathbb{Z}$	0	0	0	$\mathbb{Z}$	$\mathbb{Z}_2$	$\mathbb{Z}_2$	0
R_5	0	0	0	$\mathbb{Z}$	$\mathbb{Z}_2$	$\mathbb{Z}_2$	0	$\mathbb{Z}$
R_6	0	0	$\mathbb{Z}$	$\mathbb{Z}_2$	$\mathbb{Z}_2$	0	$\mathbb{Z}$	0
R_7	0	$\mathbb{Z}$	$\mathbb{Z}_2$	$\mathbb{Z}_2$	0	$\mathbb{Z}$	0	0

Homotopy groups for the complex and real classifying spaces, which are a list of ten topological spaces that are built out of the compact Lie groups $\mathrm{U}(N)$, $\mathrm{O}(N)$, and $\mathrm{Sp}(2N)$, the unitary, orthogonal, and symplectic matrix groups, respectively, and their quotients as is given in the last column of **Table** 8.1 for $d = 0$. The complex classifying spaces obey the periodicity condition $\pi_p(C_q) = \pi_{p+2}(C_q)$ for $q = 0, 1$. The real classifying spaces obey the periodicity condition $\pi_p(R_q) = \pi_{p+8}(R_q)$ for $q = 0, \cdots, 7$. Hence, an exhaustive list of the homotopy groups of the classifying spaces is shown in the columns "$\pi_0(V)$," "$\pi_1(V)$," $\cdots$,"$\pi_7(V)$." In each homotopy column, the three entries $\mathbb{Z}$ hold for N larger than an integer (infinity included) that depends on the order of the homotopy group and the classifying space, the two entries $\mathbb{Z}_2$ hold for N larger than an integer that also depends on the order of the homotopy group and the classifying space. The entry 0 is a shorthand for the group $\{0\}$ made of the single element 0.

interactions break the Bott periodicity of **Table** 8.3. The main result of Section 8.4 is **Table** 8.6.

8.2 Classifying Spaces of Normalized Dirac Masses

8.2.1 Overview

The phenomenon of charge fractionalization was predicted theoretically in 1976[7] by studying the spectrum of the Dirac equation in one-dimensional space when a Dirac mass hosts domains walls. The fractional charge taking precisely the value 1/2 signals the existence of a zero mode bound to a domain wall. This was the first fermionic example of an invertible topological phase of matter, whereby charge conservation and chiral symmetry are the protecting symmetries.

The experimental discovery of the integer quantum Hall effect (IQHE) in 1980[8] was soon interpreted as the second fermionic example of an invertible topological

[7] R. Jackiw and C. Rebbi, Phys. Rev. D **13**(12), 3398–3409 (1976). [Jackiw and Rebbi (1976)].

[8] K. v. Klitzing, G. Dorda, and M. Pepper, Phys. Rev. Lett. **45**(6), 494–497 (1980). [Klitzing et al. (1980)].

Table 8.3 *The tenfold way: Bott periodicity of topological insulators and superconductors*

Class	$d=0$	$d=1$	$d=2$	$d=3$	$d=4$	$d=5$	$d=6$	$d=7$
A	$\mathbb{Z}$	0	$\mathbb{Z}$	0	$\mathbb{Z}$	0	$\mathbb{Z}$	0
AIII	0	$\mathbb{Z}$	0	$\mathbb{Z}$	0	$\mathbb{Z}$	0	$\mathbb{Z}$
AI	$\mathbb{Z}$	0	0	0	$\mathbb{Z}$	0	$\mathbb{Z}_2$	$\mathbb{Z}_2$
BDI	$\mathbb{Z}_2$	$\mathbb{Z}$	0	0	0	$\mathbb{Z}$	0	$\mathbb{Z}_2$
D	$\mathbb{Z}_2$	$\mathbb{Z}_2$	$\mathbb{Z}$	0	0	0	$\mathbb{Z}$	0
DIII	0	$\mathbb{Z}_2$	$\mathbb{Z}_2$	$\mathbb{Z}$	0	0	0	$\mathbb{Z}$
AII	$\mathbb{Z}$	0	$\mathbb{Z}_2$	$\mathbb{Z}_2$	$\mathbb{Z}$	0	0	0
CII	0	$\mathbb{Z}$	0	$\mathbb{Z}_2$	$\mathbb{Z}_2$	$\mathbb{Z}$	0	0
C	0	0	$\mathbb{Z}$	0	$\mathbb{Z}_2$	$\mathbb{Z}_2$	$\mathbb{Z}$	0
CI	0	0	0	$\mathbb{Z}$	0	$\mathbb{Z}_2$	$\mathbb{Z}_2$	$\mathbb{Z}$

The ten Altland-Zirnbauer symmetry classes and their topological classification in terms of the zeroth homotopy groups of their classifying spaces define the tenfold way of strong topological insulators or superconductors.

phase of matter. A noninteracting but disordered electron gas confined to the plane normal to a strong uniform applied magnetic field carries current-carrying boundary states even though all bulk states are localized.[9,10] Charge conservation is here the only protecting symmetry, that is, the prerequisite for the IQHE is the continuity equation for the electric charge in two-dimensional space. A spin quantum Hall effect was also predicted from the general study of topological field theories in $(2+1)$-dimensional spacetime and their applications to condensed matter physics.[11,12] Hereto, it is a residual U(1) spin-rotation symmetry around a quantization axis in spin-1/2 space that is the protecting symmetry, that is, a continuity equation for the spin-1/2 component along the quantization axis holds.

Charge is not conserved in a Bogoliubov-de-Gennes superconductor, but energy is and spin can be conserved. Hence, thermal and spin counterparts to the (charge) IQHE for the symmetry classes D and C, respectively, were predicted in the late 1990s.[13,14,15,16] However, none of these proposals for realizing thermal and spin

9 R. B. Laughlin, Phys. Rev. B. **23**(10), 5632–5633 (1981). [Laughlin (1981)].
10 B. I. Halperin, Phys. Rev. B **25**(4), 2185–2190 (1982). [Halperin (1982)].
11 J. Fröhlich and U. M. Studer, Commun. Math. Phys. **148**(3), 553–600 (1992). [Fröhlich and Studer (1992)].
12 J. Fröhlich and U. M. Studer, Rev. Mod. Phys. **65**(3), 733–802 (1993). [Fröhlich and Studer (1993)].
13 T. Senthil, M. P. A. Fisher, L. Balents, and C. Nayak, Phys. Rev. Lett. **81**(21), 4704–4707 (1998). [Senthil et al. (1998)].
14 I. A. Gruzberg, A. W. W. Ludwig, and N. Read, Phys. Rev. Lett. **82**(22), 4524–4527 (1999). [Gruzberg et al. (1999)].
15 T. Senthil and M. P. A. Fisher, Phys. Rev. B **61**(14), 9690–9698 (2000). [Senthil and Fisher (2000)].
16 N. Read and D. Green, Phys. Rev. B **61**(15), 10267–10297 (2000). [Read and Green (2000)].

quantum Hall effects in two-dimensional superconductors have been realized in the laboratory so far.

Before leaving the realm of quantized Hall responses inherited from continuity equations deriving from global U(1) symmetries, it must be emphasized that it is the breaking of the time-reversal symmetry that is essential to the (charge) IQHE. Neither the presence of disorder nor of an applied magnetic field are necessary, as shown by Haldane in 1988.[17] Indeed, if one allows a spinless electron to hop with a uniform amplitude between the nearest-neighbor sites of the honeycomb lattice (the mininal model for graphene), the conduction and valence bands then touch at the two inequivalent corners of the Brillouin zone, in which case the Fermi surface at half-filling collapses to two inequivalent Fermi points. In the close vicinity of either one of these two Fermi points (valleys), the single-particle dispersion is that of a gapless Dirac Hamiltonian whose spinor (that is eigenstate) is made of two components.[18] The achievement of Haldane was to add a complex-valued next-nearest-neigbhor hopping that opens a gap at these Fermi points in such a way that the Hall conductivity is of magnitude 1 in units of e^2/h for any value of the chemical potential within this gap. In the continuum approximation by which the full lattice dispersion is replaced by that of two massive Dirac spectra at sufficiently low energies, one for each inequivalent Fermi point, each massive Dirac spectrum arises from the same massive Dirac Hamiltonian. In magnitude, the Hall conductivity of each massive Dirac Hamiltonian takes the fractional value 1/2 in units of e^2/h.[19] They add up to the value 1 in units of e^2/h, the same value that follows from computing the Hall conductivity directly from the microscopic lattice Hamiltonian. noninteracting two-dimensional lattice Hamiltonians that do not break the translation symmetry of the underlying Bravais lattice, while breaking time-reversal symmetry without a net magnetic field threading the repeat unit cell of the lattice, and have a gapped nondegenerate ground state characterized by a nonvanishing quantized Hall conductivity when periodic boundary conditions hold are called Chern band insulators.

Haldane's 1988 paper had garnered about 50 Web of Science citations by 2005 out of more than 3'000 Web of Science citations by 2020. In 2001, Kitaev wrote another sleeping-beauty paper in which he defined the Kitaev chain (5.178).[20] This paper had garnered four citations from the Web of Science by February 2010. During the next ten years, it garnered more than 2'000 Web of Science citations. The Kitaev chain was the first example of a $\mathbb{Z}_2$ topological superconductor (see Chapter 5). It is the conservation of the fermion parity that is the protecting symmetry in the Kitaev chain.

[17] F. D. M. Haldane, Phys. Rev. Lett. **61**(18), 2015–2018 (1988). [Haldane (1988)].
[18] P. R. Wallace, Phys. Rev. **71**(9), 622–634 (1947). [Wallace (1947)].
[19] S. Deser, R. Jackiw, and S. Templeton, Ann. Phys. (N.Y.) **140**(2), 372–411 (1982); *Erratum*, **185**(2), 406–406 (1988). [Deser et al. (1982, 1988)].
[20] A. Kitaev, Phys. Usp. **44**(10S), 131–136 (2001). [Kitaev (2001)].

Had the sign of the Dirac masses in the two Dirac Hamiltonians encoding the low-energy dispersions around the two inequivalent Fermi points of graphene been of opposite signs, the Hall conductivity from one massive Dirac Hamiltonian would have the opposite sign from that of the other Dirac Hamiltonian. This is what happens if the sublattice symmetry of the half-filled tight-binding Hamiltonian on the honeycomb lattice with only uniform nearest-neighbor hopping is perturbed by breaking the sublattice symmetry with a staggered chemical potential that penalizes the on-site energy to be on one sublattice relative to the other.[21] Such a perturbation to uniform nearest-neighbor hopping does not break time-reversal symmetry, unlike the one chosen by Haldane. However, there is another way to open a gap at the Fermi points of graphene without breaking time-reversal symmetry that delivers a two-dimensional $\mathbb{Z}_2$ topological insulator, as was realized by Kane and Mele, if one makes uses of the spin-1/2 degrees of freedom of electrons cleverly.

One year after the discovery of graphene in 2004,[22] Kane and Mele made a theoretical prediction that was immediately impactfull, unlike the prediction that an open Kitaev chain supports Majorana zero modes at its ends. Namely, a noninteracting tight-binding model for graphene can realize a fermionic invertible topological phase of matter without breaking time-reversal symmetry.[23,24] In this proposal, charge conservation and time-reversal symmetry are the protecting symmetries. Spin-rotation symmetry is presumed to be completely broken by spin-orbit coupling (symmetry class AII). If periodic boundary conditions are imposed, the single-particle spectrum is gapped and the many-body ground state at half-filling is nondegenerate. The latter is a gapped many-body ground state (a Slater determinant). If open boundary conditions are imposed and upon tuning of a single dimensionless parameter (the ratio of the Dresselhaus to Rashba spin-orbit couplings), one single pair of states localized along one of the disconnected components of the one-dimensional boundary displays a dispersion that connects the empty conduction bands to the filled valence bands. The two edge states making up this Kramers' degenerate pair of counterpropagating modes are helical in that their single-particle spin expectation value is proportional to their single-particle momenta along the boundary. No single-particle backscattering potential between these two counterpropagating edge states is allowed because of time-reversal symmetry. Only backscattering between helical edge states on distinct disconnected components of the one-dimensional boundary are allowed, but with an amplitude that is exponentially suppressed as a function of the dimensionless number ℓ/ξ with ℓ the distance between the two disconnected components of the boundary and ξ inversely proportional to the gap separating the conduction from the valance bands

[21] G. W. Semenoff, Phys. Rev. Lett. **53**(26), 2449–2452 (1984). [Semenoff (1984)].
[22] K. S. Novoselov, A. K. Geim, S. V. Morozov, et al., Science **306**, 666–669 (2004). [Novoselov et al. (2004)].
[23] C. L. Kane and E. J. Mele, Phys. Rev. Lett. **95**(22), 226801 (2005). [Kane and Mele (2005b)].
[24] C. L. Kane and E. J. Mele, Phys. Rev. Lett. **95**(14), 146802 (2005). [Kane and Mele (2005a)].

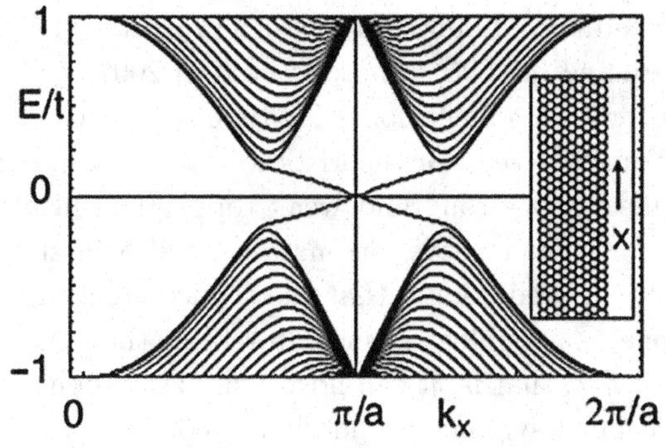

Figure 8.1 Energy eigenvalue spectrum of the tight-binding Hamiltonian used by Kane and Mele in a slab geometry with open boundary conditions along the y direction and periodic boundary conditions along the x direction, after Figure 1 from Kane and Mele (2005b). The dispersions crossing the gap are twofold degenerate. They belong to four edge states, two counterpropagating helical edge states per edge that are interchanged by reversal of time.

when periodic boundary conditions are imposed (see Figure 8.1). Unlike for the (charge) IQHE for which it is the net chirality of the current carrying boundary (edge) states that is protected by charge conservation, boundary (edge) states come in Kramers' degenerate (helical) pairs and it is the parity of the number of these helical pairs that is protected by the combination of charge conservation and time-reversal symmetry. Correspondingly, the topological index does not belong to the set of the integers that generates the group $\mathbb{Z}$ but is a two-valued index, say 0 or 1, that generates the group $\mathbb{Z}_2$. Kane and Mele also constructed a two-valued topological invariant out of all occupied Bloch states when periodic boundary conditions are imposed that plays the role of the Thouless-Kohmoto-Nightingale-Nijs formula[25] for the integer-valued Hall conductivity in units of e^2/h for the IQHE when realized for a noninteracting tight-binding Hamiltonian on a Bravais lattice. In either case, the bulk topological invariant and the number of protected edge states are related by the so-called bulk-boundary correspondence 10.[26,27] The Kane-Mele example of an invertible topological phase of matter is an example of a $\mathbb{Z}_2$ topological insulator.

However, the size of the spin-orbit coupling in graphene cannot be resolved with the lowest temperatures presently attainable when performing transport measurements. This is not the case in quantum wells (very thin layers) of mercury telluride

[25] D. J. Thouless, M. Kohmoto, M. P. Nightingale, and M. den Nijs, Phys. Rev. Lett. **49**(6), 405–408 (1982). [Thouless et al. (1982)].

[26] C. G. Callan, Jr. and J. A. Harvey, Nucl. Phys. B **250**, 427–436 (1985). [Callan and Harvey (1985)].

[27] Y. Hatsugai, Phys. Rev. Lett. **71**(22), 3697–3700 (1993). [Hatsugai (1993)].

sandwiched between cadmium telluride, for which protected one-dimensional helical edge states were predicted in 2006[28] and observed in 2007.[29]

After predictions for three-dimensional $\mathbb{Z}_2$ topological insulators were made in 2007,[30,31,32] it took another year for theorists to recognize an organization pattern that treats in a unified way strong and weak topological insulators or superconductors (the usage for the adjectives strong and weak will shortly be explained). Crucially, the theoretical prediction[33] that the semiconducting alloy $\mathrm{Bi}_{1-x}\mathrm{Sb}_x$ can be a three-dimensional $\mathbb{Z}_2$ topological insulator from the symmetry class AII depending on the value of x, that is, it can host conducting boundary states that are protected by time-reversal symmetry and charge conservation and whose (Dirac-like) dispersions connect the bulk conduction and valence bands, was verified experimentally with the help of angle-resolved photoemission spectroscopy (ARPES).[34] Angle-resolved photoemission spectroscopy is an experimental technique by which incoming light ejects electrons from a material with a subsequent measurement of the kinetic energy and momentum distributions of the emitted electrons. This technique allows to map the electronic band structure of crystalline materials, that is, the dispersion relation between the energy and momentum of sharp electronic excitations present in a crystal. Moreover, ARPES augmented with the ability to resolve the spin of the emitted electron was decisive in demonstrating that the dispersing boundary states in the semiconducting alloy $\mathrm{Bi}_{1-x}\mathrm{Sb}_x$ are Kramers' degenerate and spin polarized, that is, helical.[35] Those ARPES experiments were performed at very large synchrotron facilities that were built at the turn of the twenty-first century to deliver the necessary intensities and energy resolutions.

The organization pattern for strong and weak topological insulators or superconductors depends on the choice made for the protecting symmetries and on specifying which boundaries must support symmetry-protected boundary states.

The protecting symmetries of strong topological insulators or superconductors are in one-to-one correspondence with the symmetries defining the ten Altland-Zirnbauer symmetry classes of random matrix theory. By definition, any boundary of a strong topological insulator or superconductor must support symmetry-protected boundary states.

The protecting symmetries of weak topological insulators or superconductors in d-dimensional space are also in one-to-one correspondence with the symmetries defining the ten Altland-Zirnbauer symmetry classes of random matrix theory. However, not all $(d-1)$-dimensional boundaries of a weak topological insulators or

[28] B. A. Bernevig, T. L. Hughes, and S.-C. Zhang, Science **314**, 1757–1761 (2006). [Bernevig et al. (2006)].
[29] M. König, S. Wiedmann, C. Brüne, et al., Science **318**, 766–770 (2007). [König et al. (2007)].
[30] L. Fu, C. L. Kane, and E. J. Mele, Phys. Rev. Lett. **98**(10), 106803 (2007). [Fu et al. (2007)].
[31] J. E. Moore and L. Balents, Phys. Rev. B **75(12)**, 121306(R) (2007). [Moore and Balents (2007)].
[32] R. Roy, Phys. Rev. B **79**(19), 195322 (2009). [Roy (2009)].
[33] L. Fu and C. L. Kane, Phys. Rev. B **76**(4), 045302 (2007). [Fu and Kane (2007)].
[34] D. Hsieh, D. Qian, L. Wray, et al., Nature **452**, 970–974 (2008). [Hsieh et al. (2008)].
[35] D. Hsieh, Y. Xia, L. Wray, et al., Science **323**, 919-922 (2009). [Hsieh et al. (2009)].

superconductor must support symmetry-protected boundary states. This is so because weak topological insulators or superconductors in d-dimensional space can be thought of as the stacking of strong topological insulators or superconductors in $0 < d' < d$-dimensional space. The choice made for the stacking breaks the isotropy of space. In doing so, not all $(d-1)$-dimensional boundaries of a weak topological insulator or superconductor in d-dimensional space are equal. Only the $(d-1)$-dimensional boundaries that are "compatible" with the $(d'-1)$-dimensional boundaries of the $0 < d' < d$ strong topological insulators or superconductors support symmetry-protected boundary states. The first example of a weak topological insulator was predicted by Halperin's construction of a (charge) IQHE in three-dimensional space.[36]

Another situation besides that of weak topological insulators or superconductors for which the choice of the boundary matters, if the boundaries are to support symmetry-protected boundary modes, is that of crystalline topological insulators or superconductors. The protecting symmetries of crystalline topological insulators or superconductors in d-dimensional space are the symmetries defining the ten Altland-Zirnbauer symmetry classes of random matrix theory augmented by some crystalline symmetries. Only those boundaries that are, as a set, invariant under the protecting crystalline symmetries may support symmetry-protected boundary states in crystalline topological insulators or superconductors. The concept of crystalline topological insulators or superconductors was proposed by Fu in 2011.[37,38] Hereto, experimental validation of this theoretical prediction followed swiftly in 2012 using ARPES.[39,40,41]

Insulators or superconductors in d-dimensional space that do not support symmetry-protected states bound to their $(d-1)$-dimensional boundaries might nevertheless support symmetry-protected states on submanifolds of dimensions $d-2, \cdots, 0$, as was anticipated by Jackiw and Rossi[42] for a two-dimensional relativistic superconductor in the symmetry class C or by Hou et al.[43] for insulating graphene with a Kekulé (bond-density) gap in the symmetry class BDI. These examples were framed into a systematic organizational pattern (in essence **Table** 8.2) by making use of the theory of homotopic groups shortly after the tenfold way for strong topological insulators or superconductors was understood 3.[44] If the Altland-Zirnbauer protecting symmetries are augmented by crystalline symmetries, one arrives at the notion of higher-order topological insulators or superconductors, for example, insulators

[36] B. I. Halperin, Jpn. J. Appl. Phys. **26**, Suppl. 26-3, 1913–1919 (1987). [Halperin (1987)].
[37] L. Fu, Phys. Rev. Lett. **106**(10), 106802 (2011). [Fu (2011)].
[38] T. H. Hsieh, H. Lin, J. Liu, et al., Nature Comm. **3**, 982 (2012). [Hsieh et al. (2012)].
[39] P. Dziawa, B. Kowalski, K. Dybko, et al., Nature Mat. **11**, 1023–1027 (2012). [Dziawa et al. (2012)].
[40] Y. Tanaka, Z. Ren, T. Sato, et al., Nature Physics **8**, 800–803 (2012). [Tanaka et al. (2012)].
[41] S.-Y. Xu, C. Liu, N. Alidoust, et al., Nature Comm. **3**, 1192 (2012). [Xu et al. (2012)].
[42] R. Jackiw and P. Rossi, Nucl. Phys. B **190**, 681–691 (1981). [Jackiw and Rossi (1981)].
[43] C.-Y. Hou, C. Chamon, and C. Mudry, Phys. Rev. Lett. **98**(18), 186809 (2007). [Hou et al. (2007)].
[44] J. C. Y. Teo and C. L. Kane, Phys. Rev. B **82**(11), 115120 (2010). [Teo and Kane (2010)].

or superconductors in three-dimensional space whose two-dimensional boundaries are fully gapped when translation invariant, but may host symmetry-protected states bound to one-dimensional "defects" (edges) or zero-dimensional "defects" (corners) that break the translation invariance of the boundaries.[45,46,47,48] The bismuth-antimony alloy $Bi_{1-x}Sb_x$ was the first realization of a three-dimensional topological insulator. The composition x was used to interpolate between bismuth, an insulator that does not support symmetry-protected boundary states, and the band-inverted antimony, an insulator that does. The nominally trivial end of this interpolation, bismuth, is in fact an example of a higher-order topological insulator, as it supports symmetry-protected edge (hinge) states as was predicted theoretically and verified experimentally using scanning-tunneling microscopy and spectroscopy (STM and STS) and Josephson-interferometry measurements.[49]

The predictions of random matrix theory are universal in the limit $N \to \infty$ with N proportional to the rank of the random matrices. Similarly, some transport properties of quantum wires also become universal in the thick quantum wire limit defined in Chapter 7. Is it possible to identify a notion of universality attached to materials supporting symmetry-protected boundary states?

As in critical phenomena, universality is only expected to emerge after fixing the dimensionality of space and the protecting symmetries. Having fixed the dimension d of space and the protecting symmetries, we impose the following rules on any two d-dimensional blocks of matter A and B, each one realizing a noninteracting topological insulator or superconductor (in the sense of strong, weak, or crystalline topology). We imagine an adiabatic process by which A and B can be brought into contact along any one of their common boundaries that may host symmetry-protected boundary states so has to obtain block C. We call this process stacking.[50]

Definition 8.1 (stable stacking rules) The topological attributes of A and B as measured by the possible existence of symmetry-protected boundary states are said to be stable 4 if the following rules apply:

I If A does not host any symmetry-protected boundary states but B does, then C inherits the symmetry-protected boundary states of B.

II If A and B both host symmetry-protected boundary states, then there exists a composition rule such that the number of symmetry-protected boundary states

[45] W. A. Benalcazar, B. A. Bernevig, and T. L. Hughes, Science **357**, 61–66 (2017). [Benalcazar et al. (2017b)].

[46] W. A. Benalcazar, B. A. Bernevig, and T. L. Hughes, Phys. Rev. B **96**(24), 245115 (2017). [Benalcazar et al. (2017a)].

[47] Z. Song, Z. Fang, and C. Fang, Phys. Rev. Lett. **119**(24), 246402 (2017); J. Langbehn, Y. Peng, L. Trifunovic, F. von Oppen, and P. W. Brouwer, Phys. Rev. Lett. **119**(24), 246401 (2017). [Song et al. (2017); Langbehn et al. (2017)].

[48] F. Schindler, A. M. Cook, M. G. Vergniory, et al., Sci. Adv. **4**, eaat0346 (2018). [Schindler et al. (2018a)].

[49] F. Schindler, Z. Wang, M. G. Vergniory, et al., Nat. Phys. **14**, 918–924 (2018). [Schindler et al. (2018b)].

[50] We already encountered the notion of stacking in Section 3.12, with Eq. (5.156), with **Table** 5.8, and in Section 6.5.2.

that C inherits from A and B follows from composing the symmetry-protected boundary states of A with those of B.

III This composing rule is additive in that it forms an Abelian group, that is for any A there exists a prescription (particle-hole conjugation) that turns A into its inverse −A such that C has no symmetry-protected boundary states if B is identified with −A.

In the spirit of random matrix theory, we work in zero-dimensional space. We are going to argue, following Kitaev 4, that rules I, II, and III lead to the construction of the classifying space corresponding to the entry $d = 0$ and A of **Tables** 8.1 and 8.3. Whereas we carry this program explicitly for the symmetry class A, all remaining nine Altland-Zirnbauer symmetry classes are relegated to **Exercise** 8.1.

If we model A and B with the fermionic noninteracting Hamiltonians $\widehat{H}_{\mathsf{A}}$ and $\widehat{H}_{\mathsf{B}}$ from the symmetry class A, stability demands that we use mathematical tools that allow for the comparison of single-particle Hermitean matrices of different sizes with a gap separating all filled from all empty single-particle levels. In particular, the mathematical tools must be able to accommodate the limit $N \to \infty$ of single-particle levels.

The process by which A and B are brought into contact adiabatically so as to obtain C in such a way that rules I, II, and III apply demands that the hybridization between the single-particle levels of $\widehat{H}_{\mathsf{A}}$ and $\widehat{H}_{\mathsf{B}}$ is turned on adiabatically in such a way that the gap between occupied and empty single-particle levels never closes. As we do not preclude the possibility that this hybridization triggers level crossings among occupied (empty) single-particle levels, we may as well assume that all occupied single-particle levels are degenerate with the same energy -1 and all empty single-particle levels are degenerate with the same energy $+1$. In other words, we may replace $\widehat{H}_{\mathsf{A}}$ by the single-particle flattened Hamiltonian P_{A} and $\widehat{H}_{\mathsf{B}}$ by the single-particle flattened Hamiltonian P_{B} with the eigenvalues -1 for occupied and $+1$ for empty single-particle levels, respectively, if we are only concerned with the goal of identifying stable topological properties, that is,

$$P_{\mathsf{A}} := -\sum_{\varepsilon_{\mathsf{A},-}=-1} |\varepsilon_{\mathsf{A},-}\rangle\langle\varepsilon_{\mathsf{A},-}| + \sum_{\varepsilon_{\mathsf{A},+}=+1} |\varepsilon_{\mathsf{A},+}\rangle\langle\varepsilon_{\mathsf{A}+}| \tag{8.1}$$

say. This (square-matrix) representation of P_{A} is not unique, as we may freely and independently but without mixing transform all occupied or all empty single-particle eigenstates with norm-preserving linear transformations. These transformations span the unitary groups $\mathrm{U}(k)$ if there are k occupied single-particle levels below the gap or $\mathrm{U}(N-k)$ if there are $N-k$ empty single-particle levels above the gap. Such transformations should be quotiented from $\mathrm{U}(N)$, the Lie group generated by all $N \times N$ Hermitean matrices. Hence, we may tentatively identify $\widehat{H}_{\mathsf{A}} = \widehat{H}_{\mathsf{A}}^{\dagger}$ belonging to the symmetry class A with a point in the classifying space associated with the family

$$\mathrm{Gr}(k,N;\mathbb{K}) \equiv \frac{\mathrm{U}(N)}{\mathrm{U}(k)\times\mathrm{U}(N-k)}, \qquad \mathbb{K}=\mathbb{C} \qquad N=1,2,\cdots, \tag{8.2}$$

of Grassmannians.

The second step consists in defining equivalence classes tied to a notion of smoothness between two flattened Hamiltonians P_{A} and P_{B} from the manifold[51]

$$V_N := \bigcup_{k=0}^{N} \mathrm{Gr}(k,N;\mathbb{K}) \equiv \bigcup_{k=0}^{N} \frac{\mathrm{U}(N)}{\mathrm{U}(k)\times\mathrm{U}(N-k)}, \qquad \mathbb{K}=\mathbb{C}, \tag{8.3}$$

prior to defining some limiting procedure as $N\to\infty$. This step depends on the dimensionality d of space through the smooth dependence of the single-particle Hamiltonian on d real-valued parameters that underly the noninteracting fermionic many-body Hamiltonian $\widehat{H}_{\mathsf{A}} = \widehat{H}^{\dagger}_{\mathsf{A}}$. We momentarily made the choice $d=0$, for which classifying smoothness is the simplest.

We declare that the pair of elements P_{A} and P_{B} from V_N are homotopic,

$$P_{\mathsf{A}} \cong P_{\mathsf{B}}, \tag{8.4}$$

if and only if they can be connected by a continuous path in the manifold (8.3). As all elements of this manifold are flattened Hamiltonians with the eigenvalues ±1, P_{A} is homotopic to P_{B} if and only if P_{A} and P_{B} share the same matrix dimension and the same degeneracy of their eigenvalue -1. For example,

$$V_1 = \{-1\}\cup\{+1\} \tag{8.5a}$$

is the union of two homotopically inequivalent sets when $N=1$, while

$$V_2 = \{+\sigma_0\}\cup\{\cos\theta\,\sigma_1+\sin\theta\,\sigma_2 \mid 0\le\theta<2\pi\}\cup\{-\sigma_0\} \tag{8.5b}$$

is the union of three homotopically inequivalent sets when $N=2$ (σ_0 is the unit 2×2 matrix and $\boldsymbol{\sigma}$ the triplet of Pauli matrices). This notion of homotopy is, however, not compatible with the stacking rules in **Definition** 8.1.

The next step toward defining an equivalence relation between flattened Hamiltonians of different matrix dimensions consists in defining an equivalence relation $\sim$ between two flattened Hamiltonians in V_N by appealing to the homotopy $\cong$ in $V_{N+N'}$ for some N' through the operation of stacking. We say that any two elements P_{A} and P_{B} from the manifold V_N are equivalent,

$$P_{\mathsf{A}} \sim P_{\mathsf{B}}, \tag{8.6a}$$

if there exists a third element P_{C} from the manifold $V_{N'}$ such that[52]

$$P_{\mathsf{A}}\oplus P_{\mathsf{C}} \cong P_{\mathsf{B}}\oplus P_{\mathsf{C}} \tag{8.6b}$$

[51] In the symmetry class AI, the flattened Hamiltonian would be restricted to real-valued representations, in which case $\mathbb{K}$ is the set of real numbers. In the symmetry class AII, the flattened Hamiltonian would be restricted to quaternion-real representations, in which case $\mathbb{K}$ is the set of real quaternions.

[52] The symbol $\oplus$ produces an enlarged block diagonal matrix out of two square matrices.

in $V_{N+N'}$. We can rephrase (8.6) as declaring that P_{A} and P_{B} from the manifold V_N are $\sim$ equivalent if it is possible to stack them with the element $P_{\mathsf{C}} \in V_{N'}$ in such a way that $P_{\mathsf{A}} \oplus P_{\mathsf{C}}$ is $\cong$ homotopic in $V_{N+N'}$ to $P_{\mathsf{B}} \oplus P_{\mathsf{C}}$.

As $P_{\mathsf{A}} \sim P_{\mathsf{B}}$ holds if and only if there exists a P_{C} such that $P_{\mathsf{A}} \oplus P_{\mathsf{C}}$ and $P_{\mathsf{B}} \oplus P_{\mathsf{C}}$ share the same matrix dimension and the same number of linearly independent eigenvectors with eigenvalue -1, it follows that $P_{\mathsf{A}} \oplus P_{\mathsf{C}} \oplus (-P_{\mathsf{C}})$ and $P_{\mathsf{B}} \oplus P_{\mathsf{C}} \oplus (-P_{\mathsf{C}})$ must also share the same (enlarged) matrix-dimension and share the same degeneracy of their eigenvalue -1, that is,

$$P_{\mathsf{A}} \oplus P_{\mathsf{C}} \oplus (-P_{\mathsf{C}}) \cong P_{\mathsf{B}} \oplus P_{\mathsf{C}} \oplus (-P_{\mathsf{C}}) \tag{8.7}$$

in $V_{N+2N'}$. The flattened Hamiltonian $P_{\mathsf{C}} \oplus (-P_{\mathsf{C}}) \in V_{2N'}$ has as many linearly independent eigenvectors with eigenvalues -1 as linearly independent eigenvectors with eigenvalues $+1$. Any such flattened Hamiltonian is said to be trivial. We have thus shown that $P_{\mathsf{A}} \sim P_{\mathsf{B}} \in V_N$ if and only if there exists a trivial $P_{\mathsf{C}} \in V_{N'}$ such that

$$P_{\mathsf{A}} \oplus P_{\mathsf{C}} \cong P_{\mathsf{B}} \oplus P_{\mathsf{C}} \tag{8.8}$$

in $V_{N+N'}$.

Having defined a trivial $\sim$ equivalence class allows to define $-P_{\mathsf{A}} \in V_N$ as the inverse of $P_{\mathsf{A}} \in V_N$ under stacking (the direct sum $\oplus$).

The following definition allows to compare two pairs of flattened Hamiltonians from two pairs of manifolds V_N and $V_{N'}$ with arbitrary matrix sizes N and N', respectively. Denote with $(P_{\mathsf{A}}, P_{\mathsf{B}})$ any pair of flattened Hamiltonians P_{A} and P_{B} from the manifold V_N, thus sharing the same matrix dimension N. Denote with $(P_{\mathsf{A}'}, P_{\mathsf{B}'})$ any pair of flattened Hamiltonians $P_{\mathsf{A}'}$ and $P_{\mathsf{B}'}$ from the manifold $V_{N'}$, thus sharing the same matrix dimension N'. Observe that

$$\{(P_{\mathsf{A}}, P_{\mathsf{B}}), (P_{\mathsf{A}'}, P_{\mathsf{B}'})\} \subset (V_N \times V_N) \cup (V_{N'} \times V_{N'}). \tag{8.9}$$

These two pairs are declared to be equivalent,

$$(P_{\mathsf{A}}, P_{\mathsf{B}}) \sim (P_{\mathsf{A}'}, P_{\mathsf{B}'}), \tag{8.10a}$$

if and only if

$$P_{\mathsf{A}} \oplus P_{\mathsf{B}'} \sim P_{\mathsf{A}'} \oplus P_{\mathsf{B}}. \tag{8.10b}$$

We make the following two important observations. First, the identity

$$k(P_{\mathsf{A}} \oplus P_{\mathsf{B}'}) = k(P_{\mathsf{A}}) + k(P_{\mathsf{B}'}), \tag{8.11a}$$

where $k(P_{\mathsf{A}})$ is the degeneracy of the eigenvalue -1 of P_{A} and $k(P_{\mathsf{B}'})$ is the degeneracy of the eigenvalue -1 of $P_{\mathsf{B}'}$ implies the relation

$$k(P_{\mathsf{A}}) - k(P_{\mathsf{B}}) = k(P_{\mathsf{A}'}) - k(P_{\mathsf{B}'}) \tag{8.11b}$$

if $P_{\mathsf{A}} \oplus P_{\mathsf{B}'} \sim P_{\mathsf{A}'} \oplus P_{\mathsf{B}}$. Second, the pairs $(P_{\mathsf{A}}, P_{\mathsf{B}})$ and $\big(P_{\mathsf{A}} \oplus (-P_{\mathsf{B}}), P_{\mathsf{B}} \oplus (-P_{\mathsf{B}})\big)$ are always $\sim$ equivalent for

$$P_{\mathsf{A}} \oplus P_{\mathsf{B}} \oplus (-P_{\mathsf{B}}) \sim P_{\mathsf{A}} \oplus (-P_{\mathsf{B}}) \oplus P_{\mathsf{B}}. \tag{8.12a}$$

In other words, one may always assume that P_{B} is trivial when working with the $\sim$ equivalence class of the pair $(P_{\mathsf{A}}, P_{\mathsf{B}})$, since

$$(P_{\mathsf{A}}, P_{\mathsf{B}}) \sim \big(P_{\mathsf{A}} \oplus (-P_{\mathsf{B}}), P_{\mathsf{B}} \oplus (-P_{\mathsf{B}})\big). \tag{8.12b}$$

These $\sim$ equivalence classes are called difference classes. They are denoted

$$\begin{aligned} d(P_{\mathsf{A}}, P_{\mathsf{B}}) := \Bigg\{ & \Big[\big(P'_{\mathsf{A}} \oplus (-P'_{\mathsf{B}}), P'_{\mathsf{B}} \oplus (-P'_{\mathsf{B}})\big)\Big] \;\Bigg| \\ & (P'_{\mathsf{A}}, P'_{\mathsf{B}}) \in \bigcup_{N'=1}^{\infty} V_{N'} \times V_{N'},\ (P'_{\mathsf{A}}, P'_{\mathsf{B}}) \sim (P_{\mathsf{A}}, P_{\mathsf{B}}) \Bigg\}. \end{aligned} \tag{8.13}$$

Equation (8.11b) and the representation (8.13) of the difference classes motivates the definition of the integer

$$l := k(P_{\mathsf{A}}) - k(P_{\mathsf{B}}), \tag{8.14a}$$

where $k(P_{\mathsf{A}})$ is the degeneracy of the eigenvalue -1 of P_{A} and $k(P_{\mathsf{B}})$ is the degeneracy of the eigenvalue -1 of P_{B}. By construction, this integer has the same value for all elements of the same difference class but different values for elements of distinct difference classes, that is the integer l fully characterizes the difference class. For any $P_{\mathsf{B}} \in V_N$, the identity

$$k\Big(P_{\mathsf{B}} \oplus (-P_{\mathsf{B}})\Big) = N \tag{8.14b}$$

suggests that we may substitute the integer k in Eq. (8.2) by the integer l, thereby obtaining the family of Grassmannians

$$A^{(l)}_N := \frac{\mathrm{U}(2N)}{\mathrm{U}(N+l) \times \mathrm{U}(N-l)}, \qquad N \in \{1, 2, 3, \cdots\} \equiv I. \tag{8.14c}$$

The set I indexing the family of Grassmannians (8.14c) with the ordering relation $\leq$ is an example of a directed set. If we define for any pair $N \leq N' \in I$ the map

$$\begin{aligned} & f_{NN'} \colon A^{(l)}_N \to A^{(l)}_{N'} \\ & P_{\mathsf{A}} \mapsto f_{NN'}(P_{\mathsf{A}}) := \begin{cases} P_{\mathsf{A}}, & N' = N, \\ P_{\mathsf{A}} \oplus \Big(\mathbb{1}_{N'-N} \oplus (-\mathbb{1}_{N'-N})\Big), & N' > N, \end{cases} \end{aligned} \tag{8.15a}$$

that leaves the matrix trace invariant, one then verifies that the triplet

$$\Big\langle A^{(l)}_N,\ f_{N,N'},\ N \leq N' \in I \Big\rangle \tag{8.15b}$$

realizes a direct system over I (see **Appendix** A). The direct limit of the direct system (8.15) is then defined by

$$\varinjlim A_N^{(l)} := \left(\bigsqcup_{N\in I} A_N^{(l)} \right) \Big/ \approx . \tag{8.16a}$$

Here, the equivalence relation $\approx$ is defined by

$$\begin{aligned} &A_N^{(l)} \ni P_{\mathsf{A}} \approx P'_{\mathsf{B}} \in A_{N'}^{(l)} \\ &\iff \exists N'',\ N \le N'',\ N' \le N'',\ f_{NN''}(P_{\mathsf{A}}) = f_{N'N''}(P'_{\mathsf{B}}) \end{aligned} \tag{8.16b}$$

for any $N, N' \in I$. As elements from any finite disjoint union of the $A_N^{(l)}$ may become equal in the direct limit, it is necessary to quotient the unrestricted disjoint union by the equivalence relation $\approx$.

Finally, we arrive at the definition of the classifying space in zero-dimensional space for the symmetry class A. It is given by

$$C_0 := \bigcup_{l\in\mathbb{Z}} \varinjlim A_N^{(l)} = \bigcup_{l\in\mathbb{Z}} \varinjlim \frac{\mathrm{U}(2N)}{\mathrm{U}(N+l)\times \mathrm{U}(N-l)} . \tag{8.17}$$

The integer-valued index l, that now measures the number of occupied single-particle levels relative to that of a trivial flattened Hamiltonian, is a topological index. The prescription that leads to the index l is the counterpart to the prescription (3.39) when defining the fractional charge in polyacetylene [see also Eq. (3.626)].

We have only considered the case of noninteracting gapped fermionic Hamiltonians in the symmetry class A when space is zero dimensional so far. When $d = 1, 2, \cdots$ and the admissible noninteracting fermionic Hamiltonians are gapped, we are after the homotopy classes for the set of smooth functions from a d-dimensional manifold Ω_d to a classifying space V. The classifying space V is defined by taking the direct limit of the entries in the last column of **Table** 8.1 for each Altland-Zirnbauer symmetry class. These homotopy classes are known in algebraic topology to form an Abelian group denoted $\pi(\Omega_d, V)$, as it depends on Ω_d and V. Unlike when $d = 0$ for which Ω_0 is a set made of one element, the choice of Ω_d also matters. For example, when dealing with noninteracting insulators or superconductors,

$$\pi(\mathsf{S}^d, V) \subset \pi(\underbrace{\mathsf{S}^1 \times \cdots \times \mathsf{S}^1}_{d-\text{times}}, V) \tag{8.18}$$

with S^d denoting the d-dimensional unit sphere, as is appropriate when space is compactified, and $\mathbb{T}^d \equiv \mathsf{S}^1 \times \cdots \times \mathsf{S}^1$ denoting the d-dimensional torus, as is appropriate when the single-particle momentum space is compactified through periodic boundary conditions.

The computation of $\pi(\mathsf{S}^d, V)$ can be achieved with the tools of K-theory.[1,53] The result, that we quote without proof, is[54]

$$\pi(\mathsf{S}^d, V) = \begin{cases} K_{\mathbb{C}}^{0,q}(\mathsf{S}^d) \cong \pi_0(C_{q-d}), & q = 0, 1, \qquad V = C_0, C_1, \\ \\ K_{\mathbb{R}}^{0,q}(\mathsf{S}^d) \cong \pi_0(R_{q-d}), & q = 0, \cdots, 7, \quad V = R_0, \cdots, R_7, \end{cases} \tag{8.19}$$

from which **Tables** 8.2 and 8.3 follow. Observe that it is with Eq. (8.19) that the periodicity of two or eight appears for the complex and real classifying spaces, respectively.

The computation of $\pi(\mathbb{T}^d, V)$ is more involved than that of $\pi(\mathsf{S}^d, V)$ because both $K_{\mathbb{C}}^{0,q}(\mathbb{T}^d)$ and $K_{\mathbb{R}}^{0,q}(\mathbb{T}^d)$ are sensitive to the difference between strong and weak noninteracting topological insulators or superconductors, as was emphasized by Kitaev 4. For example,

$$K_{\mathbb{R}}^{0,q}(\mathbb{T}^d) \cong \pi_0(R_{q-d}) \bigoplus_{s=0}^{d-1} \binom{d}{s} \pi_0(R_{q-s}), \qquad q = 0, \cdots, 7. \tag{8.20}$$

A necessary condition for the existence of stable topological attributes that defines noninteracting strong topological Bloch insulators or Bloch superconductors is that the entries of **Table** 8.3 are nonvanishing, as will be explained in more detail in Section 8.2.5. What **Table** 8.3 does not do is to provide a sufficient condition that decides if the Bloch Hamiltonian $\widehat{H}$ can be assigned a quantized quantum number. However, if the Bloch Hamiltonian $\widehat{H}$ can be assigned such a quantized quantum number, the topological attribute that it encodes must then be universal, for any smooth perturbation of the Bloch Hamiltonian $\widehat{H}$ that does not close the gap cannot change the value of this quantized quantum number. Such a quantized quantum number could be the quantized Hall conductivity of a Chern insulator taking a nonvanishing integer value, in which case the bulk-boundary correspondence[27] relates the quantized Hall conductivity of $\widehat{H}$ to the existence of chiral edge states if open boundary conditions are used instead of periodic ones.

There is another approach that delivers **Table** 8.3 without assuming that the single-particle momentum is a good quantum number. 2, 3 This approach assumes three bulk energy scales. There is the bulk gap $\Delta > 0$, the characteristic strength $U > 0$ of some static disorder potential, and the magnitude $|\varepsilon_{\mathrm{bd,F}}|$ of the Fermi energy $\varepsilon_{\mathrm{bd,F}}$ of the boundary modes (that is $\varepsilon_{\mathrm{bd,F}}$ is the bulk chemical potential measured relative to the single-particle energy in the middle of the band gap, say). The hierarchy

$$|\varepsilon_{\mathrm{bd,F}}| \ll U \ll \Delta \tag{8.21}$$

[53] M. F. Atiyah, R. Bott, and A. Shapiro, Topology **3** (Supplement 1), 3–38 (1964). [Atiyah et al. (1964)].

[54] The proof involves two steps, one for each equality. The first equality makes use of K-theory and notation therein. The second equality is expressed in terms of the zeroth-homotopy group of the classifying spaces, an Abelian group labeled by the disconnected components of the classifying space.

holds by assumption. All single-particle excitations, whose energies are to be found within the bulk gap between the valence and conduction bands in the clean limit, are expected to be exponentially localized (Anderson localization) owing to the first inequality of Eq. (8.21). However, by definition, the boundary states of strong topological insulators or superconductors remain delocalized along the boundaries even though the disorder is strong for these states according to the first inequality of Eq. (8.21). To understand the mechanism by which the boundary modes can escape Anderson localization in spite of the first inequality of Eq. (8.21), one must accept as a fact that an effective theory that captures the effect of static disorder on a noninteracting gas of electrons is a nonlinear-sigma model (NLSM) with at most two coupling constants. The first coupling constant is always present. It is proportional to the average conductivity computed from a microscopic Hamiltonian within the Born approximation. It delivers the one-parameter scaling description of Anderson localization according to which all states are localized for strong disorder. The second coupling is only present when the target manifold of the NLSM has a topological attribute. When the second coupling is absent, while the first coupling is small, Anderson localization rules. However, the presence of the second coupling is conjectured to signal the absence of Anderson localization. **Table** 8.3 follows from identifying those dimensions and those target spaces (that is some combinations of the Altland-Zirnbauer protecting symmetries) for which the second coupling is permissible in the NLSM describing the effect of the static disorder on the boundary states. Hereto, this derivation of **Table** 8.3 offers no more than necessary conditions for a noninteracting insulator or superconductor to support symmetry-protected boundary states.

The third approach to deriving **Table** 8.3 that we will follow was sketched by Kitaev in 2009 4 and worked out in detail by Morimoto and Furusaki.[55] This approach relies on using massive Dirac Hamiltonians as a diagnostic of fermionic invertible topological phases of matter. This should not be a surprise given the historical role played by Dirac Hamiltonians in connection with index theorems. This approach is no less general than the one relying on Hermitean flattened Hamiltonians as justified by rules I, II, and III. This approach can accommodate static disorder as is done in Section 8.3 and will be adapted in Section 8.4 to incorporate the effects of interactions.

8.2.2 Clifford Algebras and their Classifying Spaces

The complex Clifford algebra[56]

$$C\ell_q := \mathrm{span}_{C\ell}\left\{e_1, \cdots, e_q\right\} \tag{8.22a}$$

[55] T. Morimoto and A. Furusaki, Phys. Rev. B **88**(12), 125129 (2013). [Morimoto and Furusaki (2013)].

[56] In words, $\mathrm{span}_{C\ell}$ means taking all possible linear combinations of all linearly independent products of the e's.

is a complex vector space isomorphic to $\mathbb{C}^{2^q}$ of dimension 2^q that is spanned by the basis with the basis elements

$$e_{n_1\cdots n_q} \equiv \prod_{\iota=1}^{q} (e_\iota)^{n_\iota}, \qquad n_1, \cdots, n_q = 0, 1, \tag{8.22b}$$

whereby the multiplication obeys the rule

$$\{e_\iota, e_{\iota'}\} = 2\,\delta_{\iota,\iota'} \tag{8.22c}$$

for $\iota, \iota' = 1, \cdots, q$. Owing to Eq. (8.22c), the vector space $C\ell_q$ is closed under multiplication of any two of its elements. In fact, it is an associative and unital algebra.

The real Clifford algebra

$$C\ell_{p,q} := \mathrm{span}_{C\ell}\left\{e_1, \cdots, e_p; e_{p+1} \cdots, e_{p+q}\right\} \tag{8.23a}$$

is a real vector space isomorphic to $\mathbb{R}^{2^{p+q}}$ of dimension 2^{p+q} that is spanned by the basis with the basis elements

$$e_{n_1\cdots n_{p+q}} \equiv \prod_{\iota=1}^{p+q} (e_\iota)^{n_\iota}, \qquad n_1, \cdots, n_{p+q} = 0, 1, \tag{8.23b}$$

whereby the multiplication obeys the rule

$$\begin{aligned} \{e_\iota, e_{\iota'}\} &= 2\,\eta_{\iota,\iota'}, \\ \eta_{\iota,\iota'} &= \mathrm{diag}\,(\overbrace{-1, \cdots, -1}^{p\text{-times}}, \overbrace{+1, \cdots, +1}^{q\text{-times}}), \end{aligned} \tag{8.23c}$$

for $\iota, \iota' = 1, \cdots, p+q$. Owing to Eq. (8.23c), the vector space $C\ell_{p,q}$ is closed under multiplication of any two of its elements. Hereto, it is an associative and unital algebra.

Given a representation of the complex Clifford algebra $C\ell_q$, "the extension problem" expressed as

$$C\ell_q \to C\ell_{q+1} \tag{8.24}$$

consists in identifying the "classifying space" C_q associated with the direct limit[57] of the matrix space realized by the generator e_{q+1} present in $C\ell_{q+1}$ but absent in $C\ell_q$ as the dimensionality r of the matrix representation for e_{q+1} is taken to infinity.

Similarly, given a representation of the real Clifford algebra $C\ell_{p,q}$, there are two possible extension problems.

[57] We have encountered the example (8.17) of a direct limit. The direct limit of an ordered family of structured sets is defined in all generality in **Appendix** A. The direct limit is needed to combine a family of mathematical objects of ascending "sizes" into a single "large" mathematical object. The physics rational for the direct limit is to implement the stacking rules in **Definition** 8.1.

1. There is the extension problem
$$C\ell_{p,q} \to C\ell_{p,q+1} \tag{8.25}$$
that consists in identifying the classifying space $R^{(+)}_{p,q}$ associated with the direct limit[57] of the matrix space realized by the generator e_{p+q+1} present in $C\ell_{p,q+1}$ and thus satisfying $e^2_{p+q+1} = +1$, but absent in $C\ell_{p,q}$ as the dimensionality r of the matrix representation for e_{p+q+1} is taken to infinity.
2. There is the extension problem
$$C\ell_{p,q} \to C\ell_{p+1,q} \tag{8.26}$$
that consists in identifying the classifying space $R^{(-)}_{p,q}$ associated with the direct limit[57] of the matrix space realized by the generator e_{p+1} present in $C\ell_{p+1,q}$ and thus satisfying $e^2_{p+1} = -1$, but absent in $C\ell_{p,q}$ as the dimensionality r of the matrix representation for e_{p+1} is taken to infinity.

The extension problem (8.26) is, in disguise, equivalent to the extension problem (8.25). This claim can be understood with a two-step argumentation. First, one takes advantage of the algebra isomorphism (**Exercise** 8.2)
$$C\ell_{p,q} \otimes \mathbb{R}(2) \cong C\ell_{q,p+2}, \tag{8.27a}$$
where $\mathbb{R}(2) \cong C\ell_{0,2}$, as an algebra, is generated by the two real-valued Pauli matrices
$$e'_1 \equiv \begin{pmatrix} 1 & 0 \\ 0 & -1 \end{pmatrix}, \qquad e'_2 \equiv \begin{pmatrix} 0 & 1 \\ 1 & 0 \end{pmatrix}, \tag{8.27b}$$
while $\mathbb{R}(2) \cong C\ell_{0,2}$, as a four-dimensional vector space over the field of real numbers, has the basis
$$(e'_1)^2 = (e'_2)^2 = \begin{pmatrix} 1 & 0 \\ 0 & 1 \end{pmatrix}, \; e'_1 \equiv \begin{pmatrix} 1 & 0 \\ 0 & -1 \end{pmatrix}, \; e'_2 \equiv \begin{pmatrix} 0 & 1 \\ 1 & 0 \end{pmatrix}, \; e'_1\,e'_2 = \begin{pmatrix} 0 & 1 \\ -1 & 0 \end{pmatrix}. \tag{8.27c}$$
If the extension problem (8.26) is the same as the extension problem
$$C\ell_{p,q} \otimes \mathbb{R}(2) \to C\ell_{p+1,q} \otimes \mathbb{R}(2), \tag{8.28}$$
we can bring the extension problem (8.28) to the form of the extension problem (8.25) with the help of the algebra isomorphism (8.27a), thereby establishing the claim that solving the extension problem (8.26) is equivalent to solving the extension problem (8.25). Second, the reason for which the extension problem (8.26) is the same as the extension problem (8.28) is that the choice of $\mathbb{R}(2)$ for the representation of $C\ell_{0,2}$ on both sides of the symbol $\to$ in the extension problem (8.28) is fixed, given by Eq. (8.27b) for example. The presence of $\mathbb{R}(2)$ does not enter the identification of the independent parameters needed to parameterize the generator (that we shall call e^*) present on the right-hand side of the symbol $\to$ but absent on the left-hand

side upon increasing the dimensionality r of the matrix representation for e^* as the direct limit is taken.

A priori the classifying spaces C_q and $R^{(+)}_{p,q}$ associated with the extension problem (8.24) for the complex Clifford algebras and (8.25) for the real Clifford algebras are indexed by a single integer q for the complex Clifford algebras and the pair of integers p, q for the real Clifford algebras, respectively. However, owing to the three algebra isomorphisms (see **Exercise** 8.2 or the textbooks from Karoubi[1] and Porteous[58])

$$C\ell_{q+2} \cong C\ell_q \otimes \mathbb{C}(2), \tag{8.29a}$$

$$C\ell_{p+1,q+1} \cong C\ell_{p,q} \otimes \mathbb{R}(2), \tag{8.29b}$$

$$C\ell_{p+8,q} \cong C\ell_{p,q+8} \cong C\ell_{p,q} \otimes \mathbb{R}(16), \tag{8.29c}$$

where $\mathbb{C}(2)$ is the representation of $C\ell_2$ in terms of 2×2 complex-valued matrices, $\mathbb{R}(2)$ is the representation of $C\ell_{0,2}$ in terms of 2×2 real-valued matrices, and $\mathbb{R}(16)$ is the representation of $C\ell_{0,8}$ in terms of 16×16 real-valued matrices, one deduces that C_q depends only on q modulo 2 for the complex classifying spaces, while $R^{(+)}_{p,q} \equiv R_{p-q}$ depends only on $p - q$ modulo 8 for the real classifying spaces.

We conclude that there are two families of complex classifying spaces C_0 and C_1, while there are eight families of real classifying spaces $R_0, \cdots, R_7$. In other words, the dependence on q enters modulo two for the complex classifying spaces

$$C_{q+2} \cong C_q, \tag{8.30a}$$

while it enters modulo eight for the real classifying spaces,

$$R_{q+8} \cong R_q. \tag{8.30b}$$

This periodicity is called the Bott periodicity.[1] Each classifying space V in **Table** 8.1 is obtained from taking the direct limit of a family of topological spaces labeled by the integer number N entering the rank

$$r = r_{\min} N \tag{8.31}$$

assumed for the representation of the Clifford algebras.

8.2.3 Definition of Minimal Massive Dirac Hamiltonians

We assume that space is $d = 1, 2, \cdots$ dimensional. The kinetic part of a translation-invariant Dirac Hamiltonian is

$$\mathcal{H}_{\text{kin}}(\boldsymbol{k}) = \sum_{i=1}^{d} k_i \, \alpha_i, \tag{8.32a}$$

[58] I. Porteous, *Clifford Algebras and the Classical Groups,* Cambridge Studies in Advanced Mathematics, Vol. **50**, Cambridge University Press, Cambridge 1995. [Porteous (1995)].

where $\boldsymbol{k} \equiv (k_1, \cdots, k_d) \in \mathbb{R}^d$ is the momentum and $\boldsymbol{\alpha} \equiv (\alpha_1, \cdots, \alpha_d)$ are the (Hermitean) Dirac matrices that obey the algebra

$$\{\alpha_i, \alpha_j\} = 2\delta_{i,j}, \qquad i,j = 1, \cdots, d. \tag{8.32b}$$

On the one hand, we assume that the dimensionality of the representation of the matrices $\boldsymbol{\alpha}$ is sufficiently large so that there exists at least one Hermitean matrix β such that it anticommutes with all the components of $\boldsymbol{\alpha}$ and it squares to the identity matrix,

$$\{\beta, \mathcal{H}_{\mathrm{kin}}(\boldsymbol{k})\} = 0, \qquad \beta^2 = 1. \tag{8.32c}$$

It is then possible to write the translation-invariant massive Dirac Hamiltonian

$$\mathcal{H}(\boldsymbol{k}) = \sum_{i=1}^{d} k_i\, \alpha_i + m\, \beta, \tag{8.32d}$$

where $m \in \mathbb{R}$ is a mass. However, the matrix β with its mass m (that is mass matrix in short) may not be unique. For example, if the Dirac matrices are chosen to be of rank 2, then there are two linearly independent mass matrices anticommuting with each other in $d = 1$, one possible mass matrix in $d = 2$, and none in $d = 3$. On the other hand, the translation-invariant massive Dirac Hamiltonian (8.32d) becomes decomposable for sufficiently large rank of the Dirac matrices. In any of the Altland-Zirnbauer symmetry classes, we may start from a sufficiently large matrix representation of the translation-invariant massive Dirac Hamiltonian (8.32d), which we then reduce until we reach the rank of the Dirac matrices below which we would loose all mass matrices. In this way, one obtains for each Altland-Zirnbauer symmetry class and dimension d an irreducible translation-invariant massive Dirac Hamiltonian of the form (8.32d) which is of minimum rank r_{min} (not necessarily unique in that more than one distinct mass matrix may be possible).

To which Altland-Zirnbauer symmetry class the Dirac Hamiltonian

$$\mathcal{H} = \sum_{i=1}^{d} \alpha_i \frac{\partial}{\mathrm{i}\partial x_i} + \beta\, m(\boldsymbol{x}) \tag{8.33}$$

belongs depends on whether it is possible to construct a combination from the triplet of operations

$$T \equiv \mathcal{T}\mathsf{K}, \qquad C \equiv \mathcal{C}\mathsf{K}, \qquad \Gamma, \tag{8.34a}$$

for time-reversal symmetry (TRS), particle-hole symmetry (PHS), and chiral symmetry (CHS), respectively (K denotes the operation of complex conjugation and $\mathcal{T}$, $\mathcal{C}$, and Γ are matrices sharing the same rank as the Dirac matrices $\boldsymbol{\alpha}$), such that

$$T^2 = \pm 1, \quad C^2 = \pm 1, \quad [T, C] = 0, \quad \Gamma^2 = 1, \tag{8.34b}$$

and

$$\text{TRS:} \qquad [T,\mathcal{H}]=0, \tag{8.35a}$$
$$\text{PHS:} \qquad \{C,\mathcal{H}\}=0, \tag{8.35b}$$
$$\text{CHS:} \qquad \{\Gamma,\mathcal{H}\}=0. \tag{8.35c}$$

(We have performed a global choice of gauge for which $[T,C]=0$ holds.) Equations (8.35) are equivalent to

$$\text{TRS:} \qquad [T,\beta]=\{T,\boldsymbol{\alpha}\}=0, \tag{8.36a}$$
$$\text{PHS:} \qquad \{C,\beta\}=[C,\boldsymbol{\alpha}]=0, \tag{8.36b}$$
$$\text{CHS:} \qquad \{\Gamma,\beta\}=\{\Gamma,\boldsymbol{\alpha}\}=0. \tag{8.36c}$$

Observe here that the antiunitarity of T and C interchanges the action of commutators and anticommutators when acting on the mass relative to the kinetic part of the Dirac Hamiltonian (8.33).

8.2.4 The Tenfold Way for the Clifford Algebras

We are ready to combine the Clifford algebras and their classifying spaces from Section 8.2.2 with the Altland-Zirnbauer classification of massive Dirac Hamiltonians from Section 8.2.3.

We associate with each Altland-Zirnbauer symmetry class, with each dimension d of space, and with any rank of the Dirac matrices $\boldsymbol{\alpha}\equiv(\alpha_1,\cdots,\alpha_d)$ equal to or larger than the minimal rank as defined below Eq. (8.32d), a Clifford algebra according to the following rules.

The symmetry classes A and AIII are associated with the complex Clifford algebra

$$\text{A:} \qquad C\ell_{d+1}=\operatorname{span}_{C\ell}\{\beta,\boldsymbol{\alpha}\}, \tag{8.37a}$$
$$\text{AIII:} \qquad C\ell_{d+2}=\operatorname{span}_{C\ell}\{\beta,\Gamma,\boldsymbol{\alpha}\}, \tag{8.37b}$$

respectively.

For the remaining eight symmetry classes, it is always possible to define the operation J that satisfies the relations

$$\{T,J\}=\{C,J\}=[\Gamma,J]=[\boldsymbol{\alpha},J]=[\beta,J]=0, \tag{8.38}$$

and plays the role of an imaginary unit for the real Clifford algebras as $J^2=-1$. The symmetry classes AI, BDI, D, DIII, AII, CII, C, and CI are associated with the real Clifford algebras[55]

$$\text{AI:} \qquad C\ell_{1,d+2}=\operatorname{span}_{C\ell}\{J\,\beta;T,T\,J,\boldsymbol{\alpha}\}, \tag{8.39a}$$
$$\text{BDI:} \qquad C\ell_{d+1,3}=\operatorname{span}_{C\ell}\{J\,\boldsymbol{\alpha},T\,C\,J;C,C\,J,\beta\}, \tag{8.39b}$$
$$\text{D:} \qquad C\ell_{d,3}=\operatorname{span}_{C\ell}\{J\,\boldsymbol{\alpha};C,C\,J,\beta\}, \tag{8.39c}$$

$$\text{DIII:} \qquad C\ell_{d,4} = \text{span}_{C\ell}\left\{J\,\boldsymbol{\alpha}; C, C\,J, T\,C\,J, \beta\right\}, \tag{8.39d}$$

$$\text{AII:} \qquad C\ell_{3,d} = \text{span}_{C\ell}\left\{J\,\beta, T, T\,J; \boldsymbol{\alpha}\right\}, \tag{8.39e}$$

$$\text{CII:} \qquad C\ell_{d+3,1} = \text{span}_{C\ell}\left\{J\,\boldsymbol{\alpha}, C, C\,J, T\,C\,J; \beta\right\}, \tag{8.39f}$$

$$\text{C:} \qquad C\ell_{d+2,1} = \text{span}_{C\ell}\left\{J\,\boldsymbol{\alpha}, C, C\,J; \beta\right\}, \tag{8.39g}$$

$$\text{CI:} \qquad C\ell_{d+2,2} = \text{span}_{C\ell}\left\{J\,\boldsymbol{\alpha}, C, C\,J; T\,C\,J, \beta\right\}, \tag{8.39h}$$

respectively.

8.2.5 The Tenfold Way for the Classifying Spaces V

The definition of the classifying space V associated with the Dirac Hamiltonian (8.33) depends on the Altland-Zirnbauer symmetry class to which it belongs and the dimensionality d of space. The classifying space V encodes the fact that the mass matrix β in Eq. (8.33) might not be unique for given d and the rank of the Dirac matrices $\boldsymbol{\alpha} \equiv (\alpha_1, \cdots, \alpha_d)$. The construction of the classifying space V proceeds with the following steps (**Exercises** 8.3 and 8.4).

Step 1: To each Altland-Zirnbauer symmetry class, we assign the following pair of Clifford algebras differing by one generator, holding the dimensionality d of space fixed and the rank of the Dirac Hamiltonian fixed. The Clifford algebra to the left of the arrow in the fifth column "extension" from **Table** 8.1 is obtained after removing from the tenfold list of Clifford algebras defined by Eqs. (8.37) and (8.39) one generator, namely, the mass matrix $J\,\beta$ for the symmetry classes AI and AII and the mass matrix β otherwise. This gives the tenfold list

$$\text{A:} \qquad C\ell_{d} = \text{span}_{C\ell}\left\{\boldsymbol{\alpha}\right\}, \tag{8.40a}$$

$$\text{AIII:} \qquad C\ell_{d+1} = \text{span}_{C\ell}\left\{\Gamma, \boldsymbol{\alpha}\right\}, \tag{8.40b}$$

and

$$\text{AI:} \qquad C\ell_{0,d+2} = \text{span}_{C\ell}\left\{; T, T\,J, \boldsymbol{\alpha}\right\}, \tag{8.41a}$$

$$\text{BDI:} \qquad C\ell_{d+1,2} = \text{span}_{C\ell}\left\{J\,\boldsymbol{\alpha}, T\,C\,J; C, C\,J\right\}, \tag{8.41b}$$

$$\text{D:} \qquad C\ell_{d,2} = \text{span}_{C\ell}\left\{J\,\boldsymbol{\alpha}; C, C\,J\right\}, \tag{8.41c}$$

$$\text{DIII:} \qquad C\ell_{d,3} = \text{span}_{C\ell}\left\{J\,\boldsymbol{\alpha}; C, C\,J, T\,C\,J\right\}, \tag{8.41d}$$

$$\text{AII:} \qquad C\ell_{2,d} = \text{span}_{C\ell}\left\{T, T\,J; \boldsymbol{\alpha}\right\}, \tag{8.41e}$$

$$\text{CII:} \qquad C\ell_{d+3,0} = \text{span}_{C\ell}\left\{J\,\boldsymbol{\alpha}, C, C\,J, T\,C\,J; \right\}, \tag{8.41f}$$

$$\text{C:} \qquad C\ell_{d+2,0} = \text{span}_{C\ell}\left\{J\,\boldsymbol{\alpha}, C, C\,J; \right\}, \tag{8.41g}$$

$$\text{CI:} \qquad C\ell_{d+2,1} = \text{span}_{C\ell}\left\{J\,\boldsymbol{\alpha}, C, C\,J; T\,C\,J\right\}. \tag{8.41h}$$

Step 2: For each Altland-Zirnbauer symmetry class, we seek *all* distinct Hermitean matrices sharing the same rank as the Dirac matrices $\boldsymbol{\alpha} \equiv (\alpha_1, \cdots, \alpha_d)$ such that they can be added to the list of generators entering the corresponding Clifford algebra from the tenfold list defined by Eqs. (8.40) and (8.41) so as to deliver the corresponding Clifford algebra from the tenfold list defined by Eqs. (8.37) and (8.39). These mass matrices are associated with a set V, the classifying space, needed to implement mathematically the stacking rules in **Definition** 8.1. Determining V is an example of the extension problem in K-theory. A characterization of the classifying space V can be deduced from K-theory.[1] The outcome of this exercise is listed in the last two columns of **Table** 8.1 (owing to the Bott periodicity) for the case when the rank r of the Dirac matrices $\boldsymbol{\alpha} \equiv (\alpha_1, \cdots, \alpha_d)$ is

$$r = r_{\text{min}}\, N, \tag{8.42}$$

where r_{min} is the rank of the minimal representation for the Dirac Hamiltonian (8.33) and N is an integer.

The zeroth homotopy group of the classifying spaces is given in the second column of **Table** 8.2. **Table** 8.3 follows from combining **Tables** 8.1 and 8.2. A zeroth homotopy group of cardinality larger than one has the following consequences. Imagine that d-dimensional space is divided into two halves. Both halves share a $(d-1)$-dimensional boundary. Whenever the zeroth homotopy group of the classifying space in the tenfold way is nontrivial, consider the Dirac Hamiltonian (8.33) with the mass term interpolating smoothly across the $(d-1)$-dimensional boundary between two fixed elements associated with the classifying space characterized by distinct values of the zeroth homotopy group (that is two fixed elements from different connected components of the classifying space so that it is not possible to continuously interpolate between them without leaving the classifying space). By the very definition of a homotopy group, this is only possible if the mass term vanishes along the $(d-1)$-dimensional boundary separating the two halves of d-dimensional space. As was shown by Jackiw and Rebbi for the symmetry class BDI when $d=1$ (see footnote 7), the Dirac Hamiltonian (8.33) must then support a zero mode that is extended along the $(d-1)$-dimensional boundary but exponentially localized away from it. This is the defining property of a topological insulator (superconductor). Hence, by combining the zeroth homotopy group of the classifying spaces given in the second column of **Table** 8.2 with the Bott periodicity (8.30), one infers which of the Altland-Zirnbauer symmetry classes allows a topological insulator or superconductor for any given dimensionality d of space. The periodic **Table** 8.3 for topological insulators or superconductors follows. 4, 2, 3

All higher homotopy groups of the classifying spaces are given in column three to nine of **Table** 8.2, for they obey the periodicities

$$\pi_p(C_j) = \pi_{p+2}(C_j), \qquad p = 0, 1, 2, \cdots, \tag{8.43a}$$

for the complex classes $j = 0, 1$ and

$$\pi_p(R_j) = \pi_{p+8}(R_j), \qquad p = 0, 1, 2, \cdots, \tag{8.43b}$$

for the real classes $j = 0, 1, \cdots, 7$. They also obey the translation rules

$$\pi_p(C_j) = \pi_{p+1}(C_{j+1}), \qquad p = 0, 1, 2, \cdots, \tag{8.44a}$$

for the complex classes with the integers j and $j+1$ defined modulo 2 and

$$\pi_p(R_j) = \pi_{p+1}(R_{j-1}), \qquad p = 0, 1, 2, \cdots , \tag{8.44b}$$

for the real classes with the integers j and $j-1$ defined modulo 8. If we combine these translations rules with the definition of V_d given in **Table** 8.1, we find the relation

$$\pi_D(V_d) = \pi_0(V_{d-D}) \tag{8.45}$$

for any $D = 0, 1, 2, \cdots , d$.

8.2.6 Existence and Uniqueness of Normalized Dirac Masses

We summarize the properties of d-dimensional Dirac Hamiltonians that follow from the topology of the classifying spaces V_d for the ten Altland-Zirnbauer symmetry classes[55,59] (**Exercises** 8.3 and 8.4) and that we verified explicitly by hand for the case of $d = 1$ in Section 7.8.

1. For each symmetry class, there exists a minimum rank $r_{\mathrm{min}}(d)$, an even integer equal to or larger than the integer 2, for which the d-dimensional Dirac Hamiltonian supports a mass matrix and below which either no mass matrix or no Dirac kinetic contribution are allowed by symmetry.
2. Suppose that the rank of the d-dimensional massive Dirac Hamiltonian is

$$r = r_{\mathrm{min}}(d)\, N. \tag{8.46}$$

Definition 8.2 (unique Dirac mass matrix) When there exists a Dirac mass matrix that is unique up to a sign with the following two properties

1. it squares to unity,
2. and it commutes with all other symmetry-allowed mass matrices;

we call it the unique (Dirac) mass matrix.

There are the following three cases depending on the entries in the "$\pi_0(V_d)$" column of **Table** 8.2.

1. $\pi_0(V_d) = \mathbb{Z}$: there is always a unique mass matrix for $N \geq 1$.
2. $\pi_0(V_d) = \mathbb{Z}_2$: there is a unique mass matrix β_{min} only when $N = 1$. When N is an even integer, for any given mass matrix, there exists another matrix that anticommutes with it. When N is an odd integer larger than one, the matrix $\beta_{\mathrm{min}} \otimes \mathbb{1}_N$ plays a role similar to that of the unique mass matrix in that the two matrices $\pm\beta_{\mathrm{min}} \otimes \mathbb{1}_N$ belong to different connected components of V_d, even though there exists a normalized mass matrix that neither commutes nor anticommutes with $\beta_{\mathrm{min}} \otimes \mathbb{1}_N$.

[59] T. Morimoto, A. Furusaki, and C. Mudry, Phys. Rev. B **91**(23), 235111 (2015). [Morimoto et al. (2015a)].

3. $\pi_0(V_d) = 0$: for any given mass matrix, there exists another mass matrix that anticommutes with it. That is, there is no unique mass matrix for $N \geq 1$.

When (i) $N = 1$, (ii) the Dirac Hamiltonian has a unique mass matrix that realizes topologically distinct ground states for different signs of its mass, and (iii) the mass is varied smoothly in space, then domain boundaries along which the mass vanishes are accompanied by massless Dirac fermions, whose low-energy Hamiltonians are of rank $r_{\mathrm{min}}(d)/2$. It follows that

$$r_{\mathrm{min}}(d-1) = r_{\mathrm{min}}(d) \tag{8.47}$$

since it is possible to gap these massless Dirac fermions by stacking any low-energy massless Hamiltonian of rank $r_{\mathrm{min}}(d)/2$ with its homotopic inverse.

8.2.7 Relationship to Higher Homotopy Groups

The zeroth homotopy group of a topological space indicates if it is path-connected and, if not, how to index all its distinct subspaces that are path-connected. Equation (8.45) relates the zeroth to the higher homotopy groups of the classifying spaces V_d defined in **Table** 8.2. Equation (8.45) can be given the following interpretation.

We recall that the homotopy group $\pi_n(X)$ is the set of homotopy classes of maps $f\colon \mathsf{S}^n \to X$ between the unit sphere S^n in $(n+1)$-dimensional Euclidean space and the topological space X. One may then identify D in Eq. (8.45) as the dimensionality of the sphere S^D that surrounds a defect in d-dimensional space (see **Table** 8.4).[44] For point, line, and surface defects, $D = d-1$, $D = d-2$, and $D = d-3$, respectively. In other words, $D+1$ is the codimension of such defects in d-dimensional space.[60] A homotopy group $\pi_D(V_d)$ with more than one element signals that defects of codimension $D+1$ in the normalized Dirac masses can be indexed by a topological number. For given D and d, Eq. (8.45) [both d and $d-D$ on the left- and right-hand sides of $\pi_D(V_d) = \pi_0(V_{d-D})$, respectively, are defined either modulo 2 or modulo 8 depending on the Altland-Zirnbauer symmetry class] dictates through the homotopy reduction $D \to D - D = 0$ and the dimensional reduction $d \to d-D$, which five of the ten Altland-Zirnbauer symmetry classes support topological defects in their normalized Dirac masses. In particular, point defects inherit the topological numbers from the zeroth homotopy group of V_1. Any Dirac Hamiltonian with a defective normalized Dirac mass of topological character supports eigenstates with a vanishing energy eigenvalue (zero modes) that are bound to the defect in the directions transverse to it. These zero modes are robust to any local perturbation as long as they respect the Altland-Zirnbauer symmetry class and they are not too strong.

[60] If a defect is a submanifold of dimension d_{defect} embedded in d-dimensional Euclidean space, then its codimension is $d - d_{\mathrm{defect}}$.

Table 8.4 *Point-like, line-like, and plane-like topological defects*

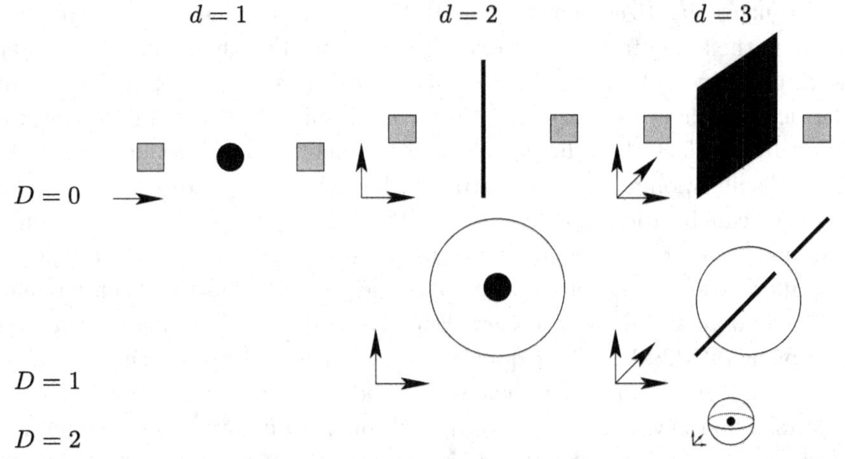

Let d be the dimensionality of space. The sphere S^D, with S^0 pictured by two squares shaded in gray, S^1 pictured by a circle, and S^2 pictured by the surface of a sphere for $D = 0, 1, 2$, respectively, surrounds a topological defect in the embedding d-dimensional space with $d > D$. This topological defect is pictured by a black disc, a thick line, or a black plane if its dimensionality is zero-, one-, or two-dimensional, respectively.

8.3 Anderson Localization and the Topology of Classifying Spaces

8.3.1 Overview

Anderson localization is the non-perturbative phenomenon by which the plane-wave solutions to a linear differential equation become exponentially localized upon the breaking of translation symmetry by a local random potential.[61,62,63,64] Anderson localization always prevails for metals in one-dimensional and for metals without spin-orbit coupling in two-dimensional space, however small the disorder strength. It requires the disorder strength to be large enough, of the order of the Fermi energy, otherwise. Until the experimental discovery of the integer quantum Hall effect (IQHE) in 1980, the most important challenge brought about by Anderson localization had been (and remains) to understand the metal-insulator transition by analytical means.

[61] P. W. Anderson, Phys. Rev. **109**(5), 1492-1505 (1958). [Anderson (1958)].

[62] A local random potential is a statistical distribution of potentials such that all m-th order cumulants of the random potential at n distinct points in space decay exponentially fast with the pairwise separation of any two points.

[63] P. A. Lee and T. V. Ramakrishnan, Rev. Mod. Phys. **57**(2), 287–337 (1985). [Lee and Ramakrishnan (1985)].

[64] F. Evers and A. D. Mirlin, Rev. Mod. Phys. **80**(4), 1355–1417 (2008). [Evers and Mirlin (2008)].

The IQHE is characterized empirically by the quantized value of the Hall conductivity at fixed filling fraction and by a sharp (quantum) transition between two consecutive quantized values of the Hall conductivity, the plateau transition in short, when the filling fraction is tuned. It is explained by the topological character of the Hall conductivity when the chemical potential is in between two consecutive Landau levels of a two-dimensional electron gas subjected to a uniform magnetic field and by the fact that this topological character is retained in the regime of Anderson localization.[9,10,25] The IQHE teaches two important experimental facts. First, there can be topologically distinct insulating phases of electronic matter. Second, direct continuous (quantum) transitions between these phases are possible. Several approaches have been used to study the plateau transition from an analytical and a computational point of view. Effective models such as quantum nonlinear sigma models (QNLSM),[65,66,67] quantum network models,[68] and Dirac fermions[69] were proposed. In parallel, the phenomenological two-parameter scaling theory of Khmelnitskii and Levine et al. was verified through numerous large-scale numerical simulations.[70] We are going to generalize the approach pioneered by Ludwig et al.[69] by which they argued that the minimal continuum model that captures the IQHE in both the clean and disordered limits is a Dirac Hamiltonian with random mass and gauge fields.

In Section 7.7, we have shown under what conditions a scaling approach to the computation of the probability distribution for the Landauer conductance applies for quantum wires at zero temperature. We saw that the effects of static disorder on a quasi-one-dimensional noninteracting gas of fermions that realizes otherwise a metallic phase at zero temperature in any one of the ten Altland-Zirnbauer symmetry classes can be described by a one-parameter scaling theory in the thick quantum wire limit. Moreover, for the special cases of the chiral symmetry classes AIII, BDI, and CII, we could demonstrate the counterpart to the plateau transition in the IQHE, namely, a direct continuous (quantum) phase transition between topologically distinct phases of matter.[71] We are going to explain why the effects of static disorder at zero temperature on a (noninteracting) strong topological insulator or superconductor in d-dimensional space are qualitatively captured by phase diagrams that depends on two parameters, the spectral gap that separates the filled single-particle levels from the empty ones in the clean limit and the strength of the static disorder. We are also going to show that these phase diagrams are of three

[65] D. E. Khmelnitskii, JETP Lett. **38**(9), 552–556 (1983). [Khmelnitskii (1983)].
[66] H. Levine, S. B. Libby, and A. M. M. Pruisken, Phys. Rev. Lett. **51**(20), 1915–1918 (1983). [Levine et al. (1983)].
[67] A. M. M. Pruisken, Nucl. Phys. B **235**, 277–298 (1984). [Pruisken (1984)].
[68] J. T. Chalker and P. D. Coddington, J. Phys. C: Solid St. Phys. **21**(14), 2665–2679 (1988). [Chalker and Coddington (1988)].
[69] A. W. W. Ludwig, M. P. A. Fisher, R. Shankar, and G. Grinstein, Phys. Rev. B **50**(11), 7526–7552 (1994). [Ludwig et al. (1994)].
[70] B. Huckestein, Rev. Mod. Phys. **67**(2), 357–396 (1995). [Huckestein (1995)].
[71] P. W. Brouwer, C. Mudry, B. D. Simons, and A. Altland, Phys. Rev. Lett. **81**(4), 862–865 (1998). [Brouwer et al. (1998)].

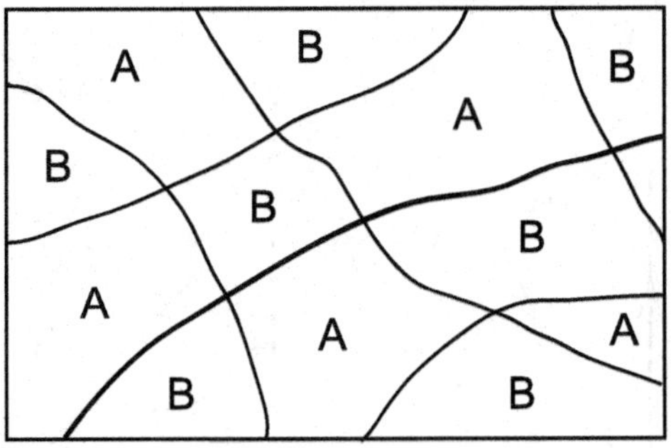

Figure 8.2 For each realization of a random potential perturbing the massless Dirac Hamiltonian defined in d-dimensional Euclidean space $\mathbb{R}^d$, we may decompose $\mathbb{R}^d$ into open sets (domains) of linear size ξ_{dis}. In each of these domains, the normalized Dirac masses correspond to a unique value of their zeroth homotopy group. At the boundary between domains differing by the values taken by their zeroth homotopy group, the Dirac masses must vanish. Such boundaries support quasi-zero-energy boundary states. When it is possible to classify the elements of the zeroth homotopy group by the pair of indices A and B, we may assign the letters A or B to any one of these domains as is illustrated. When the typical "volume" of a domain of type A equals that of type B, quasi-zero modes undergo quantum percolation through the sample and thus establish either a critical or a metallic phase of quantum matter in d-dimensional space.

types corresponding to the topological index taking values in $\mathbb{Z}$, $\mathbb{Z}_2$, or $\{0\}$, respectively. The main results from this discussion are captured in Figures 8.2 and 8.3. In Figure 8.3, we present the three typical phase diagrams when $d = 1$. We do this because Figure 8.3 can be understood with the tools introduced in Section 7.7. When $d > 1$, diffusive metallic phases are allowed in the phase diagram. These diffusive metallic phases separate Anderson localized phases that may or may not support delocalized boundary states. Understanding how these metallic phases come about theoretically when $d > 1$ requires tools that are not optimized to take advantage of the kinematics of one-dimensional space as the Dorokhov-Mello-Pereyra-Kumar (DMPK) approach used in Section 7.7 does. We refer the reader to Morimoto et al.[59] for the two- and three-dimensional counterparts to Figure 8.3.

8.3.2 Dirac Hamiltonians Perturbed by Static Disorder

We assume space to be d-dimensional. Furthermore, we assume that a microscopic lattice model with lattice spacing $\mathfrak{a}$ describing noninteracting fermions propagating in a static random environment respecting one of the ten symmetry constraints from the Altland-Zirnbauer symmetry classes is captured by the Dirac Hamiltonian

$$\mathcal{H} = \sum_{i=1}^{d} \alpha_i \frac{\partial}{\mathrm{i}\partial x_i} + V(\boldsymbol{x}) + \cdots , \tag{8.48a}$$

in the low-energy and long-wavelength limit. The rank of the Dirac matrices is

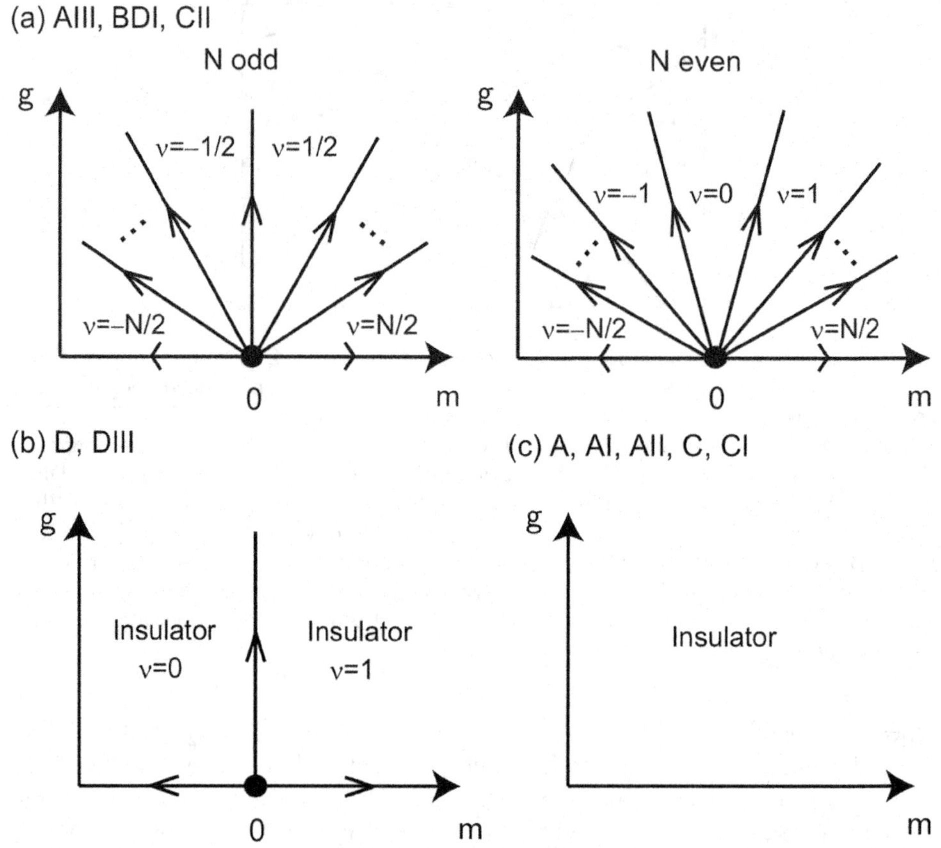

Figure 8.3 Qualitative quantum phase diagrams for 1D disordered wires with the quasi-particle energy fixed at $\varepsilon = 0$. The horizontal axis is parameterized by the characteristic value $\mathrm{m} \in \mathbb{R}$ of the disorder-averaged Dirac masses allowed by the symmetry class. The vertical axis is parameterized by the characteristic value $\mathrm{g} \geq 0$ taken by the strength of the generic static and local random mass disorder allowed by the symmetry class. Arrows on the phase boundaries and on the horizontal axis indicate flows under renormalization group transformations.

$$r = r_{\mathrm{min}}\, N, \tag{8.48b}$$

where r_{min} is the smallest rank that admits a Dirac mass matrix and $N = 1, 2, \cdots$. Hence, for any $\boldsymbol{x} \in \mathbb{R}^d$, there exists a Dirac mass matrix $V(\boldsymbol{x})$ of rank r that competes with the kinetic contribution parameterized by the Dirac matrices $\boldsymbol{\alpha} = (\alpha_1, \cdots \alpha_d)$ [that is $V(\boldsymbol{x})$ anticommutes with $\boldsymbol{\alpha}$]. The matrix elements of the Dirac mass matrix $V(\boldsymbol{x})$ in Eq. (8.48a) are random functions of $\boldsymbol{x} \in \mathbb{R}^d$. The dots represent all other static random vector and scalar potentials allowed by the Altland-Zirnbauer symmetry class. We fix the chemical potential μ to be zero.

The matrix elements of the Dirac mass matrix $V(\boldsymbol{x})$ are assumed to be random functions that change smoothly in space on the length scale of ξ_{dis} ($\gg \mathfrak{a}$). Their correlations are assumed local in that these matrix elements that are not related by the Altland-Zirnbauer symmetries are uncorrelated up to an exponential precision beyond the finite length scale $\xi_{\mathrm{dis}} \gg \mathfrak{a}$, for example, (disorder averaging is denoted by an overline)

$$\overline{V(\boldsymbol{x})} =: \mathsf{m}\,\beta_0, \tag{8.48c}$$

and

$$\frac{1}{r}\,\overline{\mathrm{tr}\,\{[V(\boldsymbol{x})-\mathsf{m}\,\beta_0]\,[V(\boldsymbol{y})-\mathsf{m}\,\beta_0]\}} =: \mathsf{g}^2\,e^{-|\boldsymbol{x}-\boldsymbol{y}|/\xi_{\mathrm{dis}}}, \tag{8.48d}$$

with all higher cumulants vanishing. The choice of the normalized Dirac mass matrix β_0 associated with the classifying space V_d that follows from the direct limit of the family of spaces $\{V_{d,r}\}$ indexed by the rank (8.48b) will be done in Sections 8.3.3, 8.3.4, and 8.3.5 in such a way that the parameter space $(\mathsf{m},\mathsf{g}) \in \mathbb{R}\times[0,\infty[$ captures the phase diagram representing the competition between delocalized and all topologically distinct localized phases of noninteracting fermions in a given d-dimensional Altland-Zirnbauer symmetry class. The former phase is favored by the Dirac kinetic contribution. The latter phases are favored by the Dirac masses. In other words, localized (insulating) phases are favored by large $|\mathsf{m}|$, whereas increasing g generates more density of states in the band (mass) gap and, in doing so, favors delocalization, if a delocalized phase exists as it does for sufficiently large dimensionality d of space.

The main predictions regarding Anderson localization follow from Figure 8.4, a consequence of **Table** 8.3 and Figure 8.2. When space is one-dimensional, the phase diagram at zero temperature that follows from these general considerations and are verified by the explicit calculations presented in Section 7.7 are summarized in Figure 8.3.

Let $V_{d,r}$ be the topological space associated with any typical realization of the random Dirac Hamiltonian (8.48) and let V_d be the associated classifying space. We then have the following phases (at the band center $\varepsilon = 0$ of the quasiparticle dispersion):

(a) If $\pi(\mathsf{S}^d, V_d) = \mathbb{Z}$ in Eq. (8.19), there are $N+1$ topologically distinct insulating phases that are separated pairwise by either a critical point or a metallic phase.
(b) If $\pi(\mathsf{S}^d, V_d) = \mathbb{Z}_2$ in Eq. (8.19), there are two topologically distinct insulating phases that are separated by either a critical point or a metallic phase.
(c) If $\pi(\mathsf{S}^d, V_d) = 0$ in Eq. (8.19), there is only one topologically trivial insulating phase.

When $d \geq 2$ (where the equality holds for classes AII, CII, D, and DIII), the ground state is metallic for sufficiently large g.[72]

[72] This claim is not obvious with the results derived so far. It follows from the one-loop

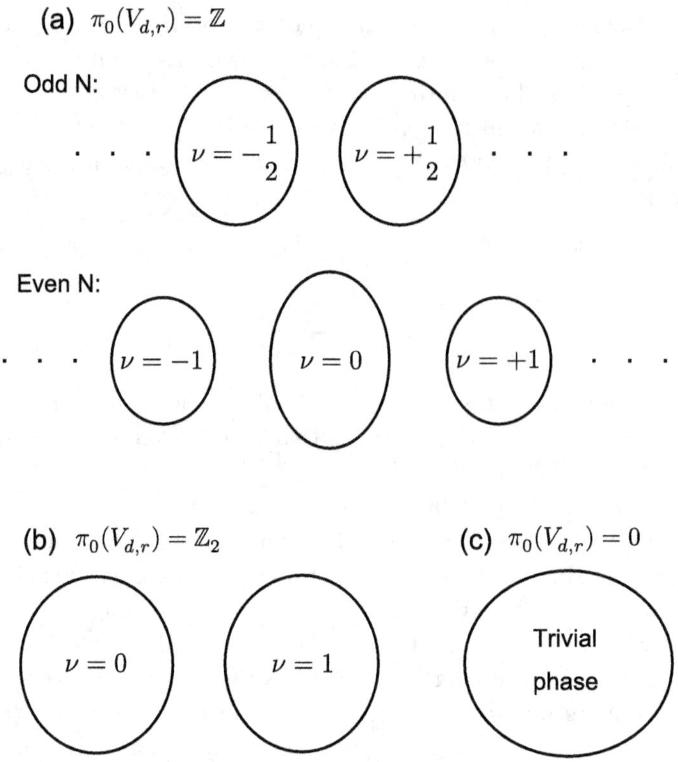

Figure 8.4 Connectedness of the compact topological space $V_{d,r}$ parameterized by the normalized Dirac masses. The zeroth homotopy groups $\pi_0(V_{d,r})$ index the disconnected parts of the compact topological space $V_{d,r}$. The zeroth homotopy group $\pi_0(V_{d,r_{\min}N})$ is either (a) $\mathbb{Z}$ (in the limit $N \to \infty$), (b) $\mathbb{Z}_2$, or (c) $\{0\} \equiv 0$ according to **Table** 8.2.

The key intuition to support items (a) and (b) is the following. Since the spatial variation of $V(\boldsymbol{x})$ is assumed to be smooth for any realization of the disorder, d-dimensional space can be decomposed into open sets (domains) with the characteristic size ξ_{dis} such that

(1) each domain can be assigned a value from $\pi(\mathsf{S}^d, V_d)$ in Eq. (8.19),
(2) and $\det[V(\boldsymbol{x})] = 0$ along the boundary of each domain.[73]

Figure 8.2 is an illustration of this decomposition of d-dimensional space when either $\pi(\mathsf{S}^d, V_d) = \mathbb{Z}$ or $\pi(\mathsf{S}^d, V_d) = \mathbb{Z}_2$ in Eq. (8.19). There are then gapless modes

renormalization group treatment of the effects of a weak static random potential on the metallic fixed point in dimensions of space larger than two as reviewed in footnotes 63 and 64.

[73] For fixed rank r in Eq. (8.48b), we must use the equivalence classes defined in Eq. (8.4) to define the notion of homotopy in $V_{d,r}$. These are indexed by the number of negative (say) eigenvalues in the mass matrix. Changing equivalence class can only take place if an eigenvalue of the mass matrix $V(\boldsymbol{x})$ smoothly changes sign as a function of $\boldsymbol{x}$, that is, $\det[V(\boldsymbol{x})]$ goes through a zero as a function of $\boldsymbol{x}$.

bound to the boundaries of these domains. We use a semiclassical picture in analogy with the Chalker-Coddington network model of the IQHE.[68]

When no connected boundaries defined by the condition $\det[V(\boldsymbol{x})] = 0$ percolate across d-dimensional space, we expect an insulating phase. However, if a connected boundary along which $\det[V(\boldsymbol{x})] = 0$ percolates across d-dimensional space, we expect departure from an insulating phase. Quantum mechanics modifies this percolating picture by turning it into that of a Chalker-Coddington-like (quantum) network model in dimension d, whereby the scattering matrix at each node of the network is fixed by the Altland-Zirnbauer symmetry class and $r = r_{\text{min}}\,N$.

8.3.3 Case of the Zeroth Homotopy Group $\mathbb{Z}$

In each dimension d of space, there are three Altland-Zirnbauer symmetry classes with $V_{d,r_{\text{min}}N}$ the unions of one of the three Grassmannian manifolds

$$\bigcup_{n=0,\cdots,N} \left\{\mathrm{U}(N)/[\mathrm{U}(n)\times\mathrm{U}(N-n)]\right\} \equiv \bigcup_{n=0,\cdots,N} \mathrm{Gr}(n,N;\mathbb{C}), \tag{8.49a}$$

$$\bigcup_{n=0,\cdots,N} \left\{\mathrm{O}(N)/[\mathrm{O}(n)\times\mathrm{O}(N-n)]\right\} \equiv \bigcup_{n=0,\cdots,N} \mathrm{Gr}(n,N;\mathbb{R}), \tag{8.49b}$$

$$\bigcup_{n=0,\cdots,N} \left\{\mathrm{Sp}(2N)/[\mathrm{Sp}(2n)\times\mathrm{Sp}(2N-2n)]\right\} \equiv \bigcup_{n=0,\cdots,N} \mathrm{Gr}(n,N;\mathbb{H}). \tag{8.49c}$$

These unions of Grassmannian manifolds are realized by the topological space of normalized Dirac masses whenever there exists a unique (up to a sign) normalized mass matrix that commutes with all other allowed normalized mass matrices. The family $\{V_{d,r_{\text{min}}N}\}$ indexed by N is a direct system with the classifying space V_d as its direct limit. The zeroth homotopy group of either one of these Grassmannian manifolds is a cyclic subgroup of $\pi(\mathrm{S}^d, V_d) = \mathbb{Z}$. It was defined by Eq. (8.4).

When the rank (8.48b) of the Dirac Hamiltonian (8.48a) belonging to any one of these Altland-Zirnbauer symmetry classes is the minimal one, $N = 1$ and there is a unique (up to a sign) normalized Dirac mass matrix β_0 of rank r_{min} such that

$$V(\boldsymbol{x}) = m(\boldsymbol{x})\,\beta_0 \tag{8.50}$$

in Eq. (8.48a). The Dirac Hamiltonian (8.48a) with the uniform mass $m > 0$ is topologically distinct from the Dirac Hamiltonian (8.48a) with the uniform mass $m < 0$. Correspondingly, the topological space $V_{d,r_{\text{min}}}$, that is, (8.49) with $N = 1$, reduces to

$$\bigcup_{n=0,1} \mathrm{U}(1)/[\mathrm{U}(n)\times\mathrm{U}(1-n)] \cong \{-1,+1\}, \tag{8.51a}$$

$$\bigcup_{n=0,1} \mathrm{O}(1)/[\mathrm{O}(n)\times\mathrm{O}(1-n)] \cong \{-1,+1\}, \tag{8.51b}$$

$$\bigcup_{n=0,1} \mathrm{Sp}(2)/[\mathrm{Sp}(2n)\times\mathrm{Sp}(2-2n)] \cong \{-1,+1\}. \tag{8.51c}$$

If the Dirac mass matrix $V(\boldsymbol{x})$ of minimal rank is random with the statistical correlation (8.48c) and (8.48d), we may decompose d-dimensional space in disjoint domains as depicted in Figure 8.2. A typical domain has the linear size ξ_{dis}. In it $m(\boldsymbol{x}) \neq 0$ with a given sign, along its boundary $m(\boldsymbol{x}) = 0$, and a sign change is only permissible across this boundary into another domain with opposite and constant sign of the mass $m(\boldsymbol{x}) \neq 0$. Any boundary separating two domains with opposite signs of $m(\boldsymbol{x}) \neq 0$ binds gapless boundary states. Whenever two boundaries approach each other within a distance much smaller than ξ_{dis}, boundary states undergo an elastic quantum scattering process dictated by the Altland-Zirnbauer symmetry class. In other words, Figure 8.2 defines a quantum network model, whereby incoming plane waves along the boundaries defined by the condition $m(\boldsymbol{x}) = 0$ scatter off each other elastically at the nodes of this network of boundaries. The mean value m of the random Dirac mass $m(\boldsymbol{x})$ dictates the relative volume occupied by the domains with $\mathrm{sgn}[m(\boldsymbol{x})] = +1$ relative to the volume occupied by the domains with $\mathrm{sgn}[m(\boldsymbol{x})] = -1$. When $\mathsf{m} = 0$, both volumes are typically equal, in which case the domain boundaries percolate across the system and the boundary states are delocalized and signal either a critical or a metallic phase.

The situation is different when $N = 2$. Indeed, the unions of Grassmannian manifolds (8.49) are now comprised of the pair of Grassmannians

$$\mathrm{U}(2)/[\mathrm{U}(2)\times\mathrm{U}(0)], \qquad \mathrm{U}(2)/[\mathrm{U}(0)\times\mathrm{U}(2)], \tag{8.52a}$$

$$\mathrm{O}(2)/[\mathrm{O}(2)\times\mathrm{O}(0)], \qquad \mathrm{O}(2)/[\mathrm{O}(0)\times\mathrm{O}(2)], \tag{8.52b}$$

$$\mathrm{Sp}(4)/[\mathrm{Sp}(4)\times\mathrm{Sp}(0)], \quad \mathrm{Sp}(4)/[\mathrm{Sp}(0)\times\mathrm{Sp}(4)], \tag{8.52c}$$

with the dimensions 0, 0, and 0, respectively, and the path-connected Grassmannians

$$\mathrm{U}(2)/[\mathrm{U}(1)\times\mathrm{U}(1)], \tag{8.53a}$$

$$\mathrm{O}(2)/[\mathrm{O}(1)\times\mathrm{O}(1)], \tag{8.53b}$$

$$\mathrm{Sp}(4)/[\mathrm{Sp}(2)\times\mathrm{Sp}(2)], \tag{8.53c}$$

with the dimensions 2, 1, and 4, respectively. Accordingly, there are three topologically distinct insulating phases when $N = 2$. To derive the dimensions of these Grassmannian manifolds, we used the fact that $\mathrm{U}(n)$, $\mathrm{O}(n)$, and $\mathrm{Sp}(2n)$ have the dimensions n^2, $n(n-1)/2$, and $n(2n+1)$, respectively.

Imagine that d-dimensional space is randomly decomposed into three types of domains according to the rule that the random Dirac mass matrix $V(\boldsymbol{x})$ is associated with only one of the three path-connected Grassmannian manifolds making up the topological space (8.49) with $N = 2$ in each domain. The domain boundaries bind gapless states, percolation of which leads to delocalization or criticality between localized phases as in the case of $N = 1$. The transitions can be induced by changing the parameter m. Unlike the $N = 1$ case, however, the ground state at $\mathsf{m} = 0$ and for any nonvanishing g not too strong is most likely a localized phase.

Indeed, localization is most likely to occur because d-dimensional space is randomly partitioned into domains such that most of the domains are characterized by a random Dirac mass matrix $V(\boldsymbol{x})$ associated with the path-connected Grassmannian manifolds of largest dimension in Eq. (8.53).

This difference between the $N = 1$ and $N = 2$ cases is not accidental. The same difference holds between the cases of odd N and even N integers, namely, that the tuning $\mathsf{m} = 0$ delivers typically a critical or metallic phase of quantum matter in d-dimensional space when N is odd, while it delivers typically a localized phase of quantum matter in d-dimensional space when N is even, as we now explain.[74]

When the topological space $V_{d,r_{\mathrm{min}}\,N}$ is any one of the three unions of Grassmannian manifolds (8.49), we can always choose to represent the Dirac Hamiltonian (8.48a) with

$$\boldsymbol{\alpha} = \boldsymbol{\alpha}_{\mathrm{min}} \otimes \mathbb{1}_N \tag{8.54a}$$

for the Dirac kinetic contribution and

$$V(\boldsymbol{x}) = \beta_{\mathrm{min}} \otimes M(\boldsymbol{x}) \tag{8.54b}$$

for the Dirac mass contribution, whereby $\boldsymbol{\alpha}_{\mathrm{min}}$ and β_{min} represent the Clifford algebra with rank r_{min}, while $\mathbb{1}_N$ is a unit $N \times N$ matrix and $M(\boldsymbol{x})$ is a random $N \times N$ Hermitean matrix which is a smooth continuous function of $\boldsymbol{x}$. For a given realization of the random matrix $M(\boldsymbol{x})$, we may partition d-dimensional space into domains whose boundaries are defined by $\det[M(\boldsymbol{x})] = 0$. Each domain can be assigned a topological index as follows. We may index each element of $\pi_0(V_{d,r_{\mathrm{min}}\,N})$ with the values of ν defined by

$$\nu := \frac{1}{2}\,\mathrm{tr}\{\mathrm{sgn}[M(\boldsymbol{x})]\} \in \begin{cases} \{-1/2, +1/2\}, & \text{if } N = 1, \\ \{-1, 0, +1\}, & \text{if } N = 2, \\ \text{and so on}, & \text{if } N > 2, \end{cases} \tag{8.55a}$$

where the Hermitean matrix $M(\boldsymbol{x})$ is diagonalized by the unitary matrix $U(\boldsymbol{x})$,

$$M =: U^{\dagger}\,\mathrm{diag}\,(\lambda_1, \cdots, \lambda_N)\,U, \tag{8.55b}$$

and

$$\mathrm{sgn}[M(\boldsymbol{x})] := U^{\dagger}\,\mathrm{diag}\left(\frac{\lambda_1}{|\lambda_1|}, \cdots, \frac{\lambda_N}{|\lambda_N|}\right) U. \tag{8.55c}$$

Each domain with $V(\boldsymbol{x})$ a smooth function of $\boldsymbol{x}$ has thereby been assigned the topological index ν.

We still need to choose the parameter space $(\mathsf{m}, \mathsf{g}) \in \mathbb{R} \times [0, \infty[$ announced in Eqs. (8.48c) and (8.48d), when the zeroth homotopy group of the classifying space is $\mathbb{Z}$. We must distinguish two cases.

[74] Observe that the direct limit (8.17) is not sensitive to this distinction between even and odd N in view of Eq. (8.15).

Case when $\mathsf{g}>0$ is a relevant perturbation to the clean critical point: We select the normalized Dirac mass matrix

$$\beta_0 := \beta_{\mathrm{min}} \otimes \mathbb{1}_N, \tag{8.56a}$$

which anticommutes with all the components of $\boldsymbol{\alpha}$ and commutes with all Dirac mass matrices allowed for given r and symmetry constraints (**Exercise** 8.3). The matrix β_0 is the unique mass matrix introduced in Section 8.2.6. We define the parameter space $(\mathsf{m},\mathsf{g}) \in \mathbb{R}\times[0,\infty[$ through the probability distribution of the Dirac mass matrix $V(\boldsymbol{x})$ given by

$$\overline{V(\boldsymbol{x})} =: \mathsf{m}\,\beta_0, \tag{8.56b}$$

and

$$\frac{1}{r}\,\overline{\mathrm{tr}\,\{[V(\boldsymbol{x})-\mathsf{m}\,\beta_0][V(\boldsymbol{y})-\mathsf{m}\,\beta_0]\}} =: \mathsf{g}^2\,e^{-|\boldsymbol{x}-\boldsymbol{y}|/\xi_{\mathrm{dis}}}, \tag{8.56c}$$

with all higher cumulants vanishing. With the definitions (8.56b) and (8.56c) for the probability distribution of $V(\boldsymbol{x})$, the point $\mathsf{m}=\mathsf{g}=0$ is a massless Dirac critical point. The critical point $\mathsf{m}=\mathsf{g}=0$ separates two insulating phases with $\nu=\pm N/2$ along the horizontal axis $\mathsf{g}=0$ in parameter space. By assumption, $\mathsf{g}>0$ is relevant in the vicinity of the clean critical point at $\mathsf{m}=\mathsf{g}=0$. This is the rule when $d=1$, in which case there appears for $\mathsf{g}>0$ $N-1$ additional localized phases, with $\nu=-(N/2)+1,-(N/2)+2,\cdots,(N/2)-1$ that are pairwise separated by lines of critical points, all of which emerge from the critical point $\mathsf{m}=\mathsf{g}=0$, as was demonstrated in Section 7.7 from the fact that the mean value of the conductance is a decreasing function of the length of the quantum wire. (It is the dimerization denoted by f in Section 7.7 that plays the role of m.) For those symmetry classes in $d=2$ for which $\mathsf{g}>0$ is (marginally) relevant in the vicinity of the clean critical point at $\mathsf{m}=\mathsf{g}=0$, it is conjectured that $N-1$ additional localized phases that we may again label with $\nu=-(N/2)+1,-(N/2)+2,\cdots,(N/2)-1$ are stabilized when $\mathsf{g}>0$. Consecutive localized phases are either separated by a line of critical points or by a metallic phase. The prescription (8.56) applies to the symmetry classes AIII, BDI, and CII in $d=1$ and A and C in $d=2$.

Case when $\mathsf{g}>0$ is an irrelevant perturbation to a clean critical point: We define the parameter space $(\mathsf{m},\mathsf{g}) \in \mathbb{R}\times[0,\infty[$ through the probability distribution of the Dirac mass matrix $V(\boldsymbol{x})$ given by

$$\overline{V(\boldsymbol{x})} =: \mathsf{m}\,\beta_0 + V_0, \tag{8.57a}$$

where β_0 again commutes with all other mass matrices permitted by the symmetry class, and

$$\frac{1}{r}\,\overline{\mathrm{tr}\,\left\{[V(\boldsymbol{x})-\overline{V(\boldsymbol{x})}][V(\boldsymbol{y})-\overline{V(\boldsymbol{y})}]\right\}} =: \mathsf{g}^2\,e^{-|\boldsymbol{x}-\boldsymbol{y}|/\xi_{\mathrm{dis}}}. \tag{8.57b}$$

Here, V_0 is any mass matrix permitted by the symmetry that satisfies the condition

$$V_0 = \beta_{\text{min}} \otimes M_0, \tag{8.57c}$$

where the $N \times N$ Hermitean matrix M_0 has N nondegenerate eigenvalues. The existence of the $r \times r$ Hermitean matrix V_0 is required to obtain $N+1$ distinct localized phases in the phase diagram by changing the parameter m in the clean limit $\mathsf{g} = 0$. The prescription (8.57) applies to the symmetry class D in $d = 2$ and all symmetry classes with $\pi_0(V) = \mathbb{Z}$ in $d \geq 3$.

The zeroth homotopy group of the topological space $V_{d,r_{\text{min}}N}$ encodes the connectedness of $V_{d,r_{\text{min}}N}$. The "volume" of each path-connected component

$$\text{Gr}(n, N; \mathbb{C}) \equiv \text{U}(N)/\big[\text{U}(n) \times \text{U}(N-n)\big], \tag{8.58a}$$
$$\text{Gr}(n, N; \mathbb{R}) \equiv \text{O}(N)/\big[\text{O}(n) \times \text{O}(N-n)\big], \tag{8.58b}$$
$$\text{Gr}(n, N; \mathbb{H}) \equiv \text{Sp}(2N)/\big[\text{Sp}(2n) \times \text{Sp}(2N-2n)\big] \tag{8.58c}$$

of $V_{d,r_{\text{min}}N}$ is measured by the dimension

$$2n(N-n) = N^2 - n^2 - (N-n)^2, \tag{8.59a}$$
$$n(N-n) = \frac{N(N-1) - n(n-1)}{2} - \frac{(N-n)(N-n-1)}{2}, \tag{8.59b}$$
$$4n(N-n) = N(2N+1) - (N-n)[2(N-n)+1] - n(2n+1), \tag{8.59c}$$

respectively, as is depicted schematically in Figure 8.4(a).

When N is odd, one can always write

$$V_{d,r_{\text{min}}N} = \text{A} \cup \text{B}, \tag{8.60a}$$

where

$$\text{A} := \bigcup_{n=0,\cdots,\frac{N-1}{2}} \{\text{U}(N)/[\text{U}(N-n) \times \text{U}(n)]\} \tag{8.60b}$$

and

$$\text{B} := \bigcup_{n=\frac{N+1}{2},\cdots,N} \{\text{U}(N)/[\text{U}(n) \times \text{U}(N-n)]\}, \tag{8.60c}$$

and similarly with the substitutions $\text{U} \to \text{O}, \text{Sp}$. The existence of a critical or metallic phase of quantum matter when $\mathsf{m} = 0$ in the parameterization (8.56) follows from repeating the argumentation for the $N = 1$ case that is captured by Figure 8.2.

When N is even, one can always write

$$V_{d,r_{\text{min}}N} = V_- \cup V_0 \cup V_+, \tag{8.61a}$$

where

$$V_{-} := \bigcup_{n=0,\cdots,\frac{N}{2}-1} \{\mathrm{U}(N)/[\mathrm{U}(N-n)\times\mathrm{U}(n)]\}, \tag{8.61b}$$

$$V_{0} := \mathrm{U}(N)/[\mathrm{U}(N/2)\times\mathrm{U}(N/2)], \tag{8.61c}$$

and

$$V_{+} := \bigcup_{n=\frac{N}{2}+1,\cdots,N} \{\mathrm{U}(N)/[\mathrm{U}(n)\times\mathrm{U}(N-n)]\}, \tag{8.61d}$$

and similarly with the substitutions $\mathrm{U}\to\mathrm{O},\mathrm{Sp}$. Since the index ν assigned to localized phases is an odd function of m, the ground state at $\mathsf{m}=0$ and g nonvanishing but not too large in the parameterization (8.56) is expected to be in the localized phase, for d-dimensional space is partitioned into domains of Dirac mass matrices which are predominantly drawn from V_0.

8.3.4 Case of the Zeroth Homotopy Group $\mathbb{Z}_2$

In each dimension d of space, there are two Altland-Zirnbauer symmetry classes with $V_{d,r_{\mathrm{min}}\,N}$ homeomorphic to either the orthogonal group $\mathrm{O}(N)$ or the coset space $\mathrm{O}(2N)/\mathrm{U}(N)$. Thus, $V_{d,r_{\mathrm{min}}\,N}$ has the zeroth homotopy group $\mathbb{Z}_2$. The former case is called the first descendant $\mathbb{Z}_2$. The second case is called the second descendant $\mathbb{Z}_2$. If so, we can always choose to represent the Dirac Hamiltonian (8.48a) with

$$\boldsymbol{\alpha} = \boldsymbol{\alpha}_{\mathrm{min}} \otimes \mathbb{1}_N \tag{8.62a}$$

for the Dirac kinetic contribution and

$$V(\boldsymbol{x}) = \rho_{\mathrm{min}} \otimes M(\boldsymbol{x}) \tag{8.62b}$$

for the Dirac mass contribution in each domain where the Dirac mass matrix $V(\boldsymbol{x})$ is continuous and invertible. Here, ρ_{min} is a $r_{\mathrm{min}}/2\times r_{\mathrm{min}}/2$ matrix such that, when tensored with the antisymmetric Pauli matrix τ_2, $\boldsymbol{\alpha}_{\mathrm{min}}$ and $\rho_{\mathrm{min}}\otimes\tau_2$ deliver a representation of the Clifford algebra of rank r_{min}. Finally, $\mathbb{1}_N$ is a unit $N\times N$ matrix and $M(\boldsymbol{x})$ is a $2N\times 2N$ Hermitean matrix which is also antisymmetric, that is

$$M(\boldsymbol{x}) = M^{\dagger}(\boldsymbol{x}), \qquad M(\boldsymbol{x}) = -M^{\mathsf{T}}(\boldsymbol{x}). \tag{8.62c}$$

Each domain in the partition of d-dimensional space into domains defined by the boundaries where $\det[M(\boldsymbol{x})]=0$ can be assigned a topological index as follows. We may index each element of $\pi_0(V_{d,r_{\mathrm{min}}\,N})$ with the values of $\nu=0,1$ defined by

$$(-1)^{\nu} := \frac{\mathrm{Pf}\,[\mathrm{i}M(\boldsymbol{x})]}{\sqrt{\det\,[\mathrm{i}M(\boldsymbol{x})]}}. \tag{8.62d}$$

Here, we have multiplied $M(\boldsymbol{x})$ by the imaginary unit so that, owing to Eq. (8.62c), $\mathrm{i}M(\boldsymbol{x})$ is an invertible real-valued antisymmetric matrix with positive determinant. Now, the Pfaffian of a $2n \times 2n$ real-valued antisymmetric matrix $A = (A_{i,j})$ is defined by (S_{2n} denotes the permutation group of $2n$ elements)

$$\mathrm{Pf}\,(A) := \frac{1}{2^n\, n!} \sum_{\sigma \in S_{2n}} \mathrm{sgn}\,(\sigma) \prod_{i=1}^{n} A_{\sigma(2i-1),\sigma(2i)}. \tag{8.63a}$$

It obeys the property

$$[\mathrm{Pf}\,(A)]^2 = \det\,(A). \tag{8.63b}$$

Hence, the index ν reflects the ambiguity when choosing the branch cut of the square root. Each domain has thereby been assigned the $\mathbb{Z}_2$-valued topological index ν.

When N is odd, we choose the parameter space $(\mathsf{m}, \mathsf{g}) \in \mathbb{R} \times [0, \infty[$ by selecting

$$\beta_0 := \rho_{\mathrm{min}} \otimes \tau_2 \otimes \mathbb{1}_N \tag{8.64a}$$

in the probability distribution of the Dirac mass matrix $V(\boldsymbol{x})$ given by

$$\overline{V(\boldsymbol{x})} =: \mathsf{m}\, \beta_0, \tag{8.64b}$$

and

$$\frac{1}{r}\, \overline{\mathrm{tr}\,\{[V(\boldsymbol{x}) - \mathsf{m}\,\beta_0][V(\boldsymbol{y}) - \mathsf{m}\,\beta_0]\}} =: \mathsf{g}^2\, e^{-|\boldsymbol{x}-\boldsymbol{y}|/\xi_{\mathrm{dis}}}, \tag{8.64c}$$

with all higher cumulants vanishing. In the clean limit $\mathsf{g} = 0$, the point $\mathsf{m} = 0$ is a massless Dirac critical point separating the two insulating phases with $\nu = 0$ ($\mathsf{m} > 0$) and $\nu = 1$ ($\mathsf{m} < 0$). Given $\mathsf{g} > 0$, the d-dimensional space is decomposed into domains, and we may identify the domains labeled by A in Figure 8.2 with $\nu = 0$ and the domains labeled by B in Figure 8.2 with $\nu = 1$ for any realization of the random Dirac mass matrix $V(\boldsymbol{x})$. Hence, tuning the mean value m to some critical value (e.g., $\mathsf{m} = 0$) realizes a situation where the domains A and B appear with equal probability and the domain boundaries percolate. This tuning stabilizes either a critical point or a metallic phase in d-dimensional space.

When N is even, we do not adopt the probability distribution (8.64), for it leads to the massless Dirac point $\mathsf{m} = \mathsf{g} = 0$ separating two insulating phases belonging to the same topological phase with $\nu = 0$ in the clean limit $\mathsf{g} = 0$. For example, when we consider $N = 2$ and the first descendant $\mathbb{Z}_2$ (**Exercise** 8.3), we may choose

$$\begin{aligned} M(\boldsymbol{x}) =& \, m_{2,0}(\boldsymbol{x})\, \tau_2 \otimes \sigma_0 + m_{2,1}(\boldsymbol{x})\, \tau_2 \otimes \sigma_1 \\ &+ m_{2,3}(\boldsymbol{x})\, \tau_2 \otimes \sigma_3 + m_{1,2}(\boldsymbol{x})\, \tau_1 \otimes \sigma_2, \end{aligned} \tag{8.65}$$

where the quadruplets τ_μ and σ_μ, $\mu = 0, 1, 2, 3$, are made of the unit 2×2 matrix and the three Pauli matrices, respectively. In this case,

$$(-1)^\nu = \mathrm{sgn}(m_{2,0}^2 + m_{1,2}^2 - m_{2,1}^2 - m_{2,3}^2). \tag{8.66}$$

If we define the controlling parameter m as in Eq. (8.64), we have $\overline{m_{2,0}} = \mathsf{m}$ and $\overline{m_{1,2}} = \overline{m_{2,1}} = \overline{m_{2,3}} = 0$, for which the localized phase with $\nu = 0$ always appear for nonvanishing m with sufficiently small g. Instead, we choose a probability distribution such that the massless Dirac point $\mathsf{m} = \mathsf{g} = 0$ in the parameter space $(\mathsf{m}, \mathsf{g}) \in \mathbb{R} \times [0, \infty[$ separates two insulating phases with $\nu = 0$ and $\nu = 1$ in the clean limit $\mathsf{g} = 0$. For the above example of $N = 2$, such a choice is given by

$$\overline{V(\boldsymbol{x})} = 0, \tag{8.67a}$$

$$\overline{m^2_{2,0}} + \overline{m^2_{1,2}} - \overline{m^2_{2,1}} - \overline{m^2_{2,3}} =: \mathsf{m}, \tag{8.67b}$$

and

$$\frac{1}{r}\,\overline{\mathrm{tr}\,[V(\boldsymbol{x})\,V(\boldsymbol{y})]} =: \mathsf{g}^2\, e^{-|\boldsymbol{x}-\boldsymbol{y}|/\xi_{\mathrm{dis}}}. \tag{8.67c}$$

For general values of N, we may adopt $\mathsf{m} := \overline{\mathrm{Pf}\,[\mathrm{i}M(\boldsymbol{x})]}$ instead of Eq. (8.64).

8.3.5 Case of the Zeroth Homotopy Group $\{0\}$

In each dimension d of space, there are five Altland-Zirnbauer symmetry classes with $V_{d,r}$ a compact and path-connected topological space. Hence $V_{d,r}$ has a vanishing zeroth homotopy group [recall that we are using Eq. (8.4) to define the homotopy group when r is fixed]. We can always choose to represent the Dirac Hamiltonian (8.48a) with

$$\boldsymbol{\alpha} = \boldsymbol{\alpha}_{\mathrm{min}} \otimes \mathbb{1}_N \tag{8.68}$$

for the Dirac kinetic contribution, and then we choose β_{min} arbitrarily from the allowed $r_{\mathrm{min}} \times r_{\mathrm{min}}$ normalized Dirac mass matrices which anticommutes with all the components of $\boldsymbol{\alpha}_{\mathrm{min}}$.

For any of these five Altland-Zirnbauer symmetry classes, we choose the parameter space $(\mathsf{m}, \mathsf{g}) \in \mathbb{R} \times [0, \infty[$ by selecting

$$\beta_0 := \beta_{\mathrm{min}} \otimes \mathbb{1}_N \tag{8.69a}$$

in the probability distribution of the Dirac mass matrix $V(\boldsymbol{x})$ given by

$$\overline{V(\boldsymbol{x})} =: \mathsf{m}\,\beta_0, \tag{8.69b}$$

$$\frac{1}{r}\,\overline{\mathrm{tr}\,\{[V(\boldsymbol{x}) - \mathsf{m}\,\beta_0][V(\boldsymbol{y}) - \mathsf{m}\,\beta_0]\}} =: \mathsf{g}^2\, e^{-|\boldsymbol{x}-\boldsymbol{y}|/\xi_{\mathrm{dis}}}, \tag{8.69c}$$

with all higher cumulants vanishing.

In these five Altland-Zirnbauer symmetry classes, the phase diagram has only a single localized phase that is adiabatically connected to a topologically trivial band insulator with reducing disorder.

Nevertheless, there can be anomalies of the conductivity and of the density of states at $\varepsilon = 0$ when higher than the zeroth homotopy group of $V_{d,r}$ have more than one element.[59]

Table 8.5 *The tenfold way: Immunity to Anderson localization*

Class	Edge of 2D TI/TSC	Surface of 3D TI/TSC
A	Ballistic	-
AIII	-	Metallic
AI	-	-
BDI	-	-
D	Ballistic	-
DIII	Ballistic	Metallic
AII	Ballistic	Metallic
CII	-	Metallic
C	Ballistic	-
CI	-	Metallic

Immunity to Anderson localization in five out of the ten Altland-Zirnbauer symmetry classes along any boundary of a strong topological insulator (TI) or a topological superconductor (TSC). This immunity is a consequence of the absence of any Dirac mass matrix with the rank $r = \widetilde{r}_{\mathrm{min}} < r_{\mathrm{min}}$ (**Exercise** 8.4) for a random Dirac Hamiltonian capturing the low-energy and long-wavelength effects of local disorder on the boundary in the corresponding Altland-Zirnbauer symmetry class. The entry ballistic refers to quantum transport without disorder for which the Landauer conductance for the conserved charge is quantized (see Figure 7.4). The entry metallic refers to the delocalized but diffusive regime of quantum transport.

8.3.6 Boundaries of Topological Insulators or Superconductors

So far, we have been concerned with the interplay between topology, local symmetries, and Anderson localization for d-dimensional random massive Dirac Hamiltonians that capture the effects of local, and smooth disorder of d-dimensional lattice models such as quantum network models at low energies and long wavelengths. We have explained in Section 8.2.5 how **Tables** 8.1, 8.2, and 8.3 can be used to identify in each dimension d of space the five Altland-Zirnbauer symmetry classes that support topologically distinct insulating phases of d-dimensional quantum matter. There is an alternative approach to the classification of topological insulators or superconductors that was explained in Section 8.2.5.

Any $(d-1)$-dimensional boundary of a d-dimensional strong topological insulator or superconductor is immune to Anderson localization.[2,3] This immunity is captured by the presence of an additional term of topological origin in the NLSM that captures the physics of Anderson localization on the $(d-1)$-dimensional boundary subject to the symmetry constraint imposed by one of the ten Altland-Zirnbauer symmetry classes.[2,3] Alternatively, the immunity against Anderson localization holds for those Altland-Zirnbauer symmetry classes that prohibit the existence of any Dirac mass matrix of rank $r = \widetilde{r}_{\mathrm{min}} < r_{\mathrm{min}}$ entering the random Dirac Hamiltonian that encodes single-particle transport at low energies and long wavelength on any $(d-1)$-dimensional boundary. In **Exercise** 8.4, the conditions for the presence (or absence) of Dirac mass matrices is formulated and solved. The conditions for the absence of normalized Dirac masses in $(d-1)$-dimensions are equivalent to the conditions for a nontrivial zeroth homotopy group of the topological spaces associated with the normalized Dirac masses in d-dimensions. This alternative derivation of the tenfold way for topological insulators or superconductors is captured by **Table** 8.5 for two- and three-dimensional spaces.

8.4 Breakdown of the Tenfold Way Due to Interactions

8.4.1 Overview

Strong topological insulating or superconducting phases of noninteracting (within the mean-field approximation) fermions are characterized by topological numbers ($\mathbb{Z}$ or $\mathbb{Z}_2$) that encode the nontrivial topology of the occupied single-particle wave functions and are accompanied by gapless excitations that are localized along any boundary. Polyacetylene and the IQHE are the first two historical examples of strong topological insulators or superconductors. Polyacetylene can be found in row BDI and column $d = 1$ of **Table** 8.3. The IQHE can be found in row A and column $d = 2$ of **Table** 8.3. It is characterized by the Hall conductivity quantized by the integer $\nu = 1, 2, \cdots$ in units of e^2/h. The topological integer ν counts the number of extended chiral edge modes propagating at the boundary of the sample.

The gapless modes appearing at the boundary in the IQHE are robust to both elastic and inelastic scattering resulting from one-body impurity potentials and many-body electron-electron interactions.[9,75] Similarly, the gapless modes in the row AII column $d = 2$ of **Table** 8.3 predicted and christened topological insulators by Kane and Mele in 2005[23,24] are immune to both backscattering resulting from one-body impurity potentials[23,24] and many-body electron-electron interactions[76,77] provided TRS is neither explicitly nor spontaneously broken.

Given the robustness to many-body fermion-fermion interactions of the edge states in the IQHE, it was a remarkable observation made by Fidkowski and Kitaev in 2010[5,6] that it is possible to gap out eight Majorana zero modes localized at the end of a one-dimensional topological superconducting wire through many-body interactions without closing the spectral gap in the bulk. In the terminology of the tenfold way,[2,4,3] it was demonstrated by Fidkowski and Kitaev that the $\mathbb{Z}$ topological classification for the noninteracting one-dimensional symmetry class BDI, when interpreted as a superconductor, is (i) unstable to quartic contact

[75] Q. Niu, D. J. Thouless, and Y.-S. Wu, Phys. Rev. B **31**(6), 3372–3377 (1985). [Niu et al. (1985)].
[76] M. Levin and A. Stern, Phys. Rev. Lett. **103**(19), 196803 (2009). [Levin and Stern (2009)].
[77] T. Neupert, L. Santos, S. Ryu, C. Chamon, and C. Mudry, Phys. Rev. B **84**(16), 165107 (2011). [Neupert et al. (2011)].

interactions that neither break explicitly nor spontaneously the TRS, and (ii) this instability reduces the noninteracting topological classification $\mathbb{Z}$ to $\mathbb{Z}_8$.

We present in Section 8.4.2 an approach[78] that allows to derive the reduction pattern of all strong noninteracting topological insulators or superconductors for any dimensionality d of space in the presence of quartic contact interactions that neither break explicitly nor spontaneously the defining symmetries. This method relies on the topology of the classifying spaces.

This approach is applied (but not limited) to the breakdown of the tenfold way for strong topological insulators or superconductors. In doing so, we prove the following properties that we report in **Table** 8.6:[79]

1. All $\mathbb{Z}_2$ entries of the periodic **Table** 8.3 irrespectively of the dimensionality of space are stable to quartic contact interactions.
2. All $\mathbb{Z}$ entries of the periodic **Table** 8.3 when the dimensionality of space is even are stable to quartic contact interactions.
3. Only the $\mathbb{Z}$ entries of the periodic **Table** 8.3 when the dimensionality of space is odd are unstable to quartic contact interactions with a reduction pattern that is computed explicitly and shown to break the Bott periodicity of two for the complex symmetry classes and of eight for the real symmetry classes.

The strategy that we use to study the robustness of ν boundary modes to quartic contact fermion-fermion interactions consists of three steps.

First, a noninteracting topological phase is represented by the many-body ground state of a massive Dirac Hamiltonian with a matrix dimension that depends on ν.

Second, a Hubbard-Stratonovich transformation is used to trade a generic quartic contact interaction in favor of dynamical Dirac mass-like bilinears coupled to their conjugate fields (that will be called Dirac masses). These dynamical Dirac masses may violate any symmetry constraint other than the PHS.

Third, the ν boundary modes that are coupled with a suitably chosen subset of dynamical masses are integrated over. The resulting dynamical theory on the $(d-1)$-dimensional boundary is a bosonic one, a QNLSM in $[(d-1)+1]$-dimensional spacetime with a target space that depends on ν (**Exercise** 5.5).

The reduction pattern is then obtained by identifying the smallest value of ν for which this QNLSM cannot be augmented by a topological term. The presence or absence of topological terms in the relevant QNLSM is determined by the topology of the spaces of boundary dynamical Dirac masses, that is the topology of classifying spaces. This is why the same approach that was used to obtain the tenfold way of noninteracting fermions can be relied on to deduce a classification of the fermionic

[78] T. Morimoto, A. Furusaki, and C. Mudry, Phys. Rev. B **92**(12), 125104 (2015). [Morimoto et al. (2015b)].

[79] As always, interactions that preserve the defining symmetries can always break spontaneously the defining symmetries when sufficiently strong. The condition of stability thus precludes the possibility of spontaneous symmetry breaking of the defining symmetries.

Table 8.6 *The tenfold way: Breakdown of the Bott periodicity due to local interactions*

Class	$d=1$	$d=2$	$d=3$	$d=4$	$d=5$	$d=6$	$d=7$	$d=8$
A	0	$\mathbb{Z}$	0	$\mathbb{Z}$	0	$\mathbb{Z}$	0	$\mathbb{Z}$
AIII	$\boxed{\mathbb{Z}_4}$	0	$\boxed{\mathbb{Z}_8}$	0	$\boxed{\mathbb{Z}_{16}}$	0	$\boxed{\mathbb{Z}_{32}}$	0
AI	0	0	0	$\mathbb{Z}$	0	$\mathbb{Z}_2$	$\mathbb{Z}_2$	$\mathbb{Z}$
BDI	$\boxed{\mathbb{Z}_8}, \boxed{\mathbb{Z}_4}$	0	0	0	$\boxed{\mathbb{Z}_{16}}, \boxed{\mathbb{Z}_8}$	0	$\mathbb{Z}_2$	$\mathbb{Z}_2$
D	$\mathbb{Z}_2$	$\mathbb{Z}$	0	0	0	$\mathbb{Z}$	0	$\mathbb{Z}_2$
DIII	$\mathbb{Z}_2$	$\mathbb{Z}_2$	$\boxed{\mathbb{Z}_{16}}$	0	0	0	$\boxed{\mathbb{Z}_{32}}$	0
AII	0	$\mathbb{Z}_2$	$\mathbb{Z}_2$	$\mathbb{Z}$	0	0	0	$\mathbb{Z}$
CII	$\boxed{\mathbb{Z}_2}, \boxed{\mathbb{Z}_2}$	0	$\mathbb{Z}_2$	$\mathbb{Z}_2$	$\boxed{\mathbb{Z}_{16}}, \boxed{\mathbb{Z}_{16}}$	0	0	0
C	0	$\mathbb{Z}$	0	$\mathbb{Z}_2$	$\mathbb{Z}_2$	$\mathbb{Z}$	0	0
CI	0	0	$\boxed{\mathbb{Z}_4}$	0	$\mathbb{Z}_2$	$\mathbb{Z}_2$	$\boxed{\mathbb{Z}_{32}}$	0

The ten Altland-Zirnbauer symmetry classes and their topological classification when (i) fermion-fermion interactions neither break explicitly their defining symmetries nor spontaneously, and (ii) the many-body ground state is nondegenerate when d-dimensional space is any compact manifold without boundary. The reduction, if any, that arises from the effects of interactions on the topological classification of noninteracting fermions for $d=1,\cdots,8$ is given in the last eight columns. Each entry with a nontrivial Abelian group defines equivalence classes of interacting topological insulators or superconductors with a nondegenerate ground state when d-dimensional space is any compact manifold without boundary. We boxed the entry corresponding to a given symmetry class and a given column of odd dimensionality d to indicate that this entry is a quotient group $\mathfrak{G}_{\mathrm{int}}$ of $\mathfrak{G}=\mathbb{Z}$. The reduction $\mathbb{Z}\to\mathfrak{G}_{\mathrm{int}}$ results from an instability of the noninteracting topological classification to fermion-fermion interactions. The double entries for the symmetry classes BDI and CII in $d=1$ and $d=5$ dimensional space, respectively, correspond to (not) imposing U(1) charge conservation symmetry for the cyclic group with the (larger) smaller order.

invertible topological phases of matter studied using the tools of group cohomology in Chapters 5 and 6 for interacting fermions (bosons) in one-dimensional space.

8.4.2 Strategy

Noninteracting fermions always belong to one of the Altland-Zirnbauer symmetry classes defined in Section 7.3 (**Exercise** 3.37). Within any one of these ten symmetry classes, the defining topological attributes of noninteracting topological insulators or superconductors are shared by equivalence classes of Hamiltonians. Any two members within a topological class can be deformed into each other by a smooth (adiabatic) deformation of the matrix elements of these Hamiltonians without closing the bulk energy gap. These equivalence classes are endowed with an Abelian group structure $\mathfrak{G}$. For any given dimensionality d of space, $\mathfrak{G}$ is non-trivial for five out of the ten Altland-Zirnbauer classes. Specifically, three of the ten Altland-Zirnbauer classes support Abelian groups $\mathfrak{G} = \mathbb{Z}$, while two of the ten Altland-Zirnbauer classes support Abelian groups $\mathfrak{G} = \mathbb{Z}_2$. The TRS, PHS, and CHS can be augmented by crystalline symmetries. noninteracting fermions obeying crystalline symmetries can also be understood as realizing topologically distinct equivalence classes, that is topological crystalline insulators.[37,80]

The topological classifications with the Abelian group $\mathfrak{G}$ for noninteracting topological insulators or superconductors can break down in the presence of many-body interactions. Namely, an Abelian group $\mathfrak{G}_{\mathrm{int}}$ that encodes the topological equivalence classes of gapped ground states for interacting fermions can be smaller than $\mathfrak{G}$ as a group (some quotient group of $\mathfrak{G}$).

In order to establish the instability of the noninteracting classification of strong topological insulators or superconductors, we choose the family of massive Dirac Hamiltonians

$$\mathcal{H}^{(0)} := -\mathrm{i}\sum_{j=1}^{d} \widetilde{\alpha}_j \otimes \mathbb{1}\,\frac{\partial}{\partial x^j} + m(\boldsymbol{x})\,\widetilde{\beta} \otimes \mathbb{1} \tag{8.70}$$

as representative single-particle Hamiltonians. Here, Dirac matrices $\widetilde{\boldsymbol{\alpha}}$ and $\widetilde{\beta}$ anti-commute with each other and have the minimal dimension (rank) r_{min} under the symmetry constraints, that is, r_{min} is the minimal rank to realize a Dirac Hamiltonian of the form (8.70). The dimension of the unit matrix $\mathbb{1}$ is $\nu = 1, 2, \cdots$. The integer $\nu \in \mathfrak{G}$ is then related to the dimension $r(\nu) = r_{\mathrm{min}}\,\nu$ of the Dirac matrices that we choose. The question that we want to address is that of the stability or instability of the boundary states of a noninteracting strong topological insulator or superconductor in the presence of many-body interactions that do not break the protecting symmetries of the noninteracting limit.[81] Here, whenever $\nu \neq 0$, the extended single-particle boundary states are governed by the massless Dirac Hamiltonian

$$\mathcal{H}^{(0)}_{\mathrm{bd}} := -\mathrm{i}\sum_{j=1}^{d-1} \alpha_j \otimes \mathbb{1}\,\frac{\partial}{\partial x^j} \equiv -\mathrm{i}\boldsymbol{\alpha} \otimes \mathbb{1} \cdot \boldsymbol{\partial}, \tag{8.71}$$

which is obtained by introducing a domain wall in the mass $m(\boldsymbol{x})$ along the x^d-direction that enters Hamiltonian (8.70). The Dirac matrices $\boldsymbol{\alpha}\otimes\mathbb{1}$ have a dimension $r(\nu)/2$ that is half that of the bulk massive Dirac Hamiltonian $\mathcal{H}^{(0)}$. The dimension of the matrices $\boldsymbol{\alpha}$ is $r_{\mathrm{min}}/2$.

[80] E. Cornfeld and S. Carmeli, Phys. Rev. Research **3**(1), 013052 (2021); and references therein. [Cornfeld and Carmeli (2021)].

[81] We consider interactions that do not break the protecting symmetries of the noninteracting limit, that are strong on the boundary, yet are not-too-strong as measured by the single-particle gap for the bulk states of insulators.

The breakdown (reduction) of the topological classification for noninteracting fermions takes place when the boundary states of topological insulators or superconductors can be gapped by many-body interactions that preserve their defining symmetries. By assumption, we consider many-body interactions that are weak relative to the bulk gap. If so, it is sufficient to treat the effects of many-body interactions for the massless Dirac fermions propagating on the $(d-1)$-dimensional boundary. To establish an instability of the noninteracting topological classification, we need not consider all possible many-body interactions. It suffices to establish that at least one family of strong (on the boundary) interactions implies the instability of the noninteracting classification $\mathfrak{G}$ by gapping out all boundary Dirac fermions. To this end, we limit ourselves to contact interactions.

Definition 8.3 (dynamical Dirac masses) Contact interactions are constructed from taking squares of local bilinears in the Dirac fermions. We have two options for these bilinears. The bilinear under consideration either commutes or anticommutes with the kinetic contribution to the Dirac Hamiltonian. We shall call the latter option a Dirac mass. In this chapter, we only consider the contact interactions obtained from taking squares of those bilinears built out of Dirac mass matrices, for only these can gap the noninteracting massless boundary Dirac fermions in a mean-field approximation. Because we assume that the protecting symmetries forbid the presence of Dirac masses on the boundary that are consistent with the protecting symmetries, the only possible Dirac masses induced by a mean-field treatment of a symmetry-preserving quartic interaction on the boundary must be odd under at least one of the protecting symmetries. We shall call such a boundary Dirac mass a boundary dynamical mass and label it with the same Greek letter β as is traditionally used when denoting a Dirac mass matrix.

We are thus led to consider the many-body interacting Dirac boundary Hamiltonian

$$\widehat{H}_{\mathrm{bd}} := \widehat{H}^{(0)}_{\mathrm{bd}} + \widehat{H}^{(\mathrm{int})}_{\mathrm{bd}}, \tag{8.72a}$$

where (the subscript "bd" stands for boundary)

$$\widehat{H}^{(0)}_{\mathrm{bd}} := \int \mathrm{d}^{d-1}\boldsymbol{x}\, \hat{\Psi}^{\dagger}(t,\boldsymbol{x})\, \mathcal{H}^{(0)}_{\mathrm{bd}}\, \hat{\Psi}(t,\boldsymbol{x}) \tag{8.72b}$$

and

$$\widehat{H}^{(\mathrm{int})}_{\mathrm{bd}} := \lambda \sum_{\{\beta\}} \int \mathrm{d}^{d-1}\boldsymbol{x}\, \left[\hat{\Psi}^{\dagger}(t,\boldsymbol{x})\, \beta\, \hat{\Psi}(t,\boldsymbol{x})\right]^{2}. \tag{8.72c}$$

We have chosen the real-valued coupling λ with the dimension of $(\text{length})^{d-2}$ to be independent of the symbol β that we use for the dynamical mass matrices for simplicity. The coupling constant λ is marginal in $d=2$ and irrelevant when $d>2$ with respect to the critical Hamiltonian corresponding to the noninteracting limit.

(Of course, it can very well be that the set $\{\beta\}$ is empty. If so, we anticipate that $\mathfrak{G} = \mathfrak{G}_{\text{int}}$ must hold. This is what happens for the strong topological insulators in the symmetry classes A, D, and C when $d = 2$.) At this stage, it is convenient to treat the many-body Hamiltonian (8.72) with the help of the path integral

$$Z_{\text{bd}} := \int \mathcal{D}[\Psi^{\dagger}, \Psi]\, e^{-S_{\text{bd}}}, \tag{8.73a}$$

where the action in Euclidean time τ is

$$S_{\text{bd}} := \int \mathrm{d}\tau \int \mathrm{d}^{d-1}\boldsymbol{x}\, \mathcal{L}_{\text{bd}}, \tag{8.73b}$$

with the Lagrangian density

$$\mathcal{L}_{\text{bd}} := \Psi^{\dagger} \left(\partial_{\tau} + \mathcal{H}^{(0)}_{\text{bd}}\right) \Psi + \lambda \sum_{\{\beta\}} \left(\Psi^{\dagger} \beta\, \Psi\right)^{2}. \tag{8.73c}$$

The path integral is over Grassmann-valued Dirac spinors.

We rewrite the quartic interaction terms by performing a Hubbard-Stratonovich transformation with respect to the bosonic fields ϕ_{β} conjugate to $\Psi^{\dagger}\, \beta\, \Psi$,[82]

$$Z_{\text{bd}} \propto \int \mathcal{D}[\Psi^{\dagger}, \Psi, \phi_{\beta}]\, e^{-S'_{\text{bd}}}. \tag{8.74a}$$

Here, the action in Euclidean time τ is

$$S'_{\text{bd}} := \int \mathrm{d}\tau \int \mathrm{d}^{d-1}\boldsymbol{x}\, \mathcal{L}'_{\text{bd}}, \tag{8.74b}$$

with the Lagrangian density

$$\mathcal{L}'_{\text{bd}} := \Psi^{\dagger} \left(\partial_{\tau} + \mathcal{H}^{(\text{dyn})}_{\text{bd}}\right) \Psi + \frac{1}{\lambda} \sum_{\{\beta\}} \phi^{2}_{\beta}, \tag{8.74c}$$

where we have introduced the dynamical one-body single-particle Hamiltonian

$$\mathcal{H}^{(\text{dyn})}_{\text{bd}}(\tau, \boldsymbol{x}) := \mathcal{H}^{(0)}_{\text{bd}}(\boldsymbol{x}) + \sum_{\{\beta\}} 2\mathrm{i}\, \beta\, \phi_{\beta}(\tau, \boldsymbol{x}). \tag{8.74d}$$

The sign of λ is selected so as to find a nonvanishing solution to the saddle-point equations. We will see that the sign of λ must be negative, that is, it corresponds to an attractive interaction.

In a saddle-point approximation, the magnitude of the vector $\boldsymbol{\phi}$ with the components ϕ_{β} can be frozen both in imaginary time and in $(d-1)$-dimensional space. Fluctuations that change this frozen magnitude are suppressed by the second term on the right-hand side of Eq. (8.74c). We will restrict the set $\{\beta\}$ to pairwise anticommuting Dirac mass matrices. If so, the direction in which the vector $\boldsymbol{\phi}$ with the components ϕ_{β} freezes in the saddle-point approximation is arbitrary.

[82] E. Fradkin, *Field Theories of Condensed Matter Physics*, Cambridge University Press, New York 2013. [Fradkin (2013)].

The saddle-point equations for $\boldsymbol{\phi}$ are given as follows. Integrating the fermionic degrees of freedom leads to the effective Lagrangian,

$$\mathcal{S}_{\text{eff}}[\boldsymbol{\phi}] := (-1)\mathrm{Tr}\,\log\left[\partial_\tau - \mathrm{i}\boldsymbol{\alpha}\cdot\boldsymbol{\partial} + 2\mathrm{i}\boldsymbol{\beta}\cdot\boldsymbol{\phi}\right] + \frac{1}{\lambda\,r}\mathrm{Tr}\,(\boldsymbol{\phi}^2), \tag{8.75}$$

where the vector of Dirac matrices $\boldsymbol{\alpha}$ for the kinetic energy and the vector of Dirac matrices $\boldsymbol{\beta}$ for the dynamical masses (with all matrices anticommuting pairwise and squaring to the unit matrix) need not have the same number of components (given by $d-1$ for $\boldsymbol{\alpha}$). The symbol Tr represents tracing over the single-particle Hilbert space of the Dirac Hamiltonian with the Dirac matrices of dimension r. The saddle-point equations

$$\begin{aligned}
0 =& -\left.\frac{\delta\mathcal{S}_{\text{eff}}[\boldsymbol{\phi}]}{\delta\boldsymbol{\phi}}\right|_{\boldsymbol{\phi}=\bar{\boldsymbol{\phi}}} \\
=&\,\mathrm{Tr}\left[\frac{2\mathrm{i}\boldsymbol{\beta}}{\partial_\tau - \mathrm{i}\boldsymbol{\alpha}\cdot\boldsymbol{\partial} + 2\mathrm{i}\boldsymbol{\beta}\cdot\bar{\boldsymbol{\phi}}} - \frac{2}{\lambda\,r}\,\bar{\boldsymbol{\phi}}\right] \\
=&\,\mathrm{Tr}\left[\frac{4\,\bar{\boldsymbol{\phi}}}{\partial_\tau^2 + \boldsymbol{\partial}^2 + 4\,\bar{\boldsymbol{\phi}}^2} - \frac{2}{\lambda\,r}\,\bar{\boldsymbol{\phi}}\right]
\end{aligned} \tag{8.76}$$

with the mean-field $\bar{\boldsymbol{\phi}}$ independent of space and time can be turned into

$$\int \mathrm{d}\omega \int \mathrm{d}^{d-1}\boldsymbol{k}\,\left(\frac{2\,\bar{\boldsymbol{\phi}}}{\omega^2 + |\boldsymbol{k}|^2 - 4\,\bar{\boldsymbol{\phi}}^2}\right) = -\frac{1}{\lambda\,r}\,\bar{\boldsymbol{\phi}}. \tag{8.77}$$

We denote with Ω_{d-1} the area of the unit sphere S^{d-1}, with k the length of the vector $(\omega,\boldsymbol{k})$, and with Λ the ultraviolet cutoff in $(\omega,\boldsymbol{k})$ space (a four-dimensional Euclidean space). The saddle-point equations reduce to the equation

$$\Omega_{d-1}\int\limits_0^\Lambda \mathrm{d}k\,k^{d-1}\,\frac{2}{k^2 - 4\,\bar{\boldsymbol{\phi}}^2} = -\frac{1}{\lambda\,r}. \tag{8.78}$$

As advertized, it has the solution

$$|\bar{\boldsymbol{\phi}}| = \mathrm{i}f(\lambda\,r) \tag{8.79}$$

if and only if $\lambda < 0$.

Fluctuations in the magnitude of the mean-field value $\bar{\boldsymbol{\phi}}$ taken by $\boldsymbol{\phi}$ cost a non-vanishing energy. For this reason, they are said to be hard. Fluctuations in the direction of the mean-field value taken by $\boldsymbol{\phi}$ can be made to cost arbitrarily small energies. For this reason, they are said to be soft. These are the Goldstone modes associated with the spontaneous breaking of a continuous symmetry.

The effective low-energy theory governing the fluctuations of these Goldstone modes is obtained from a gradient expansion of the fermion determinant

$$\mathrm{Det}\,\left(\partial_\tau + \mathcal{H}^{(\text{dyn})}_{\text{bd}}\right) := \int \mathcal{D}[\Psi,\Psi^\dagger]\,e^{-\int \mathrm{d}\tau\int \mathrm{d}^{d-1}\boldsymbol{x}\,\Psi^\dagger\left(\partial_\tau + \mathcal{H}^{(\text{dyn})}_{\text{bd}}\right)\Psi}. \tag{8.80}$$

It is captured by the partition function [see the textbook of Fradkin[82] when deducing the O(3) NLSM from the Hubbard model]

$$Z_{\mathrm{bd}} \approx \int \mathcal{D}[\boldsymbol{\phi}]\, \delta(\boldsymbol{\phi}^2 - 1)\, e^{-S_{\mathrm{QNLSM}} - S_{\mathrm{top}}}, \tag{8.81}$$

for the soft fluctuations $\boldsymbol{\phi}$ about $\bar{\boldsymbol{\phi}}$ after we have rescaled the vector $\boldsymbol{\phi}$ so that it squares to one. The Euclidean action

$$S_{\mathrm{QNLSM}} = \frac{1}{2g} \int \mathrm{d}\tau \int \mathrm{d}^{d-1}\boldsymbol{x}\, (\partial_i \boldsymbol{\phi})^2 \tag{8.82}$$

is the action of the QNLSM with the base space $\mathbb{R}^{(d-1)+1}$ in spacetime and the target space

$$\mathrm{S}^{N(\nu)-1} \tag{8.83}$$

with the integer $N(\nu)$ counting the pairwise anticommuting Dirac masses that have been retained in the set $\{\beta\}$. The effective coupling constant g is positive. The term S_{top} is present whenever any one of the homotopy groups

$$\begin{aligned} &\pi_0\left(\mathrm{S}^{N(\nu)-1}\right), \\ &\quad \pi_1\left(\mathrm{S}^{N(\nu)-1}\right), \\ &\qquad \ddots \\ &\qquad\quad \pi_d\left(\mathrm{S}^{N(\nu)-1}\right), \\ &\qquad\qquad \pi_{d+1}\left(\mathrm{S}^{N(\nu)-1}\right) \end{aligned} \tag{8.84}$$

is nontrivial. (The reason why we ignore all topological terms associated with nontrivial homotopy groups of order larger than $d+1$ is that such topological terms would modify the local equations of motion derived from S_{QNLSM} in a nonlocal way.) The topological term S_{top} signals the existence of zero modes of the Dirac Hamiltonian (8.74d) in the presence of topological defects in the order parameter $\boldsymbol{\phi}$. These zero modes can modify the Boltzmann weight $\exp(-S_{\mathrm{QNLSM}})$ in the partition function (8.81).

Conjectured topological obstruction to gapping[83] These zero modes prevent the gapping of the boundary Dirac fermions. Accordingly, we define the smallest value ν_{min} for the dimension ν of the unit matrix $\mathbb{1}$ in Eq. (8.71) for which

[83] The validity of this conjecture will be verified by working out explicitly the cases when $d = 1, 2, 3$ below and comparison with the alternative methods of Section 3.12 and Chapters 5 and 6 when $d = 1$.

$$\begin{aligned}\pi_0\left(\mathrm{S}^{N(\nu_{\mathrm{min}})-1}\right)&=0,\\ \pi_1\left(\mathrm{S}^{N(\nu_{\mathrm{min}})-1}\right)&=0,\\ &\ddots\\ \pi_d\left(\mathrm{S}^{N(\nu_{\mathrm{min}})-1}\right)&=0,\\ \pi_{d+1}\left(\mathrm{S}^{N(\nu_{\mathrm{min}})-1}\right)&=0.\end{aligned}\tag{8.85}$$

As all homotopy groups of the spheres are known, one may verify that

$$d+1<N(\nu_{\mathrm{min}})-1.\tag{8.86}$$

When Eq. (8.85) holds, the topological term S_{top} is absent, and the effective action in the partition function is simply the action (8.82) for a QNLSM on a sphere.[84] In this case, the quantum-disordered phase at the strong-coupling fixed point $g\to\infty$ is stable. In this strongly interacting phase and when $\mathfrak{G}=\mathbb{Z}$, quantum fluctuations restore dynamically and non-perturbatively all the symmetries broken by the saddle point, including any protecting symmetries. If so, all boundary Dirac fermions are gapped out. We then conclude that

$$\mathfrak{G}_{\mathrm{int}}=\mathbb{Z}_{\nu_{\mathrm{min}}}.\tag{8.87}$$

The stability

$$\mathfrak{G}_{\mathrm{int}}=\mathfrak{G}\tag{8.88}$$

when $\mathfrak{G}=\mathbb{Z}_2$ follows from the fact that one of the homotopy groups

$$\pi_D\left(\mathrm{S}^{N(\nu=1)-1}\right)\tag{8.89}$$

with $D\leq d+1$ is always nontrivial when $\mathfrak{G}=\mathbb{Z}_2$ (see Section 8.4.7).

Table 8.6 follows from the conjectured topological obstruction to gapping (8.85). It gives the equivalence classes of strong topological insulators or superconductors belonging to the ten Altland-Zirnbauer symmetry classes in the presence of interactions that realize fermionic invertible topological phases of matter. It becomes apparent that the Bott periodicity of the tenfold way, that is the periodicity of the (zeroth) homotopy groups of the classifying spaces with respect to d, is lost. It also becomes apparent that the reduction of the topologically distinct equivalence

[84] When $d=2$ and $N(\nu_{\mathrm{min}})>2$, the Mermin-Wagner theorem applied to the QNLSM describing the one-dimensional boundary prevents the spontaneous symmetry breaking on the target space $\mathrm{S}^{N(\nu_{\mathrm{min}})-1}$. The coupling constant g always flows to strong coupling, the quantum-disordered phase at $g\to\infty$. When $d>2$, the fixed point at $g=0$ of the QNLSM describing the $(d-1)$-dimensional boundary is stable. At this fixed point, one linear combination of the bilinears $\Psi^{\dagger}\,\beta\,\Psi$ acquires an expectation value. It thereby breaks spontaneously one of the protecting symmetries. In this case, interactions remove the noninteracting topological attributes by spontaneously breaking one of the protecting symmetries. The transition between the fixed point at $g=0$ and $g=\infty$ occurs at $g=g_{\star}\sim 1$. Microscopics determine if the bare value of g is smaller or larger than the unstable quantum-critical point at $g_{\star}$.

classes of noninteracting fermions for any given Altland-Zirnbauer symmetry class occurs only in odd dimensions of space. Finally, two of the Altland-Zirnbauer symmetry classes, namely, the chiral symmetry classes BDI and CII, have the particularity that they may be interpreted either as a superconductor or an insulator. Correspondingly, the reduction of their classification $\mathbb{Z} \to \mathbb{Z}_m$ for the superconductor interpretation and $\mathbb{Z} \to \mathbb{Z}_n$ for the insulator interpretation of these symmetry classes obeys

$$m = 2n, \qquad \text{class BDI}, \tag{8.90a}$$

$$m = n, \qquad \text{class CII}, \tag{8.90b}$$

when the dimensionality of space is $d = 1 \bmod 4$.

The remaining of Section 8.4 is devoted to deriving explicitly the entries of **Table** 8.6 using the conjectured topological obstruction to gapping (8.85).

8.4.3 Notation

We will use the following conventions. The operation of complex conjugation will be denoted by K. Linear maps of two-dimensional vector space $\mathbb{C}^2$ shall be represented by 2×2 matrices that we expand in terms of the unit matrix τ_0 and the three Pauli matrices τ_1, τ_2, and τ_3. Linear maps of the four-dimensional vector space $\mathbb{C}^4 = \mathbb{C}^2 \otimes \mathbb{C}^2$ will be represented by 4×4 matrices that we expand in terms of the 16 Hermitean matrices

$$X_{\mu\mu'} \equiv \tau_\mu \otimes \sigma_{\mu'}, \qquad \mu, \mu' = 0, 1, 2, 3, \tag{8.91}$$

where σ_ν is a second set comprised of the unit matrix and the three Pauli matrices. Linear maps of the 2^n-dimensional vector space $\mathbb{C}^{2^n} = \mathbb{C}^2 \otimes \cdots \otimes \mathbb{C}^2$ will be represented by $2^n \times 2^n$ matrices that we expand in terms of the 4^n Hermitean matrices

$$X_{\mu_1 \cdots \mu_n} \equiv \tau_{\mu_1} \otimes \tau_{\mu_2} \otimes \cdots \otimes \tau_{\mu_n} \tag{8.92}$$

where $\mu_1, \cdots, \mu_n = 0, 1, 2, 3$.

8.4.4 The Case of One-Dimensional Space

We want to derive the reduction $\mathbb{Z} \to \mathbb{Z}_8$ and $\mathbb{Z} \to \mathbb{Z}_2$ when the symmetry classes BDI and CII are interpreted as chains of Majorana fermions, that is, as superconductors, respectively. This derivation is complementary to that for Majorana c-chains in Section 5.5.4 and for the stability of the zero modes of polyacetylene with respect to fermion interactions in Section 3.12.

The one-dimensional chiral symmetry classes can be also realized as chains of complex fermions with sublattice symmetry and fermion-number conservation, for example, polyacetylene. For example, polyacetylene-like chains realize the symmetry class AIII when TRS is broken, the symmetry class BDI when both TRS and the SU(2) spin-rotation symmetry are present, and the symmetry class CII when TRS holds but not the SU(2) spin-rotation symmetry if spin-orbit coupling is sizable. We show that the reduction of the noninteracting topological classification for insulators is $\mathbb{Z} \to \mathbb{Z}_4$ for the symmetry classes AIII and BDI, while it is $\mathbb{Z} \to \mathbb{Z}_2$ for the symmetry class CII, provided conservation of the fermion number holds (in agreement with Section 3.12).

8.4.4.1 The Symmetry Class BDI when $d = 1$

Consider the one-dimensional bulk single-particle Dirac Hamiltonian (with Dirac matrices of dimension $r = 2 \equiv r_{\mathrm{min}}$)

$$\mathcal{H}^{(0)}(x) := -\mathrm{i}\partial_x\,\tau_3 + m(x)\,\tau_2. \tag{8.93a}$$

This single-particle Hamiltonian belongs to the symmetry class BDI, for

$$\mathcal{T}\,\mathcal{H}^{(0)}(x)\,\mathcal{T}^{-1} = +\mathcal{H}^{(0)}(x), \tag{8.93b}$$

$$\mathcal{C}\,\mathcal{H}^{(0)}(x)\,\mathcal{C}^{-1} = -\mathcal{H}^{(0)}(x), \tag{8.93c}$$

where

$$\mathcal{T} := \tau_1\,\mathsf{K}, \qquad \mathcal{C} := \tau_0\,\mathsf{K}. \tag{8.93d}$$

The Dirac mass matrix τ_2 is here the only one allowed for dimension two Dirac matrices under the constraints (8.93b) and (8.93c). As was shown by Jackiw and Rebbi, if translation symmetry is broken by the mass term supporting the domain wall

$$m(x) = m_\infty\,\mathrm{sgn}(x), \qquad m_\infty \in \mathbb{R}, \tag{8.94a}$$

at $x = 0$, then the zero mode

$$e^{-\mathrm{i}\tau_3\,\tau_2\int_0^x \mathrm{d}x'\,m(x')}\,\chi = e^{-|m_\infty\,x|}\,\chi, \tag{8.94b}$$

where

$$\tau_1\,\chi = \mathrm{sgn}\,(m_\infty)\,\chi, \tag{8.94c}$$

is the only normalizable state bound to this domain wall. This boundary state is a zero mode. It is an eigenstate of the single-particle boundary Hamiltonian

$$\mathcal{H}^{(0)}_{\mathrm{bd}} = 0. \tag{8.94d}$$

Suppose that we consider $\nu = 1, 2, \cdots$ identical copies of the single-particle Hamiltonian (8.93) by defining

$$\mathcal{H}^{(0)}_{\nu}(x) := \mathcal{H}^{(0)}(x) \otimes \mathbb{1}, \tag{8.95a}$$

and

$$\mathcal{T} := \tau_1 \otimes \mathbb{1}\,\mathsf{K}, \qquad \mathcal{C} := \tau_0 \otimes \mathbb{1}\,\mathsf{K}, \tag{8.95b}$$

where $\mathbb{1}$ is a $\nu \times \nu$ unit matrix. Observe that $\mathcal{T}$ and $\mathcal{C}$ commute with $\tau_1 \otimes \mathbb{1}$ and with each other. The domain wall (8.94a) must then support ν linearly independent boundary zero modes. They are annihilated by the boundary Hamiltonian

$$\mathcal{H}^{(0)}_{\mathrm{bd}\,\nu} = \mathcal{H}^{(0)}_{\mathrm{bd}} \otimes \mathbb{1} = 0. \tag{8.96}$$

The topological sectors for noninteracting Hamiltonians are thus labeled by the integer ν taking values in $\mathbb{Z}$ in the direct limit $\nu \to \infty$.

A generic local quartic interaction, that respects the defining BDI symmetries with the potential to gap out these boundary zero modes, reduces to a dynamical Dirac mass (that depends on imaginary time τ in addition to space x) that belongs to the symmetry class D, upon performing a Hubbard-Stratonovich transformation.[85]

Hence, we must consider the dynamical bulk single-particle Hamiltonian

$$\mathcal{H}^{(\mathrm{dyn})}_{\nu}(\tau,x) := \left[-\mathrm{i}\partial_x\,\tau_3 + m(x)\,\tau_2\right] \otimes \mathbb{1} + \mathcal{V}(\tau,x). \tag{8.97a}$$

The dynamical Dirac mass $\mathcal{V}(\tau,x)$ is here defined by the condition that it anticommutes with $\mathcal{H}^{(0)}(x) \otimes \mathbb{1}$, when independent of x, and obeys the transformation laws dictated by the symmetry class D, that is

$$\mathcal{C}\,\mathcal{V}(\tau,x)\,\mathcal{C} = -\mathcal{V}(\tau,x) \tag{8.97b}$$

with $\mathcal{C}$ defined by Eq. (8.95b). Correspondingly, it is of the form

$$\mathcal{V}(\tau,x) := \tau_1 \otimes \gamma'(\tau,x), \qquad \gamma'(\tau,x) := \mathrm{i}M(\tau,x), \tag{8.97c}$$

where

$$M(\tau,x) = M^{*}(\tau,x), \qquad M(\tau,x) = -M^{\mathsf{T}}(\tau,x) \tag{8.97d}$$

is a real-valued antisymmetric $\nu \times \nu$ matrix. Consequently, TRS is only retained for a given $\mathcal{V}(\tau,x)$ if

$$M(\tau,x) = -M(-\tau,x). \tag{8.97e}$$

On the boundary, the operations for reversal of time and charge conjugation are now represented by

$$\mathcal{T}_{\mathrm{bd}} := \mathsf{K}, \qquad \mathcal{C}_{\mathrm{bd}} := \mathsf{K}. \tag{8.98a}$$

Hence, we must consider the dynamical single-particle boundary Hamiltonian

$$\mathcal{H}^{(\mathrm{dyn})}_{\mathrm{bd}\,\nu}(\tau) \equiv \gamma'(\tau) := \mathrm{i}M(\tau), \tag{8.98b}$$

[85] Remember that, by **Definition** (8.3), the dynamical mass matrices must be odd under at least one protecting symmetry. For the symmetry class BDI, charge conjugation must remain an unbroken single-particle spectral symmetry. Hence, a dynamical mass matrix must be odd under reversal of time. As reversal of time squares to the identity, any dynamical mass matrix must belong to the symmetry class D.

where $M(\tau)$ is a real-valued antisymmetric $\nu\times\nu$ matrix. The space of boundary normalized Dirac mass matrices obtained by demanding that γ' square to the unit $\nu\times\nu$ matrix is topologically equivalent to the space

$$V_{\nu}=\mathrm{O}(\nu)/\mathrm{U}(\nu/2) \tag{8.99}$$

for the symmetry class D in zero-dimensional space, provided the rank $\nu\geq 2$ and ν is even. The direct limit $\nu\to\infty$ of these spaces is the classifying space R_2 from **Table** 8.1. In order to gap out dynamically the boundary zero modes without breaking the defining symmetries of the symmetry class BDI, we need to construct a (0+1)-dimensional QNLSM for the (boundary) dynamical Dirac masses from the zero-dimensional symmetry class D without topological obstructions. Next, we construct explicitly the spaces for the normalized boundary dynamical Dirac masses of dimension $\nu=2^n$ with $n=0,1,2,3$ of the form (8.98b).[86] The relevant homotopy groups are given in **Table** 8.7.[87]

Case $\nu=1$*:* No Dirac mass is allowed on the boundary, because the boundary is the end of a one-dimensional $\mathbb{Z}_2$ topological superconductor in the topologically nontrivial phase of the symmetry class D.

Case $\nu=2$*:* We use the representation $\mathbb{1}=\sigma_0$. There is one dynamical normalized Dirac mass on the boundary that is proportional to the matrix σ_2. A domain wall in imaginary time such as $m_{2\,\infty}\,\mathrm{sgn}(\tau)\,\sigma_2$ prevents the dynamical generation of a spectral gap on the boundary, as the single-particle fermionic kernel on the right-hand side of Eq. (8.74c) is of the form

$$\sigma_0\,\partial_\tau+\sigma_2\,m_2(\tau),\qquad m_2(\tau):=m_{2\,\infty}\,\mathrm{sgn}(\tau), \tag{8.100a}$$

with the zero mode

$$\chi(\tau):=e^{-\int_0^\tau \mathrm{d}\tau'\,m_{2\,\infty}\,\mathrm{sgn}(\tau')\,\sigma_2}\,\chi_{2\,\infty}=e^{-|m_{2\,\infty}|\,|\tau|}\,\chi_{2\,\infty}, \tag{8.100b}$$

where

$$\sigma_2\,\chi_{2\,\infty}=\mathrm{sgn}\,(m_{2\,\infty})\,\chi_{2\,\infty}. \tag{8.100c}$$

[86] In order to study the topological obstructions in the target space of the QNLSMs, it is sufficient to consider the dimensions $\nu=2^n$ with $n=0,1,2,3$ of the dynamical Dirac mass matrices. Indeed, the target space of the QNLSM is a sphere generated by a maximum number of anticommuting dynamical Dirac masses. The increase in the number of anticommuting dynamical Dirac masses $N(\nu)$ takes place if and only if the dimensions of the Dirac matrices are doubled. In other words, as $N(\nu)$ remains the same for $\nu=2^n,\cdots,2^{n+1}-1$, the same topological obstruction for the QNLSM prevents gapping out of the excitations at the boundary for $\nu=2^n,\cdots,2^{n+1}-1$. This is why, to study the breakdown of the noninteracting classification, we only focus on the cases with $\nu=2^n$ in the following.

[87] The homotopy groups for the space of $\nu\times\nu$ normalized Dirac mass matrices V_ν for finite ν can be different from those for the space R_2 (that is the direct limit $\nu\to\infty$). In fact, the latter obey the Bott periodicity, while the former do not. However, we find by an explicit enumeration of the Dirac mass matrices in the following that the nontrivial entries of the relevant homotopy groups $\pi_D(V_\nu)$ appear when $\pi_D(R_2)$ is nontrivial. It turns out that this correspondence between homotopy groups at finite ν and infinite ν always holds for any example that will be worked out later. While we do not rely on this fact for the analysis in one, two, and three dimensions, the analysis in higher dimensions made in Section 8.4.7 assumes this correspondence.

Table 8.7 *Topological obstructions for the symmetry class BDI in one-dimensional space*

D	$\pi_D(R_2)$	ν	Topological obstruction
0	$\mathbb{Z}_2$	2	Domain wall
1	0		
2	$\mathbb{Z}$	4	WZ term
3	0		
4	0		
5	0		
6	$\mathbb{Z}$	8	None
7	$\mathbb{Z}_2$		

Reduction from $\mathbb{Z}$ to $\mathbb{Z}_8$ for the topologically equivalent classes of the one-dimensional noninteracting fermionic invertible topological phases in the symmetry class BDI that arises from interactions. We denote by V_ν the space of $\nu\times\nu$ normalized Dirac mass matrices for zero-dimensional Hamiltonians belonging to the symmetry class D. The direct limit $\nu\to\infty$ of these spaces is the classifying space R_2 from **Table** 8.1. The second column shows the stable D-th homotopy groups of the classifying space R_2 as taken from **Table** 8.2. The third column gives the number ν of copies of boundary (Dirac) fermions for which a topological obstruction that is compatible with local equations of motion is permissible. The fourth column gives the type of topological obstruction that prevents the gapping of the boundary (Dirac) fermions.

Case $\nu=4$*:* We use the representation $\mathbb{1}=\sigma_0\otimes\rho_0$. A (maximum) set of pairwise anticommuting boundary dynamical Dirac mass matrices follows from the set

$$\{\sigma_2\otimes\rho_0,\ \sigma_1\otimes\rho_2,\ \sigma_3\otimes\rho_2\}. \tag{8.101}$$

This set spans the space of normalized boundary dynamical Dirac masses that is homeomorphic to S^2. Even though $\pi_{0+1}(S^2)=0$, it is possible to add a topological term that is nonlocal, yet only modifies the equations of motion of the (0+1)-dimensional QNLSM on the boundary by local terms as a consequence of the fact that $\pi_{0+1+1}(S^2)=\mathbb{Z}$. Such a term is a (0+1)-dimensional example of a Wess-Zumino (WZ) term.[88] In the presence of this WZ term, the boundary theory remains gapless. It is nothing but a bosonic representation of the gapless $S=1/2$ degrees of freedom at the end of a quantum spin-1 antiferromagnetic spin chain in the Haldane phase.

Case $\nu=8$*:* We use the representation $\mathbb{1}=\sigma_0\otimes\rho_0\otimes\lambda_0$. One set of pairwise anticommuting boundary dynamical Dirac mass matrices follows from the set

$$\{\sigma_2\otimes\rho_0\otimes\lambda_0,\ \sigma_3\otimes\rho_2\otimes\lambda_0,\ \sigma_3\otimes\rho_3\otimes\lambda_2,\ \sigma_1\otimes\rho_0\otimes\lambda_2\}. \tag{8.102}$$

[88] J. Wess and B. Zumino, Phys. Lett. B **37**(1), 95–97 (1971). [Wess and Zumino (1971)].

This set spans a manifold homeomorphic to S^3. As

$$\pi_0(S^3) = \pi_1(S^3) = \pi_2(S^3) = 0, \tag{8.103}$$

no topological term is admissible over this target manifold that delivers local equations of motion. The QNLSM over this target space endows dynamically the boundary Hamiltonian with a spectral gap. We have shown that $\nu_{\text{min}} = 8$ in Eqs. (8.85) and (8.86).

We conclude that the effects of interactions on the one-dimensional fermionic invertible topological phases of matter in the symmetry class BDI are to reduce the topological classification of the cyclic group $\mathbb{Z}$ in the noninteracting limit down to the cyclic group $\mathbb{Z}_8$ under the assumption that a Hamiltonian from the symmetry class BDI is interpreted as a mean-field description of a superconductor. The logic used to reach this conclusion is summarized by **Table** 8.7. The first input in **Table** 8.7 is the classifying space of the dynamical masses. It was determined to be R_2 as deduced from Eq. (8.99). The second and third inputs in **Table** 8.7 are the first and second columns labeled by the integer D entering the Dth homotopy group of the classifying space R_2 with the Bott periodicity of eight as given in **Table** 8.2, respectively. The fourth input in **Table** 8.7 consists, for each value of ν such that $N(\nu) - 1 \leq d + 1$ where $N(\nu)$ is defined in Eq. (8.83), in matching the first $\pi_n(S^{N(\nu)-1}) \neq 0$ as n is taking the ascending values $n = 0, 1, \cdots, d+1$ with one of the homotopy groups from the second column. Accordingly:

1. We interpret the entry corresponding to $\nu = 2$ in the third column of **Table** 8.7 as the topological obstruction brought about by the zero modes (8.100) attached to the domain wall in imaginary time. The line on which it is positioned follows from matching $\pi_0(S^0) = \mathbb{Z}_2$ with $\pi_0(R_2) = \mathbb{Z}_2$.
2. We interpret the entry corresponding to $\nu = 4$ in the third column of **Table** 8.7 as the topological obstruction brought about by the presence of a WZ term in the partition function (8.81). The line on which it is positioned follows from matching $\pi_2(S^2) = \mathbb{Z}$ with $\pi_2(R_2) = \mathbb{Z}$.
3. The next entry for D for which $\pi_D(R_2) \neq 0$ is $D = 6$. It cannot be associated with a topological obstruction delivering local equations of motion for the value $\nu = 8$, since $\pi_0(S^3) = \pi_1(S^3) = \pi_2(S^3) = 0$. The same reasoning holds for any $\nu \geq 8$ as $N(\nu) - 1 \geq 3$ and $\pi_0(S^{N(\nu)-1}) = \pi_1(S^{N(\nu)-1}) = \pi_2(S^{N(\nu)-1}) = 0$.
4. The topological obstruction preventing the gapping of the boundary states for any $\nu \geq 1$ is either of the type $\nu = 2$ modulo 8 or $\nu = 4$ modulo 8.

8.4.4.2 The Symmetry Class CII when d = 1

Consider the one-dimensional bulk single-particle Dirac Hamiltonian (with Dirac matrices of dimension $r = 4 \equiv r_{\text{min}}$)

$$\mathcal{H}^{(0)}(x) := -\mathrm{i}\partial_x X_{30} + m(x) X_{20}. \tag{8.104a}$$

This single-particle Hamiltonian belongs to the symmetry class CII, for

$$\mathcal{T}\,\mathcal{H}^{(0)}(x)\,\mathcal{T}^{-1} = +\mathcal{H}^{(0)}(x), \tag{8.104b}$$

$$\mathcal{C}\,\mathcal{H}^{(0)}(x)\,\mathcal{C}^{-1} = -\mathcal{H}^{(0)}(x), \tag{8.104c}$$

where

$$\mathcal{T} := \mathrm{i}X_{12}\,\mathsf{K}, \qquad \mathcal{C} := \mathrm{i}X_{02}\,\mathsf{K}. \tag{8.104d}$$

The Dirac mass matrix X_{20} is here the only one allowed for dimension four Dirac matrices in the symmetry class CII. If translation symmetry is broken by the Dirac mass term supporting the domain wall

$$m(x) = m_{\infty}\,\mathrm{sgn}(x), \qquad m_{\infty} \in \mathbb{R}, \tag{8.105a}$$

at $x = 0$, then the zero mode

$$e^{-\mathrm{i}X_{30}\,X_{20}\int_0^x \mathrm{d}x'\,m(x')}\,\chi = e^{-|m_{\infty}\,x|}\,\chi, \tag{8.105b}$$

where

$$X_{10}\,\chi = \mathrm{sgn}\,(m_{\infty})\,\chi \tag{8.105c}$$

is the only normalizable state bound to this domain wall. This boundary state is a zero mode. It is an eigenstate of the single-particle boundary Hamiltonian

$$\mathcal{H}^{(0)}_{\mathrm{bd}} = 0. \tag{8.105d}$$

Suppose that we consider $\nu = 1, 2, \cdots$ identical copies of the single-particle Hamiltonian (8.104) by defining

$$\mathcal{H}^{(0)}_{\nu}(x) := \mathcal{H}^{(0)}(x) \otimes \mathbb{1}, \tag{8.106a}$$

and

$$\mathcal{T} := \mathrm{i}X_{12} \otimes \mathbb{1}\,\mathsf{K}, \qquad \mathcal{C} := \mathrm{i}X_{02} \otimes \mathbb{1}\,\mathsf{K}, \tag{8.106b}$$

where $\mathbb{1}$ is a $\nu \times \nu$ unit matrix. Observe that $\mathcal{T}$ and $\mathcal{C}$ commute with $X_{10} \otimes \mathbb{1}$ and with each other. The domain wall (8.105a) must then support ν linearly independent boundary zero modes. They are annihilated by the boundary Hamiltonian

$$\mathcal{H}^{(0)}_{\mathrm{bd}\,\nu} = \mathcal{H}^{(0)}_{\mathrm{bd}} \otimes \mathbb{1} = 0. \tag{8.107}$$

The topological sectors for noninteracting Hamiltonians are thus labeled by the integer ν taking values in $\mathbb{Z}$ in the direct limit $\nu \to \infty$.

A generic local quartic interaction, that respects the defining CII symmetries with the potential to gap out the boundary zero modes, reduces to a dynamical Dirac mass (that depends on imaginary time τ in addition to space x) that belongs to the symmetry class C, upon performing a Hubbard-Stratonovich transformation.[89] Hence, we must consider the dynamical bulk single-particle Hamiltonian

$$\mathcal{H}^{(\mathrm{dyn})}_{\nu}(\tau,x) := \left[-\mathrm{i}\partial_x X_{30} + m(x)\, X_{20}\right] \otimes \mathbb{1} + \mathcal{V}(\tau,x), \tag{8.108a}$$

where the dynamical Dirac mass $\mathcal{V}(\tau,x)$ is defined by the condition that it anticommutes with $\mathcal{H}^{(0)}(x)\otimes\mathbb{1}$, when independent of x, and obeys the transformation laws dictated by the symmetry class C, that is

$$\mathcal{C}\,\mathcal{V}(\tau,x)\,\mathcal{C}^{-1} = -\mathcal{V}(\tau,x), \tag{8.108b}$$

with $\mathcal{C}$ defined in Eq. (8.106b).

On the boundary, the operations for reversal of time and charge conjugation are now represented by

$$\mathcal{T}_{\mathrm{bd}} := \sigma_2 \otimes \mathbb{1}\, \mathsf{K}, \qquad \mathcal{C}_{\mathrm{bd}} := \sigma_2 \otimes \mathbb{1}\, \mathsf{K}. \tag{8.109a}$$

Hence, we must consider the dynamical single-particle boundary Hamiltonian

$$\mathcal{H}^{(\mathrm{dyn})}_{\mathrm{bd}\,\nu}(\tau) := \gamma'(\tau), \tag{8.109b}$$

where

$$\mathcal{C}_{\mathrm{bd}}\,\gamma'(\tau)\,\mathcal{C}^{-1}_{\mathrm{bd}} = -\gamma'(\tau). \tag{8.109c}$$

The space of normalized Dirac mass matrices obtained by demanding that $\gamma'(\tau)$ squares to the unit $2\nu\times 2\nu$ matrix for all imaginary times is the space

$$V_\nu = \mathrm{Sp}(2\nu)/\mathrm{U}(\nu) \tag{8.110}$$

for the symmetry class C in zero-dimensional space. The direct limit $\nu\to\infty$ of these spaces is the classifying space R_6 from **Table** 8.1. In order to gap out dynamically the boundary zero modes without breaking the defining symmetries of the symmetry class CII, we need to construct a (0+1)-dimensional QNLSM for the (boundary) dynamical Dirac masses from the zero-dimensional symmetry class C without topological obstructions. We construct explicitly the spaces for the relevant normalized boundary dynamical Dirac masses of dimension $\nu = 2^n$ with $n = 0, 1$ in the following. The relevant homotopy groups are given in **Table** 8.8.

Case $\nu = 1$*:* The three dynamical mass matrices σ_1, σ_2, and σ_3 are allowed on the boundary. They all anticommute pairwise. A WZ term is permissible as $\pi_{0+1+1}(\mathsf{S}^2) = \mathbb{Z}$.[88] In the presence of this WZ term, the boundary theory remains gapless.

Case $\nu = 2$*:* The minimum number of dynamical anticommuting mass matrices is larger than three. Hence, the zeroth, first, and second homotopy groups over the boundary normalized dynamical Dirac masses all vanish. No topological term is

[89] Remember that, by **Definition** (8.3), the dynamical mass matrices must be odd under at least one protecting symmetry. For the symmetry class CII, charge conjugation must be a spectral symmetry. Hence, a dynamical mass matrix must be odd under reversal of time. As reversal of time squares to minus the identity, any mass matrix must belong to the symmetry class C.

Table 8.8 *Topological obstructions for the symmetry class CII in one-dimensional space*

D	$\pi_D(R_6)$	ν	Topological obstruction
0	0		
1	0		
2	$\mathbb{Z}$	1	WZ term
3	$\mathbb{Z}_2$	2	None
4	$\mathbb{Z}_2$		
5	0		
6	$\mathbb{Z}$		
7	0		

Reduction from $\mathbb{Z}$ to $\mathbb{Z}_2$ for the topologically equivalent classes of the one-dimensional noninteracting fermionic invertible topological phases in the symmetry class CII that arises from interactions. We denote by V_ν the space of $\nu \times \nu$ normalized Dirac mass matrices in zero-dimensional Hamiltonians belonging to the symmetry class C. The direct limit $\nu \to \infty$ of these spaces is the classifying space R_6 from **Table** 8.1. The second column shows the stable D-th homotopy groups of the classifying space R_6 as taken from **Table** 8.2. The third column gives the number ν of copies of boundary (Dirac) fermions for which a topological obstruction that is compatible with local equations of motion is permissible. The fourth column gives the type of topological obstruction that prevents the gapping of the boundary (Dirac) fermions.

possible according to Eq. (8.86). The (0+1)-dimensional QNLSM over this target space endows dynamically the boundary Hamiltonian with a spectral gap.

We conclude that the effects of interactions on the one-dimensional noninteracting fermionic invertible topological phases in the symmetry class CII are to reduce the topological classification $\mathbb{Z}$ in the noninteracting limit down to $\mathbb{Z}_2$ under the assumption that a Hamiltonian from the symmetry class CII is interpreted as a mean-field description of a superconductor. The logic used to reach this conclusion is summarized by **Table** 8.8 once the line corresponding to $\nu = 1$ has been identified. It is given by matching the smallest D and the smallest ν such that $\pi_D(R_6) \neq 0$ and $\pi_D(\mathrm{S}^{N(\nu)-1}) \neq 0$, where $N(\nu)$ is defined in Eq. (8.83). This exercise is then repeated for ascending values of D and ν until the entry "none" is reached.

8.4.4.3 The Chiral Symmetry Classes as One-Dimensional Insulators

So far we have interpreted the symmetry classes BDI and CII as examples of topological superconductors by focusing on the fact that their second-quantized Hamiltonian respects a unitary charge-conjugation symmetry, a PHS (**Exercise** 3.37).

As the symmetry classes BDI and CII also preserve TRS, the composition of reversal of time with charge conjugation delivers a non-unitary symmetry of their second-quantized Hamiltonian, namely, a CHS (**Exercise** 3.37). The third chiral symmetry class AIII is defined by demanding that it preserves CHS, no more and no less. Hence, any representative gapped Hamiltonian from the chiral symmetry class AIII can always be interpreted as a topological insulator with fermion number conservation.

The CHS can be implemented at the single-particle level by a unitary sublattice spectral symmetry for (complex) electrons hopping between two sublattices. From this point of view, the three chiral symmetry classes AIII, BDI, and CII, when interpreted as metals or as insulators, can be treated on equal footing. The fermion number is conserved in a metal or in an insulator, unlike in the mean-field treatment of a superconductor. This is the case when the symmetry class BDI is interpreted as an effective theory for polyacetylene, in which case the Dirac gap is induced by coupling the electrons to phonons, that is it realizes a Peierls or bond-density wave instability as covered in Chapters 2 and 3. In this interpretation of the chiral classes, it is necessary to introduce an additional particle-hole grading in order to include through dynamical Dirac masses the effects of superconducting fluctuations induced by any quartic interaction. Failure to do so can produce a distinct reduction pattern of the noninteracting topological equivalence classes arising from interactions, for it can matter whether the boundary dynamical masses belong to the classifying space associated with the symmetry class D or to the symmetry class A.

Symmetry class AIII Consider the one-dimensional bulk single-particle Dirac Hamiltonian in the symmetry class AIII:

$$\mathcal{H}^{(0)}(x) := -\mathrm{i}\partial_x \tau_3 + m(x)\tau_2. \tag{8.111a}$$

It anticommutes with the unitary operator

$$\Gamma := \tau_1 \tag{8.111b}$$

that represents the CHS. It supports the zero mode (8.94) at the boundary where it identically vanishes:

$$\mathcal{H}^{(0)}_{\mathrm{bd}}(x) = 0. \tag{8.111c}$$

Upon tensoring Hamiltonians (8.111a) and (8.111c) together with the Dirac Γ matrix by the $\nu \times \nu$ unit matrix $\mathbb{1}$, there follows $\nu = 1, 2, 3, \cdots$ boundary zero modes.

The dynamical single-particle Hamiltonian that encodes those nonsuperconducting fluctuations arising from a local quartic interaction after a Hubbard-Stratonovich transformation takes the form (8.97a) with

$$\mathcal{V}(\tau, x) := \tau_1 \otimes \gamma'(\tau, x) \tag{8.112}$$

and $\gamma'(\tau, x)$ a $\nu \times \nu$ Hermitean matrix.

On the boundary, we must consider the dynamical single-particle boundary Hamiltonian

$$\mathcal{H}^{(\mathrm{dyn})}_{\mathrm{bd}\,\nu}(\tau) = \gamma'(\tau). \tag{8.113}$$

The $\nu \times \nu$ Hermitean matrix $\gamma'(\tau)$ with no further symmetry constraints belongs to the zero-dimensional symmetry class A. Consequently, it is assigned the classifying space C_0 in the limit limit $\nu \to \infty$ with the zeroth homotopy group $\pi_0(C_0) = \mathbb{Z}$. When $\nu = 1$, γ' is a real number, and the domain wall $\gamma'(\tau) \propto \mathrm{sgn}(\tau)$ binds a zero mode at $\tau = 0$ [that is a normalizable zero mode of the operator $\partial_\tau + \mathcal{H}^{(\mathrm{dyn})}_{\mathrm{bd}\,\nu}(\tau)$]. When $\nu = 2$, we can write $\gamma'(\tau)$ as a linear combination of the matrices σ_μ with the real-valued functions $m_\mu(\tau)$ as coefficients for $\mu = 0, 1, 2, 3$, respectively. Any one of the three Pauli matrices $(\sigma_1, \sigma_2, \sigma_3)$ anticommutes with the other two Pauli matrices. Hence, the space of normalized boundary dynamical Dirac masses that anticommute pairwise is homeomorphic to S^2 with the homotopy group $\pi_{0+1+1}(\mathsf{S}^2) = \mathbb{Z}$ when $\nu = 2$. A (0+1)-dimensional QNLSM for the (boundary) dynamical Dirac masses is augmented by a WZ term.[88]

We conclude that the effects of interactions on the one-dimensional noninteracting fermionic invertible topological phases in the symmetry class AIII are to reduce the topological classification $\mathbb{Z}$ in the noninteracting limit down to $\mathbb{Z}_4$ under the assumption that only fermion-number-conserving dynamical Dirac masses are included in the single-particle Hamiltonian. The logic used to reach this conclusion is summarized by **Table** 8.9 once the line corresponding to $\nu = 1$ has been identified. It is given by matching the smallest D and the smallest ν such that $\pi_D(C_0) \neq 0$ and $\pi_D(\mathsf{S}^{N(\nu)-1}) \neq 0$, where $N(\nu)$ is defined in Eq. (8.83). This exercise is then repeated for ascending values of D and ν until the entry "none" is reached.

To include the effects of superconducting fluctuations in the single-particle Hamiltonian after a Hubbard-Stratonovich transformation, we need to consider the direct sum

$$\begin{aligned}\mathcal{H}^{(0)}_{\mathrm{BdG}}(x) &:= \left[\mathcal{H}^{(0)}(x) \otimes \mathbb{1}\right] \oplus \left[-\mathcal{H}^{(0)\,*}(x) \otimes \mathbb{1}\right]\\ &\equiv \mathcal{H}^{(0)}(x) \otimes \mathbb{1} \otimes \rho_0.\end{aligned} \tag{8.114a}$$

This Bogoliubov-de-Gennes single-particle Hamiltonian anticommutes with the operation for charge conjugation

$$\mathcal{C} := \tau_0 \otimes \mathbb{1} \otimes \rho_1\,\mathsf{K}, \tag{8.114b}$$

in addition to anticommuting with

$$\Gamma \equiv \tau_1 \otimes \mathbb{1} \otimes \rho_0 \tag{8.114c}$$

in order to represent CHS. We also observe that $\mathcal{H}^{(0)}_{\mathrm{BdG}}(x)$ commutes with

$$\mathcal{T} := \tau_1 \otimes \mathbb{1} \otimes \rho_1\,\mathsf{K}. \tag{8.114d}$$

Hence, $\mathcal{H}^{(0)}_{\mathrm{BdG}}(x)$ belongs to the symmetry class BDI. The dynamical single-particle Hamiltonian that accounts for superconducting fluctuations (dynamical Dirac

Table 8.9 *Topological obstructions for the symmetry class AIII in one-dimensional space*

D	$\pi_D(C_0)$	ν	Topological obstruction
0	$\mathbb{Z}$	1	Domain wall
1	0		
2	$\mathbb{Z}$	2	WZ term
3	0		
4	$\mathbb{Z}$	4	None
5	0		
6	$\mathbb{Z}$		
7	0		

Reduction from $\mathbb{Z}$ to $\mathbb{Z}_4$ for the topologically equivalent classes of the one-dimensional noninteracting fermionic invertible topological phases in symmetry classes AIII or BDI that arises from the fermion-number-conserving interacting channels. We denote by V_ν the space of $\nu \times \nu$ normalized Dirac mass matrices in zero-dimensional Hamiltonians belonging to the symmetry class A. The direct limit $\nu \to \infty$ of these spaces is the classifying space C_0 from **Table** 8.1. The second column shows the stable D-th homotopy groups of the classifying space C_0 as taken from **Table** 8.2. The third column gives the number ν of copies of boundary (Dirac) fermions for which a topological obstruction that is compatible with local equations of motion is permissible. The fourth column gives the type of topological obstruction that prevents the gapping of the boundary (Dirac) fermions.

masses from the symmetry class D) takes a form similar to that in Eq. (8.97c), namely,

$$\mathcal{V}(\tau, x) := \tau_1 \otimes \gamma'(\tau, x), \tag{8.115a}$$

where the $2\nu \times 2\nu$ Hermitean matrix $\gamma'(\tau, x)$ must obey

$$\gamma'(\tau, x) = -(\mathbb{1} \otimes \rho_1)\, \gamma'^{*}(\tau, x)\, (\mathbb{1} \otimes \rho_1). \tag{8.115b}$$

The stability analysis of the boundary zero modes is similar to the one performed below Eq. (8.99) except that one must replace ν in Eq. (8.99) by $\nu_{\mathrm{BdG}} = 2\nu$ and that we have a different representation of the PHS. When $\nu = 1$,

$$\gamma'(\tau) = M(\tau)\, \rho_3, \qquad M(\tau) \in \mathbb{R} \tag{8.116}$$

supports a domain wall in imaginary time on the boundary. When $\nu = 2$, we use the representation $\mathbb{1} = \sigma_0$ and introduce the notation $X_{\mu\mu'} \equiv \sigma_\mu \otimes \rho_{\mu'}$ with $\mu, \mu' = 0, 1, 2, 3$. Now,

$$\gamma'(\tau) = \sum_{\{(\mu,\mu')\}} M_{\mu\mu'}(\tau)\, X_{\mu\mu'}, \qquad M_{\mu\mu'}(\tau) \in \mathbb{R}, \tag{8.117}$$

where the sum on the right-hand side is to be performed over the six matrices X_{21},X_{22},X_{03},X_{13},X_{20},X_{33}. This set of six matrices decomposes into two triplets of pairwise anticommuting matrices, whereby all elements of one triplet commute with all the elements of the other triplet. The first triplet is given by X_{21},X_{22},X_{03}. The second triplet is given by X_{13},X_{20},X_{33}. Each triplet defines a two-sphere S^2. Hence, each triplet has the potential to accommodate a WZ term. However, we must make sure that the integrity of any one S^2 entering the decomposition $\mathsf{S}^2 \cup \mathsf{S}^2$ of the normalized dynamical masses on the boundary is compatible with maintaining the global U(1) symmetry associated with the conservation of the fermion number.

To address the fate of the fermion-number conservation, observe that Hamiltonian (8.111a) is invariant under the global U(1) transformation

$$\mathcal{H}^{(0)} \mapsto \mathcal{U}(\alpha)\,\mathcal{H}^{(0)}\,\mathcal{U}^{-1}(\alpha), \qquad \mathcal{U}(\alpha) := e^{+\mathrm{i}\alpha\,\tau_0}, \tag{8.118}$$

with $0 \leq \alpha < 2\pi$ independent of x. This symmetry is to be preserved when treating superconducting fluctuations. In the Bogoliubov-de-Gennes representation (8.114a), this symmetry becomes the symmetry under the global U(1) transformation

$$\mathcal{H}^{(0)}_{\mathrm{BdG}} \mapsto \mathcal{U}_{\mathrm{BdG}}(\alpha)\,\mathcal{H}^{(0)}_{\mathrm{BdG}}\,\mathcal{U}^{-1}_{\mathrm{BdG}}(\alpha), \tag{8.119a}$$

where

$$\mathcal{U}_{\mathrm{BdG}}(\alpha) := e^{+\mathrm{i}\alpha\,\tau_0\otimes\mathbb{1}\otimes\rho_3}. \tag{8.119b}$$

When $\nu = 1$, the boundary dynamical mass (8.116) is invariant under multiplication from the left with $\exp(+\mathrm{i}\alpha\,\rho_3)$ and multiplication from the right with $\exp(-\mathrm{i}\alpha\,\rho_3)$. When $\nu = 2$, if we normalize the boundary dynamical mass (8.117) by demanding that

$$1 = M^2_{21} + M^2_{22} + M^2_{03}, \tag{8.120a}$$

$$1 = M^2_{13} + M^2_{20} + M^2_{33}, \tag{8.120b}$$

we may then identify these boundary dynamical masses as the union of two two-spheres S^2. The global U(1) transformation defined by multiplication from the left with $\exp(+\mathrm{i}\alpha\,X_{03})$ and multiplication from the right with $\exp(-\mathrm{i}\alpha\,X_{03})$ leaves the two-sphere (8.120a) invariant as a set, for it is represented by a rotation about the north pole X_{03} that rotates the equator spanned by X_{21} and X_{22} with the angle 2α. The same transformation leaves the two-sphere (8.120b) invariant point-wise. Hence, each S^2 in $\mathsf{S}^2 \cup \mathsf{S}^2$ is compatible with the conservation of the global fermion number. Because $\pi_{0+1+1}(\mathsf{S}^2) = \mathbb{Z}$, a WZ topological term in the QNLSM for the boundary is permissible.

We may then safely conclude that the effects of interactions on the one-dimensional fermionic invertible topological phases of matter phases in the symmetry class AIII are also to reduce the topological classification $\mathbb{Z}$ in the noninteracting limit down to $\mathbb{Z}_4$ under the assumption that superconducting fluctuation channels are included in the stability analysis. The $\mathbb{Z}_4$ classification for one-dimensional fermionic invertible topological phases of matter in the symmetry class AIII agrees with the one derived using group cohomology by Tang and Wen[90] [with AIII interpreted by Tang and Wen as a time-reversal-symmetric superconductor with the full spin-1/2 rotation symmetry broken down to a U(1) subgroup].

Symmetry class BDI Dirac Hamiltonians in the symmetry class BDI are obtained from those in the symmetry class AIII by imposing the constraint of TRS, Eqs. (8.93b) and (8.93d) (note that $r_{\mathrm{min}} = 2$ for both AIII and BDI classes in $d = 1$). Since the TRS is not relevant for dynamical Dirac masses in the single-particle Dirac Hamiltonian after a Hubbard-Stratonovich transformation, the stability analysis of gapless boundary states in the symmetry class BDI in $d = 1$ follows from that of the symmetry class AIII. Consequently, the effects of interactions in the symmetry class BDI, when interpreted as realizing complex fermions as opposed to Majorana fermions, is to reduce the topological classification $\mathbb{Z}$ in the noninteracting limit down to $\mathbb{Z}_4$ under the assumption that only fermion-number preserving dynamical Dirac masses taken from the symmetry class A are included in the stability analysis. Furthermore, if dynamical superconducting fluctuations are allowed by introducing an additional particle-hole grading and dynamical Dirac masses from the symmetry class D, then the same reduction pattern $\mathbb{Z} \to \mathbb{Z}_4$ follows. The topological classification $\mathbb{Z}_8$ of the symmetry class BDI when interpreted as describing Majorana fermions is thus finer than the classification $\mathbb{Z}_4$ of the symmetry class BDI when interpreted as describing complex fermions.

Symmetry class CII We interpret the single-particle Hamiltonian (8.104a) as describing an insulator, not a superconductor. This is to say that the defining symmetries are TRS (8.104b) and the CHS

$$\Gamma\,\mathcal{H}^{(0)}(x)\,\Gamma^{-1} = -\mathcal{H}^{(0)}(x), \qquad \Gamma := X_{10}. \tag{8.121}$$

If we are after the dynamical effects of interactions that preserve the (complex) fermion number, we may use Eq. (8.108a) with the only caveat that the dynamical mass matrix is now required to belong to the symmetry class A instead of the symmetry class C.[91] The boundary dynamical Hamiltonian is then the same as for the symmetry classes AIII and BDI, that is Eq. (8.113), except for its rank being twice as large as compared to the symmetry classes AIII and BDI, for an additional grading (that of the spin-1/2 degrees of freedom) has been accounted for. This larger rank implies that the WZ term is already permissible at $\nu = 1$, that is, the reduction pattern for the noninteracting topological classification is $\mathbb{Z} \to \mathbb{Z}_2$. It remains to verify that the same reduction pattern is also obtained if superconducting fluctuations are included, as was the case for the symmetry classes AIII and BDI.

[90] E. Tang and X.-G. Wen, Phys. Rev. Lett. **109**(9), 096403 (2012). [Tang and Wen (2012)].

[91] The status of charge-conjugation symmetry in second quantization encoded by Eq. (8.104) is now replaced by U(1) charge conservation, which we do not allow to be broken dynamically. The dynamical masses must then break time-reversal symmetry, that is the dynamical masses must then necessarily belong to the symmetry class A.

To this end, we must tensor (8.104a) with ρ_0, in which case the charge conjugation symmetry is realized by $\tau_0 \otimes \sigma_0 \otimes \rho_1\,\mathsf{K}$. We can borrow the stability analysis with respect to superconducting interacting channels from the symmetry class D that we performed for the symmetry classes AIII and BDI, again with the caveat that the rank of the boundary dynamical Hamiltonian is twice as large as it was. This larger rank implies again that the WZ term is already permissible for $\nu = 1$, that is, the reduction pattern for the noninteracting topological classification is again $\mathbb{Z} \to \mathbb{Z}_2$. Thus we obtain the same topological classification $\mathbb{Z}_2$ of the symmetry class CII in $d = 1$ both when interpreted as describing Majorana fermions (superconductors) and when interpreted as describing complex fermions (insulators).

8.4.4.4 Beyond the Stability Analysis of the Tenfold Way when $d = 1$

The following argument shows that there can be fermionic invertible topological phases of matter that are not adiabatically connected to topological insulators or superconductors in one-dimensional space. Consider the quantum spin-1 nearest-neighbor antiferromagnetic Heisenberg chain

$$\widehat{H}_{\mathrm{b}} := J \sum_{j=1}^{N-b} \widehat{\boldsymbol{S}}_j \cdot \widehat{\boldsymbol{S}}_{j+1}, \qquad J > 0, \tag{8.122a}$$

$$\left[\widehat{S}^a_j, \widehat{S}^b_{j'}\right] = \mathrm{i}\epsilon^{abc}\, \widehat{S}^c_j\, \delta_{j,j'}, \qquad \widehat{\boldsymbol{S}}^2_j = 2, \qquad j, j' = 1, \cdots, N, \tag{8.122b}$$

with $b = 0$ ($b = 1$) if periodic (open) boundary conditions are selected. This Hamiltonian is the archetypal example of a bosonic invertible topological phase of matter as was discussed in Chapter 5. Its ground state is nondegenerate and gapped when $b = 0$. Its ground state is fourfold degenerate and gapped when $b = 1$, as it supports two free spin-1/2 gapless excitations on its boundaries in the thermodynamic limit $N \to \infty$. The bosonic Hamiltonian $\widehat{H}_{\mathrm{b}}$ can be turned into the fermionic Hamiltonian[92]

$$\widehat{H}_{\mathrm{f}} := \frac{J}{4} \sum_{j=1}^{N-b} \left(\hat{c}^{\dagger}_{j;\mu,\nu}\, \sigma^a_{\mu\mu'} \otimes \tau^0_{\nu\nu'}\, \hat{c}_{j;\mu',\nu'}\right) \left(\hat{c}^{\dagger}_{j+1;\rho,\lambda}\, \sigma^a_{\rho\rho'} \otimes \tau^0_{\lambda\lambda'}\, \hat{c}_{j+1;\rho',\lambda'}\right), \tag{8.123a}$$

$$\left\{\hat{c}_{j;\mu,\nu}, \hat{c}^{\dagger}_{j';\mu',\nu'}\right\} = \delta_{jj'}\, \delta_{\mu\mu'}\, \delta_{\nu\nu'}, \qquad \left\{\hat{c}^{\dagger}_{j;\mu,\nu}, \hat{c}^{\dagger}_{j';\mu',\nu'}\right\} = \left\{\hat{c}_{j;\mu,\nu}, \hat{c}_{j';\mu',\nu'}\right\} = 0, \tag{8.123b}$$

provided the local constraints

$$\hat{c}^{\dagger}_{j;\mu,\nu}\, \hat{c}_{j;\mu,\nu} = 2, \tag{8.123c}$$

$$\hat{c}^{\dagger}_{j;\mu,\nu}\, \tau^b_{\nu\nu'}\, \hat{c}_{j;\mu,\nu'} = 0, \qquad b = 1, 2, 3 \tag{8.123d}$$

hold for any site $j = 1, \cdots, N$. Here, $(\sigma^0, \sigma^1, \sigma^2, \sigma^3)$ and $(\tau^0, \tau^1, \tau^2, \tau^3)$ each denotes the 2×2 unit matrix and three Pauli matrices and the summation convention

[92] C. Itoi and H. Mukaida, J. Phys. A: Math. and Gen. **27**(13), 4695–4708 (1994). [Itoi and Mukaida (1994)].

over repeated indices is used within each parenthesis on the right-hand side of Eq. (8.123a) and on the left-hand sides of Eqs. (8.123c) and (8.123d). The four local constraints (8.123c) and (8.123d) project the 16-dimensional fermionic Fock space to the three-dimensional Fock space of a quantum spin one for each site j. As Hamiltonian (8.123a) is a sum over quartic monomials in the fermionic operators, it is purely interacting. It belongs unambiguously to the symmetry class AI once the local constraint (8.123c) has been imposed. We conclude that the symmetry class AI can support a nontrivial fermionic invertible topological phase of matter driven by strong interactions, even though this is not possible in the noninteracting limit. This possibility is accounted for by the universal data[93]

$$\begin{aligned}&(\mu,\rho,\nu)\in\mathbb{Z}_2\times C^1\left(\mathrm{G}_b,\mathbb{Z}_2\right)\times C^2\left(\mathrm{G}_b,\mathrm{U}(1)\right)\\&\text{subject to consistency conditions and stacking rules}\end{aligned}\tag{8.124}$$

describing all fermionic invertible topological phases of matter in one-dimensional space, as derived in Section 6.5.

8.4.5 The Case of Two-Dimensional Space

The notion that the chiral edge modes in the IQHE are immune to local interactions is rather intuitive. Neither backscattering nor umklapp scattering is permissible. An operational and quantitative validation for this intuition goes back to Niu, Thouless, and Wu[75], who proposed to average the Kubo Hall conductivity over all twisted boundary conditions of the many-body ground state as a signature of both the IQHE and FQHE. Hasting and Michalakis on the one hand and Koma on the other hand provided mathematically rigorous proofs of this intuition.[94,95]

This intuition readily extends to the symmetry classes D and C as they realize quantized thermal Hall effects. The robustness of chiral edge modes in the symmetry classes D, C, and A to quartic contact interactions will be derived using the conjectured topological obstruction to gapping (8.85).

Let $\boldsymbol{x}=(x_1,x_2)$ denote a point in two-dimensional space. The single-particle Dirac Hamiltonian with the smallest rank $r_{\mathrm{min}}=2$ that admits a Dirac mass can be chosen to be represented by

$$\begin{aligned}\mathcal{H}^{(0)}_{\mathrm{A}}(\boldsymbol{x}):=&\left[-\mathrm{i}\partial_1+A_1(\boldsymbol{x})\right]\sigma_3+\left[-\mathrm{i}\partial_2+A_2(\boldsymbol{x})\right]\sigma_1\\&+A_0(\boldsymbol{x})\,\sigma_0+m(\boldsymbol{x})\,\sigma_2.\end{aligned}\tag{8.125}$$

It belongs to the symmetry class A for arbitrarily chosen vector potentials $\boldsymbol{A}(\boldsymbol{x})$, scalar potential $A_0(\boldsymbol{x})$, and mass $m(\boldsymbol{x})$. When the gauge fields are vanishing,

93 The definition of n-cochains can be found in Eq. (6.113).

94 M. B. Hastings and S. Michalakis, Commun. Math. Phys. **334**(1), 433–471 (2015). [Hastings and Michalakis (2015)].

95 T. Koma, arXiv:1504.01243 (2015). [Koma (2015)].

$$\begin{aligned} \mathcal{H}^{(0)}_{\mathrm{D}}(\boldsymbol{x}) :=& -\mathrm{i}\partial_1\,\sigma_3 - \mathrm{i}\partial_2\,\sigma_1 + m(\boldsymbol{x})\,\sigma_2 \\ =& -\left[\mathcal{H}^{(0)}(\boldsymbol{x})\right]^* \end{aligned} \tag{8.126}$$

belongs to the symmetry class D (PHS is represented by $\mathcal{C} := \sigma_0\,\mathsf{K}$). Finally, the single-particle Hamiltonian with the smallest rank $r_{\mathrm{min}} = 4$ that belongs to the symmetry class C can be chosen to be represented by

$$\begin{aligned} \mathcal{H}^{(0)}_{\mathrm{C}}(\boldsymbol{x}) :=& -\mathrm{i}\partial_1\,X_{30} + \sum_{j=1}^{3} A_{1j}(\boldsymbol{x})\,X_{3j} \\ & -\mathrm{i}\partial_2\,X_{10} + \sum_{j=1}^{3} A_{2j}(\boldsymbol{x})\,X_{1j} \\ & + \sum_{j=1}^{3} A_{0j}(\boldsymbol{x})\,X_{0j} + m(\boldsymbol{x})\,X_{20} \\ =& -X_{02}\left[\mathcal{H}^{(0)}(\boldsymbol{x})\right]^* X_{02} \end{aligned} \tag{8.127}$$

(PHS is represented by $\mathcal{C} := \mathrm{i}X_{02}\,\mathsf{K}$).

In two spatial dimensions, the symmetry classes A, D, and C realize noninteracting topological insulators or superconductors with the direct systems labeled by N and defined by (see **Appendix** A)

$$V^{(\mathrm{A})}_N \equiv \bigcup_{n=0}^{N} \mathrm{U}(N)/[\mathrm{U}(n)\times\mathrm{U}(N-n)], \tag{8.128a}$$

$$V^{(\mathrm{D})}_N \equiv \bigcup_{n=0}^{N} \mathrm{O}(N)/[\mathrm{O}(n)\times\mathrm{O}(N-n)], \tag{8.128b}$$

$$V^{(\mathrm{C})}_N \equiv \bigcup_{n=0}^{N} \mathrm{Sp}(2N)/[\mathrm{Sp}(2n)\times\mathrm{Sp}(2N-2n)], \tag{8.128c}$$

associated through their direct limits to the classifying spaces C_0, R_0, and R_4, respectively. They share the same zeroth-order homotopy group $\mathbb{Z}$. This group also serves as defining the topological attributes of noninteracting topological insulators or superconductors in the symmetry classes A, D, and C.

8.4.5.1 The Symmetry Class D When $d = 2$

Let $\mathbb{1}$ denote a $\nu\times\nu$ unit matrix with $\nu = 1, 2, \cdots$. Consider the two-dimensional bulk single-particle Dirac Hamiltonian

$$\mathcal{H}^{(0)}(\boldsymbol{x}) := -\mathrm{i}\partial_1\,\sigma_3\otimes\mathbb{1} - \mathrm{i}\partial_2\,\sigma_1\otimes\mathbb{1} + m(\boldsymbol{x})\,\sigma_2\otimes\mathbb{1} \tag{8.129}$$

of rank 2ν. There is no Hermitean $(2\nu)\times(2\nu)$ matrix that anticommutes with $\mathcal{H}^{(0)}(\boldsymbol{x})$. If so, the set $\{\beta\}$ in Eq. (8.72) is empty. In other words, no dynamical mass is available to induce a dynamical instability of the ν boundary zero modes.

8.4.5.2 The Symmetry Class C When $d = 2$

The same reasoning applies to the bulk single-particle Hamiltonian

$$\mathcal{H}^{(0)}(\boldsymbol{x}) := \left[-\mathrm{i}\partial_1 X_{30} - \mathrm{i}\partial_2 X_{10} + m(\boldsymbol{x}) X_{20}\right] \otimes \mathbb{1}, \tag{8.130a}$$

of rank 4ν that realizes a topological superconductor in the symmetry class C,

$$\mathcal{H}^{(0)}(\boldsymbol{x}) = -\left(X_{02} \otimes \mathbb{1}\right) \left[\mathcal{H}^{(0)}(\boldsymbol{x})\right]^{*} \left(X_{02} \otimes \mathbb{1}\right). \tag{8.130b}$$

No dynamical mass is available to induce a dynamical instability of the ν boundary zero modes.

8.4.5.3 The Symmetry Class A When $d = 2$

If the single-particle Dirac Hamiltonian (8.129) is interpreted as describing an insulator with fermion-number conservation, then no dynamical mass that preserves the fermion number and anticommutes with $\sigma_2 \otimes \mathbb{1}$ is permissible. The same remains true if we account for superconducting fluctuations, for the Bogoliubov-de-Gennes extension of (8.129) that is given by Eq. (8.130a), whereby charge conjugation is defined by [and not by Eq. (8.130b) as $X_{02} \otimes \mathbb{1}\,\mathsf{K}$ squares to minus the identity]

$$\mathcal{C} := X_{01} \otimes \mathbb{1}\,\mathsf{K}, \qquad \mathcal{C}^2 = X_{00} \otimes \mathbb{1}, \tag{8.131}$$

fails to anticommute with any $(4\nu) \times (4\nu)$ Hermitean matrix allowed by the PHS generated by the operation of charge conjugation (8.131). Therefore, there is no reduction from $\mathbb{Z}$ to a subgroup in $d = 2$.

8.4.5.4 Beyond the Stability Analysis of the Tenfold Way When $d = 2$

Not all fermionic invertible topological phases of matter in two-dimensional space are adiabatically connected to noninteracting topological insulators or superconductors. For example, there are local electronic interactions in the symmetry class A such that the charge and thermal Hall conductance are independently quantized,[96] thereby breaking the Wiedemann-Franz law according to which the thermal and charge conductivity tensor must be proportional. As the Wiedemann-Franz law is a property of noninteracting electrons, it follows that the decoupling of the charge and thermal Hall conductance is a property driven by strong interactions in the symmetry class A.

Fermionic invertible topological phases of matter in two-dimensional space with an arbitrary fermionic on-site symmetry group G_f can be treated by methods more powerful than the stability analysis described so far.[97] Here, G_f is the central extension of an on-site bosonic symmetry group G_b by the fermion parity. This central extension is specified by the second cohomology group $H^2(\mathrm{G}_b, \mathbb{Z}_2^{\mathrm{F}})$

[96] T. Neupert, C. Chamon, C. Mudry, and R. Thomale, Phys. Rev. B **90**(20), 205101 (2014). [Neupert et al. (2014)].

[97] M. Barkeshli, Y.-A. Chen, P.-S. Hsin, and N. Manjunath, Phys. Rev. B **105**(23), 235143 (2022). [Barkeshli et al. (2022)].

(see **Property** 6.35). The outcome is a classification of fermionic invertible topological phases of matter in terms of the universal data[98]

$$(c_-, n_1, n_2, \nu_3) \in \frac{1}{2}\mathbb{Z} \times C^1(\mathrm{G}_b, \mathbb{Z}_2) \times C^2(\mathrm{G}_b, \mathbb{Z}_2) \times C^3(\mathrm{G}_b, \mathrm{U}(1)) \tag{8.132}$$
subject to consistency conditions and stacking rules.

This classification is believed to be exhaustive for all fermionic lattice Hamiltonians whose low-energy effective theories are invertible topological field theories. The quadruplet (8.132) generalizes the triplet

$$(\mu, \rho, \nu) \in \mathbb{Z}_2 \times C^1(\mathrm{G}_b, \mathbb{Z}_2) \times C^2(\mathrm{G}_b, \mathrm{U}(1)) \tag{8.133}$$
subject to consistency conditions and stacking rules

of universal data describing all fermionic invertible topological phases of matter in one-dimensional space, as derived in Section 6.5. The index c_- is the chiral central charge of the boundary theory ($2\,c_-$ is the trace of the K-matrix with determinant ± 1 from Section 3.10). It is the counterpart to the index μ measuring the parity of the number of boundary Majorana modes characterizing the boundary theory of fermionic invertible topological phases of matter in one-dimensional space. The triplet of indices (n_1, n_2, ν_3) can be interpreted as being related to the braiding and fusion properties of point-like excitations of strongly interacting bosonic phases of matter in two-dimensonal space with a symmetry group G_b that is neither explicitly nor spontaneously broken and with gapped ground states that support topological order, that is, are not invertible.

8.4.6 The Case of Three-Dimensional Space

The reduction $\mathbb{Z} \to \mathbb{Z}_{16}$ for the three-dimensional interacting topological superconductors belonging to the symmetry class DIII has been understood in the following ways after a conjecture by Kitaev.[99]

One approach[100,101,102] is to enumerate the distinct symmetric but gapped phases on the two-dimensional surface of the three-dimensional bulk with the following properties:

1. These boundary phases support point-like excitations whose exchange statistics is not fermionic.
2. These boundary phases cannot be derived from a two-dimensional lattice Hamiltonian for which all protecting symmetries are on-site ones.

[98] The definition of n-cochains can be found in Eq. (6.113).
[99] A. Kitaev, http://online.kitp.ucsb.edu/online/topomat11/kitaev (2011).
[100] L. Fidkowski, X. Chen, and A. Vishwanath, Phys. Rev. X **3**(4), 041016 (2013). [Fidkowski et al. (2013)].
[101] M. A. Metlitski, L. Fidkowski, X. Chen, and A. Vishwanath, arXiv:1406.3032 (2014). [Metlitski et al. (2014)].
[102] C. Wang and T. Senthil, Phys. Rev. B **89**(19), 195124 (2014). [Wang and Senthil (2014)].

In this approach, the breakdown of $\mathbb{Z}$ takes place when vortices (point-like defects of a symmetry-broken phase) at the surface proliferate (deconfine) so as to stabilize a gapped and fully symmetric surface phase. Using such methods, the reductions $\mathbb{Z} \to \mathbb{Z}_4$ and $\mathbb{Z} \to \mathbb{Z}_8$ for the symmetry classes CI and AIII were obtained by Wang and Senthil[102].

Another approach advocated by You and Xu[103] to study the stability of noninteracting topological insulators or superconductors to interactions is to posit that some strong attractive interactions can confine all fermions into bosonic bound states without breaking the protecting symmetries. If so, the topological classification of noninteracting topological insulators or superconductors must be unstable to the classification of bosonic invertible topological phases of matter whose symmetries are inherited from the fermionic ones.

Finally, Kapustin has proposed to classify invertible topological phases for interacting bosons or fermions by considering low-energy effective actions that are invariant under cobordism (a certain type of equivalence relation between manifolds).[104,105] A related approach was taken by Freed and Hopkins[106] who achieved an exhaustive classification of invertible topological field theories using stable homotopy theory in any dimension of space and for any on-site symmetry group (bosonic or fermionic). This amounts to an exhaustive classification of the invertible topological phases of matter supported by lattice Hamiltonians whose low-energy limits are captured by invertible topological field theories.

Our aim is to apply the conjectured topological obstruction to gapping (8.85) to the symmetry classes DIII, CI, and AIII in the presence of quartic contact interactions. We recover the reductions $\mathbb{Z} \to \mathbb{Z}_{16}$, $\mathbb{Z} \to \mathbb{Z}_4$, and $\mathbb{Z} \to \mathbb{Z}_8$ for the symmetry classes DIII, CI, and AIII, respectively. We also verify that the topological classification $\mathbb{Z}_2$ of the symmetry class AII is stable to quartic contact interactions.

We shall denote with $\boldsymbol{x} \equiv (x, y, z) \equiv (x_1, x_2, x_3)$ a point in three-dimensional space.

8.4.6.1 The Symmetry Class DIII When $d = 3$

Let $X_{\mu\mu'} \equiv \tau_\mu \otimes \rho_{\mu'}$ with $\mu, \mu' = 0, 1, 2, 3$, Consider the three-dimensional bulk single-particle Dirac Hamiltonian (with Dirac matrices of dimension $r = 4 \equiv r_{\text{min}}$),

$$\mathcal{H}^{(0)}(\boldsymbol{x}) := -\mathrm{i}\partial_1 X_{31} - \mathrm{i}\partial_2 X_{02} - \mathrm{i}\partial_3 X_{11} + m(\boldsymbol{x}) X_{03}. \tag{8.134a}$$

This single-particle Hamiltonian belongs to the three-dimensional symmetry class DIII, for

[103] Y.-Z. You and C. Xu, Phys. Rev. B **90**,(24) 245120 (2014). [You and Xu (2014)].
[104] A. Kapustin, arXiv:1403.1467 (2014); arXiv:1404.6659 (2014). [Kapustin (2014b,a)].
[105] A. Kapustin, R. Thorngren, A. Turzillo, and Z. Wang, JHEP **2015**(12), 052 (2015). [Kapustin et al. (2015)].
[106] D. S. Freed and M. J. Hopkins, Geometry and Topology **25**(3), 1165–1330 (2021). [Freed and Hopkins (2021)].

$$\mathcal{T}\,\mathcal{H}^{(0)}(\boldsymbol{x})\,\mathcal{T}^{-1} = +\mathcal{H}^{(0)}(\boldsymbol{x}), \tag{8.134b}$$

$$\mathcal{C}\,\mathcal{H}^{(0)}(\boldsymbol{x})\,\mathcal{C}^{-1} = -\mathcal{H}^{(0)}(\boldsymbol{x}), \tag{8.134c}$$

where

$$\mathcal{T} := \mathrm{i}X_{20}\,\mathsf{K}, \qquad \mathcal{C} := X_{01}\,\mathsf{K}. \tag{8.134d}$$

The multiplicative factor i in the definition of $\mathcal{T}$ is needed for $\mathcal{T}$ to commute with $\mathcal{C}$.

The Dirac mass matrix X_{03} is here the only one allowed for dimension four Dirac matrices under the constraints (8.134b) and (8.134c). Consequently, the domain wall

$$m(\boldsymbol{x}) \equiv m(y) := m_{\infty}\,\mathrm{sgn}(y), \qquad m_{\infty} \in \mathbb{R}, \tag{8.135a}$$

at $y = 0$, binds the zero mode

$$e^{-\mathrm{i}X_{02}\,X_{03}\int_{0}^{y}\mathrm{d}y'\,m(y')}\,\chi = e^{-|m_{\infty}y|}\,\chi, \tag{8.135b}$$

where

$$X_{01}\,\chi = -\mathrm{sgn}\,(m_{\infty})\,\chi \tag{8.135c}$$

with χ independent of x and z. The kinetics of the gapless boundary states is governed by the Dirac Hamiltonian

$$\mathcal{H}^{(0)}_{\mathrm{bd}}(x,z) = -\mathrm{i}\partial_x\tau_3 - \mathrm{i}\partial_z\tau_1. \tag{8.136}$$

On the boundary, the operations for reversal of time and charge conjugation are now represented by

$$\mathcal{T}_{\mathrm{bd}\,\nu} := \mathrm{i}\tau_2 \otimes \mathbb{1}\,\mathsf{K}, \qquad \mathcal{C}_{\mathrm{bd}\,\nu} := \tau_0 \otimes \mathbb{1}\,\mathsf{K}, \tag{8.137a}$$

where $\mathbb{1}$ is the $\nu \times \nu$ unit matrix. We seek the single-particle Hamiltonian on the boundary that encodes the fluctuations arising from the Hubbard-Stratonovich decoupling of quartic interactions through a generic dynamical mass that respects the PHS but is odd under the TRS on the boundary. It is given by

$$\begin{aligned}\mathcal{H}^{(\mathrm{dyn})}_{\mathrm{bd}\,\nu}(\tau,x,z) :=\; & -\mathrm{i}\partial_x\tau_3 \otimes \mathbb{1} - \mathrm{i}\partial_z\tau_1 \otimes \mathbb{1} \\ & + \tau_2 \otimes M(\tau,x,z)\end{aligned} \tag{8.137b}$$

with the $\nu \times \nu$ real-valued and symmetric matrix

$$M(\tau,x,z) = M^{*}(\tau,x,z) = M^{\mathsf{T}}(\tau,x,z). \tag{8.137c}$$

The space of normalized Dirac mass matrices of the form (8.137c) is topologically equivalent to the space

$$V_{\nu} := \bigcup_{k=1}^{\nu} \mathrm{O}(\nu)/\left[\mathrm{O}(k) \times \mathrm{O}(\nu-k)\right] \tag{8.138}$$

Table 8.10 *Topological obstructions for the symmetry class DIII in three-dimensional space*

D	$\pi_D(R_0)$	ν	Topological obstruction
0	$\mathbb{Z}$	1	Domain wall
1	$\mathbb{Z}_2$	2	Vortex line
2	$\mathbb{Z}_2$	4	Monopole
3	0		
4	$\mathbb{Z}$	8	WZ term
5	0		
6	0		
7	0		
8	$\mathbb{Z}$	16	None

Reduction from $\mathbb{Z}$ to $\mathbb{Z}_{16}$ for the topologically equivalent classes of the three-dimensional noninteracting fermionic invertible topological phases in the symmetry class DIII that arises from interactions. We denote by V_ν the space of $\nu \times \nu$ normalized Dirac mass matrices in boundary ($d = 2$) Dirac Hamiltonians belonging to the symmetry class D. The direct limit $\nu \to \infty$ of these spaces is the classifying space R_0 from **Table** 8.1. The second column shows the stable D-th homotopy groups of the classifying space R_0 as taken from **Table** 8.2. The third column gives the number ν of copies of boundary (Dirac) fermions for which a topological obstruction that is compatible with local equations of motion is permissible. The fourth column gives the type of topological obstruction that prevents the gapping of the boundary (Dirac) fermions.

for the symmetry class D in two-dimensional space. The direct limit $\nu \to \infty$ of these spaces is the classifying space R_0 from **Table** 8.1. In order to gap out dynamically the boundary zero modes without breaking the defining symmetries of the symmetry class DIII, we need to construct a (2+1)-dimensional QNLSM for the (boundary) dynamical Dirac masses from the two-dimensional symmetry class D without topological obstructions. We construct explicitly the spaces for the relevant normalized boundary dynamical Dirac mass matrices of dimension $\nu = 2^n$ with $n = 0, 1, 2, 3, 4$ in the following. The relevant homotopy groups are given in **Table** 8.10.

Case $\nu = 1$*:* There is one boundary dynamical Dirac mass matrix

$$\gamma'(\tau, x, z) \equiv \tau_2 \otimes M(\tau, x, z) \tag{8.139}$$

on the boundary that is proportional to τ_2. A domain wall in imaginary time such as $m_{2\,\infty}\,\mathrm{sgn}(\tau)\,\tau_2$ prevents the dynamical generation of a spectral gap on the boundary.

Case $\nu = 2$*:* We use the representation $\mathbb{1} = \sigma_0$. The 2×2 real-valued and symmetric matrix $M(\tau, x, z)$ is a linear combination with real-valued coefficients of the pair of anticommuting matrices σ_1 and σ_3. If $M(\tau, x, z)$ is normalized by demanding that it squares to σ_0, then the set spanned by $M(\tau, x, z)$ is homeomorphic to the one-sphere S^1. As $\pi_1(S^1) = \mathbb{Z}$, it follows that $M(\tau, x, z)$ supports vortex lines (very much as with the entry $D = 1$ and $d = 3$ from **Table** 8.4) that bind zero modes in (2+1)-dimensional space and time and thus prevent the gapping of the boundary states.[107]

Case $\nu = 4$*:* We use the representation $\mathbb{1} = \sigma_0 \otimes \sigma'_0$. The 4×4 real-valued and symmetric matrix $M(\tau, x, z)$ is a linear combination with real coefficients of $X_{\sigma_\mu \sigma'_{\mu'}} \equiv \sigma_\mu \otimes \sigma'_{\mu'}$ with $\mu, \mu' = 0, 1, 2, 3$ such that either none or two of μ and μ' equal the number 2. Of these, the three matrices X_{13}, X_{33}, and X_{01} anticommute pairwise. If $M(\tau, x, z)$ is a linear combinations with real-valued coefficients of these three matrices and if M is normalized by demanding that it squares to X_{00}, then the set spanned by $M(\tau, x, z)$ is homeomorphic to the two-sphere S^2. As $\pi_2(S^2) = \mathbb{Z}$, $M(\tau, x, z)$ supports point-like defects of the monopole type (very much as with the entry $D = 2$ and $d = 3$ from **Table** 8.4) that bind zero modes in (2+1)-dimensional space and time and thus prevent the gapping of the boundary states.[108]

Case $\nu = 8$*:* We use the representation $\mathbb{1} = \sigma_0 \otimes \sigma'_0 \otimes \sigma''_0$. The 8×8 real-valued and symmetric matrix $M(\tau, x, z)$ is a linear combination with real-valued coefficients of the matrices $X_{\mu\mu'\mu''} \equiv \sigma_\mu \otimes \sigma'_{\mu'} \otimes \sigma''_{\mu''}$ where either none or two of $\mu, \mu', \mu'' = 0, 1, 2, 3$ equal the number 2. Of these, one finds the five pairwise anticommuting matrices X_{333}, X_{133}, X_{013}, X_{001}, and X_{212}. If $M(\tau, x, z)$ is a linear combination with real-valued coefficients of these five matrices and if M is normalized by demanding that it squares to X_{000}, then the set spanned by $M(\tau, x, z)$ is homeomorphic to the four-sphere S^4. As $\pi_4(S^4) = \mathbb{Z}$, it is possible to add a topological term to the QNLSM on the boundary that is of the WZ type. This term is conjectured to prevent the gapping of the boundary states.

Case $\nu = 16$*:* We use the representation $\mathbb{1} = \sigma_0 \otimes \sigma'_0 \otimes \sigma''_0 \otimes \sigma'''_0$. The 16×16 real-valued and symmetric matrix $M(\tau, x, z)$ is a linear combination with real-valued coefficients of the matrices $X_{\mu\mu'\mu''\mu'''} = \sigma_\mu \otimes \sigma'_{\mu'} \otimes \sigma''_{\mu''} \otimes \sigma'''_{\mu'''}$ where none, two, or four of $\mu, \mu', \mu'', \mu''' = 0, 1, 2, 3$ equal the number 2. Of these, one finds the nine pairwise anticommuting matrices X_{2222}, X_{0122}, X_{0322}, X_{2012}, X_{2032}, X_{1202}, X_{3202}, X_{0001}, and X_{0003}. If $M(\tau, x, z)$ is a linear combination with real-valued coefficients of these nine matrices and if $M(\tau, x, z)$ is normalized by demanding that it squares to X_{0000}, then the set spanned by $M(\tau, x, z)$ is homeomorphic to the eight-sphere S^{9-1}. It is then impossible to add a topological term that is compatible with local equations of motion to the QNLSM on the boundary as $\pi_0(S^8) = \cdots = \pi_4(S^8) = 0$. The boundary zero modes can be gapped out.

[107] Observe that $\pi_1(S^1) = \mathbb{Z}$ whereas $\pi_1(R_0) = \mathbb{Z}_2$. This discrepancy arises because we enter the stable homotopy group $\pi_D(R_0) = \mathbb{Z}_2$ by taking the direct limit R_0 of the direct system $\{V_\nu\}$ in the second column of **Table** 8.10.

[108] Observe that $\pi_2(S^2) = \mathbb{Z}$ whereas $\pi_2(R_0) = \mathbb{Z}_2$. This discrepancy arises because we enter the stable homotopy group $\pi_D(R_0) = \mathbb{Z}_2$ by taking the direct limit R_0 of the direct system $\{V_\nu\}$ in the second column of **Table** 8.10.

We conclude that the effects of interactions on the three-dimensional noninteracting fermionic invertible topological phases in the symmetry class DIII are to reduce the topological classification $\mathbb{Z}$ in the noninteracting limit down to $\mathbb{Z}_{16}$. The logic used to reach this conclusion is summarized by **Table** 8.10 once the line corresponding to $\nu = 1$ has been identified. It is given by matching the smallest D and the smallest ν such that $\pi_D(R_0) \neq 0$ and $\pi_D(\mathsf{S}^{N(\nu)-1}) \neq 0$, where $N(\nu)$ is defined in Eq. (8.83). This exercise is then repeated for ascending values of D and ν until the entry "none" is reached.

8.4.6.2 The Symmetry Class CI When $d = 3$

Let $X_{\mu\nu\lambda} \equiv \tau_\mu \otimes \rho_\nu \otimes \sigma_\lambda$ with $\mu, \nu, \lambda = 0, 1, 2, 3$. Consider the three-dimensional bulk single-particle Dirac Hamiltonian (with Dirac matrices of dimension $r = 8 \equiv r_{\mathrm{min}}$),

$$\mathcal{H}^{(0)}(\boldsymbol{x}) := -\mathrm{i}\partial_1 X_{310} - \mathrm{i}\partial_2 X_{020} - \mathrm{i}\partial_3 X_{110} + m(\boldsymbol{x}) X_{030}. \tag{8.140a}$$

This single-particle Hamiltonian belongs to the three-dimensional symmetry class CI, for

$$\mathcal{T}\mathcal{H}^{(0)}(\boldsymbol{x})\mathcal{T}^{-1} = +\mathcal{H}^{(0)}(\boldsymbol{x}), \tag{8.140b}$$

$$\mathcal{C}\mathcal{H}^{(0)}(\boldsymbol{x})\mathcal{C}^{-1} = -\mathcal{H}^{(0)}(\boldsymbol{x}), \tag{8.140c}$$

where

$$\mathcal{T} := X_{202}\,\mathsf{K}, \qquad \mathcal{C} := \mathrm{i}X_{012}\,\mathsf{K}. \tag{8.140d}$$

The multiplicative factor i in the definition of $\mathcal{C}$ is needed for $\mathcal{T}$ to commute with $\mathcal{C}$.

The single-particle Hamiltonian (8.140a) is the direct product of the single-particle Hamiltonian (8.134a) with the unit 2×2 matrix σ_0. If we interpret the degrees of freedom encoded by σ_0 and the Pauli matrices $\boldsymbol{\sigma}$ as carrying spin-1/2 degrees of freedom, we may then interpret Eq. (8.140) as defining a spin-singlet superconductor that preserves TRS.

The Dirac mass matrix X_{030} is here the only one allowed for dimension eight Dirac matrices under the constraints (8.140b) and (8.140c). Consequently, the domain wall

$$m(\boldsymbol{x}) \equiv m(y) := m_\infty\,\mathrm{sgn}(y), \qquad m_\infty \in \mathbb{R}, \tag{8.141a}$$

at $y = 0$, binds the zero mode

$$e^{-\mathrm{i}X_{020}\,X_{030}\int_0^y \mathrm{d}y'\,m(y')}\,\chi = e^{-|m_\infty\,y|}\chi, \tag{8.141b}$$

where

$$X_{010}\,\chi = -\mathrm{sgn}(m_\infty)\,\chi \tag{8.141c}$$

with χ independent of x and z. The kinetics of the gapless boundary states is governed by the Dirac Hamiltonian

$$\mathcal{H}^{(0)}_{\text{bd}}(x,z) = -\mathrm{i}\partial_x \tau_3 \otimes \sigma_0 - \mathrm{i}\partial_z \tau_1 \otimes \sigma_0. \tag{8.142}$$

On the boundary, the operations for reversal of time and charge conjugation are now represented by

$$\mathcal{T}_{\text{bd}\,\nu} := \tau_2 \otimes \sigma_2 \otimes \mathbb{1}\mathsf{K}, \qquad \mathcal{C}_{\text{bd}\,\nu} := \mathrm{i}\,\tau_0 \otimes \sigma_2 \otimes \mathbb{1}\mathsf{K}, \tag{8.143a}$$

where $\mathbb{1}$ is the $\nu \times \nu$ unit matrix. We seek the single-particle Hamiltonian on the boundary that encodes the fluctuations arising from the Hubbard-Stratonovich decoupling of quartic interactions through a generic dynamical mass that respects the PHS but is odd under the TRS on the boundary. It is given by

$$\begin{aligned}\mathcal{H}^{(\text{dyn})}_{\text{bd}\,\nu}(\tau,x,z) := &-\mathrm{i}\partial_x \tau_3 \otimes \sigma_0 \otimes \mathbb{1} - \mathrm{i}\partial_z \tau_1 \otimes \sigma_0 \otimes \mathbb{1}\\ &+ \tau_2 \otimes M(\tau,x,z)\end{aligned} \tag{8.143b}$$

with the $2\nu \times 2\nu$ Hermitean matrix

$$M(\tau,x,z) = +(\sigma_2 \otimes \mathbb{1})\, M^{*}(\tau,x,z)\, (\sigma_2 \otimes \mathbb{1}). \tag{8.143c}$$

The space of normalized Dirac mass matrices satisfying the condition (8.143c) is topologically equivalent to the space

$$V_\nu := \bigcup_{k=1}^{\nu} \mathrm{Sp}(2\nu)/\left[\mathrm{Sp}(2k) \times \mathrm{Sp}(2\nu - 2k)\right] \tag{8.144}$$

for the symmetry class C in two-dimensional space. The direct limit $\nu \to \infty$ of these spaces is the classifying space R_4 from **Table** 8.1. In order to gap out dynamically the boundary zero modes without breaking the defining symmetries of the symmetry class CI, we need to construct a (2+1)-dimensional QNLSM for the (boundary) dynamical Dirac masses from the two-dimensional symmetry class C without topological obstructions. We construct explicitly the spaces for the relevant normalized boundary dynamical Dirac mass matrices of dimension $\nu = 2^n$ with $n = 0, 1, 2$ in the following. The relevant homotopy groups are given in **Table** 8.11.

Case $\nu = 1$*:* There is one 2×2 Hermitean matrix $M(\tau,x,z)$ on the boundary that is proportional to σ_0. A domain wall in imaginary time such as $m_{2\,\infty}\,\mathrm{sgn}(\tau)\,\tau_2 \otimes \sigma_0$ prevents the dynamical generation of a spectral gap on the boundary.

Case $\nu = 2$*:* We use the representation $\mathbb{1} = \sigma'_0$. The Hermitean 4×4 matrix $M(\tau,x,z)$ is a linear combination with real-valued coefficients of the matrices $X_{\mu\mu'} \equiv \sigma_\mu \otimes \sigma'_{\mu'}$ with $\mu, \mu' = 0, 1, 2, 3$ such that $X_{20}\, X^{*}_{\mu\mu'}\, X_{20} = +X_{\mu\mu'}$. Of these, one finds the five matrices X_{12}, X_{22}, X_{32}, X_{01}, and X_{03} that anticommute pairwise. If $M(\tau,x,z)$ is a linear combination with real-valued coefficients of these five matrices and if $M(\tau,x,z)$ is normalized by demanding that it squares to X_{00}, then the set spanned by $M(\tau,x,z)$ is homeomorphic to S^4. As $\pi_{2+1+1}(\mathrm{S}^4) = \mathbb{Z}$, it is

Table 8.11 *Topological obstructions for the symmetry class CI in three-dimensional space*

D	$\pi_D(R_4)$	ν	Topological obstruction
0	$\mathbb{Z}$	1	Domain wall
1	0		
2	0		
3	0		
4	$\mathbb{Z}$	2	WZ term
5	$\mathbb{Z}_2$	4	None
6	$\mathbb{Z}_2$		
7	0		

Reduction from $\mathbb{Z}$ to $\mathbb{Z}_4$ for the topologically equivalent classes of the three-dimensional noninteracting fermionic invertible topological phases in the symmetry class CI that arises from interactions. We denote by V_ν the space of $\nu \times \nu$ normalized Dirac mass matrices in boundary ($d = 2$) Dirac Hamiltonians belonging to the symmetry class C. The direct limit $\nu \to \infty$ of these spaces is the classifying space R_4 from **Table** 8.1. The second column shows the stable D-th homotopy groups of the classifying space R_4 as taken from **Table** 8.2. The third column gives the number ν of copies of boundary (Dirac) fermions for which a topological obstruction that is compatible with local equations of motion is permissible. The fourth column gives the type of topological obstruction that prevents the gapping of the boundary (Dirac) fermions.

then possible to add a topological term to the (2+1)-dimensional QNLSM on the boundary that is of the WZ type. This term is conjectured to prevent the gapping of the boundary states.

Case $\nu = 4$: We use the representation $\mathbb{1} = \sigma'_0 \otimes \sigma''_0$. The Hermitean 8×8 matrix $M(\tau, x, z)$ is a linear combination with real-valued coefficients of the matrices $X_{\mu\mu'\mu''} \equiv \sigma_\mu \otimes \sigma'_{\mu'} \otimes \sigma''_{\mu''}$ with $\mu, \mu', \mu'' = 0, 1, 2, 3$ such that $X_{200}\, X^*_{\mu\mu'\mu''}\, X_{200} = +X_{\mu\mu'\mu''}$. Of these, one finds the six matrices X_{120}, X_{220}, X_{320}, X_{010}, X_{031} and X_{033} that anticommute pairwise. If $M(\tau, x, z)$ is a linear combination with real-valued coefficients of these six matrices and if $M(\tau, x, z)$ is normalized by demanding that it squares to X_{000}, then the set spanned by $M(\tau, x, z)$ is homeomorphic to S^5. It is then impossible to add a topological term that is compatible with local equations of motion to the (2+1)-dimensional QNLSM on the boundary as $\pi_0(\mathrm{S}^5) = \cdots = \pi_4(\mathrm{S}^5) = 0$. The boundary zero modes can be gapped out.

We conclude that the effects of interactions on the three-dimensional noninteracting fermionic invertible topological phases in the symmetry class CI are to reduce the topological classification $\mathbb{Z}$ in the noninteracting limit down to $\mathbb{Z}_4$. The logic used to reach this conclusion is summarized by **Table** 8.11 once the line corresponding to $\nu = 1$ has been identified. It is given by matching the smallest D and the smallest ν such that $\pi_D(R_4) \neq 0$ and $\pi_D(\mathrm{S}^{N(\nu)-1}) \neq 0$, where $N(\nu)$ is defined in Eq. (8.83). This exercise is then repeated for ascending values of D and ν until the entry "none" is reached.

8.4.6.3 The Symmetry Class AIII When $d = 3$

By omitting the contributions arising from the gauge potentials, the single-particle Hamiltonian (8.134a) does not specify uniquely the symmetry class. For example, the single-particle Hamiltonian (8.134a) can also be interpreted as an insulator belonging to the symmetry class AIII, for it anticommutes with the composition

$$\Gamma := -\mathrm{i}\mathcal{T}\,\mathcal{C} = X_{21} \tag{8.145}$$

of the operations $\mathcal{T}$ and $\mathcal{C}$ for time reversal and charge conjugation, respectively, defined in Eq. (8.134d).

The tensor product of the single-particle Hamiltonian (8.134a) with the $\nu \times \nu$ unit matrix $\mathbb{1}$ supports ν zero modes bound to the boundary $y = 0$, for they are annihilated by the boundary Hamiltonian

$$\mathcal{H}^{(0)}_{\mathrm{bd}\,\nu}(x,z) := -\mathrm{i}\partial_x\,\tau_3 \otimes \mathbb{1} - \mathrm{i}\partial_z\,\tau_1 \otimes \mathbb{1} \tag{8.146a}$$

that anticommutes with

$$\Gamma^{(\mathrm{bd})} := \tau_2 \otimes \mathbb{1}. \tag{8.146b}$$

The fate of these zero modes in the presence of fermion-fermion interactions is investigated in two steps, as we did for the one-dimensional case. First, we include the effects of interactions by perturbing the boundary Hamiltonian with all boundary dynamical Dirac masses from the symmetry class A. Second, we introduce a Bogoliubov-de-Gennes (Nambu) grading to account for the interactions-driven superconducting fluctuations by perturbing the boundary Hamiltonian $\mathcal{H}^{(0)}_{\mathrm{bd}\,\nu}(x,z)\otimes\rho_0$ with all boundary dynamical Dirac masses that anticommute with $\tau_0 \otimes \mathbb{1} \otimes \rho_1\,\mathsf{K}$, that is with all boundary dynamical Dirac masses from the symmetry class D. In the first step, the boundary dynamical single-particle Hamiltonian is

$$\mathcal{H}^{(\mathrm{dyn})}_{\mathrm{bd}\,\nu}(\tau,x,z) := (-\mathrm{i}\partial_x\,\tau_3 - \mathrm{i}\partial_z\,\tau_1) \otimes \mathbb{1} + \tau_2 \otimes M(\tau,x,z), \tag{8.147a}$$

with the $\nu \times \nu$ Hermitean matrix

$$M(\tau,x,z) = M^{\dagger}(\tau,x,z). \tag{8.147b}$$

In the second step, the boundary dynamical single-particle Hamiltonian is

$$\mathcal{H}^{(\mathrm{dyn})}_{\mathrm{bd}\,\nu}(\tau,x,z) := (-\mathrm{i}\partial_x\,\tau_3 - \mathrm{i}\partial_z\,\tau_1) \otimes (\mathbb{1}\otimes\rho_0) + \tau_2 \otimes M(\tau,x,z), \tag{8.148a}$$

with the $2\nu \times 2\nu$ Hermitean matrix

$$M(\tau,x,z) = +(\mathbb{1}\otimes\rho_1)\,M^{*}(\tau,x,z)\,(\mathbb{1}\otimes\rho_1). \tag{8.148b}$$

Observe that $\tau_2 \otimes M(\tau,x,z)$ anticommutes with $\tau_0 \otimes \mathbb{1} \otimes \rho_1\,\mathsf{K}$.

The space of boundary dynamical Dirac mass matrices of the form (8.147b) that square to the unit matrix is homeomorphic to the classifying space C_0 for the symmetry class A in two-dimensional space and in the direct limit $\nu \to \infty$. In order to gap out dynamically the boundary zero modes without breaking the defining symmetries of the symmetry class AIII, we need to construct a (2+1)-dimensional QNLSM for the (boundary) dynamical Dirac masses from the two-dimensional symmetry class A without topological obstructions. When $\nu = 1$, a domain wall such as $M(\tau, x, z) = M_\infty \operatorname{sgn}(\tau)$ prevents the gapping of the boundary zero modes. When $\nu = 2$, we choose the representation $\mathbb{1} = \sigma_0$. The set spanned by $M(\tau, x, z) = \sum_{j=1}^{3} m_j(\tau, x, z)\, \sigma_j$ with the real-valued functions $m_j(\tau, x, z)$ obeying the normalization condition $\sum_{j=1}^{3} m_j^2(\tau, x, z) = 1$ supports a monopole (very much as with the entry $D = 2$ and $d = 3$ from **Table** 8.4) that binds zero modes in (2+1)-dimensional space and time, as $\pi_2(\mathsf{S}^2) = \mathbb{Z}$. When $\nu = 4$, we choose the representation $\mathbb{1} = X_{00}$ where $X_{\mu\mu'} := \sigma_\mu \otimes \sigma'_{\mu'}$ for $\mu, \mu' = 0, 1, 2, 3$. We may then write $M(\tau, x, z) = \sum_{\mu,\mu'=0}^{3} m_{\mu\mu'}(\tau, x, z)\, X_{\mu\mu'}$. Any $X_{\mu\mu'}$ other than the unit matrix X_{00} belongs to a multiplet of five pairwise anticommuting matrices of the form $X_{\nu\nu'} \neq X_{00}$. Hence, we may always construct a set of normalized $M(\tau, x, z)$ homeomorphic to S^4. Since $\pi_{2+1+1}(\mathsf{S}^4) = \mathbb{Z}$, it is possible to augment the corresponding boundary QNLSM in (2+1)-dimensional space and time by a WZ term that modifies the equations of motion in a local way. This term is conjectured to prevent the gapping of the boundary states. When $\nu = 2^n$ with $n \geq 3$, we choose the representation $\mathbb{1} = X_{00\cdots}$ where $X_{\mu\mu'\cdots} := \sigma_\mu \otimes \sigma'_{\mu'} \otimes \cdots$ for $\mu, \mu', \cdots = 0, 1, 2, 3$. Any $X_{\mu\mu'\cdots}$ other than the unit matrix $X_{00\cdots}$ belongs to a multiplet of no less than seven pairwise anticommuting matrices. It is for this reason that the boundary states are then necessarily gapped, for it is not permissible to add a topological term to the action of the boundary QNLSM for a sphere of dimension larger than four. We conclude that the effects of interactions in the three-dimensional noninteracting fermionic invertible topological phases in the symmetry class AIII is to reduce the topological classification $\mathbb{Z}$ in the noninteracting limit down to $\mathbb{Z}_8$ under the assumption that only fermion-number-conserving interacting channels are included in the stability analysis. The logic used to reach this conclusion is summarized by **Table** 8.12 once the line corresponding to $\nu = 1$ has been identified. It is given by matching the smallest D and the smallest ν such that $\pi_D(C_0) \neq 0$ and $\pi_D(\mathsf{S}^{N(\nu)-1}) \neq 0$, where $N(\nu)$ is defined in Eq. (8.83). This exercise is then repeated for ascending values of D and ν until the entry "none" is reached.

The space of boundary dynamical matrices that satisfy the condition (8.148b) and square to the unit matrix is homeomorphic to the classifying space R_0 for the symmetry class D in two-dimensional space and in the direct limit $\nu \to \infty$. Because the dimension of the boundary dynamical matrix (8.148b) is twice that of the boundary dynamical matrix (8.147b), one might have guessed that the gapping of the boundary zero modes takes place for a value of ν smaller than eight. This

Table 8.12 *Topological obstructions for the symmetry class AIII in three-dimensional space*

D	$\pi_D(C_0)$	ν	Topological obstruction
0	$\mathbb{Z}$	1	Domain wall
1	0		
2	$\mathbb{Z}$	2	Monopole
3	0		
4	$\mathbb{Z}$	4	WZ term
5	0		
6	$\mathbb{Z}$	8	None
7	0		

Reduction from $\mathbb{Z}$ to $\mathbb{Z}_8$ for the topologically equivalent classes of the three-dimensional noninteracting fermionic invertible topological phases in the symmetry classes AIII that arises from the fermion-number-conserving interacting channels. We denote by V_ν the space of $\nu \times \nu$ normalized Dirac mass matrices in boundary ($d = 2$) Dirac Hamiltonians belonging to the symmetry class A. The direct limit $\nu \to \infty$ of these spaces is the classifying space C_0 from **Table** 8.1. The second column shows the stable D-th homotopy groups of the classifying space C_0 as taken from **Table** 8.2. The third column gives the number ν of copies of boundary (Dirac) fermions for which a topological obstruction that is compatible with local equations of motion is permissible. The fourth column gives the type of topological obstruction that prevents the gapping of the boundary (Dirac) fermions.

is not so, however, because of two constraints. The first constraint is that of PHS. The second constraint restricts the target space for the boundary QNLSM that is built out of the boundary dynamical Dirac masses. The target space of the QNLSM must be invariant as a set under the action of a global gauge U(1) transformation that is generated by $\tau_0 \otimes \mathbb{1} \otimes \rho_3$ (as demanded below Eq. (8.97) when $d = 1$). This global U(1) symmetry implements conservation of the fermion number. Indeed, one verifies the following facts.

When $\nu = 1$, the boundary dynamical matrix $M(\tau, x, z)$ is a linear combination of ρ_1 and ρ_2 with real-valued functions as coefficients. Hence, the space of normalized boundary dynamical Dirac mass matrices is homeomorphic to S^1 and invariant as a set under any global gauge U(1) transformation when $\nu = 1$. Because of $\pi_1(\mathrm{S}^1) = \mathbb{Z}$, vortex lines bind zero modes (very much as with the entry $D = 1$ and $d = 3$ from **Table** 8.4) in (2+1)-dimensional space and time.

When $\nu = 2$, we represent the unit 4×4 matrix by $\sigma_0 \otimes \rho_0$ and we expand any 4×4 Hermitean matrix as a linear combination with real-valued coefficients of the 64 matrices

$$X_{\mu\mu'} = \sigma_\mu \otimes \rho_{\mu'} \tag{8.149a}$$

with $\mu,\mu' = 0,1,2,3$. The boundary dynamical matrix $M(\tau,x,z)$ is a linear combination with real-valued functions as coefficients of the ten matrices

$$\begin{array}{lll} X_{00}, & & \\ X_{10}, & X_{30}, & X_{23}, \\ X_{02}, & X_{11}, & X_{31}, \\ X_{01}, & X_{12}, & X_{32}, \end{array} \tag{8.149b}$$

that all satisfy the constraint

$$X_{\mu\mu'} = +X_{01}\,X^{*}_{\mu\mu'}\,X_{01}. \tag{8.149c}$$

Other than the unit matrix X_{00}, any one of these nine matrices belongs to a triplet of pairwise anticommuting matrices. Only the triplet (X_{10}, X_{30}, X_{23}) is made of three dymamical mass matrices, each of which is invariant under the global U(1) transformation (8.119), that is invariant under conjugation with the generator $\exp(\mathrm{i}\theta\,X_{03})$ of the global U(1) transformation parameterized by $0 \leq \theta < 2\pi$. As no other dynamical mass matrix anticommutes with this triplet of gauge-invariant dynamical masses, this triplet spans a set of normalized boundary dynamical Dirac masses that is homeomorphic to S^2, each point of which is invariant under the global U(1) transformation associated with the conservation of the fermion number. Because of $\pi_2(\mathsf{S}^2) = \mathbb{Z}$, monopoles bind zero modes (very much as with the entry $D = 2$ and $d = 3$ from **Table** 8.4) in (2+1)-dimensional space and time that prevent the gapping of the boundary states when $\nu = 2$.

When $\nu = 4$, we represent the unit 8×8 matrix by $\sigma_0\otimes\sigma'_0\otimes\rho_0$ and we expand any 8×8 matrix as a linear combination with real-valued coefficients of the 64 matrices

$$X_{\mu\mu'\mu''} := \sigma_\mu\otimes\sigma'_{\mu'}\otimes\rho_{\mu''} \tag{8.150a}$$

with $\mu,\mu',\mu'' = 0,1,2,3$. The boundary dynamical matrix $M(\tau,x,z)$ is a linear combination with real-valued functions as coefficients of those matrices:

$$X_{\mu\mu'\mu''} = +X_{001}\,X^{*}_{\mu\mu'\mu''}\,X_{001}. \tag{8.150b}$$

Other than the unit matrix X_{000}, any one of those matrices belong to a quintuplet of pairwise anticommuting matrices. Among these, each element from the quintuplet

$$X_{010},\qquad X_{030},\qquad X_{123},\qquad X_{220},\qquad X_{323} \tag{8.150c}$$

is invariant under conjugation by X_{003}, that is, it is invariant under the global gauge transformation $\exp(\mathrm{i}\theta\,X_{003})$ for any $0 \leq \theta < 2\pi$. Moreover, no other matrix, satisfying the condition $X_{\mu\mu'\mu''} = +X_{001}\,X^{*}_{\mu\mu'\mu''}\,X_{001}$, anticommutes with this quintuplet. Hence, this quintuplet spans a set of normalized boundary dynamical Dirac masses that is homeomorphic to S^4, each point of which is invariant under the global U(1) transformation associated with the conservation of the fermion number. Because of $\pi_{2+1+1}(\mathsf{S}^4) = \mathbb{Z}$, it is possible to augment the corresponding boundary QNLSM in (2+1)-dimensional space and time by a WZ term that modifies the equations of motion in a local way. This term is conjectured to prevent the gapping of the boundary states.

When $\nu = 2^{n-1}$ with $n = 4$, one verifies that any dynamical mass matrix

$$\begin{aligned} X_{\mu\mu'\mu''\cdots\mu^{(n)}} &:= \sigma_{\mu} \otimes \sigma'_{\mu'} \otimes \sigma''_{\mu''} \otimes \cdots \otimes \rho_{\mu^{(n)}} \\ &= +X_{000\cdots 1}\, X^{*}_{\mu\mu'\mu'''\cdots\mu^{(n)}}\, X_{000\cdots 1} \end{aligned} \tag{8.151a}$$

with $\mu, \mu', \mu'', \cdots, \mu^{(n)} = 0, 1, 2, 3$ belongs to a $N(\nu)$-tuplet of pairwise anticommuting permissible matrices with $N(\nu) > 5$. The $N(\nu)$-tuplet that contains the pair of anticommuting matrices

$$X_{0\cdots 010} = +X_{0\cdots 001}\, X^{*}_{0\cdots 010}\, X_{0\cdots 001}, \qquad X_{0\cdots 030} = +X_{0\cdots 001}\, X^{*}_{0\cdots 030}\, X_{0\cdots 001} \tag{8.151b}$$

has the particularity that (i) each of its elements is invariant under conjugation with $X_{000\cdots 3}$, that is it is invariant under the global gauge transformation $\exp(\mathrm{i}\theta\, X_{000\cdots 3})$ for any $0 \leq \theta < 2\pi$ and (ii) cannot be augmented by one more anticommuting $X_{\mu\mu'\mu'''\cdots\mu^{(n)}} = +X_{000\cdots 1}\, X^{*}_{\mu\mu'\mu'''\cdots\mu^{(n)}}\, X_{000\cdots 1}$. Hence, this $N(\nu)$-tuplet spans a set of normalized boundary dynamical Dirac masses that is homeomorphic to $S^{N(\nu)-1}$, each point of which is invariant under the global U(1) transformation associated with the conservation of the fermion number. Since $N(\nu) > 5$ for $\nu = 2^{n-1}$ with $n = 4$, it follows that all homotopy groups of order less than 4 for the space of the normalized boundary dynamical Dirac masses that are invariant under the global U(1) transformation are vanishing. The boundary states are then necessarily gapped by an interaction that does not break spontaneously the protecting U(1) fermion-number symmetry when $n = 4$.

We conclude that the effects of interactions on the three-dimensional noninteracting fermionic invertible topological phases in the symmetry class AIII are to reduce the topological classification $\mathbb{Z}$ in the noninteracting limit down to $\mathbb{Z}_8$.

8.4.6.4 The Symmetry Class AII When $d = 3$

We close the discussion of the stability to fermion-fermion interactions of strong noninteracting topological insulators or superconductors in three-dimensional space by illustrating why the $\mathbb{Z}_2$ noninteracting classification is stable.

To this end, consider the single-particle bulk Dirac Hamiltonian

$$\mathcal{H}^{(0)}(\boldsymbol{x}) := -\mathrm{i}\partial_x\, X_{21} - \mathrm{i}\partial_y\, X_{11} - \mathrm{i}\partial_z\, X_{02} + m(\boldsymbol{x})\, X_{03}, \tag{8.152a}$$

where $X_{\mu\mu'} := \sigma_{\mu} \otimes \tau_{\mu'}$ for $\mu, \mu' = 0, 1, 2, 3$. Because

$$\mathcal{H}^{(0)}(\boldsymbol{x}) = +\mathcal{T}\, \mathcal{H}^{(0)}(\boldsymbol{x})\, \mathcal{T}^{-1}, \qquad \mathcal{T} := \mathrm{i}X_{20}\, \mathsf{K}, \tag{8.152b}$$

we interpret this Hamiltonian as realizing a noninteracting topological insulator in the three-dimensional symmetry class AII. The domain wall in the mass

$$m(x,y,z) = m_{\infty}\,\mathrm{sgn}(z) \tag{8.153a}$$

binds a zero mode to the boundary $z = 0$ that is annihilated by the boundary single-particle Hamiltonian

$$\begin{aligned}\mathcal{H}^{(0)}_{\mathrm{bd}}(x,y) =& -\mathrm{i}\partial_x\,\sigma_2 - \mathrm{i}\partial_y\,\sigma_1\\ =& \,\mathcal{T}_{\mathrm{bd}}\,\mathcal{H}^{(0)}_{\mathrm{bd}}(x,y)\,\mathcal{T}^{-1}_{\mathrm{bd}},\end{aligned} \tag{8.153b}$$

where

$$\mathcal{T}_{\mathrm{bd}} := \mathrm{i}\sigma_2\,\mathsf{K}. \tag{8.153c}$$

The boundary dynamical Dirac Hamiltonian

$$\mathcal{H}^{(\mathrm{dyn})}_{\mathrm{bd}}(\tau,x,y) = -\mathrm{i}\partial_x\,\sigma_2 - \mathrm{i}\partial_y\,\sigma_1 + M(\tau,x,y)\,\sigma_3 \tag{8.154}$$

belongs to the symmetry class A, as the Dirac mass $M\sigma_3$ breaks TRS unless $M(-\tau,x,y) = -M(\tau,x,y)$. The space of normalized boundary dynamical Dirac mass matrices $\{\pm\sigma_3\}$ is homeomorphic to the space of normalized Dirac mass matrices for the two-dimensional system in the symmetry class A

$$V_{\nu=1} = \bigcup_{k=0}^{1} \mathrm{U}(1)/\left[\mathrm{U}(k)\times\mathrm{U}(1-k)\right]. \tag{8.155}$$

The domain wall in imaginary time $M(\tau,x,y) = M_{\infty}\,\mathrm{sgn}(\tau)$ prevents the gapping of the boundary zero modes.

We conclude that the noninteracting topological classification $\mathbb{Z}_2$ of three-dimensional insulators in the symmetry class AII is robust to the effects of interactions under the assumption that only fermion-number-conserving interacting channels are included in the stability analysis. The logic used to reach this conclusion is summarized by **Table** 8.13 once the line corresponding to $\nu = 1$ has been identified. It is given by matching the smallest D and the smallest ν such that $\pi_D(C_0) \neq 0$ and $\pi_D(\mathrm{S}^{N(\nu)-1}) \neq 0$, where $N(\nu)$ is defined in Eq. (8.83). Moreover, one verifies by introducing a Bogoliubov-de-Gennes (Nambu) grading that this robustness extends to interaction-driven dynamical superconducting fluctuations.

Homotopy groups of the dynamical boundary Dirac masses for the symmetry class CII in nine-dimensional space.

8.4.7 Higher Dimensions

The following rules can be deduced[109] by working out explicitly the effects of fermion-fermion interactions on the boundary states supported by single-particle

[109] With the usual caveat that the interactions are strong on the boundary but not too strong in the bulk.

Table 8.13 *Topological obstructions for the symmetry class AII in three-dimensional space*

D	$\pi_D(C_0)$	ν	Topological obstruction
0	$\mathbb{Z}$	1	Domain wall
1	0		

Stability to fermion-fermion interactions of the noninteracting topological classification $\mathbb{Z}_2$ for three-dimensional strong topological insulators belonging to the symmetry classes AII. We denote by V_ν the space of $\nu \times \nu$ normalized Dirac mass matrices in boundary ($d = 2$) Dirac Hamiltonians belonging to the symmetry class A. The direct limit $\nu \to \infty$ of these spaces is the classifying space C_0 from **Table** 8.1. The second column shows the stable D-th homotopy groups of the classifying space C_0 as taken from **Table** 8.2. The third column gives the number ν of copies of boundary (Dirac) fermions for which a topological obstruction that is compatible with local equations of motion is permissible. The fourth column gives the type of topological obstruction that prevents the gapping of the boundary (Dirac) fermions.

Table 8.14 *Homotopy groups of the dynamical boundary Dirac masses for the symmetry class BDI in one-dimensional space*

D	$\pi_D(R_2)$	ν	S_{QNLSM}	S_{top}
0	$\mathbb{Z}_2$	2	S^0	✓
1	0			
2	$\mathbb{Z}$	4	S^2	✓
3	0			
4	0			
5	0			
6	$\mathbb{Z}$	8	S^6	–
7	$\mathbb{Z}_2$	16	S^7	–

Application of the Bott periodicity obeyed by the homotopy groups $\pi_D(V)$ for $D = 0, 1, \cdots$ of a given classifying space V'_{d-1} of dynamical boundary Dirac masses to deduce the reduction pattern $\mathbb{Z} \to \mathbb{Z}_{\nu_{\mathrm{max}}}$ for the symmetry class BDI in dimensions $d = 1$ for which $V'_{d-1} = R_2$ and $\nu_{\mathrm{max}} = 8$. The column ν fixes the rank $r := r_{\mathrm{min}}\, \nu$ of the Dirac Hamiltonian in the symmetry class BDI. The fourth column gives the target manifold of the QNLSM with the action S_{QNLSM} that encodes the fermion-fermion interactions on the $(d-1)$-dimensional boundary. The fifth column indicates if a topological obstruction is present.

Dirac Hamiltonians representing strong topological insulators or superconductors when the dimensionality of space ranges from $d = 1$ to $d = 8$.

1. The $\mathbb{Z}_2$ topological classification of strong topological insulators or superconductors is robust to interactions in all dimensions.

Table 8.15 *Homotopy groups of the dynamical boundary Dirac masses for the symmetry class BDI in five-dimensional space*

D	$\pi_D(R_6)$	ν	S_{QNLSM}	S_{top}
0	0			
1	0			
2	$\mathbb{Z}$	1	S^2	✓
3	$\mathbb{Z}_2$	2	S^3	✓
4	$\mathbb{Z}_2$	4	S^4	✓
5	0			
6	$\mathbb{Z}$	8	S^6	✓
7	0			
8	0			
9	0			
10	$\mathbb{Z}$	16	S^{10}	–

Same as in **Table** 8.14 except for $d = 5$ for which $V'_{d-1} = R_6$ and $\nu_{\text{max}} = 16$.

2. The $\mathbb{Z}$ topological classification of strong topological insulators or superconductors is robust to interactions in all even dimensions.
3. The $\mathbb{Z}$ topological classification of strong topological insulators or superconductors is unstable to interactions in all odd dimensions.

In this section, we prove Rules 1 and 2 and we work out explicitly the reduction pattern of the noninteracting $\mathbb{Z}$ topological classification for any odd dimension.

8.4.7.1 The Case of $\mathbb{Z}_2$ Classification

The proof of Rule 1 follows the same logic as in the example of the three-dimensional symmetry class AII in Section 8.4.6.

When $d = 1$, there are two symmetry classes with $\pi_0(V) = \mathbb{Z}_2$: the symmetry classes D and DIII (see **Table** 8.6). No dynamical Dirac mass is allowed in class D, since there is no protecting symmetry to break.[110] For the symmetry class DIII, the normalized boundary dynamical Dirac masses are taken from Dirac masses in the symmetry class D with space d = 0-dimensional, that is, they belong to the classifying space R_2, according to **Tables** 8.6 and 8.1. According to **Table** 8.2, $\pi_0(R_2) = \mathbb{Z}_2$. Hence, the two noninteracting $\mathbb{Z}_2$ topological classifications are stable in one dimension.

To treat the case of $d \geq 2$, let V denote any one of the eight real classifying spaces V and observe that, according to **Table** 8.2, at least one of the homotopy

[110] Quantum mechanics posits that the total fermion parity cannot be broken (neither explicitly nor spontaneously) as fermions are always created or destroyed in pairs.

Table 8.16 *Homotopy groups of the dynamical boundary Dirac masses for the symmetry class BDI in nine-dimensional space*

D	$\pi_D(R_2)$	ν	S_{QNLSM}	S_{top}
0	$\mathbb{Z}_2$	2	S^0	✓
1	0			
2	$\mathbb{Z}$	4	S^2	✓
3	0			
4	0			
5	0			
6	$\mathbb{Z}$	8	S^6	✓
7	$\mathbb{Z}_2$	16	S^7	✓
8	$\mathbb{Z}_2$	32	S^8	✓
9	0			
10	$\mathbb{Z}$	64	S^{10}	✓
11	0			
12	0			
13	0			
14	$\mathbb{Z}$	128	S^{14}	–

Same as in **Table** 8.14 except for $d = 9$ for which $V'_{d-1} = R_2$ and $\nu_{\text{max}} = 128$.

groups $\pi_D(V)$ with $D = 0, 1, 2, 3$ is nontrivial. We specialize to any one of the two symmetry classes in d dimensions with the classifying space (the space of normalized bulk Dirac masses) V such that $\pi_0(V) = \mathbb{Z}_2$. By assumption $d + 1 \geq 3$. Let V_{bd} denote the space of the boundary dynamical Dirac masses. If this space is empty, the $\mathbb{Z}_2$ classification is stable. If this space is not empty, then we know that at least one of $\pi_D(V_{\text{bd}})$ with $D = 0, 1, \cdots, d+1$ is nonvanishing. In turn, this implies that at least one of the homotopy groups from Eq. (8.84) is nontrivial. As the sphere $\mathsf{S}^{N(1)-1}$ entering Eq. (8.84) is the target space for the QNLSM in $(d-1)+1$ space and time dimensions obtained from integrating the $\nu = 1$ boundary Dirac fermions subjected to dynamical masses, the QNLSM accommodates a topological term that prevents the gapping of the $\nu = 1$ boundary zero mode. Hence, the two noninteracting $\mathbb{Z}_2$ topological classifications are stable in any spatial dimension 79.

8.4.7.2 The Case of Even Dimensions

Because of Rule 1, we only need to consider the symmetry classes in even dimensions which have a $\mathbb{Z}$ topological classification for gapped noninteracting fermions. According to **Table** 8.6 and the Bott periodicity of two (eight) for the complex (real) symmetry classes, these are the symmetry classes (i) A for $d = 0 \bmod 2$, (ii) AI and AII for $d = 4, 8 \bmod 8$, (iii) D and C for $d = 2, 6 \bmod 8$.

Table 8.17 *Homotopy groups of the dynamical boundary Dirac masses for the symmetry class DIII in three-dimensional space*

D	$\pi_D(R_0)$	ν	S_{QNLSM}	S_{top}
0	$\mathbb{Z}$	1	S^0	✓
1	$\mathbb{Z}_2$	2	S^1	✓
2	$\mathbb{Z}_2$	4	S^2	✓
3	0			
4	$\mathbb{Z}$	8	S^4	✓
5	0			
6	0			
7	0			
8	$\mathbb{Z}$	16	S^8	–

Application of the Bott periodicity obeyed by the homotopy groups $\pi_D(V)$ for $D = 0, 1, \cdots$ of a given classifying space V'_{d-1} of dynamical boundary Dirac masses to deduce the reduction pattern $\mathbb{Z} \to \mathbb{Z}_{\nu_{\text{max}}}$ for the symmetry class DIII in dimensions $d = 3$ for which $V'_{d-1} = R_0$ and $\nu_{\text{max}} = 16$. The column ν fixes the rank $r := r_{\text{min}}\,\nu$ of the Dirac Hamiltonian in the symmetry class DIII. The fourth column gives the target manifold of the QNLSM with the action S_{QNLSM} that encodes the fermion-fermion interactions on the $(d-1)$-dimensional boundary. The fifth column indicates if a topological obstruction is present.

Proof for case (i): We start with the complex symmetry class A in even dimensions of space. In the symmetry class A, the only protecting symmetry is that of the total fermion number, which includes the total fermion parity. We proceed in two steps. First we rule out dynamical superconducting fluctuations and show the stability of the noninteracting $\mathbb{Z}$ topological classification. We then show that the inclusion of dynamical superconducting fluctuations remains harmless.

Without dynamical superconducting fluctuations, the classifying space for the normalized dynamical Dirac masses is that for the complex symmetry class A. Because (by definition) no protecing symmetry from the symmetry class is violated by such dynamical Dirac masses, dynamical Dirac masses are forbidden altogether. The noninteracting topological phase is stable to such interactions.

With dynamical superconducting fluctuations, normalized dynamical Dirac masses form the space of Dirac masses in the symmetry class D. The original single-particle Hamiltonian $\mathcal{H}^{(0)}_{\nu}$ that annihilates ν zero modes is extended to a Bogoliubov-de-Gennes (Nambu) single-particle Hamiltonian $\mathcal{H}^{(0)}_{\text{BdG}\,\nu}$ that commutes with ρ_3 and anticommutes with $\rho_1\mathsf{K}$ [recall Eq. (7.102)]. Here, ρ_0 is the unit 2×2 matrix and $\boldsymbol{\rho}$ are the Pauli matrices acting on the auxiliary particle-hole degrees of freedom. Boundary dynamical Dirac masses may then exist. However, they must anticommute with ρ_3, since no boundary Dirac mass that commutes with ρ_3 is allowed after

Table 8.18 *Homotopy groups of the dynamical boundary Dirac masses for the symmetry class DIII in seven-dimensional space*

D	$\pi_D(R_4)$	ν	S_{QNLSM}	S_{top}
0	$\mathbb{Z}$	1	S^0	✓
1	0			
2	0			
3	0			
4	$\mathbb{Z}$	2	S^4	✓
5	$\mathbb{Z}_2$	4	S^5	✓
6	$\mathbb{Z}_2$	8	S^6	✓
7	0			
8	$\mathbb{Z}$	16	S^8	✓
9	0			
10	0			
11	0			
12	$\mathbb{Z}$	32	S^{12}	–

Same as in **Table** 8.17 except for $d = 7$ for which $V'_{d-1} = R_4$ and $\nu_{\max} = 32$.

restricting $\mathcal{H}^{(0)}_{\mathrm{BdG}\,\nu}$ to the boundary as the condition for the presence of ν topologically protected boundary zero modes in the noninteractign limit. Upon integrating the boundary Dirac fermions, a QNLSM in $(d-1)+1$ space and time dimensions ensues. The target space of this QNLSM has to be closed under the action of a global U(1) gauge transformation. This is to say that a generic boundary dynamical Dirac mass must be of the form

$$\gamma' = [\cos(\theta)\rho_1 + \sin(\theta)\rho_2] \otimes M, \tag{8.156a}$$

with the $(r_{\min}\,\nu/2) \times (r_{\min}\,\nu/2)$ matrix M satisfying

$$M = M^\dagger = -M^*. \tag{8.156b}$$

We recall that $r_{\min}$ is the minimal rank of the Bogoliubov-de-Gennes Hamiltonian. The action on γ' of a global U(1) gauge transformation parameterized by the global phase α is simply the shift $\theta \mapsto \theta + \alpha$. If so, the target space of the QNLSM is homeomorphic to $\mathsf{S}^1 \times V_{\mathrm{BdG}\,\nu}$ whereby S^1 is parameterized by θ and $V_{\mathrm{BdG}\,\nu}$ is parameterized by M squaring to the unit matrix. For such a target space, we can always assign a topological term to account for the vortices supported by the parameter θ for the S^1 factor, as $\pi_1(\mathsf{S}^1) = \mathbb{Z}$. These vortices bind ν zero modes.

We conclude that the noninteracting topological classification $\mathbb{Z}$ for the symmetry class A in even dimensions survives strong interactions on the boundary provided the fermion-number conservation is neither explicitly nor spontaneously broken.

Table 8.19 *Homotopy groups of the dynamical boundary Dirac masses for the symmetry class DIII in 11-dimensional space*

D	$\pi_D(R_0)$	ν	S_{QNLSM}	S_{top}
0	$\mathbb{Z}$	1	S^0	✓
1	$\mathbb{Z}_2$	2	S^1	✓
2	$\mathbb{Z}_2$	4	S^2	✓
3	0			
4	$\mathbb{Z}$	8	S^4	✓
5	0			
6	0			
7	0			
8	$\mathbb{Z}$	16	S^8	✓
9	$\mathbb{Z}_2$	32	S^9	✓
10	$\mathbb{Z}_2$	64	S^{10}	✓
11	0			
12	$\mathbb{Z}$	128	S^{12}	✓
13	0			
14	0			
15	0			
16	$\mathbb{Z}$	256	S^{16}	–

Same as in **Table** 8.17 except for $d = 11$ for which $V'_{d-1} = R_0$ and $\nu_{\max} = 256$.

Proof for case (ii): First, we show the statement for (a) cases with dynamical Dirac masses that preserve the fermion-number U(1) symmetry. We then proceed to (b) cases with U(1)-breaking dynamical Dirac masses.

(a) We consider the massive Dirac Hamiltonian

$$\mathcal{H}^{(0)}(\boldsymbol{x}) = \sum_{j=1}^{d}(-\mathrm{i}\partial_j)\,\gamma_j + m(\boldsymbol{x})\,\gamma_0, \quad m(\boldsymbol{x}) \in \mathbb{R}, \tag{8.157}$$

obeying the TRS represented by $\mathcal{T}$ for classes AI and AII in $d = 4n$ for $n = 1, 2, \cdots$. Here, the Dirac matrices are of dimension $r \geq r_{\mathrm{min}}$. They obey the Clifford algebra $\{\gamma_\mu, \gamma_{\mu'}\} = 2\,\delta_{\mu\mu'}$ with $\mu, \mu' = 0, \cdots, d$. The Dirac matrices entering $\mathcal{H}^{(0)}(\boldsymbol{x})$ and the operator $\mathcal{T}$ that represents reversal of time can be used to define the following pair of Clifford algebras (recall Section 8.2).

For the symmetry class AI, reversal of time is represented by an element of the Clifford algebra $\mathcal{T}$ that squares to the unit matrix. Since the action of $\mathcal{T}$ includes complex conjugation K, $\mathcal{T}$ anticommutes with $\mathrm{i}\mathcal{T}$, while $\mathrm{i}\mathcal{T}$ squares to the unit matrix. Hence, $\mathcal{T}$ and $\mathrm{i}\mathcal{T}$, together with the gamma matrices $\gamma_1, \cdots, \gamma_d$, which satisfy the conditions $\gamma_1 = -\mathcal{T}\,\gamma_1\,\mathcal{T}^{-1}, \cdots, \gamma_d = -\mathcal{T}\,\gamma_d\,\mathcal{T}^{-1}$, are generators of the

Table 8.20 *Homotopy groups of the dynamical boundary Dirac masses for the symmetry class CII in one-dimensional space*

D	$\pi_D(R_6)$	ν	S_{QNLSM}	S_{top}
0	0			
1	0			
2	$\mathbb{Z}$	1	S^2	✓
3	$\mathbb{Z}_2$	2	S^3	–
4	$\mathbb{Z}_2$	4	S^4	–
5	0			
6	$\mathbb{Z}$	8	S^6	–
7	0			

Application of the Bott periodicity obeyed by the homotopy groups $\pi_D(V)$ for $D=0,1,\cdots$ of a given classifying space V'_{d-1} of dynamical boundary Dirac masses to deduce the reduction pattern $\mathbb{Z}\to\mathbb{Z}_{\nu_{\mathrm{max}}}$ for the symmetry class CII in dimensions $d=1$ for which $V'_{d-1}=R_6$ and $\nu_{\mathrm{max}}=2$. The column ν fixes the rank $r:=r_{\mathrm{min}}\,\nu$ of the Dirac Hamiltonian in the symmetry class CII. The fourth column gives the target manifold of the QNLSM with the action S_{QNLSM} that encodes the fermion-fermion interactions on the $(d-1)$-dimensional boundary. The fifth column indicates if a topological obstruction is present.

Table 8.21 *Homotopy groups of the dynamical boundary Dirac masses for the symmetry class CII in five-dimensional space*

D	$\pi_D(R_2)$	ν	S_{QNLSM}	S_{top}
0	$\mathbb{Z}_2$	2	S^0	✓
1	0			
2	$\mathbb{Z}$	4	S^2	✓
3	0			
4	0			
5	0			
6	$\mathbb{Z}$	8	S^6	✓
7	$\mathbb{Z}_2$	16	S^7	–

Same as **Table** 8.20 except with $d=5$ for which $V'_{d-1}=R_2$ and $\nu_{\mathrm{max}}=16$.

Clifford algebra. On the other hand, the Dirac mass matrix $\mathrm{i}\gamma_0$ is chosen as the generator that squares to minus the unit matrix, for $\gamma_0=+\mathcal{T}\,\gamma_0\,\mathcal{T}^{-1}$. We arrive at the Clifford algebra

$$C\ell_{1,2+d}:=\mathrm{span}_{C\ell}\,\{J\gamma_0;\mathcal{T},J\mathcal{T},\gamma_1,\cdots,\gamma_d\} \tag{8.158}$$

Table 8.22 *Homotopy groups of the dynamical boundary Dirac masses for the symmetry class CII in nine-dimensional space*

D	$\pi_D(R_6)$	ν	S_{QNLSM}	S_{top}
0	0			
1	0			
2	$\mathbb{Z}$	1	S^2	✓
3	$\mathbb{Z}_2$	2	S^3	✓
4	$\mathbb{Z}_2$	4	S^4	✓
5	0			
6	$\mathbb{Z}$	8	S^6	✓
7	0			
8	0			
9	0			
10	$\mathbb{Z}$	16	S^{10}	✓
11	$\mathbb{Z}_2$	32	S^{11}	–

Same as in **Table** 8.20 except with $d = 9$ for which $V'_{d-1} = R_6$ and $\nu_{\text{max}} = 32$.

for the symmetry class AI, where J is the generator that represents the imaginary unit "i" and satisfies the relations $J^2 = -1$ and $\{\mathcal{T}, J\} = 0$.[111]

For the symmetry class AII, reversal of time is represented by an element of the Clifford algebra $\mathcal{T}$ that squares to minus the unit matrix. It and $\mathrm{i}\mathcal{T}$ enter on equal footing with $\mathrm{i}\gamma_0$ as the generators that square to minus the unit matrix, for $\gamma_0 = +\mathcal{T}\,\gamma_0\,\mathcal{T}^{-1}$. On the other hand, the gamma matrices $\gamma_1, \cdots, \gamma_d$, which satisfy the conditions $\gamma_1 = -\mathcal{T}\,\gamma_1\,\mathcal{T}^{-1}, \cdots, \gamma_d = -\mathcal{T}\,\gamma_d\,\mathcal{T}^{-1}$, are the generators that square to the unit matrix in the Clifford algebra. We arrive at the Clifford algebra

$$C\ell_{3,d} := \operatorname{span}_{C\ell}\{J\gamma_0, \mathcal{T}, J\mathcal{T}; \gamma_1, \cdots, \gamma_d\} \tag{8.159}$$

for the symmetry class AII.

In both symmetry classes the choice of γ_0 is unique, up to a sign, as a consequence of the fact that the zeroth homotopy groups of the classifying spaces for the symmetry classes AI and AII is $\mathbb{Z}$ in $4n$ dimensions. In other words, no other Dirac mass matrix that is invariant under reversal of time anticommutes with γ_0 (recall Section 8.2). This leaves open the possibility that the Clifford algebras (8.158) and (8.159) for $d = 4n$ could admit the addition of a generator γ'_0 that anticommutes with $\mathcal{H}^{(0)}$ and is odd under reversal of time, $\gamma'_0 = -\mathcal{T}\,\gamma'_0\,\mathcal{T}^{-1}$. If so, the choice of γ_1 to γ_d in the Clifford algebras (8.158) and (8.159) would not be unique in an

[111] The notation $C\ell_{p,q} = \operatorname{span}_{C\ell}\{e_1, \cdots, e_p; e_{p+1}, \cdots, e_{p+q}\}$ for the real Clifford algebra is the span over the field $\mathbb{R}$ of all possible products obtained from the $p+q$ pairwise anticommuting generators $(e_1, \cdots, e_{p+q})$ satisfying the conditions $e_j^2 = -1$ for $j = 1, \cdots, p$ and $e_j^2 = +1$ for $j = p+1, \cdots, p+q$.

Table 8.23 *Homotopy groups of the dynamical boundary Dirac masses for the symmetry class CI in three-dimensional space*

D	$\pi_D(R_4)$	ν	S_{QNLSM}	S_{top}
0	$\mathbb{Z}$	1	S^0	✓
1	0			
2	0			
3	0			
4	$\mathbb{Z}$	2	S^4	✓
5	$\mathbb{Z}_2$	4	S^5	–
6	$\mathbb{Z}_2$	8	S^6	–
7	0			

Application of the Bott periodicity obeyed by the homotopy groups $\pi_D(V)$ for $D = 0, 1, \cdots$ of a given classifying space V'_{d-1} of dynamical boundary Dirac masses to deduce the reduction pattern $\mathbb{Z} \to \mathbb{Z}_{\nu_{\text{max}}}$ for the symmetry class CI in dimensions $d = 3$ for which $V'_{d-1} = R_4$ and $\nu_{\text{max}} = 4$. The column ν fixes the rank $r := r_{\text{min}}\,\nu$ of the Dirac Hamiltonian in the symmetry class CI. The fourth column gives the target manifold of the QNLSM with the action S_{QNLSM} that encodes the fermion-fermion interactions on the $(d-1)$-dimensional boundary. The fifth column indicates if a topological obstruction is present.

Table 8.24 *Homotopy groups of the dynamical boundary Dirac masses for the symmetry class CI in seven-dimensional space*

D	$\pi_D(R_0)$	ν	S_{QNLSM}	S_{top}
0	$\mathbb{Z}$	1	S^0	✓
1	$\mathbb{Z}_2$	2	S^1	✓
2	$\mathbb{Z}_2$	4	S^2	✓
3	0			
4	$\mathbb{Z}$	8	S^4	✓
5	0			
6	0			
7	0			
8	$\mathbb{Z}$	16	S^8	✓
9	$\mathbb{Z}_2$	32	S^9	–

Same as in **Table** 8.23 except with $d = 7$ for which $V'_{d-1} = R_0$ and $\nu_{\text{max}} = 32$.

uncountable (in a continuous) way.[112] The existence of γ'_0 is thus tied to the task of parameterizing in a continuous way the representation of the generator (e.g., γ_d)

[112] For example, the matrix γ_1 could then be replaced by the matrix $\cos\theta\,\gamma_1 + \sin\theta\gamma'_0$ for any $0 \leq \theta < 2\pi$.

Table 8.25 *Homotopy groups of the dynamical boundary Dirac masses for the symmetry class CI in 11-dimensional space*

D	$\pi_D(R_4)$	ν	S_{QNLSM}	S_{top}
0	$\mathbb{Z}$	1	S^0	✓
1	0			
2	0			
3	0			
4	$\mathbb{Z}$	2	S^4	✓
5	$\mathbb{Z}_2$	4	S^5	✓
6	$\mathbb{Z}_2$	8	S^6	✓
7	0			
8	$\mathbb{Z}$	16	S^8	✓
9	0			
10	0			
11	0			
12	$\mathbb{Z}$	32	S^{12}	✓
13	$\mathbb{Z}_2$	64	S^{13}	–

Same as in **Table** 8.23 except with $d = 11$ for which $V'_{d-1} = R_4$ and $\nu_{\max} = 64$.

present in $C\ell_{p,q+1}$ but absent in $C\ell_{p,q}$ applied to the cases $(p,q) = (1, 4n+1)$ and $(p,q) = (3, 4n-1)$ for the $4n$-dimensional symmetry classes AI and AII, respectively.[113] Both tasks are denoted by the extension problem of Clifford algebras

$$C\ell_{p,q} \to C\ell_{p,q+1}, \tag{8.160}$$

with the classifying spaces

$$R_{q-p} = \begin{cases} R_{4n}, & (p,q) = (1, 4n+1), \\ R_{4n-4}, & (p,q) = (3, 4n-1), \end{cases} \tag{8.161}$$

as solutions for the set of representations of possible γ'_0 in the symmetry classes AI and AII, respectively. Hereto, it is the zeroth homotopy group of the classifying space R_{p-q} that seals the fate of the existence of γ'_0. As $\pi_0(R_{4n}) = \pi_0(R_{4n-4}) = \mathbb{Z}$, it follows that γ'_0 does not exist, that is, no dynamical Dirac mass that breaks the TRS symmetry but preserves the global U(1) gauge symmetry is permissible for the symmetry classes AI and AII when $d = 4n$.

113 These tasks correspond to the following classification problem. How does one parameterize the generators of $C\ell_{p,q+1}$ that enter the kinetic contribution to the Dirac Hamiltonian? This classification problem is thus distinct from the one in which one seeks to parameterize the generators that enter the Dirac Hamiltonian as a Dirac mass.

(b) After having established the absence of U(1)-preserving dynamical Dirac masses, the stability analysis in the presence of dynamical superconducting fluctuations for the symmetry classes AI and AII is the same as that for the symmetry class A. The boundary dynamical Dirac mass takes the form (8.156). The target space of the QNLSM is homeomorphic to $S^1 \times V_{\mathrm{BdG}\,\nu}$ since it has to be closed under the action of a global U(1) gauge transformation. Vortices that bind boundary zero modes originate from the S^1 manifold.

Proof for case (iii): The symmetry classes D and C for $d = 2, 6 \pmod 8$ do not support dynamical Dirac masses along the boundary, because the PHS is kept as a fundamental symmetry. Their noninteracting topological classification $\mathbb{Z}$ survives strong interactions on the boundary provided the PHS is neither explicitly nor spontaneously broken.

8.4.7.3 The Case of Odd Dimensions

The topological classification $\mathbb{Z}$ of noninteracting strong topological insulators or superconductors in odd dimensions of space is reduced to the coarser classification $\mathbb{Z}_{\nu_{\mathrm{max}}}$ with ν_{max} an integer:

$$\mathbb{Z} \to \mathbb{Z}_{\nu_{\mathrm{max}}}. \tag{8.162a}$$

The label "max" stands here for maximum. The task at hand is thus to compute the integer ν_{max}. Computing ν_{max} proceeds with the following algorithm (see **Tables** 8.14–8.25).

Step 1: Choose any one of the ten Altland-Zirnbauer symmetry classes from **Table** 8.6.

Step 2: Choose any odd dimension d for which the zero-th homotopy group of the classifying space of the chosen symmetry class V_d is the set of integers $[\pi_0(V_d) = \mathbb{Z}]$. This step restricts the symmetry classes to the complex symmetry class AIII and the real symmetry classes BDI, DIII, CII, and CI.

Step 3: Identify the parent symmetry class and its classifying space V'_d that follows if CHS is broken for the complex symmetry class AIII or if TRS is broken for the real symmetry classes. This step restricts the parent symmetry classes to the complex symmetry class A if the symmetry class AIII is interpreted as realizing an insulator,[114] the real symmetry class D (A) if the symmetry classes BDI and DIII are interpreted as superconductors (insulators), and the real symmetry class C if the symmetry classes CII and CI are interpreted as superconductors.

Step 4: Assign the minimal value

$$\nu_{\mathrm{min}} := \begin{cases} 1, & \pi_0(V'_d) = 0, \\ \\ 2, & \pi_0(V'_d) \neq 0, \end{cases} \tag{8.162b}$$

[114] Gapping the boundary states through superconducting instabilities is interpreted as spontaneous symmetry breaking of the fermion-number U(1) protecting symmetry.

if the zero-th homotopy group of V'_d is trivial or nontrivial, respectively. The index ν_{min} defines the minimal number of boundary states for which a boundary dynamical mass matrix is permissible.[115]
Step 5: Identify the classifying space V'_{d-1} that determines the dynamical Dirac mass matrices induced by the fermion-fermion interactions on the boundary.
Step 6: Construct a table with lines labeled by the integer $D = 0, 1, 2, \cdots$. The first column gives the order D of the homotopy group $\pi_D(V'_{d-1})$ given in the second column. The third column is the number ν of boundary zero modes in the selected symmetry class. Enter the value ν_{min} in the third column for the smallest value of D for which $\pi_D(V'_{d-1})$ is nontrivial. The value of ν is then doubled for each successive line with $\pi_D(V'_{d-1})$ nontrivial. The fourth column denotes the target space of the QNLSM with the action S_{QNLSM} defined by integrating out all the boundary Dirac fermions when coupled to $(D+1)$ real-valued bosonic fields, each one of which couples to a Dirac mass matrix from a $(D+1)$-tuplet of pairwise anticommuting Dirac mass matrices allowed on the boundary by the parent symmetry class. The fifth column indicates when a topological term S_{top} can be added to the action S_{QNLSM}.[116]
Step 7: Let n_{WZ} be the number of lines with $\pi_D(V'_{d-1})$ nontrivial when D takes the values $D = 0, 1, \cdots, d+1$. It then follows that

$$\nu_{\mathrm{max}} = \nu_{\mathrm{min}} \times 2^{n_{\mathrm{WZ}}}. \tag{8.162c}$$

For the complex symmetry class AIII in dimension $d = 2n+1$ with $n = 0, 1, 2, \cdots$, the reduction pattern induced by the fermion-fermion interactions is

$$\mathbb{Z} \to \mathbb{Z}_{2^{n+2}}. \tag{8.163}$$

By making use of the eightfold Bott periodicity, one verifies that the reduction patterns are

(8.164)

	$d = 8n+1$	$d = 8n+3$	$d = 8n+5$	$d = 8n+7$
BDI	$\mathbb{Z} \to \mathbb{Z}_{2^{4n+3}}$	–	$\mathbb{Z} \to \mathbb{Z}_{2^{4n+4}}$	–
DIII	–	$\mathbb{Z} \to \mathbb{Z}_{2^{4n+4}}$	–	$\mathbb{Z} \to \mathbb{Z}_{2^{4n+5}}$
CII	$\mathbb{Z} \to \mathbb{Z}_{2^{4n+1}}$	–	$\mathbb{Z} \to \mathbb{Z}_{2^{4n+4}}$	–
CI	–	$\mathbb{Z} \to \mathbb{Z}_{2^{4n+2}}$	–	$\mathbb{Z} \to \mathbb{Z}_{2^{4n+5}}$

115 When $\pi_0(V'_d) = 0$, the parent symmetry class is topologically trivial, that is it is always possible to gap with dynamical masses the boundary states whose existence derive from $\pi_0(V_d) = 0$. When $\pi_0(V'_d) \neq 0$, the parent symmetry class is topologically nontrivial, that is one must double the number of boundary states in order to gap with dynamical masses the boundary states whose existence derive from $\pi_0(V_d) = 0$.

116 When identifying nontrivial homotopy groups and topological terms, we assume that the homotopy groups $\pi_D(V_\nu)$ for the space V_ν of $\nu \times \nu$ normalized Dirac mass matrices with the relevant finite ν are nontrivial whenever $\pi_D(R_q)$ is nontrivial. This is valid when ν is larger than a certain value determined by D. Here, R_q is the space of normalized Dirac mass matrices in the direct limit $\nu \to \infty$, and $\pi_D(R_q)$ obeys the Bott periodicity and is known from the mathematic literature. No proof was given for the validity of this assumption in all dimensions. It was merely observed that it always holds in one, two, and three dimensions of space.

if we interpret the symmetry classes BDI, DIII, CII, and CI as superconductors, or

	$d = 8n+1$	$d = 8n+3$	$d = 8n+5$	$d = 8n+7$
BDI	$\mathbb{Z} \to \mathbb{Z}_{2^{4n+2}}$	–	$\mathbb{Z} \to \mathbb{Z}_{2^{4n+3}}$	–
CII	$\mathbb{Z} \to \mathbb{Z}_{2^{4n+1}}$	–	$\mathbb{Z} \to \mathbb{Z}_{2^{4n+4}}$	–

(8.165)

if we interpret the symmetry classes BDI and CII as insulators. We have thus found two different patterns for the reduction of the topological classification of the symmetry class BDI depending on these two interpretations, as we have observed for $d = 1$ in Section 8.4.4.

8.5 Comments and Outlook

A noninteracting strong topological insulator or superconductor realizes a noninteracting fermionic invertible topological phase of matter that is protected by on-site symmetries, whereby the quantum-mechanical representations of the on-site symmetries act locally and faithfully.[117] In the absence of disorder, its ground state is nondegenerate with a gap Δ to all excitations when d-dimensional space is any compact manifold without boundary, while its ground state is topologically non trivial in that it supports gapless boundary states that are localized along the connected components of the boundary in the limit in which all pairs of disconnected components of the boundary are infinitely separated. In the presence of static and short-range correlated (in space) disorder with the characteristic strength $V \ll \Delta$, these gapless states remain delocalized even though the characteristic disorder strength V can be much larger than their characteristic kinetic energy measured relative to the Fermi energy. These boundary states evade Anderson localization along each connected component of the boundary because backscattering from the static and short-range correlated disorder is not permissible on it. Backscattering is only permissible between pairs of disconnected components of the boundary, a process that is suppressed if $V/\Delta \ll 1$ is held fixed as the pairwise separations between the disconnected components of the boundary is taken to infinity. The gapless boundary excitations are fractional parts of the microscopic fermions. Correspondingly, the noninteracting boundary Hamiltonians describing the boundary states of d-dimensional strong topological insulators and superconductors that are localized on a single connected component of the boundary are anomalous in that they cannot be regularized on a $(d-1)$-dimensional lattice, while retaining local and faithful representations of all their protecting on-site symmetries.

As was stressed by Haldane[118] in the context of the edge states for the quantum Hall effect, the stability of the gapless boundary states to both static and

[117] Faithful is here a shorthand for an injective group homomorphism.
[118] F. D. M. Haldane, Phys. Rev. Lett. **74**(11), 2090–2093 (1995). [Haldane (1995)].

short-range correlated disorder and many-body fermionic interactions must hold for both static and short-range correlated disorder and local interactions that are strong, when considering the boundary as an isolated quantum system. It is not sufficient to establish that the interactions are irrelevant in the sense of the renormalization group. It must be shown that the robustness to disorder or interactions holds non-perturbatively in the strength of the disorder or interactions.

From the point of view of Dirac fermions, a noninteracting strong topological insulator or superconductor in d-dimensional space can be modeled by a massive Dirac Hamiltonian such that the on-site symmetries are represented locally and faithfully and there is a special Dirac mass that binds gapless boundary states to the connected $(d-1)$-dimensional submanifold of space where this Dirac mass vanishes. The existence of this special Dirac mass is guaranteed if the classifying space of normalized Dirac masses is not path-connected. To study the robustness of these boundary states to strong local interactions, we have introduced the concept of dynamical masses (**Definition** 8.3) associated with the Hubbard-Stratonovich decoupling of local quartic interactions. If any one of these dynamical masses acquires an expectation value, it would break spontaneously the microscopic on-site symmetries. For a sufficiently strong attractive interaction, spontaneous symmetry breaking of at least one of the on-site symmetries is the expected outcome. Integrating the boundary fermions delivers a QNLSM, owing to the fine-tuned choice made for the quartic interactions. We then conjectured that, as long as each nontrivial homotopy group of the classifying space for the dynamical masses can be intrepreted as the presence of an anomalous term in the QNLSM compatible with local equations of motion, strong interactions either break spontaneously the microscopic on-site symmetries or fail to open a spectral gap. The constraint that imposes locality of the equations of motion for the QNLSM is the reason why the tenfold classification of noninteracting strong topological insulators or superconductors can be too fine, in which case it reduces to a smaller set of equivalence classes. However, the weaknesses of this approach are the following:

1. The local fermionic quartic interaction is fine-tuned. Fine-tuning is sufficient to predict the reduction of the noninteracting topological classification, but not sufficient to establish that the reduced topological classification is stable to any other strong local interaction.
2. The QNLSM is derived assuming weak fluctuations about the symmetry-breaking mean field, but applied to the regime of strong fluctuations.
3. The anomalous QNLSM has symmetries that are not present microscopically, namely, Poincaré invariance of spacetime and an internal $\mathrm{O}\left(2^{N(\nu)-1}\right)$ symmetry.[119]

[119] A priori, one cannot tell if the anomaly of the QNLSM originates from the on-site symmetries of the microscopic model or from the fine-tuned continuous $\mathrm{O}\left(2^{N(\nu)-1}\right)$ symmetry. We have taken the point of view that the restoration of the symmetry group $\mathrm{O}\left(2^{N(\nu)-1}\right)$ by the strong

4. Strong interactions can turn a noninteracting Hamiltonian in d-dimensional space with a topologically trivial gapped ground state into one with a nontrivial invertible topological phase 96.

We adopt the point of view that making use of the QNLSM is only the common thread used to decide if there are topological defects of the dynamical masses in spacetime whose dynamics are obstructions to gaping the boundary modes in the periodic table of noninteracting topological insulators or superconductors. We also observe that a complete classification in one-dimensional space of fermionic invertible topological phases of matter was derived in Chapter 6 by studying systematically anomalous representations (that is projective representations) of the on-site symmetries on the zero-dimensional boundaries, without any reference to the one-dimensional local fermionic Hamiltonian other than to its on-site symmetries. This suggests that the reduction by interactions in **Table** 8.6 could perhaps be deduced solely from revisiting index theorems for noninteracting Dirac fermions protected by on-site symmetries.

Instead of deducing the entries $\mathbb{Z}$ and $\mathbb{Z}_2$ in **Table** 8.3 from index theorems predicting the number of zero modes bound to a domain wall in the mass of the Dirac Hamiltonian for each of the ten Altland-Zirnbauer symmetry classes, the noninteracting massive Dirac Hamiltonian for each of the ten Altland-Zirnbauer symmetry classes is bosonized in the presence of a local background field that transforms nontrivially under the on-site protecting symmetries or local coordinate transformations when total energy is the only conserved observable available aside from fermion parity. The idea is that for each of the ten Altland-Zirnbauer symmetry classes, the bosonized theory can be organized into different equivalence classes with the nontrivial ones encoding an anomalous realization of the protecting symmetries and that these equivalence classes deliver the reduction by interactions in **Table** 8.6. 105 [120]

To outline this program in more detail, we begin with an observation followed by a Gedanken theory.

We argued in Section 5.5.4 that the nearest-neighbor antiferromagnetic quantum spin-1 Heisenberg Hamiltonian (5.225) is equivalent from a topological point of view to a perturbation of the Majorana 4 chain (5.198). This perturbation lifts the 16-fold degeneracy of the open Majorana 4 chain (5.198) irrespective of its length L to the minimal fourfold degeneracy that is protected by fermion parity and time-reversal symmetry in the thermodynamic limit $L \to \infty$. The dependence on β and L of the partition function $Z_{\mathrm{OBC}}(\beta, L)$ at the inverse temperature β for a chain of length L obeying open boundary conditions is given by

fluctuations of the order parameter through the anomalous terms is to be ascribed to the on-site symmetries, as the symmetry broken phase of the QNLSM necessarily breaks spontaneously at least one microscopic on-site symmetry.

[120] E. Witten, Rev. Mod. Phys. **88**(3), 035001 (2016). [Witten (2016)].

$$Z_{\mathrm{OBC}}(\beta,L)=1+3\,\exp\left(-c\,\beta\,J\,e^{-L/\xi}\right)+\cdots. \tag{8.166a}$$

Here, $0<\xi<\infty$ is the correlation length at zero temperature,[121] J is the antiferromagnetic Heisenberg exchange coupling, and c is a nonuniversal numerical constant. The first term on the right-hand side stems from the ground state with vanishing total spin for any finite L. The second term on the right-hand side stems from hybridizing the boundary quantum spin-1/2 localized on the left boundary with that localized on the right boundary into a degenerate spin-1 triplet with the nonvanishing excitation energy $c\,J\,e^{-L/\xi}$ prior to taking the thermodynamic limit. It follows that the partition function (8.166a) displays an essential singularity at

$$\beta=L=\infty. \tag{8.166b}$$

In particular, we find that

$$\lim_{L\to\infty}\lim_{\beta\to\infty}Z_{\mathrm{OBC}}(\beta,L)=1,\qquad \lim_{\beta\to\infty}\lim_{L\to\infty}Z_{\mathrm{OBC}}(\beta,L)=4. \tag{8.166c}$$

This example is a rule, not a particularity of the nearest-neighbor antiferromagnetic quantum spin-1 Heisenberg chain.

Property 8.4 The partition function with open boundary conditions of any nontrivial invertible topological phase of matter displays an essential singularity at zero temperature and in the thermodynamic limit.

The Gedanken theory is an abstraction of the observation that the quantization of the Hall conductivity in integer multiples of e^2/h is a manifestation of an underlying topological field theory.[122] Given is a quadratic form

$$\widehat{H}_{\mathrm{AZ}:A;\mathcal{M}_{\mathrm{no-bd}}}=\left(\widehat{H}_{\mathrm{AZ}:A;\mathcal{M}_{\mathrm{no-bd}}}\right)^{\dagger} \tag{8.167}$$

presented in terms of massive Dirac fermions in some background field A, some metric $\mathfrak{g}$ if space is some d-dimensional compact manifold $\mathcal{M}_{\mathrm{no-bd}}$ without boundary, and some uniform mass m for any one of the ten Altland-Zirnbauer symmetry classes that we denote with AZ. By definition of an invertible topological phase of matter, the many-body ground state $|\Psi_{\mathrm{AZ}:A;\mathcal{M}_{\mathrm{no-bd}}}\rangle$ with the energy eigenvalue $E_{\mathrm{AZ}:A;\mathcal{M}_{\mathrm{no-bd}}}$ of Hamiltonian (8.167) is gapped and nondegenerate. It is obtained by filling all the single-particle states with negative energies of the single-particle Dirac Hamiltonian, thereby defining the Dirac sea. Denote with $\mathcal{M}_{\mathrm{bd}}$ the manifold

[121] The correlation length is given by

$$\xi=c'\,\frac{\hbar\,v_{\mathrm{sw}}}{\Delta}$$

with c' another nonuniversal numerical constant and v_{sw} the spin-wave velocity of antiferromagnetic magnons. The minima of the energy dispersion of the magnons is at the wave numbers $\pm\pi$, an energy $0<\Delta<\infty$ above that of the nondegenerate ground state if periodic boundary conditions are imposed.

[122] X.-G. Wen, *Quantum Field Theory of Many-Body Systems: From the Origin of Sound to an Origin of Light and Electrons,* Oxford University Press, New York 2007. [Wen (2007)].

that follows from "cutting" open $\mathcal{M}_{\text{no-bd}}$ such that (i) $\mathcal{M}_{\text{bd}}$ is path -onnected and (ii) has two disconnected boundaries, each of which is path-connected, that are separated by the characteristic distance L.[123]

On the one hand, define the partition function

$$Z_{\text{AZ}:A;\mathcal{M}_{\text{no-bd}}}(\beta,L) := \operatorname{Tr}\left[\widehat{P}_A\, e^{-\beta\left(\widehat{H}_{\text{AZ}:A;\mathcal{M}_{\text{no-bd}}}-E_{\text{AZ}:A;\mathcal{M}_{\text{no-bd}}}\right)}\right]. \tag{8.168a}$$

Here, the unitary operator $\widehat{P}_A$ is needed to impose "twisted" boundary conditions for which the partition function becomes a complex-valued function of the complex variable $\beta+\mathrm{i}L$.[124,125,126,127] We assume that there exists an $A'\neq A$ such that

$$\widehat{H}_{\text{AZ}:A;\mathcal{M}_{\text{no-bd}}} = \widehat{U}\,\widehat{H}_{\text{AZ}:A';\mathcal{M}_{\text{no-bd}}}\,\widehat{U}^{\dagger}, \tag{8.168b}$$

$$\widehat{P}_A = \widehat{U}\,\widehat{P}_{A'}\,\widehat{U}^{\dagger}, \tag{8.168c}$$

where the unitary operator $\widehat{U}$ is related to one of the protecting symmetries that can be broken spontaneously, say the U(1) global gauge symmetry if we are considering the symmetry class A and $d=2$ (as is suited for the study of the integer quantum Hall effect), or to a local coordinate transformation when the protecting symmetry cannot be broken spontaneously,[128,129] as is the case with fermion parity in the symmetry class D. Under the assumption that the functional trace is cyclic, we then have the symmetry

$$Z_{\text{AZ}:A;\mathcal{M}_{\text{no-bd}}}(\beta,L) = Z_{\text{AZ}:A';\mathcal{M}_{\text{no-bd}}}(\beta,L). \tag{8.168d}$$

By definition of an invertible topological phase of matter and if many-body energies are measured relative to that of the ground state, we can safely take the limits of $\beta, L\to\infty$ (without paying attention to the order in which the two limits are taken) to find that

$$|Z_{\text{AZ}:A;\mathcal{M}_{\text{no-bd}}}| := \lim_{\substack{\beta\to\infty\\ L\to\infty}} |Z_{\text{AZ}:A;\mathcal{M}_{\text{no-bd}}}(\beta,L)| = 1. \tag{8.168e}$$

On the other hand, define the partition function

$$Z_{\text{AZ};A;\mathcal{M}_{\text{bd}}}(\beta,L) := \operatorname{Tr}\left[\widehat{P}_A\, e^{-\beta\left(\widehat{H}_{\text{AZ};A;\mathcal{M}_{\text{bd}}}-E_{\text{AZ};A;\mathcal{M}_{\text{bd}}}\right)}\right] \tag{8.169a}$$

[123] For the symmetry class A in $d=2$, we have in mind a 2-torus $\mathrm{S}^1\times\mathrm{S}^1$ for $\mathcal{M}_{\text{no-bd}}$ and the surface of a cylinder for $\mathcal{M}_{\text{bd}}$.

[124] C.-T. Hsieh, O. M. Sule, G. Y. Cho, S. Ryu, and R. G. Leigh, Phys. Rev. B **90**(16), 165134 (2014). [Hsieh et al. (2014)].

[125] G. Y. Cho, C.-T. Hsieh, T. Morimoto, and S. Ryu, Phys. Rev. B **91**(19), 195142 (2015). [Cho et al. (2015)].

[126] C.-T. Hsieh, G. Y. Cho, and S. Ryu, Phys. Rev. B **93**(7), 075135 (2016). [Hsieh et al. (2016a)].

[127] H. Shapourian, K. Shiozaki, and S. Ryu, Phys. Rev. Lett. **118**(21), 216402 (2017). [Shapourian et al. (2017)].

[128] S. Ryu, J. E. Moore, and A. W. W. Ludwig, Phys. Rev. B **85**(4), 045104 (2012). [Ryu et al. (2012)].

[129] S. Ryu and S.-C. Zhang, Phys. Rev. B **85**(24), 245132 (2012). [Ryu and Zhang (2012)].

under the assumption that the ground state is nondegenerate for any finite β and L. This partition function is thus a well-behaved function for finite β and L. However, whereas

$$\lim_{\substack{\beta\to\infty\\ L\to\infty}} |Z_{\mathrm{AZ};A;\mathcal{M}_{\mathrm{bd}}}(\beta,L)| = 1 \tag{8.169b}$$

if $|\Psi_{\mathrm{AZ}:A;\mathcal{M}_{\mathrm{bd}}}\rangle$ remains nondegenerate as $L\to\infty$, the limit $\beta,L\to\infty$ of Eq. (8.169a) is ill-defined if $|\Psi_{\mathrm{AZ}:A;\mathcal{M}_{\mathrm{bd}}}\rangle$ becomes degenerate as $L\to\infty$. For example, the partition function of noninteracting fermions is the fermion determinant Det $(\partial_\tau + \mathcal{H})$ with $\mathcal{H}$ the single-particle Hamiltonian acting on the single-particle Hilbert space that is selected by the twisted boundary conditions in space and imaginary time [see Eq. (3.335)]. In the limit $\beta=\infty$ with antiperiodic boundary conditions in imaginary time, zero modes of $\mathcal{H}$ are zero modes of the non-Hermitean kernel $(\partial_\tau + \mathcal{H})$ and we conclude that Det $(\partial_\tau + \mathcal{H})$ vanishes if the limit $L\to\infty$ is taken before the limit $\beta\to\infty$.

There are two difficulties with the computation of the partition function for Dirac fermions (8.168a). The first one is already present for the trivial background $A=0$ and a flat metric (for which all the eigenvalues of the Dirac Hamiltonian are known in closed form) as the spectrum of the Dirac Hamiltonian is unbounded. The second one is that, for a generic A, the eigenvalues of the Dirac Hamiltonian are not known in closed form. The first difficulty requires a regularization procedure for the product of infinitely many complex-valued eigenvalues whose magnitudes are unbounded. The second difficulty requires some perturbative expansion such as the gradient expansion of fermionic determinants from Chapter 3. The resulting approximation $Z^{\mathrm{eff}}_{\mathrm{AZ}:A;\mathcal{M}_{\mathrm{no-bd}}}$ to $Z_{\mathrm{AZ}:A;\mathcal{M}_{\mathrm{no-bd}}}$ must be a conserving approximation in that it obeys the symmetry

$$Z^{\mathrm{eff}}_{\mathrm{AZ}:A;\mathcal{M}_{\mathrm{no-bd}}} = Z^{\mathrm{eff}}_{\mathrm{AZ}:A';\mathcal{M}_{\mathrm{no-bd}}}. \tag{8.170a}$$

Here, $Z^{\mathrm{eff}}_{\mathrm{AZ}:A;\mathcal{M}_{\mathrm{no-bd}}}$, owing to the twisted boundary conditions, can take complex values subject to the condition that the partition function

$$Z^{\mathrm{tft}}_{\mathrm{AZ}:A;\mathcal{M}_{\mathrm{no-bd}}} := \exp\left(\mathrm{i}\arg\left(Z^{\mathrm{eff}}_{\mathrm{AZ}:A;\mathcal{M}_{\mathrm{no-bd}}}\right)\right) \equiv e^{\mathrm{i}I_{\mathrm{AZ}:A;\mathcal{M}_{\mathrm{no-bd}}}} \tag{8.170b}$$

defines a topological field theory (tft).

Definition 8.5 (Topological field theory) A topological field theory (tft) is a local classical field theory such that (i) it is Poincaré symmetric for the Minkowski metric and (ii) its Hamiltonian density vanishes.

Equipped with the functional dependence on A of $Z^{\mathrm{tft}}_{\mathrm{AZ}:A;\mathcal{M}_{\mathrm{no-bd}}}$, we define

$$Z^{\mathrm{tft}}_{\mathrm{AZ}:A;\mathcal{M}_{\mathrm{bd}}} := \exp\left(\mathrm{i}\arg\left(Z^{\mathrm{eff}}_{\mathrm{AZ}:A;\mathcal{M}_{\mathrm{bd}}}\right)\right) \equiv e^{\mathrm{i}I_{\mathrm{AZ}:A;\mathcal{M}_{\mathrm{bd}}}} \tag{8.170c}$$

by substituting $\mathcal{M}_{\text{no-bd}}$ with $\mathcal{M}_{\text{bd}}$ on the left-hand side of Eq. (8.170b). Does Eq. (8.170a) also hold for $Z^{\text{tft}}_{\text{AZ}:A;\mathcal{M}_{\text{bd}}}$. The answer is

$$I_{\text{AZ}:A;\mathcal{M}_{\text{bd}}} = I_{\text{AZ}:A';\mathcal{M}_{\text{bd}}} = 0, \qquad \text{if Eq. (8.169b) holds,} \tag{8.171a}$$

while

$$I_{\text{AZ}:A;\mathcal{M}_{\text{bd}}} \neq I_{\text{AZ}:A';\mathcal{M}_{\text{bd}}}, \qquad \text{otherwise.} \tag{8.171b}$$

The breakdown of the symmetry (8.170a) by Eq. (8.171b) results from the essential singularity of the partition function (8.169a) at $\beta = \infty$ and $L = \infty$ if the nondegenerate many-body ground state for finite L becomes degenerate in the limit $L \to \infty$. It is an example of a quantum anomaly. Had we computed for the $(d-1)$-dimensional gapless boundary Hamiltonian its partition function on a connected boundary of $\mathcal{M}_{\text{bd}}$ for the $(d-1)$-dimensional counterparts of the background A, we would have found a quantum anomaly that matches the quantum anomaly (8.171b).[130]

The fact that the quantum anomaly (8.171b) is the phase of an unimodular partition function is compatible with the stacking rules from Chapters 5 and 6. Indeed, multiplying the partition function (8.170c) with its complex-conjugate,

$$Z^{\text{tft}}_{\text{AZ}:A;\mathcal{M}_{\text{bd}}} \times \left(Z^{\text{tft}}_{\text{AZ}:A;\mathcal{M}_{\text{bd}}}\right)^{*} = e^{\mathrm{i}I_{\text{AZ}:A;\mathcal{M}_{\text{bd}}} - \mathrm{i}I_{\text{AZ}:A;\mathcal{M}_{\text{bd}}}} = 1 \tag{8.172}$$

is equivalent to taking the direct sum of two massive Dirac Hamiltonian in such a way that the resulting massive Dirac Hamiltonian is necessarily topologically trivial. This is to say that any topologically nontrivial massive Dirac Hamiltonian has an "inverse" with respect to the stacking rules. The topological field theories resulting from computing the partition functions of noninteracting massive Dirac fermions are examples of invertible topological field theories.

Definition 8.6 (Invertible topological field theories) Invertible topological field theories are classical field theories such that (i) they are Poincaré symmetric for the Minkowski metric, (ii) their Hamiltonian densities vanish, and (iii) heir partition functions are complex numbers of unit magnitude for any compact manifold (with or without genus or boundaries).

Remarkably, the phases of the partition functions of invertible topological field theories fall into equivalence classes labeled by the values of topological invariants that are attached to the manifolds $\mathcal{M}_{\text{no-bd}}$ associated with the protecting symmetries. For the integer quantum Hall effect in two-dimensional space, this topological invariant is the Chern number. It multiplies the Chern-Simon action for the U(1) electro-magnetic gauge field associated with the U(1) conservation of the fermionic charge. Freed and Hopkins 106 have constructed an algorithm that delivers those topological invariants for all invertible topological field theories. They also computed them for all ten Altland-Zirnbauer symmetry classes. When

[130] The U(1) gauge anomaly of Section 3.10.8 for the chiral edge states in the integer quantum Hall effect is an example thereof.

the computation of these topological invariants is restricted to the subset of all topological field theories resulting from computing the partition function of non-interacting Dirac Hamiltonians from the ten Altland-Zirnbauer symmetry classes, the computed equivalence classes match the entries of **Table** 8.6. However, the computation of Freed and Hopkins predicts the existence of fermionic invertible topological phases of matter that are driven by strong interactions in that they would otherwise be trivial in the absence of interactions. The systematic study of interaction-driven fermionic invertible topological phases of matter lie outside the scope of this book as announced in Section 1.2.

8.6 Exercises

8.1 We have derived explicitly in Section 7.3 ten distinct manifolds generated by Hermitean matrices, one for each of the ten Altland-Zirnbauer symmetry classes. Each manifold is one of the ten coset spaces entering the last column of **Table** 7.1. Show that upon spectral flattening, that is, replacing a single-particle matrix $\mathcal{H}$ in the last column of **Table** 7.1 by a Hermitean flattened Hamiltonian $\mathcal{P}$, we may identify $\mathcal{P}$ as a point in any one of the ten manifolds from the last column of **Table** 8.1.

Hint: Show that for the symmetry classes A, AI, and AII, $\mathcal{P}$ can be identified with an element from the unions of the Grassmannians given by

$$\begin{aligned}&\bigcup_{k=0}^{N}\frac{\mathrm{U}(N)}{\mathrm{U}(k)\times\mathrm{U}(N-k)},\\&\bigcup_{k=0}^{N}\frac{\mathrm{O}(N)}{\mathrm{O}(k)\times\mathrm{O}(N-k)},\\&\bigcup_{k=0}^{N}\frac{\mathrm{Sp}(2N)}{\mathrm{Sp}(2k)\times\mathrm{Sp}(2N-2k)},\end{aligned}\tag{8.173}$$

respectively. Show that imposing on the symmetry classes A, AI, and AII the chiral symmetry implies that $\mathcal{P}$ can be identified with an element of

$$\mathrm{U}(N),\qquad \mathrm{O}(N),\qquad \mathrm{Sp}(2N),\tag{8.174}$$

for the symmetry classes AIII, BDI, and CII, respectively. For the four Bogoliubov-de-Gennes symmetry classes, start from the symmetry class D. Show that

$$\exists V\in\mathrm{O}(2N),\qquad \mathcal{P}=V\,(\rho_2\otimes\mathbb{1}_N)\,V^{-1},\tag{8.175}$$

where ρ_2 is the second Pauli matrix in the particle-hole grading. Is this representation unique? Show that the subset of orthogonal matrices from $\mathrm{O}(2N)$ that commutes with $\rho_2\otimes\mathbb{1}_N$ is isomorphic to $\mathrm{U}(N)$. Conclude that $\mathcal{P}$ can be identified with an element from $\mathrm{O}(2N)/\mathrm{U}(N)$. Repeat the same strategy for the symmetry classes DIII, C, and CI.

8.2 The fields of complex and real numbers are denoted $\mathbb{C}$ and $\mathbb{R}$, respectively. The set of complex-valued $n \times n$ matrices is denoted by $\mathbb{C}(n)$. It can also be interpreted as an n^2-dimensional vector space over the field of complex numbers. The set of real-valued $n \times n$ matrices is denoted by $\mathbb{R}(n)$. It can also be interpreted as an n^2-dimensional vector space over the field of real numbers. The set of real quaternions is denoted by $\mathbb{H}$. As shown in **Exercises** 7.2, $\mathbb{H}$ can also be interpreted as a four-dimensional vector space over the field of real numbers. The symbol $\mathbb{H}(n)$ denotes the set of all $n \times n$ matrices with quaternion-valued matrix elements. The algebra isomorphisms

$$C\ell_0 \equiv \mathbb{C}, \tag{8.176a}$$
$$C\ell_1 \cong \mathbb{C} \oplus \mathbb{C}, \tag{8.176b}$$
$$C\ell_2 \cong \mathbb{C}(2), \tag{8.176c}$$
$$C\ell_q \otimes \mathbb{C}(2) \cong C\ell_{q+2}, \tag{8.176d}$$

for the complex Clifford algebras (8.22) indexed by $q = 0, 1, 2, \cdots$ and

$$C\ell_{0,0} \equiv \mathbb{R}, \tag{8.177a}$$
$$C\ell_{0,1} \cong \mathbb{R} \oplus \mathbb{R}, \tag{8.177b}$$
$$C\ell_{0,2} \cong \mathbb{R}(2), \tag{8.177c}$$
$$C\ell_{0,3} \cong \mathbb{C}(2), \tag{8.177d}$$
$$C\ell_{0,4} \cong \mathbb{H}(2), \tag{8.177e}$$
$$C\ell_{0,5} \cong \mathbb{H}(2) \oplus \mathbb{H}(2), \tag{8.177f}$$
$$C\ell_{0,6} \cong \mathbb{H}(4), \tag{8.177g}$$
$$C\ell_{0,7} \cong \mathbb{C}(8), \tag{8.177h}$$
$$C\ell_{1,0} \cong \mathbb{C}, \tag{8.177i}$$
$$C\ell_{2,0} \cong \mathbb{H}, \tag{8.177j}$$
$$C\ell_{p,q} \otimes \mathbb{R}(2) \cong C\ell_{p+1,q+1}, \tag{8.177k}$$
$$C\ell_{p,q} \otimes \mathbb{R}(2) \cong C\ell_{q,p+2}, \tag{8.177l}$$
$$C\ell_{p,q} \otimes \mathbb{H} \cong C\ell_{q+2,p}, \tag{8.177m}$$
$$C\ell_{p,q} \otimes \mathbb{H}(2) \cong C\ell_{p,q+4}, \tag{8.177n}$$
$$C\ell_{p,q} \otimes \mathbb{R}(16) \cong C\ell_{p+8,q} \cong C\ell_{p,q+8} \tag{8.177o}$$

for the real Clifford algebras (8.23) indexed by $p, q = 0, 1, 2, \cdots$ with $p+q \geq 1$ hold. The tensor product of two associative and unital algebras over the field of real or complex numbers is defined in **Appendix** A. The goal of this exercise is to provide an intuition for proving these algebra isomorphisms by verifying Eqs. (8.176b)–(8.176d), (8.177b)–(8.177j), and (8.177l) for $p = q = 1$.

8.2.1 Prove the algebra isomorphisms (8.176b)–(8.176d) and (8.177b)–(8.177j) for the Clifford algebras over the fields of complex and real numbers, respectively. *Hint:* A bijection must be constructed between the sets to the left and right of the symbol $\cong$. The addition and multiplication must be defined for elements of these sets and shown to be compatible with the bijection.

8.2.2 Let

$$\begin{aligned} \mathrm{C}\ell_{1,1} &:= \mathrm{span}_{\mathrm{C}\ell}\left\{e_1,\, e_2 \,\middle|\, -(e_1)^2=(e_2)^2=1\right\}, && (8.178a)\\ \mathrm{C}\ell_{0,2} &:= \mathrm{span}_{\mathrm{C}\ell}\left\{e'_1,\, e'_2 \,\middle|\, +(e'_1)^2=(e'_2)^2=1\right\}. && (8.178b)\end{aligned}$$

The proof of Eq. (8.177l) for $p=q=1$ is achieved in two steps. First, prove that

$$\begin{aligned}\mathrm{C}\ell_{1,1}\otimes\mathrm{C}\ell_{0,2} &:= \mathrm{span}\left\{(e_1)^{n_1}(e_2)^{n_2}\otimes(e'_1)^{n'_1}(e'_2)^{n'_2}\,\middle|\, n_1,n_2,n'_1,n'_2=0,1\right\}\\ &= \mathrm{span}_{\mathrm{C}\ell}\left\{e_2\otimes e'_1\,e'_2,\; e_1\otimes e'_1\,e'_2,\; e_1\,e_2\otimes e'_1,\; e_1\,e_2\otimes e'_2\right\}.\end{aligned} \tag{8.179a}$$

Second, prove that

$$\mathrm{C}\ell_{1,3}=\mathrm{span}_{\mathrm{C}\ell}\{e_2\otimes e'_1\,e'_2,\; e_1\otimes e'_1\,e'_2,\; e_1\,e_2\otimes e'_1,\; e_1\,e_2\otimes e'_2\}. \tag{8.179b}$$

8.3 Assume that we have chosen a dimensionality d of space and an Altland-Zirnbauer symmetry class. We may then increase the rank r of the Dirac matrices until we reach the smallest rank r_{min} for which we may write the massive Dirac Hamiltonian

$$\mathcal{H}_{\mathrm{min}}=\boldsymbol{\alpha}_{\mathrm{min}}\cdot\frac{\partial}{\mathrm{i}\partial\boldsymbol{x}}+m\,\beta_{\mathrm{min}}+\cdots, \tag{8.180}$$

where $\boldsymbol{x}\in\mathbb{R}^d$, the $(d+1)$ Hermitean matrices $\alpha_{\mathrm{min}\ 1},\cdots,\alpha_{\mathrm{min}\ d},\beta_{\mathrm{min}}$ anti-commute pairwise and square to the unit $r_{\mathrm{min}}\times r_{\mathrm{min}}$ matrix, m is a real-valued (mass) parameter, and "$\cdots$" accounts for all scalar and vector gauge contributions as well as for the possibility of additional mass terms. Let $V_{d,r_{\mathrm{min}}}$ be the compact topological space associated with the normalized Dirac masses that enter on the right-hand side of Eq. (8.180). Let V_d denote the classifying space associated with the family of manifolds $V_{d,r_{\mathrm{min}}N}$ labeled by $N=1,2,3,\cdots$.

8.3.1 Case $\pi_0(V_d)=\{0\}$

There is at least one $r_{\mathrm{min}}\times r_{\mathrm{min}}$ Hermitean matrix β'_{min} that enters "$\cdots$" in Eq. (8.180) such that it anticommutes with any one of the $(d+1)$ matrices $\alpha_{\mathrm{min}\,1},\cdots,\alpha_{\mathrm{min}\,d},\beta_{\mathrm{min}}$ and it squares to the unit $r_{\mathrm{min}}\times r_{\mathrm{min}}$ matrix. For any $0\leq\theta<2\pi$, we may then define the normalized mass matrix

$$\beta_{\mathrm{min}}(\theta):=\cos\theta\,\beta_{\mathrm{min}}+\sin\theta\,\beta'_{\mathrm{min}} \tag{8.181}$$

that provides a smooth path between $+\beta_{\mathrm{min}}$ and $-\beta_{\mathrm{min}}$. Since β_{min} can be chosen arbitrarily in the compact topological space $V_{d,r_{\mathrm{min}}}$ associated with the massive Dirac Hamiltonian (8.180), we can rule out the existence of the disconnected subspace $\{\pm\beta_{\mathrm{min}}\}$ in $V_{d,r_{\mathrm{min}}}$.

Verify on some examples such as Hamiltonian (7.447) that, for general values of $N = 1,2,3,\cdots$, the trivial zeroth homotopy group $\pi_0(V_{d,r_{\mathrm{min}}}) = \{0\}$ indicates that for any given pair $(\beta,\beta') \in V_{d,r_{\mathrm{min}}N} \times V_{d,r_{\mathrm{min}}N}$ (not necessarily anticommuting), one always finds the sequence of anticommuting pairs

$$\begin{aligned} &(\beta,\beta_1) \in V_{d,r_{\mathrm{min}}N} \times V_{d,r_{\mathrm{min}}N}, \\ &(\beta_1,\beta_2) \in V_{d,r_{\mathrm{min}}N} \times V_{d,r_{\mathrm{min}}N}, \\ &\quad\vdots \\ &(\beta_n,\beta') \in V_{d,r_{\mathrm{min}}N} \times V_{d,r_{\mathrm{min}}N}. \end{aligned} \tag{8.182}$$

8.3.2 Case $\pi_0(V_d) \neq \{0\}$

Assume that no other normalized $r_{\mathrm{min}} \times r_{\mathrm{min}}$ mass matrix than β_{min} enters the right-hand side of Eq. (8.180). It then follows that

$$V_{d,r_{\mathrm{min}}} = \{\pm\beta_{\mathrm{min}}\}. \tag{8.183}$$

Hence, the ground state of $\mathcal{H}_{\mathrm{min}}$ for $m > 0$ cannot be smoothly deformed into the ground state of $\mathcal{H}_{\mathrm{min}}$ for $m < 0$ without closing the gap proportional to $|m|$. We are going to verify that there are three possible outcomes upon increasing the rank in Eq. (8.180) from r_{min} to $r_{\mathrm{min}}N$ with $N = 2,3,\cdots$ when condition (8.183) holds.

As a warm up, verify on some examples of massive Dirac Hamiltonians in one-dimensional space that one can find more than one linearly independent mass matrices when $N = 2,3,\cdots$.

8.3.3 Case $\pi_0(V_d) = \mathbb{Z}$

We assume that there is one normalized Dirac mass matrix which commutes with all normalized Dirac mass matrices that have become available in the given Altland-Zirnbauer symmetry class upon increasing r from r_{min} to $r_{\mathrm{min}}N$ with $N = 2,3,\cdots$.

For any ν such that $2\nu = N - j$ with $j = 0,2,4,\cdots,2\lfloor N/2\rfloor$,[131] make the Ansatz

$$\beta_{\pm\nu} := \pm\beta_{\mathrm{min}} \otimes M_\nu, \tag{8.184a}$$

[131] The lower floor $\lfloor x\rfloor$ of the real number x is the largest integer smaller than or equal to x.

where the $N \times N$ matrix M_ν obeys

$$M_\nu = M_\nu^\dagger, \tag{8.184b}$$
$$2\nu = \operatorname{tr}(M_\nu), \tag{8.184c}$$
$$M_\nu^2 = \mathbb{1}_N. \tag{8.184d}$$

Verify that

8.3.3.1 the matrix $\beta_{\pm\nu}$ is Hermitean.

8.3.3.2 the matrix M_ν commutes with $\mathbb{1}_N$.

8.3.3.3 $\pm\beta_{\mathrm{min}} \otimes \mathbb{1}_N$ are the only matrices of the form (8.184) that commute with all matrices of the form (8.184).

8.3.3.4 the Ansatz

$$M_\nu = \frac{2\nu}{N}\,\mathbb{1}_N + A \tag{8.185}$$

for any traceless and Hermitean $N \times N$ matrix A solves conditions (8.184a), (8.184b), and (8.184c).

8.3.3.5 the polar decomposition Ansatz

$$\begin{aligned} M_\nu &= U_{N-n_\nu,n_\nu}\, I_{N-n_\nu,n_\nu}\, U^\dagger_{N-n_\nu,n_\nu}, \\ I_{N-n_\nu,n_\nu} &= \operatorname{diag}\left(+\mathbb{1}_{N-n_\nu} \quad -\mathbb{1}_{n_\nu}\right), \\ 2n_\nu &:= N - 2\nu, \end{aligned} \tag{8.186}$$

where the matrices U_{N-n_ν,n_ν} are arbitrarily chosen in $\mathrm{U}(N)$, $\mathrm{O}(N)$, or $\mathrm{Sp}(2N)$[132] depending on the Altland-Zirnbauer symmetry class, solves conditions (8.184a) and (8.184d).

8.3.3.6 the polar decomposition (8.186) is unique up to multiplication of U_{N-n_ν,n_ν} from the right by the block diagonal matrix

$$\operatorname{diag}\left(U_{N-n_\nu} \quad U_{n_\nu}\right), \tag{8.187}$$

where the matrix U_m belongs to either $\mathrm{U}(m)$, $\mathrm{O}(m)$, or $\mathrm{Sp}(2m)$ for $m = N - n_\nu, n_\nu$ depending on the Altland-Zirnbauer symmetry class.

8.3.3.7 the set of matrices

$$\begin{aligned} \{\beta_{\pm\nu} &\text{ of the form (8.184) with } 2\nu = N - j,\ j = 0, 2, \cdots, 2\lfloor N/2 \rfloor \\ &\text{and } M_\nu \text{ of the form (8.185) and (8.186)}\} \end{aligned} \tag{8.188}$$

spans any one of the three classifying spaces (8.49).

The representations (8.54) and (8.55) follow.

[132] The notation $\mathrm{Sp}(2N)$ is used for $N \times N$ quaternion-valued matrix elements (that is 2×2 matrices that are complex-valued). Hence, $\mathrm{Sp}(2N)$ is a $2N \times 2N$ complex-valued matrix.

8.3.4 Case of first descendant $\pi_0(V_d = R_1) = \mathbb{Z}_2$
Motivated by the explicit representations (7.424b) and (7.445b) corresponding to the one-dimensional symmetry class D with $r = r_{\mathrm{min}}$ and $r = 2r_{\mathrm{min}}$, respectively, assume that there is a pair of two anticommuting normalized Dirac mass matrices that span $V_{d,2r_{\mathrm{min}}}$ upon increasing r from r_{min} to $2r_{\mathrm{min}}$.

8.3.4.1 Construct the $r_{\mathrm{min}}/2 \times r_{\mathrm{min}}/2$ matrix ρ_{min} such that it squares to unity and $\beta_{\mathrm{min}} = \rho_{\mathrm{min}} \otimes \tau_2$ for $d = 1$ and $r_{\mathrm{min}} = 2$ using Eq. (7.424b).

8.3.4.2 Verify that there exists the Hermitean $r_{\mathrm{min}}/2 \times r_{\mathrm{min}}/2$ matrix ρ_{min} such that

$$\beta_{\mathrm{min}} =: \rho_{\mathrm{min}} \otimes \tau_2, \qquad \rho^2_{\mathrm{min}} = \mathbb{1}_{r_{\mathrm{min}}/2} \tag{8.189}$$

for $d = 1$.

8.3.4.3 Let $N = 2, 3, \cdots$ and make the Ansatz

$$\beta := \rho_{\mathrm{min}} \otimes M, \tag{8.190a}$$
$$M := \tau_2 \otimes A_2 + \tau_1 \otimes A_1, \tag{8.190b}$$
$$A_2 = A_2^{\dagger} = +A_2^{\mathsf{T}} \in \mathrm{GL}(N, \mathbb{C}), \tag{8.190c}$$
$$A_1 = A_1^{\dagger} = -A_1^{\mathsf{T}} \in \mathrm{GL}(N, \mathbb{C}), \tag{8.190d}$$
$$M^2 = \mathbb{1}_{2N}. \tag{8.190e}$$

Verify that the $2N \times 2N$ matrix M is Hermitean and antisymmetric because

- the $N \times N$ matrix A_2 is Hermitean and symmetric (A_2 commutes with the operation K for complex conjugation),
- the $N \times N$ matrix A_1 is Hermitean and antisymmetric (A_1 anticommutes with the operation K for complex conjugation).

8.3.4.4 Verify that the matrix M is a solution of the extension problem

$$\begin{aligned} C\ell_{1,2} &= \mathrm{span}_{C\ell}\,\{\mathrm{i}\tau_3 \otimes \mathbb{1}_N; \mathsf{K}\,\mathbb{1}_{2N}, \mathrm{i}\mathsf{K}\,\mathbb{1}_{2N}\} \\ &\to C\ell_{1,3} = \mathrm{span}_{C\ell}\,\{\mathrm{i}\tau_3 \otimes \mathbb{1}_N; \mathsf{K}\,\mathbb{1}_{2N}, \mathrm{i}\mathsf{K}\,\mathbb{1}_{2N}, M\}. \end{aligned} \tag{8.191}$$

Observe here that it is the matrix $\mathrm{i}\tau_3$ that enters in the kinetic contribution to the Dirac Hamiltonian in the example (7.445b) for a one-dimensional Dirac Hamiltonian in the symmetry class D. The explicit representation of Eq. (8.62) for this example thus follows from solving with Eq. (8.190) the extension problem (8.191).

It is instructive to compare Eqs. (8.189) and (8.190) with the explicit representations (7.424b) and (7.445b) corresponding to the one-dimensional symmetry class D with $r = r_{\mathrm{min}}$ and $r = 2r_{\mathrm{min}}$, respectively. This comparison suggests that the choice $A_2 = \pm\mathbb{1}_N$ and $A_1 = 0$ is special for any odd N. Indeed, one may verify that it is impossible to construct an antisymmetric matrix A_1 of odd rank N that connects smoothly, without closing a gap, between the choices $A_2 = +\mathbb{1}_N$ and $A_1 = 0$ on the one hand, and

$A_2 = -\mathbb{1}_N$ and $A_1 = 0$ on the other hand. Hence, the choices $A_2 = \pm\mathbb{1}_N$ and $A_1 = 0$ represent two topologically distinct phases when N is odd.

Finally, observe that there is no unique Dirac mass matrix that commutes with all other Dirac mass matrices for $N > 1$. For example, the Dirac mass matrix with $A_2 = \mathbb{1}_3$ and $A_1 = 0$ neither commutes nor anticommutes with the Dirac mass matrix for which

$$A_2 = \begin{pmatrix} \cos\theta & 0 & 0 \\ 0 & \cos\theta & 0 \\ 0 & 0 & 1 \end{pmatrix}, \quad A_1 = \sin\theta \begin{pmatrix} 0 & -\mathrm{i} & 0 \\ +\mathrm{i} & 0 & 0 \\ 0 & 0 & 0 \end{pmatrix}. \tag{8.192}$$

8.3.5 Case of second descendant $\pi_0(V_d = R_2) = \mathbb{Z}_2$

We start from Eq. (7.446) for the one-dimensional symmetry class DIII with $r = 4$. Needed are the normalized Dirac masses for the one-dimensional symmetry class DIII with $r = 8$.

8.3.5.1 Verify that if one imposes PHS and TRS through

$$\mathcal{H}(k) = -\mathcal{H}^*(-k), \tag{8.193a}$$

$$\mathcal{H}(k) = +\tau_2 \otimes \sigma_0 \otimes \rho_0\, \mathcal{H}^*(-k)\, \tau_2 \otimes \sigma_0 \otimes \rho_0, \tag{8.193b}$$

then

$$\begin{aligned} \mathcal{H}(k) = {} & \tau_3 \otimes \sigma_0 \otimes \rho_0\, k \\ & + \sum_{\nu=0,1,3} \tau_3 \otimes (\sigma_2 \otimes \rho_\nu\, A_{1,2,\nu} + \sigma_\nu \otimes \rho_2\, A_{1,\nu,2}) \\ & + \sum_{\nu=0,1,3} \tau_1 \otimes (\sigma_2 \otimes \rho_\nu\, M_{1,2,\nu} + \sigma_\nu \otimes \rho_2\, M_{1,\nu,2}). \end{aligned} \tag{8.193c}$$

8.3.5.2 Verify that the topological space of normalized Dirac masses obtained by adding to the Dirac kinetic contribution a mass matrix squaring to the unit matrix, obeying PHS squaring to unity and TRS squaring to minus unity is

$$\begin{aligned} V^{\mathrm{DIII}}_{d=1,r=8} &= \{\boldsymbol{M}\cdot\boldsymbol{X} | |\boldsymbol{M}|^2 = 1\} \cup \{\boldsymbol{N}\cdot\boldsymbol{Y} | |\boldsymbol{N}|^2 = 1\}, \\ \boldsymbol{M} &:= \left(M_{1,2,0}, M_{1,1,2}, M_{1,3,2}\right), \\ \boldsymbol{N} &:= \left(M_{1,0,2}, M_{1,2,1}, M_{1,2,3}\right), \\ \boldsymbol{X} &:= \left(\tau_1 \otimes \sigma_2 \otimes \rho_0, \tau_1 \otimes \sigma_1 \otimes \rho_2, \tau_1 \otimes \sigma_3 \otimes \rho_2\right), \\ \boldsymbol{Y} &:= \left(\tau_1 \otimes \sigma_0 \otimes \rho_2, \tau_1 \otimes \sigma_2 \otimes \rho_1, \tau_1 \otimes \sigma_2 \otimes \rho_3\right). \end{aligned} \tag{8.194}$$

As a topological space, $V^{\mathrm{DIII}}_{d=1,r=8} \cong S^2 \cup S^2$ is homeomorphic to $\mathrm{O}(4)/\mathrm{U}(2)$, for the homeomorphisms

$$
\begin{aligned}
O(4)/U(2) &\cong \mathbb{Z}_2 \times SO(4)/[U(1) \times SU(2)] \\
&\cong \mathbb{Z}_2 \times [SO(3) \times SU(2)]/[SO(2) \times SU(2)] \\
&\cong \mathbb{Z}_2 \times SO(3)/SO(2) \\
&\cong S^2 \cup S^2
\end{aligned}
\tag{8.195}
$$

hold between topological spaces.

Assume that there is a pair of three anticommuting normalized Dirac mass matrices that span $V_{d,2r_{\mathrm{min}}}$ upon increasing r from r_{min} to $2r_{\mathrm{min}}$.

8.3.5.3 Verify that one can choose a representation of the Dirac mass matrix as follows. If $N = 2, 3, \cdots$ and if we define the Hermitean $r_{\mathrm{min}}/2 \times r_{\mathrm{min}}/2$ matrix ρ_{min} by

$$
\beta_{\mathrm{min}} =: \rho_{\mathrm{min}} \otimes \tau_2, \tag{8.196a}
$$

we may then make the Ansatz

$$
\beta := \rho_{\mathrm{min}} \otimes M, \tag{8.196b}
$$

$$
M := \tau_2 \otimes A_2 + \tau_3 \otimes A_3 + \tau_1 \otimes A_1 + \tau_0 \otimes A_0, \tag{8.196c}
$$

for a generic Dirac mass matrix. Here, the $2N \times 2N$ matrix M is Hermitean and antisymmetric because the $N \times N$ matrix A_2 is Hermitean and symmetric (A_2 commutes with the operation K for complex conjugation), while the $N \times N$ matrices A_3, A_1, and A_0 are Hermitean and antisymmetric (A_3, A_1, and A_0 anticommute with the operation K for complex conjugation).

8.3.5.4 Verify that the matrix M is a solution of the extension problem

$$
\begin{aligned}
C\ell_{0,2} &= \mathrm{span}_{C\ell}\{; \mathsf{K}\,\mathbb{1}_{2N}, \mathrm{i}\mathsf{K}\,\mathbb{1}_{2N}\} \\
&\to C\ell_{0,3} = \mathrm{span}_{C\ell}\{; \mathsf{K}\,\mathbb{1}_{2N}, \mathrm{i}\mathsf{K}\,\mathbb{1}_{2N}, M\},
\end{aligned}
\tag{8.197}
$$

which constrains M to be pure imaginary, as can be verified for Eq. (8.196c). The representation (8.62) follows.

8.3.5.5 In order to compare the Dirac mass matrices entering Eq. (8.193) with Eq. (8.196c), notice first that $\rho_{\mathrm{min}} = \tau_1$ and second that σ_ν in Eq. (8.194) plays the role of τ_ν in Eq. (8.196c).

8.4 Assume that we are given Eq. (8.180). The question posed and solved in Section 8.2 was the following:

Q1 What is the compact topological space spanned by the normalized Dirac masses of rank $r = r_{\mathrm{min}} N$ in the limit $N \to \infty$?

The answer to question **Q1** was given in **Table** 8.1 by the extension problems (the values of p and q are taken from **Table** 8.1)

$$
\begin{aligned}
C\ell_q &= \mathrm{span}_{C\ell}\{e_1, \cdots, e_q\} \\
&\to \mathrm{span}_{C\ell}\{e_1, \cdots, e_q, e_{q+1}\} = C\ell_{q+1}
\end{aligned}
\tag{8.198a}
$$

for the complex symmetry classes A and AIII;

$$\begin{aligned} C\ell_{p,q} &= \mathrm{span}_{C\ell}\{e_1,\cdots,e_p;e_{p+1},\cdots,e_{p+q}\} \\ &\to \mathrm{span}_{C\ell}\{e_1,\cdots,e_p;e_{p+1},\cdots,e_{p+q},e_{p+q+1}\} = C\ell_{p,q+1} \end{aligned} \tag{8.198b}$$

for the real symmetry classes BDI, D, DIII, CII, C, and CI; and

$$\begin{aligned} C\ell_{p,q} &= \mathrm{span}_{C\ell}\{e_1,\cdots,e_p;e_{p+1},\cdots,e_{p+q}\} \\ &\to \mathrm{span}_{C\ell}\{e_1,\cdots,e_p,e_{p+1};e_{p+2},\cdots,e_{p+q+1}\} = C\ell_{p+1,q} \end{aligned} \tag{8.198c}$$

for the real symmetry classes AI and AII. The solutions to Eqs. (8.198) are the classifying spaces (the values of p and q are taken from **Table** 8.1)

$$V = C_q, \qquad V = R_{q-p}, \qquad V = R_{p-q+2}, \tag{8.199}$$

respectively. By definition, the Dirac Hamiltonian (8.180) realizes a topological insulator or superconductor if the compact topological space $V_{d,r_{\mathrm{min}}}$ spanned by the normalized Dirac mass β_{min} consists of no more than the set $\{\pm\beta_{\mathrm{min}}\}$. Here, the minimal rank r_{min} was defined to be the rank for which no Dirac mass is permissible upon lowering the rank of the Dirac matrices entering the Dirac Hamiltonian (8.180) holding the dimensionality of space and the Altland-Zirnbauer symmetry class fixed.

It then follows that, on any $(d-1)$-dimensional boundary of d-dimensional space, any Dirac Hamiltonian (8.180) with

$$V_{d,r_{\mathrm{min}}} \cong \{\pm\beta_{\mathrm{min}}\} \tag{8.200a}$$

must necessarily reduce to a massless Dirac Hamiltonian of the form

$$\widetilde{\mathcal{H}} = \widetilde{\boldsymbol{\alpha}} \cdot \frac{\partial}{\mathrm{i}\partial\widetilde{\boldsymbol{x}}} + \cdots, \tag{8.200b}$$

where $\widetilde{\boldsymbol{x}} \in \mathbb{R}^{d-1}$, the $(d-1)$ matrices $\widetilde{\alpha}_1,\cdots,\widetilde{\alpha}_{d-1}$ anticommute pairwise and square to the unit, their rank is

$$\tilde{r} = r_{\mathrm{min}}/2, \tag{8.200c}$$

and "$\cdots$" accounts for scalar and vector gauge contributions, but "$\cdots$" does not contain Dirac mass terms.

We now ask the following question, given the dimensionality $d-1$ of space and given an Altland-Zirnbauer symmetry class:

Q2 Is a Dirac mass of rank $\widetilde{r}_{\mathrm{min}}$ permissible or not? Here, $\widetilde{r}_{\mathrm{min}}$ denotes the minimal rank for which we may write the massless Dirac Hamiltonian

$$\widetilde{\mathcal{H}} := \widetilde{\boldsymbol{\alpha}} \cdot \frac{\partial}{\mathrm{i}\partial\widetilde{\boldsymbol{x}}} + \cdots, \tag{8.201a}$$

where $\widetilde{\boldsymbol{x}} \in \mathbb{R}^{d-1}$ and the $d-1$ Hermitean matrices $\widetilde{\alpha}_1$, $\cdots$, $\widetilde{\alpha}_{d-1}$ obey the Clifford algebra

$$\{\widetilde{\alpha}_i,\widetilde{\alpha}_j\} = \delta_{ij}, \qquad i,j=1,\cdots,d-1, \qquad d=1,2,3,\cdots. \tag{8.201b}$$

In other words, we want to know if a massless Dirac Hamiltonian belonging to some prescribed $(d-1)$-dimensional Altland-Zirnbauer symmetry class can accommodate a Dirac mass.

To answer this question, we recall that Eqs. (8.37) and (8.39) define, for any dimension d of space and for any Altland-Zirnbauer symmetry class, a Clifford algebra that supports at least one normalized Dirac mass, provided the rank r of the Dirac matrices is no less than the minimal rank r_{min}. As the Clifford algebras (8.40) and (8.41) are obtained after removing the generator associated with the normalized Dirac mass in each of the Clifford algebras (8.37) and (8.39), the extension problems defined by Eq. (8.198) aim at enumerating all the possible Dirac masses for fixed dimensionality of space and fixed Altland-Zirnbauer symmetry class.

By analogy, question **Q2** can then be rephrased as follows:
For any fixed dimensionality of space and fixed Altland-Zirnbauer symmetry class, what are the distinct ways to construct the corresponding Clifford algebra from Eqs. (8.40) and (8.41) out of one in the same symmetry class but with one fewer generator? In particular, if the answer to this extension problem is that there is only one possible additional generator (up to its sign) when going from $(d-2)$ generators to $(d-1)$ generators of the Clifford algebras, then there is no room to accommodate a mass matrix on the $(d-1)$-dimensional boundary as all $(d-1)$ generators are needed to define the kinetic energy of the Dirac Hamiltonian.

We can either take the perspective of the bulk, or that of the boundary. If there is essentially only one way to extend a Clifford algebra with $(d-2)$ generators to one with $d-1$ generators on the boundary, then there is no room to have a mass on the boundary.

8.4.1 Convince yourself that the answer to this question is given by the extension problems (the values of p and q are taken from **Table** 8.1)

$$C\ell_{q-1} = \text{span}_{C\ell}\{e_1, \cdots, e_{q-1}\} \to \text{span}_{C\ell}\{e_1, \cdots, e_{q-1}, e_q\} = C\ell_q \tag{8.202a}$$

for the complex symmetry classes A and AIII;

$$\begin{aligned} C\ell_{p,q-1} &= \text{span}_{C\ell}\{e_1, \cdots, e_p; e_{p+1}, \cdots, e_{p+q-1}\} \\ &\to \text{span}_{C\ell}\{e_1, \cdots, e_p; e_{p+1}, \cdots, e_{p+q-1}, e_{p+q}\} = C\ell_{p,q} \end{aligned} \tag{8.202b}$$

for the real symmetry classes BDI, D, DIII, CII, C, and CI; and

$$\begin{aligned} C\ell_{p-1,q} &= \text{span}_{C\ell}\{e_1, \cdots, e_{p-1}; e_p, \cdots, e_{p+q-1}\} \\ &\to \text{span}_{C\ell}\{e_1, \cdots, e_{p-1}, e_p; e_{p+1}, \cdots, e_{p+q}\} = C\ell_{p,q} \end{aligned} \tag{8.202c}$$

for the real symmetry classes AI and AII.

8.4.2 Verify that the solutions to Eq. (8.202) are the classifying spaces (the values of p and q are taken from **Table** 8.1)

$$\widetilde{V} = C_{q-1}, \qquad \widetilde{V} = R_{q-p-1}, \qquad \widetilde{V} = R_{p-q+1}, \tag{8.203}$$

respectively.

On the one hand, if $\pi_0(\widetilde{V}) = \mathbb{Z}, \mathbb{Z}_2$, then the generator e_{p+q} from the corresponding Clifford algebra is unique in that no additional generator exists that anticommutes with e_{p+q}, that is no additional generator e_{p+q+1} that plays the role of a normalized Dirac mass represented by a Hermitean matrix of rank $\widetilde{r}_{\text{min}}$ is allowed. On the other hand, if $\pi_0(\widetilde{V}) = 0$, then the generator e_{p+q} from the corresponding Clifford algebra is not unique in that there exist additional independent generators that anticommute with e_{p+q}, that is a generator e_{p+q+1} that plays the role of a normalized Dirac mass represented by a Hermitean matrix of rank $\widetilde{r}_{\text{min}}$ is allowed.

8.4.3 Deduce that the classifying space associated with **Q2** in $(d-1)$ dimensions coincides with the classifying space associated with **Q1** in d dimensions. The existence of nontrivial topological insulators in d dimensions and the gapless Dirac Hamiltonian with no allowed Dirac mass in $d-1$ dimensions are equivalent.

The final answer to the question **Q2** is as follows:

A No Dirac mass matrix of rank $\widetilde{r}_{\text{min}}$ is permissible when $\pi_0(\widetilde{V}) = \mathbb{Z}$. If N channels are added by generalizing the Dirac matrices $\widetilde{\alpha}$ to $\widetilde{\boldsymbol{\alpha}} \otimes \mathbb{1}_N$, then no Dirac mass of rank $\widetilde{r}_{\text{min}} N$ is permissible when $\pi_0(\widetilde{V}) = \mathbb{Z}$.

B No Dirac mass matrix of rank $\widetilde{r}_{\text{min}}$ is permissible when $\pi_0(\widetilde{V}) = \mathbb{Z}_2$. If N channels are added by generalizing the Dirac matrices $\widetilde{\alpha}$ to $\widetilde{\boldsymbol{\alpha}} \otimes \mathbb{1}_N$, then no Dirac mass of rank $\widetilde{r}_{\text{min}} N$ is permissible for odd N, while a Dirac mass of rank $\widetilde{r}_{\text{min}} N$ is permissible for even N, when $\pi_0(\widetilde{V}) = \mathbb{Z}_2$.

C A Dirac mass matrix of rank $\widetilde{r}_{\text{min}}$ is permissible when $\pi_0(\widetilde{V}) = 0$. If N channels are added by generalizing the Dirac matrices $\widetilde{\alpha}$ to $\widetilde{\boldsymbol{\alpha}} \otimes \mathbb{1}_N$, then a Dirac mass matrix of rank $\widetilde{r}_{\text{min}} N$ is permissible when $\pi_0(\widetilde{V}) = 0$.

Table 8.5 follows.

Appendix A
Mathematical Glossary

In order to be more or less self-contained, we review some of the mathematical concepts outside of linear algebra and analysis that we will use in this book when tools from group cohomology and topology will be needed.

Definition A.1 (equivalence relation) Let S be a non-empty set. A binary relation $\sim$ on S is called an equivalence relation on S if it satisfies

$$\textbf{reflexivity} \qquad a \sim a \qquad \forall a \in S, \tag{A.1a}$$

$$\textbf{symmetry} \qquad a \sim b \Longrightarrow b \sim a \qquad \forall a, b \in S, \tag{A.1b}$$

$$\textbf{transitivity} \qquad a \sim b,\ b \sim c \Longrightarrow a \sim c \qquad \forall a, b, c \in S. \tag{A.1c}$$

For any $a \in S$, the set

$$[a] := \{ b \in S \mid b \sim a \} \tag{A.2}$$

is called the equivalence class of a.

Definition A.2 (partially ordered set) A partially ordered set is a non-empty set S together with a binary relation $\leq$ on S also called a partial order that obeys the rules

$$\textbf{reflexivity} \qquad a \leq a \qquad \forall a \in S, \tag{A.3a}$$

$$\textbf{antisymmetry} \qquad a \leq b,\ b \leq a \Longrightarrow a = b \qquad \forall a, b \in S, \tag{A.3b}$$

$$\textbf{transitivity} \qquad a \leq b,\ b \leq c \Longrightarrow a \leq c \qquad \forall a, b, c \in S. \tag{A.3c}$$

Definition A.3 (directed set) A directed set is a non-empty set S together with a binary relation $\leq$ on S also called a preorder that obeys the rules

$$\textbf{reflexivity} \qquad a \leq a \qquad \forall a \in S, \tag{A.4a}$$

$$\textbf{transitivity} \qquad a \leq b,\ b \leq c \Longrightarrow a \leq c \qquad \forall a, b, c \in S, \tag{A.4b}$$

$$\textbf{upper bound} \qquad \forall a, b \in S, \qquad \exists c \in S,\ a \leq c,\ b \leq c. \tag{A.4c}$$

Definition A.4 (group) The pair $(\mathrm{G}, *)$ made of a non-empty set G together with a binary operation denoted $*$

$$\begin{aligned} \mathrm{G} \times \mathrm{G} &\rightarrow \mathrm{G}, \\ (a, b) &\mapsto a * b, \end{aligned} \tag{A.5a}$$

such that the three conditions

$$\textbf{associativity} \qquad (a * b) * c = a * (b * c) \qquad \forall a, b, c \in \mathrm{G}, \tag{A.5b}$$

$$\textbf{neutral element} \qquad \exists e \in \mathrm{G},\ e * a = a * e = a \qquad \forall a \in \mathrm{G}, \tag{A.5c}$$

$$\textbf{inverses} \qquad \forall a \in \mathrm{G},\ \exists a^{-1} \in \mathrm{G},\ a^{-1} * a = a * a^{-1} = e \tag{A.5d}$$

hold is called a group. It is customary to refer to a group by only referring to the set G from the pair $(\mathrm{G}, *)$ when there is no risk of confusion.

Definition A.5 (Abelian group) The pair $(\mathrm{G}, *)$ is called an Abelian or commutative group if it is a group obeying the additional condition

$$a * b = b * a \qquad \forall a, b \in \mathrm{G}, \tag{A.6}$$

in which case it is customary to denote $*$ with $+$, e with 0, and a^{-1} with $-a$.

Definition A.6 (trivial and nontrivial groups) A trivial group is made of one element, the neutral element. A nontrivial group is made of at least two elements.

Definition A.7 (subgroup) A non-empty subset of a group is a subgroup if it is a group on its own right under the same operation as defined in the group.

Definition A.8 (center of a group) The center of a group is the set of those elements from the group that commute with all elements of the group. The center of a group is always a non-empty subgroup as it must contain the neutral element.

Definition A.9 (cosets of a group) Let the subset H of a group G be a subgroup. The equivalence relation

$$g \sim g' \iff g^{-1} * g' \in \mathrm{H} \qquad (g \sim g' \iff g' * g^{-1} \in \mathrm{H}) \tag{A.7}$$

for any pair $g, g' \in \mathrm{G}$ defines the left (right) equivalence class, or coset, of H in G.

Definition A.10 (normal subgroup) The subgroup N of the group G is normal if

$$g * n * g^{-1} \in \mathrm{N} \qquad \forall g \in \mathrm{G}, \forall n \in \mathrm{N}. \tag{A.8}$$

One writes $\mathrm{N} \triangleleft \mathrm{G}$.

Property A.11 (quotient group) The collection of all the cosets of a normal subgroup H of the group G is a group called the quotient (factor) group and denoted by G/H.

Property A.12 (coset space) The collection G/H of all the cosets of the subgroup H of the group G is called a coset space when H is not a normal subgroup.

Definition A.13 (simple group) A simple group is a nontrivial group whose only normal subgroups are the trivial group and the group itself.

Property A.14 Any non-simple group can be decomposed into two smaller groups, a nontrivial normal subgroup and the corresponding quotient group. This process when iterated constitutes the Jordan–Hölder theorem. Any finite groups can thus be assigned a uniquely (up to ordering) determined sequence of simple groups. The complete classification of finite simple groups was initiated by Galois in 1832 and completed in 2004.

Definition A.15 (direct product of groups) Let $(\mathrm{G},\circ)$ denote a group with the non-empty set G equipped with the binary operation $\circ$. Let $(\mathrm{H},\bullet)$ denote a group with the non-empty set H equipped with the binary operation $\bullet$. Their direct product is the Cartesian product $\mathrm{G}\times\mathrm{H}$ with the composition law

$$\begin{aligned}&(\mathrm{G}\times\mathrm{H})\times(\mathrm{G}\times\mathrm{H})\to(\mathrm{G}\times\mathrm{H}),\\ &\Big((g,h),(g',h')\Big)\mapsto(g,h)\cdot(g',h'):=(g\circ g',h\bullet h').\end{aligned}\tag{A.9a}$$

When the groups G and H are Abelian, in which case it is customary to denote their composition law by the addition symbol $+$, their direct product is also referred to as a direct sum and is denoted $\mathrm{G}\oplus\mathrm{H}$ with the composition law

$$(g,h)+(g',h'):=(g+g',h+h').\tag{A.9b}$$

Example A.16 The group O(3) of orthogonal 3×3 matrices has the direct product decomposition

$$\mathrm{O}(3)=\{I,-I\}\times\mathrm{SO}(3),\tag{A.10}$$

where I is the 3×3 unit matrix. This property generalizes to the group $\mathrm{O}(2n+1)$ of orthogonal $(2n+1)\times(2n+1)$ matrices for any $n=0,1,2,\cdots$.

Definition A.17 (semi-direct product of groups) Let G denote a group. Let N denote a normal subgroup of G, that is, $\mathrm{N}\lhd\mathrm{G}$. Let H denote a subgroup of G. The group G is the semi-direct product of H acting on N if and only if

$$\forall g\in\mathrm{G},\ \exists!\,h\in\mathrm{H}\text{ and }\exists!\,n\in\mathrm{N}\ \text{ such that } g=h\,n.\tag{A.11a}$$

One writes

$$\mathrm{G}=\mathrm{H}\ltimes\mathrm{N}.\tag{A.11b}$$

Property A.18 The group G is the semi-direct product of H acting on N if and only if

$$\forall g\in\mathrm{G},\ \exists!\,h\in\mathrm{H}\text{ and }\exists!\,n\in\mathrm{N}\ \text{ such that } g=n\,h.\tag{A.12a}$$

One writes

$$\mathrm{G} = \mathrm{N} \rtimes \mathrm{H}. \tag{A.12b}$$

Property A.19 The group G is the semi-direct product of H acting on N if and only if the composition $\pi \circ i$ of the inclusion map $i\colon \mathrm{H} \to \mathrm{G}$ and the natural projection $\pi\colon \mathrm{G} \to \mathrm{G/N}$ is an isomorphism between H and the quotient group G/N.

Example A.20 The group O(2) of orthogonal 2×2 matrices has the semi-direct product decomposition

$$\mathrm{O}(2) = \{I, R\} \ltimes \mathrm{SO}(2), \tag{A.13}$$

where I is the 2×2 unit matrix and R is the diagonal matrix $\mathrm{diag}(-1, 1)$. This property generalizes to the group $\mathrm{O}(n)$ of orthogonal $n \times n$ matrices for any $n = 1, 2, \cdots$.

Definition A.21 Let $(\mathrm{G}, \circ)$ and $(\mathrm{G}', \bullet)$ denote a pair of groups with their composition rules $\circ$ and $\bullet$, respectively. A map $f\colon \mathrm{G} \to \mathrm{G}'$ is a (group) homomorphism if

$$f(g \circ h) = f(g) \bullet f(h) \tag{A.14}$$

for any g and h of G. A (group) isomorphism is a bijective (group) homomorphism.

Definition A.22 (ring) A ring $(\mathrm{R}, +, *)$ is a non-empty set R together with the binary operation

$$\begin{aligned} \mathrm{R} \times \mathrm{R} &\to \mathrm{R}, \\ (a, b) &\mapsto a + b, \end{aligned} \tag{A.15a}$$

denoted by the addition $+$ and the binary operation

$$\begin{aligned} \mathrm{R} \times \mathrm{R} &\to \mathrm{R}, \\ (a, b) &\mapsto a * b, \end{aligned} \tag{A.15b}$$

denoted by the multiplication $*$ such that (i) R is an Abelian group under the addition $+$ and (ii) the three conditions

associativity	$(a * b) * c = a * (b * c)$	$\forall a, b, c \in \mathrm{R}$,	(A.15c)
right distributivity	$(a + b) * c = a * c + b * c$	$\forall a, b, c \in \mathrm{R}$,	(A.15d)
left distributivity	$c * (a + b) = c * a + c * b$	$\forall a, b, c \in \mathrm{R}$,	(A.15e)

hold. It is customary to refer to a ring by only referring to the set R from the triplet $(\mathrm{R}, +, *)$ when there is no risk of confusion.

Definition A.23 (commutative ring) A ring $(\mathrm{R}, +, *)$ is commutative (Abelian) if

$$a * b = b * a \qquad \forall a, b \in \mathrm{R}. \tag{A.16}$$

Definition A.24 (unital ring) A ring $(\mathrm{R}, +, *)$ is unital (a ring with unity or identity) if there exists $e \in \mathrm{R}$ such that

$$e * a = a * e = a \qquad \forall a \in \mathrm{G}. \tag{A.17}$$

Definition A.25 (zero or trivial ring) A zero or trivial ring is any ring consisting of one and only one element. It is customary to denote this element with 0.

Definition A.26 (division ring) A nonzero unital ring $(\mathrm{R}, +, *)$ is called a division ring (skew field) if every nonzero element has a multiplicative inverse.

Definition A.27 (subring) A subset $\mathrm{S} \subseteq \mathrm{R}$ of a ring $(\mathrm{R}, +, *)$ is called a subring if it is a ring on its own right under the operations $+$ and $*$ obtained by restricting these operations defined in R to S.

Definition A.28 (center of a ring) The center $\mathrm{Z}(\mathrm{R})$ of a ring $(\mathrm{R}, +, *)$ is the subset of R such that all its elements commute with respect to the multiplication $*$ with all elements of R.

Property A.29 The center $\mathrm{Z}(\mathrm{R})$ is a commutative subring.

Definition A.30 (ideal of a ring) A subset $\mathrm{I} \subseteq \mathrm{R}$ of a ring $(\mathrm{R}, +, *)$ is called a left ideal, right ideal, or two-sided (an ideal in short) if it is an Abelian subgroup of R with respect to the addition $+$ and if it is closed under left, right, or left and right multiplication by any ring element, that is,

$$a \in \mathrm{I},\ r \in \mathrm{R} \implies r * a \in \mathrm{I}, \tag{A.18a}$$

$$a \in \mathrm{I},\ r \in \mathrm{R} \implies a * r \in \mathrm{I}, \tag{A.18b}$$

$$a \in \mathrm{I},\ r \in \mathrm{R} \implies r * a, a * r \in \mathrm{I}, \tag{A.18c}$$

respectively.

Definition A.31 (simple ring) A ring is called simple if it is a nonzero ring that has no ideal besides the zero ideal and itself.

Property A.32 (quotient ring) Let I be an ideal of the ring $(\mathrm{R}, +, *)$. The relation

$$a \sim b \iff x - y \in \mathrm{I} \tag{A.19}$$

for any pair $a, b \in \mathrm{R}$ defines an equivalence relation. The set R/I of the equivalence classes induced by this equivalence relation forms a ring under the addition

$$\begin{aligned} (\mathrm{R}/\mathrm{I}) \times (\mathrm{R}/\mathrm{I}) &\to (\mathrm{R}/\mathrm{I}), \\ ([a], [b]) &\mapsto [a] + [b] := [a + b], \end{aligned} \tag{A.20}$$

and the multiplication

$$\begin{aligned} (\mathrm{R}/\mathrm{I}) \times (\mathrm{R}/\mathrm{I}) &\to (\mathrm{R}/\mathrm{I}), \\ ([a], [b]) &\mapsto [a] * [b] := [a * b], \end{aligned} \tag{A.21}$$

called the quotient (factor) ring.

Definition A.33 (direct product of rings) The direct product of the pair of rings $(\mathrm{Q},+,*)$ and $(\mathrm{R},\dot{+},\bullet)$ is the Cartesian product $\mathrm{Q}\times\mathrm{R}$ equipped with the addition $+$ and the multiplication $\cdot$ defined for any $(q,r),(q',r')\in\mathrm{Q}\times\mathrm{R}$ by

$$(q,r)+(q',r'):=(q+q',r\dot{+}r') \tag{A.22a}$$

and

$$(q,r)\cdot(q',r'):=(q*q',r\bullet r'), \tag{A.22b}$$

respectively.

Definition A.34 (graded ring) A ring $(\mathrm{R},+,*)$ is said to be graded if there exists a unique (up to ordering) sequence of subrings $(\mathrm{R}_0,\mathrm{R}_1,\mathrm{R}_2,\cdots)$ such that for any $r\in\mathrm{R}$ there exists a unique (up to ordering) sequence $(r_0,r_1,r_2,\cdots)$ in terms of which

$$r=\sum_{n\in\mathbb{N}}r_n,\qquad r_n\in\mathrm{R}_n, \tag{A.23a}$$

whereby the condition

$$r_m\in\mathrm{R}_m,r_n\in\mathrm{R}_n\implies r_m*r_n\in\mathrm{R}_{m+n} \tag{A.23b}$$

must hold for any pair $m,n\in\mathbb{N}$. A nonzero element of R_n is said to be homogeneous of degree n. If we interpret R together with $\mathrm{R}_0,\mathrm{R}_1,\mathrm{R}_2,\cdots$ as Abelian groups with respect to the composition law $+$, then we have the direct sum decomposition

$$\mathrm{R}=\bigoplus_{n\in\mathbb{N}}\mathrm{R}_n,\qquad \mathrm{R}_m*\mathrm{R}_n\subseteq\mathrm{R}_{m+n}. \tag{A.23c}$$

Definition A.35 (field) A field $\mathbb{F}$ is a ring with the addition $+$ and multiplication $*$ such that the set $\mathbb{F}^*$ obtained from removing the neutral element 0 with respect to the addition $+$ from $\mathbb{F}$ is an Abelian group under the multiplication $*$.

Property A.36 A commutative ring is a simple ring if and only if it is a field.

Definition A.37 (module) A module M over the ring $(\mathrm{R},+,*)$ is a non-empty set equipped with two operations: the addition (denoted $\dot{+}$)

$$\begin{aligned}M\times M&\to M,\\(m,n)&\mapsto m\dot{+}n,\end{aligned} \tag{A.24a}$$

and the scalar multiplication (denoted by juxtaposition)

$$\begin{aligned}\mathrm{R}\times M&\to M,\\(\alpha,m)&\mapsto\alpha\,m,\end{aligned} \tag{A.24b}$$

such that M is an Abelian group under the addition $\dotplus$ and

$$\alpha(m \dotplus n) = \alpha\, m \dotplus \alpha\, n, \tag{A.24c}$$

$$(\alpha + \beta)\, m = \alpha\, m \dotplus \beta\, m, \tag{A.24d}$$

$$(\alpha * \beta)\, m = \alpha\, (\beta\, m) \tag{A.24e}$$

for any $\alpha, \beta \in \mathrm{R}$, $m, n \in M$. If the ring $(\mathrm{R}, +, *)$ has the identity ε with respect to the multiplication $*$ and if

$$\varepsilon\, m = m \qquad \forall m \in M, \tag{A.24f}$$

then the module is called a unital module.

Definition A.38 (submodule) A subset S of a module M is called a submodule if it is a module on its own right under the operations obtained by restricting the operations of M to S.

Definition A.39 (direct sum of modules) Given two modules M and M' over the ring $(\mathrm{R}, +, *)$, their direct sum $M \oplus M'$ is the Cartesian product $M \times M'$ with the composition law defined as the direct sum of the additions in module M and M'. (If so, the scalar multiplication inherited from M and M' is compatible with that inherited from $M \oplus M'$.)

Definition A.40 (direct sum decomposition of a module) A module M over the ring $(\mathrm{R}, +, *)$ is the direct sum over the submodules $S_1, \cdots, S_n$ if any element $m \in M$ can be written in a unique way (except for ordering) as a sum over the elements from the submodules $S_1, \cdots, S_n$. One writes

$$M = S_1 \oplus \cdots \oplus S_n. \tag{A.25}$$

Definition A.41 (graded module) A graded module is a module M over a ring $(\mathrm{R}, +, *)$ which is graded, that is,

$$\mathrm{R} = \bigoplus_{n \in \mathbb{N}} \mathrm{R}_n, \qquad \mathrm{R}_m * \mathrm{R}_n \subseteq \mathrm{R}_{m+n} \qquad \forall m, n \in \mathbb{N}, \tag{A.26a}$$

such that there exists the decomposition

$$M = \bigoplus_{n \in \mathbb{N}} M_n \qquad \text{as Abelian group}, \tag{A.26b}$$

whereby the condition

$$R_m\, M_n \subseteq M_{m+n} \qquad \forall m, n \in \mathbb{N}, \tag{A.26c}$$

for the scalar multiplication must hold.

Definition A.42 (vector space) The doublet $(V, \dot{+})$ over the field $(\mathbb{F}, +, *)$ is a vector space if V is a non-empty set equipped with two operations, the vector addition (denoted by $\dot{+}$)

$$\begin{aligned} V \times V &\to V, \\ (v, w) &\mapsto v \dot{+} w, \end{aligned} \tag{A.27a}$$

and the scalar multiplication (denoted by juxtaposition)

$$\begin{aligned} \mathbb{F} \times V &\to V, \\ (\alpha, v) &\mapsto \alpha\, v, \end{aligned} \tag{A.27b}$$

such that V is an Abelian group under the addition $\dot{+}$ and

$$\alpha\,(v \dot{+} w) = \alpha\, v \dot{+} \alpha\, w, \tag{A.27c}$$

$$(\alpha + \beta)\, v = \alpha\, v \dot{+} \beta\, v, \tag{A.27d}$$

$$(\alpha * \beta)\, v = \alpha(\beta\, v), \tag{A.27e}$$

$$\varepsilon\, v = v \tag{A.27f}$$

for any $\alpha, \beta \in \mathbb{F}$, $v, w \in V$, and with ε denoting the neutral element of the multiplication $*$ in $\mathbb{F}$.

Definition A.43 (linear subspace) A non-empty subset $W \subseteq V$ is a linear subspace of the vector space V if and only if it is closed under both the vector addition and the scalar multiplication inherited from V.

Definition A.44 (codimension) If $W \subseteq V$ is a linear subspace of the finite-dimensional vector space V with dimensions $\dim(W)$ and $\dim(V)$, respectively, then the codimension of W in V is the difference between the dimensions

$$\operatorname{codim}(W) := \dim(V) - \dim(W) \iff \dim(W) + \operatorname{codim}(W) = \dim(V). \tag{A.28}$$

Definition A.45 (graded vector space) A graded vector space is a vector space V over a field $\mathbb{F}$ that admits the direct sum decomposition

$$V = \bigoplus_{n \in \mathbb{N}} \mathrm{V}_n, \tag{A.29}$$

where each V_n is a vector space. For a given n, the elements of V_n are then called homogeneous elements of degree n.

Definition A.46 (algebra I) Let $\mathcal{A}$ denote a non-empty set. The triplet $(\mathcal{A}, \dot{+}, \bullet)$ with the doublet $(\mathcal{A}, \dot{+})$ a unital module over the unital ring $(\mathrm{R}, +, *)$ and the map

$$\begin{aligned} \bullet\colon \mathcal{A} \times \mathcal{A} &\to \mathcal{A}, \\ (v, w) &\mapsto v \bullet w \end{aligned} \tag{A.30a}$$

is an algebra if the map $\bullet$ is bilinear, that is, for any triplet $v, w, z \in \mathcal{A}$ and any $\alpha \in$ R,

$$\textbf{right distributivity} \qquad (v \dotplus w) \bullet z = v \bullet z \dotplus w \bullet z, \tag{A.30b}$$

$$\textbf{left distributivity} \qquad z \bullet (v \dotplus w) = z \bullet v \dotplus z \bullet w, \tag{A.30c}$$

$$\textbf{compatibility} \qquad \alpha(v \bullet w) = (\alpha\, v) \bullet w = v \bullet (\alpha\, w). \tag{A.30d}$$

Neither associativity nor commutativity nor existence of a neutral element with respect to the multiplication $\bullet$ were assumed. If the multiplication $\bullet$ is associative, then the algebra is called associative. An algebra is called unital if the multiplication $\bullet$ admits a neutral element.

Definition A.47 (Lie algebra) Let $\mathfrak{g}$ denote a non-empty set. The triplet $(\mathfrak{g}, \dotplus, [\cdot,\cdot])$ with the doublet $(\mathfrak{g}, \dotplus)$ a vector space over the field $(\mathbb{F}, +, *)$ and the map

$$\begin{aligned} [\cdot,\cdot]\colon \mathfrak{g} \times \mathfrak{g} &\to \mathfrak{g}, \\ (X, Y) &\mapsto [X, Y] \end{aligned} \tag{A.31a}$$

is called a Lie algebra if the map $[\cdot,\cdot]$ obeys

$$\textbf{linearity} \qquad [X, \alpha\, Y \dotplus \beta\, Z]] = \alpha\, [X, Y] \dotplus \beta\, [X, Z], \tag{A.31b}$$

$$\textbf{antisymmetry} \qquad [X, Y] = -[Y, X], \tag{A.31c}$$

$$\textbf{Jacobi identity} \qquad [X, [Y, Z]] \dotplus [Y, [Z, X]] \dotplus [Z, [X, Y]] = 0 \tag{A.31d}$$

for any $X, Y, Z \in \mathfrak{g}$ and for any $\alpha, \beta \in \mathbb{F}$.

Property A.48 A Lie algebra is a nonassociative algebra.

Property A.49 Denote with $\mathrm{gl}(n, \mathbb{K})$ the set of all $n \times n$ matrices with matrix elements taking values in the field $\mathbb{K}$. Denote with I_n the unit $n \times n$ matrix and denote with J_{2n} the matrix $\begin{pmatrix} 0 & I_n \\ -I_n & 0 \end{pmatrix}$. Denote with

$$\mathfrak{a}_n \equiv \mathfrak{sl}(n+1, \mathbb{C}) := \{X \in \mathrm{gl}(n+1, \mathbb{C}) \mid \mathrm{tr}\, X = 0\}, \tag{A.32a}$$

$$\mathfrak{b}_n \equiv \mathfrak{so}(2n+1, \mathbb{C}) := \{X \in \mathrm{gl}(2n+1, \mathbb{R}) \mid X = -X^{\mathsf{T}}\}, \tag{A.32b}$$

$$\mathfrak{c}_n \equiv \mathfrak{sp}(2n, \mathbb{C}) := \{X \in \mathrm{gl}(2n, \mathbb{C}) \mid J_{2n}\, X + X^{\mathsf{T}}\, J_{2n} = 0\}, \tag{A.32c}$$

$$\mathfrak{d}_n \equiv \mathfrak{so}(2n, \mathbb{C}) := \{X \in \mathrm{gl}(2n, \mathbb{C}) \mid X = -X^{\mathsf{T}}\} \tag{A.32d}$$

the complex classical Lie algebras. Define the Lie bracket on any of the four families $\mathfrak{a}_n$, $\mathfrak{b}_n$, $\mathfrak{c}_n$, and $\mathfrak{d}_n$ to be the commutator of two matrices. The set $\mathfrak{a}_n$ realizes a Lie algebra and as a vector space over the field $\mathbb{R}$ it has the dimension $2[(n+1)^2 - 1] = 2n(n+2)$. The set $\mathfrak{b}_n$ realizes a Lie algebra and as a vector space over the field $\mathbb{R}$ it has the dimension $2n(2n+1)$. The set $\mathfrak{c}_n$ realizes a Lie algebra and as a vector space over the field $\mathbb{R}$ it has the dimension $2n(2n+1)$. The set $\mathfrak{d}_n$ realizes a Lie algebra and as a vector space over the field $\mathbb{R}$ it has the dimension $2n(2n-1)$.

Definition A.50 (algebra II) An associative algebra over the unital ring $(\mathrm{R}, +, *)$ is a non-empty set $\mathcal{A}$ equipped with three operations: the addition (denoted by $\dot{+}$)

$$\begin{aligned} \mathcal{A} \times \mathcal{A} &\to \mathcal{A}, \\ (v, w) &\mapsto v \dot{+} w, \end{aligned} \tag{A.33a}$$

the multiplication (denoted by $\bullet$)

$$\begin{aligned} \mathcal{A} \times \mathcal{A} &\to \mathcal{A}, \\ (v, w) &\mapsto v \bullet w, \end{aligned} \tag{A.33b}$$

and the scalar multiplication (denoted by juxtaposition)

$$\begin{aligned} \mathrm{R} \times \mathcal{A} &\to \mathcal{A}, \\ (\alpha, v) &\mapsto \alpha\, v, \end{aligned} \tag{A.33c}$$

such that

(i) $\mathcal{A}$ is a unital module over the unital ring $(\mathrm{R}, +, *)$,
(ii) $\mathcal{A}$ is a ring under the addition $\dot{+}$ and the multiplication $\bullet$,
(iii) and

$$\alpha(v \bullet w) = (\alpha\, v) \bullet w = v \bullet (\alpha\, w) \tag{A.33d}$$

for any $\alpha \in \mathrm{R}$, $v, w \in \mathcal{A}$.

Definition A.51 (algebra III) The ring $(\mathcal{A}, \dot{+}, \bullet)$ together with the unital ring $(\mathrm{R}, +, *)$ form an associative algebra if the scalar multiplication

$$\begin{aligned} \mathrm{R} \times \mathcal{A} &\to \mathcal{A}, \\ (\alpha, v) &\mapsto \alpha\, v \end{aligned} \tag{A.34a}$$

obeys the compatibility condition

$$\alpha(v \bullet w) = (\alpha\, v) \bullet w = v \bullet (\alpha\, w) \tag{A.34b}$$

for any $\alpha \in \mathrm{R}$ and any pair $v, w \in \mathcal{A}$.

Definition A.52 (subalgebra) A non-empty subset $\mathcal{S} \subseteq \mathcal{A}$ of an algebra $(\mathcal{A}, \dot{+}, \bullet)$ over the unital ring $(\mathrm{R}, +, *)$ is called a subalgebra if $\mathcal{S}$ is an algebra on its own right under the restrictions to $\mathcal{S}$ of the addition $\dot{+}$, multiplication $\bullet$, and multiplication by the scalars.

Definition A.53 (center of an algebra) The center $\mathcal{C} \subseteq \mathcal{A}$ of an algebra $(\mathcal{A}, \dot{+}, \bullet)$ over the unital ring $(\mathrm{R}, +, *)$ is the subset of all elements in $\mathcal{A}$ that commute with respect to the multiplication $\bullet$ with all elements of $\mathcal{A}$.

Definition A.54 (ideal of an algebra) A subset $\mathcal{I} \subseteq \mathcal{A}$ of an algebra $(\mathcal{A}, \dot{+}, \bullet)$ over the unital ring $(\mathrm{R}, +, *)$ is called a left (right) ideal if it is a subalgebra that is closed under the left (right) multiplication $\bullet$ by any element from $\mathcal{A}$. An ideal that is simultaneously a left ideal and a right ideal is said to be two-sided, or an ideal in short. Left, right, and two-sided ideals coincide when the multiplication $\bullet$ is commutative.

Definition A.55 (simple algebra) An algebra $(\mathcal{A}, \dot{+}, \bullet)$ over the unital ring $(\mathrm{R}, +, *)$ is called simple if there is no other ideals than $\mathcal{A}$ itself or the neutral element of the multiplication $\bullet$ if the algebra is unital.

Definition A.56 (graded algebra) An algebra $(\mathcal{A}, \dot{+}, \bullet)$ over the unital ring $(\mathrm{R}, +, *)$ is graded if the ring $(\mathcal{A}, \dot{+}, \bullet)$ is graded, in which case one may interpret $\mathcal{A}$ as an Abelian group with respect to the addition $\dot{+}$ with the direct sum decomposition

$$\mathcal{A} = \bigoplus_{n \in \mathbb{N}} \mathcal{A}_n, \qquad \mathcal{A}_m \bullet \mathcal{A}_n \subseteq \mathcal{A}_{m+n} \qquad \forall m, n \in \mathbb{N}, \tag{A.35}$$

in terms of the sequence $(\mathcal{A}_0, \mathcal{A}_1, \mathcal{A}_2, \cdots)$ of subrings (each of which is also to be interpreted as an Abelian subgroup with respect to the addition $\dot{+}$). A nonzero element of $\mathcal{A}_n$ is said to be homogeneous of degree n. If the unital ring $(\mathrm{R}, +, *)$ is also graded, that is,

$$\mathrm{R} = \bigoplus_{n \in \mathbb{N}} \mathrm{R}_n, \qquad \mathrm{R}_m * \mathrm{R}_n \subseteq \mathrm{R}_{m+n} \qquad \forall m, n \in \mathbb{N}, \tag{A.36}$$

we must impose on the scalar multiplication the compatibility conditions

$$\mathrm{R}_m \mathcal{A}_n \subseteq \mathcal{A}_{m+n} \qquad \forall m, n \in \mathbb{N}, \tag{A.37}$$

as is demanded by the fact that $(\mathcal{A}, \dot{+})$ is a module over the unital ring $(\mathrm{R}, +, *)$.

Definition A.57 ($\mathbb{Z}_n$-graded algebra) A $\mathbb{Z}_n$-graded algebra is a graded algebra such that the degree n of any homogeneous element of the algebra is defined modulo n. For example, the set of complex numbers $\mathbb{C}$ can be interpreted as a $\mathbb{Z}_2$-graded algebra with even elements defining the real numbers $\mathbb{R} \subset \mathbb{C}$ and the odd elements defining the purely imaginary numbers $\mathrm{i}\,\mathbb{R} \subset \mathbb{C}$.

Definition A.58 (direct system over I) Let $\langle I, \leq \rangle$ denote a directed set. Let $\{A_i \mid i \in I\}$ denote a family of objects (sets with some additional structures so as to realize a group, a ring, a module over a fixed ring, an algebra over a fixed unital ring, etc.) indexed by I. Let $f_{ij} \colon A_i \to A_j$ be some homomorphism for all $i \leq j \in I$ with the following properties.

1 Each map f_{ii} is the identity for all $i \in I$.
2 The composition rule $f_{ik} = f_{jk} \circ f_{ij}$ holds for all $i \leq j \leq k$.

If so, the family of pairs $\langle A_i, f_{ij} \rangle$ indexed over I is called a direct system over I.

Definition A.59 (tensor product of algebras) Given the pair of associative and unital algebras $(\mathcal{A}, \bar{+}, \bar{\bullet})$ and $(\mathcal{B}, \underline{+}, \underline{\bullet})$ over the unital commutative ring $(R, +, *)$, their tensor product is the algebra $(\mathcal{A} \otimes \mathcal{B}, \dot{+}, \bullet)$ defined by assigning to any pair $(a, b) \in \mathcal{A} \times \mathcal{B}$ the element

$$a \otimes b \equiv b \otimes a \tag{A.38a}$$

with the multiplicative composition rule

$$(a \otimes b) \bullet (a' \otimes b') := (a \,\bar{\bullet}\, a') \otimes (b \,\underline{\bullet}\, b') \tag{A.38b}$$

that is compatible with the linearity under the additive rules and the multiplications by elements of the ring R in that

$$\begin{aligned}&(\alpha\, a \,\bar{+}\, \alpha'\, a') \otimes (\beta\, b \,\underline{+}\, \beta'\, b') := \\ &\qquad (\alpha * \beta)\, a \otimes b \,\dot{+}\, (\alpha * \beta')\, a \otimes b' \,\dot{+}\, (\alpha' * \beta)\, a' \otimes b \,\dot{+}\, (\alpha' * \beta')\, a' \otimes b'\end{aligned} \tag{A.38c}$$

holds for any $(a, b), (a', b') \in \mathcal{A} \times \mathcal{B}$ and for any $\alpha, \alpha', \beta, \beta' \in R$.

Definition A.60 (direct limit of a direct system over I) The direct limit of the direct system $\langle A_i, f_{ij} \rangle$ over I is defined by the disjoint union

$$\varinjlim A_i = \bigsqcup_{i \in I} A_i \Big/ \sim \tag{A.39}$$

modulo the equivalence relation defined by

$$x_i \sim x_j \iff \exists k \in I,\ i \leq k,\ j \leq k,\ f_{ik}(x_i) = f_{jk}(x_j) \tag{A.40}$$

for all $x_i \in A_i$, $x_j \in A_j$, and $i, j \in I$.

Example A.61 A collection of subsets $\{S_i,\ i \in I\}$ of a set S can be partially ordered by inclusion. If the collection is directed, its direct limit is the union

$$\bigcup_{i \in I} S_i. \tag{A.41}$$

Example A.62 Let $\mathbb{K}$ be a field. For any positive integer $n > 0$, $\mathrm{GL}(n; \mathbb{K})$ denotes the set of invertible $n \times n$ matrices with entries from $\mathbb{K}$. The map $\mathrm{GL}(n; \mathbb{K}) \to \mathrm{GL}(n + 1; \mathbb{K})$ defined by the direct sum of an element of $\mathrm{GL}(n; \mathbb{K})$ with the 1×1 matrix made of the neutral element 1 with respect to the multiplication in $\mathbb{K}$, that is, the map that enlarges matrices by adding one row and one column of vanishing entries except for the new diagonal element 1 in the lower right corner of the $(n+1) \times (n+1)$ matrix, generates all the desired homomorphisms between $\mathrm{GL}(n; \mathbb{K}) \to \mathrm{GL}(n'; \mathbb{K})$ for any $n, n' = 1, 2, \cdots$ with $n \leq n'$ needed to define a direct system over the nonvanishing and positive integers. The direct limit of this direct system over the nonvanishing and positive integers is the general linear group of $\mathbb{K}$ denoted by

$$\mathrm{GL}(\mathbb{K}). \tag{A.42}$$

Its elements can be thought of as infinite-dimensional invertible matrices that differ from the infinite-dimensional identity matrix in only finitely many entries.

Definition A.63 (topological space) A topological space is a pair $(X, \boldsymbol{T})$ with X a set and $\boldsymbol{T}$ a family of subsets of X called open sets if the following conditions hold simultaneously.

1. The set X and the empty set $\emptyset$ are open.
2. Any union of open sets is open.
3. Any finite intersection of open sets is open.

By convention, an element $x \in X$ is called a point from the topological space.

Definition A.64 (topological subspace) Let $(X, \boldsymbol{T})$ denote a topological space. For any subset $S \subseteq X$, its subspace topology is the set

$$\boldsymbol{T}_S := \{S \cap U \mid U \in \boldsymbol{T}\}. \tag{A.43}$$

The pair $(S, \boldsymbol{T}_S)$ is a topological space of its own and it realizes a subspace topology of $(X, \boldsymbol{T})$.

Definition A.65 (neighborhood of a point) Let $(X, \boldsymbol{T})$ be a topological space. A neighborhood of $x \in X$ is a subset of X that includes an open set to which x belongs to.

Definition A.66 (connected topological space) Let $(X, \boldsymbol{T})$ be a topological space. It is disconnected if it is the union of two disjoint non-empty open sets. Otherwise, it is connected. A subset of a topological space is said to be connected if it is connected under its subspace topology.

Definition A.67 (Hausdorff topological space) A topological space $(X, \boldsymbol{T})$ is called Hausdorff if any two distinct points possess disjoint neighborhoods.

Definition A.68 (homeomorphic map) Let $(X, \boldsymbol{T})$ and $(X', \boldsymbol{T}')$ denote two topological spaces. The map $f \colon X \to X'$ is homeomorphic if the following conditions hold simultaneously.

1. The map f is bijective with the inverse map f^{-1}.
2. If U is an open set in $(X, \boldsymbol{T})$, then $f(U)$ is an open set in $(X', \boldsymbol{T}')$.
3. If U' is an open set in $(X', \boldsymbol{T}')$, then $f^{-1}(U')$ is an open set in $(X, \boldsymbol{T})$.

Definition A.69 (homeomorphic topological spaces) Two topological spaces are homeomorphic if a homeomorphic map exists between them.

Definition A.70 (locally Euclidean topological space) If a topological space can be attached a nonnegative integer n such that for any point from the topological space there exists a neighborhood which is homeomorphic to $\mathbb{R}^n$, then it is called a locally Euclidean topological space.

Definition A.71 (topological manifold) An n-dimensional topological manifold is a Hausdorff topological space such that every point has a neighborhood homeomorphic to $\mathbb{R}^n$, that is, it is a topological space that is both Hausdorff and locally Euclidean.

Definition A.72 (continuous map) Let $(X, \boldsymbol{T})$ and $(X', \boldsymbol{T}')$ denote two topological spaces. The map $f \colon X \to X'$ is continuous at the point $x \in X$ if for any open set U' that contains $f(x)$, there exists an open set U that contains x such that $f(U) \subset U'$. The map f is continuous if it is continuous for all points $x \in X$.

Property A.73 Let $(X, \boldsymbol{T})$ and $(X', \boldsymbol{T}')$ denote two topological spaces and let $f \colon X \to X'$ denote a map from X to X'. The following two statements are equivalent.

1. The map f is a homeomorphism.
2. The map f is bijective and continuous, while the inverse map f^{-1} is continuous.

Definition A.74 (homotopy of two maps) Let $(X, \boldsymbol{T})$ and $(X', \boldsymbol{T}')$ denote two topological spaces. Two continuous maps $f \colon X \to X'$ and $g \colon X \to X'$ are homotopic if there exists a continuous map

$$\begin{aligned} F \colon X \times [0,1] &\to X', \\ (x,t) &\mapsto F(x,t), \end{aligned} \tag{A.44a}$$

such that

$$F(x,0) = f(x), \tag{A.44b}$$

and

$$F(x,1) = g(x). \tag{A.44c}$$

Property A.75 (homotopy of two maps) Let $(X, \boldsymbol{T})$ and $(X', \boldsymbol{T}')$ denote two topological spaces. Denote with $C(X, X')$ the set of all the continuous maps between X and X'.

1. The homotopy relation on $C(X, X')$ is an equivalence relation.
2. If X is homeomorphic to Y through the map $h \colon X \to Y$ and X' is homeomorphic to Y' through the map $h' \colon X' \to Y'$, then the maps $h' \circ f \circ h^{-1} \colon Y \to Y'$ and $h' \circ g \circ h^{-1} \colon Y \to Y'$ are homotopic if and only if the maps $f \colon X \to X'$ and $g \colon X \to X'$ are homotopic.

Definition A.76 (path between two points from a topological space) Let $(X, \boldsymbol{T})$ denote a topological space. Two points x and x' from X are path-connected if there exists a continuous map $\alpha \colon [0,1] \to X$ such that $\alpha(0) = x$ and $\alpha(1) = x'$.

Definition A.77 (path-connected topological space) A topological space $(X, \boldsymbol{T})$ is path-connected if any two points x and x' from X are path-connected.

Remark Any path-connected topological space is a connected topological space. The converse needs not be true.

Definition A.78 (path based at a point) Let $(X, \boldsymbol{T})$ denote a topological space and let x be a point from X. A continuous map $\alpha\colon [0,1] \to X$ with $\alpha(0) = x$ is based at x.

Definition A.79 (n-loop based at a point) Let $(X, \boldsymbol{T})$ denote a topological space, let x be a point from X, and let $n = 1, 2, 3, \cdots$. Let I_n denote the closed n-dimensional cube in $\mathbb{R}^n$, that is,

$$I_n := \left\{ \boldsymbol{t} \equiv (t_1, \cdots, t_n) \in \mathbb{R}^n \,\middle|\, 0 \le t_j \le 1,\ j = 1, \cdots, n \right\} \tag{A.45a}$$

and let ∂I_n denote the surface of I_n, that is,

$$\partial I_n := \left\{ \boldsymbol{t} \equiv (t_1, \cdots, t_n) \in I_n \,\middle|\, \exists j \in \{1, \cdots, n\},\ x_j = 0, 1 \right\}. \tag{A.45b}$$

1. A continuous map $\alpha\colon I_n \to X$ is an n-loop based at x if $\alpha(\partial I_n) = x$.
2. If α is an n-loop based at x, then the inverse n-loop is defined by

$$\begin{aligned} \alpha^{-1}\colon I_n &\to X, \\ \boldsymbol{t} &\mapsto \alpha^{-1}(\boldsymbol{t}) := \alpha(1 - t_1, t_2, \cdots, t_n). \end{aligned} \tag{A.45c}$$

3. The constant n-loop based at x is

$$\begin{aligned} e\colon I_n &\to X, \\ \boldsymbol{t} &\mapsto e(\boldsymbol{t}) := x. \end{aligned} \tag{A.45d}$$

Definition A.80 (product of two n-loops based at a point) Let $(X, \boldsymbol{T})$ denote a topological space, let x be a point from X, and let α and β be two n-loops based at x. Their product $\alpha \bullet \beta$ is the n-loop based at x defined by

$$\begin{aligned} \alpha \bullet \beta\colon I_n &\to X, \\ \boldsymbol{t} \mapsto \alpha \bullet \beta(t_1, t_2, \cdots, t_n) &:= \begin{cases} \alpha(2t_1, t_2, \cdots, t_n), & 0 \le t_1 \le \frac{1}{2}, \\ \beta(2t_1 - 1, t_2, \cdots, t_n), & \frac{1}{2} \le t_1 \le 1. \end{cases} \end{aligned} \tag{A.46}$$

Property A.81 Let $(X, \boldsymbol{T})$ denote a topological space and let x be a point from X. The set of all n-loops based at x form a group with respect to the composition rule (A.46) with the constant one-loop based at x as neutral element and the inverse element defined by Eq. (A.45c).

Definition A.82 (product of homotopic n-loop equivalence classes) Let $(X, \boldsymbol{T})$ denote a topological space, let x be a point from X, and denote by

$$\pi_n(X, x) := \left\{ [\alpha] \,\middle|\, [\alpha] = \{\text{all } n\text{-loops based at } x \text{ homotopic to } \alpha\} \right\} \tag{A.47}$$

the quotient space of all homotopic n-loops based at x. Define the product

$$[\alpha] \circ [\beta] := [\alpha \bullet \beta]. \tag{A.48}$$

Property A.83 Let $(X, \boldsymbol{T})$ denote a topological space, let x be a point from X, and let $\pi_n(X, x)$ be the quotient space of all homotopic n-loops based at x. The binary operation (A.48) endows $\pi_n(X, x)$ with a group structure for which the equivalence class of all maps homotopic to the constant map is the identity. The group $\pi_n(X, x)$ is called n-th homotopy group. For $n = 1$, this group is called the fundamental group. The fundamental group is not necessarily Abelian. For $n > 1$, $\pi_n(X, x)$ is Abelian.

Property A.84 Let $(X, \boldsymbol{T})$ and $(X', \boldsymbol{T}')$ denote a pair of topological spaces. Let x and x' be a pair of points from X and X', respectively. The homotopy group $\pi_n(X \times X', (x, x'))$ is then isomorphic to the homotopy group

$$\pi_n(X, x) \times \pi_n(X', x') \tag{A.49a}$$

for any $n = 1, 2, 3, \cdots$, where, for $n > 1$,

$$\pi_n(X, x) \times \pi_n(X', x') \cong \pi_n(X, x) \oplus \pi_n(X', x'). \tag{A.49b}$$

Property A.85 Let $(X, \boldsymbol{T})$ denote a path-connected topological space. Let x and x' be two points from X. The homotopy group $\pi_n(X, x)$ is then isomorphic to the homotopy group $\pi_n(X, x')$ for any $n = 1, 2, 3, \cdots$.

Remark For a path-connected topological space $(X, \boldsymbol{T})$, we may drop the reference to the base point x in the homotopy group $\pi_n(X, x)$.

References

Ablamowicz, R., and Sobczyk, G. 2004. *Lectures on Clifford (Geometric) Algebras and Applications.* Birkhäuser, Boston.

Abrahams, E., Anderson, P. W., Licciardello, D. C., and Ramakrishnan, T. V. 1979. Scaling theory of localization: Absence of quantum diffusion in two dimensions. *Phys. Rev. Lett.*, **42**(10), 673–676.

Abramowitz, M., and Stegun, I. A. 1970. *Handbook of Mathematical Functions.* Dover, New York.

Affleck, I. 1986. Universal term in the free energy at a critical point and the conformal anomaly. *Phys. Rev. Lett.*, **56**(7), 746–748.

Affleck, I., and Haldane, F. D. M. 1987. Critical theory of quantum spin chains. *Phys. Rev. B*, **36**(10), 5291–5300.

Affleck, I., and Lieb, Elliott H. 1986. A proof of part of Haldane's conjecture on spin chains. *Lett. in Math. Phys.*, **12**(1), 57–69.

Affleck, I., Kennedy, T., Lieb, E. H., and Tasaki, H. 1987. Rigorous results on valence-bond ground states in antiferromagnets. *Phys. Rev. Lett.*, **59**(7), 799–802.

Affleck, I., Kennedy, T., Lieb, E. H., and Tasaki, H. 1988. Valence bond ground states in isotropic quantum antiferromagnets. *Commun. Math. Phys.*, **115**(3), 477–528.

Aitchison, I. J. R., and Dunne, G. V. 2001. Nontopological finite temperature induced fermion number. *Phys. Rev. Lett.*, **86**(9), 1690–1693.

Aksoy, Ö. M., and Mudry, C. 2022. Elementary derivation of the stacking rules of invertible fermionic topological phases in one dimension. *Phys. Rev. B*, **106**(3), 035117.

Aksoy, Ö. M., Tiwari, A., and Mudry, C. 2021. Lieb-Schultz-Mattis type theorems for Majorana models with discrete symmetries. *Phys. Rev. B*, **104**(7), 075146.

Altland, A., and Zirnbauer, M. R. 1997. Nonstandard symmetry classes in mesoscopic normal-superconducting hybrid structures. *Phys. Rev. B*, **55**(2), 1142–1161.

Anderson, P. W. 1958. Absence of diffusion in certain random lattices. *Phys. Rev.*, **109**(5), 1492–1505.

Anderson, P. W. 1963. Plasmons, gauge invariance, and mass. *Phys. Rev.*, **130**(1), 439–442.

Anderson, P. W., Thouless, D. J., Abrahams, E., and Fisher, D. S. 1980. New method for a scaling theory of localization. *Phys. Rev. B*, **22**(8), 3519–3526.

Arovas, D. P., and Auerbach, A. 1988. Functional integral theories of low-dimensional quantum Heisenberg models. *Phys. Rev. B*, **38**(1), 316–332.

Atiyah, M. F., Bott, R., and Shapiro, A. 1964. Clifford modules. *Topology*, **3**(Supplement 1), 3–38.

Auerbach, A., and Arovas, D. P. 1988. Spin dynamics in the square-lattice antiferromagnet. *Phys. Rev. Lett.*, **61**(5), 617–620.

Baez, J. C. 2020. The tenfold way. *Notices Amer. Math. Soc.*, **67**, 1599–1601.

Barisic, S. 1983. The role of Coulomb screening in quasi one-dimensional conductors. *J. Phys. France*, **44**(2), 185–199.

Barkeshli, M., Chen, Y.-A., Hsin, P.-S., and Manjunath, N. 2022. Classification of $(2+1)$D invertible fermionic topological phases with symmetry. *Phys. Rev. B*, **105**(23), 235143.

Beenakker, C. W. J. 1997. Random-matrix theory of quantum transport. *Rev. Mod. Phys.*, **69**(3), 731–808.

Beenakker, C. W. J. 2015. Random-matrix theory of Majorana fermions and topological superconductors. *Rev. Mod. Phys.*, **87**(3), 1037–1066.

Beenakker, C. W. J., and Rejaei, B. 1993. Nonlogarithmic repulsion of transmission eigenvalues in a disordered wire. *Phys. Rev. Lett.*, **71**(22), 3689–3692.

Beenakker, C. W. J., and Rejaei, B. 1994. Exact solution for the distribution of transmission eigenvalues in a disordered wire and comparison with random-matrix theory. *Phys. Rev. B*, **49**(11), 7499–7510.

Behrends, J., and Béri, B. 2020. Supersymmetry in the standard Sachdev-Ye-Kitaev Model. *Phys. Rev. Lett.*, **124**(23), 236804.

Bell, J. S., and Rajaraman, R. 1983. On states, on a lattice, with half-integral charge. *Nucl. Phys. B*, **220**(FS8), 1–12.

Benalcazar, W. A., Bernevig, B. A., and Hughes, T. L. 2017a. Electric multipole moments, topological multipole moment pumping, and chiral hinge states in crystalline insulators. *Phys. Rev. B*, **96**(24), 245115.

Benalcazar, W. A., Bernevig, B. A., and Hughes, T. L. 2017b. Quantized electric multipole insulators. *Science*, **357**, 61–66.

Berets, D. J., and Smith, D. S. 1968. Electrical properties of linear polyacetylene. *Trans. Faraday Soc.*, **64**, 823–828.

Bernevig, B. A., Hughes, T. L., and Zhang, S.-C. 2006. Quantum spin Hall effect and topological phase transition in HgTe quantum wells. *Science*, **314**, 1757–1761.

Bethe, H. 1931. Zur Theorie der Metalle. *Z. Phys.*, **71**(3), 205–226.

Blankenbecler, R., and Boyanovsky, D. 1985. Fractional charge and spectral asymmetry in one dimension: A closer look. *Phys. Rev. D*, **31**(8), 2089–2099.

Blöte, H. W. J., Cardy, John L., and Nightingale, M. P. 1986. Conformal invariance, the central charge, and universal finite-size amplitudes at criticality. *Phys. Rev. Lett.*, **56**(7), 742–745.

Borland, R. E. 1963. The nature of the electronic states in disordered one-dimensional systems. *Proc. R. Soc. Lond. A*, **274**(1359), 529–545.

Bourne, C., and Ogata, Y. 2021. The classification of symmetry protected topological phases of one-dimensional fermion systems. *Forum of Mathematics, Sigma*, **9**(e25), 1–45.

Bravyi, S., Leemhuis, B., and Terhal, B. M. 2011. Topological order in an exactly solvable 3D spin model. *Ann. Phys.*, **326**(4), 839–866.

Bronk, B. V. 1965. Exponential ensemble for random matrices. *J. Math. Phys.*, **6**(2), 228–237.

Brouwer, P. W., Mudry, C., Simons, B. D., and Altland, A. 1998. Delocalization in coupled one-dimensional chains. *Phys. Rev. Lett.*, **81**(4), 862–865.

Brouwer, P. W., Mudry, C., and Furusaki, A. 2000a. Density of states in coupled chains with off-diagonal disorder. *Phys. Rev. Lett.*, **84**(13), 2913–2916.

Brouwer, P. W., Furusaki, A., Gruzberg, I. A., and Mudry, C. 2000b. Localization and delocalization in dirty superconducting wires. *Phys. Rev. Lett.*, **85**(5), 1064–1067.

Brouwer, P. W., Mudry, C., and Furusaki, A. 2000. Nonuniversality in quantum wires with off-diagonal disorder: A geometric point of view. *Nucl. Phys. B*, **565**, 653–663.

Brown, K. S. 1982. *Cohomology of Groups.* Springer, New York.

Bultinck, N., Williamson, D. J., Haegeman, J., and Verstraete, F. 2017. Fermionic matrix product states and one-dimensional topological phases. *Phys. Rev. B*, **95**(7), 075108.

Bundschuh, R., Cassanello, C., Serban, D., and Zirnbauer, M. R. 1998. Localization of quasiparticles in a disordered vortex. *Nucl. Phys. B*, **532**, 689–732.

Bundschuh, R., Cassanello, C., Serban, D., and Zirnbauer, M. R. 1999. Weak localization of disordered quasiparticles in the mixed superconducting state. *Phys. Rev. B*, **59**(6), 4382–4389.

Büttiker, M. 1986. Four-terminal phase-coherent conductance. *Phys. Rev. Lett.*, **57**(14), 1761–1764.

Buttiker, M. 1988. Symmetry of electrical conduction. *IBM J. Res. Dev.*, **32**(3), 317–334.

Büttiker, M. 1993. Capacitance, admittance, and rectification properties of small conductors. *J. Phys.: Condens. Matter*, **5**(50), 9361–9378.

Callan, C. G., and Harvey, J. A. 1985. Anomalies and fermion zero modes on strings and domain walls. *Nucl. Phys. B*, **250**(1), 427–436.

Caselle, M., and Magnea, U. 2004. Random matrix theory and symmetric spaces. *Phys. Rep.*, **394**(2), 41–156.

Chalker, J. T., and Coddington, P. D. 1988. Percolation, quantum tunnelling and the integer Hall effect. *J. Phys. C: Solid State Phys.*, **21**(14), 2665–2679.

Chamon, C. 2005. Quantum glassiness in strongly correlated clean systems: An example of topological overprotection. *Phys. Rev. Lett.*, **94**(4), 040402.

Chen, X., Gu, Z.-C., Liu, Z.-X., and Wen, X.-G. 2013. Symmetry protected topological orders and the group cohomology of their symmetry group. *Phys. Rev. B*, **87**(15), 155114.

Chiang, C. K., Fincher, C. R., and Park, Y. W. et al. 1977. Electrical conductivity in doped polyacetylene. *Phys. Rev. Lett.*, **39**(17), 1098–1101.

Chiang, C. K., Gau, S. C., and Fincher, C. R., Jr. et al. 1978. Polyacetylene, $(\mathrm{CH})_x$: n-type and p-type doping and compensation. *Appl. Phys. Lett.*, **33**(1), 18–20.

Chiang, C. K., Heeger, A. J., and Macdiarmid, A. G. 1979. Synthesis, structure, and electrical properties of doped polyacetylene. *Berichte der Bunsengesellschaft für physikalische Chemie*, **83**(4), 407–417.

Cho, G. Y., Hsieh, C.-T., Morimoto, T., and Ryu, S. 2015. Topological phases protected by reflection symmetry and cross-cap states. *Phys. Rev. B*, **91**(19), 195142.

Choquet-Bruhat, Y., DeWitt-Morette, C., and Dillard-Bleick, M. 1982. *Analysis, Manifolds and Physics*. Elsevier Science, Amsterdam.

Coleman, S. 1975. Quantum sine-Gordon equation as the massive Thirring model. *Phys. Rev. D*, **11**(8), 2088–2097.

Coleman, S., and Weinberg, E. 1973. Radiative corrections as the origin of spontaneous symmetry breaking. *Phys. Rev. D*, **7**(6), 1888–1910.

Cornfeld, E., and Carmeli, S. 2021. Tenfold topology of crystals: Unified classification of crystalline topological insulators and superconductors. *Phys. Rev. Res.*, **3**(1), 013052.

den Nijs, M., and Rommelse, K. 1989. Preroughening transitions in crystal surfaces and valence-bond phases in quantum spin chains. *Phys. Rev. B*, **40**(7), 4709–4734.

Deser, S., Jackiw, R., and Templeton, T. 1982. Topologically massive gauge theories. *Ann. Phys.*, **140**(2), 372–411.

Deser, S., Jackiw, R., and Templeton, T. 1988. Erratum: Topologically massive gauge theories. *Ann. Phys.*, **185**(2), 406–406.

Dijkgraaf, R., and Witten, E. 1990. Topological gauge theories and group cohomology. *Commun. Math. Phys.*, **129**(2), 393–429.

Dorokhov, O. N. 1982. Transmission coefficient and the localization length of an electron in N bound disordered chains. *JETP Lett.*, **36**(7), 318–321.

Dresselhaus, M. S., Smalley, R. E., Dresselhaus, G., and Avouris, P. 2001. *Carbon Nanotubes: Synthesis, Structure, Properties, and Applications*. Springer, Berlin Heidelberg.

Dunne, G. V., and Rao, K. 2001. Thermal fluctuations of induced fermion number. *Phys. Rev. D*, **64**(2), 025003.

Dyson, F. J. 1953. The dynamics of a disordered linear chain. *Phys. Rev.*, **92**(6), 1331–1338.

Dyson, F. J. 1962a. Statistical theory of the energy levels of complex systems. I. *J. Math. Phys.*, **3**(1), 140–156.

Dyson, F. J. 1962b. Statistical theory of the energy levels of complex systems. II. *J. Math. Phys.*, **3**(1), 157–165.

Dyson, F. J. 1962c. Statistical theory of the energy levels of complex systems. III. *J. Math. Phys.*, **3**(1), 166–175.

Dziawa, P, Kowalski, B. J., and Dybko, K. et al. 2012. Topological crystalline insulator states in $Pb_{1-x}Sn_x$.Se. *Nat. Mater.*, **11**, 1023–1027.

Edwards, J. T., and Thouless, D. J. 1972. Numerical studies of localization in disordered systems. *J. Phys. C: Solid State Phys.*, **5**(8), 807–820.

Emery, V. J. 1979. Theory of the one-dimensional electron gas, in *Highly Conducting One-Dimensional Solids*, edited by J. T. Devreese, R. P. Evrard and V. E. van Doren, Springer, Boston. 247–303.

Evers, F., and Mirlin, A. D. 2008. Anderson transitions. *Rev. Mod. Phys.*, **80**(4), 1355–1417.

Fidkowski, L., and Kitaev, A. 2010. Effects of interactions on the topological classification of free fermion systems. *Phys. Rev. B*, **81**(13), 134509.

Fidkowski, L., and Kitaev, A. 2011. Topological phases of fermions in one dimension. *Phys. Rev. B*, **83**(7), 075103.

Fidkowski, L., Chen, X., and Vishwanath, A. 2013. Non-Abelian topological order on the surface of a 3D topological superconductor from an exactly solved model. *Phys. Rev. X*, **3**(4), 041016.

Fisher, D. S., and Lee, P. A. 1981. Relation between conductivity and transmission matrix. *Phys. Rev. B*, **23**(12), 6851–6854.

Fradkin, E. 2013. *Field Theories of Condensed Matter Physics*. Cambridge University Press, New York.

Frahm, K. 1995. Equivalence of the Fokker-Planck approach and the nonlinear σ model for disordered wires in the unitary symmetry class. *Phys. Rev. Lett.*, **74**(23), 4706–4709.

Freed, D. S., and Hopkins, M. J. 2021. Reflection positivity and invertible topological phases. *Geometry and Topology*, **25**(3), 1165–1330.

Friedel, J. 1952. XIV. The distribution of electrons round impurities in monovalent metals. *The London Edinburgh Dublin Philos. Mag. & J. Sci.*, **43**(337), 153–189.

Fröhlich, J., and Studer, U. M. 1992. U(1)×SU(2)-gauge invariance of nonrelativistic quantum mechanics, and generalized Hall effects. *Commun. Math. Phys.*, **148**(3), 553–600.

Fröhlich, J., and Studer, U. M. 1993. Gauge invariance and current algebra in nonrelativistic many-body theory. *Rev. Mod. Phys.*, **65**(3), 733–802.

Fu, L. 2011. Topological crystalline insulators. *Phys. Rev. Lett.*, **106**(10), 106802.

Fu, L., and Kane, C. L. 2007. Topological insulators with inversion symmetry. *Phys. Rev. B*, **76**(4), 045302.

Fu, L., Kane, C. L., and Mele, E. J. 2007. Topological insulators in three dimensions. *Phys. Rev. Lett.*, **98**(10), 106803.

Fujikawa, K. 1979. Path-integral measure for gauge-invariant fermion theories. *Phys. Rev. Lett.*, **42**(18), 1195–1198.

Fulga, I. C., Hassler, F., and Akhmerov, A. R. 2012. Scattering theory of topological insulators and superconductors. *Phys. Rev. B*, **85**(16), 165409.

Gade, R. 1993. Anderson localization for sublattice models. *Nucl. Phys. B*, **398**, 499–515.

Gade, R., and Wegner, F. 1991. The $n = 0$ replica limit of U(n) and U(n)/SO(n) models. *Nucl. Phys. B*, **360**, 213–218.

Gamboa Saraví, R. E., Schaposnik, F. A., and Solomin, J. E. 1981. Path-integral formulation of two-dimensional gauge theories with massless fermions. *Nucl. Phys. B*, **185**, 239–253.

Georgi, H. 2000. *Lie Algebras In Particle Physics: from Isospin To Unified Theories*. CRS Press.

Goldstone, J. 1961. Field theories with "superconductor" solutions. *Il Nuovo Cimento*, **19**(1), 154–164.

Goldstone, J., and Wilczek, F. 1981. Fractional quantum numbers on solitons. *Phys. Rev. Lett.*, **47**(14), 986–989.

Gradshteyn, I. S., and Ryzhik, I. M. 1994. *Table of Integrals, Series, and Products*. Academic Press, London.

Grant, P. M., and Batra, I. P. 1983. Self-consistent crystal potential and band structure of three-dimensional trans-polyacetylene. *J. Phys. Colloques*, **44**(C3), 437–442.

Grüner, G. 1988. The dynamics of charge-density waves. *Rev. Mod. Phys.*, **60**(4), 1129–1181.

Gruzberg, I. A., Ludwig, A. W. W., and Read, N. 1999. Exact exponents for the spin quantum Hall transition. *Phys. Rev. Lett.*, **82**(22), 4524–4527.

Gu, Z.-C., and Wen, X.-G. 2009. Tensor-entanglement-filtering renormalization approach and symmetry-protected topological order. *Phys. Rev. B*, **80**(15), 155131.

Haah, J. 2011. Local stabilizer codes in three dimensions without string logical operators. *Phys. Rev. A*, **83**(4), 042330.

Hagiwara, M., Katsumata, K., and Affleck, I. et al. 1990. Observation of $S = 1/2$ degrees of freedom in an $S = 1$ linear-chain Heisenberg antiferromagnet. *Phys. Rev. Lett.*, **65**(25), 3181–3184.

Haldane, F. D. M. 1983a. Continuum dynamics of the 1-D Heisenberg antiferromagnet: Identification with the O(3) nonlinear sigma model. *Phys. Lett. A*, **93**(9), 464–468.

Haldane, F. D. M. 1983b. Nonlinear field theory of large-spin Heisenberg antiferromagnets: Semiclassically quantized solitons of the one-dimensional easy-axis Néel state. *Phys. Rev. Lett.*, **50**(15), 1153–1156.

Haldane, F. D. M. 1988. Model for a quantum Hall effect without Landau levels: Condensed-matter realization of the "parity anomaly". *Phys. Rev. Lett.*, **61**(18), 2015–2018.

Haldane, F. D. M. 1995. Stability of chiral Luttinger liquids and Abelian quantum Hall states. *Phys. Rev. Lett.*, **74**(11), 2090–2093.

Halperin, B. I. 1982. Quantized Hall conductance, current-carrying edge states, and the existence of extended states in a two-dimensional disordered potential. *Phys. Rev. B*, **25**(4), 2185–2190.

Halperin, B. I. 1987. Possible states for a three-dimensional electron gas in a strong magnetic field. *Jpn. J. Appl. Phys.*, **26**(Supplement 26-3), 1913–1919.

Halperin, B. I., Lubensky, T. C., and Ma, S. K. 1974. First-order phase transitions in superconductors and smectic-*A* liquid crystals. *Phys. Rev. Lett.*, **32**(6), 292–295.

Hanson, R., Kouwenhoven, L. P., and Petta, J. R. et al. 2007. Spins in few-electron quantum dots. *Rev. Mod. Phys.*, **79**(4), 1217–1265.

Hastings, M. B., and Michalakis, S. 2015. Quantization of Hall conductance for interacting electrons on a torus. *Commun. Math. Phys.*, **334**(1), 433–471.

Hatano, M., Kambara, S., and Okamoto, S. 1961. Paramagnetic and electric properties of polyacetylene. *J. Polym. Sci.*, **51**(156), S26–S29.

Hatsugai, Y. 1993. Chern number and edge states in the integer quantum Hall effect. *Phys. Rev. Lett.*, **71**(22), 3697–3700.

Heeger, A. J., Kivelson, S., Schrieffer, J. R., and Su, W. P. 1988. Solitons in conducting polymers. *Rev. Mod. Phys.*, **60**(3), 781–850.

Helgason, S. 2001. *Differential Geometry, Lie Groups, and Symmetric Spaces.* American Mathematical Society.

Higgs, P. W. 1964a. Broken symmetries and the masses of gauge bosons. *Phys. Rev. Lett.*, **13**(16), 508–509.

Higgs, P. W. 1964b. Broken symmetries, massless particles and gauge fields. *Phys. Lett.*, **12**(2), 132–133.

Holstein, T., Norton, R. E., and Pincus, P. 1973. De Haas-Van Alphen effect and the specific heat of an electron gas. *Phys. Rev. B*, **8**(6), 2649–2656.

Hou, C.-Y., Chamon, C., and Mudry, C. 2007. Electron fractionalization in two-dimensional graphenelike structures. *Phys. Rev. Lett.*, **98**(18), 186809.

Hsieh, C.-T., Sule, O. M., Cho, G. Y., Ryu, S., and Leigh, R. G. 2014. Symmetry-protected topological phases, generalized Laughlin argument, and orientifolds. *Phys. Rev. B*, **90**(16), 165134.

Hsieh, C.-T., Cho, G. Y., and Ryu, S. 2016a. Global anomalies on the surface of fermionic symmetry-protected topological phases in (3+1) dimensions. *Phys. Rev. B*, **93**(7), 075135.

Hsieh, D., Qian, D., and Wray, L. et al. 2008. A topological Dirac insulator in a quantum spin Hall phase. *Nature*, **452**, 970–974.

Hsieh, D., Xia, Y., and Wray, L. et al. 2009. Observation of unconventional quantum spin textures in topological insulators. *Science*, **323**, 919–922.

Hsieh, T. H., Lin, H., and Liu, J. et al. 2012. Topological crystalline insulators in the SnTe material class. *Nat. Commun.*, **3**, 982.

Hsieh, T. H., Halász, G. B., and Grover, T. 2016b. All Majorana models with translation symmetry are supersymmetric. *Phys. Rev. Lett.*, **117**(16), 166802.

Huckestein, B. 1995. Scaling theory of the integer quantum Hall effect. *Rev. Mod. Phys.*, **67**(2), 357–396.

Hüffmann, A. 1990. Disordered wires from a geometric viewpoint. *J. Phys. A: Math. Gen.*, **23**(24), 5733–5744.

Imry, Y. 1986. Physics of Mesoscopic Systems, in *Directions in Condensed Matter Physics*, edited by G. Grinstein and G. Mazenko, World Scientific, Singapore. 101–164.

Itoi, C., and Mukaida, H. 1994. Non-Abelian gauge theory for quantum Heisenberg antiferromagnetic chain. *J. Phys. A: Math. Gen.*, **27**(13), 4695–4708.

Jackiw, R. 1982. Fermion fractionization in physics, in *Quantum Structure of Space and Time*, edited by M. J. Duff and C. J. Isham, Cambridge University Press, New York. 169–184.

Jackiw, R., and Rebbi, C. 1976. Solitons with fermion number $\frac{1}{2}$. *Phys. Rev. D*, **13**(12), 3398–3409.

Jackiw, R., and Rossi, P. 1981. Zero modes of the vortex-fermion system. *Nucl. Phys. B*, **190**, 681–691.

Jackiw, R., and Semenoff, G. 1983. Continuum quantum field theory for a linearly conjugated diatomic polymer with fermion fractionization. *Phys. Rev. Lett.*, **50**(6), 439–442.

Jackiw, R., Kerman, A. K., Klebanov, I., and Semenoff, G. 1983. Fluctuations of fractional charge in soliton anti-soliton systems. *Nucl. Phys. B*, **225**(FS9), 233–246.

Jordan, P., and Wigner, E. 1928. Über das Paulische Äquivalenzverbot. *Z. Phys.*, **47**, 631–651.

Kane, C. L., and Mele, E. J. 2005a. $\mathbb{Z}_2$ Topological order and the quantum spin Hall effect. *Phys. Rev. Lett.*, **95**(14), 146802.

Kane, C. L., and Mele, E. J. 2005b. Quantum spin Hall effect in graphene. *Phys. Rev. Lett.*, **95**(22), 226801.

Kapustin, A. 2014a. Bosonic topological insulators and paramagnets: A view from cobordisms. *arXiv:1404.6659.*

Kapustin, A. 2014b. Symmetry protected topological phases, anomalies, and cobordisms: Beyond group cohomology. *arXiv:1403.1467.*

Kapustin, A., Thorngren, R., Turzillo, A., and Wang, Z. 2015. Fermionic symmetry protected topological phases and cobordisms. *JHEP*, **2015**(12), 052.

Karoubi, M. 2008. *K-Theory: An Introduction.* Springer, Berlin Heidelberg.

Kennedy, T. 1990. Exact diagonalisations of open spin-1 chains. *J. Phys.: Condens. Matter*, **2**(26), 5737–5745.

Khmelnitskii, D. E. 1983. Quantization of Hall conductivity. *JETP Lett.*, **38**(9), 552–556.

Kitaev, A. 2009. Periodic table for topological insulators and superconductors. *AIP Conference Proceedings*, **1134**, 22–30.

Kitaev, A. Y. 2001. Unpaired Majorana fermions in quantum wires. *Phys.-Uspekhi*, **44**(10S), 131–136.

Kivelson, S., and Schrieffer, J. R. 1982. Fractional charge, a sharp quantum observable. *Phys. Rev. B*, **25**(10), 6447–6451.

Kleist, F. D., and Byrd, N. R. 1969. Preparation and properties of polyacetylene. *J. Polym. Sci. Part A-1: Polymer Chemistry*, **7**(12), 3419–3425.

Klitzing, K. v., Dorda, G., and Pepper, M. 1980. New method for high-accuracy determination of the fine-structure constant based on quantized Hall resistance. *Phys. Rev. Lett.*, **45**(6), 494–497.

Koma, T. 2000. Spectral gaps of quantum Hall systems with interactions. *J. of Stat. Phys.*, **99**(1), 313–381.

Koma, T. 2015. Topological current in fractional Chern insulators. *arXiv:1504.01243.*

König, M., Wiedmann, S., and Brüne, C. et al. 2007. Quantum spin Hall insulator state in HgTe quantum wells. *Science*, **318**, 766–770.

Kramers, H. A. 1930. Théorie générale de la rotation paramagnétique dans les cristaux. *Proceedings of the Royal Netherlands Academy of Arts and Sciences*, **33**(9), 959–972.

Kronig, R. de L. 1935. Zur Neutrinotheorie des Lichtes III. *Physica*, **2**, 968–980.

Landau, L. D. 1957a. Oscillations in a Fermi liquid. *Soviet Physics JETP-USSR*, **5**(1), 101–108.

Landau, L. D. 1957b. The theory of a Fermi liquid. *Soviet Physics JETP-USSR*, **3**(6), 920–925.

Landau, L. D. 1959. On the theory of the Fermi liquid. *Soviet Physics JETP-USSR*, **8**(1), 70–74.

Landauer, R. 1957. Spatial variation of currents and fields due to localized scatterers in metallic conduction. *IBM J. Res. Dev.*, **1**(3), 223–231.

Landauer, R. 1970. Electrical resistance of disordered one-dimensional lattices. *Phil. Mag.*, **21**(172), 863–867.

Landauer, R. 1989. Conductance determined by transmission: Probes and quantised constriction resistance. *J. Phys.: Condens. Matter*, **1**(43), 8099–8110.

Langbehn, J., Peng, Y., Trifunovic, L., von Oppen, F., and Brouwer, P. W. 2017. Reflection-symmetric second-order topological insulators and superconductors. *Phys. Rev. Lett.*, **119**(24), 246401.

Laughlin, R. B. 1981. Quantized Hall conductivity in two dimensions. *Phys. Rev. B*, **23**(10), 5632–5633.

Lee, P. A., and Ramakrishnan, T. V. 1985. Disordered electronic systems. *Rev. Mod. Phys.*, **57**(2), 287–337.

Levin, M., and Stern, A. 2009. Fractional topological insulators. *Phys. Rev. Lett.*, **103**(19), 196803.

Levine, H., Libby, S. B., and Pruisken, A. M. M. 1983. Electron delocalization by a magnetic field in two dimensions. *Phys. Rev. Lett.*, **51**(20), 1915–1918.

Levinson, N. 1949. On the uniqueness of the potential in a Schrodinger equation for a given asymptotic phase. *Kgl. Danske Videnskab Selskab. Mat. Fys. Medd.*, **25**(9), 1–29.

Licciardello, D. C., and Thouless, D. J. 1975. Conductivity and mobility edges for two-dimensional disordered systems. *J. Phys. C: Solid State Phys.*, **8**(24), 4157–4170.

Licciardello, D. C., and Thouless, D. J. 1978. Conductivity and mobility edges in disordered systems. II. Further calculations for the square and diamond lattices. *J. Phys. C: Solid State Phys.*, **11**(5), 925–936.

Lieb, E., Schultz, T., and Mattis, D. 1961. Two soluble models of an antiferromagnetic chain. *Ann. Phys.*, **16**(3), 407–466.

Ludwig, A. W. W., Fisher, M. P. A., Shankar, R., and Grinstein, G. 1994. Integer quantum Hall transition: An alternative approach and exact results. *Phys. Rev. B*, **50**(11), 7526–7552.

Luther, A., and Peschel, I. 1975. Calculation of critical exponents in two dimensions from quantum field theory in one dimension. *Phys. Rev. B*, **12**(9), 3908–3917.

Luttinger, J. M. 1963. An exactly soluble model of a many-fermion system. *J. Math. Phys.*, **4**(9), 1154–1162.

MacKenzie, R., and Wilczek, F. 1984. Illustrations of vacuum polarization by solitons. *Phys. Rev. D*, **30**(10), 2194–2200.

Mandelstam, S. 1975. Soliton operators for the quantized sine-Gordon equation. *Phys. Rev. D*, **11**(10), 3026–3030.

Marshall, W. 1955. Antiferromagnetism. *Proc. R. Soc. Lond. A*, **232**(1188), 48–68.

Martin, Th., and Landauer, R. 1992. Wave-packet approach to noise in multichannel mesoscopic systems. *Phys. Rev. B*, **45**(4), 1742–1755.

Matsui, A., and Nakamura, K. 1967. Optical properties of polyacetylene. *Jpn. J. Appl. Phys.*, **6**(12), 1468–1469.

Mehta, M. L. 2004. *Random Matrices.* Elsevier Science, Amsterdam.

Mello, P. A., and Pichard, J.-L. 1991. Symmetries and parametrization of the transfer matrix in electronic quantum transport theory. *J. Phys. I France*, **1**(4), 493–513.

Mello, P. A., Pereyra, P., and Kumar, N. 1988. Macroscopic approach to multichannel disordered conductors. *Ann. Phys.*, **181**(2), 290–317.

Mermin, N. D., and Wagner, H. 1966. Absence of ferromagnetism or antiferromagnetism in one- or two-dimensional isotropic Heisenberg models. *Phys. Rev. Lett.*, **17**(22), 1133–1136.

Messiah, A. 2017. *Quantum Mechanics.* Dover, New York.

Metlitski, M. A., Fidkowski, L., Chen, X., and Vishwanath, A. 2014. Interaction effects on 3D topological superconductors: surface topological order from vortex condensation, the 16 fold way and fermionic Kramers doublets. *arXiv:1406.3032.*

Midorikawa, S. 1983. Fractional charges at finite temperature. *Prog. Theor. Phys.*, **69**(6), 1831–1834.

Midorikawa, S. 1985. Fractional fermion number and its thermal effect. *Phys. Rev. D*, **31**(6), 1499–1502.

Moore, J. E., and Balents, L. 2007. Topological invariants of time-reversal-invariant band structures. *Phys. Rev. B*, **75**(12), 121306(R).

Morimoto, T., and Furusaki, A. 2013. Topological classification with additional symmetries from Clifford algebras. *Phys. Rev. B*, **88**(12), 125129.

Morimoto, T., Furusaki, A., and Mudry, C. 2015a. Anderson localization and the topology of classifying spaces. *Phys. Rev. B*, **91**(23), 235111.

Morimoto, T., Furusaki, A., and Mudry, C. 2015b. Breakdown of the topological classification $\mathbb{Z}$ for gapped phases of noninteracting fermions by quartic interactions. *Phys. Rev. B*, **92**(12), 125104.

Moses, D., Feldblum, A., and Ehrenfreund, E. et al. 1982. Pressure dependence of the photoabsorption of polyacetylene. *Phys. Rev. B*, **26**(6), 3361–3369.

Mott, N. F., and Twose, W. D. 1961. The theory of impurity conduction. *Advan. Phys.*, **10**(38), 107–163.

Mudry, C. 2014. *Lecture Notes On Field Theory In Condensed Matter Physics.* World Scientific, Singapore.

Mudry, C., Brouwer, P. W., and Furusaki, A. 1999. Random magnetic flux problem in a quantum wire. *Phys. Rev. B*, **59**(20), 13221–13234.

Mudry, C., Brouwer, P. W., and Furusaki, A. 2000. Crossover from the chiral to the standard universality classes in the conductance of a quantum wire with random hopping only. *Phys. Rev. B*, **62**(12), 8249–8268.

Mudry, C., Brouwer, P. W., and Furusaki, A. 2001. Erratum: Crossover from the chiral to the standard universality classes in the conductance of a quantum wire with random hopping only [Phys. Rev. B 62, 8249 (2000)]. *Phys. Rev. B*, **63**(12), 129901.

Nambu, Y. 1960. Quasi-particles and gauge invariance in the theory of superconductivity. *Phys. Rev.*, **117**(3), 648–663.

Nambu, Y., and Jona-Lasinio, G. 1961. Dynamical model of elementary particles based on an analogy with superconductivity. I. *Phys. Rev.*, **122**(1), 345–358.

Natta, G., Pino, P., and Corradini, P. et al. 1955. Crystalline high polymers of alpha-olefins. *J. Am. Chem. Soc.*, **77**(6), 1708–1710.

Natta, G., Mazzanti, G., and Corradini, P. 1958. Polimerizzazione stereospecifica dell'acetilene. *Atti Accad. Naz. Lincei, Rend. Cl. Sci. Fis. Mat. e Nat.*, **25**, 3–12.

Neupert, T., Santos, L., Ryu, S., Chamon, C., and Mudry, C. 2011. Fractional topological liquids with time-reversal symmetry and their lattice realization. *Phys. Rev. B*, **84**(16), 165107.

Neupert, T., Chamon, C., Mudry, C., and Thomale, R. 2014. Wire deconstructionism of two-dimensional topological phases. *Phys. Rev. B*, **90**(20), 205101.

Ng, T.-K. 1992. Schwinger-boson mean-field theory for $S = 1$ open spin chains. *Phys. Rev. B*, **45**(14), 8181–8184.

Ng, T.-K. 1993. Edge states in Schwinger-boson mean-field theory of low-dimensional quantum antiferromagnets. *Phys. Rev. B*, **47**(17), 11575–11578.

Ng, T.-K. 1994. Edge states in antiferromagnetic quantum spin chains. *Phys. Rev. B*, **50**(1), 555–558.

Nielsen, H. B., and Ninomiya, M. 1981. Absence of neutrinos on a lattice: (I). Proof by homotopy theory. *Nucl. Phys. B*, **185**, 20–40.

Niemi, A. J. 1984. Fermion number fractionization and the Witten index – A new approach. *Phys. Lett. B*, **146**(3), 213–216.

Niemi, A. J. 1985. Spectral density and a family of Dirac operators. *Nucl. Phys. B*, **253**, 14–46.

Niemi, A. J., and Semenoff, G. W. 1984. Fractional fermion number at finite temperature. *Phys. Lett. B*, **135**(1), 121–124.

Niemi, A. J., and Semenoff, G. W. 1986. Fermion number fractionization in quantum field theory. *Phys. Rep.*, **135**(3), 99–193.

Niu, Q., Thouless, D. J., and Wu, Y.-S. 1985. Quantized Hall conductance as a topological invariant. *Phys. Rev. B*, **31**(6), 3372–3377.

Novoselov, K. S., Geim, A. K., and Morozov, S. V. et al. 2004. Electric field effect in atomically thin carbon films. *Science*, **306**, 666–669.

Ogata, Y. 2020. A $\mathbb{Z}_2$-index of symmetry protected topological phases with time reversal symmetry for quantum spin chains. *Commun. Math. Phys.*, **374**(2), 705–734.

Ogata, Y., and Tasaki, H. 2019. Lieb–Schultz–Mattis type theorems for quantum spin chains without continuous symmetry. *Commun. Math. Phys.*, **372**(3), 951–962.

Ogata, Y., Tachikawa, Y., and Tasaki, H. 2021. General Lieb–Schultz–Mattis type theorems for quantum spin chains. *Commun. Math. Phys.*, **385**(1), 79–99.

Oshikawa, M., Yamanaka, M., and Affleck, I. 1997. Magnetization plateaus in spin chains: "Haldane gap" for half-integer Spins. *Phys. Rev. Lett.*, **78**(10), 1984–1987.

Perez-Garcia, D., Verstraete, F., Wolf, M. M., and Cirac, J. I. 2007. Matrix product state representations. *Quantum Info. Comput.*, **7**(5), 401–430.

Polyakov, A. M. 1975. Interaction of Goldstone particles in two dimensions. Applications to ferromagnets and massive Yang-Mills fields. *Phys. Lett. B*, **59**(1), 79–81.

Porteous, I. R. 1995. *Clifford Algebras and the Classical Groups.* Cambridge Studies in Advanced Mathematics, Vol. **50**. Cambridge University Press, Cambridge.

Pruisken, A. M. M. 1984. On localization in the theory of the quantized Hall effect: A two-dimensional realization of the θ-vacuum. *Nucl. Phys. B*, **235**, 277–298.

Rajaraman, R., and Bell, J. S. 1982. On solitons with half integral charge. *Phys. Lett. B*, **116**(2), 151–154.

Read, N., and Green, D. 2000. Paired states of fermions in two dimensions with breaking of parity and time-reversal symmetries and the fractional quantum Hall effect. *Phys. Rev. B*, **61**(15), 10267–10297.

Reimann, S. M., and Manninen, M. 2002. Electronic structure of quantum dots. *Rev. Mod. Phys.*, **74**(4), 1283–1342.

Renard, J. P., Verdaguer, M., and Regnault, L. P. et al. 1987. Presumption for a quantum energy gap in the quasi-one-dimensional $S = 1$ Heisenberg antiferromagnet $Ni(C_2H_8N_2)_2NO_2(ClO_4)$. *Europhys. Lett.*, **3**(8), 945–952.

Rotman, J. 1999. *An Introduction to the Theory of Groups.* Springer, New York.

Roy, R. 2009. Topological phases and the quantum spin Hall effect in three dimensions. *Phys. Rev. B*, **79**(19), 195322.

Ryu, S., and Zhang, S.-C. 2012. Interacting topological phases and modular invariance. *Phys. Rev. B*, **85**(24), 245132.

Ryu, S., Schnyder, A. P., Furusaki, A., and Ludwig, A. W. W. 2010. Topological insulators and superconductors: Tenfold way and dimensional hierarchy. *New J. Phys.*, **12**(6), 065010.

Ryu, S., Moore, J. E., and Ludwig, A. W. W. 2012. Electromagnetic and gravitational responses and anomalies in topological insulators and superconductors. *Phys. Rev. B*, **85**(4), 045104.

Sattinger, D. H., and Weaver, O. L. 1986. *Lie Groups and Algebras with Applications to Physics, Geometry, and Mechanics.* Springer, New York.

Schaposnik, F. A. 1985. Fermion currents in two-dimensional models. *Z. Phys. C*, **28**, 127–131.

Schindler, F., Cook, A. M., and Vergniory, M. G. et al. 2018a. Higher-order topological insulators. *Science Advances*, **4**, eaat0346.

Schindler, F., Wang, Z., and Vergniory, M. G. et al. 2018b. Higher-order topology in bismuth. *Nature Physics*, **14**, 918–924.

Schmidt, Helmut. 1957. Disordered one-dimensional crystals. *Phys. Rev.*, **105**(2), 425–441.

Schnyder, A. P., Ryu, S., Furusaki, A., and Ludwig, A. W. W. 2008. Classification of topological insulators and superconductors in three spatial dimensions. *Phys. Rev. B*, **78**(19), 195125.

Schultz, T. D., Mattis, D. C., and Lieb, E. H. 1964. Two-dimensional Ising model as a soluble problem of many fermions. *Rev. Mod. Phys.*, **36**(3), 856–871.

Schwinger, J. 1959. Field theory commutators. *Phys. Rev. Lett.*, **3**(6), 296–297.

Schwinger, J. 1962. Gauge invariance and mass. II. *Phys. Rev.*, **128**(5), 2425–2429.

Semenoff, G. W. 1984. Condensed-matter simulation of a three-dimensional anomaly. *Phys. Rev. Lett.*, **53**(26), 2449–2452.

Senthil, T., and Fisher, Matthew P. A. 2000. Quasiparticle localization in superconductors with spin-orbit scattering. *Phys. Rev. B*, **61**(14), 9690–9698.

Senthil, T., Fisher, M. P. A., Balents, L., and Nayak, C. 1998. Quasiparticle transport and localization in high- T_c superconductors. *Phys. Rev. Lett.*, **81**(21), 4704–4707.

Shapourian, H., Shiozaki, K., and Ryu, S. 2017. Many-body topological invariants for fermionic symmetry-protected topological phases. *Phys. Rev. Lett.*, **118**(21), 216402.

Shimamura, K., Hatano, M., Kanbara, S., and Nakada, I. 1967. Electrical conduction of poly-acetylene under high pressure. *J. Phys. Soc. Jpn.*, **23**(3), 578–581.

Shuryak, E. V., and Verbaarschot, J. J. M. 1993. Random matrix theory and spectral sum rules for the Dirac operator in QCD. *Nucl. Phys. A*, **560**, 306–320.

Slevin, K., and Nagao, T. 1993. New random matrix theory of scattering in mesoscopic systems. *Phys. Rev. Lett.*, **70**(5), 635–638.

Slevin, K., and Nagao, T. 1994. Impurity scattering in mesoscopic quantum wires and the Laguerre ensemble. *Phys. Rev. B*, **50**(4), 2380–2392.

Son, W., Amico, L., and Fazio, R. et al. 2011. Quantum phase transition between cluster and antiferromagnetic states. *Europhys. Lett.*, **95**(5), 50001.

Song, Z., Fang, Z., and Fang, C. 2017. $(d-2)$-dimensional edge states of rotation symmetry protected topological states. *Phys. Rev. Lett.*, **119**(24), 246402.

Stone, A. D., and Szafer, A. 1988. What is measured when you measure a resistance? – The Landauer formula revisited. *IBM J. Res. Dev.*, **32**(3), 384–413.

Su, W. P., and Schrieffer, J. R. 1981. Fractionally charged excitations in charge-density-wave systems with commensurability 3. *Phys. Rev. Lett.*, **46**(11), 738–741.

Su, W. P., Schrieffer, J. R., and Heeger, A. J. 1979. Solitons in polyacetylene. *Phys. Rev. Lett.*, **42**(25), 1698–1701.

Su, W. P., Schrieffer, J. R., and Heeger, A. J. 1980. Soliton excitations in polyacetylene. *Phys. Rev. B*, **22**(4), 2099–2111.

Suzuki, M. 1971a. The dimer problem and the generalized X-model. *Phys. Lett. A*, **34**(6), 338–339.

Suzuki, M. 1971b. Relationship among exactly soluble models of critical phenomena. I: 2D Ising model, dimer problem and the generalized XY-model. *Prog. Theor. Phys.*, **46**(5), 1337–1359.

Takayama, H., Lin-Liu, Y. R., and Maki, K. 1980. Continuum model for solitons in polyacetylene. *Phys. Rev. B*, **21**(6), 2388–2393.

Tanaka, Y., Ren, Z., and Sato, T. et al. 2012. Experimental realization of a topological crystalline insulator in SnTe. *Nat. Phys.*, **8**, 800–803.

Tang, E., and Wen, X.-G. 2012. Interacting one-dimensional fermionic symmetry-protected topological phases. *Phys. Rev. Lett.*, **109**(9), 096403.

Tasaki, H. 2018. Lieb–Schultz–Mattis theorem with a local twist for general one-dimensional quantum systems. *J. of Stat. Phys.*, **170**(4), 653–671.

Tasaki, H. 2020. *Physics and Mathematics of Quantum Many-Body Systems.* Springer, Switzerland.

Teo, J. C. Y., and Kane, C. L. 2010. Topological defects and gapless modes in insulators and superconductors. *Phys. Rev. B*, **82**(11), 115120.

Thouless, D. J. 1983. Quantization of particle transport. *Phys. Rev. B*, **27**(10), 6083–6087.

Thouless, D. J., Kohmoto, M., Nightingale, M. P., and den Nijs, M. 1982. Quantized Hall conductance in a two-dimensional periodic potential. *Phys. Rev. Lett.*, **49**(6), 405–408.

Titov, M., Brouwer, P. W., Furusaki, A., and Mudry, C. 2001. Fokker-Planck equations and density of states in disordered quantum wires. *Phys. Rev. B*, **63**(23), 235318.

Tomonaga, S. 1950. Remarks on Bloch's method of sound waves applied to many-fermion problems. *Prog. Theor. Phys.*, **5**(4), 544–569.

Tsui, D. C., Stormer, H. L., and Gossard, A. C. 1982. Two-dimensional magneto-transport in the extreme quantum limit. *Phys. Rev. Lett.*, **48**(22), 1559–1562.

Turner, A. M., Pollmann, F., and Berg, E. 2011. Topological phases of one-dimensional fermions: An entanglement point of view. *Phys. Rev. B*, **83**(7), 075102.

Turzillo, A., and You, M. 2019. Fermionic matrix product states and one-dimensional short-range entangled phases with antiunitary symmetries. *Phys. Rev. B*, **99**(3), 035103.

Van Kampen, N. G. 2011. *Stochastic Processes in Physics and Chemistry.* Elsevier Science, Amsterdam.

van Wees, B. J., van Houten, H., and Beenakker, C. W. J. et al. 1988. Quantized conductance of point contacts in a two-dimensional electron gas. *Phys. Rev. Lett.*, **60**(9), 848–850.

Verbaarschot, J. 1994. Spectrum of the QCD Dirac operator and chiral random matrix theory. *Phys. Rev. Lett.*, **72**(16), 2531–2533.

Verresen, R., Moessner, R., and Pollmann, F. 2017. One-dimensional symmetry protected topological phases and their transitions. *Phys. Rev. B*, **96**(16), 165124.

Vijay, S., Haah, J., and Fu, L. 2015. A new kind of topological quantum order: A dimensional hierarchy of quasiparticles built from stationary excitations. *Phys. Rev. B*, **92**(23), 235136.

Wall, C. T. C. 1964. Graded Brauer Groups. *J. Reine Angew. Math.*, **213**, 187–199.

Wallace, P. R. 1947. The band theory of graphite. *Phys. Rev.*, **71**(9), 622–634.

Wang, C., and Senthil, T. 2014. Interacting fermionic topological insulators/superconductors in three dimensions. *Phys. Rev. B*, **89**(19), 195124.

Watanabe, H. 2017. Energy gap of neutral excitations implies vanishing charge susceptibility. *Phys. Rev. Lett.*, **118**(11), 117205.

Watson Jr., W. H., McMordie Jr., W. C., and Lands, L. G. 1961. Polymerization of alkynes by Ziegler-type catalyst. *J. Polym. Sci.*, **55**(161), 137–144.

Wegner, F. J. 1976. Electrons in disordered systems. Scaling near the mobility edge. *Z. Phys. B*, **25**(4), 327–337.

Wen, X. G. 1990. Topological orders in rigid states. *International Journal of Modern Physics B*, **04**(02), 239–271.

Wen, X. G. 2007. *Quantum Field Theory of Many-Body Systems: From the Origin of Sound to an Origin of Light and Electrons*. Oxford University Press, New York.

Wess, J., and Zumino, B. 1971. Consequences of anomalous ward identities. *Phys. Lett. B*, **37**(1), 95–97.

Wiegmann, P. B. 1985. Exact solution of the O(3) nonlinear σ-model. *Phys. Lett. B*, **152**(3), 209–214.

Wigner, E. P. 1951a. On a class of analytic functions from the quantum theory of collisions. *Ann. Math.*, **53**(1), 36–67.

Wigner, E. P. 1951b. On the statistical distribution of the widths and spacings of nuclear resonance levels. *Math. Proc. Cambridge Philos. Soc.*, **47**(4), 790–798.

Wigner, E. P. 1955. Characteristic vectors of bordered matrices with infinite dimensions. *Ann. Math.*, **62**(3), 548–564.

Wigner, E. P. 1957. Characteristics vectors of bordered matrices with infinite dimensions II. *Ann. Math.*, **65**(2), 203–207.

Wigner, E. P. 1958. On the distribution of the roots of certain symmetric matrices. *Ann. Math.*, **67**(2), 325–327.

Witten, E. 1982. Constraints on supersymmetry breaking. *Nucl. Phys. B*, **202**, 253–316.

Witten, E. 1984. Non-abelian bosonization in two dimensions. *Commun. Math. Phys.*, **92**(4), 455–472.

Witten, E. 2016. Fermion path integrals and topological phases. *Rev. Mod. Phys.*, **88**(3), 035001.

Wu, W. K., and Kivelson, S. 1986. Theory of conducting polymers with weak electron-electron interactions. *Phys. Rev. B*, **33**(12), 8546–8557.

Xu, S.-Y., Liu, C., and Alidoust, N. et al. 2012. Observation of a topological crystalline insulator phase and topological phase transition in $Pb_{1-x}Sn_xTe$. *Nat. Commun.*, **3**, 1192.

Yamagishi, H. 1983a. Comment on "Fractional Quantum Numbers on Solitons". *Phys. Rev. Lett.*, **50**(6), 458–458.

Yamagishi, H. 1983b. Fermion-monopole system reexamined. *Phys. Rev. D*, **27**(10), 2383–2396.

Yamanaka, M., Oshikawa, M., and Affleck, I. 1997. Nonperturbative approach to Luttinger's theorem in one dimension. *Phys. Rev. Lett.*, **79**(6), 1110–1113.

You, Y.-Z., and Xu, C. 2014. Symmetry-protected topological states of interacting fermions and bosons. *Phys. Rev. B*, **90**(24), 245120.

Youla, D. C. 1961. A Normal form for a matrix under the unitary congruence group. *Can. J. Math.*, **13**, 694–704.

Ziegler, K., Holzkamp, E., Breil, H., and Martin, H. 1955. Das Mühlheimer Normaldruck-Polyäthylen-Verfahren. *Angew. Chem.*, **67**(19-20), 541–547.

Zirnbauer, M. R. 1992. Super Fourier analysis and localization in disordered wires. *Phys. Rev. Lett.*, **69**(10), 1584–1587.

Zirnbauer, M. R. 1996. Riemannian symmetric superspaces and their origin in random-matrix theory. *J. Math. Phys.*, **37**(10), 4986–5018.

Index

Printed in the United States
by Baker & Taylor Publisher Services